Catalogue of Palaearctic Coleoptera

Volume 5

Tenebrionoidea

DEDICATION

This volume is dedicated to the late Gustav Adolf Lohse (27.XII.1910 – 30.IV.1994, Hamburg, Germany). Though he was a professional dentist, he significantly influenced the study of Central European Coleoptera, and worldwide the taxonomy of some groups. A man with sincere interests for taxonomy and natural history, he was remarkably generous to those who knew him and did not let personal gain and competition taint his compassion for the study of beetles.

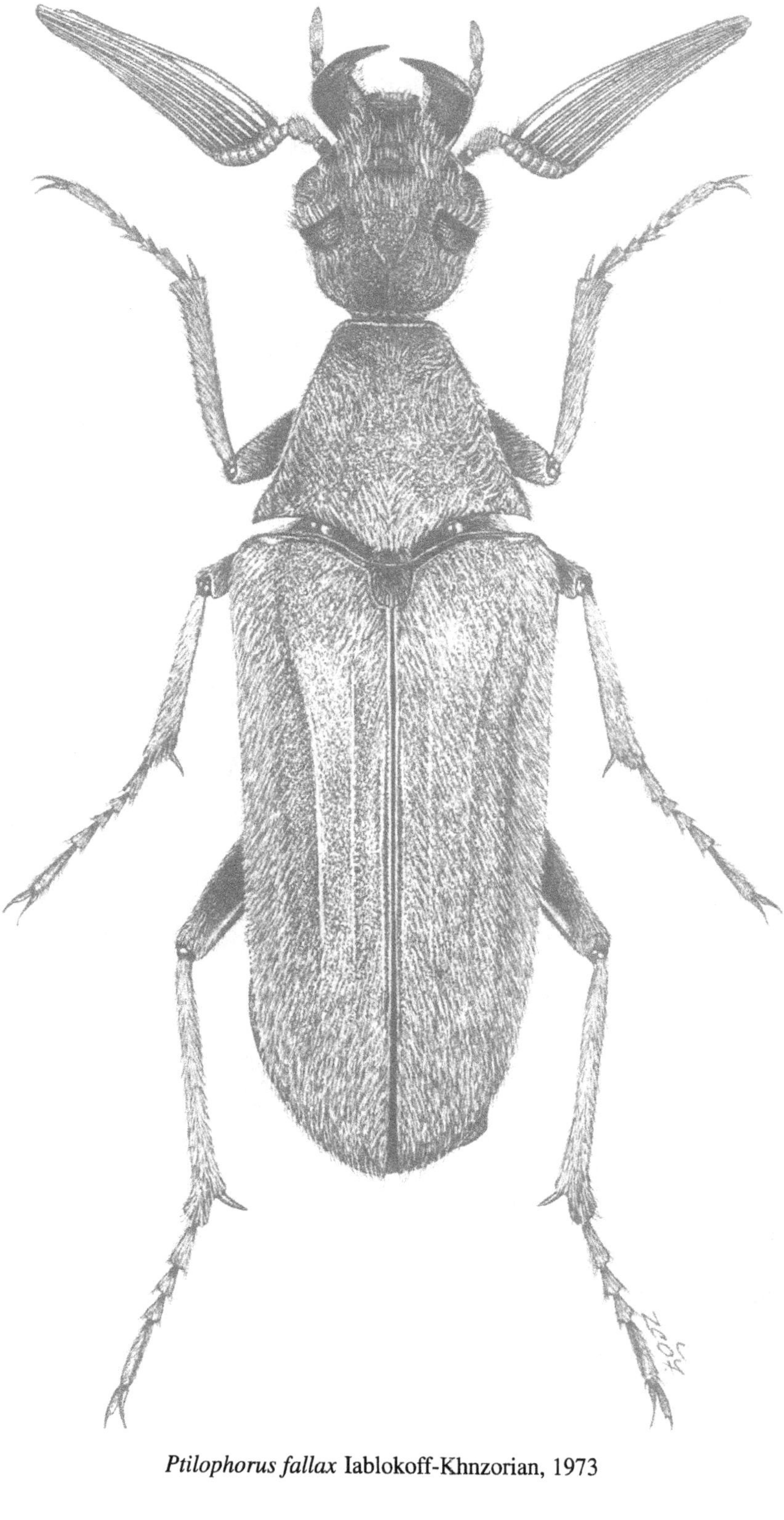

Ptilophorus fallax Iablokoff-Khnzorian, 1973

Catalogue of Palaearctic Coleoptera

Volume 5

Tenebrionoidea

Edited by
I. LÖBL & A. SMETANA

Apollo Books
Stenstrup, 2008

Cover illustration: Jan Batelka.

Edited by: I. Löbl & A. Smetana.

Printed by: Vinderup Bogtrykkeri Aps.

Published by:
Apollo Books
Kirkeby Sand 19
DK-5771 Stenstrup
Denmark
Telephone + 45 62 26 37 37
Telefax + 45 62 26 37 80
E-mail: apollobooks@vip.cybercity.dk
Web site: www.apollobooks.com

Chapters from this publication should be cited as follows:
Nikitsky, N. B.: Mycetophagidae, pp. 51-55. – In I. Löbl & A. Smetana (ed.): *Catalogue of Palaearctic Coleoptera,* Vol. 5. Stenstrup: Apollo Books, 670 pp.

ISBN 978-87-88757-69-9 (Vol. 5)
ISBN 978-87-88757-71-2 (Vols 1-8)

Date of publication: 15.04.2008.

CONTENTS

INTRODUCTION

Structured knowledge requires categorisation of observations and phenomena, and one of these phenomena is life. Structured knowledge of life, as it evolved and is present on our planet, requires cataloguing.

Taxonomy has a profound impact in many fields, including genetics, physiology, ecology, plant protection, nature conservation, and legislation. At the core of taxonomic information are taxonomic catalogues that serve as guides to the diversity of life. They summarize the work in systematics and taxonomy that enable assessment of faunal diversity and classification, and provide information on historical taxonomic actions. Catalogues are also essential in achieving a universal language in taxonomy that assures unambiguous transfer of biological information.

The ever increasing number of taxa described since the tenth edition of Linné's *Systema Naturae* was probably the main motivation for compiling catalogues in the past. Since the middle of the 19th century numerous catalogues were published, usually covering restricted parts of the world and served as checklists rather than works that we would qualify today as full catalogues. While knowledge is dynamic, catalogues are static and reflect a time restricted information. Therefore, there is always a need of newer, up-to-date, and complete catalogues. The first world catalogue of Coleoptera was Gemminger & Harold's *Catalogus Coleopterorum* issued from 1868 to 1874. It was followed by Junk & Schenkling's (editors) *Coleopterorum Catalogus*, published in separate parts from 1910 to 1940. At present, even though it is desperately obsolete, the *Coleopterorum Catalogus* remains the only source of comprehensive taxonomic information on a world basis for many families of beetles. The only catalogue covering all groups of Coleoptera occurring in the Palaearctic region is Winkler's 1924 – 1932 *Catalogus Coleopterorum regionis palaearcticae*. However, the Junk & Schenkling and Winkler catalogues differ. The former one is approaching the definition of a "full" catalogue, while the latter is rather a checklist. They both share two serious omissions. The primary sources of taxonomic acts were not consistently checked, and the information on distribution was not detailed enough to satisfy the needs of biologists. In addition, many rules subsequently adopted by the *International Code of Zoological Nomenclature* were not respected. Due to the lack of modern catalogues, the language in taxonomic literature remains inadequate in many respects. In particular, unavailable names are regularly used as available and sometimes even regarded as valid, various spellings are used for the same taxon, and references to taxonomic actions are inexact to a high degree for papers published during the 20th century.

Undoubtedly, the Coleoptera of the Palaearctic Region were over the years studied by more professional and amateur biologists than those in other parts of the World, and the results may exceed those of the study of many other arthropods. It is therefore remarkable that an overview of the present knowledge of the Palaearctic Coleoptera is badly lagging behind other groups, such as the Nearctic Coleoptera, or Palaearctic Diptera. Systematic studies of many groups of Palaearctic Coleoptera, even in view of their popularity for biologists, have to start with searches for data scattered throughout the literature of the past 250 years.

A full catalogue provides information concerning all published names within a group, their classification, taxonomic history, and so on. Ideally, the present Catalogue should include such data. However, considering the number of taxa and the low number of present day taxonomists, along with difficulties in accessing primary literature, a full catalogue of all Palaearctic beetles would not be completed within a reasonable period of time.

Having this in mind, the aim of this Catalogue is to provide a tool that meets the most urgent needs: 1) a list of available names, both valid and invalid, of taxa occurring in the Palaearctic Region, in their verified orthography and with correct publication dates, 2) a complete list of verified references to primary descriptions, and 3) informative distributions of the species and subspecies. Thus, the Catalogue is expected to respond to questions related to biodiversity and to the stability of the taxonomic nomenclature.

This Catalogue is a collective work involving many authors. For its realisation, it was necessary to achieve a reasonable consensus in very diverse fields. The Editors appreciate the understanding of the need of consistency by the contributing authors. The most controversial points concern the spelling of names, and the delimitation of the Palaearctic Region and its main subdivisions.

The first point is purely formal. The alternatives adopted by the Editors and presented below result from the consensus achieved during the discussions with colleagues. The Editors are aware of the difficulty in satisfying all opinions and appeal to the understanding of the user.

The second point pertains to the presentation of distributional data. The Palaearctic Region, as it is defined for the purpose of this Catalogue, includes some areas that are usually considered to belong to the Afrotropical, Oriental, and Pacific Regions respectively, i.e. the south of the Arabian Peninsula and Suqutra, all of Pakistan, the Himalayan part of India including Arunachal Pradesh, Bhutan, and Nepal, the south of China, and the Pacific Islands of Japan. The main reasons for inclusion of these areas are as follows:

Old records of "Arabia" and "China" may pertain to any states on the Arabian Peninsula and to Jordan, or to any part of the People's Republic of China, respectively.

1. Recent field work in the Himalaya and in the mountains of mainland China and Taiwan provides evidence of altitudinal faunal transition in Coleoptera, as well as in other insect groups. While subtropical climate with dominant Oriental taxa prevails at low elevations, significant faunal changes are found already at elevations of 1500 m, and almost "pure Palaearctic" taxa are present at elevations of 2500 m and above. Thus, it appears illusory to draw simple biogeographical frontiers in any of these parts of the world.
2. The Catalogue includes more information. This is an obvious advantage in the absence of modern catalogues covering the Afrotropical and Oriental regions.

The Catalogue of Palaearctic Coleoptera is published in a number of independent volumes, each having its own Reference section and Index. The taxonomic, geographical and bibliographical information in this volume is presented as in the previous Volumes.

The present volume treats the polyphagous superfamilies Elateroidea, Derodontoidea, Bostrichoidea, Lymexyloidea, Cleroidea, and Cucujoidea. As in the first three volumes, the adopted arrangement of families and subfamilies within the superfamilies is consistent with J. F. Lawrence & A. F. Newton's 1995 classification (in J. Pakaluk & S. A. Ślipiński (eds): *Biology, phylogeny, and classification of Coleoptera. Papers celebrating the 80th birthday of Roy A. Crowson.* Volume 2. Warszawa: Muzeum i Instytut Zoologii PAN). The more important changes concern the classification of the tenebrionines, and the treatement of colydiines as a subfamily of Zopheridae.

The volume includes about 21.290 names of taxa and 4.880 primary references to the genus- and species-groups names.

The Editors adopted the policy that the gap between the publication date of each volume of the Catalogue and the date of the corresponding entry deadline should not exceed three years. While the first Volume contains the available genus- and species-groups names in Archostemata, Myxophaga and Adephaga published before January 1, 2000, the second Volume contains the available names in Staphyliniformia published before January 1, 2002, the third Volume contains the available names in Scarabaeoidea, Dascilloidea, Buprestoidea and Byrrhoidea published before January 1, 2004, the fourth Volume contains the available names in Elateroidea, Derodontoidea, Bostrichoidea, Lymexyloidea, Cleroidea, and Cucujoidea published before January 1, 2005, and the present Volume is extended to contain the available names in Tenebrionoidea published before January 1, 2007.

An index of the species-group names was not provided for practical reasons associated with the printing and bookbinding costs. Therefore, as for the previous volumes, an electronic version of species-group names indexes is available on the web sites of the Apollo Books (www.apollobooks.com) and of the Muséum d'histoire naturelle in Geneva (www.ville-ge.ch/musinfo/mhng).

The Editors were repeatedly asked to make the Catalogue available electronically, either as an online interactive database, or as CD-Rom. After careful consideration they decided to publish the work only in book form for the following reasons: Information in taxonomy, unlike that in most other biological fields, remains useful for extremely long periods of time, extending over centuries. Taxonomic work deals commonly with sources published throughout the 19th and 20th centuries and often even earlier. Consequently, catalogues that summarize taxonomic work are consulted over a very long period of time. The simple, although often ignored, reason for the durability of taxonomic data is in the fact that taxonomy provides the language indispensable for unambiguous conveyance of biological information. The bulk of correctly recorded data in Catalogues remains informative even under a continuous inflow of additional new taxa, and even if the assignments and ranks of taxa and the validity or invalidity of names are changing, and the known distributions of species become gradually more precise. It is therefore necessary to insure access to taxonomic work, including Catalogues, not only in spatial but also in temporal dimensions. Experience shows that printed texts may remain available for centuries, while life expectancy of electronic information is unknown. Ideally, works like this Catalogue, should be available in both print- and electronic formats; however, the real world is not ideal and the production of printed Catalogues is associated with considerable costs. The consequence of providing low-cost online or CD-Rom editions would have a serious negative economic impact on the production of printed version. We believe that under the globally inadequate institutional interest for the needs of taxonomy, the livelihood of those who are willing to accept the commercially hazardous production of printed taxonomic publications should be protected. We advocate the growth and continuation of taxonomy, not its demise.

TAXONOMIC INFORMATION

The present Catalogue includes all available names, both valid and invalid, of extant beetle taxa described before January 1, 2005 and known to occur in the Palaearctic Region, as it is defined below. The higher classification, from suborder down to subfamily, is based on the work of Lawrence & Newton (1995): *Families and subfamilies of Coleoptera* (*with selected genera, notes, references and data on family-group names*). However, the classification proposed in this work is not taken as dogma, and changes are accepted when considered well founded. All taxa below subfamily rank are arranged alphabetically within the higher taxon and the synonyms follow the respective valid name alphabetically.

Extinct taxa, names rejected by the ICZN (*International Code of Zoological Nomenclature, Fourth Edition.* London: International Trust for Zoological Nomenclature, 1999), misspellings, misidentifications and other nomina nuda are not included in the Catalogue. However, concepts that are important for nomenclatural purposes, may be included. Similarly, all infrasubspecific names, such as those established as "morpha", "natio", or "race", "subvariety" and "aberration", or proposed as variety and form of a subspecies or another variety, and names published, e.g., as *A-us b-us c-us* , but specified in the text that they are actually proposed for a "natio" or "race", etc., are not considered subspecific and are therefore excluded from the Catalogue. Names proposed as varieties and forms before 1961 are included, if deemed subspecific under the provisions of the ICZN, Article 45.6. Unjustified emendations may be included.

The currently valid names of the family-group taxa include the name of the author and the year of the publication. Their synonyms are not listed.

The names of the genus-group taxa are given with the name of the author, and the year and page of publication. The page given is the page where the name and the actual description of the taxon is printed. The type species of all genus-group names are given in their original combination. If the type species is currently regarded as a junior synonym, the valid senior synonym is given in brackets in its original combination.

The names of the species-group taxa are given with the name of the author, and the year and page of publication. The page given is the page where the name and the actual description of the taxon is printed. In bi-languaged Chinese/English and Japanese/English publications both respective pages of the actual description may be given. For species-group taxa subsequently transferred to another genus, the name of the original genus is given in parentheses, following the page of publication.

Some authors (e.g., V. Apfelbeck, H. John) published the same description twice, or even more times, in separate papers. Such publications produce, de facto, primary homonyms and objective synonyms. The first publication in such cases is referred to as indicated above, followed by the mark = , the year and first page of the subsequent description/s in square brackets. This is particularly important for taxa that are erroneously associated with their junior description.

The following symbols, all given in square brackets following the page of publication, or the original combination when applicable, are used for taxonomic information : HN for homonyms, RN for replacement names, NO for nomina oblita, NP for nomina protecta, DA for doubtful assignment, and EA for erroneous assignment.

Taxa considered incertae sedis and nomina dubia are listed separately at the end of the nearest applicable taxon.

Taxonomic and nomenclatural acts published after December 31, 2006 are considered only when they concern taxa described on or before that date.

DISTRIBUTIONAL INFORMATION

The limits of the Palaearctic region, as those of other biogeographical regions, are arbitrarily defined (Map 1). For practical reasons, the boundaries of the Palaearctic Region, as they were established for the Catalogue (see above), usually follow national boundaries. The region includes Europe, Africa north of the Sahara, and Asia except for the part that is arbitrarily defined as belonging to the Oriental Region.

For each species and subspecies an outline of its present distribution is given. Fossil records are not considered. The information is given by means of symbols, presented at three levels.

The first level is the subdivision of the Palaearctic Region into three main parts, Europe (letter **E**, bold), North Africa (letter **N**, bold) and Asia (letter **A**, bold).

Europe includes the Azores, Iceland and Turkey west of the Bosporus. The eastern boundaries are a matter of controversy. In the Catalogue, Europe includes Russia west of the main ridge of the Ural mountains, the Permsk Oblast, Bashkortostan Republic and Orenburskaya Oblast, and the small part of Kazakhstan west of the Ural River. It includes the Caucasian republics of Georgia, Armenia and Azerbaijan. The south-eastern boundaries are the political boundaries of the Asian part of Turkey, Iran, Kazakhstan, and the Caspian and Black seas.

Yugoslavia has been recently split into two independent countries: Serbia and Montenegro. At the time of the split almost all distributional records for the respective species were finished, using the symbol YU. It would have been an unreasonably time consuming effort to retroactively differentiate these records, therefore, as a compromise, the term Yugoslavia is no longer used in the list of geographical symbols on page 14, but records for Serbia and Montenegro are still listed under the symbol YU in the body of the Catalogue.

North Africa includes Morocco (incl. Western Sahara), Algeria, Tunisia, Libya and Egypt west of the Suez Canal, and the Canary and Madeira islands.

Asia includes Sinai and the Arabian Peninsula (including Suqutra), Turkey east of the Bosporus, the Middle East and Central Asian countries, Russia east of the main ridge of the Ural mountains, Korea, Japan (including Ryukyu [= Nansei] Islands and the Japanese Pacific Islands), the entire People's Republic of China, Taiwan, Bhutan, Nepal, North India along the base of the Himalaya (Arunachal Pradesh, Uttarakhand (= Uttaranchal), northwestern area of former Uttar Pradesh), Himachal Pradesh), Jammu & Kashmir and all of Pakistan. Thus, India is the only state for which the strict political boundaries are not respected. Large parts of Uttar Pradesh south of Nepal are overpopulated plains. Information on Coleoptera from this North Indian state is based almost exclusively on its Himalayan districts lying west of Nepal, which are in the present state Uttaranchal.

The second level of the geographic information is provided by two-letter symbols for countries, major areas of Russia and North Indian states, and by three-letter symbols for provinces of mainland China and for Taiwan (see Table 1, Map 3).

The symbols are arranged in alphabetical order within the first-level subdivision. Some, usually older, distributional records cannot be accommodated within the structure of the Catalogue (e.g., Arabia, Caucasus, North India, Siberia). Such information is given in quotation marks (e.g., "Caucasus") behind the last symbol of the respective first-level symbol. Russia is subdivided into six major sub-regions (Map 2), each of which has its own two-letter symbol. These, as well as the symbol RU for Russia, are used only when more detailed information is not available. Similarly, the symbol CH for the People's Republic of China, as well as the seven two-letter symbols for China's major regions, are used only in the absence of more detailed geographical information.

In general, the published distributional information is based on both identified material in collections and on published records, scattered in an enormous number of taxonomic and faunistic papers that are virtually impossible to review in their entirety. Revisions of collections reveal a high proportion of misidentifications, attaining 40% of specimens in some large museums. The degree of identification reliability and of the records derived from identifications, is a function of faunal diversity and quality of systematic revisions. Thus, the *reliability in general* increases from south to north and from poorly studied groups to "popular" groups. At present, a number of modern catalogues or check-lists, covering the beetle faunas of many European and some extra-European countries or archipelagos, are available. The use of data contained in these and other faunistic works is left to the discretion of the authors, who may also add unpublished information available to them. The second-level geographic information is not necessarily exhaustive, it should rather be considered as a base for future faunal research.

The third-level geographic information concerns species and subspecies with restricted distribution. Taxa of this category may be strict endemites, or taxa comparatively widely distributed in one area but restricted in another area. For example, the distributional record of a species widely distributed in North Africa with isolated occurrence on Pantelleria would appear as follows: **E**: IT (Pantelleria) **N**: AG MO TU. The third-level information is facultative. It is given in parentheses after the respective second-level symbol. The official language of the respective state is used for records in languages using the Latin alphabet, or it is transliterated from the Cyrillic alphabet. Records in languages using non-Latin or Cyrillic characters (e.g. Chinese or Japanese pictographs) are translated into English, and the translated geographical terms are spelled as closely as possible to those used in the Times Atlas®, or in other well-known sources. Detailed geographical information may refer to natural geographical features such as islands, mountains, lakes, valleys, caves, or to administrative entities, such as districts.

The extralimital distribution of some Palaearctic species is indicated by three letter symbols in bold, cosmopolitan species by the symbol **COS**, all located at the end of the respective geographical information (see Table 1). Introductions are indicated by the letter "i" (e.g., **E**i: GB). The extralimital regions for the needs of the Catalogue are defined as follows (see Map 1): Nearctic (**NAR**): north of Mexico; Neotropical (**NTR**): south of the United States; Afrotropical (**AFR**): south of the North African states included in the Palaearctic Region; Oriental (**ORR**): areas south of the People's Republic of China and Taiwan, areas south of the Himalaya in India, the Philippines, Malaysia and Indonesia south to the Lydekker line; Australian (**AUS**): south of the Lydekker line, Pacific.

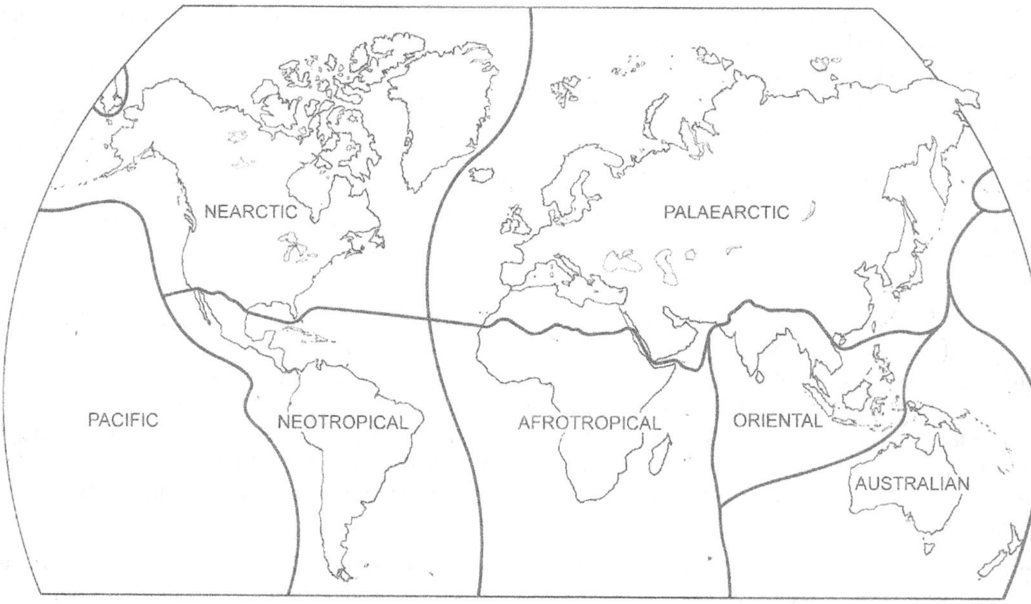

Map 1. The limits of the geographic Regions as defined for the purpose of this Catalogue

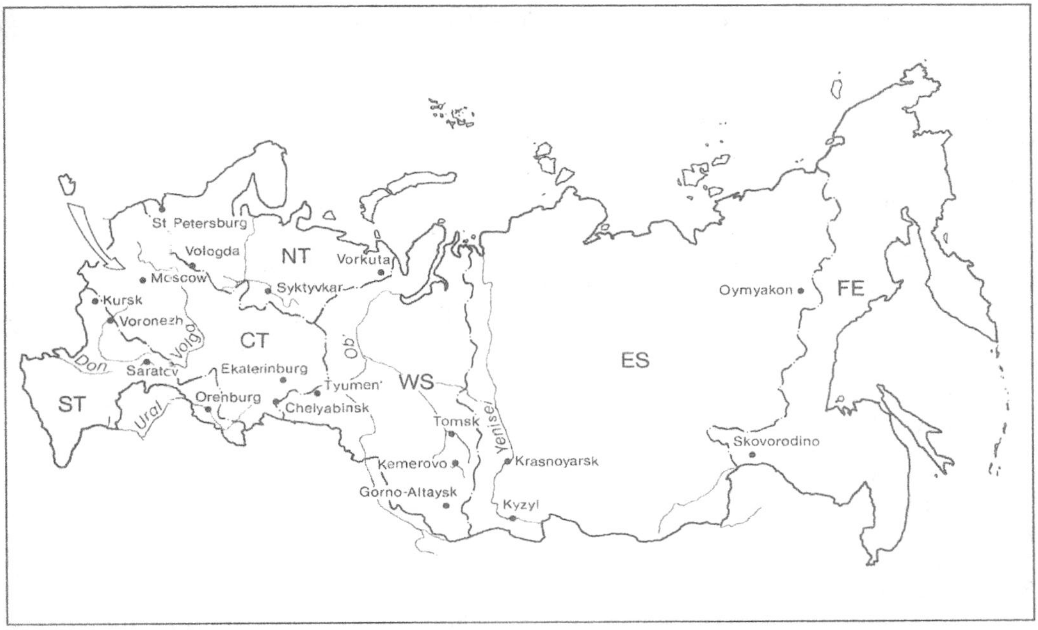

Map 2. Subdivisions of Russia

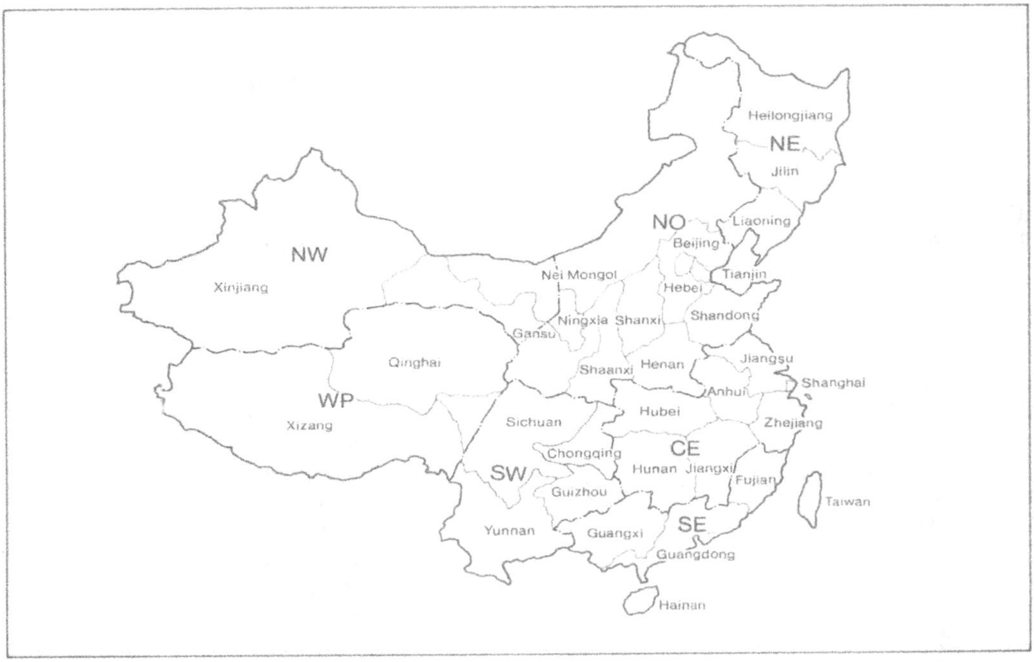

Map 3. Subdivisions and provinces of the People's Republic of China

Table 1: GEOGRAPHICAL SYMBOLS

E	**Europe**	KZ	Kazakhstan
AB	Azerbaijan	LA	Latvia
AL	Albania	LS	Liechtenstein
AN	Andorra	LT	Lithuania
AR	Armenia	LU	Luxembourg
AU	Austria	MA	Malta
AZ	Azores	MC	Macedonia
BE	Belgium	MD	Moldavia
BH	Bosnia Herzegovina	ME	Montenegro
BU	Bulgaria	NL	The Netherlands
BY	Belarus	NR	Norway
CR	Croatia	NT	Russia: North European Territory
CT	Russia: Central European Territory	PL	Poland
CZ	Czech Republic	PT	Portugal
DE	Denmark	RO	Romania
EN	Estonia	RU	Russia
FA	Faeroe Islands	SB	Serbia
FI	Finland	SK	Slovakia
FR	France (incl. Corsica, Monaco)	SL	Slovenia
GB	Great Britain (incl. Channel Islands)	SP	Spain (incl. Gibraltar)
GE	Germany	SR	Svalbard (Spitzbergen)
GG	Georgia	ST	Russia: South European Territory
GR	Greece (incl. Crete)	SV	Sweden
HU	Hungary	SZ	Switzerland
IC	Iceland	TR	Turkey
IR	Ireland	UK	Ukraine
IT	Italy (incl. Sardinia, Sicily, San Marino)	YU	Serbia and Montenegro

N	**North Africa**	LB	Libya
AG	Algeria	MO	Morocco (incl. Western Sahara)
CI	Canary Islands	MR	Madeira Archipelago
EG	Egypt	TU	Tunisia

A	**Asia**	IS	Israel
AE	Arab Emirates	JA	Japan
AF	Afghanistan	JO	Jordan
AP	India: Arunachal Pradesh	KA	India: Kashmir
BA	Bahrain	KI	Kyrgyzstan
BT	Bhutan	KU	Kuwait
CE	China: Central Territory	KZ	Kazakhstan
CH	China	LE	Lebanon
CY	Cyprus	MG	Mongolia
ES	Russia: East Siberia	NE	China: Northeast Territory
FE	Russia: Far East	NC	North Korea
HP	India: Himachal Pradesh	NO	China: Northern Territory
IN	Iran	NP	Nepal
IQ	Iraq	NW	China: Northwest Territory

OM Oman
PA Pakistan
QA Qatar (incl. United Arab Emirates)
RU Russia
SA Saudi Arabia
SC South Korea
SD India: Sikkim, Darjeeling District
SE China: Southeastern Territory (incl. Macao, Hongkong)
SI Egypt: Sinai
SW China: Southwestern Territory
SY Syria
TD Tajikistan
TM Turkmenistan
TR Turkey
UP India: Uttarakhand (= Uttaranchal), Uttar Pradesh
UZ Uzbekistan
WP China: Western Plateau
WS Russia: west Siberia
YE Yemen (incl. Suqutra)

CHINA: PROVINCES, AUTONOMOUS REGIONS OR MUNICIPALITIES, AND TAIWAN

ANH Anhui (Anhwei)
BEI Beijing (Peking or Peiping)
CHQ Chongqing
FUJ Fujian (Fukien)
GAN Gansu (Kansu)
GUA Guandong (Kwantung)
GUI Guizhou (Kweichow)
GUX Guangxi (Kwangsi)
HAI Hainan
HEB Hebei (Hopeh)
HEI Heilongjiang (Heilungkiang)
HEN Henan (Honana)
HKG Hongkong
HUB Hubei (Hupeh)
HUN Hunan
JIA Jiangsu (Kiangsu)
JIL Jilin (Kirin)
JIX Jiangxi (Kiangsi)
LIA Liaoning
MAC Macao
NIN Ningxia (Ningsia)
NMO Nei Mongol (Inner Mongolia)
QIN Qinghai (Tsinghai)
SCH Sichuan (Szechwan)
SHA Shaanxi (Shensi)
SHG Shanghai
SHN Shandong (Shantung)
SHX Shanxi (Shansi)
TAI Taiwan (Formosa)
TIA Tianjin (Tsiensin)
XIN Xinjiang (Sinkiang)
XIZ Xizang (Tibet)
YUN Yunnan
ZHE Zhejiang (Chekiang)

WORLD ZOOGEOGRAPHICAL REGIONS

AFR Afrotropical Region
AUR Australian Region
NAR Nearctic Region
NTR Neotropical Region
ORR Oriental Region

BIBLIOGRAPHIC INFORMATION

References are given to, and only to, primary sources of the genus-group and species-group names included in the Catalogue. No bibliographical reference is given to family-group names, to designation of type-species, to secondary taxonomic acts, or to sources of any other information. Names of all authors, if applicable, are given for each reference. An effort was put forth to unify the spellings of the names. For names that appeared in publications in two or more different orthographies, the one used in the Catalogue is that which was considered to be correct after thorough verification (e.g., the spelling Josef Müller for the author who also published as Giuseppe Müller). In such cases the alternative spelling follows the original one in brackets. The original name is used for authors that subsequently adopted additional names, or honorary epithets (e.g., A. Semenov, and not A. Semenov-Tian-Shanskyi). Initials are given for all authors, even if they are missing in the original publication. Names published in two or more alternative spellings, that seem to be equally correct, are spelled following Schmitt, Hübner & Gaedike (1998): *Nomina Auctorum. Auflösung von Abkürzungen taxonomischer Autoren-Namen. Nova Supplementa Entomologica* **11**: 3-189. An exception is made for F. L. Castelnau de Laporte, who is referred to as Laporte in recent major works. Therefore, the name Laporte is arbitrarily used in the Catalogue. Names in Cyrillic are transliterated as for any other Cyrillic text, unless an alternative spelling was preferred by the author, or the alternative name is generally used in the literature (e.g., Jakobson, Tschitschérine). It should be noted that lowercase prepositions such as French "de" and German "von" are not given with the author's name in the list of the taxa.

Titles of references are given in full, i.e. not abbreviated, with subtitles, as originally published. Text written in Cyrillic is transliterated. Titles in languages using non-Latin or non-Cyrillic characters are translated into English and are presented in square brackets, followed by the indication of the original language in parentheses.

Titles of periodicals are given in full. Series numbers are given in parentheses, and always precede the boldface volume number. The numbers of separate issues are only provided when they are paginated separately, in which case they are in parentheses following the volume number. Unnumbered pages are indicated in square brackets. The titles of some periodicals vary over the years. For these periodicals the titles are given as they appeared for the respective items referred to (e.g., *Coleopterologische Rundschau* / *Koleopterologische Rundschau*). Contrarily, the journal *Entomologische Blätter* had several subtitles during its existence, the last of them being *Entomologische Blätter für Biologie und Systematik der Käfer*. Since there is no danger of confusion with any other periodical, the journal is cited simply as *Entomologische Blätter*. If the year of the volume differs from the actual year of publication, the former follows the volume number and is given is square brackets. If periodicals have no volume number, the year of the volume is given.

The *Bulletin de la Société Entomologique de France* is one of the more important periodicals in the field of systematics. The title changed almost continuously for more than 60 years, appearing usually as *Bulletin des séances*, *Bulletin trimestriel* or *Bulletin entomologique*. In addition, the first numbered volume, 37, was published 100 years after the creation of the *Annales* and the *Bulletin* of the French Entomological Society in 1832. To avoid confusion, this periodical is referred to as *Bulletin de la Société Entomologique de France,* and, from 1832 to 1931, is identified by the year in which it was published. From the year 1932, the volume number is given.

The *Bulletin de la Société Entomologique de France* was published from 1873 to 1894, in addition to the well known quarterly version, in a bi-mensuel version "*Bulletin Bimensuel*". The pagination of the two versions are not identical, the publication of the bi-mensuel version preceded the quarterly one. The *Bulletin Bimensuel* is ignored by almost all coleopterists, and is not respected in the Catalogue.

For ease of access, references of the same author and year were placed in order based on the first page number of the paper, with Roman numerals being placed last, with respect to the number of the respective issue. Books were placed first for a given year, unless a page range was cited, in which case the first-page rule was followed. Papers of the same year and with the same first page numbers were placed in order of their last page number. Thus the references are not arranged chronologically within each year.

ACKNOWLEDGEMENTS

As for the previous volumes of the Catalogue, the assistance of libraries, and of colleagues who had access to such libraries, was one of the prerequisites for realisation of the present work. The most important libraries to be mentioned are: Muséum d'histoire naturelle, Genève, Switzerland (Mrs. Christele Mougin); Agriculture and Agri-Food Canada (Entomological Library) Ottawa, Canada (Mrs Patricia Madaire and Mr. Steve Gamman); Deutsches Entomologisches Institut, Müncheberg, Germany (Mrs Jutta Moebert, Monika Schiele, and Helga Wehrens, Mr. Reinhard Gaedike and Lothar Zerche), Zoologisches Museum, Berlin, Germany (Mr. Johannes Frisch), Naturhistorisches Museum, Basel, Switzerland (Daniel Burckhardt, Michel Brancucci), and Museum of Comparative Zoology, Harvard University, Cambridge, USA (Mrs Dana Fischer).

Special thanks are due to H. J. Bremer, Melle, Germany, for information on the tenebrionid genus *Amarygmus* and related taxa. Julio Ferrer, Stockholm, Sweden, provided a number of unpublished data on Tenebrionidae. David G. H. Halstead, Old Windsor, Great Britain, reviewed the tenebrionid genus *Tribolium*, *Palorus*, and allied taxa. Zbyněk Kejval, Domažlice, Czech Republic, reviewed the anthicid genera *Anthelephila* and *Stenidius*. Harold Labrique, Lyon, France, completed data on the tenebrionid genus *Scaurus*. Ottó Merkl, Budapest, Hungary, verified a large set of distributional data from Hungary and some other East European countries and assisted with references. Nikolay B. Nikitsky, Moscow, Russia, verified numerous distributional data for various Russian and Central Asian taxa. Lee H. Herman, New York, USA, and Mikael Sörensson, Lund, Sweden, continuously assisted with rare publications. Volker Puthz, Schlitz, Germany reviewed parts of the reference file.

In addition, the Editors and authors of the present volume are indebted to many individuals who commented on different parts of the Catalogue, added distributional information, completed references, and helped in various ways. These include also coauthors, and are as follows: Rolf L. Aalbu, San Francisco, USA; Roland Allemand, Lyon, France, J. Maxwell L. Barclay, London, Great Britain; Nicole Berti, Paris, France; Michel Brancucci, Basel, Switzerland; Martin J. D. Brendell, London, Great Britain; Ben Brugge, Amsterdam, The Netherlands; Hervé Brustel, Toulouse, France; Pavel Chvojka, Praha, Czech Republic; Giulio Cuccodoro, Genève, Switzerland; Mikhail L. Danilevsky, Moscow, Russia; Zachary Falin, Lawrence, USA; Claude Girard, Paris, France; Jocelyne Guglielmi, Paris, France; Matthias Hartmann, Erfurt, Germany; Jiří Háva, Praha, Czech Republic; Bernd Jaeger, Berlin, Germany; Josef Jelínek, Praha, Czech Republic; David Král, Praha Czech Republic; Vítězslav Kubáň, Brno, Czech Republic; Bernard Landry, Genève, Switzerland; John Lawrence, Gympie, Australia; Piero Leo, Italy; R. A. B. Leschen, Auckland, New Zealand; J. I. López-Colón, Madrid, Spain; late Karel Majer, Brno, Czech Republic; Noël Mal, Bruxelles, Belgium; Gleb S. Medvedev, St. Petersburg, Russia; Koji Mizota, Sapporo, Japan; Yuichi Okushima, Kurashiki, Japan; Michel Perreau, Paris, France; Roberto Poggi, Genova, Italy; Guo-dong Ren, Baoding, China; Jan Růžička, Praha, Czech Republic; Heinrich Schönmann, Wien, Austria; Rudolf Schuh, Wien, Austria; Sharon Shute, London, Great Britain; Hans Silfverberg, Helsinki, Finnland; Alexey Yu. Solodovnikov, Copenhagen, Denmark; Pavel Stoev, Sofia, Bulgaria; Petr Švácha, České Budějovice, Czech Republic; Sergei Tshernyshev, Novosibirsk, Russia; Azadeh Taghavian, Paris, France; Masatoshi Takakuwa, Iryuda, Japan; Manfred Uhlig, Berlin, Germany; Vicky Venes, London, Great Britain; Mark Volkovitsh, St. Petersburg, Russia; Petr Zahradník, Praha, Czech Republic.

The following colleagues identified errors that appeared in the previous volumes of the Catalogue: Miguel A. Alonso-Zarazaga, Madrid, Spain; Jan Batelka, Praha, Czech Republic; Yves Bousquet, Ottawa, Canada; Michel Brancucci, Basel, Switzerland; Peter Cate, Wien, Austria; Jon Cooter, Great Britain; Michael Geisthardt, Wiesbaden, Germany; Michael Geiser, Basel, Switzerland; Matthias Hartmann, Erfurt, Germany; Colin Johnson, Manchester, Great Britain; Sergei Kazantsev, Moscow, Russia; Denis Keith, Chartres, France; Vítězslav Kubáň, Praha, Czech Republic; Lucien Leseigneur, Meylan, France; Jiří Háva, Praha, Czech Republic; Chi-Feng Lee, Taipei, Taiwan; Jyrki Muona, Helsinki, Finland; Nikolai B. Nikitsky, Moscow, Russia; Giuseppe Platia, Gatteo, Italy; Paul F. Whitehead, Pershore, Great Britain.

The realisation of the Oedemeridae section was supported by the grant MK 00002327201 of the Ministry of Culture of the Czech Republic (to V. Švihla, National History Museum Kunratice, Praha).

The photography of Gustav A. Lohse to whom the present volume is dedicated was among a set kindly sent to the Editors by his son U. M. Lohse, Landkirchen, Germany, K. Hengmith, Hamburg, Germany, and H.

Meybohm, Stelle, Germany. The frontispiece illustration was made by Zuzana Čadová and provided by Jan Batelka, both from Praha, Czech Republic.

Mr. Jan Batelka, Praha, Czech Republic, thanks his wife Hana for support, and Ivan Löbl is grateful to his wife Daniela for her endless patience.

Mr. Daniel Löbl, Genève, Switzerland, developed the software tools needed to generate the Indexes, and was helpful in solving a number of computer problems.

THE EDITORS

Ivan Löbl
Muséum d'histoire naturelle

Genève, Switzerland

Aleš Smetana
Agriculture and
Agri-Food Canada
Ottawa, Canada

COORDINATOR OF THIS VOLUME

Ivan Löbl, Muséum d'histoire naturelle, Genève, Switzerland

AUTHORS

family MYCETOPHAGIDAE Leach, 1815	N. B. NIKITSKY
family ARCHEOCRYPTICIDAE Kaszab, 1964	O. MERKL
family PTEROGENIIDAE Crowson, 1953	I. LÖBL
family CIIDAE Leach, 1819	J. JELÍNEK
family TETRATOMIDAE Billberg, 1820	N. B. NIKITSKY
family MELANDRYIDAE Leach, 1815	N. B. NIKITSKY, D. A. POLLOCK
family MORDELLIDAE Latreille, 1802	J. HORÁK
family RIPIPHORIDAE Gemminger & Harold, 1870	J. BATELKA
family ZOPHERIDAE Solier, 1834	S. A. ŚLIPIŃSKI, R. SCHUH
family TENEBRIONIDAE Latreille, 1802	I. LÖBL, O. MERKL, K. ANDO, P. BOUCHARD, M. LILLIG, K. MASOMUTO, W. SCHAWALLER
tribe Lagriini Latreille, 1825	O. MERKL
tribe Asidini Solier, 1834	F. SOLDATI
tribe Blaptini Leach, 1815	I. LÖBL, M. NABOZHENKO, O. MERKL
tribe Melanimini Seidlitz, 1894	D. IWAN, I. LÖBL
tribe Opatrini Brullé, 1832	D. IWAN, I. LÖBL
tribe Pedinini Eschscholtz, 1829	D. IWAN, I. LÖBL
tribe Platyscelidini Lacordaire, 1859	L. V. EGOROV
tribe Helopini Latreille, 1802	M. NABOZHENKO, I. LÖBL
subfamily Alleculinae Laporte, 1840	V. NOVÁK, R. PETTERSSON
family PROSTOMIDAE C. G. Thomson, 1859	W. SCHAWALLER
family SYNCHROIDAE Lacordaire, 1859	I. LÖBL
family STENOTRACHELIDAE Latreille, 1810	I. LÖBL
family MELOIDAE Gyllenhal, 1810	M. A. BOLOGNA
family MYCTERIDAE LeConte, 1862	I. LÖBL
family BORIDAE C. G. Thomson, 1859	D. A. POLLOCK
family TRICTENOTOMIDAE Blanchard, 1845	D. A. POLLOCK
family PYTHIDAE Solier, 1834	D. A. POLLOCK
family PYROCHROIDAE Latreille, 1807	D. A. POLLOCK, D. K. YOUNG
family SALPINGIDAE Leach, 1915	D. A. POLLOCK, I. LÖBL
family ANTHICIDAE Latreille, 1819	D. S. CHANDLER, G. UHMANN, G. NARDI, D. TELNOV
family ADERIDAE Winkler, 1927	G. NARDI
family SCRAPTIIDAE Mulsant, 1856	P. LEBLANC, B. LEVEY, J. HORÁK

MAILING AND E-MAIL ADDRESSES OF THE AUTHORS

ANDO Kiyoshi 1, Hezuo Road, Baoding, Hebei, China; Post Code: 071002 and 100, Guilin Road, Shanghai, 200234, China. E-mail: tenebrio@entom.biz-web.jp
Private Laboratory: Kofu-dai 5-3-5, Toyono-cho, Toyono-Gun, Osaka, 563-0104 Japan (BTel&Fax. +81-72-738-1012).

BATELKA Jan, Nad vodovodem 16, 100 00 Praha 10, Czech Republic. E-mail: janbat@centrum.cz

BOLOGNA A. Marco, Università Roma Tre, Viale Marconi 446, 00146 Roma, Italy. E-mail: bologna@uniroma3.it
BOUCHARD Patrice, Canadian National Collection of Insects, Agriculture and Agri-Food Canada, 960 Carling Avenue, Ottawa, ON K1A 0C6 Canada. E-mail: bouchardpb@agr.gc.ca
CHANDLER Donald S., Department of Zoology, University of New Hampshire, Durham, NH 03824, USA. E-mail: dsc1@cisunix.unh.edu
EGOROV V. Leonid, 15, 525 Kadykova st., Cheboksary, Chuvashia, 428037 Russia. E-mail: platyscelis@rambler.ru
HORÁK Jan, K hádku 567, Dubeček, 107 00 Praha 10-Dubec, Czech Republic. E-mail: horak.mordellidae@post.cz
IWAN Dariusz, Museum and Institute of Zoology, Polish Academy of Sciences, Wilcza 64, 00-679 Warszawa, Poland. E-mail: darek@miiz.waw.pl
LEBLANC Pascal, Muséum d'Histoire Naturelle 1, rue Chrestien de Troyes, F-10000 Troyes, France. E-mail: p.leblanc@ville-troyes.fr
LEVEY Brian, National Museum of Wales, Cathay's Park, CF10 3NP Cardiff, Great Britain. E-mail:Brian.Levey@museumwales.ac.uk
LILLIG Martin, Krämersweg 55, D-66123 Saarbrücken, Germany. E-mail: martin.lillig@t-online.de
LÖBL Ivan, Muséum d'histoire naturelle, Case postale 6434, CH-1211 Genève, Switzerland. E-mail: ivan.lobl@bluewin.ch
MASOMUTO Kimio, Institute of Human Living Sciences, Otsuma Women's University, Tokyo 102-8357, Japan. E-mail: masumoto@otsuma.ac.jp
MERKL Otto, Hungarian Natural history Museum, Depart. of Zoology, Barros u. 13, H-1088 Budapest, Hungary. E-mail: merkl@zoo.zoo.nhmus.hu
NABOZHENKO Maxim, Southern Scientific Centre, Russian Academy of Sciences, 344006 Tchekhov str. 41, Rostov-on-Don, Russia. E-mail: nalassus@mail.ru
NARDI Gianluca, Centro Nazionale per lo Studio e la Conservazione della Biodiversità Forestale - Corpo Forestale dello Stato Strada Mantova 29, I-46045 Marmirolo (Mantova), Italy. E-Mail: notoxus@katamail.com
NIKITSKY Nikolay B., Zoological Museum of Moscow Lomonosov State University, Bolshaya Nikitskaya 6, Moscow 125009, Russia. E-mail: nikitsky_nb@mtu-net.ru
NOVÁK Vladimir, District Museum Praha-Východ, Masarykovo nám. 97, Brandýs n. Labem, Czech Republic. E-mail: novak@ompv.cz
PETTERSSON Roger B., Department of Animal Ecology, Swedish University of Agricultural Sciences, S-901 83 Umeå, Sweden. E-mail: roger.pettersson@szooek.slu.se
POLLOCK Darren A., Department of Biology Station #33, Eastern New Mexico University, Portales, NM 88130, USA. E-mail: darren.pollock@enmu.edu
SCHAWALLER Wolfgang, Staatliches Museum für Naturkunde, Rosenstein 1, D-70191 Stuttgart, Germany. E-mail: schawaller.smns@naturkundemuseum-bw.de
SCHUH Rudolf, Raugasse 28A/2/18, A-2700 Wiener Neustadt, Austria. E-mail: rudolf.schuh@at
ŚLIPIŃSKI S. Adam, Division of entomology, CSIRO, GOP Box 1700, Canberra, ACT 2601, Australia. E-mail: Adam.Slipinski@csiro.au
SOLDATI Fabien, France; Muséum d'Histoire Naturelle, 1, rue Chrestien de Troyes, F-10000 Troyes, France. E-mail: asida.soldati@wanadoo.fr
TELNOV Dmitry, Praulienas iela 4-35, LV-1021 Riga, Latvia. E-mail: telnovs@finieris.lv
†UHMANN Gerhard, Tannenhofstrasse 10, D-92690 Pressath, Germany.
YOUNG K. Daniel, UW Insect Research Collection, 445 Russell Laboratories, Department of Entomology, 1630 Linden Drive, University of Wisconsin, Madison, WI 53706, USA. E-mail: young@entomology.wisc.edu

ERRATA for VOLUME 1

p. 26, add: AB to distribution of *Sphaerius acaroides* Waltl, 1838
p. 27, add: *guernei* Méquignon, 1942: 9 in synonymy with *Gyrinus distinctus* Aubé, 1838
p. 35, add: SZ to distribution of *Hygrobia hermanni* Fabricius, 1775
p. 700, correct year for Linnaeus C.: *Fauna Suecica* ... to 1760
p. 715, add: Méquignon A. 1942: Voyage de MM. L. Chopard et A. Méquignon aux Açores (août-septembre 1930). XIII. Diagnoses de coléoptères nouveaux. *Bulletin de la Société Entomologique de France* **47**: 9-12.

ERRATA for VOLUME 2

p. 65, add: nomen dubium *Cercyon elegans* Betta, 1847: 25 **E**: IT
p. 130, add: **A**: FE to *Acrotrichis sjobergi*
p. 144, correct reference for *Choleva uhagoni* to 1922a: 88 and use this name as valid for the subspecies of *Choleva fagniezi*, *brevistylis* Jeannel
p, 144, place *brevistylis* Jeannel as invalid in synonymy with *Choleva fagniezi uhagoni* Jeannel
p. 145, add: (*Catops*) to *Nargus brunneus* Sturm
p. 159, change agreement in gender in *Anillochlamys* to feminine (to *subtruncata* Jeannel, and *tropica* Abeille de Perrin)
p. 163, correct year for *Adelops triangulum* Sharp to 1873c
p. 167, change agreement in gender in *Notidocharis* to feminine (to *ovoidea* Jeannel)
p. 173, correct year for *Speonomidius crotchi crotchi* Sharp to 1873c
p. 173, correct spelling of *oberthueri* to *oberthuri* (subspecies of *Speonomidius crotchi* Sharp)
p. 192, add: nomen dubium *Agathidium festivum* Betta, 1847: 33 **E**: IT
p. 209, add to *Euconnus* (*Tetramelus*)
arabiae Alonso-Zarazaga, 2004: 127 [RN] **E**: YU
apfelbecki Franz, 1968a: 531 [HN]
cetinjensis Davies, 2004: 26 [RN]
p. 227, add to nomina dubia: *Scydmaenus ullrichi* Betta, 1847: 24 **E**: IT
p. 299, correct year for *Fagniezia impressa* (Panzer) to 1803
p. 304, correct year for *Bryaxis clavicornis* (Panzer) to 1805
p. 329, correct year for *Tyrus mucronatus* (Panzer) to 1803
p. 526, add to nomina dubia: *Bledius marginalis* Betta, 1847: 17 **E**: IT and *Bledius rufus* Betta, 1847: 18 **E**: IT
p. 701, add: Alonso-Zarazaga M. A. 2004: Nombre de reemplazo en Euconnus Thomson, 1859 (Coleoptera, Scydmaenidae). *Graellsia* **60**: 127.
p. 717, add: Betta V. 1847: *De quibusdam coleopteris agri ticinensis*. Ticini Regii, 35 pp.
p. 761, correct year for Gebler F. A. von. 1830: Bemerkungen ... to 1829
p. 888, delete reference to Sharp D. S. 1872b (is given below as Sharp D. S. 1873c)

ERRATA for VOLUME 3

p. 81, add GB to distribution of *Trox scaber* (Linnaeus)
p. 146, add GB to distribution of *Psammodius nocturnus* Reitter
p. 223, correct spelling in *Onychosophrops* to: *brevisetosus* and *holosetosus*
p. 249, add: *Adoretus petrovitzi* Baraud, 1975c: 192 **A**: IN
p. 251, correct spelling in *Phaeadoretus shirazensis* to *P. shirasensis*, and correct distribution to IN (from AF)
p. 387, change agreement in gender in *Phaenops* to feminine
p. 387, add *azurea* A. G. Olivier, 1790: Errata [3] (*Buprestis*) in synonymy of *Phaenops cyaneus* Fabricius, 1775
p. 397, correct page of description of *Phaenops cyanea* Fabricius, 1775 to 223
p. 403, add under nomina dubia: *Agrilus comolli* Betta, 1847: 19 **E**: IT and *Agrilus auricollis* Betta, 1847: 20 **E**: IT
p. 408, transfer genus *Amorphosoma* with the included species under Amorphosomatina, p. 407
p. 427, correct author of *stoicus* (in synonymy with *Byrrhus fasciatus* Forster, 1771) to O. Fabricus, 1780
p. 475, add: Betta V. 1847: *De quibusdam coleopteris agri ticinensis*. Ticini Regii, 35 pp.

p. 489, correct year for Chevrolat L. A. A. 1838a to 1840 [the necrologies of Gyllenhal and J.-F. Desjardins given on p. 351 of the *Revue Entomologique* vol. 5 indicate, that the volume was published after April 1840].
p. 504, correct author of *Fauna Groenlandica* to O. Fabricius
p. 517, correct year for Gebler F. A. von. 1830: Bemerkungen ... to 1829
p. 565, correct year for Linnaeus C.: *Fauna Suecica* ... to 1760

ERRATA for VOLUME 4

p. 5, correct spelling Cerophytidae, from Cerophatidae
p. 20, add P. Audisio as the second author of Kateretidae
p. 33, correct spelling of author of Adrastini to Candèze
p. 33, correct name *Dicronychus nigromaculatus* and *Melanoxanthus nigromaculatus* to *nigropunctatus*
p. 38, add: Mardjanian, 1983 to *Cardiophorus hiekei*
p. 46, correct spelling of *Benibotarsus* to *Benototarsus*
p. 46, add to Comments under Elateridae:
Neopristilophus gougeleti Fairmaire, 1859 is a synonym of *N. insitivus* Germar, 1824 according to G. Platia (pers. comm.), who examined external characters and genitalia of material of both nominal taxa

p. 57, add year 1790 to *Attagenus unicolor* (Brahm)
p. 65, *Silpha evanescens* Marsham, 1802 is doubtful synonym [DA] of *Ptenidium pusillum* (Gyllenhal, 1808), in Vol. 2, p. 123
p. 81, replace JA by TAI for *Eurypogon granulatus* Sakai
p. 82, corrdct spelling of author of *Calyptocerus* to Guérin-Méneville
p. 82, add GB to *Farsus dubius* Piller & Mitterpacher
p. 86, correct spelling Macraulacinae, from Macrulacinae
p. 86, add tribe Macraulacini Fleutiaux, 1922 and tranfer to it genus *Dromaeolus*, from Euryptychini
p. 87 and 88, correct spelling of author of *Aulonothroscus incrassatus* and *A. intactus* to Peyerimhoff
p. 88, add p. 11 to *Aulonothoscus nikitskyi* Egorov, 1992
p. 88, correct spelling of author of *Aulonothoscus laticollis* to Rybiński
p. 88, correct page of description of *Thoscux hauseri* Cobos, 1963 (in synonymy with *Trixagus atticus* Reittter) to 53
p. 88, correct spelling of *Trixagus lesegneuri* Muona to *leseigneuri* and add distribution: BU CZ FI FR GE PL PT RO SK SP SV ST (Caucasus) SZ
p. 89, p. 88, correct page of description of *Trixagus perversus* Cobos, 1963 to 51
p. 92, add p. 44 to *Cebrio melanocephalus* (Leach, 1824)
p. 100, correct year for *Elater punctatus* Herbst, 1784 to 1779 (type species of *Lacon*)
p. 103, replace Conoderini Fleutiaux, 1919 by Oophorini Gistel, 1856; Conoderini Fleutiaux, 1919 is a junior homonym of Conoderini Schoenherr, 1833 (Curculionidae)
p. 105, add [HN] to *breviusculus* Desbrochers des Loges (in synonymy with *Drasterius bimaculatus*)
p. 107, correct spelling of author of *coomani* (in synonymy with *Cryptalaus sculptus*) to Fleutiaux
p. 111, add XIZ to *cylindriformis* Ôhira & Becker
p. 111, add **ORR** to *unicus* Platia & Gudenzi
p. 112, add **ORR** to *incognitus* Platia & Schimmel
p. 112, add XIZ to *nepalensis* Ôhira & Becker
p. 113, add SD to *umbilicatus* Platia & Schimmel
p. 114, add **N**: MO to *modestus* Kiesenwetter
p. 115, correct spelling of author of *Agriotes kirghisicus* to Iablokoff-Khnzorian
p. 115, add HUB SCH SHA SHX YUN to *hedini* Fleutiaux
p. 115, add BU to *infuscatus* Desbrochers des Loges
p. 115, add LE MG **NTRi** to *lineatus* Linnaeus
p. 116, add HEB NMO to and delete GAN SHX from *obscurus* Linnaeus
p. 116, delete *peregrinus* Gurjeva and place under synonymy of *hedini* Fleutiaux, p. 115
p. 117, add *rufus* Fleutiaux, 1918b: 266 (*Agonischius*) **A**: YUN **ORR**

p. 117, add BEI SHX to *rugipennis* Schwarz
p. 117, add GAN to *sericatus* Schwarz
p. 117, add *simplex* Candèze, 1891b: 789 (*Agonischius*) **A**: FUJ GUX HAI HUN XIZ YUN **ORR**
p. 117, add SCH to *sputator* Linnaeus
p. 117, add XIN to *squalidus squalidus* Gurjeva
p. 117, delete *soederbomi* Fleutiaux and place under synonymy of *unicolor* Koenig, p. 118
p. 118, add BEI JIX to *subvittatus subvittatus* Motschulsky
p. 118, delete *tonkinensis* Fleutiaux (occurs only in **ORR**) and transfer its synonyms *colonus* Fleutiaux, *formosanus* Miwa, *nodieri* Fleutiaux and *pallidiangulus* Miwa to *Agriotes breviusculus*, p. 114 as synonyms of this species
p. 120, add FUJ NMO to *dahuricus dahuricus* Candèze
p. 120, add JIL NMO to *piloselloides* Schwarz
p. 123, correct spelling of author of *Ampedus elongatulus* to Fabricius
p. 123, add LIA to *gagatinus* Candèze
p. 131, add to genus *Agonischius*:
lateritius Fleutiaux, 1940a: 122 **A**: NP SD **ORR**
thoracicus Fleutiaux, 1889: 145 **A**: NP SD **ORR**
p. 131, delete *breviusculus* Candèze from *Agonischius* and place as valid species under *Agriotes*, p. 114, add distribution: **A**: FUJ GUA HAI TAI **ORR**
p. 131, delete *francki* Fleutiaux from *Agonischius* and place as good species under *Agriotes*, p. 115
p. 131, delete *scapularis* Candèze from *Agonischius* and place as good species under *Agriotes*, p. 117
p. 132, correct spelling of author of *obscurum* (in synonymy with *Ludioschema sulcicolle*) to Fleutiaux
p. 133, correct spelling of author of *gagates* (in synonymy with *Ludioschema suturale*) to Fleutiaux
p. 133, add (*Ludius*) to *Mulsanteus hirtellus*, Candèze
p. 134, add to genus *Vuilletus*:
amamiensis amamiensis Ôhira, 1967d: 28 **A**: JA (Amami-Ôshima)
p. 136, correct spelling of author of *Megapenthes bengalensis* to Schimmel
p. 139, correct spelling of author of *obscurus* (in synonymy with *Xanthopenthes birmanicus*) to Fleutiaux
p. 139, correct spelling of author of *Anchastelater ornatus* to Fleutiaux
p.141, add IN to *admirabilis* Dolin
p. 141, add FUJ HUN to *Melanotus agilis* Platia & Schimmel
p. 142, add **ORR** to *Melanotus auberti* Platia & Schimmel
p. 142, add AP to *Melanotus badgleyi* Platia & Schimmel
p. 142, add NC to *Melanotus bocaki* Platia & Schimmel
p. 142, add *brevis* Candèze, 1882: 88 **A**: XIZ YUN **ORR**
p. 143, add SHA to *Melanotus cechovskyi* Platia & Schimmel
p. 143, add AP to *Melanotus coolsi* Platia & Schimmel
p. 143, add HEI XIZ to *Melanotus coomani* Platia & Schimmel, 2001
p. 143, add *cribratoides* Platia & Schimmel, 2002: 345 **A**: XIZ **ORR**
p. 144, add FUJ SHA to *Melanotus dusaneki* Platia & Schimmel
p. 144, add HAI to *Melanotus excoffieri* Platia & Schimmel
p. 145, add BEI to *Melanotus hebeiensis* Platia & Schimmel
p. 145, add PA SD to *Melanotus hirticornis* Herbst
p. 145, add AP to *Melanotus holzschuhi* Platia & Schimmel
p. 145, add HEN to *Melanotus hunanensis* Platia & Schimmel
p. 146, add **ORR** to *Melanotus melli* Platia & Schimmel
p. 147, add *novus* Platia & Schimmel, 2001: 340 **A**: HUN **ORR**
p. 148, add *pseudogigas* Platia & Schimmel, 2002: 353 **A**: YUN **ORR**
p. 148, add HAI to *Melanotus regalis* Candèze
p. 148, add *ricchiardii* Platia & Schimmel, 2001: 352 **A**: XIZ **ORR**
p. 148, add FUJ to *Melanotus rugosipennis* Fleutiaux
p. 150, add p. 2 to *Melanotus atayal* Kishii & Platia, 1991
p. 150, add **ORR** to *Melanotus villosus* Geoffroy
p. 150, add HEN SHA to *Melanotus zhilongensis* Platia & Schimmel

p. 151, add to subgenus *Spheniscosomus* the species *schuhi* Platia & Schimmel, 2002: 315 **A**: FUJ **ORR**
p. 151, add SCH to *Metriaulacus badiipennis* Candèze
p. 151, add FUJ to *Metriaulacus gobius* Candèze
p. 152, add to *Priopus* the species *quatei* Platia & Schimmel, 1996: 165 **A**: XIZ YUN **ORR**
p. 157, delete *nigricornis* C. N. F. Brisout de Barneville from subgenus *Athous* and place in synonymy with *obsoletus* Mulsant & Guillebeau, p. 157
p. 157, delete *rubrotestaceus* Desbrochers des Loges from subgenus *Athous* and place in synonymy with *obsoletus* Mulsant & Guillebeau, p. 157
p. 160, delete *angustulus* Candèze from subgenus *Orthathous* and place in synonymy with *obsoletus* Mulsant & Guillebeau, p. 157
p. 161, delete *brisouti* Sánchez-Ruiz and its synonym *lateralis* C. N. F. Brisout de Barneville from subgenus *Orthathous* and place in synonymy with *kruegeri* Reitter, p. 162 and add PT
p. 161, add AU SL to *Athous* (*Orthathous*) *caviformis* Reitter
p. 161, add SL to *Athous* (*Orthathous*) *cavus* Germar
p. 161, add AU to *Athous* (*Orthathous*) *ganglbaueri* Schwarz
p. 161, add PT to *Athous* (*Orthathous*) *gerezianus* Reitter
p. 162, delete *obsoletus* Mulsant & Guillebeau from subgenus *Orthathous* and place as valid species in subgenus *Athous* s.str., p. 157 and add SP
p. 162, add SP to *Athous* (*Orthathous*) *olbiensis* Mulsant & Guillebeau
p. 162, add AU to *Athous* (*Orthathous*) *naseri* J. Müller
p. 162, add MC YU to *Athous* (*Orthathous*) *singularis* Reitter
p. 164, correct spelling of author of *Elathous perisii* to Desbrochers des Loges
p. 167, add: *lividus* Stroem, 1783: 62 (*Elater*) in synonymy with *Denticollis linearis* (Linnaeus, 1758)
p. 168, add to genus *Denticollis*:
multiimpressus Pic, 1930g: 11 (*Lepturoides*) **A**: YUN
p. 169, correct spelling of author of Hemicrepidiina to Champion
p. 175, correct spelling of author of *Anostirus purpureus* to Poda von Neuhaus
p. 178, remove genus *Hypoganomorphus* Dolin to p. 177 before genus *Hypoganus* Kiesenwetter
p. 178, delete *Neopristilophus gougeleti* Fairmaire as valid species and add it and the synonymous *hispanicus* Sáez-Bolaño as synonyms to *insitivus* Germar. Add SP to the distribution of *insitivus* Germar
p. 179, *insularis* Platia & Gudenzi, change **E**: GR to **E**: GR (Kríti)
p. 179, add FUJ HUB to *Gnathodicrus francki* Fleutiaux
p. 179, add GUX to *Gnathodicrus tonkinensis* Fleutiaux
p. 179, add to subgenus *Gnathodicrus*:
vitalisi Fleutiaux, 1918b: 249 (*Corymbites*) A: FUJ **ORR**
p. 179, remove subgenus *Gnathodicrus* Fleutiaux from genus *Poemnites* Buysson and place it as genus before *Hypoganomorphus* Dolin on p. 177
p. 183, delete *assoi* Perez Arcas as synonym of *dima* Schaufuss and place it as valid species in *Dima* Charpentier, with distribution **E**: SP
p. 186, correct spelling of author of Hemiopinae and Hemiopini to Fleutiaux
p. 189, add FE to *Neohypdonus tumescens* LeConte
p. 190, correct spelling of author of *Quasimus chibi* to Kishii
p. 199, correct spelling of author of *Cardiophorus ovipennis* to Desbrochers des Loges
p. 202, correct spelling Grand Canaria and Gan Canaria to Gran Canaria
p. 202, in *Cardiophorus*, delete *gracilis* Gurjeva, 1966 and replace by:
gurjevae Cate, **nom. nov**. **A**: TD
gracilis Gurjeva, 1966: 90 [HN] [nec *Cardiophorus gracilis* (Wollaston, 1864)]
p. 203, add YE to *Craspedostethus buettikeri* Platia & Schimmel
p. 205, correct spelling of author of *Dicronychus variatus* to Desbrochers des Loges
p. 206, add YE to *Paracardiophorus buettikeri* Chassain
p. 206, add BEI to *Paracardiophorus sequens sequens* Candèze, 1873
p. 208, delete *Elater degener* Scopoli, it is in *Acmaeodera*, Buprestidae
p. 208, delete *Elater equestris* Fabricius, it is a synonym of *Drapetes mordelloides* (Host)
p. 209, add: *ater* Motschulsky, 1839: 81 in synonymy with *Drilus concolor* Ahrens

p. 209, correct spelling of author of *ruficollis* and *venustus* (in synonymy with *Drilus fulvicollis*) to Guérin-Méneville
p. 209, add CT ST UK to distribution of *Drilus flavescens* A. G. Olivier
p. 210, correct spelling of *Omalisus fontisbellaquei* Geoffroy to *fontisbellaquaei*
p. 214, add UK to distribution of *Benibotarus taygetanus* Pic
p. 214, correct spelling of *Benibotarus arnoldi* to *arnoldii*
p. 214, add FE (Kuril Is.) to distribution of *Dictyoptera sapporensis* Kôno
p. 215, delete A: WS from distribution of *Erotides cosnardi* Chevrolat
p. 215, change agreement in gender in *Laterialis* to feminine (*nasuta* Chevrolat)
p. 216, correct spelling of author of *pusillus* (in synonymy with *Platycis minutus*) to Gmelin
p. 216, change agreement in gender in *Punicealis* to feminine (*miranda* Barovskij)
p. 216, add JA to distribution of *Punicealis medvedevi* Kazantsev and correct distribution of *Punicealis miranda* Barovskij to: FE (Primorskiy Kray)
p. 216, add MO to distribution of *Pyropterus rubripes* Pic
p. 218, change agreement in gender in *Ponyalis* to feminine and correct spelling of *cincinata* to *cincinnata* Kazantsev
p. 218, correct spelling of *Ponyalis chifenglee* Kazantsev to *chifengleei* Kazantsev
p. 221, add p. 175 to *Atelius wittmeri* Bocák, 1997:
p. 227, add *Lampyris letourneuxi* E. Olivier, 1884: 24 **A**: IS
p. 229, add FE to distribution of *Pyrocoelia rufa* E. Olivier
p. 231, correct author of *Lampyroidea maculata* to Geisthard & Day
p. 232, add FE to distribution of *Luciola lateralis* Motschulsky and *L. mongolica* Motschulsky
p. 234, add to family Lampyridae, incertae sedis:
genus *Lamellipalpodes* Maulik, 1921: 584 type species *Lamellipalpodes annandalei* Maulik, 1921
montanus Wittmer 1995: 107 **A**: NP
nepalensis Wittmer 1995: 109 **A**: NP
pp. 235 and 236, correct spelling of authors of *Asiopodabrus kiiensis*, *A. simillimus, A. tajimensis* and *A. takaosanus* to Nakane & Makino
p. 246, correct spelling of author of *Cantharis trabeata* to Fairmaire
p. 250, delete line with *Lycocerus flavonotaticeps* (Pic, 1937) as valid species, and place *obliteratus* Pic, 1937 in synonymy with *Lycocerus fainanus* (Pic, 1910)
p. 251, add: *Lycocerus longipennis* Pic, 1947: 8 **A**: YUN
p. 253, replace *purpureus* Kazantsev by: *purpurascens* Wittmer, 1978c: 155 (*Andrathemus*) **A**: BT
p. 253, correct spelling of *Lycocerus purpurascens* Pic (in synonymy with *Lycocerus rubripenis* Hope) to *purpureopubens* Pic
p. 260, correct spelling of author of *Podistra longicornis* to Wittmer
p. 261, correct spelling of authors of *Protemus watanabei* to Okushima & Satô
p. 263, add: *Rhagonycha anceyi* Abeille de Perrin, 1909: 8 (*Nastonycha*) **N**: AG
p. 263, correct spelling of author of *delahoni* (in synonymy with *Rhagonycha fulva*) to Schilsky
p. 265, add ES FE to distribution of *Rhagonycha mongolica* Wittmer
p. 267, correct spelling of author of *florida* (in synonymy with *Rhagonycha testacea*) to Schiödte
p. 271, correct spelling of *Themus* (s.str.) *glasunovi* Barovskij to: *glazunovi*
p. 283, correct spelling of author of *Malthinus stigmatias* to Kiesenwetter
p. 300, add the author name Steven to *Dermestes dimidiatus*
p. 301, add: *laniarius subfulvicollis* Kalík, 1951c: 155 **A**: ES KZ WS
p. 302, correct spelling of author of *caninus* (in synonymy with *Dermestes undulatus*) to Gravenhorst
p. 302, add TU to *Mariouna letourneuxi* Pic
p. 314, subgenus *Nathrenus*, delete *versicolor* Reitter, 1887a: 70 from synonymy with *verbasci* (Linnaeus, 1767), and add it as valid species, with distribution: **E**: FR GR IT (Sicilia, Sardegna)
p. 316, *Globicornis hispanica* Pic, 1908, place in synonymy with *G. emarginata* (Gyllenhal, 1808)
p. 317, add: **A**: UZ to *Megatoma graeseri* (Reitter)
p. 319, add *koningsbergeri* Pic, 1933g: 71 as synonym of *Trogoderma granarium* Everts
p. 325, correct spelling of author of *Xylion adustus* to Fåhraeus
p. 355, add p. 14 to *Mesothes substriatus* Pic, 1938b

p. 359, correct type species of *Calypterus* to *Calypterus sericans* Mulsant & Godart, 1859 (= *Ptilinus bucephalus* Illiger, 1807), and of *Sternoplus* to *Ptilinus ater* Creutzer
p. 360, correct spelling of author of *sericeus* (in synonymy with *Xyletinus ornatus*) to Morawitz
p. 361, correct spelling of author of *discicollis* (in synonymy with *Xyletinus ruficollis*) to Morawitz
p. 361, correct spelling of author of *tibialis* (in synonymy with *Xyletinus subrotundatus*) to Schilsky
p. 371, correct spelling of *Bouquetoclerus* to *Bousquetoclerus*
p. 377, correct spelling of author of *schreberi* (in synonymy with *Trichodes favarius*) to Gerhardt
p. 377, correct spelling of *audouini dubreuili* Reymond to *audouinidubreuili* Reymond
p. 382, correct spelling of author of *morulus* (in synonymy with *Opetiopalpus obesus*) to Kiesenwetter
p. 384, add PT and SP to the distribution of *Acanthocnemus nigricans* Hope
p. 385, delete *Idgia curticeps* Pic, 1920 and *I. lineata* Pic, 1907; both occur only in South India
p. 386, delete as unavaialble name *tereticollis* Sturm, 1826 from synonymy with *Prionocerus caeruleipennis* Perty, 1831
p. 407, correct spelling of authors of *subdensatum*, *submicaceum* and *subnodosum* (in synonymy with *Dolichosoma lineare*) to Mulsant & Rey
p. 422, correct spelling of authors of *Attalus gracilentus* to Mulsant & Rey
p. 434, add GB to *Hypebaeus flavipes* Fabricius
p. 443, correct spelling of author of *armeniacus* (in synonymy with *Clanoptilus ambiguus*) to Ménétriés
p. 436, add (= *Attalus cyaneonotatus* Pic, 1903) to the type species of *Indiebaeus* Wittmer
p. 440, add GB to distribution of *Axinotarsus marginalis* (Laporte)
p. 447, correct spelling of authors of *signicollis* (in synonymy with *Oogynes longicollis*) to Mulsant & Rey
p. 450, add as nomen dubium: *Malachius glabrellus* Betta, 1847: 23 **E**: IT
p. 467, correct year for *Nitidula sexpustulata* Fabricus to 1792
p. 467, correct year for *Carpophilus grandis* Motschulsky to 1861
p. 499, correct year for *Silvanus surinamensis* Linnaeus to 1758
p. 500, add (= *Silvanus cribricollis* Grouvelle, 1897) to the type species of *Silvanoides* Halstead
p. 502, correct spelling of author of *brevifossum* (in synonymy with *Passandra heros*) to Newman
p. 502, correct for *grouvellei* Reitter, 1877b to Reitter, 1877e
p. 502, correct author of *siculus* (in synonymy with *Cucujus clavipes*) to Pic
p. 504, correct spelling of author of *labiatus* (in synonymy with *Laemophloeus muticus*) to Kugelann
add to *Laemophloeus*: *jureceki* Pic, 1929: 118 **A**: FE and *ussuricus* Pic, 1929: 118 **A**: FE
p. 508, correct spelling of author of *viennensis* (in synonymy with *Olibrus bimaculatus*) to Guillebeau
p. 509, correct spelling of author of *Olibrus multesimus* to Lyubarsky
p. 510, correct spelling of author of *flachi* (in synonymy with *Phalacrus caricis*) to Schilsky
p. 516, delete SP from distribution of *Cryptophagus confusus* Bruce, 1934
p. 516, add *erichsoni* Reitter, 1888a: 22 **E**: AB AR RO ST (Kavkaz), and delete *erichsoni* Reitter from nomina dubia, p. 521
p. 516, add following resurrected name and restructured synonyms:

denticulatus Heer, 1841: 426 **E**: AB AR AU BE BH BU BY CR CT CZ DE EN FI FR GB GE GG GR IR HU IT LS MC MD NL NR NT PL RO SK SL SP ST (Kavkaz) SV SZ TR UK **N**: MR **A**: ES FE IN IS TR UZ **AFR AUR NAR**

albipilus Rey, 1889b: 43
corticum Rey, 1889b: 43
obtusidens Rey, 1889b: 43
pilosus Gyllenhal, 1827: 287 [HN]
pseudodentatus Bruce, 1934d: 168
substriatus Rey, 1889b: 43
tamaricis Rey, 1889b: 43

p. 519, delete following, because of homonymy:

pilosus Gyllenhal, 1827: 287 **E**: AB AR AU BE BH BU BY CR CT CZ DE EN FI FR GB GE GG GR IR HU IT LS MC MD NL NR NT PL RO SK SL SP ST (Kavkaz) SV SZ TR UK **N**: MR A: ES FE IN IS TR UZ **AFR AUR NAR**

albipilus Rey, 1889b: 43
corticum Rey, 1889b: 43

denticulatus Heer, 1841: 426
obtusidens Rey, 1889b: 43
pseudodentatus Bruce, 1934d: 168
substriatus Rey, 1889b: 43
tamaricis Rey, 1889b: 43

p. 521, add to nomina dubia *pilosus* Herbst, 1792: 177 (*Kryptophagus*)
p. 528, correct spelling of *Atomaria fimetarius* to *fimetaria*
p. 544, correct spelling of authors of *Triplax nikritini* to Nikitsky & Kompantzev
p. 545, correct spelling of authors of *Triplax tadzhica* to Nikitsky & Kompantzev
p. 548, add *Diplocoelus talyshensis* Nikitsky, 1993: 154 **E**: AB
p. 549, add p. 41 to *Anommatus herculis* Všetečka, 1948
pp. 555, 556, *Sphaerosoma piliferum* P. W. J. Müller, 1821: 207 (*Tritoma*) is a valid name for the species listed as *S. quercus* Samuelle, 1819
place *Sphaerosoma quercus Samuelle*, 1819 in synonymy with *Sphaerosoma pilosum* (Panzer, 1793)
place *Sphaerosoma infimum* (Baudi di Selve, 1889) and *S. piligerum* (P. W. J. Müller, 1824) in synonymy with *Sphaerosoma piliferum* (P. W. J. Müller, 1821)
p. 581, correct spelling of author of *obliquus* (in synonymy with *Nephus quadrimaculatus*) to Comolli
p. 560, correct spelling of author of *scovitzi* (in synonymy with *Endomychus armeniacus*) to Faldermann
p. 561, correct spelling of author of *Leiestes menetriesi* to Faldermann
p. 563, correct spelling of author of *Eumorphus assamensis* to Gerstaecker
p. 567, change year for *Danae* Reiche, 1847 and *Danae rufula* Reiche, 1847 to 1850
p. 572, correct spelling of author of *plagiata* (in synonymy with *Coccidula rufa*) to Gerhardt
p. 609, correct spelling of author of *distincta* (in synonymy with *Ceratomegilla undecimnotata*) to L. Redtenchacher
p. 641, correct spelling of author of *microps* (in synonymy with *Thes bergrothi*) to Erichson
p. 647, correct year and page for *Melanophthalma angulata* Wollaston to 1864: 394
p. 660, add: Betta V. 1847: *De quibusdam coleopteris agri ticinensis*. Ticini Regii, 35 pp.
p. 663, correct spelling of Bonneli F. A. to Bonelli F. A.
p. 678, deleta German text in bolt characters at the end of the reference to Cherepanov A. I. 1956
p. 716, correct year for Gebler F. A. von. 1830: Bemerkungen … to 1829
p. 720, correct spelling of editor of Gmelin's 1790 work to Georg Emanuel Beer
p. 726, correct spelling of Guérin-Menéville to Guérin-Méneville
p. 727, delete "M." as inicial from Guérin-Méneville
p. 752, add: Kawashima I. & Sugaya H. 2003: An additional new species of the genus Rhagophthalmus (Coleoptera, Rhagophthalmidae) from Taiwan, with a key to the males of the Taiwanese and Japanese species. *Elytron* **31**: 353-359.
p. 776, add: Lewis G. 1888b: On the species of Cucujus found in Japan. *The Entomologist's Monthly Magazine* 25 [1888-1889]: 84.
p. 777, correct year for Linnaeus C.: *Fauna Suecica* … to 1760
p. 786, add: Maulik S. 1921: New Indian drilid beetles. *Proceedings of the Zoological Society of London* **1921**: 579-586.
p. 832, add: Pic M. 1929: Nouveaux coléoptères paléarctiques. *Časopis Československé Společnosti Entomologické* **25** [1928]: 118-120.
p. 842, change Reiche L. J. 1847 to 1850
p. 857, change Rücker H. W. 1979 to: 1979a
p. 857, add: Rücker H. W. 1979b: Beitrag zur Kenntnis Lathridiiden Vietnams (Coleoptera, Lathridiidae). *Annales Historico-Naturales Musei Nationalis Hungarici* **71**: 161-163.
p. 878, add: Stroem H. 1783: Norske insecters beskrivelse med anmaerkninger. *Nye Samling af det Kongelige Danske Videnskabers Selskabsskrifter* **2**: 49-93.
p. 888, transfer the reference: Tomaszewska K. W. 2000 from p. 888 to p. 884.

NEW NOMENCLATURAL and TAXONOMIC ACTS, and COMMENTS

Mycetophagidae

N. B. Nikitsky

New replacement name

Mycetophagus reitteri **nom. nov**. for *Mycetophagus undulatus* Reitter, 1889, nec *M. undulatus* (Marsham, 1802)

New synonyms

Atritomoides Nikitsky, 1988, **syn. nov.** of *Eulagius* Motschulsky, 1845
Litargus antennatus Miyatake, 1957, **syn. nov.** of *Litargus balteatus* LeConte, 1856
Litargus ferrantei Reitter, 1908, **syn. nov.** of *Litargus balteatus* LeConte, 1856
Litargus pilosus Wollaston, 1857, **syn. nov.** of *Litargus balteatus* LeConte, 1856
Mycetophagus sauteri Grouvelle, 1913, **syn. nov.** of *Mycetophagus antennatus* (Reitter, 1879)
Mycetophagus salicis C. Brisout de Barneville, 1862, **syn. nov.** of *Mycetophagus piceus* (Fabricius, 1777)
Typhaea decipiens Lohse, 1989, **syn. nov.** of *Typhaea haagi* Reitter, 1874

Ciidae

J. Jelínek

Comments

Anobium micans Fabricius, 1792 is a valid member of *Cis* Latreille. It was misinterpreted by subsequent workers who used that name for *Cis submicans* Abeille de Perrin, 1874.

Anobium nitidum Fabricius, 1792 is a valid member of *Sulcacis* Dury. It was misinterpreted by some workers who used that name for *Cis castaneus* Herbst, 1793. *Cis alter* Silfverberg, 1991: 22 was proposed as a replacement name for the incorrect concept and is therefore unavailable.

Tetratomidae

N. B. Nikitsky

New synonymy

M. sapporense Nakane, 1991, **syn. nov**.of *Mycetoma affine* Nikitsky, 1985

Melandryidae

N. B. Nikitsky

Changes in rank

Paralederia Nikitsky & Belov, 1982, as genus, from subgenus
Dircaea ussuriensis Nikitsky, 1985, as species, from subspecies in *Dircaea erotyloides*
Nagakuchikius Toyoshima & Y. Ishikawa 2000, as genus, from subgenus in *Phloeotrinus*
Mimoserropalpus Pic, 1922, as genus, from subgenus in *Serropalpus*
Mimoserropalpus indicus Pic, 1945, as species, from variety of *Serropalpus* (*Mimoserropalpus*) *binhanus*

New combinations

Lederina prion (Sasaji, 1987), **comb. nov**. from *Lederia*
Lederina kidoi (Sasaji, 1995), **comb. nov**. from *Lederia*
Anisoxiella laticollis (Nakane, 1963), **comb. nov**. from *Anisoxya*
Mimoserropalpus formosanus (Nomura, 1958), **comb. nov**. from *Cuphosis*
Mimoserropalpus striatus (Blair, 1931), **comb. nov**. from *Talayra*
Symphora brunnea (Marseul, 1876), **comb. nov**. from *Scraptia*
Melandrya utakoae (Sasaji, 1988), **comb. nov**. from *Prothalpia*

New synonyms

Orchesia obscuricolor Pic, 1954, **syn. nov.** of *Orchesia fusiformis* Solsky, 1871
Phloiotrya parvula (Lewis, 1895), **syn. nov.** of *Phloiotrya obscura* (Lewis, 1895)

Phloiotrya vaudoueri Mulsant, 1856, **syn. nov.** of *Serropalpus tenuis* Hampe, 1850
Xylitella paganettii Obenberger, 1917, **syn. nov.** of *Rushia pareyssii* (Mulsant, 1856)
Symphora miyakei Nomura & Hayashi, 1960, **syn. nov.** of *Symphora brunnea* (Marseul, 1876)
Bonzicomorpha Pic, 1927 (type species *Bonzicomorpha unifasciata* Pic, 1927), **syn. nov.** of *Salumia* Fairmaire, 1889 (type species *Salumia davidis* Fairmaire, 1889)
Osphya talyschensis Pic, 1902, **syn. nov.** of *Osphya aerata* Seidlitz, 1898

New replacement names
Abdera (*Adobia*) *mulsanti* Nikitsky, **nom. nov.** for *Abdera* (*Adobia*) *bifasciata* Mulsant, 1856 (nec *Abdera bifasciata* (Marsham, 1802)

Comments
Orchesia anaspioides (Motschulsky, 1845) is possibly a senior synonym of *Orchesia fusiformis* Solsky, 1871
Wanachia Schulze, 1912: the assignment of this genus to Dircaeini is doubtful

Mordellidae

J. Horák

New synonymy
Mordella ornatopallida Reitter, 1911, **syn. nov.** of *Mordella rufoaxillaris* Fairmaire & Gory, 1863

Ripiphoridae

J. Batelka

New combination
Micropelecotoides aurosericeus (Gressitt, 1941), **comb. nov.** from *Pelecotoides* Laporte, 1840.

Changes in rank
Ptilophorus afghanicus (Kaszab, 1957) as species, from subspecies of *Evaniocera fischeri*.

Designations of **type species**
Emenadia Laporte, 1840, was established for *Mordella bimaculata*, *Mordella flabellata*, *Emenadia terminata* and *Emenadia bicolor*. A type species was not designated. *Mordella bimaculata* Fabricius, 1787 is here designated as the type of *Emenadia* Laporte.
Polychronia Gistel, 1848, is currently in synonymy with *Ripiphorus* Bosc, 1791. It was established as an unnecessary replacement name of *Myodes*, and no species was included. *Ripiphorus subdipterus* Bosc, 1792 is here designated as the type species of *Polychronia* Gistel.
Ripidastes Gistel, 1848, is currently in synonymy with *Metoecus* Dejean, 1834. It was established as an unnecessary replacement name of *Metoecus*, and no species was included. *Mordella paradoxa* Linnaeus, 1761 is here designated as the type species of *Ripidastes* Gistel.

Spelling
Ripiphorus Bosc, 1791, and *Ripidius* Thunberg, 1806, are the correct original spellings. Hence the family-group names are spelled Ripiphoridae and Ripidiinae, and Rhipiphoridae, Rhipidiinae, *Rhipiphorus* and *Rhipidius* are unjustified emendations.

Comments
Rhyzostylops Silvestri, 1905 described and known from one female only may be a synonym of the genus *Blattivorus* Chobaut, 1891 of which only males are known (Švácha, 1994).
Egidyella prophetea Reitter, 1899 (= *Trichopleurias deserti* J. R. Sahlberg, 1903), listed in ripiphorids in Csiki (1913) belongs in Dermestidae.
Ripidius quadraticeps is a misspelling in Lesne (1902b) and thus unavailable. Subsequent authors incorrectly listed it as a synonym of *R. quadriceps* Abeille de Perrin, 1872.

Zopheridae

S. A. Ślipiński & R. Schuh

New synonymy

Canaricerus Oromí & Garcia, 1986, **syn. nov.** of *Xylolaemus* L. Redtenbacher, 1857 (based on description of *Canaricerus*.

Cerchanotus ogasawarensis (Nakane, 1990), **syn. nov.** of *Cerchanotus orientalis* (Ślipiński, 1985) (based on description of *C. ogasawarensis* and on examination of topotypical material).

Gempylodes lewisii Sharp, 1885, **syn. nov.** of *Gempylodes ornamentalis* (Reitter, 1878) (based on examination of type material).

Synchita hirsuta (Pic, 1922), **syn. nov.** of *Synchita crenicollis* Wollaston, 1867 (based on redescription of *S. hirsuta* by Dajoz, 1970).

New combination:

Lasconotus niponius (Lewis, 1879), **comb. nov.** from *Bitoma* (type material examined).

Comments

The Zoological Records for 1984 records *Langelandia macedonica* Franz as a new species. The name was actually listed by Franz, 1984 (*Zeitschrift der Arbeitsgemeinschaft Österreichischer Entomologen* **36**: 57-62) but is unavailable (misspelling of *Langelandia macedonia* Franz, 1980).

Tenebrionidae: Laenini

I. Löbl, W. Schawaller

New replacement name

Laena confusa Löbl & Schawaller, **nom. nov.** for *Laena chinensis* Li & Chen, 1993 (nec *Laena chinensis* Kaszab, 1965)

Tenebrionidae: Asidini

F. Soldati

Comments

Allard, 1869 and subsequent authors refer to a paper "Coléoptères Nouveaux" by Allard, 1868, that supposedly contains descriptions of numerous new species of *Asida*. It was impossible to trace this paper, including in the library of the Muséum National d'Histoire Naturelle in Paris. Allard's paper "Coléoptères Nouveaux" is therefore treated as unpublished and the respective species are considered to be published in Allard, 1869.

Viñolas & Cartagena (2005) upgraded some of the former subgenera. Their study is based only on a minor part of relevant taxa and lacks sound justification. Therefore, the classification of the Asidini is given in the present Catalogue as currently accepted anterior to their publication. The names of taxa invalided in Viñolas & Cartagena (2005) here resurrected, and the species downgraded to subspecies are again upgraded.

Resurrections

Alphasida (*Glabrasida*) *calagurriensis* (Escalera, 1943) is resurrected from synonymy with *Glabrasida brevipubens* (Reitter, 1917)

Alphasida (*Glabrasida*) *cordubensis* Reitter, 1917 is resurrected from synonymy with *Glabrasida porcata* (Solier, 1836)

Alphasida (*Glabrasida*) *heydeni* Reitter, 1917 is resurrected from synonymy with *Glabrasida dubia* (Rambur, 1838)

Asida (*Asida*) *alcarazensis* Reitter, 1917 is resurrected from synonymy with *Asida* (*Asida*) *serripes* Chevrolat, 1874

Asida (*Asida*) *heikertingeri* Schuster, 1928 is resurrected from synonymy with *Asida* (*Asida*) *spaethi* Schuster, 1928

Asida (*Planasida*) *confusa* Kraatz, 1874 is resurrected from synonymy with *Asida* (*Planasida*) *pygmaea pusillima* Kraatz, 1874

Asida (*Planasida*) *fallax* Escalera, 1921 is resurrected from synonymy with *Asida* (*Granulasida*) *esteparia* Reitter, 1917, type material examined

Asida (Planasida) latissima Wilke, 1921 is resurrected from synonymy with *Asida (Planasida) paulinoi* Perez Arcas, 1868

Changes in rank

Alphasida (Alphasida) becerrae (Escalera, 1905) is upgraded from *Alphasida (Alphasida) lorcana becerrae* (Escalera, 1905)

Alphasida (Alphasida) lazaroi Escalera, 1906 is upgraded from *Alphasida (Alphasida) solieri lazaroi* Escalera, 1906

Alphasida (Alphasida) merceti (Bolivar y Pieltain, 1914) is upgraded from *Alphasida (Alphasida) solieri merceti* (Bolivar y Pieltain, 1914)

Alphasida (Alphasida) oberthueri (Escalera, 1901) is upgraded from *Alphasida (Alphasida) solieri oberthueri* (Escalera, 1901)

Alphasida (Alphasida) rufopubescens (Escalera, 1905) is upgraded from *Alphasida (Alphasida) oberthueri* (Escalera, 1901)

Alphasida (Alphasida) typica Gebien, 1936 is upgraded from *Alphasida (Alphasida) martinezi gaditana* (Escalera, 1905)

Alphasida (Alphasida) volxemi (Escalera, 1905) is upgraded from *Alphasida (Alphasida) solieri volxemi* (Escalera, 1905)

Alphasida (Elongasida) calumniata (Escalera, 1906) is upgraded from *Elongasida grandipalpis calumniata* (Escalera, 1906)

Alphasida (Elongasida) granosa (Escalera, 1921) is upgraded from *Elongasida alonensis granosa* (Escalera, 1921)

Alphasida (Elongasida) hispaliensis (Escalera, 1906) is upgraded from *Elongasida hesperica hispaliensis* (Escalera, 1906)

Alphasida (Elongasida) rectipennis (Escalera, 1921) is upgraded from *Elongasida alonensis rectipennis* (Escalera, 1921)

Alphasida (Glabrasida) alcirensis Reitter, 1917 is upgraded from *Glabrasida dubia alcirensis* (Reitter, 1917)

Alphasida (Glabrasida) discostriata (Escalera, 1922) is upgraded from *Glabrasida parallela discostriata* (Escalera, 1922)

Alphasida (Glabrasida) frigidissima (Escalera, 1922) is upgraded from *Glabrasida zapaterii frigidissima* (Escalera, 1922)

Alphasida (Glabrasida) granadina (Escalera, 1922) is upgraded from *Glabrasida elongata granadina* (Escalera, 1922)

Alphasida (Glabrasida) javalambrensis (Escalera, 1943) is upgraded from *Glabrasida deyrollei javalambrensis* (Escalera, 1943)

Alphasida (Glabrasida) loroi (Escalera, 1922) is upgraded from *Glabrasida zapaterii loroi* (Escalera, 1922)

Alphasida (Glabrasida) marseuli (Allard, 1869) is upgraded from *Glabrasida punctipennis marseuli* (Allard, 1869)

Alphasida (Glabrasida) olmedensis (Escalera, 1922) is upgraded from *Glabrasida zapaterii olmedensis* (Escalera, 1922)

Alphasida (Glabrasida) pilosipennis (Escalera, 1923) is upgraded from *Glabrasida punctipennis pilosipennis* (Escalera, 1923)

Alphasida (Glabrasida) ruiderensis (Escalera, 1922) is upgraded from *Glabrasida toletana ruiderensis* (Escalera, 1922)

Alphasida (Glabrasida) turillensis (Escalera, 1922) is upgraded from *Glabrasida dubia turillensis* (Escalera, 1922)

Alphasida (Glabrasida) uhagoni (Escalera, 1912) is upgraded from *Glabrasida sulcata uhagoni* (Escalera, 1912)

Asida (Asida) coimbrensis Escalera, 1922 is upgraded from *Asida (Asida) pseudoreticulata coimbrensis* Escalera, 1922

Asida (Asida) spaethi Schuster, 1928 is upgraded from *Asida (Asida) serripes spaethi* Schuster, 1928

Asida (Asida) transmontana Escalera, 1922 is upgraded from *Asida (Asida) bejarensis transmontana* Escalera, 1922

Asida (Planasida) alcoyana Escalera, 1921 is upgraded from *Asida (Planasida) moroderi alcoyana* Escalera, 1921, type material examined

Asida (*Planasida*) *amplicollis* Escalera, 1921 is upgraded from *Asida* (*Planasida*) *marginicollis amplicollis* Escalera, 1921
Asida (*Planasida*) *breiti* Schuster, 1928 is upgraded from *Asida* (*Planasida*) *inquinata breiti* Schuster, 1928
Asida (*Planasida*) *breuili* Escalera, 1921 is upgraded from *Asida* (*Planasida*) *paulinoi breuili* Escalera, 1921
Asida (*Planasida*) *moczarskii* Schuster, 1928 is upgraded from *Asida* (*Planasida*) *pygmaea moczarskii* Schuster, 1928
Asida (*Planasida*) *paludosa* Escalera, 1921 is upgraded from *Asida* (*Planasida*) *moroderi paludosa* Escalera, 1921
Asida (*Planasida*) *pseudosetipennis* Escalera, 1921 is upgraded from *Asida* (*Granulasida*) *esteparia pseudosetipennis* Escalera, 1921
Asida (*Planasida*) *pusillima* Kraatz, 1874 is upgraded from *Asida* (*Planasida*) *pygmaea pusillima* Kraatz, 1874
Asida (*Planasida*) *rufa* Escalera, 1921 is upgraded from *Asida* (*Planasida*) *paulinoi rufa* Perez Arcas, 1868
Asida (*Polasida*) *levantina* Escalera, 1922 is upgraded from *Asida* (*Polasida*) *jurinei levantina* Escalera, 1922, type material examined.

Designation of type species

Aulonasida Reitter, 1917 is currently in synonymy with *Glabrasida* Escalera, 1910. It was established as a subgenus of *Alphasida* Escalera, 1905 for *Alphasida chauveneti* Solier, 1836, *A. quadricostata* Allard, 1869 and *A. lethierryi* Allard, 1869. A type species was not designated. *Asida chauveneti* Solier, 1836 is here desgnated as the type species of *Aulonasida* Reitter, 1917.

Betasida Reitter, 1917 was established as a subgenus of *Alphasida* Escalera, 1905. Viñolas and Cartagena (2005: 110) designated *Asida luctuosa* Boisduval, 1835 as the type species of this subgenus. Their designation is invalid because Reitter, 1917 conditionally included *Asida luctuosa* in *Betasida*, and this species was unknown to Reitter (see ICZN Art. 67.2.5). *Asida argenteolimbata* Escalera, 1901 was originally included in *Betasida* and was the only species well known to the author of the subgenus at the moment of its description (recommendation ICZN Art. 69A.7.). *Asida luctuosa* Boisduval, 1835 is the type species of *Subalphasida* Escalera, 1927 that is placed in synonymy with *Betasida* Reitter, 1917.

Cribrasida Reitter, 1917 is currently in synonymy with *Elongasida* Escalera, 1906. It was established as a subgenus of *Alphasida* Escalera, 1905 for *Alphasida grandipalpis* (Allard, 1869) and *A. rectipennis* Escalera, 1906. A type species was not designated. *Asida grandipalpis* is here designated as the type species of *Cribrasida* Reitter, 1917.

Dolichasida Reitter, 1917 was established as a subgenus of *Asida* Latreille, 1802, for the following species: *A. moraguesi* L. W. Schaufuss, 1879, *A. esteparia* Reitter, 1917, *A. setipennis* Allard, 1869, *A. valentina* Reitter, 1917 and *A. gibbicollis* Perez Arcas, 1865. A type species was not designated. *Asida moraguesi* is here designated as the type species of the *Dolichasida* Reitter, 1917.

Durasida Reitter, 1917 is currently in synonymy with *Glabrasida* Escalera, 1910. It was established as a subgenus in *Alphasida* Escalera, 1905 for the following species: *Alphasida villososulcata* Allard, 1869, *A. silphoides* (Linnaeus, 1767), *A. henoni Fairmaire*, 1868, *A. edithae* Reitter, 1917, *A. herminae* Reitter, 1917, *A. silphiformis* Reitter, 1917, *A. felicitana* Reitter, 1917, *A. saintpierrei* Allard, 1869, *A. himerera* Reitter, 1917, *A. opatroides* Allard, 1869, *A. tuberculosa* Reitter, 1917 and *A. quadricarinata* Reitter, 1917. A type species was not designated. *Asida silphoides* is here designated as type species of *Durasida* Reitter, 1917.

Glabrasida Escalera, 1910 was established for the following species: *Glabrasida villososulcata* (Allard, 1869), *G. punctipennis* (Perez Arcas, 1865), *G. goudoti* (Solier, 1836), *G. porcata* (Solier, 1836), *G. silphoides* (Linnaeus, 1767), *G. servillei* (Solier, 1836), *G. gracilis* (Allard, 1869), *G. vuillefroyi* (Perez Arcas, 1868), *G. conspuata* Escalera, 1910, *G. tuberculipennis* Escalera, 1910, *G. melillensis* Escalera, 1910, *G. globipennis* Escalera, 1910, *G. mazaganica* Escalera, 1910, *G. rabatica* Escalera, 1910, and *G. maroccana* Allard, 1869. The designation of *Asida elongata* Solier, 1836 as the type species in Viñolas & Cartagena (2005: 113) is invalid because that species was not originally included. *Glabrasida conspuata* Escalera, 1910 is designated here as the type species of *Glabrasida* Escalera, 1910.

Gracilasida Escalera, 1909 is currently in synonymy with *Planasida* Escalera, 1907. It was established as a subgenus of *Asida* Latreille, 1802, for *Asida ariasi* Escalera, 1909 and *A. pusillima* Kraatz, 1874. A type species was not designated. *Asida ariasi* is here designated as the type species of as the type species of *Gracilasida* Escalera, 1909.

Gymnetasida Reitter, 1917 is currently in synonymy with *Glabrasida* Escalera, 1910. It was established as a subgenus in *Alphasida* Escalera, 1905 for the following species: *Alphasida tumida* Reitter, 1917, *A. crassipes* (Allard, 1869), *A. tricostata* (Allard, 1869), *A. rugosa* (Fabricius, 1775), *A. serpiginosa* (Erichson, 1841), *A. nigerrima* (Allard, 1869), *A. laevicollis* (Allard, 1869), *A. servillei* (Solier, 1836), *A. auriculata* (Solier, 1836), *A. gracilis* (Allard, 1869), *A. algeriana* Gebien, 1910, *A. beduina* Reitter, 1917, *A. tuberculifera* (Allard, 1869), (Allard, 1869), *A. tuberculata* (Allard, 1869), *A. cylindrica* Reitter, 1917, *A. rolphii* (Fairmaire, 1867), *A. maroccana* (Allard, 1869), and *A. granifera* (Solier, 1836). A type species was not designated. *Asida tricostata* Allard, 1869 is here designated as the type species of *Gymnetasida* Reitter, 1917.

Insulasida Escalera, 1922 was placed in synonymy with *Asida* Latreille, 1802 by Viñolas and Cartagena (2005). It was established as a subgenus of *Asida* Latreille, 1802, for *Asida moraguezi* L. W. Schaufuss, 1879, *A. balearica* Escalera, 1922 and *A. planipennis* L. W. Schaufuss, 1869. A type species was not designated. *Asida moraguezi* L. W. Schaufuss, 1879 is here designated as the type species of *Insulasida* Escalera, 1922.

Leptasida Reitter, 1917 was established as a subgenus of *Asida* Latreille, 1802, for the following species: *Asida diecki* Allard, 1869, *A. planipennis* L. W. Schaufuss, 1869, *A. ricoi* Martinez y Sáez, 1873, *A. zaida* Reitter, 1917, *A. pygmaea* Rosenhauer, 1856, *A. ithae* Reitter, 1917, *A. pusillima* Kraatz, 1874, and *A. akisoides* Reitter, 1917. A type species was not designated. *Asida diecki* Allard, 1869 is here designated as the type species of *Leptasida* Reitter, 1917 (nec *Leptasida* Chatanay, 1914).

Mimelasida Reitter, 1917 is currently in synonymy with *Glabrasida* Escalera, 1910. It was established as a subgenus in *Alphasida* Escalera, 1905, for *Alphasida leperina* Reitter, 1917 and *A. puncticollis* (Solier, 1836). A type species was not designated. *Asida puncticollis* Solier, 1836 is here designated as the type species of *Mimelasida* Reitter, 1917.

Opatrasida Escalera, 1922 is currently in synonymy with *Polasida* Reitter, 1917. It was established as a subgenus in *Asida* Latreille, 1802, for the following five species: *Asida jurinei* Solier, 1836, *A. levantina* Escalera, 1922, *A. marmottani* C. Brisout de Barneville, 1863, *A. saguntina* Escalera, 1922, and *A. sericea* A. G. Olivier, 1795. A type species was not designated. *Asida jurinei* Solier, 1836 is here designated as the type species *Opatrasida* Escalera, 1922.

Pedarasida Reitter, 1917 is currently in synonymy with *Glabrasida* Escalera, 1910. It was established as a subgenus of *Alphasida* Escalera, 1905, for the following species: *Alphasida cariosicollis* (Solier, 1836), *A. tangeriana* (Solier, 1836), *A. subcostata* (Solier, 1836), *A. atrata* (Baudi di Selve, 1875), *A. asperata* (Solier, 1836), *A. melitana* (Reitter, 1894), *A. sicula* (Solier, 1836), *A. syriaca* (Allard, 1869), *A. cossyrensis* Reitter, 1917 and *A. ithana* Reitter, 1917. A type species was not designated. *Asida cariosicollis* Solier, 1836 is here designated as the type species of *Pedarasida* Reitter, 1917.

Pseudoelongasida Escalera, 1922 was established as a subgenus of *Asida* Latreille, 1802, for *Asida silvestrei* Escalera, 1922 and *A. tenuecostata* Fairmaire, 1880, both from Morocco. A type species was not designated. Only *Asida silvestrei* Escalera, 1922 was known to the author and is therefore designated here as the type species of *Pseudoelongasida* Escalera, 1922.

Pseudoplanasida Escalera, 1921 is currently in synonymy with *Planasida* Escalera, 1907. It was established as a subgenus of *Asida* Latreille, 1802, for *Asida pygmaea* Rosenhauer, 1856, *A. lindaraja* Escalera, 1921, *A. moraima* Escalera, 1921, *A. minima* Escalera, 1921, and *A. pusillima* Kraatz, 1874. A type species was not designated. *Asida pygmaea* Rosenhauer, 1856 is here designated as the type species of *Pseudoplanasida* Escalera, 1921.

Rugasida Escalera, 1922 was established as a subgenus of *Asida* Latreille, 1802, for the following species: *A. reticulata* Solier, 1836, *A. granulifera* Chevrolat, 1840, *A. pseudoreticulata* Escalera, 1922, *A. coimbrensis* Escalera, 1922, *A. bejarensis* Escalera, 1922, *A. transmontana* Escalera, 1922, *A. cazorlensis* Escalera, 1922, *A. serripes* Chevrolat, 1874, *A. segurensis* Escalera, 1922, *A. gibbicollis* Perez Arcas, 1865 and *A. diecki* Allard, 1869. A type species was not designated. *Asida reticulata* Solier, 1836 (presently *Asida sabulosa reticulata* Solier, 1836) is here designated as the type species of *Rugasida* Escalera, 1922.

Trachasida Reitter, 1917 is currently in synonymy with *Planasida* Escalera, 1907. It was established as a subgenus of *Asida* Latreille, 1802, for the following species: *Asida reichei* Allard, 1869, *A. inquinata* Rambur, 1838, *A. marginicollis* Rambur, 1838, *A. confusa* Kraatz, 1874, *A. brucki* Allard, 1869, *A. moroderi* Reitter, 1917, *A. wankai* Reitter, 1917, *A. ruficornis* Solier, 1836, *A. lefranci* Reitter, 1917, and *A. gebieni* Reitter, 1917. A type species was not designated. *Asida ruficornis* Solier, 1836 is here designated as the type species of *Trachasida* Reitter, 1917.

New synonyms

Euryasida Reitter, 1917 is currently considered as a valid subgenus of *Asida* Latreille, 1802. It is based on a single Balearic species, *Asida barceloi* Perez Arcas, 1868. *Asida barceloi* is closely related related to several Corsican and Sardinian species (e. g., *A. abrupta* Fairmaire, 1867, *A. inaequalis* Solier, 1836, and *A. solieri* Gené, 1836) all currently placed in *Asida* s. str., and shares with them also the pattern of elytral costae. No characters are known to separate clearly this species from other members of *Asida* s. str. Thus, *Euryasida* Reitter, 1917 is **syn. nov.** of *Asida* Latreille, 1802.

Granulasida Escalera, 1921 was established as a subgenus of *Asida* Latreille, 1802, and included about 15 Iberian species. It was considered by Viñolas & Cartagena (2005) as distinct from *Planasida* Escalera, 1907 because of the crenulate lateral margins of elytra in *Granulasida* while members of *Planasida* were supposed to possess smooth lateral margins of elytra. In fact about half of the North African species currently placed in *Planasida* have also crenulate lateral margins of elytra. Thus, members of *Planasida* and *Granulasida* share significant morphological characters and also ecology. As the Iberian *Granulasida* cannot be separated from *Planasida*, *Granulasida* Escalera, 1921 is **syn. nov.** of *Planasida* Escalera, 1907.

New assignments

Asida (*Globasida*) *ludovici* Perez Arcas, 1874, from *Asida* s. str.
Asida (*Planasida*) *hispidula* Pic, 1903, from *Asida* s. str.
Asida (*Planasida*) *sacculipennis* Kwieton, 1986, from *Asida* s. str.

Spelling

Alphasida interjecta var. *interstrata* Reitter, 1917 is the original, correct spelling, and not the subsequently used *A. interjecta* var. *interstriata* Reitter

Asida amorii Perez Arcas, 1868 is the original, correct spelling, and not the subsequently used *A. amori* Perez Arcas

Asida dejeanii Solier, 1836 is the original, correct spelling, and not the subsequently used *A. dejeani* Solier

Asida dufourii Baudi di Selve, 1875 is the original, correct spelling, and not the subsequently used *A. dufouri* Baudi di Selve

Asida ghilianii Baudi di Selve, 1875 is the original, correct spelling, and not the subsequently used *A. ghiliani* Baudi di Selve

Asida lusitana Escalera, 1923 is the original, correct spelling, and not the subsequently used *A. lusitanica* Escalera

Asida mahonis Boieldieu, 1865 is the original, correct spelling, and not the subsequently used *A. mahoni* Boieldieu

Asida pazii Perez Arcas, 1868 is the original, correct spelling, and not the subsequently used *A. pazi* Perez Arcas

Asida zapaterii Perez Arcas, 1872 is the original, correct spelling, and not the subsequently used *A. zapateri* Perez Arcas

Tenebrionidae: Akidini, Pedinini and Tentyriini

J. Ferrer

Resurrections, based on study of type material

Tentyria aragonica Koch, 1944 as valid species, from synonymy with *Tentyria peiroleri* Solier, 1835

Tentyria corrugata Rosenhauer, 1856 as valid species, from synonymy with *Tentyria gaditana* Rosenhauer, 1856

Tentyria gallica Solier, 1835 as valid species, from synonymy with *Tentyria curculionides interrupta* Latreille, 1807

Tentyria heydeni Haag-Rutenberg, 1870 as valid species, from synonymy with *Tentyria curculionides interrupta* Latreille, 1807

Tentyria grossa basalis L. W. Schaufuss, 1869 as valid subspecies, from synonymy with *Tentyria grossa grossa* Besser, 1832

Tentyria jordani Koch, 1944 as valid species, from synonymy with *Tentyria curculionoides curculionoides* (Herbst, 1799)

Tentyria velox Chevrolat, 1865 as valid species, from synonymy with *Tentyria curculionoides curculionoides* (Herbst, 1799)

Changes in rank

Tentyria castillana Koch, 1944 as valid species, from subspecies of *Tentyria peiroleri* Solier, 1835. Based on significanly dfifferent shape od elytra in these taxa.

New synonyms

Phylan (*Litororus*) *gadeai* Español & Viñolas, 1981, **syn. nov.** of *Phylan mulsanti* (Piochard de la Brûlerie, 1869). Based on study of type material.

Tentyria andalusiaca Kraatz, 1865, **syn. nov.** of *Tentyria corrugata* Rosenhauer, 1856

Comments

Akis spinosa (Linnaeus, 1764) is currently considered a widely distributed Mediterranean species. The study of the Linnean material of *Akis spinosa*, published as *Tenebrio spinosus*, provided evidence that it is the same taxon as "*sculptior* Koch, 1935". The latter was introduced as a variety of *Akis elevata* Solier, 1837 and is known only from Egypte. Following names placed incorrectly in synonymy with *Akis spinosa* (Linnaeus, 1764) are here placed in synonymy with *Akis trilineata* Herbst, 1799: *Akis carinata* Drapiez, 1821, *Akis nitida* Solier, 1837, and *Akis planicollis* Solier, 1837.

Tenebrionidae: Scaurini

H. Labrique

New synonymy

The following synonymies are based on study of extensive material including types, and were given in an unpublished thesis submitted in 1999.

Scaurus giganteus Küster, 1848, **syn. nov.** of *Scaurus aegyptiacus* Solier, 1838

Scaurus harbuloti Reymond, 1956, **syn. nov.** of *Scaurus mesatlanticus mesatlanticus* Peyerimhoff, 1948

Tenebrionidae: Pimeliini

M. Lillig

New assignments

Trachyderma (*Atrachyderma*) *afghanica afghanica* Kaszab, 1960 and *T.* (*Atrachyderma*) *afghanica kandaharica* Kaszab, 1970 are transferred to *Atrachyderma* from *Trachyderma* s. str., based on the characters given in the original description.

Tenebrionidae: Blaptini

M. Nabozhenko

New replacement name

Blaps neoscabiosa **nom. nov.** for *Blaps scabiosa* Faust, 1875 (nec *Blaps scabiosa* Baudi di Selve, 1874)

New synonymy

Blaps bipunctata Allard, 1880, **syn. nov.** of *Blaps lethifera lethifera* Marsham, 1802. The description of this species was based on a single female from Ekaterinoslavl' (presently Dnepropetrovsk, Ukraine). The characters and illustration provided by Allard (1880) suggest the present synonymy.

Designations of type species

Agroblaps Motschulsky, 1860 is currently placed in synonymy with *Blaps* Fabricius, 1775. It was established for 12 species, a type species was not designated. *Blaps fatidica* J. Sturm, 1807, one of the originally included species, is here designated as the type species of *Agroblaps* Motschulsky, 1860.

Blapimorpha Motschulsky, 1860 is currently placed in synonymy with *Blaps* Fabricius, 1775. It was established for four species, a type species was not designated. *Blaps reflexa* Gebler, 1832, one of the originally included species, is designated here as the type species of *Blapimorpha* Motschulsky, 1860.

Blapisa Motschulsky, 1860 is currently placed in synonymy with *Blaps* Fabricius, 1775. It was established for five species, a type species was not designated. *Blaps jaegeri* Hummel, 1827, one of the originally included species, is designated here as the type species of *Blapisa* Motschulsky, 1860.

Eupomeca Solier, 1848 was established for two species, *Blaps cylindrica* Herbst, 1799 and *Blaps attenuata* Fischer von Waldheim, 1822, both currently placed in synonymy with *Prosodes obtusa* (Fabricius, 1798).

This genus-group name was overlooked by subsequent workers and a type species was not designated. *Blaps cylindrica* Herbst is designated here as the type species of *Eupomeca* Solier, 1848.

Leptocolena Allard, 1880 is currently placed in synonymy with *Blaps* Fabricius, 1775. It was established for four species, a type species was not designated. *Blaps emoda* Allard, 1880, one of the originally included species, is designated here as the type species of *Leptocolena* Allard, 1880.

Platyblaps Motschulsky, 1860 is currently placed in synonymy with *Blaps* Fabricius, 1775. It was established for five species, a type species was not designated. *Blaps holconota* Fischer de Waldheim, 1844 (misspelled *chalconota*), one of the originally included species, is designated here as the type species of *Platyblaps* Motschulsky, 1860.

Rhizoblaps Motschulsky, 1860 is currently placed in synonymy with *Blaps* Fabricius, 1775. It was established for 11 species, a type species was not designated. *Blaps pruinosa* Eversmann, 1836, one of the originally included species, is designated here as the type species of *Rhizoblaps* Motschulsky, 1860.

Uroblaps Motschulsky, 1860 is currently placed in synonymy with *Blaps* Fabricius, 1775. It was established for 6 species, a type species was not designated. *Blaps producta* Laporte, 1840, junior synonym of *Blaps lusitanica* Herbst, 1799 is one of the originally included species and it is designated here as the type species of *Uroblaps* Motschulsky, 1860.

Tenebrionidae: Helopini

M. Nabozhenko

Resurrections

Cylindrinotina Español, 1956. Español (1956) recognized within his concept of the "Helopinae" two tribes, Helopini and Cylindronotini (= Cylindrinotini), the latter name being based on *Cylindronotus,* an unjustified emendation of *Cylindrinotus*. The two tribes were recognized as valid in several papers (Español, 1959, 1961a, 1961b, 1963, 1974, Antoine, 1958). Watt (1974) treated the Helopinae of Español as a tribe, Helopini, without having mentioned any subtribe. The name Cylindrinotini is in synonymy with Helopini in Bouchard et al. (2005). Español (1956) based the Helopini and Cylindrinotini on aedeagal characters. The larval characters studied recently (Nabozhenko, 2006) confirm the distinction of Helopina and Cylindrinotina. Consenquently, Cylindrinotina, with as type genus *Cylindrinotus* Faldermann, 1837 is resurrected, with the rank of a subtribe within Helopini.

Eustenomacidius laevicollis (Kraatz, 1882) was incorrectly synonymized with *Eustenomacidius luridus* (Ménétriés, 1848) by Nabozhenko (2006c). The latter species name, originally placed in *Helops*, is unavailable because of homonymy with *Helops luridus* Fabricius, 1775. *Eustenomacidius laevicollis* (Kraatz, 1882) has no senior synonyms and is used as valid.

New synonyms

Hedyphanes impressicollis Faldermann, 1837, **syn. nov.** of *Hedyphanes laticollis laticollis* Fischer von Waldheim in Ménétriés, 1832. The synonymy is based on study of the type material of *Hedyphanes laticollis laticollis* (Nabozhenko, 2002) and a syntype of *Hedyphanes impressicollis* labeled: "Conf. Pers.", "*Hedyphanes impressicollis* Conf. Pers. Faldm.". The syntype is preserved in the Zoological Museum of Moscow State University in Moscow.

Nalassus (*Helopocerodes*) *semiopacus* (Reitter, 1902), **syn. nov.** of *Nalassus* (*Helopocerodes*) *faldermanni* (Faldermann, 1837), based on the description and study of extensive material of *Nalassus semiopacus* and *Nalassus faldermanni* from their respective type localities.

New replacement name

Odocnemis zoltani Nabozhenko, **nom. nov.** for *Odocnemis anatolicus* Kaszab, 1961 (nec *Odocnemis anatolicus* Pic, 1899)

New combinations

Hedyphanes muminovi (Bogachev, 1963), **comb. nov.** from *Catomus*
Nalassus balchashensis (G. S. Medvedev, 1970), **comb. nov.** from *Cylindrinotus*
Nalassus halophilus (G. S. Medvedev, 1987), **comb. nov.** from *Cylindrinotus*
Nalassus kazachstanicus (G. S. Medvedev, 1964), **comb. nov.** from *Cylindrinotus*
Nalassus melchiades (Reitter, 1922), **comb. nov.** from *Cylindrinotus*

Nalassus mojynkumensis (G. S. Medvedev, 1987), **comb. nov**. from *Cylindrinotus*
Nalassus dilaticornis (Reitter, 1922), **comb. nov**. from *Cylindrinotus*
Nalassus ahngeri (G. S. Medvedev, 1998), **comb. nov**. from *Cylindrinotus*
Nalassus cambyses (Seidlitz, 1896), **comb. nov**. from *Cylindrinotus*
Nalassus cunctator (Reitter, 1922), **comb. nov**. from *Cylindrinotus*
Nalassus heres (Reitter, 1922), **comb. nov**. from *Cylindrinotus*
Nalassus ponticus (Seidlitz, 1896), **comb. nov**. from *Cylindrinotus*
Odocnemis alcides (Reitter, 1922), **comb. nov**. from *Cylindrinotus*
Odocnemis amandus (Reitter, 1922), **comb. nov**. from *Cylindrinotus*
Odocnemis amasiae (Seidlitz, 1896), **comb. nov**. from *Cylindrinotus*
Odocnemis anatolicus (Pic, 1899), **comb. nov**. from *Cylindrinotus*
Odocnemis badius (L. Redtenbacher, 1849), **comb. nov**. from *Cylindrinotus*
Odocnemis baudoni (Antoine, 1954), **comb. nov**. from *Stenomax*
Odocnemis bosphoranus (Allard, 1876), **comb. nov**. from *Cylindrinotus*
Odocnemis clypeatus (Küster, 1851), **comb. nov**. from *Cylindrinotus*
Odocnemis colasi (Español, 1954), **comb. nov**. from *Nalassus*
Odocnemis congener cedrorum (Antoine, 1962), **comb. nov**. from *Stenomax*
Odocnemis congener congener (Reiche, 1861), **comb. nov**. from *Cylindrinotus*
Odocnemis crenatostriatus (Allard, 1877), **comb. nov**. from *Cylindrinotus*
Odocnemis cretensis (Seidlitz, 1896), **comb. nov**. from *Cylindrinotus*
Odocnemis curticollis (Reitter, 1922), **comb. nov**. from *Cylindrinotus*
Odocnemis dasypus (Seidlitz, 1896), **comb. nov**. from *Cylindrinotus*
Odocnemis exaratus (Germar, 1817), **comb. nov**. from *Cylindrinotus*
Odocnemis fleischeri (Reitter, 1922), **comb. nov**. from *Cylindrinotus*
Odocnemis fundator (Reitter, 1908), **comb. nov**. from *Cylindrinotus*
Odocnemis incultus (Allard, 1877), **comb. nov**. from *Cylindrinotus*
Odocnemis intricatus (Allard, 1877), **comb. nov**. from *Cylindrinotus*
Odocnemis intrusicollis (Seidlitz, 1896), **comb. nov**. from *Cylindrinotus*
Odocnemis korbi (Reitter, 1902), **comb. nov**. from *Cylindrinotus*
Odocnemis kurdistanus (Reitter, 1902), **comb. nov**. from *Cylindrinotus*
Odocnemis mediatlantis (Antoine, 1949), **comb. nov**. from *Stenomax*
Odocnemis moabiticus (J. R. Sahlberg, 1908), **comb. nov**. from *Cylindrinotus*
Odocnemis nigropiceus (Küster, 1850), **comb. nov**. from *Cylindrinotus*
Odocnemis opertus (Reitter, 1922), **comb. nov**. from *Cylindrinotus*
Odocnemis ophonoides (Lucas, 1846), **comb. nov**. from *Nalassus*
Odocnemis pardoi (Español, 1953), **comb. nov**. from *Nalassus*
Odocnemis planivittis Allard, 1876), **comb. nov**. from *Cylindrinotus*
Odocnemis poeciloides (Seidlitz, 1896), **comb. nov**. from *Cylindrinotus*
Odocnemis praelongus (Baudi di Selve, 1876), **comb. nov**. from *Cylindrinotus*
Odocnemis protinus (Reitter, 1900), **comb. nov**. from *Cylindrinotus*
Odocnemis pseudoclarus (Reitter, 1922), **comb. nov**. from *Cylindrinotus*
Odocnemis rhaticus (Antoine, 1937), **comb. nov**. from *Stenomax*
Odocnemis ruffoi ruffoi Canzoneri, 1970), **comb. nov**. from *Cylindrinotus*
Odocnemis rufffoi osellai Gardini, 1979), **comb. nov**. from *Cylindrinotus*
Odocnemis schusteri (Reitter, 1922), **comb. nov**. from *Cylindrinotus*
Odocnemis scutellatus (Reitter, 1902), **comb. nov**. from *Cylindrinotus*
Odocnemis serbicus (J. Frivaldszky, 1894), **comb. nov**. from *Cylindrinotus*
Odocnemis seriegranatus (Seidlitz, 1896), **comb. nov**. from *Cylindrinotus*
Odocnemis steindachneri (Apfelbeck, 1907), **comb. nov**. from *Cylindrinotus*
Odocnemis strangulatus (Reitter, 1922), **comb. nov**. from *Cylindrinotus*
Odocnemis telueticus (Escalera, 1914), **comb. nov**. from *Cylindrinotus*
Odocnemis tenebrioides (Germar, 1813), **comb. nov**. from *Nalassus*
Odocnemis valgus (Baudi di Selve, 1881), **comb. nov**. from *Cylindrinotus*

Nomina dubia

Hedyphanes fovoelatusstriatus Motschulsky, 1845 was described from "Russie méridionale". The only member of *Hedyphanes* known to occur in Russia is the rare *H. laticollis nycterinoides* Faldermann, 1837, from the mountains in Dagestan. Motschulsky's description does not fit the latter species, and the type material of his *H. fovoelatostriatus* was not found in the collection of V. Motschulsky housed in the Zoological Museum of Moscow State University. Thus, it is unknown to which real taxon *H. fovoelatusstriatus* Motschulsky may refer to.

Probaticus nitidipennis Küster, 1850 was described from Sicilia. Its type material is lost and the description fits several Helopini species.

Designations of type species

Accanthopus Dejean, 1821 was incorrectly considered as a junior homonym of *Acanthopus* Klug, 1807 (Hymenoptera). *Enoplopus* Solier, 1848 with type species *Helops dentipes* Rossi, 1790 is an unnecessary replacement name for *Accanthopus* Dejean, 1821. *Enoplopus* is currently in use as valid. Silfverberg 1984: 60 used *Accanthopus* as valid and mentioned that *Enoplopus* Solier is its junior synonym. Therefore, *Enoplopus* cannot be declared nomen oblitum (ICZN, Art. 23.9) and is listed in synonymy.

Catomidius Seidlitz, 1896 is currently placed in synonymy with *Catomus* Allard, 1876. It was established for ten species, a type species was not designated. *Hedyphanes rhynchophorus* Seidlitz, 1896, one of the originally included species, is here designated as the type species of *Catomidius* Seidlitz, 1896.

Helopelius Reitter, 1922 was established for four species. A type species was not designated. *Stenomax aeneipennis* Allard, 1876, one of the originally included species, is here designated as the type species of *Helopelius* Reitter, 1922.

Nesotes Allard, 1876 was established for 21 species, a type species was not designated. *Helops asper* Küster, 1850, one of the originally included species, is here designated as the type species of *Nesotes* Allard, 1876.

Diastixus Allard, 1876 was established for 11 species, a type species was not designated. *Helops punctipennis* Lucas, 1846, one of the originally included species, is here designated as the type species of *Diastixus* Allard, 1876.

Gunarellus Reitter, 1922, currently a subgenus of *Stenohelops*, was established for six species, a type species was not designated. *Helops gratus* J. Frivaldsky, 1894, one of the originally included species, is here designated as the type species *Gunarellus* Reitter, 1922.

Helopotrichus Reitter, 1922 was established for four species, a type species was not designated. *Helops villosipennis* Lucas, 1846, one of the originally included species, is here designated as the type species of *Helopotrichus*.

Omaleis Allard, 1877 is a replacement name for the homonymous *Omalus* Allard, 1876. *Omalus* was proposed for 10 species, including *Helops congener* Reiche, 1861. A type species was not designated, including for *Omaleis*. *Helops congener* Reiche, 1861 is here designated as the type species for *Omaleis* Allard.

Tenebrionidae: Alleculinae

V. Novák

New replacement names:

Borboresthes borchmanni **nom. nov.** for *B. tibialis* Borchmann, 1941 (nec *Borboresthes tibialis* Pic, 1937)

New synonyms

Hymenorus Mulsant, 1856 **syn. nov.** of *Hymenophorus* Mulsant, 1851

Omophlus (*Odontomophlus*) *glamocensis* Obenberger, 1917 **syn. nov.** of *Megischina armillata* Brullé, 1832 (type material studied).

New combinations (type material studied):

Borboresthes acicularis (Marseul, 1876), **comb. nov.** from *Allecula*

Borboresthes elliptica (Fairmaire, 1891), **comb. nov.** from *Pseudocistela*

Borboresthes kuluensis (Pic, 1910), **comb. nov.** from *Allecula*

Borboresthes obliquefasciata (Pic, 1926), **comb. nov.** from *Allecula*

Borboresthes pieli (Pic, 1937), **comb. nov.** from

Gonodera flavida (Fairmaire, 1892), **comb. nov.** from *Cistela*

Gonodera forticornis (Fairmaire, 1892), **comb. nov.** from *Cistela*
Cistelomorpha maculicornis (Fairmaire, 1878), **comb. nov.** *Cistela*
Cistelomorpha ustiventris (Fairmaire, 1861), **comb. nov.** from *Cistela*
Cteniopinus tatsienluensis (Borchmann, 1932), **comb. nov.** from *Cistelina*
Heliotaurus splendidus (Bedel, 1905), **comb. nov.** from *Omophlus*
Hymenalia atronitens (Fairmaire, 1892), **comb. nov.** from *Gonodera*
Hymenalia minyops (Peyerimhoff, 1943), **comb. nov.** from *Mycetocharina*
Hymenalia picheyrei (Peyerimhoff, 1943), **comb. nov.** from *Mycetocharina*
Hymenalia rungsi (Peyerimhoff, 1943), **comb. nov.** from *Mycetocharina*
Isomira sericea (Drapiez, 1828), **comb. nov.** from *Cistela*
Omophlus (*Paromophlus*) *emmae* (Iablokoff-Khnzorian, 1959), **comb. nov.** from *Heliotaurus*
Paracistela rufithorax (Pic, 1915), **comb. nov.** from *Pseudocistela*
Paracistela sinensis (Pic, 1934), **comb. nov.** from *Isomira*
Pseudocistela znojkoi (Iablokoff-Khnzorian, 1959), **comb. nov.** from *Gonodera*

Authorships
Omophlus is currently credited to Solier, 1835: 246. The name was established by Dejean, 1834: 213. As type species of *Omophlus* Dejean is here designated *Cistela lepturoides* Fabricius, 1787, the type species of *Omophlus* Solier and originally included by Dejean, 1834.
Podonta is currently credited to Mulsant, 1856: 66. The name was established by Solier, 1835: 247.

Tenebrionidae: Alleculinae

J. Ferrer & V. Novák

Resurrection and new combination
Chrysomela pubescens Linnaeus, 1767 is resurrected from synonymy with *Lagris hirta* Linnaeus, 1758 and as **comb. nov.** placed as valid in *Omophlus* [see below].

New synonymy
Omophlus betulae (Herbst, 1783) **syn. nov.** of *Omophlus pubescens* (Linnaeus, 1767). Based on Sharon L. Shute's examination of the original material of *Chrysomela pubescens* Linnaeus, 1767 that was found conspecific with specimens in the Natural History Museum, London identified as *Omophlus rufitarsis* (Leske, 1785).

Tenebrionidae

K. Ando

New synonymy and combination
Aptereucyrtus shibatai Ando, 1984, **syn. nov.** of *Aptereucyrtus hemichalceus* Gebien, 1922. Based on study of type material and on Philippine specimens of *A. hemichalceus* identified by Z. Kaszab.
Derosphaerus foveolatus (Marseul, 1876), **syn. nov.** of *Derosphaerus sinensis* (Hope, 1842). The description of *D. foveolatus* was obviously based on speciens of *D. sinensis* introduced to Nagasaki from China or Vietnam.
Microcameria pygmaea Ren, 1998, the type species of *Microcameria* Ren, 1998 is transferred to *Foochounus* Pic, 1921, based on study of type material. Thus, *Microcameria* is a **syn. nov.** of *Foochounus*, and *Foochounus pygmaea* (Ren) is a **comb. nov**.
Neatus ventralis Marseul, 1876, **syn. nov.** of *Neatus picipes* (Herbst, 1797). Based on absence of distinguishing characters.
Stenophanes rubripennis (Marseul, 1876), **syn. nov.** of *Stenophanes mesostena* (Solsky, 1871)

Tenebrionidae: Opatrini

P. Bouchard & I. Löbl

Amphithrixoides Bouchard & Löbl, **nom. nov.** for *Amphithrix* Español, 1953 (nec *Amphithrix* Ragonot, 1893, in Lepidoptera)

Tenebrionidae

I. Löbl, P. Bouchard, O. Merkl & D. Iwan

New replacement names

Pimelia (*Pimelia*) *anomaloides* Löbl, Bouchard & Merkl, **nom. nov**. for *Pimelia anomala* Sénac, 1880 (nec *Pimelia anomala* Solier, 1836)

Taxa established by type species designations

Belutschistanops was introduced by Koch, 1943: 552. It was proposed as a subgenus in *Hyperops* Eschscholtz, 1831 for *Hyperops pygmaea* L. Redtenbacher, 1850, *H. schusteri* Koch, 1940, *H. nitidula* Kraatz, 1865, *H. doriae doriae* Baudi di Selve, 1874 and *H. doriae arabica* Koch, 1940, and is currently considered as valid. A type species was not designated and therefore this genus-group name is unavailable. The description given by Koch (1943) characterizes the group. *Hyperops schusteri* Koch, 1940 is here designated as the type species of *Belutschistanops*.

Erodibius was introduced by Koch, 1960: 354 as a new subgenus of *Arthrodibius* Lesne, 1915. He included in *Erodibius* 7 species, *A. plicatus* (Fairmmare, 1883), *A. cicatrix* (Fairmaire, 1879), *A. bernadellii* Koch, 1959, *A. nitidiventris* (Fairmaire, 1884), *A. cyphonotus* (Fairmaire, 1887), and as new taxa, *A. scorteccii* and *A. hemmingi*. The subgenus is currently considered valid although a type species was not designated and this genus-group name is unavailable. The description given by Koch, 1960: 354 characterizes the group. *Arthrodeis cyphonotus* Fairmaire, 1887 is here designated as the type species of *Erodibius*.

Leptoderops was introduced as a new subgenus of *Thraustocolus* Kraatz, 1865 by Koch, 1934: 96 who included two species, *T. montandoni* Schuster, 1915 and his new species *T. priesneri*. This subgenus is currently considered valid although a type species was not designated and therefore this genus-group name is unavailable. Subsequently, *T. montandoni* was placed in the subgenus *Prothraustocola* Kaszab, 1957, and *T. afghanicus* Kaszab, 1979 was added to *Leptoderops*. The description given by Koch, 1934: 96 characterizes the group. *Thraustocolus priesneri* Koch, 1934 is here designated as the type species of *Leptoderops*.

Oxypistoma was introduced by Koch, 1940: 258. It was proposed as a new subgenus of *Prochoma* Solier, 1835 for two new species, *P. bucculenta* and *P. blairi*, and for *P. clypealis* Blair, 1923. *Oxypistoma* is currently considered as valid. A type species was not designated and therefore this genus-group name is unavailable. The description given by in Koch, 1940: 258 characterizes the group. *Prochoma bucculenta* Koch, 1940 is here designated as the type species of *Oxypistoma*.

Type species designations

Abiga is currently placed in synonymy with *Scelosodis* Solier, 1835. It was established by Guérin-Méneville, 1859 for two new species, *A. humilis* and *A. cerisyi*. A type species was not designated. We designate here *Abiga humilis* Guérin-Méneville, 1859 as type species of *Abiga* Guérin-Méneville, 1859.

Acisba is currently placed in synonymy with *Pachychila* Eschscholtz, 1831. It was established by Dejean, 1834 for three species, *Tentyria cribrosa* Besser, 1832, *Tentyria brevis* Besser, 1832, and *Tentyria subovata* Besser, 1832. A type species was not designated. We designate here *Tentyria cribrosa* Besser, 1832 (= *Pimelia punctata* Fabricius, 1798) as the type species of *Acisba* Dejean, 1834.

Akis Herbst, 1799 was established for five species, including *Pimelia reflexa* Fabricius, 1775 and *P. spinosus* Linnaeus, 1764. Latreille, 1810: 429 designated *Pimelia reflexa* as the type species of *Akis*. Subsequent authors considered erroneously *Tenebrio spinosus* Linnaeus to be the type species of *Akis* Herbst, 1799.

Anebacis is currently a valid subgenus of *Pachychila* Eschscholtz, 1831. It was established by Peyerimhoff, 1927 for three species, *Tentyria cribricollis* Fairmaire, 1875, *Tentyria mairei* Peyerimhoff, 1920, and *Tentyria medioxima* Peyerimhoff, 1921. A type species was not designated. We designate here *Tentyria cribricollis* Fairmaire, 1875 as the type species *Anebacis* Peyerimhoff, 1927.

Aphanaspis is currently placed in synonymy with *Pimelia* Fabricius, 1775. It was proposed by Wollaston, 1864 who included two new species, *Pimelia granulicollis* and *Pimelia auriculata*. A type species was not designated. We designate here *Pimelia granulicollis* Wollaston, 1864 as the type species of *Aphanaspis* Wollaston, 1864.

Aprosphaena is currently placed in synonymy with *Stenosida* Solier, 1835. It was established by Reitter, 1916 for three new species, *A. adriani*, *A. crassicornis*, and *A. indicola*. A type species was not designated. We designate here *Aprosphaena adriani* Reitter, 1916 (= *Tagenia striatopunctata* Wiedemann, 1821) as the type species of *Aphanaspis* Reitter, 1916.

Catolaena is currently placed in synonymy with *Laena* Dejean, 1821. It was established by Reitter, 1900 as a subgenus of *Laena* and included originally two species, *Laena robusta* Kraatz, 1885 and *L. turkestanica* Reitter, 1897. A type species was not designated. We designate here *Laena turkestanica* Reitter, 1897 as the type species of *Catolaena* Reitter, 1900.

Chaetotoma is currently placed in synonymy with *Pimelia* Fabricius, 1775. It was established by Motschulsky, 1860 who included five species, *Pimelia kirgisica* Motschulsky, 1860, *Tenebrio cephalotes* Pallas, 1781, *Chaetotoma globulosa* Motschulsky [unavailable], *Pimelia cursor* Ménétriés, 1832, and *Chaetotoma stellifera* Motschulsky, 1860. A type species was not designated. We designate here *Tenebrio cephalotes* Pallas, 1781 as the type species of *Chaetotoma* Motschulsky, 1860.

Cnemodasus is currently placed in synonymy with *Brachyidium* Fairmaire, 1883. It was established by Gebien, 1913 for two new species, *C. rectangulus* and *C. tenuipes*. A type species was not designated. We designate here *C. rectangulus* Gebien, 1913 as the type species of *Cnemodasus* Gebien, 1913.

Coelolophus is currently in synonymy with *Strongylium* Kirby, 1819. It was established by Mäklin, 1864 for two new species, *C. bicarinatus* and *C. schaumi*. A type species was not designated. We designate here *C. bicarinatus* Mäklin, 1864 as the type species of *Coelolophus* Mäklin, 1864.

Dendarus was introduced by Dejean, 1821 who included two available species-group names, [*Helops*] *tristis* Rossi and [*Blaps*] *emarginatus* Fabricius. While *Blaps emarginatus* is placed in *Heliopathes*, *Helops tristis* Rossi cannot by reliably identified. The Naturkundemuseum in Berlin houses the Rossi collection but is it not possible to identify with certainty any of the original specimens of *Helops tristis*. Hence, *Helops tristis* Rossi is a nomen dubium and cannot be used as the type species of *Dendarus*. Seidlitz (1893) and subsequent workers considered *Dendarus tristis* Laporte, 1840 as the type species. This is incorrect, as Laporte referred to *tristis* Herbst, 1797. The characters given by Laporte for *tristis* fit *Pandarus coarcticollis* Mulsant, 1854. In order to preserve the current usage of *Dendarus*, the ICZN will be requested to choose the species *coarcticollis* Mulsant, 1854 as type species of *Dendarus* Dejean, 1821.

Dichomma was established by Solier, 1835 who included two new species-group names, *D. duponti* and *D. maillei*. Both are currently placed in synonymy with *Dichomma dardanum* (Steven, 1829). A type species was not designated. *Dichomma duponti* Solier (= *dardanum* Steven, 1829) is here designated as the type species of *Dichomma* Solier, 1835.

Dimeriseis is currently a valid subgenus of *Erodius* Fabricius, 1775. It was established by Solier, 1834 for three species, *Erodius olivieri* Solier, 1834, *E. laevigatus* A. G. Olivier, 1791 and *E. granulosus* Solier, 1834. A type species was not designated. We designate here *Erodius laevigatus* A. G. Olivier, 1791 as the type species of *Dimeriseis* Solier, 1834.

Divieta is currently placed in synonymy with *Vieta* Hope, 1840. It was established by Reitter, 1914 for two new species, *punctipennis* and *holdhausi*, and for *Vieta costata* Allard, 1874 and *V. ovalis* Allard, 1874. A type species was not designated. We designate here *Vieta costata* Allard, 1874 as the type species of *Divieta* Reitter.

Dymonus is currently placed in synonymy with *Vieta* Hope, 1840. It was established by Solier, 1844 for four species, *D. dufossei* Solier, 1844, *D. gibbicollis* Solier, 1844, *D. tuberculatus* Solier, 1844, and *D. vestitus* Solier, 1844. A type species was not designated. We designate here *Dynomus tuberculatus* Solier, 1844 as the type species of *Dynomus* Solier.

Emypsara is currently placed in synonymy with *Callicomus* Motschulsky, 1860. It was established by Pascoe, 1866 for two new species *E. adamsii* and *E. flexuosa*. Both are currently placed in synonymy with *Callicomus riederii* (Faldermann, 1833). A type species was not designated. We designate here *Emypsara adamsii* Pascoe, 1866 as the type species of *Emypsara* Pascoe, 1866.

Eustrongylium is currently placed in synonymy with *Strongylium* Kirby, 1819. It was established by Kolbe, 1894 as a subgenus of *Strongylium* for two new species, *S. ducale* and *S. episcopale*. A type species was not designated. We designate here *Strongylium episcopale* Kolbe, 1894 as the type species of *Eustrongylium* Kolbe, 1894.

Gibbostrongylium is currently placed in synonymy with *Strongylium* Kirby, 1819. It was established by Pic, 1917 as a subgenus of *Strongylium* for three new speces, *S. medanense*, *S. violaceicolle*, and *S. pallidinotatum*, and for: "ainsi que les anciennes à élytres munis de gibbosités basales ...". A type species was not designated. We designate here *Strongylium medanense* Pic, 1917 as the type species of *Gibbostrongylium* Pic, 1917.

Hegeteromorpha is currently a valid subgenus in *Pachychila* Eschscholtz, 1831. It was introduced by Escalera, 1913 as a new genus, without any species included. Peyerimhoff, 1927 listed within *Hegeteromorpha* two species, *Pachychile externecostata* Haag-Rutenberg, 1875 and *Pachychila obtusecostata* Reitter, 1900. A type species was not designated. We designate here *Pachychile externecostata* Haag-Rutenberg, 1875 as the type species of *Hegeteromorpha* Escalera, 1913.

Heliodromus is currently placed in synonymy with *Tentyria* Latreille, 1802. It was established by Brullé, 1832 for two new species, *rotundatus* and *angulatus*. A type species was not designated. We designate here *Heliodromus rotundatus* Brullé, 1832 as the type species of *Heliodromus* Brullé, 1832.

Helops was introduced by Fabricius, 1775: 257 for nine species, including *Tenebrio caeruleus* Linnaeus, 1758, the only one currently remaining in *Helops*. Latreille, 1810: 429 subsequently designated *Tenebrio lanipes* Linnaeus, 1771 as the type species of *Helops* Fabricius. The acceptance of this type species designation would affect seriously the nomenclature. An application to set away the Latreille's 1810 fixation was submitted to the ICZN (Case no. 3422).

Histeropsis is placed currently in synonymy with *Platydema* Laporte & Brullé, 1831. It was established by Chevrolat, 1878 for 13 species, a type species was not designated. We designate here *Platydema americanum* Laporte & Brullé, 1831 as type species as the type species of *Histeropsis* Chevrolat, 1831.

Leichrodomorphus is currently placed in synonymy with *Stethotrypes* Gebien, 1914. It was established by Pic, 1921 for the following eight new species: *L. brevicornis*, *L. longicornis*, *L. latifrons*, *L. gibbosus* (now in *Derispia* Lewis, 1894), *L. rufescens* (now in *Derispia* Lewis, 1894), *L. suturifer* (synonym of *Hades raffrayi* Thomson, 1878), *L. fasciatus* (synonym of *Hades raffrayi* Thomson, 1878), and *L. laticornis* (synonym of *Stethotrypes bicornutus* Gebien, 1914). A type species was not designated. We designate here *Leichrodomorphus brevicornis* Pic, 1921 as the type species of *Leichrodomorphus* Pic, 1921.

Melancrus is currently placed in synonymy with *Oxycara* Solier, 1835. It was established by Reiche & Saulcy, 1857 for three new species, *M. laevigatum*, *M. hegetericum*, and *M. pygmaeum*. A type species was not designated. We designate here *Melancrus laevigatum* Reiche & Saulcy, 1857 as the type species of *Melancrus* Reiche & Saulcy, 1857.

Melanostola is currently placed in synonymy with *Pimelia* Fabricius, 1775. It was credited to Sénac, 1887: v who introduced it as a genus-group taxon. Sénac originally included two species, *Pimelia simplex* Solier, 1836 and *Pimelia bajula* Klug, 1830. However, *Melanostola* was introduced as an available name already by Dejean, 1836: 198. Dejean included under *Melanostola* three names, "*simplex* Dej.", "*blapsoides* Dej." and "*oblonga* Dej.". A type species was not designated by subsequent workers. The first name is available. We designate here *Pimelia simplex* Solier (published in May 1836), currently in synonymy with *Pimelia gibba* Fabricius, 1792, as the type species of *Melanostola* Dejean, 1836.

Melasmocara is currently placed in synonymy with *Melanochrus* Wollaston, 1864. Reitter, 1900 established it for an unnamed "small species from Tenerife". We designate here *Melanochrus lacordairii* Wollaston, 1864 as the type species of *Melasmocara* Reitter, 1900.

Phymatiotris is a valid genus. It was established by Solier, 1836 for *Tentyria quadricollis* Brullé, 1832, *Phymatiotris obscurus* Solier, 1836, and *Pachyscelis porphyreus* Solier, 1836. A type species was not designated. We designate here *Tentyria quadricollis* Brullé, 1832 as the type species of *Phymatiotris* Solier, 1836.

Ptilonix is currently placed in synonymy with *Misolampidius* Solsky, 1876. It was established by Allard, 1877 for five species, *rubripennis* Marseul, 1876, *strigipennis* Marseul, 1876, *laevicollis* Allard, 1877, *clavicrus* Marseul, 1876, and *indicus* Allard, 1877. A type species was not designated. *Helops clavicrus* Marseul is the only species currently included in *Misolampidius* and therefore it is designated here as the type species of *Ptilonix* Allard, 1877.

Schedarosus is currently placed in synonymy with *Adelina* Dejean, 1835. It was established by Reitter, 1876 for two new species, *S. cucujiformis* and *S. scidarius*. A type species was not designated. We designate here *S. cucujiformis* Reitter, 1876 as the type species of *Schedarosus* Reitter, 1876.

Sepidium was established by Fabricius, 1775 who included two new species, *tricuspidatum* and *cristatum*. Latreille, 1810: 429 designated subsequently *Sepidium tricuspidatus* Fabricius, 1775as type species. This act was overlooked by subsequent authors who incorrectly considered *Sepidium cristatum* Fabricius, 1775as the type species.

Sicinus is currently placed in synonymy with *Gnatocerus* Thunberg, 1814. It was established by Champion, 1886 for two new species, *Sicinus guatemalensis* and *S. brevipes*. A type species was not designated. We designate here *Sicinus guatemalensis* Champion, 1886 as the type species for *Sicinus* Champion, 1886.

Tetromma was established by Dejean, 1834 for eight available species names, including "*Hegeter*" *unicolor* Herbst, 1797 (= *Upis unicolor* Herbst, 1797). We designate here *Upis unicolor* Herbst, 1797 as the type species of *Tetromma* Dejean, 1834.

Trachyderma has been attributed to Latreille, 1829: 7. However, this genus-group name was first made available one year earlier when Latreille included a short description in Bory's Dictionnaire Classique d'Histoire Naturelle (see Latreille, 1828: 576). The following species were originally included in *Trachyderma* by Latreille (1828): *Pimelia longipes* Fabricius, 1775, "*Pimelia hispida* Fabricius" (= *Tenebrio hispidus* Forsskål, 1775), *Pimelia morbillosa* Fabricius, 1801 and *Pimelia anomala* Fischer von Waldheim, 1821. The type species "*Pimelia hispida* Fabricius" (= *Tenebrio hispidus* Forsskål, 1775) was subsequently designated by Chevrolat (1849: 628).

Trigonoscelis was established by Dejean, 1834 for five species-group names including *Pimelia nodosa* Fischer von Waldheim, 1820 and *Ocnera deplanata* Krynicki, 1832, which are available. We designate here *Pimelia nodosa* Fischer von Waldheim, 1820 as the type species of *Trigonoscelis* Dejean, 1834.

Zophoserodius is currently a valid subgenus in *Erodius* Fabricius, 1775. It was originally established by Reitter, 1914 for two species, *Erodius zophosoides* Allard, 1865 and *E. planatus* Reitter, 1914. A type species was not designated. We designate here *Erodius zophosoides* Allard, 1865 as the type species of *Zophoserodius* Reitter, 1914.

Zuercheria is currently placed in synonymy with *Strongylium* Kirby, 1819. It was established by Reitter, 1908 for two new species, *Z. sulcipennis* and *Z. matthiesseni*. *Zuercheria sulcipennis* is currently placed in synonymy with *Strongylium saracenum* (Reiche & Saulcy, 1857). A type species was not designated. *Zuercheria matthiesseni* Reitter is here designated as the type species of *Zuercheria* Reitter, 1908.

New synonymy

Graecopachys Skopin, 1968, **syn. nov.** of *Phymatiotris* Solier, 1836, based on the type species shared by both genus-group names.

Iphthimus reitteri Gebien, 1913, **syn. nov.** of *Iphthiminus italicus bellardi* (Truqui, 1857), based on study of type material.

Tetromma Dejean, 1834, **syn. nov.** of *Hyperops* Eschscholtz, 1831. The synonymy is based on the type species of *Tetromma*, *Upis unicolor* Herbst, 1797, currently placed in *Curimosphena*.

Priorities

Callicomus Motschulsky, 1860 is currently in use as a valid genus. It is homonymous with *Callicomus* Agassiz, 1846 (proposed as an emendation of *Calocomus* Audinet-Serville, 1832). *Emypsara* Pascoe, 1866 placed in synonymy with *Callicomus* Motschulsky, 1860 is therefore to be used as valid.

Colposcelis is to be credited to Dejean, 1834, and not to Lacordaire, 1859 as currently stated.

Epitrichia is to be credited to Gebler, 1859, and not to Mäklin, 1872 as currently stated.

Haemerophygus Baudi di Selve, 1876: 266 (an unavailable name in Dejean, 1834: 205 and 1836: 226) was established for *Helops mucronatus* Waltl, 1838. *Idastrandiella* E. Strand, 1929, a replacement name for the homonymous *Apolites* Jacquelin du Val, 1861, is based on the same type species as *Haemerophygus* Baudi di Selve and is therefore its junior objective synonym.

Nyctipates Solier, 1848 is currently used as a valid subgenus in *Prosodes* Eschscholtz, 1829. It was established for two species, *Nictypates coriacea* Solier, 1848 and *Prosodes carinata* Fischer von Waldheim, 1844, both junior synonyms of *Prosodes angustata* (Zoubkoff, 1833). *Nyctipates* Solier, 1848 is homonymous with *Nyctipates* Gebler, 1841. Gebler, 1841: 373 established *Nyctipates* for a new species, *rugulosa*, that is at present placed in *Prosodes* s. str. Therefore, the junior synonym of *Nyctipates* Solier, *Prosodopria* Reitter, 1909, is to be used as the correct name for this subgenus.

Vieta is to be credited to Hope, 1840, and not to Laporte, 1840 as currently stated. Hope's work came out before October 1840 (see *Journal of Proceedings of the Entomological Society of London*; Additions to the library, from December 31, 1836 to September 30, 1840) while the Laporte's work came out on December 26, 1840 (see Evenhuis N. L., 1997: *Litteratura taxonomica dipterorum* (*1758-1930*). *Volume II. L-Z.* Leiden: Backhuys Publishers, iv + 427-871 pp.).

Comments

Adesmia dilatata verniculata was intrudiced by Peyerimhoff, 1935 as an aberration. It was valided by Koch, 1940 to whom the name is to be credided.

Akis spinosa (Linnaeus, 1764) is according to J. Ferrer (in litt.) misinterpreted by authors and is the same as *sculptior* Koch, 1935, currently considered as a "variety" of *Akis elevata* Solier, 1837. As this information was not published before the present volume of the Catalogue was completed, it remained unrespected.

Blaps gages (Linnaeus, 1767) was incorrectly credited to Fabricius by subsequent workers. Linnaeus, 1767 gave the name *Tenebrio gigas* to two distinct species, one from Surinam (p. 674), the second from Mediterraneous area (p. 676). He replaced his second *Tenebrio gigas* by *gages* in the Errata printed at the end of his work. This species is widely distributed and well known as *Blaps gigas*, while the *Tenebrio gigas* from the page 674 was placed by Laporte (1840: 212) in *Nyctobates* Guérin-Méneille. Therefore, the name *gages* Linnaeus, 1767 is not resurrected here.

Eumicrositus is currently used as a valid subgenus in *Phylan* Dejean, 1821, and credided to Español, 1947. Español (1947) failed to fix the name by a type species designation. Viñolas (1990) designated *Pedinus ulyssiponensis* [sic] Germar, 1824 as the type species of *Eumicrositus* and characterized the group. Although he credited *Eumicrositus* to Español, the name became available by his act and is to be credited to him.

Heterophaga is incorrectly credited to L. Redtenbacher, 1845: 127. The name *Heterophaga* was proposed by Dejean, 1834: 199 and is currently placed in synonymy with *Alphitobius* Stephens, 1829. Duponchel, 1845: 601 designated *Tenebrio mauritanicus* Fabricius, 1792 as type species of *Heterophaga*.

Oedenocera Reiche, 1862 was introduced as a replacement name for the homonymous *Pachycera* Eschscholtz, 1831 (not *Pachycera* Billberg, 1829). In fact, Reiche (1862: 372) believed, that *Pachycera* and *Pachycerus* are homonyms, and ignored the existence of the senior *Pachycera* Billberg, 1829. In addition, his concept of *Pachycera* Eschscholtz differed from the actual one because all included species, *Oedenocera biseriata* Reiche, 1862, and *minuta* Solier, 1838, *corsica* Solier, 1838, *pumila* Solier, 1838, and *subcostata* Solier, 1838 were considered to belong to the "deuxième groupe dans le genre *Tagenia* ...". Koch (1943: 529) and subsequent authors used *Oedenocera* Reiche as an invalid name within *Dichillus* Jacquelin du Val, 1861, a synonym of *Dichillocerus* Reitter, 1916 or a valid subgeneric name in *Dichillus* Jacquelin du Val, 1861.

Tentyria: *Pimelia glabra* Fabricius, 1775 is one of the two originally included species, and usually considered as the type species of *Tentyria*, although Latreille's 1810 subsequently designated *Pimelia orbiculata* Fabricius, 1775 as the type species of *Tentyria*. While the identity of *P. glabra* Fabricius is unknown, *P. orbiculata* Fabricius, 1775 is the type species of *Tentyrina* Reitter, 1900. In order to preserve the current concepts of the well-known genera *Tentyria* and *Tentyrina*, the ICZN was asked to set aside the previous fixations of the type species of *Tentyria*, and to fix *Tentyria* Latreille by *Tentyria ligurica* Solier, 1835 as type species. The case is pending on a ruling of the ICZN (case number 3442).

Unavailable genus-group names

In addition to the unavailable genus-group names treated above, and those discussed in G. S. Medvedev, 1998b, Löbl & Merkl, 2003, and Merkl & Masumoto, 2007 (*Blaptyscelis* Koch, 1961, *Catomodontus* Koch, 1935, *Cyphostethoides* Kaszab, 1979, *Lepidocnemeplatia* Kaszab, 1938, *Macradesmia* Koch, 1944, *Macropodesmia* Koch, 1944, *Oteroscelopsis* Koch, 1944, *Paramisolampidius* Nakane, 1968, *Pseudopachyscelis* Skopin, 1962, *Saxistena* Koch, 1940, *Subtentyrina* Koch, 1939, *Tetranosis* Koch, 1940), following names currently used or listed as synonyms are recognized as unavailable:

Dendarus Latreille, 1829: 20 is currently considered as an invalid but available name. It is based on an unavailable type species, *Dendarus tristis* Laporte, 1840 and is therefore unavailable.

Emsipara Reitter, 1916b: 10 is unavailable. Reitter probably referred to *Emypsara* Pascoe, 1866, and misspelled the name.

Homalopus is currently placed in synonymy with *Pimelia* Fabricius, 1775 (Kwieton, 1977b). It was used as a valid subgeneric name in *Pimelia* (Gebien, 1937: 824), and credited either to Solier, 1836: 78 who expressly stated he is not using the name, or to Chevrolat in Dejean, 1836: 422 where it is unavailable and refers to a chrysomelid.

Oogeton Miwa, 1939: 412 is an unavailable name because it was introduced without type species fixation.

Pachychila pedinoides (Eschscholtz, 1831) is a name currently used for a North African species considered valid. This name is unavailable because it was introduced without definition. Therefore, *Tentyria brevis*

Besser, 1832, incorrectly placed in synonymy with the latter name, is here used as the valid name for this species.

Pseudoammobius canariensis Franz, 1996: 131 published as a combined genus and species description for a new Canarian taxon is unavailable (ICZN, Art. 13.4) and not included in the present Catalogue.

Oedemeridae

V. Švihla

New synonyms

Chrysanthia distinctithorax Pic, 1904, **syn. nov**. of *Chrysanthia chalcochroa chalcochroa* Fairmaire, 1892
Chrysanthia cuprina [*viridissima* var.], **syn. nov**. of *Chrysanthia viridissima* (Linnaeus, 1758)
Anoncodes zimmermanni Gistel, 1857, **syn. nov**. of *Anogcodes ustulatus* (Scopoli, 1763)
Nacerda fulvicollis var. *longicollis* Roubal, 1936, **syn. nov**. of *Anogcodes fulvicollis* (Scopoli, 1763)
Oedemera abdominalis A. G. Olivier, 1795, **syn. nov**. of *Oedemera virescens* (Linnaeus, 1767)

New combinations

Asessinia iriomotoensis (Miyatake, 1985), **comb. nov**. from *Alloxacis*
Asessinia vittata (Miyatake, 1985), **comb. nov**. from *Alloxacis*
Asessinia geniculata (Kôno, 1937), **comb. nov**. from *Alloxacis*
Asessinia flavomarginata (Miyatake, 1985), **comb. nov**. from *Alloxacis*
Eobia sinensis (Gemminger, 1870), **comb. nov**. from *Sessinia*
Nacerdochroa concolor (Pic, 1942), **comb. nov**. from *Sessinia*

New replacement name

Nacerdochroa pici **nom. nov**. for *Nacerdochroa concolor* (Pic, 1942) [see above, secondary homonym of *Nacerdochroa concolor* (Brullé, 1838)]

Meloidae

I. Löbl

Comments

Mylabris Fabricus, 1775 and *Hycleus* Latreille, 1817 are species-rich and well-known genera. *Meloe cichorii* Linnaeus, 1758 was designated as the type species of *Mylabris* by Latreille, 1810, and is placed on the Official list of Specific Names in Zoology (Opinion 1809, *Bulletin of Zoological Nomenclature* **52**: 208, 1995). The same Opinion gives *Meloe cichorii* Linnaeus, 1758 as the type species of *Mylabris*. *Meloe cichorii* Linnaeus, 1758 was placed in *Hycleus*, and is still a member of that genus. Thus, both genus-group names, *Mylabris* and *Hycleus*, are in fact synonyms. Nevertheless, this synonymy is ignored in the current use of the generic names. The Linnean syntypes of *Meloe cichorii* consist of two species, one currently known as *Hycleus cichorii* (Linnaeus), the second as *Mylabris variabilis* (Pallas). The case was discussed in more detail by Bologna & Pinto, 2001: 2066-2067. A lectotype was not designated for *Meloe cichorii* Linnaeus, and therefore this name may be applied to any of the two involved species. It is in interest of the stability of nomenclature to preserve the current generic concepts and to fix the names by appropriate acts.

Priority and authors

Apalus bipunctatus is currently credited to Germar, 1831: 6. The name was introduced and made available by Megerle von Mühlfeld, 1812: 18

Cerocoma malatyensis Kaszab was introduced as an available name by Kaszab in 1951: 266, and not in his earlier paper (Kaszab, 1941: 679) where it was proposed as an infrasubspecific name.

Meloidae

M. A. Bologna

New combinations

Hycleus Latreille, 1817. Bologna (1991) and Bologna & Pinto (2002) gave evidence that all species previously placed in the subgenera *Euzonabris, Sphenabris* and *Trigrabis* (Kuzin, 1954) of *Mylabris*, or in the genera *Decapotoma* Voigts, 1902 and *Gorrizia* Pardo Alcaide, 1950, are actually members of *Hycleus*. As only some new combinations were published, the following combinations are new:

Hycleus aegyptiacus (Marseul, 1870), **comb. nov.** from *Mylabris*
Hycleus allardi (Marseul, 1870), **comb. nov.** from *Coryna*
Hycleus apicipennis (Reiche, 1865), **comb. nov.** from *Mylabris*
Hycleus arabicus (Pallas, 1781), **comb. nov.** from *Mylabris*
Hycleus belli (Borchmann, 1940), **comb. nov.** from *Mylabris*
Hycleus biguttatus (Gebler, 1841), **comb. nov.** from *Mylabris*
Hycleus bipunctatus (A. G. Olivier, 1811), **comb. nov.** from *Mylabris*
Hycleus birecurvus (Marseul, 1870), **comb. nov.** from *Coryna*
Hycleus bistillatus (Tan, 1981), **comb. nov.** from *Mylabris*
Hycleus borchmannianus (Kaszab, 1983), **comb. nov.** from *Mylabris*
Hycleus brevetarsalis (Kaszab, 1960), **comb. nov.** from *Mylabris*
Hycleus chodschenticus (Ballion, 1878), **comb. nov.** from *Mylabris*
Hycleus cingulatus (Faldermann, 1837), **comb. nov.** from *Mylabris*
Hycleus colligatus (L. Redtenbacher, 1850), **comb. nov.** from *Mylabris*
Hycleus concinnus (Marseul, 1870), **comb. nov.** from *Mylabris*
Hycleus curticornis (Pic, 1919), **comb. nov.** from *Mylabris*
Hycleus damohensis (Saha, 1979), **comb. nov.** from *Mylabris*
Hycleus diversesignatus (Pic, 1919), **comb. nov.** from *Mylabris*
Hycleus dohrni (Marseul, 1872), **comb. nov.** from *Mylabris*
Hycleus dolens (Marseul, 1870), **comb. nov.** from *Coryna*
Hycleus esfandiarii (Kaszab, 1969), **comb. nov.** from *Decapotoma*
Hycleus flavohirtus (Kaszab, 1958), **comb. nov.** from *Decapotoma*
Hycleus gratiosus (Marseul, 1870), **comb. nov.** from *Mylabris*
Hycleus hanguensis (Kaszab, 1958), **comb. nov.** from *Mylabris*
Hycleus himalayaensis (Saha 1979), **comb. nov.** from *Mylabris*
Hycleus hokumanensis (Konô, 1940), **comb. nov.** from *Mylabris*
Hycleus horai (Saha, 1972), **comb. nov.** from *Mylabris*
Hycleus infasciculatus (Pic, 1929), **comb. nov.** from *Mylabris*
Hycleus japonicus (Sumakov, 1913), **comb. nov.** from *Mylabris*
Hycleus javeti Marseul, 1870), **comb. nov.** from *Mylabris*
Hycleus kirgisicus (Aksentjev & Sadykova, 1990), **comb. nov.** from *Mylabris*
Hycleus lacteus (Marseul, 1870), **comb. nov.** from *Mylabris*
Hycleus ligatus (Marseul, 1870), **comb. nov.** from *Mylabris*
Hycleus lindbergi (Kaszab, 1973), **comb. nov.** from *Mylabris*
Hycleus linnavuorii (Pardo Alcaide, 1963), **comb. nov.** from *Mylabris*
Hycleus mannheimsi (Kaszab, 1961), **comb. nov.** from *Mylabris*
Hycleus mediozigzagus (Pic, 1924), **comb. nov.** from *Mylabris*
Hycleus mediobipunctatus Pic, 1919), **comb. nov.** from *Mylabris*
Hycleus mediofasciatellus (Pic, 1908), **comb. nov.** from *Mylabris*
Hycleus medioinsignatus (Pic, 1909), **comb. nov.** from *Mylabris*
Hycleus nigrohirtus (Kaszab, 1983), **comb. nov.** from *Decapotoma*
Hycleus ocellaris (A. G. Olivier, 1795), **comb. nov.** from *Coryna*
Hycleus ornatus (Reiche, 1865), **comb. nov.** from *Coryna*
Hycleus parvulus Frivaldszky, 1892), **comb. nov.** from *Mylabris*
Hycleus phaleratus (Pallas, 1781), **comb. nov.** from *Mylabris*
Hycleus pierrei (Kaszab, 1968), **comb. nov.** from *Mylabris*
Hycleus pitcheri (Kaszab, 1983), **comb. nov.** from *Coryna*
Hycleus postbilunulatus Pic, 1919), **comb. nov.** from *Mylabris*
Hycleus posticatus (Fairmaire, 1892), **comb. nov.** from *Mylabris*
Hycleus praeustus (Fabricius, 1793), **comb. nov.** from *Mylabris*
Hycleus pseudobrunnipes (Kaszab, 1983), **comb. nov.** from *Mylabris*
Hycleus pustulatus (Thunberg, 1791), **comb. nov.** from *Mylabris*
Hycleus quatuordecimpunctatus (Pallas, 1781), **comb. nov.** from *Mylabris*
Hycleus raphael (Marseul, 1876), **comb. nov.** from *Mylabris*

Hycleus rotroui (Pic, 1930), **comb. nov**. from *Coryna*
Hycleus rubricollis (Marseul, 1875), **comb. nov**. from *Coryna*
Hycleus scabratus (Klug, 1845), **comb. nov**. from *Mylabris*
Hycleus scapularis Klug, 1845), **comb. nov**. from *Mylabris*
Hycleus schauffelei (Kaszab, 1957), **comb. nov**. from *Mylabris*
Hycleus schoenherri (Billberg, 1813), **comb. nov**. from *Mylabris*
Hycleus sialanus (Pic, 1929), **comb. nov**. from *Mylabris*
Hycleus silbermanni (Chevrolat, 1838), **comb. nov**. from *Mylabris*
Hycleus solanensis (Saha, 1979), **comb. nov**. from *Mylabris*
Hycleus soumacovi (Pic, 1930), **comb. nov**. from *Mylabris*
Hycleus subparallelus (Pic, 1919), **comb. nov**. from *Mylabris*
Hycleus talhouki (Kaszab, 1983), **comb. nov**. from *Mylabris*
Hycleus talyshensis (Sumakov, 1929), **comb. nov**. from *Mylabris*
Hycleus tekkensis (Heyden, 1883), **comb. nov**. from *Mylabris*
Hycleus tenuepictus (Fairmaire, 1892), **comb. nov**. from *Mylabris*
Hycleus theryanus (Pic, 1940), **comb. nov**. from *Mylabris*
Hycleus thunbergi (Billberg, 1813), **comb. nov**. from *Mylabris*
Hycleus tigripennis (Marseul, 1870), **comb. nov**. from *Mylabris*
Hycleus trianguliferus (Heyden, 1883), **comb. nov**. from *Mylabris*
Hycleus varius (A. G. Olivier, 1811), **comb. nov**. from *Mylabris*
Hycleus wagneri (Chevrolat, 1838), **comb. nov**. from *Mylabris*

Spelling

Stenoria muiri Kaszab, 1956. The original spelling "*S. mairi*" is obviously incorrect and probably a typographical error. The species was named in honor of F. Muiri, and Kaszab, 1956: 632 used the correct spelling for "*Stenoria muiri* ab. *arabica*".

Comments

Deridea notata minor Kaszab, 1960. The validity of this subspecies is doubtful.
Alosimus albovillosus (Escalera, 1914) and *A. filitarsis* (Escalera, 1925) are possibly synonyms of *A. tenuicornis* (Escalera, 1914).
Alosimus crabrerai (Escalera, 1923) and *A. crassicornis* (Escalera, 1914) are possibly synonyms of *A. robustus* (Escalera, 1914).
Calydus escherichi Reitter, 1898 is possibly a synonym of *C. semenowi* (Escherich, 1896).
Lydulus uzbekistanus Kaszab, 1981 is possibly a synonym of *L. semipurpureus* Reitter, 1902.
Lydus quadrimaculatus (Tauscher, 1812) is possibly a synonym of *L. trimaculatus* (Fabricius, 1775).
Lytta flaviventris (Ballion, 1878). The subgeneric assignment of this species needs confirmation.
Lytta vesicatoria (Linnaeus, 1758). The subspecies *heydeni* (Haag-Rutenberg, 1880), *freudei* Kaszab, 1958 and *togata* Fischer von Waldheim, 1844 may be considered as distinct species.
Lytta selanderi Saha, 1979. The validity of this species needs confirmation, as well as its subgeneric assignment.
Oenas sericeus (A. G. Olivier, 1795) is possibly a synonym of *O. afer* (Linnaeus, 1767).
Sybaris flavus (Thunberg, 1791) and *S. garhwalensis* Saha, 1979. The validity of both species is uncertain.
Sybaris immunis Stephens, 1832 is not considered in the Catalogue. It was erroneously described from England. The group is Palaeotropical and this species is possibly Indian.
Cerocoma azurea Reitter, 1913 is possibly a distinct species, as suggested by Dvořák (1993).
Epicauta kwangsiensis Tan, 1958 is a possible synonym of *E. hirticornis* (Haag-Rutenberg, 1880).
Hycleus diversesignatus (Pic, 1919) is a doubtful species, possibly synonym of *H. wagneri* (Chevrolat, 1838).
Hycleus euphraticus (Marseul, 1870) is a doubtful species, possibly synonym of *H. scabiosae* (A. G. Olivier, 1811).
Hycleus heteroclytus (Pic, 1919) is not listed in the Catalogue. It was described from "Sarepta" (south Russia), but it is probably a South African species.
Hycleus infasciculatus (Pic, 1929) is a doubtful species, possibly synonym of *H. duplicatus* (Klug, 1845).
Hycleus japonicus (Sumakov, 1913) is a doubtful species, possibly synonym of *H. thunberghi* (Billberg, 1813).
Hycleus soumacovi (Pic, 1930) is possibly a synonym of *Hycleus kuschkensis* (Pic, 1919).

Mylabris Fabricius, 1775. The subgeneric names *Chalcabris* Kuzin, 1954 and *Micrabris* Kuzin, 1954 are currently valid but probably synonyms.

Mylabris crux var. *opulenta* Escherich, 1904 is placed doubtfully in synonymy with *M. crux*. It was referred to *M. undecimnotata* Heyden, 1883 by Sumakov (1930), and to *M. mannerheimii* Gebler, 1844 by Kuzin (1954).

Mylabris turkestanica Dokhtouroff, 1890 is a doubtful species. It considered as a variety and was referred to *M. undecimnotata* Heyden, 1883 by Sumakov (1930), and to *M. mannerheimii* Gebler, 1844 by Kuzin (1954).

Mylabris undecimnotata Heyden, 1883 is a doubtful species, considered by Kuzin (1954) as a variety of *M. mannerheimii* Gebler, 1844

Mylabris quadripunctata (Linnaeus, 1767). The subspecies are poorly defined. The validity of *restricta* Motschulsky, 1849, *angustata* Sumakov, 1915, and *mesoasiatica* Kuzin, 1953 needs confirmation.

Mylabris theryi Abeille de Perrin, 1894. The subgeneric assignment of this species needs confirmation.

Mylabris sairamensis Ballion, 1878. The subgeneric assignment of this species needs confirmation.

Mylabris semivittata Pic, 1947 is a doubtful species; possibly synonym of *M. splendidula* Pallas, 1781.

Mylabris staudingeri Heyden, 1881. It is considered as a synonym of *M. splendidula* Pallas, 1781, but actually it may be a synonym of *M. frolovi* Germar, 1824 or *M. lucens* Escherich, 1904.

Mylabris uhagoni Martinez y Sáez, 1873. This species should be probably placed in a new subgenus.

Mylabris beckeri Escherich, 1890 is a possible synonym of *M. sibirica* Fischer von Waldheim, 1823.

Mylabris frolovi Fischer von Waldheim, 1823. Several subspecies are currently considered as valid but need confirmation.

Mylabris inculta (Escherich, 1899) is a possible synonym of *M. modesta* Escherich, 1899.

Mylabris liliputana (Escherich, 1899) is a possible synonym of *M. steppensis* (Dokhtouroff, 1889).

Mylabris pentheri Ganglbauer, 1905 is a possibile synonym of *M. laevicollis* Marseul, 1870.

Mylabris cinctoides Kaszab, 1958. The subgeneric assignment of this species needs confirmation.

Mylabris omocrates Wellman, 1910 is a doubtful species.

Mylabris maceki Dvořák, 1985. A new subgenus should probably be established for this species (see Bologna & Pinto, 2002).

Mylabris tarbagatajensis Pic, 1919. It probably belongs to the subgenus *Micrabris* Kuzin, 1954.

Mylabris, species incertae sedis: several of them are possibly members of *Hycleus*.

Meloe proscarabaeus Linnaeus, 1758. Some of the synonyms, as *cyaneus* Mulsant, 1857, *undulatus* Baudi di Selve, 1878, *tauricus* Baudi di Selve, 1878, may represent valid subspecies.

Meloe strigosus Motschulsky, 1872 is currently placed in synonymy with *M. violacues* Marsham, 1802. It may represent a distinct taxon from FE.

Meloe specularis Gredler, 1878 and *M. stellatus* Pliginskij, 1923 are currently placed in synonymy with *M. cavensis* L. Petagna, 1819 but may represent distinct species.

Meloe, the subgenera *Coelomeloe* Reitter, 1911 and *Eurymeloe* Reitter, 1911 are currently considered valid. They may be synonyms, based on larval features.

Meloe austrinus Wollaston, 1854. Possible synonym of *M. brevicollis* Panzer, 1793. Other taxa may be referred to *M. brevicollis*, as synonyms or valid subspecies.

Meloe marianii Kaszab, 1983 is possibly a synonym of *M. saharensis* Chobaut, 1898.

Meloe bilineatus Aragona, 1830. Doubtful species; it may be a senior synonym of *M. mediterraneus* J. Müller, 1925.

Meloe sanaanus Borchmann, 1938 is currently considered valid. It may be a synonym of *M. lefevrei* Guérin-Méneville, 1849.

Meloe asperatus Tan, 1981. The subgeneric assignment of this species needs confirmation.

Oreomeloe Tan, 1981. The validity of this genus needs confirmation; its relationships are unknown.

Coriologiton Marseul, 1879 is currently in synonymy with *Apalus* Fabricius, 1775 but may be valid.

Ctenopus carinifer Semenov, 1896, *C. hauseri* Reitter, 1894, and *C. oxianus* Semenov, 1894 are currently considered valid. They may be synonyms of *C. sinuatipennis* Fairmaire, 1892.

Euzonitis Semenov, 1893 is currently considered as a genus. It may be placed as subgenus of *Zonitis* Fabricius, 1775.

Euzonitis alfierii (Pic, 1929) is possibly a synonym of *E. sexmaculata* (A. G. Olivier, 1789).

Euzonitis haroldi (Heyden, 1880) and *E. unimaculata* Pic, 1933 are possibly synonyms of *E. quadrimaculata* (Pallas, 1782).

Euzonitis varentzowi (Semenov, 1894) is possibly a synonym of *E. rubida* (Ménétriés, 1832).
Euzonitis jacobsoni Reichardt, 1934. This species should probably be placed in a separate, new genus.
Megatrachelus pallidipennis (Motschulsky, 1845), *M. politus* (Gebler, 1832), *M. quadricollis* (Fairmaire, 1892), and *M. sibiricus* (Tauscher, 1812) are possibly synonyms.
Nemognatha flavicornis Stierlin, 1876 is possibly a synonym of *N. chrysomelina* (Fabricius, 1775).
Nyadatus Aksentjev, 1981 is currently considered as a valid genus. It is possibly a synonym of *Stenoria* Mulsant, 1857.
Sitaris flavescens Betta, 1847 and *Sitaris fuliginosa* Betta, 1847 were described from Ticino (northwest Italy or south Switzerland). The type material is not available and the respective descriptions are not informative enough to assign the names, and both were not mentioned in subsequent papers. Therefore, they are not inlcuded in the taxonomic part of the Catalogue.
Sitaromorpha Dokhtouroff, 1889. The validity of this genus needs confirmation; it may be based on teratological specimens of *Sitaris* Latreille, 1802.
Stenoria nigricollis (Semenov, 1893) is possibly a synonym of *S. intermedia* (Dokhtouroff, 1889).
Stenoria tibialis (Motschulsky, 1872) is possible a synonym of *S. fasciata* (Faldermann, 1835).
Zonitis presuopraeusta Kaszab, 1958 is possible a synonym of *Z. nitritarsis* (Stierlin, 1876).

Unavailable names

Androplicata Pardo Alcaide, 1948 and *Mesolaevigata* Pardo Alcaide, 1950, are currently considered synonyms of *Micrabris* Kuzin. They are unavailable because they are not fixed by a type species designation.

Pyrochroidae

D. Pollock & D. K. Young

Designation of type species

Corphyra: Say (1835) did not specify a type species for his new *Corphyra,* originally proposed as a subgenus within *Lagria.* Herein, we designate *Anthicus lugubris* Say, 1827, a species originally assigned by Say to his *Lagria* (*Corphyra*), as the type species of *Corphyra.*
Eupyrochroa: Blair (1914) did not designate the type species for his new genus *Eupyrochroa.* Herein, we designate *Pyrochroa insignata* Fairmaire, 1894, a species originally assigned by Blair to *Eupyrochroa*, as the type species of the genus.

Salpingidae

D. Pollock

Spelling

Schizotus yamaguchii Kôno, 1936 is a present emendation of the incorrect original spelling "*yamauchii*".

Anthicidae

D. S. Chandler, G. Nardi & D. Telnov

Comments

Tenuicollis pauperculus (LaFerté-Sénectère, 1847): *Anthicus pauperculus* LaFerté-Sénectère, 1847, *Anthicus pumilus* Baudi di Selve, 1877, and *Anthicus lucasi* Marseul, 1879, all from Algeria, have long been confused. Marseul (1879), in reversed date priority, listed *A. lucasi* Bris. [sic] as the senior synonym of *A. pumilus* twice in the index, but failed to cite this synonymy under the species treatment of *A. lucasi.* Pic (1892) correctly reversed the order of Marseul's synonymy such that *A. pumilus* was now the senior synonym to *A. lucasi*, and later (1894: 69) placed *A. pauperculus* as a variety of *A. pumilus* and *A. lucasi* as a synonym of *A. olivaceus* LaFerté-Sénectère, 1847. However, *A. pauperculus* was the only one of these three names actually listed in the various Palaearctic catalogues until Pic (1911) and Winkler (1927) continued treating *A. pumilus* as the valid name, with *A. pauperculus* listed as a variety and *A. lucasi* as a questionable synonym. *Tenuicollis pauperculus* is here recognized as the senior synonym to *A. pumilus* and *A. lucasi.*

Amblyderus sabulosus (LaFerté-Sénectère, 1849): Uhmann & al. (2005) created some confusion in their treatment of *Amblyderus sabulosus* LaFerté-Sénectère, 1849. Many of the following geographic records for this species are explicitly attributed to *Tomoderus syriacus* Pic, 1907, while the taxonomic listing of *T. syriacus* was omitted. This was an error, and not to be construed as an act of synonymy.

Aderidae

G. Nardi

Comments

Mixaderus: Báguena Corella (1948, 1962) intented to establish new genus-group names but failed to designate type species. *Mixaderus* is one of those names. It was made available in the work of Collado & Alonso-Zarazaga, 1996.

The original generic placements of names introduced by Pic (1897) are here given as "(*Xylophilus*)" because they were listed in *Olotelus* which was clearly considered as a subgenus (cf. Pic, 1897i: 82).

The subfamily and tribal names obscure relationships and are not given in the Catalogue.

Scraptiidae

P. Leblanc & B. Levey

New synonymy

Anaspis bilaciniata Pic, 1918, **syn. nov**. of *Anaspis garneysi* Fowler, 1889. Based on study of type material.

Anaspis abollata Gozis, 1882, **syn. nov**. of *Anaspis frontalis* (Linnaeus, 1758). Based on original description.

Scraptia clairii Rey, 1892, **syn. nov**. of *Scraptia bifoveolata* Küster, 1853. Based on study of type material.

Scraptia oertzeni Schilsky, 1900, **syn. nov**. of *Trotommidea revelierei elongata* Pic, 1896. Based on study of type material.

Scraptia subfoveolata Reitter, 1889, **syn. nov**. of *Scraptia thoracica* Baudi de Selve, 1877. Based on study of type material.

New combinations

Anaspella desertifera (Chobaut, 1898), **comb. nov**. from *Anaspis.* Based on study of type material.

Anaspella lutea (Marseul, 1876), **comb. nov**. from *Anaspis.* Based on study of type material.

Comments

Anaspis fuscescens Stephens, 1832: Examination of the type series of *Anaspis fuscescens* Stephens has shown that it is a senior synonym of *Anaspis* (*Nassipa*) *costai* Emery, 1876 not a synonym of *Anaspis thoracica* (Linnaeus, 1758) as currently accepted. Since *Anaspis fuscescens* Stephens, 1832 has not been used as a valid name since before 1899 and *Anaspis* (*Nassipa*) *costai* Emery, 1876 has been used by all authors as the presumed valid name of this taxon since 1876, a reversal of priority is needed, and is published separately.

Anaspis humeralis sensu auct. refers to *Mordella humeralis* Fabricius, 1775 and has been commonly used for a European *Anaspis* species, while its senior synonym, *Mordella fasciata* Forster, 1771 was not used as valid. *Anaspis humeralis* (Fabricius, 1775) is unavailable as Fabricius attributes the name to *Mordella humeralis* Linnaeus, 1758 which is currently placed in the genus *Mordellistena* (Mordellidae).

Anaspis (s. str.) *maculata* (Geoffroy, 1785) is currently used as valid name for a common European species. It is a junior synonym of *Anaspis melanopa* (Forster, 1771). A reversal of priority for these names is published separately.

Anaspis (*Silaria*) *quadrimaculata* Gyllenhal, 1817 is currently used as valid name for a common European species. It is a junior synonym of *Anaspis bipustulata* (Bonelli, 1812). A reversal of priority for these names is published separately.

CATALOGUE

family MYCETOPHAGIDAE Leach, 1815
subfamily **Mycetophaginae** Leach, 1815
tribe Mycetophagini Leach, 1815

genus *Eulagius* Motschulsky, 1845a: 92 type species *Eulagius acernus* Motschulsky, 1845
Atritomoides Nikitsky, 1988b: 1847 type species *Eulagius ussuriensis* Nikitsky, 1988
Parabaptistes Ganglbauer, 1899: 838 type species *Tritoma univestris* Reitter, 1878
acernus Motschulsky, 1845a: 92 **E**: AB AR GG ST (Kavkaz)
univestris Reitter, 1878a: 166 (*Tritoma*)
chinensis Nikitsky, 1996a: 1372 **A**: SCH
dentatus Nikitsky, 1988b: 1847 **A**: FE
filicornis Reitter, 1887c: 288 (*Atritomus*) **E**: FR GBi **N**: AG TU
boissyi Caillol, 1925: 101 (*Atritomus*)
irregularis Reitter, 1888a: 174 (*Atritomus*) **E**: AB AR GG ST (Kavkaz)
lewisi Reitter, 1889d: 249 (*Atritomus*) **A**: JA TA
reitteri Lewis, 1895a: 331 (*Atritomus*) **A**: FE JA
ussuriensis Nikitsky, 1988b: 1845 **A**: FE JA
irregularis Miyatake, 1985: 288 (*Parabaptistes*) [HM]

genus *Litargops* Reitter, 1880a: 89 type species *Litargus pictus* Wollaston, 1854
maculosus Reitter, 1889d: 247 **A**: FE JA
pictus Wollaston, 1854: 200 (*Litargus*) **N**: MR

genus *Litargus* Erichson, 1846: 415 type species *Ips bifasciatus* Fabricius, 1787
subgenus *Alitargus* Casey, 1900: 136 type species *Litargus balteatus* LeConte, 1856
balteatus LeConte, 1856: 14 **E**: AU AR CZ FR GB GE GG HU IT MA NLi PL SK SL SP ST SVi SZ **N**: CI EG MO MR **A**: JA **AUR NAR NTR ORR**
antennatus Miyatake, 1957: 41
disjunctus Sharp, 1902: 639
ferrantei Reitter, 1908c: 44
infulatus LeConte, 1856: 14
pilosus Wollaston, 1857a: 71
tranisversus LeConte, 1856: 14
subgenus *Litargosomus* Motschulsky, 1858b: 52 type species *Litargosomus maculatus* Motschulsky, 1858
coloratus Rosenhauer, 1856: 105 **E**: FR GR IT MA PT SP SVi SZ **N**: AG AZ CI MO MR TU **A**: IS SY TR **AFR**
obscuripennis Pic, 1901g: 66
trifasciatus Wollaston, 1864: 154
japonicus Reitter, 1877c: 27 **A**: JA
kyushuensis Miyatake, 1985: 287 **A**: FE JA
lewisi Reitter, 1889d: 246 **A**: CH JA
sexsignatus Miyatake, 1957: 40 **A**: JA
unifasciatus Reitter, 1889d: 247 **A**: FE JA
vestitus Sharp, 1879: 88 **A**: JA **Pacific**
subgenus *Litargus* Erichson, 1846: 415 type species *Ips bifasciatus* Fabricius, 1787
caucasicus Pic, 1894h: 71 **E**: "Caucasus"
connexus Geoffroy, 1785: 138 (*Antribus*) **E**: AB AL AR AU BE BH BU BY CR CT CZ DE EN FI FR GB GE GG GR HU IR IT KZ LA LS LT MA MC MD NL NR NT PL PT RO SK SL SP ST SV SZ TR UK YU **N**: CI MO TU **A**: *CH CY ES FE IN JA JO KI KZ MG NC NE SY TR UZ*
bifasciatus Fabricius, 1787a: 47 (*Ips*)
lunatus Fabricius, 1792b: 514 (*Ips*)

mediojunctus Pic, 1903: 191
signatus Panzer, 1798: 20 (*Mycetophagus*)
leprieuri Pic, 1901g: 65 **N**: AG TU

genus *Mycetophagus* Fabricius, 1792b: 497 type species *Silpha quadrimaculata* Schaller, 1783
subgenus *Arnoldiellus* Nikitsky, 1989b: 60 type species *Tritoma tschitscherini* Reitter, 1897
tschitscherini Reitter, 1897e: 228 (*Tritoma*) **E**: CT LA NT ST SVi **A**: ES FE MG NC WS
tschitscherini Semenov, 1899b: 549 (*Tritoma*) [HN]
subgenus *Ilendus* Casey, 1900: 131 type species *Mycetophagus pluripunctatus* LeConte, 1856
irroratus Reitter, 1879: 224 (*Tritoma*) **A**: ES FE JA NE
lederi Reitter, 1897e: 228 (*Tritoma*) **A**: ES MG
multipunctatus Fabricius, 1792b: 498 **E**: AB AL AR AU BE BH BU BY CR CT CZ DE EN FI FR GB GE GG HU IT KZ LA LT MC MD NL NR NT PL RO SK ST SV SZ TR UK YU **A**: KZ ES WS
similis Marsham, 1802: 140 (*Boletaria*)
sulcatulus Roubal, 1929a: 47
subgenus *Kovalevius* Nikitsky, 1992b: 415 type species *Tritoma grandis* Reitter, 1889
grandis Reitter, 1889d: 248 (*Tritoma*) **A**: FE (S. Kuril Is., Sakhalin) JA
simplex Reitter, 1889d: 248
obsoletesignatus Miyatake, 1968: 31 **A**: JA
subgenus *Micromycetophagus* Nikitsky, 2007: 58 type species *Tritoma undulata* Reitter, 1889 (=*Mycetophagus reitteri* Nikitsky, 2008)
reitteri Nikitsky, **nom. nov**. [see New Acts] **A**: FE JA
undulatus Reitter, 1889d: 249 (*Tritoma*) [HN]
subgenus *Mycetophagoides* Nikitsky, 1988b: 1849 type species *Mycetophagus livshitzi* Nikitsky, 1988
alni Champion, 1917: 53 **A**: UP
livshitzi Nikitsky, 1988b: 1848 **A**: FE
hillerianus Reitter, 1877d: 116 **A**: FE JA NC NE
subgenus *Mycetophagus* Fabricius, 1792b: 497 type species *Silpha quadrimaculata* Schaller, 1783
Boletaria Marsham, 1802: 138 type species *Chrysomela quadripustulata* Linnaeus, 1760
Silphoides Herbst, 1783: 41 type species *Silphoides boleti* Herbst, 1783
ancora Reitter, 1884a: 6 (*Tritoma*) **E**: AB (Talysh) **A**: IN
ater Reitter, 1879: 224 (*Tritoma*) **E**: AU BH BY CR CT CZ FR GE HU LA LT NT PL RO SK SL ST SZ UK YU **A**: CH ES FE JA KZ MG NC NE WS
jaroslawensis Semenov, 1898: 547 (*Tritoma*)
quadriornatus Reitter, 1901c: 71 (*Tritoma*) **E**: AB (Talysh) **A**: IN
quadripustulatus Linnaeus, 1760: 168 (*Chrysomela*) **E**: AB AL AR AU BE BH BU BY CR CT CZ DE EN FI FR GB GE GG GR HU IT KZ LA LT MC MD NL NR NT PL PT RO SK SL SP ST SV SZ TR UK YU **N**: MO **A**: IN KZ MG TR WS
antemacularis Dalla Torre, 1879: 98 (*Tritoma*)
bipustulatus Schilsky, 1888: 184 (*Tritoma*)
boleti Herbst, 1784a: 41 pl. 21 (*Silphoides*)
impustulatus Schilsky, 1888: 184 (*Tritoma*)
quadrimaculatus Schaller, 1783: 256 (*Silpha*)
ruficollis Schilsky, 1888: 184 (*Tritoma*)
winteri Reitter, 1911d: 291
subgenus *Mycetoxides* Motschulsky, 1858c: 144 type species *Mycetophagus fulvicollis* Fabricius, 1792
Calilendus Reitter, 1911a: 94 type species *Mycetophagus fulvicollis* Fabricius, 1792
elongatus Reitter, 1889d: 247 (*Tritoma*) **A**: FE JA
fulvicollis Fabricius, 1792b: 499 **E**: AB AL AR AU BE BH BU BY CR CT CZ DE EN FI FR GB GE GG GR HU IT KZ LA LT MD NR NT PL PT RO SK SL SP ST SV SZ UK YU **A**: ES FE KZ MG TR WS
deubeli Méhely, 1890: 272 (*Tritoma*)
massanae Dajoz, 1960: 195
sexmaculatus Ragusa, 1892a: 142
subgenus *Paramycetophagus* Nikitsky, 2007: 73 type species *Mycetophagus sulcicollis* Champion, 1917
sulcicollis Champion, 1917: 53 **A**: UP

subgenus *Parilendus* Casey, 1900: 131 type species *Mycetophagus bipustulatus* Melsheimer, 1844 (= *Mycetophagus quadriguttatus* P. W. J. Müller, 1821)
quadriguttatus P. W. J. Müller, 1821: 198 **E**: AB AL AR AU BE BH BU BY CR CT CZ DE FI FR GB GE GG GR HU IT KZ LA LS LT MC MD NL NT PL RO SK SL SP ST SV SZ TR UK YU **N**: AG MO **A**: IS KI KZ SA SY TD TM TR UZ WS **AUR NAR NTR ORR**
bipustulatus Melsheimer, 1844: 114
disjunctemaculatus Roubal, 1936a: 194
pubescens Stephens, 1830: 86
variegatus C. R. Sahlberg, 1837: 168
subgenus *Philomyces* Ganglbauer, 1899: 832 type species *Mycetophagus populi* Fabricius, 1798
populi Fabricius, 1798: 175 **E**: AU BE BY CR CT CZ DE EN FI FR GB GE HU IT LA LT NL NR NT PL RO SK ST SV SZ UK YU **A**: WS
subgenus *Ulolendus* Reitter, 1911a: 93 type species *Ips piceus* Fabricius, 1777
amamianus Nakane, 1989: 6 **A**: JA
antennatus Reitter, 1879: 225 (*Tritoma*) **A**: CH ES FE JA NC NE TAI
jureceki Pic, 1921a: 1 (*Tritoma*)
sauteri Grouvelle, 1914: 75
atomarius Fabricius, 1787a: 46 (*Ips*) **E**: AL AU BE BH BU BY CR CT CZ DE EN FI FR GB GE GR HU IT LA LT MC MD NL NR NT PT PL RO SK SL SP ST SV SZ UK YU **A**: TR
bifasciatus Champion, 1919: 242 **A**: UP
ciscaucasicus Semenov, 1899b: 596 (*Tritoma*) **E**: AB AR GG ST **A**: TR
decempunctatus decempunctatus Fabricius, 1801b: 568 **E**: AB AR AU BH BU BY CR CT CZ EN FI FR GE GG HU IT LA LT MD NT PL RO SK SL ST SV SZ UK YU **A**: KZ TR
rossicus Semenov, 1898a: 86
decempunctatus syriacus Reitter, 1904c: 165 **A**: SY
hyrcanicus Nikitsky, 1989b: 59 **E**: AB (Talysh)
incognitus Nikitsky, 1988b: 1850 **A**: ES FE
infulatus Reitter, 1884a: 6 (*Tritoma*) **E**: AB (Talysh) **A**: IN
intermedius Nikitsky, 1992b: 419 **E**: FE
piceus Fabricius, 1777: 214 (*Ips*) **E**: AB AL AR AU BE BH BU BY CR CT CZ DE EN FI FR GB GE GG GR HU IT KZ LA LT LU MA MD NL NK NT PL PT RO SK SL SP SV SZ TR UK YU **N**: EG **A**: KZ MG TR UZ
bosnicus Apfelbeck, 1911: 226
brunneus Panzer, 1798: 21
decipiens Apfelbeck, 1911: 226
feliciae Ragusa, 1892a: 142
flavotinctus Roubal, 1931: 132
histrio C. R. Sahlberg, 1837: 167
humeralis Schilsky, 1888: 184 (*Tritoma*)
hungaricus Papp, 1946: 38
lunaris Fabricius, 1801b: 568
punctulatus Schilsky, 1888: 184 (*Tritoma*)
salicis C. Brisout de Barneville, 1862: 24
undulatus Marsham, 1802: 140 (*Boletaria*)
variabilis Hellwig, 1792: 397
varius Marsham, 1802: 140 (*Boletaria*)
pustulosus Reitter, 1889d: 248 (*Tritoma*) **A**: FE JA
funebris Reitter, 1889d: 248 (*Tritoma*)
ramosus Reitter, 1904c: 166 **E**: AB
tauricus Pliginskij, 1923a: 78 **E**: UK (Krym)

genus *Mycetophagus*, species incertae sedis
fraternus Grouvelle, 1911: 315 **A**: SD

genus *Pseudotriphyllopsis* Nikitsky, 1996a: 1370 type species *Pseudotriphyllopsis inseriata* Nikitsky, 1996
inseriata Nikitsky, 1996a: 1370 **A**: SCH

genus *Pseudotriphyllus* Reitter, 1880a: 88 type species *Dermestes suturalis* Fabricius, 1801
colchicus Reitter, 1876b: 293 (*Triphyllus*) **E**: AB AR GG ST
lewisianus Wollaston, 1874: 170 (*Triphyllus*) **A**: FE JA
nepalensis Nikitsky, 2003: 49 **A**: NP
ohbayashii Miyatake, 1968: 29 **A**: JA
suturalis Fabricius, 1801a: 317 (*Dermestes*) **E**: AU BH BU CR CZ FR GB GE GR HU IT RO SK SZ UK
vicarius Peyerimhoff, 1913a: 254 **N**: AG

genus *Triphyllina* Reitter, 1877h: 384 type species *Triphyllina lederi* Reitter, 1878
lederi Reitter, 1878a: 168 **E**: AB AR GG ST (Kavkaz)

genus *Triphylloides* Miyatake, 1959a: 27 type species *Triphyllus seriatus* Reitter, 1889
seriatus Reitter, 1889d: 246 (*Triphyllus*) **A**: FE JA NE

genus *Triphyllopsis* Nikitsky, 1989b: 56 type species *Triphyllopsis altaica* Nikitsky, 1989
altaica Nikitsky, 1989b: 58 **A**: KZ WS

genus *Triphyllus* Dejean, 1821: 102 type species *Mycetophagus punctatus* Fabricius, 1792
bicolor Fabricius, 1777: 216 (*Nitidula*) **E**: AB AL AR AU BE BH BU BY CR CT CZ DE FR GB GE GG GR HU IT LA MD NL PL PT RO SK SL SP ST SV SZ UK YU **A**: TR
humeralis Marsham, 1802: 123 (*Silpha*)
immaculatus Roubal, 1936a: 192
pilosus Herbst, 1792: 177, pl. 42 (*Kryptophagus*)
punctatus Fabricius, 1792b: 499 (*Mycetophagus*)

tribe Typhaeini Nikitsky, 1993

genus *Typhaea* Stephens, 1829a: 8 type species *Mycetophagus testaceus* Fabricius, 1792 (= *Dermestes stercorea* Linnaeus, 1758)
angusta Rosenhauer, 1856: 106 [EA] **E**: IT SP **N**: AG TU
haagi Reitter, 1874: 527 **E**: AU CT CZ DE FI FR GE GG HU LA PL SK ST SV UK **A**: CH FE JA **ORR**
decipiens Lohse, 1989: 145
pallidula Reitter, 1874: 527 **A**: CH FE JA NC
stercorea Linnaeus, 1758: 357 (*Dermestes*) **E**: AB AL AN AR AU AZ BE BH BU BY CR CT CZ DE EN FA FI FR GB GE GG GR HU IC IR IT KZ LA LS LT LU MA MC MD NL NR NT PL PT RO RU SK SL SP ST SV SZ TR UK YU **N**: AG CI EG LB MO MR TU **A**: AE AF BT CH CY ES FE IN IQ IS JA JO KI KZ MG NP NC PA SA SC SI SY TD TM TR UZ WS YE **COS**
crenata Melsheimer, 1844: 114 (*Atomaria*)
fumata Linnaeus, 1767: 564 (*Dermestes*)
obscura Krauss, 1911: 109
testacea Fabricius, 1792b: 499 (*Mycetophagus*)
tomentosa Stephens, 1830: 71
variabilis Herbst, 1792: 141 (*Dermestes*)

genus *Typhaeola* Ganglbauer, 1899: 836 type species *Typhaea maculata* Perris, 1865
maculata Perris, 1865: 507 (*Typhaea*) **E**: FR IT PT SP **N**: AG EG MO TU
curticollis Fairmaire, 1879b: 167 (*Triphyllus*)
umbrata Baudi di Selve, 1870: 62 (*Typhaea*)

subfamily **Bergininae** Lend, 1920

genus *Berginus* Erichson, 1846: 405 type species *Berginus tamarisci* Wollaston, 1854
tamarisci Wollaston, 1854: 195 **E**: BH CR CZ FR GE GG GR HU IT PLi PO SK SL SP ST SVi SZ TR **N**: AG EG MO MR TU **A**: TR

sub*family* **Esarcinae** Reitter, 1882

genus *Esarcus* Reiche, 1864: 238 type species *Esarcus leprieuri* Reiche, 1864
subgenus *Entoxylon* Ancey, 1869: 46 type species *Esarcus abeillei* Ancey, 1870
Atritomus Reitter, 1877h: 384 type species *Atritomus cribratus* Reitter, 1877 (= *Entoxylon abeillei* Ancey, 1870)
abeillei Ancey, 1870: 85 (*Entoxylon*) **E**: FR IT
cribratus Reitter, 1877h: 384 (*Atritomus*)
baudii Seidlitz, 1889: 149 **E**: FR IT
besucheti Dajoz, 1964: 13 **E**: SP
franzi Dajoz, 1964: 14 **E**: SP
inexpectatus Dajoz, 1964: 14 **E**: SP
martini Reitter, 1887c: 285 **N**: AG
subgenus *Esarcus* Reiche, 1864: 238 type species *Esarcus leprieuri* Reiche, 1864
fiorii Reitter, 1887c: 285 **E**: IT
iolensis Peyerimhoff, 1917a: 140 **N**: AG
leprieuri Reiche, 1864: 239 **N**: AG TU
letourneuxi Raffray, 1873: 368 (*Entoxylon*) **N**: AG
parallelus Pic, 1901a: 3

family ARCHEOCRYPTICIDAE Kaszab, 1964

genus *Sivacrypticus* Kaszab, 1964f: 384 type species *Sivacrypticus taiwanicus* Kaszab, 1964
bengalicus Kaszab, 1975a: 34 **A**: NP SD **ORR**
dilliensis Kaszab, 1979b: 200 **A**: PA
indicus Kaszab, 1964f: 386 **A**: NP PA **ORR**
kashmirensis Kaszab, 1975a: 34 **A**: KA
loebli Kaszab, 1979b: 196 **A**: NP **ORR**
sericans Fairmaire, 1896a: 24 (*Platydema*) **A**: NP **ORR**
spinifer Champion, 1920a: 73 (*Enneboeus*)
taiwanicus Kaszab, 1964f: 385 **A**: TAI
uenoi Masumoto & Yin, 1993: 241 **A**: YUN

family PTEROGENIIDAE Crowson, 1953

genus *Laenagenius* Löbl, 2005: 153 type species *Laenagenius apterus* Löbl, 2005
apterus Löbl, 2005: 155 **A**: YUN

family CIIDAE Leach, 1819

subfamily **Ciinae** Leach, 1819

genus *Acanthocis* Miyatake, 1954: 47 type species *Cis* (*Acanthocis*) *bisetosus* Miyatake, 1954 (= *Acanthocis inonoti* Miyatake, 1955)
inonoti Miyatake, 1955b: 95 [RN] **A**: JA (Shikoku)
bisetosus Miyatake, 1954: 48 (*Cis*) [HN]
quadridentatus Nobuchi & Wada, 1959: 52 **A**: JA

genus *Anoplocis* Kawanabe, 1996b: 331 type species *Ennearthron poriae* Nakane & Nobuchi, 1955
poriae Nakane & Nobuchi, 1955: 49 (*Ennearthron*) **A**: JA
ryukyuensis Kawanabe, 1996b: 333 **A**: JA (Okinawa-hontô)

genus *Atlantocis* Israelson, 1985: 80 type species *Atlantocis gillerforsi* Israelson, 1985
gillerforsi Israelson, 1985: 82 **E**: AZ
canariensis Israelson, 1985: 84 **N**: CI (Tenerife)
lauri Wollaston, 1854: 282 (*Cis*) **N**: MR

genus *Ceracis* Mellié, 1848: 375 type species *Ennearthron sallei* Mellié, 1848
cucullatus Mellié, 1848: 372 (*Ennearthron*) **E**i: FR **AFR NTR**i
bilamellatus Pic, 1916c: 20 (*Ennearthron*)
lamellatus Pic, 1939a: 8 (*Ennearthron*)
tabelliferus Mellié, 1848: 373 (*Ennearthron*)
japonus Reitter, 1878c: 36 (*Ennearthron*) **A**: JA
laminicollis Miyatake, 1982b: 77 **A**: TAI
shikokuensis Miyatake, 1954: 56 (*Ennearthron*) **A**: JA (Shikoku)

genus *Cis* Latreille, 1796: 50 type species *Dermestes boleti* Scopoli, 1763
Cisdygma Reitter, 1885b: 209 type species *Cis clavicornis* Baudi di Selve, 1874
Dimerocis Peyerimhoff, 1919b: 277 type species *Cis aurasiacus* Peyerimhoff, 1919
Eridaulus C. G.Thomson, 1859: 91 type species *Cis comptus* Gyllenhal, 1827
Macrocis Reitter, 1878c: 34 type species *Macrocis taurus* Reitter, 1878
Xestocis Casey, 1898: 85 type species *Xestocis levettei* Casey, 1898
aurasiacus Peyerimhoff, 1919b: 277 **N**: AG
aurosericeus Reitter, 1887e: 515 **A**: FE
bidentatus A. G. Olivier, 1790: 16: 11 (*Anobium*) **E**: AU BE BU BY CR CZ DE EN FI FR GB GE HU IR IT LT NL NR NT PL RO SK SV SZ UK **N**: AG
inermis Marsham, 1802: 87 (*Ptinus*)
bifasciatus Reitter, 1877g: 381 (*Cis*) **A**: JA
bilamellatus Wood, 1884: 130 **E**i: GB **AUS**
munitus Blackburn, 1888: 268
bisetiger M. Chûjô, 1939: 3 **A**: FE (Sachalin)
bituberculosus Roubal, 1937a: 39 [RN] **E**: ST (Kavkaz)
bituberculatus Roubal, 1912a: 25 [HN]
boleti Scopoli, 1763: 17 (*Dermestes*) **E**: AB AU BE BH BY CR CT CZ DE EN FI FRGB GE GG GR HU IR IT LA LS LT NL NR NT PL RO SK SL SP ST SV SZ UK **N**: AG **A**: FE IN JA (Hokkaido) MG SCH "Transcaspia"
armicollis Roubal, 1937b: 67
boletorum Marsham, 1802: 85 (*Ptinus*)
latior Pic, 1934f: 86
minor Mellié, 1848: 241
obliteratus Mellié, 1848: 241
picipes Fabricius, 1787a: 34 (*Dermestes*)
pruinosus Motschulsky, 1861: 17
substriatus Mellié, 1848: 241
brevipennis Nakane & Nobuchi, 1955: 48 **A**: JA (Honshu)
capricornis Kawanabe, 1997b: 326 **A**: JA (Ryukyus)
castaneus Herbst, 1793: 17 (*Kateretes*) **E**: AB AU BE BH BY CR CT CZ DE EN FI FR GB GE GR HU IR IT LS LT NL NR NT PL RO SK SL ST SV SZ UK YU **A**: FE "Buchara"
[*nitidus* sensu auct., nec Fabricius, 1792]
caucasicus Ménétriés, 1832: 224 **E**: "Caucasus"
cedri Peyerimhoff, 1915: 25 **N**: AG
chinensis Lawrence, 1991: 288 **A**: CH
chujoi Miyatake, 1982b: 73 **A**: TAI
clavicornis Baudi di Selve, 1874c: 338 **E**: GR UK (Krym) **A**: CY LE
comptus Gyllenhal, 1827: 625 **E**: AB AR AU BE BH BU BY CR CT CZ DE EN FI FR GE GG GR HU IT MC NL NR NT PL RO SK SL SP ST SV SZ UK **N**: MO TU **A**: ES FE KZ TM TR WS
cognatus Rey, 1892: 17
substriatus Munster, 1927: 185 [= 1928: 295] [HN]
coriaceus Baudi di Selve, 1874c: 337 **A**: CY SY
corioli Peyerimhoff, 1915: 26 (*Cisdygma*) **N**: AG
crenatus C. R. Sahlberg, 1836: 159 **E**: FI
cucullatus Wollaston, 1865: Appendix: 39 **N**: CI (Gomera)
dentatus Mellié, 1848: 324 **E**: AU BH CR CT CZ FI FR GB GE IT NR PL RO SK SV SZ
microgonus C. G. Thomson, 1868: 46

eminenticollis Nobuchi, 1955a: 56 **A**: JA (Honshu)
erinaceus Roubal, 1933a: 72 **A**: FE
fagi Waltl, 1839: 224 **E**: AU BH BY CR CT CZ DE FI FR GB GE GR HU IR IT NL NR NT PL RO SK SP SV
fuscatus Mellié, 1848: 352
fasciculosus Champion, 1922b: 151 **A**: UP
festivus Panzer, 1793a: 7 (*Anobium*) **E**: AU BE BH CT CZ DE FI FR GB GE GG HU IR IT NL NR NT PL RO SK SL ST SV SZ UK **A**: ES
fissicollis Mellié, 1848: 247 **E**: AB AU BH CR CZ FR GG HU IT PL RO SK ST UK
fissicornis Mellié, 1848: 317 **E**: AU BY CT EN FR HU IT LT PL RO ST
germanicus Abeille de Perrin, 1874a: 39
sublaminatus Wankowicz, 1869: 422
flavipes Motschulsky, 1845b: 371 **A**: FE (Kamchatka)
friebi Roubal, 1933a: 72 **A**: FE
fusciclavis Nyholm, 1953: 190 **E**: AU BH BY CT CZ DE FI FR GE IT NL NR PL SK SV
castaneus Mellié, 1848: 351 [HN]
fuscipes Mellié, 1848: 271 **Ni**: MR **NAR NTR**
gardneri Pic, 1937g: 123 **A**: UP
giganteus Pic, 1934a: 15 **A**: CH
glabratus Mellié, 1848: 249 **E**: AU BY CT CZ EN FI FR GE IT LA NR PL SK SV ST SZ
gladiator Flach, 1882: 250 **E**: GE
graecus Schilsky, 1900: 68 **E**: GR **N**: MO **A**: SY TR
hanseni A. Strand, 1965: 61 **E**: FI GE PL NR NT ST
incanus Rey, 1892: 17 **E**: FR (Lyon)
jacquemartii Mellié, 1848: 328 **E**: AB AU BH BY CR CT CZ DE FI FR GB GE GG GR HU IT LT NR NT PL RO SK SP ST SV SZ UK **A**: TR
jezoensis Nobuchi, 1960a: 38 **A**: JA (Hokkaido)
judaeus Reitter, 1901a: 55 **A**: LE
konoi M. Chûjô, 1940a: 132 **A**: FE (Sachalin) JA (Hokkaido, Honshu) JIL
laevigatus Kawanabe, 1997b: 319 **A**: JA
laminatus Mellié, 1848: 318 **E**: AU CR CT (Kaliningrad) CZ FR GE HU IT PL "Transcaucasus"
lederi Reitter, 1880c: 477 **E**: GG ST (Kavkaz)
libanicus Abeille de Perrin, 1874a: 58 **A**: LE
lineatocribratus Mellié, 1848: 336 **E**: AB AU BH BU BY CR CT CZ DE EN FI FR GE GG HU IT LA LS NR NT PL RO SK ST SV SZ UK YU
matchanus Reitter, 1915c: 66
lineatosetosus Pool, 1917: 87 **E**: GB
lineatulus Reitter, 1901a: 51 **A**: UZ
reitteri Lopes-Andrade, 2002: 54 [RN] **E**: AB (Lenkoran)
setifer Reitter, 1884a: 8 [HN]
maculatus M. Chûjô, 1939: 2 **A**: FE (Sachalin) JA (Hokkaido) LIA
manchuricus M. Chûjô, 1940b: 136 **A**: **A**: NE
maurus Peyerimhoff, 1915: 25 **N**: AG
micans Fabricius, 1792a: 238 (*Anobium*) **E**: AL AU BE BH BU BY CR CT CZ DE EN FI FR GB GE GR HU IR IT LA LS LT NL NR NT PL RO SK SL SP ST SV SZ UK **A**: ES FE MO TR
albohispidulus Reitter, 1901a: 50
hispidus Paykull, 1798: 310 (*Anobium*)
nitidicollis Abeille de Perrin, 1874a: 32
ruficornis Marsham, 1802: 87 (*Ptinus*)
savilli Donisthorpe, 1936: 220
mikagensis Nobuchi & Wada, 1955b: 106 **A**: JA (Honshu) LIA
morikawai Miyatake, 1954: 50 **A**: JA (Honshu, Shikoku)
multidentatus Pic, 1917f: 4 (*Ennearthron*) **Ei**: GE IT **A**: YUN
nigrorugosus Schilsky, 1900: 66 **A**: FE
nikkoensis Nobuchi, 1960b: 65 **A**: JA
nipponicus M. Chûjô, 1940b: 134 **A**: JA (Hokkaido, Honshu) LIA

orius Kompantsev, 1996: 587 **A**: AF KI TD UP UZ
polypori M. Chûjô, 1939: 4 **A**: JA (Honshu)
polysticti M. Chûjô, 1939: 5 **A**: TAI
pumilio Baudi di Selve, 1874c: 338 (*Cis*) **A**: CY
puncticollis Wollaston, 1860a: 360 **N**: MR
punctifer Mellié, 1848: 344 **E**: CR FR HU IT RO SZ **A**: CY
punctulatus Gyllenhal, 1827: 625 **E**: AU BE BH CR CT CZ DE EN FI FR GB GE HU IT LT NL NR NT PL RO SK SL SV
pygmaeus glutinosus Peyerimhoff, 1919a: 186 **N**: AG
pygmaeus pygmaeus Marsham, 1802: 86 (*Ptinus*) **E**: AU BE CZ DE FR GB GE HU IT NL NR PL RO SK SV
  nigricornis Marsham, 1802: 87 (*Ptinus*)
  oblongus Mellié, 1848: 341
  pandellei Abeille de Perrin, 1874a: 73
  rhododactylus Marsham, 1802: 87 (*Ptinus*)
quadridens Mellié, 1848: 270 **E**: AU BH BY CR CZ FI FR GE IT NR NT PL RO SK ST SV SZ UK FR SZ
quadridentulus Perris, 1874: liii **E**: CR FR IT
ragusai Roubal, 1916: 89 **E**: IT (Sicilia)
robustithorax Pic, 1915k: 22 **N**: AG
rufocastaneus Nakane & Nobuchi, 1955: 48 **A**: JA (Honshu, Shikoku)
rugulosus Mellié, 1848: 242 **E**: AU BE CT CZ EN FI FR GE IT LT NL NT SK ST SV SZ
  pyrrhocephalus Mellié, 1848: 244 [HN]
  rubiginosus Mellié, 1848: 243
sasajii Kawanabe, 2001: 267 **A**: JA (Hokkaidô)
sasakawai Nobuchi, 1960a: 37 **A**: Japan (Hokkaidô)
seriatocribratus Reitter, 1913b: 124 **A**: TD UZ
seriatopilosus Motschulsky, 1861: 17 **A**: FE JA
seriatulus Kiesenwetter, 1879: 320 **A**: JA
sibiricus Schilsky, 1900: 75 **A**: FE
sinensis Pic, 1917f: 4 (*Ennearthron*) **A**: YUN
striatulus Mellié, 1848: 264 **E**: AB AU BH CR FR GE GG GR HU IT PL SP ST SZ **N**: AG **A**: TR
  flavipes Lucas, 1849: 470 [HN]
  peyronis Abeille de Perrin, 1874a: 65
submicans Abeille de Perrin, 1874a: 28 **E**: AB AR AU BE BH BY CR CT CZ DE EN FI FR GB GE GG GR HU IT LA LT NL NR NT PL SK ST SV SZ UK **A**: FE TR
  [*micans* sensu auct., nec Fabricius, 1792]
taiwanus M. Chûjô, 1939: 1 **A**: JA (Amami-Ôshima) TAI
tauriensis Królik, 2002: 198 **A**: TR (Toros dağlari)
tomentosus Mellié, 1848: 253 **E**: AB AR BH CR FR GG GR IT (Sicilia) RO SP (Gibraltar) ST YU **N**: MO
  sericeus Mellié, 1848: 346
  tomentoides Reitter, 1901a: 56
vestitus Mellié, 1848: 354 **E**: AU BE BH GB CZ DE FI FR GE HU IT NL NR PL RO SK SV SZ
villosulus Marsham, 1802: 87 (*Ptinus*) **E**: AU BE BY CR CT CZ DE FR GB GE GG GR HU IR IT LS NL NR NT PL PT SK SP SV ST SZ UK **A**: ES FE TR
  flavus Stephens, 1830: 345
  petropolitanus Jakobson, 1896: 520
  plagiatus C. G. Thomson, 1863: 184
  pyrrhocephalus Marsham, 1802: 86 (*Ptinus*)
  quadricollis Schilsky, 1900: 67
  setiger Mellié, 1848: 244
  striatulus Mellié, 1848: 246 [HN]
wollastonii Mellié, 1849: lxxxvi **N**: MR
yamamotoi Kawanabe, 1997b: 321 **A**: JA

genus *Cisarthron* Reitter, 1885b: 208 type species *Cisarthron laevicolle* Reitter, 1885
laevicolle Reitter, 1885b: 209 **E**: BH ST (Kavkaz)

genus *Dichodontocis* Kawanabe, 1994a: 186 type species *Dichodontocis uncinatus* Kawanabe, 1994
uncinatus Kawanabe, 1994a: 187 **A**: JA (Shikoku, Yakushima Is.)

genus *Diphyllocis* Reitter, 1885b: 209 type species *Ennearthron opaculum* Reitter, 1878
opaculus Reitter, 1878b: 25 (*Ennearthron*) **E**: BH CR HU IT RO SK UK

genus *Dolichocis* Dury, 1919: 158 type species *Dolichocis manitoba* Dury, 1919
laricinus Mellié, 1848: 355 (*Cis*) **E**: CT FI NR RO NT "Silesia" **A**: FE WS
yuasai M. Chûjô, 1941: 85 (*Ennearthron*) **A**: JA (Hokkaido)

genus *Ennearthron* Mellié, 1847: 110 type species *Cis cornutus* Gyllenhal, 1827
abeillei Caillol, 1914: 160 **E**: FR
amamense Miyatake, 1959b: 121 **A**: JA (Amami-Ôshima)
chujoi Nakane & Nobuchi, 1955: 49 **A**: JA (Honshu, Hokkaido)
cornutum Gyllenhal, 1827: 626 (*Cis*) **E**: AB AU BE BH BY CR CT CZ DE EN FI FR GB GE GG HU IR IT LS LT NL NR NT PL RO SK SL SP ST SV SZ UK **A**: JA (Hokkaido) JIL
concinnum Marsham, 1802: 87 (*Ptinus*) [HN]
filum Abeille de Perrin, 1874c: liii (*Cis*) **E**: AU FR IT (Sicilia)
hayashii Nobuchi, 1955b: 108 **A**: JA
ishiharai Miyatake, 1954: 57 **A**: JA (Hokkaido Shikoku)
mohrii Miyatake, 1954: 56 **A**: JA (Honshu)
ondreji Roubal, 1919: 63 **E**: ST (Kavkaz)
palmi Lohse, 1966: 28 **E**: AU BY CT CZ FI GE SV
pruinosulum Perris, 1864: 291 (*Cis*) **E**: CZ FR GE HU IT SK
reichei Abeille de Perrin, 1874a: 89 **N**: EG
reitteri Flach, 1882: 249 (*Cis*) **E**: GE (Bayern)
robusticorne Kawanabe, 1996a: 105 **A**: JA (Hokkaidô)

genus *Euxestocis* Miyatake, 1954: 53 type species *Euxestocis bicornutus* Miyatake, 1954
bicornutus Miyatake, 1954: 54 **A**: JA
formosanus Miyatake, 1982b: 76 **A**: TAI

genus *Hadraule* C. G. Thomson, 1859: 91 type species *Cis elongatulus* Gyllenhal, 1827
Knablia Roubal, 1936b: 54 type species *Cis elongatulus* Gyllenhal, 1827
Maphoca Casey, 1900: 165 type species *Maphoca blaisdelli* Casey, 1900
Pityocis Peyerimhoff, 1918b: 141 type species *Pityocis coarctatus* Peyerimhoff, 1918 (= *Cis elongatulus* Gyllenhal, 1827)
elongatula Gyllenhal, 1827: 627 (*Cis*) **E**: AL AU CT CZ FI FR GB GE GR HU IT LA NR PL SK SV **N**: AG **NAR**i
coarctata Peyerimhoff, 1918b: 142 (*Pityocis*)
striata J. R. Sahlberg, 1901: 11 (*Ennearthron*)
genus *Hadraule*, species incertae sedis
fukudai M. Chûjô, 1940b: 135 (*Cis*) **A**: JA (Honshu)
hieroglyphica Reitter, 1877g: 380 (*Cis*) **A**: FE JA
japonica Nobuchi, 1955a: 57 (*Cis*): **A**: JA (Honshu, Kyushu)
sinica Pic, 1954a: 58 (*Cis*) **A**: FUJ
subrobusta Miyatake, 1955b: 95 (*Cis*) [RN] **A**: JA (Honshu, Kyushu)
robusta Miyatake, 1954: 46 (*Cis*) [HN]

genus *Hyalocis* Kawanabe, 1993: 763 type species *Hyalocis yakushimensis* Kawanabe, 1993
yakushimensis Kawanabe, 1993: 765 **A**: JA (Yakushima Is.)

genus *Lipopterocis* Miyatake, 1954: 44 type species *Lipopterocis simplex* Miyatake, 1954
simplex Miyatake, 1954: 45 **A**: JA (Shikoku)

genus *Neoennearthron* Miyatake, 1954: 58 type species *Neoennearthron bicarinatum* Miyatake, 1954
bicarinatum Miyatake, 1954: 59 **A**: JA
hisamatsui Miyatake, 1959b: 119 **A**: JA (Amami-Ôshima)

genus *Nipponapterocis* Miyatake, 1954: 42 type species *Nipponapterocis brevis* Miyatake, 1954
brevis Miyatake, 1954: 43 **A**: JA (Honshu, Kyushu, Shikoku)
hirsutus Kawanabe, 1995: 172 **A**: JA (Yakushima Is.)
inermis Kawanabe, 1995: 170 **A**: JA (Honshu)

genus *Nipponocis* Nobuchi & Wada, 1955a: 54 type species *Nipponocis longisetosus* Nobuchi, 1955
ashuensis Nobuchi, 1959: 51 **A**: JA (Honshu)
longisetosus Nobuchi, 1955a: 55 **A**: JA (Honshu) LIA
magnus Nobuchi, 1955a: 55 **A**: JA (Honshu)
unipunctatus Nakane & Nobuchi, 1956: 107 **A**: JA (Tokara Is.)

genus *Octotemnus* Mellié, 1847: 110 type species *Cis glabriculus* Gyllenhal, 1827
Orophius L. Redtenbacher, 1847: 350 type species *Cis mandibularis* Gyllenhal, 1813
aculeatus Kawanabe, 2002: 263 **A**: JA (Iriomote Is.)
chinensis J.-K. Li, 1992: 143 **A**: JIL
glabriculus Gyllenhal, 1827: 629 (*Cis*) **E**: AB AU BE BH BU BY CR CT CZ DE FI FR GB GE GG HU IR IT LA LS LT LU MC NL NR NT PL PT RO SK SL SP ST SV SZ UK **A**: JA (Hokkaido, Honshu) LIA TR
japonicus Miyatake, 1954: 64 **A**: JA
laminifrons Motschulsky, 1861: 17 (*Orophius*) **A**: JA LIA TAI
mandibularis Gyllenhal, 1813: 717 (*Cis*) **E**: AU BH BU BY CR FI FR GE GG IT LT NR PL RO SK SL SP ST SV SZ UK **A**: ES
inaequidens Chevrolat, 1835a: pl. 40 (*Cis*)
michiochujoi Kawanabe, 2005: 47 **A**: TAI
omogensis Miyatake, 1954: 61 **A**: JA (Honshû, Kyûshû, Shikoku)
opacus Mellié, 1848: 386 **N**: MR
parvulus Miyatake, 1954: 62 **A**: JA
pilosoceps Kawanabe, 2002: 249 **A**: JA (Yaku-shima Is.)
punctidorsum Miyatake, 1954: 63 **A**: JA (Honshû, Shikoku)
robustus Kawanabe, 2002: 258 **A**: JA (Tsushima)
rugosopunctatus Drogvalenko, 2002: 9 **E**: UK (Krym)

genus *Odontocis* Nakane & Nobuchi, 1955: 50 type species *Odontocis denticollis* Nakane & Nobuchi, 1955
denticollis Nakane & Nobuchi, 1955: 50 **A**: JA (Honshu, Hokkaido)

genus *Orthocis* Casey, 1898: 84 type species *Orthocis aterrima* Casey, 1898 (= *Cis punctatus* Mellié, 1848)
Mellieicis Lohse, 1964: 122 type species *Cis alni* Gyllenhal, 1813
alni Gyllenhal, 1813: 386 (*Cis*) **E**: AU BH BY CT CZ DE EN FI FR GB GE GG GR HU IR IT LT NL NR NT PL SK SP ST SV SZ UK **A**: ES
betulae Zetterstedt, 1828: 352 (*Cis*)
mitfordi Donisthorpe, 1916: 156
recticollis Abeille de Perrin, 1874a: 60 (*Cis*)
alnoides Reitter, 1884c: 120 (*Cis*) **E**: AB CR GR (Korfu) UK
volhynicus Roubal, 1912b: 71 (*Cis*)
auriculariae Lawrence, 1991: 289 **A**: CH
coluber Abeille de Perrin, 1874c: lii (*Cis*) **E**: FR GB HU IT
latifrons Pool, 1917: 89 (*Cis*)
ishiharai Kawanabe, 1994b: 187 **A**: JA (Kyûshû Shikoku, Tsushima)
juglandis Reitter, 1885b: 208 (*Cis*) **E**: AU BH FR GE HU YU
linearis J. R. Sahlberg, 1901: 10 (*Cis*) **E**: DE FI NR NT PL SV

lucasi Abeille de Perrin, 1874a: 62 (*Cis*) [RN] **E**: AU BH BY CR CZ FR GE GG HU IT PL SK SL SP ST UK YU YU **A**: FE **N**: AG
punctulatus Lucas, 1849: 470 (*Cis*) [HN]
reflexicollis Abeille de Perrin, 1874c: lii (*Cis*)
nigrosplendidus Nobuchi, 1955b: 105 (*Cis*) **A**: JA (Hokkaidô)
ornatus Reitter, 1877g: 381 (*Cis*) **A**: JA
perrisi Abeille de Perrin, 1874c: liii (*Cis*) **E**: FR IT
pseudolinearis Lohse, 1965: 179 (*Cis*) **E**: AU GE HU PL
sagittiferus Israelson, 1980: 189 (*Cis*) **N**: CI (La Palma)
schizophylli Nakane & Nobuchi, 1955: 47 (*Cis*) **A**: JA (Hokkaido, Honshu)
zoufali Reitter, 1902a: 6 (*Cis*) **E**: BH

genus *Paraxestocis* Miyatake, 1954: 51 type species *Paraxestocis unicornis* Miyatake, 1954
unicornis Miyatake, 1954: 52 **A**: JA (Shikoku)
nakanei Nobuchi, 1955a: 53 (*Xylographus*)

genus *Rhopalodontus* Mellié 1847: 109 type species *Cis perforatus* Gyllenhal, 1813
Cedrinus Abeille de Perrin, 1876: 313 type species *Rhopalodontus camelus* Abeille de Perrin, 1876
armifrons Reitter, 1913d: 655 **N**: AG
baudueri Abeille de Perrin, 1874c: lii **E**: AU DE FR GE GR HU RO SK ST
camelus berberus Peyerimhoff, 1945: 2 **N**: AG
camelus camelus Abeille de Perrin, 1876: 312 (*Rhopalodontus*) **A**: LE
harmandi Lesne, 1917: 191 **A**: JA (Hokkaido, Honshu)
novorossicus Reitter, 1901a: 58 **E**: CZ GE IT ST
perforatus Gyllenhal, 1813: 385 (*Cis*) **E**: AB AU BH BY CR CZ DE FR GB GE GG GR HU IT LA NR PL RO SK SP ST SV SZ UK **A**: JA (Hokkaido) JIL
perrini Reitter, 1878a: 221 **E**: AB GG
populi C. & H. Brisout de Barneville, 1877: cvii **E**: FR IT SP
strandi Lohse, 1969: 50 **E**: CT FI NR NT PL SV **A**: ES FE WS

genus *Strigocis* Dury, 1917: 19 type species *Strigocis opacicollis* Dury, 1917
bicornis Mellié, 1848: 356 (*Cis*) **E**: AB BH CR CZ FR GB GE HU IT RO SP SZ UK "Caucasus" **N**: AG
tokunagai Nobuchi, 1960a: 39 (*Rhopalodontus*) **A**: JA

genus *Sulcacis* Dury, 1917: 20 type species *Sulcacis lengi* Dury, 1917
Entypocis Lohse, 1964: 121 type species *Cis bidentulus* Rosenhauer, 1847
Entypus L. Redtenbacher, 1847: 350 [HN] type species *Apate fronticornis* sensu Panzer, 1805 (=*Cis affinis* Gyllenhal, 1827)
bidentulus Rosenhauer, 1847: 58 (*Cis*) **E**: AU BU CR CT CZ FR GE HU IT PL SK SZ UK **A**: ES
alpinus Mellié, 1848: 347 (*Cis*)
fronticornis Panzer, 1805b: 7 (*Apate*) **E**: AB AU BE BH BY CR CT CZ DE FI FR GE GR HU IT LA NL NR NT PL RO SK SP ST SV SZ UK YU
japonicus Nobuchi, 1960a: 39 (*Rhopalodontus*) **A**: JA (Hokkaido)
nitidus Fabricius, 1792a: 238 (*Anobium*) **E**: AB AU BE BH BU BY CR CT CZ DE EN FI FR GB GE GG HU IT LA LT NL NR NT PL PT RO SK SL SP ST SV SZ UK YU **A**: FE JA (Hokkaido, Honshu)
affinis Gyllenhal, 1827: 628 (*Cis*)
nobuchii Kawanabe, 1997a: 74 **A**: JA (Hokkaido)

genus *Syncosmetus* Sharp, 1891: 6 type species *Syncosmetus japonicus* Sharp, 1891
Omogocis Miyatake, 1954: 40 type species *Omogocis tuberculifrons* Miyatake, 1954
japonicus Sharp, 1891: 7 **A**: JA (Kyushu, Shikoku)
tuberculifrons Miyatake, 1954: 41 (*Omogocis*)

genus *Wagaicis* Lohse, 1964: 121 type species *Ennearthron wagae* Wankowicz, 1869
wagae Wankowicz, 1869: 421 (*Ennearthron*) **E**: AU BY CR CT FI FR GE PL RO NT UK

genus *Xylographella* Miyatake, 1985: 279 type species *Xylographella punctata* Miyatake, 1985
punctata Miyatake, 1985: 280 **A**: JA (Honshu, Shikoku)

genus *Xylographus* Mellié, 1847: 109 type species *Cis bostrichoides* Dufour, 1843
bostrichoides Dufour, 1843: 93 (*Cis*) **E**: AB AU BH BY CR CZ FR GR HU IT (Sardegna) PL RO SK SP ST TR UK **N**: AG MO **A**: CY TD TR "Siberia"
aubei Mellié, 1848: 232
cribratus Lucas, 1849: 469 (*Cis*)
scheerpeltzi Nobuchi & Wada, 1956: 35 **A**: JA (Hokkaido)
tomicoides Reitter, 1901a: 47 **A**: FE

family Ciidae, nomina dubia
Cis punctiger Waltl, 1839: 224 **E**: GE
Rhopalodontus gyllenhali Gistel, 1857a: 59 **E**: GE
Xylotrogus glycyrrhizae Chevrolat, 1861: 206 **N**: AG

family TETRATOMIDAE Billberg, 1820

subfamily **Tetratominae** Billberg, 1820

genus *Tetratoma* Fabricius, 1790: 217 type species *Tetratoma fungorum* Fabricius, 1790
subgenus *Abstrulia* Casey, 1900: 167 type species *Tetratoma tessellata* Melsheimer, 1844
ainu Nakane, 1963a: 30 (*Abstrulia*) **A**: FE (S. Sakhalin, S. Kuril Is.) JA
ancora Fabricius, 1790: 217 **E**: AU BE BH BY CT CZ DE EN FI FR GE GB HU IR IT LA LT MD NL NR NT PL RO SK SL SP ST SV CZ UK YU **A**: ES FE KZ MG WS
conjuncta Pic, 1901f: 58
obscurior Pic, 1901f: 58
perrisi Houlbert, 1934: 13
pfefferi Roubal, 1930: 75
picta Newman, 1835: 202
baudueri Perris, 1864: 292 **E**: FR SP PT **N**: MO
flavopicta Fairmaire, 1884f: cxviii (*Phloeophilus*)
crenicollis Baudi di Selve, 1877a: 54 **A**: CY
japonica Miyatake, 1955a: 93 **A**: FE JA
pictipennis Reitter, 1896i: 237 **A**: ES FE NC
tedaldi Reitter, 1887b: 84 **E**: IT
virgo Motschulsky, 1845b: 366 **A**: FE (Kamchatka)
subgenus *Falsoxanthalia* Pic, 1934f: 84 type species *Falsoxanthalia bicoloripes* Pic, 1934
bicoloripes Pic, 1934f: 84 (*Falsoxanthalia*) **A**: SCH
desmarestii Latrielle, 1807: 180 **E**: AU BE BH DE FR HU GB GE GR IT NL SK SP ST SV **N**: AG
pallida Curtis, 1826: pl. 123
subgenus *Paratetratoma* Nikitsky, 1998a: 16 type species *Tetratoma sakagutii* Nakane, 1955
sakagutii Nakane, 1955: 35 **A**: JA
subgenus *Tetratoma* Fabricius, 1790: 217 type species *Tetratoma fungorum* Fabricius, 1790
cyanoptera Champion, 1924c: 263 **A**: SD
fungorum Fabricius, 1790: 217 **E**: AB AR AU BE CR CZ DE EN FI FR GB GE GG HU IR IT LA LS LT MC MD ND NL NR PL RO SK SP ST SV SZ TR UK YU **A**: TR
bicolor Marsham, 1802: 141 (*Boletaria*)
collaris Quensel, 1790: 8 (*Dermestes*)
dermestoides Herbst, 1792: 88, pl. 38
fuscoguttata Nikitsky, 1998a: 21 **A**: NP
nepalensis Nikitsky, 1998a: 23 **A**: NP
hobuchii Nakane, 1955: 36 **A**: JA
talyshensis Nikitsky, 1989b: 62 **E**: AB (Talysh)
wittmeri Nikitsky, 1998a: 25 **A**: BT

subfamily **Piseninae** Miyatake, 1960

genus *Pisenus* Casey, 1900: 167 type species *Cryptophagus humeralis* Kirby, 1837
formosanus Miyatake, 1974: 60 **A**: TAI
insignis Reitter, 1889d: 245 (*Pseudotriphyllus*) **A**: JA FE
rufitarsis Reitter, 1889d: 245 (*Pseudotriphyllu*) **A**: JA FE
chujoi Miyatake, 1960: 129 **A**: JA FE

genus *Triphyllia* Reitter, 1898f: 359 type species *Triphyllia koenigi*, 1898
koenigi Reitter, 1898f: 359 **E**: GG ST (Kavkaz)

subfamily **Penthinae** Lacordaire, 1859

genus *Cyanopenthe* Nikitsky, 1998a: 29 type species *Penthe metallica* Champion, 1916
metallica Champion, 1916b: 79 (*Penthe*) **A**: BT **ORR**

genus *Penthe* Newman, 1838: 373 types species *Helops obliquatus* Fabricius, 1798
Pyrrocis Laporte, 1840: 236 type species *Helops obliquatus* Fabricius, 1798
almorensis Champion, 1917: 54 **A**: NP UP
japana Marseul, 1876d: 335 **A**: JA
kochi Mařan, 1940b: 87 **A**: FUJ JIA TAI
reitteri Nikitsky, 1998a: 32 **A**: SCH TAI YUN
rufopubens Marseul, 1876d: 336 **A**: NP SD "Indes Orientales"
similis Nikitsky, 2005: 24 **A**: "Assam" **ORR**

subfamily **Eustrophinae** Gistel, 1856
tribe Eustrophini Gistel, 1856

genus *Eustrophus* Illiger, 1802b: 301 type species *Mycetophagus dermestoides* Fabricius, 1792
dermestoides Fabricius, 1792b: 498 (*Mycetophagus*) **E**: AU BH BU BY CR CZ FI FR GE GR HU IT LA LT MC PL RO SK SL SP SZ TR UK
siculus Baudi di Selve, 1892: 121 (*Tritoma*)
niponicus Lewis, 1895a: 259 **A**: FE JA NE
yunnanensis Nikitsky, 1998a: 54 **A**: SW

genus *Synstrophus* Seidlitz, 1898b: 442 types species *Eustrophus macrophthalmus* Reitter, 1887
klapperichi Pic, 1955c: 29 **A**: FUJ
macrophthalmus Reitter, 1887d: 383 (*Eustrophus*) **A**: JA NE
rollei Pic, 1910q: 75 (*Eustrophus*) **A**: TAI

tribe Holostrophini Nikitsky, 1998

genus *Holostrophus* Horn, 1888: 36 type species *Eustrophus bifasciatus* Say, 1824
subgenus *Holostrophus* Horn, 1888: 36 type species *Eustrophus bifasciatus* Say, 1824
aureofasciatus Pic, 1954b: 63 (*Eustrophinus*) **A**: FUJ
diversefasciatus Pic, 1921a: 1: **A**: FE JA NE
katoi Nomura, 1959: 42
dux Lewis, 1895a: 259 **A**: JA
lewisi Csiki, 1924: 10 [RN] **A**: FE (Kunashir) JA
quadrimaculatus Lewis, 1895a: 260 [HN]
morimotoi Sasaji, 1974: 1 **A**: JA
unicolor Lewis, 1895a: 260 **A**: JA
subgenus *Paraholostrophus* Nikitsky, 2007: 19 type species *Holostrophus orientalis* Lewis, 1895
orientalis Lewis, 1895a: 259 **A**: JA CH NC SC TAI **ORR**
multinotatus Pic, 1911p: 188

genus *Pseudoholostrophus* Nikitsky, 1983: 37 type species *Hallomenus klapperichi* Pic, 1954
chinensis Nikitsky, 1998a: 44 **A**: SCH

klapperichi Pic, 1954a: 53 **A**: FUJ

subfamily **Hallomeninae** Mulsant, 1856

genus *Hallomenus* Panzer, 1793e: 17 type species *Hallomenus humeralis* Panzer, 1793 (= *Chrysomela binotata* Quensel, 1790)
subgenus *Hallomenus* Panzer, 1793e: 17 type species *Hallomenus humeralis* Panzer, 1793 (= *Chrysomela binotata* Quensel, 1790)
Dinophorus Illiger, 1807: 335 type species *Dinophorus axillaris* Illiger, 1807
Dryala Mulsant, 1856f: 250 [= 1856c: 58] type species *Hallomenus fuscus* Gyllenhal, 1810
Hallominus Paykull, 1799: 179 type species *Hallominus bipunctatus* Paykull, 1799
axillaris Illiger, 1807: 335 (*Dinophorus*) **E**: AU BE BY CR CT CZ DE EN FI FR GE GG HU IT LA LT MD NR NT PL RO SK ST SV SZ UK **A**: CH WS
dimidiatus Stephens, 1835: 426
dorsalis Beck, 1817: 18 pl. 5 (*Dircaea*)
fuscus Gyllenhal, 1810: 528
piceus Marsham, 1802: 494 (*Mordella*)
scapulatus Fairmaire, 1881c: xii
binotatus Quensel, 1790: 14 (*Chrysomela*) **E**: AU BE BH BY CT CZ DE EN FI FR GB GE GG HU IR IT LA LU MD NL NR NT PL RO SK SL SP ST SV SZ UK YU **A**: ES WS
bipunctatus Paykull, 1799: 179 (*Hallominus*)
boleti Kugelann, 1792: 239 (*Elateroides*)
elateroides Kugelann, 1792: 239 (*Ips*)
humeralis Panzer, 1793e: 17
chinensis Nikitsky, 1998a: 35 **A**: SCH
nipponicus Nomura & Katô, 1958: 2 **A**: FE JA
orientalis Nikitsky, 1998a: 36 **A**: KA
subgenus *Xeuxes* Champion, 1889: 85 type species *Xeuxes brevicollis* Champion, 1889
Parahallomenus Nomura & Katô, 1958: 1 type species *Hallomenus tokejii* Nomura & Katô, 1958
arimotoi Toyoshima & Y. Ishikawa, 2000: 174 **A**: TAI
tokejii Nomura & Katô, 1958: 1 **A**: FE JA

genus *Mycetoma* Dejean, 1834: 201 type species *Dryops suturalis* Panzer, 1797
affine Nikitsky, 1985c: 267 **A**: FE JA
sapporense Nakane, 1991: 7
ussuriense Nikitsky, 1985c: 266 **A**: FE
suturale Panzer, 1797: 12 (*Dryops*) **E**: AU BH CR CZ FR GE GG HU IT PL RO SK SL SP SZ RO UK YU
richteri Roubal, 1933c: 128

family MELANDRYIDAE Leach, 1815

subfamily **Melandryinae** Leach, 1815

tribe Anisoxiellini Nikitsky, 2007

genus *Anisoxiella* Nikitsky, 1989a: 42 type species *Anisoxya ocularis* Nomura & Katô, 1959
laticollis Nakane, 1963b: 222 (*Anisoxya*) **A**: JA
ocularis Nomura & Katô, 1959: 7 (*Anisoxya*) **A**: JA

tribe Dircaeini Mulsant, 1856

genus *Abdera* Stephens, 1832: 37 type species *Mordella bifasciata* Marsham, 1802
subgenus *Abdera* Stephens, 1832: 37 type species *Mordella bifasciata* Marsham, 1802
bifasciata Marsham, 1802: 493 (*Mordella*) **E**: AU BU DE EN FR GB GE IT PL SK SP **N**: AG TU
biflexuosa Curtis, 1829: 255 (*Hypulus*)
bifasciata Stephens, 1832: 38
flexuosa A. G. Olivier, 1790: 31: 53 (*Elater*)
griseoguttata Fairmaire, 1850b: 423 (*Dircaea*)
undata Perris, 1852: 44 [HN]

quadrifasciata Curtis, 1829: 255 (*Hypulus*) **E**: AU BE BU CR FR GE GB GR HU IT LU PL RO SK SL SP SZ UK YU **N**: TU **A**: TR

subgenus *Caridua* E. Strand, 1929: 24 [RN] type species *Hallomenus affinis* Paykull, 1799

Carida Mulsant, 1856c: 61 [HN]

affinis Paykull, 1799: 181 (*Hallominus*) **E**: AU BY CR CT CZ DE EN FI FR GB GE HU IT LA LT NR NT PL RO SK SL SP ST SV SZ UK **A**: ES FE KZ WS

picea Walker, 1837: 83

flexuosa Paykull, 1799: 182 (*Hallominus*) **E**: AB AL AR AU BE BH BU BY CR CT CZ DE EN FI FR GB GE GG HU IR IT LA LS LT MD NL NR NT PL RO SK SL ST SV UK **A**: TR

undata Panzer, 1799c: 23 (*Hallomenus*)

scriptipennis Lewis, 1895a: 269 **A**: FE JA NE

genus *Anisoxya* Mulsant, 1856c: 45 type species *Dircaea tenuis* Rosenhauer, 1847

conicicollis Champion, 1916b: 34 **A**: FE JA

fuscula Illiger, 1798: 132 (*Serropalpus*) **E**: AL AU BE BH BU CR CZ DE FR GB GE GG HU IT MC MD NL NR PL RO SK SL ST SV SZ UK YU

mustela Abeille de Perrin, 1869: 42

tenuis Rosenhauer, 1847: 34 (*Dircaea*)

genus *Dircaea* Fabricius, 1798: 6 type species *Dircaea quadriguttata* Fabricius, 1798

australis Fairmaire, 1856: 546 **E**: AU BH BY CR CZ FR GE HU RO SK SL SV SZ UK YU

imitata Olexa, 1955: 38

quadriguttata Duftschmid, 1812: 264 [HN]

quadrisignata Mulsant, 1856c: 71

erotyloides Lewis, 1895a: 267 **A**: JA SC

quadriguttata Paykull, 1798: 251 (*Hypulus*) **E**: BY CR CT EN FI FR IT LA NT PL SK ST SV UK **A**: ES FE JA KZ MG NE WS

guttata Stierlin, 1898: 166 [DA]

quadriguttata Fabricius, 1798: 122

shibatai Hayashi, 1960: 42

reducta Pic, 1954a: 53 **A**: FUJ

ussuriensis Nikitsky, 1985b: 57 **A**: FE

nomen dubium

Serropalpus quadrimaculata Illiger, 1798: 130

genus *Dircaeomorpha* Fairmaire, 1896b: 122 type species *Dircaeomorpha clavicornis* Fairmaire, 1896

elegans Sasaji, 1974: 4 **A**: JA SC

validicornis Lewis, 1895a: 266 (*Dircaea*) **A**: JA

genus *Paradircaea* Nikitsky, 1998b: 44 type species *Dircaea dentatomaculata* Lewis, 1895

dentatomaculata Lewis, 1895a: 267 (*Dircaea*) **A**: FE JA SC

genus *Phloiotrya* Stephens, 1832: 35 type species *Phloiotrya rufipes* Stephens, 1832

subgenus *Phloiotrya* Stephens, 1832: 35 type species *Phloiotrya rufipes* Stephens, 1832

Phloeotrya Agassiz, 1846b: 286 [unjustified emendation]

bellicosa Lewis, 1895a: 266 **A**: FE JA NC NE SC

flavitarsis Lewis, 1895a: 266 (*Dircaea*) **A**: JA NE

granicollis Seidlitz, 1898b: 556 **E**: IT **N**: TU

ishikawai Toyoshima, 1999: 98 **A**: JA

obscura Lewis, 1895a: 268 (*Dircaea*) **A**: ES FE JA SC

rufomarginella Hayashi, 1960: 43

subcostulata Pic, 1953c: 335

planiuscula Nomura & Katô, 1959: 8 **A**: JA

quercicola Champion, 1922a: 74 **A**: UP

rufipes Gyllenhal, 1810: 519 (*Dircaea*) **E**: AU BE BH BY CR CZ FI FR GB GE IT LA MD NL NR PL RO SK SL SV SZ UK YU

testaceipennis Pic, 1900g: 62 (*Dircaea*)
rufoventris Nomura, 1958: 9 **A**: JA
rugicollis Marseul, 1876d: 334 **A**: FE JA SC
shimomurai Toyoshima & Y. Ishikawa, 2000: 177 **A**: JA
subtilis Reitter, 1897g: 246 (*Dircaea*) **E**: AU BH CR CT CZ PL RO SK SL UK YU **A**: ES FE WS
tenuis Hampe, 1850: 355 (*Serropalpus*) **E**: AB AR AU BE CR CZ FR GB GE GG GR HU IT NL RO SK SP ST SZ UK
rufipes Stephens, 1832: 35 [HN]
scabra Petri, 1912: 228
stephensi Jacquelin du Val, 1863: 393
vaudoueri Mulsant, 1856c: 79
trisignata Nomura, 1958: 7 **A**: JA

genus *Stenoxylita* Nomura, 1959: 43 type species *Dircaeomorpha trialbofasciata* Hayashi & Katô, 1956
trialbofasciata Hayashi & Katô, 1956: 37 (*Dircaeomorpha*) **A**: JA

genus *Wanachia* Schulze, 1912: 11 type species *Dircaea triguttata* Gyllenhal, 1810
Caridina Seidlitz, 1898b: 520 [HN] type species *Dircaea triguttata* Gyllenhal, 1810
similis Nikitsky, 1984: 679 **A**: FE
triguttata Gyllenhal, 1810: 524 (*Dircaea*) **E**: AU BE BY CT CZ DE EN FI FR GB GG GE HU IT LA NL NR NT PL SK SP ST SV SZ UK **A**: ES FE WS
scutellaris Mulsant, 1856c: 50 (*Abdera*)
suturalis Baudi di Selve, 1877a: 62 (*Abdera*)
trisignata Champion, 1916b: 35 (*Abdera*) **A**: FE (Kunashir) JA
viberti Pic, 1905b: 98 (*Abdera*) **N**: AG

tribe Dircaeini, species incertae sedis
Hallomenus variegatus Motschulsky, 1872: 42 **A**: ST

tribe Hypulini Seidlitz, 1875

genus *Hypulus* Paykull, 1798: 251 type species *Elater quercinus* Quensel, 1790
Mystaxus Semenov, 1900e: 685 type species *Mystaxus relictus* Semenov, 1900
acutangulus Lewis, 1895a: 270 **A**: JA
bifasciatus Fabricius, 1792a: 210 (*Notoxus*) **E**: AU BH CR CT CZ DE FI FR GE GG GR HU IT LT MC MD PL RO SK SL ST SV SZ UK YU
alni D. H. Schneider, 1792: 242 (*Ptilinus*)
fasciatus Panzer, 1793a: 3 (*Helops*)
bodemeyeri Reitter, 1909b: 51 **A**:TR
cingulatus Lewis, 1895a: 269 **A**: JA
higonius Lewis, 1895a: 269 **A**: JA
maroliiformis Reitter, 1911c: 234 **A**: KZ
quercinus Quensel, 1790: 17 (*Elater*) **E**: AU BE BH BU BY CR CT CZ DE EN FR GB GE HU IT LA NL NR PL RO SK SL ST SV SZ UK YU
blekingensis D. H. Schneider, 1792: 241 (*Elater*)
dubius Fabricius, 1792a: 211 (*Notoxus*)
undulatus Pic, 1925d: 11
relictus Semenov, 1900e: 685 (*Mystaxus*) **A**: TD UZ

genus *Marolia* Mulsant, 1856c: 92 type species *Serropalpus variegatus* Bosc, 1791
grandis Peyerimhoff, 1917c: 330 **N**: AG
leseigneuri Nicolas, 1977: 282 **E**: FR IT SZ
purkynei Mařan, 1933: 112 **E**: BH BU MC RO YU
variegata Bosc, 1791: 329 (*Serropalpus*) **E**: FR GBi IT SP SZ
obscurithorax Pic, 1925a: 1

genus *Neogonus* Hampe, 1873: 165 type species *Neogonus plasoni* Hampe, 1873
subgenus *Metax* Semenov & Arnoldi, 1937: 221 type species *Neogonus hiemalis* Semenov & Arnoldi, 1937
fausti Reitter, 1880d: 546 **E**: AB
hiemalis Semenov & Arnoldi, 1937: 223 **A**: TM
subgenus *Neogonus* Hampe, 1873: 165 type species *Neogonus plasoni* Hampe, 1873
Opsigonus Baudi di Selve, 1877c: 409 type species *Opsigonus kruperi* Baudi di Selve, 1877
emgei Reitter, 1884b: 91 **E**: GR
hampei Reitter, 1897e: 221 **E**: AR
minos Reitter, 1897e: 222 **E**: GR
plasoni Hampe, 1873: 165 **E**: GR
kruperi Baudi di Selve, 1877c: 409 (*Opsigonus*)

genus *Nipponomarolia* Miyatake, 1982a: 29 type species *Nipponomarolia kobensis* Miyatake, 1982
kidoi Nakane, 1989: 9 **A**: JA
kobensis Miyatake, 1982a: 30 **A**: JA
matsuyamensis Miyatake, 1982a: 33 **A**: JA
nagaii Nakane, 1989: 9 **A**: JA
oharai Nakane, 1989: 9 **A**: JA

tribe Melandryini Leach, 1815

genus *Bonzicus* Lewis, 1895a: 270 type species *Bonzicus hypocrita* Lewis, 1895
hypocrita Lewis, 1895a: 271 **A**: JA NE
notatus Pic, 1934e: 28

genus *Ivania* Lewis, 1895a: 271 type species *Ivania coccinea* Lewis, 1895
coccinea Lewis, 1895a: 271 **A**: FE (Kunashir, Sakhalin) JA TAI

genus *Melandrya* Fabricius, 1801a: 163 type species *Helops serratus* Fabricius, 1775
subgenus *Apteromelandrya* Gusakov, 2005: 117 type species *Melandrya minshanensis* Gusakov, 2005
minshanensis Gusakov, 2005: 117 **A**: SCH
subgenus *Emmesa* Newman, 1838: 376 type species *Emmesa connectens* Newman, 1838
atricolor Lewis, 1895a: 273 **A**: JA
flavipennis Nakane & Hayashi, 1955: 81 **A**: JA
karafutona Kôno, 1930d: 50 **A**: FE JA NC
ordinaria Lewis, 1895a: 274 **A**: JA
pictipennis Lewis, 1895a: 275 **A**: JA
rausuana Nakane, 1963b: 224 **A**: ES FE JA
rufonotata Nomura, 1958: 13 **A**: JA
utakoae Sasaji, 1988: 43 (*Prothalpia*) **A**: JA
subgenus *Eumelandrya* Nomura, 1959: 45 type species *Melandrya obsoletomaculata* Nomura, 1958
duodecimmaculata Nakane & Hayashi, 1955: 81 **A**: JA
obsoletomaculata Nomura, 1958: 12 **A**: JA
subgenus *Melandrya* Fabricius, 1801a: 163 type species *Helops serratus* Fabricius, 1775
barbata Fabricius, 1787a: 213 (*Helops*) **E**: AU BE BH CR CT CZ DE FI FR GB GE HU IT NR NT PL RO SK SL SV SZ UK YU
flavicornis Duftschmid, 1812: 262
caraboides Linnaeus, 1760: 173 (*Chrysomela*) **E**: AB AR AU BE BH BU CR CZ DE FR GB GE GG HU IR IT LA LS MC MD NL NR PL RO SK SL SP ST SV SZ TR UK YU **A**: TR
coerulescens Petri, 1912: 230
luridipes Pic, 1907e: 138
rufibarbis Schaller, 1783: 324 (*Tenebrio*)
serrata Fabricius, 1775: 257 (*Helops*)
tibialis Viturat, 1895: 110
fasciata Nikitsky, 1974: 61 **A**: FE
gloriosa Lewis, 1895a: 272 **A**: FE JA NE

modesta Lewis, 1895a: 273 **A**: FE JA
quadricostata Hayashi, 1960: 46
mongolica Solsky, 1871: 378 **A**: CH ES FE JA MG NC NE SC WS
parallela Nomura & Katô, 1968: 1 **A**: FE JA
rufipes Gebler, 1829a: 130 **A**: ES FE MG WS
fulgida Motschulsky, 1845a: 81
inequalis Pic, 1927a: 2
kieberi Laporte, 1840: 249
splendida Faldermann, 1833: 58
shimoyamai Hayashi, 1960: 45 **A**: FE JA
ussuriensis Nikitsky, 1972: 342 **A**: FE
subgenus *Paramelandrya* Nikitsky, 2002: 24 type species *Tenebrio dubius* Schaller, 1783
dubia Schaller, 1783: 326 (*Tenebrio*) **E**: AU BE BU BY CZ DE EN FI FR GE HU IT LA LT NR NT PL RO SK ST SV SZ UK **A**: ES FE JA MG NC SC WS
alternans Motschulsky, 1872: 42
canaliculata Fabricius, 1787a: 213 (*Helops*)
caraboides Latreille, 1818a: 42
goryi Laporte, 1840: 249
niponica Lewis, 1895a: 273
rufipes Chevrolat, 1833a: pl. 33
subgenus *Pseudomelandrya* Nikitsky, 2002: 24 type species *Melandrya flavonotata* Pic, 1938
flavonotata Pic, 1938c: 11 **A**: FE JA NC SC
quadrisignata Nakane & Hayashi, 1955: 81
genus *Melandrya*, species incertae sedis
incostata Fairmaire, 1889b: 46 **A**: CH

genus *Phryganophilus* C. R. Sahlberg, 1833: 454 type species *Dircaea ruficollis* Fabricius, 1798
subgenus *Phryganophilus* C. R. Sahlberg, 1833: 454 type species *Dircaea ruficollis* Fabricius, 1798
Longemelandrya Pic, 1953c: 335 type species *Melandrya angustata* Pic, 1953
Phryganophiloides Morishima, 1988: 41 type species *Phryganophiloides elegans* Morishima, 1988
angustatus Pic, 1953c: 335 **E**: CT **A**: ES FE
auritus Motschulsky, 1845a: 81 **E**: AU BY CT CZ EN GE NT PL SK UK **A**: ES FE JA KZ MG WS
elegans Morishima, 1988: 42 (*Phryganophiloides*)
nigriventris Hampe, 1850: 355
pseudoauritus Nikitsky, 1988a: 1429 **E**: CT **A**: ES FE
ruficollis Fabricius, 1798: 122 (*Dircaea*) [NP] **E**: AU BH BY CR CT CZ EN FI FR GE LA NR NT PT PL RO SK SV SP UK YU **A**: ES FE JA MG NE WS
analis Fabricius, 1792a: 214 (*Cantharis*) [NO]
paradoxus Paykull, 1799: 162 (*Lymexylon*)
rosti Hubenthal, 1905: 57
ruficollis Lewis, 1895a: 274 (*Melandrya*)
subgenus *Pseudophryganophilus* Nikitsky, 2002: 27 type species *Melandrya affinis* Nikitsky, 1985
affinis Nikitsky, 1985d: 757 (*Melandrya*) **A**: FE NC SC

genus *Sallumia* Fairmaire, 1889b: 46 type species *Sallumia davidis* Fairmaire, 1889
Bonzicomorpha Pic, 1927c: 13 type species *Bonzicomorpha unifasciata* Pic, 1927
atra Pic, 1934e: 28 (*Bonzicomorpha*) **A**: CH
davidis Fairmaire, 1889b: 47 **A**: NO SCH

tribe Orchesiini Mulsant, 1856

genus *Eucinetomorphus* Perris, 1875: 7 type species *Eucinetomorphus leprieurii* Perris, 1875
asturiensis Reitter, 1881c: 232 (*Lederia*) **E**: SP
ehlersi Heyden, 1884b: 360 (*Lederia*) **E**: SP
leprieurii Perris, 1875: 8 **N**: AG TU
lucidus Peyerimhoff, 1917a: 143 (*Orchesia*) [DA] **N**: AG
trabuti Peyerimhoff, 1918a: 371 **N**: AG

genus *Lederia* Reitter, 1880c: 478 type species *Lederia suramensis* Reitter, 1880
suramensis Reitter, 1880c: 479 **E**: GG ST (Kavkaz) **A**: TR

genus *Lederina* Nikitsky & Belov, 1982: 113 type species *Lederia kaszabi* Nikitsky & Belov, 1982
angusticanalis Sasaji, 1987: 42 (*Lederia*) **A**: JA
foenilis Lewis, 1895a: 262 (*Microscapha*) **A**: JA
japonica Reitter, 1891a: 30 (*Lederia*) **A**: JA
lata Lewis, 1895a: 262 (*Microscapha*) **A**: JA
imdecorata Ishikawa & Sakai, 2001: 333 **A**: JA
indica Nikitsky & Belov, 1982: 119 (*Lederia*) **A**: SD
kidoi Sasaji, 1995: 426 (*Lederia*) **A**: JA
martensi Nikitsky, 1994b: 346 **A**: NP
mozolevskayae Nikitsky, 2001: 23 **A**: YUN
obscuripennis Nikitsky, 1994b: 347 **A**: NP
prion Sasaji, 1987: 44 (*Lederia*) **A**: JA
schawalleri Nikitsky, 1994b: 348 **A**: NP
similis Nikitsky & Belov, 1982: 119 (*Lederia*) **A**: NP

genus *Microscapha* LeConte, 1866: 152 type species *Microscapha clavicornis* LeConte, 1866
akitai Ishikawa & Sakai, 2004: 128 **A**: JA
isensis Sasaji, 1987: 35 **A**: JA
nepalensis Nikitsky, 1992a: 83 **A**: NP
yamoto Ishikawa & Sakai, 2004: 131 **A**: JA

genus *Orchesia* Latreille, 1807: 194 type species *Hallomenus micans* Panzer, 1793
subgenus *Clinocara* C. G. Thomson, 1859: 120 type species *Serropalpus fasciatus* Illiger, 1798
blandula Brancsik, 1874: 229 **E**: AU BH CR CZ IT PL RO SK SZ UK
    carpathica Reitter, 1878d: 60
    transsylvanica Reitter, 1878d: 59
diversenotata Pic, 1933f: 299 **A**: JA
duplicata Nikitsky, 1985c: 269 **A**: CH ES FE JA NC SC
elegantula Lewis, 1895a: 260 **A**: FE JA NE
fasciata Illiger, 1798: 135 (*Serropalpus*) **E**: AU BY CR CT CZ DE EN FI FR GE HU IT LA LT NR NT PL RO SK SL ST SV SZ UK **A**: ES FE KZ WS
    laterufa Pic, 1927h: 10
    trifasciata Zetterstedt, 1838: 163 (*Dircaea*)
grandicollis Rosenhauer, 1847: 33 **E**: AU BH CR CZ GE HU IT RO SL YU
    laticollis L. Redtenbacher, 1849: 611
imitans Lewis, 1895a: 261 **A**: FE JA NE SC
kamberskyi Reitter, 1888b: 213 **E**: GG ST (Kavkaz) **A**: TR
    fuscofasciata Reitter, 1888b: 213
keili Roubal, 1933c: 128 **E**: RO
maculata Mulsant & Godart, 1856: 108 **E**: IT (Sicilia) **N**: AG
minor Walker, 1837: 83 **E**: AU BE BY CT CZ DE EN FI FR GB GE HU IR IT LA LS LT NL NR NT PL RO SK SL ST SV SZ UK **A**: KZ WS
    sepicola Rosenhauer, 1847: 32
    tetratoma C. G. Thomson, 1864: 309
undulata Kraatz, 1853: 256 **E**: AB AL AR AU BE BH BU BY CR CT CZ DE ES FI FR GB GE GG HU IR IT LA MD NL NR PL RO SK SL SP ST SV SZ UK YU **N**: AG
    fasciata Gyllenhal, 1810: 531 (*Hallomenus*) [HN]
vorontsovi Nikitsky, 2001: 19 **A**: YUN
subgenus *Orchesia* Latreille, 1807: 194 type species *Hallomenus micans* Panzer, 1793
fusiformis Solsky, 1871: 377 **E**: AU BY CT CZ LA PL RO SK UK YU **A**: CH ES FE KZ NC MG WS SC
    acicularis Reitter, 1886e: 349
    nadeshdae Semenov, 1898e: 289

obscuricolor Pic, 1954a: 53
marseuli Lewis, 1895a: 262 **A**: JA
micans Panzer, 1793e: 18 (*Hallomenus*) **E**: AB AL AR AU BE BH BU BY CR CT CZ DE EN FI FR GB GE GG GR HU IR IT LA LI MD NL NR NT RO SK SL SP ST SV SZ TR UK YU **N**: AG TU **A**: WS **AFR**
abeillei Guillebeau, 1888: 14
australis Guillebeau, 1887: 3
clavicornis Latreille, 1804a: 417 (*Anaspis*)
corsica Guillebeau, 1888: 13
picea Herbst, 1792: 97 (*Megatoma*)
reyi Guillebeau, 1887: 3
subimpressa Guillebeau, 1887: 4
ocularis Lewis, 1895a: 261 **A**: FE (S. Kuril Is., Sakhalin) JA NC NE
subgenus *Orchestera* Guillebeau, 1887: 5 type species *Orchesia luteipalpis* Mulsant, 1857
arnoldii Nikitsky, 1985c: 272 **A**: ES FE
luteipalpis Mulsant & Guillebeau, 1857: 414 **E**: AU BE BH CR CZ DE FR GE GG IT LA MD NR PL RO SK SL ST SV YU

genus *Orchesia*, species incertae sedis
anaspioides Motschulsky, 1845a: 81 (*Hallomenus*) **A**: FE

genus *Paralederia* Nikitsky & Belov, 1982: 113 type species *Lederia anatolica* J. Frivaldszky, 1880
anatolica J. Frivaldszky, 1880: 262 (*Lederia*) **E**: BU **A**: TR
seidlitzi Reitter, 1916f: 290 (*Lederia*) **A**: TR

tribe Serropalpini Latreille, 1829

genus *Amblyctis* LeConte, 1879: 3 type species *Amblyctis praeses* LeConte, 1879
Dapsiloderus Fairmaire, 1887c: 157 type species *Dapsiloderus costipennis* Fairmaire, 1887
Hikohohodemia Nakane & Hayashi, 1955: 80 type species *Hikohohodemia nomurai* Nakane & Hayashi, 1955
nomurai Nakane & Hayashi, 1955: 81 (*Hikohohodemia*) **A**: JA

genus *Enchodes* LeConte, 1866: 148 type species *Dircaea sericea* Haldeman, 1848
Paramikadonius Nomura, 1959: 44 type species *Synchroa crepuscula* Lewis, 1895
crepusculus Lewis, 1895a: 263 (*Synchroa*) **A**: FE (Kunashir) JA
costulatus Pic, 1932a: 3 (*Mikadonius*)
orientalis Nikitsky, 1973b: 1729 **A**: FE SC

genus *Mikadonius* Lewis, 1895a: 264 type species *Mikadonius gracilis* Lewis, 1895
gracilis Lewis, 1895a: 264 **A**: JA SC
japonicus Hayashi, 1960: 43 **A**: JA

genus *Mimoserropalpus* Pic, 1922b: 17 type species *Serropalpus binhanus* Pic, 1922
formosanus Nomura, 1958: 10 (*Cuphosis*) **A**: TAI
indicus Pic, 1945: 4 (*Serropalpus*) **A**: HP
striatus Blair, 1931a: 200 (*Talayra*) **A**: UP

genus *Nagakuchikius* Toyoshima & Y. Ishikawa, 2000: 183 type species *Phloeotrinus tenuipes* Toyoshima & Y. Ishikawa, 2000
tenuipes Toyoshima & Y. Ishikawa, 2000: 184 (*Phloeotrinus*) **A**: JA

genus *Perakianus* Pic, 1914k: 61 type species *Perakianus atriceps* Pic, 1914
bicolor Blair, 1931a: 201 **A**: SD
hisamatsui Nakane, 1963b: 223 **A**: JA

genus *Phloeotrinus* Nikitsky, 1989a: 45 type species *Serropalpus filiformis* Marseul, 1876
subgenus *Phloeotrinops* Nikitsky, 1989a: 46 type species *Dircaea femoralis* Lewis, 1895
femoralis Lewis, 1895a: 268 (*Dircaea*) **A**: JA

subgenus *Phloeotrinus* Nikitsky, 1989a: 45 type species *Serropalpus filiformis* Marseul, 1876
filiformis Marseul, 1876d: 333 (*Serropalpus*) **A**: JA
minusculus Nomura, 1962b: 45 (*Phloeotrya*) **A**: JA TAI

genus *Serropalpus* Hellenius, 1786: 310 type species *Serropalpus striatus* Hellenius, 1786 (= *Mordella barbata* Schaller, 1783)
barbatus Schaller, 1783: 322, pl. 1 (*Mordella*) **E**: AU BE BH BY CR CT CZ DE EN FI FR GE GR HU IT LA LT MC NLi NR NT PL RO SK SL SP ST SV SZ TR UK YU **N**: MOi **A**: ES FE JA KZ MG NC SC TR WS **NAR**
biguttatus Schellenberg, 1802: 10, pl. 3 (*Lymexylon*)
niponicus Lewis, 1895a: 263
striatus Hellenius, 1786: 318
marseuli Nikitsky, 1992d: 659 **E**: CT ST (Kavkaz) **A**: ES FE JA
genus *Serropalpus*, species incertae sedis
iriei Toyoshima & Y. Ishikawa, 2000: 187 **A**: JA TAI

tribe Xylitini C. G. Thomson, 1864

genus *Dolotarsus* Jacquelin du Val, 1863: 391 type species *Dolotarsus rufipes* Jacquelin du Val, 1863
lividus C. R. Sahlberg, 1833: 448 (*Dircaea*) **E**: AU BH BU BY CR CT CZ EN FI FR GE HU IT LA LS MC NT PL RO SK SL SP SV SZ UK YU **A**: ES FE JA NE WS NAR
ephippius Schaum, 1859: 50 (*Dircaea*)
suturus Gredler, 1866: 278 (*Phryganophilus*)
rufipes Jacquelin du Val, 1863: 391

genus *Rushia* Forel, 1893: 167 type species *Carebara longula* LeConte, 1866
Carebara LeConte, 1866: 251[HN] type species *Carebara longula* LeConte, 1866
Xylitella Reitter, 1911a: 363 type species *Xylita parreyssi* Mulsant, 1856
parreyssii Mulsant, 1856c: 74 (*Xylita*) **E**: AU BH CR DE FR GG GR HU IT LA MC RO SL SP ST SV SZ UK YU **A**: CY TR
ferruginea Gredler, 1866: 279 (*Phryganophilus*)
mollis Graells, 1858: 73 pl. 12 (*Dircaea*)
paganettii Obenberger, 1917: 40 (*Xylitella*)
revelierei Mulsant, 1859: 161 (*Dircaea*)
nagaii Nakane, 1975: 168 (*Xylita*) **A**: JA

genus *Spilotus* LeConte, 1862: 251 type species *Hallomenus quadripustulatus* Melsheimer, 1846
Micromarolia Pic, 1937f: 35 type species *Micromarolia uninotata* Pic, 1937
Pseudozilora Nomura, 1959: 44 type species *Pseudozilora quadrimaculata* Nomura, 1959
quadrimaculatus Nomura, 1959: 44 (*Pseudozilora*) **A**: JA
uninotatus Pic, 1937f: 35 (*Micromarolia*) **A**: JA

genus *Xylita* Paykull, 1798: 249 type species *Elater buprestoides* Fabricius, 1792
laevigata Hellenius, 1786: 318 (*Serropalpus*) **E**: AU BE BH BU BY CR CT CZ DEi EN FI FR GB GE GR HU IT JA LA LT MC NL NR NT PL RO SK SL SV SZ UK YU **A**: ES FE KZ MG WS **NAR**
buprestoides Fabricius, 1792b: 234 (*Elater*)
decolorata Randall, 1838: 22 (*Dircaea*)
discolor Fabricius, 1798: 121 (*Dircaea*)
producta Hatch, 1965: 70
robusta Motschulsky, 1872: 43

tribe Zilorini Nikitsky, 2007

genus *Euryzilora* Lewis, 1895a: 265 type species *Euryzilora lividipennis* Lewis, 1895
lividipennis Lewis, 1895a: 265 **A**: JA
ussurica Mamaev, 1978: 47 **A**: FE

genus *Zilora* Mulsant, 1856c: 84 type species *Xylita ferruginea* Paykull, 1798
anatolica Korge, 1971: 9 **A**: TR
elongata J. R. Sahlberg, 1881: 132 **E**: BY FI CT EN NT **A**: ES FE WS
ferruginea Paykull, 1798: 250 (*Xylita*) **E**: BY EN FI GB LA NR NT PL SV
obscura Fabricius, 1794: 445 (*Parnus*) **E**: AU BH BU BY CR CZ ES FR GE HU IT LA LS LT PL RO SK SL SP UK YU
bifoveolata Dufour, 1851: 337 (*Mycetochara*)
eugeniae Ganglbauer, 1891: 132
ferruginea Mulsant, 1856c: 85
sericea J. Sturm, 1807: 261 (*Helops*)

subfamily Melandryinae, incertae sedis

genus *Adobia* Mulsant, 1856c: 56 type species *Abdera bifasciata* Mulsant, 1856
mulsanti Nikitsky, **nom. nov.** [see New Acts] **E**: FR
bifasciata Mulsant, 1856c: 56 (*Abdera*) [HN]

genus *Hira* Hayashi, 1960: 45 type species *Hira humerosignata* Hayashi, 1960
humerosignata Hayashi, 1960: 45 **A**: JA
suturalis Nomura, 1962b: 42 **A**: JA

genus *Microtonus* LeConte, 1862: 259 type species *Microtonus sericans* LeConte, 1962
dimidiatus Marseul, 1876d: 457 (*Scraptia*) **A**: JA
takaosanus Nomura, 1962b: 44 **A**: JA

genus *Symphora* LeConte, 1866: 150 type species *Scraptia flavicollis* Haldeman, 1848
atra Nomura, 1959: 45 **A**: JA
brunnea amamiana Nomura, 1962b: 44 **A**: JA
brunnea brunnea Marseul, 1876d: 457 (*Scraptia*) **A**: FE (Kunashir) JA
miyakei Nomura & Hayashi, 1960: 44

subfamily **Osphyinae** Mulsant, 1856 (1840)

genus *Conopalpus* Gyllenhal, 1810: 547 type species *Conopalpus flavicollis* Gyllenhal, 1810
brevicollis Kraatz, 1855: 374 **E**: BH CR GE FR IT NL PT SP SZ
testaceus A. G. Olivier, 1790: 21: 11 (*Melyris*) **E**: AU BE BH BU CR CZ CT DE FR GB GE GR HU IT LA NL NR PL RO SK ST SV SZ UK
flavicollis Gyllenhal, 1810: 547
nigricornis Germar, 1821: nr. 4: 4
obscuriceps Pic, 1925d: 11
ruficollis Newman, 1833: 185
thoracicus Baudi di Selve, 1877a: 67
vigorsi Stephens, 1832: 60

genus *Osphya* Illiger, 1807: 370 [RN] type species *Cantharis bipunctata* Fabricius, 1775
Nothus A. G. Olivier, 1811: 384 type species *Nothus praeustus* A. G. Olivier, 1811
Pelecina Illiger, 1807: 333 [HN] type species *Cantharis bipunctata* Fabricius, 1775
aeneipennis Kriechbaumer, 1848: 163 (*Osphya*) **E**: AU FR IT SZ
immaculata Pic, 1897b: 3
maculicollis Pic, 1897b: 3
picticollis Pic, 1897b: 3
simplicipes Seidlitz, 1898b: 674

erata Seidlitz, 1898b: 666 **E**: AB AR **A**: IN
bilineata Pic, 1922d: 18
discoidalis Pic, 1937a: 1
multiinterrupta Pic, 1937a: 1
seidlitzi Pic, 1908a: 32
subapicalis Pic, 1912g: 34
talyschensis Pic, 1902b: 9
ipunctata Fabricius, 1775: 206 (*Cantharis*) **E**: AU BH BE BU BY CT CR CZ DE FR GB GE GG GR HU IT KZ MD NR PL RO SK SL SP ST SV SZ UK YU **A**: KZ WS
apicenotata Pic, 1908a: 31
anceps Stephens, 1839: 339 (*Nothus*)
bimaculata Stephens, 1832: 59 (*Nothus*)
clavipes Olivier, 1811: 385 (*Nothus*)
clavipes Gyllenhal, 1817: 7 (*Dryops*)
impunctata Bouskell, 1899: 267
innotata Pic, 1898g: 98
maculata Bouskell, 1899: 267
obscuripennis Pic, 1897b: 2
praeusta A. G. Olivier, 1811: 385 (*Nothus*)
quadipunctata Bouskell, 1899: 267
testaceithorax Pic, 1909i: 121
uralensis Motschulsky, 1845a: 84 (*Nothus*)
vittipennis Seidlitz, 1898b: 668
cylindromorpha Abeille de Perrin, 1896: 283 **A**: SY TR
delagrangei Pic, 1902h: 64
dissimilis Champion, 1922a: 75 **A**: UP
formosana Pic, 1927i: 49 **A**: TAI
griseofasciata Pic, 1921e: 13 **A**: TR
interrupta Pic, 1921e: 13
rufithorax Pic, 1921e: 13
harmandi Pic, 1926m: 354 **A**: SD
nigriventris Champion, 1920a: 74 **A**: UP
nigroapicalis Pic, 1921a: 10 **A**: "India"
orientalis Lewis, 1895a: 276 (*Nothus*) **A**: JA
testaceipes Pic, 1914k: 61
trilineata Pic, 1910q: 75 **A**: TAI
uniformis Pic, 1921e: 13 **A**: TR
vandalitiae Kraatz, 1868b: 335 (*Nothus*) **E**: PT SP

family RIPIPHORIDAE Gemminger & Harold, 1870

subfamily **Hemirhipidiinae** Heller, 1921

genus *Heteromeroxylon* Pic, 1939h: 375 type species *Heteromeroxylon ruficeps* Pic, 1939
Ohananomia Tôyama, 1986: 43 type species *Ohananomia malayana* Tôyama, 1986
yagii Tôyama, 1987: 33 (*Ohananomia*) **A**: JA (Yonagunijima Is., Ryukyus)

genus *Nephrites* Shuckard, 1838: 512 type species *Nephrites nitidus* Schuckard, 1838
apicalis Tôyama & Hatayama, 1985: 20 **A**: SE (Lan Hsu Is.)
japonicus Tôyama & Hatayama, 1985: 19 **A**: JA (Miyako Is., Ryukyus)
kurosawai Tôyama & Hatayama, 1985: 19 **A**: JA (Ishigakijima Is., Ryukyus)
obscurus Tôyama & Hatayama, 1985: 20 **A**: JA (Nakanoshima Is., Tokama)

subfamily **Pelecotominae** Seidlitz, 1875

genus *Clinops* Gerstaecker, 1855a: 16 [= 1855b: 9] type species *Clinops badius* Gerstaecker, 1855
Scotoscopus Brenske & Reitter, 1884b: 92 type species *Scotoscopus carbonarius* Reitter, 1884
spectabilis L. W. Schaufuss, 1872: 276 **E**: GR **A**: TR (Antakya)
carbonarius Reitter, 1884b: 93 (*Scotoscopus*)

genus *Micropelecotoides* Pic, 1910g: 21 type species *Pelecotomoides sumatrensis* Pic, 1910
Micropelecotomoides Csiki, 1913: 6 [unjustified emendation]
aurosericeus Gressitt, 1941: 526 (*Pelecotoides*) **A**: HAI
japonicus Pic, 1910g: 21 (*Pelecotomoides*) **A**: JA TAI

genus *Pelecotoma* Fischer von Waldheim, 1809: 293 type species *Pelecotoma mosquense* Fischer von Waldheim, 1809
fennica Paykull, 1799: 178 (*Ripiphorus*) **E**: AU BH BY CZ FI GE HU LA PL RO RU SK UK
latreillei Fischer von Waldheim, 1823: pl. 38
mosquense Fischer von Waldheim, 1809: pl. 18
zoufali Fleischer, 1916: 120
septentrionalis Kôno, 1936b: 87 **A**: JA

genus *Trigonodera* Dejean, 1834: 217 type species *Pelecotoma leachii* Latreille, 1817
Caspyria Fairmaire, 1901a: 194 type species *Caspyria perrieri* Fairmaire, 1901
Hylacter Gistel, 1848: xi [unnecessary RN]
Pelecotoides Laporte, 1840: 263 type species *Pelecotoma leachii* Latreille, 1817
Pelecotomoides Gemminger & Harold, 1870: 2118 [unjustified emendation]
tokejii Nomura & Nakane, 1959: 115 (*Pelecotomoides*) **A**: JA

subfamily **Ptilophorinae** Gerstaecker, 1855

genus *Ptilophorus* Dejean, 1834: 218 type species *Pelecotoma dufourii* Latreille, 1817
Evaniocera Guérin-Méneville, 1835: pl. 2 type species *Pelecotoma dufourii* Latreille, 1817
afghanicus Kaszab, 1957a: 52 (*Evaniocera*) **A**: AF
dufourii Latreille, 1817b: 135 (*Pelecotoma*) **E**: AR BU CR FR GR HU IT PT RO SP ST TR "Caucasus" **N**: AG MO TU **A**: IN IS JO TR
boryi Lucas, 1847: 382 (*Evaniocera*)
frivaldjskii Guérin-Méneville, 1834a: pl. 34 (*Pelecotoma*)
stevenii Fischer von Waldheim, 1829b: 369 [= 1836: 69] (*Pelecotoma*)
striolatus Nowicki, 1873: 5 (*Evaniocera*)
fallax Iablokoff-Khnzorian, 1973: 308 **A**: TD
fischeri Ménétriés, 1848: 29 (*Evaniocera*) **A**: KI KZ TD TM UZ
fischeri Ménétriés, 1849: 245 (*Evaniocera*)
novemflabellatus J. R. Sahlberg, 1903b: 34 (*Evaniocera*)
hauseri Reitter, 1903g: 238 (*Evaniocera*) **A**: IN
plumicornis Reitter, 1898f: 349 (*Evaniocera*) **A**: TR

subfamily **Ripidiinae** Gerstaecker, 1855

tribe Eorhipidiini Iablokoff-Khnzorian, 1986

genus *Eorhipidius* Iablokoff-Khnzorian, 1986: 89 type species *Eorhipidius januschevi* Iablokoff-Khnzorian, 1986
januschevi Iablokoff-Khnzorian, 1986: 89 **A**: IN TD

genus *Pterydrias* Reitter, 1895e: 202 type species *Pterydrias debilis* Reitter, 1895
debilis Reitter, 1895e: 202 **A**: IS TR

tribe Ripidiini Gerstaecker, 1855

genus *Blattivorus* Chobaut, 1891: 237 type species *Rhipidius lusitanicus* Gerstaecker, 1855
lusitanicus Gerstaecker, 1855a: 25 [= 1855b: 15] (*Rhipidius*) **E**: PT
margaritae Richter, 1949: 360 **E**: AR

genus *Ripidius* Thunberg, 1806: 4 type species *Ripidius pectinicornis* Thunberg, 1806
Pararhipidius Coiffait, 1947: 15 type species *Pararhipidius burdigalensis* Coiffait, 1947
Rhipidius Agassiz, 1846b: 324 [unjustified emendation]
Symbius Sundevall, 1831: 1223 type species *Symbius blattarum* Sundevall, 1831
abeillei abeillei Chobaut, 1891: 235 (*Rhipidius*) **E**: FR **N**: AG MO **A**: IS

hervei Mollandin de Boissy, 1940: 94 (*Rhipidius*)
guignoti Chobaut, 1904a: 228 (*Rhipidius*)
vaulogeri Chobaut, 1893: 99 (*Rhipidius*)
abeillei bolivari Viedma, 1962: 48 (*Rhipidius*) **E**: SP
longicollis Schilder, 1923b: 203 (*Rhipidius*) **A**: TAI
pectinicornis Thunberg, 1806: 5 **E**i: AU FI FR GB GE NL SL **N**i: EG **A**i: CH **COS**
anceps Stephens, 1835: 427
blattarum Sundevall, 1831: 1223 (*Symbius*)
fairmairei Chobaut, 1894b: ci (*Rhipidius*)
quadriceps Abeille de Perrin, 1872: 36 (*Rhipidius*) **E**: AR AU CZ FI FR GE IT PL PT SK SP SZ **N**: AG MO
amori Bolívar y Pieltain, 1917: 249 (*Rhipidius*)
apicipennis Kraatz, 1891: 358 (*Rhipidius*)
boissyi Abeille de Perrin, 1909: 854 (*Rhipidius*)
burdigalensis Coiffait, 1947: 16 (*Pararhipidius*)
denisi Chobaut, 1919: 200 (*Rhipidius*)
kabylianus Chobaut, 1902: 89 (*Rhipidius*)
maroccanus Chobaut, 1922: 127 (*Rhipidius*)
parisiensis Lesne, 1902a: 182 (*Rhipidius*)
tigrani Iablokoff-Khnzorian, 1957b: 231 (*Rhipidius*)
vilarrubiai Español, 1942: 344 (*Rhipidius*)

genus *Rhyzostylops* Silvestri, 1905: 316 type species *Rhyzostylops inquirendus* Silvestri, 1905
inquirendus Silvestri, 1905: 317 **E**: IT

subfamily **Ripiphorinae** Gemminger & Harold, 1870
tribe Macrosiagonini Reitter, 1911

genus *Macrosiagon* Hentz, 1830: 462 type species *Mordella dimidiata* Fabricius, 1781
Emenadia Laporte, 1840: 261 type species *Mordella bimaculata* Fabricius, 1787
Siagonadia Reitter, 1910c: 131 type species *Macrosiagon pallidipennis* Reitter, 1898
bifasciata Marseul, 1877: 478 (*Emenadia*) **A**: CH JA NP NC SC
coreana Pic, 1955a: 2
donceeli Pic, 1908g: 60
reducta Pic, 1909m: 147
tschungseni Pic, 1955a: 2
bimaculata Fabricius, 1787a: 218 (*Mordella*) **E**: BU FR GR HU IT KZ PT RO RU SP UK YU **N**: AG EG MO **A**: IN IS KZ TR "Turkestan"
angustula Motschulsky, 1849: 128 (*Ripiphorus*)
apicalis Steven, 1832a: 429 [= 1836: 158] (*Rhipiphorus*)
bicolor A. G. Olivier, 1795: 65: 5 (*Ripiphorus*)
binotata Fischer von Waldheim, 1829a: 77 (*Rhipiphorus*)
binotata Fischer von Waldheim, 1832b: 427 [= 1836: 157] (*Rhipiphorus*)
carinthiaca Panzer, 1794a: 7 (*Ripiphorus*)
larvata Schrank von Paula, 1789: 90 (*Mordella*)
liturata Fabricius, 1798: 126 (*Ripiphorus*)
nigrithorax Steven, 1832a: 428 [= 1836: 158] (*Rhipiphorus*)
pallidipennis Reitter, 1898f: 350
quadrimaculata Gyllenhal, 1817: 36 (*Ripiphorus*)
tricuspidata Gmelin, 1790: 2024 (*Mordella*)
bipunctata Fabricius, 1801b: 120 (*Ripiphorus*) **A**: KA NP YE **AFR** **ORR**
apicalis Westwood, 1839: lvii, pl. 9 (*Ripiphorus*) [HN]
basinotata Pic, 1949b: 12
imbasalis Pic, 1949b: 12
kaschmirensis Pic, 1949b: 12
reductenotata Pic, 1949b: 12
semiobliterata Pic, 1952b: 7
tropica Nietner, 1856: 392 (*Rhipiphorus*)
elegans Marseul, 1876a: 27 (*Rhipiphorus*) **A**: SA **AFR**

ferruginea Fabricius, 1775: 262 (*Mordella*) **E**: FR GR IT MA SP **N**: EG MO **A**: AE CH CY IS JA TAI TR **AFR ORR**
flabellata Fabricius, 1781: 501 (*Mordella*)
formosicolis Kôno, 1927: 178
pectinata Villers, 1789: 408 (*Mordella*)
subattenuata Pic, 1950b: 8
uniformis Motschulsky, 1849: 127 (*Ripiphorus*)
fortieri Chobaut, 1893: 97 (*Emenadia*) **N**: EG MO
medvedevi Iablokoff-Khnzorian, 1973: 307 **E**: AB **A**: KZ MG
meridionalis A. Costa, 1859: 24 (*Emenadia*) [RN] **E**: IT **N**: AG EG TU
bipunctata A. Costa, 1853: 5 (*Emenadia*) [HN]
italica Bedel, 1895: 184 [RN]
raffrayi Fairmaire, 1873a: 342 (*Rhipiphorus*)
nasuta Thunberg, 1784: 66 (*Mordella*) **A**: CH JA NC SC TAI **ORR**
discoidalis Pic, 1950b: 8
iwatai Kôno, 1936b: 88
obscuricolor Pic, 1912m: 69
semiobscura Pic, 1923c: 15
oberthurii Fairmaire, 1879c: 200 (*Rhipiphorus*) **N**: AG TU **A**: "Arabia"
melanura Fairmaire, 1887e: 303 (*Rhipiphorus*)
praeusta Gebler, 1829a: 133 [= 1830: 133] (*Rhipiphorus*) **E**: AR BU CR FR GR IT SP ST "Caucasus" **N**: AG **A**: IS KZ SY TM TR
apicalis Küster, 1848a: 79 (*Rhipiphorus*) [HN]
fulvipennis Küster, 1846b: 71 (*Rhipiphorus*) [HN]
rufipennis Chevrolat, 1834: pl. 34 (*Rhipiphorus*)
sulcata Ménétriés, 1832: 205 (*Rhipiphorus*)
sulcata Fischer von Waldheim, 1832b: 429 [= 1836: 158] (*Rhipiphorus*) [HN]
pusilla Gerstaecker, 1855b: 32 (*Rhipiphorus*) **A**: CE FE JA NC NO NP SC SE SW UP WP **ORR**
acutipennis Gressitt, 1941: 528 [HN]
atriceps Pic, 1950b: 8
atricolor Pic, 1953a: 15
atronitida Gressitt, 1941: 529
bicoloriceps Pic, 1953a: 15
bicoloripes Pic, 1953a: 15
coomani Pic, 1953a: 15
corporaali Pic, 1950a: 12
cyanivestis Marseul, 1877: 479 (*Rhipiphorus*)
fukienensis Pic, 1955a: 2
gerstaeckeri Harold, 1878a: 82 (*Emenadia*)
pieli Pic, 1953a: 15
thibetana Pic, 1953a: 15
tonkinea Pic, 1906f: 44
variicollis Fairmaire, 1894a: 34 (*Rhipiphorus*)
spinicollis Fairmaire, 1893b: 38 (*Rhipiphorus*) **A**: GUA TAI UP **ORR**
bipartita Fairmaire, 1894a: 35 (*Rhipiphorus*)
diversipennis Pic, 1951c: 16
inapicalis Pic, 1922c: 22
leseleuci Pic, 1906f: 44
uninotaticollis M. Chûjô, 1935b: 133
vitalisi Pic, 1922c: 21
terminata Laporte, 1840: 262 (*Emenadia*) **N**: AG EG **A**: AE SA YE **AFR ORR**
bicolor Laporte, 1840: 262 (*Emenadia*) [HN]
gibbifera Abeille de Perrin, 1880: 234 (*Emenadia*)
lateapicalis Pic, 1950a: 11
nigripennis Gerstaecker, 1855b: 24 (*Rhipiphorus*)
notata Pic, 1947a: 8
perroudi Pic, 1947a: 8
scutellaris Gerstaecker, 1855b: 24 (*Rhipiphorus*)

genus *Macrosiagon,* nomina dubia
bifasciata prescutellaris Pic, 1949b: 12 **A**: CH
grombczewskii Semenov, 1891a: 377 (*Emenadia*) **A**: TD

genus *Metoecus* Dejean, 1834: 218 type species *Mordella paradoxa* Linnaeus, 1760
Cyttaroecus Schilder, 1923a: 169 type species *Mordella paradoxa* Linnaeus, 1760
Ripidastes Gistel, 1848: x type species *Mordella paradoxa* Linnaeus, 1760
morawitzi Semenov, 1891a: 376 (*Rhipiphorus*) **A**: AF XIN
paradoxus Linnaeus, 1760: 228 (*Mordella*) **E**: AB AU BE BH CZ DE FI FR GB GE GR HU IR IT LA NL PL PT SK SP SV SZ "Caucasus" **A**: SY TR
abdominalis Gradl, 1882: 326
affinis Steven, 1832a: 427 [=1836: 157] (*Rhipiphorus*)
antoniae Reitter, 1895d: 157 (*Myiodes*)
angulatus Panzer, 1803a: 3 (*Ripiphorus*)
apicalis Gradl, 1882: 326
atripennis Pic, 1916h: 22
basalis Schilder, 1924: 241
dicipiens Schilder, 1924: 244
erythrogaster Frölich, 1792: 163 (*Mordella*)
flavoniger Gradl, 1882: 326
gradli Schilsky, 1900: 35
humeralis Schilder, 1924: 241
luteonotatus Pic, 1916h: 21
macularis Gradl, 1882: 326
multinotatus Pic, 1916h: 21
nigrescens Gradl, 1882: 326
nigriventris Gradl, 1882: 326
nigrolineatus Pic, 1916h: 21
notiventris Gradl, 1882: 326
obscuripennis Schilder, 1924: 241
pictus Gistel, 1857a: 48
semiflavus Gradl, 1882: 326
theresae Pic, 1916h: 22
tricolor Gistel, 1857a: 48
trinotatus Pic, 1916h: 21
viturati Pic, 1916h: 21
satanas Schilder, 1924: 238 **A**: CH FE JA
abdominalis Matsumura, 1906: 133 (*Motoecus* [sic]) [HN]
vespae Kôno, 1927: 181 **A**: JA

tribe Ripiphorini Reitter, 1911

genus *Ripiphorus* Bosc, 1791: 327 type species *Ripiphorus subdipterus* Bosc, 1792
Dorthesia Say, 1824a: 274 type species *Dorthesia fasciata* Say, 1824
Myiodes Heyden, Reitter & Weise, 1891: 520 [unjustified emendation]
Myiodites Fabre, 1886: 221 [unjustified emendation]
Myodes Latreille, 1818b: 130 [HN] type species *Myodes dorthesii* Latreille, 1818
Myodites Latreille, 1819b: 303 [RN] type species *Myodes dorthesii* Latreille, 1818
Polychronia Gistel, 1848: x type species *Ripiphorus subdipterus* Bosc, 1792
Rhipidophorus Gemminger & Harold, 1870: 2118 type species *Ripiphorus subdipterus* Bosc, 1792
Rhipiphorus Agassiz, 1846b: 324 [unjustified emendation]
Rhipiphorus Fabricius, 1792a: xiv [HN] type species *Rhipiphorus subdipterus* Fabricus, 1792
chalcidoides Gressitt, 1941: 532 (*Rhipiphorus*) **A**: GUA HAI
davidis Fairmaire, 1878a: 124 (*Myodites*) **A**: CE HAI
minor Gressitt, 1941: 533 (*Rhipiphorus*) **A**: GUA HAI
spalatensis Obenberger, 1917: 42 (*Rhipiphorus*) **E**: CR
subdipterus Bosc, 1792: 293 **E**: AB AL AR BU CR FR GR IT PT SP YU **N**: AG MO TU **A**: CY IS KZ UZ
ambigua Giorna, 1804: 216 (*Mordella*)
antoniae Pic, 1905h: 165 (*Myiodes*) [HN]

babadjanidesi Reitter, 1912a: 43 (*Myiodes*)
caucasicus Reitter, 1906a: 453 (*Rhipiphorus*) [RN]
clermonti Chobaut, 1906: 224 (*Myiodes*) [RN]
dorthesii Latreille, 1818b: 131 (*Myodes*)
pallescens Solsky, 1881b: 57 (*Rhipidophorus*)
subdipterus Fabricius, 1792b: 109 [HN]
syriacus Pic, 1904h: 90 (*Myiodes*) **A**: TR
tenthredinoides Gressitt, 1941: 534 **A**: GUA HAI
turcicus Pic, 1914g: 42 (*Myodites*) **A**: TR

family Ripiphoridae, nomina dubia
Ripiphorus dubius Motschulsky, 1849: 128 **E**: SP
Ripiphorus fulvipennis Fonscolombe, 1845: 504 **E**: FR
Ripiphorus indicus Hope, 1831: 32 **A**: NP
Ripiphorus maurus Betta, 1847: 28 **E**: IT
Ripiphorus suturalis Betta, 1847: 28 **E**: IT
Rhipiphorus populneus Fabricius, 1792b: 112 **E**: "Germaniae"

family ZOPHERIDAE Solier, 1834
subfamily **Zopherinae** Solier, 1834
tribe Monommatini Blanchard, 1845

genus *Monomma* Klug, 1833: 182 type species *Monomma irroratum* Klug, 1833
brittoni Freude, 1955: 20 **A**: PA SD
brunneum J. Thomson, 1860b: 23 **A**: PA SD
fulvum dehradunense Freude, 1955: 24 **A**: UP
glyphysternum Marseul, 1876d: 330 **A**: JA HAI HKG SHG TAI XIN YUN
chinense Freude, 1955: 46
formosanum Pic, 1916e: 1
gressitti Freude, 1978: 221 **A**: HAI
japonicum Motschulsky, 1861: 18 [DA] **A**: JA
similifulvum Freude, 1955: 24 **A**: PA UP

genus *Inscutomonomma* Pic, 1951a: 2 type species *Monomma syriacum* J. Thomson, 1860
mycotretoides Ancey, 1881b: 413 (*Monomma*) **A**: YE
syriacum J. Thomson, 1860b: 12 (*Monomma*) **A**: SY
yemense Freude, 1958: 36 **A**: YE

tribe Phellopsini Ślipiński & Lawrence, 1999

genus *Phellopsis* LeConte, 1862: 216 type species *Boletophagus obcordatus* Kirby, 1837
Pseudonosoderma Heyden, 1885b: 305 type species *Pseudonosoderma amurense* Heyden, 1885
amurensis Heyden, 1885b: 306 (*Pseudonosoderma*) **A**: FE SC
imurai Masumoto, 1990b: 87
chinensis Semenov, 1893g: 499 (*Pseudonosoderma*) **A**: GAN NW
suberea Lewis, 1887c: 219 **A**: JA

tribe Pycnomerini Erichson, 1845

genus *Pycnomerus* Erichson, 1842: 214 type species *Ips terebrans* A. G. Olivier, 1790
Dechomus Jacquelin du Val, 1858: 180 type species *Cerylon sulcicolle* Germar, 1824
Penthelispa Pascoe, 1860b: 111 type species *Penthelispa porosa* Pascoe, 1860
Pycnomeroplesius Ganglbauer, 1899: 885 type species *Pycnomerus inexspectus* Jacquelin du Val, 1858
fuliginosus Erichson, 1842: 215 **Ei**: GB IR
inexspectus Jacquelin du Val, 1858: 179 **E**: AU BE CZ FR GE GB IT SP
italicus Ganglbauer, 1899: 887 (*Dechomus*) **E**: IT
lucidus Dajoz, 1975: 308 (*Penthelispa*) **A**: BT SW

nitidicollis Reitter, 1877f: 350 (*Penthelispa*) **A**: CH SD
sculpturatus Sharp, 1885a: 77 **A**: JA TAI
sulcicollis Germar, 1824b: 466 (*Cerylon*) **E**: AB AR BU GG GR RO ST **A**: IN TR
coniunctus Říha, 1958: 32 (*Dechomus*)
fagivorus Říha, 1958: 32 (*Dechomus*)
terebrans A. G. Olivier, 1790: 18: 5 (*Ips*) **E**: AB AU BH BU CR CT CZ FR GE GR HU IT PL PT SK SL ST SZ TR YU **A**: IN TR
vilis Sharp, 1885a: 77 **A**: JA
japonicus Reitter, 1899e: 216 (*Penthelispa*)

tribe Usechini Horn, 1867

genus *Usechus* Motschulsky, 1845a: 79 type species *Usechus lacerta* Motschulsky, 1845
chujoi Kulzer, 1960: 304 **A**: JA
ohdaiensis Sasaji, 1987: 52 **A**: JA
sasajii Saitô, 1999: 106 **A**: JA
tsushimensis Kamiya, 1963: 21 **A**: JA

subfamily **Colydiinae** Erichson, 1842

tribe Colydiini Erichson, 1842

genus *Aulonium* Erichson, 1845: 275 type species *Colydium bidentatum* Fabricius, 1801
Anoectochilus L. Redtenbacher, 1845: 124 type species *Colydium bicolor* Herbst, 1797
ruficorne A. G. Olivier, 1790: 18: 13 (*Ips*) **E**: AB AR BU CR FR GE GR HU IT MC RO SP ST TR UK YU **N**: AG MO TU **A**: IN IS LE TM TR
bicolor Herbst, 1797: 285 (*Colydium*)
sulcicolle Wollaston, 1864: 127 **N**: CI
trisulcum Geoffroy, 1785: 23 (*Dermestes*) **E**: AR AU BE BH CT CZ FR GB GE HU IT MC NL NR NT PL SK ST SV SZ UK **A**: IN
bicolor Fabricius, 1801a: 555 (*Colydium*) [HN]
sulcatum A. G. Olivier, 1790: 18: 4 (*Ips*)

genus *Colydium* Fabricius, 1792b: 495 type species *Bostrichus elongatus* Fabricius, 1787
Paschabium Gozis, 1886: 11 type species *Bostrichus elongatus* Fabricius, 1787
elongatum Fabricius, 1787a: 36 (*Bostrichus*) **E**: AB AL AR AU BE BH BU BY CR CT CZ DE FI FR GB GE GR HU IT LT MC NL PL RO SK SL SP ST SW SZ TR UK YU **N**: AG MO TU **A**: CY IN RU SY TR
filiforme Fabricius, 1792b: 496 **E**: AB AU BE BH BU BY CR CZ FR GE GG GR HU IT LA LS NR PL SK SP ST SV SZ **A**: IN RU

tribe Gempylodini Sharp, 1893

genus *Gempylodes* Pascoe, 1863c: 132 type species *Gempylodes macer* Pascoe, 1863
Mecedanops Reitter, 1878e: 120 type species *Mecedanops ornamentalis* Reitter, 1878
ornamentalis Reitter, 1878e: 121 (*Mecedanops*) **A**: JA TAI **ORR**
lewisii Sharp, 1885a: 72

genus *Pseudendestes* Lawrence, 1980: 298 type species *Pseudendestes australis* Lawrence, 1980
andrewesi Grouvelle, 1908b: 414 (*Endestes*) **A**: SD
lawrencei Pal, 1984: 32
namdaphaensis Pal, 1984: 33 **A**: AP

tribe Nematidiini Sharp, 1894

genus *Nematidium* Erichson, 1845: 275 type species *Colydium cylindricum* Fabricius, 1801
elongatum Pal & Ślipiński, 1984: 539 **A**: UP

tribe Orthocerini Blanchard, 1845

genus *Diplagia* Reitter, 1882a: 117 type species *Diplagia hellenica* Reitter, 1882
hellenica Reitter, 1882a: 118 **E**: GR

genus *Helioctamenus* Schaufuss, 1882b: 531 type species *Helioctamenus hippopotamus* Schaufuss, 1882
Acotulus Reitter, 1891f: 246 type species *Acotulus oranensis* Reitter, 1891
bedeli Dajoz, 1971: 83 **E**: PT SP
curticornis Pic, 1922e: 21 **N**: EG
espanoli Cobos, 1950: 589 **E**: SP
fernandezi Cobos, 1965b: 59 **N**: Cl
fumicornis Bedel, 1891b: cxxxv (*Orthocerus*) **N**: AG MO
aestivus Escalera, 1914: 121 (*Acotulus*)
oranensis Reitter, 189la: 247 (*Acotulus*)
gineri Español, 1948: 311 (*Acotulus*) **N**: MO
hippopotamus Schaufuss, 1882b: 531 E: SP (Baleares)
lusitanicus Reitter, 1903d: 174 **E**: PT
occidentalis Oromí, 1984a: 226 **N**: CI
pardoi Español, 1957c: 147 **N**: MO

genus *Orthocerus* Latreille, 1796: 16 type species *Tenebrio hirticornis* DeGeer, 1775
Sarrotrium Illiger, 1798: 339 type species *Hispa mutica* Linnaeus, 1767
clavicornis Linnaeus, 1758: 355 (*Dermestes*) **E**: AU BE BH BY CR CT CZ DE FI FR GE GB HU LA LS NL NR IT IR PL RO RU SK SP SV SZ YU "Caucasus" **A**: ES FE MG TR
crenulatus Motschulsky, 1845a: 80 (*Sarrotrium*)
hirticornis DeGeer, 1775: 47 (*Tenebrio*)
muticus Linnaeus, 1767: 604 (*Hispa*)
crassicornis Erichson, 1845: 260 (*Sarrotrium*) **E**: AR AU BH BY CZ FR GE HU LA IT PL PT RO SK SZ
tereticornis Erichson, 1845: 261 (*Sarrotrium*)

tribe Rhopalocerini Reitter, 1911

genus *Rhopalocerus* W. Redtenbacher, 1842: 21 type species *Rhopalocerus setosus* W. Redtenbacher, 1842 (= *Monotoma rondanii* A. Villa & J. B. Villa, 1833)
rondanii A. Villa & J. B. Villa, 1833: 36 (*Monotoma*) **E**: AU BU CR CZ GE GR HU IT PT RO SK UK YU
setosus W. Redtenbacher, 1842: 22

tribe Synchitini Erichson, 1845

genus *Acolophus* Sharp, 1885a: 66 type species *Acolophus debilis* Sharp, 1885
debilis Sharp, 1885a: 66 **A**: JA

genus *Bitoma* Herbst, 1793: 25 type species *Tritoma crenata* Fabricius, 1775
Ditoma Illiger, 1807: 320 type species *Tritoma crenata* Fabricius, 1775
Euditomum Gistel, 1856: 128 type species *Tritoma crenata* Fabricius, 1775
Synchytodes Crotch, 1873: 45 type species *Bitoma quadriguttata* Say, 1827
Xuthia Pascoe, 1863c: 128 type species *Xuthia siccana* Pascoe, 1863
crenata Fabricius, 1775: 69 (*Tritoma*) **E**: AB AL AR AU BE BH BU BY CR CT CZ DE EN FI FR GB GE GG GR HU IT LA LS KZ MA MC MD NL NR NT PL RO RU SK SL SP ST SV SZ UK YU **N**: AG TU **A**: ES FE KZ MG WS **NAR**
castanea Dalla Torre, 1879: 93 (*Synchitodes*)
ferruginea Dalla Torre, 1879: 93 (*Synchitodes*)
picipes A. G. Olivier, 1790: 18: 7 (*Ips*)
rufipennis Fabricius, 1801b: 562 (*Lyctus*)
rufithorax Pic, 1925a: 1 (*Ditoma*)
iranica Mařan, 1954a: 132 **E**: AB **A**: IN
siccana Pascoe, 1863c: 128 (*Xuthia*) **A**: BT CH HP HKG JA NP SA SD TAI **N**: AG EG **AFR NTR ORR**
lyctiformis Wollaston, 1867: 63 (*Ditoma*)
parallela Sharp, 1885a: 70
turcica Dajoz, 1973b: 80 **A**: CY TR

genus *Bolcocius* Dajoz, 1975: 297 type species *Colobicus granulosus* Sharp, 1886
bhutanensis Dajoz, 1975: 297 **A**: BT SW

formosanus Sasaji, 1984: 42 **A**: TAI
granulosus Sharp, 1885a: 65 (*Colobicus*) **A**: JA SW
shibatai Sasaji, 1984: 39 **A**: JA
yaeyamensis Sasaji, 1984: 41 **A**: JA (Ryukyus)

genus *Bhutania* Dajoz, 1975: 300 type species *Bhutania elongata* Dajoz, 1975
elongata Dajoz, 1975: 300 **A**: BT NP

genus *Cerchanotus* Erichson, 1845: 257 type species *Syntarsus asperulus* Fairmaire, 1869
Asprecodes Nakane, 1990: 65 type species *Asprecodes ogasawarensis* Nakane, 1990
Syntarsus Fairmaire, 1869b: 205 type species *Syntarsus asperulus* Fairmaire, 1869
jelineki Ślipiński, 1985: 182 (*Syntarsus*) **A**: SW
orientalis Ślipiński, 1985: 183 (*Syntarsus*) **A**: BT JA NP SD
ogasawarensis Nakane, 1990: 66 (*Asprecodes*)

genus *Colobicones* Grouvelle, 1918: 15 type species *Colobicones singularis* Grouvelle, 1918
Wittmeria Dajoz, 1975: 294 type species *Wittmeria pubscens* Dajoz, 1975
pubescens Dajoz, 1975: 295 (*Wittmeria*) **A**: BT
sakaii Okada, 2005: 428 **A**: JA (Amami-Ôshima)
tokarensis Okada, 2005: 426 **A**: JA (Tokara Isl.)

genus *Colobicus* Latreille, 1807: 9 type species *Colobicus marginatus* Latreille, 1807
gigas Grouvelle, 1908a: 119 **A**: NP
hirtus Rossi, 1790: 59 (*Nitidula*) **E**: AU BH BU CR CT CZ FR GB GE GR HU IT LS NT PL RU SK SL SP ST SV SZ UK YU **N**: MO **A**: AF FE JA NO TR WS
axillaris Duftschmid, 1825: 155 (*Monotoma*)
emarginatus Erichson, 1845: 268
hirtus Brullé, 1835: 410
marginatus Latreille, 1807: 10
parilis Pascoe, 1860b: 102 **Ei**: FR **A**: AF CH JA NP SD TAI UP

genus *Corticus* Germar, 1824b: 146 type species *Sarrotrium celtis* Germar, 1824
Ceropachys A. Costa, 1847: 150 type species *Ceropachys foveicollis* A. Costa, 1847
Horrimantus Gozis, 1881: 113 [RN]
brevipennis Reitter, 1882a: 119 **E**: AB ST **A**: IN TR
celtis Germar, 1824b: 146 (*Sarrotrium*) **E**: CR IT **N**: AG TU
foveolatus Fairmaire, 1848: 171
foveicollis A. Costa, 1847: 150

genus *Coxelus* Dejean, 1821: 67 type species *Bolitophagus pictus* J. Sturm, 1807
alinae Dajoz, 1973b: 82 **E**: GR
bituberculatus J. Frivaldszky, 1894: 85 (*Diodesma*) **E**: TR
humeridens Reitter, 1885d: 392 **A**: LE TR
luteopilosus Pic, 1901a: 3 **A**: TR
pictus J. Sturm, 1807: 161 (*Bolitophagus*) **E**: AU BH CR CZ FR GE GR HU IT LS PL RO SK SL SP SZ YU
yeti Ślipiński, 1985: 189 **A**: NP

genus *Diodesma* Latreille, 1829a: 97 type species *Diodema subterranea* Latreille, 1829
besucheti Dajoz, 1977: 156 **N**: MO
denticincta Abeille de Perrin, 1899a: 408 **E**: FR IT
parallela Reitter, 1922a: 21 **N**: AG MO
subterranea Latreille, 1829a: 97 **E**: AU BH CZ FR GE HU IT PL SK SL SZ UK YU
picea J.H.C.F. Sturm, 1849: 15

genus *Endeitoma* Sharp, 1894: 450 type species *Endeitoma mexicana* Sharp, 1894
Pabula Carter & Zeck, 1937: 193 type species *Ditoma perforata* Blackburn, 1888

bonina Nakane, 1990: 67 (*Pabula*) **A**: JA

genus *Endophloeus* Dejean, 1834: 195 type species *Eledona spinulosa* Latreille, 1807
exsculptus Germar, 1847: nr. 2 **E**: AB AR GG HU ST UK **A**: IN TR
markovichianus Piller & Mitterpacher, 1783: 105 (*Sylpha* [sic]) **E**: AU BH BU CR FR GB GR HU IT MC PL PT RO RU SK SL SP ST TR YU **N**: AG MOTU
spinulosus Latreille, 1807: 179 (*Eledona*)
serratus Sharp, 1885a: 61 **A**: JA TAI
squarrosus Germar, 1847: nr 3 **E**: GR TR YU

genus *Franzorphius* Schuh, 1998a: 228 type species *Franzorphius franzi* Schuh, 1998
franzi Schuh, 1998a: 229 **A**: SW

genus *Glyphocryptus* Sharp, 1885a: 63 type species *Glyphocryptus brevicollis* Sharp, 1885
brevicollis Sharp, 1885a: 64 **A**: JA

genus *Hyberis* Pascoe, 1860b: 112 type species *Hyberis araneiformis* Pascoe, 1860
similis Grouvelle, 1913: 106 **A**: HP SD

genus *Langelandia* Aubé, 1842: 227 type species *Langelandia anophthalma* Aubé, 1842
subgenus *Agelandia* Reitter, 1882b: 163 type species *Langelandia grandis* Reitter, 1877
gigantea Dajoz, 1968a: 986 **A**: TR
grandis Reitter, 1877e: 296 **A**: AB ST UK "Caucasus" **A**: TR
hummleri Obenberger, 1918: 58 (*Agelandia*) **A**: IT
macedonia Franz, 1980: 110 **A**: GR
merkliana Reitter, 1890f: 210 (*Agelandia*) **E**: TR **A**: IN TR
terricola Reitter, 1912c: 280 **E**: GR
wittmeri Dajoz, 1973b: 80 **A**: IN
subgenus *Fleischerella* Reitter, 1911a: 113 type species *Langelandia incostata* Perris, 1869
exigua Perris, 1869: 9 **E**: FR IT
incostata Perris, 1869: 11 **E**: FR (Corse)
caucasica Nikitsky, 1996b: 1434 **E**: ST
hypogea Normand, 1936a: 93 **N**: AG
leonhardi Reitter, 1912c: 285 **E**: IT
niticosta Schuh & Mifsud, 2000: 262 **E**: MA
nitidicollis Reitter, 1910b: 115 **E**: IT
subgenus *Langelandia* Aubé, 1842: 227 type species *Langelandia anophthalma* Aubé, 1842
aglena Reitter, 1908a: 20 **E**: BH MC
anophthalma Aubé, 1842: 228 **E**: AU CZ BE BH GB GE GR HU IT NL PL PT RO SK SP SZ YU
insularis Dajoz, 1969: 712
media Rey, 1889: 55
ausonica Obenberger, 1914: 109 **E**: IT
brachati Daffner, 1983: 22 **N**: CI
excavata Reitter, 1912c: 282 **E**: FR
gjonovici Reitter, 1912c: 281 **E**: BH CR
karamani Reitter, 1912c: 284 **E**: CR YU
mauli Franz, 1970: 3 **N**: MR
reitteri Belon, 1882: cxxv **E**: FR IT SP **N**: AG MO
galitensis Franz, 1982: 238
ibizensis Dajoz, 1969: 717
meridionalis Dajoz, 1969: 717
parallela Marquet, 1897: 107
torressalai Español, 1955d: 103
portosantoi Franz, 1970: 1 **N**: MR (Porto Santo)
thessalica Daffner, 1985: 36 **E**: GR

viennensis Reitter, 1912c: 283 **E**: AL AU GR IT LS YU
paganettii Obenberger, 1914: 109
subgenus *Normandella* Binaghi, 1937: 77 type species *Langelandia tunisina* Binaghi, 1937
antennaria Binaghi, 1937: 78 **E**: IT
tunisina Binaghi, 1937: 77 **N**: TU
zacynthia Heinze, 1943b: 127 **E**: GR
subgenus *Paganettia* Ganglbauer, 1899: 882 type species *Langelandia callosipennis* Reitter, 1881
callosipennis Reitter, 1881a: 217 **E**: BH CR
reflexipennis Reitter, 1912c: 281 **E**: CR YU

genus *Lasconotus* Erichson, 1845: 258 type species *Lasconotus complex* LeConte, 1859
Ithris Pascoe, 1863c: 134 type species *Ithris decisa* Pascoe, 1863
Lado Wankowiez [sic!], 1867: 251 type species *Bitoma jelskii* Wankowicz, 1867
Othismopteryx J. R. Sahlberg, 1871: 441 type species *Othismopteryx carinatus* J. R. Sahlberg, 1871
jelskii Wankowicz, 1867: 249 (*Bitoma*) **E**: BY CT FI LT PL RU SV **A**: KI
carinatus J. R. Sahlberg, 1871: 441 (*Othismopteryx*)
montanus Nikitsky & Ślipiński, 2003: 43 **A**: KZ KI
nepalensis Ślipiński, 1985: 184 **A**: NE
niponius Lewis, 1879: 462 (*Xuthia*) **A**: JA
saudicus Ślipiński, 1984: 253 **A**: SA
sculpturatus Sharp, 1885a: 72 (*Ithris*) **A**: JA

genus *Lastrema* Reitter, 1882a: 129 type species *Pycnomerus verrucicollis* Reitter, 1880
verrucicollis Reitter, 1880d: 543 (*Pycnomerus*) **E**: AB ST (Kavkaz) **A**: IN TR

genus *Lyreus* Aubé, 1861: 196 type species *Lyreus subterraneus* Aubé, 1861
caneparii Magrini, Fancello & Hernando, 2003: 39 **E**: SP (Girona)
septemstriatus Fancello & Leo, 1991: 129 **E**: IT (Sardegna)
subterraneus Aubé, 1861: 196 **E**: FR

genus *Microprius* Fairmaire, 1869a: 779 type species *Microprius terrenus* Fairmaire, 1868
Trionus Sharp, 1885a: 70 type species *Trionus opacus* Sharp, 1885
demissus Pascoe, 1863a: 33 (*Phormesa*) **A**: SD **ORR**
opacus Sharp, 1885a: 70 (*Trionus*) **A**: JA NP TAI UP
rufulus Motschulsky, 1863: 502 (*Bitoma*) **Ei**: GE MA SP **N**: AG EG MO TU **A**: IS JO LE SY YE
frivaldszkyi Reitter, 1877f: 326 (*Synchitodes*)
linearis Wollaston, 1867: 64 (*Ditoma*)
rufus Reitter, 1882a: 131 (*Synchitodes*)

genus *Namunaria* Reitter, 1882a: 114 type species *Coxelus guttulatus* LeConte, 1863
Sympanotus Sharp, 1885a: 62 type species *Sympanotus pictus* Sharp, 1885
bhutanensis Ślipiński, 1981a: 425 (*Tarphius*) **A**: BT
chinensis Schuh, 1999: 143 **A**: SW
mammillaris Schuh, 1999: 148 **A**: NE
picta Sharp, 1885a: 62 (*Sympanotus*) **A**: JA NP SE SW

genus *Neotrichus* Sharp, 1885a: 60 type species *Neotrichus hispidus* Sharp, 1885
afoveicollis Pal, 2003: 173 **A**: SD
hispidus Sharp, 1885a: 61 **A**: JA
lanyuensis Sasaji, 1986: 247 **A**: TAI
serraticollis Sasaji, 1986: 246 **A**: JA (Ryukyus)

genus *Niphopelta* Reitter, 1882a: 129 type species *Niphopelta imperialis* Reitter, 1882
imperialis Reitter, 1882a: 129 **E**: AB GG UK ST "Caucasus" **A**: IN

genus *Nosodomodes* Reitter, 1922a: 16 type species *Corticus rotundicollis* Pic, 1897
diabolicus Schaufuss, 1862b: 309 (*Corticus*) **E**: AR BU GR MC RO YU **A**: TR
rotundicollis Pic, 1897h: 78 (*Corticus*) **N**: AG
syriacus Fairmaire, 1892b: 144 (*Corticus*) **A**: SY TR
tauricus Germar, 1832: 4 (*Sarrotrium*) **E**: AB ST UK **A**: TR
tuberculatus Germar, 1832: 4 (*Sarrotrium*) **E**: BU GR MC RO ST UK

genus *Prosteca* Wollaston, 1860a: 254 type species *Prostheca aspera* Wollaston, 1860
Doderonymus Binaghi, 1937: 70 type species *Doderonymus lusitanicus* Binaghi, 1937
aspera Wollaston, 1860a: 255 **E**: AZ PT **N**: MR
lusitanica Binaghi, 1937: 71 (*Doderonymus*)
orientalis Vít, 1986: 453 **A**: PA

genus *Pseudotarphius* Wollaston, 1873: 1 type species *Pseudotarphius lewisii* Wollaston, 1873
bhutanensis Dajoz, 1975: 303 **A**: BT
lewisii Wollaston, 1873: 4 **A**: JA TAI

genus *Synchita* Hellwig, 1792: 401 type species *Synchita juglandis* Hellwig, 1792
Cicones Curtis, 1827: 149 type species *Cicones carpini* Curtis, 1827
Pseudocicones Fursov, 1939: 88 type species *Pseudocicones montanus* Fursov, 1939
Pseudosynchita Pic, 1922e: 21 type species *Pseudosynchita hirsuta* Pic, 1922
amurensis Mamaev, 1977: 28 **A**: FE
angustissima Nakane, 1968a: 75 (*Cicones*) **A**: JA
bitomoides Sharp, 1885a: 69 (*Cicones*) **A**: JA
bonina Nakane, 1991: 2 (*Cicones*) **A**: JA
crenicollis Wollaston, 1867: 62 (*Synchyta*) **N**: EG
hirsuta Pic, 1922e: 21 (*Pseudosynchita*)
fallax Schuh, 1998b: 321 **E**: FR GR IT TR **N**: AG MO
hayashii Sasaji, 1971: 43 (*Cicones*) **A**: JA
humeralis Fabricius, 1792a: 205 (*Elophorus*) **E**: AU BE BH BU CR CT CZ DE FI FR GB GE HU IT LA MC MD NR NT PL RO SK SL SP ST SV SZ UK YU **A**: ES FE
juglandis Hellwig, 1792: 403
obscura L. Redtenbacher, 1857: 345
striata Herbst, 1793: 23 (*Monotoma*)
iranica Dajoz, 1973b: 79 (*Cicones*) **E**: AR **A**: IN
mediolanensis A. Villa & J. B. Villa, 1833: 36 **E**: AB AU BU CZ FR GE GR HU IT PT RO SK SP ST SZ UK **N**: AG **A**: CY IN TR
berolinensis Reitter, 1922a: 34
minima Sharp, 1885a: 69 (*Cicones*) **A**: JA
montana Fursov, 1939: 88 (*Pseudocicones*) **A**: TD
nivea Sharp, 1885a: 68 (*Cicones*) **A**: JA
oculata Sharp, 1885a: 67 (*Cicones*) **A**: JA
oblonga Sharp, 1885a: 68 (*Cicones*)
separanda Reitter, 1882a: 126 (*Ditoma*) **E**: AR AU BH BU CR CT CZ FI FR GB GE HU IT NR NT PL RO SK ST SV SZ UK **A**: IN TR
angularis Abeille de Perrin, 1901a: 61
rufosignata Sasaji, 1984: 34 (*Cicones*) **A**: JA
tokarensis Nakane, 1968a: 75 (*Cicones*) **A**: JA
undata Guérin-Méneville, 1844: 189 **E**: AU BH BU CZ FR GB GE HU IT RO PT SL ST SZ
picta Erichson, 1845: 273 (*Cicones*)
ussuriensis Iablokoff-Khnzorian, 1978: 252 (*Cicones*) **A**: FE
uralensis Iablokoff-Khnzorian, 1970: 66 **E**: KZ
variegata Hellwig, 1792: 403 **E**: AU BE BH BU CR CZ FR GB GE GR HU IT LS NL PT RO RU SL SP SV SZ TR YU **N**: AG **A**: TR **NAR**i
carpini Curtis, 1827: 149 (*Cicones*)

genus *Tarphiosoma* Wollaston, 1862b: 373 type species *Tarphiodes indicus* Wollaston, 1862
Neoplatus Sharp, 1885b: 119 type species *Tarphisoma fasciata* Motschulsky, 1863
Tarphisoma Motschulsky, 1863: 504 type species *Tarphisoma fasciata* Motschulsky, 1863
bouvieri Grouvelle, 1903: 119 **A**: SD
globosa Ślipiński, 1985: 187 (*Tarphius*)
kempi Grouvelle, 1913: 104 **A**: AP

genus *Tarphius* Erichson, 1845: 256 type species *Tarphius gibbulus* Erichson, 1845
Atlantotarphius Franz, 1967: 65 type species *Tarphius canariensis* Wollaston, 1862
Caudatotarphius Franz, 1967: 64 type species*Tarphius deformis* Wollaston, 1862
Glabrotarphius Franz, 1967: 63 type species *Tarphius lutulentus* Wollaston, 1871
abbreviatus Wollaston, 1865: 20 (App.) **N**: CI
acuminatus Gillerfors, 1986a: 4 **E**: AZ
affinis Wollaston, 1862b: 383 **N**: CI
angusticollis Wollaston, 1860a: 252 **N**: MR
angustulus Wollaston, 1862d: 289 **N**: MR
azoricus Gillerfors, 1986a: 2 **E**: AZ
barbarae Gillerfors, 1991: 22 **N**: CI
besucheti Español, 1961b: 83 **E**: SP
brevicollis Wollaston, 1854: 144 **N**: MR
camelus Wollaston, 1862b: 383 **N**: CI
canariensis Wollaston, 1862b: 383 **N**: CI
caudatus Wollaston, 1862b: 386 **N**: CI
cicatricosus Wollaston, 1854: 141 **N**: MR
compactus Wollaston, 1854: 139 **N**: MR
congestus Wollaston, 1862b: 385 **N**: CI
deformis Wollaston, 1862b: 387 **N**: CI
depressus Gillerfors, 1985: 4 **E**: AZ
echinatus Wollaston, 1854: 143 **N**: MR
elongatus Dajoz, 1971: 86 **N**: CI
epinae Gillerfors, 1986b: 4 **N**: CI
ericae Gillerfors, 1986b: 4 **N**: CI
erosus Wollaston, 1862b: 384 **N**: CI
excisus Wollaston, 1857a: 47 **N**: MR
explicatus Wollaston, 1857a: 47 **N**: MR
fairmairei Marseul, 1872b: 134 **N**: AG
humerosus Fairmaire, 1866a: 20 [HN]
wollastonii Fairmaire, 1871a: 372 [HN]
fernandezlopezi Franz, 1967: 74 **N**: CI
formosus Wollaston, 1857a: 44 **N**: MR
gibbulus Erichson, 1845: 256 **E**: IT
gigas Wollaston, 1862b: 386 **N**: CI
gomerae Franz, 1967: 76 [RN] **N**: CI
affinis Wollaston, 1865: 19 (App.) [HN]
huggerti Gillerfors, 1991: 23 **N**: CI
humerosus Wollaston, 1865: 19 (App.) **N**: CI
inornatus Wollaston, 1854: 135 **N**: MR
spinipes Wollaston, 1854: 136
kiesenwetteri Heyden, 1870: 105 **E**: PT SP
lauri Wollaston, 1854: 138 **N**: MR
liliputanus Lucas, 1846: 336 (*Opatrum*) **N**: AG TU
oblongulus Fairmaire, 1866a: 21
lowei Wollaston, 1854: 134 **N**: MR
lutulentus Wollaston, 1871: 240 **N**: MR
maroccanus Franz, 1967: 65 **N**: MO

monstrosus Wollaston, 1865: 20 (App.) **N**: CI
moyanus Franz, 1967: 72 **N**: CI
nodosus Wollaston, 1854: 140 **N**: MR
oromii Franz, 1984b: 78 **N**: CI
oulmesensis Franz, 1967: 67 **N**: MO
palmensis Gillerfors, 1990: 149 **N**: CI
fraudulentus Franz, 1990a: 30
parallelus Wollaston, 1854: 134 **N**: MR
piniphilus Franz, 1967: 89 **N**: CI
pomboi Borges, 1991: 2 **E**: AZ
postcostatus Uyttenboogaart, 1937: 86 **N**: CL
quadratus Wollaston, 1862b: 384 **N**: CI
rotundatus Wollaston, 1854: 137 **N**: MR
rufonodulosus Israelson, 1984: 149 **E**: AZ
rugosus Wollaston, 1854: 144 **N**: MR
wolffi Wollaston, 1865: 21 (App.)
sculptipennis Wollaston, 1857a: 46 **N**: MR
serranoi Borges, 1991: 4 **E**: AZ
setosus Wollaston, 1865: 17 (App.) **N**: CI
simplex Wollaston, 1862b: 382 **N**: CI
constricticollis Franz, 1990a: 29
rotundicollis Franz, 1990a: 29
stagosus Gillerfors & Oromí, 1991: 85 **N**: CI
supranubis Franz, 1984b: 77 **N**: CI
sylvicola Wollaston, 1854: 137 **N**: MR
tacorontinus Franz, 1967: 71 **N**: CI
tenerifae Franz, 1967: 68 **N**: CI
testudinalis Wollaston, 1854: 141 **N**: MR
tornvalli Gillerfors, 1985: 5 **E**: AZ
truncatus Wollaston, 1854: 142 **N**: MR
wollastoni Crotch, 1867a: 386 **E**: AZ
zerchei Gillerfors, 1997: 2 **N**: MR

genus *Trachypholis* Erichson, 1845: 257 type species *Opatrum hispidum* Weber, 1801
Labromimus Sharp, 1885a: 64 type species *Labromimus variegatus* Sharp, 1885
Microvonus Sharp, 1885b: 118 type species *Microvonus squalidus* Sharp, 1885
Tarphiodes Wollaston, 1862b: 372 type species *Tarphiodes bowringii* Wollaston, 1862
chinensis Ślipiński, 1985: 186 **A**: SW
cinerea Dajoz, 1975: 307 **A**: BT
decorata Grouvelle, 1911: 126 **A**: HP
harmandi Grouvelle, 1903: 120 (*Labromimus*) **A**: NP SD
hispida Weber, 1801: 38 (*Opatrum*) **A**: AP NP SD
bowringii Wollaston, 1862b: 372 (*Tarphiodes*)
erichsoni Reitter, 1877f: 328
luteonigra Dajoz, 1975: 304 **A**: BT NP
montana Grouvelle, 1903: 121 (*Microvonus*) **A**: SD
okinawensis Nakane, 1991: 3 **A**: JA
ornata Grouvelle, 1911: 132 **A**: TAI
proxima Dajoz, 1975: 307 **A**: BT
sauteri Grouvelle, 1914: 47 (*Microvonus*) **A**: TAI
signata Grouvelle, 1911: 128 **A**: AP
variegata Sharp, 1885a: 65 (*Labromimus*) **A**: FE (Sakhalin) JA TAI

genus *Xylolaemus* L. Redtenbacher, 1857: 349 type species *Lyctus fasciculosus* Gyllenhal, 1827
Canaricerus Oromí & García, 1986: 304 type species *Canaricerus aeonii* Oromí & García, 1986
abnormis Ślipiński, 1984: 251 **A**: SA
aeonii Oromí & García, 1986: 304 (*Canaricerus*) **N**: CI
fasciculosus Gyllenhal, 1827: 632 (*Lyctus*) **E**: AU FR IT RO ST SV YU **N**: MO TU
indicus Grouvelle, 1903: 119 [DA] **A**: SD

family MORDELLIDAE Latreille, 1802
subfamily **Mordellinae** Latreille, 1802
tribe Conaliini Ermisch, 1956
genus *Conalia* Mulsant & Rey, 1858c: 313 type species *Conalia baudii* Mulsant & Rey, 1858
baudii Mulsant & Rey, 1858c: 313 **E**: AU BY CR CT CZ FR GG GR HU NT SK ST

tribe Mordellini Latreille, 1802
genus *Calycina* Blair, 1922b: 250 [RN] type species *Calycella palpalis* Blair, 1922
Calycella Blair, 1922a: 222 [HN] type species *Calycella palpalis* Blair, 1922
major Nomura, 1967a: 6 **A**: TAI
nigroapicalis Nomura, 1967a: 5 **A**: TAI

genus *Curtimorda* Méquignon, 1946: 58 type species *Mordella bisignata* L. Redtenbacher, 1849
bisignata L. Redtenbacher, 1849: 613 (*Mordella*) **E**: AU BY CT CZ FR GE HU NT PL ST SK SZ TR UK
albosignata Mulsant, 1856f: 333 (*Mordella*)
biguttata J. Sturm, 1826: 171 (*Mordella*)
maculosa Neazen, 1794: 273 (*Mordella*) **E**: AU BY CT CZ FR GE HU IT NT PL SK ST SZ UK **A**: ES FE JA SCH WS YUN
atomaria Fabricius, 1801b: 123 (*Mordella*)
guttata Paykull, 1798: 186 (*Mordella*)
guttatipennis Pic, 1927l: 183 (*Mordella*)
irrorata Trost, 1801: 27 (*Mordella*)

genus *Glipa* LeConte, 1857: 17 type species *Mordella hilaris* Say, 1835
Neoglipa Franciscolo, 1952: 332 type species *Glipa favareli* Pic, 1917
subgenus *Stenoglipa* Franciscolo, 1952: 344 type species *Mordella annulata* L. Redtenbacher, 1868
azumai Nakane, 1950: 21 **A**: CH JA
hatayamai hatayamai Takakuwa, 2000: 79 **A**: JA
hatayamai izuinsulana Takakuwa, 2000: 80 **A**: GUX HAI JA TAI YUN
hatayamai sejuncta Takakuwa, 2000: 82 **A**: SD
ishigakiana ishigakiana Kôno, 1932b: 152 **A**: JA
ishigakiana kotoensis Nomura, 1967a: 7 **A**: TAI
kurosawai Takakuwa, 1985: 9 **A**: JA
nipponica Nomura, 1957: 40 **A**: CH JA
watanabei Takakuwa, 2000: 71 **A**: JA TAI
subgenus *Macroglipa* Franciscolo, 1952: 335 type species *Mordella durvillei* Boisduval, 1835
alboscutellata Kôno, 1934a: 116 **A**: TAI
angustilineata Fan & J. Yang, 1993: 54 **A**: GUX
asahinai Nakane & Nomura, 1950: 3 **A**: JA
cinereonigra Fairmaire, 1893f: 302 (*Mordella*) **A**: CH JA
apicalis Pic, 1922c: 21
schibatai Nakane, 1960: 17
cladoda Fan & J. Yang, 1993: 51 **A**: GUX
curtopyga Fan & J. Yang, 1993: 56 **A**: YUN
fanjingana C. Yang & Fan, 1988: 109 **A**: GUI
fasciata Kôno, 1928a: 32 **A**: HUB FUJ GUA GUX HAI HUN JA JIX SCH TAI YUN ZHE
flava Fan & J. Yang, 1993: 49 **A**: YUN

formosana formosana Pic, 1911q: 191 **A**: CH JA TAI
apicepubens Pic, 1941c: 11
laterufa Pic, 1941c: 4
obscuripennis Pic, 1912i: 46
satoi Nakane & Nomura, 1950: 1
subpubens Pic, 1941c: 11
takashii Nomura, 1967a: 8
uniformis Pic, 1932d: 29
formosana yanma Takakuwa, 1986: 260 **A**: JA
fukiensis Ermisch, 1940: 166 **A**: FUJ
laosensis Pic, 1922c: 21 **A**: NP
karubei Takakuwa, 1993: 92 **A**: JA
klapperichi Ermisch, 1940: 170 **A**: FUJ HUB SCH
kusamai Takakuwa, 1999: 65 **A**: JA
masatakai M. Chûjô, 1960: 1 **A**: JA
obliquivittata Fan & J. Yang, 1993: 50 **A**: YUN TAI
obscuripennis Pic, 1912i: 46 **A**: FUJ TAI
ogasawarensis Kôno, 1928a: 32 **A**: JA
salvazai Pic, 1929c: 23
ohgushii M. Chûjô, 1957a: 2 (*Neoglipa*) **A**: JA
oshimana Nomura, 1966: 41 **A**: JA
ohgushi Nomura, 1963a: 247 [HN]
pici Ermisch, 1940: 163 **A**: FUJ GUA GUX HAI HUB HUN JA JIX SHA SCH TAI YUN ZHE
pseudofasciata Fan & J. Yang, 1993: 57 **A**: HEN
sauteri Pic, 1911p: 189 **A**: JA TAI
separata Pic, 1941c: 4 **A**: CH JA TAI
iriei Takakuwa, 1977: 9
shirozui Nakane, 1949: 29 **A**: HUB HUN JA SC TAI YUN ZHE
thoracica Horák, 1994: 102 **A**: NP
tripartita Pic, 1934b: 30 **A**: HKG NP SD
uenoi Takakuwa, 1986: 261 **A**: CH JA TAI
watanabeorum Takakuwa, 2002: 357 **A**: TAI
zhangi Fan & J. Yang, 1993: 53 **A**: YUN

genus *Glipidiomorpha* Franciscolo, 1952: 347 type species *Glipidiomorpha leucozona* Franciscolo, 1952
curticauda Ermisch, 1968a: 289 **A**: FUJ
kuantunensis Ermisch, 1968a: 289 **A**: FUJ
rufobrunnipennis Ermisch, 1968a: 291 **A**: FUJ

genus *Hoshihananomia* Kôno, 1935b: 124 type species *Mordella perlata* Sulzer, 1776
Machairorophora Franciscolo, 1943a: 34 type species *Mordella tibialis* Broun, 1880
auromaculata amamiana Nomura, 1962b: 39 **A**: JA
auromaculata auromaculata M. Chûjô, 1935a: 72 (*Mordella*) **A**: FE TAI
auromaculata nipponica Nomura, 1962b: 39 **A**: JA
auromaculata nishimurai Takakuwa, 1986a: 243 **A**: JA
borealis Nomura, 1957: 40 **A**: HAI HUB JA JIX TAI ZHE
composita composita Walker, 1858: 286 (*Mordella*) **A**: JA TAI
innotata Pic, 1926i: 67 (*Mordella*)
composita sakishimana Nomura, 1966: 42 **A**: JA
formosana Nakane & Nomura, 1950: 9 **A**: TAI
gacognei Mulsant, 1852: 49 (*Mordella*) **E**: AU BU CZ FR HU RO SL SK SP **N**: AG
brevefasciata Pic, 1925d: 11 (*Mordella*)
godarti Méquignon, 1946: 57
vauclusiana Méquignon, 1946: 57
hananomi Kôno, 1928b: 149 (*Mordella*) **A**: JA SC TAI
katoi katoi Nakane & Nomura, 1957: 3 **A**: JA TAI

katoi boninensis Nomura, 1975: 30 **A**: JA
kirai Nakane & Nomura, 1950: 6 **A**: JA TAI
kuatunensis Ermisch, 1968a: 287 **A**: FUJ
kurosai M. Chûjô & Nakane, 1955: 28 **A**: JA
kusuii Nomura, 1975: 32 **A**: JA
libanica Méquignon, 1941: 147 (*Mordella*) **A**: LE
longecaudata Fairmaire, 1891f: ccxviii (*Mordella*) **A**: SCH
michaelae Horák, 1986: 161 **A**: YE
minuscula Nomura, 1967a: 11 **A**: TAI
mitsuoi Nakane & Nomura, 1950: 7 **A**: JA
nakanei nakanei Takakuwa, 1986a: 240 **A**: JA
nakanei kuraruana Kiyoyama, 1993: 75 **A**: TAI
ochrothorax Nomura, 1975: 33 **A**: JA
oshimae Nomura, 1967b: 59 **A**: JA
perlata Sulzer, 1776: 67 (*Mordella*) **E**: AU BH BY CT CZ ES FE FR GE HU IT PL SK RO SK ST SZ UK **A**: FE GUA HEI JA KZ MG SHA WS
laticornis Stchegoleva-Barovskaya, 1927: 161 (*Tomoxia*)
multipunctata Trost, 1801: 27 (*Mordella*)
octopunctata Schrank von Paula, 1786: 80 (*Mordella*) [HN]
duodecimpunctata Rossi, 1790: 243 (*Mordella*)
sexpunctata Herbst, 1784c: 148 (*Mordella*)
pirika Kôno, 1935b: 125 **A**: JA SC
pseudoauromaculata Kiyoyama, 1993: 76 **A**: TAI
pseudohananomi Kiyoyama, 1993: 71 **A**: JA
splendens Miwa, 1933: 10 (*Mordella*) **A**: JA
transsylvanica Ermisch, 1977: 176 **E**: RO
trichopalpis Nomura, 1975: 30 **A**: JA
ussuriensis Ermisch, 1969a: 102 **A**: FE

genus *Iberomorda* Méquignon, 1946: 59 type species *Mordella sulcicauda* Mulsant, 1856
sulcicauda Mulsant, 1856f: 354 (*Mordella*) **E**: IT SP
viridipennis Mulsant, 1856f: 355 (*Mordella*) **E**: SP

genus *Klapperichimorda* Ermisch, 1968a: 279 type species *Klapperichimorda quadrimaculata* Ermisch, 1968
quadrimaculata Ermisch, 1968a: 280 **A**: FUJ
lutevittata Fan & J. Yang, 1995: 99 **A**: ZHE

genus *Macrotomoxia* Pic, 1922i: 208 type species *Macrotomoxia castanea* Pic, 1928
Higehananomia Kôno, 1935b: 123 type species *Higehananomia palpalis* Kôno, 1935
castanea Pic, 1922i: 208 **A**: FE JA SE TAI
palpalis Kôno, 1935b: 123 (*Higehananomia*)
gardneri Blair, 1931a: 203 (*Calycina*) **A**: SD

genus *Mediimorda* Méquignon, 1946: 62 type species *Mordella bipunctata* Germar, 1827
angeliquae Leblanc, 2002: 318 **E**: FR IT
attalica Schilsky, 1895: 24 (*Mordella*) **A**: TR
basicornis Schilsky, 1895: 24 (*Mordella*)
obtecta Schilsky, 1895: 24 (*Mordella*)
batteni Plaza, 1986a: 267 **E**: SP
bipunctata Germar, 1827: 12 (*Mordella*) **E**: AL AU BU CR FR GE GR IT MC PT SP SZ TR UK **N**: AG MO TU **A**: CY IS SA TR
abbreviata Schilsky, 1899: G [HN] (*Mordella*)
castellanii Franciscolo, 1949c: 84 (*Mordella*)
decora Chevrolat, 1840a: 16 (*Mordella*)
perpicilata A. Costa, 1854: 15 (*Mordella*)

brusteli Leblanc, 2002: 319 **E**: FR SP
maceki Horák, 1985: 23 **A**: IN IQ

genus *Mordella* Linnaeus, 1758: 420 type species *Mordella aculeata* Linnaeus, 1758
aculeata Linnaeus, 1758: 420 **E**: AL AU BE BH BU BY CR CT CZ DE FR GE GG GR HU IT MC PL PT RO SK SL SP ST SV SZ TR UK YU **A**: BEI ES FE GAN IN JA KZ NMO SC SHA SHG SY TAI TR WS
communis Matsumura, 1915: 111
aradasiana Patti, 1840: 291 **E**: IT
argenteoguttata Champion, 1922a: 75 **A**: UP
brachyura brachyura Mulsant, 1856f: 354 **E**: AB AL AU BH BU BY CR CT CZ FI FR GE GR GG HU IT MC PL PT RO SK SL SP ST SZ UK YU **A**: IN SC TR
brevicauda A. Costa, 1854: 15
brunneicornis Schilsky, 1895: 28
fleischeri Emery, 1876: 69
brachyura akage Nomura, 1958: 18 **A**: JA
brachyura fuscopurpurea Nomura, 1958: 18 **A**: JIL
brachyura obscurata Ermisch, 1968b: 18 **A**: FE MG
brancuccii Horák, 1995c: 168 **A**: NP SD
brevicornis Méquignon, 1946: 67 **N**: MO
cinerea Gebler, 1829a: 134 **A**: WS
cinereoplagiata Blair, 1931a: 203 **A**: UP
comptei Plaza, 1977: 173 **N**: CI
cuneiformis Ermisch, 1968a: 282 **A**: FUJ
curticornis Ermisch, 1968a: 286 **A**: FUJ
curvipalpis Stchegoleva-Barovskaya, 1930b: 751 **A**: IN PA
duplicata Schilsky, 1895: 27 **E**: AB CR TR **A**: CY JO KI KZ SY TR
korgei Ermisch, 1969b: 107
rhodiensis Ermisch, 1963c: 51
rufipalpis Schilsky, 1895: 27
funesta Faldermann, 1838: 108 **A**: UZ
gratiosa Plaza & Compte, 1980: 205 **E**: SP
himalayana Horák, 1995c: 170 **A**: NP SD
hoberlandti Horák, 1985: 19 **A**: IN
hofferi Horák, 1979: 104 **E**: BU
holomelaena holomelaena Apfelbeck, 1914: 614 **E**: AB AL AU BE BH BU CR CZ FR GE GG HU NL NT PL RO SK SL SP ST SZ UK YU **A**: KZ
holomelaena sibirica Apfelbeck, 1914: 615 **A**: ES FE MG
huetheri Ermisch, 1956: 278 **E**: AL AU BH BU CR CT BH FR GE GG GR HU IT MC NT PL SP SK ST SV SZ **A**: KZ WS
inouei Nomura, 1958: 12 **A**: TAI
insidiosa Lucas, 1847: 384 **N**: AG
izraelsoni Plaza, 1977: 175 **N**: CI
kuantunensis Ermisch, 1968a: 283 **A**: FUJ
konpira Takakuwa, 1985a: 71 **A**: JA
latipalpis Nomura, 1958: 15 **A**: JA SC
legionensis Plaza & Compte, 1980: 202 **E**: SP
leucaspis bicoloripilosa Horák, 1985: 18 **E**: AB GR **A**: IN SY TR
leucaspis leucaspis Küster, 1849a: 80 **E**: AL AU BH BU CR CZ FR GE GR FR HU IT MC PL RO SK SL SP SZ TR UK YU **A**: IS TR
adnexa Ermisch, 1969b: 104
persica Apfelbeck, 1914: 619
longicauda Roubal, 1921: 248 **E**: SL
meridionalis Méquignon, 1946: 66 **E**: AL AU FR GE HU IT MC SP SZ
micacea Pic, 1941c: 7 **A**: CH
mixta Fabricius, 1801b: 122 **A**: TAI

mongolica Ermisch, 1964b: 341 **A**: MG NMO
multilineata Champion, 1927: 51 **A**: NP SD UP
niveoscutellata Nakane & Nomura, 1950: 15 **A**: JA TAI
ochrotrichia Nomura, 1958: 8 **A**: TAI
onaga Nomura, 1958: 10 **A**: JA
oxyptera Nomura, 1958: 10 **A**. JA SC
quadriguttulata Motschulsky, 1872: 46 **A**: WS
palmae Emery, 1876: 68 **E**: FR IT SP
plagiata Mannerheim, 1849: 240 **A**: ES
purpurascens Apfelbeck, 1914: 620 **E**: AL AU BH BU CR FR GR HU IT MC PT SP TR YU
pygidialis Apfelbeck, 1914: 612 **E**: AL AU BH GE SZ UK
quelpaertensis Pic, 1941c: 5 **A**: SC
sasajii Takakuwa, 2001: 293 **A**: JA
shirozui Nomura, 1967a: 14 **A**: TAI
splendidula Faldermann, 1838: 110 **A**: UZ
strigipennis Faldermann, 1838: 109 **A**: UZ
tadjikistanica Odnosum, 2002: 21 **A**: TD UZ
tenuicauda Nomura, 1958: 15 **A**: TAI
tokejii Nomura, 1958: 13 **A**: JA JIL SC
tunisiaca Ermisch, 1969b: 105 **N**: AG TU
velutina panonica Horák, 1985: 22 **E**: CZ HU SK
velutina velutina Emery, 1876: 69 **E**: AB AL AR IT BH BU CR GR HU IT KI NT RO ST UK YU **A**: IN KZ TR UZ
vestita Emery, 1876: 69 **E**: AL AR BH BU CR FR GR IT MC SL UK YU
  horvathi Apfelbeck, 1914: 616
viridescens A. Costa, 1854: 18 **E**: AU BH BY CR CT CZ FR GE GR HU IT PL UK
yami Nomura, 1967a: 13 **A**: TAI

genus *Mordellaria* Ermisch, 1950b: 69 type species *Mordella scripta* Fairmaire & Germain, 1863
arakii Nakane & Nomura, 1950: 13 (*Mordella*) **A**: JA
aurata Kôno, 1928a: 36 (*Mordella*) **A**: FE JA
aurofasciata Comolli, 1837: 29 (*Mordella*) **E**: AU CR CZ GE GG HU IT NT PL RO SK ST YU **N**: AG **A**: IN
  conjuncta Schilsky, 1895: 23 (*Mordella*)
  sacheri Frivaldszky von Frivald, 1865: 202 (*Mordella*)
  vittata Gemminger, 1851: 52 (*Mordella*)
hananoi Nakane & Nomura, 1950: 12 (*Mordella*) **A**: JA TAI
humeralis Nomura, 1961b: 76 **A**: JA
latior Nomura, 1967a: 9 **A**: JA TAI
kanoi Kôno, 1932b: 153 (*Mordella*) **A**: TAI
triguttata Nomura, 1961b: 75 **A**: JA
zenchii Tokeji, 1953: 4 **A**: JA

genus *Neocurtimorda* Franciscolo, 1949a: 1 type species *Neocurtimorda convexa* Franciscolo, 1949
hoanensis Pic, 1941c: 4 (*Mordella*) **A**: NP SD
touzalini Pic, 1941c: 4 (*Mordella*) **A**: FUJ HUN JIX NP SD YUN

genus *Paratomoxia*, 1950: 5 type species *Tomoxia pulchella* Ermisch, 1948
subgenus *Metatomoxia* M. Chûjô, 1957a: 1 type species *Tomoxia scutellata* Kôno, 1928
crux Kôno, 1928b: 148 (*Tomoxia*) **A**: JA
nipponica Kôno, 1928b: 148 (*Tomoxia*) **A**: FE JA
scutellata Kôno, 1928b: 147 (*Tomoxia*) **A**: FE JA

genus *Paratomoxioda* Ermisch, 1954a: 173 type species *Paratomoxioda leopardina* Ermisch, 1954
evanescens Normand, 1950a: 82 (*Mordella*) **N**: AG MO TU **AFR**

fallaciosa Stchegoleva-Barovskaya, 1930b: 750 (*Mordella*) **A**: AF IN KI TD

genus *Stenomordella* Ermisch, 1941a: 115 type species *Stenomordella longeantennalis* Ermisch, 1941
longeantennalis Ermisch, 1941a: 116 **A**: FUJ HUB
ochii Kiyoyama, 1975: 46 **A**: JA JIX

genus *Tomoxia* A. Costa, 1854: 8 type species *Tomoxia bucephala* A. Costa, 1854
bucephala bucephala A. Costa, 1854: 8 **E**: AU BH BU BY CR CT CZ DE GB GE FI FR HU IT NL NR NT PL PT RO SL SK SP ST SV SZ TR UK YU **N**: AG **A**: FE KZ WS
fasciata Paykull, 1800: 455 (*Mordella*)
sericea Mulsant, 1856f: 324
formosana M. Chûjô, 1935a: 75 **A**: JA TAI
relicta Takakuwa, 1985: 8 **A**: JA
ryukyuana Takakuwa, 1985a: 69 **A**: JA
similaris Nomura, 1967b: 57 **A**: JA

genus *Tomoxioda* Ermisch, 1950: 52 type species *Tomoxioda auropubescens* Ermisch, 1950
truncatoptera Nomura, 1958: 8 (*Mordella*) **A**: CH JA TAI

genus *Variimorda* Méquignon, 1946: 62 type species *Mordella villosa* Schrank von Paula, 1781
Sulcatimorda Méquignon, 1946: 63 type species *Mordella ragusai* Emery, 1876
subgenus *Variimorda* Méquignon, 1946: 62 type species *Mordella villosa* Schrank von Paula, 1781
ainu Nomura, 1958: 6 (*Mordella*) **A**: JA
argyropleura Franciscolo, 1942a: 8 (*Mordella*) **E**: IT
basalis A. Costa, 1854: 13 (*Mordella*) **E**: AL AU BH BU BY CR CT CZ FR GR HU IT PL SK SL SP ST SZ TR UK
pseudobrachyura Franciscolo, 1949c: 81 (*Mordella*)
briantea Comolli, 1837: 28 (*Mordella*) **E**: AU BH BU BY CR CT CZ EN FR GE GR HU IT MD PL SK SL SP ST SZ UK **N**: AG **A**: TR
flavimana Marseul, 1877: 472 (*Mordella*) **A**: FUJ GUA GUI HUB JA SCH TAI SHA
holzschuhi Horák, 1985: 13 **A**: IN IQ TR
ihai boninensis Nomura, 1975: 34 **A**: JA
ihai ihai M. Chûjô, 1959: 9 **A**: JA
inomatai Takakuwa, 1985: 10 **A**: JA
ishiharai Kiyoyama, 1994: 193 **A**: TAI
kurosawai Takakuwa, 2001a: 310 **A**: JA
kuyrasha Nomura, 1962b: 41 **A**: JA
mendax Méquignon, 1946: 71 **E**: AB AL AU BH BU CR CZ FR GE GG GR HU IT PL SK SZ
chobauti Méquignon, 1946: 71
devillei Méquignon, 1946: 71
miyarabi miyarabi Nomura, 1962b: 40 **A**: JA
miyarabi chujoi Tsuru & Takakuwa, 2005: 80 **A**: JA
miyarabi michioi Tsuru & Takakuwa, 2005: 81 **A**: JA
persica Horák, 1985: 8 **A**: IN
quomoi Franciscolo, 1942a: 7 (*Mordella*) **E**: IT
ragusai Emery, 1876: 60 (*Mordella*) **E**: FR IT **N**: MO
nigrata Méquignon, 1946: 73 (*Mordella*)
obsoleta Baudi di Selve, 1878b: 345 (*Mordella*)
oertzeni Schilsky, 1895: 26 (*Mordella*)
sinensis Pic, 1917f: 13 (*Mordella*) **A**: SD YUN
truncatopyga Pic, 1938e: 16 (*Mordella*) **A**: SE TAI
taiwana Nakane & Nomura, 1950: 14 (*Mordella*)
villosa Schrank von Paula, 1781: 228 (*Mordella*) **E**: AB AL AR AU BE BH BU BY CR CT CZ DE FR GE GG GR HU IT LA MC MD NL PL RO SK SL SP ST SZ TR UK YU **A**: FE IN IQ JA KZ TR
conjucta Stchegoleva-Barovskaya, 1931: 56 (*Mordella*)

coronata A. Costa, 1854: 11 (*Mordella*)
fasciata Fabricius, 1775: 263 (*Mordella*) [HN]
fasciolata Rossi, 1792: 103 (*Mordella*)
habelmanni Emery, 1876: 66 (*Mordella*)
interrupta A. Costa, 1854: 13 (*Mordella*)
nigricornis Stchegoleva-Barovskaya, 1931: 57 (*Mordella*)
nigrosuturalis Stchegoleva-Barovskaya, 1931: 56 (*Mordella*)
ruficornis Stchegoleva-Barovskaya, 1931: 57 (*Mordella*)
seriatoguttata Mulsant, 1856f: 345 (*Mordella*)
subbasalis Stchegoleva-Barovskaya, 1931: 56 (*Mordella*)
subcoeca Mulsant, 1856f: 345 (*Mordella*)
subgenus *Galeimorda* Horák, 1985: 15 type species *Mordella fagniezi* Méquignon, 1946
caprai Franciscolo, 1951: 7 (*Mordella*) **E**: AL
fagniezi Méquignon, 1946: 74 (*Mordella*) **E**: FR PT SP
therondi Méquignon, 1946: 74 (*Mordella*)
hladili Horák, 1985: 15 **E**: BU GR MC ST TR UK **A**: KZ
krikkeni Batten, 1977: 167 **E**: CR GR
theryi Méquignon, 1946: 74 (*Mordella*) **E**: PT SP **N**: AG MO
maroccana Méquignon, 1946: 75 (*Mordella*)
nigrescens Méquignon, 1946: 75 (*Mordella*)

genus *Yukahananomia* Kôno, 1935b: 124 type species *Tomoxia yakui* Kôno, 1930
tsuyukii Takakuwa, 1978: 5 **A**: JA
uenoi Takakuwa, 1995: 433 **A**: TAI
yakui Kôno, 1930c: 104 (*Tomoxia*) **A**: CH FE JA

tribe Mordellistenini Ermisch, 1941

genus *Asiatolida* Shiyake, 2000: 26 type species *Asiatolida miyatakei* Shiyake, 2000
melana Fan & J. Yang, 1995: 95 (*Mordellistenoda*) **A**: ZHE
miyatakei Shiyake, 2000: 27 **A**: JA

genus *Dellamora* Normand, 1916: 284 type species *Dellamora palposa* Normand, 1916
palposa Normand, 1916: 285 **E**: GR PT SP **N**: LB MO TU **A**: CY TM TR

genus *Ermischiella* Franciscolo, 1930: 2 type species *Ermischiella papuana* Franciscolo, 1950
castanea Boheman, 1858: 108 (*Mordella*) **A**: JA
dodonaeae Montrouzier, 1860: 306 (*Mordella*)
chichijimana Nomura, 1975: 37 **A**: JA
fukiensis Ermisch, 1952b: 154 (*Glipostenoda*) **A**: FUJ
hahajimana Nomura, 1975: 38 **A**: JA
hasagawai Nomura, 1951: 69 (*Glipostenoda*) **A**: JA
nigriceps Nomura, 1975: 35 **A**: JA

genus *Falsomordellina* Nomura, 1966: 44 type species *Glipostenoda luteoloides* Nomura, 1961
amamiana Nomura, 1961b: 85 (*Glipostenoda*) **A**: JA
luteoloides Nomura, 1961b: 83 (*Glipostenoda*) **A**: FUJ JA TAI ZHE
luteola Nomura, 1951: 67 (*Glipostenopa*) [HN]
luteora Kôno, 1932b: 155 (*Mordellistena*)
luteorubra Ermisch, 1965c: 201 (*Mordellistena*)
nigripennis Nomura, 1967a: 31 **A**: TAI
ohsumiana Nakane, 1957: 51 (*Glipostenoda*) **A**: JA
takaosana madara Nomura, 1966: 45 **A**: JA
takaosana takaosana Kôno, 1932: 154 (*Mordellistena*) **A**: JA ZHE
vagevittata Nakane, 1957: 50 (*Falsomordellistena*) **A**: JA

genus *Falsomordellistena* Ermisch, 1941c: 724 type species *Mordellistena formosana* Pic, 1911
alpigena Tokeji, 1953: 15 **A**: JA
altestrigata Marseul, 1877: 474 (*Mordellistena*) **A**: HUB HUN JA JIX SE SHA TAI YUN
auriguttata Nomura, 1951: 64
sauteri Pic, 1926i: 67 (*Mordellistena*)
aurofasciata Nakane, 1949: 40 (*Mordellistena*) **A**: JA
auromaculata Kôno, 1928a: 38 (*Mordellistena*) **A**: JA SC
aurosuturalis Nomura, 1967a: 25 **A**: TAI
ayahime Nomura, 1967a: 21 **A**: TAI
baishanzuna Fan, 1995b: 223 **A**: ZHE
formosana boninensis Nomura, 1975: 40 **A**: JA
formosana formosana Pic, 1911q: 190 (*Mordellistena*) **A**: TAI
chrysotrichia Nomura, 1951: 49 (*Mordellistena*) **A**: JA SC
hananoi Nomura, 1951: 48 (*Mordellistena*) **A**: JA SC
hime Kôno, 1928a: 41 (*Mordellistena*) **A**: TAI
altestrigata Ermisch, 1950b: 85 [HN]
hiranoi hiranoi Nomura, 1961b: 83 **A**: JA
hiranoi takasagoana Nomura, 1967a: 26 **A**: TAI
hirasana Shiyake, 1996: 13 **A**: JA
humerosignata Nomura, 1967a: 22 **A**: TAI
inouei Nomura, 1951: 50 (*Mordellistena*) **A**: TAI
arisana Nomura, 1951: 53
capillata Nomura, 1951: 54
katoi Nomura, 1961b: 81 **A**: JA
konoi konoi Pic, 1936h: 300 (*Mordellistena*) [RN] **A**: TAI
horni Kôno, 1934a: 118 [HN]
konoi kotoensis Nomura, 1967a: 23 **A**: TAI
konoi yakushimaensis Nomura, 1951: 65 **A**: JA
kuri Nomura, 1951: 51 (*Mordellistena*) **A**: JA
loochooana Nomura, 1964: 51 **A**: JA
okamotoi Kôno, 1935b: 127 (*Mordellistena*) **A**: JA
parca Tokeji, 1953: 14 **A**: JA
pseudoalpigena Nomura, 1975: 42 **A**: JA
satoi Nomura, 1951: 51 (*Mordellistena*) **A**: JA
rosseoloides Nomura, 1975: 41 **A**: JA
shinanoensis Tokeji, 1953: 18 **A**: JA
superfusca Nomura, 1951: 48 (*Mordellistena*) **A**: JA
tokarana Nakane, 1956a: 9 **A**: JA
watanabei Nomura, 1975: 43 **A**: JA
wui Fan & J. Yang, 1995: 96 **A**: ZHE
yoshidai Nomura, 1951: 54 (*Mordellistena*) **A**: JA

genus *Glipostena* Ermisch, 1941c: 723 type species *Mordellistena pelecotomoidea* Pic, 1911
pelecotomoidea Pic, 1911p: 188 (*Mordellistena*) **A**: JA SE TAI

genus *Glipostenoda* Ermisch, 1950b: 81 type species *Glipostenoda castaneicolor* Ermisch, 1950
excisa Nomura, 1967b: 59 **A**: JA
fukiensis Ermisch, 1952b: 154 **A**: FUJ
fusciceps Nomura, 1967b: 61 **A**: JA
higashinoi Nomura, 1967b: 67 **A**: JA TAI
kaihuana Fan & J. Yang, 1995: 98 **A**: ZHE
kawasakii Nomura, 1951: 27 (*Mordellistena*) **A**: TAI
kimotoi M. Chûjô, 1957a: 17 **A**: JA
klapperichi Ermisch, 1952b: 153 **A**: FUJ
lineatisuturalis Nomura, 1967a: 26 **A**: TAI

matsumurai Kôno, 1932b: 154 (*Mordellistena*) **A**: FUJ JA TAI
melanocephala Ermisch, 1952b: 156 **A**: FUJ
pseudoexcisa Nomura, 1975: 39 **A**: JA
rarasana Nomura, 1951: 69 **A**: JA TAI
retusa Nomura, 1967b: 60 **A**: JA
rimogana Nomura, 1967a: 30 **A**: TAI
rosseola Marseul, 1877: 473 (*Mordellistena*) **A**: JA TAI SE
hisamatsui M. Chûjô, 1956: 175
phengotrichia Nomura, 1951: 66
rufobrunnea Champion, 1927a: 52 (*Mordellistena*) **A**: UP
sasajii Schiyake, 2001: 333 **A**: JA
shibatai Nomura, 1961b: 84 **A**: JA
shizuokana Kôno, 1935b: 128 (*Mordellistena*) **A**: FUJ JA TAI
taiwana Kôno, 1934a: 117 (*Mordellistena*) **A**: TAI
takashii Nomura, 1967a: 29 **A**: TAI
trichophora Nomura, 1951: 55 (*Mordellistena*) **A**: HEI JA LIA TAI

genus *Mordellina* Schilsky, 1908b: 137 type species *Mordellina gracilis* Schilsky, 1908
subgenus *Mordellina* Schilsky, 1908b: 137 type species *Mordellina gracilis* Schilsky, 1908
atrofusca atrofusca Nomura, 1951: 60 (*Mordellistena*) **A**: JA TAI
atrofusca signatelloides Nakane, 1956a: 7 (*Mordellistena*) **A**: JA
atrofusca reductemaculata Nakane, 1956a: 8 (*Mordellistena*) **A**: JA
brunneotincta Marseul, 1877: 474 (*Mordellistena*) **A**: JA ZHE
chibi Kôno, 1932b: 156 (*Mordellistena*) **A**: JA
gutianshana Fan & J. Yang, 1995: 95 **A**: ZHE
hidakai Nomura, 1963b: 41 (*Pseudomordellistena*) **A**: JA
hirayamai Kôno, 1933: 29 (*Mordellistena*) **A**: JA
koikei Tokeji, 1953: 8 (*Mordellistena*) **A**: JA
kubotai Nomura, 1951: 57 (*Mordellistena*) **A**: JA SC
longula Kôno, 1928a: 39 (*Mordellistena*) **A**: JA
marginalis Nomura, 1967a: 20 **A**: TAI
mentiens Tokeji, 1953: 10 (*Mordellistena*) **A**: JA
mikado Tokeji, 1953: 8 (*Mordellistena*) **A**: JA
nigrofusca Nomura, 1951: 58 (*Mordellistena*) **A**: JA
otohime Nomura & Katô, 1957: 84 (*Mordellistena*) **A**: JA
palleola Nomura, 1966: 43 **A**: JA
palliata Kôno, 1932b: 156 (*Mordellistena*) **A**: JA
quinquenotata Champion, 1927a: 53 (*Mordellistena*) **A**: NP UP
shimoyamai M. Chûjô, 1957a: 6 (*Mordellistena*) **A**: JA
signatella Marseul, 1877: 473 (*Mordellistena*) **A**: JA TAI
nigrofasciata M. Chûjô, 1935a: 74 (*Mordellistena*)
uruma Nomura, 1966: 42 **A**: JA
watanabei T. Tsuru, 2002: 351 **A**: JA
yamamotoi Nomura, 1951: 56 (*Mordellistena*) **A**: JA
yezoensis M. Chûjô, 1957b: 21 (*Mordellistena*) **A**: JA
subgenus *Pseudomordellistena* Ermisch, 1952b: 145 type species *Pseudomordellistena rufobrunnea* Ermisch, 1952
amamiensis Nomura, 1951: 60 (*Mordellistena*) **A**: JA TAI
callichroa callichroa Tokeji, 1953: 11 (*Mordellistena*) **A**: JA
callichroa ami Nomura, 1967a: 18 **A**: TAI
curteapicalis Pic, 1926i: 67 (*Mordellistena*) **A**: TAI
gina Nomura, 1967a: 15 **A**: TAI
kaguyahime Nomura & Katô, 1957: 85 (*Mordellistena*) **A**: JA TAI
macrophthalma Ermisch, 1952b: 147 (*Pseudomordellistena*) **A**: FUJ

paiwana Nomura, 1967a: 17 **A**: TAI
pilosovittata Nakane, 1956a: 5 (*Mordellistena*) **A**: JA TAI
pseudohumeralis Nakane, 1956a: 6 (*Mordellistena*) **A**: JA
rufobrunnea Ermisch, 1952b: 145 (*Pseudomordellistena*) **A**: FUJ
rufohumeralis Nomura, 1967a: 16 **A**: TAI
tsutsuii Nakane, 1956a: 4 (*Mordellistena*) **A**: JA TAI
vidua Nakane, 1956a: 7 (*Mordellistena*) **A**: JA

genus *Mordellistena* A. Costa, 1854: 16 type species *Mordella pumila* Gyllenhal, 1810
subgenus *Mordellokoides* Franciscolo, 1942d: 133 type species *Mordellokoides grandii* Franciscolo, 1942
grandii Franciscolo, 1942d: 134 **E**: IT
subgenus *Mordellistena* A. Costa, 1854: 16 type species *Mordella pumila* Gyllenhal, 1810
Natirrica A. Costa, 1854: 19 type species *Natirrica meridionalis* A. Costa, 1854
aegea Franciscolo, 1949c: 90 **E**: Gr
aequalica Ermisch, 1977: 156 **E**: HU
aertsi Ermisch, 1963a: 22 **E**: FR GE
ainonia M. Chûjô, 1957b: 18 **A**: JA
albocapillata Ermisch, 1965a: 70 **A**: MG
algeriensis Ermisch, 1966b: 42 **N**: AG TU
alpicola Ermisch, 1963a: 28 **E**: IT SZ
altestriatoides Horák, 1995b: 171 **A**: NP
altifrons Stchegoleva-Barovskaya, 1927: 159 **E**: ST
amurensis Horák, 1982: 48 **A**: FE
angustula Ermisch, 1977: 168 **E**: HU
anomala Ermisch, 1968b: 18 **A**: MG
apicerufa Ermisch, 1977: 158 **E**: HU
arabissa Franciscolo, 1956c: 467 **A**: YE
atrocincta lincenti Pic, 1938e: 16 **A**: GAN
atrogemellata Ermisch, 1965d: 267 **E**: BU GR MC
aureolopilosa Stchegoleva-Barovskaya, 1932: 222 **A**: KI
aureomicans Ermisch, 1965b: 70 **N**: CI
aureotomentosa Ermisch, 1966b: 40 **N**: MG
austriaca Schilsky, 1899: 77b **E**: AL AU BH CR CZ FR GE GR HU IT PL RO SK SZ UK **A**: KZ TR
micantoides Ermisch, 1954b: 176
austriacensis Ermisch, 1956: 285 **E**: AU CZ HU SK
balcanica Ermisch, 1967b: 110 **E**: BU MC
balearica Compte, 1985: 63 **E**: SP
baliani Franciscolo, 1942c: 77 **E**: IT
bavarica Ermisch, 1963a: 11 **E**: GE
batteni Horák, 1980: 281 **A**: KI TD
berbera Horák, 1983: 8 **N**: AG
bicoloripilosa Ermisch, 1967b: 113 **E**: AU BE BU BY CZ FR GE GG HU PL RO UK **A**: KZ TM UZ WS
biroi Ermisch, 1977: 166 **E**: HU
bodemeyeri Ermisch, 1956: 283 **A**: TR
boldi Ermisch, 1965a: 68 **A**: MG
bolognai Horák, 1990: 136 **A**: LE IN IQ TR SY
borogolensis Ermisch, 1964b: 348 **A**: MG
breddini Ermisch, 1963a: 26 **E**: GE PL
brevicauda Bohemann, 1849: 223 (*Mordella*) **E**: AB AR AU BH BU BY CR CZ FR GE GG GR HU IT MC PL PT RO SK SL SP ST SZ TR YU **A**: IN IS JO KI KZ SY TD TM TR UZ
brevicornis Schilsky, 1895: 45
obtusa C. Brisout de Barneville, 1859: 233
subtruncata Mulsant, 1856f: 377
brevilineata Nomura, 1961b: 77 **A**: JA
brunneispinosa Ermisch, 1963: 6 **E**: BY GE HU PL TR UK

bulgarica Ermisch, 1977: 166 **E**: BU
canariensis Ermisch, 1965b: 73 **N**: CI
caprai Franciscolo, 1942a: 8 **E**: IT
carinthiaca Ermisch, 1966a: 33 **E**: GE HU
charagolensis Ermisch, 1964b: 343 **A**: MG
chujoi Hatayama, 1985: 389 **A**: JA
shirozui M. Chûjô, 1957b: 16 (*Falsomordellistena*) [HN]
columbretensis Compte, 1970: 110 **E**: SP
comes Marseul, 1877: 473 **A**: JA SC
confinis A. Costa, 1854: 18 **E**: CR FR IT SL SP SZ **N**: AG MO
africana Roubal, 1911b: 272
connata Ermisch, 1969b: 108 **E**: AL AU BY CZ FI FR HU NT PL SK SV UK
consonbrina Ermisch, 1977: 157 **E**: HU SZ
cuneigera Champion, 1922a: 76 **A**: NP UP
curtelineata Pic, 1927k: 153 **A**: CH
curticornis Ermisch, 1977: 164 **E**: HU
cymbalistria Peyerimhoff, 1925: 8 **N**: TU
cypria Ermisch, 1963c: 62 **A**: CY IS
dalmatica Ermisch, 1956: 289 **E**: AU BH BU CR CZ GG HU IT MC SK UK **A**: TD TM TR
mostarensis Ermisch, 1977: 171
daurica Motschulsky, 1860c: 141 **A**: WS
dentata Batten, 1978: 79 **E**: GR
dieckmanni Ermisch, 1963a: 26 **E**: AU BU CR FI FR GE GR HU PL SL SK SZ
difficilis Ermisch, 1963c: 60 **A**: CY
dives Emery, 1876: 95 **E**: AR GG HU ST
dlabolai Ermisch, 1967d: 126 **A**: MG
dolini Odnosum, 2005: 53 **E**: AB
dvoraki Ermisch, 1956: 286 **E**: BU CZ FR GE HU MC PL SK TR
edashigei M. Chûjô, 1956: 174 **A**: JA
tokaraensis Nakane, 1956a: 1
echingolensis Ermisch, 1969c: 172 **A**: MG
elbrusicola Ermisch, 1969b: 111 **E**: ST
eludens Allen, 1999: 205 **E**: GB
emeryi Schilsky, 1895: 53 **E**: AL CR FR GR IT SP **N**: AG MO
episternalis Mulsant, 1856f: 371 **E**: AU BH BU CR FR GE GR HU IT PT SK SP SZ **N**: AG MO **A**: IN IS SY TM
extensa Rosenhauer, 1856: 227 (*Mordella*)
episternaloides Ermisch, 1963a: 7 **E**: FR GE
erdoesi Ermisch, 1977: 158 **E**: HU UK
ermischi Compte, 1966: 252 **E**: SP
eversi Ermisch, 1965b: 71 **N**: CI
exisa Stchegoleva-Barovskaya, 1927: 158 **A**: IN TD TM
exclusa Ermisch, 1977: 155 **E**: HU
fageli Ermisch, 1969b: 112 **E**: FR PT SP **N**: AG MO
falsoparvula Ermisch, 1956: 281 **E**: AU BU BY CZ EN FR GE GG HU NL IT PL PT SZ UK YU
falsoparvuliformis Ermisch, 1963a: 4 **E**: AL AU BY CZ GE GG HU SK ST UK **A**: TD TM TR
feigei Ermisch, 1956: 287 **E**: AU CR GE PL
fernandezi Palm, 1976: 31 **N**: CI
festiva Ermisch, 1977: 154 **E**: HU
flavella Ermisch, 1963c: 58 **A**: CY
flavospinosa Hubenthal, 1911: 186 **E**: AU GE
flavospinulosa Ermisch, 1977: 163 **E**: HU
fujiyamai Nomura, 1951: 57 **A**: JA
fulvipennis Stchegoleva-Barovskaya, 1932: 219 **A**: KZ
fusciapicalis Nomura, 1961b: 79 **A**: JA

fuscogemellata Ermisch, 1963b: 17 **E**: FR IT NL PT SP **N**: AG MO TU
fuscogemellatoides Ermisch, 1977: 165 **E**: BU GR HU **A**: KI KZ TM
fuscosuturalis Nakane, 1961: 78 **A**: JA
gallica Ermisch, 1956: 283 **E**: FR
gardneri Blair, 1930b: 180 **A**: BT NP SD
gemellata Schilsky, 1899: 79 **E**: EN GR PT SP **A**: CY
geronensis Ermisch, 1977: 169 **E**: FR SP
gfelleri Horák, 1990: 130 **E**: FR SP
ghanii Franciscolo, 1974: 71 **A**: PA
gianassoi Horák, 1992: 137 **N**: MO
goetzi Ermisch, 1969b: 114 **E**: GG ST
grisea Mulsant, 1856f: 376 **E**: CR FR SP
gutianshana Fan & C. Yang, 1995: 95 **A**: ZHE
hatorii Tokeji, 1953: 7 **A**: JA
helvetica Ermisch, 1967b: 110 **E**: SZ
hirtipes Schilsky, 1895: 46 **E**: FR GR MC TR UK **A**: CY IS JO TM TR SY
hollandica Ermisch, 1966a: 36 **E**: NL
hoberlandti Horák, 1983: 3 **E**: AR GG **A**: IN IQ IS SY TM TR UZ
horaki Piňo, 1985: 103 [RN] **A**: IQ KI TM
   bicolor Horák, 1982: 46 [HN]
horioni Ermisch, 1956: 285 **E**: AU CZ GE SK SZ
horvathi Ermisch, 1977: 173 **E**: BU CZ HU MC SK
humeralis Fabricius, 1758: 420 (*Mordella*) **E**: AB AU BY CR CZ ES FE GB GE GG HU IT MC PL SL SK SZ YU **A**: FE JA MG
   nigricollis Schilsky, 1895: 35
humeronotata Champion, 1922a: 76 **A**: NP SP UP
humeropicta Ermisch, 1963c: 63 **E**: AB AR **N**: AG **A**: CY TR
hungarica Ermisch, 1977: 163 **E**: CR HU
hypopygialis Ermisch, 1968b: 20 **A**: MO
inaequalis Mulsant, 1856f: 365 **E**: FR
incerta Ermisch, 1965a: 72 **A**: MG
inexpectata Ermisch, 1967b: 111 **E**: AU BU CZ GE HU SK ST SZ
inornata Nomura, 1961b: 80 **A**: JA
insignata Ermisch, 1965c: 200 **A**: JA
   inornata Nomura, 1961b: 80 [HN]
intersecta Emery, 1876: 93 **E**: AB AR BU CR GE GR FR MC GG GR HU IT SP ST TR UK **A**: KZ LE SY TD TM TR
irritans Franciscolo, 1991: 168 **E**: IT
istrica Ermisch, 1977: 169 **E**: FR SL
jelineki Horák, 1980: 283 **A**: TD
kaszabi Ermisch, 1965a: 65 **A**: MG
kirai Nomura, 1951: 52 **A**: JA SC
kirghizica Odnosum, 2003: 47 **A**: KI TD
kochi Ermisch, 1956: 284 **E**: SP
koelleri Ermisch, 1956: 289 **E**: AU BH BU CZ DE GE HU MC PL SK SP SV SZ YU
koma Nomura, 1951: 56 **A**: SC
korschefskyana Ermisch, 1963a: 5 [RN] **E**: GE
   korschefskyi Ermisch, 1956: 283 [HN]
kotenkoi Odnosum, 1990: 54 **E**: UK
kraatzi kraatzi Emery, 1876: 91 **E**: AB AL AR BH BU CR CZ FR GE GG GR HU PL PT SK ST SZ UK **A**: CY IN IQ KI KZ TD TM TR
kraatzi kretica Czetó, 1990: 28 **E**: GR
krauseri Plaza, 1986b: 286 **E**: SP
krujanensis Ermisch, 1963a: 14 **E**: AL BH

kuro Nomura, 1951: 51 **A**: JA
kurosai M. Chûjô, 1957b: 20 **A**: JA
labialis Motschulsky, 1860c: 142 **A**: WS
latitarsis Batten, 1983a: 88 **A**: TR
lenensis Ermisch, 1977: 173 **A**: ES
lesbia Ermisch, 1965d: 266 **E**: GR
leveyi Batten, 1989: 105 **A**: KA
lichtneckerti Ermisch, 1977: 160 **E**: AU HU RO SK TR
lindbergi Ermisch, 1963c: 59 **E**: BU GR TR **A**: CY TR
lonai Franciscolo, 1949c: 92 **E**: AL MC YU
longicornis Mulsant, 1856f: 373 **E**: GR FR
longicornoides Ermisch, 1965d: 258 **E**: BU SP TR
longipalpis Emery, 1891: 39 **N**: TU
    aypassai Batten, 1978: 78
lusitanica Ermisch, 1963b: 16 **E**: PT
luteipalpis Schilsky, 1895: 51 **E**: AU BU CZ GE GG HU PL SK
    perparvula Ermisch, 1966a: 30
maedai Nakane, 1956a: 2 **A**: JA
majuscula Ermisch, 1977: 171 **E**: IT
manteroi Franciscolo, 1942c: 76 **E**: IT
maroccana Ermisch, 1966b: 42 **N**: MO TU
mazedonica Ermisch, 1965d: 262 **E**: BU FR GR MC
mediogemellata Ermisch, 1977: 163 **E**: HU TR
meuseli Ermisch, 1956: 283 **E**: AU CZ FR GE HU SK UK **A**: KZ UZ
    minutuloides Ermisch, 1966a: 31
micans Germar, 1817: 212 (*Mordella*) **E**: AU BH CR CZ HU IT SL SP TR **A**: IS
    infima Emery, 1876: 117
microgemellata Ermisch, 1965d: 266 **E**: BU GR HU **A**: CY
michalki Ermisch, 1956: 288 **E**: AU BH CZ GE MC SK **A**: KZ TD
mihoki Ermisch, 1977: 158 **E**: CR GE HU RO UK
minima A. Costa, 1854: 18 **E**: AB BU CR FR GR HU MC IT SP TR YU **A**: CY IS KI TR
    pseudorhenana Ermisch, 1977: 164
minutula Ermisch, 1956: 285 **E**: AU HU
miyamotoi Nakane, 1956a: 3 **A**: JA TAI
mongolica Ermisch, 1964b: 344 **A**: MG
muchei Ermisch, 1965a: 75 **A**: MG
multicicatrix Kangas, 1986: 139 **E**: FI
neglecta Ermisch, 1977: 156 **E**: HU
nepalensis Horák, 1995b: 168 **A**: NP
nessebaricus Batten, 1980: 43 **E**: BU
neuwaldeggiana Panzer, 1796d: 8 (*Mordella*) **E**: AU BE BH BU BY CR CZ GR FR HU IT NL PL SK SL SP SV SZ UK YU
    brunnea Fabricius, 1801b: 125 (*Mordella*)
    ferruginea Marsham, 1802: 490 (*Mordella*)
    meridionalis A. Costa, 1854: 19 (*Nattirica*)
nigritarsis Horák, 1996: 172 **E**: CZ GE GG SK SL
nigrogemellata Ermisch, 1964b: 343 **A**: MG
nomurai Tokeji, 1953: 6 **A**: JA
obscurosuturalis Ermisch, 1964b: 345 **A**: MG
olympica Ermisch, 1965d: 268 **E**: BU GR MC YU **A**: TR
okinawana Nomura, 1963b: 40 **A**: JA
oranensis Pic, 1900l: 124 **E**: IT (Lampione Is.) **N**: AG
ozeana Nakane, *1957: 58* (*Pseudomordellina*) **A**: JA
parabrevicauda Ermisch, 1965a: 71 **A**: KI KZ MG UZ
paradohumeralis Ermisch, 1963c: 58 **A**: CY TR

paraepisternalis Ermisch, 1965d: 260 **E**: BU GR SP **A**: IQ IS SP UZ TR
paraintersecta Ermisch, 1956: 284 **E**: AR GR HU ST **A**: IN IQ SY TR
paraobscurosuturalis Ermisch, 1965a: 66 **A**: FE MG
parapentas Ermisch, 1977: 171 **E**: IT
pararhenana Ermisch, 1977: 164 **E**: CR HU
paraweisei Ermisch, 1977: 160 **E**: RO
parvicauda Ermisch, 1967b: 116 **E**: AB AL BH BU CR GG GR MC TR **A**: TR
parvula Gyllenhal, 1827: 519 (*Mordella*) **E**: AB AU BH BU BY CZ HU IT MC NL PL RO SK SL ST SZ **A**: HUB GAN KZ NMO
liliputana Mulsant, 1856f: 374
pusilla L. Redtenbacher, 1849: 614 (*Mordella*)
rectangula Roubal, 1911b: 272 [HN]
troglodytes Mannerheim, 1844a: 198 (*Mordella*)
parvuliformis Stchegoleva-Barovskaya, 1930: 57 **E**: AB BY CZ GE HU SK ST UK **A**: IN UZ WS
paucistrigosa Ermisch, 1965a: 64 **A**: MG
peloponnesensis Batten, 1980: 42 **E**: GR **A**: CY
pentas Mulsant, 1856f: 380 **E**: AL AU CZ FR GG GR HU IT MC NL PL PT RO SP SZ UK YU **N**: AG MO **A**: CY
perroudi Mulsant, 1856f: 382 **E**: AU CR FR GR HU IT MC PL SK SL SP TR **A**: IS
persica Horák, 1983: 7 **A**: IN
permira Franciscolo, 1949c: 91 **E**: GR
phaea Nomura, 1951: 46 **A**: JA
podlussanyi Czetó, 1990: 26 **E**: GR
pontica Ermisch, 1977: 169 **E**: HU
porphyria Nomura, 1951: 55 **A**: TAI
psammophila Peyerimhoff, 1943: 16 **N**: AG
praesagita Kangas, 1988: 89 **E**: FI SV
pseudobrevicauda Ermisch, 1963a: 12 **E**: AB AR AU CR CZ BH BU EN FR GE GG HU MC PL RO SK SL SP ST TR **A**: IN KZ MG TD TM TR UZ
pseudohirtipes krotosensis Czetó, 1990: 28 **E**: GR
pseudohirtipes pseudohirtipes Ermisch, 1965d: 268 **E**: AB BU CR FR GR MC SP TR UK YU **N**: AG **A**: CY TR
pseudomicans Ermisch, 1977: 169 **E**: CR MC
pseudoparvula Ermisch, 1956: 281 **E**: AU BY GB GE HU IT PL SZ
eludens Allen, 1999: 205
parvuloides Ermisch, 1956: 281
pseudopumila Ermisch, 1963a: 28 **E**: BY DE FI FR GE NL NT PL PT SP **N**: MO **A**: FE KZ ST TD WS
pseudoreichei Ermisch, 1977: 158 **E**: IT
pseudorugipennis Ermisch, 1963c: 60 **E**: BU GR MC **A**: CY IN IQ IS JO SY TR
pseudotarsata Ermisch, 1964b: 345 **A**: FUJ JA MG NE
pumila Gyllenhal, 1810: 605 (*Mordella*) **E**: AL AR AU BE BH BU BY CR CZ FI FR GE GG GR HU IT MC NL PL RO SK SL SP ST SV SZ TR UK YU **N**:AG **A**: ES GAN HUB IN IS KZ MG NC NMO SC SY TAI TD TM TR
deficiens Mulsant, 1856f: 380
elongata Kraatz, 1868a: 295 (*Mordella*)
stricta A. Costa, 1854: 16
purpurascens A. Costa, 1854: 17 **E**: CR FR IT SL SP **N**: AG MO TU **A**: KZ TD TM WS
purpureonigrans Ermisch, 1963a: 24 **E**: AL AU BU BY CZ DE EN FI GE HU IT MC NL PL SK SL SV SZ **A**: CY TR
pygmaeola Ermisch, 1956: 288 **E**: AU BU CR CZ FR GE GG HU NL NR PL SK ST SV SZ **A**: TM
pyrenaea Ermisch, 1966a: 35 **E**: FR SP
quadrisulcata Pic, 1938g: 220 **A**: JIA
reichei Emery, 1876: 91 **E**: BU CR CZ FR GE HU MC SK ST UK
mulsanti Ermisch, 1956: 283
reitteri Schilsky, 1894a: 94 **E**: AL AU BU CR FI GE GG GR HU IT RO SK SP TR YU **A**: CY

rhenana Ermisch, 1956: 285 **E**: BE EN FR GE HU **A**: KZ TM
rufifrons Schilsky, 1894a: 92 **E**: AL AU BY CR GE MG NT PL RU **A**: ES FE KZ MG
rufospina Ermisch, 1977: 159 **E**: CR
rugipennis Schilsky, 1895: 57 **E**: GR UK **A**: TD TM TR
sajoi Ermisch, 1977: 165 **E**: HU
saxonica Ermisch, 1967b: 115 **E**: GE PL
savioi Pic, 1924a: 23 **A**: JIA
schatzmayri Franciscolo, 1949c: 91 **E**: BU MC GR
pfefferi Ermisch, 1963a: 14
shimoyamai M. Chûjô, 1957a: 6 **A**: JA
secreta Horák, 1983: 10 **E**: AB AR AU BE BU CZ FR GB GG GR HU IT MC NR PL RO SK SL SP ST SZ UK YU **A**: IS TR
sejugata Ermisch, 1965a: 73 **A**: MG
semiferruginea Reitter, 1911a: 376 **E**: AU BU FR HU RO SK SZ UK **A**: TR
balazuci Ermisch, 1966a: 37
taurica Csiki, 1949: 45
semipygmaeola Ermisch, 1964b: 349 **A**: MG
semirubra Champion, 1927: 52 **A**: NP UP
sericata Wollaston, 1864: 515 **N**: CI
serraticornis Horák, 1991: 135 **N**: MO
shirozui Nomura, 1951: 53 **A**: JA
sibirica Ermisch, 1977: 156 **A**: WS
signicollis Schilsky, 1894a: 93 **A**: SY
solarii Franciscolo, 1942c: 79 **E**: IT
squamipilosa Ermisch, 1967a: 45 **A**: MG
stenidea Mulsant, 1856f: 381 **E**: AL AU BH BU BY CR FR GE GR HU IT PL SL SP SZ
flexipes Rey, 1857: 411
steppensis Ermisch, 1965a: 68 **A**: MG
stoeckleini Ermisch, 1956: 282 **E**: CZ GE HU PL SK SZ UK **A**: FE
subepisternalis Ermisch, 1965d: 262 **E**: AL BU FR HU YU **N**: MO **A**: IS TR
subincana Ermisch, 1967a: 49 **A**: MG
subsquamosa Schilsky, 1899: 80 **A**: IQ MG WS TM
syrites Ermisch, 1977: 162 **A**: SY
tadjikistanica Horák, 1980: 280 **A**: TD UZ
takizawai Kôno, 1932b: 157 **A**: JA
tamana Tokeji, 1953: 17 (*Falsomordellistena*) **A**: JA
taorminensis Ermisch, 1965a: 71 **E**: GR IT
tarsata Mulsant, 1856f: 381 **E**: AL AU BH BU CR CZ FR GE GG GR HU IT RO SK SP ST SZ UK YU **A**: KI TM KZ TR WS
teneriffensis Ermisch, 1965: 69 **N**: CI
tenuicornis Schilsky, 1899: T [RN] **E**: GR
gracilicornis Schilsky, 1895: 47 [HN]
tenuimanus Champion, 1927a: 52 **A**: NP UP
thurepalmi Ermisch, 1965c: 201 [RN] **E**: FI HU NR PL SV
palmi Ermisch, 1963a: 10 [HN]
thuringiaca Ermisch, 1963a: 23 **E**: BY FR GE HU PL **A**: FE KZ TM WS
tibiella Ermisch, 1965a: 67 **A**: MG
tokeji Nomura, 1951: 52 **A**: JA SC
tondui Horák, 1996: 176 **A**: AF
tosgoniensis Ermisch, 1969c: 174 **A**: MG
trilineata Mulsant, 1856f: 384 **E**: FR
turknenica Odnosum, 2003: 43 **A**: TM
uvsensis Ermisch, 1968b: 17 **A**: MG
variegata Fabricius, 1798: 127 (*Mordella*) **E**: AU BU BY CR CZ FR GB GE HU IT NL PL SK SL SP ST SZ UK YU **A**: ES FE KZ

bicolor Marsham, 1802: 490 (*Mordella*)
lateralis A. G. Olivier, 1795: 64: 8 (*Mordella*) [HN]
visai Horák, 1983: 5 **A**: FE
vogti Ermisch, 1963a: 8 **E**: GE
wankai Ermisch, 1966a: 33 **E**: GE HU
weisei Schilsky, 1895: 41 **E**: AU BU BY CR CZ GE HU NL PL SK ST SZ **N**: AG **A**: FE KZ TD WS
wiebesi Batten, 1977: 170 **E**: CR GR IT SP **N**: AG
yangi Fan, 1995a: 23 **A**: HEB
zaisanensis Ermisch, 1967d: 125 **A**: MG
zoltani Ermisch, 1977: 164 **E**: CR HU
zosijniensis Ermisch, 1969c: 173 **A**: MG
subgenus *Pseudomordellina* Ermisch, 1952a: 47 type species *Mordellistena nana* Motschulsky, 1860
acuticollis Schilsky, 1895: 40 **E**: AU BU BY CZ GB GE HU IT PL RO SK SZ UK **A**: KI
imitatrix Allen, 1995: 85
aliena Ermisch, 1967d: 128 **A**: MG
arabica Chobaut, 1904b: 244 **A**: SA
aritai Nomura, 1964: 50 **A**: JA
csiki Ermisch, 1977: 175 **E**: HU SZ
ferruginipes Ermisch, 1966a: 39 **E**: NL
flavospinulosa Ermisch, 1977: 163 **E**: HU
klapperichi Ermisch, 1956: 291 **E**: BH CZ FR GE HU SK SZ
lebedevi Roubal, 1929a: 48 **A**: KI TM UZ
kubani Horák, 1982: 50
luteispina Ermisch, 1977: 175 **E**: RO TR
magyarica Ermisch, 1977: 174 **E**: AU HU
mediana Ermisch, 1977: 175 **E**: CR GR HU
microscopica Ermisch, 1977: 175 **E**: CR HU
nana Motschulsky, 1860c: 142 **E**: ST
nanula Ermisch, 1967b: 118 **E**: AL AU BU BY GE PL RO SK ST UK **A**: FE KZ TM TR WS
nanuloides Ermisch, 1967b: 118 **E**: GE NL
nigrocapillata Ermisch, 1965d: 270 **E**: BU
paranana Ermisch, 1977: 174 **E**: HU
pseudonana Ermisch, 1956: 291 **E**: AU CR GE HU MC PL SK SL SP SZ
separanda Ermisch, 1965a: 76 **A**: MG
ulanbatorensis Ermisch, 1967a: 50 **A**: MG
umbra Franciscolo, 1949b: 54 **E**: IT
zavchanensis Ermisch, 1970: 174 **A**: MG
zuuncharaensis Ermisch, 1965d: 75 **A**: MG

genus *Mordellistenochroa* Horák, 1982: 52 type species *Mordellistenochroa strejceki* Horák, 1982
fallaciosa Ermisch, 1969b: 110 (*Mordellistena*) **E**: HU IT SZ
strejceki Horák, 1982: 52 **E**: GG UK **A**: TD TM

genus *Mordellistenoda* Ermisch, 1941b: 590 type species *Mordellistenoda fukiensis* Ermisch, 1941
aka Kôno, 1928a: 43 (*Mordellistena*) **A**: JA TAI SE
donan Tsuru, 2004: 408 **A**: JA
fukiensis Ermisch, 1941b: 591 **A**: FUJ
ohsumiana Nakane, 1957: 51 (*Glipostenoda*) **A**: JA

genus *Mordellistenula* Stchegoleva-Barovskaya, 1930a: 57 type species *Mordellistena planifrons* Stchegoleva-Barovskaya, 1930
anomala Ermisch, 1957: 47 **E**: FR SP
lacinicollis Czetó, 1990: 25 **E**: GR
longipalpis Ermisch, 1965d: 256 **E**: AB AR BU GR MC TR YU **A**: KZ

perrisi Mulsant, 1857b: 412 (*Mordellistena*) **E**: AB AN AR AU BE BU CR CZ DE FR GE GR HU IT KZ MC PL PT SP SV SZ UK YU **N**: MO **A**: KZ TR
engelharti Schilsky, 1910: 255 (*Mordellistena*)
rectangula C. G. Thomson, 1868: 134 (*Mordellistena*)
planifrons Stchegoleva-Barovskaya, 1930a: 56 **E**: AB AL AU BU CR GE GG GR HU IT MC RO RU SK UK **N**: AG MO
plutonica Compte, 1969: 114 **E**: SP

genus *Mordellochroa* Emery, 1876: 83 type species *Mordella abdominalis* Fabricius, 1775
abdominalis Fabricius, 1775: 264 (*Mordella*) **E**: AB AU BY BE BH BU CR CZ FR GB GE GG GR HU IT LT NT PL SK SL ST SZ UK **A**: FE TM TR
bicolor Sulzer, 1776: 67 (*Mordella*)
nigra Marsham,: 1802: 490 (*Mordella*)
ventralis Fabricius, 1792b: 114 (*Mordella*)
hasegawai Nomura & Katô, 1959: 5 **A**: JA
humerosa Rosenhauer, 1847: 38 (*Mordella*) **E**: AL BH BU FR GR HU MC PT RU SP ST SZ UK **A**: CY IS KZ MG SY TR
freyi Ermisch, 1944: 211 (*Tolida*)
milleri Emery, 1876: 83 (*Mordellistena*) **E**: AB AU CZ FR HU PL SK SL UK
pulchella Mulsant & Rey, 1859f: 232 (*Mordella*) **E**: FR IT **N**: AG MO TU
pygidialis Nomura, 1961b: 85 **A**: JA
shibatai Kiyoyama, 1987: 31 **A**: TAI
tournieri Emery, 1876: 102 (*Tolida*) **E**: AL AU BU BY CR CT CZ ES FI FE FR GE GG GR HU IT MC NT PL RU SK SP ST SZ UK **A**: FE TR SV SZ
graeca Schilsky, 1895: 56 (*Mordellistena*)
schusteri Schilsky, 1895: 55 (*Mordellistena*)
similis Stschegoleva-Barovskaya, 1930b: 752 (*Mordellistena*)
yanoi Nomura, 1951: 56 (*Mordellistena*) **A**: TAI

genus *Pseudodellamora* Ermisch, 1941c: 720 type species *Mordellistena brevicollis* Emery, 1876
brevicollis Emery, 1876: 102 (*Mordellistena*) **N**: AG
championi Schilsky, 1899: 82 (*Mordellistena*) **E**: SP **M**: MO
distinguenda Ermisch, 1963b: 15 **E**: CR FR PT SP **A**: TR
grossepalpalis Ermisch, 1942a: 87 **E**: AB

genus *Pseudotolida* Ermisch, 1950b: 86 type species *Mordellistena ephippiata* Champion, 1891
awana Kôno, 1932b: 157 (*Mordellistena*) **A**: HUB JA TAI
mitsuhashii Nomura, 1951: 68 (*Glipostenoda*)
tokyoensis Nomura & Katô, 1959: 6
menoko Kôno, 1932b: 158 (*Mordellistena*) **A**: JA
morimotoi Nomura, 1967a: 32 **A**: TAI
multisulcata Nomura, 1966: 45 **A**: JA
sinica Fan & J. Yang, 1995: 96 **A**: ZHE

genus *Pulchrimorda* Ermisch, 1968: 37 type species *Mordellistena grelaki* Pic, 1923
savioi Pic, 1924a: 23 (*Mordellistena*) **A**: JIA

genus *Tolida* Mulsant, 1856f: 387 type species *Mordellistena artemisiae* Mulsant, 1856
artemisiae Mulsant, 1856f: 385 (*Mordellistena*) **E**: BE FR GE IT SP

genus *Tolidopalpus* Ermisch, 1952b: 144 type species *Tolidopalpus castaneicolor* Ermisch, 1952
castaneicolor Ermisch, 1952b: 149 **A**: FUJ JA
galloisi Kôno, 1932b: 158 (*Mordellistena*) **A**: FE JA TAI

genus *Tolidostena* Ermisch, 1942b: 674 type species *Tolidostena tarsalis* Ermisch, 1942
subgenus *Tolidostena* Ermisch, 1942b: 674 type species *Tolidostena tarsalis* Ermisch, 1942
atripennis Nakane, 1956a: 8 **A**: FE JA TAI
ermischi Nakane, 1956a: 9 **A**: JA
fusei Tokeji, 1953: 14 (*Mordellochroa*) **A**: JA
japonica Tokeji, 1953: 13 (*Mordellochroa*) **A**: JA
montana Kiyoyama, 1991: 54 **A**: TAI
similator Kiyoyama, 1991: 56 **A**: TAI
tarsalis Ermisch, 1942b: 675 **A**: FUJ GUI SHA
subgenus *Neotolidostena* Kiyoyama, 1991: 53 type species *Mordellochroa taiwana* Kiyoyama, 1987
hayashii Kiyoyama, 1991: 59 **A**: TAI
taiwana Kiyoyama, 1987: 34 (*Mordellochroa*) **A**: TAI

genus *Uhligia* Horák, 1990: 139 type species *Mordellistena schilskyi* Csiki, 1915
schilskyi Csiki, 1915: 50 (*Mordellistena*) **E**: AR **A**: TM UZ
deserticola Horák, 1981: 337 (*Mordellistenula*)
murina Schilsky, 1903: 100 (*Mordellistena*) [HN]

tribe Stenaliini Franciscolo, 1955

genus *Brodskyella* Horák, 1989: 37 type species *Stenalia angustata* Pic, 1923
holzschuhi Horák, 1989: 42 **A**: NP SD

genus *Stenalia* Mulsant, 1856f: 387 type species *Mordella testacea* Fabricius, 1787
araxicola Khnzorian, 1957a: 163 **E**: AB AR GG GR TR **N**: TU **A**: IN IQ IS KZ TD TM TR UZ
muchei Ermisch, 1965d: 251
ascaniaenovae Lazorko, 1975: 110 **E**: AR UK **A**: KZ UZ
atra Perris, 1875: 8 (*Mordella*) **N**: AG EG TU
argentata Schilsky, 1899: L
balcanica Franciscolo, 1949c: 88 **E**: AL
bilyi Horák, 1977: 400 **A**: KI TD UZ
bisecta Baudi di Selve, 1883a: 5 **E**: IT
biskrensis Schilsky, 1903: 99 **N**: AG TU
brunneipennis brunneipennis Mulsant, 1856f: 389 **E**: AR GR **N**: EG **A**: IN IQ IS JO LE SY TR
brunneipennis unistrigosa Chobaut, 1924d: 182 **A**: SY
dolini Lazorko, 1974: 112 **E**: UK
dzhulfae Odnosum, 2001: 22 **E**: AB
ermolenkoi Odnosum, 2000: 46: **E**: AB
escherichi Schilsky, 1899: 76 **E**: AR HU RO TR SK **N**: MO **A**: KZ TD TR UZ WS
uralensis Csiki, 1915: 45 (*Mordellistena*)
singularis Reitter, 1911a: 377 (*Mordellistena*) [HN]
gracilicornis Baudi di Selve, 1878b: 348 **E**: AB AR **A**: CY
graeca Ermisch, 1965d: 254 **E**: GR
gridellii Franciscolo, 1949c: 87 **E**: GR
hispana Schilsky, 1895: 32 **E**: SP
indica Horák, 1995a: 97 **A**: SD
iranica Horák, 1981: 335 **A**: IN
jordanensis Ermisch, 1965d: 254 **A**: IS JO
lindbergi Ermisch, 1963c: 53 **E**: GR **A**: CY IS
merkli Schilsky, 1895: 33 **E**: BU TR **A**: IQ IS JO LE SY TR
rufohumeralis Pic, 1926a: 2 **A**: FE
sefrensis peyerimhofferi Chobaut, 1924d: 183 **N**: AG
sefrensis sefrensis Chobaut, 1924d: 183 **N**: AG
stipae Chobaut, 1924d: 183 **E**: SP **N**: AG

testacea testacea Fabricius, 1787a: 218 (*Mordella*) **E**: AR BU CR FR GE IT PT SL SP ST SZ UK **N**: AG TU MO

chiragra Dufour, 1843: 72 (*Mordella*)
dolini Lazorko, 1974: 112
flavipennis J. Sturm, 1826: 171 (*Mordella*)
luteicornis Schilsky, 1895: 30

testacea tristrigosa Chobaut, 1924d: 181 **N**: AG
vladimiri Odnosum, 2001: 21 **E**: AB

family TENEBRIONIDAE Latreille, 1802
subfamily **Lagriinae** Latreille, 1825
tribe Belopini Reitter, 1917

genus *Centorus* Mulsant, 1854: 272 type species *Calcar procerus* Mulsant, 1854
subgenus *Belopomerus* Reitter, 1920a: 3 type species *Calcar zoufali* Reitter, 1915
lucasi Mulsant & Godart, 1859b: 243 **N**: AG LB
zoufali Reitter, 1915e: 89 (*Calcar*) **N**: LB TU
subgenus *Belopus* Gebien, 1911: 459 [RN] type species *Tenebrio elongatus* Herbst, 1797
Calcar Dejean, 1821: 67 [HN] type species *Tenebrio elongatus* Herbst, 1797
Calcar Latreille, 1829: 25 [HN] type species *Trogosita calcar* Fabricius, 1801
aegyptiacus Zoufal, 1893: 116 (*Calcar*) **N**: EG LB **AFR**
elongatus ecalcaratus Seidlitz, 1896: 655 (*Calcar*) **E**: IT (Sicilia) MA **N**: AG TU **A**: CY **AFR**
siculus Canzoneri, 1977: 279 (*Belopus*)
elongatus elongatus Herbst, 1797: 259 (*Tenebrio*) **E**: FR IT (Sardegna) PT SP **N**: CI MO MR TU
calcar Fabricius, 1801a: 153 (*Trogosita*)
heydeni Zoufal, 1893: 116 (*Calcar*) **N**: EG **A**: CY IN IQ IS SY TR
humeralis Champion, 1891d: 387 (*Calcar*) **N**: AG MO
truncaticollis Zoufal, 1893: 117 (*Calcar*)
parvulus G. S. Medvedev, 1964a: 55 (*Belopus*) **A**: TM (Kugitang)
syriacus Zoufal, 1893: 117 (*Calcar*) **A**: IS "Syria"
subgenus *Centorus* Mulsant, 1854: 272 type species *Calcar procerus* Mulsant, 1854
alashanicus Skopin, 1974a: 90 (*Belopus*) **A**: NMO
bucharicus Obenberger, 1914: 141 **A**: KZ TD
porrecticollis Reitter, 1915e: 90
calcaroides calcaroides Reitter, 1920a: 6 (*Belopus*) **E**: ST **A**: AF KI KZ TM UZ
calcaroides gobiensis Kaszab, 1964g: 399 (*Belopus*) **A**: ES (Tuva) MG
calcaroides intermedius Skopin, 1974a: 93 (*Belopus*) **A**: KZ
calcaroides tsharynensis Skopin, 1974a: 95 (*Belopus*) **A**: KZ
chappuisi Peyerimhoff, 1933: 346 (*Belopus*) **N**: AG
crassipes borealis Skopin, 1974a: 93 (*Belopus*) **A**: KZ
crassipes crassipes Fischer von Waldheim, 1844: 123 (*Calcar*) **E**: GR IT MC ST UK **A**: KZ TR
reitteri Zoufal, 1893: 118
sulcatus Fischer von Waldheim, 1844: 124 (*Calcar*)
csikii bagdadensis Reitter, 1920a: 9 (*Belopus*) **A**: CY IQ IS JO SA YE
csikii csikii Reitter, 1920a: 6 (*Belopus*) **E**: ST **A**: AF IN KZ TM
ferganensis Skopin, 1974a: 92 (*Belopus*) **A**: UZ (Fergana)
filiformis Motschulsky, 1872: 40 **E**: AB AR ST **A**: AF GAN IN KZ TM XIN
araxidis Reitter, 1920a: 10 (*Belopus*)
americaus [sic] Motschulsky, 1872: 40
tenuicornis Motschulsky, 1872: 40
helanensis Ren & Wu, 1994: 351 (*Belopus*) **A**: NIN
luculentus Ren, 1999: 177 (*Belopus*) **A**: XIN
lucalentus Ren, 1999: 393 (*Belopus*)
microceps Motschulsky, 1872: 39 **N**: EG
ocularis Reitter, 1920a: 7 (*Belopus*) **A**: "Turkestan"

pakistanus Kaszab, 1961d: 226 (*Belopus*) **A**: PA
pilosus Skopin, 1974a: 92 (*Belopus*) **A**: IQ
proceroides Leo, 1984: 165 (*Belopus*) **E**: IT (Sardegna)
procerus moldaviensis Reitter, 1920a: 7 (*Belopus*) **E**: MD RO ST UK **A**: KZ UZ
procerus procerus Mulsant, 1854: 272 (*Calcar*) **E**: FR SP ST **N**: TU
procerus Rosenhauer, 1856: 217 (*Calcar*) [HN]
raffrayi Fairmaire, 1873a: 340 (*Calcar*) **N**: TU
reymondi Antoine, 1951a: 76 (*Belopus*) **N**: MO
rufipes Gebler, 1833: 288 (*Calcar*) **E**: ST UK **A**: KZ WS
saccardyi Peyerimhoff, 1939a: 25 (*Belopus*) **N**: AG
sefiranus Reitter, 1920a: 11 (*Belopus*) **A**: AF KZ
sublaevis Reitter, 1920a: 11 **A**: AF KZ TM
steppensis Kaszab, 1964g: 401 (*Belopus*) **A**: ES MG
tibialis Zoufal, 1893: 118 **E**: AB UK (Krym) **A**: KZ
trogosita Motschulsky, 1872: 39 **E**: AR ST **A**: AF KZ TM UZ
tumotensis Ren & Zheng, 1993d: 52 (*Belopus*) **A**: NMO
turcicus Kaszab, 1959d: 81 (*Belopus*) **A**: TR
tuvensis Knor, 1976: 106 (*Belopus*) **A**: ES (Tuva)
subgenus *Nanocalcar* Skopin, 1974a: 102 type species *Nanocalcar nanus* Skopin, 1974
nanus Skopin, 1974a: 95 (*Belopus*) **A**: KZ (Almaatinskaya Oblast)

tribe Cossyphini Latreille, 1802

genus *Cossyphus* A. G. Olivier, 1791: 119 type species is *Lampyris depressa* Fabricius, 1781
subgenus *Acontodactylus* Desbrochers des Loges, 1894: 5 type species *Cossyphus minutissimus* Laporte, 1840
Acanthodactylus [incorrect subsequent spelling]
Paracossyphus Viñolas & Cartagena, 2005: 29 [unnecessary RN]
minutissimus Laporte, 1840: 228 **E**: PT SP **N**: MO
pygmaeus Brême, 1846: 29
tuberculatus Brême, 1846: 27
punctatostriatus Kocher, 1953: 67 **N**: MO
pusillus Pascoe, 1887: 14 **A**: NP **ORR**
subgenus *Cossyphus* A. G. Olivier, 1791: 119 type species *Lampyris depresas* Fabricius, 1781
algiricus Laporte, 1840: 228 **N**: AG MO TU
algyricus Laporte, 1840: 228 [incorrect original spelling]
ovatus Brême, 1846: 22
dejeani Brême, 1846: 21 **E**: PT SP **N**: AG MO
depressus Fabricius, 1781: 254 (*Lampyris*) **A**: NP SD **ORR**
asperatus Fairmaire, 1896a: 59
planus Fabricius, 1801b: 98
subgranulatus Reitter, 1917b: 138
hoffmanseggii Herbst, 1797: 229 **E**: PT SP **N**: AG MO
bletoni Antoine, 1954: 211
rifensis Kocher, 1963c: 198
sublaevis Kocher, 1953: 66
laevis Laporte, 1840: 228 **E**: SP **N**: AG MO
barbarus Brême, 1846: 24
incostatus Brême, 1846: 25
substriatus Laporte, 1840: 229
moniliatus Reitter, 1917b: 135 **E**: GR **A**: SY
moniliferus moniliferus Chevrolat, 1833a: pl. 31 **E**: GR IT (Sardegna) MA SP (Baleares) **N**: AG CI EG LB MO TU **A**: SY **AFR**
insularis Laporte, 1840: 229
ruffoi indicus Scupola, 2000: 241 **A**: NP **ORR**
rugulosus Peyron, 1854: 224 **A**: CY IN IS SY TR
subgranulatus Reitter, 1917b: 138 **A**: SD **ORR**
tauricus Steven, 1832: 425 **E**: AL AR BU GR IT ST TR UK **A**: AF CY IN IQ IS SY TD TM TR UZ

tribe Goniaderini Lacordaire, 1859

genus *Anaedus* Blanchard, 1843: 198 type species *Anaedus punctatissimus* Blanchard, 1843
buricus Schawaller, 1994: 266 **A**: NP
nepalicus Schawaller, 1994: 267 **A**: NP
mroczkowskii Kaszab, 1968a: 10 **A**: NC SC
spinicornis Kaszab, 1973a: 39 **A**: NP TAI **ORR**
substriatus Pic, 1938a: 17 **A**: SCH

tribe Laenini Seidlitz, 1896

genus *Hypolaenopsis* Masumoto, 2001a: 45 type species *Hypolaenopsis uenoi* Masumoto, 2001 (=*Hypolaenopsis nanpingica* Schawaller, 2001)
benesi Schawaller, 2001a: 7 (*Laena*) **A**: SCH
honggyuanica Schawaller, 2001a: 21 (*Laena*) **A**: SCH
hualongica Schawaller, 2001a: 22 (*Laena*) **A**: SCH
nanpingica Schawaller, 2001a: 29 (*Laena*) **A**: SCH
  uenoi Masumoto, 2001a: 46
nomurai Schawaller, 2001a: 29 (*Laena*) **A**: SCH

genus *Laena* Dejean, 1821: 64 type species *Scaurus viennensis* J. Sturm, 1807
  Catolaena Reitter, 1900h: 282 type species *Laena turkestanica* Reitter, 1897
  Ebertius Jedlička, 1965: 98 type species *Ebertius nepalensis* Jedlička, 1965
  Laena Latreille, 1829: 39 [HN] type species *Scaurus viennensis* J. Sturm, 1807
  Psilolaena Heller, 1923: 70 type species *Psilolaena schusteri* Heller, 1923
acuticollis Kaszab, 1978a: 175 **A**: SD
aenea Schuster, 1926a: 49 **A**: SD
  darjeelingiana Kaszab, 1938e: 226
affinis Schuster, 1935d: 451 **A**: SD
alaiensis Reitter, 1906h: 446 **A**: KI UZ (Fergana)
  edda Reitter, 1906h: 447
almorensis Schuster, 1926a: 44 **A**: UP
alpina Kaszab, 1977b: 264 **A**: NP
alticola Blair, 1923b: 284 **A**: XIZ
angorensis Reitter, 1900h: 284 **A**: TR
angulifemoralis Musumoto, 1996: 169 **A**: YUN
angusta Weise, 1878: 233 **E**: UK (Krym) "Caucasus"
annapurna Schawaller, 2002: 41 **A**: NP
apfelbecki Schuster, 1915c: 358 **E**: GR **A**: TR
arun Schawaller, 2002: 41 **A**: NP
atripennis Pic, 1919c: 6 **E**: GR (Taygetos)
badakschanica Kaszab, 1960a: 170 **A**: AF
baishuica Schawaller, 2001a: 5 **A**: YUN
barypithoides Schuster, 1916: 623 **A**: KA
baudii Weise, 1878: 239 **E**: GG
becvari Schawaller, 2001a: 6 **A**: YUN
beesoni Schuster, 1935d: 459 **A**: HP
bhatiai Schuster, 1935d: 447 **A**: HP UP
bhutanensis Kaszab, 1975e: 329 **A**: BT
bifoveolata Reitter, 1889k: 709 **A**: GAN SHA SCH
  imurai Masumoto, 1996d: 179
  shaanxiica Masumoto, 1996d: 178
blairi Schuster, 1926a: 47 **A**: UP
bogatschevi Iablokoff-Khnzorian, 1984: 43 **E**: AR
bowaica Schawaller, 2001a: 8 **A**: SCH
brahmae Masumoto, 1990f: 190 **A**: SD
  gomcheyensis Masumoto, 1990f: 194

breiti Schuster, 1916: 588 **E**: GR (Taygetos)
brendelli Schawaller, 2001a: 9 **A**: YUN
brevipennis Reitter, 1901b: 63 **A**: KZ
broscosomoides Kaszab, 1977b: 268 **A**: NP
businskyorum Schawaller, 2001a: 9 **A**: YUN
byzantina Apfelbeck, 1901: 466 **E**: TR **A**: TR
bodemeyeri Reitter, 1903e: 224
cameroni Schuster, 1926a: 36 **A**: UP
cardiothorax Kaszab, 1978a: 175 **A**: SD
carinata Schuster, 1926a: 38 **A**: UP
carinipennis Schuster, 1935d: 448 **A**: HP
carolinae Schuster, 1935d: 450 **A**: KA
caucasica Motschulsky, 1845a: 75 **E**: "Caucasus"
championi Schuster, 1926a: 41 **A**: UP
chatkalica Schawaller, 1995a: 69 **A**: KI
chatterjeei Schuster, 1935d: 452 **A**: HP
chaukiensis Masumoto, 1990f: 192 **A**: NP
chejuana Mt. Chûjô, 1992: 41 **A**: SC (Chejudo Is.)
chetri Schawaller, 2002: 42 **A**: NP
chinensis Kaszab, 1965c: 284 **A**: SCH YUN
youcida Masumoto, 1996d: 171
cholanica Schawaller, 2001a: 11 **A**: XIZ
chomolungma Schawaller, 2002: 43 **A**: NP
clivinoides Baudi di Selve, 1876c: 243 **E**: BH **A**: CY SY
clypealis Fairmaire, 1896b: 98 **A**: HP PA
confusa Löbl & Schawaller, **nom. nov.** [see New Acts] **A**: ANH
chinensis J.-K. Li & Chen, 1993: 166 [HN]
coniceps Kaszab, 1973a: 56 **A**: NP
consimilis Kaszab, 1973a: 45 **A**: NP
mandli Kaszab, 1975b: 125
constricta Iablokoff-Khnzorian, 1957a: 166 **E**: AR
convexicollis Reitter, 1908k: 275 **A**: HP
corallipes Reitter, 1908k: 280 **A**: KA
crenulicollis Kaszab, 1977b: 265 **A**: NP
cribella Reitter, 1908k: 279 **A**: HP KA
curvipes Desbrochers des Loges, 1881: 134 **A**: CY
cylindrica Schuster, 1926a: 52 **A**: XIZ
daliensis Masumoto & Yin, 1994: 168 **A**: YUN
tumidoculata Masumoto, 1996d: 173
xiaoi Masumoto & Yin, 1994: 169
yunnanensis Masumoto & Yin, 1994: 167
dampaensis Kaszab, 1975b: 125 **A**: NP
daxueica Schawaller, 2001a: 13 **A**: SCH
denticollis Schuster, 1926a: 35 **A**: KA UP
denticrus Fairmaire, 1896b: 97 **A**: HP
dentipennis Reitter, 1908k: 279
dentipes Schuster, 1935d: 458 **A**: NP SD
adriani Schawaller, 2001a: 34 [RN]
flavicincta Kaszab, 1938e: 228
schusteri Kaszab, 1938e: 228 [HN]
dentitibia Reitter, 1901b: 62 **A**: UZ (Bukhara)
denudata Kaszab, 1975e: 330 **A**: BT
deplanata Weise, 1878: 240 **A**: TR
deqenica Schawaller, 2001a: 13 **A**: YUN
dhorpatanica Kaszab, 1977b: 267 **A**: NP

diancangica Schawaller, 2001a: 14 **A**: YUN
dickorei Schawaller, 2001a: 14 **A**: XIZ
dilutella Solsky, 1881b: 55 **A**: AF KI KZ TD TM UZ
dumialensis Nakane, 1966: 244 **A**: PA
edmundi Schuster, 1915c: 359 **A**: KA
emeishana Masumoto, 1996d: 175 **A**: SCH
espagnoli Kaszab, 1965a: 128 **A**: SD
ezimai Masumoto & Akita, 2001a: 15 **A**: JA
fangjingshanana Ren & Hua, 2006: 267 **A**: GUI
fengileana Masumoto, 1996d: 180 **A**: SCH SHA
ferruginea Küster, 1846b: 68 **E**: AL BH BU CR GR YU **A**: IS LE TR
krueperi Reitter, 1881a: 222
marthae Reitter, 1881a: 224
minima Motschulsky, 1858b: 112
parvula Reitter, 1885e: 402
villosula Motschulsky, 1851a: 594
formaneki Schuster, 1916: 561 **A**: XIZ
tibetana Schuster, 1916: 563
franzi Kaszab, 1973a: 44 **A**: NP
franziana Kaszab, 1973a: 61 **A**: NP
freudei Kaszab, 1961d: 228 **A**: NP
fulunga Kaszab, 1973a: 63 **A**: NP
strigosa Kaszab, 1973a: 62
gairibasensis Masumoto, 1990f: 177 **A**: NP SD
ganglbaueri Reitter, 1891b: 33 **E**: BU TR **A**: TR
turcica Reitter, 1900h: 282
ganzica Schawaller, 2001a: 17 **A**: SCH
gardneri Schuster, 1935d: 454 **A**: UP
gebieni Reitter, 1906h: 443 **A**: KA
kaschmirensis Reitter, 1906h: 443
gigantea Schuster, 1940b: 864 **A**: SCH
glabriuscula J. R. Sahlberg, 1908: 87 **A**: LE
goetzi Kaszab, 1970c: 427 **A**: NP
gracilis Schuster, 1926a: 46 **A**: XIZ
graeca Weise, 1878: 231 **E**: BU GR TR
grandis Schuster, 1935d: 465 **A**: UP
gyamdaica Schawaller, 2001a: 18 **A**: XIZ
habashanica Schawaller, 2001a: 19 **A**: YUN
haigouica Schawaller, 2001a: 19 **A**: SCH
hartmanni Schawaller, 2002: 43 **A**: NP
hauseri Reitter, 1906h: 445 **A**: TD (Hissar Alai)
heinzi Schawaller, 2001a: 20 **A**: XIZ
hengduanica Schawaller, 2001a: 20 **A**: SCH YUN
herbertfranzi Kaszab, 1973a: 60 **A**: NP
augur Kaszab, 1973a: 59
magarkensis Masumoto, 1990f: 188
schusteriana Kaszab, 1973a: 58
heydeni Weise, 1880a: 102 **E**: BU GR
himalayana Schuster, 1915c: 357 **A**: HP
hingstoni Schuster, 1926a: 42 **A**: XIZ
hiranoi Masumoto & Akita, 2001a: 16 **A**: JA (Ryukyus)
hirtella Solsky, 1881b: 54 **A**: KI
auliensis Reitter, 1902c: 189
hirtipes Reitter, 1881a: 222 **E**: AB **A**: IN
hopffgarteni Weise, 1878: 230 **E**: BH BU CR MC RO YU

holdhausi Schuster, 1916: 622 **A**: KZ (Transili Alatau)
houzhenzica Schawaller, 2001a: 22 **A**: SHA
hubeica Schawaller, 2001a: 23 **A**: HUB
incomperta Kaszab, 1973a: 50 **A**: NP
indica Fairmaire, 1896b: 96 **A**: HP
insularis Kaszab, 1964a: 5 **A**: JA (Ryukyus)
interrupta Schuster, 1935d: 460 **A**: SD
irregularis Schuster, 1935d: 464 **A**: UP
jalaorana Reitter, 1908k: 278 **A**: KA
jizushana Masumoto, 1996d: 168 **A**: YUN
jumlana Kaszab, 1975b: 123 **A**: NP
justinae Reitter, 1887e: 523 **E**: GG
kaliensis Schuster, 1935d: 445 **A**: UP
kangchendzongensis Masumoto, 1990f: 182 **A**: NP SD
kangdingica Schawaller, 2001a: 24 **A**: SCH
karakorumensis Kaszab, 1961f: 354 **A**: PA (Karakoram)
kaszabi Masumoto, 1990f: 181 **A**: NP SD
kaufmanni Reitter, 1881a: 222 **E**: AL BH CR YU **A**: TR
khumbuana Kaszab, 1977b: 266 **A**: NP
korbi Heyden, 1890d: 374 **A**: TR
kraatzii Weise, 1878: 236 **E**: TR
  alternata Reitter, 1885e: 401
kubani Schawaller, 2001a: 25 **A**: YUN
kuluana Reitter, 1908k: 278 **A**: HP KA PA
lacordairei Marseul, 1876c: 133 **A**: UP
laevigata Schuster, 1926a: 48 **A**: UP
laevipennis Schuster, 1926a: 45 **A**: SD
  sandakphuensis Masumoto, 1990f: 189
lamjurensis Masumoto, 1990f: 180 **A**: NP
langmusica Schawaller, 2001a: 25 **A**: SCH SHA
lebedevi Roubal, 1929c: 193 **A**: KZ
leonhardi Schuster, 1916: 612 **A**: KI XIN
  spaethi Schuster, 1916: 613
lederi Weise, 1878: 229 **E**: GG
libanotica Reitter, 1885e: 401 **A**: LE
lilliputana Kaszab, 1968e: 462 **A**: TR
limbu Schawaller, 2002: 44 **A**: NP
lindbergi Kaszab, 1959c: 55 **A**: TR
longipilis Schuster, 1926a: 52 **A**: NP SD
  nepalensis Jedlička, 1965: 98 (*Ebertius*)
longula Marseul, 1876c: 133 **A**: CY IS LE SY
loricera Kaszab, 1973a: 54 **A**: NP
ludingica Schawaller, 2001a: 26 **A**: SCH
luguica Schawaller, 2001a: 27 **A**: SCH
luhuoica Schawaller, 2001a: 27 **A**: SCH
luprops Kaszab, 1973a: 57 **A**: NP
magar Schawaller, 2002: 45 **A**: NP
manang Schawaller, 2002: 46 **A**: NP
martensi Kaszab, 1973a: 47 **A**: NP
  dedita Kaszab, 1973a: 48
  zurstrasseni Kaszab, 1977b: 263
martinhauseri Schawaller, 2002: 46 **A**: NP
merditana Schuster, 1916: 505 **E**: AL
merkli Weise, 1883: 313 **E**: TR **A**: TR
merklottoi Masumoto, 1990f: 184 **A**: NP SD

minuta Fairmaire, 1896b: 97 **A**: HP
mirabilis Kaszab, 1970b: 121 **A**: SCH
mulica Schawaller, 2001a: 28 **A**: SCH
newar Schawaller, 2002: 47 **A**: NP
nigritissima Reitter, 1906h: 444 **A**: KA
nishikawai Masumoto, 1990f: 176 **A**: SD
nitida Schuster, 1935d: 453 **A**: KA
nyingchica Schawaller, 2001a: 30 **A**: XIZ
obscuricornis Reitter, 1900h: 283 **A**: TR
ocys Kaszab, 1973a: 54 **A**: NP
oedipus Kaszab, 1977b: 261 **A**: NP
oertzeni Reitter, 1885e: 401 **E**: GR (Kríti)
opaca Kaszab, 1970b: 120 **A**: NP SD
  bikhebhanjangensis Masumoto, 1990f: 178
opacicollis Kaszab, 1970c: 425 **A**: NP
  bembidion Kaszab, 1973a: 50
orbicollis Schuster, 1926a: 43 **A**: UP
osmanlis Reitter, 1906h: 447 **E**: TR **A**: TR
ovipennis Schuster, 1926a: 41 **A**: CH "near Hong Kong"
pahakhola Schawaller, 2002: 47 **A**: NP
pakistanica Kaszab, 1962c: 316 **A**: PA
paomaica Schawaller, 2001a: 30 **A**: SCH
parallelocollis Schuster, 1926a: 40 **A**: XIZ
parateneta Kaszab, 1973a: 57 **A**: NP
parkeri Schuster, 1935d: 446 **A**: UT
parvidens Reitter, 1890i: 373 **A**: TR
phithangensis Masumoto, 1990f: 175 **A**: SD
  shunichii Masumoto, 1990f: 187
pici Reitter, 1900c: 87 **A**: TR
piligera Weise, 1878: 235 **E**: AR
  longicollis Weise, 1878: 237
pilosissima Reitter, 1906h: 448 **E**: RO
planipennis Schuster, 1926a: 50 **A**: UP
pokharana Kaszab, 1973a: 64 **A**: NP
  silvicola Kaszab, 1973a: 63
prehimalayaica Kaszab, 1977b: 269 **A**: NP
  moodlungensis Masumoto, 1990f: 193
pseudofranzi Kaszab, 1975b: 124 **A**: NP
puetzi Schawaller, 2001a: 31 **A**: SCH
pulchella Fischer von Waldheim, 1823: 201 **E**: BU
  pubella Solier, 1848: 190
  rubiginosa Solier, 1848: 190
punctatissima Schuster, 1926a: 44 **A**: UP
puncticollis Schuster, 1935d: 463 **A**: UP
punctiventris Schuster, 1926a: 51 **A**: KA
qinlingica Schawaller, 2001a: 32 **A**: SCH SHA
quadricollis Weise, 1878: 232 **E**: "Caucasus"
rai Schawaller, 2002: 48 **A**: NP
reitteri Weise, 1877: 27 **E**: AU BH BU CR HU PL RO SK SL UK
  ormayi Reitter, 1887e: 523
rhododendri Kaszab, 1977b: 269 **A**: NP
robusta Reitter, 1885e: 401 **A**: UZ (Samarkand)
rosti Reitter, 1906h: 444 **A**: HP KA
rotundicollis Marseul, 1876c: 134 **A**: JA
rubripes Reitter, 1908k: 279 **A**: KA

rugosa Schuster, 1916: 553 **A**: HP KA
safraneki Schawaller, 2001a: 33 **A**: XIZ
sagarmatha Schawaller, 2002: 49 **A**: NP
sakaii Masumoto, 1990f: 188 **A**: SD
schmidti Schawaller, 2002: 50 **A**: NP
schulkei Schawaller, 2001a: 33 **A**: SCH
schusteri Heller, 1923: 70 (*Psilolaena*) **A**: SCH
schwarzi Reitter, 1885e: 402 **E**: AL GR YU
shaluica Schawaller, 2001a: 35 **A**: SCH
sherpa Schawaller, 2002: 50 **A**: NP
simillima Schuster, 1935d: 462 **A**: KA
similis Schuster, 1926a: 37 **A**: UP
transversicollis Schuster, 1926a: 37
singalilensis Masumoto, 1990f: 180 **A**: SD
smetanai Schawaller, 2001a: 35 **A**: SCH
sparsepunctata Kaszab, 1970c: 428 **A**: NP
starcki Reitter, 1887e: 523 **E**: ST (Kavkaz)
striolata Reitter, 1900h: 285 **A**: TR
studiosa Kaszab, 1973a: 46 **A**: NP
subalpina Kaszab, 1977b: 262 **A**: NP
subcoeca Kaszab, 1973a: 52 **A**: NP
sulcata Schuster, 1915c: 360 **A**: KA
syriaca Baudi di Selve, 1876c: 243 **A**: CY SY TR
tabanai Masumoto, 1998b: 316 **A**: SCH
tachysoides Kaszab, 1973a: 55 **A**: NP
takara Mt. Chûjô, 1996: 136 **A**: JA (Ryukyus)
takolana Kaszab, 1973a: 50 **A**: NP
tamang Schawaller, 2002: 51 **A**: NP
tamur Schawaller, 2002: 57 **A**: NP
thameogensis Masumoto, 1990f: 183 **A**: NP
theana Reitter, 1900h: 283 **A**: TR
thodunga Kaszab, 1973a: 60 **A**: NP
tibialis Schuster, 1926a: 39 **A**: "India"
tryznai Schawaller, 2001a: 36 **A**: SCH XIZ
tuntalica Schawaller, 2001a: 37 **A**: SCH XIZ
turkestanica Reitter, 1897e: 221 **A**: KI UZ (Margilan)
turnai Schawaller, 2001a: 38 **A**: SCH
viennensis J. Sturm, 1807: 180 (*Scaurus*) **E**: AU BH CR CZ HU IT SK SL
pimelia Duftschmid, 1812: 289 (*Scaurus*)
villosa Schuster, 1915c: 357 **A**: KA UT
vishnua Masumoto, 1990f: 186 **A**: SD
kephakensis Masumoto, 1990f: 193
wanensis Schuster, 1940b: 863 **A**: TR (Van)
watanabei Masumoto & Yin, 1993: 240 **A**: YUN
weisei Reitter, 1878a: 238 **E**: GR TR
wittmeri Kaszab, 1973a: 72 **E**: GR (Kríti)
wolongica Schawaller, 2001a: 39 **A**: SCH
wrasei Schawaller, 2001a: 39 **A**: SCH
xuerenensis Masumoto, 1996d: 172 **A**: YUN
yajiangica Schawaller, 2001a: 40 **A**: SCH
yasuakii Masumoto, 1996d: 166 **A**: YUN
yodai Kaszab, 1970c: 430 **A**: NP
yufengsi Masumoto, 1996d: 167 **A**: YUN
yulongica Schawaller, 2001a: 42 **A**: YUN

yuzhuensis Masumoto & Yin, 1994: 165 **A**: YUN
zogqenica Schawaller, 2001a: 43 **A**: SCH

genus *Nepalolaena* Schawaller, 2001d: 277 type species *Nepalolaena kira* Schawaller, 2001
kira Schawaller, 2001d: 278 **A**: NP

tribe Lagriini Latreille, 1825
subtribe Lagriina Latreille, 1825

genus *Adynata* Fåhraeus, 1870: 330 type species *Adynata tricolor* Fåhraeus, 1870
Microlagria Seidlitz, 1898b: 336 type species *Lagria poupillieri* Reiche, 1864
poupillieri Reiche, 1864: 242 (*Lagria*) **N**: AG MO

genus *Aulonogria* Borchmann, 1929: 9 type species *Lagria rugosa* Fabricius, 1801
chinensis Borchmann, 1936: 149 **A**: HAI
concolor Blanchard, 1853: 184 (*Lagria*) **A**: NP PA **ORR**

genus *Bothynogria* Borchmann, 1915a: 128 type species *Bothynogria calcarata* Borchmann, 1915
bicolor L. Redtenbacher, 1844: 534 (*Lagria*) **A**: NP SD UP
bhutanica Merkl, 1990: 292 **A**: BT
calcarata Borchmann, 1915a: 129 **A**: FUJ GUI HUB HUN JIX SCH TAI YUN **ORR**
himalayana Borchmann, 1936: 116 **A**: NP UP
ruficollis Hope, 1831: 32 (*Lagria*) **A**: NP

genus *Cerogria* Borchmann, 1909a: 210 type species *Lagria anisocera* Wiedemann, 1823
Aeschrocera Chen & Chou, 1996: 265 type species *Cerogria brunneocollis* Chen & Chou, 1996
Cerogriodes Borchmann, 1941: 25 type species *Cerogria klapperichi* Borchmann, 1941
abnormicornis Borchmann, 1942a: 8 **A**: SD YUN **ORR**
anisocera Wiedemann, 1823: 81 (*Lagria*) **A**: SCH TAI YUN **ORR**
corporaali Borchmann, 1929: 8
basalis Hope, 1831: 32 (*Lagria*) **A**: BH NP SD YUN **ORR**
dichroa Reitter, 1880b: 257 (*Lagria*)
birmana Borchmann, 1942a: 7 **A**: YUN **ORR**
castaneipennis Borchmann, 1936: 124 **A**: SCH YUN **ORR**
chinensis Fairmaire, 1886c: 349 (*Lagria*) **A**: HUB FUJ SCH TAI XIZ YUN
crassa Borchmann, 1911: 221 **A**: NP SD **ORR**
distincticornis Reitter, 1880b: 256 (*Lagria*) **A**: SD
diversicornis Pic, 1933b: 9 **A**: GAN SCH
brunneocollis Chen & Chou, 1996: 266
harpacorna Chen & Chou, 1996: 267
flavicornis Borchmann, 1911: 214 **A**: YUN **ORR**
glabricollis Chen & Xia, 2001: 5 (*Schevodera*)
forteimpressa Pic, 1955c: 31 **A**: FUJ
hajimei Masumoto, 1988: 37 **A**: TAI
hauseri Borchmann, 1936: 125 **A**: YUN
janthinipennis Fairmaire, 1886c: 349 (*Lagria*) **A**: FUJ GUI JIX HUN HUB SC SCH SHA ZHE
antennata Borchmann, 1909b: 714 (*Lagria*) [RN]
distincticornis Heyden, 1887a: 269 (*Lagria*) [HN]
kikuchii Kôno, 1929a: 28 (*Lagria*) **A**: FUJ SCH TAI
praecox Borchmann, 1941: 24
klapperichi Borchmann, 1941: 25 **A**: FUJ JIX SHG TAI ZHE
miwai Kôno, 1930a: 18 **A**: FUJ JIX SCH TAI YUN
gardneri Borchmann, 1936: 127
montana Merkl, 1991a: 10 **A**: BH NP SD
nepalensis Hope, 1831: 32 (*Lagria*) **A**: BT NP SD UP
aenea L. Redtenbacher, 1844: 553 (*Lagria*)

nigrita Borchmann, 1915b: 180 (*Lagria*) **A**: NP
nigrosparsa Pic, 1928g: 248 (*Lagria*) **A**: YUN **ORR**
odontocera Fairmaire, 1886c: 348 (*Lagria*) **A**: SCH SHA YUN
pachycera Fairmaire, 1886c: 347 (*Lagria*) **A**: JIX SCH
popularis Borchmann, 1936: 121 **A**: FUJ GUI GUX HEN SCH SHN YUN
quadrimaculata Hope, 1831: 32 (*Lagria*) **A**: FUJ GAN GUX HP HUB NP PA SCH SD UP YUN **ORR**
cardonii Fairmaire, 1894a: 31 (*Lagria*)
fukienensis Borchmann, 1941: 24
variabilis L. Redtenbacher, 1844: 534 (*Lagria*)
sikkimensis Borchmann, 1936: 128 **A**: HP NP PA SD UP

genus *Chrysolagria* Seidlitz, 1898b: 339 type species *Lagria viridipennis* Fabricius, 1798
viridipennis Fabricius, 1798: 118 (*Lagria*) **E**: SP **N**: AG MO TU

genus *Lagria* Fabricius, 1775: 124 type species *Chrysomela hirta* Linnaeus, 1758
Lachna Billberg, 1820: 35 type species *Chrysomela hirta* Linnaeus, 1758
subgenus *Apteronympha* Seidlitz, 1898b: 336 type species *Lagria rubida* Graells, 1855
rubida Graells, 1858: 74 **E**: SP
elliptica Fairmaire, 1880c: 242
parvula Perris, 1864: 291
rugosula Rosenhauer, 1856: 223 **E**: FR IT PT SP SZ **N**: MO
glabrata A. G. Olivier, 1797: 446 [HN]
melichari Reitter, 1890e: 173
pubescens Latreille, 1804a: 351 [HN]
tristis Bonelli, 1812: 172 [HN]
tenenbaumi Pic, 1929a: 2 **A**: IS
subgenus *Lagria* Fabricius, 1775: 124 type species *Chrysomela hirta* Linnaeus, 1758
atripes Mulsant & Guillebeau, 1855a: 74 **E**: AB AR AU BE BH BU CR CT CZ FR GE GG GR HU IT MC MD PL PT RO SK SL SP ST SZ TR UK YU **A**: IN TM TR
puncticollis Seidlitz, 1898b: 337
tenuicollis Seidlitz, 1898b: 338 [HN]
babai Nakane, 1994b: 243 **A**: JA
brevipilis Desbrochers des Loges, 1881: 143 **N**: AG
carinulata Fairmaire, 1891f: ccxvii **A**: HUB
comosella Fairmaire, 1894a: 30 **A**: NP SD **ORR**
conspersa Reitter, 1880b: 256 **A**: GUX NP SD XIZ YUN
multipunctata Fairmaire, 1894a: 31
formosensis Borchmann, 1912a: 6 **A**: TAI
grenieri C. Brisout de Barneville, 1867: 181 **E**: FR PT SP
hirta Linnaeus, 1758: 377 (*Chrysomela*) **E**: AB AL AN AR AU BE BH BU BY CR CT CZ DE EN ES FI FR GB GE GG GR HU IR IS IT KZ LA LS LT LU MC MD NL NR NT PL PT RO SK SL SP ST SV SZ TR UK YU **A**: ES CY IN IQ KZ TD TM TR UZ WS
caucasica Motschulsky, 1860c: 144
depilis Mulsant, 18756: 21
flava Geoffroy, 1758: 155 (*Cantharis*)
fuscata Motschulsky, 1860: 144
glabrata Fabricius, 1775: 125
limbata Desbrochers des Loges, 1881: 142
lurida Krynicki, 1832: 137
nudipennis Mulsant, 1856a: 20
pontica Motschulsky, 1860c: 144
seminuda Reitter, 1889e: 259
spadicea Scopoli, 1763: 43 (*Cantharis*)
villosa DeGeer, 1775: 44 (*Tenebrio*)
indicola Bates, 1879: 483 **A**: HP NP PA
costatula Borchmann, 1909b: 714 [RN]
subcostata Fairmaire, 1896a: 59 [HN]

klapperichi Pic, 1955c: 32 **A**: FUJ
semiobscurior Pic, 1955c: 32
kondoi Masumoto, 1988: 35 **A**: TAI
lameyi Fairmaire, 1893g: 325 **A**: FUJ GUI HUB HUN JIX NMO SCH **ORR**
annulipes PIC, 1955c: 32 [HN]
geniculata Seidlitz, 1898b: 339
lata Fabricius, 1801b: 70 **E**: SP **N**: AG MO
rugata Desbrochers des Loges, 1881: 145
laticollis Motschulsky, 1860c: 143 **A**: ES MO WS FE
moutoni Pic, 1928d: 4 **A**: SHG
notabilis Lewis, 1896: 341 **A**: JA (Ryukyus)
nigricollis Hope, 1843: 63 **A**: FE FUJ HEN HUB HUN JA JIX LIA NC SC SCH XIN ZHE
picea Brancsik, 1914: 58
subtilipunctata Seidlitz, 1898b: 340
oharai Masumoto, 1988: 34 **A**: TAI
okinawana Mt. Chûjô, 1985: 88 (*Cerogria*) **A**: JA (Ryukyus)
ophthalmica Fairmaire, 1891f: ccxvi **A**: HUB
pallidipennis Borchmann, 1936: 33 **A**: GUI FUJ SHN
paracomosella Merkl, 1991a: 6 **A**: NP
pici Normand, 1936a: 161 **N**: TU
picta Borchmann, 1911: 203 **A**: HKG YUN **ORR**
coadunata Borchmann, 1932: 96
rubella Borchmann, 1932: 95 **A**: SCH XIZ YUN
rubiginea Fairmaire, 1886c: 347 **A**: CH
rufipennis Marseul, 1876d: 337 **A**: BEI FE HUB GAN JA JIX NC NIN SC SCH SHA
vervex Marseul, 1876d: 338
rufiventris Pic, 1928d: 4 **A**: XIZ
schawalleri Merkl, 1991a: 8 **A**: NP SD UP
scutellaris Pic, 1910v: 259 **A**: TAI
sakaii Masumoto, 1988: 36
theryi Pic, 1938d: 14 **N**: MO
tibetana Blair, 1927: 249 **A**: NP XIZ
tigrina Fairmaire, 1893g: 325 **A**: GUI GUX YUN **ORR**
tristicula Fairmaire, 1886c: 346 **A**: CH
ventralis Reitter, 1880b: 255 **A**: GUI HP SD NP UP YUN **ORR**
villosa Fabricius, 1781: 160 **A**: YE **AFR NTR**i
confusa Reiche, 1850: 374
obscura Fabricius, 1792b: 79

genus *Neogria* Borchmann, 1911: 222 type species *Neogria sulcipennis* Borchmann, 1911
cyanipennis Borchmann, 1911: 226 **A**: NP **ORR**

genus *Schevodera* Borchmann, 1936: 59 type species *Lagria hirticollis* Borchmann, 1911
gracilicornis Borchmann, 1911: 204 (*Lagria*) **A**: YUN **ORR**
inflata Borchmann, 1925: 348 (*Lagria*) **A**: YUN **ORR**

genus *Xanthalia* Fairmaire, 1894e: 395 [RN] type species *Xanthia curticollis* Fairmaire, 1893
Diversogria Pic, 1954c: 229 type species *Heterogria lepersonnei* Pic, 1954
Heterogria Fairmaire, 1896a: 42 type species *Heterogria punctatissima* Fairmaire, 1896
Lagriocera Fairmaire, 1896a: 41 type species *Lagriocera cavicornis* Fairmaire, 1896
Lagriodes Borchmann, 1930d: 432 type species *Heterogria armigera* Borchmann, 1930
Pachylagria Borchmann, 1912b: 17 type species *Pachylagria ovata* Borchmann, 1912
Xanthia Fairmaire, 1893b: 31 [HN] type species *Xanthia curticollis* Fairmaire, 1893
Wallardilagria Pic, 1910q: 74 type species *Heterogria pallidicolor* Pic, 1910
atricolor Pic, 1920f: 15 (*Heterogria*) **A**: SD
cavicornis Fairmaire, 1896a: 41 (*Lagriocera*) **A**: NP **ORR**

clavata Merkl, 1991a: 13 **A**: NP
franzi Kaszab, 1973a: 40 **A**: NP
nepalica Kaszab, 1973a: 42 **A**: NP
nigrovittata Pic, 1910q: 75 (*Lagriocera*) **A**: NP SCH SHA YUN **ORR**
brunneovittata Pic, 1918d: 15 (*Heterogria*)
martensi Merkl, 1991a: 12
reducta Pic, 1955a: 2 (*Heterogria*) **A**: FUJ
obliterata Pic, 1955a: 3 (*Heterogria*)
serrifera Borchmann, 1930d: 440 (*Heterogria*) **A**: FUJ NC SCH
sinensis Pic, 1918d: 15 (*Heterogria*) **A**: CH
spinosa Kaszab, 1965d: 289 **A**: FUJ **ORR**

genus *Xenocerogria* Merkl, 2007: 269 [RN] type species *Lagriocera feai* Borchmann, 1911
Xenocera Borchmann, 1936: 116 [HN] type species *Lagriocera feai* Borchmann, 1911
feai Borchmann, 1911: 209 (*Lagriocera*) **A**: YUN **ORR**
ignota Borchmann, 1941: 22 (*Xenocera*) **A**: FUJ **ORR**
ruficollis Borchmann, 1912a: 7 (*Lagriocera*) **A**: FUJ TAI

subtribe Statirina J. L. LeConte, 1862

genus *Anisostira* Borchmann, 1915c: 296 type species *Anisostira varicolor* Borchmann, 1915
elegans M. Chûjô, 1959: 8 **A**: JA (Ryukyus)
rugipennis Lewis, 1896: 341 (*Macrolagria*) **A**: FUJ GUA GUI HUB JA (Ryukyus) SC SCH TAI ZHE
abdominalis Kôno, 1929a: 33
abnormipes Borchmann, 1912a: 10 (*Nemostira*)
cognata Borchmann, 1912a: 11 (*Nemostira*)
lucidicollis Borchmann, 1915c: 298
ferriei Pic, 1914r: 304 (*Nemostira*)
flavipes Kôno, 1929a: 33
kikuchii Kôno, 1929a: 34
nigripes Pic, 1911q: 190 (*Nemostira*)
rufoscutellaris Pic, 1914n: 76 (*Nemostira*)
similaris Borchmann, 1915c: 297
sinuatipes Pic, 1911b: 7 (*Nemostira*)
testaceithorax Pic, 1914r: 305 (*Nemostira*)
varicolor Borchmann, 1915c: 298

genus *Arthromacra* Kirby, 1837: 238 type species *Arthromacra donacioides* Kirby, 1837
Macrarthra Agassiz, 1846b: 219 [unjustified emendation]
amamiana Nakane, 1963a: 30 **A**: JA (Ryukyus)
bicolor Saitô, 1995: 178 **A**: TAI
chifengi Merkl, 2004: 300 **A**: TAI
cuprina Borchmann, 1942a: 11 **A**: YUN **ORR**
decora Marseul, 1876d: 339 (*Lagria*) **A**: JA
apicalis Kôno, 1929a: 31
bicoloripes Pic, 1917f: 16
obscuripes Pic, 1917f: 16
donckieri Pic, 1910s: 86 **A**: YUN
entypodera Chen & X. Yang, 1997: 2 **A**: FUJ
flavipes Nakane, 1994a: 85 **A**: JA
formosana Sasaji, 1986: 11 **A**: TAI
gracilis Borchmann, 1915a: 133 **A**: FUJ
higoniae Lewis, 1895a: 278 **A**: JA
imasakai Saitô, 1995: 180 **A**: TAI
kinodai Nakane, 1997: 5 **A**: JA
kyushuensis Nakane, 1997: 4 **A**: JA
majuscula Nakane, 1994a: 84 **A**: JA

minuta Masumoto, 1988: 39 **A**: TAI
narukawai Imasaka, 2005: 91 **A**: JA (Honshu)
okiana Imasaka, 2005: 100 **A**: JA (Oki Is.)
oodaigahara Imasaka, 2005: 80 **A**: JA (Honshu)
pretiosa Reiche, 1862c: 544 **E**: GG ST (Kavkaz) **A**: TR
rubidorsalis Chen & X. Yang, 1997: 3 **A**: ZHE
shiraishii Imasaka, 2005: 97 **A**: JA (Shikoku)
subopaca Pic, 1910s: 86 (*Chlorophila*) **A**: YUN
sumptuosa Lewis, 1895a: 277 **A**: JA
takahashii Imasaka, 2005: 81 **A**: JA (Honshu)
tibetana Chen & X. Yang, 1997: 3 **A**: XIZ
tsuyukii Masumoto, 1988: 38 **A**: TAI
viridissima Lewis, 1895a: 277 **A**: JA
yakushimana Imasaka, 2005: 93 **A**: JA (Yakushima)

genus *Arunogria* Merkl, 1991a: 14 type species *Arunogria pubescens* Merkl, 1991
pubescens Merkl, 1991a: 17 **A**: NP

genus *Casnonidea* Fairmaire, 1882c: 264 type species *Casnonidea holomelaena* Fairmaire, 1882
Synatractus W. J. Macleay, 1887: 312 type species *Synatractus variabilis* W. J. Macleay, 1887
occipitalis Borchmann, 1912a: 8 (*Nemostira*) **A**: JA (Ryukyus) TAI
patruelis Borchmann, 1936: 384 **A**: SD
rufipennis Borchmann, 1912a: 9 (*Nemostira*) **A**: FUJ TAI
terminata Fairmaire, 1889a: 365 **A**: HAI **ORR**
tibialis Fairmaire, 1893f: 301 **A**: YUN
viridimetallica Pic, 1911b: 7 (*Nemostira*) **A**: TAI

genus *Chlorophila* Semenov, 1891a: 374 type species *Lagria* (*Chlorophila*) *portschinskii* Semenov, 1891
campestris Fairmaire, 1894b: 221 **A**: SCH YUN
cyanea Pic, 1915g: 15 **A**: FUJ
davidi Fairmaire, 1898d: 399 **A**: SCH
portschinskii Semenov, 1891a: 374 **A**: FUJ GAN NIN SCH SD SHA YUN
basipennis Fairmaire, 1897j: 229
carolina Fairmaire, 1899: 632
immarginata Fairmaire, 1897j: 231
ingeniculata Pic, 1917f: 15
nitidicollis Fairmaire, 1897j: 230
obscuripennis Fairmaire, 1894b: 220
semenowi Fairmaire, 1899: 633 **A**: FUJ HUB

genus *Donaciolagria* Pic, 1914b: 14 type species *Donaciolagria impressipennis* Pic, 1914
admirabilis Sasaji, 1986: 9 (*Chlorophila*) **A**: TAI
femoralis Borchmann, 1915a: 135 (*Arthromacra*) **A**: GUI FUJ SCH
flavitibialis Chen, 1997a: 125 **A**: SCH
kurosawai Masumoto, 1988: 40 (*Chlorophila*) **A**: TAI
pici Borchmann, 1936: 179 **A**: NP SD
sikkimensis Borchmann, 1915a: 132 **A**: SD
viridimetallica Chen, 1997a: 124 **A**: ANH SCH

genus *Exostira* Borchmann, 1925: 353 type species *Exostira sellata* Borchmann, 1925
punctator Borchmann, 1930d: 523 **A**: TAI
schroederi Borchmann, 1936: 421 **A**: FUJ GUA GUI JIX TAI
genus *Hosohamudama* Masumoto, 1988: 43 type species *Hosohamudama sasajii* Masumoto, 1988
abnormalis Kôno, 1929a: 31 (*Arthromacra*) **A**: TAI

miyakei Masumoto, 1988: 45 **A**: TAI
sasajii Masumoto, 1988: 44 **A**: TAI

genus *Lagriogonia* Fairmaire, 1891: ccxvii type species *Lagriogonia humerosa* Fairmaire, 1891
humerosa Fairmaire, 1891: ccxvii **A**: HUB
genus *Macrolagria* Lewis, 1895a: 422 type species *Statyra rufobrunnea* Marseul, 1876
denticollis Fairmaire, 1891: ccxviii (*Casnonidea*) **A**: FUJ HUB ZHE
fujisana Lewis, 1895a: 422 **A**: JA
hirsuta Lewis, 1895a: 423 **A**: FE (Kuril Is.) JA
robusticeps Lewis, 1895a: 276 (*Eutrapela*) **A**: FE (Kuril Is.) JA
rufobrunnea Marseul, 1876d: 340 (*Statyra*) **A**: JA
shibatai Masumoto, 1987a: 51 **A**: JA

genus *Merklia* Chen, 1997b: 307 type species *Merklia bimaculata* Chen, 1997
bimaculata Chen, 1997b: 307 **A**: YUN **ORR**

genus *Mimoborchmania* Pic, 1934e: 31 type species *Nemostira coloripes* Pic, 1922
Mimoborchmannia Borchmann, 1936: 490 type species *Nemostira coloripes* Pic, 1922
yangi Merkl & Chen, 1997: 113 **A**: YUN **ORR**

genus *Odontocerostira* Merkl, 2007: 269 [RN] type species *Odontocera quinlingensis* Chen & Yuan, 1996
Odontocera Chen & Yuan, 1996: 183 [HN] type species *Odontocera quinlingensis* Chen & Yuan, 1996
qinlingensis Chen & Yuan, 1996: 184 (*Odontocera*) **A**: SHA TAI
hexamaculata Chen & Yuan, 1996: 185 (*Odontocera*) **A**: SHA

genus *Pachystira* Chen, 1997b: 308 type species *Pachystira impressipennis* Chen, 1997
impressipennis Chen, 1997b: 309 **A**: YUN

genus *Splichalia* Reitter, 1913d: 664 type species *Splichalia tigrinella* Reitter, 1913
andrewesi Borchmann, 1936: 340 **A**: KA
nigrofasciata Pic, 1912l: 63 (*Nemostira*) **A**: KA PA UP
tigrinella Reitter, 1913d: 665 **A**: KA

genus *Sora* Walker, 1859a: 259 type species *Sora marginata* Walker, 1859
Nemostira Fairmaire, 1869a: 815 type species *Nemostira coquerelii* Fairmaire, 1868
aeneipennis Pic, 1927f: 5 (*Nemostira*) **A**: CH
angulicollis Pic, 1920f: 15 (*Nemostira*) **A**: CH
cavaleriei Pic, 1937h: 148 **A**: GUI
crenulipennis Fairmaire, 1896b: 120 (*Nemostira*) **A**: HP
inopaca Pic, 1923c: 16 (*Nemostira*) **A**: CH
humeralis Pic, 1907m: 182 (*Nemostira*) **A**: SCH
klapperichi Pic, 1955c: 31 **A**: FUJ
longissima Pic, 1920f: 15 (*Nemostira*) **A**: CH
mimica Pic, 1912f: 31 (*Nemostira*) **A**: YUN **ORR**
rugulicollis Fairmaire, 1899: 633 (*Nemostira*) **A**: CH
sinensis Pic, 1912f: 30 (*Nemostira*) **A**: YUN
subcordicollis Pic, 1937f: 35 **A**: CH
subrecticollis Pic, 1914h: 44 (*Nemostira*) **A**: HUB
thibetana Pic, 1914r: 304 (*Nemostira*) **A**: XIZ

genus *Strongylagria* Pic, 1915c: 4 type species *Strongylagria metallica* Pic, 1915
metallica Pic, 1915c: 5 **A**: XIZ
genus *Taiwanolagria* Masumoto, 1988: 41 type species *Taiwanolagria merkli* Masumoto, 1988
merkli Masumoto, 1988: 41 **A**: FUJ GUI TAI

tribe Lupropini Ardoin, 1958

genus *Bolitrium* Gebien, 1914d: 390 type species *Bolitrium crenulicolle* Gebien, 1914
crenulicolle Gebien, 1914d: 391 **A**: TAI **ORR**
chinensis Kaszab, 1940c: 157 (*Lorelus*)
sasajii Masumoto & Akita, 2001c: 248 (*Lorelus*) **A**: JA (Ryukyus)

genus *Falsotithassa* Pic, 1934a: 18 type species *Falsotithassa sumatrana* Pic, 1934
sumatrana Pic, 1934a: 19 **A**: BT GUX NP **ORR**

genus *Indenicmosoma* Ardoin, 1964: 688 type species *Enicmosoma indochinensis* Kaszab, 1940
ardoini Kaszab, 1970a: 176 **A**: AF
indochinensis Kaszab, 1940c: 158 (*Enicmosoma*) **A**: NP **ORR**
nathani Ardoin, 1964: 689 **A**: BT NP **ORR**
pocsi Kaszab, 1965d: 291 **A**: NP **ORR**

genus *Luprops* Hope, 1833: 63 type species *Luprops chrysophthalmus* Hope, 1833
Etazeta Fairmaire, 1889a: 358 type species *Etazeta aeneicolor* Fairmaire, 1888
Oligorus Dejean, 1834: 206 type species *Tagenia indica* Wiedemann, 1823
Syggona Fåhraeus, 1870: 330 type species *Syggona concinna* Fåhraeus, 1870
Syngona Rye, 1893: 293 [unjustified emendation]
aeneicolor Fairmaire, 1889a: 359 (*Etazeta*) **A**: FUJ **ORR**
afghanicus Kaszab, 1973b: 57 **A**: AF
bruyanti Pic, 1917d: 18 (*Lyprops*) **A**: YE **AFR**
cribratellus Fairmaire, 1896b: 102 (*Lyprops*) **A**: SD
cribrifrons Marseul, 1876c: 125 (*Lyprops*) **A**: JA SC TAI **ORR**
curticollis Fairmaire, 1896a: 28 (*Lyprops*) **A**: PA SD **ORR**
gracilior Fairmaire, 1896a: 28 (*Lyprops*) **A**: AP BT PA SD **ORR**
horni Gebien, 1914b: 34 (*Lyprops*) **A**: TAI
indicus Wiedemann, 1823: 42 (*Tagenia*) **A**: NP
irregularis Gebien, 1914b: 35 (*Lyprops*) **A**: TAI
kaszabi Schawaller, 1997: 296 **A**: NP
luzonicus Gebien, 1914d: 405 (*Lyprops*) **A**: TAI **ORR**
orientalis Motschulsky, 1868: 195 (*Anaedus*) **A**: BT FE FUJ GAN HAI HEB HEI HEN HUB JA JIA JIL JIX LIA MG NC NIN NMO NP SC SCH SHA SHX TAI ZHE **ORR**
sinensis Marseul, 1876c: 126 (*Lyprops*)
tristis Fabricius, 1801b: 70 (*Lagria*) **A**: SD **ORR**
piceus Fairmaire, 1894a: 24 (*Lyprops*)
yunnanus Fairmaire, 1887b: 128 (*Lyprops*) **A**: BT GUA HAI NP YUN

genus *Micropedinus* Lewis, 1894: 379 type species *Micropedinus algae* Lewis, 1894
Notoprataeus Carter, 1924a: 37 type species *Notoprataeus litoralis* Carter, 1924
pallidipennis Lewis, 1894: 380 **A**: FE JA TAI "Korea" "East China" **ORR**
maderi Kaszab, 1940c: 154 (*Microlyprops*)
pullulus Boheman, 1858: 94 (*Heterophaga*) **A**: FE HKG JA NC SC TAI
algae Lewis, 1894: 379
rotundicollis Mt. Chûjô, 1996: 135 **A**: JA (Ryukyus)

genus *Sphingocorse* Gebien, 1921: 110 type species *Sphingocorse angulicollis* Gebien, 1921
nepalica Kaszab, 1975b: 122 **A**: NP UP

genus *Spinolyprops* Pic, 1917c: 12 type species *Spinolyprops rufithorax* Pic, 1917
himalayicus Kaszab, 1965a: 124 **A**: BT SD UP **ORR**
pakistanus Schawaller, 1996c: 692 **A**: PA

subfamily **Cossyphodinae** Wasmann, 1899

tribe Cossyphodini Wasmann, 1899

genus *Cossyphodes* Westwood, 1851: 168 type species *Cossyphodes wollastonii* Westwood, 1851
wollastonii Westwood, 1851: 170 **N**: CI MR **AFR**

genus *Mimocossyphus* Pic, 1923f: 5 type species *Mimocossyphus minor* Pic, 1923
minor Pic, 1923f: 5 **N**: EG

genus *Paramellon* Waterhouse, 1882: iv type species *Paramellon sociale* Waterhouse, 1882
socotraense Schawaller, 2006b: 248 **A**: YE (Suqutra)

subfamily **Phrenapatinae** Solier, 1834

tribe Archaeoglenini Watt, 1974

genus *Archaeoglenes* Broun, 1893a: 188 type species *Archaeoglenes costipennis* Broun, 1893
orientalis Sasaji, 1983: 47 **A**: JA

tribe Penetini Lacordaire, 1859

genus *Clamoris* Des Gozis, 1886: 25 [RN] type species *Phtora crenata* Mulsant, 1854
Phtora Mulsant, 1854: 228 [HN] type species *Phtora crenata* Mulsant, 1854
canalicollis Lewis, 1894: 466 (*Phthora*) **A**: FE JA
crenatus Mulsant, 1854: 229 (*Phtora*) **E**: FR IT MA PT SP **N**: AG TU
insurgens Des Gozis, 1886: 25
formosanus Masumoto, 1982a: 143 (*Phthora*) **A**: TAI
naranus Nakane, 1988b: 82 (*Phthora*) **A**: JA (Honshu)

genus *Dioedus* J. F. LeConte, 1862: 238 type species *Dioedus punctatus* J. F. LeConte, 1862
Arrhabaeus Champion, 1886a: 144 type species *Arrhabaeus convexus* Champion, 1886
Brachycilibe Carter, 1911: 207 type species *Brachycilibe antennata* Carter, 1911
Tagalus Gebien, 1914d: 388 type species *Tagalus impressicollis* Gebien, 1913
miyakensis Nakane, 1963a: 27 (*Tagalus*) **A**: JA (Izu Is.)
tokaranus Nakane, 1963a: 28 (*Tagalus*) **A**: JA

genus *Taiwanotagalus* Masumoto, 1982a: 143 type species *Taiwanotagalus klapperichi* Masumoto, 1982
klapperichi Masumoto, 1982a: 143 **A**: TAI
nepalicus Schawaller, 1999: 425 **A**: NP

subfamily **Pimeliinae** Latreille, 1802

tribe Adelostomini Solier, 1834

genus *Adelostoma* Duponchel, 1827: 338 type species *Adelostoma sulcatum* Duponchel, 1827
subgenus *Adelostoma* Duponchel, 1827: 338 type species *Adelostoma sulcatum* Duponchel, 1827
Polyscopus Waltl, 1853: 73 type species *Polyscopus costatus* Waltl, 1835
bedeli Reitter, 1885c: 317 **N**: AG
bicarinatum Waterhouse, 1881: 475 **A**: YE (Suqutra)
cordatum Solier, 1837b: 169 **N**: EG LB **A**: IS SI
granti Gahan, 1900: 10 **A**: YE (Abd al-Kuri, Suqutra)
granulithorax Escalera, 1914: 292 **N**: MO
hauseri Reitter, 1895e: 203 **A**: LE
longiceps Reitter, 1895e: 203 **N**: TU
metatarsale Carl, 1991a: 25 **A**: IQ
ovalipenne Chevrolat, 1877a: 117 **A**: TR
subtile arabicum Kaszab, 1981c: 366 **A**: SA
subtile sabulosum Kaszab, 1981c: 366 **A**: SA
subtile subtile Reitter, 1900d: 193 **A**: IN IS JO SY TR

sulcatum crassicorne Peyerimhoff, 1931b: 97 **N**: AG MO
sulcatum sulcatum Duponchel, 1827: 338 **E**: CR SP **N**: AG EG LB MO **A**: CY IQ IS JO SI SY TR
carinatum Solier, 1837b: 168
costatum Waltl, 1835: 74 (*Polyscopus*)
cristatum Eschscholtz, 1831: 11 (*Nyctoporis*)
deplanatum Haag-Rutenberg, 1875d: 422
grandiforme Koch, 1935a: 39
nitidum Haag-Rutenberg, 1875d: 421
parallelum Haag-Rutenberg, 1875d: 421
parallelum Chevrolat, 1877a: 117 [HN]
scabrum Chevrolat, 1877a: 118
waterhousei Schawaller, 2004c: 447 **A**: YE (Samha, Suqutra)
subgenus *Zarudnionymus* Semenov & Bogatchev, 1947: 175 type species *Zarudnionymus persis* Semenov & Bogatchev, 1947 (= *Adelostoma grande* Haag-Rutenberg, 1879)
Adelostomoides Carl, 1991a: 24 type species *Adelostoma grande* Haag-Rutenberg, 1879
Falsaspila Koch, 1952: 28 type species *Adelostoma batesi* Haag-Rutenberg, 1872
abyssinicum hirsutum Koch, 1935a: 39 **N**: EG
batesi Haag-Rutenberg, 1875d: 426 **A**: OM SA YE
grande Haag-Rutenberg, 1879a: 294 **A**: IN IS IQ SI
intermedium J. R. Sahlberg, 1913b: 164
perissinottoi Canzoneri, 1965: 168 (*Falsaspila*)
persis Semenov & Bogatchev, 1947: 176 (*Zarudnionymus*)

genus *Machlopsis* Pomel, 1871: 236 type species *Eurychora levaillantii* Lucas, 1850
subgenus *Hidrosella* Koch, 1952: 32 type species *Hidrosis incostata* Haag-Rutenberg, 1875
incostata Haag-Rutenberg, 1875d: 425 (*Hidrosis*) **N**: MO (Western Sahara) **AFR**
subgenus *Machlopsis* Pomel, 1871: 236 type species *Eurychora levaillantii* Lucas, 1850
Hidrosis Haag-Rutenberg, 1875c: 120 type species *Steira crenatocostata* L. Redtenbacher, 1868
alata Fairmaire, 1875a: 523 (*Eurychora*) **N**: TU
crenatocostata L. Redtenbacher, 1868: 120 (*Steira*) **N**: EG LB TU **A**: IS SI SY **AFR**
aegyptiaca Kirsch, 1871: 389 (*Steira*)
squalida Baudi di Selve, 1875c: 65 (*Eurychora*)
doderoi Gridelli, 1930: 235 **E**: IT (Lampedusa) **N**: LB
elongatula G. Quedenfeldt, 1890: 395 (*Hidrosis*) **N**: LB TU
levaillantii Lucas, 1850b: vii (*Eurychora*) **N**: AG
pseudoalata Gridelli, 1930: 236 **N**: EG LB
subfusca Pic, 1910m: 42 **N**: AG

genus *Pogonobasis* Solier, 1837b: 161 type species *Pogonobasis ornata* Solier, 1837
Peristeptus Haag-Rutenberg, 1875d: 382 type species *Pogonobasis laevigata* Gerstaecker, 1854
opaca Haag-Rutenberg, 1875d: 388 **A**: SA YE
ornata Solier, 1837b: 163 **N**: EG **AFR**

genus *Serrichora* Koch, 1952: 15 type species *Eurychora crenata* Solier, 1837
arabica Kaszab, 1981c: 364 **A**: OM

tribe Adesmiini Lacordaire, 1859

genus *Adesmia* Fischer von Waldheim, 1822: 153 type species *Adesmia longipes* Fischer von Waldheim, 1822 (= *Pimelia anomala* Fischer von Waldheim, 1820)
subgenus *Adesmia* Fischer von Waldheim, 1822: 153 type species *Adesmia longipes* Fischer von Waldheim, 1822 (= *Pimelia anomala* Fischer von Waldheim, 1820)
anomala anomala Fischer von Waldheim, 1820: pl. 14 (*Pimelia*) **A**: UZ
longipes Fischer von Waldheim, 1822: 153 [HN]
anomala dejeanii Gebler, 1841b: 589 **A**: ES KZ MG XIN
aweinsis Ren & Wang, 1993a: 3 **A**: XIN
austera Baudi di Selve, 1881: 275 **A**: YE

tuberculosa Champion, 1895: 250 [RN]
tuberculifera Gahan, 1895: 289
beduina Reitter, 1916a: 21 **N**: TU
biskreensis asperrima Peyerimhoff, 1931b: 94 **N**: MO
biskreensis biskreensis Lucas, 1844: 264 **N**: AG TU
immargo Reitter, 1916a: 25
biskreensis recticollis Peyerimhoff, 1931b: 92 **N**: AG **AFR**
fischeri caspia Bogatchev, 1964b: 177 **E**: AB
fischeri fischeri Faldermann, 1837: 22 **E**: AR **A**: IN TR
ehrenbergi Faldermann, 1837: 24
tatjanae Bogatchev, 1941a: 97
fischeri interposita Bogatchev, 1964b: 176 **A**: IN (Azarbayjan)
jugalis conveniens Bogatchev, 1941a: 97 **A**: AF
kabuliensis Kaszab, 1959f: 400
jugalis gracilenta Reitter, 1916a: 29 **A**: TD UZ (Fergana)
jugalis gridellii Kaszab, 1959e: 243 [= 1959f: 399] **A**: AF
jugalis jugalis Reitter, 1916a: 29 **A**: PA XIN
jugalis kasyi Kaszab, 1970a: 10 **A**: AF
jugalis kulzeri Kaszab, 1959f: 401 **A**: AF
jugalis septentrionalis Kaszab, 1973b: 40 **A**: AF
karelini karelini Fischer von Waldheim, 1835: 312 **A**: IN TM UZ XIN
karelini qalatensis Kaszab, 1970a: 12 **A**: AF
lehmanni Ménétriés, 1849: 224 **A**: KZ
biseriata Reitter, 1916a: 26
intersticta Reitter, 1916a: 26
septemcostata Reitter, 1916a: 27
maillei Solier, 1835c: 530 **E**: AB AR **A**: IQ IN TR
pulcherrima Fischer von Waldheim, 1835: 315
strophium Fischer von Waldheim, 1838: 316
montana acervata Klug, 1830: nr. 26, pl. xii **N**: AG EG LB TU
dubia Solier, 1835c: 527
klugii Solier, 1835c: 539
montana hamadae Koch, 1937: 393 **N**: LB (Hamada)
montana minor Gridelli, 1930: 232 **N**: EG LB
montana montana Klug, 1830: nr. 25, pl. xii (*Pimelia*) **A**: IQ IS JO SA SI
perpolita Reitter, 1916a: 24
montana parallela Miller, 1861: 171 **N**: EG **AFR**
panderi iranica Bogatchev, 1947a: 15 **A**: IN
panderi lindbergi Kaszab, 1970a: 11 **A**: AF
panderi panderi Fischer von Waldheim, 1835: 313 **A**: AF KZ UZ XIN
laevicollis Kraatz, 1882c: 107
panderi pupillata Kaszab, 1970a: 11 **A**: AF
planidorsis Reitter, 1916a: 28 **A**: TD TM UZ
qapqalensis Ren & Wang, 1993a: 4 **A**: XIN
tenebrosa Solier, 1835c: 536 **A**: IN
punctipennis Haag-Rutenberg, 1875a: 18
subgenus *Adesmina* Reitter, 1916a: 29 type species *Adesmia arabica* Reitter, 1916
arabica arabica Reitter, 1916a: 30 **A**: OM
arabica guichardi Kaszab, 1981c: 362 **A**: OM
arabica thesigeri Gridelli, 1953: 32 **A**: OM YE
arabica wittmeri Kaszab, 1981c: 362 **A**: OM
persiana Reitter, 1916a: 30 **A**: IN PA
subgenus *Macradesmia* Löbl & Merkl, 2003: 244 type species *Pimelia cancellata* Klug, 1830
aberrans Kwieton, 1978a: 8 **N**: AG (Tassili-n'Ajjer)
anatolica Ganglbauer, 1905: 275 **A**: TR SY
anthracina Klug, 1830: nr. 28, pl. xiii (*Pimelia*) **A**: IS LE SY

libana Baudi di Selve, 1875c: 30
olivieri Reiche & Saulcy, 1857: 224
procera Miller, 1861: 170
ramburi Solier, 1835c: 529
antiqua Klug, 1830: nr. 35, pl. xiii (*Pimelia*) **A**: IQ **AFR**
aethiopia Allard, 1885a: 182
asperata Baudi di Selve, 1875c: 29 **A**: YE
assimilis Gahan, 1895: 290 **A**: YE
cancellata abbreviata Klug, 1830: nr. 27, pl. xiii (*Pimelia*) **A**: SI
aspera Reitter, 1916a: 20
hieroglyphica Baudi di Selve, 1875c: 30
latreillei Solier, 1835c: 540
sinaitica Crotch, 1872: 267
cancellata cancellata Klug, 1830: nr. 37, pl. xiii (*Pimelia*) **A**: BA IS JO KU OM PA SA SY YE
cancellata clathrata Solier, 1835c: 541 **A**: IN IQ OM SA SY
babylonica Reitter, 1916a: 14
cancellata hadramautina Kaszab, 1981c: 360 **A**: YE
conjungens Koch, 1951: 130 **A**: IQ TR
gibbula Reitter, 1916a: 17 **A**: TR
interrupta interrupta Klug, 1830: nr. 36, pl. xiii (*Pimelia*) **A**: SA YE
interrupta pyriformis Schuster, 1938a: 57 **A**: SA YE
mesopotamica Chevrolat, 1877a: 113 **A**: IN IQ
mimica Reitter, 1916a: 15 **A**: IQ
monilis fougieri Koch, 1937: 390 **N**: LB
monilis monilis Klug, 1830: nr. 38, pl. xiii (*Pimelia*) **N**: EG LB
monilis pluriseriata Schuster, 1928e: 375 **N**: LB
monilis reducta Schuster, 1928e: 375 **N**: LB
sculptilis Reitter, 1916a: 17 **N**: LB
servillei chaldaea Koch, 1951: 124 **A**: IN
servillei fagergreenii Baudi di Selve, 1874a: 92 **A**: IN
servillei holdhausi Reitter, 1916a: 13 **A**: IN
servillei parisii Koch, 1940d: 287 **A**: AF
servillei provincialis Kaszab, 1959f: 399 **A**: AF PA
servillei schatzmayri Koch, 1940d: 286 **E**: ST **A**: AF IN KZ TM UZ
servillei schusteri Reitter, 1916a: 13 **A**: IN PA
servillei servillei Solier, 1835c: 538 **A**: IN
immarginicollis Koch, 1951: 124
ulcerosa arca Reiche & Saulcy, 1857: 222 **A**: IS LE
ulcerosa confusa Ardoin, 1978b: 310 **A**: IS
ulcerosa ulcerosa Klug, 1830: nr. 34, pl. xiii (*Pimelia*) **A**: IS JO SY
perplexa Schaum, 1864: 671 [RN]
reticulata Solier, 1835c: 528 [HN]
wahabita Koch, 1951: 129 **A**: YE
elateralis Koch, 1951: 129
subgenus *Oteroscelis* Solier, 1835c: 546 type species *Pimelia metallica* Klug, 1830
aenescens Kühnelt, 1951: 245 **A**: IN KU
allardi allardi Gebien, 1910a: 83 [RN] **A**: SA
fossulata Allard, 1885a: 165 [HN]
allardi interpunctata Schuster, 1938a: 56 **A**: YE
audouini aegrota Faldermann, 1837: 25 **A**: "Transcaucasus"
audouini audouini Solier, 1835c: 550 **E**: AR **A**: IN
faldermanni Fischer von Waldheim, 1835: 314
nodulosa Faldermann, 1837: 26
pulcherrima Solier, 1835c: 546
audouini kandaharica Kaszab, 1959f: 397 **A**: AF
carinata carinata Solier, 1835c: 547 **A**: IN IQ JO SA SY TR

aenea L. Redtenbacher, 1850: 48
elevata Solier, 1835c: 549
carinata persica Koch, 1940d: 284 **A**: IN
cothurnata cothurnata Forskål, 1775: 80 (*Tenebrio*) **N**: EG **A**: AE BA IQ IS JO OM SA SI YE **AFR**
bicarinata Klug, 1830: nr. 31, pl. xiii (*Pimelia*)
cursor Chevrolat, 1877a: 113
extensa Klug, 1830: nr. 29, pl. xiii (*Pimelia*)
glabrior Schatzmayr & Koch, 1934b: 23
lefebvrii Fischer von Waldheim, 1835: 315
omanensis Kaszab, 1981c: 353
pariseti Solier, 1835c: 552
subserrata Chevrolat, 1877a: 113
villae Osculati, 1844: 72
cothurnata montisatri Peyerimhoff, 1929b: 156 **N**: AG (Hoggar) LB
cothurnata rathjensi Schuster, 1938a: 56 **A**: AE SA YE
coucyi Lucas, 1850c: xi **N**: AG
devecchii Osculati, 1844: 72 **E**: AR **A**: IN SY
filippovi Bogatchev, 1941a: 98 **A**: YE
gebleri elliptica Reitter, 1916a: 6 **A**: AF TM
gebleri gebleri Gebler, 1844: 102 **A**: AF KZ TD TM
oschana Reitter, 1916a: 6
khaliensis Blair, 1931b: 270 **A**: OM SA
lacunosa fossulata Allard, 1885a: 191 **A**: SA YE
lacunosa lacunosa Klug, 1830: nr. 33, pl. xiii (*Pimelia*) **A**: SA YE
polita Baudi di Selve, 1874a: 95
semiglabra Chevrolat, 1877a: 113
legatoria Bogatchev, 1941a: 99 **A**: AF
lii Ren & Wang, 1993a: 1 **A**: XIZ
metallica atlantica Escalera, 1935a: 296 **N**: MO (Ifni)
metallica besnardi Kaszab, 1981c: 356 **A**: SA
metallica brozai Ardoin, 1978b: 300 **A**: IS SI
metallica laevior Ardoin, 1978b: 299 **A**: IS SI
metallica mediocostata Koch, 1940d: 273 **N**: AG MO
metallica metallica Klug, 1830: nr. 23, pl. xii (*Pimelia*) **N**: AG EG LB MO TU
andresi Böhm, 1908: 165
chakouri Reitter, 1916a: 6
faremontii Lucas, 1844: 264
ganglbaueri Reitter, 1916a: 10
metallica syriaca Baudi di Selve, 1874a: 95 **A**: IN IS JO
carmelitana Reitter, 1916a: 8
fortesculpta Reitter, 1916a: 8
microgranulosa Carl, 1992b: 332 [RN] **A**: IQ
microgranulata Carl, 1990a: 73 [HN]
rathjensi Schuster, 1938a: 56 **A**: YE
scorteccii Koch, 1965a: 117 **A**: YE
sodalis elegantissima Koch, 1940d: 282 **A**: IN (Beluzhistan)
sodalis ghazniana Kaszab, 1974: 194 **A**: AF
sodalis laghmanica Kaszab, 1959f: 396 **A**: AF
sodalis sodalis Waterhouse, 1889: 125 (*Adesmia*) **A**: AF PA (Balochistan)
afghanica Koch, 1940d: 280
belutschistana Schuster, 1930b: 237
stoeckleini stoeckleini Koch, 1940d: 271 **A**: JO SA
stoeckleini rasalkhaymana Kaszab, 1981c: 355 **A**: AE
subgenus *Oteroscelopsis* Löbl & Merkl, 2003: 244 type species *Pimelia dilatata* Klug, 1830
dilatata affinis Solier, 1835c: 535 **N**: AG TU
duponchelii Allard, 1885a: 179
reitteri Koch, 1940d: 297

solieri Lucas, 1844: 266

tibialis Reitter, 1916a: 21

dilatata dilatata Klug, 1830: nr. 32, pl. xiii (*Pimelia*) **N**: EG **A**: IS JO SI

convergens Walker, 1871: 14

drakii Crotch, 1872: 267

insignis Miller, 1861: 172

macropus Solier, 1835c: 542

dilatata douei Lucas, 1844: 265 **N**: AG TU

dilatata getula Peyerimhoff, 1935b: 24 **N**: AG MO

dilatata tripolitana Reitter, 1916a: 22 **N**: LB TU

dilatata tunisia Reitter, 1916a: 22 **N**: TU

dilatata vermiculata Koch, 1940d: 300 **N**: AG MO

longipes Fabricius, 1775: 251 (*Pimelia*) **N**: EG

concisa Walker, 1871: 15

maroccana Haag-Rutenberg, 1877b: 283 **N**: MO

maroccana Haag-Rutenberg, 1877b: 283 **N**: MO

microcephala Solier, 1835c: 533 **N**: AG

zanoni Schuster, 1922b: 16 **N**: LB

genus *Adesmia*, species incertae sedis

baudii Kraatz, 1882c: 106 **A**: IN

tribe Akidini Billberg, 1820

genus *Akis* Herbst, 1799: 124 type species *Pimelia reflexa* Fabricius, 1775

Acis Billberg, 1820: 32 [unjustified emendation]

acuminata Fabricius, 1787a: 209 (*Pimelia*) **E**: PT SP **N**: MO

andromeda Bogatchev, 1947a: 17 **A**: IN

antoinei antoinei Escalera, 1934b: 306 **N**: MO

antoinei midletensis Antoine, 1957: 50 **N**: MO

antoinei sabasi Antoine, 1957: 50 **N**: MO

bacarozzo Schrank von Paula, 1786: 22 (*Pimelia*) **E**: FR IT SP (Baleares)

plicata Latreille, 1816: 280

punctata Thunberg, 1787: 49 (*Pimelia*)

bernhaueri Schuster, 1919a: 29 **N**: LB

biskrensis Reitter, 1904a: 43 **N**: AG

bremeri Ardoin, 1979: 155 **E**: SP (Formentera)

cordicollis Fairmaire, 1884c: 446 **N**: MO

cyrenaica Schuster, 1928e: 376 **N**: EG LB

discoidea Quensel, 1806: 138 **E**: SP

dorsigera Reitter, 1904a: 45 **E**: PT

duplicata Reitter, 1904a: 41 **E**: SP (Cuenca)

elegans Charpentier, 1825: 216 (*Acis*) **E**: SP **N**: AG MO

carinata Solier, 1837a: 674

kobelti Heyden, 1883b: 236

otoes Fischer von Waldheim, 1837b: 14

pachecoi Escalera, 1934b: 303

elevata Solier, 1837a: 671 **N**: EG **A**: SA SI "Arabia"

elongata elongata Brullé, 1832: 194 **E**: AL BU CR GR MC TR **A**: CY TR

anatolica Reitter, 1904a: 40

deplanata Waltl, 1838: 461

taygetana Koch, 1948a: 326

elongata opaca Heyden, 1897: 31 **A**: TR

genei Solier, 1837a: 668 **E**: SP

gestroi Schuster, 1922b: 18 **N**: LB (Cyrenaica)

granulifera C. R. Sahlberg, 1823: 11 **E**: PT SP **N**: CIi

bayardi Solier, 1837a: 670

triseriata zur Strassen, 1957: 54

heydeni araneipes Antoine, 1957: 53 **N**: MO
heydeni heydeni Haag-Rutenberg, 1876c: 76 **N**: MO
longipennis Escalera, 1934b: 309
lozanoi Escalera, 1934b: 307
obtusangula Théry, 1932a: 190
penoti Antoine, 1957: 51
rhatica Antoine, 1957: 47
susica Escalera, 1934b: 309
hispanica Solier, 1837a: 667 **E**: SP
ilonca zur Strassen, 1957: 54 **E**: PT SP
latreillei Solier, 1837a: 675 **E**: GR **N**: EG **A**: CY IQ IS SY TR
lusitanica Solier, 1837a: 670 **E**: SP
maresi Ardoin, 1971a: 49 **N**: TU
nevadensis Español, 1954c: 114 **E**: SP (Granada)
perissinottoi Canzoneri, 1971: 155 **A**: IN
reflexa goryi Solier, 1837a: 676 **N**: AG EG LB MO TU
reflexa reflexa Fabricius, 1775: 252 (*Pimelia*) **N**: EG **A**: IS SI
hispida Herbst, 1799: 131
pterygodes Pallas, 1781: 41 (*Tenebrio*)
reflexa tatae Théry, 1932a: 191 **N**: MO
richteri G. Quedenfeldt, 1889: 354 **N**: MO
rotundicollis Escalera, 1934b: 301
salzei Solier, 1837a: 663 **E**: SP (Cartagena)
sansi Solier, 1837a: 660 **E**: SP
spinosa Linnaeus, 1764: 101 (*Tenebrio*) **N**: EG
sculptior Koch, 1935a: 51
stricticollis Théry, 1931a: 26 **N**: MO
subterranea Solier, 1837a: 656 **E**: IT MA
melitana Reitter, 1904a: 40
subtricostata L. Redtenbacher, 1850: 48 **A**: IN IQ SY
tingitana tingitana Lucas, 1859b: cxiv **N**: MO
tingitana venerata Antoine, 1926: 256 **N**: MO
trilineata Herbst, 1799: 130 **E**: GR IT RO TR **N**: AG LB TU **A**: SY
algeriana Solier, 1837a: 663
angusticollis Baudi di Selve, 1875c: 76
barbara Solier, 1837a: 673
carinata Drapiez, 1821: 276
italica Solier, 1837a: 674
nitida Solier, 1837a: 666
olivieri Solier, 1837a: 665
planicollis Solier, 1837a: 664
ragusae Reitter, 1904b: 97
romana Luigioni, 1921: 61
sardea Solier, 1837a: 669
speluncaria Koch, 1939: 266
terricola Ménétriés, 1838: 36
tuberculata Kraatz, 1865a: 249 **E**: SP (Baleares)

genus *Cyphogenia* Solier, 1837a: 677 type species *Tenebrio auritus* Pallas, 1781
subgenus *Cyphogenia* Solier, 1837a: 677 type species *Tenebrio auritus* Pallas, 1781
Eocyphogenia G. S. Medvedev, 1968c: 897 type species *Akis rugipennis* Faldermann, 1835
aurita aurita Pallas, 1781: 40 (*Tenebrio*) **E**: ST **A**: AF IN KZ TD TM UZ XIN
brevicostata Kraatz, 1882b: 83
truncata Gebler, 1844: 101 (*Akis*)
unicostata Kraatz, 1882b: 84
aurita cratii A. F. Morawitz, 1865: 40 **A**: IN TM
cassiopea Bogatchev, 1952b: 284

chinensis Faldermann, 1835b: 392 (*Akis*) **A**: BEI GAN KZ LIA MG NIN NMO SHA XIN
intermedia Bogatchev, 1962: 45
gibba gibba Fischer von Waldheim, 1820: pl. 15 (*Akis*) **A**: AF IN KI KZ PA TM
angustata Zubkov, 1833: 329 (*Akis*)
zablotzkii Zubkov, 1837: 67 (*Akis*)
gibba persica Baudi di Selve, 1874a: 108 [= 1875c: 78] (*Akis*) **A**: AF IN
humeralis Bates, 1879: 471 **A**: GAN KA NMO XIN XIZ
semicarinata Reitter, 1887d: 360
rugipennis Faldermann, 1835b: 389 (*Akis*) **A**: BEI HEB HEN GAN LIA NMO SHA SHN SHX TIA
subgenus *Lechriomus* A. F. Morawitz, 1865: 21 type species *Akis lucifuga* Adams, 1817
bienerti A. F. Morawitz, 1865: 27 (*Akis*) **A**: AF IN TD TM
bogatshevi G. S. Medvedev, 2004b: 560 **A**: KZ
funesta Faldermann, 1835b: 390 (*Akis*) **A**: "China borealis" "Mongolia"
sepulchralis Faldermann, 1835b: 391 (*Akis*)
limbata Fischer von Waldheim, 1820: pl. 15 (*Akis*) **A**: KZ TM UZ
lucifuga Adams, 1817: 304 (*Akis*) **E**: AB AR GG ST UK **A**: IN IQ KZ SY TM TR
depressa Küster, 1848b: 74 (*Akis*)

genus *Morica* Dejean, 1834: 182 type species *Akis planata* Fabricius, 1801
Morica Solier, 1837a: 646 type species *Akis planata* Fabricius, 1801
costitubera Marseul, 1883b: 182 (*Akis*) **N**: EG LB
pharao Reitter, 1904a: 38
schweinfurthi M. Quedenfeldt, 1890: 139 (*Akis*)
favieri Lucas, 1859b: cxiii **E**: SP **N**: MO
grossa Linnaeus, 1767: 676 (*Tenebrio*) **N**: AG EG LB TU
octocostata Solier, 1837a: 649
hybrida Charpentier, 1825: 215 (*Acis*) **E**: SP **N**: AG
obtusa Solier, 1837a: 650
jevini Lucas, 1850a: iv **N**: AG
planata Fabricius, 1801a: 134 (*Akis*) **E**: SP **N**: MO
tingitana Baudi di Selve, 1875c: 75

genus *Sarothropus* Kraatz, 1865a: 268 type species *Akis depressa* Zubkov, 1837
depressus Zubkov, 1837: 68 (*Akis*) **A**: AF KZ TM (Karakumy)
fallax G. S. Medvedev, 1961: 365 **A**: KZ
kiritshenkoi G. S. Medvedev, 1961: 366 **A**: TD TM UZ
medvedevi Bogatchev, 1962: 45 (*Sarathropus*) **A**: TD

genus *Solskyia* Solsky, 1881b: 48 type species *Solskyia peregrina* Solsky, 1882
afghanica Kaszab, 1970a: 48 (*Solskya*) **A**: AF
caporiaccoi Gridelli, 1934: 53 **A**: CH KA
concavidorsis Reitter, 1909f: 296 (*Solskia*) **A**: TD UZ
depressiuscula Fairmaire, 1891d: xcii (*Cyphogenia*) **A**: KA
angusticollis Della Beffa, 1931: 185
inermis Della Beffa, 1931: 185
grombczewskii Semenov, 1889a: 221 (*Akis*) **A**: AF KA PA TD
kaszabi Español, 1961c: 125 **A**: KA
lhasana Ren & Yu, 2000: 325 **A**: XIZ
lopatini G. S. Medvedev, 1968c: 893 **A**: TD
parvicollis Kraatz, 1865a: 251 (*Akis*) **A**: KA XIN XIZ "Himalaya"
kuenluna Kaszab, 1965c: 282
morawitzi Semenov, 1891a: 363
plana Bates, 1879: 471 (*Cyphogenia*)
peregrina Solsky, 1881b: 49 **A**: TD UZ (Samarkand)
schmidi Español, 1961c: 128 **A**: HP KA
subperegrina Reitter, 1904a: 49 **A**: "Turkestan"

tribe Asidini Fleming, 1821

genus *Alphasida* Escalera, 1905a: 380 type species *Asida holosericea* Germar, 1824
subgenus *Alphasida* Escalera, 1905a: 380 type species *Asida holosericea* Germar, 1824
becerrae Escalera, 1905a: 387 (*Asida*) **E**: SP
holosericea Germar, 1824b: 139 (*Asida*) **E**: SP
bicostata Escalera, 1905a: 386 (*Asida*)
ramburi Solier, 1836c: 438 (*Asida*)
lazaroi Escalera, 1906c: 235 **E**: SP
lorcana almeriensis Escalera, 1901: 174 (*Asida*) **E**: SP
lorcana lopezi Escalera, 1906e: 383 **E**: SP
depilata Escalera, 1906e: 384
lorcana lorcana Perez Arcas, 1865: 45 (*Asida*) **E**: SP
sanchezgomezi Escalera, 1901: 173 (*Asida*)
lorcana mendizabali Koch, 1944b: 234 **E**: SP
martinezi Escalera, 1901: 175 (*Asida*) **E**: SP
merceti Bolívar y Pieltain, 1914: 237 (*Asida*) **E**: SP
oberthueri Escalera, 1901: 175 (*Asida*) **E**: SP
rufopubescens Escalera, 1905a: 388 (*Asida*) **E**: SP
calva Escalera, 1905a: 388 (*Asida*) [HN]
solieri escalerae Oberthür, 1903: 74 (*Asida*) **E**: SP
alpujarrensis Escalera, 1905a: 386 (*Asida*)
leonhardi Reitter, 1917a: 12
solieri fuscopepla Koch, 1940d: 302 **E**: SP
solieri martini Escalera, 1903: 75 (*Asida*) **E**: SP
bolivari Escalera, 1906e: 382
solieri solieri Rambur, 1838: pl. 19 (*Asida*) **E**: SP
clementei Perez Arcas, 1865b: 432 (*Asida*)
solieri ziziphi Cobos, 1988a: 9 **E**: SP
typica Gebien, 1937: 717 [RN] **E**: SP
gaditana Escalera, 1905a: 385 (*Asida*) [HN]
volxemi Escalera, 1905a: 387 (*Asida*) **E**: PT
subgenus *Betasida* Reitter, 1917a: 11 type species *Asida argenteolimbata* Escalera, 1901
Subalphasida Escalera, 1927a: 136 type species *Asida luctuosa* Boisduval, 1835
argenteolimbata Escalera, 1901: 172 (*Asida*) **E**: SP
luctuosa Rambur, 1838: pl. 19 [HN]
espanoli Cobos, 1962: 39 **E**: SP
ferreri Cobos, 1988b: 54 **E**: SP
luctuosa Boisduval, 1835: 250 (*Asida*) **E**: SP
luctuosa Rosenhauer, 1856: 193 (*Asida*) [HN]
minor Rosenhauer, 1856: 194 (*Asida*) [HN]
subgenus *Elongasida* Escalera, 1906d: 306 type species *Asida grandipalpis* Allard, 1869
Cribrasida Reitter, 1917a: 38 type species *Asida grandipalpis* Allard, 1869
alonensis Martinez y Sáez, 1873b: 409 (*Asida*) **E**: SP
asperata Solier, 1836c: 450 (*Asida*) **E**: SP
rugosula Rambur, 1838: pl. 19 (*Asida*)
calumniata Escalera, 1906d: 311 (*Asida*) **E**: SP
grandipalpis Allard, 1869a: 288 (*Asida*) **E**: SP
granosa Escalera, 1921a: 118 (*Asida*) **E**: SP
hesperica Rambur, 1838: pl. 20 (*Asida*) **E**: SP
hispalensis Escalera, 1906d: 309 (*Asida*) **E**: SP
inesperata Escalera, 1921a: 117 (*Asida*) **E**: SP
rectipennis Escalera, 1906d: 312 (*Asida*) **E**: SP
rufomarginalis Escalera, 1906d: 310 (*Asida*) **E**: SP
subgenus *Glabrasida* Escalera, 1910c: 408 type species *Glabrasida conspuata* Escalera, 1910
Aplanasida Reitter, 1917a: 11 type species *Alphasida brevicosta* Solier, 1836 (= *Asida depressa* Solier, 1836)

Aulonasida Reitter, 1917a: 19 type species *Asida chauveneti* Solier, 1836
Durasida Reitter, 1917a: 14 type species *Tenebrio silphoides* Linnaeus, 1767
Gymnetasida Reitter, 1917a: 22 type species *Asida tricostata* Allard, 1869
Melambasida Reitter, 1917a: 21 type species *Alphasida interjecta* Reitter, 1917
Mimelasida Reitter, 1917a: 21 type species *Asida puncticollis* Solier, 1836
Pedarasida Reitter, 1917a: 11 type species *Asida cariosicollis* Solier, 1836
alcirensis Reitter, 1917a: 38 **E**: SP
alcirensis Escalera, 1923a: 466 (*Asida*) [HN]
alluaudi alluaudi Escalera, 1925a: 331 (*Glabrasida*) **N**: MO
alluaudi maritima Antoine, 1934: 117 (*Gymnetasida*) **N**: MO
altomirana Escalera, 1923b: 38 (*Asida*) **E**: SP
amorii Perez Arcas, 1868: 510 (*Asida*) **E**: SP
gaditana Rambur, 1838: pl. 20 (*Asida*) [HN]
gaditana Reitter, 1917a: 37 [HN]
annamariae annamariae Antoine, 1934: 118 (*Gymnetasida*) **N**: MO
annamariae hantzi Antoine, 1937: 186 **N**: MO
annina Reitter, 1917a: 37 **E**: SP
ardoisi Reitter, 1917a: 31 **E**: SP
ardoisi Escalera, 1922e: 391 (*Asida*) [HN]
asturica Schuster, 1928f: 946 [RN] **E**: SP
heydeni Escalera, 1923c: 87 (*Asida*) [HN]
auriculata Solier, 1836c: 451 (*Asida*) **N**: EG LB
bengasiana Wilke, 1921: 254
mystica Wilke, 1921: 255
azruensis Escalera, 1925a: 326 (*Glabrasida*) **N**: MO
baezensis Escalera, 1923a: 467 (*Asida*) **E**: SP
balboana Koch, 1937: 411 **N**: LB
bifoveata Allard, 1869a: 294 (*Asida*) **N**: MO
bonacherai Español, 1958a: 16 **E**: SP
boscai Escalera, 1922e: 389 (*Asida*) **E**: SP
breuiliana Escalera, 1923c: 86 (*Asida*) **E**: SP
brevicosta Solier, 1836c: 449 (*Asida*) **E**: SP (Baleares)
crenata Schaufuss, 1882c: 623 (*Asida*)
depressa Solier, 1836c: 466 (*Asida*) [HN]
laevigata Rambur, 1838: pl. 19 (*Asida*)
brevipubens Reitter, 1917a: 32 **E**: SP
calagurriensis Escalera, 1943a: 203 (*Asida*) **E**: SP
cariosicollis Solier, 1836c: 446 (*Asida*) **N**: AG
castellana Graells, 1858: 69 (*Asida*) **E**: SP
centralis centralis Escalera, 1925a: 330 (*Glabrasida*) **N**: MO
centralis ouezzania Kocher, 1968: 109 **N**: MO
centralis volubilitana Escalera, 1929: 108 (*Glabrasida*) **N**: MO
cerebellum cerebellum Antoine, 1933a: 211 (*Gymnetasida*) **N**: MO
cerebellum gundafica Antoine, 1933a: 213 (*Gymnetasida*) **N**: MO
cereocosta Kwieton, 1986a: 8 **N**: AG
chauveneti chauveneti Solier, 1836c: 440 (*Asida*) **N**: AG TU
lapidaria Lucas, 1846: 323 (*Asida*)
quadricarinata Reitter, 1917a: 19
chauveneti quadricostata Allard, 1869a: 233 (*Asida*) **N**: AG TU
appulsa Reitter, 1917a: 20
conjuncta Escalera, 1923b: 40 (*Asida*) **E**: SP
conspuata Escalera, 1910c: 409 (*Glabrasida*) **N**: MO
cordubensis Escalera, 1923c: 82 (*Asida*) **E**: SP
cortesensis cortesensis Reitter, 1917a: 33 **E**: SP
grossepunctata Escalera, 1923a: 473 (*Asida*)
cortesensis millarensis Escalera, 1943a: 205 (*Asida*) **E**: SP

cortesensis sacanetensis Escalera, 1943a: 205 (*Asida*) **E**: SP
costulata Solier, 1836c: 455 (*Asida*) **E**: PT SP
scabrosa Allard, 1869a: 270 (*Asida*)
ventricosa Solier, 1836c: 458 (*Asida*)
crassicollis crassicollis Fairmaire, 1868: 487 (*Asida*) **N**: AG
silphiformis Reitter, 1917a: 17
crassicollis felicitana Reitter, 1917a: 18 **N**: AG
cylindrica cylindrica Reitter, 1917a: 27 **N**: MO
cylindrica rehamnensis Antoine, 1937: 184 **N**: MO
dantini Escalera, 1923a: 472 (*Asida*) **E**: SP
deyrollei Reitter, 1917a: 34 **E**: SP
terolensis Escalera, 1923a: 473 (*Asida*)
discostriata Escalera, 1923a: 468 (*Asida*) **E**: SP
djebiletensis Antoine, 1937: 185 **N**: MO
dubia Rambur, 1838: pl. 19 (*Asida*) **E**: SP
edithae Reitter, 1917a: 16 **N**: AG
tuberculosa Reitter, 1917a: 19
elongata Solier, 1836c: 463 (*Asida*) **E**: SP
embergeri embergeri Antoine, 1936b: 71 (*Glabrasida*) **N**: MO
embergeri infrasedens Antoine, 1937: 241 (*Machlasida*) **N**: MO
embergeri supersedens Antoine, 1937: 241 (*Machlasida*) **N**: MO
exilata Escalera, 1925b: 153 (*Glabrasida*) **E**: SP
flens Antoine, 1933a: 200 (*Gymnetasida*) **N**: MO
frigidissima Escalera, 1923a: 470 (*Asida*) **E**: FR SP
gigas Dufour, 1820: 309 **E**: SP
globipennis Escalera, 1910c: 412 (*Glabrasida*) **N**: MO
goudoti Solier, 1836c: 457 (*Asida*) **E**: SP PT
atrata Baudi di Selve, 1875c: 107 (*Asida*)
perezi Chevrolat, 1865: 390 (*Asida*)
gracilis gracilis Allard, 1869a: 295 (*Asida*) **E**: SP
gracilis mariae Cobos, 1988a: 9 **E**: SP
gracilis occulta Cobos, 1988a: 9 **E**: SP
granadina Escalera, 1922e: 391 (*Asida*) **E**: SP
granifera Solier, 1836c: 460 (*Asida*) **E**: SP
granipennis breiti Antoine, 1937: 188 **N**: MO
granipennis granipennis Escalera, 1925a: 328 (*Glabrasida*) **N**: MO
granipennis punctibasis Escalera, 1925a: 330 (*Glabrasida*) **N**: MO
granipennis schusteri Antoine, 1937: 188 **N**: MO
granipunctipennis Escalera, 1923c: 86 (*Asida*) **E**: SP
granulata Fabricius, 1787a: 49 (*Silpha*) **N**: TU
lethierryi Allard, 1869a: 236 (*Asida*)
variolosa Fabricius, 1792a: 108 (*Blaps*)
grassellyi Antoine, 1937: 189 **N**: MO
gregoirei Antoine, 1933a: 213 (*Pedarasida*) **N**: MO
grossa grossa Solier, 1836c: 453 (*Asida*) **E**: IT
cadamuroi Canzoneri, 1963b: 36
cossyrensis Reitter, 1917a: 30
otrantina Leoni, 1929: 151 (*Asida*)
grossa melitana Reitter, 1894b: 82 (*Asida*) **E**: MA
grossa sicula Solier, 1836c: 454 (*Asida*) **E**: IT (Sicilia)
halophila halophila Koch, 1937: 406 **N**: LB
halophila syrtica Koch, 1939: 265 **N**: LB
herminae Reitter, 1917a: 17 **N**: AG
heydeni Reitter, 1917a: 36 **E**: SP
himerera Reitter, 1917a: 18 **E**: IT (Sicilia)

hispanica Solier, 1836c: 461 (*Asida*) **E**: SP
sibirica Solier, 1838c: 462 (*Asida*)
ibicensis ibicensis Perez Arcas, 1868: 507 (*Asida*) **E**: SP (Baleares)
ibicensis medae Español, 1940: 9 **E**: SP (Baleares)
ibicensis ovalaris Escalera, 1925e: 409 (*Asida*) **E**: SP (Baleares)
iblanensis iblanensis Antoine, 1939: 148 **N**: MO
iblanensis reymondi Antoine, 1952: 113 **N**: MO
indecisa indecisa Escalera, 1923c: 80 (*Asida*) **E**: SP
indecisa vivesi Viñolas & Cartagena, 2005: 35 (*Glabrasida*) [RN] **E**: SP
calva Escalera, 1923c: 80 (*Asida*) [HN]
interjecta brachycera Antoine, 1937: 191 **N**: MO
interjecta interjecta Reitter, 1917a: 22 **N**: AG
integra Reitter, 1917a: 22
interstrata Reitter, 1917a: 22
ithana Reitter, 1917a: 30 **E**: SP
javalambrensis Escalera, 1943a: 206 (*Asida*) **E**: SP
jumillensis Escalera, 1923a: 467 (*Asida*) **E**: SP
kocheri Antoine, 1934: 120 (*Gymnetasida*) **N**: MO
koltzei Reitter, 1917a: 36 **E**: SP
insularis Escalera, 1922e: 391 (*Asida*)
lacunosa Escalera, 1923c: 83 (*Asida*) **E**: SP
laevis approximata Escalera, 1923a: 464 (*Asida*) **E**: SP
laevis laevis Solier, 1836c: 464 (*Asida*) **E**: SP
laevis salinatrix Koch, 1944b: 238 **E**: SP
lapini babberretiana Cobos, 1968: 913 **N**: MO
lapini lapini Kocher, 1966a: 45 **N**: MO
lassallei J. Ferrer, 1983: 137 **E**: SP
lecerfi ighrezranensis Koch, 1940d: 304 **N**: MO
lecerfi lecerfi Koch, 1940d: 302 **N**: MO
lecerfi timeliltensis Koch, 1940d: 304 **N**: MO
leonensis gallaeciana Escalera, 1923c: 84 (*Asida*) **E**: SP
leonensis leonensis Reitter, 1917a: 32 **E**: PT SP
leonensis Escalera, 1923c: 84 (*Asida*) [HN]
leperina Reitter, 1917a: 21 **N**: AG
loroi Escalera, 1923a: 471 (*Asida*) **E**: SP
louati Antoine, 1933a: 198 (*Gymnetasida*) **N**: MO
lucens Escalera, 1923c: 87 (*Asida*) **E**: SP
lusitana Escalera, 1923c: 82 (*Asida*) **E**: PT SP
magnifica Escalera, 1923b: 41 (*Asida*) **E**: SP
maroccana Allard, 1869a: 260 (*Asida*) **N**: MO
blattiformis Reitter, 1917a: 27
marseuli Allard, 1869a: 274 (*Asida*) **E**: SP
mazaganica baudoni Antoine, 1957: 43 **N**: MO
mazaganica mazaganica Escalera, 1910c: 413 (*Glabrasida*) **N**: MO
mazaganica microgranifera Escalera, 1922b: 173 **N**: MO
melillensis Escalera, 1910c: 411 (*Glabrasida*) **N**: MO
meridionalis meridionalis Escalera, 1923b: 40 (*Asida*) **E**: SP
meridionalis plana Escalera, 1923b: 41 (*Asida*) **E**: SP
multicostata Escalera, 1923c: 84 (*Asida*) **E**: PT
multigranulosa Escalera, 1923c: 83 (*Asida*) **E**: SP
nemethi ganayi Antoine, 1942b: 46 (*Glabrasida*) **N**: MO
nemethi nemethi Théry, 1932a: 189 **N**: MO
nigroopaca G. Quedenfeldt, 1890: 398 (*Asida*) **N**: LB TU
obsoleta obsoleta Fairmaire, 1867: 405 (*Asida*) **N**: AG
obsoleta tuberculifera Allard, 1869a: 259 (*Asida*) **N**: AG

algeriana Gebien, 1910a: 126 (*Asida*) [RN]
tournieri Allard, 1869a: 264 (*Asida*)
tuberculata Allard, 1869a: 263 (*Asida*)
olmedensis Escalera, 1923a: 470 (*Asida*) **E**: SP
opatroides Allard, 1869a: 249 (*Asida*) **N**: AG
sulcipennis Fairmaire, 1868: 485 (*Asida*)
ovalis Escalera, 1923b: 38 (*Asida*) **E**: SP
orientalis Escalera, 1929: 108 (*Glabrasida*) **N**: MO
rotroui Escalera, 1929: 108 (*Glabrasida*)
tazekensis Antoine, 1939: 149
parallela almagrerensis Cobos, 1988a: 10 **E**: SP
parallela parallela Solier, 1836c: 459 (*Asida*) **E**: SP
montana Rambur, 1838: pl. 19 (*Asida*)
morenoi Koch, 1944b: 239
vuillefroyi Perez Arcas, 1868: 505 (*Asida*)
parallela tijolensis Escalera, 1923a: 468 (*Asida*) **E**: SP
pernitida Antoine, 1942b: 45 (*Glabrasida*) **N**: MO
pilosipennis Escalera, 1923c: 85 (*Asida*) **E**: SP
politicollis gracilitarsis Antoine, 1957: 43 **N**: MO
politicollis politicollis Fairmaire, 1868: 489 (*Asida*) **N**: AG
porcata Solier, 1836c: 448 (*Asida*) **E**: SP
pozuelensis Escalera, 1923b: 41 (*Asida*) **E**: SP
calva Escalera, 1923b: 42 (*Asida*) [HN]
primoti Antoine, 1933a: 204 (*Gymnetasida*) **N**: MO
proconsularis Kwieton, 1986a: 20 **N**: TU
pseudogoudoti Escalera, 1923b: 39 (*Asida*) **E**: SP
puncticollis moltonii Canzoneri, 1972b: 291 **E**: IT (Sicilia)
puncticollis puncticollis Solier, 1836c: 439 (*Asida*) **N**: AG TU
vagecostata Fairmaire, 1875a: 528 (*Asida*)
puncticollis tirellii Leoni, 1929: 149 (*Asida*) **E**: IT (Sicilia)
ragusae Leoni, 1929: 151 (*Asida*)
punctigranosa Escalera, 1923b: 39 (*Asida*) **E**: SP
punctipennis Perez Arcas, 1865b: 435 **E**: SP
querensis Escalera, 1923b: 39 (*Asida*) **E**: SP
rabatica Escalera, 1910c: 414 **N**: MO
raymondae Antoine, 1937: 190 **N**: MO
renata Koch, 1937: 416 **N**: LB
riffensis Antoine, 1933a: 210 (*Gymnetasida*) **N**: MO
robusta Escalera, 1923a: 471 (*Asida*) **E**: SP
rolphii Fairmaire, 1867: 404 (*Asida*) **N**: MO
rotundicollis Escalera, 1913: 40 (*Glabrasida*) **N**: MO
rugosa Fabricius, 1775: 252 (*Pimelia*) **N**: MO
fabricii Allard, 1869a: 245 (*Asida*)
ruiderensis Escalera, 1923a: 472 (*Asida*) **E**: SP
sagrensis Escalera, 1923a: 471 (*Asida*) **E**: SP
saharensis Koch, 1937: 408 **N**: LB
saintpierrei Allard, 1869a: 239 (*Asida*) **N**: AG
salmantina Escalera, 1923b: 38 (*Asida*) **E**: SP
schatzmayri Koch, 1940d: 305 **N**: MO
scobina otini Antoine, 1937: 188 **N**: MO
scobina scobina Antoine, 1933a: 206 (*Gymnetasida*) **N**: MO
semiresecta Antoine, 1933a: 208 (*Gymnetasida*) **N**: MO
serpiginosa dermatodes Fairmaire, 1868: 488 (*Asida*) **N**: AG
clypeata Allard, 1869a: 256 (*Asida*)
serpiginosa laevicollis Allard, 1869a: 243 (*Asida*) **N**: AG
serpiginosa nigerrima Allard, 1869a: 242 (*Asida*) **N**: AG

serpiginosa scabrata Fairmaire, 1868: 485 (*Asida*) **N**: AG
allardi Baudi di Selve, 1875c: 112 (*Asida*)
opaca Allard, 1869a: 235 (*Asida*)
tumida Reitter, 1917a: 22
serpiginosa serpiginosa Erichson, 1841: 180 (*Asida*) **N**: AG
servillei nitidicollis Reitter, 1917a: 35 **N**: TU
servillei servillei Solier, 1836c: 443 (*Asida*) **N**: AG
affinis Lucas, 1846: 323 (*Asida*)
depressa Solier, 1836c: 444 (*Asida*) [HN]
dufourii Baudi di Selve, 1875c: 110 (*Asida*)
minor Solier, 1836c: 444 (*Asida*) [HN]
miliaris Erichson, 1841: 179 (*Asida*)
pseudotuberculifera Reitter, 1917a: 24
subcostata Solier, 1836c: 446 (*Asida*)
silphoides silphoides Linnaeus, 1767: 677 (*Tenebrio*) **N**: AG MO
laevigata Fabricius, 1792a: 108 (*Blaps*)
silphoides henoni Fairmaire, 1868: 486 (*Asida*) **N**: AG TU
dissimilis Allard, 1869a: 241 (*Asida*)
squalida Allard, 1869a: 289 (*Asida*) **E**: SP
strangulata Escalera, 1923a: 465 (*Asida*) **E**: SP
subgracilis Escalera, 1914: 304 (*Glabrasida*) **N**: MO
sulcata Allard, 1869a: 271 (*Asida*) **E**: SP
syriaca Allard, 1869a: 287 (*Asida*) **A**: SY
tangeriana Solier, 1836c: 456 (*Asida*) **N**: MO
barbara Allard, 1869a: 262 (*Asida*)
timaditensis timaditensis Escalera, 1925a: 324 (*Glabrasida*) **N**: MO
hebriensis Antoine, 1933a: 202 (*Gymnetasida*)
mrirtensis Antoine, 1933a: 202 (*Gymnetasida*)
pulvisculosa Antoine, 1953b: 211
toletana Escalera, 1923a: 472 (*Asida*) **E**: SP
tricostata Allard, 1869a: 238 (*Asida*) **N**: AG MO
crassipes Allard, 1869a: 256 (*Asida*)
tuberculipennis sarroensis Antoine, 1951b: 314 (*Glabrasida*) **N**: MO
tuberculipennis tuberculipennis Escalera, 1910c: 410 (*Glabrasida*) **N**: MO
turrillensis Escalera, 1923a: 466 (*Asida*) **E**: SP
uhagoni Escalera, 1912: 166 (*Asida*) **E**: SP
vetusta Antoine, 1937: 187 **N**: MO
vidali Antoine, 1937: 192 **N**: MO
villososulcata Allard, 1869a: 247 (*Asida*) **N**: AG
vinalopensis Viñolas & Cartagena, 2005: 23 (*Glabrasida*) **E**: SP
zapaterii Perez Arcas, 1872: 103 (*Asida*) **E**: SP
granulosa Escalera, 1923a: 470 (*Asida*)
minor Escalera, 1923a: 470 (*Asida*) [HN]
subgenus *Granasida* Reitter, 1917a: 14 type species *Asida granulifera* Chevrolat, 1840
granulifera Chevrolat, 1840a: 16 (*Asida*) **E**: PT SP
subgenus *Machlasida* Escalera, 1907: 338 type species *Asida muleyhafidi* Escalera, 1907
Protomachlasida Escalera, 1927a: 137 type species *Machlasida liouvillei* Escalera, 1925
acuticosta Fairmaire, 1880b: 190 (*Asida*) **N**: MO
lucusi Escalera, 1922b: 171 (*Machlasida*)
antoinei antoinei Escalera, 1922b: 172 (*Machlasida*) **N**: MO
latior Antoine, 1937: 184
antoinei bremondi Antoine, 1934: 108 (*Machlasida*) **N**: MO
azilalensis Escalera, 1925c: 315 (*Machlasida*) **N**: MO
bletoni Antoine, 1937: 182 **N**: MO
boissangeri Antoine, 1942b: 44 (*Machlasida*) **N**: MO
bordei Théry, 1931b: 122 **N**: MO

corvus Antoine, 1934: 117 (*Machlasida*) **N**: MO
finei Kocher, 1964: 43 **N**: MO
gattefossei Antoine, 1934: 115 (*Machlasida*) **N**: MO
hachtamii Escalera, 1910a: 283 (*Machlasida*) **N**: MO
joliveti Cobos, 1968: 911 **N**: MO
kraatzi kraatzi Allard, 1869a: 293 (*Asida*) **N**: MO
kraatzi occidentalis Escalera, 1922b: 170 (*Machlasida*) **N**: MO
kraatzi parallelissima Escalera, 1927a: 138 (*Machlasida*) **N**: MO
liouvillei Escalera, 1925c: 313 (*Machlasida*) **N**: MO
muleyhafidi Escalera, 1907: 336 (*Machlasida*) **N**: MO
nitidicosta Antoine, 1934: 114 (*Machlasida*) **N**: MO
olcesei crassimargo Antoine, 1937: 184 **N**: MO
olcesii granipuncticollis Escalera, 1927a: 146 (*Machlasida*) **N**: MO
olcesii olcesii Fairmaire, 1867: 405 (*Asida*) **N**: MO
olcesei safiana Antoine, 1937: 184 **N**: MO
segonzaci Escalera, 1914: 302 (*Machlasida*) **N**: MO
secsaui Escalera, 1913: 39 (*Machlasida*) **N**: MO
singularis Reitter, 1917a: 14 **N**: MO
surcoufi Antoine, 1934: 106 (*Machlasida*) **N**: MO
telueti Escalera, 1910a: 283 (*Machlasida*) **N**: MO
theryi Escalera, 1925c: 314 (*Machlasida*) **N**: MO
variesculpta tazultensis Kocher, 1953: 63 (*Machlasida*) **N**: MO
variesculpta variesculpta Antoine, 1951b: 314 (*Machlasida*) **N**: MO
zemensis meskina Antoine, 1934: 109 (*Machlasida*) **N**: MO
zemensis zemensis Escalera, 1925c: 316 (*Machlasida*) **N**: MO

genus *Asida* Latreille, 1802: 269 type species *Opatrum griseum* Fabricius, 1792 (= *Opatrum griseum* Rossi, 1790)
subgenus *Asida* Latreille, 1802: 269 type species *Opatrum griseum* Fabricius, 1792 (= *Opatrum griseum* Rossi, 1790)
Dolichasida Reitter, 1917a: 40 type species *Asida moraguesi* L. W. Schaufuss, 1879
Euryasida Reitter, 1917a: 40 type species *Asida barceloi* Perez Arcas, 1868
Insulasida Escalera, 1922a: 64 type species *Asida moraguezi* L. W. Schaufuss, 1879
Leptasida Reitter, 1917a: 60 [HN] type species *Asida diecki* Allard, 1869
Rugasida Escalera, 1922a: 65 type species *A. reticulata* Solier, 1836
abrupta Fairmaire, 1867: 406 **N**: AG
acuticollis acuticollis Allard, 1869a: 205 **E**: AL BH CR
acuticollis ganglbaueri J. Müller, 1917: 15 **E**: BH CR
alcarazensis Reitter, 1917a: 44 **E**: SP
argentierae Leo, 1980c: 189 **E**: IT (Sardegna)
ascoensis F. Soldati & L. Soldati, 2001b: 180 **E**: FR (Corse)
australis Baudi di Selve, 1875b: 27 **E**: IT (Sardegna)
baguenai Escalera, 1943a: 201 **E**: SP
banatica Frivaldszky von Frivald, 1865: 200 **E**: BU RO
barceloi Perez Arcas, 1868: 494 **E**: SP (Baleares)
bayardi bayardi Solier, 1836c: 423 **E**: IT
bayardi blaptoides Leoni, 1909: 222 **E**: IT
bayardi leosinii Leoni, 1909: 226 **E**: IT
bejarensis Escalera, 1922a: 68 **E**: SP
calabra Leoni, 1909: 222 **E**: IT
cardonae Perez Arcas, 1868: 496 **E**: SP (Baleares)
horrens L. W. Schaufuss, 1869: 24
reichei Allard, 1869a: 211
carinata carinata Solier, 1836c: 426 **E**: FR (Corse)
carinata lepidoptera Allard, 1869a: 194 **E**: FR (Corse)

squamigera Reitter, 1917a: 58
cazorlensis Escalera, 1922a: 68 **E**: SP
cephalonica Reitter, 1917a: 48 **E**: GR
christinae F. Soldati & Leo, 2005: 342 **E**: FR (Corse)
coachei F. Soldati & Leo, 2005: 343 **E**: FR (Corse)
cocquempoti F. Soldati & L. Soldati, 2001c: 186 **E**: AL MC GR
coimbrensis Escalera, 1922a: 67 **E**: PT
combae Gené, 1839: 71 **E**: IT (Sardegna)
proxima Leoni, 1910: 74
corsica corsica Laporte, 1833: 33 **E**: FR (Corse) IT (Sardegna)
corsica genei Solier, 1836c: 435 **E**: IT (Sardegna)
dominula Reitter, 1917a: 53
ignorata Reitter, 1917a: 53
curta Allard, 1869a: 220 **N**: AG TU
defricta Peyerimhoff, 1920c: 245 **N**: AG
dejeanii Solier, 1836c: 420 **E**: FR IT
crispata Rey, 1892: 42
detrita Rey, 1892: 42
sinuans Rey, 1892: 42
devillei Leoni, 1910: 54 **E**: FR (Corse)
diecki Allard, 1870: 50 **E**: SP
meridionalis Escalera, 1922a: 69
doderoi Leoni, 1910: 49 **E**: IT (Sardegna)
dorgaliensis dorgaliensis Leoni, 1911: 3 **E**: IT (Sardegna)
dorgaliensis montalbica Reitter, 1917a: 56 **E**: IT (Sardegna)
fairmairei fairmairei Boieldieu, 1865: 8 **E**: GR
fairmairei graeca Allard, 1869a: 290 **E**: GR
fairmairei moriasica F. Soldati & L. Soldati, 2001c: 193 **E**: GR
fairmairei parnassiana F. Soldati & L. Soldati, 2001c: 192 **E**: GR
fascicularis brattiensis J. Müller, 1917: 9 **E**: CR
fascicularis fascicularis Germar, 1817: 190 (*Platynotus*) **E**: CR IT SL
fascicularis fiorii Leoni, 1909: 224 **E**: IT
fascicularis lineatocollis Küster, 1849a: 26 **E**: AL BH CR GR
setulifera Küster, 1849a: 25
terricola Küster, 1849a: 27
fascicularis meledana J. Müller, 1917: 10 **E**: CR
fascicularis pubipennis J. Müller, 1917: 12 **E**: CR
gestroi capraiensis Gridelli, 1972: 196 **E**: IT
gestroi gardinii Lo Cascio, 2000: 159 **E**: IT
gestroi gestroi Leoni, 1910: 45 **E**: IT
obliterata Leoni, 1910: 46
gestroi lanzai Leo, 1998: 74 **E**: IT
gestroi thyrrena Leoni, 1910: 46 **E**: IT
gibbicollis Perez Arcas, 1865b: 423 **E**: PT
angusta Allard, 1869a: 217
fuentei Fairmaire, 1896e: 345
glacialis barbaricina Leoni, 1911: 2 **E**: IT (Sardegna)
glacialis glacialis Gené, 1839: 69 **E**: IT (Sardegna)
solarii Leoni, 1910: 77
glacialis rustica Gené, 1839: 70 **E**: IT (Sardegna)
exculpta Baudi di Selve, 1875b: 30
piriensis Leoni, 1910: 79
undulata Leoni, 1910: 81
goryi Solier, 1836c: 424 **E**: IT
angustata Leoni, 1910: 15
atriventris Reitter, 1917a: 57

cribricollis Leoni, 1910: 15
grisea Rossi, 1790: 57 (*Opatrum*) **E**: IT
grisea Fabricius, 1792a: 88 (*Opatrum*)
heikertingeri Schuster, 1928f: 949 [RN] **E**: SP
segurensis Escalera, 1922a: 68 [HN]
inaequalis carthagina Kwieton, 1986b: 11 **N**: TU
inaequalis inaequalis Solier, 1836c: 428 **N**: AG TU
foveicollis Allard, 1883: 27
rhytirrhina Reitter, 1917a: 50
incerta Leoni, 1910: 87 [DA] **E**: IT
ligurica Baudi di Selve, 1875b: 27 **E**: FR IT
longicollis Solier, 1836c: 425 **E**: FR (Corse) IT
squamulata Leoni, 1910: 88
lostiae Allard, 1888: 333 **E**: IT (Sardegna)
luigionii doriae Leoni, 1910: 43 **E**: IT
luigionii insularis Leoni, 1910: 94 **E**: IT
luigionii luigionii Leoni, 1910: 40 **E**: IT
lulensis Reitter, 1917a: 57 **E**: IT (Sardegna)
lutosa Solier, 1836c: 422 **E**: BU MD RO UK ST
maltinii Ardoin, 1971a: 47 **N**: TU
massiliensis Baudi di Selve, 1875c: 115 [DA] **E**: FR
melisellensis melisellensis J. Müller, 1946: 115 **E**: CR
melisellensis pomoensis J. Müller, 1950: 13 **E**: CR
minima Reitter, 1917a: 46 **E**: IT (Sicilia)
minuta Rambur, 1838: pl. 20 **E**: SP
hebes Rosenhauer, 1856: 200
moraguezi L. W. Schaufuss, 1879: 534 **E**: SP (Baleares)
novasiriensis Grimm, 1985: 10 **E**: IT
nuragica Leo, 1985a: 73 **E**: IT (Sardegna)
ochsi Ardoin, 1958b: 37 **E**: FR
piligera Leoni, 1909: 221 **E**: IT
pirazzolii pirazzolii Allard, 1869a: 200 **E**: IT
intermedia Leoni, 1909: 184
pirazzolii sardiniensis Allard, 1869a: 291 **E**: IT
baudii Leoni, 1909: 186
planipennis cabrerensis Viñolas & Cartagena, 2005: 24 **E**: SP (Baleares)
planipennis minoricensis Español, 1954a: 37 **E**: SP (Baleares)
planipennis planipennis L. W. Schaufuss, 1869: 23 **E**: SP (Baleares)
balearica Escalera, 1922a: 65
anodonta Wilke, 1921: 258
pseudoreticulata Escalera, 1922a: 67 **E**: PT
punica Kwieton, 1986b: 10 **N**: TU
sabulosa reticulata Solier, 1836c: 419 **E**: FR PT SP
convexicollis Allard, 1883: 27
reitteri Gebien, 1906b: 248 [RN]
sabulosa sabulosa Fuessly, 1775: 20 (*Tenebrio*) **E**: BE FR GE IT LU SZ
catenulata Mulsant, 1854: 92
consanguinea Allard, 1869a: 292
glabricosta Solier, 1836c: 416
helvetica Solier, 1836c: 418
insidiosa Mulsant, 1854: 92
morbillosa Fabricius, 1801a: 140 (*Platynotus*)
obesa Allard, 1869a: 179
rugosa Geoffroy, 1785: 157 (*Tenebrio*)
undata Fabricius, 1801a: 139 (*Platynotus*)
vicina Solier, 1836c: 417
sardoa Leoni, 1910: 82 **E**: IT (Sardegna)

schawalleri F. Soldati & L. Soldati, 2003b: 43 **E**: GR
schusteri Reitter, 1917a: 52 **E**: FR (Corse) IT (Sardegna)
serripes Chevrolat, 1874: 157 **E**: SP
solieri fancelloi Leo, 1985a: 76 **E**: IT (Sardegna)
solieri solieri Gené, 1836b: iv [= 1836a: 197] **E**: IT (Sardegna)
spaethi Schuster, 1928f: 947 [RN] **E**: SP
lanceocollis Reitter, 1917a: 44 [HN]
stierlini Reitter, 1917a: 56 **E**: FR (Corse)
transmontana Escalera, 1922a: 68 **E**: PT
subgenus *Globasida* Escalera, 1905a: 430 type species *Asida oblonga* Rambur, 1838
almeriana almeriana Escalera, 1905a: 437 **E**: SP
almeriana zarcoi Koch, 1944b: 241 **E**: SP
barrosi Escalera, 1921a: 119 **E**: PT
cartagenica Escalera, 1905a: 433 **E**: SP
cincta Rambur, 1838: pl. 20 **E**: SP
cincta Rosenhauer, 1856: 196 [HN]
curvatipennis curvatipennis Escalera, 1905a: 440 **E**: SP
curvatipennis dubiosa Escalera, 1905a: 440 **E**: SP
curvatipennis mateui Viñolas & Cartagena, 2005: 25 **E**: SP
setosa Escalera, 1905a: 441 [HN]
curvatipennis quadrata Escalera, 1905a: 439 **E**: SP
curvatipennis schrammi Escalera, 1906b: 193 **E**: SP
deformis Escalera, 1905a: 437 **E**: SP
gambeyi Allard, 1869a: 184 **N**: AG
ludovici Perez Arcas, 1874: 136 **E**: SP (Baleares)
minorata Escalera, 1925e: 408
monchiquensis Escalera, 1921a: 118 **E**: PT
mollicoma Reitter, 1917a: 70 **E**: SP
nerjensis Escalera, 1905a: 438 **E**: SP
oblonga bacaresensis Escalera, 1905a: 435 **E**: SP
oblonga frigida Escalera, 1905a: 436 **E**: SP
oblonga intermedia Escalera, 1905a: 435 **E**: SP
oblonga novissima Escalera, 1905a: 436 **E**: SP
oblonga oblonga Rambur, 1838: pl. 20 **E**: SP
bonvouloiri Allard, 1869a: 210
elongata Rambur, 1838: pl. 20
oblonga rotunda Escalera, 1905a: 436 **E**: SP
segurensis Escalera, 1905a: 439 **E**: SP
setosa Perez Arcas, 1874: 139 **E**: SP
setigera Gebien, 1910a: 137 [RN]
sinuaticollis cristata Cobos, 1963a: 19 **N**: MO
sinuaticollis sinuaticollis Solier, 1836c: 433 **N**: AG MO
mauritana Escalera, 1909a: 135
subgenus *Peltasida* Reitter, 1917a: 39 type species *Asida favieri* Fairmaire, 1880
favieri confluens Antoine, 1933a: 195 **N**: MO
favieri favieri Fairmaire, 1880b: 190 (*Asida*) **N**: MO
subgenus *Planasida* Escalera, 1907: 337 type species *Asida paulinoi* Perez Arcas, 1868
Gracilasida Escalera, 1909a: 135 type species *Asida ariasi* Escalera, 1909
Granulasida Escalera, 1921c: 427 type species *Asida setipennis* Allard, 1869
Pseudoplanasida Escalera, 1921b: 360 type species *Asida pygmaea* Rosenhauer, 1856
Trachasida Reitter, 1917a: 62 type species *Asida ruficornis* Solier, 1836
akisoides Reitter, 1917a: 62 **E**: SP
akisoides Escalera, 1921b: 358 [HN]
alcoyana Escalera, 1921c: 436 **E**: SP
amplicollis Escalera, 1921b: 356 **E**: SP

basiplicata Heyden, 1884b: 360 **E**: SP
bereai africana Escalera, 1925a: 333 (*Pseudoplanasida*) **N**: MO
bereai bereai Escalera, 1907: 337 **N**: MO
bereai vaucheri Escalera, 1907: 339 **N**: MO
breiti Schuster, 1928f: 949 [RN] **E**: SP
acuticollis Escalera, 1921b: 355 [HN]
breuili Escalera, 1921b: 352 **E**: SP
brucki Allard, 1869a: 292 **E**: SP
ghilianii Baudi di Selve, 1875b: 25
approximata Escalera, 1921c: 429
garruchensis Escalera, 1921c: 429
larrinuae Escalera, 1921c: 430
nuda Escalera, 1921c: 430
candidoi Escalera, 1912: 167 **E**: SP
confusa Kraatz, 1874: 114 **E**: SP
cossyphoides Antoine, 1951b: 315 **N**: MO
curvicollis Escalera, 1921b: 353 **E**: PT
esteparia Reitter, 1917a: 59 **E**: SP
fallax Escalera, 1921c: 431 **E**: SP
granularis Escalera, 1921c: 431
granatensis Escalera, 1921c: 428 **E**: SP
hispidula Pic, 1903g: 114 **N**: TU
pilosula Gebien, 1910a: 134 [RN]
humeralis Escalera, 1943a: 199 **E**: SP
inquinata Rambur, 1838: pl. 20 **E**: SP **N**: MO
inquinata Rosenhauer, 1856: 195 [HN]
inquinatoides Escalera, 1921b: 356 **E**: PT
ithae Reitter, 1917a: 62 **E**: SP
lanceocollis Escalera, 1914: 300 (*Planasida*) **N**: MO
latissima Wilke, 1921: 259 **E**: PT
lauzannei arribasi Cobos, 1961: 87 **N**: MO
lauzannei freyi Koch, 1940d: 308 **N**: MO
lauzannei lauzannei Rotrou, 1938: 119 **N**: MO
marginicollis Rambur, 1838: pl. 20 **E**: SP
marginicollis Rosenhauer, 1856: 197 [HN]
mater cunicularia Escalera, 1925e: 406 **E**: SP (Baleares)
mater gasulli Español, 1951a: 16 **E**: SP (Baleares)
mater inmarginata Escalera, 1925e: 407 **E**: SP (Baleares)
mater josefinae Español, 1951a: 17 **E**: SP (Baleares)
mater mater Escalera, 1925e: 404 **E**: SP (Baleares)
morae Perez Arcas, 1868: 502 (*Asida*) **E**: SP
moczarskii Schuster, 1928f: 950 [RN] **E**: SP
minima Escalera, 1921b: 363 [HN]
moroderi Reitter, 1917a: 65 **E**: SP
moroderi Escalera, 1921c: 435 [HN]
normandi Koch, 1940d: 309 **N**: TU
oertzeni Escalera, 1921c: 434 **E**: SP
paludosa Escalera, 1921c: 436 **E**: SP
paulinoi Perez Arcas, 1868: 499 **E**: PT
planidorsis Escalera, 1921b: 357 **E**: SP
pseudosetipennis Escalera, 1921c: 431 **E**: SP
pusillima Kraatz, 1874: 112 **E**: SP
pygmaea filabrica Cobos, 1988a: 8 **E**: SP
pygmaea lencinai Viñolas & Cartagena, 2005: 26 **E**: SP
pygmaea lindaraja Escalera, 1921b: 361 **E**: SP

pygmaea moraima Escalera, 1921b: 362 **E**: SP
pygmaea pygmaea Rosenhauer, 1856: 199 **E**: SP
pygmaea suarezi Cobos, 1988a: 7 **E**: SP
radulicollis Kwieton, 1986b: 4 **N**: TU
ricoi cobosi Viñolas, 1989: 33 **E**: SP
ricoi laufferi Escalera, 1921c: 434 **E**: SP
ricoi murciana Escalera, 1921c: 433 **E**: SP
ricoi ricoi Martinez y Sáez, 1873b: 412 **E**: SP
ricoi salinosa Escalera, 1921c: 433 **E**: SP
rufa Escalera, 1921b: 351 **E**: PT
ruficornis antoinei Cobos, 1963a: 35 **N**: MO
ruficornis ariasi Escalera, 1909a: 135 **N**: MO
ruficornis kocheri Cobos, 1965b: 55 **N**: MO
ruficornis lefrancei Reitter, 1917a: 66 **N**: AG
ruficornis pardoi Cobos, 1963a: 35 **N**: MO
ruficornis rotroui Cobos, 1963a: 35 **N**: MO
ruficornis ruficornis Solier, 1836c: 421 **N**: AG MO TU
  complanata Lucas, 1846: 322
  gebieni Reitter, 1917a: 66
sacculipennis Kwieton, 1986b: 8 **N**: TU
setipennis Allard, 1869a: 218 **E**: SP
timida Escalera, 1921b: 350 **E**: SP
valentina Reitter, 1917a: 59 **E**: SP
  valentina Escalera, 1921c: 432 [HN]
wankai Reitter, 1917a: 65 **E**: SP
zaida Reitter, 1917a: 61 **E**: SP
  zaida Escalera, 1921c: 437 [HN]
zoraida elegans Cobos, 1988a: 7 **E**: SP
zoraida zoraida Escalera, 1921c: 437 **E**: SP
subgenus *Polasida* Reitter, 1917a: 41 type species *Opatrum sericeum* A. G. Olivier, 1795
  Opatrasida Escalera, 1922a: 69 type species *Asida jurinei* Solier, 1836
jurinei jurinei Solier, 1836c: 429 **E**: FR SP
  bigorrensis Solier, 1836c: 430
  mahonis Boieldieu, 1865: 9
  pazii Perez Arcas, 1865b: 425
  subcylindrica Leoni, 1909: 178
jurinei pyrenaea Baudi di Selve, 1875c: 117 **E**: FR
levantina Escalera, 1922a: 70 **E**: SP
marmottani C. Brisout de Barneville, 1863: 88 **E**: AN FR SP
poneli F. Soldati & L. Soldati, 2001a: 103 **E**: FR IT
saguntina Escalera, 1922a: 71 **E**: SP
sericea A. G. Olivier, 1795: 56: 5 (*Opatrum*) **E**: FR SP
  difficilis Allard, 1869a: 200
subgenus *Pseudoelongasida* Escalera, 1922b: 173 type species *Asida silvestrei* Escalera, 1922
silvestrei Escalera, 1922b: 174 (*Asida*) **N**: MO

tribe Asidini, species incertae sedis
Alphasida beduina Reitter, 1917a: 26 **N**: MO
Asida anceyi Allard, 1870: 50 **A**: "Syria"
Asida nitida Bedel, 1918: 363 **N**: MO
Asida tenuecostata Fairmaire, 1880b: 190 **N**: MO

tribe Boromorphini Skopin, 1978

genus *Boromorphus* Wollaston, 1854: 492 type species *Boromorphus maderae* Wollaston, 1854
aegyptiacus Reitter, 1908h: 217 **N**: EG

armeniacus Reitter, 1889a: 33 **E**: AB AR
libanicus Baudi di Selve, 1881: 286 **A**: IS LE "Syria"
maderae Wollaston, 1854: 493 **N**: MR
opaculus Reitter, 1887e: 521 **E**: AB AR **A**: AF IQ KZ TM UZ
parvus Wollaston, 1864: 502 **N**: CI
tagenioides Lucas, 1846: 338 (*Boros*) **E**: SP **N**: AG MO MR TU

tribe Ceratanisini Gebien, 1937

genus *Ceratanisus* Gemminger, 1870b: 1818 [RN] type species *Anisocerus tristis* Faldermann, 1837
Anisocerus Faldermann, 1837: 39 [HN] type species *Anisocerus tristis* Faldermann, 1837
Seidlitzellus Reitter, 1920b: 221 [RN]
funebris Reitter, 1898d: 106 (*Anisocerus*) **A**: TR
tristis Faldermann, 1837: 40 (*Anisocerus*) **E**: AR **A**: TR
turkestanicus Reitter, 1898d: 106 (*Anisocerus*) **A**: UZ (Tashkent)

genus *Haemerophygus* Baudi di Selve, 1876c: 266 type species *Helops mucoreus* Waltl, 1838
Apolites Jacquelin du Val, 1861: 324 [HN] type species *Helops mucoreus* Waltl, 1838
Idastrandiella E. Strand, 1929: 23 [RN] type species *Helops mucoreus* Waltl, 1838
allardi Reitter, 1884b: 88 (*Apolites*) **E**: BU GR **A**: TR
graecus Kraatz, 1877: 304 (*Apolites*) **E**: GR
guerroumii F. Soldati & L. Soldati, 2002b: 177 (*Idastrandiella*) **E**: GR
mucoreus Waltl, 1838: 464 (*Helops*) **E**: GR **A**: TR
osellai Scupola, 1984: 69 (*Idastrandiella*) **E**: GR
taygetanus Reitter, 1898d: 105 (*Apolites*) **E**: GR

genus *Tenebriocephalon* Pic, 1925b: 10 type species *Tenebriocephalon piceum* Pic, 1925
Klapperichia Kaszab, 1954: 249 type species *Klapperichia mirabilis* Kaszab, 1954
piceum Pic, 1925b: 10 **A**: YUN
rufum Pic, 1925b: 10
thoracicum Pic, 1925b: 10 **A**: FUJ
mirabilis Kaszab, 1954: 250 (*Klapperichia*)

tribe Cnemeplatiini Jacquelin du Val, 1861

genus *Cnemeplatia* A. Costa, 1847: 146 type species *Cnemeplatia atropos* A. Costa, 1847
Autocera Wollaston, 1857a: 154 type species *Autocera laticeps* Wollaston, 1857
atropos africana Kaszab, 1938c: 81 **E**: PT SP **N**: AG MO TU
atropos atropos A. Costa, 1847: 148 **E**: AB AR CR GR HU IT MA PT RO SP TR **N**: EG MO TU **A**: AF IQ IS KZ SA SY TD TM TR UZ YE **AFR**
anticipes Wollaston, 1857a: 155 (*Autocera*)
rufa Tournier, 1874: lxxxvi
indica indica Fairmaire, 1896a: 16 **A**: AF NP **ORR**
laticeps Wollaston, 1857a: 155 (*Autocera*) **N**: CI MR
oxyana G. S. Medvedev, 1982: 40 **A**: UZ (Karakalpakistan)
theryi Kaszab, 1938c: 82 **E**: SP (Algecirás)
zarudnyi G. S. Medvedev, 1991: 568 **A**: IN

genus *Lepidocnemeplatia* Löbl & Merkl, 2003: 245 type species *Cnemeplatia sericea* Horn, 1870
szekessyi Kaszab, 1942a: 41 **A**: BT **ORR**

genus *Philhammus* Fairmaire, 1871a: 393 type species *Philhammus sericans* Fairmaire, 1871
subgenus *Philhamellus* Kaszab, 1962a: 83 type species *Philhammus myrmecophilus* Kaszab, 1960
leei Kaszab, 1962a: 81 **A**: XIN
myrmecophilus Kaszab, 1960a: 137 **A**: AF IN
tadzhikistanicus G. S. Medvedev, 1991: 569 **A**: TD

subgenus *Philhammus* Fairmaire, 1871a: 393 type species *Philhammus sericans* Fairmaire, 1871
Canariella Uyttenboogaart, 1929: 342 type species *Canariella arenapta* Uyttenboogaart, 1929
aharonii Reitter, 1910d: 197 **N**: EG MO **A**: IS JO **AFR**
andresi Koch, 1935a: 89 (*Philhammus*)
sericans Fairmaire, 1871a: 393 **N**: CI EG MO
arenapta Uyttenboogaart, 1929: 343 (*Canariella*)
parallelus Théry, 1931b: 124 (*Cnemoplatia*)
zaitzevi G. S. Medvedev, 1979b: 43 **A**: KZ TM

genus *Psilachnopus* Reitter, 1901e: 161 type secies *Psilachnopus cribratellus* Reitter, 1901
cribratellus Reitter, 1901e: 162 **E**: AR

tribe Elenophorini Solier, 1837

genus *Leptoderis* Billberg, 1820: 31 type species "*Akis* auct." *collare* Linnaeus, 1767
Elenophorus Dejean, 1821: 64 type species *Tenebrio collaris* Linnaeus, 1767
Elenophorus Latreille, 1829: 10 [HN] type species *Tenebrio collaris* Linnaeus, 1767
collaris Linnaeus, 1767: 677 (*Tenebrio*) **E**: CR FR IT MA SP **N**: MO

tribe Erodiini Billberg, 1820

genus *Ammodoides* Lesne, 1915: 233 type species *Arthrodeis lateripunctatus* Fairmaire, 1890
franchettii Gridelli, 1931: 220 **A**: "Arabia" **AFR**
lateripunctatus lateripunctatus Fairmaire, 1891a: 553 (*Arthrodeis*) **A**: YE
lateripunctatus yemenesis Gebien, 1938: 53 (*Spyrathus*) **A**: YE
yemenensis Gebien, 1938: 53 (*Spyrathus*) **A**: YE

genus *Ammozoides* Kaszab, 1979a: 91 type species *Ammozoum validicorne* Reitter, 1900
bulla Semenov, 1896c: 353 (*Ammozoum*) **A**: AF TM UZ
hauseri Reitter, 1894a: 48 (*Arthrodeis*) **A**: TM (Karakumy)
validicorne Reitter, 1900k: 301 (*Ammozoum*) **A**: AF TM UZ

genus *Ammozoum* Semenov, 1891a: 352 type species *Ammozoum hyalinum* Semenov, 1891
hyalinum Semenov, 1891a: 353 **A**: AF TM

genus *Amnodeis* Miller, 1858: 117 type species *Amnodeis grandis* Miller, 1858
asiaticus Miller, 1858: 122 **A**: SY TR
wagneri Chevrolat, 1877a: 113 [DA]
confluens confluens Miller, 1858: 123 **A**: IQ IS JO SI
arabicus Reitter, 1914b: 83
diffluens Reitter, 1914b: 83
nitidus Chevrolat, 1877a: 113
confluens freyi Schuster, 1937a: 46 **A**: SY
gebieni Reitter, 1914b: 82 **A**: IS JO
giganteus Reiche & Saulcy, 1857: 187 (*Anodesis*) **A**: IQ SY
grandis Miller, 1858: 120 **A**: IN IQ SY TR
intermedius Reitter, 1914b: 82 **A**: SY TR
milleri Reitter, 1914b: 81 **A**: SY TR

genus *Apentanodes* Reitter, 1914b: 53 type species *Arthrodeis globosus* Reiche, 1857
subgenus *Apentanodes* Reitter, 1914b: 53 type species *Arthrodeis globosus* Reiche, 1857
arabicus arabicus Kirchsberg, 1877b: 204 (*Arthrodeis*) **A**: AE OM SA
arabicus edomitus Koch, 1940c: 87 **A**: IS JO
buettikeri Kaszab, 1979c: 270 **A**: OM SA
globosus globosus Reiche & Saulcy, 1857: 186 (*Arthrodeis*) **E**: GR (Kríti) **A**: CY LE

syriacus Kraatz, 1865a: 11 (*Arthrodeis*) [RN]
globosus reductepleuralis Koch, 1935e: 309 **E**: GR (Ródos)
longulus Kulzer, 1956: 635 **A**: IN
morgani Kulzer, 1956: 636 **A**: IN
mesopotamicus Blair, 1923a: 119 **A**: IQ
pakistanicus Kwieton, 1981a: 2 **A**: PA (Balochistan)
persicus Koch, 1940c: 87 **A**: IN
philbyi Blair, 1933: 4 **A**: SA
subgenus *Rasphytus* Kulzer, 1956: 639 type species *Rasphytus freyi* Kulzer, 1956
freyi Kulzer, 1956: 640 **A**: PA **ORR**
globatus Kulzer, 1956: 642 **A**: PA **ORR**

genus *Arthrodeis* Solier, 1834: 513 type species *Arthrodeis rotundatus* Solier, 1834
Arthrodes Wollaston, 1864: 437 [injustified emendation]
subgenus *Apentanes* Reitter, 1914b: 45 type species *Arthrodeis occidentalis* Fairmaire, 1868
bolivari Español, 1951b: 364 **N**: MO (Tan-Tan)
murati Koch, 1943a: 500 **N**: MO
gattefossei Antoine, 1946a: 26
occidentalis erodioides Fairmaire, 1875c: 543 **N**: MO
occidentalis occidentalis Fairmaire, 1868: 484 (*Arthrodeis*) **N**: MO
oblongior Escalera, 1914: 266 **N**: MO
opaculus Fairmaire, 1884f: cxix **N**: MO
subgenus *Arthrodeis* Solier, 1834: 513 type species *Arthrodeis rotundatus* Solier, 1834
atlanticus Escalera, 1922d: 373 **N**: MO
costifrons Wollaston, 1864: 445 (*Arthrodes*) **N**: CI (Fuerteventura, Isla de Lobos, Lanzarote)
emarginatus Wollaston, 1864: 447 (*Athrodes*) **N**: CI (Fuerteventura)
geotrupoides Wollaston, 1864: 447 (*Arthrodes*) **N**: CI (Fuerteventura)
globulosus Escalera, 1922d: 375 **N**: MO
glomeratus Fairmaire, 1871a: 384 **N**: MO
fairmairei Reitter, 1914b: 52
prosternalis Reitter, 1914b: 53
subasperatus Reitter, 1914b: 52
hartungii Wollaston, 1864: 442 (*Arthrodes*) **N**: CI (Fuerteventura, Lanzarote)
heydeni Reitter, 1914b: 53 **N**: MO
densepunctatus Escalera, 1914: 268
impressifrons Baudi di Selve, 1881: 274 **A**: SA YE
kocheri Español, 1963e: 379 **N**: MO
longicornis Ecalera, 1914: 269 **N**: MO
malleatus Wollaston, 1864: 446 (*Arthrodes*) **N**: CI (Fuerteventura, Lanzarote, Graciosa)
montanus Escalera, 1914: 265 **N**: MO
parcepunctatus Wollaston, 1864: 443 (*Arthrodes*) **N**: CI (Gran Canaria)
punctator Reitter, 1914b: 53 **N**: MO
punctatulus Wollaston, 1864: 443 (*Arthrodes*) **N**: CI (Fuerteventura, Lanzarote, Isla de Lobos, Graciosa)
rotundatus confalonierii Gridelli, 1930: 196 **N**: LB
rotundatus rotundatus Solier, 1834: 515 **N**: EG **A**: SY YE
cruciatus Solier, 1834: 517
gebieni Reitter, 1914b: 53
obliteratus Solier, 1834: 516
uamarensis Escalera, 1922d: 376 **N**: MO
subgenus *Arthrodinus* Reitter, 1900j: 299 type species *Erodius obesus* Brullé, 1838
byrrhoides Wollaston, 1864: 441 (*Arthrodes*) **N**: CI (Fuerteventura)
curtus Brullé, 1838: 63 (*Erodius*) **N**: CI (Gran Canaria)
inflatus Wollaston, 1864: 439 (*Arthrodes*) **N**: CI (Graciosa, Lanzarote, Montaña Clara)
laticollis Brullé, 1838: 63 (*Erodius*) **N**: CI (Fuerteventura, Isla de Lobos, Graciosa, Lanzarote)
obesus crassus Wollaston, 1865: 389 (*Arthrodes*) **N**: CI (Gran Canaria)

perraudieri Wollaston, 1865: 58 [Appendix] (*Arthrodes*)
obesus gomerensis Oromí, 1977: 15 **N**: CI (La Gomera)
obesus obesus Brullé, 1838: 63 (*Erodius*) **N**: CI (Tenerife)
obesus simillimus Wollaston, 1864: 440 (*Arthrodes*) **N**: CI (El Hierro, La Palma)
subciliatus Wollaston, 1864: 444 (*Arthrodes*) **N**: CI
subcostatus Brullé, 1838: 64 (*Erodius*) **N**: CI (Gran Canaria)
subgenus *Kocheria* Antoine, 1946a: 25 type species *Arthrodeis rungsi* Español, 1943
rungsi Español, 1943c: 121 **N**: MO (Western Sahara) **AFR**
bulloni Español, 1943c: 122
mauritanicus Koch, 1943a: 501
genus *Arthrodeis*, nomen dubium
crypticoides Fairmaire, 1879a: 4 **A**: SA **N**: EG

genus *Arthrodibius* Lesne, 1915: 235 type species *Arthrodeis laxepunctatus* Fairmaire, 1884
subgenus *Erodibius* Löbl, Bouchard & Merkl, 2008 [see New Acts] type species *Arthrodeis cyphonotus* Fairmaire, 1887
cicatrix Fairmaire, 1879a: 4 (*Arthrodeis*) **A**: SA YE
wissmanni Gebien, 1938: 52

genus *Arthrodosis* Reitter, 1900j: 299 type species *Erodius globosus* Faldermann, 1837
afghanicus Kaszab, 1959f: 332 **A**: AF
bogatshevi G. S. Medvedev, 1976: 898 **A**: IN
bucharicus Reitter, 1915i: 131 **A**: KZ UZ
carmanicus G. S. Medvedev, 1976: 899 **A**: IN
castaneus Vauloger de Beaupré, 1897: 294 (*Arthrodes*) **A**: AF TM UZ
costatus Kaszab, 1959f: 329 **A**: AF
globosus Faldermann, 1837: 3 (*Erodius*) **E**: AB AR **A**: IN
intermedius Reitter, 1889a: 27 (*Arthrodeis*) **A**: TM
kawirensis Schuster, 1935b: 22 **A**: IN
klapperichi Kaszab, 1959f: 332 **A**: AF
lindbergi Kaszab, 1968b: 36 **A**: AF
lucidus Kaszab, 1968b: 37 **A**: AF
orientalis Faust, 1875: 165 (*Arthrodeis*) **A**: KZ TM
lobicollis Reitter, 1914b: 57
pappi Kaszab, 1979a: 111 **A**: AF
planosternum Reitter, 1915i: 130 **A**: KZ UZ
pleuralis Kaszab, 1979a: 112 **A**: AF
politus Waterhouse, 1889: 124 (*Spirathus*) **A**: AF PA
povolnyi Kaszab, 1968b: 40 **A**: AF
puncticollis Reitter, 1902f: 214 **A**: KZ UZ
pusztae Kaszab, 1979a: 108 **A**: IN
rhomboidalis Kwieton, 1981a: 6 **A**: PA (Balochistan)
schusteri Reitter, 1914b: 57 **A**: AF KZ TM UZ
fuscocastaneus Reitter, 1914b: 57
tuxeni shvrtshinorum Kwieton, 1981a: 3 **A**: PA (Quetta)
tuxeni tuxeni Kaszab, 1959e: 238 [= 1959f: 331] **A**: AF
waziristanicus Kaszab, 1979a: 115 **A**: PA

genus *Arthrohyalosis* Kaszab, 1979a: 75 type species *Arthrodosis mostofii* Pierre, 1975
mostofii Pierre, 1975: 299 (*Arthrodosis*) **A**: IN

genus *Bulbulus* Lesne, 1915: 240 type species *Arthrodeis byrrhiformis* Fairmaire, 1892
bidentulus Fairmaire, 1879a: 4 (*Arthrodeis*) **A**: SA YE
immarginatus Gebien, 1938: 51 (*Arthrodibius*)
pinguis Ancey, 1881b: 509 (*Arthrodes*)

genus *Capricephalius* Koch, 1943a: 508 type species *Arthrodibius bazmanicus* Schuster, 1938
bazmanicus Schuster, 1938b: 79 (*Arthrodibius*) **A**: AE IN OM
zarudnyi Bogatchev, 1950: 235 (*Spyrathus*) **A**: IN

genus *Diaphanidus* Reitter, 1900j: 299 type species *Arthrodeis antennatus* Reitter, 1894
subgenus *Diaphanidus* Reitter, 1900j: 299 type species *Arthrodeis antennatus* Reitter, 1894
Globularthrodosis Kaszab, 1979a: 95 type species *Diaphanidus semenowi* Reitter, 1900
antennatus Reitter, 1894a: 48 (*Arthrodeis*) **A**: TM (Karakumy)
ferrugineus Fischer von Waldheim, 1821: 12 (*Erodius*) **E**: ST **A**: KZ TM TR UZ
volganus Semenov & Bogatchev, 1940: 208 (*Diaphanidius*)
globosus Skopin, 1961b: 381 **A**: TD UZ
granulatus Bogatchev, 1950: 233 **A**: UZ
narynensis G. S. Medvedev & Kaltaev, 1979: 59 **A**: KI
semenowi Reitter, 1900l: 302 **A**: KZ TM
subgenus *Pseudodiaphanidus* Bogatchev, 1950: 234 type species *Diaphanidus robustus* Bogatchev, 1950
khashensis Kaszab, 1979a: 101 **A**: IN
lutensis Kaszab, 1979a: 100 **A**: IN
robustus Bogatchev, 1950: 235 (*Diaphanidus*) **A**: IN
talatabensis G. S. Medvedev, 1976: 900 (*Arthrodosis*) **A**: IN

genus *Erodiontes* Reitter, 1914b: 79 type species *Erodiontes vermiculatus* Reitter, 1914
Iranarthrodosis Kaszab, 1959f: 334 type species *Arthrodosis pfaundleri* Schuster, 1935
aelleni Kaszab, 1968b: 33 (*Arthrodosis*) **A**: IN
pfaundleri Schuster, 1935b: 21 (*Arthrodosis*) **A**: IN
menetriesi Bogatchev, 1950: 232 (*Arthrodosis*)
vermiculatus Reitter, 1914b: 80 **A**: IN

genus *Erodius* Fabricius, 1775: 258 type species *Erodius gibbus* Fabricius, 1775
subgenus *Dimeriseis* Solier, 1834: 530 type species *Erodius laevigatus* Olivvier, 1791
bicostatus bicostatus Solier, 1834: 536 **N**: AG EG LB MO
tripolitanus Mader, 1943: 41
bicostatus coccinelloides Koch, 1937: 330 **N**: LB
bicostatus costatulus Koch, 1939: 253 **N**: LB (Tripoli)
bicostatus leptis Koch, 1937: 330 **N**: LB
bicostatus vittatus Peyerimhoff, 1943: 22 **N**: AG MO
bicostatus pulvereus Reitter, 1914b: 60 **N**: AG TU
bilineatus bungemensis Koch, 1937: 328 **N**: LB (Tripolitana)
bilineatus mocciai Koch, 1937: 327 **N**: LB (Tripolitana, Fezzan)
bilineatus rohlfi Koch, 1941c: 258 **N**: LB (Cyrenaica)
exilipes exilipes Lucas, 1858a: clxxx **N**: AG LB MO
exilipes thiebaultii Fairmaire, 1877: 141 **N**: AG (Grand Erg Oriental)
glabratus Solier, 1834: 544 **A**: SA
graniventris Peyerimhoff, 1921c: 233 **N**: AG **AFR**
kneuckeri kneuckeri Andres, 1920: 61 **N**: EG (Ismailia) **A**: IS SI
kneuckeri semisculptus Koch, 1935a: 12 **N**: EG (Gebel Asfar)
latreillei Solier, 1834: 537 **N**: AG LB **AFR**
mateui Español, 1946: 108 **N**: MO
opacus Kraatz, 1865a: 25 **N**: EG **A**: IS SI
peyerimhoffi Koch, 1939: 258 **N**: AG (In Salah)
piestognathoides Koch, 1941c: 257 **N**: AG TU
subgenus *Dirosis* Miller, 1858: 115 type species *Dirosis nervosus* Miller, 1858 (= *Erodius servillei* Solier, 1834)
antonei Kocher, 1951a: 214 **N**: MO
barbarus agedabicus Koch, 1939: 255 **N**: LB (Cyrenaica)
barbarus barbarus Solier, 1834: 538 **N**: LB TU

barbarus syrticus Koch, 1937: 334 **N**: LB
bicarinatus bicarinatus Erichson, 1841: 175 **N**: AG TU
bicarinatus oranensis Koch, 1941c: 260 **N**: AG (Oran)
bicarinatus tunesius Koch, 1941c: 263 **N**: TU
brevicollis Kraatz, 1865a: 64 **N**: AG MO
chauveneti Solier, 1834: 568 **N**: AG MO
subnitidus Solier, 1834: 579 [DA]
costatus costatus Solier, 1834: 539 **N**: EG LB (Cyrenaica)
costatus iforax Peyerimhoff, 1931b: 92 **N**: AG LB (Fezzan) **AFR**
dejeanii Solier, 1834: 543 **A**: IS SI SY
duponchelii Allard, 1865a: 388 **A**: IS SI
lineolatus Reitter, 1914b: 73 (*Derosis*)
octolineatus Reitter, 1914b: 73 (*Derosis*)
edomitus Koch, 1940c: 94 **A**: IS (Negev) JO
elegans elegans Kraatz, 1865a: 51 **N**: AG TU
semicostatus Allard, 1873: 160
elegans hadjouensis Koch, 1941c: 263 **N**: AG (Oran)
elegans oujdensis Koch, 1941c: 264 **N**: MO (Oujda)
escalerai Blair, 1935a: 103 [RN] **E**: SP **N**: MO
rabaticus Koch, 1941c: 266
reticulatus Escalera, 1914: 270 [HN]
rugosus Kraatz, 1865a: 32 [HN]
festae Schuster, 1922a: 4 **N**: LB (Tripolitania, Cyrenaica)
goryi goryi Solier, 1834: 564 **E**: PT SP (Cadiz, Andalusia)
faroensis Reitter, 1914b: 68 (*Derosis*)
inconstatus Reitter, 1914b: 68 (*Derosis*)
subobtusu Koch, 1944b: 219
goryi obtusus Allard, 1873: 221 **E**: PT SP
bolivari Koch, 1944b: 220
sevillanus Koch, 1944b: 220
granipennis fauconnieri Kocher, 1957: 372 **N**: MO (Meknès)
granipennis granipennis Fairmaire, 1871a: 384 **N**: MO
allardi Kocher, 1951a: 208
deplanatus Kocher, 1951a: 210
maximus Allard, 1873: 206
minimus Kocher, 1951a: 211
reymondi Kocher, 1951a: 211
testudo Reitter, 1914b: 69 (*Derosis*)
granipennis reitteri Kocher, 1951a: 213 **N**: MO
granipennis validus Baudi di Selve, 1874b: 192 **N**: MO
hebraicus Lillig, 1997: 55 **A**: IS SI
henoni Allard, 1873: 147 [RN] **N**: AG
antennarius Vauloger de Beaupré, 1897: 307
deserticola Vauloger de Beaupré, 1897: 307
granulosus Allard, 1865b: 391 [HN]
impressicollis impressicollis Vauloger de Beaupré, 1897: 307 **N**: AG TU
impressicollis zavattarii Koch, 1939: 254 **N**: LB
klugii Allard, 1865a: 388 **A**: IS LE
kruegeri Schuster, 1928e: 374 **N**: EG LB
laticollis laticollis Solier, 1834: 558 **N**: AG
laticollis lucescens Peyerimhoff, 1949b: 277 **N**: MO
lefrancii Kraatz, 1865a: 26 **N**: AG LB TU
maroccanus Koch, 1941c: 264 **N**: MO
multicostis Reitter, 1914b: 73 (*Derosis*) **A**: "Palestina"
nanus Vauloger de Beaupré, 1897: 308 **N**: LB TU
opaciventris garamas Peyerimhoff, 1931b: 92 **N**: AG (Hoggar)

opaciventris opaciventris Peyerimhoff, 1931b: 91 **N**: AG (Hoggar)
parvus Solier, 1834: 557 **E**: SP
pseudogoryi Koch, 1944b: 221
puncticollis maillei Solier, 1834: 546 [DA] **A**: TR
puncticollis puncticollis Solier, 1834: 556 **N**: EG LB **AFR**
contractus Kraatz, 1865a: 31
ecostatus Crotch, 1872: 267
puncticollis sinaiticus Crotch, 1872: 267 **A**: IS SI
reticulatus Reitter, 1914b: 69 (*Derosis*) **N**: MO
retrangulus Reitter, 1914b: 66 (*Derosis*) **N**: AG
rugosus Allard, 1865a: 387 **N**: AG TU
rugicollis Allard, 1873: 145
semenovi Bogatchev, 1950: 231 **A**: IQ
servillei Solier, 1834: 540 **A**: AF IN IQ SA SY
nervosus Miller, 1858: 116 (*Dirosis*)
scaber Solier, 1834: 542
singularis Kocher, 1951a: 227 **N**: AG MO
theryi Koch, 1940c: 97 **A**: SY
wagneri Erichson, 1841: 175 **N**: AG
subgenus *Eodirosis* Kwieton, 1980b: 25 type species *Erodius quadrilineatus* Kraatz, 1865
acutangulus Bogatchev, 1957: 530 **A**: IN
fabricii Solier, 1834: 552 **A**: CY
freyi Schuster, 1934b: 181 **A**: JO (Dead Sea)
subpuncticollis Schuster, 1934b: 183
matthiasi Lillig, 1994: 153 **A**: SA (Hail Prov.)
octocostatus octocostatus Peyerimhoff, 1907a: 28 **A**: IQ JO
octocostatus verruculiferus Kaszab, 1981c: 308 **A**: SA (Hail Prov.)
orientalis boyeri Solier, 1834: 553 **E**: GR (Kríti)
orientalis brevicostatus Solier, 1834: 551 **E**: GR
orientalis oblongus Solier, 1834: 555 **E**: GR (Kríti, Dodecanese Is.) **A**: TR
orientalis orientalis Brullé, 1832: 192 **E**: GR
duponti Solier, 1834: 562
longus Solier, 1834: 553
orientalis Solier, 1834: 553 [HN]
pavliceki Lillig, 1994: 152 **A**: IS (Negev)
quadrilineatus Kraatz, 1865a: 31 **N**: EG **A**: IS
gibbus Solier, 1834: 547 [HN]
reichei besnardi Kaszab, 1981c: 313 **A**: SA (Eastern Prov.)
reichei curvipes Kaszab, 1981c: 313 **A**: SA (Riyad Prov.)
reichei reichei Allard, 1875: 164 **A**: IS (Negev) JO
rothi Koch, 1940c: 90 **A**: IS (Negev) JO
wohlberedti Koch, 1940c: 93
rubalkhalianus Kaszab, 1981c: 314 **A**: OM SA
sauditus boreocentralis Kaszab, 1981c: 312 **A**: SA (Hail Prov.)
sauditus costatissimus Kaszab, 1981c: 312 **A**: AE OM SA YE
sauditus granularis Kaszab, 1981c: 311 **A**: SA (Gasim Prov.)
sauditus insularis Kaszab, 1981c: 312 **A**: BA
sauditus occidentalis Kaszab, 1981c: 311 **A**: SA (Tabuk Prov.)
sauditus sauditus Kaszab, 1981c: 310 **A**: KU SA
sauditus simplex Kaszab, 1981c: 311 **A**: SA

subgenus *Erodius* Fabricius, 1775: 258 type species *Erodius gibbus* Fabricius, 1775
audouini audouini Solier, 1834: 569 **N**: MO
audouini destefanii Failla-Tebaldi, 1887: 102 **E**: IT (Lampedusa)
audouini peiroleri Solier, 1834: 590 **E**: IT (Sardegna)

belutschistanicus Kaszab, 1957d: 63 **A**: IN
carinatus carinatus Solier, 1834: 549 **E**: SP **N**: MO
  latus Solier, 1834: 574
carinatus mendizabali Koch, 1944b: 221 **E**: SP (Almería)
chouberti Kocher, 1951a: 199 **N**: MO
convexus Grimm & Lillig, 2000: 32 **N**: AG MO
dentifer Peyerimhoff, 1920b: 21 **N**: AG (Grand Erg Occidental)
dilatatus Reitter, 1914b: 79 **N**: AG
emondi emondi Solier, 1834: 585 **N**: AG
  ambiguus Solier, 1834: 586
  marginicollis Solier, 1834: 587
emondi laevis Solier, 1834: 588 **E**: SP **N**: AG MO TU
  affinis Solier, 1834: 589
  africanus Solier, 1834: 578 [DA]
  imberbitus Reitter, 1914b: 79
  mittrei Solier, 1834: 591
  nitidicostis Allard, 1873: 220
externus duplex Escalera, 1914: 271 **N**: MO
externus externus Fairmaire, 1875c: 543 **N**: MO
gibbus cyrenaicus Schuster, 1926b: 130 **E**: LB (Cyrenaica)
gibbus gibbus Fabricius, 1775: 258 **N**: EG **A**: IS SI
  syriacus Solier, 1834: 592
lusitanicus Solier, 1834: 565 **E**: PT SP **N**: MO
nitidicollis Solier, 1834: 583 **N**: AG
  pulverulentus Allard, 1873: 208
  solieri Allard, 1873: 212
proximus Solier, 1834: 575 **E**: SP
siculus dalmatinus Kraatz, 1865a: 44 **E**: AL CR GR IT
siculus melitensis Reitter, 1914b: 77 **E**: MA
siculus neapolitanus Solier, 1834: 571 **E**: IT
siculus siculus Solier, 1834: 570 **E**: IT (Sicilia, Favignana)
  pyriformis Allard, 1873: 203
  rotundatus Kraatz, 1865a: 43 [DA]
  vicinus Solier, 1834: 582
subparallelus Solier, 1834: 584 **N**: AG
tangerianus Solier, 1834: 581 **N**: MO
  nitidiventris Solier, 1834: 576
  pseudocarinatus Kocher, 1951a: 218
tibialis Linnaeus, 1766: 678 (*Tenebrio*) **E**: PT SP **N**: MO (Tanger, Ceuta)
  bilineatus Herbst, 1799: 174
  curvipes Solier, 1834: 567
  distinctus Allard, 1865b: 394
  europaeus Solier, 1834: 560
subgenus *Zophoserodius* Reitter, 1914b: 58 type species *Erodius zophosoides* Allard, 1865
planatus Reitter, 1914b: 61 **N**: AG
  melillensis Kocher, 1951a: 222
zophosoides marrakensis Kocher, 1951a: 223 **N**: MO (Marrakesh)
zophosoides nunicus Kocher, 1956b: 248 **N**: MO
zophosoides schatzmayri Koch, 1937: 331 **N**: LB (Tripolitania)
zophosoides subbicostatus Koch, 1937: 333 **N**: LB (Agedabia)
zophosoides zophosoides Allard, 1865a: 387 **N**: AG EG LB TU **A**: SI
  dimidiatipennis Kraatz, 1865a: 30
genus *Erodius*, nomen dubium
kraatzi Reitter, 1914b: 66 (*Derosis*) **E**: SP

genus *Farsarthrosis* Kaszab, 1979a: 86 type species *Farsarthrosis benardi* Kaszab, 1979
benardi Kaszab, 1979a: 87 **A**: IN

genus *Foleya* Peyerimhoff, 1916a: 71 type species *Foleya brevicornis* Peyerimhoff, 1916
brevicornis Peyerimhoff, 1916a: 71 **N**: AG

genus *Histeromimus* Gahan, 1895: 288 type species *Histeromimus arabicus* Gahan, 1895
arabicus Gahan, 1895: 288 **A**: YE

genus *Histeromorphus* Kraatz, 1865a: 11 type species *Histeromorphus plicatus* Kraatz, 1865
plicatus plicatus Kraatz, 1965: 12 **A**: YE (Suqutra)
plicatipennis Waterhouse, 1881: 473
socotranus Lesne, 1915: 229
undatus Gahan, 1900: 8 **A**: YE (Suqutra)

genus *Hyalarthrodosis* Kaszab, 1979a: 74 type species *Arthrodosis monodi* Pierre, 1974
monodi Pierre, 1974: 230 (*Arthrodosis*) **A**: IN

genus *Hyalerodius* Kaszab, 1979a: 80 type species *Hyalerodius jirofti* Kaszab, 1979
jirofti Kaszab, 1979a: 80 **A**: IN

genus *Iranerodius* Kaszab, 1959f: 334 type species *Arthrodosis richteri* Kaszab, 1957
iranicus Kaszab, 1979a: 84 **A**: IN
menetriesi Bogatchev, 1950: 232 (*Arthrodosis*) **A**: IN
richteri Kaszab, 1957d: 62 (*Arthrodosis*) **A**: IN
beludzhistanicus G. S. Medvedev, 1976: 896 (*Arthrodosis*)
semenovi Bogatchev, 1949a: 40 [= 1950: 233] (*Arthrodosis*) **A**: IN

genus *Leptonychoides* Schawaller, 1990b: 51 type species *Leptonychoides juengeri* Schawaller, 1990
juengeri Schawaller, 1990b: 52 **A**: SA

genus *Leptonychus* Chevrolat, 1833b: 26 type species *Leptonychus erodioides* Chevrolat, 1833
curvicornis Peyerimhoff, 1916a: 73 **N**: AG
laevisternus Fairmaire, 1883l: clxxviii **N**: AG LB TU
reymondi Pierre, 1954: 24 **N**: MO
rufipennis Guérin-Méneville, 1859: clxxxvii **N**: AG
pellucidus iguidiense Pierre, 1954: 28 **A**: AF
pellucidus pellucidus Mulsant & Godart, 1865: 450 (*Erodius*) **N**: AG
convexiventris Fairmaire, 1875a: 517
pellucidus tunisiana Dajoz, 1983: 65 **N**: TU
sabulicola Chobaut, 1898a: 81 **N**: AG

genus *Piestognathoides* Kaszab, 1981c: 305 type species *Piestognathoides bahrainicus* Kaszab, 1981
bahrainicus Kaszab, 1981c: 306 **A**: BA

genus *Piestognathus* Lucas, 1858b: clxxxviii type species *Piestognathus douei* Lucas, 1858
douei Lucas, 1858b: clxxxviii **N**: AG TU
asperipennis Fairmaire, 1875a: 516

genus *Spyrathus* Kraatz, 1865a: 9 type species *Spyrathus indicus* Kraatz, 1865
assimilis Kaszab, 1957f: 290 **A**: PA
fabrii Fairmaire, 1883f: lv **A**: "Hindoustan"
gebieni Reitter, 1914b: 55 **A**: PA **ORR**
indicus Kraatz, 1865a: 10 **A**: "India orientale anteriore"

maindroni Vauloger de Beaupré, 1897: 294 (*Arthrodes*) **A**: PA **ORR**

tribe Idisiini G. S. Medvedev, 1973

genus *Idisia* Pascoe, 1866: 452 type species *Idisia ornata* Pascoe, 1866
ornata Pascoe, 1866: 452 **A**: FE JA NE NMO SC "Northeast China"
vestita Marseul, 1876c: 100 **A**: JA

tribe Kuhitangiini G. S. Medvedev, 1962

genus *Kuhitangia* G. S. Medvedev, 1962: 1184 type species *Kuhitangia kryzhanovskii* G. S. Medvedev, 1962
kryzhanovskii G. S. Medvedev, 1962: 1185 **A**: TM

tribe Lachnogyini Seidlitz, 1894

subtribe Klewariina Gebien, 1910

genus *Klewaria* Reitter, 1910a: 20 type species *Klewaria colydiiformis* Reitter, 1910
colydiiformis Reitter, 1910a: 21 **A**: TD TM

subtribe Lachnodactylina Reitter, 1904

genus *Lachnodactylus* Seidlitz, 1898a: 838 [RN] type species *Lachnopus digitalis* Seidlitz, 1893
Lachnopus Seidlitz, 1894: 476 [HN] type species *Lachnopus digitalis* Seidlitz, 1893
digitalis Seidlitz, 1894: 476 (*Lachnopus*) **A**: KZ TM

subtribe Lachnogyina Seidlitz, 1894

genus *Lachnogya* Ménétriés, 1849: 228 type species *Lachnogya squamosa* Ménétriés, 1849
orientalis Kaszab, 1982a: 57 **A**: "Bengal" **ORR**
squamosa Ménétriés, 1849: 229 **E**: AR ST **A**: AF KZ PA TM UZ
skopini J. Ferrer & Yvinec, 2004: 45

subtribe Netuschiliina J. Ferrer & Yvinec, 2004

genus *Netuschilia* Reitter, 1904a: 35 type species *Lachnopus hauseri* Reitter, 1897
hauseri Reitter, 1897e: 217 (*Lachnopus*) **E**: GG ST **A**: AF GAN HEB KZ LIA NMO PA TD TM UZ ZIN

tribe Leptodini Lacordaire, 1859

genus *Leptodes* Dejean, 1934: 181 type species *Sepidium boisduvalii* Zubkov, 1833
Leptodes Solier, 1838b: 191 [HN] type species *Sepidium boisduvalii* Zubkov, 1833
Leptodinopsis Kaszab, 1959g: 352 type species *Sepidium boisduvali* Zubkov, 1833
Leptodopsis Haag-Rutenberg, 1879b: 408 type species *Leptodopsis insignis* Haag-Rutenberg, 1879
andrejevae G. S. Medvedev, 1990a: 110 **A**: TD
asanovitshae asanovitshae G. S. Medvedev, 1990a: 107 **A**: UZ (Uramskiy Khrebet)
asanovitshae kassansaicus G. S. Medvedev, 1990a: 108 **A**: KI (Chatkalsiy Khrebet)
belousovi G. S. Medvedev, 1997a: 180 **A**: KI
boisduvalii Zubkov, 1833: 328 (*Sepidium*) **A**: KZ (Ustyurt) TM
brevicarina Ren, 1999: 48 **A**: XIN
cavicola Kaszab, 1959g: 355 **A**: AF
chinensis Kaszab, 1962a: 78 **A**: GAN NIN NMO XIN
fedtschenkoi Semenov, 1893b: 233 **A**: TD TM UZ
quadricostatus Kaszab, 1959g: 358
gurjevae G. S. Medvedev, 1990a: 105 **A**: KI
heydeni Reitter, 1892b: 95 **A**: TM
insignis Haag-Rutenberg, 1879b: 409 (*Leptodopsis*) **A**: KI XIN
kabakovi G. S. Medvedev, 1997a: 176 **A**: AF (Herat)
karatavicus G. S. Medvedev, 1990a: 109 **A**: KZ (Karatau)
kaszabi G. S. Medvedev, 1972a: 309 **A**: TD
kulzeri Kaszab, 1959g: 356 **A**: IN
lederi Reitter, 1889a: 32 **E**: AB AR

medvedevi Kaszab, 1970a: 21 **A**: KZ (Altyn-Emel Mts)
mongolicus G. S. Medvedev, 1990b: 136 **A**: MG
olegi G. S. Medvedev, 1997a: 178 **A**: AF (Herat)
ovtshinnikovi G. S. Medvedev, 1990a: 106 **A**: KI
panfilovi G. S. Medvedev, 1967: 356 **A**: KI
reimovi G. S. Medvedev & B. P. Pirnazarov, 1972: 125 **A**: UZ (Ustyurt)
rimicola Bogatchev & Kryzhanovskiy, 1960: 265 **A**: TM (Kopet Dag)
sagarensis G. S. Medvedev, 1997a: 179 **A**: AF (Ghor)
schusteri Reitter, 1916d: 134 **A**: UZ
semenowi Reitter, 1892b: 95 **E**: AB AR GG
shurabensis Bogatchev, 1967c: 359 **A**: TD (Shurab)
solieri Semenov, 1909: 302 [RN] **A**: IN TM UZ
boisduvali Solier, 1838b: 193 [HN]
stackelbergi G. S. Medvedev, 1967: 357 **A**: KI
sulcicollis Reitter, 1889k: 687 **A**: GAN QIN XIN XIZ
susamyrica G. S. Medvedev, 1997a: 179 **A**: KI
suworowi Reitter, 1907f: 205 (*Leptodopsis*) **A**: KZ (Almaatinskaya Oblast)
szekessyi Kaszab, 1962a: 79 **A**: NIN NMO SHA SHX
tadzhikistanicus G. S. Medvedev, 1979c: 71 **A**: TD
tjanschanicus egorovi G. S. Medvedev, 1990a: 113 **A**: KI
tjanschanicus tjanschanicus Semenov, 1893b: 232 **A**: KI KZ
transcaspicus Kaszab, 1959g: 352 **A**: UZ
tschitscherini Semenov, 1895a: 139 **A**: IN
turkestanicus Semenov, 1889a: 219 (*Leptodopsis*) **A**: TD UZ
zubkovi Semenov, 1909: 300 **A**: TM
subgenus *Paraleptodes* G. S. Medvedev, 1967: 354 type species *Leptodes lindbergi* Kaszab, 1959
badakhshanicus G. S. Medvedev, 1990a: 104 **A**: TD
kryzhanovskiji Kaszab, 1970a: 24 **A**: AF
lindbergi Kaszab, 1959g: 354 **A**: AF TM UZ
tadzhibaevi G. S. Medvedev, 1979c: 69 **A**: TD
terminassianae G. S. Medvedev, 1970: 404 **A**: TM
subgenus *Proleptodes* G. S. Medvedev, 1967: 354 type species *Leptodes sulcicollis* Reitter, 1889
reitteri Semenov, 1891b: 271 **A**: XIN
sulcicollis Reitter, 1889k: 687 **A**: GAN XIZ

genus *Tapenopsis* Solier, 1844: 217 type species *Tapenopsis costata* Solier, 1844
costata Solier, 1844: 218 **A**: AF IN

tribe Pimeliini Latreille, 1802

genus *Afghanopachys* Kwieton, 1978b: 30 type species *Pachyscelis haarlovi* Gridelli, 1954
freyi freyi Kaszab, 1960a: 35 (*Pachyscelis*) **A**: AF
freyi kabuliensis Kaszab, 1960a: 37 (*Pachyscelis*) **A**: AF
haarlovi Gridelli, 1954: 230 [= 1955: 41] (*Pachyscelis*) **A**: AF
zhenzhuristi Bogatchev, 1939a: 124 (*Pachyscelis*) **A**: AF
afghanica Gridelli, 1954: 228 [= 1955: 40] (*Pachyscelis*)

genus *Allotadzhikistania* Bogatchev, 1960a: 43 type species *Allotadzhikistania comata* Bogatchev, 1960
comata Bogatchev, 1960a: 44 **A**: TD
setosa G. S. Medvedev, 1970: 403 **A**: TD (Zaalaskij Khrebet)

genus *Apatopsis* Semenov, 1891a: 368 type species *Apatopsis grombczewskii* Semenov, 1891
conradti Semenov, 1891a: 370 **A**: XIN
grombczewskii Semenov, 1891a: 368 **A**: XIN

genus *Argyradelpha* G. S. Medvedev, 2005b: 306 type species *Argyradelpha lopatini* G. S. Medvedev, 2005
lopatini G. S. Medvedev, 2005b: 307 **A**: AF (Registan)

genus *Argyrophana* Semenov, 1889a: 222 type species *Argyrophana deserti* Semenov, 1889
caspia Semenov, 1910c: 43 **A**: TM
deserti Semenov, 1889a: 224 **A**: TM
diaphana Semenov, 1910c: 44 **A**: UZ (Kyzylkum)
vicaria Semenov, 1910c: 43 **A**: UZ (Kyzylkum)

genus *Astorthocnemis* Lillig & Pavlíček, 2002: 98 type species *Storthocnemis saudita* Koch, 1965
becvarorum Lillig & Pavlíček, 2002: 99 **A**: JO SI
saudita Koch, 1965a: 130 (*Storthocnemis*) **A**: SA YE

genus *Balachowskya* Peyerimhoff, 1928: 61 type species *Balachowskya portentosa* Peyerimhoff, 1928
portentosa Peyerimhoff, 1928: 62 **N**: AG

genus *Bogatshevia* G. S. Medvedev & Iwan, 2007: 618 [RN] type species *Achaemenes villosus* Bogatchev, 1949
Achaemenes Bogatchev, 1949a: 39 [HN] type species *Achaemenes villosus* Bogatchev, 1949
bogatchevi Kwieton, 1982b: 23 (*Thriptera*) [RN] **A**: IN PA
villosa Bogatchev, 1949a: 40 [HN]

genus *Diesia* Fischer von Waldheim, 1820: pl. xiv type species *Diesia sexdentata* Fischer von Waldheim, 1820
Diesiola Skopin, 1961b: 397 type species *Diesia quadridentata* Fischer von Waldheim, 1820
aralensis Skopin, 1960: 301 (*Idiesa*) **A**: KZ
quadridentata iliensis Skopin, 1961b: 401 **A**: KZ (Alamtinskaya oblast)
quadridentata quadridentata Fischer von Waldheim, 1820: pl. xiv **A**: AF KI KZ TM
sexdentata auliensis Reitter, 1901c: 78 (*Trigonoscelis*) **A**: KZ
sexdentata sexdentata Fischer von Waldheim, 1820: pl. xiv **A**: KZ TM UZ
sexdentata termesiana Skopin, 1960: 302 **A**: KZ TM
sexdentata vachshiana Skopin, 1960: 302 **A**: TD

genus *Dietomorpha* Kühnelt, 1957: 86 type species *Dietomorpha pardalis* Kühnelt, 1957
pardalis Kühnelt, 1957: 86 **A**: IN PA
szekessyi Kaszab, 1957f: 295 (*Kawiria*)

genus *Earophanta* Semenov, 1903b: 172 [RN] type species *Platyope serrata* Semenov, 1893
Earophila Semenov, 1903a: 9 [HN] type species *Platyope serrata* Semenov, 1893
Earophilina E. Strand, 1917: 99 [RN] type species *Platyope serrata* Semenov, 1893
beludzhistana Bogatchev, 1957: 531 **A**: PA (Balochistan)
collaris Fischer von Waldheim, 1844: 62 (*Platyope*) **A**: KZ XIN
regeli Ballion, 1878: 318 (*Platyope*)
holdhausi Reymond, 1937: 247 **A**: KZ (Dzungaria)
pilosissima Reitter, 1895d: 155 (*Przewalskia*) **A**: AF TM
loudoni Semenov, 1903a: 11
planidorsis Reitter, 1889h: 285 (*Platyope*) **A**: AF TM
autumnalis Semenov, 1903a: 9 (*Platyope*)
pubescens Skopin, 1960: 306 **A**: KZ (Zhambylskaya oblast)
serrata Semenov, 1893c: 254 (*Platyope*) **A**: TM (Karakumy)
tomentosa Semenov, 1893c: 256 (*Platyope*) **A**: UZ

genus *Euryostola* Reitter, 1893b: 207 type species *Pachyscelis minor* Baudi di Selve, 1875
minor Baudi di Selve, 1875d: 689 (*Pachyscelis*) **A**: IN
zophosioides Baudi di Selve, 1875d: 691 (*Leucolaephus*)

genus *Euthriptera* Reitter, 1893b: 229 type species *Thryptera grisescens* Fairmaire, 1875
delicata Bedel, 1893: 102 (*Thriptera*) **N**: TU **AFR**
grisescens Fairmaire, 1875a: 528 (*Thryptera*) **N**: AG LB TU
piceola Desbrochers des Loges, 1881: 86 [= 1898: 81] (*Ocnera*)

genus *Gedeon* Reiche & Saulcy, 1857: 219 type species *Gedeon hierichonticus* Reiche & Saulcy, 1857
hierichonticus Reiche & Saulcy, 1857: 221 **A**: IS JO SA SY
arabicus Solier, 1836a: 126 (*Pimelia*) [HN]
magniceps Reitter, 1915a: 60 **A**: AF UZ
oblongoovatus Schuster, 1937a: 48 (*Pimelia*) **A**: IN IQ
parallelus Solier, 1836a: 125 (*Pimelia*) **A**: IN IQ SY
borrei Haag-Rutenberg, 1878: 91
persicus Baudi di Selve, 1875d: 694 **A**: IN SY
baudii Ganglbauer, 1905: 277 (*Pimelia*)
remaudierei Pierre, 1964c: 1045
werneri Ganglbauer, 1905: 277 (*Pimelia*) **A**: TR

genus *Habrobates* Semenov, 1903a: 11 type species *Habrobates vernalis* Semenov, 1903
agnesae Schuster, 1938b: 86 **A**: IN
vejisovi Kelejnikova, 1977: 653 **A**: TM (Karakumy)
vernalis Semenov, 1903a: 12 **A**: TM (Karakumy)

genus *Habrochiton* Semenov, 1907a: 179 type species *Habrochiton vernus* Semenov, 1908
primaeverris Semenov, 1907a: 181 **A**: XIN
vernus Semenov, 1907a: 179 **A**: XIN

genus *Homopsis* Semenov, 1893c: 258 type species *Homopsis grumi* Semenov, 1893
grumi Semenov, 1893c: 258 **A**: XIN

genus *Idiesa* Reitter, 1893b: 245 type species *Diesia fischeri* Ménétriés, 1849
eversmanni afghanica Kaszab, 1970a: 51 **A**: AF
eversmanni eversmanni Kraatz, 1865a: 276 (*Diesia*) **A**: KZ UZ
fischeri Ménétriés, 1849: 225 (*Diesia*) **A**: TM KZ UZ
trigonosceloides Lebedev, 1931: 78 **A**: TM (Karakum)

genus *Iranolasiostola* Pierre, 1968: 1020 type species *Iranolasiostola davatchii* Pierre, 1968
davatchii Pierre, 1968: 1020 **A**: IN

genus *Iranopachyscelis* Pierre, 1968: 1027 type species *Iranopachyscelis eghbali* Pierre, 1968
persica lahlezarica Kwieton, 1982a: 37 **A**: IN
persica persica L. Redtenbacher, 1850: 48 (*Thriptera*) **A**: IN
eghbali Pierre, 1968: 1027
ispahansis Pierre, 1968: 1028

genus *Kawiria* Schuster, 1935b: 26 type species *Kawiria gabrieli* Schuster, 1935
gabrieli Schuster, 1935b: 28 **A**: IN

genus *Lasiostola* Dejean, 1834: 179 type species *Tenebrio pubescens* Pallas, 1781
subgenus *Centrocnemita* E. Strand, 1935: 284 [RN] type species *Centrocnemis mollis* Kraatz, 1882
Centrocnemis Kraatz, 1882d: 330 [HN] type species *Centrocnemis mollis* Kraatz, 1882
bendamira Kaszab, 1973b: 42 **A**: AF
lindbergi Kaszab, 1970a: 59 **A**: AF
linnei Kaszab, 1973b: 43 **A**: AF
lopatini Skopin, 1960: 304 **A**: TD

mollis Kraatz, 1882d: 330 (*Centrocnemis*) **A**: UZ (Samarkand)
seminuda pimela Kaszab, 1970a: 58 **A**: AF
seminuda seminuda Reitter, 1900g: 277 **A**: AF TD TM
tomentosa Reitter, 1900g: 278 **A**: UZ
subgenus *Lasiocnema* G. S. Medvedev, 1993: 110 type species *Lasiostola heterogena* Fischer von Waldheim, 1844
amseli Kaszab, 1962c: 312 [RN] **A**: AF
afghanica Kaszab, 1961e: 237 [HN]
carinata Kraatz, 1882b: 89 **A**: KZ UZ
costulifera Heyden, 1892b: 107 **A**: UZ (Margilan)
espanoli Skopin, 1960: 304 **A**: TD
gemmata Reitter, 1889h: 286 **A**: AF IN PA TM
grandis aideriensis G. S. Medvedev, 1993: 117 **A**: TM
grandis elongata Kraatz, 1883b: 359 **A**: TM
gracilis Waterhouse, 1889: 127 (*Ocnera*)
grandis grandis Kraatz, 1883b: 358 **A**: IN TM
grisescens Kraatz, 1883a: 350 **A**: AF KZ TM
heterogena Fischer von Waldheim, 1844: 58 **A**: IN TM
minuta Kraatz, 1865a: 285
kabakovi G. S. Medvedev, 1993: 116 **A**: AF
minima Ménétriés, 1849: 223 **A**: KZ UZ (Samarkand)
montana Kaszab, 1970a: 62 **A**: AF
nephelidis Reitter, 1893b: 223 **A**: AF IN TM
pulla Reitter, 1893b: 224 **A**: TM
stackelbergiana Bogatchev, 1949b: 279 **A**: TD
zarudnyi G. S. Medvedev, 1993: 115 **A**: IN
subgenus *Lasiograna* G. S. Medvedev, 1993: 111 type species *Lasiostola interrupta* Reitter, 1901
interrupta Reitter, 1901e: 161 (*Centrocnemis*) **A**: TM
machnovskii Skopin, 1960: 305 **A**: TD
rigrevana Kaszab, 1970a: 62 **A**: AF
seriegranosa Kaszab, 1970a: 60 **A**: AF
boreoafghanica G. S. Medvedev, 1970: 401
subgenus *Lasiostola* Dejean, 1834: 179 type species *Tenebrio pubescens* Pallas, 1781
affinis Kraatz, 1882b: 89 **A**: UZ (Margilan)
simillima Kraatz, 1882b: 89
afghanica Kaszab, 1960a: 26 **A**: AF
ashkhabadensis Bogatchev & Kryzhanovskiy, 1960: 267 **A**: TM (Kopet Dag)
crassa G. S. Medvedev, 1993: 111 **A**: TD
granulata Kraatz, 1882b: 90 **A**: UZ (Margilan)
hirta alakuliensis Skopin, 1961a: 188 **A**: KZ (Alakol)
hirta hirta Fischer von Waldheim, 1820: pl. XIV (*Pimelia*) **E**: ST **A**: UZ
nigrohirta Reitter, 1893b: 222
hirta robusta G. S. Medvedev, 1964a: 61 **A**: TM
hirtipes Kraatz, 1885: 331 **A**: UZ
laticollis Kraatz, 1882b: 90 **A**: UZ (Margilan)
piligera Kraatz, 1882d: 331 **A**: KZ UZ (Samarkand)
plustschevskyi Reitter, 1893a: 112 **E**: ST
pubescens Pallas, 1781: 53 (*Tenebrio*) **E**: ST **A**: KI KZ TM
scabricollis Reitter, 1907f: 206 **A**: KZ (Almaatinskaya Oblast)
setifera Kaszab, 1970a: 64 **A**: AF
variolaris Gebler, 1841a: 373 **A**: KZ
zarjanovi G. S. Medvedev, 1993: 113 **A**: KZ

genus *Leucolaephus* Lucas, 1859a: xxii type species *Leucolaephus perrisii* Lucas, 1859
Mecopisthopus Karsch, 1881: 46 type species *Mecopisthopus rohlfsi* Karsch, 1881
liliputanus Lucas, 1857: lvi (*Pimelia*) **N**: AG
perrisii Lucas, 1859a: xxiii
nigropunctatus Lucas, 1858a: clxxx (*Pimelia*) **N**: AG TU
rohlfsi Karsch, 1881: 47 (*Mecopisthopus*) **N**: LB
tripolitanus G. Quedenfeldt, 1889: 353 **N**: LB

genus *Mantichorula* Reitter, 1889k: 695 type species *Mantichorula semenowi* Reitter, 1889
grandis Semenov, 1893c: 263 **A**: GAN NIN NMO
mongolica Schuster, 1940a: 23 **A**: NMO
semenowi Reitter, 1889k: 695 **A**: GAN MG NIN NMO SHA XIN

genus *Meladiesia* Reitter, 1909g: 309 type species *Meladiesia miritarsis* Reitter, 1909
miritarsis Reitter, 1909g: 309 **A**: TM

genus *Ocnera* Fischer von Waldheim, 1822: 169 type species *Pimelia imbricata* Fischer von Waldheim, 1820
imbricata Fischer von Waldheim, 1820: pl. xiv (*Pimelia*) **A**: AF IN TM UZ
lepidacantha Fischer von Waldheim, 1830: 187 **A**: AF KZ TD TM
raddeana Reitter, 1889b: 116
longicollis Solsky, 1881b: 51 **A**: IN
pilicollis Faldermann, 1836: 378 (*Trachyderma*) **A**: AF IN KA KI KZ SY TM UZ
sublaevigata przewalskyi Reitter, 1887d: 380 **A**: KI XIN
sublaevigata schusteri Gridelli, 1954: 210 **A**: AF
sublaevigata sublaevigata Bates, 1879: 477 **A**: KA XIN

genus *Pachylodera* G. Quedenfeldt, 1890: 399 type species *Pachylodera brevicornis* G. Quedenfeldt, 1890
brevicornis G. Quedenfeldt, 1890: 400 **N**: EG LB TU

genus *Pachyscelina* Kwieton, 1978b: 30 type species *Pachyscelis micros* Kaszab, 1970
micros Kaszab, 1970a: 74 (*Pachyscelis*) **A**: AF

genus *Pachyscelis* Solier, 1836a: 54 [NP] type species *Pachyscelis depressa* Solier, 1836 (= *Pimelia musiva* Faldermann, 1832)
subgenus *Pachyscelis* Solier, 1836a: 54 [NP] type species *Pachyscelis depressa* Solier, 1836 (= *Pimelia musiva* Faldermann, 1832)
Brachyscelis Dejean, 1834: 179 [NO] type species *Pimelia mussiva* Ménétriés, 1832
banghaasi Reitter, 1893b: 208 **A**: AF TD TM UZ
cycladica Reitter, 1893b: 212 **E**: GR **A**: TR
euphratica Kraatz, 1865a: 310 **A**: IN IQ TR
klapperichi centralis Kaszab, 1970a: 72 **A**: AF (Saleha-lamg)
klapperichi imitata Kaszab, 1973b: 46 **A**: AF (Naurak, Terbolaq)
klapperichi klapperichi Kaszab, 1960a: 33 **A**: AF (Bande-e-Mir)
metopotapha Fischer von Waldheim, 1832a: 192 (*Pimelia*) **E**: AB **A**: IN TM
gemmans Baudi di Selve, 1875d: 690 [= 1876a: 14]
matilei Pierre, 1968: 1027
mollis Kraatz, 1882d: 329 (*Stalagmoptera*) **A**: UZ (Samarkand)
musiva achaemenia Bogatchev, 1949b: 278 **A**: IN
descampsi Pierre, 1968: 1025
escalerai Pierre, 1968: 1025
musiva divisa Reitter, 1893b: 210 **E**: AR **A**: IQ TR
musiva leprosa Faldermann, 1837: 14 (*Brachyscelis*) **E**: AB AR **A**: TR
gastridula Faldermann, 1837: 16 (*Brachyscelis*)
musiva mammillata Faldermann, 1837: 17 (*Brachyscelis*) **E**: AB **A**: IN IQ TM TR
musiva musiva Faldermann, 1832: 193 (*Pimelia*) **E**: AB AR ST **A**: IN IQ TR

bilineata Baudi di Selve, 1876a: 12
clavaria Faldermann, 1832: 193 (*Pimelia*)
depressa Solier, 1836a: 57
ordinata Solier, 1836a: 58
narynensis G. S. Medvedev, 1992: 793 **A**: KI
pauxilla Hampe, 1852: 312 (*Pimelia*) **A**: IN
remaudierei Pierre, 1968: 1026 **A**: IN (Elborz)
subgenus *Parapachyscelis* Kwieton, 1978b: 29 type species *Pimelia villosa* Drapiez, 1820
chrysomeloides A. G. Olivier, 1795: 59: 12 (*Pimelia*) **E**: GR **A**: LE
coriacea Klug, 1830: nr. 1, pl. xi (*Pimelia*)
kraatzi Baudi di Selve, 1875d: 689
piochardi Desbrochers des Loges, 1881: 95
rotundata Kraatz, 1865a: 313 **A**: IS JO LE SY
villosa Drapiez, 1820b: 118 (*Pimelia*) **E**: GR "Transcaucasus" **N**: EG **A**: IQ IS SY TR
crinita Solier, 1836a: 61
granulosa Solier, 1836a: 60
hirtella Solier, 1836a: 62
tenebrosa Solier, 1836a: 60
wettsteini Schuster, 1936b: 72

genus *Pachyscelodes* Sénac, 1887b: 189 type species *Pachyscelis henoni* Sénac, 1887
ferdinandi ferdinandi Escalera, 1911: 300 **N**: MO
orientalis Antoine, 1937: 241
ferdinandi reymondi Antoine, 1957: 55 **N**: MO
henoni Sénac, 1887b: 191 (*Pachyscelis*) **N**: AG TU
malleata Wollaston, 1875: 219 (*Pimelia*) **N**: MO (Grand Atlas)
occidentalis Escalera, 1914: 309
semiasperula Fairmaire, 1876: 38 (*Pimelia*) **N**: AG
tuberculifera debduensis Antoine, 1957: 54 **N**: MO
tuberculifera kocheri Antoine, 1934: 122 **N**: MO (Moyen Atlas)
tuberculifera lyauteyi Escalera, 1925c: 312 **N**: MO (Moyen Atlas)
tuberculifera tuberculifera Lucas, 1858c: ccxxi (*Pimelia*) **N**: AG
dayensis Mulsant & Godart, 1865: 451 (*Pimelia*)
serieperlata Fairmaire, 1876: 38 (*Pimelia*)

genus *Paraplatyope* Gridelli, 1953: 45 type species *Leucolaephus arabicus* Blair, 1930
arabica arabica Blair, 1931b: 270 (*Leucolaephus*) **A**: AE KU SA
arabica gridellii Koch, 1965a: 133 **A**: SA
arabica thesigeri Gridelli, 1953: 47 **A**: OM
arabica tihamana Koch, 1965a: 133 **A**: SA
popovi Koch, 1965a: 134 **A**: SA YE

genus *Pelorocnemis* Solsky, 1876: 283 type species *Pimelia punctigera* Ménétriés, 1849
darwini Faust, 1875: 214 (*Ocnera*) **A**: IN TM
hesperia Semenov & Bogatchev, 1940: 203
kraatzi Solsky, 1876: 285 **A**: KZ UZ
punctigera Ménétriés, 1849: 219 (*Pimelia*) [RN] **A**: AF KZ TD TM UZ
punctata Gebler, 1844: 101 (*Pimelia*) [HN]
spectabilis Kraatz, 1882d: 329 (*Pimelia*)

genus *Phymatiotris* Solier, 1836a: 63 type species *Tentyria quadricollis* Brullé, 1832
Graecopachys Skopin, 1968: 99 type species *Pachyscelis quadricollis* Brullé, 1832
parallela Solier, 1836a: 64 (*Pachyscelis*) **E**: GR
cycladica Reitter, 1893b: 212 (*Pachyscelis*)
quadricollis cretica Koch, 1948a: 327 (*Pachyscelis*) **E**: GR (Kríti)
quadricollis obscura Solier, 1836a: 65 (*Pachyscelis*) **E**: GR

porphyrea Solier, 1836a: 65 (*Pachyscelis*)
quadricollis quadricollis Brullé, 1832: 199 (*Tentyria*) **E**: GR MC **A**: TR
quadricollis smyrnensis Kraatz, 1865a: 309 (*Pachyscelis*) **A**: TR
quadraticollis thessalica Reitter, 1893b: 212 (*Pachyscelis*) **E**: GR

genus *Pimelia* Fabricius, 1775: 251 type species *Pimelia angulata* Fabricius, 1775
subgenus *Amblyptera* Solier, 1836a: 195 type species *Pimelia scabrosa* Solier, 1836
fornicata bathyglypta Antoine, 1949a: 56 **N**: MO
fornicata fornicata Herbst, 1799: 79 **E**: SP **N**: CI
obesa Solier, 1836a: 191
sculptilis Reitter, 1915a: 57
sparsegranulata Koch, 1941a: 110
fornicata rifensis Antoine, 1951c: 96 **N**: MO (Rif)
fornicata subris Koch, 1941a: 111 **N**: MO
incostata Escalera, 1914: 321 **N**: MO
intermedia Fischer von Waldheim, 1837b: 13 **A**: TM
marginata Fischer von Waldheim, 1844: 53 **A**: "Turkomania"
plinthota Fischer von Waldheim, 1837b: 17 **A**: TM
rotundipennis mogadora Fairmaire, 1871a: 387 **N**: MO
pseudorugosa Koch, 1941a: 113
subglabra Koch, 1941a: 112
rotundipennis rotundipennis Kraatz, 1865a: 370 **N**: MO
glabra Koch, 1941a: 112
simonis Sénac, 1887a: 124
rugosa Fabricius, 1792a: 103 **N**: MO
curticollis Haag-Rutenberg, 1875a: 53
fairmairii Kraatz, 1865a: 369
haagi Koch, 1941a: 105
haroldi Sénac, 1887a: 127
laevisulcata Kraatz, 1887: 363
monilis Haag-Rutenberg, 1875a: 55
tumidipennis Haag-Rutenberg, 1875a: 54
scabrosa costulata Schuster, 1929: 40 **N**: MO
scabrosa scabrosa Solier, 1836a: 188 **E**: SP **N**: MO
fornicantiformis Koch, 1941a: 105
ifranensis Antoine, 1949a: 52
scabrosiformis Reitter, 1915a: 57 **N**: MO
tristis apicesculpta Koch, 1941a: 109 **N**: MO (Mechra ben Abbou)
tristis ifniensis Antoine, 1949a: 54 **N**: MO
tristis rhatica Antoine, 1949a: 54 **N**: MO
tristis tristis Haag-Rutenberg, 1875a: 54 **N**: MO
uniformis Koch, 1941a: 110
zaerensis Antoine, 1949a: 54
ventricosa Faldermann, 1837: 7 **E**: "Transcaucasus"
subgenus *Camphonota* Solier, 1836a: 195 type species *Tenebrio subglobosus* Pallas, 1781
Eurypimelia Reitter, 1915a: 9 type species *Tenebrio subglobosus* Pallas, 1781
akbesiana akbesiana Fairmaire, 1884a: 170 **A**: SY TR
akbesiana evanescens Kwieton, 1982a: 3 **A**: TR
anatolica Reitter, 1915a: 51 **A**: TR
latitibia Reitter, 1915a: 51
asperula Solier, 1836a: 182 **E**: GR
exanthematica Solier, 1836a: 183 [HN]
monilifera Solier, 1836a: 184
lendli Kaszab, 1940a: 4 **A**: TR
monikovae Kwieton, 1982a: 3 **A**: TR
phymatodes Solier, 1836a: 187 **E**: GR (Kríti)
minos Lucas, 1853: 575

praetermissa Sénac, 1887a: 62 **E**: GR
puncticollis Reitter, 1915a: 55 **E**: GR
bittneri Reitter, 1915a: 56
repleta Reitter, 1915a: 50 **A**: TR
sericella Solier, 1836a: 185 **E**: GR (Kríti)
calculosa Solier, 1836a: 186
trachyderma Solier, 1836a: 186
subglobosa cephalenica Kraatz, 1865a: 348 **E**: GR
subglobosa graeca Brullé, 1832: 192 **E**: GR
subglobosa krueperi Reitter, 1915a: 54 **E**: GR (Morea)
subglobosa mongeneti Solier, 1836a: 177 **E**: GR
balcanica Kirchsberg, 1876: 30
granipenna Solier, 1836a: 178
subglobosa polita Solier, 1836a: 176 **E**: BU GR RO TR **A**: TR
declivis Reitter, 1915a: 52
euboica Boieldieu, 1865: 7
lesbicola J. R. Sahlberg, 1913b: 168
subglobosa salaria Reitter, 1915a: 51 **A**: TR
subglobosa separanda Reitter, 1915a: 54 **E**: GR
subglobosa subglobosa Pallas, 1781: 50 (*Tenebrio*) **E**: BU GR RO ST UK **A**: KZ
fulvopuberula Reitter, 1915a: 53
inaequalis Fischer von Waldheim, 1837b: 18
uralensis Reitter, 1915a: 53
subglobosa verruculifera Solier, 1836a: 180 **E**: BU GR MC TR **A**: TR
coordinata Fischer von Waldheim, 1837b: 17
diffusa Solier, 1836a: 180
phymatoptera Solier, 1836a: 177
varicosa Ménétriés, 1836a: 150
tomentifera Reitter, 1915a: 53
varicellosa J. R. Sahlberg, 1913b: 169
testudo Kraatz, 1865a: 362 **A**: TR
timarchoides Ménétriés, 1836a: 150 **A**: TR
intersticta Fischer von Waldheim, 1837b: 16
subgenus *Pimelia* Fabricius, 1775: 251 type species *Pimelia angulata* Fabricius, 1775
Agelarches Gistel, 1848: x [unjustified emendation]
Aphanaspis Wollaston, 1864: 478 type species *Pimelia granulicollis* Wollaston, 1864
Chaetotoma Motschulsky, 1860d: 436 type species *Tenebrio cephalotes* Pallas, 1781
Doderoella Schuster, 1926b: 133 type species *Doderoella cyrenaica* Schuster, 1926
Ecphoroma Solier, 1836a: 195 type species *Pimelia hemisphaerica* Solier, 1836
Melanostola Dejean, 1836: 198 type species *Pimelia simplex* Solier, 1836 (= *Pimelia gibba* Fabricius, 1792)
Pimidia Rafinesque, 1815: 113 [unjustified emendation]
abnormis Reitter, 1915a: 23 **A**: AF TM
adramantina Koch, 1965a: 142 **A**: YE
adriani Reitter, 1915a: 36 **A**: JO
aegyptiaca Andres, 1929: 15 **N**: EG
afghanistanica Gridelli, 1954: 236 [= 1955: 44] **A**: AF
albinea Reitter, 1915a: 11 **A**: YE
alboepisternalis Bogatchev, 1953: 302 **A**: IN
amseli Kaszab, 1961e: 238 **A**: AF
angulata angulata Fabricius, 1775: 352 **N**: EG **A**: SI
aculeata Klug, 1830: nr. 17, pl. xii
alternata Klug, 1830: nr. 16 nota
alternata Sénac, 1884a: 24 [HN]
asperrima Pallas, 1781: 55 (*Tenebrio*)
spinosa Forskål, 1775: 80 (*Tenebrio*)
angulata antiaegypta Koch, 1937: 446 **N**: EG MO (Western Sahara)
angulata expiata Peyerimhoff, 1923: 319 **N**: AG LB MO **AFR**

angulata nilotica Sénac, 1884a: 27 **N**: EG **AFR**
angulata sinaitica Schatzmayr & Koch, 1934b: 26 **A**: SI
angulata syriaca Sénac, 1884a: 24 **A**: IS
angulosa A. G. Olivier, 1795: 59: 11 **N**: LB **AFR**
angusticollis angusticollis Solier, 1836a: 163 **E**: FR (Corse)
angusticollis sulcitana Leo & Pisano, 1978: 80 **E**: IT (Sardegna)
anomaloides Löbl, Bouchard & Merkl, **nom. nov**. [see New Acts] **N**: AG
anomala Sénac, 1880: xxiv [= 1880: 261] [HN]
arabica arabica Klug, 1830: nr. 18, pl. xii **A**: AE BA IQ JO OM SA YE
serricosta Solier, 1836a: 102
arabica edomita Koch, 1940c: 102 **A**: IS JO SI
arabica emiri Kaszab, 1982b: 149 **A**: AE BA YE
arabica omanica Kaszab, 1982b: 148 **A**: AE OM
arenacea Solier, 1836a: 114 **N**: AG TU
arenosa Reitter, 1915a: 21 **N**: AG (Oran
ariasi Escalera, 1914: 311 **N**: MO
ascendens Wollaston, 1864: 473 **N**: CI (Tenerife)
atlantis atlantis Solier, 1836a: 139 **N**: MO (Moyen Atlas)
atlantis frigidoides Koch, 1941a: 85 **N**: MO (Moyen Atlas)
baetica Solier, 1836a: 170 **E**: PT SP
bajula bajula Klug, 1830: nr 8, pl. xi **N**: EG **A**: IQ IS JO LE SY
bajula solieri Mulsant & Wachanru, 1853: 8 **A**: CY IN TR
cylindrica Solier, 1836a: 124
mulsanti Reiche, 1861b: 88 [RN]
balachowskyi Pierre, 1968: 1029 **A**: IN
barthelemyi Solier, 1836b: 350 **N**: EG LB **A**: SI
bengasiana bengasiana Schuster, 1922b: 20 **N**: LB
bengasiana plurigranula Koch, 1939: 277 **N**: LB
bipunctata bipunctata Fabricius, 1781: 316 **E**: FR
aspera Germar, 1824b: 134
bipunctatoides Reitter, 1915a: 40
muricata A. G. Olivier, 1795: 59: 9
sulcata Geoffroy, 1785: 159 (*Tenebrio*)
bipunctata cajetana Sénac, 1887a: 96 **E**: IT
bipunctata papii Canzoneri, 1963a: 14 **E**: IT
boehmi Reitter, 1915a: 34 **N**: EG
bottae Sénac, 1887a: 48 **A**: SA SI YE
carinosa Sénac, 1887a: 48
boyeri boyeri Solier, 1836a: 143 **N**: AG TU
asphaltina Baudi di Selve, 1875c: 100
granifera Solier, 1836a: 147
graniventris Baudi di Selve, 1876a: 25
rugifera Solier, 1836a: 144
boyeri cedrorum Koch, 1941a: 87 **N**: MO (Rif)
boyeri rifatlanta Koch, 1941a: 86 **N**: MO (Rif)
boyeri septemcostata Antoine, 1949a: 45 **N**: MO
brevicollis Solier, 1836a: 172 **E**: SP
buqueti Lucas, 1858c: ccxxi **N**: AG
canariensis Brullé, 1838: 67 **N**: CI (Tenerife)
canescens arabicola Schatzmayr & Koch, 1934b: 26 **N**: EG
canescens canescens Klug, 1830: nr. 3, pl. xi **N**: EG LB **A**: IS SI
depilata Solier, 1836a: 129
capito Krynicki, 1832: 131 **E**: AB AR GG ST
anomala Solier, 1836a: 118
neglecta Fischer von Waldheim, 1832a: 194
schonherri Solier, 1836a: 117

cephalotes bactriana Bogatchev, 1964a: 44 **A**: UZ (Surkhandarinskaya oblast)
cephalotes cephalotes Pallas, 1781: 49 (*Tenebrio*) **E**: ST **A**: IN KZ TD TM
kirgisica Motschulsky, 1859c: 472 [= 1860e: 533]
menetriesi Sénac, 1884a: 93
oxysterna Solier, 1836a: 121
ciliaticornis Koch, 1941c: 307 **A**: YE
claudia Buquet, 1840: 242 **N**: AG
spectabilis Haag-Rutenberg, 1875a: 52
comata Klug, 1830: nr. 4, pl. xi **N**: EG LB **A**: "Syria"
ornata Miller, 1861: 179
confusa Sénac, 1884c: xxiv **N**: TU
conjuncta Kwieton, 1976: 1 **A**: IQ
consonbrina confalonierii Gridelli, 1930: 274 **N**: AG LB
consonbrina consonbrina Lucas, 1858c: ccxx **N**: AG
densata Reitter, 1915a: 15
lesnei Peyerimhoff, 1911: 347
sparsidens Reitter, 1915a: 13
cordata cordata Kraatz, 1865a: 368 **N**: MO
athlela Antoine, 1949a: 29
maroccana Fairmaire, 1875c: 543
cordata desertorum Escalera, 1914: 313 **N**: MO (Western Sahara)
cordata mackenziei Escalera, 1914: 312 **N**: MO (Cabo Juby)
cordata zarcoi Español, 1943c: 133 **N**: MO (Tiznit, Tan-Tan)
costata Waltl, 1835: 69 **E**: PT SP
gadium Solier, 1836a: 169
graphica Baudi di Selve, 1876a: 22
hesperica Solier, 1836a: 167
lineata Solier, 1836a: 169
crenata crenata Fabricius, 1798: 44 **N**: MO
acroterica Antoine, 1949: 31
denisi Antoine, 1949: 31
grandicollis Kraatz, 1865a: 368
discicollis Fairmaire, 1875c: 544
intermittens Reitter, 1915a: 15 (*Pieterotarsa*)
gracilenta Haag-Rutenberg, 1875a: 52
granulicosta Escalera, 1914: 313
pilosa Escalera, 1914: 314
crenata taroudantica Koch, 1941a: 62 **N**: MO (Taroudant)
cribra Solier, 1836a: 151 **E**: SP (Baleares)
cribripennis Solier, 1836a: 117 **N**: AG MO
cursor Ménétriés, 1832: 192 **E**: AB **A**: AF IN
davletschinae Bogatchev, 1964a: 42 **A**: UZ (Kyzylkum)
damasci Sénac, 1880: xlviii [= 1880: 265] **A**: SY
depressa Solier, 1836a: 116 **N**: AG
papulosa Solier, 1836a: 113
subquadrata Solier, 1836a: 113 [HN]
derasa barcana Koch, 1941c: 311 [RN] **N**: LB
tricostata Koch, 1937: 452 [HN]
derasa derasa Klug, 1830: nr. 7, pl. xi **N**: EG LB
derasa derasoides Schuster, 1922b: 20 **A**: IS JO SY
discoides Kwieton, 1978a: 10 **A**: IN
distincta Solier, 1836a: 172 **E**: PT SP
doderoi deckeli Koch, 1939: 276 **N**: LB
doderoi doderoi Schuster, 1925a: 30 **N**: LB
doderoi doloresina Koch, 1937: 453 **N**: LB
dubia bellica Bogatchev, 1953: 305 **A**: IN (Teheran)
dubia dubia Faldermann, 1837: 8 **E**: AB AR **A**: IN TR

furcicosta Reitter, 1915a: 41
indistincta Reitter, 1915a: 41
persica Faldermann, 1837: 10
duponti Solier, 1836a: 146 **N**: AG
echidna densa Koch, 1941a: 78 **N**: MO
echidna echidna Fairmaire, 1875c: 544 **N**: MO
echidna schatzmayri Koch, 1941a: 78 **N**: MO
echidna tafrautensis Antoine, 1952: 113 **N**: MO
edura Reitter, 1915a: 34 **A**: JO
elevata Sénac, 1887a: 75 **E**: SP (Baleares)
errans errans Miller, 1861: 181 **A**: SY
errans kraatzi Sénac, 1887a: 17 **A**: IN IQ SA SY
estevezi Oromí, 1990: 246 **N**: CI (Gran Canaria)
fernandezlopezi Machado, 1979: 123 **N**: CI (Gomera)
fortecarinata Schuster, 1929: 39 **N**: MO
freyi Schuster, 1937a: 46 **A**: IQ
fulvovillosa Schuster, 1935a: 17 **N**: LB
gestroana Gridelli, 1930: 279 **N**: LB
gestroi Sénac, 1884a: 94 **A**: IN
gibba oasis Koch, 1941a: 75 **N**: AG TU
gibba gibba Fabricius, 1792a: 100 **N**: AG LB TU
simplex Solier, 1836a: 123
gibba subtriseriata Koch, 1941a: 76 **N**: MO
goryi cassolai Ardoin, 1973: 305 **E**: IT (Sardegna)
goryi goryi Solier, 1836a: 162 **E**: IT (Sardegna)
corsica Solier, 1836a: 164
sardea Solier, 1836a: 164
gracilipes Sénac, 1886: xlvi **A**: AF IN
gracilis Schuster, 1935a: 20 **N**: LB
grandis amplipennis Pierre, 1978: 200 **N**: MO (Tizi n'Tazazert)
grandis depressipennis Pierre, 1978: 201 **N**: MO
grandis echidniformis Reitter, 1915a: 18 (*Piesterotarsa*) **N**: AG
politidorsum Koch, 1941a: 73
grandis grandis Klug, 1830: nr. 5, pl. xi **N**: EG **AFR**
denticostata Koch, 1935a: 70
sudanica Fairmaire, 1882b: 66
grandis latastei Sénac, 1884b: x **N**: AG LB MO
mixta Koch, 1941a: 73
grandis minor Pierre, 1978: 198 **N**: MO (Ouarzazate)
granulata doumeti Sénac, 1882a: xxxi **N**: TU
granulata granulata Solier, 1836a: 103 **N**: AG
intertuberculata Lucas, 1858c: ccxx
lucasii Reiche, 1861b: 87
granulata imitata Reitter, 1915a: 20 (*Piesterotarsa*) **N**: TU
granulata papulenta Reiche, 1861b: 88 **N**: AG
granulata prophettei Sénac, 1884b: x **N**: AG
granulata realii Koch, 1937: 441 **N**: LB (Tripolis)
granulata rectangularis Kwieton, 1987: 13 **N**: AG
granulata sefrensis Pic, 1894p: 247 **N**: AG
granulata tenuicornis Solier, 1836a: 100 **N**: LB
dohrni Reitter, 1915a: 12 (*Piesterotarsa*)
tripolitana Sénac, 1884a: 39
granulicollis Wollaston, 1864: 478 **N**: CI (Gran Canaria)
auriculata Wollaston, 1864: 479
hybrida Uyttenboogaart, 1930: 233
granulithorax Escalera, 1914: 317 **N**: MO

gridellii Schuster, 1928a: 26 **N**: LB
grossa Fabricius, 1792a: 101 **E**: IT **N**: TU
inflata Herbst, 1799: 98
vestita Sénac, 1884a: 54
haarlovi Gridelli, 1954: 240 [= 1955: 46] **A**: AF
hayekae Kaszab, 1957f: 297 **A**: PA
hemisphaerica Solier, 1836a: 193 **N**: AG MO
capillata Solier, 1836a: 194
insignis Fairmaire, 1867: 407
hirtella Sénac, 1887a: 47 **N**: EG **A**: SA SI
horni Schuster, 1936c: 195 **A**: AF
incerta Solier, 1836a: 166 **E**: PT SP
evorensis Reitter, 1915a: 39
indica Sénac, 1882b: lvi **A**: PA **ORR**
inexspectata Sénac, 1882a: xxxi **A**: PA
convexicollis Sénac, 1882b: lvii
integra Rosenhauer, 1856: 190 **E**: PT SP
castellana Perez Arcas, 1865a: 176
interjecta Solier, 1836a: 152 **E**: FR SP (Baleares)
interpunctata interpunctata Klug, 1830: nr. 2, pl. xi (*Pimelia*) **N**: EG LB
andreinii Gridelli, 1929a: 10 (*Doderoella*)
carinata Solier, 1836a: 97
cyrenaica Schuster, 1925a: 29 (*Doderoella*)
interstitialis interstitialis Solier, 1836a: 105 **N**: AG LB MO TU
dejeanii Solier, 1836a: 132
intervallaris Kwieton, 1981b: 405 **A**: OM
iscariota Reitter, 1915a: 35 **A**: IQ
jansseni Reitter, 1915a: 35 **A**: JO
kruegeri Schuster, 1928e: 377 **N**: LB
kumshensis Antoine, 1954: 211 **N**: MO (Moyen Atlas)
laevidorsis Reitter, 1915a: 33 **A**: JO
laevigata costipennis Wollaston, 1864: 476 **N**: CI (El Hierro)
ambigua Wollaston, 1864: 475
laevigata laevigata Brullé, 1838: 67 **N**: CI (La Palma)
laevigata validipes Wollaston, 1864: 476 **N**: CI (Gomera)
alluaudi Sénac, 1892: 107
laeviuscula brachythorax Escalera, 1914: 311 **N**: MO
laeviuscula laeviuscula Kraatz, 1865a: 367 **N**: MO
externeserrata Fairmaire, 1875c: 544
parvula Antoine, 1949: 28
latreillei Solier, 1836a: 93 **N**: EG **AFR**
denticulata Solier, 1836a: 95
permixta Sénac, 1881: xx [RN]
letourneuxi gridelliana Koch, 1939: 278 **N**: LB
letourneuxi letourneuxi Sénac, 1880: xxiv [= 1880: 263] **N**: EG LB
letourneuxi platytubera Koch, 1939: 280 **N**: LB
longula Kwieton, 1981b: 405 **A**: BA SA
lutaria Brullé, 1838: 68 **N**: CI (Lanzarote, Fuerteventura)
lutulenta Wollaston, 1865: 403
marseuli asirensis Koch, 1965a: 140 **A**: SA
marseuli marseuli Sénac, 1887a: 50 **A**: SA
matthiesseni Reitter, 1915a: 26 **A**: IN
matutinalis Antoine, 1954: 205 **N**: MO
maura Solier, 1836a: 137 **E**: PT SP **N**: MO
obscura Reitter, 1915a: 37
mauritanica breiti Koch, 1941a: 89 **N**: AG MO

rotroui Antoine, 1949a: 41
mauritanica hollandei Pierre, 1985: 290 **N**: MO (Haut Atlas)
mauritanica mauritanica Solier, 1836a: 139 **N**: AG
mauritanica moltonii Koch, 1941a: 90 **N**: MO (Melilla)
pseudotimarcha Antoine, 1949a: 42
mauritanica parisii Koch, 1941a: 90 **N**: MO (Midelt)
bletoni Antoine, 1949a: 42
lapidicola Antoine, 1949a: 42
mauritanica sarroensis Antoine, 1949a: 43 **N**: MO (Jbel Sarro)
mauretanica verecostata Koch, 1941a: 89 **N**: AG
mittrei Solier, 1836a: 134 **A**: IS LE SI
modesta Herbst, 1799: 96 **E**: PT SP
monticola Rosenhauer, 1856: 192 **E**: SP
strobli Reitter, 1915a: 47
moralesi Español, 1943c: 135 **N**: MO (Western Sahara)
multigranifera Escalera, 1913: 46 **N**: MO
nazarena Miller, 1861: 178 **A**: IS IQ SY
nazarenoides Kaszab, 1982b: 150 **A**: SA YE
nesterovi Bogatchev, 1953: 305 **A**: IN
oblonga mediatlanta Koch, 1941a: 79 **N**: MO (Moyen Atlas)
oblonga oblonga Sénac, 1885: lxxxi **N**: MO
obsoleta giorgii Koch, 1937: 448 **N**: LB (Tripolis)
obsoleta insinuata Kwieton, 1977a: 21 **N**: AG TU
obsoleta latipes Solier, 1836a: 109 **N**: TU
amicta Baudi di Selve, 1876a: 20
obsoleta nitida Schuster, 1935a: 16 **N**: LB
obsoleta obsoleta Solier, 1836a: 104 **N**: LB TU
wohlberedti Schuster, 1928d: 154
obsoleta pilifera Sénac, 1884b: xi **N**: AG MO TU
rugosipennis Escherich, 1896d: 276
orientalis Sénac, 1886: xlv **A**: IS JO SY
nabathaea Peyerimhoff, 1921a: 328
pakistana Kaszab, 1961d: 222 **A**: PA
payraudi payraudi Latreille, 1829: 7 **E**: FR (Corse) IT
latericosta Reitter, 1915a: 42
rugatula Solier, 1836a: 159
payraudi punctatorugosa Reitter, 1915a: 42 **E**: IT (Sardegna)
payraudi subalpina Ardoin, 1973: 306 **E**: IT (Sardegna)
perezi Sénac, 1887a: 69 **E**: SP
peyerimhoffi Pierre, 1976: 19 **N**: AG (Beni Abbés)
platynota Fairmaire, 1875c: 544 **N**: MO
prolongata Miller, 1861: 180 **A**: IS
pseudointerstitialis Pierre, 1982: 2 **N**: AG
pseudopilifera Reitter, 1915a: 20 (*Piesterotarsa*) **N**: TU
puberula Chevrolat, 1873: 203 **A**: LE
punctata Solier, 1836a: 148 **E**: SP
crassipes Solier, 1836a: 190
manchega Lauffer, 1905: 406
radula oromii Viñolas, 1994: 140 **N**: CI (Tenerife)
asperata Har. Lindberg, 1950: 10 [HN]
granulata Wollaston, 1864: 474 [HN]
radula radula Solier, 1836a: 136 **N**: AG
pseudoradula Har. Lindberg, 1950: 10
remaudieri Pierre, 1964c: 1045 **A**: SY
retrospinosa Lucas, 1858a: clxxix **N**: AG
semihispida Fairmaire, 1874b: 407

robusta Kraatz, 1865a: 362 **E**: AR **A**: TR
roseni Koch, 1940c: 103 **A**: JO
rotundata Solier, 1836a: 149 **E**: SP
hispanica Solier, 1836a: 150
rugulosa apula Gridelli, 1950: 162 **E**: IT
rugulosa melitana Reitter, 1915a: 43 **E**: MA
rugulosa pelagosana J. Müller, 1912: 291 **E**: GR (Ionioi)
rugulosa rugulosa Germar, 1824b: 134 **E**: IT
bifurcata Solier, 1836a: 157
calabrica Reitter, 1915a: 44
oblita Reitter, 1915a: 44
rugulosa sublaevigata Solier, 1836a: 154 **E**: IT (Sicilia)
subscabra Solier, 1836a: 160
ruida Solier, 1836a: 153 **A**: SP
ryssos Herbst, 1799: 90 **N**: AG
rugosa A. G. Olivier, 1795: 59: 10 [HN]
salebrosa Solier, 1836a: 142
scabricollis Sénac, 1886: xlv **N**: AG
schoenherri Faldermann, 1837: 11 **E**: AR **A**: IN
schusteri Reitter, 1915a: 33 **A**: JO SA
scorteccii Gridelli, 1939d: 439 **N**: AG
semenovi balkhensis Kaszab, 1974: 197 **A**: AF
semenovi klapperichi Kaszab, 1960a: 41 **A**: AF PA
semenovi lindbergi Kaszab, 1970a: 80 **A**: AF
semenovi saroubiana Kaszab, 1973b: 47 **A**: AF
semenovi semenovi Bogatchev, 1953: 305 **A**: AF
afghanica Gridelli, 1954: 235 [= 1955: 42]
seminuda capannii Koch, 1937: 442 **N**: LB (Misurata)
seminuda seminuda Schuster, 1935a: 19 **N**: LB (Benghazi, Derna)
seminuda subpustulata Koch, 1937: 442 **N**: LB
semiopaca Sénac, 1884c: xxiv **N**: AG
senegalensis A. G. Olivier, 1795: 59: 7 **N**: MO **AFR**
cultrimargo Sénac, 1884c: xxv
sericea A. G. Olivier, 1795: 59: 8 **N**: EG
asperata Klug, 1830: nr. 10
seriegranulata Escalera, 1913: 43 **N**: MO
servillei Solier, 1836a: 112 **N**: AG
similis Kwieton, 1978a: 9 **A**: IN
simplicior medialticola Pierre, 1978: 204 **N**: MO
simplicior simplicior Escalera, 1914: 318 **N**: MO
sparsa albohumeralis Har. Lindberg, 1950: 11 **N**: CI (Gran Canaria)
sparsa serrimargo Wollaston, 1864: 477 **N**: CI (Gran Canaria)
parvula Har. Lindberg, 1950: 10 [HN]
sparsa sparsa Brullé, 1838: 67 **N**: CI (Gran Canaria)
fritschii Heyden, 1875: 141
stellifera Motschulsky, 1860e: 533 (*Chaetotoma*) **A**: "déserts méridionaux des Kirghises"
subcanescens Pierre, 1982: 5 **N**: AG LB TU
subquadrata chudeaui Koch, 1940d: 312 **N**: MO (Western Sahara) **AFR**
subquadrata subquadrata J. Sturm, 1826: 68 **N**: EG **A**: IS SI
irrorata Solier, 1836a: 99
subquadrata valdanii Guérin-Méneville, 1859: clxxxvii **N**: AG LB MO TU **AFR**
malzyi Peyerimhoff, 1938: 178 (*Euthriptera*)
superba Schuster, 1928b: 123 **N**: LB
tenuitarsis Reitter, 1915a: 33 **N**: EG
theveneti Sénac, 1880: xlix [= 1880: 267] **N**: EG **A**: SI
thomasi nefudana Koch, 1965a: 138 **A**: SA

thomasi thomasi Blair, 1931b: 270 **A**: SA
thomsoni brunneipedes Escalera, 1913: 44 [= 1914: 316] **N**: MO (Grand Atlas)
thomsoni thomsoni Nevinson, 1889: 255 **N**: MO (Haut Atlas)
tatae Théry, 1934: 47
thomsoni tirsalensis Antoine, 1937: 242 **N**: MO (Grand Atlas)
thomsoni villiersi Koch, 1941a: 92 **N**: MO (Grand Atlas)
titan Bogatchev, 1953: 304 **A**: IN
tizintesti Escalera, 1913: 45 **N**: MO
torquata Baudi di Selve, 1875d: 692 **A**: IN
atarnites Baudi di Selve, 1875d: 692
areolata Reitter, 1915a: 24
tricostata kandaharica Kaszab, 1960a: 42 **A**: AF
tricostata mukurensis Kaszab, 1974: 198 **A**: AF
tricostata paktiana Kaszab, 1974: 199 **A**: AF
tricostata tricostata Gebien, 1913: 4 **A**: IN PA (Balochistan)
tricostata vartiani Kaszab, 1970a: 80 **A**: AF
tuberculata Fischer von Waldheim, 1832a: 194 **E**: AB **A**: IN
tunisea Fairmaire, 1879c: 191 **N**: TU
undulata Solier, 1836a: 161 **E**: IT (Sardegna)
humericosta Reitter, 1915a: 44
urticata Klug, 1830: nr. 11, pl. xi **N**: EG LB
exanthematica Klug, 1830: nr. 13, pl. xi
tuberosa Klug, 1830: nr. 12, pl. xi
valida amblyptericollis Koch, 1941a: 94 **N**: AG
valida valida Erichson, 1841: 176 **N**: AG
scabra Fabricius, 1792a: 101
variolosa Solier, 1836a: 130 **E**: SP
verrucosa Fischer von Waldheim, 1820: pl. xiv **A**: AF KI TM UZ
verrucosa Brullé, 1838: 67 [HN]
villanovae Sénac, 1887a: 78 **E**: SP (Granada)
xauenensis Escalera, 1923d: 130 **N**: MO
zarudnyi Bogatchev, 1953: 304 **A**: IN
zhenzhuristi Bogatchev, 1953: 303 **A**: SA YE
subgenus *Pseudoamblyptera* Pierre, 1985: 295 type species *Pimelia brisouti frigida* Escalera, 1925
brisouti brisouti Sénac, 1885: lxxxii **N**: AG MO
brisouti frigida Escalera, 1925f: 498 **N**: MO
antoinei Schuster, 1929: 38

genus *Pimelia*, nomen dubium
rugosicollis Levrat, 1859b: 41 **E**: "Sicile, Espagne"

genus *Pimeliocnera* Reitter, 1909g: 310 type species *Pimeliocnera gebieni* Reitter, 1909
gebieni Reitter, 1909g: 310 **A**: TD TM UZ

genus *Pimelipachys* Skopin, 1962: 232 type species *Pachyscelis laevicollis* Reitter, 1893
laevicollis laevicollis Reitter, 1893b: 211 (*Pachyscelis*) **A**: KZ TM
laevicollis mesophilus Skopin, 1960: 303 (*Pachyscelis*) **A**: KZ (Yuzhnyy Kazakhastan)
rickmersi Reitter, 1898a: 13 (*Pachyscelis*) **A**: UZ

genus *Pisterotarsa* Motschulsky, 1860e: 532 [= 1860d: 436] type species *Pimelia gigantea* Fischer von Waldheim, 1820
Piesterotarsa Sénac, 1884b: xi [unjustified emendation]
Sympiezocnemis Solsky, 1876: 290 type species *Sympiezocnemis kessleri* Solsky, 1876
gigantea gigantea Fischer von Waldheim, 1820: pl. xiv (*Pimelia*) **A**: AF KZ TM UZ
gigantea zoubkoffi Reitter, 1915a: 61 (*Sympiezocnemis*) **A**: TM UZ

kiritshenkoi Semenov, 1914: 506 (*Sympiezocnemis*) **A**: UZ (Bukhara)
kessleri Solsky, 1876: 291 (*Sympiezocnemis*) **A**: AF TD TM
pavlovskii Bogatchev, 1959: 63 **A**: TD

genus *Platyesia* Skopin, 1971: 335 type species *Diesia karelini* Fischer von Waldheim, 1844
iduna Reitter, 1902b: 179 (*Trigonoscelis*) **A**: KZ (Almaatinskaya Oblast) XIN
karelini Fischer von Waldheim, 1844: 63 (*Diesia*) **A**: KZ TD TM UZ
medvedevi Skopin, 1971: 344 [= 1972: 122] **A**: KZ TM
sefirana Reitter, 1894d: 244 (*Diesia*) **A**: AF IN TM
barchanica Sumakov, 1902: 42 (*Diesia*)
sericata Zubkov, 1833: 326 (*Pimelia*) **E**: ST (Astrakhanskaya Oblast) **A**: KZ
argentata Bogatchev, 1961b: 369 (*Diesia*)

genus *Platyope* Fischer von Waldheim, 1820: pl. xv type species *Tenebrio leucographa* Pallas, 1773 (= *Tenebrio leucogramma* Pallas, 1773)
altaiensis Ren & Wang, 1993b: 46 **A**: XIN
bairinana Ren & Dong, 2001: 299 **A**: NMO
balteiformis Ren & Wang, 1993b: 44 **A**: NIN
granulata Fischer von Waldheim, 1820: pl. xv **A**: XIN
korgasica Wu & Huang, 2005: 285 **A**: XIN
korlaensis Fan & Huang, 2005: 286 **A**: XIN
leucogramma Pallas, 1773: 719 (*Tenebrio*) **E**: ST **A**: KZ
bassii Solier, 1836a: 14
dichotoma Fischer von Waldheim, 1830: 187
leucographa Pallas, 1781: 54 (*Tenebrio*)
lineata Fabricius, 1781: 319 (*Akis*)
obliterata Fischer von Waldheim, 1844: 60
mongolica Faldermann, 1835b: 388 **A**: ES (Tuva) JIL LIA MG NIN NMO
ordossica Semenov, 1907a: 183 **A**: GAN NIN NMO
pointi Schuster & Reymond, 1937: 234 **A**: NMO
proctoleuca chinensis Kaszab, 1962c: 311 **A**: MG XIN
proctoleuca proctoleuca Fischer von Waldheim, 1820: pl. xv **A**: KZ (Zaisan)
qitaiensis Wu & Huang, 2005: 282 **A**: XIN
trichophora Ren & Dong, 2001: 297 **A**: NMO
unicolor Zubkov, 1829: 160 **E**: ST **A**: KZ
victor Schuster & Reymond, 1937: 235 **A**: GAN

genus *Podhomala* Solier, 1836a: 72 type species *Podhomala suturalis* Solier, 1836
subgenus *Podhomala* Solier, 1836a: 72 type species *Podhomala suturalis* Solier, 1836
Podomala Agassiz, 1846b: 300 [unjustified emendation]
lucidula Krynicki, 1832: 132 (*Pimelia*) **E**: ST **A**: KZ TM
suturalis Solier, 1836a: 74 **E**: ST **A**: KZ
subgenus *Pterocomodes* Reitter, 1901e: 159 type species *Pterocomodes acutus* Reitter, 1901
acuta Reitter, 1901e: 160 (*Pterocomodes*) **A**: KZ
serrata Fischer von Waldheim, 1830: 187 (*Pimelia*) **A**: KI KZ
bicarinata Gebler, 1841a: 373 (*Pimelia*)
subgenus *Uriela* Reitter, 1887e: 518 type species *Podhomala fausti* Kraatz, 1881
bicostata Sénac, 1884a: 3 (*Pimelia*) **A**: "Siberia"
fausti Kraatz, 1881d: 332 **A**: UZ (Margilan) "Kurdistan"
cristata Sénac, 1882a: xxx (*Pimelia*)
heydeni Reitter, 1907h: 414 **A**: UZ

genus *Prionotheca* Dejean, 1834: 179 type species *Pimelia coronata* A. G. Olivier, 1795
coronata coronata A. G. Olivier, 1795: 59: 4 (*Pimelia*) **N**: AG EG LB MO TU **A**: SI **AFR**
coronata frey Bythinski-Salz, 1955: 313 **A**: IS (Negev)

coronata ovalis Ancey, 1881a: 397 **A**: BA IN IQ SA YE

genus *Przewalskia* Semenov, 1893c: 262 type species *Platyope dilatata* Reitter, 1887
dilatata Reitter, 1887d: 375 (*Platyope*) **A**: QIN XIZ
lineata Reitter, 1887d: 375 (*Platyope*) **A**: XIZ
trinkleri Gebien, 1940a: 1 **A**: XIZ

genus *Pseudopachyscelis* Löbl & Merkl, 2003: 249 type species *Trigonoscelis pygmaea* Ménétriés, 1849
galinae G. S. Medvedev, 1964c: 659 (*Pachyscelis*) **A**: AF TM
pygmaea Faust, 1875: 218 (*Pachyscelis*) [HN]
bucharensis Sumakov, 1902: 41 (*Pachyscelis*) **A**: UZ
kaszabi Pierre, 1964c: 1048 (*Pachyscelis*) **A**: IN
pygmaea major Kraatz, 1882b: 86 (*Pachyscelis*) **A**: AF UZ (Margilan)
pygmaea pygmaea Ménétriés, 1849: 222 (*Trigonoscelis*) **A**: TD TM UZ
nitidula Kraatz, 1882b: 86 (*Pachyscelis*)
pygmaea tadzika Skopin, 1960: 303 **A**: TD

genus *Pseudoplatyope* Pierre, 1964b: 867 type species *Storthocnemis antoinei* Español, 1951
niger Pierre, 1964b: 867 **N**: AG

genus *Pseudopodhomala* Schuster, 1938b: 88 type species *Pseudopodhomala gabrieli* Schuster, 1938
Gedrosia Bogatchev, 1961a: 111 [HN] type species *Gedrosia monstrosa* Bogatchev, 1961
Pseudopodhomalina Kaszab, 1960a: 22 type species *Diesia costifera* Waterhouse, 1889
barclayi G. S. Medvedev, 2004b: 561 **A**: IN
costifera Waterhouse, 1889: 127 (*Diesia*) **A**: PA (Balochistan)
gabrieli eremita Pierre, 1968: 1019 **A**: IN
gabrieli gabrieli Schuster, 1938b: 89 **A**: AF IN
monstrosa Bogatchev, 1961a: 112 (*Gedrosia*)

genus *Pseudostorthocnemis* Gridelli, 1952: 83 type species *Storthocnemis patrizii* Gridelli, 1933
patrizii longicornis Koch, 1937: 428 (*Storthocnemis*) **N**: LB (Fezzan)
patrizii patrizii Gridelli, 1933: 232 (*Storthocnemis*) **N**: LB
renatae Koch, 1937: 429 (*Storthocnemis*) **N**: LB

genus *Pterocoma* Dejean, 1834: 178 type species *Pimelia piligera* Gebler, 1830
subgenus *Dzungaropterocoma* Skopin, 1974b: 150 type species *Pterocoma subnuda* Reitter, 1897
autumnalis Semenov, 1908: 123 **A**: QIN XIN
kashtshenkoi Semenov, 1908: 124 **A**: QIN
pachyscelis Skopin, 1974b: 141 **A**: QIN XIN
submetallica Semenov, 1908: 124 **A**: QIN XIN
subnuda Reitter, 1897e: 218 **A**: QIN XIN
subgenus *Eupterocoma* Skopin, 1974b: 152 type species *Pterocoma ganglbaueri* Reitter, 1890
beicki Skopin, 1974b: 141 **A**: GAN XIN XIZ
ganglbaueri ganglbaueri Reitter, 1890h: 362 **A**: KI UZ
ganglbaueri punctipleuris Skopin, 1974b: 142 **A**: KZ XIN
iliensis Semenov, 1907c: 264 **A**: KZ XIN
grigorievi Semenov, 1908: 122 **A**: XIN
lepechini Semenov, 1908: 121 **A**: XIN
suworowi Reitter, 1907f: 206 **A**: KZ
subgenus *Hemipterocoma* Skopin, 1974b: 159 type species *Pterocoma nikolskii* Semenov, 1910
nikolskii Semenov, 1910d: 151 **A**: KZ
subgenus *Mesopterocoma* Skopin, 1974b: 147 type species *Pterocoma semicarinata* Bates, 1879
amandana amandana Reitter, 1887d: 374 **A**: NIN XIN XIZ
amandana edmundi Skopin, 1974b: 141 **A**: XIN
loczyi J. Frivaldszky, 1889: 209 **A**: XIN

semicarinata Bates, 1879: 477 **A**: KA XIN (Kashgar)
subgenus *Mongolopterocoma* Skopin, 1974b: 146 type species *Pterocoma reitteri* Frivaldszky, 1889
parvula J. Frivaldszky, 1889: 209 **A**: GAN MG NMO
reitteri J. Frivaldszky, 1889: 208 **A**: GAN MG NIN NMO
obesa J. Frivaldszky, 1889: 209
subgenus *Neopterocoma* Skopin, 1974b: 158 type species *Pterocoma balchachense* Skopin, 1974
balchashense Skopin, 1974b: 143 **A**: KZ
subgenus *Pachypterocoma* Skopin, 1974b: 157 type species *Pterocoma pallasi* Semenov, 1910
pallasi Semenov, 1910d: 152 **A**: KZ
valida Skopin, 1974b: 143 **A**: KZ
subgenus *Parapterocoma* Skopin, 1974b: 149 type species *Pterocoma vittata* J. Frivaldszky, 1889
vittata J. Frivaldszky, 1889: 208 **A**: GAN NIN NMO XIN
hedini Schuster, 1936a: 5
subgenus *Poopterocoma* Skopin, 1974b: 145 type species *Pterocoma zaidamica* Skopin, 1974
zaidamica Skopin, 1974b: 141 **A**: QIN
subgenus *Propterocoma* Skopin, 1974b: 144 type species *Pterocoma tibialis* Bates, 1879
convexa Bates, 1879: 476 **A**: KI TD XIN
plicicollis Reitter, 1902d: 191 **A**: XIN (Kashgar)
tibialis Bates, 1879: 476 **A**: KA XIN XIZ
alutacea Reitter, 1902d: 190
chan Reitter, 1897e: 218
fuscopilosa Reitter, 1902d: 191
serrimargo Bates, 1879: 476
subgenus *Pseudopimelia* Motschulsky, 1860e: 536 type species *Pterocoma tuberculata* Motschulsky, 1845
mongolica Kaszab, 1968c: 371 **A**: MG
tuberculata Motschulsky, 1845a: 62 **A**: KZ NMO XIN
subgenus *Pseudopterocoma* Skopin, 1974b: 145 type species *Pterocoma trapezicollis* Skopin, 1974
trapezicollis Skopin, 1974b: 140 **A**: KI
subgenus *Pterocoma* Dejean, 1834: 178 type species *Pimelia piligera* Gebler, 1830
baeckmanni Semenov, 1910d: 151 **A**: KZ
brevicollis Ménétriés, 1849: 220 **A**: KZ
costata Pallas, 1781: 52 (*Tenebrio*) **E**: ST **A**: KI KZ
aucta Semenov, 1910d: 150
dubianskii Semenov, 1910d: 150
gracilicornis Solier, 1836a: 47
sarpae Solier, 1836a: 46
subarmata Motschulsky, 1860e: 535 [= 1860d: 440]
denticulata denticulata Gebler, 1829a: 114 (*Pimelia*) **A**: KZ XIN
subrugosa Motschulsky, 1860e: 535
denticulata piligera Gebler, 1829a: 115 (*Pimelia*) **A**: KZ
denticulata reducta Skopin, 1974b: **A**: XIN
gebleri Skopin, 1974b: 142 **A**: XIN (Tarbag)

genus *Pterolasia* Solier, 1836a: 66 type species *Pterolasia squalida* Solier, 1836
multicostata Blair, 1931b: 271 **A**: OM
squalida Solier, 1836a: 68 **N**: EG LB (Cyrenaica) MO **A**: IS SI **AFR**
boehmi Reitter, 1908c: 48 (*Thryptera*)
distincta Solier, 1836a: 69
lanata Peyerimhoff, 1907a: 35

genus *Spectrocnera* Kwieton, 1981b: 402 type species *Spectrocnera anguliceps* Kwieton, 1981
anguliceps Kwieton, 1981b: 403 **A**: YE

genus *Stalagmoptera* Solsky, 1876: 286 type species *Stalagmoptera tuberculatocostata* Solsky, 1876
Arnoldiola Semenov & Bogatchev, 1940: 201 type species *Arnoldiola peculiaris* Semenov & Bogatchev, 1940 (= *Stalagmoptera ruginota* Reitter, 1896)
alaica Skopin, 1972: 309 **A**: UZ (Fergana)
clypeata Skopin, 1972: 305 **A**: TD
confusa confusa Solsky, 1876: 289 **A**: IN TD UZ
confusa deserticola Skopin, 1972: 306 **A**: TD
confusa incostata Kraatz, 1882d: 327 **A**: UZ (Samarkand)
confusa tuberculosa Kraatz, 1882d: 327 **A**: KZ TD UZ
tomentosa Reitter, 1893b: 226
gurievae Skopin, 1972: 309 **A**: TD
hybrida hybrida Skopin, 1969: 886 **A**: TM (Kughitangay)
hybrida tadzhika Skopin, 1972: 304 **A**: TD UZ
intermedia Kraatz, 1882d: 328 **A**: TD UZ (Samarkand)
dubia Kraatz, 1882d: 328
modesta Kraatz, 1882d: 328
laticollis fernanensis Skopin, 1972: 304 **A**: UZ (Fergana)
laticollis laticollis Solsky, 1876: 289 **A**: KI KZ TD TM UZ
striata Kraatz, 1882d: 328
oblonga Skopin, 1972: 309 **A**: TD
pachyscelis alticola Skopin, 1972: 303 **A**: TD
pachyscelis pachyscelis Kraatz, 1882b: 85 (*Pimelia*) **A**: UZ (Fergana)
ruginota Reitter, 1896a: 39 **A**: IN TM (Kopet Dag)
peculiaris Semenov & Bogatchev, 1940: 202 (*Arnoldiola*)
seriatogemmata Solsky, 1876: 288 **A**: TD
heydeni Kraatz, 1882d: 326
sordida costipennis Skopin, 1972: 305 **A**: UZ (Fergana)
sordida sordida Kraatz, 1881d: 331 (*Pimelia*) **A**: UZ (Fergana)
staudingeri badakschanica Kaszab, 1960a: 39 **A**: AF TD
staudingeri humeralis Skopin, 1972: 308 **A**: TD
staudingeri soluta Skopin, 1972: 308 **A**: TD
staudingeri staudingeri Reitter, 1893b: 227 **A**: TD
tuberculatocostata pterocomoides Reitter, 1893b: 225 **A**: KI (Kirghiz Range) KZ (Karatau)
tuberculatocostata simulatrix Kraatz, 1882b: 85 (*Pimelia*) **A**: UZ (Margilan)
tuberculatocostata solskyi Semenov & Skopin, 1951: 127 **A**: KZ
tuberculatocostata tienshanica Skopin, 1972: 302 **A**: KI UZ
tuberculatocostata tuberculatocostata Solsky, 1876: 287 **A**: KZ TD UZ
variabilis Kraatz, 1881d: 331 (*Pimelia*)

genus *Sternocnera* Skopin, 1964: 394 type species *Sternocnera lindti* Skopin, 1964
lindti Skopin, 1964: 396 **A**: TD

genus *Sternodes* Fischer von Waldheim, 1837b: 10 type species *Tenebrio caspicus* Pallas, 1781
caspicus caspicus Pallas, 1781: 47 (*Tenebrio*) **A**: IN TM UZ
pallasi Semenov & Bogatchev, 1940: 205
caspicus eous Semenov & Bogatchev, 1940: 204 **A**: AF TM

genus *Sternoplax* J. Frivaldszky, 1889: 207 type species *Trigonoscelis szechenyii* J. Frivaldszky, 1889
subgenus *Mesosternoplax* Skopin, 1973: 142 type species *Trigonoscelis laeviuscula* Kraatz, 1882
belutschinstanica Skopin, 1973: 148 **A**: PA (Balochistan)
blairi Skopin, 1974c: 259 **A**: "Baluchistan"
bogatshevi Skopin, 1960: 300 **A**: TM UZ
laeviuscula Kraatz, 1882b: 87 (*Trigonoscelis*) **A**: UZ
matthiesseni Reitter, 1907b: 87 **A**: IN
persica Schuster, 1935b: 25 (*Trigonoscelis*) **A**: IN

planiuscula Kraatz, 1882b: 88 (*Trigonoscelis*) **A**: AF TM UZ
sexseriata Reitter, 1913d: 659 **A**: IN
souvorowiana Reitter, 1907b: 88 **A**: AF GAN KZ UZ XIN
subgenus *Pachysternoplax* Skopin, 1973: 154 type species *Trigonoscelis armeniaca* Faldermann, 1837
armeniaca Faldermann, 1837: 21 (*Trigonoscelis*) **E**: AB AR
nicomedia Reitter, 1908g: 214 (*Trigonoscelis*) **A**: TR
subgenus *Parasternoplax* Skopin, 1973: 148 type species *Pimelia deplanata* Krynicki, 1832
affinis affinis Zubkov, 1833: 327 (*Pimelia*) **A**: KZ
abbreviata Motschulsky, 1860e: 533 [= 1860d: 437] (*Ocnera*)
affinis australis Skopin, 1973: 152 **A**: AF IN TM UZ
ballioni Skopin, 1973: 153 **A**: GAN XIN
deplanata deplanata Krynicki, 1832: 130 (*Pimelia*) **E**: ST **A**: KZ TM
perevostchikovii Zubkov, 1833: 326 (*Pimelia*)
deplanata kuldzhana Skopin, 1973: 149 **A**: XIN
fossor Kaszab, 1967c: 320 **A**: MG
subgenus *Pseudosternoplax* Skopin, 1973: 141 type species *Trigonoscelis lacerta* Bates, 1879
lacerta Bates, 1879: 475 (*Trigonoscelis*) **A**: KA XIN XIZ
pustulosa Reitter, 1887d: 378 (*Diesia*)
subgenus *Sternoplax* J. Frivaldszky, 1889: 207 type species *Trigonoscelis szechenyii* J. Frivaldszky, 1889
Pseudeuthriptera Bogatchev & Kryzhanovskiy, 1955: 240 type species *Trigonoscelis uzboica* Bogatchev & Kryzhanovskiy, 1955
costatissima Reitter, 1899d: 204 (*Trigonoscelis*) **A**: QIN XIN
echinata Fischer von Waldheim, 1844: 57 (*Trigonoscelis*) **E**: AB ST (Dagestan) **A**: KZ TM
apscheronica Bogatchev, 1934: 70 (*Trigonoscelis*)
sinuatocollis Desbrochers des Loges, 1881: 93 (*Trigonoscelis*)
kughitangi G. S. Medvedev, 1979a: 40 **A**: TM
schusteri Günther, 1937: 132 **A**: XIN
steinbergi Bogatchev & Kryzhanovskiy, 1960: 266 (*Trigonoscelis*) **A**: TM UZ
szechenyii J. Frivaldszky, 1889: 207 **A**: GAN XIN
uzboica Bogatchev & Kryzhanovskiy, 1955: 240 (*Trigonoscelis*) **A**: TM

genus *Sternotrigon* Skopin, 1973: 109 type species *Trigonoscelis setosa* Bates, 1879
boldi Kaszab, 1964b: 17 [= 1964f: 385] **A**: MG
grandis Faldermann, 1835b: 387 (*Platyope*) **A**: GAN MG
hiekei Skopin, 1973: 176 **A**: XIN
kraatzi J. Frivaldszky, 1889: 206 (*Trigonoscelis*) **A**: GAN MG NMO XIN
mongolica Reitter, 1889k: 896 (*Trigonoscelis*) **A**: GAN MG NMO XIN
opaca Reitter, 1907b: 91 (*Sternoplax*) **A**: NMO
setosa juvencus Reitter, 1893b: 241 (*Trigonoscelis*) **A**: UZ XIN XIZ
kashgarensis Reitter, 1907b: 90 (*Sternoplax*)
seidlitzi Reitter, 1893b: 243 (*Trigonoscelis*)
setosa setosa Bates, 1879: 475 (*Trigonoscelis*) **A**: KA NIN XIN
impressicollis Reitter, 1893b: 242 (*Trigonoscelis*)
niana Reitter, 1887d: 377 (*Diesia*)
zichyi Csiki, 1901a: 110 (*Trigonoscelis*) **A**: MG

genus *Storthocnemis* Karsch, 1881: 47 type species *Storthocnemis steckeri* Karsch, 1881
antoinei Español, 1953c: 298 **N**: MO
kulzeri Koch, 1969: 484 **A**: YE
steckeri dufouri Peyerimhoff, 1935c: 90 **N**: LB MO
steckeri steckeri Karsch, 1881: 48 **N**: LB

genus *Tadzhikistania* Bogatchev, 1960b: 35 type species *Tadzhikistania mystacea* Bogatchev, 1960
mystacea Bogatchev, 1960b: 35 **A**: TD
stoljarovi G. S. Medvedev, 1970: 402 **A**: AF

genus *Thriptera* Solier, 1836a: 48 type species *Thriptera maillei* Solier, 1836 (= *Pimelia crinita* Klug, 1830)
afghanica Kaszab, 1970a: 70 **A**: AF
ballionis Kraatz, 1882b: 87 **A**: TM UZ
crinita crinita Klug, 1830: nr. 22, pl. xii (*Pimelia*) **N**: EG **A**: IN JO SA SI YE **AFR**
maillei Solier, 1836a: 50
pygmaea Miller, 1861: 181 (*Ocnera*)
dubia Kaszab, 1973b: 44 (*Thryptera*) **A**: AF
foucauldi Peyerimhoff, 1929c: 285 **N**: AG (Hoggar)
freyi freyi Gridelli, 1954: 223 **A**: AF
freyi gracillima Kaszab, 1960a: 32 (*Thryptera*) **A**: AF **ORR**
jaffana Miller, 1861: 182 (*Ocnera*) **A**: IS SY
asphaltidis Reiche & Saulcy, 1875: 218
kashmirensis Kaszab, 1957f: 297 **A**: KA
klapperichi Kaszab, 1960a: 24 (*Lasiostola*) **A**: AF
kraatzi Haag-Rutenberg, 1876b: 75 **A**: IN JO OM PA SA SI YE **ORR**
groschkei Kaszab, 1957d: 65
sorocula Reitter, 1894e: 303
yemenensis Gridelli, 1940: 236
laperrinei laperrinei Peyerimhoff, 1929c: 284 **N**: AG (Hoggar)
longipilis Reitter, 1895d: 156 **A**: AF TD TM
sparsicollis Reitter, 1909g: 311 (*Tryptera*)
malyi Obenberger, 1914: 139 **A**: UZ (Bukhara)
maroccana Théry, 1932a: 191 **N**: MO **AFR**
minuta Pic, 1897m: 124 **N**: LB
murina murina Baudi di Selve, 1881: 280 **A**: IN SA YE
olegi G. S. Medvedev, 2004b: 565 **A**: AF
pastor Reitter, 1893b: 216 **A**: KZ
protensa Fairmaire, 1891d: xciv (*Ocnera*) **A**: KA
varvasi pilipes Kraatz, 1865a: 299 **N**: EG **A**: SI
guyoti Andres, 1920: 70
heydeni Reitter, 1893b: 214
varvasi varvasi Solier, 1836a: 52 **N**: AG LB TU
bedeli Allard, 1886: (3)
debilicornis Kraatz, 1865a: 300
hamadensis Koch, 1937: 436
lybica Koch, 1937: 435
phazanica Koch, 1937: 436
villosa Solier, 1836a: 53 **N**: LB
alternans Baudi di Selve, 1876a: 10 (*Ocnera*)
tunicensis Desbrochers des Loges, 1898a: 82 (*Ocnera*)
zarudnyi G. S. Medvedev, 2004b: 567 **A**: IN

genus *Trachyderma* Latreille, 1828: 576 type species *Pimelia hispida* Fabricius (= *Tenebrio hispidus* Forskål, 1775)
subgenus *Atrachyderma* Skopin, 1962: 228 type species *Pimelia setosa* Fischer von Waldheim, 1832
afghanica afghanica Kaszab, 1960a: 29 **A**: AF
afghanica kandaharica Kaszab, 1970a: 69 **A**: AF
allardi Gridelli, 1928: 17 (*Ocnera*) [RN] **N**: EG
grandis adriani Kaszab, 1974: 196 **A**: AF
grandis grandis Schuster, 1930b: 238 (*Ocnera*) **A**: IN PA (Beluzhistan)
hispida Forskål, 1775: 79 (*Tenebrio*) **N**: EG LB MO TU **A**: IN IQ IS SA SI SY YE **AFR**
depressa Baudi di Selve, 1875d: 685 (*Ocnera*)
latreillei Solier, 1836a: 36
leprieuri Allard, 1886: (4) (*Ocnera*)
major Baudi di Selve, 1875d: 684 (*Ocnera*)
minor J. R. Sahlberg, 1913b: 170

sparsispina Böhm, 1916b: 67 (*Ocnera*)
maroccana Schuster, 1921b: 46 (*Ocnera*) **N**: MO
menetriesi afghanimorpha Skopin, 1968: 98 (*Brachyscelis*) **A**: TD
menetriesi hafizi Kwieton, 1981a: 14 **A**: PA (Beluchistan)
menetriesi macrogranulata Gridelli, 1954: 218 [= 1955: 38] (*Trachyderma*) **A**: AF
menetriesi menetriesi Kraatz, 1865a: 296 (*Ocnera*) **A**: AF KZ TM
asperula Ménétriés, 1849: 221 [RN]
muricata Fischer von Waldheim, 1837b: 15 (*Brachyscelis*)
menetriesi robusta Faust, 1875: 213 (*Ocnera*) **A**: AF IN
perlata Baudi di Selve, 1875d: 685 (*Ocnera*)
setosa Fischer von Waldheim, 1832a: 192 (*Pimelia*) **E**: AB AR ST **A**: AF IN TM TR
beckeri Desbrochers des Loges, 1881: 90 (*Ocnera*)
granulata Fischer von Waldheim, 1844: 56 (*Ocnera*)
setosa Faldermann, 1837: 20 [HN]
triangularis australis Medvedev, 1964c: 658 (*Ocnera*) **A**: TM (Badkhyz)
triangularis triangularis Faust, 1875: 212 (*Ocnera*) **A**: AF KZ TM UZ
subgenus *Trachyderma* Latreille, 1829: 7 type species *Pimelia hispida* Fabricius (= *Tenebrio hispidus* Forskål, 1775)
andresi Gridelli, 1928: 16 (*Ocnera*) **A**: SI
christophi benardi Pierre, 1968: 1023 **A**: IN
christophi christophi Faust, 1875: 208 (*Ocnera*) **E**: AR ST **A**: AF IN TM
longicollis Baudi di Selve, 1875d: 687 (*Ocnera*)
cyrenaica Gridelli, 1930: 270 (*Ocnera*) **N**: LB
emondi Solier, 1836a: 38 **N**: EG
longicollis Desbrochers des Loges, 1881: 87 (*Ocnera*) [HN]
genei Solier, 1836a: 38 **N**: EG LB **A**: SI
habelmanni Kraatz, 1865a: 292 (*Ocnera*)
gomorrhana Reiche & Saulcy, 1857: 215 **A**: IS
lima L. Petagna, 1819: 29 (*Pimelia*) **E**: GR IT MA **N**: AG EG LB TU **A**: CY IS LE SY TR
angustata Solier, 1836a: 37
persea farahense Kaszab, 1974: 196 **A**: AF
persea persea Baudi di Selve, 1875d: 687 (*Ocnera*) **A**: IN PA
persea khashensis Pierre, 1968: 1022 **A**: IN
philistina Reiche & Saulcy, 1857: 214 **E**: GR (Ródos) **N**: EG **A**: AE BA CY IN IQ IS JO OM SA SI SY TR YE **ORR**
graeca Desbrochers des Loges, 1881: 89 (*Ocnera*)
ovalipennis Baudi di Selve, 1875d: 685 [= 1876: 5] (*Ocnera*)
parvicollis Baudi di Selve, 1875d: 686 (*Ocnera*)
seurati Peyerimhoff, 1929c: 283 (*Ocnera*) **N**: AG (Hoggar) LB
striatogranosa Fairmaire, 1882a: 67 (*Thriptera*) **A**: PA SA YE **AFR**

genus *Trigonocnera* Reitter, 1893b: 202 type species *Trigonoscelis pseudopimela* Reitter, 1889
pseudopimelia pseudopimelia Reitter, 1889k: 697 (*Trigonoscelis*) **A**: GAN
amitina Kolbe, 1908: 90 (*Trigonoscelis*)
pseudopimelia reitteri Csiki, 1901a: 111 (*Trigonoscelis*) **A**: MG NMO

genus *Trigonopachys* Skopin, 1968: 300 type species *Trigonopachys michailovi* Skopin, 1968
michailovi Skopin, 1968: 302 **A**: UZ

genus *Trigonoscelis* Dejean, 1834: 179 type species *Pimelia nodosa* Fischer von Waldheim, 1820
subgenus *Chinotrigon* Skopin, 1973: 170 type species *Trigonoscelis sublaevigata* Reitter, 1887
holdereri Reitter, 1900e: 161 **A**: XIN
sublaevigata granicollis Kaszab, 1964b: 15 **A**: MG
sublaevigata sublaevigata Reitter, 1887e: 519 **A**: NMO XIN
subgenus *Echinotrigon* Skopin, 1973: 170 type species *Trigonoscelis granulata* Reitter, 1915
granulata Reitter, 1915e: 90 **A**: TD

tadzhikistana Bogatchev, 1963b: 94 **A**: TD
subgenus *Trigonoscelis* Dejean, 1834: 179 type species *Pimelia nodosa* Fischer von Waldheim, 1820
apicalis Reitter, 1907b: 83 **A**: TM
borosi Kaszab, 1951a: 2 **A**: TM
ceromatica Bogatchev, 1963b: 95 **A**: UZ (Surkhandarinska oblast)
cordata Allard, 1886: (2) **E**: AB
fasciculitarsis fasciculitarsis Reitter, 1893b: 236 **A**: KI UZ
fasciculitarsis saeptiarius Skopin, 1973: 168 **A**: UZ (Fergana)
gemmulata callosa Motschulsky, 1860e: 534 **A**: KZ
gemmulata gemmulata Ménétriés, 1849: 221 **A**: KZ UZ
gemmulata intermedia Skopin, 1973: 160 **A**: KZ
gemmulata kabuliensis Kaszab, 1974: 195 **A**: AF
gemmulata oxiana Skopin, 1973: 160 **A**: UZ
gemmulata punctipleuris Reitter, 1893b: 235 **A**: TM
gemmulata seriatula Reitter, 1907b: 84 **A**: AF TM
  aequalis Reitter, 1907b: 84
hayekae Skopin, 1973: 165 **A**: PA
muricata abnormis Skopin, 1973: 157 **A**: KZ
muricata mirabilis Faldermann, 1836: 380 **A**: KZ
muricata muricata Pallas, 1781: 48 (*Tenebrio*) **E**: ST **A**: KZ TM
  spinigera Motschulsky, 1860d: 439 [RN]
muricata sparsa Reitter, 1893b: 233 **E**: ST
nodosa gigas Reitter, 1893b: 231 **A**: TM
nodosa grandis Kraatz, 1865a: 279 **A**: TM
  contraria Desbrochers des Loges, 1881: 91
  corallifera Reitter, 1893b: 234
  elegans Allard, 1886: [1]
nodosa mixta Skopin, 1973: 162 **A**: AF KZ
nodosa nodosa Fischer von Waldheim, 1820: pl. xiv (*Pimelia*) **A**: AF KZ TM
nodosa pustulifera Skopin, 1973: 164 **A**: TM (Karakumy)
nodosa schrenkii Gebler, 1844: 101 **A**: KZ XIN
seriata Ménétriés, 1849: 222 **A**: TD TM UZ
  pecirkana Reitter, 1909g: 308
sublaevicollis Reitter, 1893b: 236 **A**: IN TM UZ
submuricata Kraatz, 1882b: 95 **A**: UZ (Fergana)
zoufali Reitter, 1893b: 237 **A**: KZ UZ

genus *Urielina* Reitter, 1888c: 331 type species *Podhomala nitida* Baudi di Selve, 1876
nitida Baudi di Selve, 1876a: 30 (*Podhomala*) **A**: AF IN TM UZ

genus *Waterhousia* Skopin, 1973: 109 type species *Trigonoscelis longipes* Waterhouse, 1889
  Heinrichesia Carl, 2000: 258 type species *Heinrichesia schaeferi* Carl, 2000
longipes Waterhouse, 1889: 128 (*Trigonoscelis*) **A**: AF PA (Balochistan)
lopatini G. S. Medvedev, 2005b: 310 **A**: AF (Regestan)
schaeferi Carl, 2000: 258 (*Heinrichesia*) **A**: IN (Sistan)
warei Skopin, 1973: 179 **A**: PA

tribe Pimeliini, species incertae sedis
Trigonoscelis occidentalis Motschulsky, 1860d: 438 **A**: LE (Beirut)

tribe Sepidiini Eschscholtz, 1829

subtribe Molurina Solier, 1834

genus *Psammophanes* Lesne, 1922: 690 type species *Psammodes duodecimcostatus* Lesne, 1922
subgenus *Somalarabes* Koch, 1953: 155 type species *Psammodes gracilentus* Fairmaire, 1882
arabicus Gebien, 1938: 58 (*Psammodes*) **A**: YE

benardellii Koch, 1965a: 126 **A**: YE

subtribe Sepidiina Eschscholtz, 1829

genus *Sepidiostenus* Fairmaire, 1884e: lxxv type species *Sepidiostenus erinaceus* Fairmaire, 1884
Sepidiacis Fairmaire, 1884h: cxlvi type species *Sepidiacis compressa* Fairmaire, 1884
pradierii Guérin-Méneville, 1858: 128 (*Sepidium*) **A**: SA

genus *Sepidium* Fabricius, 1775: 250 type species *Sepidium tricuspidatum* Fabricius, 1775
aliferum Erichson, 1841: 178 **E**: SP **N**: AG MO TU
aitagiae Escalera, 1913: 41 **N**: MO
bathelemyi Solier, 1844: 236 **N**: EG
barbarum Solier, 1844: 235 **N**: AG TU
pallens Allard, 1874: 137
solieri Desbrochers des Loges, 1881: 101
bicaudatum Fairmaire, 1871a: 388 **N**: MO
bidentatum Solier, 1844: 227 **E**: PT SP **N**: MO
perforatum Allard, 1874: 130
cristatum Fabricius, 1775: 250 **A**: YE
notoceros Pallas, 1781: 59 (*Tenebrio*)
cyrenaicum Schuster, 1928b: 122 **N**: LB
dathan Crotch, 1872: 268 **N**: EG LB **A**: IS LE SI SY
abiram Crotch, 1872: 268
vietaeformis Reitter, 1914h: 385
douei Solier, 1844: 230 **N**: AG
flexuosum Solier, 1844: 237 **N**: EG
fusiforme Kwieton, 1980a: 17 **N**: MO
gypsicola Escalera, 1913: 42 **N**: MO
hoseini Escalera, 1911: 301 **N**: MO (Tiznit)
hystrix desertica Español, 1944: 12 **N**: MO (Western Sahara)
hystrix hystrix Antoine, 1932: 185 **N**: MO (Sous)
hystrix ifniensis Escalera, 1940: 5 **N**: MO (Tiznit)
subdesertica Escalera, 1940: 8
inaequale Reitter, 1914h: 386 **N**: LB
lusitanicum Kaszab & Pinheiro, 1972: 6 **E**: PT SP
elongatum Mal, 1984: 200
marraquense Escalera, 1911: 302 **N**: MO
mesopotamicum Reitter, 1914h: 386 **A**: IQ
mittrei Solier, 1844: 228 **N**: AG
bicorne Solier, 1844: 229
mskalicum Escalera, 1914: 307 **N**: MO
penicilligerum Karsch, 1881: 49 **N**: LB
peyerimhoffi Antoine, 1932: 183 **N**: MO (Sous)
reichei Allard, 1870: 49 **N**: EG LB TU
bispinicollis Reitter, 1914h: 389
requieni Solier, 1844: 241 **N**: AG LB TU
siculum siculum Solier, 1844: 231 **E**: IT (Sicilia)
genei Solier, 1844: 230
tricuspidatum mogadoricum Escalera, 1914: 306 **N**: MO
guisseri Kocher, 1968: 111
schrammi Kocher, 1968: 110
tricuspidatum multispinosum Solier, 1844: 241 **N**: AG TU
laghoatense Baudi di Selve, 1875d: 695
tricuspidatum tomentosum Erichson, 1841: 178 **E**: MA **N**: EG TU
maillei Solier, 1844: 239
remotum J. R. Sahlberg, 1903f: 50
serratum Solier, 1843: 240

tricuspidatum tricuspidatum Fabricius, 1775: 250 **N**: EG LB **A**: IS SI SY
alexandrinus Forskål, 1775: 78 (*Tenebrio*)
cerisyi Solier, 1844: 238
immundum Reitter, 1914h: 388
korah Crotch, 1872: 268
tualensis Escalera, 1940: 9 **N**: MO (Ifni)
uncinatum Erichson, 1841: 178 **N**: AG LB MO TU
capricorne Desbrochers des Loges, 1881: 96
sefrense Rotrou, 1943: 235
subcornutum Escalera, 1925d: 376
variegatum Fabricius, 1792a: 112 (*Tenebrio*) **N**: AG MO TU
angustatum Solier, 1844: 233
dispar Desbrochers des Loges, 1881: 101
dufouri Solier, 1844: 234
integrum Desbrochers des Loges, 1881: 100
laterale Allard, 1874: 133
subfurcatum Desbrochers des Loges, 1881: 100
wagneri macrotys Antoine, 1951c: 94 **N**: MO
wagneri wagneri Erichson, 1841: 179 **N**: AG TU
confusum Allard, 1874: 141 [HN]

genus *Vieta* Hope, 1840: 116 type species *Sepidium vestitum* Guérin-Méneville, 1831
Divieta Reitter, 1914h: 390 type species *Vieta costata* Allard, 1874
Dymonus Solier, 1844: 219 type species *Dymonus tuberculatus* Solier, 1844
Vieta Laporte, 1840: 196 [HN]
algeriana Allard, 1870: 50 [= 1874: 146] **N**: AG
clypeata Gahan, 1896b: 454 **A**: YE
costata Allard, 1874: 149 **N**: EG **AFR**
gracilenta Ancey, 1881a: 397 **A**: YE
luxorii Allard, 1874: 150 **N**: EG
millingenii Kirchsberg, 1877a: 203 **A**: SA YE
tuberculata Solier, 1844: 223 (*Dymonus*) **N**: EG **A**: SA SI YE
gibbicollis Solier, 1844: 224 (*Dymonus*)

genus *Vietomorpha* Fairmaire, 1887a: 186 type species *Vietomorpha foveipennis* Fairmaire, 1887
foveipennis Fairmaire, 1887a: 186 **A**: YE **AFR**
arabica Gebien, 1938: 59

tribe Stenosini Lacordaire, 1859

genus *Afghanillus* Kaszab, 1960a: 1 type species *Afghanillus klapperichi* Kaszab, 1960
klapperichi Kaszab, 1960a: 2 **A**: AF

genus *Aspidocephalus* Motschulsky, 1839: 63 type species *Aspidocephalus desertus* Motschulsky, 1839
desertus Motschulsky, 1839: 64 **E**: AB AR GG ST **A**: KZ

genus *Dichillus* Jacquelin du Val, 1861: 253 type species *Tagenia minuta* Solier, 1838
subgenus *Dichillesthes* Reitter, 1916d: 156 type species *Dichillus cordicollis* Reitter, 1886
cordicollis Reitter, 1886a: 134 **A**: KZ TD TM UZ
kuhitangi G. S. Medvedev, 1994b: 854 **A**: TM (Kuhitanga)
subgenus *Dichillinus* Reitter, 1916d: 161 type species *Tagenia pusilla* Ménétriés, 1849
afghanicus afghanicus Kaszab, 1960a: 11 **A**: AF
afghanicus badakschanicus Kaszab, 1960a: 12 **A**: AF
afghanicus khinjanensis Kaszab, 1960a: 12 **A**: AF
afghanicus klapperichi Kaszab, 1960a: 12 **A**: AF
buettikeri Kaszab, 1982b: 127 **A**: SA

cylindricus Baudi di Selve, 1875a: 298 **A**: IS LE TR
unistriatus Desbrochers des Loges, 1881: 80
desioi Gridelli, 1934: 58 **A**: KA
dolosus Reitter, 1886a: 138 **A**: IS LE "Syria"
epipleuralis Reitter, 1916d: 163 **A**: AF TD UZ
explanatus explanatus Reitter, 1900b: 54 **A**: KZ TD UZ
explanatus kataghanicus Kaszab, 1960a: 13 **A**: AF
farsistanicus G. S. Medvedev, 1976: 894 **A**: IN
fusicornis Reitter, 1916d: 166 **A**: KZ TM UZ
karatavicus Skopin, 1966: 341
ganquanensis Ren, 1999: 333 **A**: SHA
heydeni Reitter, 1886a: 137 **A**: "Syria"
iranicus Kaszab, 1963a: 74 **A**: IN
kryzhanovskii G. S. Medvedev, 1975: 602 **A**: TM (Kopet Dag)
laeviusculus Kraatz, 1862a: 99 **E**: SP **N**: AG TU
distinguendus Fairmaire, 1884d: lxvi
modestus Fairmaire, 1884d: lxvi
lisae Carl, 1990c: 105 **A**: IQ
nitidulus nitidulus Reitter, 1886a: 137 **A**: IS
nitidulus transjordanicus Koch, 1936: 296 **A**: JO
nitidus Kraatz, 1862a: 99 **A**: IQ SY
ocellaris Kaszab, 1970a: 36 **A**: AF TM
pauxillus filum Reitter, 1916d: 163 **A**: AF TM
pauxillus pauxillus Reitter, 1900b: 54 **A**: AF TD TM UZ
pusillus Ménétriés, 1849: 228 (*Tagenia*) **A**: AF KI KZ TD UZ
brunneus Kraatz, 1882d: 321
fulvescens Reitter, 1916d: 162
nigritulus Reitter, 1916d: 162
reitteri Semenov, 1891a: 361 **A**: KZ TM UZ
sanguinipes sanguinipes Fairmaire, 1888b: 126 **A**: BEI
sanguinipes shantungensis G. S. Medvedev, 1994a: 309 **A**: SHN
schmidi Kaszab, 1961a: 116 **A**: AF
semipolitus Reitter, 1916d: 164 **A**: JO
simeki Kaszab, 1970a: 33 **A**: AF TD UZ
simillimus Kaszab, 1960a: 13 **A**: AF
sinensis G. S. Medvedev, 1994a: 308 **A**: SCH YUN
tenebrosus Reitter, 1886a: 136 **A**: KZ TM
subgenus *Dichillocerus* Reitter, 1916d: 167 type species *Hyperops pertusa* Kiesenwetter, 1861
araxidis Reitter, 1889a: 30 **E**: AB AR **A**: TR
crassicornis Reitter, 1916d: 167 **E**: GR **N**: LB **A**: IQ LE IS TR
cyrenaicus Gridelli, 1930: 242 **N**: EG LB
damascenus Reitter, 1916d: 167 **A**: SY
obenbergeri Mařan, 1935b: 147 **E**: GR (Kríti)
pertusus Kiesenwetter, 1861a: 224 (*Hyperops*) **E**: GR IT MA **A**: TR
biseriatus Reiche, 1862a: 371 (*Oedenocera*)
subgenus *Dichillodontus* Reitter, 1916: 166 type species *Dichillus adriani* Reitter, 1916
adriani Reitter, 1916d: 167 **A**: TD TM UZ
aruktavicus G. S. Medvedev, 1991: 561 **A**: TD
dentipes G. S. Medvedev, 1964a: 61 **A**: TM (Kuhitang)
subgenus *Dichillomessor* Reitter, 1916d: 155 type species *Dichillus himalayanus* Fairmaire, 1896
himalayanus Fairmaire, 1896a: 57 **A**: HP
indicus Wasmann, 1897: 239 **A**: SD **ORR**
kashmirensis Kaszab, 1975a: 15 **A**: KA
lindemannae Kaszab, 1961d: 219 **A**: PA
pakistanus Kaszab, 1961d: 221 **A**: PA

piffli Kaszab, 1961f: 344 **A**: PA (Karakoram)
subgenus *Dichillus* Jacquelin du Val, 1861: 253 type species *Tagenia minuta* Solier, 1838
alfierii Koch, 1935a: 46 **A**: SI
algiricus Lucas, 1846: 118 (*Tagenia*) **N**: AG TU
bicarinatus Baudi di Selve, 1875a: 299
castanescens Fairmaire, 1884d: lxvi
henoni Allard, 1869b: 468 (*Stenosis*)
angelicae Reitter, 1897e: 217 **E**: AB AR **A**: IN TR
angusticollis Kaszab, 1959e: 244 [= Kaszab, 1960a: 10] **A**: AF
carinatus Küster, 1848b: 89 (*Tagenia*) **E**: BH BU CR GR MC TR
corsicus Solier, 1838a: 33 (*Tagenia*) **E**: FR IT
pumilus Solier, 1838a: 33 (*Tagenia*)
formicophilus Breit, 1916: 61 **E**: GR RO UK
laevicornis Reitter, 1886a: 132 **N**: MO
libanicus Koch, 1936: 295 **A**: LE
lindbergi Kaszab, 1970a: 31 **A**: AF
michailovi G. S. Medvedev, 1975: 598 **A**: TD
minutus Solier, 1838a: 32 (*Tagenia*) **E**: FR IT SP
myrmecophilus Fairmaire, 1884d: lxvii **N**: AG
palaestinus Baudi di Selve, 1875a: 296 **A**: "Palestina"
rugatus bogatshevi G. S. Medvedev, 1975: 597 **E**: AB (Talish)
rugatus rugatus Baudi di Selve, 1874a: 105 [= 1875a: 298, 1875c: 70] **E**: AR **A**: IN IQ
scheibei Kaszab, 1960a: 7 **A**: AF
schusteri Reitter, 1916d: 160 **A**: IN TM
seminitidus Solsky, 1881b: 47 **A**: AF KZ TD TM UZ
inhumeralis Pic, 1906b: 18
socius Rottenberg, 1871: 247 **E**: IT (Sicilia)
stackelbergi G. S. Medvedev, 1975: 595 **A**: UZ (Tian Shan
stoeckleini Kaszab, 1960a: 9 **A**: AF
strigiceps Reitter, 1886a: 141 **A**: SY
subcostatus Solier, 1838a: 34 (*Tagenia*) **E**: SP **N**: MO
andulusicus Rosenhauer, 1856: 202 (*Tagenia*)
elevatus Reitter, 1886a: 142
valentinus Reitter, 1916d: 159
subsetulosus Reitter, 1886a: 140 **E**: GR (Rhodos) **A**: CY TR
subtilis Kraatz, 1862a: 98 **E**: IT (Sicilia)
weberi J. Müller, 1916: 93 **E**: CR
subgenus *Myrmecodichillus* Kaszab, 1960a: 6 type species *Dichillus reichenspergeri* Kaszab, 1960
reichenspergeri Kaszab, 1960a: 5 **A**: AF
subgenus *Pushtunillus* G. S. Medvedev, 1994b: 855 type species *Dichillus skopini* G. S. Medvedev, 1994
skopini G. S. Medvedev, 1994b: 855 **A**: AF
genus *Dichillus*, species incertae sedis
duricornis Reitter, 1886a: 139 **A**: TR

genus *Ethas* Pascoe, 1862: 324 type species *Ethas carbonarius* Pascoe, 1862
anomalus Fairmaire, 1896a: 56 **A**: HP
semicribrosus Fairmaire, 1896a: 56 **A**: "North India"

genus *Eutagenia* Reitter, 1886a: 125 type species *Stenosis smyrnensis* Reitter, 1889 (= *Eutagenis cribricollis* Reitter, 1916)
aegyptiaca aegyptiaca Reitter, 1889h: 285 **N**: EG LB
aegyptiaca tunisea Normand, 1936b: 359 **E**: IT (Lampedusa) **N**: BL TU
annae Grimm, 1991: 20 **A**: CY
cribricollis Reitter, 1916d: 152 [RN] **A**: IS TR
smynensis Reitter, 1889h: 284 [HN]

elvirae Marcuzzi & Turchetto, 1982: 235 **E**: IT (Lucania)
minutissima Pic, 1903m: 162 **E**: GR (Kríti, Kos, Ródos, Dodecanese Is.) **A**: TR
smyrnensis Solier, 1838a: 31 (*Tagenia*) **E**: BU GR **A**: JO SY TR
hellenica Reitter, 1889h: 285
longicollis Rey, 1892: 41 (*Stenosis*)
syriaca fortepunctata Koch, 1935d: 273 **A**: JO SA
syriaca ovaliceps Koch, 1935d: 272 **A**: IS
syriaca syriaca Reitter, 1889h: 284 **A**: IS SY
turcomania Reitter, 1889h: 285 **A**: AF KZ TD TM UZ

genus *Gebieniella* Koch, 1940e: 736 type species *Ethas stenosides* Pascoe, 1862
hainanensis G. S. Medvedev, 1994a: 307 **A**: HAI

genus *Herbertfranzia* Kaszab, 1973a: 26 type species *Herbertfranzia nepalica* Kaszab, 1973
nepalica Kaszab, 1973a: 28 **A**: NP

genus *Herbertfranziella* Kaszab, 1973a: 26 type species *Herbertfranzia eutagenoides* Kaszab, 1973
eutagenoides Kaszab, 1973a: 29 (*Herbertfranzia*) **A**: NP
janushevi G. S. Medvedev, 1991: 560 **A**: TD
kumaona G. S. Medvedev, 1994b: 854 **A**: UP (Kumaon)

genus *Induchillus* Koch, 1941c: 300 type species *Induchillus cristatus* Koch, 1941
convexigena Ren & Shi, 2006b: 182 **A**: XIZ

genus *Indostola* G. S. Medvedev, 1991: 557 type species *Indostola pulchella* G. S. Medvedev, 1991
pulchella G. S. Medvedev, 1991: 558 **A**: SD

genus *Microblemma* Semenov, 1889a: 213 type species *Microblemma simplex* Semenov, 1889
afghanica Kaszab, 1960a: 18 **A**: AF
cordicollis Kaszab, 1960a: 19 **A**: AF
kasyi Kaszab, 1970a: 42 **A**: AF
lindbergi Kaszab, 1970a: 40 **A**: AF
quadraticollis Kaszab, 1970a: 41 **A**: AF
simpex Semenov, 1889a: 213 **A**: TM

genus *Microtelus* Solier, 1838a: 9 type species *Microtelus asiaticus* Solier, 1838
afghanicus Kaszab, 1970a: 37 **A**: AF
angulatus Carl, 1992b: 334 **A**: IN
asiaticus Solier, 1838a: 10 **E**: GR **A**: TR
cariniceps binodiceps Reitter, 1907c: 115 **N**: EG **A**: SI
sinaiticus Koch, 1935a: 48
cariniceps cariniceps Reiche & Saulcy, 1857: 227 **N**: EG **A**: IQ IS JO LE SY
cariniceps cordicollis Kaszab, 1982b: 128 **A**: SA
insolitus Carl, 1992a: 78 **A**: IQ
interstitialis Reitter, 1914a: 14 **A**: IQ IS JO SA SY
lethierryi Reiche, 1860: 334 **E**: FR (Corse) **N**: AG EG IS LB MO TU
torretassoi Schatzmayr & Koch, 1934: 24
limitis Reitter, 1914a: 14 **A**: IS SY
nejdanus Kaszab, 1982b: 128 **A**: SA
persis Baudi di Selve, 1874a: 106 **A**: IN IQ KU
schusteri Carl, 1992a: 75 **N**: EG

genus *Mitotagenia* Reitter, 1916d: 153 type species *Stenosis arabs* Baudi di Selve, 1881
aegyptiaca Koch, 1941b: 40 **N**: EG **A**: IS SA SI
afghanica G. S. Medvedev, 2004b: 558 **A**: AF
arabs Baudi di Selve, 1881: 279 (*Stenosis*) **A**: YE **AFR**
biroi Koch, 1941b: 40 **A**: OM
pohli Schawaller, 2004c: 448 **A**: YE (Suqutra)
priesneri Andres, 1931: 56 **N**: EG

genus *Oogaster* Faldermann, 1837: 30 type species *Oogaster menetriesii* Faldermann, 1837
lehmanni Ménétriés, 1849: 226 **A**: AF IN KZ TM UZ
piceus Ménétriés, 1832: 196 (*Tagenia*) **E**: AB AR **A**: IN
doriae Baudi di Selve, 1874a: 106
menetriesii Faldermann, 1837: 31

genus *Platamodes* Ménétriés, 1849: 233 type species *Platamodes dentipes* Ménétriés, 1849
dentipes dentipes Ménétriés, 1849: 233 **E**: AB AR **A**: AF KZ TM UZ
dentipes orientalis Kaszab, 1970a: 43 **A**: AF

genus *Pseudethas* Fairmaire, 1896a: 57 type species *Pseudethas quadraticeps* Fairmaire, 1896
Schizillus Wasmann, 1899: 166 type species *Schizillus rogersi* Wasmann, 1899
Stenillus Blair, 1927: 245 type species *Stenillus monticola* Blair, 1927
afghanicus Kaszab, 1960a: 3 **A**: AF
antennalis Kaszab, 1981b: 298 **A**: UP (Kumaon)
costatus Kaszab, 1981b: 303 **A**: UP
incostatus Kaszab, 1981b: 302 **A**: PA (Swat, Chitral)
jaegeri Schawaller, 2001c: 143 **A**: XIZ
ladakhensis Kaszab, 1978b: 223 **A**: KA (Ladakh)
monticola Blair, 1927: 245 (*Stenillus*) **A**: XIN
nepalicus Kaszab, 1981b: 299 **A**: NP
pakistanus Kaszab, 1981b: 301 **A**: PA (Hazara)
quadraticeps Fairmaire, 1896a: 57 **A**: HP
rogersi Wasmann, 1899: 166 (*Schizillus*) **A**: HP UP **ORR**

genus *Reitterella* Semenov, 1891a: 362 type species *Reitterella fusiformis* Semenov, 1891
fusiformis Semenov, 1891a: 363 **A**: TM
gracillima Semenov, 1903c: 356 **A**: TM
turkestanica Pic, 1906e: 33 **A**: UZ (Fergana)

genus *Stenosis* Herbst, 1799: 160 type species *Stenosis angustata* Herbst, 1799 (= *Stenosis intermedia* Küster, 1848)
subgenus *Stenosis* Herbst, 1799: 160 type species *Stenosis angustata* Herbst, 1799 (= *Stenosis intermedia* Küster, 1848)
Tagenia Latreille, 1802: 170 type species *Tenebrio filiformis* Fabricius, 1792
affinis Solier, 1838a: 26 (*Tagenia*) **N**: EG **A**: SI
algirica Lucas, 1846: 318 (*Tagenia*) **N**: AG
alluaudi Antoine, 1936a: 14 **N**: MO
altiatlantis Antoine, 1936a: 14 **N**: MO (Grand Atlas)
angustata Fabricius, 1792a: 105 (*Pimelia*) **E**: SP **N**: TU
angusticollis angusticollis Reiche, 1861c: 209 (*Tagenia*) **E**: FR (Var, Corse) IT
angusticollis elongatissima Koch, 1940e: 702 **E**: IT (Sardegna)
antoinei Koch, 1937: 398 **N**: LB
bengala Fairmaire, 1894a: 18 **A**: UP **ORR**
brenthoides brenthoides Rossi, 1790: 135 (*Rhinomacer*) **E**: CR IT
italica Kraatz, 1862a: 99
maroccana Desbrochers des Loges, 1881: 73

brenthoides carfii Ardoin, 1971a: 46 **N**: TU
brentoides solieri Koch, 1940e: 706 **E**: PT SP
brignonei Koch, 1935e: 314 **E**: IT (Sicilia)
canaliculata Miller, 1861: 176 **A**: LE SY
canariensis Franz, 1990b: 93 **N**: CI (Gran Canaria)
cephalonica Koch, 1940e: 726 **E**: GR (Kephalonia)
cobosi Español, 1958b: 46 **E**: SP (Málaga)
colfalonierii Gridelli, 1933: 224 **N**: LB
comata comata Reiche & Saulcy, 1857: 230 (*Tagenia*) **E**: AR **A**: IS SA SY
comata longiceps Koch, 1935d: 271 **A**: IS LE
consiglioi Canzoneri, 1976: 10 **E**: IT (Palmarola)
costulata Baudi di Selve, 1881: 278 **A**: SA YE
  ecostata Gridelli, 1953: 42
cretica Koch, 1940e: 726 **E**: GR
crivellarii Koch, 1935e: 313 **E**: GR (Kastellorhizo)
cugiae Gridelli, 1930: 238 **N**: EG LB
dianae J. R. Sahlberg, 1908: 77 **A**: SY TR
dilutipes Reitter, 1887a: 77 **E**: AR **A**: IS LE TR
  reitteri Marseul, 1887: 311 [RN]
duplex Antoine, 1948: 284 **N**: MO
elongata Solier, 1838a: 24 (*Tagenia*) **N**: EG
elvirae arganeticola Antoine, 1957: 42 **N**: MO
elvirae elvirae Antoine, 1936a: 14 **N**: MO
elvirae reymondi Kocher, 1965a: 26 **N**: MO
elvirae sarhoensis Kocher, 1965a: 26 **N**: MO
esau J. R. Sahlberg, 1908: 77 **E**: GR (Samos) **A**: TR
espanoli Antoine, 1948: 289 **N**: MO
fausti Reitter, 1888d: 430 **A**: KZ
figuerasi Español, 1958c: 111 **N**: MO
filiformis Fabricius, 1792a: 114 (*Tenebrio*) **N**: AG MO
foveiventris Desbrochers des Loges, 1881: 77 **E**: UK (Krym)
freyi Koch, 1940e: 700 **E**: IT MA
frioli Solier, 1838a: 19: (*Tagenia*) **N**: AG TU
fulvipennis fulvipennis Pic, 1912o: 90 **N**: AG MO
  safiana Antoine, 1948: 283
fulvipennis ramburi Antoine, 1936a: 13 **E**: SP
fulvipes Reiche & Saulcy, 1857: 232 (*Tagenia*) **A**: IS JO LE SY
  canaliculata Miller, 1861: 176
  planicollis Reitter, 1916d: 147
gattefossei Antoine, 1936a: 15 **N**: MO
grandis grandis Solier, 1838a: 23 (*Tagenia*) **N**: TU
grandis ravasinii Gestro, 1880: 417 **N**: TU
hesperica Solier, 1838a: 29 (*Tagenia*) **N**: EG
hispanica abti Antoine, 1949c: 215 **N**: MO (Moyen Atlas)
hispanica hispanica Solier, 1838a: 25 (*Tagenia*) **E**: PT SP **N**: EG MO
  elongata Solier, 1838a: 24 (*Tagenia*)
  rabatica Théry, 1921: 53
  tangeriana Desbrochers des Loges, 1881: 72
  tazekensis Antoine, 1949c: 215
insignita Antoine, 1949c: 226 [RN] **N**: MO
  peyerimhoffi Antoine, 1948: 286 [HN]
intermedia dalmatina Reitter, 1916d: 145 **E**: AL CR GR
intermedia intermedia Solier, 1838a: 17 (*Tagenia*) **E**: FR IT SP
  angustata Herbst, 1799: 160 [HN]
  coarctata Gemminger, 1870a: 121 [RN]

fiorii Antoine, 1949c: 231
sicula Solier, 1838a: 18 (*Tagenia*)
intermedia muelleri Koch, 1940e: 710 **E**: BH CR
intermedia pelagosana J. Müller, 1912: 290 **E**: GR (Ionioi) **A**: TR
intricata Reitter, 1886a: 120 **E**: SP (Baleares, Catalán)
keosana Koch, 1948a: 322 **E**: GR (Kyklades)
kochi Antoine, 1948a: 288 **N**: MO
kuntzeni Koch, 1941c: 295 **A**: AF TD TM
laevicollis Solier, 1838a: 19 (*Tagenia*) **E**: SP **N**: TU
hipponensis Desbrochers des Loges, 1881: 75
impunctipennis Chevrolat, 1877a: 118
lafanecherei Antoine, 1951c: 94 **N**: MO (Rif)
lateralis bogosensis Reitter, 1886a: 105 **N**: EG **AFR**
lateralis lateralis Reitter, 1886a: 104 **N**: EG **AFR**
leprieuri Antoine, 1949c: 232 **N**: AG
lindbergi Schuster, 1933a: 16 **N**: MO
longipilis Reitter, 1916d: 151 **E**: PT SP
mamorensis Théry, 1924: 149 **N**: MO
maxima Desbrochers des Loges, 1881: 70 **N**: AG MO
alticola Antoine, 1949c: 212
mazetieri Antoine, 1936a: 13 **N**: MO
mediatlantis kocheri Antoine, 1957: 42 **N**: MO
mediatlantis mediatlantis Antoine, 1936a: 15 **N**: MO (Moyen Atlas)
mediatlantis robustior Antoine, 1957: 42 **N**: MO
melitana Reitter, 1894b: 82 **E**: IT MA
mendizabali Español, 1960a: 109 **E**: SP (Almería)
milosana Koch, 1948a: 323 **E**: GR (Milos)
mimeuri Antoine, 1936a: 16 **N**: MO
mogadorica Antoine, 1936a: 16 **E**: SP **N**: MO
normandi Koch, 1940e: 696 **N**: AG
vaulogeri Koch, 1940e: 697
obliterata chobauti Antoine, 1949c: 219 **N**: AG
obliterata obliterata Solier, 1838a: 30 (*Tagenia*) **E**: SP **N**: MO TU
bathyglypta Antoine, 1949c: 218
ferruginea Desbrochers des Loges, 1881: 71
obsoleta Desbrochers des Loges, 1881: 71
occidentalis Théry, 1921: 53 **N**: MO
oertzeni Koch, 1948a: 323 **E**: GR (Nikaria)
olcesii jacentipilis Antoine, 1949c: 221 **N**: MO
errans Antoine, 1949c: 221
olcesii olcesii Fairmaire, 1871a: 388 (*Tagenia*) **N**: AG
opaca Reitter, 1886a: 112 **N**: AG TU
opacoides Normand, 1939: 508 **N**: TU
orientalis jonica Koch, 1940e: 728 **E**: GR (Ionioi)
orientalis orientalis Brullé, 1832: 194 (*Tagenia*) **E**: AL GR TR **A**: TR
oteroi Español, 1981: 58 **E**: SP (Pontevedra)
otini Antoine, 1948: 283 **N**: MO
pardoi Español, 1966: 174 **N**: MO
parvicollis Desbrochers des Loges, 1881: 76 **N**: AG MO
peyerimhoffi Koch, 1940e: 714 **A**: JO
pilifera Solier, 1838a: 21 (*Tagenia*) **E**: GR (Morea)
pubescens Solier, 1838a: 22 (*Tagenia*) **N**: EG
pueli Antoine, 1949c: 219 **N**: AG
punctiventris Eschscholtz, 1831: 12 (*Tagenia*) **E**: UK **A**: TR
oblongicollis Desbrochers des Loges, 1881: 73

quadraticollis Desbrochers des Loges, 1881: 79
punica Pic, 1905b: 98 **N**: LB TU
quedenfeldti Reitter, 1916d: 146
rhodica profundepunctata Koch, 1935d: 269 **A**: IS
rhodica rhodica Koch, 1935e: 311 **E**: GR (Ródos) **A**: LE
rungsi Antoine, 1936a: 14 **N**: MO
saintjeani Normand, 1939: 510 **N**: TU
sardoa ardoini Canzoneri, 1970: 63 **E**: IT (Sicilia)
sardoa laeviventris Desbrochers des Loges, 1881: 78 **E**: UK (Krym) TR **A**: TR
brevicollis Reiche, 1857: 232 [RN]
sardoa sardoa Küster, 1848b: 83 (*Tagenia*) **E**: FR IT SP **N**: AG TU
duchoni Antoine, 1949c: 231
schembrii Canzoneri, 1979: 20 **E**: MA
schrammi Antoine, 1933b: 225 **N**: MO
silvestrii schusteri Koch, 1940e: 726 **A**: TR
silvestrii silvestrii Koch, 1935e: 312 **E**: GR (Ródos)
striolata Escalera, 1914: 291 **N**: MO
sulcata Miller, 1861: 177 **A**: CY TR
sulcicollis Ménétriés, 1849: 227 (*Tagenia*) **A**: KZ TM UZ
syrensis naxica Koch, 1940e: 724 **E**: GR (Kyklades)
syrensis syrensis Koch, 1936: 293 **E**: GR (Syra)
teluetica Antoine, 1949c: 223 **N**: MO (Grand Atlas)
tenuicornis Baudi di Selve, 1874a: 104 [= 1875c: 69, 1875a: 295] **A**: IN
theryi macra Antoine, 1936a: 16 **N**: MO
theryi theryi Antoine, 1936a: 15 **N**: MO
torretassoi Koch, 1935a: 42 **N**: EG
villosa andalusica Español, 1958b: 50 **E**: SP
villosa villosa C. Brisout de Barneville, 1866: 382 **E**: SP
pilifera Reitter, 1886a: 115
webbi ramburi Antoine, 1936a: 13 **E**: SP
webbii webbii Guérin-Méneville, 1844: 113 (*Tagenia*) **N**: AG MO TU
pici Antoine, 1936a: 15
rotroui Antoine, 1936a: 13
temporalis Reitter, 1916d: 142
zaiana Antoine, 1936a: 15 **N**: MO
zarcoi Español, 1947b: 186 **E**: SP (Castellón)

genus *Tagenostola* Reitter, 1916d: 151 type species *Stenosis turkestanica* Reitter, 1886
pilosa Motschulsky, 1839: 65 (*Tagenia*) **E**: AB AR **A**: IN KZ TD TM UZ
seriepilosa muelleri Reitter, 1886a: 123 (*Stenosis*) **N**: EG **A**: SA **AFR**
turkestanica albovillosa Koch, 1940e: 735 **A**: BT NP UP
turkestanica turkestanica Reitter, 1886a: 123 (*Stenosis*) **E**: AB **A**: KZ TM UZ

genus *Tetranillus* Wasmann, 1899: 167 type species *Tetranillus costatus* Wasmann, 1899
costatus Wasmann, 1899: 167 **A**: IN **ORR**
kobmanni Koch, 1940e: 744 **A**: IN
longicarinatus Ren & Shi, 2006b: 181 **A**: XIZ

genus *Tetranosis* G. S. Medvedev, 1994b: 858 type species *Tetranosis clypeolobus* Koch, 1940
subgenus *Hypermicrotelopsis* Koch, 1940e: 743 type species *Microtelopsis thibetana* Koch, 1940
thibetanus Koch, 1940e: 743 (*Microtelopsis*) **A**: HP XIZ
subgenus *Microtelopsis* Koch, 1940e: 742 type species *Tetranillus simplicifrons* Gridelli, 1934
simplicifrons Gridelli, 1934: 55 (*Tetranillus*) **A**: KA
topali Kaszab, 1975a: 10 **A**: KA

subgenus *Tetranosis* G. S. Medvedev, 1994b: 858 type species *Tetranosis clypeolobus* Koch, 1940
ciliaticostis Koch, 1940e: 742 **A**: SD
clypeolobus Koch, 1940e: 741 **A**: KA
franzi Kaszab, 1973a: 25 **A**: NP
wittmeri Kaszab, 1981b: 295 **A**: HP NP

tribe Tentyriini Eschscholtz, 1831

genus *Abigopsis* Escalera, 1914: 277 type species *Scelosodis ustus* Fairmaire, 1879
atlantica Escalera, 1914: 279 **N**: MO (Sous)
usta Fairmaire, 1879c: 189 (*Scelosodis*) **N**: MO (Mogador)

genus *Alcinoeta* E. Strand, 1929: 23 [RN] type species *Alcinoe helopioides* Ménétriés, 1849
Alcinoe Ménétriés, 1849: 230 [HN] type species *Alcinoe helopioides* Ménétriés, 1849
Allodengitha Bogatchev, 1963a: 57 type species *Allodengitha deserta* Bogatchev, 1963
deserta Bogatchev, 1963a: 58 (*Allodengitha*) **A**: KZ UZ
microderoides Skopin, 1964: 292 (*Alcinoe*)
helopioides helopioides Ménétriés, 1849: 231 (*Alcinoe*) **A**: AF KZ TM UZ
helopioides spectabilis Kraatz, 1882b: 82 (*Alcinoe*) **A**: TM UZ
iranica G. S. Medvedev, 1988b: 800 **A**: IN

genus *Amblycarenum* Gebien, 1910a: 77 [RN] type species *Amblycara biskrensis* Fairmaire, 1893 (= *Melancrus alutaceum* Fairmaire, 1875)
Amblycara Fairmaire, 1893i: cxlvii [HN] type species *Amblycara biskrensis* Fairmaire, 1893 (= *Melancrus alutaceum* Fairmaire, 1875)
alutaceum Fairmaire, 1875b: 495 (*Melancrus*) **N**: AG TU
biskrensis Fairmaire, 1893i: cxlviii (*Amblycara*)

genus *Ammogiton* Peyerimhoff, 1920a: 326 type species *Ammogiton viberti* Peyerimhoff, 1920
buettikeri buettikeri Kaszab, 1979c: 281 **A**: SA (Prov. Riyadh, Prov. Eastern)
buettikeri major Kaszab, 1981c: 348 **A**: SA (Prov. Riyadh)
buettikeri medinae Kaszab, 1981c: 349 **A**: SA (Prov. Medina)
jolyi Español, 1953c: 296 **N**: MO (Baja Daoura)
omanicum Schawaller, 1990a: 44 **A**: OM
peltieri Peyerimhoff, 1924: 75 **N**: AG (Grand Erg Occidental)
sonyae Kaszab, 1979c: 279 **A**: SA
viberti Peyerimhoff, 1920a: 326 **N**: TU

genus *Anatolica* Eschscholtz, 1831: 7 type species *Tentyria subquadrata* Tauscher, 1812
subgenus *Anatolica* Eschscholtz, 1831: 7 type species *Tentyria subquadrata* Tauscher, 1812
abbreviata Gebler, 1832: 116 **E**: ST UK **A**: KZ WS
gnathosioides Motschulsky, 1870: 389
amoena Faldermann, 1835b: 397 (*Tentyria*) **A**: ES (Tuva) MG NMO
auctiformis Skopin, 1964: 376
iduna Reitter, 1900d: 122
typonota Ménétriés, 1836b: 181
amoenula Reitter, 1889k: 683 **A**: GAN MG NIN NMO
dschungarica Kaszab, 1967a: 21
angustata Steven, 1829: 85 (*Tentyria*) **E**: ST **A**: KZ
audouini Solier, 1835b: 385
conica Motschulsky, 1845a: 73
elongata Gebler, 1829a: 119 (*Tentyria*)
glaberrima Baudi di Selve, 1875c: 38
mallei Solier, 1835b: 386
arethusa Reitter, 1900d: 119 **A**: UZ
arnoldii Skopin, 1964: 398 **A**: KZ
atshitnura Kaszab, 1969b: 308 **A**: MG

aucta Faldermann, 1835b: 401 (*Tentyria*) **A**: ES (Tuva) MG NMO
boldi Kaszab, 1965e: 303 **A**: MG
cecchiniae Bogdanov-Katjkov, 1915: 4 **A**: MG
cellicola Faldermann, 1835b: 398 (*Tentyria*) **A**: ES (Tuva) HEB LIA MG NMO
chogsonzhavi Kaszab, 1967a: 18 **A**: MG
crassiuscula Bogatchev, 1961a: 117 **A**: MG
dashidorzsi dashidorzsi Kaszab, 1965e: 306 **A**: MG
dashidorzsi fortepunctata Kaszab & Knor, 1976: 295 **A**: ES (Tuva)
dashidorzsi subalpina Kaszab & Knor, 1976: 296 **A**: ES (Tuva) WS (Altay)
dashidorzsi temporalis Kaszab, 1965e: 308 **A**: ES (Tuva) MG
deserticola deserticola Skopin, 1960: 296 **A**: KZ
deserticola orientalis Skopin, 1964: 397 **A**: KZ
elegans Skopin, 1961b: 389 **A**: KZ (Almatynskaya oblast)
eremita Steven, 1829: 87 (*Tentyria*) **E**: ST UK
  besseri Krynicki, 1829: 193 (*Tentyria*)
  besseri Solier, 1835b: 390 [HN]
  ventralis Motschulsky, 1845a: 73
externecostata Fairmaire, 1888b: 125 **A**: BEI
fausti Reitter, 1900d: 118 **A**: KI
gibbosa gibbosa Steven, 1829: 87 (*Tentyria*) **E**: ST **A**: KZ UZ
  balassogloi Dohrn, 1882: 245
  depressa Besser, 1832: 8 (*Tentyria*)
  lineata Steven, 1829: 87 (*Tentyria*)
  punctipleuris Bogdanov-Katjkov, 1915: 3
  saisanensis Motschulsky, 1870: 396
gibbosa thoracica Besser, 1832: 8 (*Tentyria*) **A**: XIN
gobialtaica altaica Kaszab & G. S. Medvedev, 1972: 217 **A**: MG
gobialtaica gobialtaica Kaszab, 1965e: 314 **A**: MG
gobialtaica ovalis Kaszab & G. S. Medvedev, 1976: 109 **A**: MG
granulipleuris Reitter, 1900d: 124 **A**: MG NMO
gravidula J. Frivaldszky, 1889: 204 **A**: GAN
grebenscikovi Kaszab, 1964b: 6 **A**: MG
guentheri Kaszab, 1967a: 20 **A**: MG
hammarstromi Poppius, 1904: 12 **A**: ES (Tuva) MG NMO
hauseri Reitter, 1900d: 117 **A**: KI (Tian Shan)
holdereri Reitter, 1900d: 123 **A**: XIN XIZ
humerangula Kaszab, 1965e: 318 **A**: MG
immarginata Reitter, 1889k: 681 **A**: GAN MG NMO XIN
  hummeli Schuster, 1936a: 4
impressa Tauscher, 1812a: 30 (*Tentyria*) **E**: ST **A**: KI KZ XIN
integra Reitter, 1887d: 355 **A**: MG QIN XIZ
knori Kaszab, 1976d: 294 **A**: ES (Tuva)
koltzei Reitter, 1900d: 117 **A**: KZ (Kirghiz Range)
korunolena Skopin, 1964: 400 **A**: KZ (Almaatinskaya Oblast)
kozlovi Bogatchev, 1961a: 117 **A**: GAN
kozloviana Skopin, 1964: 379 **A**: XIN
kulzeri Kaszab, 1965c: 280 **A**: XIN
lacustris Kaszab, 1967a: 291 **A**: MG
lata Steven, 1829: 86 (*Tentyria*) **E**: ST **A**: KZ MG NMO WS XIN
  albovittis Motschulsky, 1845a: 73
  oblonga Allard, 1883: 19
  tristis Solier, 1835b: 388
lepida Faldermann, 1835b: 394 (*Tentyria*) **A**: ES MG NMO
  atramentaria Faldermann, 1835b: 395 (*Tentyria*)
  damascena Kraatz, 1865a: 98
longithorax Reitter, 1900d: 118 **A**: KZ (Kirghiz Range)

lucidula Bogatchev, 1961a: 116 **A**: GAN MG
malkovskii Skopin, 1961b: 390 **A**: KZ (Muyunkum)
medusa Reitter, 1900d: 126 **A**: XIN
midas Reitter, 1900d: 116 **A**: KZ (Vostochnyy Kazakhstan)
mirabilis Kaszab, 1962c: 307 **A**: XIN
modesta Bogatchev, 1961a: 118 **A**: ES (Tuva) MG
uljasutaja Kaszab, 1965e: 311
muchei Kaszab, 1965e: 309 **A**: MG
mucronata Reitter, 1889k: 682 **A**: GAN MG NIN NMO QIN XIZ
mustacea Kolbe, 1908: 88 **A**: GAN QIN
nureti Schuster & Reymond, 1937: 237 **A**: GAN MG NIN NMO
chauveti Schuster & Reymond, 1937: 236
omnoensis Skopin, 1964: 374 **A**: MG NIN NMO
pandaroides Reitter, 1889k: 680 **A**: GAN NIN
paphia Reitter, 1900d: 117 **A**: TD (Alai)
paradoxa Reitter, 1900d: 120 **A**: ES (Tuva) MG NMO
planata J. Frivaldszky, 1889: 203 **A**: GAN
polita borealis Kaszab, 1964b: 3 **A**: GAN MG NIN NMO
polita polita J. Frivaldszky, 1889: 204 **A**: GAN NMO
potanini Reitter, 1889k: 683 **A**: GAN MG NIN NMO XIZ
psammophila Skopin, 1964: 377 **A**: MG
montagui Kaszab, 1967c: 308
pseudaucta Reitter, 1900d: 124 **A**: ES MG NMO
pseudiduna Kaszab, 1965c: 281 **A**: XIN
pusilla Kaszab, 1967c: 297 **A**: MG
quadricollis Ballion, 1878: 297 (*Colposcelis*) **A**: KZ XIN
lucina Reitter, 1900d: 116
relicta Kaszab, 1964b: 8 **A**: MG
salinicola Kaszab, 1967c: 295 **A**: MG
scythisoides Kaszab, 1968c: 359 **A**: MG
semenowi Reitter, 1889k: 680 **A**: GAN
splendida G. S. Medvedev, 1990b: 134 **A**: MG
sternalis gobiensis Kaszab, 1964g: 371 **A**: MG
sternalis sternalis Reitter, 1889k: 681 **A**: GAN NMO XIN
loczyi J. Frivaldszky, 1889: 205
strigosa redikorzevi Bogdanov-Katjkov, 1916b: xxx **A**: "Altai"
strigosa strigosa Germar, 1824b: 138 (*Tentyria*) **A**: ES (Tuva) MG WS (Altay) XIN XIZ
suavis J. Frivaldszky, 1889: 205 **A**: GAN
subtrapezicollis Kaszab, 1965e: 309 **A**: MG
subquadrata Tauscher, 1812a: 31 (*Tentyria*) **E**: ST **A**: KZ
coriacea Motschulsky, 1845a: 73
sulcipennis laevior Kaszab, 1965e: 316 **A**: MG
sulcipennis sulcipennis Reitter, 1900d: 125 **A**: MG NMO
suturalis Reitter, 1889k: 682 **A**: NMO
syrtensis Kaszab, 1962e: 306 **A**: KI
tibialis Reitter, 1900d: 127 **A**: ES (Tuva) MG XIN
tsharynensis Skopin, 1964: 399 **A**: KZ (Almaatinskaya Oblast)
tsendsureni Skopin, 1964: 372 **A**: MG XIN
colposcina Skopin, 1964: 374
undulata Gebler, 1832: 55 (*Tentyria*) **A**: ES MG NMO
implana Faldermann, 1835b: 400 (*Tentyria*)
propinqua Faldermann, 1835b: 396 (*Tentyria*)
torulosa Fischer von Waldheim, 1844: 65
subgenus *Eurepipleura* Bogdanov-Katjkov, 1915: 2 type species *Anatolica minima* Bogdanov-Katjkov, 1915
ebenina Fairmaire, 1887f: 323 **A**: BEI

minima Bogdanov-Katjkov, 1915: 2 **A**: "China borealis"

genus *Ascelosodis* L. Redtenbacher, 1868: 117 type species *Ascelosodis serripes* L. Redtenbacher, 1868
assimilis Bates, 1879: 468 **A**: KA
punctatissima Reitter, 1915b: 62 (*Gnathosia*)
ciliatis Bates, 1879: 468 **A**: KA
concinna Bates, 1879: 468 **A**: TD
semenowi Reitter, 1897h: 303 (*Orocina*)
emarginata Ren & Shi, 2006a: 4 **A**: XIN
emarginiventris Koch, 1949: 76 **A**: KA XIZ
everestina Blair, 1923b: 278 **A**: SD XIZ
forsteri Kaszab, 1961d: 214 **A**: NP
grandis Bates, 1879: 469 **A**: KA
ovoidea Fairmaire, 1891d: xcii
haagi Bates, 1879: 469 **A**: KA UZ XIZ
inermis Fairmaire, 1891d: xciii **A**: KA
intermedia Bates, 1879: 469 **A**: HP KA
kochi Kaszab, 1959f: 357 **A**: AF
lindbergi Kaszab, 1968b: 52 **A**: AF
longstaffi Blair, 1923b: 279 **A**: XIZ
minor Kaszab, 1978b: 219 **A**: KA
nitida Blair, 1923b: 281 **A**: SD XIZ
punctulata Ren & Shi, 2006a: 4 **A**: XIN
reinigi Koch, 1949: 78 **A**: HP XIZ
rugulosa Ren & Shi, 2006a: 4 **A**: XIN
schmidi Kaszab, 1965a: 108 **A**: SD XIZ
serripes L. Redtenbacher, 1868: 118 **A**: KA XIZ
thibetana Blair, 1923b: 281 **A**: XIZ
waltoni Blair, 1923b: 280 **A**: XIZ

genus *Calyptopsis* Solier, 1835b: 269 type species *Calyptopsis emondi* Solier, 1835 (= *Hegeter caraboides* Brullé, 1832)
Choristopsis Kraatz, 1965: 228 type species *Choristopsis caucasica* Kraatz, 1865
amaroides Baudi di Selve, 1874a: 96 [= 1875c: 32] **A**: IN
bazmanica Schuster, 1938b: 82 **A**: IN
capnisiformis Reitter, 1903b: 44 **A**: TR
capnisoides Reitter, 1897i: 307 **E**: GR TR **A**: TR
caraboides Brullé, 1832: 201 (*Hegeter*) **E**: GR
emondi Solier, 1835b: 271
caucasica Kraatz, 1865a: 228 (*Choristopsis*) **E**: AB AR **A**: TR
morawitzi Faust, 1877a: 37
clypeata Faust, 1875: 179 **A**: AF IN TM
convexicollis Desbrochers des Loges, 1881: 65 **E**: AB GG
antoniae Reitter, 1896a: 29
deplanata Faust, 1875: 177 **A**: IN TM TR
emarginata Reitter, 1889a: 28 **E**: AB **A**: TR
escherichi Reitter, 1900d: 133 **A**: TR
gigas Kaszab, 1962c: 309 **A**: IN
goliath Kaszab, 1959e: 242 [= 1959f: 385] **A**: AF
harpaloides Baudi di Selve, 1874a: 96 [= 1875c: 33] **E**: AR **A**: IN
armeniaca Baudi di Selve, 1874a: 97 [= 1875: 34]
incerta Faust, 1875: 179 **E**: IN TM
jeremias Reiche & Saulcy, 1857: 197 **A**: IS SY
laevipennis Reitter, 1897i: 310 **A**: KZ
lineimargo Reitter, 1897i: 310 **A**: SY

nitescens Reitter, 1897i: 310 **E**: AR
pandaroides Reitter, 1897i: 308 **E**: BU GR
equalis Reitter, 1897i: 308
ovalis Reitter, 1897i: 308
pulchella Faldermann, 1837: 36 (*Gnathosia*) **E**: AB AR **A**: TR
rugipennis Faldermann, 1837: 37 (*Gnathosia*)
punctiventris Baudi di Selve, 1874a: 100 [= 1875c: 34] (*Tentyria*) **A**: TM (Kopet Dag)
rosti Reitter, 1897i: 306 **A**: IN
schach Kaszab, 1963a: 70 **A**: IN
solieri libanica Koch, 1936: 291 **A**: LE
solieri solieri Reiche & Saulcy, 1857: 199 **A**: IS SY TR
sulcimargo Reitter, 1897i: 306 **A**: IN LE
theodoridesi Kaszab, 1962c: 308 **A**: IN

genus *Cantopipleurus* Koch, 1943a: 578 type species *Tentyria mesostenoides* Baudi di Selve, 1881
mesostenoides Baudi di Selve, 1881: 276 (*Tentyria*) **A**: YE

genus *Capnisiceps* Chatanay, 1914c: 215 type species *Capnisiceps maindroni* Chatanay, 1914
brunneus Schawaller, 1993a: 103 **A**: OM
lilligi G. Wagner, 2005: 178 **A**: OM
maindroni Chatanay, 1914c: 215 **A**: AE OM
masirahensis G. Wagner, 2005: 179 **A**: OM (Masirah)

genus *Catomulus* Reitter, 1897h: 302 type species *Catomulus subaeneus* Reitter, 1897 (= *Oxycara olcesii* Fairmaire, 1883)
olcesii olcesii Fairmaire, 1883g: cxi (*Oxycara*) **N**: MO
minimus Escalera, 1914: 280
subaeneus Reitter, 1897h: 302
olcesii rugosicollis Español, 1944: 9 **N**: MO (Western Sahara)

genus *Cimipsa* Peyerimhoff, 1911: 346 type species *Cimipsa sergenti* Peyerimhoff, 1911
peyerimhoffi Théry, 1931b: 120 **N**: MO
hamadensis Kocher, 1951b: 129
reymonti Kocher, 1951b: 126 **N**: MO
sergenti Peyerimhoff, 1911: 346 **N**: AG
zarcoi rungsi Kocher, 1966b: 136 **N**: MO (Western Sahara)
zarcoi zarcoi Español, 1944: 8 **N**: MO (Western Sahara)

genus *Colposcelis* Dejean, 1834: 185 type species *Tentyria longicollis* Zubkov, 1833
subgenus *Colposcelis* Dejean, 1834: 185 type species *Tentyria longicollis* Zubkov, 1833
afghanica Gridelli, 1954: 192 [= 1955: 32] **A**: AF
clypealis Kaszab, 1968b: 110 **A**: AF
ferghanensis Kaszab, 1962c: 305 **A**: TD
globicollis Kaszab, 1968b: 108 **A**: AF
humerangula Reitter, 1900d: 108 **A**: KZ
hybrida Reitter, 1900d: 109 **A**: KI KZ
lindbergi Kaszab, 1968b: 112 **A**: AF
longicollis Zubkov, 1833: 328 (*Tentyria*) **E**: AB **A**: TM
microderoides microderoides Reitter, 1900d: 106 **A**: MG NMO XIN
mongolica Skopin, 1965: 375
microderoides strigipleuris Kaszab, 1967c: 286 **A**: MG
minor Kaszab, 1968b: 107 **A**: AF
minuta Reitter, 1900d: 108 **A**: TM
oschensis Kaszab, 1962c: 305 **A**: TD
regeli G. S. Medvedev, 1990b: 132 **A**: XIN

simillima Kaszab, 1968b: 111 **A**: AF
skopini Kaszab, 1962c: 304 **A**: KZ
subgenus *Colposceloides* Schuster, 1940a: 20 type species *Colposcelis licenti* Schuster, 1940
licenti Schuster, 1940a: 20 **A**: NMO
subgenus *Colposcythis* Reitter, 1889b: 113 type species *Colposcythis walteri* Reitter, 1889
walteri Reitter, 1889b: 114 (*Colposcythis*) **A**: TM
zarudnyi G. S. Medvedev, 1978c: 857 **A**: IN
subgenus *Scelocolpis* Reitter, 1900d: 105 type species *Colposcelis damone* Reitter, 1900
amanda Reitter, 1900d: 107 **A**: XIN
balchashense Skopin, 1964: 407 **A**: KZ
brunnea G. S. Medvedev, 1989b: 371 **A**: MG
bulganica Kaszab, 1967c: 288 **A**: MG
damone Reitter, 1900d: 107 **A**: MG XIN
elegans Kaszab, 1968c: 352 **A**: MG
foersteri Reitter, 1900d: 107 **A**: XIN
kamalovi G. S. Medvedev, 1989b: 373 **A**: MG
montivaga Bates, 1879: 470 (*Anatolica*) **A**: KA
pavlovskii Bogatchev, 1961a: 115 **A**: XIN (Turfan)
sulcatipleuris Skopin, 1961b: 385 **A**: KZ
tentyrioides Skopin, 1961b: 385 **A**: KZ
trisulcata Reitter, 1900d: 106 **A**: XIN
subgenus *Turcmenicola* Bogatchev, 1952a: 44 type species *Turcmenicola jachontovi* Bogatchev, 1952
jachontovi Bogatchev, 1952a: 45 (*Turcmenicola*) **A**: KZ TM UZ
kerzhneri G. S. Medvedev, 1970: 406 **A**: UZ
lopatini G. S. Medvedev, 1970: 406 **A**: TM
selevini Skopin, 1951: 124 **A**: KZ (Betpak-Dala)

genus *Colposphena* Semenov, 1891a: 360 type species *Sphenaria brevicollis* Solsky, 1882
brevicollis Solsky, 1881b: 45 (*Sphenaria*) **A**: KZ (Kyzylkum) TM
  breviceps Semenov, 1889a: 216 (*Sphenaria*)
  obtusa Semenov, 1889a: 214 (*Sphenaria*)

genus *Cyphostethe* Marseul, 1867: xxxix type species *Himatismus ferrugineus* Marseul, 1867
subgenus *Apterocyphostethe* Kaszab, 1962c: 302 type species *Cyphostethe koltzei* Reitter, 1895
amseli Kaszab, 1961e: 236 **A**: AF
iranica Kaszab, 1962c: 302 **A**: IN
koltzei Reitter, 1895d: 155 (*Himatismus*) **A**: AF XIN
mongolica Kaszab, 1968c: 348 **A**: MG
subgenus *Cyphostethoides* Löbl & Merkl, 2003: 244 type species *Cyphostethe brunnea* Kaszab, 1962
angulosicolle Chobaut, 1924c: 166 **N**: AG
brunnea Kaszab, 1962c: 303 **A**: IN
brunneoides Kaszab, 1979c: 274 **A**: PA
grombczewskii Semenov, 1891a: 354 (*Himatismus*) **A**: XIN
heydeni Haag-Rutenberg, 1877a: 282 (*Himatismus*) **N**: AG EG LB MO TU **A**: IS SA SI YE **AFR**
  saharensis Chobaut, 1897d: 205 (*Himatismus*)
jelineki Merkl, 1991b: 63 **A**: IN
komarowii Reitter, 1889b: 115 (*Sphenaria*) **A**: IN KZ TM
semenovi Bogatchev, 1947b: 80 **E**: AB AR **A**: IN
subgenus *Cyphostethe* Marseul, 1867: xxxix type species *Himatismus ferrugineus* Marseul, 1867
  Asphena Semenov, 1889a: 218 type species *Aphena komarowi* Semenov, 1889
antonowi Semenov, 1889b: 310 **A**: KZ TM UZ
belutschistanica Kaszab, 1957d: 64 **A**: IN PA SA
ferruginea Marseul, 1867: xxxviii (*Himatismus*) **N**: AG LB TU **A**: IS SA
komarowi Semenov, 1889a: 218 (*Asphena*) **A**: TM

seidlitzi Reitter, 1916e: 142 **A**: KZ TD TM UZ
stehliki Kaszab, 1968b: 44 **A**: AF
wittmeri Kaszab, 1979: 270 **A**: SA

genus *Dailognatha* Steven, 1829: 88 type species *Tentyria aequalis* Tauscher, 1812
Delognatha Agassiz, 1846b: 118 [unjustified emendation]
aequalis Tauscher, 1812a: 311 [40] (*Tentyria*) **E**: AB ST (Dagestan)
depressicollis Kraatz, 1865a: 89
depressicornis Faldermann, 1837: 35 (*Gnathosia*)
levigata Brullé, 1832: 199 (*Tentyria*)
arnoldii Bogatchev, 1947d: 460 **A**: TM UZ
bogatshevi afghanica Kaszab, 1968b: 102 **A**: AF
bogatshevi bogatshevi G. S. Medvedev, 1964a: 63 **A**: TD TM
caraboides Eschscholtz, 1831: 9 (*Hegeter*) **E**: AB AR BU GG GR TR **A**: IN IQ SY TR
audouini Solier, 1835b: 266
caucasica Desbrochers des Loges, 1881: 67 (*Gnathosia*)
crenata Reiche & Saulcy, 1857: 196 **A**: CY IS JO LE SI SY TR
cylindritarsis cylindritarsis Koch, 1948a: 304 **E**: GR (Kríti)
cylindritarsis probsti Picka, 1984: 20 **E**: GR (Kríti)
gracilicornis Reitter, 1897k: 317 **A**: KZ
gracilitarsis Kaszab, 1968b: 104 **A**: TM UZ
granimentum Reitter, 1897k: 317 **A**: KZ UZ
hellenica Reitter, 1897k: 316 **E**: GR
humeridens Kaszab, 1968b: 100 **A**: AF
impressicollis Solier, 1835b: 263 **A**: TR
marginicollis Kaszab, 1968b: 97 **A**: AF
minuscula Desbrochers des Loges, 1881: 68 (*Gnathosia*) **A**: TR
nasuta Ménétriés, 1849: 231 (*Anatolica*) **A**: AF KZ TD TM UZ
planata Miller, 1861: 173 (*Gnathosia*) **A**: JO SY TR
pumila Baudi di Selve, 1874b: 197 [= 1875c: 31] (*Gnathosia*) **E**: AR **A**: TR
humeralis Desbrochers des Loges, 1881: 69 (*Gnathosia*)
quadricollis anaphiana Koch, 1948a: 301 **E**: GR (Cyclades)
quadricollis carceli Solier, 1835b: 265 **E**: AR CR GR MC **A**: TR
quadricollis laticollis Besser, 1832: 9 (*Tentyria*) **E**: AL GR (Kerkyra) TR
chevrolati Allard, 1883: 17 (*Gnathosia*)
quadricollis montana Koch, 1948a: 302 **E**: GR (Kríti)
quadricollis obtusangula Reitter, 1897k: 314 **E**: GR
quadricollis plicata Koch, 1948a: 303 **E**: GR (Kríti)
quadricollis pseudobtusangla Koch, 1948a: 302 **E**: GR
quadricollis quadricollis Brullé, 1832: 193 (*Pimelia*) **E**: BU GR RO **A**: LE TR
crassicornis Desbrochers des Loges, 1881: 66 (*Gnathosia*)
sinuatocollis Desbrochers des Loges, 1881: 66 (*Gnathosia*)
variabilis Solier, 1835b: 262
quadricollis rhodica Koch, 1948a: 301 **E**: GR (Ródos)
quadricollis rugata Solier, 1835b: 263 **E**: GR (Kríti)
quadricollis samosana Koch, 1948a: 302 **E**: GR (Samos)
quadricollis sporadica Koch, 1948a: 301 **E**: GR (Sporades)
rugipleuris Reitter, 1897k: 316 **E**: GR (Kastellorhizo) **A**: TR
vicina Brullé, 1832: 200 (*Tentyria*) **E**: AL GR
opaca Allard, 1883: 16 (*Gnathosia*)

genus *Dengitha* Reitter, 1887d: 516 type species *Dengitha lutea* Reitter, 1887
crystallina Semenov, 1896b: 49 **A**: TM UZ
hyalina Semenov, 1896b: 52 **A**: TM
lutea Reitter, 1887e: 517 **A**: UZ

symmetrica G. S. Medvedev, 1964b: 160 **A**: KZ

genus *Dichomma* Solier, 1835b: 271 type species *Dichomma duponti* Solier, 1835 (= *Tentyria dardana* Steven, 1829)
dardanum Steven, 1829: 89 (*Tentyria*) **E**: GR **A**: SY TR
chevrolatii Reiche & Saulcy, 1857: 200
duponti Solier, 1835b: 274
foraminosum Küster, 1851: 64 (*Helops*)
glabrum Brullé, 1832: 201 (*Hegeter*)
maillei Solier, 1835b: 273

genus *Epitrichia* Gebler, 1859: 475 type species *Helops tomentosus* Gebler, 1843
fusca Ren & Zheng, 1993c: 51 **A**: NIN
kerzhneri Kaszab & G. S. Medvedev, 1972: 215 **A**: MG
intermedia Kaszab & G. S. Medvedev, 1976: 100
knori Kaszab, 1976b: 99 **A**: ES (Tuva)
mongolica Kaszab, 1965e: 299 **A**: ES (Tuva) MG
ningsiana Kaszab, 1965c: 279 **A**: NIN
semenovi semenovi Bogatchev, 1949b: 277 **A**: KZ MG NMO
semenovi tsendsureni Kaszab, 1968c: 343 **A**: MG
tomentosa Gebler, 1843: 39 (*Helops*) **A**: KZ
zaissanica Skopin, 1960: 295 **A**: KZ

genus *Eschatostena* Kelejnikova, 1977: 654 type species *Eschatostena kuznetzovi* Kelejnikova, 1977
kuznetzovi Kelejnikova, 1977: 655 **A**: TM

genus *Eulipus* Wollaston, 1864: 448 type species *Tentyria elongata* Brullé, 1838
Mogadoria Escalera, 1905b: 467 type species *Tentyria subelegans* Fairmaire, 1871
elongatus elongatus Brullé, 1838: 66 (*Tentyria*) **N**: CI
brullaei Wollaston, 1865: 393 (*Tentyria*) [RN]
subelegans Fairmaire, 1871a: 386 (*Tentyria*) **N**: MO
elongatus punctidorsis Reitter, 1900d: 184 **N**: MO
foveifrons Escalera, 1935a: 302
quirogai Escalera, 1905b: 468
kruegeri Schuster, 1926b: 132 **N**: LB (Cyrenaica)
vittembeschii Koch, 1937: 378 **N**: LB (Tripolitania)

genus *Eusyntelia* Waterhouse, 1881: 473 type species *Eusyntelia balfouri* Waterhouse, 1881
balfouri Waterhouse, 1881: 474 **A**: YE (Darsa, Suqutra)
ebenina Waterhouse, 1881: 474 **A**: YE (Suqutra)
exigua Gahan, 1900: 8 (*Rhytidonota*) **A**: YE (Suqutra)
glabra glabra Waterhouse, 1881: 475 **A**: YE (Suqutra)
glabra infraplicata Koch, 1970: 104 **A**: YE (Samha, Suqutra)
opacicollis Gahan, 1900: 9 **A**: YE (Suqutra)
socia Gahan, 1900: 9 (*Rhytidonota*) **A**: YE (Samha, Suqutra)

genus *Falsocatomulus* Pic, 1914e: 10 type species *Falsocatomulus euphraticus* Pic, 1914
Microhionthis Blair, 1923a: 120 type species *Microhionthis patriciae* Blair, 1923
euphraticus Pic, 1914e: 11 **A**: IN IQ
patriciae Blair, 1923a: 121 (*Microhionthis*)

genus *Freudeia* Kaszab, 1961d: 216 type species *Freudeia nepalica* Kaszab, 1961
granulipleuris Kaszab, 1977b: 245 **A**: NP
martensi Kaszab, 1977b: 245 **A**: NP
nepalica Kaszab, 1961d: 217 **A**: NP

punctipleuris Kaszab, 1977b: 243 **A**: NP

genus *Gnathosia* Fischer von Waldheim, 1821: 13 type species *Gnathosia glabra* Fischer von Waldheim, 1821
Capnisa Dejean, 1836: 197 type species *Bradyus karelini* Faldermann, 1836
afganica Bogatchev, 1947d: 461 **A**: AF
agaboides Kaszab, 1968b: 86 **A**: UZ
andersoni Bogdanov-Katjkov, 1915: 1 **A**: TD UZ (Bukhara)
aphodina Kaszab, 1959f: 370 **A**: TM
apicepilosa Gridelli, 1954: 186 [= 1955: 28] **A**: AF
aphodina Kaszab, 1959f: 370 **A**: AF
balachana Koch, 1949: 80 **A**: TM
barschewskyi Reitter, 1896c: 130 (*Capnisa*) **A**: UZ
interrupta Reitter, 1896c: 133 (*Capnisa*)
biconiger Kaszab, 1968b: 91 **A**: AF
blapoides Kaszab, 1959f: 365 **A**: AF
bogatschevi Kaszab, 1968b: 74 **A**: UZ
brincki Kaszab, 1973b: 36 **A**: AF
compressa Reitter, 1896c: 134 (*Capnisa*) **A**: TM (Kopet Dag)
cribratella Fairmaire, 1892e: cliii (*Capnisa*) **A**: TM
crypticola Reitter, 1896c: 132 (*Capnisa*) **A**: AF KZ TD TM UZ
declivis Reitter, 1896c: 131 (*Capnisa*) **A**: TD UZ
effugens Koch, 1949: 78 **A**: TD UZ (Fergana)
elliptica Ménétriés, 1849: 219 (*Capnisa*) **A**: UZ
elongata Reitter, 1900d: 98 **A**: KZ TD UZ
fausti Reitter, 1889a: 29 **A**: IN
glabra Fischer von Waldheim, 1821: 14 **A**: KZ TD UZ
gridellii Kaszab, 1959e: 241 [= 1959f: 368] **A**: AF
haarlovi Gridelli, 1954: 187 [= 1955: 30] **A**: AF
humerosa Reitter, 1900d: 100 **A**: UZ
hygrobiiformis Reitter, 1896c: 133 (*Capnisa*) **A**: TM UZ
jacobsoni Bogdanov-Katjkov, 1916a: 69 **A**: AF IN
kandaharica Kaszab, 1959f: 373 **A**: AF
karelini Faldermann, 1836: 375 (*Bradyus*) **A**: AF KZ TD TM UZ
depressiuscula Kraatz, 1882b: 81 (*Capnisa*)
kashmirensis Kaszab, 1961f: 343 **A**: PA (Karakoram)
klapperichi Kaszab, 1959f: 367 **A**: AF
kuhitangi G. S. Medvedev, 1964a: 62 **A**: TM (Kuhitang)
kulzeri Kaszab, 1959f: 363 **A**: AF
kushkensis G. S. Medvedev, 1964c: 660 **A**: TM (Kushka)
lindbergi Kaszab, 1968b: 83 **A**: AF
lopatini Kaszab, 1968b: 72 **A**: TD
lucida Kaszab, 1968b: 70 **A**: AF
minutissima G. S. Medvedev, 1979d: 78 **A**: TD
modesta Faldermann, 1837: 28 (*Bradytus*) **E**: AB AR **A**: IN TM
montana Bogatchev, 1947d: 462 **A**: UZ
nuristanica Kaszab, 1959f: 371 **A**: AF
orozgana Kaszab, 1968b: 93 **A**: AF
paghmatica Kaszab, 1968b: 81 **A**: AF
piffli Kaszab, 1961f: 342 **A**: PA (Karakoram)
praeoccupata Kaszab, 1968b: 78 [RN] **A**: AF
afghanica Gridelli, 1954: 183 [= 1955: 27] [HN]
proxima Reitter, 1915b: 61 **A**: IN
puncticeps Kaszab, 1968b: 79 **A**: AF
schrenkii Gebler, 1844: 100 (*Capnisa*) **A**: AF KZ TM UZ
caudata Gridelli, 1954: 189 [= 1955: 31]

semenovi Bogatchev, 1946c: 393 **A**: AF
simillima Kaszab, 1968b: 92 **A**: AF
skobelewi Starck, 1889: 53 (*Capnisa*) **A**: AF TM
somocoelioides Kaszab, 1968b: 77 **A**: AF
stehliki Kaszab, 1968b: 89 **A**: AF
sublaevigata Bogatchev & Kryzhanovskiy, 1960: 264 **A**: TM (Kopet Dag)
suturalis Reitter, 1896c: 130 (*Capnisa*) **A**: UZ
tadzhikistanica G. S. Medvedev & Kompantseva, 1989: 98 **A**: TD
trimarginata Kaszab, 1968b: 87 **A**: AF
turcomanica Kaszab, 1968b: 73 **A**: TM

genus *Gnophota* Erichson, 1843: 237 type species *Gnophota antracina* Erichson, 1843
inaequalis Wollaston, 1864: 466 **N**: CI (Gran Canaria)

genus *Hegeter* Latreille, 1802: 172 type species *Blaps elongata* A. G. Olivier, 1795 (= *Blaps tristis* Fabricius, 1792)
subgenus *Hegeter* Latreille, 1802: 172 type species *Blaps elongata* A. G. Olivier, 1795 (= *Blaps tristis* Fabricius, 1792)
abbreviatus Brullé, 1838: 66 **N**: CI (Gran Canaria)
amaroides Solier, 1835b: 378 **N**: CI
brevicollis Brullé, 1838: 65 **N**: CI (Tenerife)
brevicornis Har. Lindberg, 1953: 9 **N**: CI (Gomera)
costipennis Wollaston, 1864: 457 **N**: CI (Gran Canaria)
excisus Har. Lindberg, 1950: 9 **N**: CI (Tenerife)
glaber Brullé, 1838: 65 **N**: CI (La Palma)
gomerensis Wollaston, 1864: 456 **N**: CI (Gomera)
grancanariensis Har. Lindberg, 1950: 8 **N**: CI
impressus Brullé, 1838: 64 **N**: CI (Gran Canaria)
intercedens Har. Lindberg, 1950: 9 **N**: CI (Tenerife)
latebricola Wollaston, 1854: 510 **N**: MR (Salvages)
lateralis Brullé, 1838: 65 **N**: CI (Tenerife)
proximus Har. Lindberg, 1950: 9 **N**: CI (Tenerife)
subrotundatus Wollaston, 1864: 459 **N**: CI (Gran Canaria)
tenuipunctatus Brullé, 1838: 65 **N**: CI (Tenerife)
transversus Brullé, 1838: 65 **N**: CI (Tenerife)
tristis Fabricius, 1792a: 108 (*Blaps*) **E**: AZ **N**: CI MO MR **AFR**
elongatus A. G. Olivier, 1795: 60: 9 (*Blaps*)
picicornis Fabricius, 1792a: 121 (*Helops*)
striatus Latreille, 1804a: 276
webbianus Heineken, 1835: 40 **N**: CI (Gran Canaria)
subgenus *Homalapipleurus* Español, 1957d: 166 type species *Hegeter gonzalezi* Español, 1957
deyrollei Wollaston, 1864: 462 (*Thalpophila*) **N**: CI
gonzalezi Español, 1957d: 169 **N**: CI (Fuerteventura)
subgenus *Pseudotalpophila* Reitter, 1900d: 95 type species *Thalpophila plicifrons* Wollaston, 1864
fernandezi Español, 1957a: 12 **N**: CI (Fuerteventura)
plicifrons Wollaston, 1864: 461 (*Thalpophila*) **N**: CI (Fuerteventura)

genus *Hegeterocara* Reitter, 1900d: 190 type species *Hegeterocara arabica* Reitter, 1900
Fourtaus Pic, 1921g: 47 type species *Fourtaus brevicornis* Pic, 1921
arabica Reitter, 1900d: 190 **A**: IS JO SI
brevicornis Pic, 1921g: 48 (*Fourtaus*)

genus *Herlesa* Reitter, 1897h: 298 type species *Herlesa globicollis* Reitter, 1897 (= *Micipsa cavifrons* Fairmaire, 1863)
cavifrons Fairmaire, 1863: 642 (*Micipsa*) **N**: AG LB MO

gastonis Fairmaire, 1871a: 387 (*Micipsa*)
globicollis Reitter, 1897h: 299
grandis Peyerimhoff, 1933: 348 **N**: AG
semiopaca Peyerimhoff, 1916b: 83 (*Tentyronota*) **N**: MO

genus *Hionthis* Miller, 1861: 174 type species *Hionthis tentyroides* Miller, 1861
angusticollis Schuster, 1920b: 235 **N**: EG
heydeni Reitter, 1900d: 139 **A**: IS
mesopotamica Schuster, 1920b: 234 **A**: IQ
saulcyi Reitter, 1900d: 139 **A**: IS SY
tentyrioides Miller, 1861: 176 **N**: EG LB (Cyrenaica) **A**: SI SY "Palestina"

genus *Homoeonota* Fairmaire, 1882a: 63 type species *Homoeonota subopaca* Fairmaire, 1882
Isonota Fairmaire, 1887a: 171 type species *Isonota opaca* Fairmaire, 1887
dailognathoides Kaszab, 1981c: 350 **A**: OM
omanica Kaszab, 1981c: 351 **A**: OM

genus *Hyonthosoma* Reitter, 1900d: 142 type species *Hionthis occidentalis* Fairmaire, 1897
balboi Koch, 1937: 354 (*Hionthisoma*) **N**: LB (East Tripolitania)
bernardi Pierre, 1958: 136 (*Hionthisoma*) **N**: AG (Hoggar, Tassili n'Ajjer)
coutayari coutayari Chobaut, 1918: 74 (*Hionthisoma*) **N**: LB (West Tripolitania) TU
coutayari lybica Koch, 1937: 354 (*Hionthisoma*) **N**: LB (Tripolitania)
occidentalis grandis Español, 1953c: 288 **N**: AG MO
occidentalis occidentalis Fairmaire, 1897h: 180 (*Hionthis*) **N**: AG
sicardi Peyerimhoff, 1921b: 62 (*Hionthisoma*) **N**: MO

genus *Hyperops* Eschscholtz, 1831: 9 type species *Hyperops tagenoides* Eschscholtz, 1831
subgenus *Belutschistanops* Löbl, Bouchard & Merkl, 2008 [see New Acts] type species *Hyperops schusteri* Koch, 1940
doriae Baudi di Selve, 1874a: 102 [= 1875c: 61] **A**: IN
pygmaea afghanica Kaszab, 1959f: 387 (*Pachycera*) **A**: AF PA
pygmaea arabica Koch, 1940e: 686 **A**: AE BA OM SA
pygmaea pygmaea L. Redtenbacher, 1850: 48 **A**: IN PA
schusteri Koch, 1940e: 686 **A**: PA
subgenus *Hyperops* Eschscholtz, 1831: 9 type species *Hyperops tagenoides* Eschscholtz, 1831
Oedenocera Reiche, 1862a: 372 [RN] type species *Akis laevigata* sensu Eschscholtz, 1831 (= *Tenebrio buprestoides* Fabricius, 1781)
Pachycera Eschscholtz, 1831: 7 [HN] type species *Akis laevigata* sensu Eschscholtz, 1831 (= *Tenebrio buprestoides* Fabricius, 1781)
Tetromma Dejean, 1834: 183 type species "*Hegeter*" *unicolor* Herbst, 1797 (= *Upis unicolor* Herbst, 1797)
masirahana Kaszab, 1981c: 328 (*Pachycera*) **A**: OM
nitidula Kraatz, 1865a: 237 **A**: PA **ORR**
pfeifferi pfeifferi Koch, 1940e: 689 **A**: PA
tagenoides aegyptiaca Kraatz, 1865a: 236 **N**: EG **AFR**

genus *Hypsosoma* Ménétriés, 1854: 30 type species *Hypsosoma mongolica* Ménétriés, 1854
juxtalacum Skopin, 1964: 383 (*Scythis*) **A**: MG
mongolicum Ménétriés, 1854: 31 **A**: BEI HEB LIA NMO SHA SHX
extrema Reitter, 1889k: 679 (*Anatolica*)
rotundicolle Fairmaire, 1888b: 125 **A**: BEI SHX
sulciceps Gebler, 1841b: 592 (*Anatolica*) **A**: ES MG NMO
rectangulum Poppius, 1904: 13 (*Scythis*)

genus *Imatismus* Dejean, 1834: 202 type species *Helops fasciculatus* Fabricius, 1798
subgenus *Imatismus* Dejean, 1834: 202 type species *Helops fasciculatus* Fabricius, 1798
Curimosphena Gebien, 1920: 43 type species *Himatismus villosus* Haag-Rutenberg, 1870
arabicus Kaszab, 1981c: 324 (*Himatismus*) **A**: SA YE
chinensis Reitter, 1916e: 149 (*Himatismus*) **A**: YUN
fasciculatus Fabricius, 1798: 52 (*Helops*) **A**: PA SA **N**: MR **AFR ORR**
luctuosus Fairmaire, 1895a: 444 (*Himatismus*) **N**: MR **AFR**
ovipennis Koch, 1950b: 191 (*Himatismus*) **A**: YE **AFR**
sublinearis Walker, 1871: 15 (*Himatismus*) **A**: SA
villosus Haag-Rutenberg, 1870b: 90 (*Himatismus*) **E**: GR (Kríti) IT (Linosa) SP **N**: EG LB MO TU **A**: CY IN IQ IS LE OM PA SA SI SY YE **AFR ORR**
forticornis Baudi di Selve, 1874a: 103 (*Himatismus*)

genus *Kokeniella* Reitter, 1906c: 41 type species *Kokeniella mesostenoides* Reitter, 1906
afghanistana Kaszab, 1968b: 116 (*Mesostena*) **A**: AF
convergens Koch, 1940a: 87 (*Mesostena*) **A**: IN
freyi Koch, 1940a: 92 (*Mesostena*) **A**: IN (Bushire)
gracilicornis boeckelei Kaszab, 1974: 194 (*Mesostena*) **A**: AF
gracilicornis gracilicornis Reitter, 1915h: 128 **A**: IN
lineatopunctata Kraatz, 1865a: 173 (*Rhytinota*) **A**: IN PA
major interjecta Koch, 1940a: 91 (*Mesostena*) **A**: PA (Beluchistan)
major major Baudi di Selve, 1874a: 101 [= 1875c: 56] (*Mesostenopa*) **A**: IN PA
major persica Schuster, 1938b: 83 (*Mesostenopa*) **A**: IN
major pfaundleri Koch, 1940a: 89 (*Mesostena*) **A**: IN (Bijnabad)
matthiesseni Reitter, 1915h: 127 **A**: IN
mesostenoides afghanica Koch, 1940a: 85 (*Mesostena*) **A**: AF
mesostenoides mesostenoides Reitter, 1906c: 42 **A**: PA
perforalva luristana Koch, 1943a: 570 **A**: IN (Lorestan)

genus *Leptosphena* Semenov, 1891a: 356 type species *Sphenaria rubripes* Reitter, 1889
rubripes Reitter, 1889b: 115 (*Sphenaria*) **A**: TM (Karakumy)
tomentosa Semenov, 1889a: 214 (*Sphenaria*)

genus *Megagenius* Solier, 1835c: 513 type species *Megagenius frioli* Solier, 1835
frioli frioli Solier, 1835c: 514 **N**: AG MO TU
frioli troglophilus Kocher, 1960: 233 **N**: MO

genus *Melanochrus* Wollaston, 1864: 467 type species *Melanochrus lacordairii* Wollaston, 1864
Melasmocara Reitter, 1900d: 95 type species *Melanochrus lacordairii* Wollaston, 1864
blairi Uyttenboogaart, 1929: 347 **N**: CI
lacordairii Wollaston, 1864: 468 **N**: CI

genus *Melaxumia* Reitter, 1895f: 280 type species *Melaxumia acutangula* Reitter, 1895 (= *Tentyria angulosa* Gebler, 1832)
angulosa Gebler, 1832: 53 (*Tentyria*) **A**: ES MG NMO SHX
acutangula Reitter, 1895f: 280

genus *Mesostena* Eschscholtz, 1831: 9 type species *Mesostena punctata* Eschscholtz, 1831 (=*Pimelia angustata* Fabricius, 1775)
subgenus *Mesostena* Eschscholtz, 1831: 9 type species *Mesostena punctata* Eschscholtz, 1831 (= *Pimelia angustata* Fabricius, 1775)
angustata angustata Fabricius, 1775: 253 (*Pimelia*) **N**: EG LB TU **A**: IN IQ IS JO LE SI SY **AFR**
elegans Solier, 1835b: 399
laevicollis Solier, 1835b: 402
oblonga Solier, 1835b: 401

politipennis Fairmaire, 1877: 141
punctata Eschscholtz, 1831: 9 (*Mesostena*)
punctipennis Solier, 1835b: 403
angustata longicollis Lucas, 1858c: ccxxii **N**: AG LB
angustata praesahariana Koch, 1940a: 78 **N**: AG LB MO TU
klugii Solier, 1835b: 404 **N**: EG
parvula Reiche & Saulcy, 1857: 211 **A**: LE
puncticollis Solier, 1835b: 405 **E**: AR **A**: IN IQ IS JO SA SY TM YE **AFR**
armeniaca Motschulsky, 1870: 397
avenae J. R. Sahlberg, 1908: 76 (*Mesostenopa*)
bramina Motschulsky, 1870: 397
yemenensis Scupola, 1994: 25 **A**: YE
subgenus *Mesostenopa* Kraatz, 1865a: 179 type species *Mesostenopa picea* Kraatz, 1865
aegyptiaca Koch, 1940a: 44 **N**: EG
arabica arabica Gestro, 1881: 661 (*Mesostenopa*) **A**: YE
arabica titschaki Gebien, 1938: 55 (*Mesostenopa*) **A**: OM YE
caeciliae G. Wagner, 2003: 114 **A**: YE
dentrix Baudi di Selve, 1874a: 102 [= 1875c: 57] (*Micipsa*) **A**: IN
gridellii Koch, 1940a: 53 **A**: SA YE
habbanensis G. Wagner, 2003: 112 **A**: YE
habessinica Kraatz, 1865a: 181 (*Mesostenopa*) **N**: EG **AFR**
kuriamuriaca Schawaller, 1993a: 104 **A**: OM (Is. Kuria Muria)
nabathaea Peyerimhoff, 1907a: 31 (*Mesostenopa*) **A**: JO SY
perimensis Koch, 1943a: 563 (*Mesostena*) **A**: YE (Perim Is.)
peyerimhoffi Gridelli, 1938: 109 [RN] **A**: SI
gracilis Peyerimhoff, 1907a: 31 (*Mesostenopa*) [HN]
picea agilis Gestro, 1881: 661 (*Mesostenopa*) **N**: EG **A**: SI **AFR**
picea caporiaccoi Gridelli, 1938: 115 (*Mesostena*) **N**: LB
picea kulzeri Koch, 1940a: 45 **A**: SA YE
picea picea Kraatz, 1865a: 180 (*Mesostenopa*) **N**: EG
picea sinaitica Schatzmayr & Koch, 1934b: 17 (*Mesostenopa*) **A**: SI
rathjensi rathjensi Gebien, 1938: 54 (*Mesostenopa*) **A**: YE
sericea moltonii Schatzmayr, 1938: 173 **N**: LB
sericea sericea Fairmaire, 1884i: cccxlvi (*Mesostenopa*) **A**: YE **AFR**
socotraensis Koch, 1970: 95 (*Mesostenopa*) **A**: YE (Samha, Suqutra)
transnilotica Koch, 1937: 350 (*Mesostenopa*) **N**: LB
subgenus *Platystena* Koch, 1940a: 95 type species *Mesostena blairi* Koch, 1940
blairi Koch, 1940a: 96 **A**: OM
subgenus *Saxistena* Löbl & Merkl, 2003: 251 type species *Mesostena longicornis* Kraatz, 1865
cavatica Andres, 1926: 287 (*Mesostenopa*) **N**: EG
darsiana Koch, 1970: 100 **A**: YE (Darsa, Suqutra, Samba)
ibnsaudi Koch, 1940a: 73 **A**: SA
longicornis longicornis Kraatz, 1865a: 181 **A**: IS JO
longicornis rothi Koch, 1940c: 100 **A**: JO
mequignoni Koch, 1940a: 69 **A**: YE
tibialis Gahan, 1900: 9 (*Rhytidonota*) **A**: YE (Abd al-Kuri, Suqutra)

genus *Micipsa* Lucas, 1855b: xxxiv type species *Micipsa rufitarsis* Lucas, 1855 (= *Pimelia mulsanti* Levrat, 1853)
subgenus *Cirsa* Lucas, 1857: lvi type species *Cirsa striaticollis* Lucas, 1857
chobauti Peyerimhoff, 1926b: 194 **N**: AG MO
instriata Pic, 1922g: 29 **N**: MO
striaticollis Lucas, 1857: lvi (*Cirsa*) **N**: AG

subgenus *Micipsa* Lucas, 1855b: xxxiv type species *Micipsa rufitarsis* Lucas, 1855 (= *Pimelia mulsanti* Levrat, 1853)
agnesae Schuster, 1935b: 24 **A**: IN
angustipennis Fairmaire, 1875a: 520 **N**: TU
rufescens Pic, 1922g: 29
substrigosa Reitter, 1900d: 187
arabica Kaszab, 1981c: 343 **A**: SA (Eastern Prov.)
batesi Haag-Rutenberg, 1875b: 92 **A**: "Syria"
burtoni Baudi di Selve, 1881: 277 **A**: SY
douei distincticornis Peyerimhoff, 1933: 347 **N**: AG LB
garamantium Koch, 1937: 381
douei douei Lucas, 1856: xlvi **N**: AG EG **A**: JO LB SI **AFR**
grandis Kraatz, 1865a: 188
longiceps Pic, 1922g: 30
elliptipennis Schuster, 1925a: 25 **N**: LB
gracilipes Fairmaire, 1875a: 522 **N**: AG TU
kerimii Fairmaire, 1875a: 520 **N**: TU
mulsanti Levrat, 1853: 1 (*Pimelia*) **N**: AG TU
cursor Guérin-Méneville, 1859: clxxxviii (*Cirta*)
rufitarsis Lucas, 1855b: xxxiv
velox Guérin-Méneville, 1859: clxxxix (*Cirta*)
ovoidea Fairmaire, 1883d: 460 **N**: LB
persica Baudi di Selve, 1874a: 102 [= 1875c: 58] **A**: IN
pubescens Haag-Rutenberg, 1875b: 90
philistina Reiche & Saulcy, 1857: 212 **A**: IQ IS JO SI "Syria"
bysantica Motschulsky, 1858d: 189
poripennis Fairmaire, 1875a: 523 **N**: AG
princeps occidentalis Español, 1953c: 291 **N**: MO
princeps princeps Peyerimhoff, 1934: 106 **N**: LB MO **AFR**
rufitarsis Lucas, 1855b: xxxiv **N**: AG
schaumii Kraatz, 1865a: 111 **N**: EG **A**: SI
similis Haag-Rutenberg, 1875b: 91 **A**: IN
variabilis balbus Koch, 1937: 388 **N**: LB (Fezzah)
variabilis granulipennis Koch, 1937: 388 **N**: LB
variabilis variabilis Koch, 1937: 385 **N**: LB

genus *Microdera* Eschscholtz, 1831: 6 type species *Tentyria deserta* Tauscher, 1812
subgenus *Amicrodera* Kaszab, 1966b: 292 type species *Microdera lindbergi* Kaszab, 1966
lindbergi Kaszab, 1966b: 293 **A**: AF
subgenus *Dordanea* Reitter, 1887d: 357 type species *Dordanea elegans* Reitter, 1887
Adordanea Reitter, 1897f: 229 type species *Microdera subseriata* Reitter, 1889
duplicatipunctata Ren, 1999: 116 **A**: GAN
elegans Reitter, 1887d: 358 (*Dordanea*) **A**: GAN LIA NIN NMO QIN SHX XIN
globata Faldermann, 1835b: 402 (*Tentyria*) **A**: GAN NIN MG NMO SHX
interrupta Reitter, 1897f: 230 **A**: MG NMO
jurganovae Skopin, 1964: 385 **A**: ES (Tuva) MG
kanssuana Kaszab, 1957f: 292 **A**: GAN
kraatzi alashanica Skopin, 1964: 384 **A**: GAN NIN NMO
kraatzi kraatzi Reitter, 1889k: 685 (*Dordanea*) **A**: MG NIN NMO
lampabilis Ren, 1999: 113 **A**: GAN GUA NIN
luoshanica Ren, 1999: 109 **A**: NIN
luoshanana Ren, 1999: 391
ordossica Schuster, 1940a: 21 **A**: NMO
rotundithorax Ren, 1999: 118 **A**: NIN
shenmuana Ren, 1999: 111 **A**: SHA

shenmana Ren, 1999: 391
subseriata Reitter, 1889k: 686 **A**: ES (Tuva) GAN NMO
subfoveolata Skopin, 1964: 387
subgenus *Falsomicrodera* Kaszab, 1966b: 294 type species *Microdera turkerstanica* Schuster, 1915
turkestanica Schuster, 1915b: 86 (*Microdera*) **A**: XIN
subgenus *Iliodera* Skopin, 1961b: 396 type species *Microdera desertoides* Skopin, 1961
desertoides Skopin, 1961b: 396 **A**: KZ (Alamatinskaya oblast)
subgenus *Microdera* Eschscholtz, 1831: 6 type species *Tentyria deserta* Tauscher, 1812
Rhostax Fischer von Waldheim, 1844: 67 type species *Rhostax menetriesii* Fischer von Waldheim, 1844
aciculata aciculata Reitter, 1897f: 232 **A**: KZ XIN
aciculata penkinae Kaszab, 1966b: 298 **A**: KZ
afghanica Gridelli, 1954: 198 [= 1955: 34] **A**: AF
aurita Reitter, 1889k: 686 (*Dordanea*) **A**: GAN NMO XIN
badghysi G. S. Medvedev & Nepesova, 1985b: 358 **A**: TM
balchaschensis balchaschensis Skopin, 1960: 297 **A**: KZ
balchaschensis wernoyensis Kaszab, 1966b: 298 **A**: KZ
campestris Steven, 1829: 93 **A**: KZ TM
margelanica Kraatz, 1882b: 81
chan Kaszab, 1963a: 71 **A**: IN
charpentieri Kaszab, 1973b: 39 **A**: AF
convexa convexa Tauscher, 1812a: 39 (*Tentyria*) **E**: AB ST **A**: KZ TM
convexa kaszabi Skopin, 1964: 409 **A**: KZ (Almaatinska Oblast)
deserta deserta Tauscher, 1812a: 38 (*Tentyria*) **E**: ST **A**: KZ
deplanata Gebler, 1841a: 373 (*Tentyria*)
deserta skopini Kaszab, 1966b: 296 **A**: KZ
dzhungarica G. S. Medvedev, 1964b: 161 **A**: KZ
extabesa Skopin, 1961b: 393 **A**: KZ TM UZ
fausti Kraatz, 1888b: 294 **A**: TM
ferghanensi Kaszab, 1957f: 290 **A**: KZ UZ
fortepunctata Reitter, 1915f: 94 **A**: KZ
gigas G. S. Medvedev, 1964b: 162 **A**: KZ XIN
globulicollis Ménétriés, 1849: 230 **A**: TM
globithorax Steven, 1829: 93 (*Tentyria*) **A**: KZ "Siberia"
gorochovi G. S. Medvedev, 2004b: 555 **A**: TM
gracilis Eschscholtz, 1831: 6 **A**: KI KZ
macrops Reitter, 1915f: 93
menetriesii Fischer von Waldheim, 1844: 68 (*Rostrax*)
grandicollis Bogatchev, 1946a: 211 **A**: AF KZ TD
badakshanica Kaszab, 1959f: 390
grandipunctata Ren, 1999: 122 **A**: XIN
habaensis Ren, 1999: 123 **A**: XIN
heydeni Kraatz, 1882b: 82 **A**: KZ TD TM UZ
hybrida Skopin, 1966: 69 **A**: KZ
iliensis Skopin, 1961b: 394 **A**: KZ
iranica Kaszab, 1966b: 299 **A**: IN
karelini Fischer von Waldheim, 1844: 67 (*Rhostrax*) **A**: "Songaria"
kaszabi Skopin, 1961a: 179 **A**: KZ (Amatinskaya oblast)
kazenasi Skopin, 1960: 297 **A**: KZ
keramana Ren, 1999: 124 **A**: XIN
karamiensis Ren, 1999: 390
kermanica Kaszab, 1966b: 294 **A**: IN
laevigatula Reitter, 1915f: 95 **A**: AF IN UZ
laticollis Bates, 1879: 470 **A**: KA KZ XIN
przewalskyi Reitter, 1887d: 359 (*Tentyria*)
minax Reitter, 1897f: 233 **A**: AF KZ TM UZ

mongolica kozlovi Kaszab, 1966b: 295 **A**: MG NMO
mongolica mongolica Reitter, 1889k: 686 (*Dordanea*) **A**: GAN NMO QIN XIN
obesitas Ren, 1999: 127 **A**: XIN
obsitas Ren, 1999: 390
ocularis Reitter, 1915f: 95 **E**: GG
parvicollis Bates, 1879: 470 **A**: KA KZ XIN
basalis Reitter, 1887d: 360 (*Tentyria*)
semenowi Reitter, 1897f: 231
pleuralis Reitter, 1887d: 360 (*Tentyria*) **A**: KZ NMO XIN
pulla Reitter, 1897f: 234 **A**: IN
punctipennis Kaszab, 1967c: 311 **A**: MG
reitteri Kaszab, 1966b: 280 [RN] **E**: AB
parvicollis Reitter, 1897f: 235 [HN]
schusteriana Kaszab, 1957f: 292 **A**: KZ
scyta Baudi di Selve, 1875c: 55 **A**: TM
semenoviana Bogatchev, 1947d: 463 [RN] **A**: AF TD UZ
semenovi Bogatchev, 1946a: 210 [HN]
shasenema G. S. Medvedev & Nepesova, 1985b: 360 **A**: TM
strigipleuris Reitter, 1897f: 232 **A**: IN
strigiventris Reitter, 1900d: 157 **A**: MG XIN
halophila Skopin, 1961b: 392
subglobosa Kraatz, 1865a: 187 **A**: KZ
subsulcata Kraatz, 1865a: 186 **A**: KZ
thoracica Kaszab, 1966b: 300 **A**: IN
transversicollis Reitter, 1887e: 517 **E**: AR **A**: AF IN TM UZ
tscharynensis Kaszab, 1966b: 297 **A**: KZ
vachshiana Skopin, 1964: 411 **A**: TD
subgenus *Tentyrodera* Koch, 1943a: 572 type species *Microdera marginata* Baudi di Selve, 1874
marginata brandti Koch, 1943a: 573 **A**: IN
marginata deserticola Blair, 1933: 5 (*Rhytinota*) **A**: AE OM SA YE
perezi Koch, 1943a: 574
marginata evansi Blair, 1923a: 121 (*Rhytinota*) **A**: IQ SA
marginata marginata Baudi di Selve, 1874a: 97 [= 1874a: 278, 1875c: 53] **A**: IN

genus *Microdera*, species incertae sedis
convexicollis Motschulsky, 1870: 379 (*Rhostax*) **A**: KI
excavata Motschulsky, 1870: 382 **A**: KI

genus *Neognathosia* Kaszab, 1959f: 383 type species *Gnathosia pseudanemia* Reitter, 1915
pseudanemia Reitter, 1915b: 62 (*Gnathosia*) **A**: UZ

genus *Oterophloeus* Desbrochers des Loges, 1881: 52 type species *Oterophloeus picipes* Desbrochers des Loges, 1881 (=*Pachychila humerosus* Fairmaire, 1875)
Tynteria Reitter, 1897h: 301 type species *Pachychila humerosa* Fairmaire, 1875
alveatus alveatus Peyerimhoff, 1931b: 95 **N**: AG LB MO **AFR**
alveatus peyerimhoffi Koch, 1935a: 23 **N**: EG **A**: SI
deflexangulus deflexangulus Reitter, 1900d: 143 **N**: LB (Tripolitania) TU
deflexangulus syrticus Koch, 1937: 359 **N**: LB (Cyrenaica, Tripolitania)
densepunctatus Escalera, 1935a: 298 (*Pachychila*) **N**: MO **AFR**
desioi desioi Gridelli, 1930: 210 **N**: LB
desioi divinus Koch, 1935a: 25 **N**: EG (Siwa)
haagii haagii Kraatz, 1865a: 155 (*Tentyria*) **N**: EG
tenuimarginatus Böhm, 1916a: 50 (*Tentyria*)
haagii orientalis Koch, 1935a: 25 **N**: EG **A**: SI
humerosus Fairmaire, 1875a: 518 (*Pachychila*) **N**: AG LB TU **AFR**

picipes Desbrochers des Loges, 1881: 53
scuticollis Fairmaire, 1879c: 188 (*Tentyria*) **N**: AG
torretassoi Koch, 1937: 360 **N**: LB (Fezzan)

genus *Oxycara* Solier, 1835b: 254 type species *Oxycara blapsoides* Solier, 1835
subgenus *Oxycara* Solier, 1835b: 254 type species *Oxycara blapsoides* Solier, 1835
Crypticoides Fairmaire, 1898d: 389 type species *Crypticoides mellyi* Fairmaire, 1898
Emmenastus Motschulsky, 1845a: 75 type species *Emmenastus rugosus* Motschulsky, 1845
Melancrus Reiche & Saulcy, 1857: 190 type species *Melancrus laevigatum* Reiche & Saulcy, 1857
aethiopium aethiopium Gredler, 1878: 515 **N**: AG (Tassili n'Ajjer) LB (Fezzan, Tibesti) **AFR**
ardoini Kaszab, 1979c: 285 **A**: IS (Negev) JO SI
balachowskyi Pierre, 1958: 138 **N**: AG (Tassili n'Ajjer)
becharense Chobaut, 1923c: 165 **N**: AG
buettikeri Kaszab, 1979c: 286 **A**: AE BA OM SA
chudeaui Peyerimhoff, 1942c: 276 **N**: MO (Western Sahara) **AFR**
cyrenaicum Gridelli, 1933: 222 **N**: LB
deronectinum Reitter, 1900d: 190 **N**: EG
gastonis Reitter, 1903b: 45 **N**: AG LB MO **AFR**
chobauti Gebien, 1910a: 76 [RN]
trapezicolle Chobaut, 1898a: 82 [HN]
laevigatum Reiche & Saulcy, 1857: 192 (*Melancrus*) **A**: IS JO SY
lavocati Español, 1953c: 294 **N**: AG MO
llorensi Escalera, 1935a: 306 **N**: AG MO **AFR**
mandli Kaszab, 1963a: 72 **A**: IN
mellyi Fairmaire, 1898d: 390 (*Crypticoides*) **A**: AF PA XIZ **ORR**
tharense Kulzer, 1956: 645
pierrei Girard, 1966b: 204 **A**: IQ SY
murati Peyerimhoff, 1942b: 217 **N**: MO (Western Sahara)
productum Peyerimhoff, 1907a: 33 **A**: SI
pygmaeum Reiche & Saulcy, 1857: 194 (*Melancrus*) **N**: EG LB **A**: SI
sanaaense forte Kaszab, 1981c: 346 **A**: SA
sanaaense sanaaense Gridelli, 1953: 30 **A**: SA YE
saudarabicum Kaszab, 1979c: 286 **A**: SA
subgenus *Symphoxycara* Koch, 1943a: 577 type species *Oxycara breviusculum* Fairmaire, 1892
evae Lillig, 2001: 379 **A**: OM
ferreri Lillig, 2001: 372 **N**: EG (Red See: Isl. Zabargad)
gallagheri Lillig, 2001: 375 **A**: OM
grimmi Lillig, 2001: 378 **A**: OM
girardi Lillig, 2001: 382 **A**: IN
grande Kaszab, 1981c: 345 **A**: OM YE
hansbremeri Lillig, 2001: 380 **A**: AE BA IN OM YE
hegetericum Reiche & Saulcy, 1857: 193 (*Melancrus*) **N**: EG **A**: "Dead Sea" **AFR**
laticolle Lillig, 2001: 383 **A**: IN
minutum Lillig, 2001: 382 **A**: IN
occidentale Lillig, 2001: 369 **N**: EG
peyerimhoffi Lillig, 2001: 369 **A**: IS JO SI
pumilum Reitter, 1900d: 189 **A**: EG (Red See: Is. Homan)
schawalleri Lillig, 2001: 377 **A**: OM
subcostatum Guérin-Méneville, 1862: 375 (*Melancrus*) **A**: YE
ulfbremeri Lillig, 2001: 373 **A**: SA

genus *Oxycarops* Reitter, 1900d: 94 type species *Hegeter fuscipes* Brullé, 1832
fuscipes Brullé, 1838: 66 (*Hegeter*) **N**: CI (Fuerteventura, Lanzarote)
submetallicus Wollaston, 1864: 464 (*Thalpophila*)

genus *Pachychila* Eschscholtz, 1831: 5 type species *Pimelia punctata* Fabricius, 1798
subgenus *Anebacis* Peyerimhoff, 1927a: 4 type species *Tentyria cribricollis* Fairmaire, 1875
cribricollis Fairmaire, 1875a: 519 (*Tentyria*) **N**: AG TU
elegans Escalera, 1935a: 296 **N**: MO (Tiznit)
mairei Peyerimhoff, 1920a: 327 (*Tentyria*) **N**: MO (region of Figuid)
rouleaui Antoine, 1952: 112 **N**: MO
subgenus *Hegeteromorpha* Escalera, 1913: 38 type species *Pachychile externecostata* Haag-Rutenberg, 1875
externecostata Haag-Rutenberg, 1875a: 48 (*Pachychile*) **N**: MO
obtusecostata Reitter, 1900d: 152 (*Pachychile*) **N**: MO
trisulcata Antoine, 1942b: 65 **N**: MO
subgenus *Neocisba* Reitter, 1900d: 154 type species *Pachychile dissecta* Kraatz, 1865
Neacisba Peyerimhoff, 1927a: 53 [unjustified emendation]
dissecta Kraatz, 1865a: 204 (*Pachychile*) **N**: AG
latevagans Antoine, 1952: 110 **N**: MO
mediatlantis Antoine, 1939: 148 **N**: MO (Moyen Atlas)
medioxima Peyerimhoff, 1921b: 64 (*Tentyria*) **N**: AG
muluyensis Antoine, 1942b: 60 **N**: MO (Moyenne Moulouya)
prosternalis ayachiensis Antoine, 1942b: 59 **N**: MO (Grand Atlas)
prosternalis prosternalis Bedel, 1903: 369 **N**: MO (Moyen Atlas)
rotroui perissa Antoine, 1942b: 59 **N**: MO
rotroui rotroui Antoine, 1932: 181 **N**: MO
sicardi interrupta Antoine, 1957: 40 **N**: MO (Moyenne Moulouya)
sicardi reymondi Antoine, 1957: 40 **N**: MO (Grand Atlas)
sicardi sicardi Peyerimhoff, 1927b: 218 **N**: MO
subgenus *Pachychila* Eschscholtz, 1831: 5 type species *Pimelia punctata* Fabricius, 1798
Acisba Dejean, 1834: 185 type species *Tentyria cribrosa* Besser, 1832 (=*Pimelia punctata* Fabricius, 1798
Lophoma Solier, 1835b: 285 type species *Pimelia punctata* Fabricius, 1798
Pachychile Lacordaire, 1859: 46 [unjustified emendation]
acuminata Erichson, 1841: 177 **N**: MO
angulicollis angulicollis Fairmaire, 1875c: 543 **N**: MO
fritschi Haag-Rutenberg, 1875a: 45 (*Pachychile*)
angulicollis haagi Antoine, 1932: 179 **N**: MO (Sous)
angulicollis macra Antoine, 1932: 178 **N**: MO (Grand Atlas)
angulicollis ventriosa Antoine, 1932: 177 **N**: MO
antiatlantis Antoine, 1952: 111 **N**: MO (Anti Atlas)
aulonoscelis Reitter, 1900d: 148 (*Pachychile*) **E**: SP
bifida Rosenhauer, 1865: 184 **E**: SP (Andalucía)
bremondi Koch, 1941c: 286 **N**: MO
bergevini bergevini Peyerimhoff, 1925: 12 **N**: MO (Atlantic)
bergevini theryi Peyerimhoff, 1925: 12 **N**: MO (Sebou)
brevis Besser, 1832: 11 (*Tentyria*) **N**: AG MO
cognata Haag-Rutenberg, 1875a: 46 (*Pachychile*)
[*pedinoides* Eschscholtz, 1831: 6; unavailable]
pedinoides Solier, 1835b: 294
breviuscula breviuscula Haag-Rutenberg, 1875a: 47 (*Pachychile*) **N**: MO
breviuscula discrepans Antoine, 1942b: 61 **N**: MO
calida calida Escalera, 1925d: 375 **N**: MO
calida subangulicollis Antoine, 1942b: 61 **N**: MO (Moyen Atlas)
chobauti Antoine, 1926: 253 **N**: MO
compacta compacta Antoine, 1942b: 41 **N**: MO (Grand Atlas)
compacta subaspera Antoine, 1942b: 42 **N**: MO (Moyen Atlas)
cornuta Peyerimhoff, 1925: 10 **N**: MO
crassicollis cossyrensis Ragusa, 1875: 252 (*Pachychile*) **E**: IT (Pantelleria) **N**: TU
parallela Schuster, 1919b: 42 (*Pachychilina*)
crassicollis crassicollis Kraatz, 1865a: 210 **N**: AG

cursitans Escalera, 1925d: 374 **N**: MO
denticollis Grimm, 1987: 22 **E**: SP (Almería)
desbrochersi Reitter, 1900d: 150 (*Pachychile*) **N**: AG
forticornis Escalera, 1923d: 129 **N**: MO (Rif)
foveipennis Kraatz, 1865a: 201 (*Pachychile*) **N**: MO
soligena Antoine, 1932: 182
freyi Koch, 1944b: 223 **E**: SP (Almería)
frioli Solier, 1835b: 301 **E**: GR IT SP **N**: AG LB TU
quadricollis Reiche, 1860: 729
germari Solier, 1835b: 302 **E**: IT **N**: AG MO TU
haroldi Kraatz, 1865a: 211 (*Pachychile*)
sardea Kraatz, 1865a: 209
glabella Herbst, 1799: 61 (*Pimelia*) **E**: SP
grandis derduensis Antoine, 1942b: 61 **N**: MO
grandis grandis Desbrochers des Loges, 1881: 54 **N**: AG
rugatula Desbrochers des Loges, 1881: 55
grandis gravida Antoine, 1933b: 223 **N**: MO
grandis kocheri Antoine, 1934: 106 **N**: MO (Grand Atlas)
grandis montana Peyerimhoff, 1925: 13 **N**: MO (Moyen Atlas)
grandis strenua Antoine, 1932: 180 **N**: MO
hispanica Solier, 1835b: 292 **E**: SP
impressifrons Solier, 1835b: 299 **N**: AG
breviuscula Desbrochers des Loges, 1881: 61
conformis Desbrochers des Loges, 1881: 62
continua Desbrochers des Loges, 1881: 58
curtula Gebien, 1910a: 58 [RN]
emarginata Desbrochers des Loges, 1881: 59
rugatipennis Desbrochers des Loges, 1881: 59
subcylindrica Solier, 1835b: 300
impunctata Fairmaire, 1860: 632 (*Lophoma*) **N**: MO
ifniensis Escalera, 1935a: 300
intermedia Haag-Rutenberg, 1875: 45 (*Pachychile*)
incrassata Rosenhauer, 1856: 183 **E**: SP
insculpta Escalera, 1914: 288 **N**: MO
kunzei Solier, 1835b: 298 **N**: AG
discedens Desbrochers des Loges, 1881: 56
lesnei Peyerimhoff, 1925: 13 **N**: AG MO
maroccana Haag-Rutenberg, 1875a: 50 (*Pachychile*) **N**: MO
mskalica Escalera, 1914: 281 **N**: MO (Grand Atlas)
mscalica Peyerimhoff, 1927a: 42 [unjustified emendation]
otini otini Antoine, 1939: 146 **N**: MO (Moyen Atlas)
otini ranarum Antoine, 1939: 148 **N**: MO (Moyen Atlas)
parallelipennis Escalera, 1914: 284 **N**: MO
plasoni beidensis Antoine, 1926: 255 **N**: MO
plasoni plasoni Haag-Rutenberg, 1875a: 50 (*Pachychile*) **N**: MO
prujai Antoine, 1962: 51 **N**: MO
punctata mamorensis Antoine, 1942b: 60 **N**: MO
punctata punctata Fabricius, 1798: 45 (*Pimelia*) **E**: SP **N**: AG MO
cribrosa Besser, 1832: 11 (*Tentyria*)
punctulata Lucas, 1846: 314 **E**: SP **N**: AG
refleximargo Desbrochers des Loges, 1881: 59 **N**: AG MO
reini asniensis Antoine, 1942b: 63 **N**: MO (Grand Atlas)
reini reini Haag-Rutenberg, 1875a: 49 (*Pachychile*) **N**: MO
fairmairei Haag-Rutenberg, 1875a: 49 (*Pachychile*)
rotunda Escalera, 1925d: 372 **N**: MO
sabulosa Lucas, 1846: 314 **N**: AG

salzmanni jordanai Escalera, 1925d: 373 **N**: MO (Rif)
salzmanni salzmanni Solier, 1835b: 297 **E**: SP **N**: MO
salzmanni simoni Escalera, 1914: 285 **N**: MO
sarroensis Antoine, 1947: 148 **N**: MO (Sarro)
schrammi Antoine, 1942b: 63 **N**: MO
secsaui imperfecta Peyerimhoff, 1927a: 23 **N**: MO (Grand Atlas)
secsaui insidiosa Antoine, 1942b: 63 **N**: MO (Azilal)
secsaui secsaui Escalera, 1913: 39 **N**: MO (Western Grand Atlas)
servillei Solier, 1835b: 309 (*Microdera*) **E**: FR (Corse) IT (Sardegna)
pygmaea Solier, 1835b: 310 (*Microdera*)
simytae Antoine, 1934: 104 **N**: MO
sublunata Solier, 1835b: 308 (*Microdera*) **E**: SP (Baleares)
opaca Breit, 1908: 63
sulcifrons Solier, 1835b: 295 **N**: MO
tetuanica Escalera, 1914: 287
susiana Escalera, 1914: 289 **N**: MO
susana Peyerimhoff, 1927a: 48 [unjustified emendation]
tazmaltensis Desbrochers des Loges, 1881: 56 **E**: IT (Lampedusa) **N**: AG TU
lampedusae Kraatz, 1896: 101
transversithorax Escalera, 1914: 284 **N**: MO
schenklingi Schuster, 1919b: 34
tripoliana Solier, 1835b: 303 **N**: AG MO
tumidifrons Kraatz, 1865a: 200 (*Pachychile*) **N**: AG MO
ariasi Escalera, 1914: 286
cardaillaci Peyerimhoff, 1913b: 520 (*Pachychile*)
orientalis Escalera, 1914: 286
subgenus *Pachychilina* Reitter, 1900d: 145 type species *Tentyria dejeani* Besser, 1832
alluaudi alluaudi Peyerimhoff, 1925: 9 **N**: MO (Grand Atlas)
alluaudi escalerai Peyerimhoff, 1925: 9 **N**: MO (Grand Atlas)
andreinii andreinii Dodero, 1916: 350 **N**: LB TU
gracilis Schuster, 1919b: 38 (*Pachychilina*)
andreinii djerbensis Normand, 1938a: 302 [= 1938b: 365] (*Pachychilina*) **N**: TU (Djerba)
bolivari Antoine, 1926: 249 **N**: MO (Moyen Atlas)
cedreticola Antoine, 1947: 147 **N**: MO (Moyen Atlas)
dejeani dejeani Besser, 1832: 11 (*Tentyria*) **E**: IT **N**: AG TU
steveni Solier, 1835b: 296
subovata Besser, 1832: 11 (*Tentyria*)
wiedemanni Fischer von Waldheim, 1837b: 14 (*Acispa*)
dejeani doderoi Peyerimhoff, 1927a: 17 **E**: IT (Lampedusa)
dejeani infrastriata Escalera, 1914: 283 **N**: MO (Moyen Atlas)
dejeani longipennis Kraatz, 1865a: 212 (*Pachychile*) **N**: TU
doriae Haag-Rutenberg, 1875a: 51
leonhardi Schuster, 1919b: 38 (*Pachychilina*)
oxyophthalmus Reitter, 1900d: 146
dejeani montana Koch, 1937: 364 **N**: LB
dejeani nitens Solier, 1835b: 293 **N**: AG
brevis Kraatz, 1865a: 202 (*Pachychile*)
violantis Gestro, 1880: 416

subgenus *Tentyromorpha* Escalera, 1913: 37 type species *Tentyromorpha telueti* Escalera, 1913
Tentyriomorpha Peyerimhoff, 1927a: 53 [unjustified emendation]
antoinei Peyerimhoff, 1927a: 53 [RN] **N**: MO
mairei Antoine, 1926: 252 [HN]
densepunctata Escalera, 1935a: 298 **N**: MO
fallaciosa Antoine, 1933b: 221 **N**: MO
gattefossei Antoine, 1933b: 220 **N**: MO

telueti Escalera, 1913: 38 (*Tentyromorpha*) **N**: MO

genus *Paivaea* Wollaston, 1864: 449 type species *Tentyria hispida* Brullé, 1838
hispida Brullé, 1838: 66 (*Tentyria*) **N**: CI
mogadorica Escalera, 1914: 274 **N**: MO

genus *Parabigopsis* Español, 1946: 111 type species *Parabigopsis peyerimhoffi* Español, 1946
peyerimhoffi Español, 1946: 115 **N**: MO (Tan-Tan)

genus *Paracirta* Schuster, 1930a: 66 type species *Paracirta schatzmayri* Schuster, 1930
peyerimhoffi daourensis Español, 1953c: 293 **N**: AG
peyerimhoffi peyerimhoffi Español, 1944: 10 **N**: MO (Western Sahara)
schatzmayri algirica Koch, 1941c: 293 **N**: AG
schatzmayri schatzmayri Schuster, 1930a: 67 **N**: LB

genus *Phaeotribon* Kraatz, 1865a: 242 type species *Phaeotribon pulchellus* Kraatz, 1865
pulchellus Kraatz, 1865a: 243 **N**: EG **A**: SA YE **AFR**

genus *Prochoma* Solier, 1835b: 393 type species *Prochoma audouini* Solier, 1835
subgenus *Oxypistoma* Löbl, Bouchard & Merkl, 2008 [see New Acts] type species *Prochoma bucculenta* Koch, 1940
blairi Koch, 1940d: 262 **A**: PA **ORR**
bucculenta Koch, 1940d: 260 **A**: AE IN OM SA
clypealis Blair, 1923a: 119 **A**: BA IQ OM SA
margolateralis Carl, 1994: 86 **A**: OM
subgenus *Prochoma* Solier, 1835b: 393 type species *Prochoma audouini* Solier, 1835
audouini Solier, 1835b: 395 **A**: IN IQ IS JO SI SY
membranea Carl, 1994: 83 **A**: IN
pierrei Girard, 1966b: 204 **A**: IN IQ
pilosa Carl, 1994: 81 **A**: IN

genus *Psammocryptus* Kraatz, 1865a: 239 type species *Tentyria minuta* Tauscher, 1812
bayeri Koch, 1943a: 583 **A**: TM
minutus bergi Kuzin, 1934: 160 **A**: KZ
minutus minutus Tauscher, 1812a: 33 (*Tentyria*) **E**: AB ST **A**: KZ TM
prosternalis Koch, 1943a: 581 **A**: AF PA
    semenovi Bogatchev, 1946c: 393

genus *Psammoica* Solier, 1835b: 307 type species *Microdera lucida* Solier, 1835
    Psammoeca Agassiz, 1846b: 311 [unjustified emendation]
laticollis Schuster, 1926b: 131 **N**: LB
lucida lucida Solier, 1835b: 307 (*Microdera*) **N**: EG
lucida schusteri Gridelli, 1933: 217 **N**: LB

genus *Rhomaleus* Chatanay, 1915: 64 type species *Rhomaleus scauroides* Chatanay, 1915
scauroides Chatanay, 1915: 64 **N**: MR **AFR**

genus *Rhytinota* Eschscholtz, 1831: 7 type species *Rhytinota scabriuscula* Eschscholtz, 1831
subgenus *Sphenariopsis* Kraatz, 1865a: 175 type species *Sphenariopsis tristis* Kraatz, 1865
subfossulata Solier, 1835b: 369 (*Tentyria*) **A**: SD **ORR**
    morio Baudi di Selve, 1875c: 52
    pilicollis Kraatz, 1865a: 173
    pubens Fairmaire, 1896a: 8
    rugipennis Kraatz, 1880b: 92
tristis belutschistana Koch, 1943b: 773 **A**: PA (Balochistan)

tristis kokeni Reitter, 1906c: 42 **A**: PA (Punjab)
tristis tristis Kraatz, 1865a: 176 (*Sphenariopsis*) **A**: HP UP

genus *Scelosodis* Solier, 1835b: 283 type species *Cratopus castaneus* Eschscholtz, 1831
Abiga Guérin-Méneville, 1859: clcccix type species *Abiga humilis* Guérin-Méneville, 1859
Cratopus Eschscholtz, 1831: 8 [HN] type species *Cratopus castaneus* Eschscholtz, 1831
Sceleodis Gemminger & Harold, 1870: 1836 [unjustified emendation]
Sceleoides Agassiz, 1846b: 333 [unjustified emendation]
alluaudi Peyerimhoff, 1933: 345 **N**: MO (Western Sahara) **AFR**
besnardi aindarensis Kaszab, 1981c: 335 **A**: SA (Eastern Prov.)
besnardi besnardi Kaszab, 1981c: 334 **A**: SA
castaneus castaneus Eschscholtz, 1831: 8 (*Cratopus*) **N**: AG EG LB **A**: SI
cerisyi Guérin-Méneville, 1859: cxc (*Abiga*)
humilis Guérin-Méneville, 1859: clxxxix (*Abiga*) **N**: AG TU
laticollis Chatanay, 1914a: 19 **N**: AG

genus *Schweinfurthia* Andres, 1922: 26 type species *Schweinfurthia sinaitica* Andres, 1922
alfierii Schatzmayr & Koch, 1934b: 20 **A**: SI
nana Fairmaire, 1893a: 28 (*Psammodes*) **A**: YE **AFR**
schusteri Gridelli, 1930: 211 **N**: LB
sinaitica Andres, 1922: 26 **A**: SI

genus *Scythis* Schaum, 1865: 102 type species *Tentyria macrocephala* Tauscher, 1812
Megascythis Kelejnikova, 1963: 622 type species *Megascythis panfilovi* Kelejnikova, 1963
Semenovonymus Bogatchev, 1946c: 391 type species *Semenovonymus tenuis* Bogatchev, 1946
affinis Ballion, 1878: 299 **A**: KZ (Almaatinskaya Oblast) XIN
altaica G. S. Medvedev, 1990b: 135 **A**: XIN
alticola Reinig, 1931: 901 **A**: TD (Karategin)
angusticollis Gebler, 1829a: 117 (*Tentyria*) **A**: KZ XIN
integrina Reitter, 1900d: 165
kraatzi Reitter, 1900d: 164
laticeps Reitter, 1915d: 67
longipes Kraatz, 1865a: 104
ardoini Skopin, 1960: 299 **A**: KZ (Alamatinskaya oblast)
arenaria arenaria Faldermann, 1835b: 403 (*Tentyria*) **A**: KZ MG XIN
arenaria australis Skopin, 1979: 179 **A**: KZ
athelea Reitter, 1915d: 72 **A**: KI KZ XIN
gorgone Reitter, 1915d: 72
substriatula Reitter, 1915d: 72
ballioni Reitter, 1900d: 166 **A**: KZ TD (Mogol-Tau)
banghaasi Reitter, 1900d: 166 **A**: ES (Tuva) GAN MG NMO XIN
bulganica Kaszab, 1967a: 31 **A**: MG XIN
constricta Steven, 1829: 86 (*Tentyria*) **A**: ES UZ
tenebricosa Faldermann, 1835b: 399 (*Tentyria*)
denticulata Gebler, 1841b: 591 (*Anatolica*) **A**: ES (Altai)
dschungarica Kaszab, 1967a: 23 **A**: MG
edda Reitter, 1915d: 69 **A**: KZ (Almaatinskaya Oblast)
grossepunctata Kaszab, 1965e: 326 **A**: MG
intermedia intermedia Ballion, 1878: 300 **A**: KZ (Almaatinskaya Oblast) XIN
anthoini Skopin, 1960: 298
intermedia scythiformis Reitter, 1915f: 93 (*Microdera*) **A**: XIN
ivanovi G. S. Medvedev, 2004b: 553 **A**: ES (Tuva)
latithorax Skopin, 1979: 175 **A**: XIN
lehri Skopin, 1961a: 182 [= 1964c: 404] **A**: KZ
macrocephala Tauscher, 1812a: 32 (*Tentyria*) **E**: ST **A**: KZ
panfilovi Kelejnikova, 1963: 623 (*Megascythis*) **A**: KI

piechockii Kaszab, 1964b: 13 **A**: MG
pusilla Skopin, 1964: 381 **A**: ES (Tuva) MG
  septentrionalis Kaszab, 1965e: 328
rotundicollis alexandrae Skopin, 1960: 299 **A**: KZ (Vostochnyy Kazakhstan)
rotundicollis rotundangula Skopin, 1979: 176 **A**: KZ
rotundicollis rotundicollis Reitter, 1915d: 69 **A**: KI KZ (Zhambylskaya oblast)
sculptilis Reitter, 1900e: 163 **A**: BEI NMO SHX
selivanovskii Skopin, 1961b: 391 **A**: TD (Pamir)
sithon Reitter, 1915e: 83 (*Anatolica*) **A**: KZ
skopini Kaszab, 1964b: 12 **A**: MG
tatarica gracilis Ballion, 1878: 298 **A**: KI KZ (Alamtinskaya oblast) XIN
tatarica pseudoscythis Reitter, 1900d: 119 (*Anatolica*) **A**: KI KZ XIN
tatarica suturalis Reitter, 1915d: 70 **A**: KI KZ
tatarica tatarica Gebler, 1844: 102 (*Anatolica*) **A**: KI KZ
  aglaja Reitter, 1915d: 71
  dominula Reitter, 1915d: 70
  prosternalis Reitter, 1915d: 69
  thetis Reitter, 1915d: 70
tatarica weichboldi Melichar, 1912: 33 (*Anatolica*) **A**: KI KZ
tenuis Bogatchev, 1946c: 391 (*Semenovonymus*) **A**: MG NMO
tuvae Kaszab & Knor, 1976: 296 **A**: ES (Tuva)
vitabunda Skopin, 1961a: 181 [= 1964c: 403] **A**: KZ (Almaatinskaya Oblast)
vtorovianus Skopin, 1979: 177 **A**: KI

genus *Scythodonta* Reitter, 1897h: 300 type species *Scythis humeridens* Reitter, 1887
humeridens Reitter, 1887d: 356 (*Scythis*) **A**: NMO XIN
inhumeridens Reitter, 1915e: 83 (*Scytodonta*) **A**: XIN

genus *Scytosoma* Reitter, 1895f: 281 type species *Scytosoma arcibasis* Reitter, 1895 (= *Tentyria pygmaea* Gebler, 1832)
constrictum Ren & Zheng, 1993b: 40 **A**: QIN
fascia Ren & Zheng, 1993b: 38 **A**: NIN
funebre Reitter, 1900d: 162 **A**: KZ (Dzungaria)
obesum Ren & Zheng, 1993b: 35 **A**: NMO
opacum Reitter, 1889k: 684 (*Scytis*) **A**: NMO XIN
ovale Ren & Zheng, 1993b: 37 **A**: GAN
pygmaeum Gebler, 1832: 54 (*Tentyria*) **A**: ES MG NMO NIN
  aequale Reitter, 1895f: 281
  arcibasis Reitter, 1895f: 281
  bellum Faldermann, 1835b: 393 (*Tentyria*)
rufiabdominum Ren & Zheng, 1993b: 39 **A**: NMO
scalare Ren & Zheng, 1993b: 35 **A**: GAN

genus *Sinoecia* Chatanay, 1914c: 220 type species *Sinoecia puncticollis* Chatanay, 1914
puncticollis Chatanay, 1914c: 221 **A**: HEB

genus *Sphenaria* Ménétriés, 1849: 240 type species *Sphenaria elongata* Ménétriés, 1849
barchanica Skopin, 1961b: 384 **A**: TM
elongata Ménétriés, 1849: 241 **A**: IN TM UZ
hauseri Reitter, 1894a: 48 **A**: TM UZ
  acuta Semenov, 1894c: 532 (*Himatismus*)
karelini Ménétriés, 1849: 242 **A**: TD TM UZ
lubricula Reitter, 1916e: 144 **A**: TM UZ
menetriesi Semenov, 1891a: 360 (*Himatismus*) **A**: TD TM UZ
  glabra Semenov, 1889b: 309

longula Reitter, 1916e: 144
reitteri Brancsik, 1899: 99
moritzi Shestoperov, 1930: 101 **A**: IN
nana Skopin, 1961b: 385 **A**: UZ
obtusa Semenov, 1889a: 214 **A**: UZ
pierrei Peyerimhoff, 1949a: 70 **N**: AG
sewertzowi Semenov, 1894c: 530 (*Himatismus*) **A**: UZ

genus *Stegastopsis* Kraatz, 1865a: 176 type species *Stegastopsis babylonica* Kraatz, 1865
Ohyonthis Reitter, 1898f: 347 type species *Ohyonthis microderoides* Reitter, 1898
arabica Allard, 1883: 23 (*Stegatopsis*) **A**: OM
babylonica Kraatz, 1865a: 177 **A**: IQ LE
crassicornis Baudi di Selve, 1874a: 101 [= 1875c: 60] **A**: IN
microderoides Reitter, 1898f: 347 (*Ohyonthis*) **A**: TR
persica Baudi di Selve, 1874a: 101 [= 1875c: 59] **A**: IN

genus *Stenosida* Solier, 1835b: 281 type species *Stenosida tenuicollis* Solier, 1835 (=*Tagenia striatopunctata* Wiedeman, 1821)
Aprosphaena Reitter, 1916e: 142 type species *Aprosphaena adriani* Reitter, 1916 (= *Tagenia striatopunctata* Wiedemann, 1821)
Notioscythis Fairmaire, 1883b: 31 type species *Notioscypthis punctoseriata* Fairmaire, 1883
indica Haag-Rutenberg, 1877a: 278 (*Himatismus*) **A**: GUA HAI **ORR**
indicola Reitter, 1916e: 143 (*Aprosphaena*)
striatopunctata Wiedemann, 1821: 149 (*Tagenia*) **A**: NP YUN **ORR**
adriani Reitter, 1916e: 143 (*Aprosphaena*)
tenuicollis Solier, 1835b: 282

genus *Syachis* Bates, 1879: 467 [= 1886: 55] type species *Syachis himalaicus* Bates, 1879
Orocina Reitter, 1897h: 303 type species *Orocina capnisiceps* Reitter, 1897
afghanicus Schuster, 1936c: 193 **A**: AF
ajmonis Gridelli, 1934: 44 **A**: KA XIN
badakschanicus Kaszab, 1959f: 352 **A**: AF
capnisiceps Reitter, 1897h: 303 (*Orocina*) **A**: AF IN TD (Pamir)
marginatus Reinig, 1931: 899 (*Ascelosodis*)
cugiae Gridelli, 1934: 45 **A**: KA
darvazianus Skopin, 1964: 408 **A**: TD (Pamir)
globuliformis Della Beffa, 1931: 184 **A**: KA
gnathosoides Kaszab, 1959f: 350 **A**: AF
haafi Kaszab, 1959f: 348 **A**: AF TD
himalaicus Bates, 1879: 467 **A**: KA XIZ
klapperichi Kaszab, 1959f: 351 **A**: AF
marussii Gridelli, 1957: 49 **A**: KA XIN
mediocris Fairmaire, 1891d: xciii (*Capnisa*) **A**: KA TM
osellai Kaszab, 1985: 450 **A**: KA
paludani Kaszab, 1959f: 239 **A**: AF
picicornis Bates, 1879: 467 **A**: KA XIN
proximus Kaszab, 1959f: 346 **A**: AF
rugosipennis Della Beffa, 1931: 184 **A**: KA
xizangana Ba & Ren, 2004: 763 **A**: XIZ

genus *Tamena* Reitter, 1900d: 143 type species *Psammocryptus rugiceps* Reitter, 1897
rugiceps Reitter, 1897j: 311 (*Psammocryptus*) **A**: XIN

genus *Tentyria* Latreille, 1802: 170 type species: see Comments in New Acts
subgenus *Subtentyrina* Löbl & Merkl, 2003: 251 type species *Tentyria elongata* Waltl, 1835
aragonica Koch, 1944b: 231 **E**: SP
bassii Solier, 1835b: 356 **E**: PT SP
cantabrica Koch, 1944b: 230
gredosana Koch, 1944b: 230
meridionalis Koch, 1944b: 231
castillana Koch, 1944b: 226 **E**: SP
corrugata Rosenhauer, 1856: 189 **E**: PT SP
andalusiaca Kraatz, 1865a: 142
elongata elongata Waltl, 1835: 70 **E**: SP (Andalucía oriental)
arenaria Kraatz, 1865a: 140
elongata sinuatocollis Rosenhauer, 1856: 185 **E**: SP (Andalucía occidental)
rugosostriata Kraatz, 1865a: 140
elongata subcostata Solier, 1835b: 325 **N**: AG MO
acuminipennis Lucas, 1855a: 291
beidensis Antoine, 1926: 248
godartiana Lucas, 1855a: 291
marocana Solier, 1835b: 333
maura Erichson, 1841: 177
eulipoides Koch, 1944b: 231 **E**: SP
gaditana Rosenhauer, 1856: 186 **E**: SP
heydeni Haag-Rutenberg, 1870a: 130 **E**: SP
jordani Koch, 1944b: 229 **E**: SP
levis Solier, 1835b: 358 **E**: SP
leptidea Fairmaire, 1879c: 188
sublaevis Kraatz, 1865a: 144
ligurica confusa Ardoin, 1973: 304 **E**: IT (Sardegna)
ligurica ligurica Solier, 1835b: 324 **E**: FR IT
cribrata Gené, 1836b: iii
dentibasis Reitter, 1900d: 168
ottii Lucas, 1855a: 291
ligurica pseudorugosa Ardoin, 1973: 304 **E**: IT (Sardegna)
mucronata mucronata Steven, 1829: 90 **E**: FR SP
glabra Herbst, 1799: 60 [HN]
oblonga Solier, 1835b: 321
nitida Solier, 1835b: 323
mucronata ophiusae Codina, 1918: 265 **E**: SP (Formentera)
schaumi Kraatz, 1865a: 142 **E**: SP (Baleares)
pazi Español, 1958a: 13 **E**: SP (Islas Columbretes)
peiroleri Solier, 1835b: 357 **E**: PT SP
subrugosa Solier, 1835b: 352 **E**: SP
velox Chevrolat, 1865: 390 **E**: SP
subgenus *Tentyria* Latreille, 1802: 170 type species: see Comments in New Acts
Heliodromus Brullé, 1832: 196 type species *Heliodromus rotundatus* Brullé, 1832
acuticollis Reitter, 1900d: 182 **A**: KI KZ
ganglbaueri Reitter, 1900d: 182
arabica arabica Schuster, 1938a: 55 **A**: SY YE
arabica scorteccii Koch, 1965a: 115 **A**: YE
atlantis Antoine, 1942b: 43 **N**: MO (Grand Atlas)
balachowskyi Girard, 1965: 139 **N**: AG
barbara Solier, 1835b: 340 **N**: AG
bogdanovi Gebien, 1937: 634 [RN] **A**: AF IN TM UZ
[*robusta* sensu Reitter, 1900. 183]
castrogironai Escalera, 1923d: 128 **N**: MO
clavata Kraatz, 1865a: 151 **E**: "Caucasus"

collatina Reiche & Saulcy, 1857: 207 **A**: IS
latiuscula Kraatz, 1865a: 159
confalonierii confalonierii Gridelli, 1930: 222 **N**: LB
confalonierii coniophthalma Koch, 1937: 376 **N**: LB
curculionoides Herbst, 1799: 58 (*Pimelia*) **E**: PT
curta Steven, 1829: 89
cylindrica Solier, 1835b: 353 **A**: CY SY TR
acuminata Reiche & Saulcy, 1857: 209
lata Baudi di Selve, 1875a: 284
cypria Kraatz, 1865a: 160 **A**: CY TR
cyrenaica Schuster, 1919a: 27 **N**: LB
subpunctata Schuster, 1922b: 15
discicollis Reiche & Saulcy, 1857: 207 **A**: IS JO SI SY
duponti Solier, 1835b: 323 **N**: AG EG
elegans Schuster, 1919a: 28 **N**: LB
emarginata Kraatz, 1865a: 141 **E**: PT SP
euphratica escalerai Girard, 1966b: 204 **A**: SY
euphratica euphratica Girard, 1966a: 84 **A**: IQ
euphratica kursistana Girard, 1966a: 84 **A**: "Kurdistan"
excavata Solier, 1835b: 364 **N**: AG TU
scabriuscula Solier, 1835b: 365
fossulata C. Brisout de Barneville, 1888: cxcii **N**: AG
frivaldskii Kraatz, 1865a: 150 **E**: YU
banatica J. Frivaldszky, 1865: 43
geliae Carl, 1989: 25 **A**: IQ
gigas Faldermann, 1836: 376 **A**: KZ TD TM UZ XIN
asiatica Skopin, 1966: 5
laevicollis Gebler, 1844: 102
robusta Motschulsky, 1870: 383
valida Motschulsky, 1870: 397 (*Mesostena*)
gibbicollis Lucas, 1855a: 291 **N**: AG TU
strauchii Kraatz, 1865a: 166
grossa angustata Kraatz, 1896: 103 **E**: IT (Pantelleria)
grossa basalis L. W. Schaufuss, 1869: 21 **E**: SP
grossa grossa Besser, 1832: 11 **E**: GR IT MA SP **N**: AG MO TU
dejeanii Solier, 1835b: 345
grandis Solier, 1835b: 343
sicula Solier, 1835b: 342
tristis Solier, 1835b: 344
grossa sardiniensis Ardoin, 1973: 305 **E**: IT (Sardegna)
grossa sommieri Baudi di Selve, 1874a: 99 [= 1875c: 47] **E**: IT (Linosa, Lampione)
herculeana Reiche & Saulcy, 1857: 204 **A**: IS LE SY
saulcyi Reiche, 1861d: 212 [RN]
solieri Reiche, 1857: 206
incerta Solier, 1835b: 359 **E**: SP
intermedia Skopin, 1968: 106 [RN] **A**: KZ UZ
monticola Skopin, 1966: 4 [HN]
interrupta Latreille, 1807: 155 **E**: FR PT SP
gallica Solier, 1835b: 333
italica Solier, 1835b: 348 **E**: CR GR IT
calabrica Reitter, 1906e: 130
romana Baudi di Selve, 1875a: 283
tarquinia Baudi di Selve, 1875c: 44
kantarae Fairmaire, 1879c: 187 **N**: AG
kriegi Koch, 1940d: 264 **A**: SA YE
laevigata laevigata Steven, 1829: 92 **E**: IT
laevigata leachii Baudi di Selve, 1875c: 42 **E**: MA

lateritia Reitter, 1900d: 169 **E**: PT SP
laticollis Kraatz, 1865a: 158 **A**: CY IS JO SY
latreillei brachythorax Gridelli, 1930: 219 **N**: LB
latreillei latreillei Solier, 1835b: 339 **E**: IT
sardea Solier, 1835b: 340
latreillei oblongipennis Fairmaire, 1875a: 518 **N**: TU
latreillei syrtica Koch, 1937: 374 **N**: LB
mauritanica Baudi di Selve, 1875c: 45 **N**: EG LB
mesopotamica Holdhaus, 1919: 53 **A**: IQ
morgani Girard, 1968: 873 **A**: IN
mulsanti Lucas, 1855a: 291 **N**: AG LB TU
nomas coerulescens Bogdanov-Katjkov, 1915: 6 **E**: ST (Kavkaz)
nomas nomas Pallas, 1781: 43 (*Tenebrio*) **E**: AB RO ST UK **A**: KZ "Siberia"
gebleri Besser, 1832: 15 (*Tenebrio*)
podolica Steven, 1829: 91
rugulosa Germar, 1824b: 138
saisanensis Motschulsky, 1845a: 74 (*Anatolica*)
salzmanni Solier, 1835b: 351
sibirica Gebler, 1829a: 120
nomas taurica Tauscher, 1812a: 37 **E**: ST UK (Krym)
fischeri Solier, 1835b: 355
occidentalis Peyerimhoff, 1925: 14 **N**: MO (Moyen Atlas)
ocularis calcophila Koch, 1937: 368 **N**: LB (Tripolitania)
ocularis gusbatensis Koch, 1937: 369 **N**: LB (Tripolitania)
ocularis ocularis Reitter, 1900d: 174 **N**: LB
olgae Bogdanov-Katjkov, 1915: 5 **E**: ST (Kavkaz)
parallela Baudi di Selve, 1874a: 98 [= 1875c: 49] **A**: IN
pierrei Girard, 1965: 138 **N**: AG
platyceps Steven, 1829: 92 **E**: PT SP
calcarata Reitter, 1900d: 170
goudoti Solier, 1835b: 360
modesta Rosenhauer, 1856: 188
prolixa Rosenhauer, 1856: 187
pseudogigas Skopin, 1968: 107 **A**: KI KZ UZ
pseudolaevis Koch, 1944b: 231 **E**: SP
punctatostriata Solier, 1835b: 337 **N**: EG LB **A**: SI
puncticeps puncticeps Miller, 1861: 173 **A**: CY JO SY
puncticeps variabilis Girard, 1966b: 205 **A**: IQ SY
ramburi cassolai Ardoin, 1973: 304 **E**: IT (Sardegna)
ramburi floresii Gené, 1836b: iii [= 1836a: 194] **E**: FR (Corse) IT (Sardegna)
monticola Gené, 1839: 69
ramburi ramburi Solier, 1835b: 327 **E**: FR IT
angusticollis Solier, 1835b: 328
levicollis Solier, 1835b: 329
maillei Solier, 1835b: 330
substriata Solier, 1835b: 332
ramburi rugosa Gené, 1836b: iii [= 1836a: 193] **E**: IT (Sardegna)
robustoides Reitter, 1900d: 183 **A**: AF IN TM
rotundata angulata Brullé, 1832: 197 (*Heliodromus*) **E**: BU GR MC ST **A**: IN TR
kindermanni Fischer von Waldheim, 1844: 64
subdepressa Solier, 1835b: 346
rotundata convexipennis Koch, 1948a: 316 **A**: TR
rotundata daghestris Koch, 1948a: 316 **A**: TR
rotundata ionica Koch, 1948a: 313 **E**: GR (Ionioi)
rotundata mittrei Solier, 1835b: 341 **E**: GR (Chios, Kos, Nisyros, Ródos, Samos) **A**: TR
subsulcata Koch, 1948a: 316

rotundata orbicollis Solier, 1835b: 362 **E**: GR
rotundata paganettii Schuster, 1915a: 6 **E**: GR (Kríti)
rotundata rotundata Brullé, 1832: 197 **E**: GR
grossa Solier, 1835b: 361 [HN]
rotundata sulcatipennis Schuster, 1936b: 69 **E**: GR
rotundata winkleri C. Koch, 1936: 293 **A**: IN TR
sinaitica Peyerimhoff, 1907a: 32 **A**: SI
solieri Lucas, 1846: 312 **N**: AG TU
sporadica gavdosana Koch, 1948a: 307 **E**: GR (Gavsos)
sporadica sporadica Reitter, 1900d: 179 **E**: GR (Sporades)
sporadica zimmermanni Koch, 1948a: 308 **E**: GR (Kufonisi)
striatopunctata Ménétriés, 1832: 195 **E**: AB AR GG ST **A**: IN TR
punctipleuris Reitter, 1900d: 182
syrtana Normand, 1936b: 357 **N**: TU
tessulata incusa Ménétriés, 1832: 195 **E**: AB
tessulata rugipleuris Bogdanov-Katjkov, 1916: xxix **A**: IN (Adzerbaijan)
tessulata scabripennis Solier, 1835b: 366 **A**: IN TR
tessulata tessulata Tauscher, 1812a: 316 **E**: AB AR GG ST **A**: IN TR
alpina L. Redtenbacher, 1850: 48
olivieri Solier, 1835b: 368
tessulata villiersi Girard, 1968: 871 **A**: IN
thunbergii Steven, 1829: 90 **N**: AG MO
affinis Lucas, 1846: 311
bipuncta Solier, 1835b: 336
transversicollis Fairmaire, 1879c: 187 **N**: AG
valentinae Bogdanov-Katjkov, 1915: 7 **E**: ST (Kavkaz)
varioculata Carl, 1991b: 173 **A**: IQ
vieta Faldermann, 1835b: 402 **A**: NMO
wiedemanni Ménétriés, 1838: 38 (*Heliodromus*) **A**: TR
bisinuatibasalis Koch, 1936: 292
sinuatocollis Kaszab, 1940a: 3 [HN]

genus *Tentyria*, nomina dubia
glabra Fabricius, 1775: 253 (*Pimelia*) distribution unknown
pygmaea Gené, 1836b: iii **E**: IT (Sardegna)

genus *Tentyrina* Reitter, 1900d: 166 type species *Akis orbiculata* Fabricius, 1775
Tentyriina Peyerimhoff, 1907b: 186 [unjustified emendation]
deserta deserta Kaszab, 1981c: 338 **A**: SA (Eastern Prov.)
deserta hadraumatica Kaszab, 1981c: 339 **A**: YE
deserta sparsepunctata Kaszab, 1981c: 339 **A**: AE
deserta thesigeri Kaszab, 1981c: 339 **A**: OM
duplicata boehmi Reitter, 1909a: 29 **N**: EG LB
duplicata brachythorax Gridelli, 1929b: 55 (*Tentyriina*) **N**: LB (Cyrenaica)
duplicata duplicata Reitter, 1900d: 172 (*Tentyria*) **N**: LB TU
duplicata manzonii Gridelli, 1929b: 58 (*Tentyriina*) **A**: SA YE **AFR**
duplicata schusteri Schatzmayr & Koch, 1934b: 22 (*Tentyriina*) **N**: EG
gracilis Schuster, 1930b: 236 (*Tentyriina*) **A**: IN PA SA
granulicauda Koch, 1940a: 123 **A**: IN
longicollis Lucas, 1855a: 291 (*Tentyria*) **N**: AG LB MO TU
orbiculata orbiculata Fabricius, 1775: 253 (*Pimelia*) **N**: EG
aegyptiaca Solier, 1835b: 334 (*Tentyria*)
orbiculata subsulcata Reiche & Saulcy, 1857: 203 (*Tentyria*) **N**: EG **A**: CY IS LE SI SY
palmeri palmeri Crotch, 1872: 267 (*Tentyria*) **N**: EG LB **A**: IN IQ JO SA SI SY
giraffa Allard, 1883: 22 (*Tentyria*)

palmeri persicola Schuster, 1938b: 84 (*Tentyria*) **A**: IN PA
palmeri thomasi Blair, 1931b: 269 (*Tentyria*) **A**: OM SA
sculptissima Koch, 1940c: 101 (*Tentyria*) **A**: JO
senegalensis hircipes Peyerimhoff, 1916a: 75 **N**: AG LB MO
genus *Tentyrina*, nomen dubium
reichii Besser, 1832: 14 (*Tentyria*) distribution unknown

genus *Tentyronota* Reitter, 1900d: 188 type species *Micipsa rotundicollis* Kraatz, 1865
atrimembris Pic, 1922g: 30 (*Micipsa*) **N**: MO
ifnica Escalera, 1935a: 302 (*Tentyrionota*) **N**: MO
rotundicollis lucida Peyerimhoff, 1931b: 96 **N**: AG
rotundicollis rotundicollis Kraatz, 1865a: 108 (*Micipsa*) **N**: AG LB

genus *Thalpobia* Fairmaire, 1871a: 385 type species *Thalpobia laevipennis* Fairmaire, 1871
Micipsina Reitter, 1900d: 188 type species *Micipsina rolphi* Reitter, 1900
fonti fonti Escalera, 1935a: 305 **N**: MO
fonti korsakovi Español, 1946: 117 **N**: MO (Tan-Tan)
laevipennis Fairmaire, 1871a: 386 **N**: MO
gracilitarsis Antoine, 1933b: 224
meridionalis clermonti Español, 1946: 116 **N**: MO (Tan-Tan)
rolphi meridionalis Escalera, 1914: 280 **N**: MO
alluaudi Kocher, 1957: 272
maxima Kocher, 1957: 272
rolphi rolphi Reitter, 1900d: 189 (*Micipsina*) **N**: MO
gattefossei Antoine, 1946b: 37
reitteri Escalera, 1913: 36 **N**: MO
zemensis Antoine, 1946b: 36 **N**: MO

genus *Thraustocolus* Kraatz, 1865b: 414 [RN] type species *Calobamon leptoderus* Kraatz, 1865
subgenus *Leptoderops* Löbl, Bouchard & Merkl, 2008 [see New Acts] type species *Thraustocolus priesneri* Koch, 1934
afghanicus Kaszab, 1979c: 278 **A**: AF
priesneri Koch, 1934a: 97 **N**: EG **AFR**
subgenus *Prothraustocola* Kaszab, 1957f: 293 type species *Ibnsaudia belutschistanicus* Kaszab, 1957
belutschistanicus Kaszab, 1957f: 294 (*Ibnsaudia*) **A**: IN
montandoni Schuster, 1915b: 87 **N**: EG
pitcheri Johnson, 1989: 126 **A**: SA (Eastern Prov.)
subgenus *Thraustocolus* Kraatz, 1865b: 414 [RN] type species *Calobamon leptoderus* Kraatz, 1865
Calobamon Kraatz, 1865a: 105 [HN] type species *Calobamon leptoderus* Kraatz, 1865
Ibnsaudia Koch, 1941c: 290 type species *Ibnsaudia propheta* Koch, 1941
arabicus Kaszab, 1979c: 277 **A**: SA
catomoides Kaszab, 1981c: 342 **A**: SA
klapperichi Kaszab, 1959f: 393 **A**: AF
leptoderus Kraatz, 1865a: 106 (*Calobamon*) **N**: EG **A**: IS SI "Dead Sea"
propheta Koch, 1941c: 290 (*Ibnsaudia*) **A**: SA
rugosus Kaszab, 1979c: 276 **A**: AE OM SA

genus *Trichosphaena* Reitter, 1916e: 145 type species *Sphenaria olgae* Semenov, 1889
arabica Kaszab, 1961c: 170 **A**: OM SA YE
chogsonzhavi Kaszab, 1968c: 346 **A**: MG
chotanica Semenov, 1891a: 354 (*Himatismus*) **A**: NMO XIN
compactilis Merkl, 1991b: 63 **A**: IN
dunhangensis Ren & Zheng, 1993c: 52 **A**: GAN
gobica Kaszab & G. S. Medvedev, 1972: 216 **A**: MG
iranica Bogatchev, 1967b: 350 (*Epitrichia*) **A**: IN SA

olgae Semenov, 1889b: 309 (*Sphenaria*) **A**: AF TM UZ
perraudieri perraudieri Marseul, 1867: xxxix (*Himatismus*) **N**: AG EG MO TU **A**: IS OM SA SI YE **AFR**
persica Schuster, 1935b: 23 **A**: IN
quadrate [sic!] Ren & Zheng, 1993c: 55 **A**: GAN
reitteri Semenov, 1891a: 353 (*Himatismus*) **A**: MG NMO
sahlbergi Reitter, 1916e: 147 **A**: TM
schusteri Reitter, 1916e: 147 **A**: UZ
suturalis Semenov, 1889b: 309 (*Sphenaria*) **A**: TM UZ
ulanbuhensis Ren & Zheng, 1993c: 55 **A**: NMO
vestita Reitter, 1889b: 116 (*Sphenaria*) **A**: MG NMO UZ XIN
turkestanica Semenov, 1891a: 357 (*Himatismus*)
vestita Semenov, 1889b: 215 (*Sphenaria*) [HN]
zarudnyana Schuster, 1938b: 80 **A**: IN
zarudnyi Bogatchev, 1949b: 277

genus *Uyttenboogaartia* Koch, 1943a: 595 type species *Hegeter cribricollis* Brullé, 1838
cribricollis Brullé, 1838: 66 (*Hegeter*) **N**: CI (Gran Canaria)
punctipennis Wollaston, 1864: 467 (*Gnophota*) **N**: CI (Gran Canaria)

tribe Tentyriini, nomina dubia
Dailognatha hispana Solier, 1835b: 261 **E**: PT SP
Tentyria oblonga Tauscher, 1812a: 314 **E**: "Iberia"
Tentyria punctata Tauscher, 1812a: 313 [41] **E**: "regionibus Caucasicis"

tribe Zophosini Solier, 1834

genus *Zophosis* Latreille, 1802: 167 type species *Erodius testudinarius* Fabricius, 1781
subgenus *Cheirosis* Deyrolle, 1867: 220 type species *Zophosis ovata* Faldermann, 1837 (= *Pedinus acuminatus* Fischer von Waldheim, 1832)
acuminata Fischer von Waldheim, 1832a: 202 (*Pedinus*) **E**: AB **A**: IN
ovata Faldermann, 1837: 5 [HN]
subgenus *Gahanosis* Penrith, 1983b: 379 type species *Zophosis undulata* Gahan, 1900
undulata Gahan, 1900: 8 **A**: YE (Suqutra)
subgenus *Hologenosis* Deyrolle, 1867: 82 type species *Hologenosis lacerata* Deyrolle, 1867
angustula Reitter, 1916c: 88 **N**: TU
atlantica Escalera, 1913: 36 **N**: MO
emiliana Reitter, 1916c: 88 **N**: MO
nigroaenea Deyrolle, 1867: 181 **N**: AG EG LB MO **A**: SI **AFR**
aelanitica Peyerimhoff, 1907a: 26
convergens Koch, 1941c: 269
mozabita Fairmaire, 1897f: 120
theryi Koch, 1941c: 270
picipennis arabica Lesne, 1920: 15 **A**: YE
oxycaroides Koch, 1935a: 6 **N**: EG
subgenus *Oculosis* Penrith, 1977: 126 type species *Zophosis boei* Solier, 1834
abbreviata Solier, 1834: 606 **N**: EG **A**: SA **AFR**
reichei Guérin-Méneville, 1849: 315
semilineata Deyrolle, 1867: 107
aequalis Waterhouse, 1881: 471 **A**: YE (Darsa, Suqutra)
dilatata Deyrolle, 1876: 208 **E**: GR **A**: LE SY TR
persis Chatanay, 1914d: 379 **A**: IN
punctata alborana Baudi di Selve, 1883c: 760 **E**: SP (Alboran Is.)
punctata punctata Brullé, 1832: 191 **E**: AB AL AR GR IT SP **N**: AG LB MO TU **A**: AF CY IN IQ IS JO KZ LE PA SA SI SY TD TM TR UZ XIN
algeriana Solier, 1834: 613
barbara Solier, 1834: 608

curta Solier, 1834: 611
deflexa Reitter, 1916c: 87
faldermanni Deyrolle, 1867: 116
gibbicollis Solier, 1834: 612
kollari Deyrolle, 1867: 112
maillei Solier, 1834: 613
mediocris Deyrolle, 1867: 114
nitida Gebler, 1844: 100 [= 1860: 10]
nitidula Motschulsky, 1860e: 526 [= 1860d: 426] [HN]
orientalis Deyrolle, 1867: 115
osmanlis Deyrolle, 1867: 114
ovata Brullé, 1832: 190
polita Lucas, 1854: 28
steppensis Gemminger, 1870a: 121 [RN]
subinflata Brullé, 1832: 191
truquii Deyrolle, 1867: 116
variepunctata Deyrolle, 1867: 113
vescoi Deyrolle, 1867: 113
punctata pygmaea Solier, 1834: 611 **N**: EG **A**: IS KU
quadrilineata A. G. Olivier, 1791: 426 (*Erodius*) **N**: AG LB **AFR**
subgenus *Septentriophosis* Penrith, 1982a: 167 type species *Erodius planus* Fabricius, 1775
acuticosta Fairmaire, 1886a: 448 **A**: YE **AFR**
aciculaticollis C. Müller, 1887: 297
bottegoi Chatanay, 1917b: 602
gestroi Chatanay, 1917b: 600
pinardi Théry, 1892: clxxx
nitidicostis Chatanay, 1917b: 598
ogadenica Gestro, 1892: 762
subcostulata Fairmaire, 1887a: 166
aegresinuata Penrith, 1986: 429 **A**: IN
afghanica Kaszab, 1959f: 339 **A**: AF
alticola Peyerimhoff, 1929b: 154 **N**: AG
asiatica Miller, 1861: 169 **A**: IN SY TR
puncticeps Baudi di Selve, 1874a: 90
bicarinata bicarinata Solier, 1834: 617 **N**: CI
clarkii Deyrolle, 1864: 435
ecostata Håk. Lindberg, 1962: 25
gomerae Håk. Lindberg, 1962: 25
nitidula Solier, 1834: 619
quadricostata Deyrolle, 1864: 433
vagans Brullé, 1838: 64
verrucosa Håk. Lindberg, 1962: 25
bicarinata cognata Miller, 1861: 170 **A**: IS LE
orbiculata Deyrolle, 1867: 209
bicarinata dinauxi Peyerimhoff, 1929b: 154 **N**: AG MO
aerifer Antoine, 1933b: 218
bicarinata plicata Brullé, 1838: 64 **N**: CI (Fuerteventura, Lanzarote)
bicarinata renatae Penrith, 1982a: 197 **N**: MO
bicarinata quadricostata Solier, 1834: 628 **N**: EG **A**: JO SA YE **AFR**
intermedia Chatanay, 1917b: 619
bicarinata supposita Peyerimhoff, 1931b: 90 **N**: AG
bicarinata susica Escalera, 1914: 272 **N**: MO **AFR**
graphita Koch, 1941c: 274
tricarinata Escalera, 1914: 273
clypeosuturata Carl, 1990b: 80 **A**: JO
complanata Solier, 1834: 626 **N**: EG **A**: IN IQ IS JO SA SI
deplanata Böhm, 1909: 112
humberti Deyrolle, 1867: 177
miliaris Deyrolle, 1867: 179

pulverulenta Deyrolle, 1867: 178
sulphurea Deyrolle, 1867: 180
deserticola Kulzer, 1956: 643 **A**: AF PA **ORR**
dimorpha Carl, 1990b: 79 **A**: IQ
farinosa Blair, 1914b: 483 [RN] **A**: IN IQ JO SA SI
latissima Bogatchev, 1957: 529
testudinaria Solier, 1834: 622 [HN]
finzii Koch, 1937: 343 **N**: LB
globulipennis Escalera, 1914: 271 **N**: MO
hydrobiiformis hydrobiiformis Reitter, 1916c: 98 **A**: SY
hydrobiiformis khuzistanica Kaszab, 1959a: 2 **A**: IN
inlustris Penrith, 1984: 411 **A**: IN IQ
kaszabi Penrith, 1984: 407 **A**: PA
klapperichi Kaszab, 1959f: 341 **A**: AF PA **ORR**
leonardii Kaszab, 1972: 370 **A**: SA YE
lethierryi Deyrolle, 1867: 211 **N**: AG LB TU **A**: SI
maindroni Chatanay, 1914d: 381 **A**: PA
circumducta Kwieton, 1881: 6
robusta Kwieton, 1981a: 8
marseuli Deyrolle, 1867: 210 **N**: AG TU
migneauxi Deyrolle, 1867: 171 **A**: SA YE
minuta Fabricius, 1775: 259 (*Erodius*) **E**: PT SP **N**: MO
errans Solier, 1834: 614
minutecostulata Koch, 1945: 393
suborbicularis Solier, 1834: 616
unicostulata Koch, 1945: 393
oblonga Solier, 1834: 625 **A**: LE IQ SA SY TR YE
cursor Deyrolle, 1867: 170
persica ardistana Reitter, 1916c: 95 **A**: IN TM
persica gridellii Kaszab, 1959e: 239 [= 1959f: 343] **A**: AF IN
persica persica Kraatz, 1882b: 94 **A**: AF IN TM
personata Erichson, 1841: 176 **N**: AG EG LB TU **A**: SI
curta Deyrolle, 1867: 211
depressipennis Lucas, 1858c: ccxxii
tunisea Reitter, 1916c: 91
wollastoni Deyrolle, 1867: 212
zuberi Deyrolle, 1867: 213
pharaonis pharaonis Reitter, 1916c: 90 [RN] **N**: EG **A**: IS SA SI
rotundata Deyrolle, 1867: 215 [HN]
saudita Koch, 1965a: 112
pharaonis simplex Kaszab, 1981c: 319 **A**: AE
plana Fabricius, 1775: 259 (*Erodius*) **N**: AG EG LB **A**: SA SI **AFR**
approximata Deyrolle, 1867: 195
astacura Peyerimhoff, 1931b: 90
beraudi Peyerimhoff, 1920a: 325
carinata Solier, 1834: 630
elongata Koch, 1935a: 8
germari Solier, 1834: 631
phazanica Koch, 1937: 346
praedeserta Koch, 1937: 348
reaumuri Solier, 1834: 630
schoenherri Solier, 1834: 629
posticalis Deyrolle, 1867: 199 **N**: AG EG LB **AFR**
lapruni Chatanay, 1914d: 381
niegeri Peyerimhoff, 1929b: 155
punctatissima Peyerimhoff, 1929b: 153
rotundata Ménétriés, 1849: 218 **A**: IN KZ TM

rugosa Faldermann, 1837: 4 **E**: AB AR **A**: IN
armeniaca Deyrolle, 1867: 172
sabaea Baudi di Selve, 1881: 273 **A**: YE **AFR**
aenea Chatanay, 1917b: 586
blairi Chatanay, 1917b: 586
costulata Chatanay, 1917b: 584
longa Chatanay, 1917b: 584
tenuitarsis Chatanay, 1917b: 586
scabriuscula lata Kraatz, 1882b: 94 **A**: AF KZ UZ
blattoides Reitter, 1916c: 98
scabriuscula scabriuscula Ménétriés, 1849: 218 **A**: KZ TM UZ
nivosa Dohrn, 1884: 40
scortecciana Kaszab, 1979c: 259 [RN] **A**: SA YE
scorteccii Kaszab, 1972: 370 [HN]
stilla Penrith, 1986: 431 **A**: IN
sulcata sulcata Deyrolle, 1867: 189 **N**: EG **A**: SA YE **AFR**
alternans Deyrolle, 1867: 190
arabs Chatanay, 1917b: 606
aromatum Gestro, 1895: 358
bicostis Chatanay, 1917b: 607
bohemanni Deyrolle, 1867: 188
chalybaea Chatanay, 1917b: 611
crispata Fairmaire, 1891g: ccxci
dakhlensis Andres, 1928: 48
lineata Chatanay, 1917b: 606
maeklini Deyrolle, 1867: 186
opacipennis Chatanay, 1917b: 605
planidorsis Chatanay, 1917b: 606
plicatipennis Deyrolle, 1867b: 190
pulchra Chatanay, 1917b: 613
rodolphi Chatanay, 1917b: 608
strigipleuris Chatanay, 1917b: 605
testudo Chatanay, 1917b: 589
zhenzhuristi Bogatchev, 1957: 529
viridilimbata Chobaut, 1899: 409 **N**: EG LB TU
cyrenaica Schuster, 1928e: 375
scorteccii Gridelli, 1939d: 421

subfamily **Tenebrioninae** Latreille, 1802

tribe Alphitobiini Reitter, 1917

genus *Alphitobius* Stephens, 1829a: 19 type species *Helops picipes* Panzer, 1794 (= *Opatrum laevigatum* Fabricius, 1781)

Cryptops Solier, 1851: 235 type species *Cryptops ulomoides* Solier, 1851

Heterophaga Dejean, 1834: 199 type species *Tenebrio mauritanicus* Fabricius, 1792 (= *Opatrum laevigatum* Fabricius, 1781)

Latetribolium Lepesme, 1943: 45 type species *Latetribolium risbeci* Lepesme, 1943

Microphyes W. J. MacLeay, 1873b: 286 type species *Microphyes rufipes* W. J. MacLeay, 1873

diaperinus Panzer, 1796e: 16 (*Tenebrio*) **E**: AU AZ BE BY CR CT CZ DE FI GB GE GR HU IR IT LU MA NL PL RO SA SK SL SP ST SZ UK **N**: EG CI LB MR TU **A**: AF ANH BA BT CY FE FUJ GUA GUX HAI HEB HEI HKG HUB HUN IQ IS JA JIA JIX KZ LIA MG NE NP SA SCH SHA SHX SI TAI TIA TM YE YUN ZHE "Korea" **COS**
cafer Fåhraeus, 1870: 266 (*Proselytus*)
longipennis Walker, 1858: 284 (*Crypticus*)
opatroides Brullé, 1838: 70 (*Uloma*)
ovatus Herbst, 1799: 16 (*Tenebrio*)
rufipes Walker, 1858: 284 (*Phaleria*)

laevigatus Fabricius, 1781: 90 (*Opatrum*) **E**: AR AZ BE GB GE GR HU IT LU MA NL PL SK SP ST SZ UK **N**: CI EG LB MR TU **A**: AF BA BT CY FE FUJ GUA GUI GUX HAI HEB HEI HEN HUB HUN IQ JA JIA JIL JIX KZ LIA NMO SA SCH SHA TAI YE YUN "Korea" **COS**

granivorus Mulsant & Godart, 1868: 288
mauritanicus Fabricius, 1792a: 113 (*Tenebrio*)
piceus A. G. Olivier, 1795: 58: 17 (*Helops*)
picipes Panzer, 1794c: 4 (*Helops*)
ruficolor Pic, 1925d: 11
rufipes W. J. MacLeay, 1873b: 286 (*Microphyes*)
striatulus Fairmaire, 1869b: 231 (*Cataphronetis*)

viator Mulsant & Godart, 1868: 290 **Ei**: FR SP **AFR**

genus *Diaclina* Jacquelin du Val, 1861: 296 type species *Tenebrio chrysomelinus* Herbst, 1799 (= *Tenebrio testudinea* Piller & Mitterpacher, 1783)

ceylonica Kaszab, 1980a: 185 **A**: NP **ORR**

fagi Panzer, 1799a: 3 (*Tenebrio*) **E**: AB AL AU CZ FR GE GG GR HU IT RO SK SL ST SZ TR UK YU **A**: CY IS TR

gracilipes Baudi di Selve, 1876c: 231 (*Alphitobius*)

nakatai Akita & Masumoto, 2003: 178 **A**: JA (Ishigaki)

plagiata Marseul, 1876c: 108 (*Alphitophagus*) **A**: JA SC

quadrimaculata Gebien, 1914d: 394 (*Alphitobius*) **A**: AP JA (Ryukyus) NP TAI **ORR**

testudinea Piller & Mitterpacher, 1783: 83 (*Tenebrio*) **E**: AB AL AR AU BH BU CR CZ GE GR HU GG IT MC RO SK SL ST UK YU **A**: IN TR

chrysomelina Rossi, 1790: 233 (*Tenebrio*)
chrysomelina Herbst, 1799: 17 (*Tenebrio*) [HN]
glabrata Fabricius, 1801b: 567 (*Mycetophagus*)
picea L. Redtenbacher, 1874: 113 (*Alphitobius*)

yaeyamaensis Masumoto & Akita, 2002: 150 **A**: JA (Ryukyus)

genus *Hoplopeltis* Fairmaire, 1894a: 22 type species *Hoplopeltis tricornis* Fairmaire, 1894

tricornis Fairmaire, 1894a: 23 **A**: NP **ORR**

genus *Metaclisa* Jacquelin du Val, 1861: 296 [NP] type species *Platydema parallela* Fairmaire, 1855

Amarantha Motschulsky, 1860: 141 [NO] type species *Amarantha viridis* Motschulsky, 1860

atrocyanea Lewis, 1891a: 70 (*Amarantha*) **A**: CH JA SC

azurea Waltl, 1838: 463 (*Diaperis*) **E**: GG HU RO SL ST (Kavkaz) TR **A**: IS SY TR

gracilis M. Chûjô, 1968b: 395 **A**: TAI

hamai Shibata, 1978: 24 **A**: JA (Ryukyus)

nagaii Nakane, 1968b: 79 **A**: JA (Ryukyus)

ornata Shibata, 1978: 26 **A**: TAI

viridis Motschulsky, 1860b: 141 (*Amarantha*) **E**: AB **A**: IN

chlorodia Baudi di Selve, 1876c: 247 (*Platydema*)

genus *Peltoides* Laporte, 1832: 401 type species *Peltoides senegalensis* Laporte, 1832

Oopiestus Chevrolat, 1833c: 30 type species *Oopiestus ovalis* Chevrolat, 1833

capensis longulus Fairmaire, 1897e: 126 **A**: SA YE **AFR**

tribe Amarygmini Gistel, 1848

genus *Amarygmus* Dalman, 1823: 60 type species *Chrysomela micans* Fabricius, 1794

subgenus *Amarygmus* Dalman, 1823: 60 type species *Chrysomela micans* Fabricius, 1794

Anacycus Fairmaire, 1896a: 33 type species *Dietysus navicularis* Fairmaire, 1894
Aphyllocerus Fairmaire, 1881a: 282 type species *Aphyllocerus decipiens* Fairmaire, 1881
Dietysus Pascoe, 1866: 486 type species *Dietysus confusus* Pascoe, 1866
Elixota Pascoe, 1866: 475 type species *Elixota cuprea* Pascoe, 1866
Eurypera Pascoe, 1870: 106 type species *Eurypera cuprea* Pascoe, 1870
Platolenes Gebien, 1914d: 420 type species *Platolenes rufipes* Gebien, 1913

Podamarygmus Carter, 1928: 287 type species *Podamarygmus alternans* Carter, 1928
Pseudamarygmus Pic, 1915f: 9 type species *Pseudamarygmus testaceipes* Pic, 1915
adonis Pic, 1922b: 12 **A**: NE **ORR**
ardoini Bremer, 2001: 57 [RN] **A**: SCH YUN
viridiscens Pic, 1926j: 75 (*Anacycus*) [HN]
bremeri Masumoto, Akita & Lee, 2005: 248 **A**: TAI
carbonarius Hope, 1843: 63 (*Cnodulon*) **A**: FUJ
cuprarius cuprarius Weber, 1801: 40 (*Helops*) **A**: JA TAI **AUR ORR**
callichromus Fairmaire, 1897b: 70
iridipennis Fairmaire, 1893c: 58
laosensis Pic, 1922j: 304
curvus Marseul, 1876d: 316 **A**: JA SC ZHE
pellegrini Pic, 1922f: 24 (*Anacycus*)
cyaneipennis Pic, 1938a: 10 **A**: TAI **ORR**
diversiceps Pic, 1926b: 21 **A**: TAI **ORR**
bicoloriceps Pic, 1923b: 20 [HN]
sakaii Masumoto, 1981: 29
falsicurva Masumoto, 1982b: 51 (*Elixota*) **A**: TAI
familiaris Masumoto, 1985c: 122 **A**: HKG
filicornis Gravely, 1915: 532 (*Dietysus*) **A**: AP **ORR**
atricolor Pic, 1922j: 304 (*Dietysus*)
hoanus Pic, 1929f: 44 **A**: TAI **ORR**
duporti Pic, 1938a: 9
hsiaohsingus Masumoto, 1984: 16
hopei Bremer, 2001: 57 [RN] **A**: FUJ GUX
cupreus Pascoe, 1866: 475 (*Elixota*) [HN]
iridicollis Nakane, 1968b: 83 (*Elixota*) **A**: JA (Ryukyus)
izumii Masumoto, 1983a: 81 (*Elixota*) **A**: JA (Ryukyus) TAI
latifrons Gravely, 1915: 534 (*Dietysus*) **A**: AP
nakanei Masumoto, 1982b: 53 (*Elixota*) **A**: TAI
navicularis Fairmaire, 1894a: 26 (*Dietysus*) **A**: AP NP UP **ORR**
nepalensis Kaszab, 1977b: 281 (*Platolenes*) **A**: NP **ORR**
brancuccii Bremer, 2003: 51
nodicornis Gravely, 1915: 533 (*Dietysus*) **A**: AP
vitalisi Pic, 1922j: 305
parvus Pic, 1926b: 20 **A**: HAI GUA **ORR**
picitarsis Fairmaire, 1882c: 250 (*Dietysus*) **A**: JA (Ryukyus) TAI **ORR**
aeneus Wiedemann, 1821: 154 (*Cnodalon*) [HN]
blaisei Pic, 1923b: 21
inadai Masumoto & Akita, 2001a: 21
rouyeri Pic, 1951a: 18
pilipes Gebien, 1914a: 42 **A**: AP FUJ JIX NP SD TAI YUN **ORR**
formosanus Pic, 1915g: 21
longipillis Pic, 1922j: 303
punctatus Pic, 1922b: 11 (*Anacycus*) **A**: CH
purpureofossus Fairmaire, 1896a: 32 **A**: AP NP SD **ORR**
darjilingensis Saha, 1988: 434 (*Platolenes*)
rufipennis Pic, 1922b: 11 (*Anacycus*) **A**: AP
sikkimensis Saha, 1988: 435 (*Platolenes*) **A**: SD
sinensis Pic, 1922b: 11 **A**: ZHE
speciosus Dalman, 1823: 61 **A**: SD YUN **ORR**
taiwanus Masumoto, 1981c: 31 **A**: TAI
trichopus Kaszab, 1941c: 69 **A**: TAI
vicinus Pic, 1922b: 10 (*Anacycus*) **A**: HAI SHG
subgenus *Oogeton* Kaszab, 1941c: 69 type species *Oogeton nigrocoeruleum* Kaszab, 1941 (=*Amarygmus makii* Miwa, 1939)

makii Miwa, 1939: 412 **A**: TAI
nigrocoerulus Kaszab, 1941c: 70
subgenus *Pyanirygmus* Pic, 1915f: 9 type species *Amarygmus corinthus* Pic, 1915
Inspinogeton Pic, 1937i: 174 type species *Cyriogeton impressipenne* Pic, 1937
mizusawai Yamazaki, 1968: 27 (*Plesiophthalmus*) **A**: JA (Amami-Ôshima)

genus *Eumolpamarygmus* Pic, 1923h: 11 type species *Eumolpamarygmus bigibbosus* Pic, 1923
becvari Masumoto, 1999c: 367 **A**: YUN
paotao Masumoto, 1989a: 107 **A**: TAI

genus *Gonocnemis* C. G. Thomson, 1858: 101 type species *Gonocnemis strigipennis* C. G. Thomson, 1858
senegalensis Fairmaire, 1894f: 673 **A**: YE **AFR**
surcoufi Pic, 1928h: 258 **A**: YE **AFR**

genus *Hoplobrachium* Fairmaire, 1886b: 74 type species *Hoplobrachium asperipenne* Fairmaire, 1886 (= *Helops dentipes* Fabricius, 1781)
Cephaloplonyx Pic, 1922b: 12 type species *Cephaloplonyx opacus* Pic, 1922 (= *Helops dentipes* Fabricius, 1781)
dentipes Fabricius, 1781: 326 (*Helops*) **A**: YUN **AFR ORR**
asperipenne Fairmaire, 1886b: 74
ebeninus Walker, 1858: 285 (*Helops*)
opacus Pic, 1922b: 12 (*Cephaloplonyx*)

genus *Javamarygmus* Pic, 1928b: 22 type species *Javamarygmus tristis* Pic, 1928
chingshanus Masumoto, 1982b: 49 (*Eumolpocyriogeton*) **A**: TAI
maculosus Pic, 1927g: 8 (*Eumolpamarygmus*) **A**: HAI **ORR**

genus *Plesiophthalmus* Motschulsky, 1858a: 34 type species *Plesiophthalmus nigrocyaneus* Motschulsky, 1858
Cyriogeton Pascoe, 1871: 356 type species *Cyriogeton insignis* Pascoe, 1871
antrax Fairmaire, 1903a: 17 **A**: YUN **ORR**
subnitidipennis Pic, 1934a: 17
ater Pic, 1930b: 34 **A**: FUJ GUI SCH YUN
aurichalceus M. Chûjô, 1978a: 75 **A**: JA (Ryukyus)
borchmanni Kaszab, 1957c: 56 **A**: FUJ
brevipennis brevipennis Lewis, 1896: 340 **A**: JA (Amami-Ôshima)
brevipennis makiharai M. Chûjô, 1977: 16 **A**: JA (Tokunoshima Is.)
caeruleus Pic, 1914d: 14 **A**: "South China" **ORR**
chujoi Masumoto, 2005a: 76 **A**: TAI
colossus Kaszab, 1957c: 59 **A**: FUJ GUA
convexus Pic, 1914d: 13 (*Cyriogeton*) **A**: GUI GUX HKG YUN **ORR**
auricolle Pic, 1952a: 3 (*Coriogeton*)
pulchrum Pic, 1926d: 28 (*Cyriogetum*)
cruralis Fairmaire, 1903a: 17 **A**: FUJ GUI **ORR**
davidis Fairmaire, 1878a: 121 **A**: BEI FE HUB HEN NC SC **ORR**.
lentus Heyden, 1887a: 268
donckieri Pic, 1914d: 14 (*Cyriogeton*) **A**: YUN
excellens Gebien, 1914b: 41 (*Cyriogeton*) [RN] **A**: BT SD
formosanus Miwa, 1939: 412 **A**: TAI
fujitai Masumoto, 1981c: 24 (*Cyriogeton*) **A**: TAI
fukiensis Masumoto, 1990g: 717 **A**: FUJ GUA
fuscoaenescens Fairmaire, 1898a: 14 **A**: JA (Ryukyus)
gaoligongensis Masumoto, 2000a: 148 **A**: YUN
gibbosulus Fairmaire, 1893c: 41 (*Eucyrtus*) **A**: FUJ **ORR**
guizhouensis Masumoto, 1999c: 355 **A**: GUI
gracilis M. Chûjô, 1979: 45 **A**: JA (Iriomote)

hainanensis Masumoto, 1990d: 268 **A**: HAI
impressipennis Pic, 1937i: 174 (*Cyriogeton*) **A**: FUJ ZHE
impressus Pic, 1927j: 132 (*Cyriogeton*) **A**: FUJ
inexpectatus Masumoto, 1989d: 764 **A**: SCH YUN
jessopi Masumoto, 1990d: 269 **A**: BH **ORR**
kanoi Masumoto, 1981c: 26 **A**: TAI
kaszabi Masumoto, 1990a: 40 **A**: FUJ
kikutii Yamazaki, 2002: 377 **A**: JA
kondoi Masumoto, 1981c: 18 (*Cyriogeton*) **A**: TAI
kulzeri Kaszab, 1954: 263 **A**: FUJ GUA
kumejimensis Masumoto, 1987b: 87 **A**: JA (Ryukyus)
kurosawai Masumoto, 1991a: 30 **A**: SD
laevicollis Harold, 1878a: 79 **A**: JA
latus Pic, 1914d: 13 (*Cyriogeton*) **A**: NP **ORR**
semilatus Pic, 1952a: 3 (*Cyriogeton*)
lineipunctatus Fairmaire, 1891c: xx **A**: HUB
longipes Pic, 1938a: 8 **A**: FUJ GUI SCH YUN XIZ
longulus Fairmaire, 1893b: 33 (*Rehumius*) **A**: FUJ
mayumiae Masumoto, 1981c: 21 (*Cyriogeton*) **A**: TAI
michitakai Masumoto, 1990f: 716 **A**: TAI
morio Pic, 1938a: 8 **A**: HUB HUN JIA ZHE
nakamurai Masumoto, 1987b: 89 **A**: JA (Amami-Ôshima)
nanshanchiense Masumoto, 1981c: 25 (*Cyriogeton*) **A**: TAI
nigroaeneus Gebien, 1914b: 40 (*Cyriogeton*) **A**: TAI
nigrocyaneus aeneus Motschulsky, 1862: 19 **A**: JA
yakushimanus Yamazaki, 1964: 24
nigrocyaneus nigrocyaneus Motschulsky, 1858a: 35 **A**: FE JA NE "Korea"
aenescens Marseul, 1876d: 318
mikuraensis Nakane, 1968b: 83
nigratus Motschulsky, 1862: 20
nitidior Pic, 1952a: 3
olivaceus Pic, 1939g: 342
sericeifrons Marseul, 1876d: 317
nigrocyaneus oyamai Nakane, 1956b: 169 **A**: JA
nishikawai Masumoto, 1981c: 22 (*Cyriogeton*) **A**: TAI
oblongus Masumoto, 1991a: 10 **A**: YUN
pallidicrus Fairmaire, 1889b: 46 **A**: FUJ GUX HUN SCH YUN
bicoloripes Pic, 1914d: 14
pernitidus Fairmaire, 1897k: 251 (*Cyriogeton*) **A**: GUI SCH
perpulchrus Pic, 1930b: 34 (*Cyriogetom*) **A**: YUN **ORR**
pulchrus Pic, 1927m: 187 (*Cyriogeton*) [HN]
piceus M. Chûjô, 1978a: 76 **A**: JA (Okinawa-Honto Is.)
pieli Pic, 1937i: 175 **A**: FUJ GUA ZHE **ORR**
niger Kaszab, 1957c: 58
pseudometallicus Masumoto, 1990e: 478 **A**: YUN
punctatus M. Chûjô, 1979: 50 **A**: JA (Ryukyus)
puncticollis Yamazaki, 1964: 26 **A**: JA
rubrofasciatus Ren, 1998: 233 **A**: ZHE
sameshimai Masumoto, 2001b: 58 **A**: JA (Kyushu)
semipurpureus Pic, 1917c: 9 (*Coriogeton*) **A**: SD
shibatai Masumoto, 1990g: 711 **A**: TAI
shigeoi Masumoto, 1981c: 19 (*Cyriogeton*) **A**: TAI
sichuanicus Masumoto, 1999c: 356 **A**: SCH
spectabilis okinawanus Nomura, 1964: 49 **A**: JA (Okinawa)
spectabilis spectabilis Harold, 1875: 293 **A**: BEI HEI JA LIA NC NMO SC TAI ZHE

arciferens Fairmaire, 1878a: 120
obesus Marseul, 1876d: 319
subparallelus Pic, 1916d: 11
taiwanus Nomura, 1964: 49
splendens Masumoto, 1990e: 494 **A**: BT **ORR**
yunnanus Pic, 1952a: 3 (*Cyriogeton*) **A**: YUN

genus *Pseudoogeton* Masumoto, 1989b: 304 type species *Plesiophthalmus amplipennis* Fairmaire, 1897
amplipennis Fairmaire, 1897k: 252 (*Plesiophthalmus*) **A**: SCH
gebieni Masumoto, 1989b: 312 **A**: SCH
kimurai Masumoto, 1996a: 211 **A**: JA (Ryukyus)
ovipennis Fairmaire, 1891c: xx (*Plesiophthalmus*) **A**: GUA HUB JIA
uenoi Masumoto, 1981: 93 (*Plesiophthalmus*) [RN] **A**: TAI
taiwanus Masumoto, 1981c: 28 (*Plesiophthalmus*) [HN]

tribe Apocryphini Lacordaire, 1859

genus *Diplocyrtus* G. Quedenfeldt, 1887: 257 type species *Diplocyrtus floccosus* G. Quedenfeldt, 1887
baudoni Kocher, 1962a: 200 **N**: MO
beidensis Antoine, 1936c: 141 (*Diplocyrthus*) **N**: AG
floccosus G. Quedenfeldt, 1887: 258 **N**: MO
formicoides Kocher, 1963b: 114 **N**: MO
hiemalis Escalera, 1914: 356 **N**: MO
hirtus Escalera, 1914: 357 **N**: MO
longithorax Escalera, 1914: 356 **N**: MO
marmoratus Kocher, 1962b: 306 **N**: MO
minimus Antoine, 1952: 118 **N**: MO (Anti Atlas)
panousei Kocher, 1962a: 201 **N**: MO
schrammi Antoine, 1936c: 139 (*Diplocyrthus*) **N**: MO
semihispidus Antoine, 1951b: 319 **N**: MO (Sous)
susicus Escalera, 1913: 50 **N**: MO
thamii Kocher, 1962a: 202 **N**: MO
vaulogeri Escalera, 1914: 358 **N**: MO

genus *Physohelops* Schuster, 1937a: 50 type species *Physohelops freyi* Schuster, 1937
freyi Schuster, 1937a: 51 **A**: IQ (Bacuba) SY

tribe Blaptini Leach, 1815

subtribe Blaptina Leach, 1815

genus *Ablapsis* Reitter, 1887d: 368 type species *Blaps compressipes* Reitter, 1887
compressipes Reitter, 1887d: 368 (*Blaps*) **A**: "Central China"

genus *Blaps* Fabricius, 1775: 254 type species *Tenebrio mortisagus* Linnaeus, 1758
subgenus *Arenoblaps* G. S. Medvedev, 1999a: 400 type species *Blaps hiemalis* Semenov & Bogatchev, 1940
hiemalis Semenov & Bogatchev, 1940: 206 **A**: TM UZ
subgenus *Blaps* Fabricius, 1775: 254 type species *Tenebrio mortisagus* Linnaeus, 1758
Acanthoblaps Reitter, 1889k: 687 type species *Blaps dentitibia* Reitter, 1889
Agroblaps Motschulsky, 1860e: 531 type species *Blaps fatidica* J. Sturm, 1807 (= *Blaps lethifera* Marsham, 1802)
Blapidurus Fairmaire, 1891d: xcvi type species *Blapidurus crassicornis* Fairmaire, 1891
Blapimorpha Motschulsky, 1860e: 531 type species *Blaps reflexa* Gebler, 1832
Blapisa Motschulsky, 1860e: 530 type species *Blaps jaegeri* Hummel, 1827 (= *Blaps mortisaga* Linnaeus, 1758)
Leptocolena Allard, 1880: 320 type species *Blaps emoda* Allard, 1880
Leptomorpha Faldermann, 1835b: 406 type species *Leptomorpha chinensis* Faldermann, 1835
Lithoblaps Motschulsky, 1860e: 532 type species *Tenebrio gigas* Linnaeus, 1767
Nanoblaps Semenov & Bogatchev, 1936: 565 type species *Blaps jakovlevi* Semenov & Bogatchev, 1936
Platyblaps Motschulsky, 1860e: 531 type species *Blaps holconota* Fischer von Waldheim, 1844

Rhizoblaps Motschulsky, 1860e: 532 type species *Blaps pruinosa* Eversmann, 1836
Uroblaps Motschulsky, 1860e: 530 type species *Blaps producta* Laporte, 1840 (= *Blaps lusitanica* Herbst, 1799)
abbreviata abbreviata Ménétriés, 1836a: 150 **E**: AU BU CR GR HU MC RO SK SL ST YU **A**: IN SY TR
ecaudata Küster, 1845b: 45 [HN]
ecaudata Seidlitz, 1893: 306 [HN]
ovata Solier, 1848: 305
rotundicollis Reiche & Saulcy, 1857: 240
abbreviata indagator Reiche & Saulcy, 1857: 238 **A**: IS
acuminata acuminata Fischer von Waldheim, 1820: pl. 16 **A**: GAN MG NMO QIN "Siberia"
przewalskyi Reitter, 1887: 370
acuminata kulszhana Skopin, 1973: 872 **A**: XIN
acutangula Ren & Wang, 2001: 19 **A**: NIN SHA
adelae L. Soldati & F. Soldati, 1999: 29 **A**: TR
afghanistanica Gridelli, 1954: 266 [= 1955: 61] **A**: AF
altaica Knor, 1970: 85 **A**: WS (Altay)
alternans Brullé, 1838: 68 **N**: CI
amurensis Allard, 1880: 318 **A**: NMO
angusta Seidlitz, 1893: 307 **A**: IN
anura Seidlitz, 1893: 306 **A**: AF TD UZ
appendiculata Motschulsky, 1851b: 653 [RN] **N**: AG MO TU
caudata Solier, 1848: 332 [HN]
caudigera Gemminger, 1870a: 121 [RN]
approximans Seidlitz, 1893: 256 **N**: LB TU
aruktavikus G. S. Medvedev, 2004b: 572 **A**: TD UZ
aspericollis Schuster, 1923b: 159 **A**: BEI
badakschanica Kaszab, 1960a: 73 **A**: AF TD
balashovi Bogatchev & G. S. Medvedev, 1974: 645 **A**: TM (Kopet Dag)
ballioni Skopin, 1977: 151 **A**: KI (Tien Shan) XIN
batesi Allard, 1880: 304 **A**: IQ SA YE
rathjensi Schuster, 1938a: 61
bedeli bedeli Chatanay, 1914b: 75 **N**: AG
bedeli torresallai Español, 1961f: 408 **E**: SP (Menorca)
bengalensis Hope, 1831: 31 **A**: AF IN **ORR**
spathulata Solier, 1848: 328
berezowskii G. S. Medvedev, 1998b: 206 **A**: SCH
bifurcata bifurcata Solier, 1848: 348 **N**: EG LB **A**: IS (Negev) SI
andresi Koch, 1935a: 27
dimidiata Solier, 1848: 349
bifurcata strauchii Reiche, 1861b: 88 **N**: AG LB MO TU
binominata Escalera, 1914: 324 [RN] **N**: AG MO TU
[*caudata* sensu Allard, 1880: 314]
bogatshevi G. S. Medvedev, 1964a: 60 **A**: TD TM
brevicollis Fairmaire, 1891e: cxxxi (*Leptomorpha*) **A**: KA
brevis Ren & Wang, 2001: 21 **A**: XIN
brunnea Allard, 1880: 316 **A**: "Himalaya"
bucharensis Seidlitz, 1893: 308 **A**: UZ
bushirensis Kaszab, 1959c: 54 **A**: IN
caraboides caraboides Allard, 1882a: 135 **A**: AF GAN KI KZ NIN QIN SHA TD XIN XIZ
aberrans Reinig, 1931: 878
alaiensis Reinig, 1931: 877
chinensis Reinig, 1931: 878 [HN]
emarginata Reinig, 1931: 878
licinoides Seidlitz, 1893: 308
ovata Reinig, 1931: 879
schusteri Reinig, 1934: 162 [RN]
caraboides intermittens Kaszab, 1962c: 313 **A**: KZ

chinensis Faldermann, 1835b: 407 (*Leptomorpha*) **A**: BEI GAN HEB HEN HUB JIA LIA NIN NMO SHN SHX
choui Ren & Wang, 2001: 21 **A**: SHA
clotzeri Fischer von Waldheim, 1844: 105 **A**: IN TM
cognata Solier, 1848: 336 **N**: EG
concii Koch, 1965a: 143 **A**: YE
conradtii Seidlitz, 1893: 308 **A**: KI KZ TD TM UZ
cordicollis Solier, 1848: 346 **N**: EG **A**: SY
crassicornis Fairmaire, 1891d: xcvi (*Blapidurus*) **A**: KA
cretensis Koch, 1948a: 332 **E**: GR (Kríti)
cribrosa Solier, 1848: 304 **A**: IS SY TR
  angulata Reiche & Saulcy, 1857: 247
cychroides Fairmaire, 1887b: 127 **A**: YUN
davidis Deyrolle, 1878: 119 **A**: BEI GAN HEI HUB MG NIN NMO SHA SHN "Korea"
  davidea Allard, 1886: [6] (*Blapimorpha*)
debdouensis Obenberger, 1914: 138 **N**: MO
dehaani Baudi di Selve, 1875c: 100 **A**: AF IN TM
dentitibia Reitter, 1889k: 687 **A**: GAN NIN QIN XIN
deplanata Ménétriés, 1832: 199 **E**: AB **A**: IN KZ TM UZ
  cordata Seidlitz, 1893: 288
  curvipes Baudi di Selve, 1874a: 111 [= 1875c: 91]
  muricata Fischer von Waldheim, 1832a: 199
dispar Seidlitz, 1893: 308 **A**: "Turkestan"
divergens Fairmaire, 1875a: 527 **N**: TU
doderoi Schuster, 1922b: 21 **N**: LB
dorsogranata Fairmaire, 1887b: 128 **A**: YUN
eleodes Kaszab, 1962c: 313 **A**: HUB
elliptica Seidlitz, 1893: 309 **A**: "Turkestan"
emoda Allard, 1880: 320 **A**: HP XIZ
emondi Solier, 1848: 331 **N**: AG TU
  nitidula Solier, 1848: 330
evanida Seidlitz, 1893: 285 **A**: KZ
eximia Semenov & Bogatchev, 1936: 561 **A**: KZ XIN
fascinosa Seidlitz, 1893: 305 **A**: IQ
faustii bactriana Bogatchev, 1959: 65 **A**: TD
faustii faustii Seidlitz, 1893: 305 **A**: AF KZ TD TM UZ
felix Waterhouse, 1889: 126 **A**: AF IN TM
  regia Seidlitz, 1893: 308
femoralis femoralis Fischer von Waldheim, 1844: 141 (*Pandarus*) **A**: GAN HEB MG NIN NMO SHA SHX
femoralis medusula Skopin, 1964: 372 **A**: ES (Tuva) MG NMO
femoralis rectispina Kaszab, 1968c: 377 **A**: MG
ferganica Bogatchev, 1959: 65 **A**: KI
fortesculpta Gridelli, 1953: 55 **A**: YE
freibergi Skopin, 1969: 881 **A**: TM (Kutitang-Tay)
furcala Ren & Wang, 2001: 16 **A**: NMO
garzica Ren & Wang, 2001: 15 **A**: SCH
gibba Laporte, 1840: 200 **E**: CR FR GR IT SP UK "Caucasus"
  australis Solier, 1848: 322
  impressicollis Solier, 1848: 322
  italica Baudi di Selve, 1875c: 97
  planicollis Solier, 1848: 322
  rectangularis Solier, 1848: 316
gigas Linnaeus, 1767: 676 (*Tenebrio*) **E**: AL AUi AZ BU CR CZ FR GEi GR IC IT MA PT SL SP SZ UK **N**: AG CI LB MR TU **A**: IQ KZ TR **AFR**
  avenae J. R. Sahlberg, 1903d: 13
  azorica Seidlitz, 1893: 258

gages Linnaeus, 1767: [Errata] (*Tenebrio*) [RN]
gigantea L. Petagna, 1819: 30
obtusangula Rey, 1892: 41
occulta Seidlitz, 1893: 312
producta Brullé, 1833: 302 [HN]
glabrata Seidlitz, 1893: 310 **A**: CH
glazunovi Semenov & Bogatchev, 1936: 562 **A**: IN (Elburs)
graeca Solier, 1848: 312 **E**: GR **A**: TR
ledereri Fairmaire, 1866b: 263 (*Prosodes*)
granulata altynemelis Skopin, 1966: 337 **A**: KZ (Almaatinskaya Oblast)
granulata granulata Gebler, 1825: 47 **A**: GAN KI KZ XIN
seidlitzi Schuster, 1936a: 5
granulata granulosa Ménétriés, 1854: 34 **A**: KI XIN
granulata kirgisica Skopin, 1966: 336 **A**: KI
granulata obliterata Ménétriés, 1849: 236 **A**: KZ
granulata stackelbergi Bogatchev, 1952b: 284 **A**: KZ (Vostochnyy Kazakhstan)
granulata turcomana Fischer von Waldheim, 1843: 16 [= 1844: 88] **A**: KI KZ NMO UZ XIN
caudata Gebler, 1844: 103
granulipennis Skopin, 1966: 337 **A**: KZ (Almaatinskaya Oblast)
gressoria Reitter, 1889k: 689 **A**: GAN NIN NMO QIN
haarlovi Gridelli, 1954: 264 [= 1955: 59] **A**: AF TD
haberti Peyerimhoff, 1931b: 102 **N**: AG (Hoggar) LB
helopioides Seidlitz, 1893: 309 **A**: GAN
heydeni heydeni Allard, 1880: 305 **N**: MO
heydeni taciturna Peyerimhoff, 1949b: 279 **N**: MO
hispanica Laporte, 1840: 200 **E**: SP
holconota Fischer von Waldheim, 1844: 71 **A**: AF KZ TM UZ
corrosa Fischer von Waldheim, 1844: 74
quinquecostata Fischer von Waldheim, 1844: 76
stenothorax Fischer von Waldheim, 1844: 73
indicola Bates, 1879: 472 **A**: AF IN KA XIN XIZ
lucens Fairmaire, 1891d: xcv
inflata Allard, 1880: 305 **N**: MO
inflatipennis J. R. Sahlberg, 1908: 75 **A**: LE
inflexa Zubkov, 1833: 331 **A**: KZ TM
carbonaria Kraatz, 1888a: 197
haberhaueri Seidlitz, 1893: 308
iraquensis Kaszab, 1959c: 53 **A**: IQ
jakovlevi Semenov & Bogatchev, 1936: 565 **A**: AF TM
japonensis japonensis Marseul, 1879c: 99 **A**: BEI JA LIA "Korea"
japonensis szetschwana Schuster, 1923b: 156 **A**: ANH FUJ GUI JIA SCH YUN
jeannei J. Ferrer & L. Soldati, 1999: 66 **A**: TR
judaeorum Miller, 1861: 178 **A**: IQ IS LE SY
judaica Seidlitz, 1893: 265
juliae Allard, 1880: 311 **A**: IS
kabuliensis Kaszab, 1960a: 66 **A**: AF
kadyrbekovi G. S. Medvedev, 2004b: 570 **A**: KZ
kaifensis Seidlitz, 1893: 305 **A**: IQ IS
kashgarensis gobiensis J. Frivaldszky, 1889: 207 **A**: HEB SHX
mongolica Seidlitz, 1893: 274
kashgarensis kashgarensis Bates, 1879: 473 **A**: XIN
eusoma Kraatz, 1888a: 195
lobnoriana Reitter, 1887d: 371
kiritshenkoi Semenov & Bogatchev, 1936: 555 **A**: GAN MG NIN NMO
klapperichi Kaszab, 1960a: 71 **A**: AF
kolbei Schuster, 1914b: 142 **A**: XIZ

kollari kollari Seidlitz, 1893: 263 **N**: EG **A**: IQ JO OM SA YE
kollari zhenzhuristi Bogatchev, 1939a: 125 **A**: YE
ladakensis Bates, 1879: 472 **A**: KA XIZ
lata Seidlitz, 1893: 309 **A**: TR
laticollis Solier, 1848: 324 **N**: EG
lethifera lethifera Marsham, 1802: 479 **E**: AB AL AR AU AZ BE BH BU CR CT CZ EN FI FR DE FI GB GE GG GR HU IR IT LS MC NL NR PL PT RO SK SP ST SV SZ UK WS **N**: AG MR CI XIN **A**: ES IN JO KZ TM TR WS **NARi**
abdita Picka, 1978: 100
asiatica Solier, 1848: 310
bipunctata Allard, 1880: 320
brevis Fischer von Waldheim, 1844: 96
carinula Seidlitz, 1893: 306
convexa Fischer von Waldheim, 1844: 92
crassa Reiche & Saulcy, 1857: 245
damascena Fischer von Waldheim, 1844: 101
dorsata Fischer von Waldheim, 1844: 98
fatidica J. Sturm, 1807: 205
laevicollis Rey, 1892: 41
milleri Seidlitz, 1893: 306
orbicollis Motschulsky, 1845a: 68
planicollis Motschulsky, 1845a: 65
plicaticollis Ménétriés, 1836a: 151
proxima Solier, 1848: 315
reflexicollis Solier, 1848: 314 [HN]
similis Latreille, 1804a: 279
striatopunctata Allard, 1880: 314
subquadrata Brullé, 1832: 204
lethifera pterotapha Fischer von Waldheim, 1832a: 201 **E**: AB AR GG ST **A**: IN TM TR
angulicollis Motschulsky, 1860e: 528 (*Agroblaps*)
anthracina Faldermann, 1837: 50
araxicola Seidlitz, 1893: 280
menetriesii Kraatz, 1881a: 56
pallasii Seidlitz, 1893: 306
pulvinata Ménétriés, 1849: 235
tarda Motschulsky, 1860e: 530 (*Agroblaps*)
lethifera subalpina Ménétriés, 1832: 200 **E**: ST
lindbergi Kaszab, 1959c: 52 **A**: TR
lineata Laporte, 1840: 200 **N**: AG
longicornis Kraatz, 1882b: 95 **A**: KA KI UZ
longula Reiche & Saulcy, 1857: 234 **A**: IS TR
lucidula G. S. Medvedev, 1998b: 204 **A**: SCH
luctuosa Ménétriés, 1832: 201 **E**: AB AR GG **A**: TR
araxicola Seidlitz, 1893: 306
pudica Ballion, 1888b: 704
lugens Seidlitz, 1893: 304 **E**: PT
lusitanica ceballosi Koch, 1944b: 245 **E**: SP (Madrid)
lusitanica espanoli Koch, 1944b: 246 **E**: SP (Baleares)
lusitanica lusitanica Herbst, 1799: 197 **E**: FR PT SP
kordofana Baudi di Selve, 1875c: 92
mequignoni Koch, 1944b: 246
producta Laporte, 1840: 200
magica Erichson, 1841: 183 **N**: AG
marginicollis Fairmaire, 1891e: cxxxi (*Blapidurus*) **A**: KA
maroccana Seidlitz, 1893: 257 **N**: MO
martensi Kaszab, 1978b: 227 **A**: KA
medusa Reitter, 1900e: 161 **A**: GAN MG NMO
medvedevi Bogatchev, 1961a: 107 **A**: TD TM

megalatlantica Koch, 1945: 409 **N**: MO (Haut Atlas)
menetrisiana Bogatchev, 1939: 152 **E**: AB
miliaria Fischer von Waldheim, 1844: 103 **A**: ES (Tuva) GAN MG NIN NMO SHX XIN
moerens Allard, 1880: 319 **A**: HP SCH UP XIZ
mortisaga Linnaeus, 1758: 418 (*Tenebrio*) **E**: AB AR AU BE BU BY CR CT CZ DE EN FI GB GE GG GR HU IC LA LS LT NL NR NT PL RO SK SL SP ST SV SZ UK **A**: IN IQ TM TR **NAR**i
acuminata DeGeer, 1775: 31 (*Tenebrion*)
brevicornis Seidlitz, 1893: 307
canaliculata Faldermann, 1837: 48
carbo Fischer von Waldheim, 1844: 87
confusa Fischer von Waldheim, 1844: 81
dahlii Solier, 1848: 317
elongata Ménétriés, 1832: 199
encifer Motschulsky, 1845a: 67
jaegeri Hummel, 1827: 40
koenigii Seidlitz, 1893: 307
reflexicollis Fischer von Waldheim, 1844: 82
motschulskiana Bogatchev, 1947e: 515 [RN] **A**: AF TM
gigantea Motschulsky, 1845a: 65 [HN]
mucronata Latreille, 1804a: 278 **E**: AU BE BU CR CZ DEi FIi FR GE GB GR IR IT LS MA NL NRi PL RO SK SP SVi SZ UK **N**: AG **A**: TR **NAR**i
chevrolatii Solier, 1848: 298
dilatata Laporte, 1840: 200
foveicollis Allard, 1880: 320
obtusa J. Sturm, 1807: 206 [HN]
reflexa Solier, 1848: 299
stricticollis A. Villa & J. B. Villa, 1835: 48
striolata Küster, 1846c: 46
sublineata Brullé, 1832: 204
subovata Solier, 1848: 299
murati Peyerimhoff, 1943: 23 **N**: MO **AFR**
murgabensis Bogatchev, 1961a: 109 **A**: TD TM
mutata Gemminger, 1870a: 122 [RN] **A**: IQ IS LE TR SY
convexa Reiche & Saulcy, 1857: 243 [HN]
nadaii G. S. Medvedev, 2004a: 182 **A**: UP
nanshanica Semenov & Bogatchev, 1936: 556 **A**: GAN NMO QIN
nefzauensis nefzauensis Seidlitz, 1893: 257 **N**: TU
nefzauensis vespertina Koch, 1937: 456 **N**: LB
neoscabiosa Nabozhenko, **nom. nov.** [see New Acts] **A**: IN
scabiosa Faust, 1875: 229 [HN]
nitens barbara Solier, 1848: 343 **N**: MO
nitens brachyura Küster, 1848a: 66 **E**: SP
abbreviata Solier, 1848: 339
nitens cyrenaia Seidlitz, 1893: 264 **N**: LB
nitens laportei Ardoin, 1973: 292 [RN] **N**: EG LB **A**: IS (Negev) SI
nitens mercatii Canzoneri, 1969: 44 **E**: IT (Sardegna)
nitens nitens Laporte, 1840: 200 **N**: AG TU
stygia Erichson, 1841: 182
subteres Solier, 1848: 341
tunisia Seidlitz, 1893: 264
vicina Solier, 1848: 342 [HN]
nitens praedeserta Koch, 1944b: 249 **N**: AG TU
nitens requieni Solier, 1848: 340 **N**: AG MO TU
nitens substriata Solier, 1848: 345 **N**: LB
nitida Fischer von Waldheim, 1844: 96 **E**: "Caucasus" **A**: KZ WS
amoena Fischer von Waldheim, 1844: 99
nuristanica Kaszab, 1959e: 250 [= 1960a: 75] **A**: AF

oberthuerii Seidlitz, 1893: 299 **A**: XIZ
oblonga Kraatz, 1883a: 349 **A**: GAN TD UZ WS XIN
gracilicollis Kraatz, 1885: 292
major Kraatz, 1883a: 350
ocreata Allard, 1880: 314 **E**: GR (Kríti) **A**: TR
oertzenii Seidlitz, 1893: 309 **E**: GR **A**: TR
oglobini Semenov & Bogatchev, 1936: 558 **A**: UZ (Bukhara)
ominosa Ménétriés, 1832: 198 **E**: AB AR **A**: AF TR
armeniaca Faldermann, 1837: 42
opaca Reitter, 1889k: 691 **A**: GAN NIN NMO XIN
orientalis Solier, 1848: 327 **A**: PA **ORR**
ovipennis Seidlitz, 1893: 254 **N**: MO
paludani Kaszab, 1959e: 249 [= 1960a: 73] **A**: AF
parvicollis Zubkov, 1829: 160 **E**: ST **A**: KZ
affinis Seidlitz, 1893: 307
krynickii Krynicki, 1829: 195
scabra Fischer von Waldheim, 1842: 15
pauliani Koch, 1945: 408 **N**: MO
pavlovskii Bogatchev, 1959: 64 **A**: TD
perlonga Bates, 1879: 472 **A**: KA
prolongata Fairmaire, 1887g: clxxix (*Leptomorpha*)
persica Seidlitz, 1893: 305 **A**: IN
persicola Seidlitz, 1893: 306 **A**: IN
petiti Pierre, 1964a: 408 **A**: LE
peyerimhoffi Koch, 1945: 410 **N**: MO (Moyen Atlas)
pharao Seidlitz, 1893: 266 **N**: EG **A**: SI
pilosa Ren & Wang, 2001: 19 **A**: XIZ
pinguis Allard, 1880: 309 **N**: MO
plana Solier, 1848: 338 **E**: FR SP SVi YU **N**: AG EG TU **A**: TR
planicollis Laporte, 1840: 200 **N**: EG
platythorax Gemminger, 1870a: 121 [RN] **A**: IN
laticollis L. Redtenbacher, 1850: 49
polychresta Forskål, 1775: 79 (*Tenebrio*) **N**: EG LB (Cyrenaica) **A**: IS SI **AUR**i
kuesteri Ballion, 1888a: 276
maura Solier, 1848: 347
minor Solier, 1848: 347
sulcata Fabricius, 1775: 254 [HN]
potanini Reitter, 1889k: 690 **A**: GAN NIN NMO QIN XIZ
povolnyi Kaszab, 1970a: 113 **A**: AF
prodigiosa Erichson, 1841: 182 **N**: AG
multicosta Solier, 1848: 326
propheta fiorii Español, 1967b: 179 **N**: LB (Tripolitania)
propheta propheta Reiche, 1861b: 89 **N**: AG LB MO TU
algirica Ballion, 1888b: 695
pruinosa Eversmann, 1836: 53 **E**: ST **A**: KZ TM
fischeri Fischer von Waldheim, 1844: 94
longipes Zubkov, 1833: 331
rorulenta Motschulsky, 1845a: 65
pseudocaudata Kaszab, 1960a: 69 **A**: AF
pterosticha Fischer von Waldheim, 1844: 93 **A**: KI KZ MG XIN
akinina Allard, 1882a: 135 (*Agroblaps*)
multistriata Ballion, 1878: 302
quadricollis Ballion, 1878: 304
pubescens Allard, 1880: 307 **N**: TU "Bou-Saada"
puella Allard, 1880: 317 **E**: AB AR **A**: TR
punctostriata Solier, 1848: 329 **A**: IN

putrida Motschulsky, 1845a: 68 **A**: ES KZ WS
quedenfeldtii Seidlitz, 1893: 254 **N**: MO
ratalaria Ren & Wang, 2001: 22 **A**: QIN XIZ
reflexa Gebler, 1832: 55 **A**: ES HEB MG NMO XIN
vermiculata Allard, 1882a: 93
reichardti Semenov & Bogatchev, 1936: 558 **A**: KI TD UZ (Bukhara)
reitteri Allard, 1885b: 180 (*Uroblaps*) **E**: AB
rhynchoptera Fairmaire, 1886c: 344 **A**: YUN **ORR**
inermis Fairmaire, 1887b: 127 (*Coelocnemodes*)
rimskii Semenov & Bogatchev, 1936: 560 **A**: QIN
robusta Motschulsky, 1845a: 68 **E**: ST
rotundata Solier, 1848: 325 **A**: IS SY TR
sodalis Reiche & Saulcy, 1857: 249
rougemonti Kaszab, 1986: 148 **A**: YE
rugosa Gebler, 1825: 48 **A**: ES GAN HEB JIL LIA MG NIN NMO SHA
scabripennis Faldermann, 1835b: 405
variolaris Gemminger, 1870a: 122 [RN]
variolosa Fischer von Waldheim, 1844: 104 [HN]
variolota Gemminger, 1870a: 122 [RN]
rugulipennis Fairmaire, 1891d: xcvii **A**: KA
ruhmeni Seidlitz, 1893: 260 **N**: LB
rybalovi Tschernyshev & Mordkovitsh, 2002: 385 **A**: WS
scabiosa Baudi di Selve, 1874a: 111 **A**: UZ (Tashkent)
scabriuscula montana Motschulsky, 1839: 62 **E**: GG ST (Dagestan)
scabriuscula scabriuscula Ménétriés, 1832: 200 **E**: AB
schweinfurthii Seidlitz, 1893: 263 **N**: EG
scutellata Fischer von Waldheim, 1844: 72 **A**: AF KZ TM
semenovi Bogatchev, 1939a: 127 **A**: KI (Arslanbob, Ak-Terek)
semenoviana Bogatchev, 1941b: 109 **A**: YE
seriata Fischer von Waldheim, 1820: pl. 16 **A**: IN TM UZ
acuta Seidlitz, 1893: 307
lajoyei Allard, 1883: 26
seriatimpunctata Fischer von Waldheim, 1844: 86
striola Motschulsky, 1860e: 528
shach Kaszab, 1970a: 121 **A**: AF
simplex Kaszab, 1960a: 72 **A**: AF
sinuatocollis sinuatocollis Solier, 1848: 314 **E**: ST
sinuatocollis suecica J. Ferrer & Picka, 1990: 30 **E**: SV
sonamarga Skopin, 1978: 228 **A**: KA
spinosa Allard, 1880: 304 **A**: IQ
splichali Gebien, 1913: 1 **A**: CY TR
srinagaricus Kaszab, 1975a: 17 **A**: KA
stoetzneri Schuster, 1923b: 158 **A**: QIN SCH XIZ
stoliczkana Bates, 1879: 471 **A**: TD
strandi Semenov, 1936: 557 **A**: GAN NMO
subcordata Seidlitz, 1893: 306 **A**: KZ
sulcatipennis Schuster, 1920a: 200 **A**: XIN
sulcifera Seidlitz, 1893: 264 **N**: LB
superstitiosa icosiensis Peyerimhoff, 1925: 14 **N**: AG
superstitiosa superstitiosa Erichson, 1841: 183 **N**: AG TU
taeniolata Ménétriés, 1832: 198 **E**: AB AR GR TR **N**: EG **A**: CY IN IQ LE SY TM TR
hians Fischer von Waldheim, 1844: 78
taiyuanica Ren & Wang, 2001: 17 **A**: SHX
tatsienlua Schuster, 1923a: 31 **A**: SCH XIZ
tenuicauda Seidlitz, 1893: 307 **A**: KZ

tenuicollis Solier, 1848: 301 **E**: GR **A**: IS LE
tenuepunctata Motschulsky, 1860e: 528 (*Agroblaps*) **A**: KZ
  vicina Ménétriés, 1849: 235 [HN]
theodoridis Pierre, 1964a: 408 **A**: IS
thibetana Blair, 1922c: 559 **A**: XIZ
tianshanica Semenov & Bogatchev, 1936: 559 **A**: XIN (Dzhungaria)
tibialis Reiche & Saulcy, 1857: 241 **E**: BU GR ST TR UK **A**: IS TR
  strigicollis Baudi di Selve, 1875b: 20
tibiella Seidlitz, 1893: 306 **A**: TR
tingitana antennalis Allard, 1880: 305 **N**: MO
tingitana tingitana Allard, 1880: 305 **N**: MO
  desertica Escalera, 1913: 47
titanus Ménétriés, 1849: 234 **E**: AB **A**: IN TM
transversalis Fischer von Waldheim, 1844: 105 **A**: KZ XIN
trapezoidalis Kaszab, 1970a: 119 **A**: AF
tridentata Waterhouse, 1889: 125 **A**: AF
tripolitanica Karsch, 1881: 48 **N**: LB
tristiciae Bogatchev, 1949a: 39 **A**: KA
tschitscherini Semenov & Bogatchev, 1936: 564 **A**: SCH
tsharynensis balchashensis Skopin, 1966: 338 **A**: KZ (Almaatinskaya Oblast)
tsharynensis tsharynensis Skopin, 1961a: 190 **A**: KZ (Almaatinskaya Oblast)
turcomanorum Seidlitz, 1893: 305 **A**: UZ
turkestanica Seidlitz, 1893: 309 **A**: UZ
tuxeni Gridelli, 1954: 262 [= 1955: 57] **A**: AF
umbilicata Seidlitz, 1893: 310 **A**: GAN QIN
urophora Fairmaire, 1891d: xcv **A**: KA
varicosa Seidlitz, 1893: 308 **A**: HEB
variolosa Faldermann, 1835b: 404 **A**: BEI ES (Tuva) GAN MG NIN NMO TM WS (Altay)
  tschiliana Wilke, 1921: 256 [RN]
verrucosa Adams, 1817: 305 **E**: GG
vialattei Peyerimhoff, 1920a: 328 **N**: AG (Hoggar) LB MO
virgo Seidlitz, 1893: 307 **A**: KI XIN
  sagitata Seidlitz, 1893: 307
waltlii Seidlitz, 1893: 304 **E**: PT SP
waschana Schuster, 1923a: 31 **A**: SCH SHA XIZ
wiedemannii Solier, 1848: 350 **N**: EG LB **A**: IS SA SI YE
wolinskii Bodemeyer, 1927: 24 **A**: IN
zarudniana chorassanica Semenov & Bogatchev, 1936: 555 **A**: IN (Chorassan)
zarudniana persis Semenov & Bogatchev, 1936: 555 **A**: IN (Kuhistan)
zarudniana zarudniana Semenov & Bogatchev, 1936: 554 **A**: IN
zugmayeri Schuster, 1935c: 239 **A**: IN (Baluchestan) PA (Balochistan)
subgenus *Dineria* Motschulsky, 1860e: 530 type species *Blaps confusa* Ménétriés, 1832 (= *Blaps halophila* Fischer von Waldheim, 1820)
  Laraliprosodes Bogatchev, 1947e: 513 type species *Prosodes lar* Bogatchev, 1947.
halophila Fischer von Waldheim, 1820: pl. 16 **E**: AB AU BU CR CT CZ GG HU MC MO PL RO SK SL ST UK YU **A**: KZ WS
  confusa Ménétriés, 1832: 201
  confluens Fischer von Waldheim, 1844: 84
  convexicollis Motschulsky, 1845a: 68
  coriacea Fischer von Waldheim, 1842: 15
  intrusa Fischer von Waldheim, 1844: 95
  longicollis Fischer von Waldheim, 1842: 83
  reflexicollis Fischer von Waldheim, 1844: 82
  solieri Allard, 1880: 317
  songorica Fischer von Waldheim, 1844: 97
lar Bogatchev, 1947e: 513 (*Prosodes*) **A**: UZ (Tashkentskaya oblast)

subgenus *Prosoblapsia* Skopin & Kaszab, 1978: 208 type species *Blaps allardiana* Reitter, 1889
allardiana alaiensis Seidlitz, 1893: 309 **A**: UZ (North Alaisky Ridge)
sunpantinga Schuster, 1923a: 30
allardiana allardiana Reitter, 1889k: 691 **A**: GAN QIN SCH XIZ
znoikoi Semenov & Bogatchev, 1936: 563
apicecostata Blair, 1922c: 559 **A**: BT NP QIN SD XIZ
breiti Reitter, 1913d: 658 **A**: HP XIZ
trapezicollis Gebien, 1913: 3
brevipes Seidlitz, 1893: 310 **A**: XIZ "Himalaya"
gentilis gentilis Fairmaire, 1887b: 127 **A**: BT GAN SCH YUN
gentilis punctithorax Kaszab, 1977b: 253 **A**: NP
gentilis semistriatomorpha Skopin & Kaszab, 1978: 211 **A**: NP
gentilis transversithorax Skopin & Kaszab, 1978: 211 **A**: NP
helleri Schuster, 1923b: 158 **A**: SCH QIN
himalaica Blair, 1923b: 283 **A**: NP SD XIZ
latericosta Reitter, 1889k: 688 **A**: GAN XIN
maeander Kraatz, 1885: 291 **A**: KI UZ (Tian Shan)
rudesculpta Semenov & Bogatchev, 1936: 560
socia socia Seidlitz, 1893: 309 **A**: NP SD
gentiloides Kaszab, 1977b: 254
radula Kaszab, 1977b: 254
semistriata Kaszab, 1977b: 255
tentyroides Seidlitz, 1893: 310
tenuicornis Gebler, 1847: 476 **A**: ES WS
altaica Kelejnikova, 1970: 27
depressa Gebler, 1829a: 121 [HN]
transversimsulcata meanderoides Skopin, 1977: 151 **A**: KI KZ
transversimsulcata transversimsulcata Ballion, 1878: 301 **A**: XIN

genus *Caenoblaps* König, 1906: 24 type species *Caenoblaps difformis* König, 1906
baeckmanni Schuster, 1928c: 125 **E**: ST (Dagestan)
difformis König, 1906: 24 **A**: TR (Kars)
kulzeri Schuster, 1928c: 123 **A**: TR (Van)
kulzeriana Pierre, 1964c: 1049 [RN] **A**: IN
kulzeri Pierre, 1964c: 1049 (*Blaps*) [HN]
nitida Schuster, 1920a: 199 **A**: IN TR

genus *Coelocnemodes* Bates, 1879: 474 [= 1886: 64] type species *Coelocnemodes stoliczkanus* Bates, 1879
aspericollis Fairmaire, 1886c: 341 **A**: YUN
stoliczkanus Bates, 1879: 475 **A**: PA

genus *Dila* Fischer von Waldheim, 1844: 111 type species *Blaps laevicollis* Gebler, 1841
afghanica Kaszab, 1960a: 45 **A**: AF
alaica Semenov & Bogatchev, 1940: 207 **A**: KI (Alai Range)
angustata Baudi di Selve, 1875c: 102 (*Blaps*) **A**: IN
laticollis Baudi di Selve, 1875c: 103 (*Blaps*)
mniszechi Allard, 1880: 277
armata Blair, 1913: 56 (*Blaps*) **A**: PA
bomiana Ren & Li, 2001: 311 **A**: XIZ
caudata Kaszab, 1970a: 86 **A**: AF
daghestanica G. S. Medvedev & Abdurachmanov, 1984: 291 **E**: ST (Dagestan)
kuntzeni Schuster, 1914c: 182 **A**: IN
laevicollis Gebler, 1841a: 373 (*Blaps*) **A**: AF KI KZ TD TM UZ
leptoscelis Reitter, 1900a: 239 **A**: IN
lindbergi Kaszab, 1970a: 84 **A**: AF

persiana Reitter, 1900i: 295 **A**: IN
sedecimstriata Reitter, 1900i: 295 **A**: KI UZ
seriata Reitter, 1894d: 242 **A**: TD UZ
 bucharica Reitter, 1900i: 296
transversecordata Reitter, 1899d: 203 **A**: TR

genus *Dilablaps* Bogatchev, 1976: 98 type species *Dilablaps paradoxa* Bogatchev, 1976
paradoxa Bogatchev, 1976: 99 **A**: TD

genus *Hoplitoblaps* Fairmaire, 1888a: 26 type species *Hoplitoblaps fallaciosa* Fairmaire, 1888
fallaciosa Fairmaire, 1888a: 27 **A**: YUN

genus *Nalepa* Reitter, 1887d: 366 type species *Blaps cylindracea* Reitter, 1887
cylindracea Reitter, 1887d: 366 (*Blaps*) **A**: "Central China"

genus *Neoblaps* Ren & Li, 2001: 310 type species *Neoblaps huizensis* Ren & Li, 2001
huizensis Ren & Li, 2001: 311 **A**: YUN

genus *Protoblaps* G. S. Medvedev, 1998b: 200 type species *Protoblaps kashkarovi* G. S. Medvedev, 1998
kashkarovi G. S. Medvedev, 1998b: 202 **A**: SCH

genus *Thaumatoblaps* Kaszab & G. S. Medvedev, 1984: 79 type species *Thaumatoblaps marikovskiji* Kaszab & G. S. Medvedev, 1984
marikovskiji Kaszab & G. S. Medvedev, 1984: 80 **A**: TM
zhengi Ren & Luo, 1995: 28 **A**: XIN

subtribe Blaptina, species incertae sedis
Blaps indica Hope, 1831: 31 **A**: NP
Blaps aurichalcea Adams, 1817: 307 **E**: GG

subtribe Gnaptorina G. S. Medvedev, 2001

genus *Gnaptor* Brullé, 1832: 203 type species *Tenebrio spinimanus* Pallas, 1781
 Petrobius Brullé, 1832: 202 [HN] type species *Tenebrio spinimanus* Pallas, 1781
prolixus Fairmaire, 1866b: 264 **A**: TR
spinimanus Pallas, 1781: 55 (*Tenebrio*) **E**: AL AU BH BU GR HU MC RO SK SL ST TR UK YU
 boryi Laporte, 1840: 208 (*Petrobius*)
 glabratus Fabricius, 1792a: 100 (*Pimelia*)
 graecus Küster, 1848a: 64 (*Blaps*)
 laevigatus Fabricius, 1792a: 100 (*Pimelia*)

subtribe Gnaptorinina G. S. Medvedev, 2001

genus *Agnaptoria* Reitter, 1887d: 372 type species *Agnaptoria rubripes* Reitter, 1887
abdita G. S. Medvedev, 2002: 861 **A**: GAN
amdoensis G. S. Medvedev, 2006a: 197 **A**: GAN
anthracina anthracina G. S. Medvedev, 2002: 863 **A**: GAN
anthracina labrana G. S. Medvedev, 2002: 865 **A**: GAN
belousovi G. S. Medvedev, 2002: 847 **A**: SCH
crassa G. S. Medvedev, 2006a: 180 **A**: SCH
danbaensis G. S. Medvedev, 2006a: 195 **A**: SCH
dzaensis G. S. Medvedev, 2002: 853 **A**: SCH
gansuensis G. S. Medvedev, 2002: 849 **A**: GAN
grummi G. S. Medvedev, 2002: 859 **A**: QIN
kabaki G. S. Medvedev, 2002: 845 **A**: SCH
kamensis G. S. Medvedev, 2002: 853 **A**: QIN

kozlovi G. S. Medvedev, 2002: 855 **A**: QIN
kunda G. S. Medvedev, 2006a: 186 **A**: GAN
lauta G. S. Medvedev, 2006a: 184 **A**: SCH
minuta G. S. Medvedev, 2006a: 178 **A**: SCH
markana Shi & Ren, 2005: 279 **A**: XIZ
miroshnikovi G. S. Medvedev, 1998b: 183 **A**: SCH
proba G. S. Medvedev, 2006a: 190 **A**: GAN
rubripes Reitter, 1887d: 372 **A**: QIN SCH
seidlitzi Reitter, 1893d: 315 **A**: XIN
shamaevi shamaevi G. S. Medvedev, 1999a: 401 **A**: GAN
shamaevi sitshuana G. S. Medvedev, 1999a: 403 **A**: SCH
shangmengi G. S. Medvedev, 2006a: 192 **A**: SCH
zamotailovi G. S. Medvedev, 1998b: 180 **A**: SCH

genus *Asidoblaps* Fairmaire, 1886c: 342 type species *Asidoblaps davidis* Fairmaire, 1886
altimontana G. S. Medvedev, 2002: 73 **A**: SCH
angustitarsis G. S. Medvedev, 2005c: 545 **A**: SCH
annae G. S. Medvedev, 2002: 85 **A**: SCH
belousovi G. S. Medvedev, 2002: 68 **A**: SCH
berezowskii G. S. Medvedev, 1998b: 196 **A**: SCH
bianera G. S. Medvedev, 2005c: 543 **A**: SCH
bifida G. S. Medvedev, 2005c: 565 **A**: SCH
confinis G. S. Medvedev, 2005c: 534 **A**: SCH
danba G. S. Medvedev, 2005c: 531 **A**: SCH
davidiani G. S. Medvedev, 2002: 70 **A**: SCH
davidis Fairmaire, 1886c: 342 **A**: YUN
expressa G. S. Medvedev, 2005c: 552 **A**: GAN
friedrichi G. S. Medvedev & Merkl, 2001: 625 **A**: SCH
galinae G. S. Medvedev, 2002: 80 **A**: SCH
gansuensis G. S. Medvedev, 2002: 82 **A**: GAN
glyptoptera Fairmaire, 1886c: 343 **A**: SCH YUN
gorgneri G. S. Medvedev & Merkl, 2001: 623 **A**: SCH
gracilis G. S. Medvedev, 2002: 87 **A**: SCH
ignorata G. S. Medvedev, 2005c: 541 **A**: SCH
igori G. S. Medvedev, 2002: 91 **A**: SCH
inermis G. S. Medvedev, 1998b: 198 **A**: SCH
inopinata G. S. Medvedev, 2005c: 558 **A**: GAN
kabaki G. S. Medvedev, 2002: 94 **A**: SCH
longicornis G. S. Medvedev, 2001: 83 **A**: SCH
miroshnikovi G. S. Medvedev, 1998b: 191 **A**: SCH
modica G. S. Medvedev, 2002: 77 **A**: SCH
multa G. S. Medvedev, 2005c: 548 **A**: GAN
munda G. S. Medvedev, 2005c: 537 **A**: GAN
polita Schuster, 1935c: 162 **A**: SCH
potanini G. S. Medvedev, 1998c: 583 **A**: XIZ
rilonga G. S. Medvedev, 2005c: 539 **A**: SCH
silvicola G. S. Medvedev, 2005c: 560 **A**: SCH
sinensis G. S. Medvedev, 1998b: 194 **A**: SCH
siniaevi G. S. Medvedev, 2002: 65 **A**: SCH
smetanai G. S. Medvedev, 2001: 81 **A**: TAI
subopaca Schuster, 1935c: 162 **A**: GAN SHA SCH
yaanensis G. S. Medvedev, 2005c: 563 **A**: SCH
yuana G. S. Medvedev, 2002: 75 **A**: SCH
zamotailovi G. S. Medvedev, 1998b: 189 **A**: YUN

genus *Blaptogonia* G. S. Medvedev, 1998b: 186 type species *Tagonoides costulata* Fairmaire, 1901
costulata Fairmaire, 1901b: 267 (*Tagonoides*) **A**: SD
subcarinata Blair, 1927: 243 (*Blaps*) **A**: XIZ
tshernjachovskii G. S. Medvedev, 2004a: 177 **A**: NP
yini Ren, X.-P. Wang & Y.-Z. Yu 2000: 27 **A**: XIZ
zurstrasseni Kaszab, 1977b: 246 (*Blaptyscelis*) **A**: NP

genus *Colasia* Koch, 1965b: 131 type species *Colasia akisoides* Koch, 1965
akisoides Koch, 1965b: 131 **A**: SCH YUN

genus *Gnaptorina* Reitter, 1887d: 364 type species *Gnaptorina felicitana* Reitter, 1887
brucei Blair, 1923b: 282 **A**: NP SD XIZ
cordicollis G. S. Medvedev, 1998c: 555 **A**: QIN XIZ
cylindricollis Reitter, 1889k: 693 **A**: GAN SCH XIZ
fairmairei Koch, 1965b: 130 **A**: GAN
felicitana Reitter, 1887d: 365 **A**: NMO XIZ
kashkarovi G. S. Medvedev, 1998c: 564 **A**: QIN XIZ
kozlovi G. S. Medvedev, 1998c: 560 **A**: QIN XIZ
lii Z. Li & Ren, 2004: 256 **A**: NIN
longicornis Z. Li & Ren, 2004: 261 **A**: XIZ
media G. S. Medvedev, 1998c: 558 **A**: QIN XIZ
miroshnikovi G. S. Medvedev, 1998c: 562 **A**: QIN XIZ
montana G. S. Medvedev, 1998c: 557 **A**: QIN XIZ
quadratus Z. Li & Ren, 2004: 260 **A**: XIZ
platytarsia Z. Li & Ren, 2004: 259 **A**: XIZ
potamini Reitter, 1889k: 694 **A**: GAN SCH
proxima Reitter, 1889k: 694 **A**: GAN SCH
reitteri Koch, 1965b: 129 **A**: SCH
rufipes Z. Li & Ren, 2004: 257 **A**: QIN
rugosipennis G. S. Medvedev, 1998c: 565 **A**: QIN XIZ
sikkimensis Kaszab, 1965a: 110 **A**: SD
tenuicorpra Z. Li & Ren, 2004: 258 **A**: QIN
tishkovi G. S. Medvedev, 1998c: 567 **A**: XIZ

genus *Itagonia* Reitter, 1887d: 362 type species *Itagonia gnaptorinoides* Reitter, 1887
bisetosa G. S. Medvedev, 1998c: 568 **A**: SCH
deplanata Reinig, 1931: 866 **A**: TD UZ (Bukhara)
elegans G. S. Medvedev, 1998c: 574 **A**: SCH
ganglbaueri Schuster, 1914a: 58 **A**: HEB NMO SHA
gnaptorinoides Reitter, 1887d: 362 **A**: TD UZ
kabaki G. S. Medvedev, 1998c: 578 **A**: TD
lopatini lopatini G. S. Medvedev, 1998c: 576 **A**: TD
lopatini turkestanica G. S. Medvedev, 1998c: 578 **A**: TD
mera G. S. Medvedev, 1998c: 572 **A**: XIZ
provostii Fairmaire, 1889c: cci (*Platyscelis*) **A**: BEI HEB NIN NMO
    ganglbaueri Schuster, 1914a: 58
semenowi Reitter, 1889k: 694 **A**: QIN SCH
shamaevi G. S. Medvedev, 2004a: 168 **A**: SCH
szetschwana Schuster, 1923b: 162 (*Asidoblaps*) **A**: SCH
trisetosa G. S. Medvedev, 1998c: 570 **A**: SCH

genus *Montagona* G. S. Medvedev, 1998b: 176 type species *Tagonoides asperula* Fairmaire, 1897
asperula Fairmaire, 1896b: 95 (*Tagonoides*) **A**: HP XIN XIZ

pustulosa Fairmaire, 1886c: 344 (*Tagonoides*) **A**: YUN
sculpturata Gebien, 1913: 2 (*Blaps*) **A**: KA

genus *Nepalindia* G. S. Medvedev, 1998b: 187 type species *Taganoides* [sic] *alpina* Kaszab, 1965
alpina Kaszab, 1965a: 109 (*Tagenoides*) **A**: NP SD
dreuxi Pierre, 1961: 213 (*Blaptyscellis*) **A**: NP
nepalica Kaszab, 1973a: 30 (*Tagenoides*) **A**: NP
volkovitshi infida G. S. Medvedev, 2004a: 173 **A**: NP
volkovitshi volkovitshi G. S. Medvedev, 2004a: 170 **A**: NP

genus *Pseudognaptorina* Kaszab, 1977b: 250 type species *Pseudognaptorina nepalica* Kaszab, 1977
exsertogena Shi, Ren & Merkl, 2005: 164 **A**: XIZ
nepalica Kaszab, 1977b: 251 **A**: NP
obtusa Shi, Ren & Merkl, 2005: 166 **A**: XIZ

genus *Sintagona* G. S. Medvedev, 1998c: 585 type species *Sintagona miranda* G. S. Medvedev, 1998
miranda G. S. Medvedev, 1998c: 586 **A**: QIN

genus *Tagonoides* Fairmaire, 1886c: 343 type species *Tagonoides delavayi* Fairmaire, 1886
ampliata Fairmaire, 1887b: 126 **A**: YUN
delavayi Fairmaire, 1886c: 343 **A**: YUN
kabaki G. S. Medvedev, 2004a: 164 **A**: SCH
skopini G. S. Medvedev & Merkl, 2001: 620 **A**: YUN
volkovitshi G. S. Medvedev, 2004a: 166 **A**: YUN
zamotailovi G. S. Medvedev, 1998b: 173 **A**: YUN

subtribe Prosodina Skopin, 1960

genus *Prosodes* Eschscholtz, 1829: 9 type species *Blaps attenuata* Fischer von Waldheim, 1820 (= *Blaps obtusa* Fabricius, 1798)
subgenus *Altiprosodes* G. S. Medvedev, 1997b: 580 type species *Prosodes kuhistanica* G. S. Medvedev, 1996
janushevi G. S. Medvedev, 1996: 616 **A**: TD
kuhistanica G. S. Medvedev, 1996: 624 **A**: TD
merkli G. S. Medvedev, 1996: 619 **A**: TD
oschanini Semenov, 1891a: 366 **A**: UZ (Bukhara)
pseudalaiensis Skopin, 1971: 496 **A**: UK (Samarkand)
vashgirda G. S. Medvedev, 1996: 618 **A**: TD
subgenus *Dilopersina* Reitter, 1909d: 117 type species *Prosodes jakowlewi* Semenov, 1894
chorassanica G. S. Medvedev, 1996: 603 **A**: IN (Khorasan)
cribrella cribrella Baudi di Selve, 1874a: 110 [= 1875c: 104] (*Blaps*) **A**: IN "Transcaspia"
  pustulata Faust, 1875: 237
cribrella consanguinea Allard, 1885b: 180 **A**: IN
cribrella vestita Allard, 1880: 289 **A**: IN TM (Kopet Dag)
  neopersis Semenov, 1909: 435 [RN]
emiri Sumakov, 1902: 39 **A**: TM
exilis G. S. Medvedev, 1996: 599 **A**: IN (Khorasan)
jakowlewi Semenov, 1894a: 192 **A**: IN UZ
  grouvellei Reitter, 1909d: 132
mithras Reitter, 1904d: 258 **A**: IN
parthenica G. S. Medvedev, 1996: 596 **A**: TM (Kopet Dag)
transversosulcata Allard, 1880: 292 **A**: "Inde orientale"
subgenus *Diprosodes* Reitter, 1909d: 115 type species *Prosodes biformis* Semenov, 1894
biformis Semenov, 1894a: 181 **A**: TD UZ
biformiopaca biformiopaca Reitter, 1909d: 127 **A**: UZ (Bukhara)
biformiopaca occidentalis G. S. Medvedev, 1997b: 571 **A**: UZ

biformiopaca saganiana G. S. Medvedev, 1997b: 574 **A**: UZ
elateroides Semenov, 1894a: 185 **A**: UZ (Bukhara)
callosa Semenov, 1894a: 188
gymnandra Semenov, 1894:a 189 **A**: TD UZ (Bukhara)
juliae Bogatchev, 1960c: 82
kunduzensis Kaszab, 1970a: 104 **A**: AF
lopatini Bogatchev, 1960c: 81 **A**: TD (Vakhshkiy Khrebet)
psammophila G. S. Medvedev, 1997b: 579 **A**: TD
sphodroides Semenov, 1894a: 190 **A**: UZ (Bukhara)
stackelbergi G. S. Medvedev, 1997b: 557 **A**: TD
subgenus *Euryprosodes* Reitter, 1909d: 122 type species *Prosodes areolata* Reitter, 1893
areolata Reitter, 1893c: 269 **A**: TD UZ (Bukhara)
plicata Reitter, 1909d: 166
brevipes Reitter, 1893c: 269 **A**: UZ (Bukhara)
steppensis Kaszab, 1960a: 64 **A**: AF
trifossa Reitter, 1909d: 165 **A**: TD
subgenus *Ferganoprosodes* G. S. Medvedev, 1997b: 595 type species *Prosodes angulicollis* Kraatz, 1883
angulicollis Kraatz, 1883a: 347 **A**: UZ (Fergana)
nitida Kraatz, 1885: 289
pleuralis Reitter, 1893c: 266 **A**: UZ
subgenus *Gebleria* Motschulsky, 1860e: 529 type species *Dila philacoides* Fischer von Waldheim, 1844
Aulonoscelis Reitter, 1896g: 173 type species *Platynoscelis hauseri* Reitter, 1895
Aulonoscelis Reitter, 1909d: 114 [HN] type species *Prosodes phylacoides* [sic] Fischer von Waldheim, 1844
Phylacoprosodes Schuster, 1934a: 75 [RN] type species *Prosodes philacoides* Fischer von Waldheim, 1844
philacoides Fischer von Waldheim, 1844: 118 (*Dila*) **A**: ES KI KZ XIN
brevis Gebler, 1844: 103 [= 1860: 22] (*Tagona*)
gnaptorina Reitter, 1909d: 123
hauseri Reitter, 1895d: 156 (*Platynoscelis*) [HN]
regeli Semenov, 1894a: 202 **A**: KZ XIN
subgenus *Hypoprosodes* Reitter, 1909d: 122 type species *Prosodes minuta* Kraatz, 1881
minuta Kraatz, 1881b: 62 **A**: KZ UZ
angulata Reitter, 1890h: 361
sulcipubens Reitter, 1909d: 165
subgenus *Indoprosodes* G. S. Medvedev, 2003: 690 type species *Prosodes boorpi* Kaszab, 1956
boorpi Kaszab, 1956d: 420 **A**: UP
subgenus *Iranosodes* G. S. Medvedev, 1996: 605 type species *Prosodes kaszabi* G. S. Medvedev & Kabakov, 1996
diversa Waterhouse, 1889: 126 **A**: AF TM
quadriimpressa Reitter, 1909d: 149
jakesi Kaszab, 1970a: 105 **A**: AF
kaszabi G. S. Medvedev & Kabakov, 1996: 607 **A**: AF
laticauda Reitter, 1896a: 38 **A**: IN TM (Kopet Dag)
subgenus *Lyprosodes* Reitter, 1909d: 116 type species *Prosodes quadricostata* Reitter, 1893
aciculata Reitter, 1893c: 274 **A**: TD UZ (Bukhara)
fallax Kraatz, 1886: 189 **A**: UZ (Margilan)
pavlovskii Bogatchev, 1960c: 80 **A**: TD
pseudaciculata Reitter, 1909d: 129 **A**: UZ (Bukhara)
quadricostata Reitter, 1893c: 278 **A**: UZ
subgenus *Megaprosodes* Reitter, 1909d: 118 type species *Prosodes striata* Reitter, 1893
alaiensis alaiensis Kraatz, 1885: 288 **A**: UZ
sparsa Reitter, 1909d: 135
alaiensis transmontana G. S. Medvedev, 1996: 612 **A**: TD
baratovi G. S. Medvedev, 1996: 613 **A**: TD
jardanus Reitter, 1909d: 135 **A**: UZ (Bukhara)
latitibia Kaszab, 1960a: 62 **A**: AF
praelonga Reitter, 1893c: 286 **A**: UZ (Bukhara)

pubistriata Reitter, 1893c: 307 **A**: UZ (Bukhara)
lateralis Reitter, 1893c: 276
striata Reitter, 1893c: 289 **A**: UZ (Bukhara)
obliterata Semenov, 1894a: 206
svetlanae G. S. Medvedev & Michailov, 1978: 61 **A**: TD
subgenus *Meropersina* Reitter, 1909d: 117 type species *Prosodes cordicollis* Allard, 1883
amica G. S. Medvedev, 2001: 86 **A**: IQ
banghaasi Reitter, 1893c: 293 **A**: TR
cordicollis Allard, 1883: 25 **A**: IN
fabiani G. S. Medvedev & Merkl, 2005: 172 **A**: IN (Zagros)
laevigata Baudi di Selve, 1874a: 109 [= 1875c: 101] (*Blaps*) **A**: IN
persica Faust, 1875: 236
vermiculosa Reitter, 1909d: 131 **A**: IN
vigi G. S. Medvedev & Merkl, 2005: 172 **A**: IN (Zagros)
subgenus *Mesoprosodes* G. S. Medvedev, 1995: 827 type species *Prosodes nuratensis* Semenov, 1894
fausti Semenov, 1894a: 217 **A**: UZ
foveicollis Semenov, 1894a: 213 **A**: TD
karatavica Skopin, 1966: 339 **A**: KZ (Karatay)
kyzylkumensis G. S. Medvedev, 1995: 828 **A**: UZ
nuratensis Semenov, 1894a: 214 **A**: UZ
turkestanica Semenov, 1894a: 211 **A**: TD UZ
undulata Kraatz, 1882d: 322 **A**: UZ (Samarkand)
subgenus *Oliprosodes* Reitter, 1909d: 118 type species *Prosodes trisulcata* Bates, 1879
kuluana G. S. Medvedev, 1999b: 857 **A**: HP
laevicostata G. S. Medvedev, 1999b: 853 **A**: AF
rufosulcata Fairmaire, 1891d: xciv **A**: KA
trisulcata Bates, 1879: 473 **A**: KA PA XIZ
vicina Bates, 1879: 474
subgenus *Peltarium* Fischer von Waldheim, 1844: 106 type species *Peltarium marginatum* Fischer von Waldheim, 1844 (=*Blaps baerii* Fischer von Waldheim, 1842)
baerii Fischer von Waldheim, 1842: 16 (*Blaps*) **A**: KZ
marginata Fischer von Waldheim, 1842: 28 (*Blaps*)
marginata Fischer von Waldheim, 1844: 108 (*Peltarium*)
karelini Gebler, 1841b: 593 (*Blaps*) **A**: KZ XIN
altaica Reitter, 1909d: 147
kataevi G. S. Medvedev, 1995: 831 **A**: KI (Alaiyskiy Khrebet)
staudingeri Kraatz, 1882d: 305 **A**: UZ (Samarkand)
subgenus *Planoprosodes* G. S. Medvedev, 2005a: 68 type species *Prosodes reitteri* Reitter, 1893
reitteri Reitter, 1893c: 285 **A**: KZ
asperipennis Reitter, 1893c: 311
subgenus *Prosodella* Reitter, 1909d: 120 type species *Prosodes bactriana* Semenov, 1894
Paraprosodes Reitter, 1909d: 119 type species *Prosodes triangulipes* Reitter, 1909
bactriana Semenov, 1894a: 199 **A**: AF TD UZ
bactrianiformis Reitter, 1901c: 77 **A**: UZ (Bukhara)
bogatshevi G. S. Medvedev, 1996: 631 **A**: TD
convergens Reitter, 1893c: 283 **A**: UZ (Bukhara)
divergens Reitter, 1893c: 283 **A**: UZ (Bukhara)
singularis Reitter, 1909d: 145 **A**: UZ (Bukhara)
triangulipes Reitter, 1893c: 282 **A**: "Turkestan" (Tien Shan)
subgenus *Prosodes* Eschscholtz, 1829: 9 type species *Blaps attenuata* Fischer von Waldheim, 1820 (= *Blaps obtusa* Fabricius, 1798)
Blaptoprosodes Reitter, 1909d: 120 type species *Prosodes mucronata* Reitter, 1893
Eupomeca Solier, 1848: 289 type species *Blaps cylindrica* Herbst, 1799 (= *Blaps obtusa* Fabricius, 1798)
Lioprosodes Reitter, 1909d: 121 type species *Prosodes dilaticollis* Motschulsky, 1860
Nyctipates Gebler, 1841a: 373 type species *Nyctipates rugulosa* Gebler, 1841
Platyprosodes Reitter, 1909d: 121 type species *Nyctipates rugulosa* Gebler, 1841

Prosodila Reitter, 1909d: 121 type species *Prosodes strigiventris* Reitter, 1893
Pseudoprosodes Reitter, 1909d: 120 type species *Prosodes transfuga* Reitter, 1893
angustipleuris angustipleuris Reitter, 1893c: 292 **A**: KZ UZ
angustipleuris boroldaica Skopin, 1966: 341 **A**: (Boroldai Tau)
angustipleuris ugamica Skopin, 1966: 341 **A**: UZ (Ugamsky Khrebet)
blapoides Reitter, 1893c: 292 **A**: KZ UZ (Margilan, Tashkent)
cribripennis Reitter, 1909d: 155
brevicauda brevicauda Reitter, 1909d: 156 **A**: UZ (Margilan, Tashkent)
brevicauda laevicauda Skopin, 1966: 341 **A**: UZ (Ugamsky Khrebet)
edmundi Semenov, 1894a: 220 **A**: XIN
deplanata Ballion, 1878: 314 **A**: KI
dilaticollis Motschulsky, 1859a: 302 **A**: KI XIN
lucida Ballion, 1878: 316
fulminans Reitter, 1893c: 303 **A**: KI
galinae G. S. Medvedev, 1995: 830 **A**: UZ (Nuratau)
gracilis Faust, 1875: 241 **A**: KZ UZ XIN
granulipleuris G. S. Medvedev, 1995: 822 **A**: UZ
hiekei G. S. Medvedev, 1995: 819 **A**: TD (Kuramin Mts)
kassansaica G. S. Medvedev, 1995: 821 **A**: KI
kumbeliana G. S. Medvedev, 1995: 826 **A**: UZ (Fergana)
longicornis Kraatz, 1883a: 349 **A**: KI
tenuicornis G. S. Medvedev, 1995: 825
ludmilae G. S. Medvedev, 1995: 816 **A**: TD (Kuramin Mts)
mogoltavica G. S. Medvedev, 1995: 813 **A**: TD (Mogoltay)
mucronata Reitter, 1893c: 291 **A**: UZ
composita Reitter, 1893c: 291
obtusa Fabricius, 1798: 46 (*Blaps*) **E**: ST UK **A**: KZ TM
attenuata Fischer von Waldheim, 1820: pl. 16 (*Blaps*)
cylindrica Herbst, 1799: 185 (*Blaps*)
cylindropastica Krynicki, 1832: 133 (*Blaps*)
pastica Solier, 1848: 290 (*Nyctipates*)
punctata Fischer von Waldheim, 1844: 109 (*Peltarium*)
pangazica G. S. Medvedev, 1995: 814 **A**: TD (Kuramin Mts)
parallelocollis Kraatz, 1882c: 108 **A**: UZ (Margilan)
rudesculpta Reitter, 1909d: 158
pecirkana Reitter, 1909d: 157 **A**: KI (Koksuiskiy Khrebet)
pekinensis Fairmaire, 1887f: 323 **A**: BEI HEB GAN NIN NMO QIN SHA SHX XIN
kreitneri J. Frivaldszky, 1889: 206
przewalskii Semenov, 1894a: 195 **A**: XIN
rugosicollis padshaatinus Skopin, 1971: 500 **A**: KI
rugosicollis rugosicollis Kraatz, 1885: 290 **A**: UZ (Namanganskaya Oblast)
rugosicollis sarytshelekis Skopin, 1971: 500 **A**: KI
rugulosa granulosa Ballion, 1878: 307 **A**: KZ (Almaatinskaya Oblast)
rugulosa degenerata Semenov, 1894a: 198 **A**: XIN
rugulosa hetschkoi Reitter, 1909d: 163 **A**: KI
rugulosa nitidula Motschulsky, 1859b: 309 [= 1859a: 300] **A**: KI XIN
costipennis Ballion, 1878: 312
dubia Ballion, 1878: 316
similis Ballion, 1878: 309
rugulosa rugulosa Gebler, 1841a: 373 (*Nyctipates*) **A**: KI KZ GAN UZ
bicostata Fischer von Waldheim, 1844: 108 (*Peltarium*)
costata Fischer von Waldheim, 1844: 120 (*Nyctipates*)
foveata Fischer von Waldheim, 1842: 15 (*Blaps*)
sobolevae G. S. Medvedev, 1995: 812 **A**: TD (Kuraminsky Khrebet)
strigiventris Reitter, 1893c: 277 **A**: KI TD UZ
sulcicauda Reitter, 1893c: 290 **A**: UZ

transfuga cariosa Skopin, 1961a: 193 **A**: KI KZ
transfuga transfuga Reitter, 1893c: 304 **A**: KI KZ UZ
sulcatissima Reitter, 1901c: 78
tshistjakovae G. S. Medvedev, 1995: 824 **A**: UZ (Chatkalskiy Khrebet)
valida Reitter, 1893c: 304 **A**: KI UZ (Margilan)
zarudnyi G. S. Medvedev, 1995: 818 **A**: UZ XIN
subgenus *Prosodestes* Reitter, 1909d: 114 type species *Prosodes grandicollis* Kraatz, 1883
bucharica Reitter, 1915f: 92 **A**: UZ (South West)
dolini Kh. Dadabaev, 1986: 80 **A**: TD (Khodzha-Maston)
grandicollis grandicollis Kraatz, 1883a: 348 **A**: KI (Osh)
grandicolli centralis G. S. Medvedev, 2005a: 68 **A**: TD (Shartuskiyv Rayon)
grandicolli interga G. S. Medvedev, 2005a: 69 **A**: TD (South West)
lindbergi Kaszab, 1970a: 95 **A**: AF
malyi malyi Obenberger, 1914: 139 **A**: TD (South West)
malyi tupalangi G. S. Medvedev, 2005a: 77 **A**: UZ (Tupalangdarya Valley)
subgenus *Prosodinia* Reitter, 1909d: 115 type species *Prosodes calcarata* Reitter, 1893
afghanica Schuster, 1936c: 197 **A**: AF
badakschanica Kaszab, 1960a: 51 **A**: AF
baghlana G. S. Medvedev, 2003a: 374 **A**: AF
boeckeleri Kaszab, 1974: 199 **A**: AF
brunnea Kaszab, 1960a: 50 **A**: AF
calcarata Reitter, 1893c: 309 **A**: IN TM
chazariensis G. S. Medvedev, 2003a: 386 **A**: AF
ferozensis G. S. Medvedev, 2003a: 378 **A**: AF
ghourbandensis Kaszab, 1970a: 99 **A**: AF
haarlovi Gridelli, 1954: 256 [= 1955: 54] **A**: AF
higginsi Kaszab, 1956d: 418 **A**: KA PA
kabakovi G. S. Medvedev, 1999b: 862 **A**: AF
kabulensis G. S. Medvedev, 1999b: 878 **A**: AF
klapperichi Kaszab, 1960a: 55 **A**: AF
kraatzi Reitter, 1893c: 309 **A**: AF IN
refleximargo Gridelli, 1954: 251 [= 1955: 50]
kunarensis G. S. Medvedev, 2003a: 366 **A**: AF
olegi G. S. Medvedev, 1999b: 874 **A**: AF
oruzgana G. S. Medvedev, 2003a: 394 **A**: AF
provincialis Kaszab, 1959e: 246 [= 1960a: 48] **A**: AF
quadraticollis Kaszab, 1960a: 47 **A**: AF
scheibei Schuster, 1936c: 199 **A**: AF
shahranensis Medvedev, 2003a: 382 **A**: AF
simplicithorax Gridelli, 1954: 255 [= 1955: 52] **A**: AF
speciosa Kaszab, 1970a: 102 **A**: AF
subbrunnea Kaszab, 1970a: 100 **A**: AF
tuxeni Kaszab, 1959e: 247 [= 1960a: 52] **A**: AF
undulatipennis Gridelli, 1954: 257 [= 1955: 55] **A**: AF
zoltani G. S. Medvedev, 1997b: 564 **A**: AF (Hindu Kush)
subgenus *Prosodopria* Reitter, 1909d: 116 type species *Blaps angustata* Zubkov, 1833
Nyctipates Solier, 1848: 285 [HN] type species *Nyctipates coriacea* Solier, 1848 (= *Blaps angustata* Zubkov, 1833)
angustata Zubkov, 1833: 329 (*Blaps*) **A**: KI
asperata Zubkov, 1833: 330 (*Blaps*)
carinata Fischer von Waldheim, 1844: 119 (*Nyctipates*)
coriacea Solier, 1848: 289 (*Nyctipates*)
mammillata Faldermann, 1836: 382 (*Blaps*)
subgenus *Prosodoscelis* Reitter, 1909d: 117 type species *Prosodes solskyi* Faust, 1875
dentimana Reitter, 1909d: 130 **A**: IN

irinae Skopin, 1964: 284 **A**: KZ (Mangistauskaya Oblast)
solskyi Faust, 1875: 239 **A**: TM (Kopet Dag)
subgenus *Prosodura* Reitter, 1909d: 118 type species *Prosodes semenowi* Reitter, 1893
peloroides Reitter, 1909d: 134 **A**: UZ (Bukhara)
prosternalis Reitter, 1893c: 284 **A**: UZ (Bukhara)
punctimana Reitter, 1893c: 284 **A**: UZ (Bukhara)
semenowi darvazica G. S. Medvedev, 1996: 611 **A**: TD
semenowi semenowi Reitter, 1893c: 273 **A**: TD UZ
semenowi sexualis Kaszab, 1960a: 61 **A**: AF
semenowi vachshiana G. S. Medvedev, 1996: 609 **A**: TD
subgenus *Tangiprosodes* G. S. Medvedev, 2005a: 68 type species *Prosodes pygmaea* Kraatz, 1882
kuhitangiana G. S. Medvedev, 1964a: 56 **A**: TM (Kuhitang)
pygmaea pygmaea Kraatz, 1882d: 325 **A**: AF KZ UZ
digitula Reitter, 1915e: 91
kelessiana Skopin, 1971: 495
minima Kraatz, 1882d: 326
subgenus *Uroprosodes* Reitter, 1909d: 119 type species *Prosodes costifera* Kraatz, 1886
altimontana G. S. Medvedev, 1995: 848 **A**: TD (Zheranshanskiy Khrebet)
amathia Reitter, 1893c: 275 **A**: TD
arnoldii G. S. Medvedev, 1995: 852 **A**: UZ (Fergana)
ballionis Kraatz, 1885: 288 [RN] **A**: UZ
belousovi G. S. Medvedev, 1996: 627 **A**: TD
brevitarsis Kraatz, 1882d: 305 **A**: UZ (Samarkand)
georgii Reitter, 1909d: 153
catenulata Kraatz, 1883a: 346 **A**: KI
chodzhaksaris Skopin, 1969: 885 **A**: TM (Kuhitang-Tay)
costifera Kraatz, 1886: 187 **A**: KI (Alai) TD XIN
noethi Reinig, 1931: 872
crassitibia G. S. Medvedev, 1995: 839 **A**: UZ
danijarovi G. S. Medvedev, 1995: 847 **A**: TD
diloides Kraatz, 1882d: 323 **A**: UZ (Samarkand)
dushenkoi G. S. Medvedev, 1996: 629 **A**: TD
fleischeri Reitter, 1894d: 243 **A**: UZ (Samarkand)
glasunowi Semenov, 1894a: 208 **A**: UZ (Bukhara)
semenoviana Bogatchev, 1941b: 110
grombczewskii Semenov, 1894a: 203 **A**: UZ (Bukhara)
hauseri Reitter, 1893c: 300 **A**: UZ
heydeni Kraatz, 1882d: 324 **A**: UZ
gebieni Reitter, 1909d: 152
parumpunctata Kraatz, 1882d: 321
pubibasis Reitter, 1909d: 142
hissarica hissarica G. S. Medvedev, 1995: 837 **A**: UZ
hissarica infida G. S. Medvedev, 1995: 837 **A**: UZ
hissarica inopinata G. S. Medvedev, 1997b: 584 **A**: TD
kondarensis H. R. Dadabaev, 1986: 127 **A**: TD
korotyaevi G. S. Medvedev, 1997b: 588 **A**: TD
kryzhanovskii G. S. Medvedev, 1995: 845 **A**: TD
michailovi G. S. Medvedev, 1995: 846 **A**: TD
monticola G. S. Medvedev, 1964a: 58 **A**: TM (Kuhitang)
nalepoides Reitter, 1893c: 299 **A**: UZ (Bukhara)
novemcostata Semenov, 1891a: 367 **A**: UZ
acutimarga Skopin, 1971: 497
obliquesulcata Kraatz, 1882d: 322 **A**: UZ (Samarkand, Tashkent)
oblonga G. S. Medvedev, 1995: 841 **A**: UZ
opaca Bogatchev, 1960c: 84 **A**: TD

ovtshinnikovi G. S. Medvedev, 1995: 843 **A**: TD
rectilineata Bogatchev, 1960c: 83 **A**: TD
reflexicollis Reitter, 1893c: 290 **A**: UZ (Bukhara)
rugicostata Reitter, 1900b: 55 **A**: UZ (Bukhara)
rugosipennis G. S. Medvedev, 1995: 850 **A**: TD
sogdiana G. S. Medvedev, 1995: 840 **A**: TD (Zerafshanskiy Khrebet)
spectabilis Reitter, 1893c: 287 **A**: UZ (Arslanbovskiy Massiv)
inconstans Reitter, 1909d: 139
kiritchenkoi Semenov & Bogatchev, 1939: 125
subpilosa G. S. Medvedev, 1964a: 59 **A**: TM (Kuhitang)
freibergi Skopin, 1969: 882
suturangula Reitter, 1900b: 55 **A**: UZ (Bukhara)
suturidens Reitter, 1909d: 142 **A**: UZ
tachtaensis tachtaensis G. S. Medvedev, 1995: 834 **A**: UZ
tachtaensis umnovi G. S. Medvedev, 1995: 837 **A**: UZ
tshikatunovi Kh. Dadabaev, 1986: 78 **A**: TD
tudaensis G. S. Medvedev, 1995: 832 **A**: UZ
vereshtshaginae G. S. Medvedev, 1997b: 591 **A**: UZ (Kutikangay)
vincens Reitter, 1900b: 55 **A**: UZ (Samarkand)
ypsilon Reitter, 1893c: 296 **A**: UZ (Samarkand)
genus *Prosodes*, species incertis sedis
motschulskyi Motschulsky, 1853: 78 **A**: XIN

genus *Tagona* Fischer von Waldheim, 1820: pl. 16 type species *Tagona macrophthalma* Fischer von Waldheim, 1820
acuminata Fischer von Waldheim, 1820: pl. 16 **A**: KZ TM
lata Sumakov, 1908b: 245
macrophthalma macrophthalma Fischer von Waldheim, 1820: pl. 16 **A**: AF KZ TM UZ
longicollis Brancsik, 1899: 100
tinodactyla Fischer von Waldheim, 1844: 121
macrophthalma rugipleuris Reitter, 1901c: 76 **A**: KZ (Muyumkum)
robusta Skopin, 1970: 166 **A**: UZ

subtribe Remipedellina Semenov, 1907

genus *Remipedella* Semenov, 1907b: 257 type species *Remipedella deserti* Semenov, 1907
deserti Semenov, 1907b: 259 **A**: TM (Karakumy)
semenovi G. S. Medvedev, 1989a: 90 **A**: TM

tribe Bolitophagini Kirby, 1837

subtribe Bolitophagina Kirby, 1837

genus *Atasthalomorpha* Miyatake, 1964: 77 type species *Atasthalus dentifrons* Lewis, 1894
dentifrons Lewis, 1894: 385 (*Atasthalus*) **A**: ES FE JA

genus *Boletoxenus* Motschulsky, 1858b: 63 type species *Boletoxenus gibber* Motschulsky, 1858
assamicus Kaszab, 1965a: 115 (*Bolitoxenus*) **A**: AP
bellicosus Lewis, 1894: 386 (*Atasthalus*) **A**: FE JA SC
galloisi Pic, 1932a: 3 (*Bolitonaeus*)
formosanus Masumoto, 1982c: 31 **A**: TAI
incurvatus Lewis, 1894: 386 (*Atasthalus*) **A**: FE JA
kawadai Masumoto & Akita, 2003: 176 **A**: JA (Ryukyus)
sasajii Masumoto & Akita, 2001c: 247 **A**: JA (Amami-Ôshima)

genus *Bolitonaeus* Lewis, 1894: 387 type species *Bolitonaeus mergae* Lewis, 1894
mergae Lewis, 1894: 387 **A**: JA

quadridentatus Candèze, 1861: 368 (*Bolitotherus*) **A**: NP **ORR**

genus *Bolitophagiella* Miyatake, 1964: 71 type species *Bolitophagus pannosus* Lewis, 1894
pannosa Lewis, 1894: 384 (*Bolitophagus*) **A**: JA

genus *Bolitophagus* Illiger, 1798: 100 type species *Silpha reticulata* Linnaeus, 1767
Boletophagus Zetterstedt, 1828: 265 type species *Silpha reticulata* Linnaeus, 1767
granulatus Fischer von Waldheim, 1842: 16 [= 1844: 129] **A**: KZ (Dzungaria)
interruptus Illiger, 1800: 112 **E**: AU BH CR CZ FR GE HU IT PL RO SK SL
goedeni Panzer, 1803b: 1 (*Bolitophagus*)
reticulatus Linnaeus, 1767: 572 (*Silpha*) **E**: AL AR AU BE BH BU BY CR CT CZ DE EN FI FR GB GE GR HU IT LS LT NL NR NT PL RO SK SP ST SV SZ UK YU **A**: FE ES JA KZ SC TR WS
boleti Piller & Mitterpacher, 1783: 113 (*Sylpha*)
cornutus Thunberg, 1784: 15 (*Hispa*)
crenatus Fabricius, 1792a: 89 (*Opatrum*)
subinteger Reitter, 1896b: 74 **E**: AB (Talysh) **A**: IN

genus *Bolitotrogus* Miyatake, 1964: 80 type species *Bolitotrogus kurosonis* Miyatake, 1964
formosanus Masumoto, 1982c: 30 **A**: TAI
kurosonis Miyatake, 1964: 82 **A**: JA
ohdaiensis Miyatake, 1964: 83 **A**: JA

genus *Byrsax* Pascoe, 1860a: 42 type species *Byrsax coenosus* Pascoe, 1860
fukiensis Kaszab, 1954: 251 **A**: FUJ
kaszabi Miyatake, 1970: 120 **A**: JA
kimurai Miyatake, 1970: 118 **A**: FE JA SC
niponicus Lewis, 1894: 388 **A**: JA
shibatai Masumoto, 1982c: 26 **A**: TAI
kawadai Masumoto, 1982c: 28
spiniceps Lewis, 1894: 388 **A**: SC
tsushimensis Miyatake, 1970: 122 **A**: JA (Tsushima)
tuberculatus Gravely, 1915: 522 **A**: AP NP **ORR**

genus *Eledona* Latreille, 1796: 19 type species *Opatrum agricola* Herbst, 1783
agricola Herbst, 1783: 35 (*Opatrum*) **E**: AB AR AU BE BH BY CR CT CZ DE EN FI FR GB GE GG GR HU IT LA LS LT MA NL NR PL PT SK SL SP ST SV SZ TR UK **N**: TU **A**: TM TR **NAR**
fungicola Horn, 1870: 389
sulcata Thunberg, 1784: 2 (*Dermestes*)
turcica Seidlitz, 1891: 553
hellenica Reitter, 1885d: 381 (*Heledona*) **E**: GR TR

genus *Eledonoprius* Reitter, 1911a: 338 type species *Opatrum armatum* Panzer, 1799
armatus Panzer, 1799a: 2 (*Opatrum*) **E**: AB AU BU CR CZ DE FR GE GR HU IT PL RO SK SL SP SV SZ UK **N**: MO **A**: CY **NAR**
serrifrons Reitter, 1890e: 172 (*Bolitophagus*) **E**: AB IT

genus *Lanhsia* Shibata, 1980: 64 type species *Lanhsia bucca* Shibata, 1980
bucca Shibata, 1980: 64 **A**: TAI

genus *Parabolitophagus* Miyatake, 1964: 70 type species *Bolitophagus felix* Lewis, 1894
felix Lewis, 1894: 384 (*Bolitophagus*) **A**: JA SC

subtribe Rhipidandrina LeConte, 1862

genus *Rhipidandrus* J. F. LeConte, 1862: 236 type species *Xylotinus flabellicornis* J. Sturm, 1826
Bolitopertha Gebien, 1910b: 379 type species *Bolitopertha novemcostata* Gebien, 1910

Cherostus Waterhouse, 1894: 68 type species *Cherostus walkeri* Waterhouse, 1894
Eutomus Lacordaire, 1866: 369 type species *Eutomus micrographus* Lacordaire, 1866
Heptaphylla Friedenreich, 1883: 375 type species *Heptaphylla fungicola* Friedenreich, 1883
speculifrons Gebien, 1922a: 288 (*Cherostus*) **A**: JA (Okinawa) TAI **AFR AUR ORR**
dybasi Kulzer, 1957: 204
scolytoides M. Chûjô, 1985: 61
sodalis Kulzer, 1957: 206

tribe Dissonomini G. S. Medvedev, 1968

genus *Bradyus* Dejean, 1834: 190 type species *Aphaleria capnisoides* Reitter, 1896 (= *Erodius pygmaeus* Fischer von Waldheim, 1821)
Aphaleria Reitter, 1896i: 235 type species *Erodius pygmaeus* Fischer von Waldheim, 1821
brodskyi Picka, 1987: 33 (*Aphaleria*) **A**: TM (Kurtlinsk)
oblonga G. S. Medvedev, 1964c: 657 (*Aphaleria*) **A**: TM UZ
pygmaea Fischer von Waldheim, 1821: 12 (*Erodius*) **A**: AF KZ TM UZ
capnisoides Reitter, 1896i: 236 (*Aphaleria*)

genus *Dissonomus* Jacquelin du Val, 1861: 280 type species *Heliopates picipes* Faldermann, 1837
subgenus *Dissonomus* Jacquelin du Val, 1861: 280 type species *Heliopates picipes* Faldermann, 1837
Heterophylus Mulsant & Rey, 1859b: 69 [= 1859e: 7] [HN] type species *Helipathes picipes* Faldermann, 1837
badghysi G. S. Medvedev, 1964a: 656 **A**: TM
cavicola Kaszab, 1970: 137 **A**: AF
ellipticus Desbrochers des Loges, 1881: 120 (*Heterophylus*) **E**: "Russia mer."
iranicus G. S. Medvedev, 1968a: 226 **A**: IN
kermanicus G. S. Medvedev, 1968a: 228 **A**: IN
picipes Faldermann, 1837: 55 (*Helipathes*) **E**: AB AR ST **A**: IN TR
parvulus Reitter, 1896e: 159 (*Heterophylus*)
tibialis Reitter, 1904a: 70 (*Heterophylus*) **A**: KZ TM
subgenus *Eudissonomus* G. S. Medvedev, 1968a: 218 type species *Heterophylus substriatus* Reitter, 1898
luristanicus G. S. Medvedev, 1968a: 219 **A**: IN
ovatus G. S. Medvedev, 1968a: 222 **A**: IN
politus Kaszab, 1968e: 460 **A**: IN TR
substriatus Reitter, 1898f: 348 (*Heterophylus*) **A**: IQ TR
zarudnyi G. S. Medvedev, 1968a: 220 **A**: IN
subgenus *Neodissonomus* G. S. Medvedev, 1968a: 235 type species *Heterophylus angustitarsis* Reitter, 1896
angustitarsis Reitter, 1896e: 160 (*Heterophylus*) **A**: TM
amplicollis Reitter, 1896e: 160 (*Heterophylus*)
desertus Skopin, 1966: 332 **A**: KZ TD UZ
latiusculus Mulsant & Rey, 1859b: 72 (*Heterophylus*) **A**: AF KZ TD TM UZ
subgenus *Paradissonomus* G. S. Medvedev, 1968a: 229 type species *Dissonomus longulus* Bogatchev & Kryzhanovskiy, 1960
badghysi G. S. Medvedev, 1964c: 656 **A**: IN TM
franzi Kaszab, 1960a: 131 **A**: AF
gracilis G. S. Medvedev, 1968a: 232 **A**: TD
kazachstanicus Skopin, 1968: 80 **A**: KZ (Karatau)
longulus Bogatchev & Kryzhanovskiy, 1960: 268 **A**: TM
rosae Schuster, 1915b: 89 **A**: KZ
validior G. S. Medvedev, 1968a: 234 **A**: IN
genus *Dissonomus*, species incertae sedis
seriepunctatus Reitter, 1904a: 69 (*Heterophylus*) **N**: AG

tribe Heleini Fleming, 1821

subtribe Cyphaleina Lacordaire, 1859

genus *Lepispilus* Westwood, 1841: 44 type species *Lepispilus sulcicollis* Hope, 1841 (= *Pachycaelia sulcicollis* Boisduval, 1835)

Lepidospilus Agassiz, 1846b: 205 [unjustified emendation]
Pachycaelia Boisduval, 1835: 248 [HN] type species *Pachycaelia sulcicollis* Boisduval, 1835
Tyndarisus Pascoe, 1869: 294 type species *Tyndarisus longitarsis* Pascoe, 1869 (= *Lepispilus stygianus* Pascoe, 1896)
punctatolineatus Allard, 1896: 318 [EA] **A**: SD

tribe Helopini Latreille, 1802
subtribe Cylindrinotina Español, 1956

genus *Armenohelops* Nabozhenko, 2002a: 42 type species *Armenohelops armeniacus* Nabozhenko, 2002
armeniacus Nabozhenko, 2002a: 42 **E**: AR
kagyzmanicus Nabozhenko, 2002a: 44 **A**: TR (Kars)

genus *Cylindrinotus* Faldermann, 1837: 73 type species *Helops femoratus* Faldermann, 1837
Cylindronotus Agassiz, 1846b: 111 [unjustified emendation]
acutangulus Seidlitz, 1896a: 797 (*Hedyphanes*) **E**: GG
bellator Reitter, 1902g: 218 (*Helops*) **E**: GG
batesi Allard, 1876a: 38 **A**: TR
constrictus Seidlitz, 1896a: 722 (*Helops*) **E**: TR
erivanus Reitter, 1902g: 219 (*Helops*) **E**: AR
femoratus Faldermann, 1837: 71 (*Helops*) **E**: AB AR
lugubris Faldermann, 1837: 74
funestoides Reitter, 1900a: 159 (*Helops*) **A**: TR
funestus Faldermann, 1837: 76 **E**: AR
gibbicollis Faldermann, 1837: 78 **E**: AR GG **A**: TR
flavipes Allard, 1877a: 219
corallipes Reitter, 1902g: 219 (*Helops*)
umbrinus Faldermann, 1837: 77
gibbosus Seidlitz, 1896a: 721 (*Helops*) **E**: "Caucasus"
hoberlandti Kaszab, 1959d: 81 (*Cylindronotus*) **A**: TR
nitidus Seidlitz, 1896a: 721 (*Helops*) **A**: SY TR
genus *Cylindrinotus,* species incertae sedis
pekinensis Fairmaire, 1888b: 130 (*Helops*) **A**: BEI

genus *Ectromopsis* Antoine, 1949b: 145 type species *Catomus politicollis* Allard, 1876
bogatschevi Iablokoff-Khnzorian, 1957a: 169 (*Helops*) **E**: AR
bulgarica G. S. Medvedev & Angelov, 1981: 312 (*Gunarus*) **E**: BU GR
mendizabali Cobos, 1953: 127 **E**: SP (Almería)
politicollis Allard, 1876a: 31 (*Catomus*) **N**: AG MO
tantilla Ménétriés, 1848: 28 (*Helops*) **E**: ST **A**: KZ
monilicornis Baudi di Selve, 1876d: 273 (*Helops*)

genus *Eustenomacidius* Nabozhenko, 2006b: 807 type species *Helops luridus* Ménétriés, 1848 (= *Eustenomacidius laevicollis* Kraatz, 1882)
subgenus *Caucasohelops* Nabozhenko, 2006b: 816 type species *Eustenomacidius svetlanae* Nabozhenko, 2006
svetlanae araxi Nabozhenko, 2006b: 819 **E**: AB (Nakhichevan)
svetlanae svetlanae Nabozhenko, 2006b: 817 **E**: AB (Talysh)
subgenus *Eustenomacidius* Nabozhenko, 2006b: 807 type species *Helops luridus* Ménétriés, 1848 (= *Stenomax laevicollis* Kraatz, 1882)
hirtipennis Seidlitz, 1896a: 792 (*Hedyphanes*) **A**: KI KZ
laevicollis Kraatz, 1882d: 333 (*Stenomax*) **A**: KI KZ TD TM UZ
lucidicollis Kraatz, 1882d: 333 (*Stenomax*)
luridus Ménétriés, 1848: 27 (*Helops*) [HN]
provocator Reitter, 1922b: 9 (*Catomus*)
mongolicus Kaszab, 1968c: 395 (*Catomus*) **A**: MG
turcmenicus G. S. Medvedev, 1964c: 652 (*Catomus*) **A**: TM (Kopet Dag)

wagnae Ren, 1999: 312 (*Catomus*) **A**: GAN

genus *Nalassus* Mulsant, 1854: 323 type species *Helops dryadophilus* Mulsant, 1854
subgenus *Caucasonotus* Nabozhenko, 2000: 107 type species *Cylindrinotus dombaicus* Nabozhenko, 2000
adriani Reitter, 1922b: 143 (*Cylindronotus*) **E**: GG
alanicus Nabozhenko, 2000: 111 (*Cylindronotus*) **E**: ST (Severnaya Ossetia)
colchicus Nabozhenko, 2001b: 651 **E**: GG
diteras Allard, 1876a: 66 (*Helops*) **E**: AB AR GG ST
dombaicus Nabozhenko, 2000: 108 (*Cylindronotus*) **E**: GG ST (Kavkaz)
ludmilae Nabozhenko, 2001b: 648 **E**: GG (Abkhazia)
pharnaces Allard, 1876a: 41 **E**: GG ST (Kavkaz)
aeneipennis Reitter, 1888b: 212 (*Stenomax*) [HN]
glorificus Seidlitz, 1896a: 729 (*Helops*) [RN]
svaneticus Nabozhenko & Dzhambazishvili, 2001: 90 **E**: GG
zakatalensis Nabozhenko, 2001b: 646 **E**: AB
subgenus *Helopocerodes* Reitter, 1922b: 144 type species *Helops faldermanni* Faldermann, 1837
balchashensis G. S. Medvedev, 1970: 396 **A**: KZ
cyrensis Bogatchev, 1946b: 246 (*Cylindronotus*) **E**: AB GG
dilaticornis Reitter, 1922b: 146 (*Cylindronotus*) **A**: TR
dissonus Nabozhenko, 2001b: 635 **E**: ST (Severnaya Osetiya)
faldermanni Faldermann, 1837: 70 (*Helops*) **E**: AB AR GG ST UK **A**: IN TM TR
amaroides Küster, 1850b: 69 (*Helops*)
ambrosius Reitter, 1922b: 146 (*Cylindronotus*)
confluens Seidlitz, 1896a: 740 (*Helops*)
eligius Reitter, 1922b: 145 (*Cylindronotus*)
impressicollis Faldermann, 1837: 69 (*Helops*)
semiopacus Reitter, 1902g: 222 (*Xanthomus*)
zabroides Küster, 1850b: 67 (*Helops*)
halophilus G. S. Medvedev, 1987b: 101 (*Cylindronotus*) **A**: KZ
kalashiani Nabozhenko, 2001b: 640 **E**: ST (Daghestan, Chechnya)
kaszabi Nabozhenko, 2001b: 643 **A**: TR (Van)
kazachstanicus G. S. Medvedev, 1964b: 159 (*Cylindronotus*) **A**: KZ
lutshniki Nabozhenko, 2001b: 636 **E**: ST
marteni Español, 1953a: 73 (*Stenomax*) **N**: MO
medvedevi Nabozhenko, 2006a: 248 **A**: IN
melchiades Reitter, 1922b: 147 (*Cylindronotus*) **A**: KZ
mojynkumensis G. S. Medvedev, 1987b: 100 (*Cylindronotus*) **A**: KZ
romashovi Nabozhenko, 2001b: 639 **E**: UK (Krym)
zaratustrai Nabozhenko, 2006a: 246 **A**: IN
subgenus *Helopondrus* Reitter, 1922b: 153 type species *Stenomax sareptanus* Allard, 1876
abkhasicus Nabozhenko, 2001b: 644 **E**: GG (Abkhazia)
adzharicus Nabozhenko & Dzhambazishvili, 2001: 92 **E**: GG **A**: TR
ahngeri G. S. Medvedev, 1998a: 138 (*Cylindronotus*) **A**: TM
arcanus Nabozhenko, 2001b: 657 **E**: GG
assimilis Küster, 1850b: 36 (*Helops*) **E**: FR IT SP
aeneoniger Küster, 1850b: 35 (*Helops*)
convexicollis Küster, 1850b: 34 (*Helops*)
cambyses Seidlitz, 1896a: 738 (*Helops*) **A**: IN
clavicornis Allard, 1876a: 20 (*Nalassus*) **A**: TR
cunctator Reitter, 1922b: 155 (*Cylindronotus*) **A**: IN
gloriosus Faldermann, 1837: 67 (*Helops*) **E**: AB AR GG ST (Nalchik, Adler)
pulcher Allard, 1877a: 256 (*Helops*)
heres Reitter, 1922b: 154 (*Cylindronotus*) **A**: IN
lineatus Allard, 1877a: 259 (*Stenomax*) **E**: AB GG ST **A**: IN
propheta Seidlitz, 1896a: 731 (*Helops*)
vexator Reitter, 1887e: 524 (*Stenomax*)

planivittis Allard, 1876a: 39 (*Stenomax*) **E**: GR **A**: TR
ponticus Seidlitz, 1896a: 737 (*Helops*) **A**: IN TR
sareptanus Allard, 1876a: 54 (*Stenomax*) **E**: GG (Abkhazia) KZ (Uralsk) MD ST UK
subgenus *Nalassus* Mulsant, 1854: 323 type species *Helops dryadophilus* Mulsant, 1854
abeillei Seidlitz, 1896a: 736 (*Helops*) **E**: FR (Corse)
aemulus aemulus Küster, 1850b: 61 (*Helops*) **E**: IT MA SP **N**: LB MO TU
fusculus Allard, 1876a: 40 (*Omalus*)
aemulus calaritanus Leo, 1985b: 144 **E**: IT (Sardegna)
alpigradus Fairmaire, 1883k: clxix (*Helops*) **E**: FR IT
brevicollis Steven, 1832: 137 (*Helops*) **E**: CT GG ST UK
brevicollis Küster, 1850b: 80 (*Helops*) [HN]
brevis J. Sturm, 1843: 163 (*Helops*)
gilvipes Ménétriés, 1848: 27 (*Helops*)
calpensis Champion, 1891d: 389 (*Helops*) **E**: SP
hispanus Seidlitz, 1896a: 753 (*Helops*)
colasi cobosi Español, 1974: 215 **E**: SP (Sierra Filabres)
colasi colasi Español, 1954c: 119 (*Cylindronotus*) **E**: SP (Sierra Nevada)
convexus Comolli, 1837: 26 (*Helops*) **E**: AU GE HU IT LS SZ
convexus Küster, 1850b: 74 (*Helops*) [HN]
laevigatus Küster, 1850b: 79 (*Helops*)
dermestoides Illiger, 1798: 120 (*Helops*) **E**: AL AU BH BY CR CT CZ EN GE HU IT LA LT MD PL RO SK SV SZ UK YU
brevis Allard, 1876a: 22 (*Cylindronotus*) [HN]
fabricii Gemminger, 1870a: 123 (*Helops*) [RN]
picipes Küster, 1850b: 78 (*Helops*)
quisquilius Panzer, 1794c: 3 (*Helops*) [HN]
dryadophilus Mulsant, 1854: 337 (*Helops*) **E**: AL BU FR GR IT RO SP SZ TR **A**: TR
ecoffeti Küster, 1850: 70 (*Helops*) **E**: FR SZ
picinus Küster, 1850: 81 (*Helops*)
rufocastaneus Pic, 1919c: 5 (*Helops*)
schaeferi Ardoin, 1958a: 41
temperei Ardoin, 1958a: 41
estrellensis Kraatz, 1870: 133 (*Helops*) **E**: PT SP
genei genei Gené, 1839: 74 (*Helops*) **E**: FR (Corse)
genei melonii Leo, 1982a: 1 **E**: IT (Sardegna)
graecus Seidlitz, 1896a: 745 (*Helops*) **E**: BU GR
harpaloides Küster, 1850b: 73 (*Helops*) **E**: FR
laevioctostriatus Goeze, 1777: 683 (*Tenebrio*) **E**: BE CZ FR GE GB IR IT NL PL PT SK SP SZ **A**: TR
caraboides Panzer, 1794c: 3 (*Helops*)
latiusculus Kraatz, 1880d: 300
lusitanus Kraatz, 1870: 134 (*Helops*)
ruficollis Fabricius, 1787a: 214 (*Helops*)
striatus Geoffroy, 1785: 157 (*Tenebrio*)
longipennis Küster, 1850b: 68 (*Helops*) **E**: SP
phaecus J. R. Sahlberg, 1903c: 33 (*Helops*) **E**: GR
planipennis Küster, 1850b: 83 (*Helops*) **E**: IT
ruficornis Baudi di Selve, 1876c: 259 (*Helops*)
plebejus Küster, 1850b: 84 (*Helops*) **E**: AL CR GR IT **A**: TR
distinguendus Küster, 1850b: 82 (*Helops*)

genus *Odocnemis* Allard, 1876a: 4 type species *Odocnemis caudatus* Allard, 1876
subgenus *Heloponotus* Reitter, 1922b: 126 type species *Helops perplexus* Ménétriés, 1848
douei Allard, 1876a: 39 (*Stenomax*) **E**: UK (Krym)
excavatus Seidlitz, 1896a: 752 (*Helops*)
perplexus Ménétriés, 1848: 26 (*Helops*) **E**: ST UK **A**: KZ (Uralsk)
grandicollis Küster, 1851: 62 (*Helops*)
sturmi Küster, 1851: 60 (*Helops*)

tauricus Mulsant & Godart, 1870: 49 (*Helops*)

subgenus *Odocnemis* Allard, 1876a: 4 type species *Odocnemis caudatus* Allard, 1876
Homaleis Rye, 1879: 62 [unjustified emendation]
Homalus Rye, 1878: 69 [unjustified emendation, HN]
Odontocnemis Rye, 1878: 69 [unjustified RN]
Omaleis Allard, 1877a: 151 [RN] type species *Helops congener* Reiche, 1861
Omalus Allard, 1876a: 4 [HN] type species *Helops congener* Reiche, 1861
alcides Reitter, 1922b: 140 (*Cylindronotus*) **E**: GR
amandus Reitter, 1922b: 132 (*Cylindronotum*) **E**: "Pyrenees"
amasiae Seidlitz, 1896a: 732 (*Helops*) **A**: TR
anatolicus Pic, 1899p: 411 (*Helops*) **A**: TR
badius L. Redtenbacher, 1849: 601 (*Helops*) **E**: BU GR RO
asphaltinus Küster, 1850b: 55 (*Helops*)
baudoni Antoine, 1954: 213 (*Stenomax*) **N**: MO (Moyen Atlas)
bosphoranus Allard, 1876a: 56 (*Stenomax*) **E**: TR
caudatus Allard, 1876a: 36 **A**: TR SY
clarus Allard, 1876a: 37 **N**: EG
clypeatus Küster, 1851: 68 (*Helops*) **E**: IT (Sicilia)
colasi Español, 1954c: 119 (*Nalassus*) **E**: SP (Sierra Nevada)
congener cedrorum Antoine, 1962: 55 (*Stenomax*) **N**: MO (Rif)
congener congener Reiche, 1862a: 372 (*Helops*) **N**: MO
crenatostriatus Allard, 1877a: 146 (*Stenomax*) **E**: GR **A**: CY TR
pindicus Apfelbeck, 1901: 465
cretensis Seidlitz, 1896a: 733 (*Helops*) **E**: GR (Kríti)
curticollis Reitter, 1922b: 149 (*Cylindronotus*) **E**: BH
dasypus Seidlitz, 1896a: 725 (*Helops*) **A**: TR
exaratus Germar, 1817: 192 (*Helops*) **E**: AL BH CR GR IT MA SL
brunnitarsis Küster, 1851: 69 (*Helops*)
fleischeri Reitter, 1922b: 130 (*Cylindronotum*) **E**: GR (Heraklea)
fundator Reitter, 1908g: 215 (*Stenomax*) **A**: TR
incultus Allard, 1877a: 132 (*Stenomax*) **A**: TR
lordiscelis Reitter, 1900a: 157 (*Helops*)
intricatus Allard, 1877a: 257 (*Stenomax*) **E**: BU
intrusicollis Seidlitz, 1896a: 733 (*Helops*) **A**: CY
korbi Reitter, 1902g: 219 (*Odocnemus*) **A**: TR
kurdistanus Reitter, 1902g: 221 (*Stenomax*) **A**: TR (Erzurum)
mediatlantis Antoine, 1949b: 129 (*Stenomax*) **N**: MO (Moyen Atlas)
moabiticus J. R. Sahlberg, 1908: 88 (*Helops*) **A**: JO
nigropiceus Küster, 1850b: 56 (*Helops*) **E**: BU GR
atticus Allard, 1881b: ciii (*Omalois*)
opertus Reitter, 1922b: 132 (*Cylindronotus*) **A**: TR
ophonoides Lucas, 1846: 352 (*Helops*) **E**: SP **N**: TU
pardoi Español, 1953: 75 (*Cylindronotus*) **N**: MO
perarmatus Nabozhenko & Tichý, 2006: 183 **A**: TR
poeciloides Seidlitz, 1896a: 739 (*Helops*) **N**: AG TU
praelongus Baudi di Selve, 1876c: 250 [= 1876b: 101] (*Helops*) **A**: LE SY "Kurdistan"
protinus Reitter, 1900a: 167 (*Helops*) **A**: TR
pseudoclarus Reitter, 1922b: 134 (*Cylindronotus*) **A**: LE
punctatus Allard, 1876a: 38 **E**: GR MC **A**: IS
rhaticus Antoine, 1936: 75 (*Helops*) **N**: MO (Grand Atlas)
recticollis Allard, 1876a: 38 (*Stenomax*) **E**: AR **A**: TR
rufffoi osellai Gardini, 1979: 75 (*Cylindronotus*) **E**: IT (Is. Montecristo)
ruffoi ruffoi Canzoneri, 1970: 81 (*Cylindronotus*) **E**: IT (Sicilia, Is. Marettimo)
schusteri Reitter, 1922b: 133 (*Cylindronotus*) **E**: AL GR

scutellatus Reitter, 1902g: 221 (*Stenomax*) **A**: TR
serbicus J. Frivaldszky, 1894: 86 (*Helops*) **E**: YU
seriegranatus Seidlitz, 1896a: 733 (*Helops*) **E**: TR
steindachneri Apfelbeck, 1907: 1671 (*Helops*) **E**: AL
strangulatus Reitter, 1922b: 130 (*Cylindronotum*) **A**: TR
telueticus Escalera, 1914: 350 (*Helops*) **N**: MO
tenebrioides Germar, 1813: 123 (*Helops*) **E**: PT SP
tuberculatus Küster, 1851: 67 (*Helops*) **E**: AL GR
tuberculiger Reiche & Saulcy, 1857: 265 (*Helops*) **E**: BU GR MC TR
valgus Baudi di Selve, 1881: 291 (*Helops*) **A**: IS "Syria"
zoltani Nabozhenko, **nom. nov.** [see New Acts] **A**: TR
  anatolicus Kaszab, 1961c: 173 [HN]

genus *Pseudoprobaticus* Nabozhenko, 2001a: 513 type species *Helops granipennis* Allard, 1876
granipennis Allard, 1876a: 51 (*Helops*) **E**: TR **A**: TR

genus *Reitterohelops* Skopin, 1960: 308 type species *Zophohelops lazarus* Reitter, 1922
ahngeri G. S. Medvedev, 1964c: 655 (*Zophohelops*) **A**: TM (Kopet Dag)
badghysi G. S. Medvedev, 1964c: 654 (*Zophohelops*) **A**: TM
kocheri Skopin, 1960: 309 (*Zophohelops*) **A**: KI KZ UZ
kulzeri Skopin, 1960: 310 (*Zophohelops*) **A**: TD UZ
lazarus Reitter, 1922b: 118 (*Zophohelops*) **A**: AF TM
steinbergi G. S. Medvedev, 1964c: 653: (*Zophohelops*) **A**: TM (Kopet Dag)
striatipennis G. S. Medvedev, 1970: 395 (*Zophohelops*) **A**: TD

genus *Stenomax* Allard, 1876a: 4 type species *Tenebrio lanipes* Linnaeus, 1771 (= *Stenomax aeneus* Scopoli, 1763)
subgenus *Asyrmatus* Canzoneri, 1959b: 149 type species *Helops piceus* J. Sturm, 1826
foudrasii Mulsant, 1854: 361 (*Helops*) **E**: FR IT
meridianus Mulsant, 1854: 370 (*Helops*) **E**: FR
piceus J. Sturm, 1826: 70 (*Helops*) **E**: FR IT
  cordatus Küster, 1850b: 53 (*Helops*)
  subaeneipennis Pic, 1919c: 5 (*Helops*)
subgenus *Stenomax* Allard, 1876a: 4 type species *Tenebrio lanipes* Linnaeus, 1771: 82 (= *Stenomax aeneus* Scopoli, 1763)
aeneus Scopoli, 1763: 82 (*Tenebrio*) **E**: AU BH BU BY CR CZ FR GE HU IT LA LS PL RO SK SL ST SZ TR UK YU
  arboreus Schrank von Paula, 1781: 219 (*Tenebrio*)
  cupreus Geoffroy, 1785: 158 (*Tenebrio*)
  incurvus Küster, 1850b: 54 (*Helops*)
  lanipes Linnaeus, 1771: 533 (*Tenebrio*)
litigiosus Vauloger de Beaupré, 1900: 701 (*Helops*) **N**: AG
martini Vauloger de Beaupré, 1900: 700 (*Helops*) **N**: AG

genus *Stygohelops* Leo & Liberto, 2003: 299 type species *Probaticus kalavriticus* Schawaller, 2001
kalavriticus Schawaller, 2001b: 44 (*Probaticus*) **E**: GR (Pelopponisos)
genus *Turkmenohelops* G. S. Medvedev, 1987a: 102 type species *Zophohelops balchanicus* G. S. Medvedev & Nepesova, 1985
arvatensis G. S. Medvedev, 1964c: 654 (*Zophohelops*) **A**: TM (Kopet Dag)
balchanicus G. S. Medvedev & Nepesova, 1985b: 362 (*Zophohelops*) **A**: IN TM (Kopet Dag)

genus *Xanthohelops* Nabozhenko, 2006b: 822 type species *Xanthohelops karakumicus* Nabozhenko & G. S. Medvedev, 2006
karakumicus Nabozhenko & G. S. Medvedev, 2006: 823 **A**: TM

genus *Xanthomus* Mulsant, 1854: 302 type species *Helops pallidus* Curtis, 1830
cyprius Grimm, 1991: 41 **E**: GR (Ródos) **N**: EG **A**: CY TR
graecus Dajoz, 1984: 89 **E**: GR
grimmi J. Ferrer & Whitehead, 2002: 393 **E**: IT
interstitialis Grimm, 1991: 42 **A**: CY
israelsoni J. Ferrer & Whitehead, 2002: 388 **N**: CI (Gran Canaria)
ovulus Seidlitz, 1896a: 739 (*Helops*) **E**: GR (Kríti)
pallidus Curtis, 1830: 298 (*Helops*) **E**: BE FR GB GE IR IT MA NL PT SP SZi **N**: AG CI EG MO MR **A**: CY
ghidinii Canzoneri, 1959a: 146
residuus Canzoneri, 1959a: 145
testaceus Küster, 1850b: 63 (*Helops*)
pellucidus Mulsant & Rey, 1856: 15 (*Helops*) **E**: FR GR IT SP **N**: AG
subsulcatus Rey, 1892: 65 (*Helops*)
viklundi J. Ferrer & Whitehead, 2002: 393 **N**: TU

genus *Zophohelops* Reitter, 1902h: 221 [RN] type species *Helops tiro* Reitter, 1902
Euryhelops Reitter, 1902g: 214 [HN] type species *Helops tiro* Reitter, 1902
bozbutavicus G. S. Medvedev, 1987a: 115 **A**: KI (Mts Bozbutay)
cylindronotoides Reitter, 1902g: 216 (*Helops*) **A**: KI KZ
gebieni Reitter, 1922b: 120
davletshinae G. S. Medvedev, 1987a: 106 **A**: KZ (Ugamskiy Khrebet)
demetrius Reitter, 1922b: 119 **A**: UZ (Ferganskiy Khrebet)
diabolinus Brancsik, 1899: 101 (*Helops*) **A**: TM (Ashkhabat)
ferganensis G. S. Medvedev, 1987a: 119 **A**: UZ (Ferganskiy Khrebet)
gurjevae G. S. Medvedev, 1987a: 106 **A**: KI (Mts Bozbutay)
hirtus G. S. Medvedev, 1987a: 121 **A**: KI UZ
pullus Seidlitz, 1896a: 705: 750 (Helops) **E**: AB (Nakhichevan) AR
humeridens Reitter, 1902g: 223 (*Xanthomus*)
karatavicus G. S. Medvedev, 1987a: 108 **A**: KZ (Karatau Khrebet)
kaszabi G. S. Medvedev, 1987a: 108 **A**: UZ (Ferganskiy Khrebet)
kuramensis G. S. Medvedev, 1987a: 104 **A**: UZ
michajlovi Nabozhenko, 2001a: 512 **A**: TD
nadari Reitter, 1902g: 215 (*Helops*) **A**: KI (Alai)
pangazensis G. S. Medvedev, 1987a: 105 **A**: TD
pilosus G. S. Medvedev, 1987a: 120 **A**: KZ
protzenkoi Skopin, 1964: 389 **A**: KI
pyriformis G. S. Medvedev, 1987a: 110 **A**: KZ (Ugamskiy Khrebet)
sergei G. S. Medvedev, 1987a: 112 **A**: KZ (Ugamskiy Khrebet)
skopini G. S. Medvedev, 1987a: 113 **A**: KZ (Karatau Khrebet)
spiridion Reitter, 1922b: 121 **A**: KI KZ UZ
stephanus Reitter, 1922b: 121 **A**: KI (Osch)
tar G. S. Medvedev, 1987a: 118 **A**: UZ (Ferganskiy Khrebet)
tiro Reitter, 1902g: 215 (*Helops*) **A**: KI UZ
ugamicus Skopin, 1966: 325 **A**: KZ (Ugamskiy Khrebet)
zarudnyi G. S. Medvedev, 1987a: 114 **A**: KI (Chatkalskiy Khrebet)
zolotarewi Reitter, 1902c: 190 (*Helops*) **A**: UZ (Tashkent)

subtribe Helopina Latreille, 1802

genus *Accanthopus* Dejean, 1821: 71 type species *Blaps caraboides* sensu Germar, 1817 (= *Tenebrio velikensis* Piller & Mitterpacher, 1783)
Enoplopus Solier, 1848: 158 type species *Helops dentipes* Rossi, 1790 (= *Tenebrio velikensis* Piller & Mitterpacher, 1783)
reitteri Brenske, 1884: 88 (*Acanthopus*) **E**: AL GR MC
velikensis Piller & Mitterpacher, 1783: 83 (*Tenebrio*) **E**: AL BH BU CR FR GR HU IT MC RO SL SZ TR YU
caraboides V. Petagna, 1786: 15 [= 1787: 15] (*Tenebrio*) [HN]
dentipes Rossi, 1790: 236 (*Helops*)

latissimus Stierlin, 1902: 426 (*Dendarus*)
rufescens Pic, 1919c: 5

genus *Adelphinus* Fairmaire & Coquerel, 1866: 44 type species *Eutrapela suturalis* Lucas, 1846
subgenus *Adelphinus* Fairmaire & Coquerel, 1866: 44 type species *Eutrapela suturalis* Lucas, 1846
ciliatus Seidlitz, 1898a: 853 **N**: AG MO
rotundicollis Kocher, 1958a: 124 **N**: MO
suturalis Lucas, 1847: 363 (*Eutrapela*) **N**: AG TU
maculatus Pic, 1899h: 115
pallidipennis Pic, 1899h: 114
rasus Seidlitz, 1898: 799 (*Nephodes*)
subgenus *Adelphinops* Reitter, 1922b: 168 type species *Adelphinus ordubadensis* Reitter, 1890
afghanicus Kaszab, 1960a: 174 **A**: AF
ordubadensis Reitter, 1890d: 150 **E**: AB AR
fulvipennis Reitter, 1906f: 242
fulvovittatus Reitter, 1890d: 150

genus *Allardius* Ragusa, 1898: 130 type species *Parablops oculatus* Baudi di Selve, 1876
Parablops Allard, 1877a: 230 [HN]
Pseudoparablops Heyden, 1908: 132 type species *Paraplops sardiniensis* Allard, 1877
oculatus Baudi di Selve, 1876c: 264 [= 1876b: 135] (*Paraplops*) **E**: IT (Sicilia)
sardiniensis Allard, 1877a: 261 (*Paraplops*) **E**: IT (Sardegna)

genus *Apterotarpela* Kaszab, 1954: 262 type species *Apterotarpela klapperichi* Kaszab, 1954
klapperichi Kaszab, 1954: 262 **A**: FUJ

genus *Catomus* Allard, 1876a: 4 type species *Catomus persicus* Allard, 1876
subgenus *Catomodontus* Löbl & Merkl, 2003: 251 type species *Catomus coronatus* Koch, 1935
coronatus Koch, 1935a: 108 **N**: LB
subgenus *Catomus* Allard, 1876a: 4 type species *Catomus persicus* Allard, 1876
Catomidius Seidlitz, 1896a: 792 type species *Hedyphanes rhynchophorus* Seidlitz, 1896
acutipennis Reiche & Saulcy, 1857: 269 (*Helops*) **A**: IS JO
amabilis Vauloger de Beaupré, 1900: 716 (*Helops*) **N**: AG TU
antennatus Bogatchev, 1963b: 100 **A**: TD
anthicoides Vauloger de Beaupré, 1900: 710 (*Helops*) **N**: AG
antoniae Reitter, 1890e: 172 **E**: AB AR
arabicus Kaszab, 1982b: 183 **A**: OM
batnensis Fairmaire, 1883e: xliv (*Helops*) **N**: AG
biskrensis Vauloger de Beaupré, 1900: 692 (*Helops*)
soricinus Vauloger de Beaupré, 1900: 692 (*Helops*)
bengasianus Schuster, 1925b: 88 **N**: LB
capilatus Vauloger de Beaupré, 1900: 693 **N**: AG TU
ciliatus Vauloger de Beaupré, 1900: 710 (*Helops*) **N**: AG
comatus Vauloger de Beaupré, 1900: 710 (*Helops*) **N**: AG
consentaneus Küster, 1851: 63 (*Helops*) **E**: AL FR GR IT MC SP **N**: AG EG TU **A**: CY SY TR
convexifrons Fairmaire, 1877: 141 (*Hedyphanes*) [sic] **N**: AG TU
dichrous Reitter, 1902g: 224 **A**: TR
echinatus Vauloger de Beaupré, 1900: 692 (*Helops*) [RN] **N**: AG
fallax Vauloger de Beaupré, 1900: 694 (*Helops*) **N**: LB TU
flavus Allard, 1876a: 28 **E**: AL
fragilis Ménétriés, 1848: 27 (*Helops*) **A**: TM UZ
subniger Reitter, 1902b: 181
frivaldszkyi Seidlitz, 1896a: 754 (*Helops*) **E**: TR **A**: TR
gracilis Küster, 1850b: 43 (*Helops*) [HN]
fulvipes Reiche & Saulcy, 1857: 267 (*Helops*) **A**: IS TR
gossypiatus Reiche, 1862a: 373 (*Helops*) **N**: AG TU

grosclaudei Normand, 1936b: 367 **N**: AG TU
henoni Allard, 1876a: 49 **N**: AG
hirtus Mulsant & Godart, 1870: 52 (*Hedyphanes*) **N**: AG
hesperides Reiche, 1861a: 7 (*Helops*) **E**: AR **A**: CY IS JO LE TR
berytensis Kraatz, 1880d: 299 (*Isopedes*)
improbulus Normand, 1950a: 85 **N**: TU
indubitatus Nabozhenko, 2006b: 838 **A**: TD UZ
karakalensis G. S. Medvedev, 1964c: 652 **A**: IN TM
dolini G. S. Medvedev, 2004b: 575
laenoides Reitter, 1922b: 11 **A**: SY
lepidus Reitter, 1922b: 10 **A**: SY
liliputanus Reitter, 1906g: 375 (*Helops*) **E**: SP
longiceps longiceps Chobaut, 1918: 75 (*Helops*) **N**: TU
longiceps minutepunctatus Chobaut, 1918: 76 (*Helops*) **N**: TU
niger Kraatz, 1882d: 332 (*Hedyphanes*) **A**: KZ TD UZ
minor Kraatz, 1882d: 332 (*Hedyphanes*)
nitidior Antoine, 1949b: 143 **N**: MO (Grand Atlas)
noctivagus Nabozhenko, 2006b: 840 **A**: UZ
obsoletus biasiolii Koch, 1937: 495 **N**: LB
obsoletus obsoletus Allard, 1876a: 30 **N**: AG LB TU
ovatus Koch, 1935a: 109 **N**: LB
persicus Allard, 1876a: 30 **A**: IN
pilosus Allard, 1876a: 49 **N**: AG (Oran)
puber Allard, 1876a: 50 **N**: AG
pullatus Peyerimhoff, 1917b: 318 (*Helops*) **N**: AG (Djurdjura)
rhynchophorus Seidlitz, 1896a: 792 (*Hedyphanes*) **N**: AG
rotundicollis Guérin-Méneville, 1825b: 112 (*Helops*) **E**: FR IT MA SP **N**: TU
agonus Mulsant, 1854: 377 (*Helops*)
angustatus Lucas, 1846: 354 (*Helops*)
curvipes Reitter, 1922b: 12
juncorum Küster, 1850b: 41 (*Helops*)
obtusangulus Küster, 1850b: 37 (*Helops*)
pygmaeus Küster, 1850b: 38 (*Helops*)
rotundicollis Küster, 1850b: 40 (*Helops*) [HN]
siculus Küster, 1850b: 39 (*Helops*)
tagenioides Küster, 1850b: 42 (*Helops*)
rungsi Antoine, 1949b: 143 **N**: MO
schusteri Normand, 1936b: 366 **N**: LB TU
seidlitzi Gebien, 1911: 552 (*Helops*) [RN] **A**: IS "Syria"
hierochonticus J. R. Sahlberg, 1913b: 182 [HN]
semiruber Allard, 1876a: 29 **A**: IN
senussianus Koch, 1937: 495 **N**: LB
sicardi Vauloger de Beaupré, 1900: 690 (*Helops*) **N**: AG TU
submetallicus Vauloger de Beaupré, 1900: 713 (*Helops*) **N**: AG
subparallelus Koch, 1935a: 109 **N**: EG
sulcatus G. S. Medvedev, 1964c: 652 **A**: UZ
testaceipes Fairmaire, 1884d: lxvii (*Helops*) **N**: AG MO TU
thamii Antoine, 1955: 364 **N**: MO
torretassoi Koch, 1935a: 109 **N**: EG
ursus Antoine, 1949b: 143 **N**: MO
vaulogeri Théry, 1931b: 126 **N**: MO (Haute Moulouya)
virgilius Reitter, 1922b: 11 **N**: AG (Dra el Mizan)
walkeri Champion, 1891d: 390 (*Helops*) **E**: SP (Gibraltar)
wangae Ren, 1999: 312 **A**: GAN

subgenus *Montanocatomus* Nabozhenko, 2006b: 842 type species *Catomus grandis* G. S. Medvedev, 1978
badachshanicus G. S. Medvedev, 1970: 393 **A**: TD
fabiani Nabozhenko, 2006b: 849 **A**: KZ
grandis G. S. Medvedev, 1978b: 50 **A**: KI (Alayskiy ridge)
pilosulus Kraatz, 1886: 190 **A**: UZ (Fergana)
gracilicollis Kraatz, 1886: 189 (*Stenomax*)
reinigi Shuster, 1931: 903 (*Stenomacidius*) **A**: KI TD
alaensis G. S. Medvedev, 1970: 394
subgenus *Sinocatomus* Nabozhenko, 2006b: 852 type species *Catomus solitarius* Nabozhenko, 2006
solitarius Nabozhenko, 2006b: 852 **A**: SCH

genus *Ceratopelius* Antoine, 1962: 52 type species *Ceratopelius mussardi* Antoine, 1962
mussardi Antoine, 1962: 52 **N**: MO (Rif)

genus *Deretus* Gahan, 1900: 10 type species *Deretus denticollis* Gahan, 1900
denticollis Gahan, 1900: 10 **A**: YE (Suqutra)
spinicollis Schawaller, 2004c: 456 **A**: YE (Suqutra)
wraniki Schawaller, 2004c: 457 **A**: YE (Suqutra)

genus *Entomogonus* Solier, 1848: 155 type species *Entomogonus barthelemyi* Solier, 1848
subgenus *Delonurops* Reitter, 1922b: 24 type species *Entomogonus clavimanus* Reitter, 1903
Macrophanes Iablokoff-Khnzorian, 1957a: 165 type species *Hedyphanes corax* Iablokoff-Khnzorian, 1957
amandanus Reitter, 1902g: 211 (*Helops*) **E**: AB AR
corax Iablokoff-Khnzorian, 1957a: 163
clavimanus Reitter, 1903a: 18 **E**: AR
duchoni Reitter, 1903a: 20 **A**: IQ SY TR
egregius Seidlitz, 1896a: 703 (*Helops*) **A**: TR
obtusus Seidlitz, 1896a: 704 (*Helops*) **A**: CY TR
peyronis Reiche, 1861a: 5 (*Helops*) **A**: TR "Syria"
fusiformis Allard, 1877a: 76 (*Helops*)
saphyrinus Allard, 1876a: 43 [= 1877b: 69] (*Helops*) **A**: IQ TR "Syria"
fausti Allard, 1877a: 255 (*Helops*)
subgenus *Entomogonus* Solier, 1848: 155 type species *Entomogonus barthelemyi* Solier, 1848
angulicollis aleppensis Reitter, 1922b: 28 **A**: SY TR
angulicollis angulicollis Mulsant & Wachanru, 1853: 11 (*Hediphanes*) **A**: SY TR
gassneri Reitter, 1908b: 30
helopinus Reitter, 1922b: 28
angulicollis gebieni Reitter, 1922b: 28 **A**: TR
barthelemyi Solier, 1848: 157 **A**: TR "Syria"
subgenus *Eutelogonus* Reitter, 1922b: 24 type species *Helops davidis* Fairmaire, 1884
davidis Fairmaire, 1884a: 171 (*Helops*) **A**: TR
nylanderi J. Ferrer & L. Soldati, 1999: 74 **A**: TR (Antalya)

genus *Erionura* Reitter, 1903a: 18 type species *Helops gigantea* Kraatz, 1862
gigantea Kraatz, 1862b: 124 (*Helops*) **E**: GR (Taygetos)

genus *Erulipus* Fairmaire, 1903a: 14 type species *Erulipus fruhstorferi* Fairmaire, 1903
multicolor Pic, 1927c: 18 **A**: FUJ **ORR**

genus *Euboeus* Boieldieu, 1865: 10 type species *Euboeus minonti* Boieldieu, 1865
minonti Boieldieu, 1865: 11 **E**: BU GR **A**: TR

genus *Gunarus* Des Gozis, 1886: 25 type species *Helops hirtulus* Reiche, 1861
arenicola Antoine, 1949b: 140 **N**: MO
bremondi Antoine, 1949b: 140 **N**: MO

hirtulus Reiche, 1862a: 374 (*Helops*) **E**: SP **N**: AG MO
minutus Mulsant & Godart, 1870: 51 (*Helops*)
lapidicola Küster, 1850b: 62 (*Helops*) **E**: BU CR GR YU
rufescens Küster, 1851: 59 (*Helops*)
splendidulus L. W. Schaufuss, 1862a: 92 (*Helops*)
kaszabi Grimm, 1981: 10 **E**: GR (Santorin)
nodicornis Reitter, 1922b: 117 **E**: TR
ovipennis Allard, 1877a: 123 (*Xanthomus*) **E**: GR
parvulus Lucas, 1846: 355 (*Helops*) **E**: GR IT MA SP **N**: MO TU
heydeni Allard, 1876a: 42 (*Nalassus*)
intersparsus Küster, 1850b: 58 (*Helops*)
nanus Küster, 1850b: 60 (*Helops*)
tingitanus Allard, 1877a: 116 (*Xanthomus*) **E**: SP (Granada) **N**: MO
laeviusculus Kraatz, 1883c: 395 (*Nalassus*)

genus *Hedyphanes* Fischer von Waldheim, 1820: pl. 15 type species *Hedyphanes coerulescens* Fischer von Waldheim, 1820
subgenus *Hedyphanes* Fischer von Waldheim, 1820: pl. 15 type species *Hedyphanes coerulescens* Fischer von Waldheim, 1820
Coelophanes Iablokoff-Khnzorian, 1964: 304 type species *Hedyphanes impressicollis* Faldermann, 1837
albertus Reitter, 1922b: 16 **A**: TM
besseri Faldermann, 1837: 95 **A**: TM
bodemeyeri Reitter, 1914e: 185 **A**: IN
chalybaeus Faldermann, 1837: 96 **E**: AR
coerulescens Fischer von Waldheim, 1820: pl. 15 **E**: ST **A**: AF KZ TD TM UZ
cordicollis Seidlitz, 1896a: 797 **A**: TR
cribripennis Lucas, 1854: 32 **E**: GR (Kríti)
dejeani Faldermann, 1837: 88 **E**: AR
desertus Motschulsky, 1845a: 82 **E**: AB
elongatus Allard, 1876a: 42 (*Entomogonus*) **E**: AB **A**: TR
europs Reitter, 1914e: 187 **A**: IN
gebieni Reitter, 1914e: 186 **A**: IN
hegeteroides Faldermann, 1837: 92 **E**: AR
helopinus Gemminger, 1870a: 123 **E**: GR (Kríti)
helopioides Lucas, 1854: 33 [HN]
helopioides Faldermann, 1837: 83 **E**: AR
iranicus G. S. Medvedev, 1976: 890 **A**: IN
kiritshenkoi G. S. Medvedev, 1978b: 51 **A**: TD
koltzei Heyden, 1892b: 108 **A**: KI
kuschkensis Kaszab, 1960a: 172 **A**: AF TM
laticollis laticollis Fischer von Waldheim, 1832a: 196 **E**: AB AR GG
impressicollis Faldermann, 1837: 94
laticollis nycterinoides Faldermann, 1837: 93 **E**: ST (Dagestan)
nycteroides Reitter, 1914e: 185
mannerheimi Faldermann, 1837: 91 **E**: AR
matthiesseni Reitter, 1914e: 186 **A**: IN
mesopotamicus Nabozhenko, 2005: 351 **A**: IQ
muminovi Bogatchev, 1963b: 98 **A**: UZ (Surkhandar'inskaya Oblast')
ocularis Reitter, 1914e: 187 **A**: IN
parvicollis Seidlitz, 1896a: 797 **A**: TM
seidlitzi Reitter, 1914e: 184 **A**: IN TM
semnanicus Nabozhenko, 2005: 352 **A**: IN
tagenoides Faldermann, 1832: 197 **E**: AB AR GG **A** IN
cruralis Fischer von Waldheim, 1837a: 153
menetriesii Fischer von Waldheim, 1832a: 197
pubiventris Reitter, 1922b: 18

tentyrioides Faldermann, 1837: 85
tuxeni Kaszab, 1959e: 253 [= 1960a: 172] **A**: AF IN
upioides Faldermann, 1837: 87 **A**: TR
angustus Marseul, 1879c: 75 (*Apolites*)
lutosus Allard, 1877a: 229 [= 1877b: 97]
zarudnyi G. S. Medvedev, 1976: 890 **A**: IN

genus *Hedyphanes*, nomen dubium
fovoelatusstriatus Motschulsky, 1845a: 82 **E**: ST

genus *Helopelius* Reitter, 1922b: 152 type species *Stenomax aeneipennis* Allard, 1876
aeneipennis Allard, 1876a: 18 (*Stenomax*) **E**: GR (Ródos) **A**: TR "Syria"
disgregus Reitter, 1922b: 152 (*Cylindronotus*) **E**: GR (Ródos)
nodifer Kraatz, 1880d: 299 (*Isopedes*) **E**: SP
otini Antoine, 1949b: 141 **N**: MO (Moyen Atlas)
subsinuatus Antoine, 1951b: 318 **N**: MO (Moyen Atlas)
verrucosus Vauloger de Beaupré, 1900: 683 (*Helops*) **E**: PT **N**: AG
zaianus Antoine, 1949b: 142 **N**: MO

genus *Helops* Fabricius, 1775: 257 type species *Tenebrio caeruleus* Linnaeus, 1758
subgenus *Helops* Fabricius, 1775: 257 type species *Tenebrio caeruleus* Linnaeus, 1758
Anteros Laporte, 1840: 235 type species *Helops chalybaeus* Latreille, 1804 (= *Tenebrio caeruleus* Linnaeus, 1758)
caeruleus caeruleus Linnaeus, 1758: 418 (*Tenebrio*) **E**: AB AL BH BU CR FR GB GE GR IT NL RO SL SP ST SZ UK **A**: IN TR
chalibaeus Rossi, 1790: 236
chalybaeus Latreille, 1804a: 346 [unjustified emendation]
violaceus Marsham, 1802: 480 (*Blaps*)
caeruleus stevenii Krynicki, 1834: 167 **E**: GG ST UK (Krym)
caeruleus talyshensis Bogatchev, 1949a: 38 **A**: AB (Talysh)
glabriventris glabriventris Reitter, 1885d: 382 **E**: GR **A**: CY TR
glabriventris jelineki Picka, 1984: 27 **E**: GR (Kríti)
insignis insignis Lucas, 1846: 348 **N**: AG TU
insignis maroccanus Fairmaire, 1873a: 341 **N**: MO
magnificus Escalera, 1925a: 334
rossii Germar, 1817: 191 **E**: AL BH CR FR GR IT LS MA RO SL TR **A**: TR
marloysi Laporte, 1840: 235
subgenus *Mesohelops* Reitter, 1922b: 31 type species *Helops cyanipes* Allard, 1877
cyanipes Allard, 1877a: 256 **A**: IS LE TR
carinimentum Reitter, 1885d: 383
thoracicus Grimm, 1991: 41 **A**: CY
genus *Helops*, species incertae sedis
Helops caucasicus Allard, 1877b: 97 **E**: "Caucasus"
Helops gracilis Fischer von Waldheim, 1823: 200 **E**: "Rossia meridionalis"
Helops arboreus Fischer von Waldheim, 1823: 200 **E**: "Rossia meridionalis"

genus *Italohelops* Español, 1961e: 295 type species *Parablops subchalybaeus* Reitter, 1907
subchalybaeus Reitter, 1907e: 204 (*Parablops*) **E**: IT

genus *Mamorina* Antoine, 1951c: 98 type species *Mamorina sulcaticeps* Antoine, 1951
sulcaticeps Antoine, 1951c: 99 **N**: MO

genus *Nephodinus* Gebien, 1943: 900 [RN] type species *Nephodes villiger* Rosenhauer, 1856
subgenus *Nephodinus* Gebien, 1943: 900 [RN] type species *Nephodes villiger* Rosenhauer, 1856
Nephodes Blanchard, 1845: 34 [HN] type species *Nephodes villiger* Rosenhauer, 1856
gracilior Fairmaire, 1884d: lxvii (*Nephodes*) **N**: AG

metallescens Küster, 1846c: 47 (*Helops*) **E**: FR (Corse) IT (Sardegna)
corsicus Allard, 1877a: 222 (*Nephodes*)
pubescens modestus Kraatz, 1870: 142 (*Nephodes*) **E**: SP
infoveicollis Pic, 1910c: 10 (*Nephodes*)
pubescence pubescence Küster, 1850b: 88 (*Helops*) **E**: SP
foveicollis Küster, 1851: 72 (*Helops*)
villiger barbarus Kraatz, 1872: 182 (*Nephodes*) **N**: AG MO TU
villiger villiger Rosenhauer, 1856: 218 (*Nephodes*) **E**: SP **N**: MO
incanus Fairmaire, 1879c: 199 (*Nephodes*)
subdepressus Fairmaire, 1868: 492 (*Nephodes*) **N**: AG
subgenus *Paranephodes* Antoine, 1954: 208 type species *Paranephodes camusi* Antoine, 1954
camusi Antoine, 1954: 208 (*Nephodes*) **N**: MO

genus *Nesotes* Allard, 1876a: 5 type species *Helops asper* Küster, 1850
subgenus *Helopogonus* Reitter, 1922b: 122 type species *Helops viridicollis* L. W. Schaufuss, 1869
viridicollis ibicensis Español, 1946: 105 **E**: SP (Ibiza)
viridicollis viridicollis L. W. Schaufuss, 1869: 25 (*Helops*) **E**: SP (Baleares)
rugipennis Tenenbaum, 1915: 841 (*Helops*)
subgenus *Nesotes* Allard, 1876a: 5 type species *Helops asper* Küster, 1850
Diastixus Allard, 1876a: 5 type species *Helops punctipennis* Lucas, 1846 (= *Helops heteromorphus* Lucas, 1846)
aenescens Reiche, 1861b: 90 (*Helops*) **N**: AG MO
acutangulus Fairmaire, 1879c: 199 (*Diastixus*)
lucens Escalera, 1914: 352 (*Helops*)
aethiops Wollaston, 1864: 509 (*Helops*) **N**: CI
afer Erichson, 1841: 184 (*Helops*) **N**: AG
altivagans Wollaston, 1864: 503 (*Helops*) **N**: CI (Tenerife)
andreinii Schuster, 1925b: 89 (*Diastixus*) **N**: LB
arboricola Wollaston, 1862d: 338 (*Helops*) **N**: MR
asper asper Küster, 1850b: 65 (*Helops*) **N**: MR
vulcanus Wollaston, 1854: 513 (*Helops*)
asper obliteratus Wollaston, 1871: 280 (*Helops*) **N**: MR
aterrimus Wollaston, 1864: 506 (*Helops*) **N**: CI (Gomera)
azoricus Crotch, 1867a: 390 (*Helops*) **N**: AZ
carbo Küster, 1850b: 33 (*Helops*) **E**: SP
carbunculus Wollaston, 1854: 519 (*Helops*) **N**: CI
catomoides bibersoni Antoine, 1954: 208 **N**: MO
catomoides catomoides Español, 1952a: 74 **N**: MO (Western Sahara)
cinnamomeus Wollaston, 1854: 520 (*Helops*) **N**: MR
confertus colasi Ardoin, 1960a: 118 **N**: MR
confertus confertus Wollaston, 1854: 515 (*Helops*) **N**: MR
conformis conformis Gemminger, 1870a: 123 [RN] **N**: CI
congener Wollaston, 1864: 504 (*Helops*) [HN]
conformis grancanariensis Español, 1962b: 339 **N**: CI (Gran Canaria)
conformis turgiticollis Wollaston, 1864: 504 (*Helops*) **N**: CI
congestus Wollaston, 1864: 505 **N**: CI
congregatus Wollaston, 1854: 518 (*Helops*) **N**: MR
crassicollis Küster, 1850b: 32 (*Helops*) **E**: SP
ghilianii Baudi di Selve, 1876c: 253 [= 1877b: 48] (*Helops*)
ibericus Allard, 1877a: 182 (*Diastixus*)
effeminatus Antoine, 1949b: 135 **N**: MO
elliptipennis Wollaston, 1864: 503 (*Helops*) **N**: CI (Tenerife)
fusculus Wollaston, 1864: 511 (*Helops*) **N**: CI
rufipes Har. Lindberg, 1953: 10
futilis Wollaston, 1854: 520 (*Helops*) **N**: MR
gagatinus Küster, 1850b: 66 (*Helops*) **N**: MR
pluto Wollaston, 1854: 516 (*Helops*)

gomerensis Wollaston, 1965: 64 [Appendix] (*Helops*) **N**: CI (Gomera)
graniger Küster, 1850b: 64 (*Helops*) **N**: MR
validus Reitter, 1922b: 162 (*Cylindronotus*)
helleri Reitter, 1922b: 163 (*Cylindronotus*) **N**: CI (Tenerife)
heteromorphus Lucas, 1846: 351 (*Helops*) **N**: AG
punctipennis Lucas, 1846: 352 (*Helops*)
infernus infernus Wollaston, 1854: 517 (*Helops*) **N**: MR (Porto Santo)
infernus wollastoni Ardoin, 1960a: 121 **N**: MR (Porto Santo)
leacockianus Wollaston, 1854: 517 (*Helops*) **N**: MR
lindbergi Español, 1962b: 342 **N**: CI (Gran Canaria)
lucasi Méquignon, 1937: 194 [RN] **N**: AG MO TU
rotundicollis Lucas, 1846: 350 (*Helops*) [HN]
lucifugulus Reitter, 1922b: 519 (*Cylindronotus*) **N**: CI
lucifugus Wollaston, 1854: 518 (*Helops*) **N**: MR (Porto Santo)
marseulii Wollaston, 1865: 63 [Appendix] (*Helops*) **N**: CI
mogadoricus Escalera, 1914: 351 (*Helops*) **N**: MO
monodi Alluaud, 1935: 41 (*Helops*) **N**: MR (Salvages)
nigroaeneus Küster, 1850b: 31 (*Helops*) **E**: FR SP
alicantinus Kraatz, 1883c: 395 (*Nalassus*)
subaeneus Baudi di Selve, 1876c: 258 [= 1877b: 49] (*Helops*)
nitens Wollaston, 1864: 506 (*Helops*) **N**: CI (Tenerife)
nitidicollis Lucas, 1846: 354 (*Helops*) **N**: AG MO
nigricans Vauloger de Beaupré, 1900: 688 (*Helops*)
thalassinus Allard, 1877a: 181 (*Diastixus*)
obtusatus Fairmaire, 1883e: xliv (*Helops*) **N**: AG TU
deserticola Vauloger de Beaupré, 1900: 688 (*Helops*)
picescens Wollaston, 1864: 509 (*Helops*) **N**: CI
piger Vauloger de Beaupré, 1900: 705 (*Helops*) **N**: TU
piliger Har. Lindberg, 1953: 10 (*Cylindronotus*) **N**: CI (Gran Canaria)
porrectus Wollaston, 1864: 508 (*Helops*) **N**: CI
portosanctanus Wollaston, 1854: 521 (*Helops*) **N**: MR (Porto Santo)
pulvinatus Vauloger de Beaupré, 1900: 704 (*Helops*) **N**: AG
puncticollis Lucas, 1846: 349 (*Helops*) **N**: AG TU
quadratus Brullé, 1838: 70 (*Helops*) **N**: CI (Gran Canaria)
rimosus Wollaston, 1864: 508 (*Helops*) **N**: CI (Fuerteventura)
rufulus Vauloger de Beaupré, 1900: 708 **N**: AG
sabulicola Israelson, 1980: 192 **N**: CI (Lanzarote)
sefruensis Antoine, 1949b: 137 **N**: MO (Moyen Atlas)
subdepressus Wollaston, 1857a: 158 (*Helops*) **N**: MR
tenuipunctatus Har. Lindberg, 1950: 12 (*Cylindronotus*) **N**: CI (Tenerife)
transversus Brullé, 1838: 70 (*Helops*) **N**: CI
tuberculipennis ramirezi Español, 1961e: 304 **E**: SP (Cádiz, Gibraltar)
tuberculipennis schrammi Antoine, 1949b: 135 **N**: MO
tuberculipennis tuberculipennis Lucas, 1846: 349 (*Helops*) **E**: SP (Gibraltar) **N**: AG MO
tuberculipennis vilarrubiai Español, 1943c: 144 (*Cylindronotus*) **N**: MO (Tiznit)

genus *Nipponohelops* Masumoto, Ando & Akita, 2006: 33 type species *Nipponohelops ishikawai* Masumoto, Ando & Akita, 2006
ishikawai Masumoto, Ando & Akita, 2006: 34 **A**: JA (Honshu)

genus *Probaticus* Seidlitz, 1896a: 764 type species *Helops mori* Brullé, 1832
subgenus *Helopidoxus* Reitter, 1922b: 44 type species *Helops superbus* Mulsant & Godart, 1855
superbus Mulsant & Godart, 1855b: 153 [= 1855a: 83] (*Helops*) **E**: FR (Corse)
subgenus *Helopostygnus* Antoine, 1949b: 133 type species *Helops atlantis* Antoine, 1926
atlantis atlantis Antoine, 1926: 258 (*Helops*) **N**: MO (Grand Atlas)

atlantis secedens Antoine, 1933b: 230 (*Helops*) **N**: MO
confinalis confinalis Antoine, 1949b: 133 **N**: MO
confinalis kertensis Español, 1953a: 78 **N**: MO
longulus Reiche, 1862a: 373 (*Helops*) **N**: AG
subgenus *Helopotrichus* Reitter, 1922b: 44 type species *Helops villosipennis* Lucas, 1846
gibbithorax Gemminger, 1870a: 123 (*Helops*) [RN] **E**: IT (Sardegna)
gibbicollis Küster, 1850b: 86 (*Helops*) [HN]
sphaericollis Küster, 1850b: 87 (*Helops*) **E**: IT
tomentosus Reitter, 1906g: 376 (*Helops*) **E**: IT (Sicilia)
siculus Canzoneri, 1960b: 50 (*Nephodinus*)
villosipennis Lucas, 1846: 350 (*Helops*) **N**: AG TU
normandi Vauloger de Beaupré, 1900: 681
subgenus *Pelorinus* Vauloger de Beaupré, 1900: 675 type species *Helops anthracinus* Germar, 1813
adimonius Allard, 1876a: 35 (*Helops*) **A**: TR
anthracinus Germar, 1813: 124 (*Helops*) **E**: FR PT SP **N**: AG MO
coriaceus Küster, 1851: 66 (*Helops*)
anthrax Seidlitz, 1896a: 711 (*Helops*) **E**: IT
balearicus Español, 1980: 108 **E**: SP (Baleares)
bodemeyeri Reitter, 1900a: 156 (*Helops*) **A**: TR
chorasanicus G. S. Medvedev, 1976: 893 (*Probaticus*) **A**: IN
corrugatus Seidlitz, 1896a: 707 (*Helops*) **A**: TR
corrugosus Seidlitz, 1896a: 707 (*Helops*) **E**: GR **A**: TR
dieckii Kraatz, 1870: 132 (*Helops*) **E**: SP
castilianus Reitter, 1922b: 40
dorsalis dorsalis Allard, 1877a: 97 (*Helops*) **A**: TR
dorsalis violaceus Pic, 1899p: 411 **A**: TR
ebeninus cassolai Ardoin, 1973: 306 **E**: IT (Sardegna)
ebeninus ebeninus A. Villa & J. B. Villa, 1838: 63 (*Helops*) **E**: FR IT
germari Küster, 1850b: 20 (*Helops*)
puncticollis Rey, 1892: 65 (*Helops*)
robustus Mulsant, 1854: 320 (*Helops*)
euboeuicus Reitter, 1885d: 382 (*Helops*) **E**: GR
foveolatus Seidlitz, 1896a: 710 (*Helops*) **E**: SP
obesus Allard, 1876a: 32 (*Helops*) [HN]
tardus Vauloger de Beaupré, 1900: 682 [RN]
freyi Español, 1956a: 109 **E**: SP
fulei Küster, 1850b: 21 (*Helops*) **E**: BU GR
glyptus Antoine, 1949b: 132 **N**: MO
gorganicus G. S. Medvedev, 1976: 892 (*Probaticus*) **A**: IN
grajus Allard, 1876a: 34 (*Helops*) **E**: GR
granicollis Seidlitz, 1896a: 716 (*Helops*) **E**: TR
granosus Seidlitz, 1896a: 716 (*Helops*) **A**: TR
granulatus Allard, 1876a: 32 (*Helops*) **E**: PT SP
cordicollis Baudi di Selve, 1876c: 245 (*Helops*)
ehlersi Kraatz, 1882a: 47 (*Helops*)
granulifer Seidlitz, 1896a: 711 (*Helops*) **E**: PT SP
immarginatus Reitter, 1902g: 212 (*Helops*) **A**: TR
interstitialis Küster, 1850b: 25 (*Helops*) **E**: SP
elongatus Español, 1956a: 100
recticollis Español, 1956a: 100
kobeltii Seidlitz, 1896a: 717 **N**: TU
lacertosus Küster, 1845a: 30 (*Helops*) **E**: BU GR MC **A**: TR
laticollis Küster, 1850b: 76 (*Helops*) **E**: FR (Pyrénées) SP
cerberus Mulsant, 1854: 316 (*Helops*)
ponferradanus Reitter, 1922b: 38
linearis Vauloger de Beaupré, 1900: 681 (*Helops*) [RN] **N**: AG

[*cribripennis* sensu Allard, 1877]
maurus Waltl, 1838: 464 (*Helops*) **E**: BU GR MC MD RO TR UK **A**: TR
gibbus Küster, 1851: 60 (*Helops*)
melas Küster, 1850b: 22 (*Helops*) **E**: SP
micantipennis Allard, 1876a: 55 (*Stenomax*) **E**: PT
myops Allard, 1876a: 50 (*Helops*) **A**: TR "Kurdistan"
obesus Frivaldszky von Frivald, 1835: 262 (*Helops*) **E**: BU GR MC TR **A**: TR
oliveirae Seidlitz, 1896a: 709 **E**: PT SP
parthorum G. S. Medvedev, 1976. 891 (*Probaticus*) **A**: IN
pentheri Reitter, 1905e: 278 (*Helops*) **A**: TR
peropacus Reitter, 1902g: 214 (*Helops*) **A**: TR (Erzurum)
picianus Reitter, 1907a: 30 (*Helops*) [RN] **A**: TR
carinatus Pic, 1899p: 411 (*Helops*) [HN]
priapus Seidlitz, 1896a: 751 (*Helops*) **A**: TR
prometheus Reitter, 1902g: 213 (*Helops*) **E**: AB **A**: IN (Talysh)
quadricollis Baudi di Selve, 1876c: 246 **E**: AB (Talysh)
quadraticollis Ménétriés, 1832: 196 (*Hedyphanes*) [HN]
subrugosus Duftschmid, 1812: 284 (*Blaps*) **E**: BU CR GR HU KZ MD MC RO SK ST UK
arboreus Germar, 1824b: 160 (*Helops*) [HN]
damascenus Küster, 1850b: 28 (*Helops*)
rugosus Küster, 1850b: 26 (*Helops*)
schmidti Ahrens, 1814, 2: 3 (*Helops*)
strigicollis Seidlitz, 1896a: 712 (*Helops*) [HN]
strigithorax Reitter, 1906a: 497 (*Helops*) [RN]
sulcatus Fischer von Waldheim, 1844: 124 (*Helops*)
tarsatus Küster, 1850b: 29 (*Helops*) **E**: BU MD RO
tazekensis Antoine, 1952: 116 **N**: MO
tenebricosus Brullé, 1832: 223 (*Helops*) **E**: AL BU GR RO YU **A**: TR
theanus Reitter, 1902g: 213 (*Helops*) **A**: TR
tripolitanus Seidlitz, 1896a: 717 **N**: LB
valdanii Guérin-Méneville, 1859: cxc (*Helops*) **N**: AG
vicinus Allard, 1876a: 34 (*Helops*) **E**: AB (Nakhichevan) AR
zoroaster Seidlitz, 1896a: 714 (*Helops*) **A**: TM (Kopet Dag)
subgenus *Probaticus* Seidlitz, 1896a: 764 type species *Helops mori* Brullé, 1832
huedepohli Kulzer, 1964: 228 **A**: IN
mori Brullé, 1832: 222 (*Helops*) **E**: GR
angulicollis Fairmaire, 1883k: clxx (*Helops*)
terrenii Frivaldszky von Frivald, 1845: 183 (*Helops*)
tentyrioides Küster, 1851: 71 (*Helops*) **E**: GR **A**: TR

genus *Probaticus*, nomina dubia
nitidipennis Küster, 1850b: 23 (*Helops*) **E**: IT (Sicilia)
rufipes Allard, 1876a: 33 (*Helops*) **A**: TR

genus *Raibosceles* Allard, 1876a: 5 type species *Helops corvinus* Küster, 1850
Rhaebosceles Rey, 1878: 69 [unjustified emendation]
angustitarsis Reitter, 1902g: 210 (*Helops*) **E**: AR **A**: TR
azureus azureus Brullé, 1832: 224 (*Helops*) **E**: GR IT (Sicilia)
tumidicollis Küster, 1850b: 45 (*Helops*)
azureus stichi Mader, 1943: 42 (*Raiboscelis*) **E**: GR (Is. Thera)
coelestinus coelestinus Waltl, 1838: 464 (*Helops*) **E**: GR **A**: TR
quadraticollis Küster, 1850b: 48 (*Helops*)
coelestinus rugipleuris Reitter, 1922b: 30 **E**: GR (Ródos)
corvinus bilyi Picka, 1984: 26 **E**: GR (Gaudos Is.)
corvinus brodskyi Picka, 1984: 24 **E**: GR (Kríti: Matala)
corvinus corvinus Küster, 1850b: 44 (*Helops*) **E**: GR (Kríti) **A**: TR

corvinus slamai Picka, 1984: 25 **E**: GR (Kríti: Nida-Ida)
cyprius Allard, 1876a: 44 **A**: CY
eleodinus Reiche, 1861a: 5 (*Helops*) **A**: IS TR (Antakya)
latimargo Seidlitz, 1896a: 705 (*Helops*) **A**: TR
obliteratus Allard, 1876a: 56 **E**: GR
syriacus immargo Reitter, 1922b: 30 **A**: TR
syriacus syriacus Reiche, 1861a: 6 (*Helops*) **A**: SY TR
iconiensis Pic, 1900d: 31 (*Helops*)

genus *Sabularius* Escalera, 1914: 353 type species *Helops fossor* Escalera, 1914
fossor Escalera, 1914: 354 (*Helops*) **N**: MO
hoplites Antoine, 1937b: 177 **N**: MO
kocheri Antoine, 1953b: 215 **N**: MO
cephalotes Antoine, 1953b: 217
mimeuri Antoine, 1937b: 179 **N**: MO
punctatostriatus Antoine, 1953b: 214 **N**: MO
serrulifer Antoine, 1937b: 178 **N**: MO
talpa Antoine, 1937b: 175 **N**: MO

genus *Stenohelops* Reitter, 1922b: 113 type species *Isopedus plicatulus* Kraatz, 1880
subgenus *Gunarellus* Reitter, 1922b: 114 type species *Helops gratus* J. Frivaldszky, 1894
calvus Seidlitz, 1896a: 719 (*Helops*) **N**: MO TU
carlofortinus Leo, 1980b: 36 **E**: IT (Sardegna)
gratus J. Frivaldszky, 1894: 85 (*Helops*) **E**: RO YU
macellus Kraatz, 1870: 136 (*Helops*) **E**: SP
mauritanicus Seidlitz, 1896a: 719 (*Helops*) **N**: MO
piligerus Kraatz, 1870: 135 (*Helops*) **E**: SP
silvestrei Escalera, 1914: 349 (*Helops*) **N**: MO
tronqueti F. Soldati & L. Soldati, 2002a: 139 **E**: SP
villosus L. W. Schaufuss, 1861: 92 (*Helops*) **E**: SP
subgenus *Stenohelops* Reitter, 1922b: 113 type species *Isopedus plicatulus* Kraatz, 1880
berettai Marcuzzi, 2001: 1 **N**: TU (Djerba)
filum Reitter, 1922b: 114 **E**: SP
montanus Kraatz, 1870: 137 (*Helops*) **E**: PT
plicatulus Kraatz, 1880d: 299 (*Isopedus*) **E**: SP
prolixus Seidlitz, 1896a: 748 (*Helops*) **E**: SP
sublinearis Kraatz, 1870: 139 (*Helops*) **E**: SP
subgenus *Stenomaleis* Español, 1957b: 22 type species *Stenohelops ardoini* Español, 1957
ardoini Español, 1957b: 22 **E**: SP
effeminatus Español, 1963f: 392 **N**: MO (Rif)
protensulus Seidlitz, 1896a: 735 (*Helops*) **E**: SP
pyrenaeus Mulsant, 1854: 374 (*Helops*) **E**: FR SP
genus *Stenohelops*, species incertae sedis
carinatus Seidlitz, 1896a: 719 (*Helops*) **E**: "Caucasus"
narcissus Reitter, 1922b: 114 **E**: IT (Sicilia)

genus *Tarpela* Bates, 1870: 272 type species *Tarpela brownii* Bates, 1870
Lamperos Allard, 1876a: 4 type species *Helops micans* Fabricius, 1798
akitai Masumoto, 1998b: 317 **A**: JA (Ryukyus)
amamiensis Kaszab, 1964a: 7 **A**: JA (Ryukyus)
andoi Masumoto, 1993d: 134 **A**: JA (Yakushima)
brunnea brunnea Marseul, 1876c: 140 (*Helops*) **A**: JA
japonica Allard, 1876a: 46 (*Lamperos*)
brunnea konoi Nakane, 1963a: 28 **A**: JA (Ryukyus)

clypealis Kaszab, 1954: 260 **A**: FUJ
cordicollis Marseul, 1876c: 141 (*Helops*) **A**: JA SC
elegantula Lewis, 1894: 478 (*Lamperos*) **A**: JA SC
formosana Masumoto, 1981c: 35 **A**: TAI
kimurai Masumoto, 1996a: 213 **A**: JA (Ryukyus)
lewisi Masumoto, 1993d: 132 **A**: JA
magyari Kaszab, 1968a: 12 **A**: NC
subasperipennis Kaszab, 1954: 261 **A**: FUJ
todai Ando, 2006: 29 **A**: JA (Ishigaki Is.)
tokunoshimana Masumoto & Akita, 2001a: 20 **A**: JA (Ryukyus)
tsushimana Nakane, 1979: 114 **A**: JA (Tsushima)
zoltani Masumoto, 1981c: 34 **A**: TAI

tribe Helopinini Lacordaire, 1859

genus *Drosochrus* Erichson, 1843: 243 type species *Drosochrus brunnipes* Erichson, 1843
subgenus *Helopinus* Solier, 1848: 197 type species *Helopinus costatus* Solier, 1848
costatus aegyptiacus Gridelli, 1939b: 198 (*Helopinus*) **N**: EG **A**: SI **AFR**
costatus costatus Solier, 1848: 199 (*Helopinus*) **N**: EG **A**: JO SA **AFR**
  misolampoides Lacordaire, 1876: pl. 58 (*Helopinus*)
pilosus Gebien, 1921: 122 (*Helopinus*) **A**: YE
psalidiformis Ancey, 1881a: 397 (*Helopinus*) **A**: YE

genus *Micrantereus* Solier, 1848: 175 type species *Acanthomerus anomalus* Guérin-Méneville, 1834
  Solenomerus Fåhraeus, 1870: 306 type species *Solenomerus longipes* Fåhraeus, 1870
arabicus Blair, 1933: 6 **A**: SA
desertus Koch, 1965a: 146 **A**: SA
nitidus nitidus Gahan, 1896b: 455 **A**: SA YE
nitidus yemensis Koch, 1965a: 145 **A**: SA YE
szalaymarzsoi Kaszab, 1972: 382 **A**: YE
tentyrioides Pascoe, 1882: 29 **A**: YE

tribe Melanimini Seidlitz, 1894 (1854)

genus *Cheirodes* Gené, 1839: 73 type species *Cheirodes sardous* Gené, 1839
subgenus *Cheirodes* Gené, 1839: 73 type species *Cheirodes sardous* Gené, 1839
  Anemia Laporte, 1840: 218 type species *Anemia granulata* Laporte, 1840
  Chirodes Agassiz, 1846b: 81 [unjustified emendation]
aethiopia Ardoin, 1971b: 376 (*Anemia*) **A**: YE **AFR**
chobauti Reitter, 1898f: 349 (*Anemia*) **N**: AG LB MO TU **A**: IS SA YE **AFR**
  reitteri Pic, 1899n: 278 (*Anemia*) [RN]
dentipes Ballion, 1878: 332 (*Microzoum*) **E**: AB ST **A**: AF IN KZ MG SA TM UZ XIN
  reichardti Kontkanen, 1956: 57 (*Anemia*)
drurei Pic, 1923f: 5 (*Anemia*) **A**: IQ
granulatus Laporte, 1840: 218 (*Anemia*) **A**: SA **AFR**
sardous denticulatus Wollaston, 1867: 197 (*Anemia*) **A**: OM YE **AFR**
sardous sardous Gené, 1839: 73 **E**: AB AR FR GR IT PT SP **N**: CI EG LB MO TU **A**: CY IN IQ IS JO SI TR **AUR**i
  caulobioides Carter, 1920: 224 (*Anemia*)
sinuatifrons Reitter, 1887d: 389 (*Anemia*) **A**: XIN
subgenus *Histiaea* Fairmaire, 1892a: 107 type species *Histiaea bidentula* Fairmaire, 1892
ardoini Kaszab, 1972: 378 (*Anemia*) **A**: YE
asperulus Reitter, 1884f: 260 (*Anemia*) **N**: EG LB TU **A**: IN IS JO SA SI SY YE **AFR**
  aegyptiacus Pic, 1936h: 300 (*Anemia*) [RN]
  denticulatus Pic, 1923f: 5 (*Anemia*) [HN]
  seriesetosus Baudi di Selve, 1894: 8 (*Anemia*)
bidentulus Fairmaire, 1892a: 108 (*Histiaea*) **A**: SA **AFR**

subgenus *Pseudanemia* Wollaston, 1864: 492 type species *Pseudanemia brevicollis* Wollaston, 1864
Ammidanemia Reitter, 1904a: 132 type species *Anemia fausti* Solsky, 1882
brevicollis Wollaston, 1864: 493 **E**: IT (South, Sicilia) MA SP **N**: AG CI EG LB MO TU **A**: AF BA IN IS IQ JO KZ MG OM PA SA SI TM UZ XIN YE **AFR**
aphodioides Walker, 1871: 16 (*Anemia*)
curticollis Pic, 1923f: 5 (*Anemia*)
fausti Solsky, 1881b: 52 (*Anemia*)
hauseri Reitter, 1894a: 50 (*Anemia*)
pharao Reitter, 1897b: 43 (*Anemia*)
rotundicollis Desbrochers des Loges, 1881: 126 (*Anemia*)
striolatus Fairmaire, 1894c: 311 (*Anemia*)
submetallicus Raffray, 1873: 377 (*Anemia*) **E**: SP (Gibraltar) **N**: AG EG LB MO TU **A**: JO SA **AFR**
palaestinus Pic, 1899n: 278 (*Anemia*)
zhengi Ren & Yu, 1994: 88 **A**: NIX
subgenus *Spinanemia* Ardoin, 1971b: 378 type species *Anemia cornuta* Pic, 1898
cornutus cornutus Pic, 1898h: 121 (*Anemia*) **A**: AF IN SA
cornutus panelii Ardoin, 1971b: 379 (*Anemia*) **A**: SA **AFR**
pilosus Tournier, 1868: 146 (*Anemia*) **N**: AG CI EG LB MO TU **A**: IS (Negev) OM SA SI YE **AFR**
fenyesi Reitter, 1897b: 42 (*Anemia*)
fissidens Reitter, 1898f: 348
thoracicus Fairmaire, 1891b: 250 (*Anemia*) **A**: SA **AFR**

genus *Dolamara* Reichardt, 1935: 251 type species *Melanimon cupreomicans* Reitter, 1915
cupreomicans Reitter, 1915e: 88 (*Melanimon*) **A**: KI MG TD UZ

genus *Melanimon* Steven, 1829: 18 type species *Opatrum tibiale* Fabricius, 1781
Fungulus Gistel, 1848: 125 [RN]
Microzoom Hope, 1840: 112 [unjustified emendation]
Microzoum Dejean, 1834: 193 type species *Opatrum tibiale* Fabricius, 1781
inermus Picka, 1983b: 265 **E**: BU
kiritshenkoi Reichardt, 1936a: 68 **E**: AB (Talysh) **A**: TR
tibialis subaereus Reitter, 1904a: 129 **A**: KI KZ
tibialis tibialis Fabricius, 1781: 90 (*Opatrum*) **E**: AU BE BH BU BY CR CZ DE EN FI FR GB GE GR HU IR IT LA LT NL NR NT PL RO SK SL SP ST SV SZ UK **N**: MO **A**: KZ TM UZ WS

tribe Opatrini Brullé, 1832

subtribe Heterocheirina Koch, 1956

genus *Diphyrrhynchus* Fairmaire, 1849: 445 type species *Diphyrrhynchus chalcens* Fairmaire, 1849
Abantiades Fairmaire, 1894e: 395 [RN] type species *Abantis aenescens* Fairmaire, 1892
Abantis Fairmaire, 1892a: 109 [HN] type species *Abantis aenescens* Fairmaire, 1892
Acanthosternus Montrouzier, 1860: 289 type species *Acanthosternus halorageos* Montrouzier, 1860
Neoabantis Gebien, 1910a: 341 [RN] type species *Abantis aenescens* Fairmaire, 1892
aenescens Fairmaire, 1892a: 110 (*Abantis*) **A**: SA SI YE **AFR**
dilatipennis Schawaller, 1991b: 290 **A**: IN OM YE (Suqutra)
iriomotensis M. Chûjô, 1966: 4 (*Diphyrrhinchus*) **A**: JA TAI **ORR**
oharensis botelensis Nakane, 1973b: 105 **A**: TAI
oharensis ohariensis Nakane, 1973b: 104 **A**: JA (Ryukyus) **ORR**
shibatai Kaszab, 1964c: 43 (*Diphyrrhinchus*) **A**: JA (Amami, Asani)

tribe Heterocheirini, species incertae sedis
Diphyrrhynchus geminatus Allard, 1896: 320 **A**: SD

subtribe Heterotarsina Blanchard, 1845

genus *Heterotarsus* Latreille, 1829: 26 type species *Heterotarsus tenebrioides* Guérin-Méneville, 1831
Helopimorphus Desbrochers des Loges, 1881: 141 type species *Helopimorphus angulipennis* Desbrochers des Loges, 1881

Hopatropteron Reitter, 1889k: 701 type species *Hopatropteron subcostatum* Reitter, 1889
Oubanghinum Pic, 1933d: 4 type species *Oubanghinum atrum* Pic, 1933
aphodioides Fairmaire, 1886c: 346 **A**: CH
auricularis Marseul, 1876c: 128 **A**: CH TAI **ORR**
subcrenatus Fairmaire, 1886c: 345
carinula Marseul, 1876c: 127 **A**: ANH FE FUJ GAN GUI HAI HUB JA JIA NC SC SCH TAI ZHE **ORR**
subcostatus Reitter, 1889k: 701 (*Hopatropteron*)
crenulifer Kaszab, 1941c: 58 **A**: JA (Ryukyus) TAI
indicus indicus Marseul, 1876c: 128 **A**: NP SD YUN **ORR**
inflatus Lacordaire, 1859: 399 **A**: NP YUN **ORR**
subinflatus Pic, 1923j: 75
metallifer Kaszab, 1976a: 50 **A**: YUN
pustulifer Fairmaire, 1889a: 361 **A**: ANH FUJ GAN JIA **ORR**
topali Kaszab, 1976a: 48 **A**: NP **ORR**
urbani Kaszab, 1941c: 59 **A**: TAI

subtribe Neopachypterina Bouchard, Löbl & Merkl, 2007

genus *Neopachypterus* Bouchard, Löbl & Merkl, 2007: 386 [RN] type species *Pachypterus mauritanicus* Lucas, 1846
Pachypterus Lucas, 1846: 325 [HN] type species *Pachypterus mauritanicus* Lucas, 1846
gebieni Kaszab, 1941b: 18 (*Pachypterus*) **A**: SD **ORR**
mauritanicus Lucas, 1846: 325 (*Pachypterus*) **E**: GR SP **N**: AG MO TU
niloticus niloticus Miller, 1861: 205 (*Pachypterus*) **N**: EG
serrulatus Reitter, 1904a: 75 (*Pachypterus*) **E**: AB AR **A**: AF KZ TD TM UZ
zarudnyi G. S. Medvedev, 1968a: 251 (*Pachypterus*) **A**: IN

subtribe Opatrina Brullé, 1832

genus *Adavius* Mulsant & Rey, 1859e: 138 [= 1860: 46] type species *Adavius clavipes* Mulsant & Rey, 1859
Udebra Reitter, 1896i: 236 type species *Udebra hauseri* Reitter, 1896
clavipes Mulsant & Rey, 1859e: 138 [= 1860: 46] **A**: XIZ **ORR**
fimbriatus Ménétriés, 1849: 217 (*Erodius*) **E**: ST **A**: AF IN KZ TM UZ
hauseri Reitter, 1896i: 236 (*Udebra*)

genus *Ammobius* Guérin-Méneville, 1844: 121 type species *Trachyscelis rufus* Latreille (= *Trachyscelis rufus* Lucas, 1846)
Ammophtorus Lacordaire, 1859: 284 type species *Trachyscelis rufus* Lucas, 1846
asperatus Champion, 1894b: 448 (*Ammophthorus*) **A**: GUA (Nam'ao Is.)
cyprius Grimm, 1991: 29 **A**: CY TR
rufus Lucas, 1846: 339 (*Trachyscelis*) **E**: AL BU CR FR GG GR IT MA MC RO SP ST UK **N**: AG LB MO TU **A**: CY IS LE SI SY TR
insularis Reitter, 1893e: 315
jakovlevi Semenov, 1905: 123
minotaurus Obenberger, 1922: 41
rugosus Rosenhauer, 1856: 214
genus *Amphithrixoides* Bouchard & Löbl, nom. nov. [see New Acts] type species *Amphithrix peyerimhoffi* Español, 1953
Amphithrix Español, 1953c: 306 [HN] type species *Amphithrix peyerimhoffi* Español, 1953
peyerimhoffi Español, 1953c: 309 **N**: AG

genus *Anatrum* Reichardt, 1936a: 84 type species *Anatrum songoricum* Reichardt, 1936
shandanicum Ren, 1999: 226 **A**: GAN
songoricum Reichardt, 1936a: 85 **A**: MG XIN

genus *Asiocaedius* G. S. Medvedev & Nepesova, 1985a: 138 [RN] type species *Pseudocaedius kiseritzkii* Medvedev, 1966

Pseudocaedius G. S. Medvedev, 1966: 98 [HN] type species *Pseudocaedius kiseritzkii* Medvedev, 1966

kiseritzkii G. S. Medvedev, 1966: 98 (*Pseudocaedius*) **A**: TM

kuznetzovi G. S. Medvedev, 1970: 400 (*Pseudocaedius*) **A**: TM UZ

genus *Brachyesthes* Fairmaire, 1868: 490 type species *Melanesthes pilosella* Marseul, 1867

brevior Fairmaire, 1883h: cxiii [RN] **N**: AG LB TU

chrysomelina A. Costa, 1878: 23 (*Caedius*) **N**: EG **A**: TR **ORR**

indica Kaszab, 1938d: 230

gastonis gastonis Fairmaire, 1873b: 392 **N**: AG EG LB

gastonis kebiliensis Picka, 1988: 111 (*Brachyestes*) **N**: TU

olexai Picka, 1988: 111 **N**: AG

peyerimhoffi Koch, 1937: 473 **N**: LB

pilosella pilosella Marseul, 1867: xxxvi (*Melanesthes*) **N**: AG MO TU

approximans Fairmaire, 1873b: 391

indentata Pic, 1929d: 6

pilosella salina Koch, 1937: 473 **N**: LB

genus *Brachyidium* Fairmaire, 1883b: 33 type species *Brachyidium breviusculum* Fairmaire, 1883

Cnemodasus Gebien, 1914d: 374 type species *Cnemodasus rectangulus* Gebien, 1913

iwojimae Masumoto, 1991b: 27 **A**: JA (Iwojima Is., Volcano Is.)

palauense Kulzer, 1957: 196 (*Caedius*) **A**: JA (Volcano Is.) **AUR**

genus *Caediexis* Lebedev, 1932: 125 type species *Caediexis arenicola* Lebedev, 1932

arenicola Lebedev, 1932: 125 **A**: TD TM UZ

genus *Caedius* Blanchard, 1845: 13 type species *Opatrum sphaeroides* Hope, 1842

Caedius Mulsant & Rey, 1859: 126 [HN] type species *Caedius latipes* Mulsant & Rey, 1859

Isarida Pascoe, 1866: 456 type species *Isarida testacea* Pascoe, 1866

cassidoides Fairmaire, 1894f: 661 **N**: EG

chinensis Kaszab, 1942a: 34 **A**: FUJ **ORR**

ciliger Mulsant & Rey, 1859e: 135 [= 1860: 43] **A**: XIZ

eberti Kaszab, 1980a: 145 **A**: AF **ORR**

fluviatilis Nakane & Masumoto, 1979: 112 **A**: FE JA

formosanus Kaszab, 1942a: 39 **A**: TAI

fulvus Mulsant & Rey, 1859e: 133 [= 1860: 41] **A**: NP PA UP XIZ

asper Fairmaire, 1896a: 18

testaceus Pascoe, 1866: 456

himalayensis Kaszab, 1941d: 119 **A**: BT HP NP

indicus Fairmaire, 1896a: 18 **A**: NP **ORR**

maderi maderi Kaszab, 1942a: 35 **A**: JA "Korea" **ORR**

maderi nakanei Ferrer, 2003: 303 [RN] **A**: JA (Ryukyus)

minor Nakane, 1963a: 26 [HN]

marinus Marseul, 1876c: 96 **A**: JA NC

obesus Kulzer, 1964: 266 **A**: JA "S. China" **AUR ORR**

orientalis Fairmaire, 1893g: 316 (*Cyptus*) **A**: TAI **ORR**

shoitii Mt. Chûjô, 1992: 34 **A**: SC

yangi Ren, 1995a: 24 **A**: HEB

genus *Clitobius* Mulsant & Rey, 1859e: 141 [= 1860: 49] type species *Clitobius sabulicola* Mulsant & Rey, 1859 (= *Opatrum ovatum* Erichson, 1843)

subgenus *Apteroclitobius* Koch, 1960: 391 type species *Clitobius grandis* Fairmaire, 1896

cribricollis Allard, 1882b: lxxxvi (*Halonomus*) **A**: YE **AFR**

subgenus *Clitobius* Mulsant & Rey, 1859e: 141 [= 1860: 49] type species *Clitobius sabulicola* Mulsant & Rey, 1859 (= *Opatrum ovatum* Erichson, 1843)
Halonomus Wollaston, 1861a: 201 type species *Halonomus grayii* Wollaston, 1861 (= *Opatrum ovatus* Erichson, 1843)
oblongiusculus oblongiusculus Fairmaire, 1875b: 495 (*Halonomus*) **E**: AB **N**: AG EG LB MO TU **A**: CY IN IQ IS JO SA SI TD TM UZ
lineicollis Fairmaire, 1879a: 12 (*Halonomus*)
oblongiusculus schneideri Allard, 1882b: lxxxvi (*Halonomus*) **N**: EG **AFR**
ovatus ovatus Erichson, 1843: 249 (*Opatrum*) **E**: IT MA **N**: AG CI EG LB MO TU **A**: TR **AFR ORR**
grayii Wollaston, 1861a: 203 (*Halonomus*)
immarginatus Fairmaire, 1888c: 195
sabulicola Mulsant & Rey, 1859e: 141 [= 1860: 49] (*Clitobius*)
ovatus salinicola Wollaston, 1861a: 203 (*Halonomus*) **N**: CI MO
opacus Har. Lindberg, 1950: 12

genus *Cyptus* Gerstaecker, 1871: 61 type species *Cyptus scabrosus* Gerstaecker, 1871
Neocaedius Pierre, 1972: 960 type species *Caedius halli* Kaszab, 1949
aegyptiacus Mulsant & Rey, 1859e: 136 [= 1860: 44] (*Caedius*) **N**: EG **A**: AF IS **AFR**
alternatus Fairmaire, 1894d: 321 (*Coedius*) **N**: AG TU **A**: SA YE **AFR**
auriculatus Chobaut, 1924b: 154 (*Caedius*) **N**: AG **AFR**

genus *Dilamus* Jacquelin du Val, 1861: 279 type species *Boros rufipes* Lucas, 1846
subgenus *Choresmolamus* G. S. Medvedev, 1978a: 150 type species *Dilamus oxianus* G. S. Medvedev, 1978
oxianus G. S. Medvedev, 1978a: 150 **A**: UZ
subgenus *Dilamus* Jacquelin du Val, 1861: 279 type species *Boros rufipes* Lucas, 1846
arabicus Kaszab, 1979c: 286 **A**: SA
aralensis G. S. Medvedev, 1978a: 149 **A**: UZ (Karapalkistan)
caspicus G. S. Medvedev, 1978a: 147 **A**: TM
gnom Skopin, 1961b: 405 **A**: KZ MG
grandiceps Bogatchev & Kryzhanovskiy, 1960: 269 **A**: TM UZ
gurjevae G. S. Medvedev, 1972b: 491 **A**: MG
iranicus G. S. Medvedev, 1968a: 259 **A**: IN
laticollis Baudi di Selve, 1876c: 241 **A**: "Palestina"
mandli Kaszab, 1963a: 75 **A**: IN
mongolicus Kaszab, 1968c: 380 **A**: MG
myrmecophilus Skopin, 1966: 331 **A**: KZ
obsoletus obsoletus Baudi di Selve, 1876c: 240 **A**: IS
obsoletus transjordanicus Koch, 1936: 300 **A**: JO
pardilloi Español, 1956b: 277 **N**: MO (Western Sahara)
planicollis laticeps Peyerimhoff, 1931b: 103 **N**: AG (Hoggar)
planicollis planicollis Fairmaire, 1883h: xliv **E**: IT **N**: AG EG LB MO TU
rufipes Lucas, 1846: 338 (*Boros*) **E**: IT SP **N**: AG TU
andalusiae Baudi di Selve, 1876c: 240
congener Rottenberg, 1871: 250
rungsi Antoine, 1955: 363 **N**: MO
scoparius Antoine, 1952: 114 **N**: MO (Anti Atlas)
tangerianus Desbrochers des Loges, 1881: 132 **N**: MO
volkonskyi battareli Peyerimhoff, 1943: 24 **N**: AG
volkonskyi volkonskyi Peyerimhoff, 1943: 24 **N**: AG
zarudnyi G. S. Medvedev, 1968a: 262 **A**: IN
subgenus *Ochrolamus* Reitter, 1904a: 73 type species *Dilamus pictus* Baudi di Selve, 1881
pictus Baudi di Selve, 1881: 285 **N**: EG **AFR**

genus *Eumylada* Reitter, 1904a: 170 type species *Myladina punctifera* Reitter, 1889
glandilosa X.-J. Yang & Ren, 2004: 305 **A**: NMO
obenbergeri Schuster, 1933b: 97 (*Myladina*) **A**: NMO

ordosana Reitter, 1889k: 708 (*Myladina*) **A**: NMO
potanini Reitter, 1889k: 708 (*Myladina*) **A**: GAN NIN NMO
punctifera amaroides Reichardt, 1936a: 171 **A**: MG
punctifera punctifera Reitter, 1889k: 707 (*Myladina*) **A**: MG NMO

genus *Eurycaulus* Fairmaire, 1868: 492 type species *Eurycaulus marmottani* Fairmaire, 1868
subgenus *Ammotrypes* Fairmaire, 1879c: 194 type species *Ammotrypes crenulicollis* Fairmaire, 1879
Scleronimon Reitter, 1904a: 127 type species *Eurycaulus granulatus* Reitter, 1904
crenulicollis Fairmaire, 1879c: 195 (*Ammotrypes*) **N**: AG TU
koestlini Español & Viñolas, 1981a: 2
granulatus Reitter, 1904a: 127 **N**: EG
hellmanni Grimm, 2000: 90 **N**: AG
henoni Fairmaire, 1897f: 121 **N**: EG **A**: IS
peyerimhoffi Reitter, 1904a: 127 **A**: IS (Negev) SI
subgenus *Eurycaulinus* Koch, 1937: 468 type species *Eurycalus quedenfeldti* Heyden, 1890
quedenfeldti Heyden, 1890b: 75 (*Eurycalus*) **N**: LB TU
subgenus *Eurycaulus* Fairmaire, 1868: 492 type species *Eurycaulus marmottani* Fairmaire, 1868
buettikeri Schawaller, 1993a: 106 **A**: OM
dubius dubius Gridelli, 1930: 295 (*Scleron*) **N**: EG LB
dubius scleronoides Koch, 1937: 465 **N**: LB
marmottani Fairmaire, 1868: 492 **N**: AG LB TU
kabakovi G. S. Medvedev, 1974: 946 **A**: AF
nasreddinovi G. S. Medvedev, 1970: 947 **A**: AF TD
pachecoi Escalera, 1943b: 221 **N**: MO (Western Sahara)
reymondi Español, 1953c: 303 **N**: AG MO
subgenus *Scleronopsis* Koch, 1935a: 87 type species *Scleron hirsutus* Miller, 1861
hirsutus heliopolos Koch, 1935a: 86 **N**: EG
hirsutus hirsutus Miller, 1861: 204 (*Scleron*) **N**: EG **A**: OM

genus *Falsammidium* Koch, 1960: 395 type species *Clitobius laevipennis* Fairmaire, 1892
rugulipennis Fairmaire, 1896c: 224 (*Clitobius*) **A**: SA YE **AFR**

genus *Falsocaedius* Español, 1943c: 140 type species *Clitobius fossulatus* Escalera, 1914
fossulatus Escalera, 1914: 340 (*Clitobius*) **N**: CI MO

genus *Falsolobodera* Kaszab, 1967a: 24 type species *Falsolobodera skopini* Kaszab, 1967
skopini Kaszab, 1967a: 25 **A**: MG

genus *Gonocephalum* Solier, 1834: 498 type species *Opatrum pygmaeum* Steven, 1829
subgenus *Gonocephalum* Solier, 1834: 498 type species *Opatrum pygmaeum* Steven, 1829
Dasus Motschulsky, 1845a: 78 type species *Opatrum fuscum* "Fabricius" [= *Opatrum rusticum* A. G. Olivier, 1811]
Megadasus Reitter, 1904a: 146 type species *Opatrum lefranci* Fairmaire, 1863
acoriaceum M. Chûjô, 1975: 15 **A**: JA (Ryukyus)
adpressiforme Kaszab, 1951c: 182 **A**: JA (Volcano Is.) TAI GUA **AUR ORR**
aequatoriale Blanchard, 1853: 152 (*Opatrum*) **A**: HAI **ORR**
alaticolle Fairmaire, 1893b: 23 (*Bradymerus*) **A**: SD **ORR**
andrewesi Kaszab, 1952d: 649 **A**: TAI **AUR ORR**
annamita Chatanay, 1917a: 236 **A**: ANH AP FUJ GUA HAI HEN HKG JA NMO YUN **AUR ORR**
arisi Reitter, 1905a: 92 **A**: AF IN TM
assimile Küster, 1849a: 52 (*Opatrum*) **E**: IT
belli Kaszab, 1952d: 513 **A**: NP **ORR**
besnardi besnardi Kaszab, 1982b: 163 **A**: AE OM SA
besnardi muehlei J. Ferrer, 1993a: 91 **A**: SA YE
besnardi pierrei Ferrer, 2004: 87 **A**: IN

bilineatum Walker, 1858: 284 (*Opatrum*) **A**: BT FE FUJ GUA HAI HKG JA NP SCH SD UP YUN "Korea" **AUR NAR ORR**
kamtschaticum Motschulsky, 1860c: 139
latifrons LeConte, 1874a: 70 (*Blapstinus*)
orarium Lewis, 1894: 380 (*Opatrum*)
birmanicum Kaszab, 1952d: 484 **A**: NP TAI UP **ORR**
biseriatum Kaszab, 1975b: 121 **A**: NP
brachelytra Kaszab, 1952d: 632 **A**: AF PA UP **ORR**
calcaripes Karsch, 1881: 59 (*Opatrum*) **E**: FR PT SP **N**: AG MO **AFR**
yelamosi Español & Viñolas, 1983a: 37
catenulatum Fairmaire, 1896a: 19 (*Hopatrum*) **A**: SD **ORR**
chinense Gebien, 1910a: 322 [RN] **A**: HAI YUN **ORR**
subsetosum Reitter, 1904a: 147 [HN]
civicum Kaszab, 1952d: 612 **A**: NP SD UP **ORR**
coarticolle Kaszab, 1952d: 500 **A**: "Indus" **ORR**
coenosum Kaszab, 1952d: 643 **A**: JA FUJ GUA HKG HUB JIA SC SCH TAI XIN ZHE
consobrinum Blair, 1923a: 122 **A**: AF IN IQ PA SA UP YE **AFR ORR**
controversum Gridelli, 1948: 11 [RN] **N**: EG **A**: BA JO SA SY YE **AFR**
coriaceum Motschulsky, 1858a: 34 **A**: FE FUJ GUA GUX JA NMO NP SC SCH SHN TAI ZHE
costatum costatum Brullé, 1832: 218 (*Opatrum*) **E**: AB AR CR GR IT SP TR **A**: CY IN IS SY TR
lineare Küster, 1849a: 54 (*Opatrum*)
costatum rugulosum Küster, 1849a: 55 (*Opatrum*) **E**: AR **A**: CY IN IS SY TR
crassepunctatum Kaszab, 1952d: 526 **A**: BT NP **ORR**
curiosum Kaszab, 1952d: 460 **A**: NP **ORR**
curvicolle Reitter, 1889k: 705 **A**: GAN HP KA NMO SCH XIN
deliense Kaszab, 1952d: 505 **A**: GUX **ORR**
dentipes Kaszab, 1952d: 655 **A**: TAI **ORR**
depressum Fabricius, 1801a: 116 (*Opatrum*) **A**: AF AP BT GUA GUX HAI HP KA NP PA SD TAI YUN **AUR ORR**
dilatatum Wollaston, 1854: 501 (*Opatrum*) **N**: MR (Salvages)
dorsogranosum Fairmaire, 1896a: 19 (*Hopatrum*) **A**: AF PA SD UP **ORR**
foveolatus Kaszab, 1963b: 343
ermischi Kaszab, 1960a: 139 **A**: AF PA
espagnoli Kaszab, 1965a: 111 **A**: AP
ficifolium Mt. Chûjô, 1995a: 113 **A**: JA
formosanum Gebien, 1914b: 4 **A**: TAI
freudei Kaszab, 1960a: 143 **A**: AF PA
gallagheri J. Ferrer, 1997b: 214 **A**: OM
gebienianum Kaszab, 1952d: 470 **A**: SD **ORR**
geneirotundum Ren, 1998: 111 **A**: ZHE
gracile Bates, 1879: 482 (*Penthicus*) **A**: HP KA NP SD XIN XIZ
granulatipenne Kaszab, 1978a: 173 **A**: BT SD
granulatum afghanicum Gridelli, 1954: 283 [= 1955: 72] **A**: AF TM TR
granulatum granulatum Fabricius, 1792a: 90 (*Opatrum*) **N**: MO
granulatum meridionale Küster, 1849a: 57 (*Opatrum*) **E**: IT SP **N**: AG TU
granulatum nigrum Küster, 1849a: 58 (*Opatrum*) **E**: FR IT SP
granulatum pusillum Fabricius, 1792a: 91 (*Opatrum*) **E**: AB AL AR AU BU CR CZ HU GR IT MC PL RO SK SP ST UK YU **N**: MO **A**: AF IN KZ MG NIN SY TM TR XIZ
interstitiale Küster, 1849a: 59 (*Opatrum*)
muricatum Brullé, 1832: 215 (*Opatrum*)
nitidulum Küster, 1849a: 56 (*Opatrum*)
parvulum Lucas, 1846: 334 (*Opatrum*)
pruinosum Fischer von Waldheim, 1844: 128 (*Opatrum*)
ragusanum Küster, 1849c: 52 (*Opatrum*)
subnodosum Rey, 1892: 42
viennense Duftschmid, 1812: 293

guerryi Chatanay, 1917a: 244 **A**: AF HKG HUB PA SCH SD YUN **ORR**
potanini Reichardt, 1936a: 108
guignoti Kaszab, 1960a: 144 **A**: AF PA
hadroide Fairmaire, 1888b: 129 (*Hopatrum*) **A**: BEI SCH
reichardti Gebien, 1939: 447 [RN]
hauschildi Kaszab, 1952d: 498 **A**: GUX NP SD UP "Thibet" **ORR**
helopoide Fairmaire, 1894a: 19 (*Hopatrum*) **A**: NP SD UP **ORR**
himalayense Kaszab, 1952d: 528 **A**: BT NP SD **ORR**
hingstoni Kaszab, 1952d: 524 **A**: SD **ORR**
hintoni Kaszab, 1952d: 668 **A**: XIZ
hiranoi Akita & Masomoto, 2003: 173 **A**: JA (Honshu)
hispidulum Kaszab, 1952d: 657 **A**: FUG HKG ZHE
hoffmannseggii Steven, 1829: 96 (*Opatrum*) **A**: BT NP SD **ORR**
japanum Motschulsky, 1861: 16 **A**: JA "East China" "Korea"
expansicolle Lewis, 1894: 380 (*Opatrum*)
kandaharicum Kaszab, 1960a: 148 **A**: AF
karakorumense Kaszab, 1961f: 353 **A**: KA
klapperichi Kaszab, 1952d: 550 **A**: FUJ JA SHA
kochi Kaszab, 1952d: 578 **A**: FUJ
kondoi Masumoto, 1983b: 173 **A**: JA (Ryukyus) TAI
konoi Kaszab, 1952d: 482 **A**: TAI
koreanum Kaszab, 1952d: 548 **A**: SC
kuenelti Kaszab, 1961f: 351 **A**: KA
kuluanum Kaszab, 1952d: 659 **A**: AF HP NP PA
kuntzeni Kaszab, 1952d: 580 **A**: HAI
lefranci Fairmaire, 1863: 643 (*Opatrum*) **E**: IT (Sardegna) **N**: AG TU
longitarse Kaszab, 1952d: 465 **A**: YUN **ORR**
macrophthalmum Kaszab, 1952d: 582 **A**: JA (Shikoku) **ORR**
malayanum Gebien, 1935: 53 **A**: HAI HEN SC **ORR**
martensi Kaszab, 1977b: 258 **A**: NP
merensi estevezi J. Ferrer, 1993a: 97 **N**: CI (Gran Canaria)
merensi merensi Uyttenboogaart, 1929: 348 **N**: CI (Gran Canaria)
mesopotamicum Blair, 1923a: 123 **A**: IQ
minusculum Fairmaire, 1894a: 20 (*Hopatrum*) **A**: UP **ORR**
miyakense Nakane, 1963a: 26 **A**: JA (Ryukyus.)
modestum Reiche, 1850: 368 (*Opatrum*) **A**: IN SA YE **AFR**
inornatum Schuster, 1938a: 61
sulcipenne C. G. Thomson, 1858: 84 (*Opatrum*)
moluccanum Blanchard, 1853: 153 (*Opatrum*) **A**: FUJ GUA HAI JA TAI **AUR ORR**
muelleri Gridelli, 1945: 23 **A**: SA YE **AFR**
nepalicum Kaszab, 1973a: 32 **A**: BT NP SD
niloticum Ardoin, 1965a: 73 **A**: SA **AFR**
oblitum Wollaston, 1864: 489 (*Opatrum*) **N**: CI
oblongum Fabricius, 1801a: 117 (*Opatrum*) **A**: NP SD **ORR**
obscurum aethiops J. Ferrer, 1993a: 78 **A**: OM SA **AFR**
obscurum obscurum Küster, 1849a: 46 (*Opatrum*) **E**: FR (Alpes-Maritimes, Corse) IT SP **N**: AG EG LB MO **AFR**
parvulum Lucas, 1846: 33 (*Opatrum*)
okinawanum M. Chûjô, *1963b: 149* **A**: JA (Ryukyus.)
outreyi Chatanay, 1917a: 240 **A**: GUA HKG SC TAI YUN **ORR**
parcesetosum Kaszab, 1952d: 516 **A**: NP "North India" **ORR**
papulosum Fairmaire, 1896a: 19 (*Hopatrum*) **A**: PA **ORR**
parallelum Kaszab, 1952d: 606 **A**: AF IQ KA PA **ORR**
patricium Kaszab, 1952d: 604 **A**: PA SD UP **ORR**

patruele patruele Erichson, 1843: 248 (*Opatrum*) **N**: AG CI EG LB MO TU **A**: AE AF IN IS OM SA SI SY YE **AFR**

lutosum Wollaston, 1864: 486 (*Opatrum*)

sericeum Baudi di Selve, 1875d: 701 [= 1876b: 228] (*Opatrum*)

patruele turchestanicum Gridelli, 1948: 9 **A**: AF KZ TM XIN

perplexum Lucas, 1846: 334 (*Opatrum*) [NP] **E**: CR GR IT (Lampedusa, Pantelleria) SP **N**: AG EG LB MO TU **A**: CY IS SA SI SY YE **AFR**

famelicum A. G. Olivier, 1811: 497 (*Opatrum*) [NO]

persimile Lewis, 1894: 381 (*Opatrum*) **A**: FE HEI JA MG NC NMO SC

platipenne Kaszab, 1960a: 149 **A**: AF

pottsi Kulzer, 1957: 194 **A**: JA (Bonin Is.) AUR

prolixum Erichson, 1843: 248 (*Opatrum*) **E**: IT SP **N**: AG CI EG LB MO TU **A**: CY IN SA SY **AFR**

clavipes Wollaston, 1867: 188 (*Opatrum*)

hespericum Motschulsky, 1849: 126 (*Dasus*)

lugens Küster, 1849a: 50 (*Opatrum*)

pseudopubens Kaszab, 1952d: 592 **A**: HAI HKG TAI ZHE **ORR**

pubens Marseul, 1876c: 97 (*Opatrum*) **A**: AP HAI JA NE SC SHN

obtusicolle Kolbe, 1886: 95 (*Opatrum*)

pubiferum Reitter, 1904a: 142 **E**: AB AR ST **A**: AF IN IQ KZ MG OM PA SA TD TM UZ XIN

pygmaeum Steven, 1829: 98 (*Opatrum*) **E**: AU BU CR FR GE GG HU IT SK SL ST SZ UK **A**: KZ TR XIN

pedestre Rosenhauer, 1847: 31 (*Opatrum*)

pesthiense Besser, 1832: 16 (*Opatrum*)

sturmii Küster, 1849a: 62 (*Opatrum*)

recticolle Motschulsky, 1866: 173 **A**: FE GAN HEB HEI JA JIL LIA NIN NMO QIN SC TAI "Korea"

reticulatum Motschulsky, 1854: 47 **A**: BEI ES GAN HEB HEI HEN JIA JIL MG NC NIN NMO QIN SHA SHN SHX TIA

mongolicum Reitter, 1889k: 706

rileyi Kaszab, 1952d: 473 **A**: NO **ORR**

roseni Kaszab, 1952d: 546 **A**: AP BT NP SD UP **ORR**

rusticum A. G. Olivier, 1811: 498 (*Opatrum*) [NP] **E**: AB AL AR AZ CR FR GR IT MA MC SP ST UK **N**: CI EG LB MO MR TU **A**: AF ES HEN IN IQ IS KZ LIA MG NMO OM SA SHN SI TD TM TR UZ XIN

errans Wollaston, 1854: 501 (*Opatrum*)

famelicum Küster, 1849a: 48 (*Opatrum*)

fuscum Herbst, 1793: 225 (*Opatrum*) [NO]

modestum Küster, 1849a: 53 (*Opatrum*)

obductum Gebler, 1860: 38 (*Opatrum*)

pulverulentum A. G. Olivier, 1811: 500 (*Opatrum*)

rusticum Brullé, 1832: 218 (*Opatrum*) [HN]

setulosum Küster, 1849a: 47

substriatum Besser, 1832: 18 (*Opatrum*)

terrosum Küster, 1845a: 28 (*Opatrum*)

trichopterum Gemminger, 1870a: 122 (*Hopatrum*) [RN]

tricostatum Fischer von Waldheim, 1844: 128 (*Bolitophagus*)

vestitum Küster, 1849a: 45 (*Opatrum*)

vicinum Gemminger, 1870a: 122 (*Hopatrum*)

villosum Steven, 1829: 97 (*Opatrum*)

sauteri Kaszab, 1952d: 529 **A**: TAI

sawadai Masumoto, 1985b: 28 **A**: JA (Honshu)

schneideri Reitter, 1898a: 14 **A**: KZ TM UZ

semipatruele Kaszab, 1952d: 608 **A**: IN PA UP YE **ORR**

senkakuense M. Chûjô, 1973: 13 **A**: JA (Ryukyus)

setulosum angustum Har. Lindberg, 1950: 11 **E**: IT (Sicilia) **N**: AG CI EG MO TU

lindbergi J. Ferrer, 1993a: 75 [RN]

setulosum kashgarense Bates, 1879: 481 (*Opatrum*) **A**: IN KA ZU XIZ

setulosum setulosum Faldermann, 1837: 60 (*Opatrum*) **E**: AR GR IT MA SP ST **N**: CI EG LB MO TU **A**: AF CY IN IQ IS KZ LE SA SI SY TD TM TR UZ XIN YE **AFR**

demaisonis Allard, 1883: 32

minutum Ménétriés, 1849: 239 (*Opatrum*)
parallelum Normand, 1936b: 363
sexuale Marseul, 1876c: 98 (*Opatrum*) **A**: BEI FE JA NC SC SCH
sikkimense Kaszab, 1952d: 517 **A**: SD
simplex yemenicum J. Ferrer, 1993a: 103 **A**: YE
simulatrix Fairmaire, 1891d: c (*Pseudoblaps*) **A**: AF GAN HP KA NP PA UP **ORR**
infimum Fairmaire, 1896a: 57 (*Scleropatrum*)
sinicum Reichardt, 1936a: 109 **A**: BT GAN SCH **ORR**
somalicum Gridelli, 1945: 19 **A**: YE (Suqutra) **AFR**
soricinum insidiosum Fairmaire, 1880a: 16 (*Opatrum*) **N**: AG MO
soricinum soricinum Reiche & Saulcy, 1857: 259 (*Opatrum*) **A**: AF IN IS JO PA SA SI SY YE **AFR**
spinicolle Fairmaire, 1896a: 22 (*Bradymerus*) **A**: HP **ORR**
stoeckleini Kaszab, 1952d: 662 **A**: KA PA UP **ORR**
strangulatum Fairmaire, 1888b: 128 (*Hopatrum*) **A**: BEI HEI
strigosum Reiche, 1850: 369 **N**: AG EG TU **A**: IS **AFR**
subrugulosum Reitter, 1887d: 388 **A**: KI MG NMO XIN XIZ
subspinosum Fairmaire, 1894a: 19 (*Hopatrum*) **A**: AP BT FUJ GUA GUI HUB HUN JIA NP SCH SD TAI XIZ YUN **ORR**
takara Nakane, 1963a: 26 **A**: JA (Ryukyus)
tellinii Gridelli, 1945: 20 **A**: OM SA YE **AFR**
tenuicorne Kaszab, 1952d: 590 **A**: SD **ORR**
tenuipes Kaszab, 1952d: 587 **A**: NP **ORR**
terminale Reichardt, 1936a: 110 **A**: FE JA NC SC
tibetanum Kaszab, 1952d: 511 **A**: XIZ
titschacki Kaszab, 1952d: 509 **A**: FUJ JA
tonkinense Kaszab, 1952d: 532 **A**: NP **ORR**
tschilianum Kaszab, 1952d: 630 **A**: SHX
tuberculatum Hope, 1831: 31 (*Opatrum*) **A**: AF HAI HKG HP HUB KA NP PA SCH SD UP TAI YUN **ORR**
vagum Steven, 1829: 96 (*Opatrum*) **A**: NP SD YUN **ORR**
wittmeri Kaszab, 1975e: 320 **A**: BT NP
woynarovichi Kaszab, 1973a: 33 **A**: NP
wui Ren, 1995b: 234 **A**: ZHE
subgenus *Myladanesthes* Skopin, 1961a: 202 type species *Myladina fortidens* Reitter, 1915
fortidens Reitter, 1915e: 88 (*Myladina*) **A**: KZ (Almaatinskaya Oblast) XIN
kalidii Skopin, 1964: 281 **A**: KZ
karatalica Skopin, 1966: 330 [= 1967: 209] **A**: KZ
kazenasi Skopin, 1961a: 203 **A**: KZ
subgenus *Opatropis* Reitter, 1904a: 159 type species *Opatrum hispidum* Brullé, 1838 (= *Opatrum affine* Billberg, 1815)
affine Billberg, 1815: 275 (*Opatrum*) **E**: GR SP **N**: CI EG MO MR **A**: IS **AFR**
hispidum Brullé, 1838: 68 (*Opatrum*)
virgatum Erichson, 1843: 249 (*Opatrum*)
blairi Gebien, 1922a: 276 **A**: YE (Suqutra) **AFR**
oculare Kaszab, 1952d: 622 (*Opatropis*) **A**: NP UP **ORR**
pilosum Ferrer, 2000b: 124 **N**: EG
genus *Gonocephalum*, species incertae sedis
tomentosum Walker, 1871: 16 (*Hopatrum*) **A**: "Arabia"

genus *Gonocephalum*, nomen dubium
spoliatum Rey, 1892: 42 **A**: TR

genus *Hadrus* Wollaston, 1854: 502 type species *Hadrus alpinus* Wollaston, 1854
alpinus Wollaston, 1854: 502 **N**: MR
carbonarius carbonarius Quensel, 1806: 142 (*Platynotus*) **N**: MR
cinerascens Wollaston, 1854: 503

verrucatus Desbrochers des Loges, 1881: 124
carbonarius paivae Wollaston, 1860b: 50 **N**: MR
carbonarius sousai Ardoin, 1960a: 108 **N**: MR
illotus Wollaston, 1854: 503 **N**: MR
europaeus Motschulsky, 1858d: 189
subellipticus Desbrochers des Loges, 1881: 124
subplumbeus Fairmaire, 1856: 533 (*Platydema*)

genus *Jintainum* Ren, 1999: 228 type species *Jintainum sulcatum* Ren, 1999
sulcatum Ren, 1999: 228 **A**: GAN NIN

genus *Mateuina* Español, 1944: 22 type species *Mateuina kaszabi* Español, 1944
kaszabi Español, 1944: 24 **N**: MO (Western Sahara)

genus *Melanesthes* Dejean, 1834: 191 type species *Opatrum sibiricum* Faldermann, 1833
subgenus *Lesbidana* Reitter, 1904a: 173 type species *Melanesthes simplex* Reitter, 1897
coriaria Reitter, 1904a: 173 **A**: UZ
chinganica Reichardt, 1936a: 182 **A**: GAN MG NMO
kasachstanica Kaszab, 1962c: 315 **A**: KZ
simplex Reitter, 1897e: 220 **A**: KI XIN
subcoriaria Reitter, 1897e: 220 **A**: KI XIN
subgenus *Melanesthes* Dejean, 1834: 191 type species *Opatrum sibiricum* Faldermann, 1833
Hemitrichestes Reitter, 1904a: 169 type species *Penthicus hirsutus* Reitter, 1896
Miglica Reitter, 1904a: 171 type species *Pedinus laticollis* Gebler, 1830
ciliata Reitter, 1889k: 703 **A**: MG NIN NMO XIN
psammophila Kaszab, 1964g: 397
conica Kaszab, 1965c: 283 **A**: XIN
davadshamsi Kaszab, 1964g: 394 **A**: MG NIN
desertora Ren, 1993b: 487 [486] **A**: NIN
exilidentata Ren, 1993a: 30 **A**: NIN
faldermanni Mulsant & Rey, 1859b: 98 [= 1859e: 36] **A**: ES (Tuva) MG NMO XIN
laticollis Faldermann, 1835a: 167 [HN]
hirsuta balchashensis Skopin, 1966: 328 [= 1967: 207] **A**: KZ
hirsuta hirsuta Reitter, 1896f: 170 (*Penthicus*) **A**: KZ TM
hirsuta seminuda Skopin, 1966: 329 [= 1967: 207] **A**: KZ
iliensis Skopin, 1966: 329 [= 1967: 208] **A**: KZ (Almaatinskaya Oblast)
jenseni jenseni Schuster, 1922d: 164 **A**: ES MG WS XIN
furva Kontkanen, 1956: 58
jenseni lacustris G. S. Medvedev, 1989b: 385 **A**: MG
jenseni meridionalis Kaszab, 1968c: 390 **A**: MG NMO
jintaiensis Ren, 1992: 330 **A**: NIN
kaszabi Bogatchev, 1972: 629 **A**: KZ (Karagiye)
laticollis Gebler, 1829a: 123 (*Pedinus*) **E**: ST **A**: KI KZ
lingwuensis Ren, 1993b: 487 **A**: NIN
maowusuensis Ren, 1992: 329 **A**: NMO
maxima borealis Kaszab, 1968c: 391 **A**: MG
maxima maxima Ménétriés, 1854: 33 **A**: MG NIN NMO
bielawskii Kaszab, 1964g: 391
medvedevi Kaszab, 1973c: 103 **A**: MG
mongolica Csiki, 1901a: 112 **A**: MG NIN NMO
opaca Reitter, 1889k: 703 **A**: NMO
parvula Kaszab, 1967c: 348 **A**: MG
sibirica Faldermann, 1833: 55 (*Opatrum*) **A**: ES MG NMO
alutacea Ménétriés, 1836b: 181
undentata Ren, 1993b: 488 **A**: NIN

subgenus *Mongolesthes* Reitter, 1904a: 174 type species *Melanesthes heydeni* Csiki, 1901
altaica altaica Kaszab, 1967a: 26 [= 1967c: 344] **A**: MG
altaica dschungarica Kaszab, 1967c: 345 **A**: MG
csikii Kaszab, 1965e: 343 **A**: GAN MG NIN NMO
heydeni australis G. S. Medvedev, 1989b: 383 **A**: MG
heydeni heydeni Csiki, 1901a: 112 **A**: MG NMO
multidentata Ren, 1993a: 29 **A**: XIN
tridentata Ren & Meng, 1995: 55 **A**: XIN
subgenus *Opatronesthes* Reitter, 1904a: 174 type species *Melanesthes punctipennis* Reitter, 1889
monatroides Reichardt, 1936a: 185 **A**: GAN
ningxianensis Ren, 1993a: 31 **A**: NIN
punctipennis Reitter, 1889k: 704 **A**: GAN
rugipennis Reitter, 1889k: 704 **A**: GAN NIN
tuberculosa Reitter, 1889k: 704 **A**: GAN

genus *Mesomorphus* Miedel, 1880: 140 type species *Opatrum murinum* Baudi di Selve, 1876 (= *Opatrinus setosus* Mulsant & Rey, 1853)
Hopatromorpha Blackburn, 1907: 286 [RN]
Pentholasius Reitter, 1904a: 178 type species *Penthicus variolatus* Allard, 1884
annamitus Kaszab, 1963b: 348 **A**: BT **ORR**
apterus Kaszab, 1963c: 364 **A**: SA **AFR**
belutschistanicus Kaszab, 1963b: 341 **A**: AF NP PA **ORR**
birmanicus Kaszab, 1963b: 343 **A**: BT NP YUN **ORR**
bremeri J. Ferrer, 1997a: 75 **A**: PA **ORR**
brevis Kaszab, 1963b: 344 **A**: YUN **ORR**
feai Kaszab, 1963b: 351 **A**: BT YUN **ORR**
fulvoscutatus Fairmaire, 1898c: 233 **N**: MR
gabrielae Schawaller, 2000a: 42 **A**: NP
hispidus Español, 1956b: 279 **N**: MO (Western Sahara) **AFR**
kulzeri Kaszab, 1963b: 349 **A**: NP **ORR**
latiusculus Chatanay, 1917a: 234 **A**: NP YUN **ORR**
longulus Reiche & Saulcy, 1857: 263 (*Crypticus*) **A**: CY IN IS JO LE SY TR
cyprius Kaszab, 1963b: 338
pellitus maroccanus Español, 1963e: 381 **N**: MO
rugulosus Chatanay, 1917a: 234 **A**: HAI NP **ORR**
maquilingius Gebien, 1921: 440
siamicus Kaszab, 1963b: 350 **A**: YUN **ORR**
setosus Mulsant & Rey, 1853c: 88 [= 1853d: 313] (*Opatrinus*) **N**: EG **A**: IS SI **AFR**
murinus Baudi di Selve, 1876b: 227 (*Opatrum*)
striolatus Fairmaire, 1896a: 21 (*Penthicus*) **A**: AF KA PA **ORR**
variolatus Allard, 1884: 248 (*Halonomus*) **A**: IS SY
subpubescens Reitter, 1896f: 163 (*Penthicus*)
villiger Blanchard, 1853: 154 (*Opatrum*) **A**: AF ANH FE FUJ GUA GUI GUX HAI HEB HEI HEN HKG HUB HUN JA JIA KA LIA NIN NMO NP SC SCH SD SHA SHN SHX TAI YUN **AFR AUR ORR**
asperulus Fairmaire, 1898c: 234
dermestoides Reitter, 1904a: 74
dispersus Champion, 1894a: 361 (*Opatrum*)
mustelinus Fairmaire, 1882c: 221 (*Opatrum*)
puberulus Fauvel, 1867: 187 (*Gonocephalum*)
ussuriensis Solsky, 1871: 374 (*Gonocephalum*)
wittmeri Kaszab, 1960a: 133 **A**: AF

genus *Moragacinella* Español, 1954d: 162 [RN] type species *Moralesia longepilosa* Kaszab, 1944
Moralesia Kaszab, 1944: 18 [HN] type species *Moralesia longepilosa* Kaszab, 1944
lindbergi G. S. Medvedev, 2004b: 574 **N**: MO (south)

longepilosa Kaszab, 1944a: 19 **N**: MO (Western Sahara)

genus *Myladina* Reitter, 1889k: 706 type species *Myladina unguiculina* Reitter, 1889
unguiculina Reitter, 1889k: 707 **A**: NMO SHA

genus *Nesocaedius* Kolbe, 1915: 262 type species *Nesocaedius schultzei* Kolbe, 1915
minimus M. Chûjô, 1966: 9 (*Caedius*) **A**: JA (Kyushu, Ryukyus)
pardoi Español, 1956b: 280 **N**: MO (Western Sahara)
taiwanus Shibata, 1979: 72 **A**: TAI
vermiculus Shibata, 1979: 71 **A**: TAI

genus *Opatroides* Brullé, 1832: 219 type species *Opatroides punctulatus* Brullé, 1832
Penthicus Faldermann, 1837: 61 type species *Penthicus parvulus* Faldermann, 1837 (=*Opatroides punctulatus* Brullé, 1832)
curtulus Fairmaire, 1892b: 149 **A**: IS LE SY TR
asiaticus Desbrochers des Loges, 1881: 122 (*Penthicus*)
judaicus Baudi di Selve, 1881: 283 **A**: IS SY TR
punctulatus punctulatus Brullé, 1832: 220 **E**: GR IT MA **N**: AG EG LB MO TU **A**: BA CY ES IQ IS JO KA KZ LE PA SA SI SY TR UZ WS YE **AFR**
minutus Mulsant & Rey, 1859b: 78 [= 1859e: 16]
parvulus Faldermann, 1837: 62 (*Penthicus*)
punctulatus subcylindricus Ménétriés, 1849: 238 (*Penthicus*) **E**: AB AR **A**: AF IN OM TD TM TR UZ
thoracicus Rosenhauer, 1856: 205 **E**: SP **N**: AG MO
vicinus angulatus Baudi di Selve, 1876b: 215 **A**: IQ
vicinus vicinus Fairmaire, 1896a: 21 (*Penthicus*) **A**: AE AF BA IN KU NP OM PA SA YE **ORR**

genus *Opatrum* Fabricius, 1775: 76 type species *Silpha sabulosa* Linnaeus, 1760
subgenus *Colpopatrum* Reitter, 1904a: 148 type species *Opatrum asperipenne* Reitter, 1897
asperipenne Reitter, 1897e: 219 **A**: GAN MG NIN NMO
subgenus *Colpophorus* Mulsant & Rey, 1859d: 106 [= 1859e: 44] type species *Opatrum emarginatum* Lucas, 1846
Colpophorinus Escalera, 1914: 336 type species *Opatrum gonocephaloides* Escalera, 1914
arrundarum Koch, 1945: 427 **N**: MO (Grand Atlas)
baeticum almeriense Español, 1954e: 323 **E**: SP (Andalucía)
baeticum baeticum Rosenhauer, 1856: 206 **E**: SP **N**: AG MO
granatum Fairmaire, 1871a: 392
baeticum gregarium Rosenhauer, 1856: 208 **E**: SP (Andalucía)
baeticum orientale Español, 1954e: 322 **E**: SP (Andalucía)
brevepilosum Antoine, 1942a: 22 **N**: MO (Grand Atlas)
emarginatum Lucas, 1846: 333 **E**: IT MA **N**: AG TU
inaequale Reitter, 1904a: 149
freudei Español, 1963b: 73 **N**: AG
gonocephaloides Escalera, 1914: 336 **N**: MO
hookeri hookeri Wollaston, 1875: 219 (*Hopatrum*) **N**: MO
hookeri malandriosum Antoine, 1942a: 23 **N**: MO (Grand Atlas)
lanzai Ardoin, 1971a: 51 **N**: TU
moralesi Español, 1959: 247 **N**: MO
nivale Gené, 1839: 72 (*Philax*) **E**: IT (Sardegna)
otini mediatlantis Antoine, 1942a: 27 **N**: MO
otini otini Antoine, 1942a: 25 **N**: MO
porcatum Fabricius, 1792a: 89 **N**: AG MO TU
granuliferum Lucas, 1846: 332
rhaticum Antoine, 1942a: 24 **N**: MO
ualdunum Koch, 1945: 426
rifense Koch, 1945: 426 **N**: MO (Rif)

rotundatum ifrense Español, 1956b: 280 **N**: MO (Ifra)
rotundatum rotundatum Escalera, 1914: 338 **N**: MO
simplicior antoinei Koch, 1945: 426 **N**: MO (Grand Atlas)
simplicior megalatlanticum Koch, 1945: 427 **N**: MO (Grand Atlas)
simplicior rugultense Koch, 1945: 428 **N**: MO (Tizi n'Rougoult)
simplicior simplicior Escalera, 1914: 337 **N**: MO
simplicior simplicissimum Antoine, 1942a: 24 **N**: MO
talboti Koch, 1945: 430
tangerianum Koch, 1945: 426 **N**: MO (Tanger)
thalense Mayet, 1890: civ **N**: TU
trisinuatum Koch, 1945: 429 **N**: MO
validum marcuzzii Canzoneri, 1972a: 190 **E**: IT (Ustica)
validum rottenbergi Canzoneri, 1972a: 188 **E**: IT (Lapione)
validum schickli Gebien, 1906a: 213 [RN] **E**: IT (Pantelleria) **N**: AG MO TU
validum validum Rottenberg, 1871: 250 **E**: IT (Sicilia) **N**: AG
venustulum Antoine, 1942a: 24 **N**: MO
subgenus *Opatrum* Fabricius, 1775: 76 type species *Silpha sabulosa* Linnaeus, 1760
alternatum Küster, 1849a: 39 **E**: GR RO TR **A**: TR
asperidorsum Fairmaire, 1878b: lvi **E**: FR IT (Sardegna)
cypraeum Reitter, 1904a: 155 **A**: CY
dahli Küster, 1849a: 34 **E**: FR (Corse) IT (Sardegna)
excisum Seidlitz, 1894: 429 **N**: AG
geminatum Brullé, 1832: 213 **E**: AB AR GR **A**: CY LE SY TR
asiaticum Reitter, 1904a: 153
granulosum Fischer von Waldheim, 1842: 16 **A**: KZ
grenieri Perris, 1870: 19 **E**: FR (Corse)
italicum Reitter, 1904a: 156 **E**: IT
libanii Baudi di Selve, 1876a: 64 [= 1876b: 222] **A**: IS LE SY
melitense Küster, 1849a: 38 **E**: MA
obesum A. G. Olivier, 1811: 497 **E**: GR IT TR **A**: IQ SY TR
curtum Miller, 1861: 205
dardanum Steven, 1829: 94
elevatum Brullé, 1832: 215
gibbum Besser, 1832: 17 [HN]
messeniacum Rottenberg, 1871: 251
sulcatum Küster, 1849c: 49
perlatum Germar, 1824b: 144 **E**: PT SP
riparium W. Scriba, 1865: 50 **E**: AU BY CR CT CZ DE EN FI FR HU GE HU LT NR NT PL RO SK ST SZ UK **A**: WS
sabulosum amauropterum Reichardt, 1936a: 131 **E**: ST (Kavkaz)
sabulosum guttifer Seidlitz, 1898a: 830 **E**: SP
sabulosum lucifugum Küster, 1849a: 41 **E**: AU CR GR IT
sabulosum pyrenaeum Español, 1963c: 480 **E**: SP (Pirineos centrales)
sabulosum reitteri Schuster, 1915b: 90 **E**: AB AR GG **A**: IN
sabulosum sabulosum Linnaeus, 1760: 150 (*Silpha*) **E**: AL AR AU BE BH BU BY CR CT CZ DE EN CZ FI FR GB GE GG GR HU IR IT LA LS LT NL NR NT PL RO SK SL ST SV SZ TR UK YU **A**: ES GAN HEN KZ NMO TD TM TR WS XIZ
distinctum A. Villa & J. B. Villa, 1835: 49
intermedium Fischer von Waldheim, 1844: 127
rugosum DeGeer, 1775: 43 (*Tenebrio*)
tricarinatum Motschulsky, 1859b: 307
sabulosum sculptum Rey, 1892: 42 **E**: FR IT
subdilatatum Reitter, 1908a: 23
sculpturatum caprariense Razzauti, 1919: 5 **E**: IT (Capraia)
sculpturatum igiliense Razzauti, 1919: 6 **E**: IT (Giglio)

sculpturatum ilvense Razzauti, 1919: 7 **E**: IT (Elba)
sculpturatum sculpturatum Fairmaire, 1860: 630 **E**: FR IT
sculpturatum urgonense Razzauti, 1919: 6 **E**: IT (Gorgona)
subaratum Faldermann, 1835b: 413 (*Opatrum*) **A**: ANH ES FE GAN GUI GUX HEI HEB HEN HUB HUN JIL JIX HEI JA LIA MG NC NIN NMO QIN SC SCH SHA SHN SHX TAI
tebessanum Reitter, 1904a: 153 **N**: AG TU
triste Steven, 1829: 94 **E**: UK (Krym)
areolatum Küster, 1849a: 40
granulatostriatum Küster, 1849a: 43
politum Besser, 1832: 17
verrucosum Germar, 1817: 188 **E**: AL BH CR GR IT TR **A**: TR
graniger Brullé, 1832: 214
granigerum Besser, 1832: 16
hispidosum Brullé, 1832: 217
trilobatum Baudi di Selve, 1875b: 221
verrucosum Waltl, 1835: 72 [HN]

genus *Penthicinus* Reitter, 1896f: 161 type species *Penthicinus koltzei* Reitter, 1896
ghilarovi G. S. Medvedev, 1988a: 55 **A**: KZ
koltzei Reitter, 1896f: 171 **A**: UZ XIN
netshaevae G. S. Medvedev, 1970: 397 **A**: TM
pedinoides Reitter, 1896f: 172 **A**: UZ

genus *Penthicus* Faldermann, 1836: 386 type species *Penthicus pinguis* Faldermann, 1836
subgenus *Allomyladion* Bogatchev, 1972: 626 type species *Penthicus poricollis* Reitter, 1896
poricollis Reitter, 1896f: 165 **A**: AF TD UZ
zoltani G. S. Medvedev & Iwan, 2006: 617 [RN] **A**: TD
kaszabi Bogatchev, 1972: 626 [HN]
subgenus *Aulonolcus* Reitter, 1904a: 168 type species *Pedinus altaicus* Gebler, 1830
altaicus Gebler, 1829a: 123 (*Pedinus*) **A**: ES (Tuva) MG NMO WS (Altay) XIN
sulcibasis Reitter, 1904a: 169 (*Lobothorax*)
tenebrioides Faldermann, 1835b: 411 (*Heliophilus*)
subgenus *Discotus* Reitter, 1904a: 161 type species *Opatrum dilectans* Faldermann, 1836
dilectans Faldermann, 1836: 389 (*Opatrum*) **E**: AB AR ST **A**: AF IN KI KZ MG TM UZ
echingolensis Kaszab & G. S. Medvedev, 1984: 82 (*Lobodera*) **A**: MG
kaszabi Skopin, 1960: 307 (*Lobodera*) **A**: KZ
semenovi Reichardt, 1936a: 141 (*Lobodera*) **A**: KZ MG TD TM UZ XIN
viberti Reitter, 1906f: 240 (*Lobothorax*) **N**: AG EG **A**: IS **AFR**
subgenus *Myladion* Reitter, 1887d: 385 type species *Myladion acuticolle* Reitter, 1887
Penthicomelanesthes Bogatchev, 1972: 631 type species *Heliophilus gibbulus* Faldermann, 1836
acuticollis acuticollis Reitter, 1887d: 386 (*Myladion*) **A**: XIN XIZ
acuticollis pevtzovi Bogatchev, 1972: 627 **A**: TD
alashanicus Reichardt, 1936a: 163 (*Lobodera*) **A**: MG NIN NMO
beicki beicki Reichardt, 1936a: 164 (*Lobodera*) **A**: MG XIN
beicki cerberus Reichardt, 1936a: 165 (*Lobodera*) **A**: MG
brutus Reichardt, 1936a: 160 (*Lobodera*) **A**: MG
bulganicus G. S. Medvedev, 1990b: 137 **A**: MG QIN XIN
davadshamsi davadshamsi Kaszab, 1965e: 336 (*Lobodera*) **A**: MG XIN
davadshamsi kobdoensis Kaszab, 1977c: 278 (*Lobodera*) **A**: MG
davadshamsi vulgaris Kaszab, 1967c: 334 (*Lobodera*) **A**: MG
dschungaricus Kaszab, 1967c: 339 (*Lobodera*) **A**: MG
explanatus explanatus Reitter, 1896f: 164 **A**: ES (Tuva) WS (Altay)
explanatus laesus Reichardt, 1936a: 159 (*Lobodera*) **A**: MG
explanatus reichardti Kaszab, 1969b: 325 (*Lobodera*) **A**: MG
frater Kaszab, 1967c: 336 (*Lobodera*) **A**: MG NMO

ferganensis Schuster, 1919a: 32 (*Lobothorax*) **A**: KI (Alai Range)
gibbulus Faldermann, 1835b: 412 (*Heliophilus*) **A**: "Altayskiy Kray"
hypolithus Gebler, 1836: 341 (*Heliophilus*)
gobiensis Kaszab, 1964b: 24 (*Lobodera*) **A**: MG
heydeni Reitter, 1896f: 168 **A**: UZ (Fergana)
humeridens Reitter, 1896f: 164 **A**: MG XIN "Altay"
kerzhneri G. S. Medvedev, 1989b: 380 **A**: MG
kiritshenkoi Reichardt, 1936a: 160 (*Lobodera*) **A**: MG NIN
kochi kochi Skopin, 1960: 308 **A**: KZ (Vostochnyy Kazakhstan)
kochi tarbagataicus Skopin, 1964: 393 **A**: KZ (Vostochnyy Kazakhstan)
obsoletus G. S. Medvedev, 1964b: 163
korotyaevi G. S. Medvedev, 1984: 298 **A**: ES (Tuva)
laelaps Reichardt, 1936a: 161 (*Lobodera*) **A**: MG
lycaon Reichardt, 1936a: 161 (*Lobodera*) **A**: NMO
marginalis Reichardt, 1936a: 160 (*Lobodera*) **A**: MG NMO
nanshanicus Reichardt, 1936a: 164 (*Lobodera*) **A**: GAN MG
nojonicus Kaszab, 1968c: 385 (*Lobodera*) **A**: GAN MG NIN NMO
oblongipennis Bogatchev, 1961a: 113 (*Lobodera*) **A**: TD (Turkestanskiy Khrebet)
obtusangulus bajanus G. S. Medvedev, 1989b: 379 **A**: MG
obtusangulus obtusangulus Reitter, 1889k: 709 (*Opatroides*) **A**: ES MG NMO
parvulus Kaszab, 1967c: 340 (*Lobodera*) **A**: MG
pavlovskyi Reichardt, 1935: 253 (*Lobodera*) **A**: AF TD UZ
reitteri Csiki, 1901a: 111 (*Gonocephalum*) **A**: MG NMO
serpens Reitter, 1896f: 168 **A**: KI
schusteri Reichardt, 1936a: 157 (*Lobodera*) **A**: GAN
sequensi Reitter, 1901c: 79 (*Lobothorax*) **A**: MG
tannuolensis G. S. Medvedev & Mordkovich, 1970: 30 **A**: ES (Tuva)
teter major Kaszab, 1969b: 322 (*Lobodera*) **A**: MG
teter teter Reichardt, 1936a: 158 (*Lobodera*) **A**: MG
truncatus Reitter, 1896f: 167 **A**: UZ (Bukhara)
tuberculosus Heyden, 1887c: 319 **A**: KI
tzaidamicus Bogatchev, 1951: 72 (*Lobodera*) **A**: QIN
vulneratus Kraatz, 1885: 293 **A**: KI UZ
waziristanica Kaszab, 1957f: 298 (*Lobodera*) **A**: PA
subgenus *Penthicus* Faldermann, 1836: 384 type species *Penthicus pinguis* Faldermann, 1836
Lobodera Mulsant & Rey, 1859b: 80 [= 1859e: 18] type species *Lobodera rufescens* Mulsant & Rey, 1859
Lobothorax Gemminger, 1870a: 124 [RN] type species *Lobodera rufescens* Mulsant & Rey, 1859
Penthomegus Reitter, 1904a: 161 type species *Penticus corpulentus* Reitter, 1896
Stonavus Reitter, 1904a: 161 type species *Penticus alaiensis* Reitter, 1896
adustus Bogatchev, 1951: 71 (*Lobodera*) **A**: AF
alaiensis Reitter, 1896f: 171 **A**: UZ (Alai Mts)
antoinei Español, 1945a: 228 (*Lobodera*) **N**: MO (Western Sahara)
ardoini Kaszab, 1969d: 454 (*Lobodera*) **A**: IN
auliensis Reitter, 1904a: 163 (*Lobothorax*) **A**: KZ UZ
belutschistanicus Kaszab, 1961a: 119 (*Lobodera*) **A**: PA
bogatshevi Kaszab, 1970a: 164 **A**: AF
corpulentus Reitter, 1896f: 163 **A**: IN
curtus Kaszab, 1970a: 163 (*Lobodera*) **A**: AF
fartilis Reitter, 1899d: 205 **A**: TM (Kopet Dag)
freyi Schuster, 1937b: 79 (*Lobodera*) **A**: IN
granulifera Reichardt, 1936a: 145 (*Lobodera*) **A**: UZ
granulosus Ménétriés, 1849: 239 **A**: KI KZ TD TM UZ XIN
semigranosus Reitter, 1904a: 162 (*Lobothorax*)
horni Schuster, 1922e: 212 (*Lobothorax*) **A**: IN
iners Ménétriés, 1832: 202 (*Heliophilus*) **E**: AB **A**: IN KI TR XIN

molestus Faldermann, 1836: 388
sagitatus Ménétriés, 1832: 203 (*Opatrum*)
kirgizicus Bogatchev, 1961a: 114 (*Lobodera*) **A**: KZ (Bednak-Dala)
lenczyi Kaszab, 1968c: 383 (*Lobodera*) **A**: GAN MG
lindbergi Kaszab, 1970a: 161 (*Lobodera*) **A**: AF
medvedevorum Bogatchev, 1972: 628 **A**: TD
netuschili Reitter, 1904a: 162 (*Lobothorax*) **A**: XIN
obesus Kulzer, 1964: 223 (*Lobodera*) **A**: IN
oblongopunctatus Reitter, 1904a: 164 (*Lobothorax*) **A**: AF IN IQ JO PA SA TM
persicus Schuster, 1919a: 31 (*Lobothorax*) **A**: IN
pinguis arnoldianus Bogatchev, 1951: 70 (*Lobodera*) **A**: UZ
pinguis kughitangi G. S. Medvedev & Nepesova, 1985b: 361 **A**: TM (Kugitangtay)
pinguis medvedevi Iwan & Löbl, 2007: 734 [RN] **A**: TM (Badkhyz)
granulifera G. S. Medvedev, 1964c: 658 (*Lobodera*) [HN]
pinguis nitescens Reichardt, 1935: 255 **A**: UZ
pinguis pinguis Faldermann, 1836: 386 **E**: AB AR **A**: AF TM
remotus Reitter, 1896f: 170 **A**: AF IN TM
rufescens rufescens Mulsant & Rey, 1859b: 80 [= 1859e: 18] (*Lobodera*) **E**: AB AR GG ST **A**: KZ TM UZ
rufescens tenuepunctatus Bogatchev, 1951: 72 (*Lobodera*) **A**: KZ
rugosicollis Bogatchev, 1951: 69 (*Lobodera*) **A**: UZ
samgarensis Reitter, 1896f: 168 **A**: UZ
similis Schuster, 1937b: 80 (*Lobodera*) **A**: IN
turcomanicus G. S. Medvedev, 1964c: 658 (*Lobodera*) **A**: TM
villiersi Kaszab, 1969d: 452 (*Lobodera*) **A**: IN
winkleri Schuster, 1915b: 92 (*Lobodera*) **A**: KZ XIN
zhenzhuristi afghana Gridelli, 1954: 289 (*Lobodera*) **A**: AF PA
zhenzhuristi zhenzhuristi Bogatchev, 1951: 70 (*Lobodera*) **A**: AF
subgenus *Pseudopenthicinus* Bogatchev, 1972: 628 type species *Penthicus medvedevi* Bogatchev, 1972
medvedevi Bogatchev, 1972: 628 **A**: TD
genus *Penthicus*, species incertae sedis
cribellatus Fairmaire, 1888b: 127 (*Opatroides*) **A**: BEI
gracilis Bates, 1879: 482 **A**: KA XIN
granulatus Desbrochers des Loges, 1881: 121 **E**: "South Russia"

genus *Perithrix* Fairmaire, 1879c: 193 type species *Perithrix granidorsum* Fairmaire, 1879
granidorsum Fairmaire, 1879c: 193 **N**: AG TU
tunisiensis Dajoz, 1983: 65 **N**: AG TU

genus *Phelopatrum* Marseul, 1876c: 100 type species *Hadrus scaphoides* Marseul, 1876
Pseudadrus Fairmaire, 1897a: 39 type species *Hadrus scaphoides* Marseul, 1876
scaphoides Marseul, 1876c: 99 (*Hadrus*) **A**: JA

genus *Platynosum* Mulsant & Rey, 1859b: 143 [= 1859e: 81] type species *Platynosum paulinae* Mulsant & Rey, 1859
collare Motschulsky, 1839: 60 (*Microzoum*) **E**: AR **A**: SA
paulinae Mulsant & Rey, 1859b: 143 [= 1859e: 81] **E**: GR (Ródos) **A**: CY IS KU SA SY
sabulorum Chobaut, 1900a: 31 (*Melanimon*) **N**: LB TU
zacheus J. R. Sahlberg, 1908: 84 **A**: IS

genus *Polycoelogastridion* Reichardt, 1936a: 85 type species *Scleron sexcostatum* Motschulsky, 1858
sexcostatum Motschulsky, 1858d: 188 (*Scleron*) **E**: AB GR **A**: SY TR
scleroide Baudi di Selve, 1876b: 234 (*Opatrum*)
vinciguerrae Gridelli, 1930: 297 (*Scleropatrum*) **N**: LB

genus *Prodilamus* Ardoin, 1969: 258 type species *Dilamus brevicollis* Fairmaire, 1894

boehmi Reitter, 1904a: 71 (*Dilamus*) **N**: EG **AFR**
fausti fausti Reitter, 1890d: 150 (*Dilamus*) **E**: AR **A**: CY IN IQ TM
fausti major Kaszab, 1982b: 162 **A**: SA
ferrantei Reitter, 1908c: 49 (*Dilamus*) **N**: EG **AFR**

genus *Proscheimus* Desbrochers des Loges, 1881: 127 type species *Proscheimus arabicus* Desbrochers des Loges, 1881
arabicus Desbrochers des Loges, 1881: 128 **A**: IS JO SA SI **N**: EG
fulvipes Ménétriés, 1849: 240 (*Microzoum*) **A**: TM

genus *Psammestus* Reichardt, 1936a: 194 type species *Ammobius dilatatus* Reitter, 1893
dilatatus Reitter, 1893e: 315 (*Ammobius*) **A**: KZ MG NMO TM UZ XIN
caspicus Semenov, 1905: 124 (*Ammobius*)
kozlovi Semenov, 1905: 124 (*Ammobius*)
panfilovi G. S. Medvedev, 1970: 398 **A**: TM
pseudocaedius Reitter, 1895d: 156 (*Ammobius*) **A**: AF TM

genus *Pseudolamus* Fairmaire, 1874a: 388 type species *Pseudolamus seriatoporus* Fairmaire, 1874
schatzmayri Koch, 1937: 459 **N**: LB
seriatoporus Fairmaire, 1874a: 389 **E**: SP **N**: MO
pusillus Baudi di Selve, 1876b: 219 (*Pachypterus*)

genus *Reichardtiellina* Kaszab, 1982a: 79 [RN] type species *Reichardtiella armata* Kaszab, 1942
Reichardtiella Kaszab, 1942a: 18 [HN] type species *Reichardtiella armata* Kaszab, 1942
armata Kaszab, 1942a: 19 (*Reichardtiella*) **A**: HP
tibetana Kaszab, 1942a: 21 (*Reichardtiella*) **A**: XIN

genus *Sclerum* Dejean, 1834: 193 type species *Opatrum orientale* Fabricius, 1775
Anticlia Gistel, 1848: x [RN]
Chlamydion Gistel, 1848: x [RN]
Scleron Hope, 1840: 111 [HN] type species *Opatrum orientale* Fabricius, 1775
aequale Reitter, 1904a: 124 **A**: SA
andreinii Gridelli, 1930: 295 **N**: LB
armatum Waltl, 1835: 72 (*Opatrum*) **E**: FR IT SP **N**: AG CI EG LB MO TU **A**: SY
algiricum Lucas, 1846: 335 (*Opatrum*)
damascenum Baudi di Selve, 1876a: 60 (*Sclerum*)
fernandezi Har. Lindberg, 1950: 11
asperulum Wollaston, 1864: 486 (*Sclerum*) **N**: CI (Gran Canaria)
bengalensis Kaszab, 1942a: 10 **A**: AF **ORR**
bodenheimeri Koch, 1935d: 282 **A**: IS
carinatum Baudi di Selve, 1875d: 700 [= 1876a: 60] **E**: AB **A**: AF CY IN IQ SA SY TD TM TR UZ
dubium Gridelli, 1930: 295 **N**: EG LB
evansi Blair, 1923a: 122 **A**: IN IQ SA
ferrugineum Fabricius, 1801a: 118 (*Opatrum*) **A**: AP TAI "S. China" **ORR**
fossulatum Mulsant & Wachanru, 1852: 9 (*Sclerum*) **A**: CY IQ IS LE SY TR
angustum Miller, 1861: 203 (*Scleron*)
humerosum Miller, 1861: 202 (*Scleron*) **A**: CY SY TR
mariae Mulsant & Rey, 1859b: 140 [= 1859e: 78] (*Sclerum*)
kandaharicum Kaszab, 1960a: 135 **A**: AF
multistriatum Forskål, 1775: 77 (*Silpha*) **E**: GR IT MA ST TR **N**: EG LB **A**: IS SY TR
abbreviatum Reiche & Saulcy, 1857: 260
foveolatum A. G. Olivier, 1811: 500 (*Opatrum*)
orientale orientale Fabricius, 1775: 76 (*Opatrum*) **N**: EG **A**: IS JO SA SI YE **AFR**
subclathratum Reitter, 1904a: 125 [RN]
orientale yemense Kaszab, 1972: 377 **A**: YE
reitteri Gebien, 1906a: 234 [RN] **A**: NP UP **ORR**

socotranum Schawaller, 2004c: 449 **A**: YE (Suqutra)
sudanicum Koch, 1935a: 84 (*Scleron*) **N**: EG **AFR**
sulcatum lomii Gridelli, 1953: 59 **A**: SA **AFR**
sulcatum saharense Peyerimhoff, 1931b: 103 **N**: AG LB MO
sulcatum scotti Gridelli, 1953: 58 **A**: YE
sulcatum sulcatum Baudi di Selve, 1876a: 59 (*Sclerum*) **A**: SA

genus *Scleropatroides* Löbl & Merkl, 2003: 250 type species *Opatrum hirtulum* Baudi di Selve, 1876
Scleropatrum Reitter, 1890d: 149 [HN] type species *Opatrum hirtulum* Baudi di Selve, 1876
breviusculus Reitter, 1889a: 32 (*Scleropatrum*) **E**: AB AR **A**: AF KZ TM UZ
gallagheri Ferrer, 2000a: 37 (*Scleropatrum*) **A**: OM
hirtulus Baudi di Selve, 1876a: 69 (*Opatrum*) **E**: AB AR ST UK **A**: AF IN IQ KZ SA SI SY TD TM UZ
seidlitzi Reitter, 1898b: 39 (*Scleropatrum*) **E**: AB AR ST **A**: AF GAN IN KZ NMO TM UZ
strigatus Fabricius, 1798: 41 (*Opatrum*) **A**: OM **AFR ORR**
turanicus Reitter, 1904a: 139 (*Scleropatrum*) **A**: KZ TM

genus *Scleropatrum* Reitter, 1887d: 388 type species *Scleropatrum tuberculatum* Reitter, 1887
Monatrum Reichardt, 1936a: 81 type species *Opatrum carinatum* Gebler, 1830
carinatum carinatum Gebler, 1829a: 126 (*Opatrum*) **A**: KI
carinatum relictum Skopin, 1964: 394 (*Monatrum*) **A**: KZ
csikii Kaszab, 1967c: 332 (*Monatrum*) **A**: GAN XIN
elongatum Guérin-Méneville, 1834b: 32 (*Opatrum*) **A**: HAI HP KA **ORR**
quadrinodosum Reitter, 1904a: 146 (*Gonocephalum*)
horridum horridum Reitter, 1898b: 37 **A**: GAN NIN NMO SHX XIN
horridum humeralis Kaszab, 1964g: 390 (*Monatrum*) **A**: MG
mongolicum Kaszab, 1967c: 329 (*Monatrum*) **A**: MG
prescottii Faldermann, 1833: 54 (*Opatrum*) **A**: ES (Tuva) MG NMO XIN
carinatum Csiki, 1901a: 91
tuberculatum Reitter, 1887d: 388 **A**: GAN NMO QIN XIN XIN
tuberculiferum Reitter, 1890d: 148 **A**: GAN NIN NMO
striatogranulatum Reitter, 1890d: 149

genus *Sinorus* Mulsant & Revelière, 1860: 153 type species *Sinorus ciliaris* Mulsant & Revelière, 1860 (= *Opatrum colliardi* Fairmaire, 1860)
colliardi Fairmaire, 1860: 630 (*Opatrum*) **E**: FR (Corse) IT (Sardegna)
ciliaris Mulsant & Revelière, 1860: 154

genus *Socotropatrum* Koch, 1970: 109 type species *Opatrum costiferum* Waterhouse, 1881
costiferum Waterhouse, 1881: 476 (*Opatrum*) **A**: YE (Samha, Suqutra)
guichardi Koch, 1970: 113 **A**: YE (Suqutra)

genus *Tidiguinia* Español, 1959: 249 type species *Tidiguinia bolivari* Español, 1959
bolivari Español, 1959: 252 **N**: MO (Rif)

genus *Trigonopoda* Gebien, 1914b: 2 type species *Trigonopoda crassipes* Gebien, 1914
carmanicus Reichardt, 1936a: 199 (*Caedius*) **A**: IN
crassipes Gebien, 1914b: 3 **A**: TAI
ovalipennis Shibata, 1979: 68 **A**: TAI
piffli Kaszab, 1961f: 348 **A**: AF PA

genus *Weisea* Semenov, 1891a: 370 type species *Weisea sabulicola* Semenov, 1890
sabulicola Semenov, 1891a: 371 **A**: TM (Karakumy)

tribe Opatrini, species incertae sedis
Gonocephalum angustatum Chevrolat, 1873: 204 **A**: "Syria"

Hopatrum opacum Walker, 1871: 15 **N**: EG
Heliopates damascenus Mannerheim, 1825: 30 **E**: ST
Heliopates rugipennis Faldermann, 1837: 56 **A**: IN

tribe Palorini Matthews, 2003

genus *Palorinus* Blair, 1930a: 135 type species *Palorus humeralis* Gebien, 1914
bicolor Blair, 1930a: 143 **A**: NP **ORR**
nepalicus Grimm, 2003: 4 **A**: NP

genus *Paloropsis* Masumoto & Grimm, 2004: 127 type species *Paloropsis irei* Masumoto & Grimm, 2004
irei Masumoto & Grimm, 2004: 127 **A**: JA

genus *Palorus* Mulsant, 1854: 250 type species *Hypophloeus depressus* Fabricius, 1790
Caenocorse C. G. Thomson, 1859: 117 type species *Hypophlaeus depressus* Fabricius, 1790
Circomus Fleischer, 1900: 236 type species *Hypophloeus subdepressus* Wollaston, 1864
Eba Pascoe, 1863c: 129 type species *Eba cerylonoides* Pascoe, 1863
Platyotus Gerstaecker, 1871: 62 type species *Platyotus glabratus* Gerstaecker, 1871
Stenopalorus Blair, 1930a: 136 type species *Palorus hypophloeoides* Blair, 1930
cerylonoides Pascoe, 1863c: 129 (*Eba*) **A**: FUJ GUA GUI GUX HAI HEB HEN HUB HUN IN JA JIA JIX NMO NP SCH SHA SHX TAI ZHE **AFR AUR ORR**
exilis Marseul, 1876c: 116 (*Hypophloeus*)
minor Waterhouse, 1894: 71
praslinensis Gebien, 1922a: 304
papuanus Kaszab, 1939b: 218
zimmermanni Kaszab, 1955c: 657
depressus Fabricius, 1790: 223 (*Hypophlaeus*) **E**: AL AU BH BY CR CT CZ EN FI FR GE GR HU IT NL NR NT PL RO SK SL SP SV SZ UK **N**: EG **A**: JA KZ TAI TD TR **ORR**
formiceticola Munster, 1928: 296 (*Caenocorse*)
unicolor A. G. Olivier, 1790: 18: 12 (*Ips*)
euphorbiae Wollaston, 1862a: 183 (*Hypophloeus*) **N**: CI
ficicola Wollaston, 1867: 205 (*Hypophloeus*) **N**: AG EG LB MO **A**: PA SA **AFR ORR**
deserticola J. R. Sahlberg, 1913a: 51 (*Caenocorse*)
subfilum A. Fleischer, 1900: 237
hypophloeoides Blair, 1930a: 136 **A**: UP **ORR**
longifoliae Blair, 1930a: 138 **A**: UP **ORR**
orientalis Fleischer, 1900: 237 **E**: AB **A**: IN
ratzeburgii Wissmann, 1848: 77 (*Hypophloeus*) **E**: AR AU AZ BE BY CR CZ DE FR GB GE GR HU IR IT LU NL PL SK SL SP SV SZ UK **N**: CI EG MR **A**: GAN GUI GUX HEB HEI HEN HUB HUN FE IQ IS JA JIA JIL JIX LIA NMO SCH SHA SHN SY TR YUN **COS**
ambiguus Wollaston, 1857a: 152 (*Hypophloeus*)
floricola Marseul, 1876c: 115 (*Hypophloeus*)
galilaeus J. R. Sahlberg, 1913a: 49 (*Caenocorse*)
shoreae Blair, 1930a: 138 **A**: NP **ORR**
sinuaticollis Blair, 1930a: 137 **A**: TAI **ORR**
fuhoshoanus Kaszab, 1941c: 56
subdepressus Wollaston, 1864: 499 (*Hypophloeus*) **E**: AZ BH CR CZ FR GB GE GR HU IR IT MA PL SK SP SZ UK **N**: EG CI LB MO MR TU **A**: AF ANH FE FUJ GUI GUX HEB HEN HUB IS JA JIA JIX KZ MG NE NMO SA SCH SHA SHN TAI TM TR YE YUN ZHE **COS**
bifoveolatus Baudi di Selve, 1876c: 235

genus *Platycotylus* Olliff, 1883b: 182 type species *Platycotylus inusitatus* Olliff, 1883
ferrugineus Kaszab, 1939a: 105 (*Doliema*) **A**: NP TAI **ORR**
nitidulus W. J. MacLeay, 1873a: 168 (*Ipsaphes*) **A**: NP TAI **AUR ORR**
coloratus Blackburn, 1903: 147
inusitatus Olliff, 1883b: 183
laticollis Pic, 1923d: 25 (*Doliema*)

nigripes Olliff, 1883a: 60
spinicollis Fairmaire, 1893b: 27 (*Sitophagus*)
suturalis Fairmaire, 1893b: 28 (*Sitophagus*)

genus *Ulomina* Baudi di Selve, 1876c: 235 type species *Ulomina carinata* Baudi di Selve, 1876
Coelopalorus Blair, 1930a: 135 type species *Palorus foveicollis* Blair, 1930 (= *Ulomina carinata* Baudi di Selve, 1876)
carinata Baudi di Selve, 1876c: 236 **E**i: GB IT **A**: HAI HEN GUA GUX JA NP SD TAI YUN **AFR AUR NAR NTR ORR**
foveicollis Blair, 1930a: 136 (*Palorus*)
martinae Scupola, 2002: 188 [RN] **A**: HAI **ORR**
carinata Blair, 1930a: 143 (*Palorus*) [HN]

tribe Pedinini Eschscholtz, 1829
subtribe Dendarina Seidlitz, 1889
genus *Bioplanes* Mulsant, 1854: 144 type species *Bioplanes meridionalis* Mulsant, 1854
meridionalis Mulsant, 1854: 145 **E**: FR IT

genus *Dendarus* Dejean, 1821: 65 type species *Helops tristis* sensu Laporte, 1840 (= *Pandarus coarcticollis* Mulsant, 1854)
subgenus *Dendaroscelis* Reitter, 1904a: 79 type species *Dendarus serripes* Reitter, 1904
serripes Reitter, 1904a: 92 (*Dendarus*) **E**: SP **N**: MR
subgenus *Dendarus* Dejean, 1821: 65 type species *Helops tristis* sensu Laporte, 1840 (= *Pandarus coarcticollis* Mulsant, 1854)
Microdendarus Escalera, 1944: 84 type species *Dendarus schusteri* Español, 1937
Pandarus Dejean, 1834: 191 [unjustified emendation] [HN]
Phylax Brullé, 1832: 211 type species *Phylax emarginatus* Brullé, 1832
Reitterellus Escalera, 1944: 91 type species *Dendarus depressus* Reitter, 1915
aubei aubei Mulsant & Rey, 1854a: 64 [= 1854b: 208] (*Pandarus*) **E**: SP (Andalucía)
aubei cazorlensis Koch, 1944b: 254 **E**: SP (Sierrade Cazorla)
seguranus Escalera, 1944: 89
calcaratus Baudi di Selve, 1881: 281 **A**: IS LE
carinatus Mulsant & Rey, 1854a: 58 [= 1854b: 202] (*Pandarus*) **E**: FR (Corse) IT (Sardegna)
corsicus Perris, 1875: 5 (*Pandarinus*)
tricostatus Mulsant & Rey, 1854a: 60 = 1854b: 204] (*Pandarus*)
castilianus Piochard de la Brûlerie, 1869: 31 (*Pandarus*) **E**: PT SP
cedrorum F. Soldati & L. Soldati, 2000: 68 **A**: LB
coarcticollis Mulsant, 1854: 142 (*Pandarus*) **E**: AU CR FR IT SZ
[*tristis* sensu Laporte, 1840]
crenulatus Ménétriés, 1832: 202 (*Heliophilus*) **E**: AB AR ST **A**: TR
cribratus Waltl, 1838: 462
depressus Reitter, 1915e: 86 **E**: SP (Baleares)
disparicus Clermont, 1925: 183
hildti Tenenbaum, 1915: 837
emarginatus Brullé, 1832: 211 (*Phylax*) **E**: GR **A**: TR
sinuatus Mulsant & Rey, 1854a: 69 [= 1854b: 213] (*Pandarus*)
tibialis Seidlitz, 1893: 398
graecus Brullé, 1832: 207 **E**: GR (Kríti)
montanus Koch, 1948a: 343
insidiosus alcojonensis Español, 1961a: 62 **E**: SP (Málaga)
insidiosus bolivari Escalera, 1944: 87 **E**: SP (Cádiz)
insidiosus insidiosus Mulsant & Rey, 1854a: 66 [= 1854b: 210] (*Pandarus*) **E**: SP **N**: MO
mohgrebicus Escalera, 1944: 88
insidiosus riffensis Escalera, 1944: 88 **N**: MO
pardoi Escalera, 1944: 89 **N**: MO
pectoralis bejarensis Español, 1961a: 45 **E**: SP
pectoralis ilerdensis Español, 1937: 49 **E**: SP

pectoralis pectoralis Mulsant & Rey, 1854a: 62 [= 1854b: 206] (*Pandarus*) **E**: PO SP **N**: MO
coriaceus Baudi di Selve, 1876a: 36
piochardi Español, 1937: 51 **E**: PT SP
stygius kochi G. S. Medvedev, 1968a: 100 [RN] **E**: GR (Evvoia)
oertzeni Koch, 1948a: 349 [HN]
stygius stygius Waltl, 1838: 462 **E**: GR **A**: TR
schusteri Español, 1937: 54 **E**: SP
tauricus Koch, 1948a: 350 **A**: SY TR
werneri Koch, 1948a: 342 **E**: GR (Kyklades)
zariquieyi almeriensis Escalera, 1944: 90 **E**: SP
zariquieyi vivesi Español, 1961a: 58 **E**: SP (Granada)
zariquieyi zariquieyi Español, 1937: 62 **E**: SP **N**: MO
subgenus *Pandarinus* Mulsant & Rey, 1854a: 103 [= 1854b: 247] type species *Pandarinus tenellus* Mulsant & Rey, 1854
alleonis Seidlitz, 1893: 399 **E**: TR **A**: TR
ananensis Chatzimanolis, Engel & Trichas, 2002: 260 **E**: GR (Ananes Is.)
anaphianus Koch, 1948a: 344 **E**: GR (Anaphis Is.)
angulitibia Koch, 1948a: 345 **E**: GR (Tinos Is.)
antikythirensis Chatzimanolis, Engel & Trichas, 2002: 264 **E**: GR (Antythira Is.)
armeniacus Baudi di Selve, 1876c: 266 [= 1976b: 140] **E**: AR **A**: AF IN TM TR
coelatus Brullé, 1832: 208 **E**: AL GR IT **A**: TR
crenulatus Ménétriés, 1832: 202 (*Heliophilus*) **E**: AB AR GG ST **A**: IN TR
dardanus Faldermann, 1837: 51 (*Pandarus*)
cribratus Waltl, 1838: 462
dalmatinus Germar, 1824b: 142 (*Pedinus*) **E**: AL BH CR GR IT SL **A**: TR
dentitibia Koch, 1948a: 344 **E**: GR (Andros)
dragonadanus Koch, 1948a: 348 **E**: GR (Dragonada)
extensus Faldermann, 1837: 52 (*Pandarus*) **E**: AR **A**: IN TM
falassamensis Chatzimanolis, Engel & Trichas, 2002: 263 **E**: GR (satellite West Kríti islet)
foveolatus Seidlitz, 1893: 399 **E**: AR **A**: TR
grampusanus Koch, 1948a: 347 **E**: GR (Grampusa)
leonhardi Schuster, 1940b: 861 **A**: IN TM (Kopet Dag)
maximus Koch, 1948a: 348 **E**: GR (Gavdos)
messenius Brullé, 1832: 211 (*Phylax*) **E**: AL GR MC TR YU **A**: TR
gravidus Brullé, 1832: 210 (*Phylax*)
ottomanus Mulsant & Rey, 1854a: 101 [=1854b: 245] (*Pandarus*)
tentyrioides Brullé, 1832: 212 (*Phylax*)
moesiacus Mulsant & Rey, 1854a: 87 [= 1854b: 231] (*Pandarus*) **E**: AL BU GR MC **A**: TR
mylonasi Chatzimanolis, Engel & Trichas, 2002: 262 **E**: GR (Augo Is.)
opacus Koch, 1948a: 347 **E**: GR (Kríti)
orientalis Seidlitz, 1893: 399 **A**: CY SY
plicatulus jonicus Koch, 1948a: 352 **E**: GR
plicatulus paganettii Koch, 1948a: 353 **E**: GR
plicatulus plicatulus Brullé, 1832: 210 (*Phylax*) **E**: GR TR
simius Mulsant & Rey, 1854a: 76 [= 1854b: 220] (*Pandarus*)
plicatulus victoris Mulsant & Rey, 1854a: 84 [= 1854b: 228] (*Pandarus*) **E**: CR GR
seidlitzi Reitter, 1904a: 80
punctatus Audinet-Serville, 1825: 26 (*Pedinus*) **E**: AR ST UK **A**: TR
puncticollis Koch, 1948a: 348 **E**: GR (Kríti)
schatzmayri Koch, 1948a: 344 **E**: GR (Syra, Kythera)
scoparipes Reitter, 1904a: 86 **E**: GR (Kythera)
scrobiculatus Seidlitz, 1893: 399 **A**: TR
simplex Seidlitz, 1893: 399 **E**: AR **A**: IQ IN SY TR
sporadicus Koch, 1948a: 347 **E**: GR (Sporades)
stampalicus Koch, 1948a: 347 **E**: GR (Sporades)

subopacus Baudi di Selve, 1881: 282 (*Pandarinus*) **A**: LE SY
tenellus Mulsant & Rey, 1854a: 106 [= 1854b: 250] (*Pandarinus*) **E**: GR **A**: IQ TR
torpidus Mulsant & Rey, 1854a: 82 [= 1854b: 226] (*Pandarus*) **A**: TR
transcaspicus afghanicus Kaszab, 1973b: 53 **A**: AF
transcaspicus medvedevi Kaszab, 1973b: 55 **A**: AF
transcaspicus transcaspcius Brancsik, 1899: 101 **A**: TM (Kopet Dag)
vagabundus Reitter, 1904a: 84
vagans Reitter, 1902b: 180
wernerianus Koch, 1948a: 344 **E**: GR (Sikinos)
wettsteini Koch, 1948a: 348 **E**: GR (Kríti)
subgenus *Paroderus* Mulsant & Rey, 1854a: 111 [= 1854b: 255] type species *Pandarinus elongatus* Mulsant & Rey, 1854
Dichromma Mulsant & Rey, 1855c: 194 type species *Pandarinus foraminosus* Mulsant & Rey, 1855
corcyrensis Koch, 1948a: 352 **E**: GR (Kerkyra)
elongatus Mulsant & Rey, 1854a: 111 [= 1854b: 255] (*Pandarinus*) **E**: SP (Andalucía)
nevadensis Piochard de la Brûlerie, 1869: 33 (*Pandarinus*)
creticus Motschulsky, 1858d: 188 (*Bioplanes*)
foraminosus Mulsant & Rey, 1855b: 95 [= 1855c: 195] **E**: GR (Kríti)
punctulatus Lucas, 1854: 30 (*Philax*)
lugens Mulsant & Rey, 1854a: 77 [= 1854b: 221] (*Pandarus*) **E**: IT MA
libanicus Desbrochers des Loges, 1881: 116 [HN]
politus Reitter, 1904a: 89 **E**: GR (Kríti)
rhodius Baudi di Selve, 1876a: 39 **E**: GR (Ródos, Sporades)
subgenus *Rhizalemus* Reitter, 1904a: 79 type species *Dendarus reitteri* Seidlitz, 1893
calcaroides Peyerimhoff, 1907a: 36 (*Phylax*) **A**: SI
crassiusculus Mulsant & Rey, 1854a: 246 [= 1855a: 134] (*Bioplanes*) **A**: IS
impressus Reiche & Saulcy, 1857: 255 (*Bioplanes*) **A**: IS "Syria"
oertzenii Seidlitz, 1893: 393 **A**: IS
pauper Mulsant & Rey, 1854a: 114 [= 1854b: 258] (*Pandarinus*) **A**: IS JO
plorans Mulsant & Rey, 1854a: 245 [= 1855a: 133] (*Bioplanes*) **A**: IS
saginatus Baudi di Selve, 1876a: 41
reitteri Seidlitz, 1893: 398 **A**: LE "Syria"
libanicus Baudi di Selve, 1875d: 697
vagans Seidlitz, 1893: 399 **A**: LE
subgenus *Rizalus* Mulsant & Rey, 1854a: 104 [= 1854b: 248] type species *Opatrum piceus* A. G. Olivier, 1811
piceus A. G. Olivier, 1811: 501 (*Opatrum*) **N**: EG **A**: CY IS LE SY
tenuicornis Miller, 1861: 201 (*Pandarus*)
syriacus Reiche & Saulcy, 1857: 257 (*Bioplanes*) **N**: EG **A**: IS JO SI
nitidulus Seidlitz, 1893: 394
viduus Reiche, 1857: 256 (*Bioplanes*)

genus *Dendarus*, nomen dubium
obscuripennis Brullé, 1832: 212 (*Phylax*) **E**: GR

genus *Heliopates* Dejean, 1834: 191 type species *Tenebrio lusitanicus* Herbst, 1797
subgenus *Heliocrates* Reitter, 1904a: 98 type species *Heliophilus humerangulus* Reitter, 1904
humerangulus Reitter, 1904a: 99 (*Heliophilus*) **E**: SP
strigosus Reitter, 1904a: 98 (*Heliophilus*) **E**: SP
sulcipennis Reitter, 1904a: 99 (*Heliophilus*) **E**: SP
transversalis Mulsant & Rey, 1854a: 232 [= 1855a: 120] **E**: SP
subgenus *Heliopates* Dejean, 1834: 191 type species *Tenebrio lusitanicus* Herbst, 1797
Heliocaes Bedel, 1906b: 92 [RN]
Heliopathes Mulsant, 1854: 157 type species *Pedinus luctuosus* Le Peletier de Saint-Fargeau & Audinet-Serville, 1825
agrestis agrestis Mulsant & Rey, 1854a: 239 [= 1855a: 127] **E**: SP
agrestis forcadelli Español, 1958a: 23 **E**: SP (Columbrete)
angustipennis Español, 1958c: 115 **E**: SP

atlasicus Escalera, 1925f: 501 (*Heliocaes*) **N**: MO (Moyen Atlas, Grand Atlas)
altiatlantis Antoine, 1936b: 72 (*Heliophilus*)
bergmanni Antoine, 1951b: 318 (*Heliophilus*)
avarus avarus Mulsant & Rey, 1854a: 234 [= 1855a: 122] **E**: IT MA **N**: AG
ambiguus Baudi di Selve, 1876b: 164
avarus donatellae Canzoneri, 1970: 72 (*Heliocaes*) **E**: IT (Pantelleria)
avarus dwejrensis Scupola & Mifsud, 2001: 1 **E**: MA
balearicus Español, 1951a: 21 [= 1954a: 56] **E**: SP (Ibiza)
batnensis Mulsant & Godart, 1865: 455 **E**: IT (Sicilia) **N**: AG MO TU
conjuctus Reitter, 1904a: 101 (*Heliophilus*) **E**: SP (Huesca, Zaragoza, Lérida)
cribratostriatus Mulsant & Rey, 1854a: 234 [= 1855a: 122] **E**: PT SP **N**: AG MO
cribratus Chevrolat, 1865: 391 **E**: PT SP
parcefoveatus Reitter, 1904a: 100 (*Heliophilus*)
emarginatus Fabricius, 1792a: 108 (*Blaps*) **E**: PT SP **N**: AG MO
subvariolosus Lucas, 1846: 330 (*Dendarus*)
escalerai Español, 1943b: 106 (*Heliophilus*) **E**: SP (Navarra)
heydeni Reitter, 1904a: 102 (*Heliophilus*) **E**: SP
ibericus Mulsant & Rey, 1854a: 235 [= 1855a: 123] **E**: PT SP
immarginatus Reitter, 1904a: 101 (*Heliophilus*) **E**: SP
interstitialis Mulsant & Rey, 1854a: 231 [= 1855a: 119] **N**: AG MO
liebei Schuster, 1926c: 243 (*Heliophilus*) **E**: SP (Pontevedra)
littoralis Español, 1958a: 20 **E**: FR SP
luctuosus Le Peletier de Saint-Fargeau & Audinet-Serville, 1825: 27 (*Pedinus*) **E**: FR SP
simulans Chevrolat, 1865: 391
lusitanicus Herbst, 1797: 244 (*Tenebrio*) **E**: PT SP
franzi Español, 1955c: 97
obsoletus Marsham, 1802: 475 (*Tenebrio*)
montivagus Mulsant & Rey, 1854a: 233 [= 1855a: 121] **E**: SP (Granada)
morandi Español, 1969: 81 **E**: SP (Salinas, Torrevieja)
neptunius Baudi di Selve, 1875b: 164 **E**: IT (Sicilia: Mti Madone)
rex Antoine, 1942c: 48 (*Heliophilus*) **N**: MO (Rif)
rectangularis Antoine, 1951c: 97 (*Heliophilus*) **N**: MO
rotundicollis Lucas, 1846: 329 (*Dendarus*) **E**: PT SP **N**: AG LB TU
rotundicollis Mulsant & Rey, 1854a: 237 [= 1855a: 125] (*Heliopathes*) [HN]
tripolianus Baudi di Selve, 1876a: 50
tunisiensis Dajoz, 1985: 193
sculpturatus Reitter, 1904a: 102 (*Heliophilus*) **E**: SP
strigicollis Baudi di Selve, 1876b: 163 **N**: MO
subimpressus Reitter, 1904a: 103 (*Heliophilus*) **E**: SP
tazakensis Antoine, 1942c: 48 (*Heliophilus*) **N**: MO
vivesi Español & Viñolas, 1983b: 216 **E**: SP

genus *Heliopathes*, nomen dubium
meridianus Brullé, 1832: 208 (*Heliophilus*) **E**: GR

genus *Neoisocerus* Bouchard, Lawrence, Davies & Newton, 2005: 499 [RN] type species *Tenebrio purpurascens* Herbst, 1799
Isocerus Dejean, 1821: 66 [HN] type species *Tenebrio purpurascens* Herbst, 1799
Isocerus Laporte, 1840: 210 [HN] type species *Helops ferrugineus* Fabricius, 1798
balearicus L. W. Schaufuss, 1879: 535 (*Isocerus*) **E**: SP (Baleares)
barbieri Ardoin, 1972: 239 (*Isocerus*) **N**: AG
ferrugineus Fabricius, 1798: 53 (*Helops*) **E**: PT SP **N**: MO
cameroni Reitter, 1904a: 78 (*Isocerus*)
lusitanicus Reitter, 1886d: 152 (*Isocerus*)
purpurascens Herbst, 1799: 20 (*Tenebrio*)
tunisiensis E. Strand, 1935: 294 (*Isocerus*) [RN] **N**: AG TU

ferrugineus Reitter, 1886d: 152 (*Isocerus*) [HN]

genus *Micrositus* Mulsant & Rey, 1854a: 148 [= 1854b: 292] type species *Micrositus orbicularis* Mulsant & Rey, 1854
opacus Reitter, 1904a: 116 **A**: TR
orbicularis Mulsant & Rey, 1854a: 155 [= 1854b: 299] **E**: GR **A**: TR

genus *Phylan* Dejean, 1821: 65 type species *Opatrum gibbum* Fabricius, 1775
subgenus *Dendarophylan* Español, 1945b: 312 type species *Phylan pardoi* Español, 1945
pardoi Español, 1945b: 349 **E**: SP **N**: MO
subgenus *Eumicrositus* Viñolas, 1990: 62 type species *Pedinus ulissiponensis* Germar, 1824
emmanueli Español, 1958c: 108 (*Micrositus*) **E**: SP (Guadalupe)
laufferi Reitter, 1915e: 87 (*Micrositus*) **E**: SP (Granada)
martinezi Español, 1947a: 34 (*Micrositus*) **E**: SP (Ávila)
miser almeriensis Español, 1947a: 30 (*Micrositus*) **E**: SP (Almeriá)
miser miser Mulsant & Rey, 1854a: 168 [= 1854b: 312] (*Micrositus*) **E**: PT SP **N**: MO
obesus agricola Mulsant & Rey, 1854a: 164 [= 1854b: 308] (*Micrositus*) **E**: SP
obesus obesus Waltl, 1835: 71 (*Heliophilus*) **E**: PT SP
baeticus Mulsant & Rey, 1854b: 310 [= 1854a: 166] (*Micrositus*)
saxicola Mulsant & Rey, 1854a: 193 [= 1855a: 81] (*Omocrates*) **E**: PT SP
saxeticola Graells, 1858: 71 (*Phylan*)
ulissiponensis conveximargo Español, 1947a: 24 (*Micrositus*) **E**: PT SP
ulissiponensis montanus Mulsant & Rey, 1854a: 160 [= 1854b: 304] (*Micrositus*) **E**: SP
ulissiponensis paivae Perez Arcas, 1865b: 439 (*Micrositus*) **E**: SP
ulissiponensis ulissiponensis Germar, 1824b: 143 (*Pedinus*) **E**: PT SP
obesus Mulsant & Rey, 1854a: 165 [= 1854b: 309] (*Micrositus*) [HN]
ulyssiponensis Le Peletier de Saint-Fargeau & Audinet-Serville, 1825: 26 (*Pedinus*)
ulissiponensis ventralis Marseul, 1868b: 20 (*Micrositus*) [RN] **E**: SP
levis Perez Arcas, 1865a: 178 (*Micrositus*)
vivesi Español, 1963d: 185 (*Micrositus*) **E**: SP
subgenus *Litoboriolus* Español, 1945b: 310 type species *Omocrates collaris* Mulsant & Rey, 1854
collaris Mulsant & Rey, 1854a: 195 [= 1855a: 83] (*Omocrates*) **E**: SP (Guadarrama)
companyae Viñolas, 1997a: 98 **E**: SP (Zamora)
mulsanti Piochard de la Brûlerie, 1869: 36 (*Heliopathes*) **E**: SP
gadeai Español & Viñolas, 1981b: 108
reyi Piochard de la Brûlerie, 1869: 37 (*Heliopathes*)
subgenus *Litororus* Reitter, 1904a: 96 type species *Phylax balearicus* Schaufuss, 1882 (= *Micrositus semicostatus* Mulsant & Rey, 1854)
nitidicollis Perez Arcas, 1872: 107 (*Micrositus*) **E**: SP (Mallorca)
cabrerensis Tenenbaum, 1915: 839 (*Dendarus*)
melas Tenenbaum, 1915: 838 (*Dendarus*)
semicostatus curtulus Breit, 1908: 63 **E**: SP (Cabrera)
semicostatus semicostatus Mulsant & Rey, 1854a: 178 [= 1854b: 322] (*Micrositus*) **E**: SP (Baleares)
balearicus Schaufuss, 1882c: 623 (*Phylax*)
subgenus *Meladeras* Mulsant & Rey, 1854a: 219 [= 1855a: 107] type species *Meladeras quadratulus* Mulsant & Rey, 1854
amoenus Mulsant & Rey, 1854a: 224 [= 1855a: 112] (*Meladeras*) **N**: AG
barbarus Lucas, 1846: 329 (*Dendarus*) **N**: AG MO
marocanus Obenberger, 1914: 140 (*Meladeras*)
quadratulus Mulsant & Rey, 1854a: 220 [= 1855a: 108] (*Meladeras*)
obscurus Mulsant & Rey, 1854a: 223 [= 1855a: 111] (*Meladeras*) **N**: AG MO
subgenus *Meladocrates* Reitter, 1904a: 96 type species *Omocrates planiusculus* Mulsant & Rey, 1854
costulatus Escalera, 1914: 327 **N**: MO
orientalis Dajoz, 1985: 195 **N**: TU
planiusculus Mulsant & Rey, 1854a: 216 [= 1855a: 104] (*Omocrates*) **E**: SP (Cádiz) **N**: MO
lacertosus Antoine, 1942c: 56

latipennis Champion, 1891d: 384 (*Heliopathes*)
mediatlantis Antoine, 1942c: 57
tenuicornis Antoine, 1942c: 55 **N**: MO
viaticus Mulsant & Rey, 1854a: 218 [= 1855a: 106] (*Omocrates*) **E**: SP **N**: MO
subgenus *Phylan* Dejean, 1821: 65 type species *Opatrum gibbum* Fabricius, 1775
Heliophilus Dejean, 1821: 65 [HN] type species *Pedinus hybridus* Latreille, 1804 (= *Tenebrio abbreviatus* A. G. Olivier, 1795)
Olocrates Mulsant, 1854: 383 [RN] type species *Tenebrio abbreviatus* A. G. Olivier, 1795
Olocrates Mulsant & Rey, 1854a: 248 [= 1855a: 136] [RN] type species *Tenebrio abbreviatus* A. G. Olivier, 1795
Omocrates Mulsant, 1854: 150 type species *Tenebrio abbreviatus* A. G. Olivier, 1795
Omocrates Mulsant & Rey, 1854a: 191 [= 1855a: 79] type species *Tenebrio abbreviatus* A. G. Olivier, 1795
Phylan Stephens, 1832: 20 [HN] type species *Opatrum gibbum* Fabricius, 1775
abbreviatus abbreviatus A. G. Olivier, 1795: 57: 17 (*Tenebrio*) **E**: FR SP
dubius Baudi di Selve, 1875b: 161 (*Heliopathes*)
hybridus Latreille, 1804a: 284 (*Pedinus*)
intermedius Dajoz, 1965: 131
tristis Herbst, 1797: 243 (*Tenebrio*)
abbreviatus chossoni Pic, 1937d: 10 **N**: AG
abbreviatus italicus Reitter, 1904a: 93 (*Olocrates*) **E**: IT
abreviatus hamilcaris Reitter, 1904a: 94 (*Olocrates*) **E**: SP
abbreviatus wankai Reitter, 1915e: 87 **E**: SP
ceballosi Español, 1945b: 335 **E**: SP (Granada, Murcia)
difficilis Antoine, 1942c: 54 **N**: MO (Moyen Atlas)
minutus Antoine, 1942c: 54
foveipennis foveipennis Mulsant & Rey, 1854a: 203 [= 1855a: 91] (*Omocrates*) **E**: SP
dendaroides Baudi di Selve, 1876a: 46 (*Heliopates*)
foveostriatus Baudi di Selve, 1875b: 160 (*Heliopathes*)
foveipennis kochi Español, 1945b: 319 **E**: PT
foveipennis lineatopunctatus Mulsant & Rey, 1854a: 205 [= 1855a: 93] (*Omocrates*) **E**: SP
foveolatus Graells, 1858: 71 (*Heliopates*) **E**: SP
fossulatus Mulsant & Rey, 1854a: 200 [= 1855a: 88] (*Omocrates*)
milleporus Baudi di Selve, 1875b: 160 (*Heliopathes*)
gibbus Fabricius, 1775: 76 (*Opatrum*) **E**: BE DE EN FI FR GB GE IR LA LT LU NL NR NT PL PT SP SV
arenosus Marsham, 1802: 475 (*Tenebrio*)
gemellatus Marsham, 1802: 475 (*Tenebrio*)
paludicola Chevrolat, 1865: 391 (*Olocrates*)
pilipes Herbst, 1797: 260 (*Tenebrio*)
ilerdensis Español & Viñolas, 1981b: 105 **E**: SP
indiscretus indiscretus Mulsant & Rey, 1854a: 208 [= 1855a: 96] (*Omocrates*) **E**: SP
indiscretus mendizabali Español, 1945b: 332 **E**: SP (Almería, Jaén)
indiscretus nivalis Baudi di Selve, 1876a: 50 (*Heliopathes*) [= 1876b: 161] **E**: SP (Granada)
mahjoubi Antoine, 1942c: 52 **N**: MO (region of Fés)
mediterraneus Piochard de la Brûlerie, 1869: 37 (*Heliopathes*) **E**: SP (Baleares)
nitidior Antoine, 1942c: 53 **N**: MO (Meknès)
ribesi ardoini Español & Viñolas, 1981b: 102 **E**: SP (Zamora)
ribesi ribesi Español, 1969: 83 **E**: SP
subheliopates Reitter, 1904a: 95 (*Olocrates*) **E**: SP
subgenus *Platyolus* Mulsant & Rey, 1854a: 153 [= 1854b: 297] type species *Micrositus longulus* Mulsant & Rey, 1854
furvus filabricus Español, 1963e: 387 (*Micrositus*) **E**: SP (Almería)
furvus furvus Mulsant & Rey, 1854a: 181 [= 1854b: 324] (*Micrositus*) **E**: SP
furvus stenopterus Español, 1947a: 49 (*Micrositus*) **E**: SP (Almería)
gibbulus Motschulsky, 1849: 124 (*Pandarus*) **E**: SP
gustavii Mulsant & Rey, 1854a: 244 (*Micrositus*)
heeri Mulsant & Rey, 1854a: 171 [= 1854b: 315] (*Micrositus*) **E**: SP (Almería)
lapidarius lapidarius Reitter, 1904: 121 (*Micrositus*) **E**: SP (Almería, Granada)
lapidarius strasseni Español, 1954d: 161 (*Micrositus*) **E**: SP (Granada)

longulus Mulsant & Rey, 1854a: 185 [= 1854b: 329] (*Micrositus*) **E**: PT SP
melancholicus Mulsant & Rey, 1854a: 176 [= 1854b: 320] (*Micrositus*) **E**: PT SP **N**: MO
milleri Reitter, 1904a: 120 (*Micrositus*) [DA] **E**: SP (Cádiz)
subcylindricus Motschulsky, 1849: 123 (*Pandarus*) **E**: SP
tenebrioides Reitter, 1904a: 118 (*Micrositus*) **E**: PT SP

subtribe Leichenina Mulsant, 1854

genus *Apsheronellus* Bogatchev, 1967a: 157 type species *Apsheronellus arenarius* Bogatchev, 1967
Microleichenum G. S. Medvedev, 1973: 648 type species *Microleichenum choresmensis* G. S. Medvedev, 1973
arenarius Bogatchev, 1967a: 159 **E**: AB
choresmensis G. S. Medvedev, 1973: 648 (*Microleichenum*) **A**: KZ TM UZ

genus *Leichenum* Dejean, 1834: 194 type species *Opatrum pictum* Fabricius, 1801
Endothina Carter, 1924b: 536 type species *Endothina squamosa* Carter, 1924
Lichenum Agassiz, 1846b: 209 [unjustified emendation]
canaliculatum canaliculatum Fabricius, 1798: 42 (*Opatrum*) **E**: GG GR **A**: FUJ IQ JA NP TAI TM UP YE **AUR ORR**
argillaceum Motschulsky, 1863: 475
foveistrium Marseul, 1876c: 102 (*Lichenum*)
seriehispidum Marseul, 1876c: 101 (*Lichenum*)
squamosum Carter, 1924b: 537 (*Endothina*)
variegatum Klug, 1833: 88 (*Opatrum*)
mucronatum Küster, 1849a: 67 **E**: AB GR **A**: AF IN IQ LE KZ SY TR
caucasicum Reitter, 1899a: 84
caudatum Reitter, 1899a: 85
muelleri muelleri Gridelli, 1939c: 222 **A**: SA YE **AFR**
pictum Fabricius, 1801a: 117 (*Opatrum*) **E**: AL AU BH BU CR HU GR IT RO SK ST SZ UK
heyrovskyi Obenberger, 1922: 41 (*Lichenum*)
pulchellum pulchellum Lucas, 1846: 336 (*Opatrum*) **E**: FR IT SP **N**: AG MO
gebieni Reitter, 1906a: 489 (*Lichenum*) [RN]
pulchellum Küster, 1849a: 65 [HN]
variegatum Küster, 1849a: 66 [HN]
pulchellum pumilum Baudi di Selve, 1876a: 73 **E**: AB GR **N**: EG LB TU **A**: BA CY IQ IN IS LE PA SA SI SY TR YE **AFR**
incisum Reitter, 1899a: 85

subtribe Melambiina Mulsant & Rey, 1854

genus *Allophylax* Bedel, 1906c: 177 [RN] type species *Phylax littoralis* Mulsant, 1854 (= *Opatrum picipes* A. G. Olivier, 1811)
subgenus *Allophylax* Bedel, 1906c: 177 [RN] type species *Phylax littoralis* Mulsant, 1854 (= *Opatrum picipes* A. G. Olivier, 1811)
Melambiophylax Schuster, 1922c: 48 type species *Phylax sardous* Baudi di Selve, 1875
Neophylax Bedel, 1906b: 92 [HN] type species *Phylax littoralis* Mulsant, 1854 (= *Opatrum picipes* A. G. Olivier, 1811)
brevicollis Baudi di Selve, 1876a: 55 [= 1876b: 214] (*Phylax*) **E**: IT (Sardegna)
picipes melitensis Baudi di Selve, 1876a: 54 (*Phylax*) **E**: MA
picipes picipes A. G. Olivier, 1811: 500 (*Opatrum*) **E**: CR FR IT **N**: AG TU
dalmatinus Reitter, 1904: 109 (*Phylax*)
fraternus Mulsant & Rey, 1854: 287 (*Phylax*)
littoralis Mulsant, 1854: 148 (*Phylax*)
sardous Baudi di Selve, 1876a: 55 [= 1876b: 214] (*Phylax*) **E**: IT (Sardegna)
subgenus *Phylaximon* Koch, 1948b: 414 type species *Opatrum variolosus* A. G. Olivier, 1811
costatipennis costatipennis Lucas, 1846: 328 (*Philax*) **E**: IT (Linosa, Lampedusa) **N**: AG TU
sicardi Reitter, 1904a: 111 (*Phylax*)
costatipennis godenigoi Canzoneri, 1970: 75 **E**: IT (Marettimo)
ignavus Mulsant & Rey, 1854a: 147 [= 1854b: 291] (*Phylax*) **N**: AG

ingratus Mulsant & Rey, 1854a: 143 [= 1854b: 287] (*Phylax*) **N**: AG TU
segnis Mulsant & Rey, 1854a: 145 [= 1854b: 289] (*Phylax*) **N**: AG
undulatus Mulsant & Rey, 1854a: 136 [= 1854b: 280] (*Phylax*) **N**: AG TU
ovipennis Fairmaire, 1871a: 391 (*Phylax*) **N**: AG
variolosus A. G. Olivier, 1811: 497 (*Opatrum*) **N**: AG TU
subgenus *Litoboromimus* Koch, 1948b: 413 type species *Litoborus parallelus* Schuster, 1919
parallelus Schuster, 1919a: 30 (*Litoborus*) **N**: LB
rotundicollis Chobaut, 1925: 228 **N**: TU
genus *Allophylax*, species incertae sedis
insignis Reitter, 1904a: 110 (*Phylax*) **A**: "Syria"

genus *Bermejoina* Español, 1944: 12 type species *Bermejoina aiunica* Español, 1944
aiunica Español, 1944: 15 **N**: MO (Western Sahara)

genus *Guildia* Antoine, 1956: 346 type species *Micrositus punctistriatus* Escalera, 1925
punctistriata Escalera, 1925d: 379 (*Micrositus*) **N**: MO (Moyen Atlas)

genus *Hoplarion* Mulsant & Rey, 1854a: 150 [= 1854b: 294] type species *Hoplarion tumidum* Mulsant & Rey, 1854
subgenus *Atlasion* Koch 1948: 428 type species *Micrositus bedeli* Escalera, 1914
Megatlasion Koch, 1948b: 427 type species *Micrositus atlantis* Escalera, 1914
angulatum angulatum Escalera, 1913: 48 (*Micrositus*) **N**: MO (Grand Atlas)
angulatum gundaficum Antoine, 1942b: 37 (*Melambius*) **N**: MO (Grand Atlas)
angulatum kochi Antoine, 1946c: 55 (*Hoplariobius*) **N**: MO (Grand Atlas)
atlantis atlantis Escalera, 1914: 333 (*Micrositus*) **N**: MO (Grand Atlas)
atlantis gattefossei Antoine, 1936b: 74 (*Micrositus*) **N**: MO (Grand Atlas)
atlantis subtile Antoine, 1956: 355 **N**: MO (Grand Atlas)
bedeli baudoni Antoine, 1956: 358 **N**: MO (Grand Atlas)
bedeli bedeli Escalera, 1914: 334 (*Micrositus*) **N**: MO (Moyen Atlas)
globosum Escalera, 1925a: 374 (*Micrositus*)
lecerfi Antoine, 1956: 358 (*Hoplariobius*)
bedeli escalerai Koch, 1945: 417 (*Micrositus*) **N**: MO (Moyen Atlas)
nitens Antoine, 1956: 359 (*Hoplariobius*)
bedeli insolens Antoine, 1942b: 46 (*Melambius*) **N**: MO (Grand Atlas)
bedeli major Antoine, 1942b: 40 (*Melambius*) **N**: MO (Grand Atlas)
bedeli remotum Antoine, 1956: 359 **N**: MO (Grand Atlas)
bedeli septentrionale Antoine, 1942b: 38 (*Melambius*) **N**: MO (Moyen Atlas)
bedeli subcostatum Koch, 1945: 418 (*Hoplariobius*) **N**: MO (Moyen Atlas)
latissimum Antoine, 1942b: 38 (*Melambius*) **N**: MO (Moyen Atlas)
theryi sarroense Antoine, 1956: 354 **N**: MO (Sarhro)

theryi theryi Antoine, 1933b: 227 (*Micrositus*) **N**: MO (Grand Atlas)
rugosium Antoine, 1956: 354
subgenus *Hoplarion* Mulsant & Rey, 1854a: 150 [= 1854b: 294] type species *Hoplarion tumidum* Mulsant & Rey, 1854
attritum Bedel, 1918: 364 **N**: MO
humile Antoine, 1942b: 52 (*Melambius*) **N**: MO
kocheri Antoine, 1934: 122 (*Hoplarion*) **N**: MO
torretassoae Koch, 1948b: 429 **N**: AG
tumidum Mulsant & Rey, 1854a: 150 [=1854b: 294] (*Micrositus*) **N**: AG
subgenus *Glyptariobius* Koch, 1948b: 423 type species *Glyptariobius excavatus* Koch, 1948
excavatum Koch, 1948b: 423 **N**: AG
subgenus *Hoplariobius* Reitter, 1904a: 115 type species *Micrositus decurtatus* Fairmaire, 1884
declive Antoine, 1951b: 317 (*Melambius*) **N**: MO (Grand Atlas)
decurtatum Fairmaire, 1884f: cxx (*Micrositus*) **N**: MO

tenuepunctatum Escalera, 1914: 333 (*Micrositus*) **N**: MO
granulosum Billberg, 1815: 276 (*Opatrum*) **N**: AG
subgenus *Mentariobius* Koch, 1948b: 425 type species *Hoplariobius pueli* Koch, 1948
pueli Koch, 1948b: 425 (*Hoplariobius*) **N**: AG
distinguendum Mulsant & Rey, 1854a: 159 [= 1854b: 303] (*Micrositus*) **N**: AG
subgenus *Saharoplarion* Koch, 1948b: 431 type species *Micrositus compactus* Fairmaire, 1880
compactum Fairmaire, 1880a: 16 (*Micrositus*) **N**: AG LB TU

genus *Litoborus* Mulsant & Rey, 1854a: 126 [= 1854b: 270] type species *Philax moreleti* Lucas, 1846
subgenus *Litoborus* Mulsant & Rey, 1854a: 126 [= 1854b: 270] type species *Philax moreleti* Lucas, 1846
clermonti Antoine, 1931: 187 **N**: AG MO
maroccanus Escalera, 1914: 329 **N**: MO (Haut Atlas)
moreleti mediatlantis Antoine, 1942a: 22 **N**: MO (Moyen Atlas)
moreleti moreleti Lucas, 1846: 327 (*Philax*) **N**: AG MO
  baudii Reitter, 1904a: 113
moreleti peyerimhoffi Antoine, 1931: 185 **N**: MO
subgenus *Paralitoborus* Antoine, 1931: 190 type species *Litoborus sternalis* Fairmaire, 1884
chobauti Antoine, 1931: 190 **N**: MO
escalerai Antoine, 1931: 192 **E**: SP
forticostis forticostis Escalera, 1914: 330 **N**: MO
forticostis saharensis Español, 1946: 120 **N**: MO (Western Sahara)
lindbergi Español, 1967a: 39 **N**: MO
olcesii atlantis Antoine, 1931: 203 **N**: MO (Haut Atlas)
olcesii olcesii Fairmaire, 1871a: 392 (*Phylax*) **N**: MO
planicollis bolivari Antoine, 1931: 195 **N**: MO
planicollis planicollis Waltl, 1835: 71 (*Phylan*) **E**: PT SP **N**: AG MO
  algiricus Baudi di Selve, 1875b: 158
  planicollis Mulsant & Rey, 1854a: 128 [= 1854b: 272] [HN]
sternalis Fairmaire, 1884c: 446 **N**: MO
subtilimargo Reitter, 1904a: 113 **N**: AG

genus *Melambius* Mulsant & Rey, 1854a: 124 [= 1854b: 268] type species *Opatrum barbarum* Erichson, 1841
subgenus *Melambatlasus* Koch, 1948b: 433 type species *Micrositus hebes* Antoine, 1933
androgynus Antoine, 1942b: 31 **N**: MO (Grand Atlas)
  antoinei Koch, 1945: 421 (*Micrositus*)
anemophilus Antoine, 1942b: 30 **N**: MO (Grand Atlas)
asniensis Antoine, 1942b: 28 **N**: MO (Grand Atlas)
cacuminorum Antoine, 1936b: 73 (*Micrositus*) **N**: MO (Grand Atlas)
cultriger Antoine, 1942b: 33 **N**: MO (Grand Atlas)
hebes Antoine, 1933b: 228 (*Micrositus*) **N**: MO (Grand Atlas)
hassani hassani Antoine, 1942b: 26 **N**: MO (Grand Atlas)
hassani jejunus Antoine, 1956: 362 **N**: MO (Grand Atlas)
lasciniosus Antoine, 1956: 364 **N**: MO (Grand Atlas)
meridionalis Antoine, 1951b: 317 **N**: MO (Sarhro)
mirus Antoine, 1956: 361 **N**: MO (Moyen Atlas)
pauliani Koch, 1945: 422 (*Micrositus*) **N**: MO (Grand Atlas)
rugatithorax Antoine, 1951b: 316 **N**: MO (Grand Atlas)
venustus Antoine, 1942b: 32 **N**: MO (Moyen Atlas)
subgenus *Hadromelambius* Koch, 1948b: 432 type species *Melambius telueticus* Escalera, 1914
telueticus Escalera, 1914: 331 **N**: MO
subgenus *Hoplambius* Reitter, 1904a: 114 type species *Hoplarion melambioides* Fairmaire, 1893
melambioides Fairmaire, 1893i: cxlviii (*Hoplarion*) **N**: AG TU
subgenus *Melambius* Mulsant & Rey, 1854a: 124 [= 1854b: 268] type species *Opatrum barbarum* Erichson, 1841

barbarus Erichson, 1841: 181 (*Opatrum*) **N**: AG
bidens Antoine, 1932: 188 **N**: MO (Grand Atlas)
breviusculus breviusculus Fairmaire, 1871a: 390 **N**: AG
breviusculus syrtanus Normand, 1938a: 303 [= 1938b: 366] **N**: TU
denticollis Escalera, 1913: 48 **N**: MO (Grand Atlas)
mideltensis Antoine, 1932: 188 **N**: MO
otini Antoine, 1942b: 42 **N**: MO
rotroui Antoine, 1951b: 315 **N**: MO (Rif)
rungsi Antoine, 1956: 364 **N**: MO (Grand Atlas)
schrammi Antoine, 1956: 365 **N**: MO
sulcatus Antoine, 1956: 366 **N**: MO (Moyen Atlas)
teinturieri Mulsant & Godart, 1865: 454 **N**: AG TU
    asperocostatus Fairmaire, 1871a: 391
tuniseus Levrat, 1859a: 34 (*Philax*) **N**: TU

genus *Melansis* Wollaston, 1864: 491 type species *Phylax costatus* Brullé, 1838
angulata angulata Wollaston, 1864: 492 **N**: CI (La Palma)
angulata hierroensis Español, 1962a: 206 **N**: CI
costata Brullé, 1838: 69 (*Phylax*) **N**: CI (Gran Canaria)
kaszabi J. Ferrer, 1992: 83 **N**: CI (Gran Canaria)
reyesi Gancía & Oromí, 2000: 53 **N**: CI (Tenerife)
tibicena Liberto & Leo, 2002: 20 **N**: CI (Gran Canaria)

genus *Melasmana* E. Strand, 1935: 297 [RN] type species *Phylax lineatus* Brullé, 1838
subgenus *Heliomelasma* Koch, 1948b: 408 type species *Melasmana appenhageni* Koch, 1948
appenhageni Koch, 1948b: 408 **N**: CI (Fuerteventura)
subgenus *Melasmana* E. Strand, 1935: 297 [RN] type species *Phylax lineatus* Brullé, 1838
    Melasma Wollaston, 1864: 484 [HN] type species *Phylax lineatus* Brullé, 1838
lineatum Brullé, 1838: 69 (*Phylax*) **N**: CI

genus *Oreomelasma* Español, 1975b: 238 type species *Oreomelasma oromii* Español, 1975
oromii Español, 1975b: 241 **N**: CI (Fuerteventura)

genus *Otinia* Antoine, 1942b: 22 type species *Otinia iblanensis* Antoine, 1942
subgenus *Antoineius* Koch, 1948b: 418 type species *Micrositus jeanneli* Koch, 1945 (= *Melambius inermis* Antoine, 1942)
espanoli Antoine, 1956: 350 **N**: MO (Rif)
inermis Antoine, 1942b: 35 (*Melambius*) **N**: MO (Moyen Atlas)
    jeanneli Koch, 1945: 420 (*Micrositus*)
kocheri Antoine, 1946c: 54 (*Melambius*) **N**: MO (Grand Atlas)
vidali Antoine, 1942b: 27 (*Melambius*) **N**: MO (Rif)

subgenus *Orophylaxus* Koch, 1948b: 417 type species *Philax incertus* Mulsant & Godart, 1865
incerta disparis Escalera, 1925d: 377 (*Micrositus*) **N**: MO (Moyen Atlas)
    occidentalis Antoine, 1956: 350
incerta incerta Mulsant & Godart, 1865: 453 (*Philax*) **N**: AG
    oxyholma Fairmaire, 1876: 38 (*Phylax*)
    serripes Desbrochers des Loges, 1881: 119 (*Micrositus*)
lacouri debduensis Antoine, 1942b: 52 **N**: MO
lacouri lacouri Antoine, 1942b: 41 (*Melambius*) **N**: MO
subgenus *Otinia* Antoine, 1942b: 22 type species *Otinia iblanensis* Antoine, 1942
iblanensis Antoine, 1942b: 23 **N**: MO (Moyen Atlas)
    embergeri Koch, 1945: 419 (*Micrositus*)

genus *Peyerimhoffius* Koch, 1948b: 420 type species *Philax plicatus* Lucas, 1846

plicatus Lucas, 1846: 326 (*Philax*) **N**: AG
plicatus Mulsant & Rey, 1854a: 157 [= 1854b: 301] (*Micrositus*) [HN]

genus *Psammoardoinellus* Leo, 1980a: 34 type species *Isocerus sardiniensis* Ardoin, 1972
sardiniensis Ardoin, 1972: 240 (*Isocerus*) **E**: IT (Sardegna)

subtribe Pedinina Eschscholtz, 1829

genus *Blindus* Mulsant & Rey, 1853c: 206 [= 1853e: 122] type species *Pedinus strigosus* Faldermann, 1835
fulvicornis Reitter, 1889k: 700 (*Pedinus*) **A**: NMO SCH
japonicus Seidlitz, 1893: 376 (*Pedinus*) **A**: JA
potanini nudiventris G. S. Medvedev, 1968a: 162 **A**: SCH
potanini potanini G. S. Medvedev, 1968a: 160 **A**: SCH
reichardti G. S. Medvedev, 1968a: 163 **A**: SCH
strigosus Faldermann, 1835b: 410 (*Pedinus*) **A**: BEI FE HEB HUB JA LIA MG NC NMO SC TAI TIA
faldermanni Baudi di Selve, 1876a: 46 (*Colpotus*)
thibetanus Fairmaire, 1897i: 217 (*Pedinus*) **A**: SCH XIZ

genus *Cabirutus* E. Strand, 1929: 24 [RN] type species *Cabirus minutissimus* Mulsant & Rey, 1853
subgenus *Asiobirus* G. S. Medvedev, 1968a: 179 type species *Cabirus validipes* Reitter, 1891
ferganensis G. S. Medvedev, 1968a: 201 **A**: KI
glazunovi G. S. Medvedev, 1968a: 193 **A**: UZ
gurjevae G. S. Medvedev, 1968a: 199 **A**: KI
kirghizicus G. S. Medvedev & Kaltaev, 1979: 57 **A**: KI
kuhitangi G. S. Medvedev, 1968a: 195 **A**: TM
kuramensis G. S. Medvedev, 1968a: 191 **A**: TD
obtusicollis Reitter, 1891d: 224 (*Cabirus*) **A**: KI KZ UZ
oxianus G. S. Medvedev, 1968a: 203 **A**: TM
procerulus Reitter, 1904a: 65 (*Cabirus*) **A**: KI KZ
puncticollis Reitter, 1904a: 66 (*Cabirus*) **A**: TD
ruficornis G. S. Medvedev, 1968a: 185 **A**: KI TD UZ
tibialis Reitter, 1891d: 224 (*Cabirus*) **A**: UZ
validipes Reitter, 1891d: 224 (*Cabirus*) **A**: KI KZ TD UZ
zaslavskii G. S. Medvedev, 1968a: 189 **A**: "Tian-Shan"
subgenus *Cabirutus* E. Strand, 1929: 24 [RN] type species *Cabirus minutissimus* Mulsant & Rey, 1853
Cabirus Mulsant & Rey, 1853c: 223 [= 1853e: 139] [HN] type species *Cabirus minutissimus* Mulsant & Rey, 1853
ampliatus Desbrochers des Loges, 1881: 109 (*Cabirus*) **A**: SY
castaneus Reitter, 1904a: 67 (*Cabirus*) **A**: IS
cavimanus J. R. Sahlberg, 1908: 81 (*Cabirus*) **A**: LE TR
cilicicus J. R. Sahlberg, 1908: 82 (*Cabirus*) **A**: TR
convexiusculus Desbrochers des Loges, 1881: 110 (*Cabirus*) **A**: "Syria"
cordicollis Kaszab, 1970b: 117 **A**: IN

cribricollis Baudi di Selve, 1876b: 156 (*Cabirus*) **E**: GR (Ródos) **A**: CY TR
cribricollis Desbrochers des Loges, 1881: 115 [HN]
cyrenaicus Gridelli, 1930: 290 (*Cabirus*) **N**: EG LB
desbrochersi Reitter, 1904a: 68 **A**: IS LE
elenae G. S. Medvedev, 1978c: 855 **A**: IN
elongatus J. R. Sahlberg, 1908: 79 (*Cabirus*) **A**: SY
gracilis Reitter, 1904a: 66 (*Cabirus*) **A**: SY TR
iranicus G. S. Medvedev, 1968a: 176 **A**: IN
libanicus Desbrochers des Loges, 1881: 110 (*Cabirus*) **A**: LE
minutissimus Mulsant & Rey, 1853c: 224 [= 1853e: 140] (*Cabirus*) **A**: IS LE
mulsanti Desbrochers des Loges, 1881: 111 (*Cabirus*) **A**: IS LE
obsoletus Baudi di Selve, 1881: 281 (*Cabirus*) **A**: IS LE

latiusculus Desbrochers des Loges, 1881: 111 (*Cabirus*)
persis Baudi di Selve, 1875d: 697 (*Cabirus*) **A**: IN
persicus Baudi di Selve, 1876b: 156 (*Cabirus*)
rotundicollis Miller, 1861: 202 (*Cabirus*) **A**: IS LE
pubescens Desbrochers des Loges, 1881: 108 (*Cabirus*)
simonis Reitter, 1904a: 68 (*Cabirus*) **A**: IN IS SY
strictulus Desbrochers des Loges, 1881: 110 (*Cabirus*) **A**: IS
subpubescens Reitter, 1904a: 67 (*Cabirus*) **A**: LE
thoracicus J. R. Sahlberg, 1908: 83 (*Cabirus*) **A**: LE
turcmenicus G. S. Medvedev, 1968a: 178 **A**: TM (Kopet Dag)
vicinus Desbrochers des Loges, 1881: 112 (*Cabirus*) **A**: "Syria"
subgenus *Dentibirus* G. S. Medvedev, 1968a: 204 type species *Heliopathes pusillus* Ménétriés, 1849
pusillus Ménétriés, 1849: 327 (*Heliopathes*) **A**: AF KZ TD TM UZ
dentimanus Reitter, 1904a: 64 (*Cabirus*)
femoralis Reitter, 1915e: 85 (*Cabirus*)
pusillus Mulsant & Rey, 1853c: 225 [= 1853e: 141] (*Cabirus*) [HN

genus *Colpotinus* Fairmaire, 1891c: xvii type species *Colpotinus simulator* Fairmaire, 1891
simulator Fairmaire, 1891c: xvii **A**: HUB SCH

genus *Colpotus* Mulsant & Rey, 1853c: 208 [= 1853e: 124] type species *Pandarus strigosus* A. Costa, 1847
Pandarus A. Costa, 1847: 144 [HN] type species *Pandarus strigosus* A. Costa, 1847
adebratti J. Ferrer & L. Soldati, 1999: 70 **A**: TR
byzantinicus Waltl, 1838: 462 (*Pedinus*) **E**: BU GG GR SY TR **A**: TR
angustulus Reitter, 1887e: 520
cobosi Español, 1953b: 139 **E**: SP
dentatus Grimm & Schawaller, 2000: 6 **A**: TR
godarti Mulsant & Rey, 1853c: 214 [= 1853e: 130] **E**: FR (Corse) IT (Sardegna)
magnicollis Fairmaire, 1892b: 148 [DA] **A**: TR
ovoideus Grimm & Schawaller, 2000: 8 **E**: GR
pectoralis Mulsant & Rey, 1853c: 220 [= 1853e: 136] **E**: AL GR
corcyrensis Koch, 1948a: 358
damorei Koch, 1948a: 358
ruficornis Reitter, 1884b: 94 (*Pandarinus*)
punctipleuris Reitter, 1904a: 63 **A**: TR (Boz Dagh)
reitteri Zürcher, 1911: 292 **A**: CY
schusteri Koch, 1936: 298 **A**: TR
lasallei J. Ferrer & L. Soldati, 1999: 71
similaris nevadensis Español, 1953b: 138 **E**: SP
similaris similaris Mulsant & Rey, 1853c: 212 [= 1853e: 128] **E**: PT SP
strigosus ganglbaueri D'Amore-Fracassi, 1907: 191 **E**: IT
strigosus oglasensis Gardini, 1975: 44 **E**: IT (Montecristo Is.)
strigosus ragusai D'Amore-Fracassi, 1907: 190 **E**: IT (Sicilia)
strigosus strigosus A. Costa, 1847: 144 (*Pandarus*) **E**: IT
strigicollis Mulsant & Rey, 1853c: 210 [= 1853e: 126]
sulcatus Ménétriés, 1838: 39 (*Pedinus*) **E**: GR **A**: TR
asiaticus Koch, 1948a: 356
rhodius Koch, 1948a: 355
sulcatus Mulsant, 1853c: 218 [= 1853e: 134] (*Colpotus*)[HN]
vogti Koch, 1948a: 357 **E**: GR (Samos) **A**: TR

genus *Pedinus* Latreille, 1796: 20 type species *Tenebrio femoralis* Linnaeus, 1767
subgenus *Pedinulus* Seidlitz, 1893: 373 type species *Pedinus ragusae* Baudi di Selve, 1875
ionicus Kiesenwetter, 1880: 66 **E**: GR
ragusae Baudi di Selve, 1876a: 44 [= 1876b: 151] **E**: GR IT
subgenus *Pedinus* Latreille, 1796: 20 type species *Tenebrio femoralis* Linnaeus, 1767

aetolicus Apfelbeck, 1901: 464 **E**: GR
affinis affinis Brullé, 1833: 206 **E**: GR
gibbosus Mulsant & Rey, 1853c: 165 [=1853e: 81] [HN]
affinis alziari Ardoin, 1978a: 134 **E**: GR (Peloponnisos)
affinis krueperi Seidlitz, 1893: 375 **E**: GR
affinis reitteri Seidlitz, 1893: 375 **E**: GR TR
aliquoi F. Soldati & L. Soldati, 2002b: 178 **E**: GR
balcanicus Apfelbeck, 1901: 465 **E**: AL BH GR YU
bodemeyeri Reitter, 1904a: 56 **A**: TR
borysthenicus Reichardt, 1936b: 676 **E**: ST UK
knechteli Panin, 1943: 179
cimmerius caucasicus G. S. Medvedev, 1968a: 149 **E**: ST
cimmerius cimmerius G. S. Medvedev, 1968a: 147 **E**: UK (Krym)
cimmerius znoikoi G. S. Medvedev, 1968a: 150 **E**: ST UK
circassicus Reitter, 1887e: 520 (*Micrositus*) **E**: GG ST **A**: TR
clementis Seidlitz, 1893: 373 **E**: GR
curvipes Mulsant & Rey, 1853c: 189 [= 1853e: 105] **E**: BU GR RO TR **A**: TR
tibipinnus Koch, 1936: 296
dilaticollis J. R. Sahlberg, 1908: 78 **A**: SY
fallax fallax Mulsant & Rey, 1853c: 168 [=1853e: 84] **E**: AL BH BU CR IT SK SL ST UK YU
fallax gracilis Mulsant & Rey, 1853c: 171 [=1853e: 87] **E**: CR HU RO
ulrichii Seidlitz, 1893: 371
fatuus Mulsant & Rey, 1853c: 178 [= 1853e: 94] **E**: GR
debilis Kiesenwetter, 1880: 68
femoralis femoralis Linnaeus, 1767: 679 (*Tenebrio*) **E**: AR AU BU BY CT CZ FR GE GR HU IT MD PL RO SK ST SZ TR UK **A**: ES KZ WS XIN
aequalis Faldermann, 1837: 57
curtulus Mulsant & Rey, 1853c: 196 [= 1853e: 112]
dermestoides Fabricius, 1787a: 212 (*Tenebrio*)
fornicatus Faldermann, 1837: 58
gemellatus Geoffroy, 1785: 157 (*Tenebrio*)
ovatus Faldermann, 1837: 59
striopunctatus Goeze, 1777: 687 (*Tenebrio*)
fodori Kaszab, 1941e: 161 **E**: GR
helopioides Germar, 1814: 2 **E**: AL BH BU CR GR IT RO SK SL TR **A**: TR
gibbosus Brullé, 1832: 206
inflatus Reiche & Saulcy, 1857: 262 (*Crypticus*)
laticollis Herbst, 1799: 198 (*Blaps*)
hungaricus Seidlitz, 1893: 376 **E**: BH HU YU
intermedius Seidlitz, 1893: 375 **A**: "Syria"
kiesenwetterii Seidlitz, 1893: 375 **E**: TR **A**: TR
longulus Rottenberg, 1871: 249 **E**: IT (Sicilia)
marani Kaszab, 1959d: 80 **A**: TR
meridianus Mulsant & Rey, 1853c: 175 [= 1853e: 91] **E**: CR FR IT
androgyne Reitter, 1904: 57
natolicus Mulsant & Rey, 1853c: 187 [= 1853e: 103] **A**: TR
oblongus Mulsant & Rey, 1853c: 179 [= 1853e: 95] **E**: AL BU GR
olivieri Mulsant & Rey, 1853c: 157 [= 1853e: 73] **E**: GR
punctulatus Kiesenwetter, 1880: 70
olympicus Kiesenwetter, 1880: 67 **E**: GR
paradoxus Reitter, 1904a: 59 **A**: SY
peristericus Apfelbeck, 1901: 464 **E**: AL GR
podager Seidlitz, 1893: 375 **A**: TR
punctatostriatus Mulsant & Rey, 1853c: 172 [= 1853e: 88] **E**: IT (Sicilia)
quadratus Brullé, 1832: 205 **E**: AL CR GR TR
schaumii Mulsant & Rey, 1853c: 181 [= 1853e: 97] **E**: GR **A**: "Syria Turc"

sicanus Canzoneri, 1984: 3 **E**: IT (Sicilia)
siculus Seidlitz, 1893: 367 **E**: IT (Sicilia)
simplex Seidlitz, 1893: 376 **E**: GR
strabonis Seidlitz, 1893: 375 **E**: AB AR **A**: IN TR
strigicollis Reitter, 1904a: 61 **E**: UK (Krym)
subdepressus Brullé, 1832: 207 **E**: GR
cylindricus Waltl, 1838: 462
szekessyi Kaszab, 1941e: 163 **E**: GR (Zakynthos)
tauricus Mulsant & Rey, 1853c: 197 [= 1853e: 113] **E**: RO UK (Krym) **A**: TR
problematicus Reichardt, 1936b: 674
taygetanus Kiesenwetter, 1880: 67 **E**: GR
tibipinnus Koch, 1936: 296 **A**: TR
tieffenbachi Reitter, 1915e: 84 **E**: GR
volgensis Mulsant & Rey, 1853c: 203 [= 1853e: 119] **E**: ST UK
subgenus *Vadalus* Mulsant & Rey, 1854e: 66 [= 1854c: 150] type species *Pedinus punctulatus* Mulsant & Rey, 1853
circassicus Reitter, 1887: 520 (*Micrositus*) **E**: GG ST **A**: TR
punctulatus Mulsant & Rey, 1853c: 150 [= 1853e: 66] **A**: TR
genus *Pedinus*, nomen dubium
sibiricus Gebler, 1833: 287 **A**: "Montis Altaicis"

subtribe Platynotina Mulsant & Rey, 1853

genus *Adamus* Iwan, 1997: 255 type species *Platynotus micrositoides* Kaszab, 1975
sikkimensis Kaszab, 1975d: 315 (*Platydendarus*) **A**: SD **ORR**

genus *Notocorax* Dejean, 1834: 191 type species *Opatrum javanum* Wiedemann, 1819
Platydendarus Kaszab, 1975d: 312 [HN] type species *Opatrum javanum* Wiedemann, 1819
javanus Wiedemann, 1819: 163 (*Opatrum*) **A**: HAI YUN **ORR**
westermanni Mannerheim, 1844b: 862 (*Notocorax*)

genus *Penthicoides* Fairmaire, 1896a: 20 type species *Penthicoides seriatoporus* Fairmaire, 1896
seriatoporus Fairmaire, 1896a: 20 **A**: SD **ORR**

genus *Platynotus* Fabricius, 1801a: 138 type species *Blaps excavata* Fabricius, 1775
perforatus Mulsant & Rey, 1853c: 44 [= 1853d: 269] **A**: UP XIZ **ORR**

genus *Pseudoblaps* Guérin-Méneville, 1834b: 28 type species *Pseudoblaps substriata* Guérin-Méneville, 1834
Nyctalops Gistel, 1848: 125 [unjustified emendation]
becvari Iwan, 2001: 149 **A**: PA
deserticola Kaszab, 1975d: 329 **A**: PA
indica Kaszab, 1975d: 328 **A**: KA
iranica Kaszab, 1975d: 327 **A**: IN
kabuliensis kabuliensis Kaszab, 1970a: 134 **A**: AF
kabuliensis lindemannae Kaszab, 1975d: 327 **A**: PA
kazakhstaniensis Iwan, 2001: 151 **A**: KZ
medvedevi Kaszab, 1975d: 325 **A**: KA
nuristanica Kaszab, 1960a: 128 **A**: AF KA
pakistana Kaszab, 1961d: 223 **A**: IN PA
substriata Guérin-Méneville, 1834b: 29 **A**: KA KZ **ORR**
ambigua Mulsant & Rey, 1853c: 55 [= 1853d: 280] (*Notocorax*)
mellyi Mulsant & Rey, 1853c: 54 [= 1853d: 279] (*Notocorax*)
parallela Mulsant & Rey, 1853c: 56 [= 1853d: 281] (*Notocorax*)
strigipennis Mulsant & Rey, 1853c: 61 [= 1853d: 286] (*Notocorax*)
zarudnyi G. S. Medvedev, 1968b: 178 **A**: IN

genus *Zidalus* Mulsant & Rey, 1853c: 71 [= 1853d: 296] type species *Opatrinus corvinus* Mulsant & Rey, 1853
Apterozidalus Ardoin, 1965b: 1315 type species *Apterozidalus royi* Ardoin, 1965
Zodinus Mulsant & Rey, 1853c: 90 [= 1853d: 315] type species *Opatrinus servus* Mulsant & Rey, 1853
corvinus Mulsant & Rey, 1853: 71 [= 1853: 296] (*Opatrinus*) **N**: EG **AFR**
pinkey Koch, 1956: 96 (*Opatrinus*)
erythraeus Gridelli, 1940: 239 (*Opatrinus*) **N**: EG **AFR**
niloticus Mulsant & Rey, 1853c: 87 [= 1853d: 312] (*Opatrinus*) **N**: CI (Fuerteventura) LB **A**: YE **AFR**
angusticollis Fairmaire, 1887e: 283 (*Opatrinus*)
edentatus Koch, 1956: 99 (*Opatrinus*)
olivensis Wollaston, 1864: 501 (*Tenebrio*)
zolotarevskyi Español, 1943c: 138 (*Opatrinus*)

tribe Platynotini, nomen dubium
Platynotus pekinensis Fairmaire, 1888b: 127 **A**: "Montagnes de Pékin"

tribe Platyscelidini Lacordaire, 1859

genus *Bioramix* Bates, 1879: 478 [= 1890: 69] type species *Bioramix ovalis* Bates, 1879
subgenus *Bioramix* Bates, 1879: 478 [= 1890: 69] type species *Bioramix ovalis* Bates, 1879
Botiras Fairmaire, 1891d: xcviii type species *Botiras striatellus* Fairmaire, 1891
Euryhelops Reitter, 1902g: 209 type species *Helops championi* Reitter, 1891
espanoli Kaszab, 1961a: 118 (*Platynoscelis*) **A**: PA
freudei Kaszab, 1970a: 126 (*Platynoscelis*) **A**: AF
glacialis Kaszab, 1975a: 19 (*Platynoscelis*) **A**: KA
kashmirensis Kaszab, 1940b: 178 (*Platynoscelis*) **A**: HP KA
klapperichi Kaszab, 1960a: 85 (*Platynoscelis*) **A**: AF
laeviuscula Fairmaire, 1891d: xcvii (*Faustia*) **A**: KA
oblongula Fairmaire, 1898d: 389 (*Botiras*) **A**: HP KA NP XIZ
himalajensis Kaszab, 1940b: 179 (*Platynoscelis*)
ovalis Bates, 1879: 478 **A**: HP NP KA XIZ
paghmanica Kaszab, 1960a: 86 (*Platynoscelis*) **A**: AF
puncticeps Bates, 1879: 479 **A**: HP KA NP SD
princeps Bogatchev, 1952b: 286 (*Platynoscelis*)
rotundicollis Kaszab, 1940b: 175 (*Platynoscelis*) **A**: NP UP XIZ
stoeckleini Kaszab, 1960a: 87 (*Platynoscelis*) **A**: AF
subgenus *Cardiobioramix* Kaszab, 1940b: 183 type species *Bioramix asidioides* Bates, 1879
asidioides Bates, 1879: 479 **A**: KA PA XIN
punctatella Fairmaire, 1891d: xcix (*Botiras*)
bechynei Kaszab, 1960a: 89 (*Platynoscelis*) **A**: AF
blairi Kaszab, 1940b: 183 (*Platynoscelis*) **A**: KA
championi Reitter, 1891e: 226 (*Helops*) [RN] **A**: GAN
subaenea Reitter, 1889k: 710 (*Helops*) [HN]
chinensis Kaszab, 1940b: 197 (*Platynoscelis*) **A**: SCH
dubiosa Kaszab, 1940b: 207 (*Platynoscelis*) **A**: KA
fairmairei Kaszab, 1940b: 184 (*Platynoscelis*) **A**: HP KA
gridellii Kaszab, 1940b: 186 (*Platynoscelis*) **A**: KA
haafi Kaszab, 1960a: 88 (*Platynoscelis*) **A**: AF
kabaki L. V. Egorov, 2006: 794 **A**: GAN
korschefskyi Kaszab, 1940b: 196 (*Platynoscelis*) **A**: SCH
kulzeri Kaszab, 1960a: 91 (*Platynoscelis*) **A**: SCH
ovata Kaszab, 1940b: 189 (*Platynoscelis*) **A**: KA
punjabensis L. V. Egorov, 2004: 597 [RN] **A**: HP KA
championi Kaszab, 1940b: 194 (*Platynoscelis*) [HN]
schusteri Kaszab, 1940b: 188 (*Platynoscelis*) **A**: KA
sikkimensis Kaszab, 1940b: 191 (*Platynoscelis*) **A**: SD

splendida L. V. Egorov, 2006: 789 **A**: SCH
subaenescens Schuster, 1923b: 160 (*Platyscelis*) **A**: SCH
szetschuana Kaszab, 1940b: 192 (*Platynoscelis*) **A**: SCH
tshartymensis L. V. Egorov, 1990a: 120 **A**: TD
wittmeri L. V. Egorov, 1990a: 121 **A**: PA
zabriformis Fairmaire, 1896a: 55 (*Tagonoides*) **A**: KA
subgenus *Cardiochianalus* Kaszab, 1940b: 202 type species *Botiras sculptipennis* Fairmaire, 1891
batesi Kaszab, 1940b: 203 (*Platynoscelis*) **A**: HP KA
cordicollis Kaszab, 1940b: 202 (*Platynoscelis*) **A**: HP KA
schawalleri L. V. Egorov, 2004: 599 [RN] **A**: HP KA
kaszabi Koch, 1965b: 133 (*Platynoscelis*) [RN] [HN]
sculptipennis Fairmaire, 1891d: xcix (*Botiras*) **A**: KA
sommerssmithi Fairmaire, 1896a: 54 (*Tagonoides*)
subgenus *Chianalus* Bates, 1879: 479 [= 1890: 72] type species *Chianalus costipennis* Bates, 1879
costipennis Bates, 1879: 480 (*Chianalus*) **A**: KA PA
falsa Kaszab, 1961f: 347 (*Platynoscelis*) **A**: KA (Karakoram)
striatella Fairmaire, 1891d: xcviii (*Botiras*) **A**: KA PA
subcostipennis Gridelli, 1934: 64 (*Chianalus*)
subgenus *Faustia* Kraatz, 1882b: 92 type species *Faustia modesta* Kraatz, 1882
lopatini L. V. Egorov, 1990a: 117 **A**: TD
modesta Kraatz, 1882b: 93 (*Faustia*) **A**: KI TD
subgenus *Leipopleura* Seidlitz, 1893: 354 type species *Faustia integra* Reitter, 1887
aenescens Blair, 1923b: 284 (*Platyscelis*) **A**: NP SD XIZ
crypticoides Reitter, 1887d: 384 (*Faustia*) **A**: GAN QIN SCH XIN XIZ
darbukensis Kaszab, 1940b: 160 (*Platynoscelis*) **A**: KA
frivaldszkyi Kaszab, 1940b: 167 (*Platynoscelis*) **A**: GAN QIN
integra Reitter, 1887d: 382 (*Faustia*) **A**: GAN QIN SCH
kochi Kaszab, 1940b: 169 (*Platynoscelis*) **A**: NP SD XIZ
micans Reitter, 1889k: 699 (*Faustia*) **A**: GAN QIN XIZ
brevipilosa Meng & Ren, 2005: 106 (*Myatis*)
siningensis J. Frivaldszky, 1889: 210 (*Faustia*)
nagquana Meng & Ren, 2005: 107 (*Myatis*) **A**: XIZ
picipes Gebler, 1833: 287 (*Platyscelis*) **A**: ES MG WS
angustata Faldermann, 1835b: 409 (*Platyscelis*)
reitteri Seidlitz, 1893: 354 (*Platyscelis*)
politicollis Kaszab, 1940b: 171 (*Platynoscelis*) **A**: NP XIZ
reinigi Kaszab, 1940b: 163 (*Platynoscelis*) **A**: QIN XIZ
tibetana Kaszab, 1940b: 162 (*Platynoscelis*)
rubripes Reitter, 1889k: 700 (*Faustia*) **A**: QIN XIZ
rufipalpis Reitter, 1887d: 383 (*Faustia*) **A**: QIN
subgenus *Nudoplatyscelis* Kaszab, 1940b: 222 type species *Platynoscelis turanica* Reitter, 1896 (= *Bioramix turanica* Reitter, 1896)
kaszabi Gridelli, 1954: 270 [= 1955: 64] (*Platynoscelis*) **A**: AF
ghorana Kaszab, 1974: 200 (*Platynoscelis*)
turanica Reitter, 1896g: 176 (*Platynoscelis*) **A**: IN TM
subgenus *Ovalobioramix* L. V. Egorov, 2004: 603 type species *Platynoscelis molesta* Bogatchev, 1947
molesta Bogatchev, 1947c: 87 (*Platynoscelis*) **A**: TD UZ
hissarica L. V. Egorov, 1992: 796 **A**: KI TD
subgenus *Planoplatyscelis* Kaszab, 1940b: 206 type species *Platyscelis margelanica* Kraatz, 1882 (= *Bioramix pamirensis* Bates, 1879)
Pleioplatyscelis Kaszab, 1940b: 220 type species *Platynoscelis lederi* Seidlitz, 1893
andreevae L. V. Egorov, 1990b: 891 **A**: KI (Talasskiy Khrebet)
aruktavica L. V. Egorov, 1990a: 119 **A**: TD (Mts. Aruktau)
caraboides Kaszab, 1960a: 92 (*Platynoscelis*) **A**: AF
conradti Seidlitz, 1893: 354 (*Platyscelis*) **A**: TD

pseudohelops Bogatchev, 1947c: 84 (*Platynoscelis*)
faldermanni Seidlitz, 1893: 354 (*Platyscelis*) **A**: TD UZ
auliensis Kaszab, 1940b: 212 (*Platynoscelis*)
blaptiformis Bogatchev, 1946d: 437 (*Platynoscelis*)
duplicata Kaszab, 194b0: 218 (*Platynoscelis*)
haberhaueri Seidlitz, 1893: 354 (*Platyscelis*) **A**: KI
humeralis Reitter, 1896g: 175 (*Platynoscelis*)
latipennis Kaszab, 1940b: 217 (*Platynoscelis*) **A**: TD
lederi Seidlitz, 1893: 355 (*Platyscelis*) **A**: KI TD
pamirensis Bates, 1879: 478 **A**: KI TD
caroli Kaszab, 1940b: 221 (*Platynoscelis*)
difficilis Kraatz, 1883a: 351 (*Platyscelis*)
latipes Kraatz, 1886: 189 (*Platyscelis*)
margelanica Kraatz, 1882b: 84 (*Platyscelis*)
rotundangula Kraatz, 1883a: 351 (*Platyscelis*)
simplex Kraatz, 1883a: 352 (*Platyscelis*)
subgenus *Platynoscelis* Kraatz, 1882b: 91 type species *Platynoscelis helopioides* Kraatz, 1882
afghanica Schuster, 1936c: 200 **A**: AF
afghanistana Gridelli, 1954: 275 [= 1955: 65] (*Platynoscelis*) **A**: AF
bogatshevi Kaszab, 1970a: 128 (*Platynoscelis*) [RN] **A**: AF
rotundicollis Kaszab, 1960a: 93 (*Platynoscelis*) [HN]
cylindrica Kaszab, 1960a: 104 (*Trichomyatis*) **A**: AF
cylindricollis L. V. Egorov, 1990a: 130 **A**: AF
delerei Kaszab, 1960a: 94 (*Platynoscelis*) **A**: AF
gazni L. V. Egorov, 1990a: 127 **A**: AF
gigantea Kaszab, 1974: 202 (*Platynoscelis*) **A**: AF
gracilioides Kaszab, 1970a: 130 (*Platynoscelis*) **A**: AF
gracilipenis Kaszab, 1960a: 98 (*Platynoscelis*) **A**: AF
gracilis Seidlitz, 1893: 355 (*Platyscelis*) **A**: TD
granulipennis Kaszab, 1960a: 95 (*Platynoscelis*) **A**: AF
gurjevae L. V. Egorov, 1990a: 123 **A**: TD
haarlovi Kaszab, 1959e: 251 [= 1960a: 99] (*Platynoscelis*) **A**: AF
helopioides Kraatz, 1882b: 92 (*Platynoscelis*) **A**: KI TD
hirtipennis Kaszab, 1960a: 103 (*Trichomyatis*) **A**: AF
kabakovi L. V. Egorov, 1990a: 131 **A**: AF
kapisensis L. V. Egorov, 1990a: 133 **A**: AF
lindbergi Kaszab, 1973b: 49 (*Platynoscelis*) **A**: AF
lucida Gridelli, 1954: 278 [= 1955: 68] (*Platynoscelis*) **A**: AF
merkli L. V. Egorov, 1990a: 129 **A**: AF
montana Kaszab, 1960a: 97 (*Platynoscelis*) **A**: AF
subalpestris Kaszab, 1973b: 48 (*Platynoscelis*)
oruzganensis L. V. Egorov, 1990a: 132 **A**: AF
ovipennis Kaszab, 1960a: 106 (*Trichomyatis*) **A**: AF
parvula L. V. Egorov, 1990a: 127 **A**: TD
rectangularis Kaszab, 1960a: 107 (*Trichomyatis*) **A**: AF
rufipes Kaszab, 1940b: 230 (*Platynoscelis*) **A**: AF TD
scheerpeltzi Kaszab, 1960a: 100 (*Platynoscelis*) **A**: AF
setosa L. V. Egorov, 1990a: 126 **A**: "Ost Bukhara"
shugnanensis L. V. Egorov, 1990a: 125 **A**: TD
sistensis L. V. Egorov, 1990a: 122 **A**: TD
tuxeni Gridelli, 1954: 280 [= 1955: 70] (*Platynoscelis*) **A**: AF
waziristanica Kaszab, 1940b: 233 (*Platynoscelis*) **A**: AF PA
paktiana Kaszab, 1974: 201
subgenus *Trichochianalis* Kaszab, 1940b: 201 type species *Platynoscelis monticola* Kaszab, 1940
monticola Kaszab, 1940b: 201 (*Platynoscelis*) **A**: KA

subgenus *Tricholeipopleura* Kaszab, 1940b: 223 type species *Platynoscelis lucidicollis* Kraatz, 1882
constricta Seidlitz, 1893: 355 (*Platyscelis*) **A**: TD
horni Kaszab, 1940b: 227 (*Platynoscelis*) **A**: TD
lucidicollis Kraatz, 1882b: 92 (*Platynoscelis*) **A**: KI XIN
psalidium Seidlitz, 1893: 355 (*Platyscelis*) **A**: TD
sinuatocollis Reitter, 1902b: 180 (*Platyscelis*) **A**: TD UZ
szekessyi Kaszab, 1938a: 1 (*Platyscelis*) **A**: TD
subgenus *Trichoplatyscelis* Reinig, 1931: 895 type species *Trichoplatyscelis pamirensis* Reinig, 1931 (= *Bioramix lapidicola* Kaszab, 1940)
Pseudotrichoplatyscelis Kaszab, 1960a: 82 type species *Platynoscelis badakschanica* Kaszab, 1960 (= *Bioramix lapidicola* Kaszab, 1940)
lapidicola Kaszab, 1940b: 166 (*Platynoscelis*) [RN] **A**: AF (Badakhshan) TD (West Pamir)
badakschanica Kaszab, 1960a: 83 (*Platynoscelis*)
pamirensis Reinig, 1931: 895 (*Trichoplatyscelis*) [HN]
tadzhika Bogatchev, 1947c: 85 (*Platynoscelis*)
paludani Gridelli, 1954: 269 [= 1955: 63] (*Platynoscelis*) **A**: AF (Kunar)
genus *Bioramix*, species incertae sedis
gebieni Kaszab, 1940b: 211 (*Platynoscelis*) **A**: KZ ("Karabutak")

genus *Microplatyscelis* Kaszab, 1940b: 144 type species *Microplatyscelis seriepunctata* Reitter, 1890
seriepunctata Reitter, 1890h: 362 (*Faustia*) **A**: AF KZ TD TM UZ

genus *Myatis* Bates, 1879: 480 [= 1890: 73] type species *Myatis humeralis* Bates, 1879
humeralis Bates, 1879: 480 **A**: KA TD XIN XIZ
quadraticollis Bates, 1879: 481
tenuissima Reitter, 1896g: 174 (*Leipopleura*)
variabilis Bates, 1879: 481
schaeferi Kaszab, 1940d: 900 **A**: XIZ

genus *Oodescelis* Motschulsky, 1845a: 76 type species *Blaps polita* J. Sturm, 1807
subgenus *Acutoodescelis* Kaszab, 1940d: 951 type species *Platyscelis punctatissima* Fairmaire, 1886
emmerichi Kaszab, 1940d: 953 **A**: SHA SHX
punctatissima Fairmaire, 1886c: 345 (*Platyscelis*) **A**: BEI NMO SCH TIA
pyripenis Ren, 1999: 293 [393] **A**: SHA
subgenus *Clavatoodescelis* Kaszab, 1940d: 974 type species *Platyscelis melas* Fischer von Waldheim, 1823
acutanguloides Kaszab, 1940d: 976 **A**: XIN
melas Fischer von Waldheim, 1823: pl. 20 [= 1824: 194] (*Platyscelis*) **E**: CT HU KZ PL RO ST UK
oblonga Ballion, 1878: 327 (*Platyscelis*) **A**: KZ XIN
gebieni Kaszab, 1940d: 978
kuntzeni Kaszab, 1940d: 977
regeli Ballion, 1878: 326 (*Platyscelis*)
subgenus *Convexoodescelis* L. V. Egorov, 2004: 593 type species *Platyscelis brevipennis* Kaszab, 1938
brevipennis brevipennis Kaszab, 1938d: 92 (*Platyscelis*) **A**: KI KZ XIN
brevipennis wernoyensis Kaszab, 1940d: 950 **A**: KI KZ
subgenus *Longuloodescelis* Kaszab, 1940d: 957 type species *Platyscelis hirta* Seidlitz, 1893
Trichoodescelis Kaszab, 1940d: 954 type species *Platyscelis acutangula* Kraatz, 1884
acutangula Kraatz, 1884: 224 (*Platyscelis*) **A**: KI
hirta Seidlitz, 1893: 355 (*Platyscelis*) **A**: KI
medvedevi L. V. Egorov, 1989a: 102 **A**: KZ
pilosa L. V. Egorov, 1987: 800 **A**: KZ UZ
songariensis Kaszab, 1940d: 960 **A**: KI
ballioni Skopin, 1973: 873
subgenus *Montanoodescelis* L. V. Egorov, 2004: 592 type species *Platyscelis sahlbergi* Reitter, 1900
femoralis Kaszab, 1938d: 91 (*Platyscelis*) **A**: KZ (Zailiyskiy Alatau)
ketmeniana Skopin, 1961a: 198 [= 1965: 66] **A**: KZ (Khrebet Ketmen)

sahlbergi Reitter, 1900f: 220 (*Platyscelis*) **A**: KZ (Zailiyskiy Alatau)
transiliensis L. V. Egorov, 1987: 801 **A**: KZ (Zailiyskiy Alatau)
subgenus *Ovaloodescelis* Kaszab, 1940d: 947 type species *Platyscelis affinis* Seidlitz, 1893
adriani Kaszab, 1940d: 948 **A**: KZ
affinis Seidlitz, 1893: 355 (*Platyscelis*) **A**: KI KZ XIN
heydeni Seidlitz, 1893: 355 (*Platyscelis*) **A**: KZ XIN
xerophila Skopin, 1965: 68
karaganae Skopin, 1965: 67 **A**: KZ
ovalis Ballion, 1878: 329 (*Platyscelis*) **A**: XIN
chinensis Kaszab, 1962a: 84
przewalskii Bogatchev, 1946d: 436
pseudotibialis Skopin, 1973: 874 **A**: KI KZ XIN
similis similis Kaszab, 1938b: 55 (*Platyscelis*) **A**: KI KZ XIN
clavatipes Kaszab, 1938b: 55 (*Platyscelis*)
iliensis Skopin, 1958: 146
similis blattiformis Kaszab, 1938b: 54 (*Platyscelis*) **A**: KI KZ UZ
subattenuata L. V. Egorov, 1991: 146 [RN] **A**: KZ XIN
attenuata Kaszab, 1940d: 982 [HN]
tibialis Ballion, 1878: 331 (*Platyscelis*) **A**: KI KZ XIN
curvipes Skopin, 1966: 335
heptapotamica Skopin, 1966: 335
sachtlebeni Kaszab, 1940d: 980
subgenus *Oodescelis* Motschulsky, 1845a: 76 type species *Blaps polita* J. Sturm, 1807
depilata Skopin, 1966: 333 **A**: KZ
polita J. Sturm, 1807: 208 (*Blaps*) **E**: AB AU BU CT CZ GG HU KZ LT PL RO SK SL ST UK **A**: ES KZ WS
arnoldii Skopin, 1964: 390
gages Fischer von Waldheim, 1823: pl. 20, fig. 5 [= 1824: 195] (*Platyscelis*)
globosa Seidlitz, 1893: 355 (*Platyscelis*)
minima Motschulsky, 1860e: 527
subgenus *Planoodescelis* L. V. Egorov, 2004: 591 type species *Oodescelis kansouensis* Kaszab, 1940
kansouensis Kaszab, 1940d: 954 **A**: GAN
subgenus *Spinoodescelis* Kaszab, 1940d: 966 type species *Platyscelis somocoeloides* Seidlitz, 1893
acuta Kaszab, 1940d: 968 **A**: KI KZ UZ
dispar Skopin, 1965: 65
semenoviana Bogatchev, 1946d: 435
hirtipennis Kaszab, 1940d: 972 **A**: UZ TD
longisterna Kaszab, 1940d: 971 **A**: "Turkestan"
necopinata Skopin, 1965: 66 **A**: KZ
ovulum Seidlitz, 1893: 355 (*Platyscelis*) **A**: KI TD UZ
latipleura Kaszab, 1940d: 969
punctolineata Kaszab, 1940d: 973 **A**: KI
somocoeloides grandis Skopin, 1965: 69 **A**: KZ
alticola Skopin, 1965: 69
somocoeloides somocoeloides Seidlitz, 1893: 356 (*Platyscelis*) **A**: KZ UZ
blapoides Reitter, 1896g: 175 (*Platynoscelis*)
transcaspica Kaszab, 1940d: 960 **A**: KZ
subgenus *Splenoodescelis* L. V. Egorov, 2004: 593 type species *Platyscelis turkestanica* Seidlitz, 1893
turkestanica Seidlitz, 1893: 355 (*Platyscelis*) **A**: KI
hirsuta Reitter, 1896g: 176
turul Kaszab, 1940d: 966
subgenus *Truncatoodescelis* Kaszab, 1940d: 962 type species *Platyscelis longicollis* Kraatz, 1884
longicollis Kraatz, 1884: 224 (*Platyscelis*) **A**: KI
kiritchenkoi Bogatchev, 1939a: 123 (*Platyscelis*)
kirghizica kirghizica L. V. Egorov, 1989a: 104 **A**: KI (Talasskiy Alatay)
kirghizica ferghanensis L. V. Egorov, 1989a: 105 **A**: KI (Ferganskiy Khrebet)
schusteri schusteri Kaszab, 1940d: 961 **A**: KI (Kirgizskiy Khrebet)

schusteri truncatoides Skopin, 1968: 83 **A**: KI KZ
genus *Oodescelis*, species incertae sedis
attenuata Motschulsky, 1860e: 527 **A**: KZ

genus *Platyscelis* Latreille, 1818c: 23 type species *Tenebrio hypolithus* Pallas, 1781
subgenus *Oblongoplatyscelis* Kaszab, 1940d: 916 type species *Platyscelis ganglbaueri* Seidlitz, 1893
ganglbaueri Seidlitz, 1893: 354 **A**: KI KZ
seidlitzi Reitter, 1896g: 173
subgenus *Paraplatyscelis* Kaszab, 1940d: 936 type species *Platyscelis sinuata* Seidlitz, 1893
sinuata Seidlitz, 1893: 354 **A**: KI
subgenus *Platyscelis* Latreille, 1818c: 23 type species *Tenebrio hypolithus* Pallas, 1781
Kaszaboscelis Löbl & Merkl, 2003: 246 type species *Tenebrio hypolithos* Pallas, 1781
amdoensis L. V. Egorov, 1989b: 347 **A**: GAN
angusticollis angusticollis Kaszab, 1940d: 931 **A**: KI KZ XIN
skopini Kaszab, 1962c: 314
angusticollis platytarsis Kaszab, 1940d: 932 **A**: KZ XIN
ballioni Reitter, 1889k: 698 **A**: SHX
bogatshevi L. V. Egorov, 1989b: 341 **A**: SCH
brevis Baudi di Selve, 1876a: 35 **A**: ES NMO MG
rugifrons Germar, 1824b: 139 (*Blaps*) [HN]
confusa Schuster, 1934a: 75 [RN] **A**: GAN NIN QIN
hauseri Reitter, 1899d: 205 [HN]
freyi Kaszab, 1940d: 925 **A**: GAN
gebieni Schuster, 1915b: 88 **A**: HEB NMO SHA SHX
hungarica Frivaldszky von Frivald, 1865: 200 **E**: HU
hypolitha Pallas, 1781: 44 (*Tenebrio*) [NP] **E**: CT GG KZ ST UK **A**: KZ WS
laevis Forskål, 1775: 78 (*Tenebrio*) [NO]
spinolae Solier, 1848: 211
ketmeniana L. V. Egorov, 1989b: 343 **A**: KZ (Mts. Ketmen)
kirghisica Reitter, 1896g: 174 **E**: KZ **A**: KZ
aralensis Reitter, 1896g: 174
rugifrons Fischer von Waldheim, 1823: pl. 20 (*Trachyscelis*) [= 1824: 194 (*Platyscelis*)] **A**: KI KZ WS
intermedia Motschulsky, 1859c: 481
potanini Bogatchev, 1961a: 110
subcordata Seidlitz, 1893: 354 **A**: BEI HEB LIA SHX SHN
strigicollis Lewis, 1895a: 447
suiyuana Kaszab, 1940d: 928 **A**: NMO SHA SHX
subgenus *Pleiopleura* Seidlitz, 1893: 343 type species *Platyscelis striatus* Motschulsky, 1859
ovata Ballion, 1878: 324 **A**: KI KZ XIN
striata Motschulsky, 1859b: 307 [= 1859a: 299] **A**: KZ XIN
przewalskii Bogatchev, 1961a: 109
sulcata Ballion, 1878: 322
genus *Platyscelis*, species incertae sedis
obovata Ren, 1999: 307 (393) [DA] **A**: QIN

genus *Somocoelia* Heyden & Kraatz, 1882d: 331 type species *Somocoelia pinguis* Heyden & Kraatz, 1882
arnoldii L. V. Egorov, 1992: 797 **A**: TM
pinguis pinguis Heyden & Kraatz, 1882d: 332 **A**: TD UZ
globulus Seidlitz, 1893: 354 (*Platyscelis*)
pinguis gracilipes Kaszab, 1940d: 907 **A**: TD
pinguis kuhitangi L. V. Egorov, 1998: 213 **A**: TM (Mts. Kugitangtau)

genus *Somocoeloplatys* Skopin, 1968: 82 type species *Platynoscelis boroldaica* Skopin, 1965
boroldaica Skopin, 1965: 64 (*Platynoscelis*) **A**: KZ

genus *Trichomyatis* Schuster, 1931: 893 type species *Trichomyatis conradti* Schuster, 1931
Trichoplatynoscelis Kaszab, 1940d: 896 type species *Trichoplatynoscelis pamirensis* Kaszab, 1940
conradti Schuster, 1931: 894 **A**: TD
michailovi L. V. Egorov, 1992: 798 **A**: TD
nuristanensis L. V. Egorov, 1992: 798 **A**: AF
pamirensis Kaszab, 1940d: 897 (*Trichoplatynoscelis*) **A**: TD
rugicollis Kaszab, 1960a: 108 (*Trichoplatynoscelis*) **A**: AF TD
tadzhika Bogatchev, 1952b: 285 (*Myatis*) **A**: TD

tribe Praeugenini De Moor, 1970

genus *Miltoprepes* Gerstaecker, 1871: 65 type species *Miltoprepes laetus* Gerstaecker, 1871
erythraeus Gridelli, 1939a: 78 **A**: YE **AFR**

genus *Praeugena* Laporte, 1840: 241 type species *Praeugena rubripes* Laporte, 1840
Ergenna Fairmaire, 1897e: 139 type species *Ergenna caerulescens* Fairmaire, 1897
Lamprobothris Fairmaire, 1887e: 302 type species *Lamprobothris attenuata* Fairmaire, 1887 (= *Praogena fossulata* C. Müller, 1887)
Tactoderus Fairmaire, 1892a: 112 type species *Tactoderus subopacus* Fairmaire, 1892
gagatina Mäklin, 1863: 579 (*Praogena*) **A**: SA YE **AFR**
subopaca Fairmaire, 1892a: 112 (*Tactoderus*)
tenuicollis Fairmaire, 1894d: 328 (*Dysgena*) **A**: SA **AFR**

tribe Rhysopaussini Wasmann, 1896

genus *Azarelius* Fairmaire, 1892c: vii type species *Azarelius sculpticollis* Fairmaire, 1892
singularis Wasmann, 1896b: 151 **A**: SD

genus *Rhysopaussus* Wasmann, 1896a: 617 type species *Rhysopaussus dohertii* Wasmann, 1896
dohertii Wasmann, 1896a: 617 **A**: NP **ORR**

genus *Ziaelas* Fairmaire, 1892d: cx type species *Ziaelas insolitus* Fairmaire, 1892
formosanus Hozawa, 1914: 484 **A**: TAI

tribe Scaurini Billberg, 1820

genus *Cephalostenus* Solier, 1838b: 184 type species *Cephalostenus dejeanii* Solier, 1836 (= *Scaurus elegans* Brullé, 1832)
alziari Grimm, 1991: 24 **A**: CY
elegans Brullé, 1832: 195 (*Scaurus*) **E**: GR **A**: TR
dejeanii Solier, 1838b: 186
demaisoni Reitter, 1903c: 132 **A**: TR
orbicollis Ménétriés, 1836a: 151 **E**: BU GR TR **A**: SY TR

genus *Scaurus* Fabricius, 1775: 253 type species *Scaurus atratus* Fabricius, 1775
aegyptiacus Solier, 1838b: 170 **E**: GR (Castellorizo) IT (Sardegna, Sicilia, Stromboli) MA **N**: AG EG LB MO TU **A**: CY IQ LE SI SY TR
giganteus Küster, 1848a: 54
alticola Escalera, 1914: 296 **N**: MO
angustus Reiche, 1864: 242 **N**: AG TU
araxinus Richter, 1945: 103 **E**: AR
atratus Fabricius, 1775: 253 **E**: FR IT (Sicilia, Sardegna) **N**: AG TU
breviatus Desbrochers des Loges, 1881: 81
morio Fabricius, 1777: 241 (*Helops*)
planidorsis Fairmaire, 1879c: 190
quadraticollis Fairmaire, 1875a: 526
vagecostatus Fairmaire, 1879c: 189
barbarus barbarus Solier, 1838b: 165 **N**: LB TU

gracilicornis Fairmaire, 1875a: 524
interruptus Baudi di Selve, 1875b: 11
pellegrinii Koch, 1937: 424
bougonii bougonii Fairmaire, 1883i: cxxv **N**: EG LB TU
bougonii schusteri Gridelli, 1930: 260 **N**: LB
calcaratus Fabricius, 1798: 42 **N**: MO
distinctus Solier, 1838b: 173
latipennis Baudi di Selve, 1875b: 13
sublaevis Desbrochers des Loges, 1881: 84
camelus Kocher, 1958b: 195 **N**: MO
carinatus carinatus Solier, 1838b: 176 **N**: EG LB **A**: SI
carinatus vicinoides Schuster, 1935a: 15 **N**: LB
contractus Fairmaire, 1880a: 13 **N**: MO
incostulatus Escalera, 1935a: 308
dubius Solier, 1838b: 181 **N**: AG MO TU
amplicollis Fairmaire, 1875a: 526
barbarus Erichson, 1841: 181 (*Scaurus*) [HN]
elongatus Mulsant & Godart, 1865: 452 **N**: AG TU
gigas Waltl, 1835: 70 **E**: PT SP **N**: AG MO
ferreri Español, 1943c: 130
hespericus Solier, 1838b: 178
maroccanus Fairmaire, 1875a: 525
hypogeus Antoine, 1954: 203 **N**: MO (Tleta des Akhsas)
jahandiezi Théry, 1931b: 123 **N**: MO
kocheri Peyerimhoff, 1948b: 171 **N**: MO (Grand Atlas)
kraatzi Haag-Rutenberg, 1876a: 10 **N**: MO (Atlantic)
mairei Peyerimhoff, 1929c: 282 **N**: AG (Hoggar)
manueli Labrique, 1992: 123 **N**: MO
mesatlanticus horstensis Labrique & Chavanon, 2005: 41 **N**: MO (Moyen Atlas: Horst)
mesatlanticus mesatlanticus Peyerimhoff, 1948b: 170 **N**: MO (Moyen Atlas: Taza)
harbuloti Reymond, 1956b: 266 [= 1956a: 40]
microcephalus Escalera, 1914: 297 **N**: MO
lopezi Español, 1943c: 131
pierrei Antoine, 1954: 204
quirogai Escalera, 1914: 299
puncticollis dlabolai Kaszab, 1959d: 79 **A**: TR
puncticollis getula Peyerimhoff, 1946: 97 **N**: MO
puncticollis macricollis Allard, 1882b: lxxxvii [= 1883: 24] **A**: IN IQ TR
puncticollis puncticollis Solier, 1838b: 172 **N**: AG EG LB TU **A**: IS SA SI SY **AFR**
parvicollis Fairmaire, 1875a: 525
rugicollis Reitter, 1914g: 374
puncticollis syriacus Reitter, 1914g: 374 **A**: CY SY TR
rugulosus Solier, 1838b: 174 **E**: SP
eleonorae Tenenbaum, 1915: 840
gracilis Küster, 1848a: 61
sanctiamandi Solier, 1838b: 177 **N**: LB MO TU
lucidulus Desbrochers des Loges, 1881: 82
ovipennis Fairmaire, 1875: 525
striatus Fabricius, 1792a: 93 **E**: FR GR IT MA SP
maremmanus Solier, 1838b: 180
sardous A. Costa, 1884: 13
tingitanus antarensis Labrique, 1995: 124 **N**: AG (Dj. Antar)
tingitanus gimeli Antoine, 1953a: 59 **N**: MO
tingitanus tingitanus Peyerimhoff, 1948b: 169 **E**: SP **N**: MO
sabasi Kocher, 1965b: 34
tristis A. G. Olivier, 1795: 62: 4 **E**: FR GR IT MA SP SZ **N**: AG MO TU
abbreviatus Reitter, 1914g: 371

algericus Küster, 1848a: 56
gracilipes Solier, 1838b: 167
lugens Küster, 1848a: 57
uncinus Forster, 1771: 60 (*Tenebrio*) **E**: FR GR IT (Sardegna) PT SP **N**: AG CI MO **AFR**
asperulus Fairmaire, 1875a: 525
punctatus Fabricius, 1798: 43
mendax Reitter, 1914g: 379
punctatus Herbst, 1799: 138 [HN]
sticticus Gemminger, 1870a: 121 [RN]
variolosus Wollaston, 1867: 178
varvasi Solier, 1838b: 175 **N**: AG MO
porcatus Erichson, 1841: 182
vicinus Solier, 1838b: 171 **E**: SP **N**: MO
balearicus Reitter, 1914g: 372
balearicus Tenenbaum, 1915: 840 [HN]

genus *Scaurus*, nomina dubia
carinatus V. Petagna, 1787: 26 (*Pimelia*) **E**: IT
dentipes Villers, 1789: 396 (*Tenebrio*) **E**: "Galia"
sulcatus Fabricius, 1792a: 93 **E**: FR

tribe Tenebrionini Latreille, 1802

genus *Ariarathus* Fairmaire, 1891f: ccxi type species *Ariarathus ulomoides* Fairmaire, 1891 (=*Tenebrio atronitens* Fairmaire, 1891)
Teneatopus Reitter, 1920a: 23 type species *Tenebrio atronitens* Fairmaire, 1891
atronitens Fairmaire, 1891c: xvii (*Tenebrio*) **A**: BEI FUJ GUI HUB SCH SHG
ulomoides Fairmaire, 1891f: ccxi (*Ariarathus*)

genus *Bius* Dejean, 1834: 205 type species *Trogosita thoracica* Fabricius, 1792
Bia Hope, 1840: 132 [unjustified emendation]
Dendroscopius Gistel, 1848: 125 [unnecessary RN]
thoracicus Fabricius, 1792a: 116 (*Trogosita*) **E**: AU BY CR CT ES FI FR GE NR NT PL RO SK SV SZ **A**: ES FE JA KZ MG WS
laevis Stephens, 1835: 426 (*Tenebrio*)

genus *Neatus* J. L. LeConte, 1862: 233 type species *Helops tenebrioides* Palisot de Beauvois, 1811
noctivagus Mulsant & Rey, 1853a: 7 [= 1853b: 9] (*Tenebrio*) **E**: AL CR IT
picipes Herbst, 1797: 245 (*Tenebrio*) **E**: AU BU BY CR CT CZ DE ES FR GB GE GR HU IT LA PL RO SK ST SZ UK **A**: ANH BEI ES FE GAO HUB IN JA KZ NC NE SC SHA SHX TR WS **NAR**
badius Say, 1824a: 265 (*Tenebrio*)
intermedius Fiori, 1903: 226 (*Tenebrio*)
laticollis Stephens, 1832: 8 (*Tenebrio*)
loripes J. Sturm, 1807: 223 (*Tenebrio*)
transversalis Duftschmid, 1812: 300 (*Tenebrio*)
ventralis Marseul, 1876c: 123 (*Tenebrio*)
subaequalis Reitter, 1920a: 22 **E**: AB AL AR GG GR MC ST **A**: IN TR

genus *Tenebrio* Linnaeus, 1758: 417 type species *Tenebrio molitor* Linnaeus, 1758
Menedrio Motschulsky, 1872: 27 type species *Tenebrio obscurus* Fabricius, 1792
Tenebrionellus Crotch, 1873: 105 [unnecessary RN]
alternicostis Marseul, 1876c: 123 **A**: JA (Kyushu)
angustus Zoufal, 1892: 284 **E**: AB **A**: AF IN KZ TD TM UZ
cedrorum Pic, 1925c: 6 **N**: AG
fuliginosus Fairmaire, 1874a: 388 **N**: AG MO
medius Desbrochers des Loges, 1881: 131 **N**: AG
molitor Linnaeus, 1758: 417 **E**: AB AL AU BE BH BU BY CR CT CZ DE EN FI FR GB GE GR HU IC IR IT LA LS LT LU MA NL NR NT PL RO SK SP ST SV SZ UK YU **N**: CI EG MR **A**: FE FUJ GAN GUA GUX

HAI HEB HEI HEN HKG HUB JA JIA JIL JIX LIA NIN NMO SCH SHN SHX TAI TD TM YUN "Korea" **AFR AUR NAR NTR**

obscurus Fabricius, 1792a: 111 **E**: AB AL AU AZ BE BH BU BY CR CT CZ DE EN FI FR GB GE GR HU IR IT LT LU MA NL NT PL RO SK SL SP ST SV SZ UK YU **N**: CI EG MO MR TU **A**: AF ANH CY FE FUJ GUA GUI GUX HAI HEB HEI HEN HUN IN IQ JA JIA JIL JIX KZ LIA NMO QIN SC SCH SHA SHN SHX TAI TD TM TR UZ XIN ZHE **COS**

codinae Reitter, 1912d: 310

longicollis Motschulsky, 1872: 37 (*Menedrio*)

pollens Casey, 1924: 321

opacus Duftschmid, 1812: 299 **E**: AU BU CR CT CZ DE FR GE GG GR HU IT MC PL RO SK SL SV SZ UK

punctipennis Seidlitz, 1896a: 633 **E**: BU GR IT **N**: TU

punctatissimus Denisova, 1940: 243 **A**: AF TD UZ

syriacus Zoufal, 1892: 284 **N**: EG **A**: SY TR

tribe Toxicini Lacordaire, 1859

genus *Cryphaeus* Klug, 1833: 177 type species *Cryphaeus aries* Klug, 1833

Anthracias Dejean, 1834: 177 type species *Uloma cornuta* Fischer von Waldheim, 1823

Anthracias L. Redtenbacher, 1857: 617 [HN] type species *Anthracias bicornis* L. Redtenbacher, 1857 (= *Uloma cornuta* Fischer von Waldheim, 1823)

albopilosus Mt. Chûjô & Lee, 1993: 116 **A**: SC SHA

amurensis Kulzer, 1950: 41 **A**: FE

bicornutus Pic, 1921c: 28 (*Toxicum*) **A**: FUJ YUN

boleti Lewis, 1894: 471 (*Anthracias*) **A**: JA

cavifrons Kulzer, 1950: 38 **A**: FUJ YUN

chengi Masumoto, Akita & Lee, 2005: 252 **A**: TAI

cornutus Fischer von Waldheim, 1823: 199 (*Uloma*) **E**: AB BH BU FR GG GR HU RO SL ST UK **N**: CI **A**: CY SY TR

bicornis L. Redtenbacher, 1857: 617 (*Anthracias*)

dissolutus Kulzer, 1950: 39 **A**: SCH

duellicus Lewis, 1894: 470 (*Anthracias*) **A**: FE HAI JA SC

formosanus Masumoto, 1982d: 62 **A**: TAI

inomatai Masumoto, 1982d: 65 **A**: TAI

laticeps Lillig, 2006: 49 **A**: IS

punctatulus Lewis, 1894: 470 (*Anthracias*) **A**: JA

fagi Lewis, 1894: 471 (*Anthracias*)

punctipennis Gravely, 1915: 529 **A**: AF AP BT NP PA SD **ORR**

rotundicollis Mt. Chûjô & Lee, 1993: 117 **A**: SC

satoi Kaszab, 1964a: 1 **A**: JA (Ryukyus) TAI

incurvatus Masumoto, 1982d: 63

iriomotensis M. Chûjô, 1979: 42

taiwanus Masumoto, 1982d: 64 **A**: TAI

taurus Fabricius, 1801a: 153 (*Trogosita*) **A**: YE **AFR**

tenuis Fairmaire, 1896a: 26 (*Toxicum*) **A**: NP **ORR**

genus *Epitoxicum* Bates, 1873a: 46 type species *Epitoxicum haplandroides* Bates, 1873

haplandroides Bates, 1873a: 46 **A**: NP **ORR**

genus *Toxicum* Latreille, 1802: 174 type species *Toxicum richesianum* Latreille, 1804

Trestonia Rafinesque, 1815: 113 [unnecessary RN]

angustatum Pic, 1921c: 27 **A**: AP BT SCH

assamense Pic, 1913c: 18 **A**: AP NP SD XIZ YUN **ORR**

cavifrons Pic, 1913c: 17 **A**: TAI

fangjingshananum Ren & Hua, 2006: 271 **A**: GUI

formosanum Kulzer, 1950: 37 **A**: TAI

funginum Lewis, 1894: 469 **A**: JA TAI

subtricornutum M. Chûjô, 1978a: 70 **A**: JA (Ryukyus)
tricornutum Waterhouse, 1874: 126 **A**: JA SC
tuberculifrons Lewis, 1894: 469
umbrosum Harold, 1880: 165

genus *Taiwanocryphaeus* Masumoto, 1996c: 67 type species *Taiwanocryphaeus rhinoceros* Masumoto, 1996
rhinoceros Masumoto, 1996c: 68 **A**: TAI

tribe Triboliini Gistel, 1848

genus *Latheticus* Waterhouse, 1880: 147 type species *Latheticus oryzae* Waterhouse, 1880
oryzae Waterhouse, 1880: 148 **E**i: AU BE CZ DE FI FR GB GE HU IT LA LU NL NR PL SK SP ST SV SZ UK **N**i: CI LB MO TU **A**: FE HEB HEN HUB GUA GUX IN IQ IS JA JIA JIX KZ MG NE NMO SA SCH SHA SHX TM YE "Korea" **COS**
striolatus Fairmaire, 1892a: 111 (*Lyphia*)

genus *Lyphia* Mulsant & Rey, 1859d: 166 type species *Lyphia ficicola* Mulsant & Rey, 1859 (= *Bius tetraphylla* Fairmaire, 1856)
Lindia Blackburn, 1888: 275 [HN] type species *Lindia angusta* Blackburn, 1888
abyssinica Fairmaire, 1893a: 29 **A**: YE **AFR**
angusta Lucas, 1846: 345 (*Hypophloeus*) **E**: FR (Corse) SP (Mallorca) **N**: AG CI
exigua Marseul, 1876c: 113 **A**: JA (Honshu)
formosana Masumoto, 1982d: 67 **A**: TAI
subopaca Blair, 1930a: 150 **A**: NP UP **ORR**
tetraphylla Fairmaire, 1856: 534 (*Bius*) **E**i: CR CZ FR GR IT SP **A**i: CY IS TR **NAR**
ficicola Mulsant & Rey, 1859d: 167
rugosa Dury, 1902: 171 (*Hypophloeus*)

genus *Platybolium* Blair, 1938: 222 type species *Platybolium alvearium* Blair, 1938
alvearium Blair, 1938: 223 **A**: FUJ **ORR**

genus *Tribolium* W. S. MacLeay, 1825: 47 type species *Colydium castaneum* Herbst, 1797
Aphanotus J. L. LeConte, 1862: 233 type species *Eulabis brevicornis* J. L. LeConte, 1859
Eusemostene Gebien, 1940b: 764 [RN] type species *Tribolium confusum* Jacquelin du Val, 1861
Leanum Uyttenbenboogaart, 1934: 29 type species *Tribolium myrmecophilum* Lea, 1904
Margus Dejean, 1834: 200 type species *Colydium castaneum* Herbst, 1797
Stene Stephens, 1832: 9 type species *Stene ferruginea* Stephens, 1832
Stene Seidlitz, 1890: 518 [HN] type species *Tribolium confusum* Jacquelin du Val, 1861
Tenebrioloma Gebien, 1910b: 386 type species *Tenebrioloma semicostata* Gebien, 1910
castaneum Herbst, 1797: 282 (*Colydium*) **E**: AL AU AZ BE BU BY CR CT CZ EN DE FI FR GB GE GR HU IC IR IT LA LS LT LU MA NL NR NT PL RO SK SL SP SV SZ UK **N**: CI EG LB MO MR TU **A**: AF ANH BT CY FE FUJ GUA GUI GUX HAI HEN HKG HUB HUN IQ IS JA JIA JIX KZ MG NE NIN SA SCH SHX SI TAI TM TR YE YUN ZHE **COS**
bifoveolatum Duftschmid, 1812: 304 (*Tenebrio*)
ferrugineum Fabricius, 1787a: 212 (*Tenebrio*) [HN]
testaceum Fabricius, 1798: 179 (*Ips*)
confusum Jacquelin du Val, 1861: 181 **E**: AL AU AZ BE BU BY CR CT CZ DE EN FI FR GB GE GR HU IC IR IT LT LU NL NR NT PL SK SP ST SV SZ UK **N**: AG CI EG LB MO TU **A**: ANH CY ES FE FUJ GUI GUX HAI HEI HEN HUB HUN IS JA JIA JIL JIX KZ LIA MG NIN SA SCH SHA SHN SHX SI TM YUN ZHE "Korea" **COS**
destructor Uyttenboogaart, 1933: xlii [= 1934: 26] **E**: BE BY CT CZ DE EN FI FR GB GE HU IC LT LU NL NR PL SK SV SZ UK **N**: CI **A**: FE MG TM YE **COS**
freemani Hinton, 1948: 49 **A**: JA GAN KA NIN SHX XIN
indicum Blair, 1930a: 148 **A**: IN SA YE **AFR ORR**
ares Hinton, 1948: 44
seres Hinton, 1948: 44

madens Charpentier, 1825: 218 (*Tenebrio*) **E**: AR AU CZ FR GE GR HU IC IT NT PL RO SK ST SV SZ UK **N** EG **A**: NIN TD TM WS **NAR**
obscurum W. Redtenbacher, 1842: 17 (*Margus*)
uezumii Nakane, 1963a: 28 **A**: JA (Honshu)

tribe Ulomini Blanchard, 1845

genus *Apithesis* Waterhouse, 1881: 476 type species *Apithesis obesa* Waterhouse, 1881
obesa Waterhouse, 1881: 477 **A**: YE (Suqutra)

genus *Basanopsis* Gebien, 1914b: 21 type species *Basanopsis curvipes* Gebien, 1914
curvipes Gebien, 1914b: 22 **A**: TAI

genus *Brachypohloeus* Fairmaire, 1897d: 113 type species *Hypophloeus dimidiatipennis* Fairmaire, 1880
Leptoscapha Fairmaire, 1886b: 73 [RN, HN] type species *Stenoscapha spissicornis* Fairmaire, 1885
Stenoscapha Fairmaire, 1884b: 234 [HN] type species *Stenoscapha spissicornis* Fairmaire, 1884
lignicola Gravely, 1915: 531 (*Leptoscapha*) **A**: AP NP
pulcher Gravely, 1915: 530 (*Leptoscapha*) **A**: AP BT
unifasciatus Kaszab, 1964c: 47 (*Leptoscapha*) **A**: JA (Ryukyus)

genus *Cenoscelis* Wollaston, 1867: 179 type species *Cenoscelis tibialis* Wollaston, 1867 (= *Uloma pulla* Erichson, 1843)
atra Pic, 1926c: 29 (*Eutochia*) **A**: JA TAI **AUR ORR**
lateralis Boheman, 1858: 94 (*Heterophaga*) **A**: HAI HKG JA TAI "Korea" **AUR ORR**
picescens Fairmaire, 1885b: cix (*Holaniara*)
pulla Erichson, 1843: 253 (*Uloma*) **A**: SA YE **AFR ORR**
tibialis Wollaston, 1867: 180

genus *Cneocnemis* Gebien, 1914a: 32 type species *Uloma haemorrhoa* Fairmaire, 1893
chetri Schawaller, 1998: 464 **A**: NP
laminipes Gebien, 1914b: 26 **A**: JA TAI "Korea"
newar Schawaller, 1998: 465 **A**: NP

genus *Uloma* Dejean, 1821: 67 type species *Tenebrio culinaris* Linnaeus, 1758
subgenus *Uloma* Dejean, 1821: 67 type species *Tenebrio culinaris* Linnaeus, 1758
Melasia Perroud & Mulsant, 1856: 160 type species *Melasia gagatina* Perroud & Mulsant, 1856
Prioscelida White, 1846: 11 type species *Prionoscelida tenebrionoides* White, 1846
bonzica Marseul, 1876c: 114 **A**: JA NC
robustior Nakane, 1956b: 167
bhutanensis Kaszab, 1975e: 327 **A**: BT
castanea Ren & Liu, 2004: 299 **A**: GUX YUN
contracta Fairmaire, 1882c: 226 **A**: GUX HAI YUN **ORR**
culinaris Linnaeus, 1758: 417 (*Tenebrio*) **E**: AB AL AU BE BH BU BY CR CT CZ DE EN FR GB GE GG GR HU IT LA LS LT NR PL RO SK SL SP ST SV SZ TR UK YU **A**: IN WS
cucullata Fischer von Waldheim, 1832a: 204
ferruginea Panzer, 1793c: 2 (*Tenebrio*)
gagatina Perroud & Mulsant, 1856: 161 (*Melasia*)
rubens Laporte, 1840: 220
cypraea Kraatz, 1873: 196 **E**: GR **A**: CY TR
crenulata Baudi di Selve, 1876c: 230
cypria Allard, 1876b: 21
cypriotes Marseul, 1876b: 40
euphorbiae Blair, 1930b: 177 **A**: NP SD
sherpa Schawaller, 1996a: 121
excisa excisa Gebien, 1914b: 24 **A**: FUJ GUX HAI TAI **ORR**
tschungseni Kaszab, 1954: 254
excisa nanseiensis Masumoto & Nishikawa, 1986: 27 **A**: JA (Ryukyus)

formosana Kaszab, 1941c: 54 **A**: TAI
fukiensis Kaszab, 1954: 254 **A**: FUJ TAI
gongshanica Ren & Liu, 2004: 300 **A**: YUN
hikosana Nakane, 1956b: 167 **A**: JA
hiranoi Masumoto & Akito, 2002: 148 **A**: JA (Ryukyus)
ichoi Nakane, 1956b: 168 **A**: JA (Ryukyus)
ishigakiensis M. Chûjô, 1983: 49 **A**: JA (Ryukyus)
kondoi Nakane, 1968b: 77 **A**: JA (Yakushima)
latimanus Kolbe, 1886: 202 **A**: JA NC SC
lewisi Nakane, 1956b: 167 **A**: JA
liangi Ren & Liu, 2004: 298 **A**: YUN
marseuli fujitai Masumoto, 1985a: 21 **A**: JA (Ryukyus)
marseuli marseuli Nakane, 1956b: 168 **A**: JA SC
meifengensis Masumoto, 1982c: 25 **A**: TAI
miyakei Masumoto & Nishikawa, 1986: 23 **A**: TAI
mulidenta Ren & Liu, 2004: 296 **A**: YUN
nakanei Masumoto & Nishikawa, 1986: 25 **A**: TAI
nanshanchica Masumoto & Nishikawa, 1986: 30 **A**: TAI
nepalica Kaszab, 1977b: 259 **A**: NP
nomurai Masumoto, 1982c: 25 **A**: TAI
picea Küster, 1846a: 77 **E**: CR
polita Wiedemann, 1821: 149 (*Phaleria*) **A**: BT GUX JA NP SD TAI **AFR ORR**
anomala Herbst, 1799: 23 (*Tenebrio*) [HN]
latipes Fairmaire, 1903b: 208 (*Eutochia*)
retenta Walker, 1858: 283 (*Tenebrio*)
reitteri Kaszab, 1941d: 123 **A**: SCH
rubripes minor Gebien, 1914b: 28 **A**: AP **ORR**
rubripes rubripes Hope, 1831: 31 (*Tenebrio*) **A**: AP BT NP TAI **AUR ORR**
prehimalayana Kaszab, 1975e: 325
rufa Piller & Mitterpacher, 1783: 39 (*Tenebrio*) **E**: AB AU BH BY CR CT CZ EN FI FR GE GR HU IT LA LT NR NT PL RO SK SP ST SV SZ UK **A**: ES FE KZ
perroudi Mulsant & Guillebeau, 1855b: 421
ruficornis Allard, 1896: 319 **A**: SD
kulzeri Ardoin, 1960b: 273
rufilabris Fairmaire, 1882c: 226 **A**: NP SD UP **ORR**
sakuraii Masumoto, 1983a: 89 **A**: JA (Ryukyus)
sauteri Kaszab, 1941c: 55 **A**: TAI
scita Walker, 1858: 284 **A**: BT NP PA SD **ORR**
splendida Ren & Liu, 2004: 301 **A**: YUN
spinipes Kaszab, 1975e: 328 **A**: BT NP
tadafumii Masumoto & Akita, 2003: 179 **A**: JA (Ryukyus)
takagii Masumoto & Nishikawa, 1986: 28 **A**: TAI
takarai M. Chûjô, 1983: 51 **A**: JA (Iriomote)
tsugeae Masumoto, 1982c: 23 **A**: TAI
versicolor Ren & Liu, 2004: 297 **A**: YUN

subfamily Tenebrioninae, genus incertae sedis
genus *Anophthalmolamus* Ferrer, 1993b: 122 type species *Anophthalmolamus fuerteventurae* J. Ferrer, 1993
fuerteventurae J. Ferrer, 1993b: 122 **N**: CI (Fuerteventura)

subfamily **Diaperinae** Latreille, 1802

tribe Crypticini Brullé, 1832

genus *Cechenosternum* Gebien, 1921: 9 type species *Cechenosternum nigromaculatum* Gebien, 1921
bengalense Kaszab, 1960a: 165 **A**: PA "Bengale"

klapperichi Kaszab, 1960a: 164 **A**: AF
rufulum Motschulsky, 1873: 480 (*Neomida*) **N**: EG **AFR**

genus *Crypticus* Latreille, 1817a: 298 [= 1817b: 525] type species *Helops glaber* Fabricius, 1775 (=*Tenebrio quisquilius* Linnaeus, 1760)
subgenus *Crypticopsis* Antoine, 1945: 270 type species *Crypticus corticeus* Fairmaire, 1871
corticeus Fairmaire, 1871a: 389 **N**: MO
proximus Escalera, 1914: 342
meridionalis meridionalis Escalera, 1914: 343 **N**: MO
meridionalis validus Español, 1963e: 382 **N**: MO
subgenus *Crypticus* Latreille, 1817a: 298 [= 1817b: 525] type species *Helops glaber* Fabricius, 1775
aksuensis Español, 1955a: 14 **E**: AB **A**: AF KZ XIN
antoinei Español, 1950a: 144 **E**: SP
arandai arandai Escalera, 1914: 344 **N**: MO (Grand Atlas)
gundaficus Antoine, 1945: 266
major Antoine, 1945: 266
arandai faucium Antoine, 1933b: 229 **N**: MO
atlasicus atlasicus Théry, 1931b: 125 **N**: MO (Moyen Atlas)
atlasicus culminicola Antoine, 1936b: 74 **N**: MO (Grand Atlas)
azruensis Escalera, 1925a: 335 **N**: MO
beitunensis Ren & Zheng, 1993a: 14 **A**: XIN
burqinensis Ren & Zheng, 1993a: 17 **A**: XIN
capsianus capsianus Antoine, 1945: 268 **N**: MO
capsianus littoralis Español, 1950a: 141 **N**: LB TU
cyrenaicus Koch, 1937: 477 **N**: LB
dactylispinus Marseul, 1870a: 381 **N**: AG LB MO TU
melancroides Fairmaire, 1879c: 196
fairmairei Desbrochers des Loges, 1881: 101 **N**: AG
fernandezi Español, 1954b: 105 **N**: CI
gibbulus Quensel, 1806: 163 (*Helops*) **E**: FR IT MA PT SP **N**: AG MO TU **A**: CY JO SY
pubescens Dufour, 1820: 311 (*Pedinus*)
gomerensis Oromí, 1978: 105 **N**: CI (Gomera)
iblanensis Antoine, 1945: 265 **N**: MO
impunctatus Théry, 1932a: 193 **N**: MO
kochi Español, 1950a: 146 **E**: SP
latiusculus Ménétriés, 1849: 327 (*Heliopathes*) **A**: GAN HUB JIA KZ TM XIN
substriatus Desbrochers des Loges, 1881: 166
leptodentatus Ren & Zheng, 1993a: 13 **A**: XIN
lecerfi Théry, 1932a: 194 **N**: MO (Grand Atlas)
minutus Brullé, 1838: 69 **N**: CI (Gran Canaria)
navicularis amaroides Har. Lindberg, 1953: 9 **N**: CI (Tenerife)
navicularis calvus Wollaston, 1865: 59 [Appendix] **N**: CI (El Hierro)
navicularis canariensis Wollaston, 1864: 481 **N**: CI (Tenerife)
navicularis latihumeralis Har. Lindberg, 1953: 9 **N**: CI (Tenerife)
navicularis navicularis Brullé, 1838: 69 **N**: CI (Tenerife)
navicularis punctatissimus Wollaston, 1864: 480 **N**: CI (La Palma)
nitidulus Wollaston, 1865: 60 [Appendix] **N**: CI (Gomera)
nondentatus Ren & Zheng, 1993a: 12 **A**: XIN
obesus Lucas, 1846: 331 **N**: AG
oblongus oblongus Wollaston, 1864: 482 **N**: CI (Tenerife)
oblongus subnitens Har. Lindberg, 1950: 12 **N**: CI (Tenerife)
obtusus Reitter, 1896b: 74 **A**: KZ (Aral sea)
pentheri Reitter, 1905e: 278 **A**: TR
petraeus Escalera, 1913: 49 **N**: MO
pettersoni J. Ferrer & L. Soldati, 1999: 73 **A**: TR

platystaturalis Ren & Zheng, 1993a: 16 **A**: XIN
platytibialis Ren & Zheng, 1993a: 10 **A**: XIN
pubens balearicus Español, 1950a: 144 **E**: SP (Baleares)
pubens pubens Fairmaire, 1881b: cxxxii **E**: SP **N**: MO
quisquilius aprutianus Gridelli, 1949: 272 **E**: IT
quisquilius pyrenaeus Baudi di Selve, 1876a: 33 **E**: FR SP
quisquilius quisquilius Linnaeus, 1760: 226 (*Tenebrio*) **E**: AL AR AU BE BH BU BY CR CT CZ DE EN FI FR GB GE GR HU IR IT LA LT NL NR NT PL RO SK SP ST SV SZ TR UK YU **A**: ES FE KI KZ MG TR WS XIN
glaber Fabricius, 1775: 258 (*Helops*)
laevis Gmelin, 1790: 2010 (*Pimelia*)
laticollis Panzer, 1796d: 1 (*Helops*)
luctuosus Schrank von Paula, 1781: 219 (*Tenebrio*)
niger Geoffroy, 1785: 158 (*Tenebrio*)
pusillus Herbst, 1799: 199 (*Blaps*)
opacus Brancsik, 1899: 100
strauchi Roubal, 1911a: 133
rhaticus Antoine, 1945: 263 **N**: MO
schrammi Antoine, 1945: 269 **N**: MO
sibiricus Solsky, 1871: 373 **A**: ES
subpunctatus Walker, 1871: 15 **N**: EG
tachdirtensis Antoine, 1945: 263 **N**: MO
tauri Pic, 1907c: 123 **A**: TR
tenietensis Desbrochers des Loges, 1890: clxxxv **N**: AG
thessalicus Reitter, 1896d: 150 **E**: GR
yosiii Nakane, 1966: 243 **A**: AF
zuberi Marseul, 1870a: 382 **E**: AB ST UK **A**: AF KZ MG UZ XIN
ovalis Ballion, 1878: 321
subgenus *Platycrypticus* Español, 1952b: 117 [RN] type species *Crypticus viaticus* Fairmaire, 1851
Ulomoides Escalera, 1927b: 501 [HN]
castaneus Baudi di Selve, 1876a: 34 **E**: GR **A**: TR
frigidus Escalera, 1927b: 503 **E**: SP (Andaluzía)
pardoi Español, 1952b: 118 **E**: SP (Ciudad Real)
punctatolineatus Fairmaire, 1880d: 251 **N**: MO
vaucheri Bedel, 1906a: 272 **N**: MO
viaticus Fairmaire, 1851: 528 **E**: SP
ulomoides Fairmaire, 1852: 85
subgenus *Seriscius* Motschulsky, 1845a: 77 type species *Seriscius pubescens* Motschulsky, 1845 (= *Crypticus rufipes* Gebler, 1830)
paradoxus Reitter, 1896d: 149 **N**: MO
rufipes Gebler, 1829a: 125 **E**: ST **A**: BEI ES FE GAN MG NIN NMO SCH WS XIN "Korea"
asiaticus Reitter, 1889k: 698
fuscovariegatus Reitter, 1889k: 698
ovatulus Reitter, 1896d: 150
pubescens Motschulsky, 1845a: 77 (*Seriscius*)
soricinus Fairmaire, 1887f: 324

genus *Ellipsodes* Wollaston, 1854: 485 type species *Sphaeridium glabratum* Fabricius, 1792
subgenus *Anthrenopsis* Koch, 1950a: 74 type species *Platydema scriptipenne* Fairmaire, 1875 (= *Basites ziczac* Motschulsky, 1873)
hieroglyphicus Fairmaire, 1896a: 25 (*Platydema*) **A**: NP **ORR**
scriptus Lewis, 1894: 396 (*Platydema*) **A**: ANH FUJ GUA GUI GUX HEB HEN HUB HUN JA JIA JIX NMO NP SCH SD SHA SHN SHX TAI UP YUN ZHE **AUR NAR ORR**
minutissimus Gebien, 1914b: 6 (*Crypticus*)
submaculatus Champion, 1894a: 362 (*Crypticus*)
ziczac nuristanicus Kaszab, 1960a: 163 **A**: AF

ziczac ziczac Motschulsky, 1873: 475 (*Basides*) **A**: HEN HUB JIX NP QIN SD **AFR AUR ORR**
bifasciatus Chevrolat, 1878a: 222 (*Platydema*)
scriptipennis Fairmaire, 1875d: xxxiii (*Platydema*)
subgenus *Ellipsodes* Wollaston, 1854: 485 type species *Sphaeridium glabratum* Fabricius, 1792
glabratus Fabricius, 1781: 78 (*Sphaeridium*) **N**: MR
oblongior Wollaston, 1857a: 150 **N**: MR

genus *Lamprocrypticus* Español, 1950b: 127 type species *Crypticus alpinus* Comolli, 1837
alpinus Comolli, 1837: 26 (*Crypticus*) **E**: IT SZ (Valais)
baguenai Español, 1948b: 488 (*Crypticus*) **E**: SP
kraatzii C. Brisout de Barneville, 1866: 383 (*Crypticus*) **E**: PT SP
minutissimus Escalera, 1934a: 211 (*Crypticus*) **E**: SP (Sierra de Segura)
pusillus Rosenhauer, 1856: 212 (*Crypticus*) **E**: SP
tonsilis Bedel, 1893: 103 (*Crypticus*) **N**: AG MO
zophosioides Heyden, 1870: 131 (*Crypticus*) **E**: PT SP

genus *Oochrotus* Lucas, 1852: xxix type species *Oochrotus unicolor* Lucas, 1852
glaber glaber Demaison, 1905: 287 **E**: GR (Ródos) **A**: CY SY TR
boyadjiani Lokay, 1907: 91
glaber rhodicus Koch, 1935e: 319 **E**: GR (Ródos)
unicolor ardoini Canzoneri, 1961a: 21 **E**: IT
unicolor lusitanicus Canzoneri, 1961a: 16 **E**: PT
unicolor moltonii Canzoneri, 1961a: 20 **E**: IT (Sicilia)
unicolor unicolor Lucas, 1852: xxix **E**: FR IT (Sardegna) SP **N**: AG MO TU
chilivanii Rallo, 1974: 62
espagnoli Canzoneri, 1961a: 20
hispanus Canzoneri, 1961a: 17
laurae Canzoneri, 1961: 23
meridionalis Canzoneri, 1961a: 19
sardous Canzoneri, 1961a: 23

genus *Pseudoseriscius* Español, 1950b: 125 type species *Pedinus pruinosus* Dufour, 1820
adspersus adspersus Küster, 1851: 51 (*Crypticus*) **E**: SP
adspersus nebulosus Fairmaire, 1871a: 389 (*Crypticus*) **N**: AG LB MO TU
uniformis Reitter, 1896d: 150 (*Crypticus*)
adspersus timaditensis Théry, 1932a: 194 (*Crypticus*) **N**: MO (Moyen Atlas)
cameroni Reitter, 1902e: 194 (*Crypticus*) **E**: MA
fonti Escalera, 1923d: 132 (*Crypticus*) **N**: CI MO (Western Sahara)
alluaudi Peyerimhoff, 1942a: 11 (*Crypticus*)
gridellii Español, 1949: 211 (*Crypticus*) **E**: IT **N**: LB
griseovestis Fairmaire, 1879c: 192 (*Crypticus*) **E**: IT (Linosa) **N**: AG EG LB TU **A**: CY JO KU
pelitus Reitter, 1896d: 150 (*Crypticus*)
helvolus adriaticus Español, 1949: 208 (*Crypticus*) **E**: AL GR IT
helvolus helvolus Küster, 1852a: 83 (*Crypticus*) **E**: IT **N**: TU
ibicensis Español, 1955b: 27 **E**: SP (Ibiza)
maculosus maculosus Fairmaire, 1871a: 390 (*Crypticus*) **A**: IN IS LE SA SY
maculosus murinus Allard, 1882b: lxxxvii [= 1883: 30] (*Crypticus*) **N**: EG **A**: SI
mus Reitter, 1896d: 149 (*Crypticus*)
mollis Reitter, 1896d: 150 (*Crypticus*) **N**: AG LB TU
munyozi Viñolas, 1997b: 97 **E**: SP (Almería)
normandi normandi Español, 1949: 209 (*Crypticus*) **N**: TU
normandi pacificii Leo, 1982b: 104 **E**: IT
olivierii olivierii Desbrochers des Loges, 1881: 103 (*Crypticus*) **N**: AG LB MO TU
olivierii sardiniensis Leo, 1982b: 102 **E**: IT (Sardegna)
pruinosus Dufour, 1820: 310 (*Pedinus*) **E**: FR PT SP **N**: LB MO

richteri Kaszab, 1959a: 3 **A**: IN

tribe Diaperini Latreille, 1802

genus *Adelina* Dejean, 1835: 315 type species *Cucujus planus* Fabricius, 1801

Doliema Pascoe, 1860a: 50 type species *Doliema platisoides* Pascoe, 1860

Schedarosus Reitter, 1876a: 42 type species *Schedarosus cucujiformis* Reitter, 1876

turcica Reitter, 1877b: 8 (*Sitophagus*) **E**: GR RO TR **A**: IS

genus *Alphitophagus* Stephens, 1832: 12 type species *Alphitophagus quadripustulatus* Stephens, 1832 (= *Diaperis bifasciatus* Say, 1824)

Phyletes L. Redtenbacher, 1845: 128 type species *Phylethus populi* L. Redtenbacher, 1849 (= *Diaperis bifasciatus* Say, 1824)

bifasciatus Say, 1824a: 268 (*Diaperis*) **E**: AB AL AR AU BE BY CR CT CZ DE EN FI FR GB GE GR HU IT LT MC NL NR PL SK SP ST SV UK **N**: EG TU **A**: AF CY GAN GUA HAI HEB HEN HUB IS JA JIA JIL JIX KZ LIA NIN NMO SCH SHA SHN SHX SY TM TR XIN "Korea" **COS**

iudaeus Roubal, 1929b: 97

lilliputanus Carter, 1937: 130 (*Platydema*)

pictus Ménétriés, 1832: 203 (*Diaperis*)

populi L. Redtenbacher, 1849: 589 (*Phylethus*)

quadripustulatus Stephens, 1832: 12

convexicollis Reitter, 1911d: 268 **A**: TR

nigrotrabeatus Fairmaire, 1879c: 197 **N**: MO (West Sahara)

obtusangulus J. Müller, 1904: 209 **E**: CR (Meleda) GR FR

socotranus Schawaller, 2004c: 454 **A**: YE (Suqutra)

xaxarsi Reitter, 1914b: 82 (*Alphitobius*) **E**: SP (Baleares)

genus *Ceropria* Laporte & Brullé, 1831: 396 type species *Helops indutus* Wiedemann, 1819

Dictysomorphus Pic, 1921c: 24 type species *Dictysomorphus dentaticornis* Pic, 1921

Epilamprus Gistel, 1848: xiv [unjustified emendation]

Epilampus Dejean 1834: 198 type species *Helops indutus* Wiedemann, 1819

Epilampus Blanchard, 1845: 30 [HN]

chinensis Masumoto, 1995a: 3 **A**: FUJ

formosana Gebien, 1914b: 20 **A**: TAI

induta induta Wiedemann, 1819: 164 (*Helops*) **A**: AP BT FUJ HAI JA NP TAI YUN XIZ "Korea" **ORR**

kinugasai Masumoto, 1982a: 151

subocellata Laporte & Brullé, 1831: 398

induta purpurina Gebien, 1925a: 277 **A**: AP HP NP SD **ORR**

krausei Masumoto, 1994b: 772 **A**: YUN

laticollis Fairmaire, 1903a: 13 **A**: AP FUJ JA SC SD SW YUN **ORR**

opacipennis Pic, 1921c: 26

lii Ando & Ren, 2006: 82 **A**: YUN

schenklingi Gebien, 1914b: 19 **A**: TAI

serripes Gebien, 1925a: 284 **A**: NP YUN **ORR**

striata Lewis, 1894: 399 **A**: JA SC

subocellata Laporte & Brullé, 1831: 398 **A**: AP NP SD **ORR**

sulcifrons Harold, 1878c: 353 **A**: JA SC (Jejudo Is.)

superba Wiedemann, 1823: 43 (*Cnodalon*) **A**: "South China" **ORR**

chrysosticta Hope, 1843: 65 [= 1845: 52] (*Epilampus*)

festiva Laporte & Brullé, 1831: 400

taiwana Masumoto, 1995a: 1 **A**: TAI

thailandica Masumoto, 1995b: 724 **A**: YUN **ORR**

versicolor Laporte & Brulle, 1831: 401 **A**: YUN **ORR**

concolor Pic, 1923c: 20

impressifrons Fairmaire, 1882c: 222

genus *Cynaeus* J. L. LeConte, 1862: 233 type species *Platydema angustum* J. L. LeConte, 1851
angustus J. L. LeConte, 1851: 149 (*Platydema*) **E**i: GE SV **NAR**
opacus Champion, 1886a: 156 **E**i: FI **NTR**

genus *Diaperis* Geoffroy, 1762: 337 type species *Chrysomela boleti* Linnaeus, 1758
Diaperis O. F. Müller, 1764: xv [HN] type species *Chrysomela boleti* Linnaeus, 1758
boleti boleti Linnaeus, 1758: 371 (*Chrysomela*) **E**: AR AU BE BU BY CT CZ DE EN FI FR GB GE GR HU IT LA LS LT NL NR NT PL PT SK SP ST SV SZ UK **N**: AG MO TU **A**: CY FE ES IS KI KZ SY TR WS
bipustulata Laporte & Brullé, 1831: 337
cypria Baudi di Selve, 1876c: 229
fasciata Scopoli, 1763: 79 (*Coccinella*)
fagniezi Méquignon, 1947: 60
fungi Motschulsky, 1873: 466
interrupta Heyden, 1890b: 78
morio Baudi di Selve, 1876c: 228
posticalis Heyden, 1890b: 78
quercana Baudi di Selve, 1876c: 228
lewisi intersecta Gebien, 1914b: 15 **A**: HKG JA (south, Ryukyus) TAI **ORR**
lewisi lewisi Bates, 1873a: 14 **A**: ES FE GUA HKG HUB JA NC SC TAI "North China" **ORR**
rubrofasciata Reitter, 1879: 226
sinensis Gebien, 1925a: 155
niponensis Lewis, 1887b: 217 **A**: FE JA
quadrimaculata Pic, 1926c: 22 **A**: XIZ

genus *Espagnolina* Kaszab, 1965a: 117 type species *Espagnolina assamica* Kaszab, 1965
assamica Kaszab, 1965a: 117 **A**: AP

genus *Falsocosmonota* Kaszab, 1962a: 75 type species *Falsocosmonota cheni* Kaszab, 1962
cheni Kaszab, 1962a: 77 **A**: GUA HAI

genus *Gnatocerus* Thunberg, 1814: 47 type species *Gnatocerus ruber* Thunberg, 1814 (= *Trogosita cornuta* Fabricius, 1798)
subgenus *Echocerus* Horn, 1870: 366 type species *Trogossita maxillosa* Fabricius, 1801
maxillosus Fabricius, 1801a: 155 (*Trogossita*) **E**: BE FR GB IT LU NL **N**: CI MR **A**: CH JA **AFR NAR NTR ORR**
subgenus *Gnatocerus* Thunberg, 1814: 47 type species *Gnatocerus ruber* Thunberg, 1814
Cerandria Dejean, 1834: 200 type species *Trogosita cornuta* Fabricius, 1798
Gnathocerus Agassiz, 1846a: 73 [unjustified emendation]
Sicinus Champion, 1886a: 146 type species *Sicinus guatemalensis* Champion, 1886
cornutus Fabricius, 1798: 51 (*Trogosita*) **E**: AU AZ BE CR CT CZ DE EN FI FR GB GE HU IC IR IT LA LS LT LU MA NL NR NT PL SK SP ST SV SZ UK **N**: CI LB MO MR TU **A**: FE FUJ GUA GUI GUX HEN HKG JA JIA KZ MG NMO SCH TAI TM YUN **COS**
cucullatus Montrouzier, 1860: 294 (*Neomida*)
laeviusculus Stephens, 1832: 10 (*Uloma*)
maxillosus Palisot de Beauvois, 1805: 125 (*Trogossita*) [HN]
ruber Thunberg, 1814: 47

genus *Ischnodactylus* Chevrolat, 1877d: 173 type species *Ischnodactylus quadridentatus* Chevrolat, 1877
bisbifasciatus Gebien, 1925b: 446 **A**: HKG TAI
ruficornis Pic, 1925g: 16 [Pic, 1925j: 437] (*Basides*)
formosanus Gebien, 1925b: 445 **A**: FUJ TAI
iriomotensis Ando, 1978: 81 **A**: JA (Iriomote)
loripes Lewis, 1894: 392 **A**: HEI JA SC
parallelicornis Nakane, 1956b: 164 (*Platydema*) **A**: JA (Honshu)
rubromarginatus yunnanus Kaszab, 1965c: 284 **A**: YUN

genus *Neomida* Latreille, 1829: 29 type species *Ips haemorrhoidalis* Fabricius, 1787
Arrhenoplita Kirby, 1837: 235 type species *Diaperis bicornis* A. G. Olivier, 1791
Evoplus J. L. LeConte, 1866: 128 type species *Evoplus ferruginea* [sic] J. L. LeConte, 1860
Hoplocephala Agassiz, 1846: 185 [unjustified emendation]
Oplocephala Laporte & Brullé, 1831: 338 type species *Ips haemorrhoidalis* Fabricius, 1787
haemorrhoidalis Fabricius, 1787b: 378 (*Ips*) **E**: AL AU BH BU BY CR CT CZ DE EN FI FR HU GE GR IT LA LT NR NT PL RO SK SL SP ST SV SZ UK **A**: AF ES CY FE IN JA KI TR WS
asiatica Lewis, 1894: 392 (*Arrhenoplita*)
cornifrons Rossi, 1792: 92 (*Tenebrio*)
quadricornis Motschulsky, 1873: 466 (*Oplocephala*) **E**: AB GG **A**: IN
quadricornis Reitter, 1884a: 8 (*Hoplocephala*) [HN]

genus *Pelleas* Bates, 1872: 98 type species *Tenebrio crotchii* Wollaston, 1865
crotchii Wollaston, 1865: 62 (*Tenebrio*) **N**: CI

genus *Pentaphyllus* Dejean, 1821: 68 type species *Mycetophagus testaceus* Hellwig, 1792
Iphicorynus Jacquelin du Val, 1861: 299 type species *Pentaphyllus melanophthalmus* Mulsant, 1854 (= *Nitidula chrysomeloides* Rossi, 1792)
chrysomeloides Rossi, 1792: 24 (*Nitidula*) **E**: AB AL AU BU CR DE EN FI FR GE GG GR HU IT LA PL RO SK SL SP ST SV SZ UK **N**: TU **A**: AF CY IN TD TM TR
melanophthalmus Mulsant, 1854: 197
dilatipes Shibata, 1978: 22 **A**: JA (Ryukyus) TAI
izuanus Nakane, 1978: 161 **A**: JA (Izu Is.)
lewisi Kaszab, 1982a: 63 **A**: JA (Kyushu)
nitidulus Reitter, 1884a: 8 **E**: AB
oblongus Lewis, 1894: 398 **A**: JA (Honshu)
philippinensis Kaszab, 1956a: 94 **A**: TAI **ORR**
quadricornis Gebien, 1914: 23 **A**: JA (Ryukyus) TAI **ORR**
testaceus Hellwig, 1792: 400 (*Mycetophagus*) **E**: AL AU BE BY CR CT CZ DE FI FR GB GE GG GR HU IT MA NL PL RO SK SP ST SV SZ UK **N**: AG TU **A**: TD
barbarus Motschulsky, 1873: 481
tokarensis Nakane, 196a3: 27 **A**: JA (Ryukyus) TAI **ORR**

genus *Platydema* Laporte & Brullé, 1831: 350 type species *Platydema dejeanii* Laporte & Brullé, 1831
Anisochara Gebien, 1925d: 101 type species *Anisochara gynadromorpha* Gebien, 1925
Histeropsis Chevrolat, 1878a: 221 type species *Platydema americanum* Laporte & Brullé, 1831
Typhobia Pascoe, 1869: 279 type species *Typhobia fuliginea* Pascoe, 1869
alticorne Gravely, 1915: 524 **A**: AP NP TAI **ORR**
gyandromorpha Gebien, 1925d: 102 (*Anisochara*)
aurimaculatum Gravely, 1915: 523 **A**: AP TAI **ORR**
cederholmi Kaszab, 1980a: 167
monoceratoides Masumoto, 1982a: 145
brahma Schawaller, 2003b: 271 **A**: NP
caesifrons Marseul, 1876b: 38 **N**: EG
capreolum Chevrolat, 1877d: 170 (*Hoplocephala*) **A**: NP **ORR**
celatum Nakane, 1983b: 146 **A**: JA (Iriomote)
coeruleum Gebien, 1925c: 570 **A**: TAI
dejeanii Laporte & Brullé, 1831: 351 **E**: AU BH BY CR CZ FR HU GE GR PL RO SK SL ST SZ UK YU **A**: ES FE WS
detersum Walker, 1858: pl. IV **A**: AP NP TAI YUN **AUR ORR**
endoi Masumoto, 1984: 22 **A**: TAI
europaeum Laporte & Brullé, 1831: 365 **E**: CR FR GR IT SP ST SZ UK **N**: AG TU **A**: TR
atrum Motschulsky, 1873: 477 (*Neomida*)
subsericeum Motschulsky, 1873: 478 (*Neomida*)
flavopictum Gebien, 1914b: 17 **A**: TAI **ORR**
fumosum formosanum Gebien, 1925c: 588 **A**: TAI

morimotoi Nakane, 1968a: 76
fumosum fumosum Lewis, 1894: 395 **A**: FUJ HUB JA SC ZHE
fumosum kawaii Nakane, 1968a: 76 **A**: JA (Ryukyus)
ganesha Schawaller, 2003b: 271 **A**: NP
guangxicum Schawaller, 2004b: 24 **A**: GUX
haemorrhoidale Gebien, 1914b: 16 **A**: FUJ NP TAI **ORR**
higonium Lewis, 1894: 394 **A**: JA
kochi Schawaller, 2004c: 452 **A**: YE (Suqutra)
koreanum Mt. Chûjô, 1992: 37 **A**: SC
krishna Schawaller, 2003b: 271 **A**: NP
kulzerianum Nakane, 1977: 158 **A**: JA (Bonin Is.)
kurama Nakane, 1963a: 26 **A**: JA SC
lynceum Lewis, 1894: 395 **A**: JA SC
marseuli Lewis, 1894: 393 **A**: JA TAI **ORR**
benakatense Masumoto & Makihara, 1997: 119
nigroaeneum Motschulsky, 1861: 18 **A**: JA "Korea"
musiva Harold, 1878a: 78
nigropictum Nakane, 1963a: 27 **A**: FUJ JA TAI **ORR**
pallidicolle Lewis, 1894: 398 (*Alphitophagus*) **A**: CH JA TAI **AUR ORR**
sodale Waterhouse, 1894: 70
parachalceum Masumoto, 1982a: 147 **A**: FUJ GUI HEN HUB SHA TAI
zoltani Masumoto, 1985c: 121
planicorne Mt. Chûjô, 1978: 3 **A**: JA (Ryukyus)
recticorne Lewis, 1894: 394 **A**: JA SC **ORR**
reitteri Pic, 1934f: 85 (*Ceropria*) **A**: SCH
sakishimense Nakane, 1973b: 106 **A**: JA TAI
satoi Kaszab, 1964c: 45 **A**: JA (Tsushima Is.)
sauteri omoto Nakane, 1973b: 107 **A**: JA (Ryukyus)
sauteri sauteri Gebien, 1914b: 15 **A**: TAI
sawadai Masumoto, 1991b: 25 **A**: JA (Ryukyus)
semimetallicum Blair, 1930b: 177 **A**: NP SD XIZ
martensi Schawaller, 1994: 262
subfascia subfascia Walker, 1858: 284 (*Alphitophagus*) **A**: FUJ JA NP SC TAI **AFR AUR ORR**
celebum Chevrolat, 1877d: 177 (*Hoplocephala*)
diversidens Fairmaire, 1893b: 24 (*Hoplocephala*)
hoanum Pic, 1929c: 33
japanum Marseul, 1876c: 109 (*Alphitophagus*)
picicolle Motschulsky, 1873: 474 (*Basides*)
subfascia yaeyama Nakane, 1973b: 105 **A**: JA (Ryukyus)
sylvestre Lewis, 1894: 394 **A**: JA
takeii Nakane, 1956b: 165 **A**: JA
terusane Masumoto, 1984: 21 **A**: TAI
triste Laporte & Brullé, 1831: 359 **E**: AR GG ST UK **A**: IN TM
armeniacum Desbrochers des Loges, 1881: 129 (*Scaphidema*)
cophosioides Faldermann, 1837: 64 (*Neomida*)
tuchinlongi Masumoto, 1982a: 149 **A**: TAI
umbratum Marseul, 1876c: 107 **A**: JA (Kyushu)
ussurianum Kaszab, 1977a: 118 **A**: FE
violaceum Fabricius, 1790: 219 (*Diaperis*) **E**: AL AR AU BE BH BU BY CR CT CZ DE EN FI FR GB GE GR HU IT LA LS LT NL NT PL RO SK SL SP ST SV SZ UK
yangmingense Masumoto, 1982a: 148 **A**: TAI
yunnanicum Schawaller, 2004b: 46 **A**: YUN

genus *Sitophagus* Mulsant, 1854: 264 type species *Sitophagus solieri* Mulsant, 1854
hololeptoides Laporte, 1840: 220 (*Uloma*) **N**: MR **A**: TR
solieri Mulsant, 1854: 265

genus *Ulomoides* Blackburn, 1888: 274 type species *Ulomoides humeralis* Blackburn, 1888
Martianus Fairmaire, 1893h: 540 type species *Martianus castaneus* Fairmaire, 1893
Palembus Casey, 1891: 65 type species *Palembus ocularis* Casey, 1891
Tenebriomimus Kolbe, 1901: 342 type species *Tenebriomimus adansoniarum* Kolbe, 1901
castaneus Fairmaire, 1893h: 541 (*Martianus*) **A**: SA **AFR**
dermestoides Chevrolat, 1878a: 242 (*Histeropsis*) **E**: SVi **A**: FUJ GUA GUX HAI HUB JA JIA TAI "Korea" **AUR ORR**

tribe Gnathidiini Gebien, 1921

genus *Enanea* Lewis, 1894: 467 type species *Enanea testacea* Lewis, 1894
chujoi Ando & Mt. Chûjô, 2005: 72 **A**: JA (Amami-Ôshima)
testacea Lewis, 1894: 467 **A**: JA (Kyushu)
yakushimana Nakane, 1984: 597 **A**: JA (Yakushima)

genus *Menimus* Sharp, 1876: 73 type species *Menimus batesi* Sharp, 1876
Ceramba Fauvel, 1904: 206 type species *Ceramba hydrovatina* Fauvel, 1904
Microcilibe Carter, 1919: 147 type species *Microcilibe castaneum* Carter, 1919
Neomenimus Kaszab, 1939b: 190 type species *Neomenimus clavatus* Kasazab, 1939
Paita Fauvel, 1904: 173 type species *Paita setulosa* Fauvel, 1904
Tjikoraia Pic, 1921c: 18 type species *Tjikoraia javana* Pic, 1921
niponicus Lewis, 1894: 398 **A**: JA
wittmeri Kaszab, 1982a: 64 **A**: SD

genus *Sakaiomenimus* Ando, 2003c: 135 type species *Sakaiomenimus todai* Ando, 2003
todai Ando, 2003c: 139 **A**: JA

tribe Hypophlaeini Billberg, 1820

genus *Corticeus* Piller & Mitterpacher, 1783: 87 type species *Corticeus unicolor* Piller & Mitterpacher, 1783
subgenus *Corticeus* Piller & Mitterpacher, 1783: 87 type species *Corticeus unicolor* Piller & Mitterpacher, 1783
Hypophlaeus Fabricius, 1790: 222 type species *Ips bicolor* A. G. Olivier, 1790
Paraphloeus Seidlitz, 1894: 553 type species *Hypophloeus longulus* Gyllenhal, 1827
amamiensis Kaszab, 1964c: 48 (*Hypophloeus*) **A**: JA (Honshu, Ryukyus)
basalis Reitter, 1884a: 9 **E**: AB
becvari Bremer, 1999: 29 **A**: YUN
bicolor A. G. Olivier, 1790: 18: 12 (*Ips*) **E**: AL AR AU BE BH BU BY CR CT CZ DE EN FI FR GB GE GR HU IT LA LT NL NR NT PL RO SK SP ST SV SZ UK **N**: AG **A**: ES FE WS
bicoloroides Roubal, 1933b: 78 (*Hypophloeus*) **E**: AU CZ FR GE HU IT PL RO SK SZ
colydioides Lewis, 1894: 468 **A**: JA SC
engelhardtiae Blair, 1937: 36 **A**: SD TAI **ORR**
zoltani Masumoto, 1981e: 92 (*Hypophloeus*)
fasciatus Fabricius, 1790: 223 (*Hypophlaeus*) **E**: AL AU BY CR CZ DE EN FR GE GR HU IT PL RO SK SP ST SV SZ UK **N**: AG TU
pusillus Rossi, 1790a: 50 (*Ips*)
unicolor Ragusa, 1892b: 31 [HN]
fraxini Kugelann, 1794: 527 (*Hypophloeus*) **E**: AU BE BU BY CR CT CZ EN FI GB GE GR HU IT LA LT NL NR NT PL RO SK SP ST SV SZ UK **A**: CY ES FE KZ MG TR WS
ferrugineus Mulsant, 1854: 254 (*Hypophloeus*)
fusciventris Reitter, 1884e: 256 **E**: BH CR GR
haoi J.-K. Li, 1992: 165 (*Hypophloues*) **A**: LIA
krueperi Obenberger, 1922: 42 (*Hypophloues*) **E**: GR
leonhardi Reitter, 1906f: 241 (*Hypophloeus*) **E**: SP
linearis Fabricius, 1790: 223 (*Hypophlaeus*) **E**: AU BE BY CT CZ DE EN FI FR GB GE GR HU IT LA LT NL NR NT PL RO SK SP ST SV SZ **N**: TU **A**: ES FE WS
perrisi Pic, 1917g: 5 (*Hypophloeus*)

longicornis Wollaston, 1867: 206 **A**: YE **AFR**
longulus Gyllenhal, 1827: 517 (*Hypophloeus*) **E**: AL AU BY CR CT CZ EN FI FR GE GR HU LA LT NR NT PL RO SK SL SP ST SV SZ UK YU **A**: ES FE MG WS
maehleri mayumiae Masumoto, 1983b: 175 (*Hypophloeus*) **A**: JA (Ryukyus)
nepalicus Bremer, 1999: 27 **A**: BT NP
okuezonis Kôno, 1939: 85 (*Hypophloeus*) **A**: FE JA
pini Panzer, 1799b: 19 (*Hypophloeus*) **E**: AU BE BY CR CT CZ EN FR GE GR HU IT LS PL PT RO SK SP ST SZ UK **N**: CI TU **A**: ES FE IS LE MG SY TR
ferrugineus Creutzer, 1799: 126 (*Hypophloeus*)
leonardi Reitter, 1906f: 241 (*Hypophlaeus*)
nocivus Wollaston, 1862: 442 (*Hypophloeus*)
praenuntius Bremer, 1999: 23 **A**: HP NP PA UP
rufithorax Pic, 1903m: 162 **E**: GR SP
rufobasalis Pic, 1917g: 5 (*Hypophloeus*) **A**: TR
stebbingi Blair, 1921: 5 (*Hypophloeus*) **A**: HP UP
suberis Lucas, 1846: 345 (*Hypophloeus*) **E**: AU CR CZ FR GE HU IT PL RO SK ST **N**: AG **A**: CY IS TR
insidiosus Mulsant, 1854: 384 (*Hypophloeus*)
rufulus Rosenhauer, 1847: 32 (*Hypophloeus*)
suturalis Paykull, 1800: 323 (*Hypophlaeus*) **E**: BY CT CZ EN FI LA LT NR NT PL RO SK ST SV **A**: ES FE JA
bivittatus Reitter, 1875: 362 (*Hypophloeus*)
ulomoides Blair, 1921: 4 (*Hypophloeus*) **A**: KA PA
unicolor Piller & Mitterpacher, 1783: 87 **E**: AR AU BE BU BY CR CT CZ DE EN FR GB GE GR HU IT LA LS LT MA NL NR PL RO SK SP ST SV SZ UK **A**: IS TR
castaneus Fabricius, 1790: 223 (*Hypophlaeus*)
nitidus Melsheimer, 1846: 62 (*Hypophloeus*)
pseudocastaneus Reitter, 1906f: 242 (*Hypophlaeus*)
purkynei Roubal, 1932b: 130 (*Hypophloeus*)
taxicornis Rossi, 1790a: 49 (*Ips*)
versipellis Baudi di Selve, 1876c: 234 (*Hypophloeus*) **E**: AU CZ GE HU IT PL RO SK SZ
hopffgarteni Reitter, 1877a: 26
subgenus *Stenophloeus* Blair, 1921: 1 type species *Hypophloeus filum* Fairmaire, 1893
Cnemophloeus Bremer, 1998: 10 type species *Hypophloeus filum* Fairmaire, 1893
analis Gebien, 1914d: 395 (*Hypophloeus*) **A**: JA (Ryukyus) TAI **ORR**
andamanus Pic, 1945: 7 (*Hypophloeus*)
attenuicollis Pic, 1946: 3 (*Hypophlaeus*)
borneensis Pic, 1914d: 16 (*Hypophloeus*)
javanus Pic, 1946: 3 (*Hypophlaeus*)
obscurior Pic, 1946: 3 (*Hypophlaeus*)
cephalotes Gebien, 1914b: 28 (*Hypophloeus*) **A**: AP UP TAI **ORR**
andaiensis Pic, 1946: 3 (*Hypophlaeus*)
cornutus Pic, 1914d: 16 (*Hypophloeus*)
palawanus Pic, 1945: 4 (*Hypophloeus*)
subcastaneus Pic, 1945: 8 (*Hypophloeus*)
filum Fairmaire, 1893: 29 (*Hypophloeus*) **A**: CH NP UP TAI **AFR AUR ORR**
niasensis Pic, 1914d: 16 (*Hypophloeus*) **A**: TAI **ORR**
similis Bremer, 1999: 45 **A**: TAI **ORR**
subgenus *Tylophloeus* Bremer, 1998: 10 type species *Hypophloeus flavipennis* Motschulsky, 1860
curtithorax Pic, 1924b: 14 (*Hypophlaeus*) **A**: GUA HAI NP SD YUN **ORR**
beesoni Blair, 1930a: 146 (*Hypophloeus*)
dorsalis Blair, 1930a: 145 (*Hypophloeus*) **A**: UP
flavipennis Motschulsky, 1860a: 99 (*Hypophlaeus*) **A**: AP HP NP PA UP YUN **ORR**
brunneicollis Pic, 1924a: 29 (*Hypophloeus*)
gentilis Lewis, 1894: 468 **A**: HKG JA NP TAI YUN **ORR**
robustus Gebien, 1914b: 29 (*Hypophloeus*)

tribe Leiochrinini Lewis, 1894

genus *Ades* Guérin-Méneville, 1857: 277 type species *Ades hemisphericus* Guérin-Méneville, 1857
Leiochrodes Westwood, 1883: 69 type species *Leiochrodes discoidalis* Westwood, 1883
Leiochrota Westwood, 1883: 70 type species *Leiochrinus uniformis* Westwood, 1883
assimilis Kaszab, 1961i: 461 (*Leiochrodes*) **A**: NP SD
bengalicus Kaszab, 1975a: 36 (*Leiochrodes*) **A**: SD
convexus Lewis, 1894: 391 (*Leiochrodes*) **A**: JA SD TAI **ORR**
subovatus Pic, 1921b: 8 (*Leiochrodes*)
coomani Pic, 1927b: 14 (*Leichrodes*) **A**: SCH SD **ORR**
diaphanus Fabricius, 1798: 76 (*Coccinella*) **A**: NP UP **ORR**
emeicus Schawaller, 1993b: 357 (*Leiochrodes*) **A**: SCH
formosanus Kaszab, 1946: 157 (*Leiochrodes*) **A**: GUI SCH TAI
glabratus Walker, 1859b: 219 (*Lycoperdina*) **A**: BT FUJ SD TAI **ORR**
obscuricollis Fairmaire, 1893b: 25 (*Liochrinus*)
tibialis Gebien, 1914b: 23 (*Leiochrodes*)
tonkineus Pic, 1922h: 100 (*Leiochrodes*)
himalayensis Kaszab, 1946: 158 (*Leiochrodes*) **A**: BT GUA HP **ORR**
lanceolatus Kaszab, 1961i: 456 (*Leiochrodes*) **A**: NP SCH YUN ZHE
masidai Nakane, 1963a: 26 (*Leiochrodes*) **A**: JA
nigronotatus Pic, 1934f: 84 (*Leiochrodes*) **A**: SCH ZHE
politus Kaszab, 1961i: 455 (*Leiochrodes*) **A**: YUN
rufofulvus Westwood, 1883: 71 (*Leiochrinus*) **A**: SD **ORR**
sheduwicus Schawaller, 1992: 9 (*Leiochrodes*) **A**: NP
sikkimensis Kaszab, 1961i: 464 (*Leiochrodes*) **A**: NP SD
subaeneus Pic, 1918a: 14 (*Leiochrodes*) **A**: SD **ORR**
taiwanus Masumoto, 1981d: 25 (*Leiochrodes*) **A**: TAI
testaceicollis Kaszab, 1946: 155 (*Leiochrodes*) **A**: HUN TAI

genus *Crypsis* Waterhouse, 1877: 73 type species *Crypsis violaceipennis* Waterhouse, 1877
Leiochrotina Westwood, 1883: 76 type species *Leiochrinus indicus* Westwood, 1883
bimaculatus Kaszab, 1946: 191 **A**: NP SD
blairi Kaszab, 1946: 192 **A**: NP SD
chinensis Kaszab, 1946: 196 **A**: FUJ GUX
gebieni Kaszab, 1946: 195 **A**: TAI
minutus Kaszab, 1946: 197 **A**: NP SD **ORR**
purpuripennis Neervoort van de Poll, 1886: 34 (*Leiochrotina*) **A**: SD **ORR**
ruficollis Kaszab, 1946: 193 **A**: SD **ORR**
rufomarginatus Kaszab, 1946: 195 **A**: BT NP SD
migliaccioi J. Ferrer, 1983: 140
szekessyi Kaszab, 1946: 194 **A**: GUI
violaceipennis Waterhouse, 1877: 74 **A**: SD UP **ORR**
braeti Pic, 1918a: 15 (*Leiochrotina*)
vitalisi Kaszab, 1961h: 375 **A**: YUN **ORR**
wrasei Schawaller, 2005: 222 **A**: HUB SCH SHA
yunnanus Kaszab, 1961h: 376 **A**: YUN **ORR**
genus *Derispia* Lewis, 1894: 389 type species *Diaperis maculipennis* Marseul, 1876
aeneonigra Kaszab, 1946: 65 **A**: SD **ORR**
akitai Masumoto, 1994a: 42 **A**: JA (Ryukyus)
amamiana Miyatake, 1961: 37 **A**: JA (Amami-Ôshima)
bisexnotata Kaszab, 1961b: 177 **A**: GUX HKG **ORR**
bisquinquemaculata Blair, 1937: 36 **A**: SD
bistrimaculata bistrimaculata Blair, 1937: 36 **A**: NP SD
bistrimaculata championi Kaszab, 1961b: 178 **A**: BT UP
bhutanensis Kaszab, 1975e: 323 **A**: BT
chinensis Kaszab, 1946: 108 **A**: GUI HKG YUN

confluens Kaszab, 1946: 112 **A**: NP SD **ORR**
cooteri Schawaller, 2005: 216 **A**: JIX
diaperoides Kaszab, 1946: 105 **A**: NP SD
flavicornis Kaszab, 1946: 79 **A**: SD
formosana Masumoto, 1981d: 20 **A**: TAI
fukiensis Kaszab, 1961b: 181 **A**: FUJ GUA ZHE
indica Kaszab, 1946: 79 **A**: BT NP SD **ORR**
janetscheki Kaszab, 1970c: 432 **A**: NP
japonica japonica Kaszab, 1961b: 181 **A**: JA (Okinawa)
japonica yaeyamana Mt. Chûjô, 1997b: 200 **A**: JA (Ryukyus)
japonicola Miyatake, 1961: 36 **A**: JA
jizushanica Schawaller, 2005: 218 **A**: YUN
klapperichi Kaszab, 1942c: 216 **A**: FUJ SCH
kondoi Masumoto, 1981d: 23 **A**: TAI
kryzhanovskii Kaszab, 1961b: 182 **A**: YUN
lineolata Pic, 1922f: 23 (*Leiochrodes*) **A**: FUJ GUA HUN SCH **ORR**
maculipennis Marseul, 1876c: 105 (*Diaperis*) **A**: JA FUJ GUA GUX HUN SCH SHA
  klapperichiana Kaszab, 1954: 253
melli Kaszab, 1946: 106 **A**: GUA
miyatakei Masumoto, 1994a: 42 **A**: JA (Okinawa-Honto)
nanshanchiensis Masumoto, 1981d: 21 **A**: TAI
notata Kaszab, 1946: 81 **A**: NP **ORR**
ocellata Kaszab, 1946: 109 **A**: SD "Himalaya" **ORR**
sauteri Kaszab, 1946: 108 **A**: TAI
shibatai M. Chûjô & Miyatake, 1961: 38 **A**: JA (Amami-Ôshima)
sichuanensis Schawaller, 1993b: 354 **A**: SCH
sikkimensis Kaszab, 1946: 103 **A**: SD
similis Kaszab, 1961b: 176 **A**: YUN
tenuipunctata Kaszab, 1946: 105 **A**: SD
titschacki Kaszab, 1946: 101 **A**: GUA HKG JIA SHG **ORR**
tricolor Kaszab, 1942c: 216 **A**: FUJ GUA
truncata Kaszab, 1961b: 152 **A**: BT NP SD
undulata Pic, 1922h: 101 (*Leiochrodes*) **A**: YUN
walkeri Kaszab, 1961b: 180 **A**: HKG
wittmeri Kaszab, 1975e: 321 **A**: BT NP
zoltani Masumoto, 1981d: 22 **A**: TAI

genus *Derispiella* Kaszab, 1961h: 364 type species *Derispiella hingstoni* Kaszab, 1961
bhutanensis Kaszab, 1975e: 324 **A**: BT
hingstoni Kaszab, 1961h: 364 **A**: BT SD

genus *Derispiola* Kaszab, 1946: 115 type species *Derispiola fruhstorferi* Kaszab, 1946
blairi Kaszab, 1946: 119 **A**: NP SD YUN **ORR**
darjeelingiana Kaszab, 1946: 117 **A**: NP SD
fruhstorferi Kaszab, 1946: 118 **A**: SCH **ORR**
unicornis Kaszab, 1946: 116 **A**: FUJ GUA GUX HUB HUN JIX SCH ZHE **ORR**

genus *Leiochrinus* Westwood, 1883: 68 type species *Leiochrinus fulvicollis* Westwood, 1883
bifurcatus Kaszab, 1946: 132 **A**: SHG "North India" **ORR**
lutescens Westwood, 1883: 71 **A**: NC **ORR**
metallicus Schawaller, 2005: 221 **A**: SD
satzumae Lewis, 1894: 390 **A**: JA **ORR**
sauteri minor Kaszab, 1961h: 371 **A**: HP UP
sauteri sauteri Kaszab, 1946: 132 **A**: GUI NP SD TAI "South China" **ORR**

genus *Leiochrodontes* Kaszab, 1946: 200 type species *Leiochrodes madurensis* Pic, 1918
madurensis Pic, 1918a: 15 (*Leiochrodes*) **A**: SD **ORR**

genus *Pimplema* Pascoe, 1887: 16 type species *Pimplema ampliata* Pascoe, 1887
Hades J. Thomson, 1860: 13 [HN] type species *Hades tenebrosus* Thomson, 1860
hemisphaerica Laporte & Brullé, 1831: 367 (*Platydema*) **A**: NP SD **ORR**
ampliata Pascoe, 1887: 17
rufolimbatus Fairmaire, 1882c: 257 (*Hades*)
tenebrosus J. Thomson, 1860: 13 (*Hades*)

genus *Stethotrypes* Gebien, 1914: 26 type species *Stethotrypes bicornutus* Gebien, 1914
Leichrodomorphus Pic, 1921b: 6 type species not designated *Leichrodomorphus brevicornis* Pic, 1921
korschefskyana Kaszab, 1942c: 215 (*Derispia*) **A**: FUJ

tribe Myrmechixenini Jacquelin du Val, 1858

genus *Myrmechixenus* Chevrolat, 1835b: 267 type species *Myrmechixenus subteranneus* Chevrolat, 1835
Myrmecoxenus Agassiz, 1846b: 243 [unjustified emendation]
picinus Aubé, 1850: 330 (*Myrmechoxenus*) **E**: FR IT MA SK **N**: AG EG MR
subterraneus Chevrolat, 1835b: 268 **E**: AU BY CZ DE FI FR GE HU IT LS NR NT PL SK SV SZ
vaporariorum Guérin-Méneville, 1843a: 70 **E**: AU CZ DE FI FR GE HU IT PL SK SV SZ **AFR ORR**

tribe Phaleriini Blanchard, 1845

genus *Emypsara* Pascoe, 1866: 460 type species *Emypsara adamsii* Pascoe, 1866 (= *Diaperis riederii* Faldermann, 1833)
Callicomus Motschulsky, 1860c: 138 [HN] type species *Diaperis riederii* Faldermann, 1833
Obenbergeria E. Strand, 1929: 24 [RN]
riederii Faldermann, 1833: 57 (*Diaperis*) **A**: FE JA NE SC "Korea"
adamsii Pascoe, 1866: 461 (*Emypsara*)
flexuosus Pascoe, 1866: 461 (*Emypsara*)
hilgendorfi Harold, 1878a: 76 (*Phaleria*)

genus *Halammobia* Semenov, 1901a: 91 type species *Tenebrio pellucidus* Herbst, 1799
pellucida Herbst, 1799: 19 (*Tenebrio*) **E**: FR IT
hemisphaerica Küster, 1852b: 67 (*Phaleria*)
pallens Latreille, 1802: 162 (*Phaleria*)

genus *Paranemia* Heyden, 1892a: 103 type species *Paranemia schroederi* Heyden, 1892
Taclamacanius J. Ferrer & Yvinec, 2005: 121 [RN] type species *Taklamakania lepetzi* J. Ferrer & Yvinec, 2004
Taklamakania J. Ferrer & Yvinec, 2004: 42 [HN] type species *Taklamakania lepetzi* J. Ferrer & Yvinec, 2004
bicolor Reitter, 1895d: 157 **A**: ES (Tuva) MG NMO XIN
lepetzi J. Ferrer & Yvinec, 2004: 42 (*Taklamakania*) **A**: XIN
schroederi Heyden, 1892a: 103 **E**: AB **A**: KZ TM UZ
argiropuloi Bogatchev, 1965: 372
argyropuloi Bogatchev, 1967a: 159

genus *Phaleria* Latreille, 1802: 162 type species *Tenebrio cadaverinus* Fabricius, 1792
subgenus *Epiphaleria* Lewis, 1894: 382 type species *Epiphaleria atriceps* Lewis, 1894
atriceps Lewis, 1894: 384 (*Epiphaleria*) **A**: JA TAI **ORR**
prolixa Fairmaire, 1869a: 797 **N**: EG **A**: AE BA IN KU OM SA SI YE **AFR**
aegyptiaca Seidlitz, 1894: 482
fuscata Fairmaire, 1886a: 437
munda Walker, 1871: 16
pusilla Boheman, 1858: 93 **A**: HKG
subgenus *Eremophaleria* Español, 1951a: 32 type species *Phaleria bedeli* Chobaut, 1900
bedeli Chobaut, 1900b: 340 **N**: AG

subgenus *Phaleria* Latreille, 1802: 162 type species *Tenebrio cadaverina* Fabricius, 1792
Atahualpina Español, 1960b: 113 type species *Atahualpina peruviana* Español, 1960
Halophalerus Crotch, 1873: 107 [unnecessary RN]
Phalerisida Kulzer, 1959: 563 type species *Phalerisida maculata* Kulzer, 1959
Sepedonastes Gistel, 1856: 382 type species *Sepedonastes bimaculatus* Herbst, 1799 (= *Dytiscus bimaculatus* Linnaeus, 1767)
acuminata acuminata Küster, 1852b: 69 **E**: CR FR GR IT MA SP **N**: AG EG LB MO TU **A**: IS LE TR
incerta Rey, 1891a: 86
limbata Baudi di Selve, 1876c: 226
maculosa Seidlitz, 1894: 481
seductor Rey, 1891a: 85
submaculata Rey, 1891a: 85
tunisia Reitter, 1916b: 8
acuminata nigriceps Mulsant & Wachanru, 1853: 10 **A**: TR
acuminata syriaca Reitter, 1916b: 7 **N**: EG **A**: IS LE
atlantica Fauvel, 1899: 77 [RN] **E**: FR PT SP **N**: MR
divisa Pic, 1923a: 3
bimaculata adriatica Rey, 1891a: 86 **E**: CR GR (Dodecanese) IT **A**: TR
bimaculata bimaculata Linnaeus, 1767: 667 (*Dytiscus*) **E**: AL BU FR GR IT MA RO SP UK **N**: EG LB TU **A**: TR
abdominalis Seidlitz, 1894: 484
bisignata Rey, 1892: 53
calabrica Canzoneri, 1961b: 25
concii Canzoneri, 1961b: 25
conicicollis Reitter, 1918: 154
cordata O. Schneider, 1902: 10
delata Rey, 1891a: 85
dorsalis Fabricius, 1792a: 85 (*Byrrhus*)
dorsigera Fabricius, 1787a: 201 (*Carabus*)
giordanii Canzoneri, 1960a: 44
irregularis Pic, 1923a: 3
italica Reitter, 1916b: 8
kochi Canzoneri, 1961b: 26
ligurica O. Schneider, 1902: 10
longicollis Rey, 1892: 53
nigricollis J. R. Sahlberg, 1903c: 3
nigricordis O. Schneider, 1902: 11
obscuriceps O. Schneider, 1902: 11
ornata O. Schneider, 1902: 11
schatzmayri Canzoneri, 1961b: 25
signata O. Schneider, 1902: 11
bimaculata jonica Canzoneri, 1968: 155 **E**: IT (Calabria)
bimaculata marcuzzii Aliquò, 1993: 113 **E**: IT (Isole Eolie: Vulcano)
cadaverina cadaverina Fabricius, 1792a: 113 (*Tenebrio*) **E**: AZ BE DE FR GB GE IR NL PL PT SP SV **N**: CI
apicalis O. Schneider, 1902: 12
cava Herbst, 1799: 15 (*Tenebrio*)
fuscata O. Schneider, 1902: 12
leegei O. Schneider, 1902: 12
cadaverina maroccana Pic, 1923a: 3 **N**: CI MO
ciliata Wollaston, 1854: 488 **N**: MR
insulana Rey, 1890: 329 **E**: FR IT (Sardegna)
ornata Wollaston, 1864: 494 **N**: CI
biplagiata Heyden, 1909: 486
boeckeri Heyden, 1909: 486
giebeleri Heyden, 1909: 486
nigrothoracica Heyden, 1909: 486
picta Wollaston, 1861a: 246 [HN]
perissinottoi Canzoneri, 1973: 41 **A**: TR
pontica Semenov, 1901a: 90 **E**: BU GG RO ST UK

provincialis cypria Grimm, 1991: 30 **A**: CY
provincialis ghidinii Canzoneri, 1961c: 61 **E**: IT
provincialis intermedia Schuster, 1930a: 65 **E**: IT
provincialis provincialis Fauvel, 1901: 253 **E**: FR SP
pueli Pic, 1902h: 64
pujeti Español, 1951a: 27 [= 1954a: 65] **E**: SP (Baleares)
revellierii Mulsant & Rey, 1858b: 310 [= 1859a: 52] **E**: FR IT SP **N**: AG TU
beloni Pic, 1923a: 3
cinctella Rey, 1891a: 84
circumducta Rey, 1891b: 238
heydenii Seidlitz, 1894: 482
lineolata Rey, 1891b: 238
oblonga Küster, 1852b: 70
reyi Seidlitz, 1894: 482
seidlitzi O. Schneider, 1902: 11
sublaevicollis Rey, 1891b: 236

genus *Phaleromela* Reitter, 1916b: 4 type species *Phaleria subhumeralis* Marseul, 1876
humeralis Laporte, 1840: 219 (*Phaleria*) **A**: NMO **NAR**
subhumeralis Marseul, 1876c: 102 (*Phaleria*) **A**: FE JA SC

genus *Phtora* Germar, 1836c: 11 type species *Phtora crenata* Germar, 1836
subgenus *Clypeophtora* F. Soldati & L. Soldati, 2003a: 4 type species *Phtora tronqueti* F. Soldati & L. Soldati, 2003
tronqueti F. Soldati & L. Soldati, 2003a: 4 **E**: SP
subgenus *Phtora* Germar, 1836c: 11 type species *Phtora crenata* Germar, 1836
Cataphronetis Lucas, 1846: 342 type species *Cataphronetis levaillantii* Lucas, 1846 (= *Phtora crenata* Germar, 1836)
Pseudostene Wollaston, 1861a: 247 type species *Pseudostene angusta* Wollaston, 1861
aegyptiaca Reitter, 1899c: 159 (*Phthora*) **N**: EG LB
alashanensis Ren & Zheng, 1993e: 77 **A**: NMO
[*alanshanensis* Ren & Zheng, 1993e: 77]
angusta Wollaston, 1861a: 249 (*Pseudostene*) **E**: SP **N**: AG CI MO **AFR**
fossoria Wollaston, 1861a: 250 (*Pseudostene*)
apicilaevis Marseul, 1876b: 36 (*Cataphronetis*) **N**: EG LB TU **A**: SI
bicolor Reitter, 1899c: 159 (*Phthora*) **N**: AG TU
confluens Reitter, 1894c: 104 **E**: PT SP **N**: LB TU
crenata Germar, 1836c: 11 **E**: FR IT MA SP **N**: AG LB MO TU
attenuata Motschulsky, 1849: 121 (*Cataphronetes*)
brunnea Mulsant, 1854: 241 (*Cataphronetis*)
levaillanti Lucas, 1846: 342 (*Cataphronetis*)
hauseriana Reitter, 1895c: 149 (*Cataphronetis*) **E**: "Caucasus" **A**: TM UZ
kandaharica Kaszab, 1960a: 168 **A**: AF
millingeni Reitter, 1899c: 159 (*Phthora*) **A**: IQ
plagiocnema Iablokoff-Khnzorian, 1956b: 41 (*Cataphronetis*) **E**: AR
priva G. S. Medvedev, 2004b: 577 **A**: TM
prolixa Fairmaire, 1879c: 198 (*Cataphronetis*) **N**: AG TU
quadricollis Reitter, 1895c: 148 (*Cataphronetis*) **E**: ST **A**: AF KI KZ MG TM UZ XIN
reitteri cypria Grimm, 1991: 34 **A**: CY
reitteri reitteri Seidlitz, 1894: 572 (*Cataphronetis*) **E**: GR UK **A**: TR
hellenica Reitter, 1895c: 148 (*Cataphronetis*)
subclavata Wollaston, 1861a: 250 (*Pseudostene*) **N**: EG **A**: IS SI SA
tenuis Marseul, 1876b: 36 (*Cataphronetis*)
tenuicornis Reitter, 1895c: 148 (*Cataphronetis*) **E**: ST **A**: AF KZ TM

tribe Scaphidemini Reitter, 1922

genus *Basanus* Lacordaire, 1859: 306 type species *Basanus javanus* Chevrolat, 1878
amamianus M. Chûjô, 1966: 13 **A**: JA (Ryukyu, Yakushima) SC (Cheju-do)
annamitus Gebien, 1925a: 147 **A**: FUJ SHA ZHE **ORR**
erotyloides Lewis, 1891a: 71 **A**: JA
flaviventris Blair, 1937: 35 **A**: SD **ORR**
formosanus Masumoto, 1982d: 68 **A**: TAI
fukudai Nakane, 1968b: 77 **A**: JA (Ryukyus)
himalayanus Kaszab, 1965a: 119 **A**: AP
lui Masumoto & Merkl, 2003: 169 **A**: TAI
manangensis Schawaller, 1995b: 332 **A**: NP
tsushimensis kompancevi Kaszab & G. S. Medvedev, 1984: 84 **A**: FE "Korea"
tsushimensis tsushimensis M. Chûjô, 1963a: 17 **A**: JA SC

genus *Pseudoscaphidema* Pic, 1926a: 2 type species *Pseudoscaphidema rufonotata* Pic, 1926
rufonotata Pic, 1926a: 3 **A**: FE (Kamtchatka)

genus *Scaphidema* L. Redtenbacher, 1849: 591 type species *Mycetophagus metallicus* Fabricius, 1792
Microbasanus Pic, 1921a: 1 type species *Microbasanum jureceki* Pic, 1921
Nelites J. L. LeConte, 1850: 232 type species *Nelites aeneolum* J. L. LeConte, 1850
aikoae M. Chûjô, 1983: 47 **A**: JA (Yakushima)
angustatum Pic, 1935a: 25 **A**: CH
discale Lewis, 1894: 397 **A**: JA
formosanum Masumoto, 1982a: 150 **A**: TAI
haruhiae M. Chûjô, 1985: 63 **A**: JA (Tsushima Is.)
hiranoi Masumoto & Akita, 2001b: 75 **A**: JA (Okinawa)
insularis G. S. Medvedev & Kompantseva, 1989: 100 **A**: FE (Sachalin)
jureceki Pic, 1921a: 1 (*Microbasanus*) **A**: FE HEI
khnzoriani Kaszab, 1975c: 127
kayokoae Mt. Chûjô, 1992: 36 **A**: SC (Chejodo Is.)
kondoi Nakane, 1968a: 76 **A**: JA (Amami-Ôshima)
metallicum bicolor Fabricius, 1798: 179 (*Scaphidium*) **E**: HU UK
metallicum metallicum Fabricius, 1792b: 499 (*Mycetophagus*) **E**: AR AU BE BH BY CT CZ DE EN FI FR GB GE HU IT LA LS LT NL NR NT PL SK SP ST SV SZ UK YU **A**: WS
aeneuum Panzer, 1793b: 2 (*Scolytus*)
aheneum Marsham, 1802: 176 (*Chrysomela*)
sulcipenne Petri, 1926: 194
michihidei Mt. Chûjô & Lee, 1993: 111 **A**: SC
nigricorne Lewis, 1894: 397 **A**: JA (Honshu)
ornatellum Lewis, 1894: 396 **A**: JA
pictipenne Lewis, 1894: 397 **A**: JA
rousi Picka, 1983a: 262 **E**: GG
schmidti Schawaller, 2003a: 2 **A**: NP
sichuanum Schawaller, 2003a: 3 **A**: SCH
trimaculatum M. Chûjô, 1968b: 393 **A**: TAI

genus *Spiloscapha* Bates, 1873b: 202 type species *Spiloscapha crassicornis* Bates, 1873 (= *Platydema thallioides* Pascoe, 1869)
bipunctata Schawaller, 1997: 4 **A**: YUN **ORR**
cooteri Schawaller, 2004a: 3 **A**: JIX
hiraii Masumoto & Akita, 2001b: 76 **A**: JA (Okinawa)
ichihashii Nakane, 1956b: 166 (*Platydema*) **A**: JA
elegans M. Chûjô, 1968c: 398 (*Basanus*)
kobayashii Shibata, 1978: 19 **A**: TAI
nepalica Schawaller, 1997: 7 **A**: NP **ORR**

taiwana Masumoto & Merkl, 2003: 169 **A**: TAI
unicolor Blair, 1937: 35 **A**: SD

tribe Trachyscelini Blanchard, 1845

genus *Trachyscelis* Latreille, 1809: 379 type species *Trachyscelis aphodioides* Latreille, 1809
aphodioides aphodioides Latreille, 1809: 379 **E**: AL AZ BU CR FR GG GR IT MA MC SP ST UK **N**: CI EG LB MO TU **A**: CY KZ TR **AFR**
lopadusae Koch, 1935a: 98
subopacus Pic, 1923a: 1 (*Tachyscelis*)
chinensis Champion, 1894b: 448 **A**: GUA (Nam'ao Is.) JA (Ryukyus)
sabuleti Lewis, 1894: 383 **A**: JA SC
tenuestriatus Fairmaire, 1886a: 449 **A**: JO SA SI YE **AFR**

subfamily **Alleculinae** Laporte, 1840
tribe Alleculini Laporte, 1840
subtribe Alleculina Laporte, 1840

genus *Allecula* Fabricius, 1801b: 21 type species *Allecula morio* Fabricius, 1801
subgenus *Allecula* Fabricius, 1801b: 21 type species *Allecula morio* Fabricius, 1801
aeneipennis Harold, 1878a: 80 **A**: JA
akitai Hanatsuka, Masumoto & Kon, 2006: 179 **A**: JA
amamiensis amamiensis Maeda & Nakane, 1988: 1 **A**: JA (Amami-Ôshima)
amamiensis okinawaensis Maeda & Nakane, 1988: 2 **A**: JA (Okinawa)
angusta Borchmann, 1941: 26 **A**: FUJ
angustatissima Pic, 1930b: 29 **A**: CH
arthritica Fairmaire, 1896a: 36 **A**: SD **ORR**
balthasari Mařan, 1940c: 170 **A**: NO
bicoloripes Pic, 1920a: 23 **A**: FUJ
bilamellata Marseul, 1876d: 323 **A**: CH FE JA TAI
cincta Pic, 1929b: 22 [RN] **A**: SW YUN
cinctipennis Pic, 1915b: 15 [HN]
cinctipennis Pic, 1909a: 19 **A**: SW TAI YUN
coreana Kolbe, 1886: 211 **A**: SC
cryptomeriae Lewis, 1895a: 250 **A**: JA
densaticollis Fairmaire, 1891f: ccxv **A**: HEB HUB
divisa Reitter, 1883a: 115 **E**: AR "Caucasus" **A**: TM UZ
elongatissima Pic, 1926d: 30 **A**: YUN
endemica Muche, 1979b: 269 **A**: PA
estriata Seidlitz, 1896b: 37 **A**: TR
femorata Borchmann, 1915b: 183 **A**: SD **ORR**
formosana Pic, 1910u: 94 **A**: TAI
guerryi Pic, 1934b: 26 **A**: YUN
himalaica Blair, 1927: 248 **A**: XIZ
impunctaticollis Borchmann, 1932: 91 **A**: SCH
indica Borchmann, 1909b: 714 [RN] **A**: SD **ORR**
brachydera Fairmaire, 1896a: 38 [HN]
ishigakiensis Maeda & Nakane, 1988: 3 **A**: JA
klapperichi Pic, 1955c: 31 **A**: FUJ
kuatunensis Pic, 1955c: 31 **A**: FUJ
kurinoensis Maeda & Nakane, 1988: 5 **A**: JA
longeimpressa Pic, 1955c: 30 **A**: FUJ
mausonensis Borchmann, 1938a: 296 **A**: TAI **ORR**
maxima Pic, 1910u: 94 **A**: FUJ SW TAI YUN
metallicipennis Pic, 1926d: 29 **A**: YUN
mongolica Muche, 1972b: 223 **A**: MG

morio Fabricius, 1787a: 86 (*Cistela*) **E**: AU BE BH CR CZ DE FI FR GE HU IT LA NL PL RO SK SL SV SZ UK
opaca Illiger, 1794: 610 (*Cistela*)
rufipes Fabricius, 1792b: 44 (*Cistela*)
semilivida Pic, 1891a: 51
moupinea Fairmaire, 1891f: ccxv **A**: SCH
nipponica Miyatake, 1985: 347 **A**: JA
nitidipennis Borchmann, 1932: 93 **A**: SCH
noctivaga Lewis, 1895a: 251 **A**: JA "Korea"
ohshimana Maeda & Nakane, 1988: 4 **A**: JA
ontakensis Maeda & Nakane, 1988: 2 **A**: JA
oronthea Baudi di Selve, 1881: 292 **A**: LE TR
paulonotata Pic, 1936a: 29 **A**: JA
platicera Dubrovin, 1992b: 513 **A**: FE NW
rhenana Bach, 1856: 228 **E**: AU BH CR CZ DE FR GE GG IT PL RO RU SK SL ST SV SZ UK YU
loevendali Reitter, 1886c: 140
satsumae Fairmaire, 1896b: 116 **A**: JA
savioi Pic, 1926l: 302 **A**: CH
sculpturata Borchmann, 1932: 92 **A**: SCH
semeti Pic, 1909l: 142 **A**: SW YUN
seminitens Blair, 1927: 249 **A**: XIZ
shibatai Nomura, 1964: 49 **A**: JA
simiola Lewis, 1895a: 251 **A**: JA
sobrina Borchmann, 1915b: 186 **A**: SD
striatipennis Pic, 1924b: 7 **A**: CH
subbinotata Pic, 1934c: 23 **A**: CH
tenuis Marseul, 1876d: 326 **A**: JA
touzalini Pic, 1936a: 30 **A**: YUN
tschungseni Pic, 1955c: 31 **A**: FUJ
umbilicata Seidlitz, 1896b: 38 **A**: CH JA TAI
subgenus *Dietopsis* Solier, 1840: 243 type species *Allecula solieri* Laporte, 1840
costipennis Bates, 1879: 482 **A**: PA
subgenus *Upinella* Mulsant, 1856b: 17 type species *Allecula aterrima* Rosenhauer, 1847
aterrima Rosenhauer, 1847: 122 (*Allecula*) **E**: AR AU BH BU CR GR HU IT RO RU ST UK YU
frankenbergeri Mařan, 1940c: 168 **A**: FUJ
fuliginosa Mäklin, 1875b: 666 (*Allecula*) **A**: GAN JA SC TAI
obscura Harold, 1878b: 132 (*Allecula*)
velutina Marseul, 1876d: 322 (*Allecula*)
melanaria Mäklin, 1875b: 669 (*Allecula*) **A**: FUJ JA SC
rufipes Marseul, 1876d: 322 (*Allecula*)
ussuriensis Borchmann, 1937: 210 **A**: FE ES NE NC SC
mandshurica Mařan, 1940c: 169 (*Allecula*)

genus *Anthracula* Fairmaire, 1896d: 236 type species *Anthracula latifrons* Fairmaire, 1896
latifrons Fairmaire, 1896d: 236 **A**: HP

genus *Balassogloa* Semenov, 1891a: 372 type species *Balassogloa sphenaroides* Semenov, 1891
minor Semenov, 1891a: 373 **A**: TM
sphenarioides Semenov, *1891a: 372* **A**: TM

genus *Bolbostetha* Fairmaire, 1896b: 117 type species *Bolbostetha soleata* Fairmaire, 1896
Alleculodes Borchmann, 1925: 335 type species *Alleculodes discrepans* Borchmann, 1925
maxima Pic, 1910u: 94 (*Allecula*) **A**: FUJ SW TAI YUN
opaca Borchmann, 1925: 342 (*Alleculodes*) **A**: HKG
proavia Borchmann, 1925: 343 (*Alleculodes*) **A**: HKG

sauteri oshimana Nakane, 1968c: 85 (*Alleculodes*) **A**: JA (Ryukyus)
sauteri sauteri Borchmann, 1925: 341 (*Alleculodes*) **A**: JA TAI

genus *Borboresthes* Fairmaire, 1897k: 253 type species *Allecula cruralis* Marseul, 1876
acicularis Marseul, 1876d: 325 (*Allecula*) **A**: JA SC
amamianus Maeda & Nakane, 1988: 5 **A**: JA
bertrandi Pic, 1934b: 20 **A**: "Himalaya"
borchmanni Novák, **nom. nov.** [see New Acts] **A**: FUJ
tibialis Borchmann, 1941: 27 [HN]
castaneus Pic, 1928c: 15 **A**: YUN
cruralis Marseul, 1876d: 324 (*Allecula*) **A**: JA FE TAI
elliptica Fairmaire, 1891f: ccxiv (*Cistela*) **A**: HUB SCH
fainanensis fainanensis Pic, 1922h: 102 **A**: FUJ TAI ZHE
fainanensis diversicollis Pic, 1937i: 173 **A**: CH
flavicornis Borchmann, 1941: 28 **A**: NE
fokienensis Pic, 1922f: 24 **A**: FUJ
formosensis Pic, 1934b: 21 **A**: TAI
fuliginosus Fairmaire, 1897k: 253 **A**: JA SCH TAI WP
hovactyliformis hovactyliformis Pic, 1934d: 22 **A**: CH
hovactyliformis lemoulti Pic, 1934d: 23 **A**: CH
impressithorax Pic, 1922f: 24 **A**: CH
insulcatus Pic, 1922f: 24 **A**: SW YUN
jaegeri Novák, 2005: 126 **A**: NP
kabirensis Maeda & Nakane, 1988: 6 **A**: JA
klapperichi klapperichi Pic, 1955c: 30 **A**: FUJ
klapperichi obscurior Pic, 1955c: 30 **A**: CH
klapperichi rufescens Pic, 1955c: 30 **A**: CH
kuluensis Pic, 1910u: 94 (*Allecula*) **A**: HP NP
major Pic, 1934b: 21 **A**: CH SCH
minor Pic, 1930b: 29 **A**: FUJ SHA YUN
minutulus Mařan, 1944b: 76 **A**: CE FUJ
obliquefasciata Pic, 1926d: 29 (*Allecula*) **A**: YUN
obscurithorax Pic, 1922f: 24 **A**: TAI
piceus Borchmann, 1941: 27 **A**: FUJ
pieli Pic, 1937i: 174 (*Allecula*) **A**: CE JIX
ruficollis Borchmann, 1940a: 155 **A**: TAI
rufina rufina Borchmann, 1940a: 156 **A**: TAI
rufina piceicollis Borchmann, 1940a: 156 **A**: TAI
rufosuturalis Pic, 1934b: 21 **A**: CH
sanensis Pic, 1934b: 20 **A**: CH
subapicalis Pic, 1934b: 22 **A**: CH
submetallicus Pic, 1934b: 20 **A**: CH
subsulcatus Pic, 1922f: 24 **A**: FUJ SW YUN
suturalis Borchmann, 1915b: 186 **A**: SD
thibetanus Pic, 1934b: 20 **A**: XIZ
tsinensis Pic, 1934b: 21 **A**: CH
ussirianus Dubrovin, 1992b: 513 **A**: FE
vicinus Pic, 1930b: 28 **A**: YUN

genus *Cistelopsis* Fairmaire, 1896a: 39 type species *Cistelopsis rufina* Fairmaire, 1896
aborensis Borchmann, 1915b: 187 **A**: AP
klapperichi Mařan, 1944b: 77 **A**: FUJ
rufimembris Pic, 1930e: 76 **A**: NMO **ORR**
sinensis Pic, 1955c: 30 **A**: FUJ

genus *Hymenalia* Mulsant, 1856b: 48 type species *Cistela fusca* Illiger, 1794 (= *Cistela rufipes* Fabricius, 1792)
subgenus *Hymenalia* Mulsant, 1856b: 48 type species *Cistela fusca* Illiger, 1794 (= *Cistela rufipes* Fabricius, 1792)
asahiensis Maeda & Nakane, 1991: 6 **A**: JA
atronitens Fairmaire, 1892b: 151 (*Gonodera*) **A**: IS SY TR
badia Kiesenwetter, 1861a: 234 (*Cistela*) **E**: AB GR TR
basalis Faust, 1877b: 320 (*Allecula*) **E**: AB AR "Caucasus" **A**: IN
brignolii Muche, 1974: 219 (*Allecula*) **A**: TR
castaneipennis Fairmaire, 1884a: 172 **A**: TR
crassicollis crassicollis Fairmaire, 1866a: 47 **N**: AG MO
bispilosa Desbrochers des Loges, 1884: 169 (*Isomira*)
crassicollis obscurimembris Pic, 1938d: 14 **N**: MO
denticulata Muche, 1982: 122 (*Prionychus*) **A**: AE OM SA
elongata Pic, 1925a: 2 **E**: GR
graeca Seidlitz, 1896b: 80 **E**: BH BU CR GR SL TR YU **A**: TR
gravida Küster, 1850a: 77 (*Cistela*) **E**: BH CR GR **A**: TR
klapperichi klapperichi Pic, 1955c: 30 **A**: FUJ SCH
klapperichi tschungseni Pic, 1955c: 30 **A**: CH
minuta Pic, 1910d: 14 **A**: YUN
minyops Peyerimhoff, 1943: 17 (*Mycetocharina*) **N**: AG MO TU
morio L. Redtenbacher, 1849: 602 **E**: AU HU SK UK
amplicollis Linder, 1864: 251 (*Cistella*)
obscuriceps Pic, 1925a: 2 **E**: GR
obscuripennis Pic, 1905g: 162 **A**: TR
pallidipennis Pic, 1926d: 31 **A**: YUN
picheyrei Peyerimhoff, 1943: 20 (*Mycetocharina*) **N**: AG
purkynei Obenberger, 1917: 38 **E**: BU RO
reticulata Seidlitz, 1896b: 77 **A**: IQ
rufipennis Marseul, 1876d: 328 (*Cistela*) **A**: CH ES FE JA SC TAI
rufipes Fabricius, 1792b: 44 (*Cistela*) **E**: AR AU BE BH BU CR CZ DE FR GE GR HU IT LA NL NR PL RO RU SK SL SP ST SV SZ TR UK YU "Caucasus" **A**: TR
fusca Illiger, 1794: 610 (*Cistela*)
pauliani Pic, 1931e: 13 (*Gonodera*)
sericans Rey, 1892: 65
rungsi Peyerimhoff, 1943: 19 (*Mycetocharina*) **N**: MO
sasajii Saitô, 2001: 347 **A**: JA
smirnovi Dubrovina, 1978: 54 **E**: UK
unicolor Nakane, 1963a: 30 **A**: FE JA SC
zoufali Mařan, 1935a: 141 **E**: BH BU
subgenus *Nikomenalia* Dubrovina, 1975: 166 type species *Hymenalia kaszabi* Muche, 1972
impunctaticollis Dubrovina, 1975: 171 **A**: GAN NMO
kaszabi Muche, 1972b: 224 (*Hymenalia*) **A**: MG
medvedevi Dubrovina, 1975: 170 **A**: MG NMO

genus *Hymenophorus* Mulsant, 1851: 201 [= 1852: 68] type species *Hymenophorus doublieri* Mulsant, 1851
Hymenorus Mulsant, 1856e: 20 type species *Hymenorus rugicollis* Mulsant, 1856
baudii Seidlitz, 1896b: 52 (*Hymenorus*) **A**: CY
candeli Pardo Alcaide, 1978: 90 (*Hymenorus*) **N**: MO
doublieri Mulsant, 1851: 202 [= 1852: 68] **E**: AU BH BU CR CT CZ FI FR GG GR HU IT NT RO RU SK SL SP ST SV SZ UK "Caucasus" **A**: ES FE MG RU SHA TR WS
andalusiacus Cobos, 1954: 36 (*Hymenorus*)
avajewi Semenov, 1901b: 168 (*Hymenorus*)
rugicollis Mulsant, 1856e: 20 (*Hymenorus*)

scutellatus Pic, 1901o: 254 (*Hymenorus*)
evae Novák, 2006c: 318 **A**: IN
gerdae Novák, 2006c: 322 **A**: IN
indicus Fairmaire, 1896a: 58 (*Hymenorus*) **A**: HP IN
veterator Lewis, 1895a: 252 (*Hymenorus*) **A**: JA FE

genus *Mycetocharina* Seidlitz, 1891: 136 type species *Allecula orientalis* Faust, 1877
subgenus *Alleculopsis* Semenov, 1893c: 361 [= 1894d: 609] type species *Alleculopsis deserticola* Semenov, 1893
deserticola Semenov, 1893c: 362 [= 1894d: 610] **A**: KZ TM UZ
macrophthalma Gebler, 1859: 341 (*Allecula*) **A**: KI KZ MG TM
puncticollis Reitter, 1898c: 65 **A**: UZ
ruficeps Pic, 1923g: 10 **A**: YE
subgenus *Mycetocharina* Seidlitz, 1891: 136 type species *Allecula orientalis* Faust, 1877
Caristela Fairmaire, 1894c: 311 type species *Caristela megalops* Fairmaire, 1894
adusta Reitter, 1898c: 65 **A**: TR
afghanica Muche, 1986: 181 **A**: AF IN
arabica Muche, 1982: 119 **A**: SA
braaschi Muche, 1982: 117 **A**: AE SA
castanea Faust, 1877b: 322 (*Allecula*) **E**: AR CT ST **A**: WS
jelineki Novák, 2006a: 87 **A**: IN
kabateki Novák, 2006a: 90 **A**: IN
konardanica Novák, 2006a: 94 **A**: IN
marocana Pic, 1951e: 75 **N**: MO
megalops longipennis Pic, 1923g: 10 **A**: SI
megalops megalops Fairmaire, 1894c: 311 (*Caristela*) **N**: AG EG MO TU **A**: IS
megalops pallidior Pic, 1938d: 14 **N**: MO
mogadorica Escalera, 1914: 360 **N**: MO
orientalis Faust, 1877b: 318 (*Allecula*) **E**: AB AR "Caucasus" **A**: IS JO TM TR
beckeri Kiesenwetter, 1878: 245 (*Hymenalia*)
riabovi Iablokoff-Khnzorian, 1959c: 141 **E**: AB AR
rufofusca Muche, 1982: 119 **A**: SA
rufotestacea Reitter, 1898c: 65 **A**: TR
seminigra Pic, 1925i: 228 **N**: EG
syriaca Baudi di Selve, 1881: 293 (*Cistela*) **A**: IS SY
volkonskyi Peyerimhoff, 1943: 21 **N**: "Sahara centralis"
wittmeri Muche, 1982: 116 **A**: SA

genus *Netopha* Fairmaire, 1893f: 299 type species *Netopha pallidipes* Fairmaire, 1893
pallidipes Fairmaire, 1893f: 300 **A**: TAI YUN **ORR**

genus *Prionychus* Solier, 1835a: 237 type species *Helops ater* Fabricius, 1775
Eryx Stephens, 1832: 27 [HN] type species *Eryx nigra* Stephens, 1832
anthracina Mulsant, 1856e: 23 (*Eryx*) **N**: AG TU
asiatica Fairmaire, 1892b: 151 (*Eryx*) **E**: AB **A**: TR
ater Fabricius, 1775: 258 (*Helops*) **E**: AU CR CZ DE EN FI FR GB GE GR HU IT LA LT NR NT PL RO RU SK SV SZ UK **A**: WS
ater Gmelin, 1790: 2011 (*Pimelia*)
*niger*DeGeer, 1775: 25 (*Pyrochroa*)
subsulcatus Fairmaire, 1862: 583 (*Eryx*)
cisteloides Seidlitz, 1896b: 60 **A**: CY IN LE SY TR
delagrangei Fairmaire, 1892b: 150 (*Gonodera*) **A**: TR
fairmairii Reiche, 1860: 731 (*Eryx*) **E**: FR IT
lugens Küster, 1850a: 70 (*Cistela*) **E**: FR (Corse) IT (Sicilia) **N**: AG MO
bellieri Reiche, 1860: 730 (*Eryx*)

mauritanica Mulsant, 1856e: 23 (*Eryx*)
melanarius Germar, 1813: 123 (*Helops*) **E**: AU BE BH CR CZ DE FI FR GB GE GR HU IT NR NT PL RO RU SK SL SV UK YU
ater L. Redtenbacher, 1849: 602 [HN]
ater Baudi di Selve, 1877c: 385 (*Cistela*) [HN]
laevis Rosenhauer, 1847: 122 (*Cistela*)
nitidula C. G. Thomson, 1864: 286 (*Cistela*)
nitidissimus Pic, 1905l: 286 **A**: TR
striatipennis Pic, 1909f: 114 **E**: BU RO
montandoni Reitter, 1913a: 26

subtribe Gonoderina Seidlitz, 1896

genus *Copistethus* Seidlitz, 1890: 524 type species *Copistethus spadix* Kiesenwetter, 1861
spadix abdominalis Pic, 1934d: 26 **E**: GR
spadix spadix Kiesenwetter, 1861a: 235 (*Cistela*) **E**: BU GR MC

genus *Cornucistela* Campbell, 1980: 133 type species *Cornucistela serrata* Campbell, 1980
serrata Campbell, 1980: 135 **A**: SA

genus *Gerandryus* Rottenberg, 1873: 217 [RN] type species *Parablops aetnensis* Rottenberg, 1871
Parablops Rottenberg, 1871: 254 [HN] type species *Parablops aetnensis* Rottenberg, 1871
aetnensis Rottenberg, 1871: 256 (*Parablops*) **E**: FR GR IT

genus *Gonodera* Mulsant, 1856b: 41 type species *Cistela fulvipes* Fabricius, 1792
bicolor Reitter, 1884b: 89 **E**: GR
obscuripennis Pic, 1902g: 57
obscuripes Pic, 1902g: 56
corinthia Fairmaire, 1884a: 172 **A**: SY TR
fulvipennis Reitter, 1895b: 84
luperoides Reitter, 1895b: 83
flavida Fairmaire, 1892b: 150 (*Cistela*) **A**: TR
forticornis Fairmaire, 1892b: 150 (*Cistela*) **A**: TR
jureceki Mařan, 1944d: 193 **A**: TR
luperus hellenica Mařan, 1944d: 191 **E**: GR
luperus luperus Herbst, 1783: 65 (*Cistela*) **E**: AU BH BU CR CZ DE EN FR GB GE GR HU IR IT MC PL PT RO SK SL SP ST SV SZ YU
castanea Marsham, 1802: 223 (*Crioceris*)
erythropa Marsham, 1802: 223 (*Crioceris*)
ferruginea Fabricius, 1792b: 45 (*Cistela*)
fulvipes Fabricius, 1792b: 44 (*Cistela*)
obscuriceps Pic, 1926f: 9
macrophthalma Reitter, 1884b: 89 **E**: AR ST "Caucasus" **A**: TR
latior Pic, 1903m: 163
metallica Küster, 1850a: 73 (*Cistela*) **E**: IT SZ
pulcherrima Faldermann, 1837: 103 (*Cistela*) **E**: AR GG "Caucasus"
viridis Allard, 1877a: 234 (*Euboeus*)
rufoaenea Reitter, 1900a: 159 **A**: SY TR
subaenea Küster, 1850a: 72 (*Cistela*) **E**: GR MC

genus *Isomira* Mulsant, 1856b: 52 type species *Cistela antennata* Panzer, 1796
subgenus *Apteromira* Weise, 1974: 123 type species *Cistela ovulum* Kiesenwetter, 1863
ovulum Kiesenwetter, 1863: 427 (*Cistela*) **E**: SP **N**: MO
acuminata Fairmaire, 1866a: 46 (*Isomera*)
subgenus *Asiomira* Dubrovina, 1973b: 367 type species *Isomira opthalmica* Seidlitz, 1896
Kirgisomira Weise, 1974: 71 type species *Isomira opthalmica* Seidlitz, 1896
bicolorata Muche, 1982: 121 **A**: SA

firjusana Dubrovina, 1973b: 371 **A**: TD TM
kabadiana Dubrovina, 1973b: 372 **A**: TD
keleinikovae Dubrovina, 1973b: 373 **A**: TD
muchei Dubrovina, 1982: 133 **E**: GG ST
monticola Muche, 1972a: 127
ophthalmica Seidlitz, 1896b: 107 **A**: IN KI TM UZ
rufescens Dubrovina, 1982: 142 **A**: TD
stoetzneri Muche, 1981: 157 **A**: GUX SCH
subgenus *Danielomira* Weise, 1974: 119 type species *Danielomira cantabrica* Weise, 1974
cantabrica Weise, 1974: 120 **E**: SP
schedli Muche, 1984: 207 **E**: SP
scutellaris Baudi di Selve, 1877c: 388 (*Cistela*) **E**: IT
subgenus *Heteromira* Hölzel, 1958: 19 type species *Heteromira moroi* Hölzel, 1958
costessii Bertolini, 1868: 119 (*Cistela*) **E**: AU IT
moroi Hölzel, 1958: 19 **E**: AU IT
subgenus *Isomira* Mulsant, 1856b: 52 type species *Cistela antennata* Panzer, 1798
alleuculoides Faldermann, 1837: 102 (*Cistela*) **E**: ST
anaspiformis Weise, 1974: 102 **E**: IT
antennalis Reitter, 1884a: 9 **E**: AB AR GG "Caucasus" **A**: IN TR
antennalis Reitter, 1890j: 394 [HN]
antennata Panzer, 1798: 8 (*Cistela*) **E**: AU BU CZ FR GR HU IT MC PT RO SK SP UK YU
budensis Csiki, 1940b: 919
euboica Pic, 1903l: 146
tristicula Reitter, 1890j: 393
armena Iablokoff-Khnzorian, 1976: 324 **E**: AR
atlantis Escalera, 1914: 359 **N**: MO
avula Seidlitz, 1896b: 107 **E**: AU GE
bodemeyeri Reitter, 1903b: 45 **A**: TR
caucasica Reitter, 1890j: 394 **E**: AB AR GG ST **A**: IN
atriceps Reitter, 1896b: 75
convexior Weise, 1974: 100 **E**: SP
discithorax Pic, 1917b: 20 **A**: SD
estrellana Kiesenwetter, 1870: 144 **E**: PT SP
eustrophoides Pic, 1930b: 30 **A**: CH
excavata Dubrovina, 1982: 138 **E**: AB GG ST
formosana Pic, 1917b: 20 **A**: TAI
funerea Kiesenwetter, 1861a: 236 (*Cistela*) **E**: GR
genistae Rottenberg, 1871: 256 (*Cistela*) **E**: IT
georgica Iablokoff-Khnzorian, 1976: 321 **E**: GG
gracilicornis Roubal, 1919: 63 **E**: AR GG ST
fallax Iablokoff-Khnzorian, 1976: 323
granifera Kiesenwetter, 1878: 245 **E**: AR GG "Caucasus"
hispanica Kiesenwetter, 1870: 143 **E**: PT SP **N**: AG MO
hypocrita Mulsant, 1856b: 60 **E**: AU FR GE IT SL SZ
icteropa H. Wagner, 1917: 346
oblonga Küster, 1850a: 76 (*Cistela*)
icteropa Küster, 1852b: 72 (*Cistela*) **E**: AU BH BU CR CT FR GE IT PL RO SP ST SZ UK **A**: WS
intrusa Seidlitz, 1896b: 107 **E**: AB GG "Caucasus"
murina Kiesenwetter, 1878: 246
ishigakiensis Hanatsuka, Masumoto & Kon, 2005: 414 **A**: JA (Ishigaki Is.)
khnzoriani Dubrovina, 1982: 139 **E**: AR
kompantzevae Dubrovin, 1992a: 14 **A**: TM
konoi Pic, 1931b: 171 **A**: TAI
formosana Kôno, 1930b: 96
marcida Kiesenwetter, 1863: 425 **E**: AU IT SL SZ

matsumurai Kôno, 1930b: 95 (*Isomia*) **A**: TAI
melanophthalma Lucas, 1846: 356 (*Cistela*) **E**: BH CR FR IT PT **N**: AG MO
corsica Mulsant, 1856e: 22
ferruginea Küster, 1850a: 78 (*Cistela*)
meybohmi Weise, 1974: 110 **E**: SP
minuta Pic, 1934b: 24 **A**: CH
mitis Bonadona, 1979: 2 **E**: FR (Corse)
murina ciscaucasica Dubrovina, 1982: 135 **A**: FE WS
murina murina Linnaeus, 1758: 377 (*Chrysomela*) **E**: AR AU BH CR CT CZ FR DE EN FI GB GE GR HU IT LA LT NR NT PL RO RU SK SL SP ST SV SZ UK **A**: FE WS
aemiliae Bedel, 1904: 211
arenaria Gerhardt, 1904: 79
ecchelii Bertolini, 1892: 350 (*Gonodera*)
evonymi Fabricius, 1792b: 45 (*Cistela*)
fulva Geoffroy, 1785: 162 (*Mordella*)
fusca Marsham, 1802: 223 (*Crioceris*)
galii Brahm, 1790: 100 (*Chrysomela*)
reppensis Herbst, 1783: 65 (*Cistela*)
semiflava Küster, 1852b: 71 (*Cistela*)
subnitida Rey, 1892: 65
nitida Reitter, 1890j: 394 **E**: GR **A**: TR
impressithorax Pic, 1903l: 146
nitidula Kiesenwetter, 1861a: 237 (*Cistela*) **E**: GR ST TR **A**: TR
pallidior Pic, 1901a: 3
rhodius Pic, 1901a: 3
ochropus Küster, 1850b: 92 (*Cistela*) **E**: BH CR IT
murina Kiesenwetter, 1863: 424 (*Cistela*) [HN]
parvula Rottenberg, 1871: 257 (*Cistela*)
oertzeni Reitter, 1889e: 257 **E**: GR
parvuloides Weise, 1974: 103 **E**: IT
pilosicollis Escalera, 1913: 51 **N**: AG MO
radula Weise, 1974: 106 **E**: SP
sanquinicollis Reitter, 1911a: 354 **A**: IS SY TR
sericea Drapiez, 1820a: 351 (*Cistela*) **E**: GE
striatipennis Dubrovina, 1982: 141 **E**: GR **A**: TR
stricta Iablokoff-Khnzorian, 1976: 319 **E**: AR
subelongata Pic, 1917c: 20 **A**: TAI
testacea Seidlitz, 1896b: 121 **E**: AU BH CR IT
paupercula Baudi di Selve, 1883a: 3 (*Cistela*)
testaceicornis obscuricolor Pic, 1904d: 27 **A**: FE
testaceicornis testaceicornis Pic, 1904d: 26 **A**: ES FE
thibetana Pic, 1917c: 20 **A**: XIZ
thoracica Fabricius, 1792b: 45 (*Cistela*) **E**: CZ GE HU SK
hungarica Csiki, 1901b: 488
rubricollis Panzer, 1796b: 9 (*Cistela*)
tonkinea Pic, 1917c: 19 **A**: TAI **ORR**
umbellatarum Kiesenwetter, 1863: 426 (*Cistela*) **E**: AU BH CR IT
unicolor Iablokoff-Khnzorian, 1976: 325 **E**: AR GG ST
viridimetallicus Pic, 1909l: 142 **A**: PA
yunnana Pic, 1930b: 30 **A**: YUN
subgenus *Paraisomira* Dubrovina, 1982: 138 type species *Cistela oculata* Marseul, 1876
oculata Marseul, 1876d: 327 (*Cistela*) **A**: JA RU

genus *Microcistela* Pic, 1904d: 26 type species *Microcistela rosinae* Pic, 1904
rosinae Pic, 1904d: 26 **A**: ES FE SHA
infernalis Pic, 1904d: 26

genus *Paracistela* Borchmann, 1942a: 30 type species *Paracistela variabilis* Borchmann, 1942
rufithorax Pic, 1913e: 19 (*Pseudocistela*) **A**: JIX **ORR**
sinensis Pic, 1934b: 24 (*Isomira*) **A**: CH

genus *Pseudocistela* Crotch, 1873: 108 type species *Cistela brevis* Say, 1823
ceramboides ceramboides Linnaeus, 1758: 377 (*Chrysomela*) **E**: AU BU CR CZ DE EN FI FR GB GE GR HU IR IT LA LT NR PL RO RU SK SP ST SV SZ UK "Caucasus"
rufa DeGeer, 1775: 23 (*Pyrochroa*)
saperdoides Küster, 1850a: 71 (*Cistela*)
striata Geoffroy, 1785: 161 (*Mordella*)
ceramboides serrata Chevrolat, 1844: 125 (*Cistela*) **E**: HU SK
haagi Harold, 1878a: 80 **A**: FE JA
impressicollis Chevrolat, 1873: 205 (*Cistela*) **A**: SY
muhldorfi Borchmann, 1930c: 244 **A**: ES FE
pamirica Plavilshchikov, 1927: 133 **A**: TD
rubroflava Fairmaire, 1878a: 123 (*Cistela*) **A**: CE
semirubra Pic, 1910q: 75 **A**: TAI
watanabei Kôno, 1934b: 37 **A**: FE
znojkoi Iablokoff-Khnzorian, 1959a: 60 (*Gonodera*) **E**: AB

subtribe Mycetocharina Gistel, 1848

genus *Mycetochara* Berthold, 1827: 371 [RN] type species *Cistela linearis* Illiger, 1794
subgenus *Ernocharis* C. G. Thomson, 1859: 118 type species *Cistela brevis* Illiger, 1794
Stigmatoma LeConte, 1862: 244 type species *Cistela fraterna* Say, 1823
abschasica Pic, 1925c: 6 **E**: "Caucasus"
angustifrons Reitter, 1899b: 155 **E**: ST
brenskei Seidlitz, 1896b: 136 **E**: GR
cretica Mařan, 1954b: 169 **E**: GR
excelsa Reitter, 1884d: 246 **E**: AB GG ST "Caucasus" **A**: IN
flavicornis Miller, 1883: 265 (*Mycetochares*) **E**: GR
flavipennis Reitter, 1908e: 129 **E**: IT
gracilicornis Roubal, 1935a: 24 **E**: ST "Caucasus"
graciliformis Reitter, 1899b: 156 **E**: CR RO
gracilis Faldermann, 1837: 98 (*Mycetochares*) **E**: AR GG ST "Caucasus"
croceipes Weise, 1880b: 478 (*Mycetochares*)
hirta J. R. Sahlberg, 1908: 89 **A**: TR
humeralis Fabricius, 1787a: 85 (*Cistela*) **E**: AU BE BH CR CZ DE EN FI FR GB GE GG HU IT LA LT NL NR PL RO RU SK SL SV SZ UK YU
bipustulata Illiger, 1794: 606 (*Cistela*)
scapularis Illiger, 1805: 90 (*Cistela*)
impressa Pic, 1925a: 2 **N**: AG
jonica Obenberger, 1917: 38 **E**: GR
koltzei Reitter, 1896b: 75 **A**: ES FE JA (Hokkaido) MG
satanula Reitter, 1899b: 155
laticornis Reitter, 1884d: 249 **A**: IS LE
longior Fairmaire, 1892b: 149 **A**: IS TR
maculipes Reitter, 1899b: 158 **A**: TR
maura Fabricius, 1792b: 45 (*Cistela*) **E**: AB AR AU BE CR CZ DE EN FR GE GG GR HU IT LA NL NR PL PT SK SP SV SZ UK "Caucasus" **N**: MO **A**: TR
barbata Laporte, 1840: 244
barbatus Latreille, 1804a: 348 (*Helops*)
brevis Illiger, 1794: 608 (*Cistela*)
dalmatina Baudi di Selve, 1877d: 588
hirsuta Pic, 1925a: 1

linearis Illiger, 1794: 607 (*Cistela*)
rufipennis Pic, 1925a: 1
melandryina Roubal, 1934b: 35 (*Mycetochares*) **E**: SK
metallica Pic, 1934c: 22 **A**: XIZ
myrmecophila Obenberger, 1917: 37 **E**: GR
netolitzkyi Penecke, 1912: 242 **E**: BU
obscura Zetterstedt, 1840: 162 (*Mycetocharis*) **E**: CZ FI LA NR NT PL RU SV
obtusicollis Reitter, 1899b: 157 **E**: AR
ocularis Reitter, 1884d: 245 **E**: AB "Caucasus"
orientalis Dubrovin, 1992b: 516 **A**: FE
pygmaea L. Redtenbacher, 1874: 133 (*Mycetochares*) **E**: AU BH CR CZ HU IT PL RO RU SK SL YU "Caucasus"
quadrimaculata Latreille, 1804a: 349 (*Helops*) **E**: AU BH CR CZ FR GR HU IT RO RU SK UK YU **A**: TR
schwarzi Reitter, 1888d: 431
quadrispilota Mařan, 1954b: 170 **A**: NC FE SC
roubali Mařan, 1935a: 141 **E**: CZ HU PL RO SK UK
rudis Küster, 1850b: 97 (*Mycetochares*) **E**: GR TR **A**: CY
ruficollis Baudi di Selve, 1877d: 589 **A**: CY LE SY
ruficornis Pic, 1925a: 1 **A**: TR
seidlitzi Reitter, 1899b: 156 **E**: ST
straussii Seidlitz, 1896b: 166 **E**: AU CR
striatipennis Pic, 1904f: 74 **A**: TR
thoracica Gredler, 1854: 15 (*Mycetochares*) **E**: AU FR IT
fasciata E. Mulsant & V. Mulsant, 1855: 155 (*Mycetochares*)
zolotareffi Reitter, 1896b: 76 **E**: GG ST "Caucasus" **A**: TR
subgenus *Mycetochara* Berthold, 1827: 371 [RN] type species *Cistela linearis* Illiger, 1794
Mycetocharis Gyllenhal, 1827: 510 [RN] type species *Cistela linearis* Illiger, 1794
Mycetochara Laporte, 1840: 244 [HN] type species *Mycetochara barbata* Laporte, 1840
Mycetocharis J. R. Sahlberg, 1833: 456 [HN] type species *Mycetophila linearis* Gyllenhal, 1810
Mycetocharus Stephens, 1829b: 245 type species *Mycetophila scapularis* Gyllenhal, 1810
Mycetophila Gyllenhal, 1810: 541 [HN] type species *Cistela linearis* Illiger, 1794
aomoriensis Nakane, 1991: 6 **A**: JA
axillaris axillaris Paykull, 1799: 123 (*Cistela*) **E**: AU BE BU CZ DE EN FI FR GE GR HU IT LA LT NL NT NR PL RO RU SK SV SZ UK **A**: "Siberia"
linearis Gyllenhal, 1810: 541 (*Mycetophila*)
maurina Mulsant, 1856b: 33 (*Mycetochares*)
morio L. Redtenbacher, 1849: 605
collina Lewis, 1895a: 253 (*Mycetochares*) **A**: JA
elongata Miyatake, 1985: 348 **A**: JA
flavipes Fabricius, 1792b: 45 (*Cistela*) **E**: AU BY CZ DE EN FI FR GE GG HU IT LA LT NR NT PL RO RU SK SV SZ YU "Caucasus" **A**: FE MG "South Siberia"
bimaculata Mannerheim, 1844a: 197 (*Mycetochares*)
bipustulata Thunberg, 1784: 17 (*Leptura*)
grandicollis Fairmaire, 1892b: 149 **A**: TR
kimotoi Hanatsuka, Masumoto & Kon, 2006: 181 **A**: JA
mimica Lewis, 1895a: 253 (*Mycetochares*) **A**: JA
nigripes Petri, 1926: 193 **E**: AU RO
scutellaris Lewis, 1895a: 253 (*Mycetochares*) **A**: JA
sulcipennis Reitter, 1896b: 76 **E**: CZ RO
subgenus *Pterna* Seidlitz, 1896b: 137 type species *Pterna auricoma* Reitter, 1884
arnoldii Dubrovina, 1976: 1408 **A**: TD
auricoma Reitter, 1884d: 249 (*Ernocharis*) **E**: CT MD RO ST UK
retowskyi Reitter, 1889j: 373 (*Ernocharis*)

tribe Cteniopodini Solier, 1835

genus *Brachycryptus* Quedenfeldt, 1891: 129 type species *Brachycryptus tripolitanus* Quedenfeldt, 1891

tripolitanus G. Quedenfeldt, 1891: 130 **N**: LB

genus *Cistelina* Seidlitz, 1896b: 195 type species *Cistela davidis* Fairmaire, 1873
crassicornis Borchmann, 1916: 103 **A**: TAI
davidis Fairmaire, 1878a: 123 (*Cistela*) **A**: CE MG
rufa Borchmann, 1930a: 118 **A**: SCH
sulphurea Seidlitz, 1896b: 196 **A**: SCH
tatsienluensis Borchmann, 1932: 94 **A**: SCH
tokaraensis Nomura, 1961a: 38 (*Ctenioplus*) **A**: JA
tricolor Borchmann, 1916: 101 **A**: TAI **ORR**

genus *Cistelomorpha* L. Redtenbacher, 1868: 134 type species *Cistelomorpha haematica* L. Redtenbacher, 1868
alternans Fairmaire, 1894a: 40 **A**: SD
apicipalpis Fairmaire, 1889b: 47 (*Cistela*) **A**: FUJ GUI SCH
atripes Fairmaire, 1899: 631 **A**: FUJ
bina Fairmaire, 1899: 631 **A**: FUJ TAI
bisbinotata bisbinotata Pic, 1909k: 134 **A**: "Himalaya"
bisbinotata subobliterata Pic, 1909k: 134 **A**: "Himalaya"
costatipennis Pic, 1908c: 38 **A**: YUN
costulata Pic, 1908h: 62 **A**: YUN
densestriata Fairmaire, 1896b: 119 **A**: SD "Himalaya"
diversestriata Pic, 1908e: 47 **A**: CH
donckieri Pic, 1908e: 48 **A**: YUN
grandis Pic, 1908e: 47 **A**: CH
holoxanta Pic, 1908e: 48 **A**: JIX YUN
hummeli Pic, 1933b: 8 **A**: GAN NW
infranigra Fairmaire, 1899: 632 **A**: CH
lameyi Pic, 1908c: 39 **A**: SCH **ORR**
luteicolor Pic, 1915c: 7 **A**: XIZ
maculicornis Fairmaire, 1878a: 121 (*Cistela*) **A**: CE **ORR**
mausonensis Borchmann, 1938a: 296 **A**: CH "Kanshirei" **ORR**
  flavipilis Borchmann, 1938a: 297
melanopyga Fairmaire, 1893g: 322 **A**: TAI **ORR**
nigripilis Borchmann, 1940a: 154 **A**: TAI
nigrosparsa nigrosparsa Fairmaire, 1899: 630 **A**: YUN
nigrosparsa separata Pic, 1908c: 40 **A**: YUN
nigrotibialis Fairmaire, 1893g: 301 **A**: CH **ORR**
obscuriceps Pic, 1909a: 19 **A**: YUN
rufina Fairmaire, 1893b: 322 **A**: TAI **ORR**
savioi Pic, 1924a: 31 **A**: SHG
simillima Fairmaire, 1891c: xxi (*Cteniopus*) **A**: HEB HUB
spurcaticornis Fairmaire, 1891c: xxi (*Cteniopus*) **A**: JIX HUB
straminea L. Redtenbacher, 1868: 135 **A**: CH
thibetana Pic, 1915c: 7 **A**: XIZ
ustiventris Fairmaire, 1878a: 122 (*Cistela*) **A**: CE

genus *Cnecosochara* Reitter, 1913d: 660 type species *Cnecosochara petriiformis* Reitter, 1913
filia Znojko, 1936: 74 **A**: KZ UZ
petriiformis Reitter, 1913d: 661 **A**: TD

genus *Cteniopinus* Seidlitz, 1896b: 200 type species *Cistela altaica* Gebler, 1830
altaicus altaicus Gebler, 1829a: 128 (*Cistela*) **A**: CE ES FE MG ST WS
  longicornis Borchmann, 1930b: 155

altaicus kaszabi Muche, 1965a: 207 **A**: MG
andrewesii Fairmaire, 1896a: 58 (*Cistelomorpha*) **A**: ES HP MG NP WS XIZ **ORR**
andrewesii annuligera Fairmaire, 1896b: 118 (*Cistelomorpha*) **A**: HP
andrewesii confluens Borchmann, 1930b: 151 **A**: KA **ORR**
ater Borchmann, 1930b: 160 **A**: TAI
borchmanni Znojko, 1950: 43 **A**: XIZ
brevithoracus Bai & Ren, 2003: 682 **A**: YUN
brunneicapilus Yu & Ren, 1997: 9 **A**: SHX
diversipes Pic, 1937i: 173 **A**: ZHE
diversipunctatus Yu & Ren, 1997: 10 **A**: NMO
elegans Nomura, 1961a: 39 **A**: TAI
flavifemur Borchmann, 1930b: 161 **A**: CH SD
foveicollis Borchmann, 1930b: 154 **A**: TAI
glabratus Yu & Ren, 1997: 8 **A**: GAN NIN
hypocrita Marseul, 1876d: 329 (*Cteniopus*) **A**: BEI FUJ GAN GUA GUX HUN JA JIA NE SC SCH SHA SHG XIZ
potanini Heyden, 1889b: 677 (*Cteniopus*)
impressithorax Pic, 1910w: 272 **A**: TAI
iriomotensis Nakane, 1968b: 84 **A**: JA
iyonis Kôno, 1930b: 97 **A**: JA
koltzei Heyden, 1884a: 295 (*Cteniopus*) **A**: FE NE NO SC
koreanus atricornis Pic, 1904d: 26 **A**: "Sibérie"
koreanus koreanus Seidlitz, 1896b: 203 **A**: ES FE NE NO SC
koreanus subelongata Pic, 1908k: 75 **A**: "Corée"
kwanhsienensis Borchmann, 1930b: 159 **A**: GUX NE NO SCH WP XIZ
nigricornis Borchmann, 1930b: 153 **A**: FUJ JIX
nigritarsis Borchmann, 1930b: 156 **A**: NC
nigrocapitis Bai & Ren, 2003: 682 **A**: YUN
nigroventrus Bai & Ren, 2003: 684 **A**: SCH
parvus Yu & Ren, 1997: 10 **A**: NIN
protuberans Yu & Ren, 1997: 8 **A**: GAN SHA
ruber Pic, 1923j: 76 **A**: CH YUN
rubipes Bai & Ren, 2003: 686 **A**: SCH
scapulatus Borchmann, 1930b: 162 **A**: SCH XIZ YUN
semirufus Pic, 1923j: 76 **A**: XIZ YUN
similis Borchmann, 1930b: 158 **A**: JIL LIA MG NE
subobscurus Pic, 1910w: 271 **A**: TAI
sulcatus Pic, 1925a: 2 **E**: "Caucasus"
taiwanus Kôno, 1930b: 98 **A**: TAI
tenuitarsis Borchmann, 1930b: 157 **A**: NC
tschiliensis Borchmann, 1930b: 157 **A**: HEB JIL MG NE NMO
unicolor Kôno, 1930b: 98 **A**: TAI
varicolor Heyden, 1889b: 676 (*Cteniopus*) **A**: GAN NO SCH SHA SHX "Korea"
subgenus *Lechinius* Borchmann, 1930b: 151 type species *Cistelomorpha fossulata* Pic, 1913
fossulatus Pic, 1913m: 142 (*Cistelomorpha*) **A**: NP SD XIZ **ORR**
catenulatus Blair, 1922c: 561
semicoccineus Blair, 1922c: 560 **A**: NP XIZ

genus *Cteniopus* Solier, 1835a: 246 type species *Cistela sulphurea* Fabricius, 1775
subgenus *Ctenioposomus* Reitter, 1906d: 131 type species *Cteniopus frater* Reitter, 1903
bodemeyeri Reitter, 1903h: 257 (*Cteniopus*) **A**: TR
frater Reitter, 1903h: 258 (*Cteniopus*) **A**: TR
gibbosus Baudi di Selve, 1877c: 394 (*Cteniopus*) **A**: IS SY **N**: EG
intermedius Fairmaire, 1895b: cx (*Cteniopus*) **A**: TR

persicus Ogloblin, 1950: 72 **A**: AF IN IQ SY TR
priesneri luteiceps Pic, 1939f: 146 **A**: SI
priesneri priesneri Wittmer, 1936: 17 **N**: EG
turcicus Pic, 1913h: 121 **A**: TR
subgenus *Cteniopus* Solier, 1835a: 246 type species *Cistela sulphurea* Fabricius, 1775
Sarandonyx Des Gozis, 1881: cxiii [unnecessary RN]
anatolicus Pic, 1903i: 123 **E**: TR **A**: TR
angustatus Pic, 1905j: 177 **A**: IQ
expulsus Roubal, 1917b: 47 **E**: GG
impressicollis Fairmaire, 1892b: 152 **A**: TR
neapolitanus Baudi di Selve, 1877c: 393 **E**: IT
trifossus Pic, 1903i: 123
nigrifrons Fairmaire, 1895b: cxi **A**: TR
pallidus Küster, 1850a: 82 (*Cistela*) **A**: CY IQ SY TR
intrusus Seidlitz, 1896b: 210
persimilis Reitter, 1890g: 258 **E**: AB AR **A**: IN TR
purkynei Mařan, 1936: 205 **E**: MC
sulphureus Linnaeus, 1758: 376 (*Chrysomela*) **E**: AU CR CT CZ DE FI FR GB GE GR HU IT KZ LA LT NL NR PL RO SK SP ST SV SZ UK YU "Caucasus" **A**: ES FE MG TR WS
altaicus Reitter, 1906d: 133
analis Seidlitz, 1896b: 212
bicolor Fabricius, 1794: 447 (*Cistela*)
flavus Scopoli, 1763: 84 (*Tenebrio*)
gilvus Seidlitz, 1896b: 212
luteus Geoffroy, 1785: 159 (*Tenebrio*)
murinus Herbst, 1783: 64 (*Cistela*)
palpalis Seidlitz, 1896b: 212
sulphuratus Gmelin, 1790: 1717 (*Cryptocephalus*)
varians Fabricius, 1787a: 85 (*Cistela*)
tokatensis Pic, 1904d: 26 **A**: TR
subgenus *Rhinobarus* Reitter, 1906d: 131 type species *Cistela sulphuripes* Germar, 1824
elegans Faldermann, 1837: 104 (*Cistela*) **E**: AR GG "Caucasus"
punctatissimus Baudi di Selve, 1877c: 395 **E**: GR
graecus Heyden, 1883c: 312
sulphuripes Germar, 1824b: 162 (*Cistela*) **E**: AB AU BU CR CZ FR GR HU IT MC MD PL RO RU SK ST UK YU **A**: ES TR
collaris Küster, 1850a: 75 (*Cistela*)
notatus Pic, 1925a: 2
sulfuripes W. Redtenbacher, 1842: 18 (*Cistela*)

genus *Falsomophlus* Pic, 1925e: 11 type species *Falsomophlus niger* Pic, 1925
niger Pic, 1925e: 12 **A**: "Himalaya"

genus *Gastrhaema* Jacquelin du Val, 1863: 353 type species *Cistela rufiventris* Waltl, 1835
abdominalis Laporte, 1840: 247 (*Omophlus*) **E**: SP **A**: AL MO
boiteli Peyerimhoff, 1931a: 5 (*Heliotaurus*) **N**: MO
gharbensis Théry, 1932b: 133 **N**: MO
hoemorroidale Fairmaire, 1866a: 50 **N**: AG
martini Bedel, 1894: 164 **N**: AG TU
nigripennis discipennis Reitter, 1906d: 136 **N**: AG
nigripennis erythrocephala Solier, 1835a: 248 (*Megichia*) **N**: AG
nigripennis nigripennis Fabricius, 1792b: 44 (*Cistela*) **N**: AG TU
crassicornis Desbrochers des Loges, 1881: 135 (*Omophlus*)
nigripennis nigrocephala Reitter, 1906d: 137 **N**: AG
nigripennis subrubrica Reitter, 1906d: 137 **N**: AG
ovalis Laporte, 1840: 247 (*Omophlus*) **N**: AG MO TU

rufiventris anceps Mulsant, 1856b: 41 **N**: MO
rufiventris rufiventris Waltl, 1835: 75 (*Cistela*) **E**: SP **N**: AG MO
tingitana Théry, 1932b: 135 (*Heliotaurus*) **N**: MO

genus *Heliomophlus* Reitter, 1906d: 147 type species *Heliotaurus scabriusculus* Fairmaire, 1866
albopilosus Kocher, 1956a: 240 **N**: MO
dasytoides Fairmaire, 1871a: 395 (*Heliotaurus*) **N**: AG
epipleuricus Reitter, 1908a: 22 **N**: MO
mimeuri Kocher, 1956a: 236 **N**: MO
scabriusculus Fairmaire, 1866a: 54 (*Heliotaurus*) **N**: AG TU
gracilior Fairmaire, 1871a: 394 (*Heliotaurus*)
kirschii Reitter, 1872: 174 (*Omophlus*)
mulsanti Kirsch, 1869: 103 (*Omophlus*)
theryi mairei Kocher, 1956a: 239 **N**: MO
theryi occipitalis Kocher, 1956a: 239 **N**: MO
theryi theryi Escalera, 1926: 184 **N**: MO

genus *Heliosthraema* Reitter, 1890b: 34 type species *Heliotaurus griseolineatus* Reitter, 1890
griseolineata alterneseriatus Escalera, 1911: 303 **N**: MO
griseolineata griseolineata Reitter, 1890b: 36 **N**: MO
poweli Théry, 1932b: 136 **N**: MO
reymondi Pic, 1953b: 62 **N**: MO
rolphii Fairmaire, 1867: 408 (*Heliotaurus*) **N**: MO
susica Escalera, 1913: 52 (*Heliotaurus*) **N**: MO
vegepunctata Escalera, 1913: 53 (*Heliotaurus*) **N**: MO

genus *Heliotaurus* Mulsant, 1856b: 73 type species *Cistela distincta* Laporte, 1840
subgenus *Heliotaurus* Mulsant, 1856b: 73 type species *Cistela distincta* Laporte, 1840
aestivalis Escalera, 1914: 363 **N**: MO
analis analis Desbrochers des Loges, 1881: 137 (*Heliotaurus*) **N**: AG MO TU
analis brevicrinitus Muche, 1967: 23 **N**: TU
analis henoni Pic, 1936c: 22 **N**: AG
angusticollis Mulsant, 1856b: 48 **N**: AG EG LB MO TU **AFR**
oberthuri Reitter, 1890b: 38
brisouti Bedel, 1891a: 794 (*Omophlus*) **N**: AG
chalybaeus Kirsch, 1869: 103 **N**: EG LB
chobauti Bedel, 1894: 169 **N**: AG
coeruleus Fabricius, 1787a: 85 (*Cistela*) **N**: AG LB MO TU
coerulescens A. G. Olivier, 1791: 6 (*Cistela*)
confusus Reitter, 1890b: 40 **N**: AG MO TU
corallinus Reitter, 1890b: 51 **N**: LB
crassidactylus crassidactylus Seidlitz, 1896b: 228 **E**: PT SP
crassidactylus stenodactylus Reitter, 1906d: 140 **E**: SP
distincta distincta Laporte, 1840: 246 (*Cistela*) **E**: IT (Sicilia) SP **N**: AG MO LB TU
distincta kabylianus Pic, 1896h: 93 **N**: AG
distincta perroudi Mulsant, 1856b: 45 **N**: AG
distincta plenifrons Fairmaire, 1866a: 51 **N**: AG
distincta riffensis Escalera, 1922c: 360 **N**: MO
distincta seidlitzi Reitter, 1906d: 141 **N**: AG
distincta variventris Desbrochers des Loges, 1881: 136 **N**: AG TU
doriae doriae Bedel, 1891a: 795 (*Omophlus*) **N**: AG LB TU
doriae obscuripes Pic, 1936c: 22 **N**: TU
erythrogaster Lucas, 1846: 358 (*Omophlus*) **N**: AG
testacea Fabricius, 1787a: 85 (*Cistela*)
gasthraemoides Reitter, 1890b: 52 **N**: MO

gestroi Peyerimhoff, 1918c: 257 **N**: AG
goedeli Reitter, 1890b: 38 **A**: SY **N**: EG **AFR**
gracilior Fairmaire, 1871a: 394 **N**: AG
larbii Escalera, 1911: 304 **N**: MO
longitarsis Escalera, 1914: 365 **N**: MO
maroccanus Lucas, 1846: 357 (*Omophlus*) **N**: AG MO
janthinus Raffray, 1873: 378
menticornis ciliatus Reitter, 1890b: 37 **N**: AG
menticornis menticornis Reitter, 1872: 172 (*Omophlus*) **N**: AG MO TU
anthracinus Fairmaire, 1875a: 530 (*Omophlus*)
moderatus Peyerimhoff, 1939b: 120 **N**: AG
monticola Normand, 1955: 85 **N**: TU
parvicollis Reitter, 1872: 144 **N**: EG LB MO TU
parvicollis Reitter, 1890b: 52 [HN]
productus Rosenhauer, 1856: 220 (*Omophlus*) **E**: SP **N**: MO
rasalis Normand, 1955: 86 **N**: TU
ruficollis incertus Mulsant, 1856e: 48 **E**: SP
ruficollis punctosulcatus Fairmaire, 1879d: 242 **E**: SP
ruficollis ruficollis Fabricius, 1781: 147 (*Cistela*) **E**: PT SP **N**: AG MO
maculicollis Desbrochers des Loges, 1881: 92 (*Omophlus*)
ruficollis tangerianus Escalera, 1922c: 359 **N**: MO
rufithorax cincticollis Escalera, 1933: 130 **N**: MO
rufithorax nigritulus Escalera, 1914: 365 **N**: MO
rufithorax rufithorax Reitter, 1890b: 36 **E**: SP **N**: MO
sanchezperezi nigricolor Escalera, 1933: 131 **N**: MO
sanchezperezi sanchezperezi Escalera, 1933: 131 **N**: MO
sanguinicollis nigricollis Reitter, 1906d: 143 **E**: SP
sanguinicollis rubronotatus Fuente, 1907: 319 **E**: SP
sanguinicollis sanguinicollis Reitter, 1906d: 143 **E**: SP
scabriusculus Fairmaire, 1866a: 54 **N**: MO
splendidus Bedel, 1905: 272 (*Omophlus*) **N**: MO
pulchellus Reitter, 1908a: 21
subpilosus Seidlitz, 1896b: 232 **N**: TU
tenuipes Seidlitz, 1896b: 229 **E**: SP
tournieri scultipennis Escalera, 1933: 130 **N**: MO
tournieri tournieri Pic, 1896f: 61 **N**: MO
tuniseus Fairmaire, 1875a: 529 **N**: AG LB TU
subgenus *Julogenius* Reitter, 1906d: 138 type species *Julogenius reichei* Mulsant, 1856
longipilus Fairmaire, 1866a: 54 **N**: AG MO TU
quedenfeldti Reitter, 1890b: 52 (*Heliotaurus*) **N**: MO
reichii Mulsant, 1856b: 50 (*Heliotaurus*) **N**: AG LB MO
grilati Mulsant & Godart, 1876b: 255 (*Heliotaurus*)
oranensis Reitter, 1872: 173 (*Omophlus*)

genus *Holdhausia* Reitter, 1906d: 125 type species *Cteniopus crassus* Fairmaire, 1892
crassus Fairmaire, 1892b: 152 (*Cteniopus*) **A**: IS SY TR
genus *Hypocistela* Bates, 1879: 482 type species *Hypocistela tenuipes* Bates, 1879
tenuipes Bates, 1879: 483 **A**: NW

genus *Megischia* Solier, 1835a: 247 type species *Cistela curvipes* Brullé, 1832
curvimanus Reitter, 1890b: 40 **A**: CY SY
curvipes corcyrea Mařan, 1936: 207 **E**: GR (Corfu)
curvipes curvipes Brullé, *1832: 226* (*Cistela*) **E**: FR GR IT SP
prosternalis Reitter, 1890b: 40
galbanatus Kiesenwetter, 1861a: 238 (*Omophlus*) **E**: BH BU CR GR **A**: TR

curvipes Reitter, 1890b: 40
laticollis Mařan, 1936: 207
schmalfussi Muche, 1987: 135 **E**: GR

genus *Megischina* Reitter, 1906d: 171 type species *Cistela armillata* Brullé, 1832
adaliae Reitter, 1890b: 41 (*Omophlus*) **A**: TR
armillata Brullé, 1832: 225 (*Cistela*) **E**: BH CR GR IT RO TR YU **A**: TR
epipleuralis Seidlitz, 1896b: 241 (*Odontomophlus*)
glamocensis Obenberger, 1917: 39 (*Odontomophlus*)
interstitialis Reitter, 1906d: 172
armillata merthae Reitter, 1898f: 349 (*Omophlus*) **E**: RO
bozdaglariensis Novák, 2006b: 100 **A**: TR
cypria Mařan, 1936: 206 **A**: CY
rosinae Seidlitz, 1896b: 263 (*Odontomophlus*) **E**: SP

genus *Omophlina* Reitter, 1890b: 34 type species *Omophlina podontoides* Reitter, 1890
alpina Mulsant, 1856e: 30 (*Podonta*) **A**: TR UZ
arcuata Gebler, 1829a: 129 (*Cistela*) **A**: AF IN KI KZ MG TD TM UZ WS
heydeni Reitter, 1890b: 35
corva Solsky, 1881a: 253 (*Omophlus*) **A**: KI KZ TD TM UZ
matthiesseni Reitter, 1907f: 207
diversipilosa Muche, 1973: 39 **A**: UZ
hirtipennis Solsky, 1881a: 251 (*Podonta*) **A**: KZ TD TM UZ
pubifer Reitter, 1890b: 35
rubricollis Reitter, 1898a: 14
tenuis Kraatz, 1882c: 114 (*Podonta*)
kirgisica Muche, 1973: 37 **A**: KI
muminovi Muche, 1980: 291 **A**: TD
pentheri Ganglbauer, 1905: 279 **A**: TR
podontoides podontoides Reitter, 1890b: 35 **A**: IN SY
podontoides syriaca Muche, 1963b: 63 **A**: SY
reitteri Ogloblin, 1950: 34 **A**: TD
seidlitzi Reitter, 1898a: 15 **A**: KI UZ
turcmenica Iablokoff-Khnzorian, 1970: 64 **A**: TM
willbergi Reitter, 1892c: 135 (*Omophlus*) **A**: TM UZ
bicolorata Reitter, 1906d: 128

genus *Omophlus* Dejean, 1834: 213 type species *Cistela lepturoides* Fabricius, 1787
subgenus *Euomophlus* Iablokoff-Khnzorian, 1983: 134 type species *Cistela rugosicollis* Brullé, 1832
caucasicus caucasicus Kirsch, 1869: 123 (*Omophlus*) **E**: AB AR GG ST TR "Caucasus" **A**: IN TR
somcheticus Seidlitz, 1896b: 265
caucasicus svaneticus Znojko, 1950: 107 **E**: GG
caucasicus talyschensis Reitter, 1906d: 160 **E**: AR
dilatatus Faldermann, 1837: 99 **E**: AB AR "Caucasus" **A**: IN TR
pallitarsis Reitter, 1890b: 50
dubitatus Reitter, 1906d: 164 **E**: AB AR

laevigatus Seidlitz, 1896b: 265 **E**: AR **A**: IN SY TR
politipennis Reitter, 1906d: 161
nitidicollis Seidlitz, 1896b: 265 **E**: AR **A**: IN IQ TR
oblongus Znojko, 1950: 117 **E**: AR GG **A**: TM
pilicollis filitarsis Reitter, 1890b: 45 **A**: TR
pilicollis fortipes Reitter, 1906d: 168 **A**: UZ
pilicollis longipilis Reitter, 1906d: 168 **A**: UZ
pilicollis marginatus Kirsch, 1869: 110 **E** "Caucasus"

pilicollis pilicollis Faldermann, 1832: 204 (*Cistela*) **E**: AB AR CT GG KZ RU ST UK **A**: IN KI KZ TM TR UZ WS

mitis Reitter, 1906d: 165

volgensis Kirsch, 1869: 115

pilicollis zolotarewi Reitter, 1906d: 168 **A**: UZ

pruinosus Reitter, 1890b: 45 **E**: AR GG ST "Caucasus"

rugosicollis Brullé, 1832: 225 (*Cistela*) **E**: AN AU BH BU CR CZ FR GR HU IT RO SK SZ TR YU **A**: TR

brevicollis Mulsant, 1856b: 91

rugicollis Küster, 1849c: 60

subgenus *Micromophlus* Znojko, 1950: 124 type species *Omophlus subtilis* Solsky, 1881

subtilis Solsky, 1881a: 255 (*Omophlus*) **A**: IN TM UZ

subgenus *Odontomophlus* Seidlitz, 1896b: 263 type species *Cistela armillatus* Brullé, 1932

afghanus Muche, 1965b: 96 **A**: AF PA

anatolicus Reitter, 1906d: 155 **A**: TR

atticus Reitter, 1906d: 153 **E**: GR

baudueri Baudi di Selve, 1877c: 401 **A**: SY TR

blumenthali Muche, 1979a: 173 **A**: IN

bodemeyeri Reitter, 1906d: 154 **A**: TR

compressus Seidlitz, 1896b: 263 **E**: GR

crinifer Seidlitz, 1896b: 263 **A**: TM

cyrenaicus Gridelli, 1930: 190 **N**: LB

dispar A. Costa, 1847: 154 **E**: IT

emgei Reitter, 1891c: 199 **E**: BU GR

fallaciosus Rottenberg, 1871: 258 **E**: IT (Sicilia)

falsarius Kirsch, 1869: 109 **E**: AR BU RO TR **A**: IN IQ SY TR

pilifer Seidlitz, 1896b: 248

flavipennis Küster, 1849c: 57 E: AB AR BU CR CT GG GR MD RO ST TR UK **A**: IQ SY TR

atripes Küster, 1850a: 61

basicornis Reitter, 1890b: 48

dalmatinus Kirsch, 1869: 116

quadricollis Brullé, 1832: 224 (*Cistela*)

forficula Seidlitz, 1896b: 263 **A**: IS JO SY TR

foveicollis Fairmaire, 1892b: 153 **A**: LE SY TR

furca Seidlitz, 1896b: 264 **A**: IS LE SY

furcula Seidlitz, 1896b: 264 **A**: TR

hirsutus Seidlitz, 1896b: 264 **A**: IS SY TM TR UZ

infirmus Kirsch, 1869: 119 **E**: GR IT

kalalae Mařan, 1935a: 145 **A**: IQ

kruegeri Gridelli, 1930: 193 **N**: LB

kurda Znojko, 1950: 97 **A**: IQ

laciniatus Seidlitz, 1896b: 263 **E**: AB AR GG **A**: KI IN SY TR

latipleuris Reitter, 1890b: 47 **E**: AR **A**: TM TR

lepturoides Fabricius, 1787a: 85 (*Cistela*) **E**: AU BH BU CR CZ FR GE GR IT RO SK SP ST SZ TR UK **A**: IN TR

alpinus Miller, 1851: 112

betulae Küster, 1850a: 63

elongatus Küster, 1850a: 59

libanonensis Muche, 1964b: 592 **A**: LE

longipleuris Znojko, 1950: 89 **E**: AR **A**: TR

lucidus lucidus Kirsch, 1869: 125 **E**: GR **N**: EG **A**: IS JO SY

lucidus validiusculus Muche, 1965b: 93 **A**: JO

luciolus Seidlitz, 1896b: 264 **E**: ST **A**: TR

medius Muche, 1964b: 591 **A**: IN TR

melitensis Baudi di Selve, 1877c: 400 **E**: MA

championi Reitter, 1891g: 260

nadayi Muche, 1965b: 85 **A**: TR
obscurus Reitter, 1890b: 46 **E**: AB AR
ochraceipennis Faldermann, 1837: 100 **E**: AB AR GG **A**: IN TR
kuesteri Kirsch, 1869: 105
latitarsis Reitter, 1890b: 41
tarsalis Kirsch, 1869: 115
pilifer Seidlitz, 1896b: 264 **A**: TR
pilosellus Kirsch, 1869: 120 **E**: CR GR IT **A**: IQ TR
pseudolucidus Muche, 1965b: 95 **A**: AF
sandneri Reitter, 1906d: 157 **A**: IN IQ SY TR
schmidi Muche, 1965b: 89 **A**: PA
silifkeensis Muche, 1964b: 593 **A**: TR
sulcipleuris Seidlitz, 1896b: 263 **E**: GR
syriacus Mulsant, 1856e: 57 **A**: IS SY TR
varicolor Kirsch, 1869: 112
terminatus Fairmaire, 1884a: 171 **A**: TR
tibialis Reitter, 1890b: 46 **A**: IS JO SY
tumidipes Kirsch, 1878: 247 **E**: AB AR **A**: TM TR
turcicus Kirsch, 1869: 117 **E**: GR MC TR
foveola Seidlitz, 1896b: 264
subgenus *Omophlus* Dejean, 1834: 213 type species *Cistela lepturoides* Fabricius, 1787
anthracinus Znojko, 1950: 102 **E**: ST
brulléi Kirsch, 1869: 109 **E**: GR
hirtellus Kirsch, 1869: 111
candiota Obenberger, 1917: 40 **E**: GR
caspius Znojko, 1950: 105 **E**: AB **A**: IN TM
coriaceus Seidlitz, 1896b: 264 **E**: AB GG
cribricollis Fairmaire, 1892b: 153 **A**: TR
deserticola Kirsch, 1869: 125 **E**: AB CT KZ ST **A**: KI KZ TM WS
eckerleini Muche, 1964b: 594 **A**: LE
ganglbaueri Reitter, 1890b: 51 **A**: IN
gracilipes Kirsch, 1869: 108 **N**: EG **A**: IS JO SY TR
hirtipennis Seidlitz, 1896b: 265 **A**: TR
iranicus Muche, 1964a: 238 **A**: IN
kaszabi Muche, 1965b: 88 **A**: IN
klapperichi Muche, 1965b: 91 **A**: JO
kuznetzovi Znojko, 1950: 120 **A**: IN
lepidus Reitter, 1906d: 166 **A**: TR
lindti Muche, 1978: 358 **A**: TM
longicornis Bertolini, 1868: 119 **E**: AU CR CZ BH IT RO SK SL "Caucasus" **A**: TR
luristanicus Muche, 1965b: 94 **A**: IN
nanulus Muche, 1964a: 235 **E**: RO
nasreddini Reitter, 1890b: 51 **A**: IN IQ SY TR
conicicollis Reitter, 1906d: 164
nigrinus Reitter, 1889e: 257 **E**: GR
nigropilosellus Muche, 1964a: 239 **A**: TR
ocularis Kirsch, 1869: 128 **A**: IS JO
orestias Muche, 1964b: 596 **A**: TR
orientalis Mulsant, 1856b: 53 **E**: AB AR GR ST TR UK **A**: TR
pollinosus Znojko, 1950: 104 **E**: AB **A**: IN
propagatus Kirsch, 1869: 111 **A**: CY
proteus proteus Kirsch, 1869: 123 **E**: AU CZ BU GR HU IT RO SK ST TR UK YU "Caucasus" **A**: IN SY TM TR
lepturoides L. Redtenbacher, 1849: 604
longicornis Reitter, 1890b: 50

proteus megista Muche, 1964b: 597 **A**: TR
proteus prometheus Reitter, 1906d: 160 **A**: TM
reitteri König, 1906: 25 **E**: AR **A**: TR
rugipennis Seidlitz, 1896b: 265 **E**: AB AR "Caucasus"
salbachi Reitter, 1906d: 163 **A**: TR
scutellaris Mulsant, 1856b: 56 **N**: EG
subcostatus Reitter, 1908d: 56 **A**: TR
tenellus tenellus Kirsch, 1869: 126 **N**: EG LB **A**: IS
tenellus tripolitanus Muche, 1965b: 87 **N**: LB
transitus Muche, 1965b: 98 **A**: AF IN
turanicus Reitter, 1906d: 167 **E**: AR **A**: TM UZ
turcomanicus Znojko, 1950: 118 **A**: TM
subgenus *Paromohlus* Znojko, 1950: 125 type species *Cistela subalpina* Ménétriés, 1832
agrapha Reitter, 1890b: 42 **E**: AB AR GR
armeniacus Znojko, 1950: 126 **E**: AR
pubescens Linnaeus, 1758: 601 (*Chrysomela*) **E**: AB AU BE BH CZ EN FR GB GE GR HU IT LA LT NT PL RU SK SP SV SZ UK
amerinae Kirsch, 1869: 107
armeriae Curtis, 1836: 622
betulae Herbst, 1783: 65 (*Cistela*)
frigidus Mulsant, 1856b: 81
pinicola L. Redtenbacher, 1849: 604
pubescens Mulsant, 1856b: 84 [HN]
rufitarsis Leske, 1785: 15 (*Chrysomela*)
curtellus Seidlitz, 1896b: 265 **A**: SY
curtulus Kirsch, 1878: 246 **E**: AB AR GG RU ST "Caucasus" **A**: TR
curtus Küster, 1850a: 60 **E**: GG GR **A**: TR
densepunctatus Fairmaire, 1892b: 153 **A**: TR
emmae Iablokoff-Khnzorian, 1959b: 191 (*Heliotaurus*) **E**: AR
excavatus Seidlitz, 1896b: 265 **A**: IN
gracilis Znojko, 1950: 123 **E**: AB TR
hirtus Seidlitz, 1896b: 265 **E**: GR IT (Sicilia)
irrasus Seidlitz, 1896b: 265 **A**: TR
irregularis Reitter, 1906d: 170 **A**: TR
lividipes Mulsant, 1856b: 87 **E**: AU BH CT CZ FR GE GR HU IT KZ MD PL RU SK ST UK **A**: WS
picipes L. Redtenbacher, 1849: 604
nigricolor Reitter, 1913c: 290 **A**: IN
nigripes Küster, 1850a: 65 **E**: GR **A**: IQ SY TR
picipes Fabricius, 1792b: 43 (*Cistela*) **E**: AU CZ FR GR HU IT SK SP
sericeicollis Küster, 1850a: 64
tibialis A. Costa, 1847: 156 (*Omophlus*)
subalpinus Ménétriés, 1832: 204 (*Cistela*) **E**: ST "Caucasus"
subgenus *Paurodontomophlus* Muche, 1979a: 171 type species *Paurodontomophlus wittmeri* Muche, 1979
wittmeri Muche, 1979a: 171 **A**: TR

genus *Petria* Semenov, 1893f: 363 [= 1894c: 611] type species *Petria tachyptera* Semenov, 1893
ales Znojko, 1936: 77 **A**: UZ
antennata Semenov, 1893f: 366 [= 1894c: 614] **A**: TM
nitidicollis Semenov, 1896a: 29 **A**: TM
tachyptera Semenov, 1893f: 365 [= 1894c: 613] **A**: TM

genus *Podonta* Solier, 1835a: 247 type species *Cistela nigrita* Fabricius, 1794
Pododonta Agassiz, 1846: 300 [unjustified emendation]
ambigua Kiesenwetter, 1873: 11 **A**: SY TR
anatolica Muche, 1965c: 86 **A**: TR

antennata Muche, 1965c: 83 **E**: GR
atrata Kiesenwetter, 1873: 15 **A**: CY IS SY TR
aubei Mulsant, 1856e: 29 **E**: GR
biformis Reitter, 1889j: 374 **E**: AR **A**: IQ SY TR
ruficollis Seidlitz, 1896b: 189
bodemeyeri Reitter, 1906d: 123 **A**: TR
brevicornis Seidlitz, 1896b: 189 **E**: BU TR
carbonaria Kiesenwetter, 1873: 16 **A**: IN IQ IS SY
convexicollis Küster, 1850a: 81 (*Cistela*) **E**: TR
corvina Kiesenwetter, 1873: 21 **E**: GR RO TR
daghestanica daghestanica Reitter, 1885d: 383 **E**: BU MC RU ST UK YU **A**: IN
daghestanica macedonica Mařan, 1935a: 143 **E**: BU MC
dalmatina Baudi di Selve, 1877c: 395 **E**: BH CR RO RU ST TR UK YU
dentella Muche, 1962b: 145 **A**: TR
eckerleini Muche, 1963b: 61 **A**: TR
elongata Ménétriés, 1832: 205 **E**: AB AR GG ST **A**: SY TR
flecki Reitter, 1906d: 122 **E**: BU RO
frater Seidlitz, 1896b: 190 **A**: TR
graeca Seidlitz, 1896b: 190 **E**: GR
heydeni Kiesenwetter, 1873: 19 **E**: BU GR RO TR **A**: TR
italica Baudi di Selve, 1877c: 397 **E**: IT
kesanensis Muche, 1965c: 85 **A**: TR
lugubris Küster, 1850a: 80 (*Cistela*) **E**: BU CR RO **A**: TR
milleri Kiesenwetter, 1873: 19 **E**: GR
morio Kiesenwetter, 1873: 21 **E**: AB GG GR TR
nigrita Fabricius, 1794: 447 (*Cistela*) **E**: AU BH BU CR CZ GR HU IT PL RO RU SK ST SZ TR UK YU "Caucasus" **A**: TR
paradoxa Schilsky, 1889: 354
patagiata Seidlitz, 1896b: 190 **A**: TR
purkynei Mařan, 1935a: 143 **E**: MC
rambouseki Mařan, 1935a: 144 **E**: MC
rotundangula Reitter, 1906d: 120 **A**: TR
simplex Seidlitz, 1896b: 190 **E**: GR
soror Seidlitz, 1896b: 189 **A**: SY TR
turcica Kiesenwetter, 1873: 14 **E**: RO **A**: TR
korbii Seidlitz, 1896b: 190

genus *Podontinus* Seidlitz, 1896b: 197 type species *Cteniopus punctatissimus* Kiesenwetter, 1861
punctatissimus Kiesenwetter, 1861a: 238 (*Cteniopus*) **A**: SY

genus *Proctenius* Reitter, 1890g: 256 type species *Cistela granatensis* Rosenhauer, 1856
chamaeleon chamaeleon Reitter, 1894e: 305 (*Cteniopus*) **E**: SP
chamaeleon ictericus Viedma, 1959: 262 **E**: SP
chamaeleon sanguinosus Viedma, 1959: 262 **E**: SP
gaditanus canescens Viedma, 1959: 265 **E**: SP
gaditanus gaditanus Viedma, 1959: 264 **E**: SP
granatensis granatensis Rosenhauer, 1856: 221 (*Cistela*) **E**: SP
granatensis incanus Viedma, 1959: 263 **E**: SP
luteus Küster, 1847c: 82 (*Cistela*) **E**: SP
obscurithorax Pic, 1899o: 300

genus *Stenerophlina* Reitter, 1906d: 130 type species *Omophlina hauseri* Reitter, 1894
hauseri Reitter, 1894a: 50 (*Omophlina*) **A**: TM UZ
levipes Dubrovina, 1973a: 284 **A**: UZ

genus *Stenerxy* Reitter, 1890g: 256 type species *Cistela dejeanii* Faldermann, 1836
angustatus Pic, 1938g: 221 **A**: BEI
dejeanii Faldermann, 1836: 391 (*Cistela*) **A**: KZ MG TM UZ
gussakovskyi Znojko, 1936: 73 **A**: UZ
hauseri Seidlitz, 1896b: 205 **A**: KZ NW TM UZ
semenovi Znojko, 1936: 73 **A**: TM

subfamily **Stenochiinae** Kirby, 1837

tribe Cnodalonini Gistel, 1856

genus *Ainu* Lewis, 1894: 479 type species *Ainu tenuicornis* Lewis, 1894
angustus Ren & Yuan, 2005: 98 **A**: GUX SCH
fukudai Masumoto, 1981a: 41 **A**: TAI
grandis Ren & Yuan, 2005: 98 **A**: GUX
masumotoi Nakane, 1968b: 83 **A**: JA (Ryukyus)
medoganus Ren & Yuan, 2005: 100 **A**: XIZ
tenuicornis Lewis, 1894: 480 **A**: JA "Korea"

genus *Andocamaria* Masumoto, 1993b: 230 type species *Campsiomorpha formosana* Pic, 1930
andoi Masumoto, 1992: 89 (*Campsiomorpha*) **A**: TAI
formosana Pic, 1930b: 31 (*Campsiomorpha*) **A**: TAI
ryukyuensis Masumoto, 1985a: 25 (*Campsiomorpha*) **A**: JA (Ryukyus)

genus *Androsus* Gebien, 1920: 386 type species *Chariotheca violacea* Pascoe, 1887
adonis Pic, 1927c: 18 (*Chariotheca*) **A**: NP **ORR**
amaroides Gebien, 1927: 41 **A**: TAI **ORR**
cyanescens Pic, 1934f: 85 **A**: SCH
sinensis Pic, 1934b: 33 (*Chariotheca*) **A**: SCH

genus *Aptereucyrtus* Gebien, 1922b: 477 type species *Aptereucyrtus hemichalceus* Gebien, 1922
hemichalceus Gebien, 1922b: 477 **A**: TAI **ORR**
shibatai Ando, 1984: 69

genus *Asbolodomimus* Pic, 1921c: 20 type species *Asbolodomimus subcarinatus* Pic, 1921
Allopezus Gebien, 1922b: 504 type species *Allopezus miritarsis* Gebien, 1922
bhutanicus Masumoto, 2002: 373 (*Allopezus*) **A**: BT

genus *Augolesthus* Motschulsky, 1872: 34 type species *Augolesthus purpureofasciatus* Motschulsky, 1872
Chrysomaia Kulzer, 1952: 755 type species *Eucyrtus carbunculus* Fairmaire, 1885
subgenus *Falsoaugolesthus* Masumoto, 1993c: 38 type species *Eucyrtus pulcher* Pic, 1927
kurosawai Masumoto, 1986a: 251 **A**: TAI

genus *Bradymerus* Perroud, 1864: 110 type species *Bradymerus tuberculatus* Perroud, 1864
Bradynocerus Fairmaire, 1883b: 36 type species *Bradynocerus aulacopterus* Fairmaire, 1883
Isostira Pascoe, 1870: 97 type species *Isostira crenata* Pascoe, 1870
Pseudobradymerus Pic, 1926g: 16 type species *Bradymerus simplithorax*Pic, 1926
clathratus L. W. Schaufuss, 1887: 134 **A**: JA (Ryukyus) TAI **ORR**
aequecostatus Fairmaire, 1893c: 20
junctus Shibata, 1980: 66
kanchenjungicus Schawaller, 2006a: 47 **A**: NP
kondoi Kulzer, 1957: 199 **A**: JA (Bonin Is.)
masumotoi Schawaller, 2006a: 53 **A**: TAI

genus *Camarimena* Motschulsky, 1863: 473 type species *Camarimena ovicauda* Motschulsky, 1863
Espitomorphus Pic, 1921c: 24 type species *Espitomorphus multicolor* Pic, 1921
Sinopium Pascoe, 1866: 487 type species *Strongylium variabile* Walker, 1858

viridans Kulzer, 1954: 56 **A**: SD

genus *Campsiomorpha* Pic, 1917f: 19 type species *Camaria lata* Pic, 1915
lata Pic, 1915g: 17 (*Camaria*) **A**: FUJ JIX SCH
corinthia Heller, 1923: 71 (*Eucamaria*)
siemsseni Gebien, 1919b: 150 (*Eucamaria*)
subangulata Pic, 1920b: 2

genus *Catapiestus* Perty, 1831: xxxviii type species *Catapiestus piceus* Perty, 1831
Plateia Laporte, 1840: 215 type species *Plateia orientalis* Laporte, 1840
crenulicollis Fairmaire, 1889a: 357 **A**: FUJ YUN **ORR**
indicus Fairmaire, 1896a: 28 **A**: AP SD **ORR**
rugipennis M. Chûjô, 1984: 1 **A**: JA (Ryukyus)
subrufescens Pic, 1911i: 134 **A**: TAI
tonkineus Pic, 1912b: 17 **A**: YUN **ORR**

genus *Chariophenus* Blair, 1929: 239 type species *Chariophenus wasmanni* Blair, 1929
schmidti Schawaller, 2000b: 354 **A**: NP
tibialis Blair, 1930b: 178 **A**: BH SD

genus *Cleomis* Fairmaire, 1893c: 54 type species *Cleomis violaceipes* Fairmaire, 1893
Oyanus Pic, 1921c: 23 type species *Oyanus curticornis* Pic, 1921
Pseudeucyrtus Pic, 1916g: 14 type species *Pseudeucyrtus niasensis* Pic, 1916
marginicollis Gebien, 1914b: 38 (*Psydus*) **A**: TAI **ORR**
substriatus Pic, 1930b: 30 (*Psydus*)

genus *Coelometopus* Solier, 1848: 278 type species *Blaps clypeata* Germar, 1813
clypeatus Germar, 1813: 122 (*Blaps*) **E**: PT SP
cobosi Español, 1963a: 52 **E**: SP (Granada)

genus *Cybopiestes* Reitter, 1917b: 150 type species *Cybopiestes csikii* Reitter, 1917
csikii Reitter, 1917b: 150 **N**: MO

genus *Derosphaerus* C. G. Thomson, 1858: 99 type species *Derosphaerus globicollis* C. G. Thomson, 1858
Cholipus Pascoe, 1866: 471 type species *Cholipus brevicornis* Pascoe, 1866
Encyalesthus Motschulsky, 1860c: 139 type species *Encyalesthus subviolaceus* Motschulsky, 1860
Falsoencyalesthus Pic, 1923b: 29 type species *Falsoencyalesthus latipennis* Pic, 1923
Neandrosus Pic, 1921e: 12 type species *Neandrosus singularipes* Pic, 1921
Notiolesthus Motschulsky, 1872: 25 type species *Notiolesthus natalensis* Motschulsky, 1872
Pachyurgus J. L. LeConte, 1862: 230 type species *Iphthinus aeraeus* Melsheimer, 1846
aeruginosus Fabricius, 1787a: 213 (*Helops*) **A**: NP **ORR**
asperipennis Allard, 1896: 320 **A**: BT NP SD UP XIZ
angustatus Pic, 1922c: 24 (*Leprocaulus*)
brevicornis Mäklin, 1863: 553 **A**: SD YUN **ORR**
crassicrus Fairmaire, 1903a: 14 **A**: FUJ **ORR**
cribripennis Fairmaire, 1891f: ccxi **A**: CH
exularis Gebien, 1914b: 31 (*Encyalesthus*) **A**: AP JA (Ryukyus) NP SD TAI **ORR**
foveoseriatus Fairmaire, 1889a: 358 **A**: CH **ORR**
grandipes Fairmaire, 1893f: 297 **A**: CH **ORR**
coomani Pic, 1926d: 26
leseleuci Pic, 1923b: 32
instriatus Pic, 1922c: 27 **A**: XIN
niger Wiedemann, 1819: 164 (*Helops*) **A**: "Bengal"
nitidus Pic, 1922c: 27 **A**: CH FUJ TAI
rotundicollis Laporte, 1840: 213 (*Upis*) **A**: TAI **ORR**
rugosus Gravely, 1915: 528 **A**: AP NP SD UP **ORR**

sinensis Hope, 1842: 52 (*Upis*) **A**: CH
foveolatus Marseul, 1876c: 119 (*Upis*)
foveolatus Pic, 1923b: 31 (*Encyalesthus*) [HN]
inaequalis Fairmaire, 1889a: 357 (*Nyctobates*)
stevensi Gravely, 1915: 527 (*Encyalesthus*) **A**: UP
subviolaceus Motschulsky, 1860c: 139 (*Encyalesthus*) **A**: FE JA NC NE SC
violaceipennis Marseul, 1876c: 118 (*Upis*)
thibetanus Pic, 1922c: 26 **A**: XIZ

genus *Dioscoridemus* Koch, 1970: 115 type species *Dioscoridemus vonhayekae* Koch, 1970
vonhayekae Koch, 1970: 116 **A**: YE (Suqutra)

genus *Eucyrtus* Pascoe, 1866: 473 type species *Scotaeus pretiosus* Lacordaire, 1859
annulipes Kraatz, 1880c: 103 **A**: GUX HAI **ORR**
anthracinus anthracinus Kraatz, 1880c: 102 **A**: HAI **ORR**
opacus Kraatz, 1880c: 102 **A**: "Himalaya" **ORR**
rondoni Ando, 2003a: 96 **A**: YUN **ORR**
septentrionalis Ando, 2003a: 91 **A**: BT YUN **ORR**
splendens elatus Ando, 2003a: 54 **A**: BT NP SD **ORR**
splendens splendens Lacordaire, 1859: 417 (*Scotaeus*) **A**: AP BT SD **ORR**
coeruleicollis Pic, 1924e: 27 (*Hemicera*)
semiviolacea Pic, 1923c: 19 (*Hemicera*)

genus *Euhemicera* Ando, 1996: 189 type species *Epilampus pulcher* Hope, 1842
aeneovirgulata Ando, 2003a: 200 **A**: FUJ GUI
alternata Gebien, 1914b: 37 (*Hemicera*) **A**: TAI **ORR**
chapaensis Pic, 1923c: 19 (*Hemicera*)
chinensis Kulzer, 1954: 42 (*Hemicera*) **A**: HUB
chujoi Ando, 2003a: 233 **A**: FUJ
formosana Ando, 1997: 110 **A**: TAI
foveoseriata Kaszab, 1941c: 63 (*Hemicera*) **A**: TAI
parafoveoseriata Masumoto, 1981c: 37 (*Hemicera*)
gebieni Kaszab, 1941c: 61 (*Hemicera*) **A**: FUJ TAI **ORR**
fukiensis Kaszab, 1954: 258 (*Hemicera*)
hajimei Masumoto, 1983a: 86 (*Hemicera*) **A**: JA (Ryukyus)
himalaica Blair, 1930b: 178 (*Hemicera*) **A**: NP SD **ORR**
inexpecta Ando, 2003: 307 **A**: NP **ORR**
japonica Masumoto, 1983a: 83 (*Hemicera*) **A**: JA (Ryukyus)
nodokai Nakane, 1963a: 28 (*Hemicera*) **A**: JA (Ryukyus)
pingtita Masumoto, 1981e: 84 (*Hemicera*) **A**: ANH TAI
pulchra Hope, 1842: 52 [= 1845: 16] (*Epilampus*) **A**: GUA JA TAI
zigzaga Marseul, 1876c: 111 (*Hemicera*)
sakaii Masumoto, 1981c: 38 (*Hemicera*) **A**: TAI
sakishimensis M. Chûjô, 1978b: 8 (*Hemicera*) **A**: JA (Ryukyus) TAI
shantita Masumoto, 1981e: 86 (*Hemicera*) **A**: TAI
shibatai Ando, 1997: 107 **A**: TAI
theresae Pic, 1921c: 30 (*Hemicera*) **A**: YUN **ORR**
undulata Pic, 1923k: 271 (*Hemicera*) **A**: FUJ GUA **ORR**
viridiapicalis Ando, 1994: 111 (*Hemicera*) **A**: NP **ORR**
vittatipennis Pic, 1923c: 19 (*Hemicera*) **A**: TAI **ORR**
tenuestriata Kaszab, 1941c: 61 (*Hemicera*)

genus *Euphloeus* Pascoe, 1887: 15 type species *Euphloeus verrucosus* Pascoe, 1887
verrucosus Pascoe, 1887: 16 **A**: HP **ORR**
bigranosus Fairmaire, 1897i: 221

granosus Allard, 1894: 104 (*Osdara*)

genus *Falsocamaria* Pic, 1917f: 19 type species *Falsocamaria obscura* Pic, 1917
Eucamaria Gebien, 1919b: 149 type species *Camaria spectabilis* Pascoe, 1860
microdera Fairmaire, 1899: 630 (*Camaria*) **A**: FUJ SCH
obscura Pic, 1917f: 19
subopaca Pic, 1920b: 2 (*Camaria*)
muelleri Kaszab, 1954: 259 (*Campsiomorpha*) **A**: FUJ
spectabilis Pascoe, 1860a: 52 (*Camaria*) **A**: "North China"

genus *Foochounus* Pic, 1921c: 22 type species *Foochounus convexipennis* Pic, 1921
Anobriomaia Kaszab, 1941c: 67 type species *Anobriomaia sulcata* Kaszab, 1941
Microcameria Ren, 1998: 108 type species *Microcameria pygmaea* Ren, 1998
assamicus Kaszab, 1965a: 127 (*Anobriomaia*) **A**: AP NP
convexipennis Pic, 1921c: 22 **A**: CH
pygmaeus Ren, 1998: 108 (*Microcameria*) **A**: ZHE
sulcatus Kaszab, 1941c: 68 (*Anobriomaia*) **A**: TAI **ORR**

genus *Gauromaia* Pascoe, 1866: 473 type species *Gauromaia dives* Pascoe, 1866
subgenus *Gauromaia* Pascoe, 1866: 473 type species *Gauromaia dives* Pascoe, 1866
Cephaleucyrtus Pic, 1923b: 22 type species *Cephaleucyrtus viridicollis* Pic, 1923
formosana Masumoto, 1981c: 41 **A**: TAI
himalaica Schawaller, 2006c: 479 **A**: NP
tenuestriata Fairmaire, 1891e: cxxxii **A**: KA

genus *Gnesis* Pascoe, 1866: 477 type species *Gnesis helopioides* Pascoe, 1866
Tromosternus Harold, 1878b: 130 type species *Tromosternus haagi* Harold, 1878
formosanus M. Chûjô, 1973: 20 (*Phaedis*) **A**: TAI
helopioides helopioides Pascoe, 1866: 478 **A**: JA NE "Korea" "Mantchuria"
haagi Harold, 1878b: 131 (*Tromosternus*)
helopioides okinawanus Nakane, 1968b: 80 (*Phaedis*) **A**: JA (Okinawa-honto)
helopioides kentingensis Masumoto, 1981e: 87 **A**: TAI
helopioides purpurascens Nakane, 1968b: 79 (*Phaedis*) **A**: JA (Ryukyus)
amamianus M. Chûjô, 1977: 12 (*Phaedis*)
helopioides senkakuensis M. Chûjô, 1973: 19 (*Phaedis*) **A**: JA (Senkaku Is.)
helopioides yakuanus Nakane, 1968b: 80 (*Phaedis*) **A**: JA (Yakushima)
liukueiensis Masumoto, 1981c: 45 **A**: TAI
magnipunctatus M. Chûjô, 1966: 23 (*Phaedis*) **A**: JA (Ryukyus)

genus *Hemicera* Laporte & Brullé, 1831: 393 type species *Cnodalon splendens* Wiedemann, 1823
subgenus *Hemicera* Laporte & Brullé, 1831: 393 type species *Cnodalon splendens* Wiedemann, 1823
consonbrina Ando, 2003a: 388 **A**: SCH
gilvipes gilvipes Ando, 2003a: 422 **A**: NP **ORR**
kurosawai Masumoto, 1981c: 40 **A**: SCH TAI
nakamurai Masumoto, 1982d: 56
masumotoi dilatilineata Ando, 2003a: 383 **A**: GUI **ORR**
masumotoi masumotoi Ando, 2003a: 383 **A**: GUX **ORR**
nakanei Ando, 2003a: 467 **A**: NP **ORR**
oblita Ando, 2003a: 394 **A**: NP SD YUN **ORR**
tsuyukii Masumoto, 1981e: 83 **A**: TAI

genus *Hexarhopalus* Fairmaire, 1891c: xix type species *Hexarhopalus sculpticollis* Fairmaire, 1891
subgenus *Hexarhopalus* Fairmaire, 1891c: xix type species *Hexarhopalus sculpticollis* Fairmaire, 1891
Apteroleprocaulus Kaszab, 1983b: 182 type species *Leprocaulus montanus* Kaszab, 1983
Hexaroptrum Fairmaire, 1894a: 38 [unnecessary RN] type species *Hexarhopalus sculpticollis* Fairmaire, 1891
Laosocryptobates Pic, 1928b: 25 type species *Laosocryptobates tuberculatus* Pic, 1928

sculpticolle Fairmaire, 1891c: xix **A**: HUB **ORR**

subgenus *Leprocaulus* Fairmaire, 1896b: 95 type species *Leprocaulus clavipes* Fairmaire, 1896

Pseudoderosphaerus Pic, 1922c: 24 type species: *Leprocaulus rotundicollis* Pic, 1922

attenuatus Pic, 1922c: 24 **A**: YUN

genus *Iphthiminus* Spilman, 1973: 42 type species *Iphthimus italicus* Truqui, 1857

italicus bellardi Truqui, 1857: 94 (*Iphthimus*) **E**: GR **A**: CY IS SY TR

reitteri Gebien, 1913: 7 (*Iphthimus*)

italicus croaticus Truqui, 1857: 93 (*Iphthimus*) **E**: AL BH CR GR

italicus italicus Truqui, 1857: 93 (*Iphthimus*) **E**: FR IT

genus *Iphthimulus* Reitter, 1920a: 17 type species *Iphthimus truquii* Marseul, 1869

truquii Marseul, 1869: 274 (*Iphthimus*) **E**: GR **A**: SY TR

genus *Macrostethus* Wollaston, 1854: 504 type species *Macrostethus tuberculatus* Wollaston, 1854

tuberculatus Wollaston, 1854: 505 **N**: MR

genus *Menephilus* Mulsant, 1854: 291 type species *Tenebrio curvipes* Fabricius, 1792 (= *Tenebrio cylindricus* Herbst, 1784)

aborensis Gravely, 1915: 529 **A**: AP

arciscelis Marseul, 1876c: 119 **A**: JA

atronitens Kulzer, 1957: 235 (*Setenis*) **A**: JA (Bonin)

cylindricus cylindricus Herbst, 1784c: 144 (*Tenebrio*) **E**: AL AR AU BH BU CR CZ FR GE GR HU IT PL PT RO SK SL SP ST SV SZ UK **A**: CY SY TR **NAR**

affinis Rossi, 1790a: 232 (*Tenebrio*)

curvipes Fabricius, 1792a: 111 (*Tenebrio*)

loripes Illiger, 1798: 113 (*Tenebrio*)

cylindricus maroccanus Théry, 1931b: 127 **N**: AG MO

formosanus Masumoto, 1981e: 91 **A**: TAI

lucens Marseul, 1876c: 121 **A**: JA "Korea"

medius Marseul, 1876c: 120 **A**: FUJ JA **ORR**

striatifrons Fairmaire, 1896b: 101 (*Nyctobates*) **A**: JIX

taiwanus Masumoto, 1981e: 92 **A**: TAI

genus *Misolampidius* Solsky, 1876: 292 type species *Misolampidius tentyrioides* Solsky, 1876

Ptilonix Allard, 1877a: 62 type species *Helops clavicrus* Marseul, 1876

adachii Mt. Chûjô & Imasaka, 1982: 128 **A**: JA (Tsushima)

changbaiensis J.-K. Li, 1992: 166 **A**: JIL

chejudoensis Mt. Chûjô & Imasaka, 1982: 124 **A**: SC (Chejudo)

clavicrus Marseul, 1876c: 139 (*Helops*) **A**: JA (Kyushu)

entomogonoides Allard, 1896: 319 (*Ptilonix*) **A**: NP SD

foveipenne Fairmaire, 1894a: 38 (*Hexaroptrum*) **A**: SD

gotoensis M. Chûjô & Imasaka, 1982: 130 **A**: JA (Goto)

hanshungliengensis J.-K. Li, 1992: 166 **A**: JIL

indicus Allard, 1877a: 63 (*Ptilonix*) **A**: "Himalaya"

iriei M. Chûjô, 1975: 21 **A**: JA (Yakushima)

kawadai Nakane, 1983b: 147 **A**: JA (Izu Is., Mikura Is.)

koreanus Mt. Chûjô & Imasaka, 1982: 123 **A**: SC

laevicollis Allard, 1877a: 62 (*Ptilonix*) **A**: "Himalaya"

maoershanensis J.-K. Li, 1992: 165 **A**: HEI LIA

molytopsis Marseul, 1876c: 135 (*Heliophygus*) **A**: JA "Korea"

hikosanus Nakane, 1963a: 28

morio Kolbe, 1886: 209 **A**: "Korea"

ohminesanus Masumoto & Akita, 2001a: 19 **A**: JA (Honshu)

okiensis Nakane, 1983b: 147 **A**: JA (Oki Is.)

okumurai Nakane, 1968b: 81 **A**: JA
paratentyrioides J.-K. Li, 1992: 165 **A**: HEI
qianshanensis J.-K. Li, 1992: 166 **A**: LIA
    liaoningensis J.-K. Li, 1992: 166
rugipennis rugipennis Lewis, 1894: 476 **A**: JA
rugipennis subnitidus Nakane, 1968b: 81 **A**: JA
satoi Masumoto, Akita & Utsunomiya, 2003: 327 **A**: SCH
sobosanus Mt. Chûjô & Imasaka, 1982: 132 **A**: JA (Kyushu)
tentyrioides Solsky, 1876: 293 **A**: FE JIL NC SC
tsurugisanus Nakane, 1963a: 28 **A**: JA (Shikoku)
tsushimanus Nakane, 1968b: 80 **A**: JA (Tsushima)
yakushimanus Nakane, 1968b: 81 **A**: JA (Yakushima)

genus *Misolampomorphus* Kaszab, 1941a: 6 type species *Misolampomorphus kochi* Kaszab, 1941 (= *Lyprocaulus reitteri* Pic, 1934)
reitteri Pic, 1934f: 84 (*Lyprocaulus*) **A**: SCH
    kochi Kaszab, 1941a: 7

genus *Misolampus* Latreille, 1807: 160 type species *Misolampus hoffmannseggii* Latreille, 1807 (= *Pimelia gibbula* Herbst, 1799)
gibbulus Herbst, 1799: 51 (*Pimelia*) **E**: PT SP
    hoffmannseggii Latreille, 1807: 161
goudotii erichsoni Vauloger de Beaupré, 1900: 674 **E**: SP **N**: AG
    goudotii Erichson, 1841: 184 [HN]
goudotii goudotii Guérin-Méneville, 1834b: 28 **N**: MO
goudotii peyerimhoffi Antoine, 1926: 257 **N**: MO (Grand Atlas)
lusitanicus Brème, 1842: 83 **E**: PT SP
ramburii Brème, 1842: 83 **E**: PT SP
scabricollis Graells, 1849: 621 **E**: PT SP
subglaber Rosenhauer, 1856: 204 **E**: SP

genus *Morphostenophanes* Pic, 1925b: 7 type species *Morphostenophanes aenescens* Pic, 1925
aenescens Pic, 1925b: 7 **A**: YUN
    diversus Pic, 1925b: 7
    subparallelus Pic, 1925b: 7
jendeki jendeki Masumoto, 1998b: 307 **A**: YUN
jendeki similis Masumoto, 1998b: 308 **A**: YUN
papillatus Kaszab, 1941a: 11 **A**: GUI SCH
tanikadoi Masumoto, 1998b: 305 **A**: SCH

genus *Necrobioides* Fairmaire, 1882c: 234 type species *Necrobioides coeruleatus* Fairmaire, 1882
kabakovi Kaszab, 1980b: 199 **A**: TAI **ORR**

genus *Neoplamius* Masumoto, 1981d: 15 type species *Neoplamius zoltani* Masumoto, 1981
endoi Masumoto, 1981d: 18 **A**: TAI
zoltani Masumoto, 1981d: 15 **A**: TAI
genus *Oedemutes* Pascoe, 1860a: 51 type species *Oedemutes tumidus* Pascoe, 1860
subgenus *Tamdaous* Pic, 1923h: 11 type species *Oedemutes impressus* Pic, 1923
chinensis Ando & Ren, 2006: 85 **A**: YUN
formosanus Masumoto, 1981c: 42 **A**: TAI
hirashimai hirashimai M. Chûjô, 1973: 17 (*Oedemutes*) **A**: JA (Ryukyus)
hirashimai inadai Ando, 2000: 309 **A**: JA (Ryukyus)
itoi Ando, 1989: 108 **A**: TAI

genus *Paramisolampidius* Merkl & Masumoto, 2007: 127 type species *Paramisolampidius kagoshimensis* Nakane, 1968
formosanus Masumoto, 1981c: 32 **A**: TAI
kagoshimensis Nakane, 1968b: 82 **A**: JA
kinugasai Masumoto, 1984: 17 **A**: TAI
shirozui M. Chûjô, 1967: 379 (*Misolampidius*) **A**: TAI
alishanus Masumoto, 1981e: 79
wufengus Masumoto, 1981e: 80
tenghsiensis Masumoto, 1984: 19 **A**: TAI
tokarensis Nakane, 1963a: 29 (*Misolampidius*) **A**: JA

genus *Phaedis* Pascoe, 1866: 474 type species *Phaedis elysius* Pascoe, 1866
subgenus *Phaedis* Pascoe, 1866: 474 type species *Phaedis elysius* Pascoe, 1866
Pseudeumolpus Kraatz, 1880c: 111 type species *Pseudeumolpus bicolor* Kraatz, 1880
iriei M. Chûjô, 1980: 5 **A**: JA (Ryukyus)
marmoratus M. Chûjô, 1978b: 11 **A**: JA (Ryukyus)
mushanus Masumoto, 1981c: 44 **A**: TAI
oshimensis Nakane, 1979: 113 **A**: JA (Ryukyus)
taiwanus Masumoto, 1981c: 42 **A**: TAI

genus *Pigeus* Gebien, 1919b: 153 type species *Camarimena nitidipes* Fairmaire, 1893
Pseudocamarimena Pic, 1923d: 21 type species *Pseudocamarimena striata* Pic, 1923
striatus Pic, 1923d: 22 (*Pseudocamarimena*) **A**: SHG

genus *Plamius* Fairmaire, 1896a: 30 type species *Plamius tenuestriatus* Fairmaire, 1896
Cnemandrosus Gebien, 1927: 42 type species *Eucyrtus semipurpureus* Fairmaire, 1896
convexus M. Chûjô, 1973: 15 **A**: HKG
fukiensis Picka, 1990: 110 **A**: FUJ
kaszabi Picka, 1990: 111 **A**: ZHE
quadrimaculatus Kaszab, 1941c: 65 (*Cnemandrosus*) **A**: TAI
quadrinotatus Pic, 1925b: 11 (*Rehumius*) **A**: FUJ TAI **ORR**
yaeyamensis Nakane, 1968b: 78 **A**: JA (Ryukyus)

genus *Planibates* Kaszab, 1939b: 221 type species *Planibates papuanus* Kaszab, 1939
fukiensis Kaszab, 1954: 257 **A**: FUJ

genus *Platycrepis* Lacordaire, 1859: 418 type species *Platycrepis violaceus* Lacordaire, 1859
schmidi Kaszab, 1965a: 125 **A**: UP (Kumaon)
violaceus Lacordaire, 1859: 418 **A**: BT FUJ TAI **ORR**
impressithorax Pic, 1924a: 26 (*Hemicera*)
latitarsis Fairmaire, 1882c: 239 (*Eucyrtus*)
tonkineus Pic, 1922a: 23
yangi Masumoto, 1986b: 64 **A**: TAI

genus *Promethis* Pascoe, 1869: 148 type species *Upis angulatus* Erichson, 1842
Mederis Motschulsky, 1872: 24 type species *Upis angulata* Erichson, 1842
Pediris Motschulsky, 1872: 24 type species *Pediris longipes* Motschulsky, 1872
Pseudobates Fairmaire, 1882c: 231 type species *Nyctobates coracina* Fairmaire, 1882
Setenis Motschulsky, 1872: 24 type species *Tenebrio valgus* Wiedemann, 1823
almorensis Kaszab, 1988: 101 **A**: UP (Almora)
amanoi Masumoto, 1987b: 92 **A**: JA
angulicollis Kaszab, 1988: 113 **A**: YUN
barbereti Fairmaire, 1888a: 27 (*Nyctobates*) **A**: ZHE (Zhoushan)
chinensis Kaszab, 1988: 106 **A**: YUN **ORR**
confusa Fairmaire, 1896a: 27 (*Nyctobates*) **A**: NP **ORR**

semivalga Blair, 1913: 57 (*Setenis*)
cordicollis Kaszab, 1988: 95 **A**: FUJ GUA YUN **ORR**
crenatostriata Motschulsky, 1872: 30 (*Setenis*) **A**: AP BT HAI SCH SD YUN **ORR**
foveicollis Fairmaire, 1893b: 29 (*Nyctobates*)
sinuatipes Pic, 1928i: 377 (*Setenis*)
evanescens Gebien, 1918: 124 [= 1919a: 9] (*Setenis*) **A**: FUJ HAI SHN **ORR**
quadraticollis Fairmaire, 1893f: 297 (*Nyctobates*) [HN]
exigua M. Chûjô, 1980: 1 (*Setenis*) **A**: JA (Yakushima)
formosana Masumoto, 1981e: 88 (*Setenis*) **A**: TAI
gebieni Kaszab, 1941d: 124 (*Setenis*) **A**: SCH
glabra Hope, 1831: 31 (*Tenebrio*) **A**: BT NP SD **ORR**
glabricula Motschulsky, 1872: 31 (*Nyctobates*) **A**: NP UP XIZ
cribrifrons Fairmaire, 1894a: 38 (*Nyctobates*)
semiopaca Blair, 1913: 57 (*Setenis*)
harmandi Allard, 1896: 318 **A**: BT NP PA SD YUN XIZ **ORR**
heros Gebien, 1918: 125 (*Setenis*) **A**: YUN **ORR**
insomnis Lewis, 1894: 472 (*Setenis*) **A**: JA "Korea"
higonia Lewis, 1894: 472 (*Setenis*)
iriomotensis M. Chûjô, 1979: 40 (*Setenis*) **A**: JA (Ryukyus)
kaoshana Masumoto, 1982d: 60 (*Setenis*) **A**: TAI
kempi kempi Gravely, 1915: 526 (*Setenis*) **A**: AP YUN **ORR**
kurosawai Masumoto, 1982d: 58 (*Setenis*) **A**: TAI
laevis laevis Fairmaire, 1896d: 230 (*Nyctobates*) **A**: AP BT SD TAI **ORR**
nitidula M. Chûjô, 1981: 133 (*Setenis*)
laevis vietnamica Kaszab, 1980b: 198 (*Setenis*) **A**: HAI **ORR**
microcephala Fairmaire, 1899: 629 (*Nyctobates*) **A**: SCH
microthorax Kaszab, 1988: 101 **A**: SCH
nitouana Kaszab, 1988: 108 **A**: SCH
noctivigila Lewis, 1894: 473 (*Setenis*) **A**: JA
okinawana M. Chûjô, 1978a: 66 (*Setenis*) **A**: JA (Ryukyus)
oshimana Miwa, 1935: 21 (*Setenis*) **A**: JA (Amami-Ôshima)
parallela cheni Kaszab, 1988: 95 **A**: SCH XIZ YUN
parallela parallela Fairmaire, 1897k: 251 (*Nyctobates*) **A**: SCH YUN ZHE **ORR**
pauperula Gebien, 1918: 124 (*Setenis*) **A**: HAI YUN **ORR**
persimilis M. Chûjô, 1980: 3 (*Setenis*) **A**: JA (Yakushima)
piluensis Masumoto, Akita & Lee, 2005: 249 **A**: TAI
punctatostriata Motschulsky, 1872: 29 (*Setenis*) **A**: AP BT HAI NP SD YUN "Korea" **ORR**
foveiceps Pic, 1929b: 17 (*Setenis*)
indosinica Fairmaire, 1893f: 296 (*Nyctobates*)
pilipes Pic, 1929b: 17 (*Setenis*)
puncticollis Motschulsky, 1872: 29 (*Setenis*) **A**: CH **ORR**
rectangula Motschulsky, 1872: 28 (*Setenis*) **A**: HAI SC YUN **ORR**
biangulata Gebien, 1918: 122 (*Setenis*)
semisulcata Fairmaire, 1882c: 228 (*Nyctobates*)
sinuatocollis Kaszab, 1988: 103 **A**: HAI SCH YUN **ORR**
striatipennis Lewis, 1894: 473 (*Setenis*) **A**: JA SC
subbiangulata subbiangulata M. Chûjô, 1978a: 68 (*Setenis*) **A**: JA (Ryukyus)
subrobusta indochinensis Kaszab, 1988: 109 **A**: HAI SCH YUN **ORR**
subrobusta subrobusta Motschulsky, 1872: 32 (*Nyctobates*) **A**: "Korea" **ORR**
coracina Fairmaire, 1882c: 230 (*Nyctobates*) [HN]
fairmairei Kolbe, 1900: 74 (*Setenis*) [RN]
sulcicollis Kaszab, 1988: 119 **A**: YUN **ORR**
sulcigera Boisduval, 1835: 256 (*Upis*) **A**: TAI **AUR ORR**
sulcitibia Ren & Hua, 2006: 268 **A**: GUI GUX
szetchuanica Kaszab, 1988: 104 **A**: SCH YUN **ORR**
taiwana Masumoto, 1981e: 89 (*Setenis*) **A**: TAI

similis M. Chûjô, 1981: 136 (*Setenis*)
tatsienlua Kaszab, 1988: 101 **A**: SCH YUN
tokunoshimana Mt. Chûjô, 1995b: 119 **A**: JA (Ryukyus)
tonkinensis Gebien, 1918: 124 [= 1919a: 8] (*Setenis*) **A**: TAI YUN **ORR**
aritai M. Chûjô, 1981: 131 (*Setenis*)
transversicollis Motschulsky, 1872: 29 (*Setenis*) **A**: BT NP SD YUN **ORR**
laevis Fairmaire, 1897j: 230 (*Nyctobates*)
valgipes subvalgipes M. Chûjô, 1981: 138 (*Setenis*) **A**: CH TAI
valgipes valgipes Marseul, 1876c: 117 (*Nyctobates*) **A**: GUI HAI HUB HUN JA JIX SC **ORR**
davidis Fairmaire, 1878a: 119 (*Nyctobates*)
villosipes Marseul, 1876c: 118 (*Nyctobates*)
yunnanica Kaszab, 1988: 96 **A**: YUN

genus *Promorphostenophanes* Kaszab, 1960e: 277 type species *Promorphostenophanes atavus* Kaszab, 1960
atavus Kaszab, 1960e: 279 **A**: YUN

genus *Pseudonautes* Fairmaire, 1893c: 52 type species *Pseudonautes vagevittatus* Fairmaire, 1893
Thydemus Lewis, 1894: 475 type species *Scotaeus purpurivittatus* Marseul, 1876
purpurivittatus Marseul, 1876c: 130 (*Scotaeus*) **A**: JA TAI **ORR**
unicolor Pic, 1923b: 24 (*Thydemus*) **A**: FUJ NP **ORR**
viridiaeneum Allard, 1896: 321 (*Strongylium*) **A**: NP SD

genus *Psydomorphus* Pic, 1921c: 24 type species *Psydomorphus diversipes* Pic, 1921
diversipes Pic, 1921c: 24 **A**: YUN

genus *Rhopalobates* Fairmaire, 1897j: 230 type species *Rhopalobates villardi* Fairmaire, 1897
villardi Fairmaire, 1897j: 231 **A**: FUJ GUA SD **ORR**
klapperichi Kaszab, 1954: 255 (*Encyalesthus*)

genus *Schizomma* Gebien, 1920: 392 type species *Schizomma cucumericola* Gebien, 1920
kondoi Kulzer, 1957: 245 **A**: JA (Bonin Is.)

genus *Scotaeus* Hope, 1834a: 14 type species *Scotaeus corallipes* Hope, 1834
focalis Gebien, 1935: 70 **A**: FUJ JIX

genus *Simalura* Gebien, 1914c: 71 type species *Simalura jacobsoni* Gebien, 1914
Microhemicera Pic, 1921c: 30 type species *Hemicera humeralis* Pic, 1921
coerulea Lewis, 1894: 475 (*Eucyrtus*) **A**: JA
splendens Ando & Ren, 2006: 87 **A**: GUX

genus *Steneucyrtus* Fairmaire, 1896a: 31 type species *Steneucyrtus pexicollis* Fairmaire, 1896
Pseudochariotheca Pic, 1934b: 32 type species *Pseudochariotheca minutissima* Pic, 1934
major Pic, 1934b: 33 (*Pseudochariotheca*) **A**: CH **ORR**

genus *Stenophanes* Solsky, 1876: 294 type species *Hedyphanes mesostena* Solsky, 1871
mesostena Solsky, 1871: 376 (*Hedyphanes*) **A**: FE JA (Tsushima) JIL NC SC NE
rubripennis Marseul, 1876c: 137 (*Helops*)
strigipennis Marseul, 1876c: 138 (*Helops*) **A**: FE JA NMO "Korea"
doii Kôno, 1935a: 89

genus *Taichius* Ando, 1996: 196 type species *Platycrepis hemiceroides* Blair, 1929
forticornis Pic, 1922a: 21 (*Steneucyrtus*) **A**: GUX ZHE **ORR**
frater Ando, 1998: 370 **A**: FUJ ZHE
fukudai Masumoto, 1981e: 81 (*Hemicera*) **A**: TAI

genus *Taiwanomenephilus* Masumoto, 1986b: 61 type species *Taiwanomenephilus chui* Masumoto, 1986
chui Masumoto, 1986b: 62 **A**: TAI

genus *Tearchus* Kraatz, 1880c: 110 type species *Tearchus annulipes* Kraatz, 1880
Heteromerotylus Pic, 1921e: 11 type species *Heteromerotylus bicoloripes* Pic, 1921
annulipes Kraatz, 1880c: 110 **A**: AP BT **ORR**
geniculatus Pic, 1925h: 76 (*Heteromerotylus*) **A**: GUX **ORR**
vitalisi Pic, 1922i: 209 (*Heteromerotylus*) **A**: YUN **ORR**

genus *Teles* Mulsant & Godart, 1876a: 181 type species *Teles eutymi* Mulsant & Godart, 1876
eutymi Mulsant & Godart, 1876a: 182 **A**: TR

genus *Tetragonomenes* Chevrolat, 1878c: clii type species *Tetragonomenes semiviridis* Chevrolat, 1878
Obriomaia Gebien, 1927: 45 type species *Eucyrtus subcostatus* Fairmaire, 1893
Tetragonomecus Rye, 1880: 87 [unjustified emendation]
boninensis Nakane, 1977: 159 (*Obriomaia*) **A**: JA (Bonin Is.)
hirasawai Masumoto, 1984: 20 **A**: TAI
japonicus Masumoto, 1985a: 23 **A**: JA (Ryukyus)
komiyai Masumoto, 1981b: 31 (*Obriomaia*) **A**: JA (Ryukyus)
palpalis Kaszab, 1941c: 64 (*Obriomaia*) **A**: TAI
palpaloides Nakane, 1963a: 28 (*Obriomaia*) **A**: JA
pseudorufiventris Masumoto, 1981c: 46 (*Obriomaia*) **A**: TAI
rufiventris Kaszab, 1964a: 4 (*Obriomaia*) **A**: JA (Ryukyus)
semiviolaceus Nakane, 1968b: 80 (*Obriomaia*) **A**: JA (Ryukyus)

genus *Tetraphyllus* Laporte & Brullé, 1831: 404 type species *Tetraphyllus latreillei* Laporte & Brullé, 1831
Addia Lewis, 1894: 465 type species *Addia scatebrae* Lewis, 1894
Hybonotus Dejean, 1834: 211 type species *Tetraphyllus formosus* Laporte & Brullé, 1831
brunneipes brunneipes Kaszab, 1941c: 60 **A**: TAI
latior Nakane, 1963a: 27 (*Addia*) **A**: JA (Ryukyus)
amamiensis Kaszab, 1964a: 2
monticola Nakakita, 1987: 63 (*Addia*) **A**: TAI
nakanei Masumoto, 1982c: 22 (*Addia*) **A**: TAI
paykullii laevis M. Chûjô, 1977: 11 **A**: JA (Ryukyus)
paykullii paykullii Dalman, 1823: 60 (*Amarygmus*) **A**: JA
lunuliger Marseul, 1876c: 129 (*Artactes*)
punctatus yunnanus Kaszab, 1944b: 57 **A**: YUN
satoi Ando & Sakai, 1999: 157 **A**: TAI
scatebrae Lewis, 1894: 466 (*Addia*) **A**: JA
unicolorata Nakane, 1991: 4
shibatai Nakakita, 1987: 61 (*Addia*) **A**: TAI
tsaii Masumoto, 1986b: 66 (*Addia*) **A**: TAI

genus *Tonkinius* Fairmaire, 1903a: 13 type species *Tonkinius sculptilis* Fairmaire, 1903
sculptilis Fairmaire, 1903a: 13 **A**: FUJ **ORR**
striatipennis Gravely, 1915: 527 **A**: AP
thibetanus Kaszab, 1943: 72 **A**: XIZ **ORR**

genus *Upis* Fabricius, 1792b: 515 type species *Attelabus ceramboides* Linnaeus, 1758
ceramboides Linnaeus, 1758: 388 (*Attelabus*) **E**: BY CT EN FI LT NR NT PL SV SZ UK **A**: CH ES FE HEI KZ MG TR WS **NAR**
reticulata Say, 1824a: 279 (*Zophosis*)
variolosa DeGeer, 1775: 32 (*Tenebrio*)

genus *Zabroideus* Fairmaire, 1894b: 219 type species *Zabroideus pinguis* Fairmaire, 1894
pinguis Fairmaire, 1894b: 220 **A**: XIZ

tribe Stenochiini Kirby, 1837

genus *Eucrossoscelis* Nakane, 1963a: 29 type species *Eucrossoscelis broscosomoides* Nakane, 1963
araneiformis Allard, 1876a: 67 (*Helops*) **A**: JA (Kyushu)
helopioides Lewis, 1894: 482 (*Strongylium*)
broscosomoides Nakane, 1963a: 29 **A**: JA (Amami-Ôshima)
maruyamai Masumoto, 1999a: 121 **A**: JA (Ryukyus)
michioi M. Chûjô, 1978a: 78 **A**: JA (Ryukyus)

genus *Stenochinus* Motschulsky, 1860a: 102 type species *Stenochinus reticulatus* Motschulsky, 1860
Brachypilium Fairmaire, 1896a: 23 type species *Brachypilium sculpturatum* Fairmaire, 1896
Dicraeosis Gebien, 1911: 355 [RN] type species *Dicraeus bacillus* Marseul, 1876
Dicraeus Marseul, 1876c: 103 [HN] type species *Dicraeus bacillus* Marseul, 1876
amplus Gebien, 1914b: 8 (*Dicraeosis*) **A**: TAI
bacillus Marseul, 1876c: 103 (*Dicraeus*) **A**: JA
capucinus Gravely, 1915: 520 (*Dicraeosis*) **A**: NP SD
apterus Kaszab, 1978a: 174 (*Dicraeosis*)
carinatus carinatus Gebien, 1914b: 12 (*Dicraeosis*) **A**: JA (Ryukyus) NP TAI **ORR**
carinatus shirozui Nakane, 1956b: 164 (*Dicraeosis*) **A**: JA (Kyushu)
cylindricus Gebien, 1914b: 9 (*Dicraeosis*) **A**: JA (Ryukyus) TAI
furcifer Shibata, 1980: 73 (*Dicraeosis*) **A**: TAI
meshimanus Nakane, 1985: 163 (*Dicraeosis*) **A**: JA (Meshima Is., Danjo-Gunto Is.)
miyakoensis Nakane, 1985: 163 (*Dicraeosis*) **A**: JA (Ryukyus)
oshimanus Nakane, 1985: 162 (*Dicraeosis*) **A**: JA (Amami-Ôshima)
pokharicus Schawaller, 2006c: 477 **A**: NP
reticulatus Motschulsky, 1860a: 102 **A**: NP **ORR**
sauteri Gebien, 1914b: 11 (*Dicraeosis*) **A**: TAI
subparallelus Pic, 1926b: 21 (*Dicraeosis*) **A**: CH
unicornis Shibata, 1980: 69 (*Dicraeosis*) **A**: TAI
wakoi Masumoto & Akita, 2002: 147 **A**: JA (Ryukyus)
yonagunianus Nakane, 1985: 163 (*Dicraeosis*) **A**: JA (Ryukyus: Yonakuni Is.)

genus *Strongylium* Kirby, 1819: 417 [NP] type species *Strongylium chalconatum* Kirby, 1819
Anomoearthrum Mäklin, 1864: 482 type species *Anomoearthrum debile* Mäklin, 1864
Coelolophus Mäklin, 1864: 394 type species *Coelolophus bicarinatus* Mäklin, 1864
Crossoscelis Gebien, 1914b: 52 type species *Crossoscelis clauda* Gebien, 1914
Ebenolus Fairmaire, 1897j: 227 type species *Ebenolus vernicatus* Fairmaire, 1897
Eustrongylium Kolbe, 1894b: 366 type species *Strongylium episcopale* Kolbe, 1894
Falsolophocnemis Pic, 1917c: 13 type species *Falsolophocnemis sinuatipes* Pic, 1917
Gentinadis Laporte, 1840: 240 type species *Stenochia (Gentinadis) caerulea* Laporte, 1840
Gibbostrongylium Pic, 1917c: 18 type species *Strongylium medanense* Pic, 1917
Massalia Pascoe, 1883: 442 type species *Massalia varians* Pascoe, 1883
Microstrongylium Pic, 1917c: 13 type species *Microstrongylium cyanicolle* Pic, 1917
Notostrongylium Carter, 1915: 523 type species *Notostrongylium rugosicolle* Carter, 1915
Pedostrongylium Pic, 1916f: 11 type species *Strongylium luteonotatum* Pic, 1916
Poecilesthostrongylium Pic, 1918b: 12 type species *Strongylium amazonicum* Pic, 1918
Pseudocaelophus Pic, 1922c: 28 type species *Strongylium difforme* Pic, 1922
Reminius Casey, 1924: 321 type species *Reminius ocularis* Casey, 1924
Saerangodes Gistel, 1848: 126 type species *Saerangodes rubripennis* Gistel, 1848
Stenochia Kirby, 1819: 423 type species *Stenochia rufipes* Kirby, 1819
Strongyliastrum Fairmaire, 1894a: 41 type species *Strongyliastrum braetii* Fairmaire, 1894
Styrax Westwood, 1875: 227 type species *Styrax tricondyloides* Westwood, 1875
Xanthothopeia Mäklin, 1864: 223 type species *Xanthothopeia rufipennis* Mäklin, 1864
Zuercheria Reitter, 1908f: 134 type species *Zuercheria matthiesseni* Reitter, 1908
akitai Masumoto, 1998a: 195 **A**: JA (Ryukyus)

albopilosum Gebien, 1914b: 44 **A**: TAI
alishanum Masumoto, 1981c: 15 **A**: TAI
amamianum Nomura, 1963b: 38 **A**: JA (Ryukyus)
andoi Masumoto, 1996b: 138 **A**: TAI
anhuiense Masumoto, 2000b: 168 **A**: ANH
angustissimum Pic, 1922c: 27 **A**: YUN
antracinum Mäklin, 1864: 306 **A**: "China boreale"
araii M. Chûjô, 1980: 7 [RN] **A**: JA (Sendaku Is.)
costatum M. Chûjô, 1979: 54 [HN]
atritarse Pic, 1916e: 13 **A**: YUN
basifemoratum Mäklin, 1864: 326 **A**: JIA
beesoni Blair, 1937: 37 **A**: SD
bhutanicum Masumoto, 1999b: 337 **A**: BH SD
binhense Pic, 1922h: 101 **A**: YUN **ORR**
nodieri Pic, 1936b: 19
brevicorne Lewis, 1894: 482 **A**: JA
brunneum Yuan & Ren, 2005: 399 **A**: GUX HAI
carbonarium Gebien, 1914b: 47 **A**: TAI
cariosipenne Fairmaire, 1896b: 112 **A**: SD
castanescens Fairmaire, 1896b: 112 **A**: SD
chihpenense Masumoto, 2005b: 205 **A**: TAI
chinense Fairmaire, 1891f: ccxiii **A**: HUB
chutungense Masumoto, 2005b: 202 **A**: TAI
claudum Gebien, 1914b: 53 (*Crossoscelis*) **A**: TAI
clavipes Mäklin, 1864: 331 **A**: JIA
cultellatum cultellatum Mäklin, 1864: 345 **A**: JA HKG SC
cultellatum taiwanum Nomura, 1963b: 40 **A**: CH TAI
culteatum yuwanum Nomura, 1963b: 40 **A**: JA (Amami-Ôshima)
curvicornis Gravely, 1915: 535 **A**: AP
dimidiatum Fairmaire, 1891f: cciv **A**: SCH SHN
endoi Masumoto, 1982d: 55 **A**: TAI
erythrocephalum Fabricius, 1801a: 156 (*Helops*) **A**: HAI TAI **ORR**
flavilabre Fairmaire, 1891f: ccxii **A**: HUB HUN
foochouense Masumoto, 2004: 380 **A**: FUJ
formosanum Gebien, 1914b: 45 **A**: TAI
fujitai Masumoto, 1981a: 46 **A**: TAI
fugianense Masumoto, 2000b: 170 **A**: FUJ
fuscum Yuan & Ren, 2005: 402 **A**: CHQ
gaoliense Masumoto, 2004: 378 **A**: YUN
gardneri Blair, 1930b: 180 **A**: SD
gibbosipenne Nakane, 1963a: 29 **A**: JA
gibbosulum Fairmaire, 1891f: ccxii **A**: GUA JIX
guizhouense Masumoto, 2000b: 166 **A**: GUI
habashanense habashanense Masumoto, 1999a: 114 **A**: YUN
habashanense lijiangense Masumoto, 1999a: 115 **A**: YUN
holcopterum Fairmaire, 1893g: 320 **A**: BT **ORR**
hyacinthinum Kaszab, 1957c: 61 **A**: FUJ
impigrum Lewis, 1894: 481 **A**: JA
impressipenne Pic, 1916d: 12 **A**: FUJ
japanum japanum Marseul, 1876d: 320 **A**: JA
japanum naraense Nakane, 1968b: 84 **A**: JA (Nara Pref.)
japanum tanegashimae Nomura, 1963b: 38 **A**: JA (Tanegashima Is.)
japanum yakushimae Nomura, 1963b: 37 **A**: JA (Yakushima)
jizushanense Masumoto, 1999a: 116 **A**: YUN

jucundum Mäklin, 1864: 353 **A**: HKG
katsumii Masumoto, 1999b: 347 **A**: TAI
kawadai Masumoto, 1981b: 29 **A**: JA (Ryukyus)
kentingense Masumoto, 1981a: 47 **A**: TAI
klapperichi Kaszab, 1957c: 60 **A**: FUJ
kuantouense Masumoto, 2005b: 195 **A**: TAI
kubani Masumoto, 2004: 387 **A**: YUN
laszlorum Masumoto, 2005b: 201 **A**: TAI
lishanum Masumoto, 1981a: 43 **A**: TAI
longipenne Fairmaire, 1903c: 318 (*Coelolophus*) **A**: FUJ
longissimum Gebien, 1914b: 50 **A**: TAI
lutaoense Masumoto, 1992: 93 **A**: TAI
macrops Wiedemann, 1823: 41 (*Helops*) **A**: NP **ORR**
marseuli marseuli Lewis, 1894: 481 **A**: JA NC
apterum Nomura & Yamazaki, 1960: 14
marseuli nigripes Nomura & Yamazaki, 1960: 15 **A**: JA (Hachijo-jima)
marseuli watanabai Nomura & Yamazaki, 1960: 15 **A**: JA (Shikoku)
matthiesseni Reitter, 1908f: 135 (*Zuercheria*) **A**: IN
mesopotamicum Gebien, 1911: 598 [RN] **A**: IQ
cancellatum Abeille de Perrin, 1894a: 94 (*Stenochia*) [HN]
fossum Allard, 1896: 321
metallescens Dohrn, 1880: 377
miwai Masumoto, 1996b: 139 [RN] **A**: TAI
insolitum Miwa, 1939: 413 [HN]
nigripes Nomura & Yamazaki, 1960: 15
watanabei Nomura & Yamazaki, 1960: 15
multiimpressum Pic, 1917e: 23 **A**: CH
multipunctatum Pic, 1936b: 19 **A**: YUN
nakanei Masumoto, 1981a: 49 **A**: TAI
nanfangum Masumoto, 1982d: 53 **A**: TAI
niponicum Lewis, 1894: 480 **A**: JA "Korea"
ochii Masumoto, 1992: 92 **A**: TAI
ochrolatum Ren, 1998: 111 **A**: ZHE
okumurai Masumoto, 1981a: 44 **A**: TAI
opacicolle Fairmaire, 1891f: ccxiv **A**: SCH
osawai Masumoto, 2005b: 196 **A**: TAI
oshimanum Fairmaire, 1897g: 167 **A**: JA (Ryukyus)
palingense Masumoto, 2005b: 200 **A**: TAI
phedongense Masumoto, 1998a: 174 **A**: BH SD
pilimarginum Yuan & Ren, 2005: 403 **A**: GUX
pilosulum Fairmaire, 1891f: ccviii **A**: SCH
pinfaense Masumoto, 2000c: 405 **A**: GUI
pseudogibbosipenne Masumoto, 1981a: 43 **A**: TAI
quadrimaculatum Yuan & Ren, 2005: 401 **A**: SCH
rufipenne L. Redtenbacher, 1844: 533 **A**: NP SD UP **ORR**
braetii Fairmaire, 1894a: 40 (*Strongyliastrum*)
semirufum Pic, 1917c: 15
rufitarse Pic, 1916d: 12 **A**: YUN
saracenum Reiche & Saulcy, 1857: 270 (*Stenochia*) **E**: GR **A**: IS TR
perdubium Gebien, 1911: 599 [RN]
sulcipenne Reitter, 1908f: 134 (*Zuercheria*) [HN]
schawallerianum Masumoto, 2004: 371 **A**: NP
schenklingi Gebien, 1914b: 48 **A**: TAI
shibatai Nomura, 1963b: 39 **A**: JA (Ryukyus)
simulator Dohrn, 1880: 373 **A**: SD **ORR**

interruptum Blair, 1930b: 179
sinuatipenne Miwa, 1939: 414 **A**: TAI
sobrinum Dohrn, 1880: 376 **A**: SD UP
stanislavium Masumoto, 2004: 381 **A**: YUN
stevensi Gravely, 1915: 534 **A**: AP
subaeneum Pic, 1917h: 12 **A**: YUN
sulcielytratum Yuan & Ren, 2005: 400 **A**: GUX
szentivanyi Kaszab, 1941c: 71 **A**: TAI
tabanai Masumoto, 1998a: 196 **A**: SCH
taliense Pic, 1940b: 13 **A**: YUN
taliopacium Masumoto, 2004: 374 **A**: YUN
tanakai Ando, 2003b: 79 **A**: JA
tanikadoi Masumoto, 1998a: 197 **A**: SCH
tehuashense Masumoto, 2005b: 204 **A**: TAI
thibetanum Pic, 1916d: 13 **A**: XIZ YUN
yunnanum Pic, 1936a: 36
uedai Nomura, 1964: 48 **A**: JA (Yakushima)
undulatum kuatunense Kaszab, 1957c: 60 **A**: FUJ
wadai Masumoto, 2005b: 193 **A**: TAI
westermanni Mäklin, 1864: 341 **A**: "Himalaya"
yakushimanum Nakane, 1963a: 29 **A**: JA
yasumatsui M. Chûjô, 1968a: 16 **A**: TAI
yasuhikoi Masumoto, 1996b: 137 **A**: TAI
yokoyamai Masumoto, 1981a: 48 **A**: TAI
yunnanatrum Masumoto, 2004: 373 **A**: YUN
yunnanicum Masumoto, 1999b: 338 **A**: YUN
zoltani Masumoto, 1981c: 16 **A**: TAI

family Tenebrionidae, nomina dubia
Diaperis cornigera Gistel, 1857: 12 **E**: HU
Erelus Mulsant & Rey, 1853f: 185 and *Erelus sulcipennis* Mulsant & Rey, 1853f: 188 from Sicily
Heliopates rugipennis Faldermann, 1837: 56 from Iran
Helops lapidicola Gistel, 1857: 14 **E**: AU
Opatrum glabrum Gistel, 1857: 18 "Taurien"
Opatrum lutosum Gistel, 1857: 47 **E**: PT
Opatrum politum Gistel, 1857: 51 "Taurica chersonesus"
Pedinus sibiricus Gebler, 1833: 287 **A**: Altai
Pygidiphorus Mulsant, 1856f: 520 [= 1856a: unnumb. pp.] with *P. caroli* Mulsant, 1856f: 521 [= 1856a: unnumb. pp.] from France: Bordeaux
Tenebrio tristis Herbst, 1797: 243

family PROSTOMIDAE C. G. Thomson, 1859

genus *Prostomis* Latreille, 1819a: 171 type species *Trogosita mandibularis* Fabricius, 1801
Megagnathus Dejean, 1821: 103 [HN] type species *Trogosita mandibularis* Fabricius, 1801
beatae Schawaller, 1991a: 6 **A**: NP
edithae Schawaller, 1991a: 9 **A**: BT NP SCH YUN **ORR**
grandis Fursov, 1935: 90 **E**: ST "Caucasus"
latoris Reitter, 1889i: 315 **A**: JA SC
mandibularis Fabricius, 1801a: 155 (*Trogosita*) **E**: AB AR AU FR GR GG HU IT PT RO SP ST UK **A**: IN TR
brunneus Fursov, 1935: 90
elburica Fleischer, 1919b: 211
maxillosus P. W. J. Müller, 1805: 205 (*Cucujus*)
strandi Roubal, 1936b: 58
mordax Reitter, 1887d: 287 **A**: FE SC

morsitans Pascoe, 1860b: 98 **A**: NP SD
subtilis Szallies, 1994: 259 **A**: TR
susannae Schawaller, 1991a: 9 **A**: NP **ORR**

family SYNCHROIDAE Lacordaire, 1859

genus *Synchroa* Newman, 1838: 378 type species *Synchroa punctata* Newman, 1838
Phaiona Haldeman, 1848: 99 type species *Phaiona murina* Haldeman, 1848
chinensis Nikitsky, 2000: 45 **A**: SCH
melanotoides Lewis, 1895a: 263 **A**: FE JA JIL "Korea"

family OEDEMERIDAE Latreille, 1810

subfamily **Calopodinae** A. Costa, 1852

genus *Calopus* Fabricius, 1775: 20 type species *Cerambyx serraticornis* Linnaeus, 1758
serraticornis Linnaeus, 1758: 395 (*Cerambyx*) **E**: AL AN AU BH BU BY CR CT CZ EN FI FR GE GG GR HU IT LA LT NR NT PL RO SK SP ST SV SZ UK **A**: ES TR WS
pretneri J. Müller, 1929: 63

genus *Sparedrus* Dejean, 1821: 72 type species *Calopus testaceus* Andersch, 1797
Ocularium Pic, 1922b: 18 type species *Ocularium rufum* Pic, 1922
Polacus Sasaji, 1985: 10 type species *Polacus longicornis* Sasaji, 1985 (= *Sparedrus sasajii* Švihla, 2007
Saloninus Fairmaire, 1891e: cxxxiii type species *Saloninus nebulosus* Fairmaire, 1891
Sparedropsis Heyden, 1886a: 191 type species *Sparedropsis fuscus* Heyden, 1886
albipilis Švihla, 1983a: 108 (*Sparedropsis*) **A**: IN
affinis Švihla, 2006a: 20 **A**: PA
alesivani Švihla, 2006a: 6 **A**: NP
bulgaricus Švihla, 2006b: 124 **E**: BU
championi Švihla, 2006a: 19 **A**: HP UP
concolor Švihla, 2006a: 13 **A**: AF
davidis Fairmaire, 1888b: 131 **A**: HEB
elegans Švihla, 2006a: 17 **A**: PA
fuscus Heyden, 1886a: 191 (*Sparedropsis*) **A**: KI TD UZ
holzschuhi Švihla, 2006a: 13 **A**: AF
hurkai Švihla, 2006a: 22 **A**: PA
kazakhstanicus Švihla, 2006a: 21 **A**: KZ
klapperichi Švihla, 2006a: 14 **A**: AF
kutiauensis Švihla, 2006a: 14 **A**: AF
lencinae Vázquez, 1988: 259 **E**: SP
lepidus Švihla, 2006a: 18 **A**: KA PA
longicollis Švihla, 2006a: 6 **A**: IN
longicornis Iablokoff-Khnzorian, 1970: 62 (*Sparedropsis*) **A**: TD
murinus Švihla, 2006a: 9 **A**: PA
nebulosus Fairmaire, 1891e: cxxxiii (*Saloninus*) **A**: KA
orsinii A. Costa, 1852: 5 **E**: IT
pakistanicus Švihla, 2006a: 16 **A**: PA
proximus Švihla, 2006a: 9 **A**: AF
sasajii Švihla, 2007: 160 [RN] **A**: TAI
longicornis Sasaji, 1985: 10 [HN]
sepiaceus Švihla, 2006a: 21 **A**: PA UP
similis Švihla, 2006a: 12 **A**: AF
subserratus Gressitt, 1939: 218 (*Sparedropsis*) **A**: HAI
testaceus Andersch, 1797: 166 (*Calopus*) **E**: AL AU BH BU CR CZ GR HU IT RO SK SL SZ YU
unicolor Švihla, 1982: 381 (*Sparedropsis*) **A**: KA

uzbekistanicus Švihla, 2006a: 11 **A**: UZ

subfamily **Oedemerinae** Latreille, 1810
tribe Asclerini Semenov, 1894

genus *Alloxantha* Seidlitz, 1899: 828 type species *Alloxantha lutea* Seidlitz, 1899
Alloxanthina Švihla, 1983a: 117 type species *Alloxanthina flava* Švihla, 1983
Ananconiomorpha Švihla, 1983a: 120 type species *Ananconiomorpha platytarsia* Švihla, 1983
albopubens Švihla, 1987: 23 **A**: IN
carinata Karsch, 1881: 50 (*Nacerdes*) **N**: AG EG LB **A**: SI
sinaita Pic, 1929g: 109 (*Colobostomus*)
fernandezi Švihla, 1996: 86 **N**: CI (Gran Canaria)
flava Švihla, 1983a: 117 (*Alloxanthina*) **A**: AE IN OM
flavoides Švihla, 1995: 18 **A**: IN PA
fulva Wollaston, 1854: 523 (*Ditylus*) **E**: PT (Salvages: Pitao Grande) **N**: CI (Gran Canaria, Tenerife)
fuscostrigosa Fairmaire, 1879c: 203 (*Nacerdes*) **N**: AG MO TU **AFR**
lutea Seidlitz, 1899: 830 **N**: CI (Tenerife, El Hierro, La Palma, Gomera, Gran Canaria)
novellae Vázquez, 2002: 24 **N**: CI (Gran Canaria)
obsoleta Fairmaire, 1894c: 312 (*Ananca*) **N**: AG TU **AFR**
ochracea Seidlitz, 1899: 829 **N**: CI (Tenerife, El Hierro)
platytarsia Švihla, 1983a: 120 (*Ananconiomorpha*) **A**: IN OM PA SA
*sahariana*Vázquez, 1991: 107 **N**: MO
seidlitzi Švihla, 1988b: 376 **N**: CI (Tenerife)
svihlai Vázquez, 1993: 127 **N**: AG
talhouki Švihla, 1984: 265 **A**: AE IN OM SA
wollastoni Švihla, 1988b: 375 **N**: CI (Tenerife)

genus *Anacerda* Champion, 1927b: 249 type species *Anacerda leptidioides* Champion, 1927
leptidioides Champion, 1927b: 249 **A**: AF HP UP

genus *Anisochroa* Semenov, 1900d: 652 type species *Anisochroa zarudnyi* Semenov, 1900
semenovi Švihla, 1983a: 111 **A**: IN
volkovitshi Švihla, 1995: 22 **A**: TD TM UZ
zarudnyi Semenov, 1900d: 653 **A**: IN

genus *Anisochrodes* Švihla, 1983a: 108 type species *Anisochrodes jelineki* Švihla, 1983
escalerai Švihla, 1999a: 81 **A**: IN
holzschuhi Švihla, 1983b: 337 **A**: AF
jelineki Švihla, 1983a: 110 **A**: IN PA

genus *Ascleropsis* Seidlitz, 1899: 860 type species *Asclera maculicollis* Ganglbauer, 1889
jakowlewi Semenov, 1894c: 542 (*Asclera*) **A**: KI KZ
kolibaci Švihla, 1997b: 422 **A**: SCH
maculicollis Ganglbauer, 1889: 42 (*Asclera*) **A**: GAN SCH
maculicollis Pic, 1907j: 174 (*Probosca*) [HN]
miladae Švihla, 2005b: 74 **A**: SCH
oudai Švihla, 1999a: 82 **A**: SCH
similis Švihla, 1997b: 421 **A**: SHA SHX
volkovitshi Švihla, 1997b: 421 **A**: SCH

genus *Asessinia* Švihla, 1986a: 196 type species *Sessinia subopaca* Pic, 1924
flavomarginata Miyatake, 1985: 407 (*Anoxacis*) **A**: JA (Shikoku, Ryukyus)
geniculata Kôno, 1937a: 10 (*Alloxacis*) **A**: JA (Ryukyus)
iriomotoensis Miyatake, 1985: 407 (*Anoxacis*) **A**: JA (Iriomote Is.)
vittata Miyatake, 1985: 407 (*Anoxacis*) **A**: JA (Kyushu, Ryukyus)

genus *Chitona* W. L. E. Schmidt, 1844: 46 type species *Stenostoma variegatum* Germar, 1824 (= *Leptura connexa* Fabricius, 1798)
Dolichopyga Seidlitz, 1899: 832 type species *Probosca acuminata* Reitter, 1890 (= *Probosca cretica* Fairmaire, 1861)
baulnyi Fairmaire, 1863: 646 **N**: AG TU
callida Kocher, 1968: 109 **N**: MO
connexa Fabricius, 1798: 154 (*Leptura*) **E**: PT SP **N**: AG MO
variegata Germar, 1824b: 167 (*Stenostoma*)
cretica Fairmaire, 1861: 199 **E**: GR "Balkans" **A**: TR
acuminata Reitter, 1890d: 152 (*Probosca*)
balcanica Pic, 1913r: 169 (*Dolichopyga*)
fucata fucata Faldermann, 1837: 139 (*Nacerdes*) **E**: AB ST (Dagestan) **A**: IN TM
cinerea Motschulsky, 1849: 134 (*Probosca*)
fucata tristis Faldermann, 1837: 140 (*Nacerdes*) **E**: AB (Naxcivan) AR **A**: TR
sieversi Kiesenwetter, 1878: 257
ganglbaueri Reitter, 1889a: 38 **E**: AB (Naxcivan)
gracilicollis Fairmaire, 1879c: 205 **N**: AG
incana W. L. E. Schmidt, 1846: 132 (*Probosca*) **A**: TR
innotaticollis Pic, 1913r: 170 **N**: TU
kocheri Pic, 1949c: 74 **N**: MO
macedonica Švihla, 2006c: 114 **E**: BU GR MC
semividua Fairmaire, 1876: 50 **N**: AG
suturalis A. G. Olivier, 1811: 446 (*Oedemera*) **E**: SP **N**: CI (Fuerteventura) MO
ornata Küster, 1847c: 88
strigilata Suffrian, 1848: 170

genus *Chitonoidea* Švihla, 1983a: 116 type species *Chitonoidea persica* Švihla, 1983
persica Švihla, 1983a: 116 (*Chitonoidea*) **A**: IN IS

genus *Colobostomoides* Švihla, 1983a: 113 type species *Colobostomoides longepubens* Švihla, 1983
longepubens Švihla, 1983a: 115 **A**: IN OM PA

genus *Colobostomus* Fairmaire, 1886a: 452 type species *Colobostomus griseovestitus* Fairmaire, 1886
Isoloxantha Semenov, 1902b: 353 [RN] type species *Ananconia handlirschi* Seidlitz, 1899
Xanthomima Semenov, 1900d: 645 [HN] type species *Ananconia handlirschi* Seidlitz, 1899
arabicus arabicus Švihla, 1984: 260 **A**: SA
arabicus persicus Švihla, 2006d: 3 **A**: IN
griseovestitus Fairmaire, 1886a: 453 **N**: EG **A**: SA YE **AFR**
handlirschi Seidlitz, 1899: 832 (*Ananconia*) **N**: AG EG
boisyi Ancey, 1900: 140 (*Ananca*)
obscuricolor Pic, 1925i: 228
nitidus Pic, 1919e: 14 **N**: EG

genus *Copidita* J. L. LeConte, 1866: 164 type species *Probosca quadrimaculata* Motschulsky, 1853
testaceopyga Pic, 1923c: 32 [DA] **A**: "China"

genus *Dainsclera* Švihla, 1997b: 423 type species *Asclera obsoleta* Ganglbauer, 1889
obscura Švihla, 1980: 51 (*Indasclera*) **A**: BT
obscuroides Švihla, 1987: 18 (*Ascleropsis*) **A**: NP SD
obscuroviridis Švihla, 1997b: 426 **A**: SCH YUN
obsoleta obsoleta Ganglbauer, 1889: 43 (*Asclera*) **A**: GAN SCH SHX
obsoleta sulcicollis Pic, 1929d: 6 (*Asclera*) **A**: FE HEI

genus *Dentostomus* Švihla, 1984: 261 type species *Ananca anceyi* Pic, 1920
Hypascleroides Švihla, 1986a: 197 type species *Hypascleroides socotrensis* Švihla, 1986
anceyi Pic, 1920c: 5 (*Ananca*) **A**: SA YE **AFR**

guichardi Švihla, 1987: 25 (*Paroxacis*) **A**: YE (Abdelkuri I., Suqutra)
socotrensis Švihla, 1986a: 198 (*Hypascleroides*) **A**: YE (Suqutra, Darsah Is.)

genus *Eobia* Semenov, 1894b: 455 type species *Asclera cinereipennis* Motschulsky, 1866
subgenus *Eobia* Semenov, 1894b: 455 type species *Asclera cinereipennis* Motschulsky, 1866
chinensis ambusta Lewis, 1895a: 438 **A**: JA TAI
chinensis chinensis Hope, 1843: 63 (*Nacerdes*) **A**: ZHE (Zhousan Is.)
chinensis kotoensis Kôno, 1937b: 138 **A**: TAI (Lanxu Is.)
cinereipennis cinereipennis Motschulsky, 1866: 173 (*Asclera*) **A**: JA SC
japonica Harold, 1878b: 133 (*Ananca*)
cinereipennis ishigakiana Kôno, 1937a: 9 **A**: JA (Ryukyus)
cinereipennis ogasawarensis Matsumura, 1915: 227 (*Xanthochroa*) **A**: JA (Bonin Is.)
curticornis Pic, 1911p: 188 **A**: TAI
diversithorax Pic, 1927i: 48
formosana Pic, 1923c: 31 **A**: TAI
fuscipennis Nakane, 1954: 178 **A**: JA (Honshu, Ryukyus)
magna M. Chûjô & Satô, 1972: 26 **A**: JA (Iriomote Is.)
matsumurai Kôno, 1937b: 138 **A**: JA (Bonin Is.)
nigriceps Pic, 1923c: 30 **A**: JA
pallidipennis Gressitt, 1939: 219 **A**: HAI
piceopyga Pic, 1923c: 31 **A**: "China"
sinensis Gemminger, 1870a: 124 (*Ananca*) [RN] **A**: HKG
chinensis Boheman, 1858: 111 (*Nacerdes*) [HN]
testaceiceps Pic, 1923c: 30 **A**: "China"
subgenus *Pareobia* Švihla, 1986a: 189 type species *Eobia florilega* Lewis, 1895
florilega Lewis, 1895a: 438 **A**: JA TAI "China"

genus *Ganglbaueria* Semenov, 1891a: 378 type species *Ganglbaueria collaris* Semenov, 1891
subgenus *Ganglbaueria* Semenov, 1891a: 378 type species *Ganglbaueria collaris* Semenov, 1891
collaris Semenov, 1891a: 379 **A**: MG XIN
hedini Pic, 1933b: 8 (*Nacerdochroa*)
turkestanica Švihla, 1995: 10 **A**: KZ
subgenus *Pseudananconia* Švihla, 1983a: 121 type species *Ananconia mesopotamica* Reitter, 1908 (= *Ananca heluanensis* Pic, 1901)
heluanensis Pic, 1901c: 34 (*Ananca*) **N**: EG **A**: IN IQ IS JO SY
mesopotamica Reitter, 1908i: 246 (*Ananconia*)
kermanica Švihla, 1983a: 124 (*Pseudananconia*) **A**: IN
richteri Švihla, 1987: 24 (*Pseudananconia*) **A**: IN
vazquezi Švihla, 1995: 14 **A**: TD
subgenus *Zubkovia* Semenov, 1894b: 467 type species *Zubkovia turcomanica* Semenov, 1894
iranica Švihla, 1995: 12 **A**: IN
turcomanica Semenov, 1894b: 468 (*Zubkovia*) **A**: TD TM

genus *Gilotia* Pardo Alcaide, 1968: 42 type species *Gilotia perroti* Pardo Alcaide, 1968
perroti Pardo Alcaide, 1968: 43 **N**: MO

genus *Hyperopselaphus* Mizota, 1999b: 287 type species *Hyperopselaphus ikedai* Mizota, 1999
ikedai Mizota, 1999b: 288 **A**: JA (Yaeyama Is.) TAI

genus *Indasclera* Švihla, 1980: 48 type species *Asclera indica* Fairmaire, 1894
binotata Pic, 1927c: 13 (*Asclera*) **A**: YUN
bocaki Švihla, 1997b: 443 **A**: SCH
brancuccii Švihla, 1997b: 440 **A**: BT NP SD
brevicollis Gressitt, 1939: 221 (*Asclera*) **A**: FUJ GUA GUI GUX
brunneipennis Lewis, 1895a: 440 (*Asclera*) **A**: JA

carinicollis Lewis, 1895a: 439 (*Oxacis*) **A**: JA (Hokkaido, Honshu)
chinensis Švihla, 1997b: 456 **A**: "China"
diluta Gressitt, 1939: 222 (*Asclera*) **A**: GUA
formosana Pic, 1910u: 95 (*Asclera*) **A**: TAI
haemorrhoidalis Pic, 1907j: 174 (*Probosca*) **A**: YUN
himalaica Švihla, 1980: 50 **A**: NP SD
godawariensis Švihla, 1997b: 437
igai Nakane, 1954: 177 (*Asclera*) **A**: JA
incostata Pic, 1926c: 19 (*Asclera*) **A**: GAN SCH SHA XIZ **ORR**
indica Fairmaire, 1894a: 42 (*Asclera*) **A**: AP BT NP SD
excellens Fleischer, 1919a: 168 (*Ascleropsis*)
japonica amamiana Miyatake, 1985: 405 (*Asclera*) **A**: JA (Ryukyus)
japonica japonica Pic, 1910u: 95 (*Asclera*) **A**: JA (Honshu, Kyushu)
satana Nakane, 1954: 175 (*Asclera*)
loi Gressitt, 1939: 223 (*Asclera*) **A**: HAI
nakanei Mizota, 1999a: 82 **A**: JA (Honshu, Kyushu)
nepalensis lubosi Švihla, 2002: 336 **A**: SD **ORR**
nepalensis nepalensis Švihla, 1987: 17 (*Ascleropsis*) **A**: NP
peculiaris Pic, 1914b: 16 (*Asclera*) **A**: YUN **ORR**
puncticollis Švihla, 2006d: 5 **A**: AP
ruficollis Lewis, 1895a: 434 (*Ditylus*) **A**: JA
konoi Nakane, 1973a: 7 (*Asclera*)
rugosipennis Pic, 1910u: 95 (*Asclera*) **A**: TAI
similis Švihla, 2002: 335 **A**: YUN **ORR**
strangulata Fairmaire, 1889b: 50 (*Anoncodes*) **A**: SCH
subrugosa Kôno, 1937b: 137 (*Asclera*) **A**: JA (Kyushu, Ryukyus)
kyushuensis Nakane, 1954: 176 (*Asclera*)
thibetana Pic, 1915c: 5 (*Asclera*) **A**: "Tibet"
uenoi Švihla, 1997b: 434 **A**: TAI
unicostata Pic, 1914l: 68 (*Asclera*) **A**: "Tibet"
wittmeri Švihla, 1997b: 429 **A**: TAI

genus *Ischnomera* Stephens, 1832: 53 type species *Cantharis caerulea* Linnaeus, 1758
Asclera Dejean, 1834: 228 type species *Necydalis sanguinicollis* Fabricius, 1787
Ascleronia Seidlitz, 1899: 859 type species *Ischnomera semiflava* Reitter, 1891
Chromasclera Seidlitz, 1899: 859 type species *Asclera partitipennis* Fairmaire, 1892
abdominalis Heyden, 1887b: 304 (*Asclera*) **A**: BEI FE GAN JA (Honshu)
analis Fairmaire, 1888b: 131 (*Oedemera*) [HN]
flavipes Reitter, 1891a: 31
apicipennis Švihla, 1992: 40 **A**: KZ
auripennis Reitter, 1903b: 45 (*Asclera*) **E**: GR **A**: TR
bajsunensis Švihla, 2004: 64 **A**: UZ
besucheti Švihla, 1993a: 53 **A**: PA
bilyi Švihla, 1992: 41 **A**: TD UZ
caerulea Linnaeus, 1758: 403 (*Cantharis*) **E**: AB AL AR AU BH BU CR CZ FR GB GE GG GR HU IT MC NR PL RO SK SL SP ST SV SZ UK YU **A**: IN TM TR
carinithorax Roubal, 1936a: 259 (*Asclera*)
coerulescens Fabricius, 1775: 208 (*Necydalis*)
nigripes A. G. Olivier, 1811: 446 (*Oedemera*)
obliterata Pic, 1926e: 6 (*Asclera*)
opacocoerulea Reitter, 1911a: 408
quadrinervosa Latreille, 1804b: 12 (*Oedemera*)
unicolor Melsheimer, 1846: 54
cinerascens bodemeyeri Ganglbauer, 1900: 161 **A**: TR
cinerascens cinerascens Pandellé, 1867: 182 (*Asclera*) **E**: AU BH CR CZ DE FR GB GE HU IT NR PL RO SK SL SP SV SZ YU

cinerascens kurdistana Švihla, 1988a: 219 **A**: IQ TR
cinerascens reitteri Ganglbauer, 1881: 106 **E**: AB AL BU GR **A**: TR
cyanea Fabricius, 1792b: 351 (*Necydalis*) **E**: AL AU BH BU CR CZ DE FR GB GE GR HU IT MC MD NL PL RO SK SL SP SV SZ TR UK YU **A**: TR
graeca Dahlgren, 1976: 22 (*Asclera*)
viridescens Pic, 1920c: 5 (*Asclera*)
danilevskiji Švihla, 1992: 39 **A**: TD
deyrollei Pic, 1895h: 243 (*Asclera*) **E**: GG ST (Caucasus) **A**: SY TR
milleri Pic, 1898c: 38 (*Asclera*)
dzhungarica Švihla, 1986b: 216 **A**: KZ MG XIN
erdemliensis Švihla, 1998c: 239 **A**: SY TR
fulvicollis Reitter, 1897a: 38 (*Asclera*) **E**: GR (Kríti, Ikaria) **A**: CY TR
fuscipennis Švihla, 1988a: 209 **A**: TR
girardi Švihla, 1992: 37 **A**: YUN
haemorrhoidalis W. L. E. Schmidt, 1846: 43 (*Asclera*) **E**: GR TR **A**: TR
impressithorax Pic, 1897e: 42 (*Asclera*)
hauseri ferganica Švihla, 1992: 41 **A**: KI
hauseri hauseri L. Heyden, 1887c: 322 (*Asclera*) **A**: KZ TD UZ
holzschuhi Švihla, 1988a: 207 **A**: IN
jezeki Švihla, 1988a: 219 **A**: IN
klapperichi klapperichi Švihla, 1997a: 86 **A**: AF
klapperichi wittmeri Švihla, 1997a: 87 **A**: PA
kopetzi Švihla, 2004: 64 **A**: NP
lutescens Abeille de Perrin, 1892: 64 (*Asclera*) **A**: TR
maceki Švihla, 1988a: 208 **A**: IN IQ TR
nigricans Fairmaire, 1897j: 238 (*Asclera*) **A**: HP NP **ORR**
viator Champion, 1927b: 250 (*Asclera*)
nigrocyanea koreana Fleischer, 1919a: 168 **A**: FE "Korea"
nigrocyanea nigrocyanea Lewis, 1895a: 440 (*Asclera*) **A**: FE (Kuriles) JA
okushimai Švihla, 1997a: 85 **A**: JA (Okinawa)
partitipennis Fairmaire, 1892b: 158 (*Asclera*) **A**: IN IQ TR
iranica Zaitzev, 1941: 247 (*Asclera*)
persica Švihla, 1992: 43 **A**: IN
rosti Pic, 1914j: 57 (*Asclera*) **A**: KA PA
russowi Seidlitz, 1899: 851 (*Asclera*) **A**: KI KZ TD UZ
turkestanica Pic, 1912e: 26 (*Asclera*)
sanguinicollis Fabricius, 1787a: 170 (*Necydalis*) **E**: AL AU BE BH BU CR CZ DE FR GB GE GR HU IR IT LA LT MC NR PL RO SK SL SP SV SZ UK YU
flavicollis Panzer, 1794c: 18 (*Necydalis*)
semiflava Reitter, 1891a: 31 **A**: KZ UZ XIN
nigripennis Reitter, 1897a: 37 (*Asclera*)
similis Švihla, 1982: 384 (*Asclera*) **A**: IN TR
sveci Švihla, 2004: 62 **A**: TR
taiwana Švihla, 1992: 38 **A**: TAI
tenietensis Abeille de Perrin, 1896: 283 (*Asclera*) **N**: AG TU
testaceipes testaceipes Pic, 1914j: 57 (*Asclera*) **A**: KA
testaceipes viridipennis Švihla, 1986b: 215 **A**: PA
xanthoderes Mulsant, 1858a: 232 (*Asclera*) **E**: FR PT SP **N**: MO

genus *Nacerdochroa* Reitter, 1893a: 113 type species *Nacerdochroa plutschevskyi* Reitter, 1893 (= *Nacerdes caspia* Faldermann, 1836)
Holoxantha Semenov, 1894b: 470 type species *Ditylus concolor* Brullé, 1838
carinatopyga Švihla, 1983a: 113 **A**: AE IN OM SA YE
caspia Faldermann, 1836: 392 (*Nacerdes*) **E**: AB ST **A**: KZ TD TM UZ

glasunovi Semenov, 1894b: 466
plutschevskyi Reitter, 1893a: 114
concolor Brullé, 1838: 70 (*Ditylus*) **E**: PT (Salvages) **N**: CI
hierrensis Franz, 1985: 102 (*Alloxantha*)
pici Švihla, **nom. nov**. [see New Acts] **N**: MO
concolor Pic, 1942c: 11 (*Sessinia*) [HN]

genus *Oschaninia* Semenov, 1898: 175 type species *Ganglbaueria wilkinsi* Heyden, 1894
subgenus *Homomorpha* Semenov, 1898: 177 type species *Homomorpha crucifera* Semenov, 1898
crucifera Semenov, 1898: 177 (*Homomorpha*) **A**: XIN
subgenus *Oschaninia* Semenov, 1898: 175 type species *Ganglbaueria wilkinsi* Heyden, 1894
semenovi Kuznetzova, 1935: 298 (*Homomorpha*) **A**: TM
wilkinsi Heyden, 1894a: 52 (*Ganglbaueria*) **A**: MG TM
zhelochovtzevi Kuznetzova, 1935: 300

genus *Paroxacis* Arnett, 1951: 334 type species *Probosca lucana* LeConte, 1866
sahariana Švihla, 1995: 22 **N**: AG

genus *Probosca* W. L. E. Schmidt, 1846: 130 type species *Probosca viridana* W. L. E. Schmidt, 1846 (= *Callidium virens* Fabricius, 1792)
subgenus *Ananconia* Seidlitz, 1899: 831 type species *Chitona martini* Fairmaire, 1896
martini Fairmaire, 1896c: 224 (*Chitona*) **A**: SA YE **AFR**
spurcaticollis Fairmaire, 1886a: 452 (*Ananca*) **A**: SA YE **AFR**
subgenus *Asclerella* Semenov, 1900d: 648 type species *Asclerella tenera* Semenov, 1900
tenera Semenov, 1900d: 651 (*Asclerella*) **A**: IN PA
elongatipennis Pic, 1935e: 101 (*Ananconia*)
subgenus *Probosca* W. L. E. Schmidt, 1846: 130 type species *Probosca viridana* W. L. E. Schmidt, 1846 (= *Callidium virens* Fabricius, 1792)
infima Fairmaire, 1879c: 204 **N**: AG
letorneuxi Pic, 1894j: 93 **N**: EG **A**: CY
metallescens Fairmaire, 1866a: 57 (*Chitona*) **N**: MO
pallipes A. G. Olivier, 1811: 446 (*Oedemera*) **N**: EG
unicolor Küster, 1847c: 87 (*Chitona*) **E**: SP **N**: MO
plumbea Suffrian, 1848: 169
virens Fabricius, 1792b: 338 (*Callidium*) **E**: IT (Sardegna, Sicilia) SP **N**: AG TU
boiteli Pic, 1936f: 148
hispanica Pic, 1920c: 5
nigrofemorata Pic, 1898h: 122
purpureomicans Pic, 1913r: 170
viridana W. L. E. Schmidt, 1846: 130
subgenus *Proboxantha* Švihla, 1995: 16 type species *Ananconia comata* Blair, 1923
afroarabica Švihla, 1984: 262 (*Ananconia*) **N**: EG **A**: IS YE **AFR**
batesi Švihla, 1984: 263 (*Ananconia*) **A**: IQ SA
comata Blair, 1923a: 125 (*Ananconia*) **A**: AE IN IQ JO SA
fuscipennis Blair, 1923a: 124 (*Isoloxantha*) **A**: AE IN IQ KU SA
persica Pic, 1929d: 6 (*Zubkovia*)
maindroni Pic, 1935e: 101 (*Ananconia*) **A**: PA YE (Suqutra)
marginata Walker, 1871: 17 **N**: EG LB **A**: SA YE **AFR**
notatithorax Pic, 1906c: 26

genus *Schellia* Reitter, 1914c: 49 type species *Schellia sitaroides* Reitter, 1914
sitaroides Reitter, 1914c: 50 **A**: KZ TM

genus *Schistopselaphus* Fairmaire, 1896b: 126 type species *Schistopselaphus apicatus* Fairmaire, 1896
sonani Kôno, 1937b: 140 **A**: TAI (Lanxu Is.)

genus *Xanthochroina* Ganglbauer, 1881: 105 type species *Xanthochroa auberti* Abeille de Perrin, 1876
Anancosessinia Kôno, 1937b: 139 type species *Anancosessinia tarsalis* Kôno, 1937
Asclerostoma Fleischer, 1919a: 169 type species *Asclerostoma reitteri* Fleischer, 1919
auberti Abeille de Perrin, 1876: clxvi (*Xanthochroa*) **E**: CR FR GR IT SP **A**: LE TR
latetestacea Pic, 1920g: 18
reitteri Fleischer, 1919a: 170 (*Asclerostoma*)
tarsalis Kôno, 1937b: 140 (*Anancosessinia*) **A**: JA (Ryukyus) TAI

tribe Ditylini Mulsant, 1858

genus *Ascleranoncodes* Pic, 1915i: 10 type species *Ascleranoncodes distincticornis* Pic, 1915
Indanerces Pic, 1923c: 24 type species *Indanerces nitidicollis* Pic, 1923
suturalis Švihla, 1998b: 163 (*Pseudonerdanus*) **A**: NP SD

genus *Chrysanthia* W. L. E. Schmidt, 1844: 46 type species *Cantharis viridissima* Linnaeus, 1758
bhutanica Švihla, 1980: 54 **A**: BT NP
canaliculata Iablokoff-Khznorian, 1974: 190 **A**: TD
chalcochroa chalcochroa Fairmaire, 1892b: 159 **A**: TR
distinctithorax Pic, 1904e: 57
tauricola Pic, 1920c: 5
chalcochroa holzschuhi Pardo Alcaide, 1972: 214 **A**: TR (Prov. Antalya)
cyprica Pic, 1920c: 5 **A**: CY
flavipes Reitter, 1889f: 266 **E**: GR (Sporades) **A**: IS JO SY TR
oralis Fairmaire, 1892b: 158
semicuprea Pic, 1920c: 5
fuscimembris fuscimembris Fairmaire, 1891e: cxxxiv **A**: KA PA UP
cyanescens Pic, 1926c: 19
fuscimembris kaganensis Švihla, 2001a: 96 **A**: PA
geniculata chinensis Švihla, 1999a: 80 **A**: HEB HUB SHA SCH
geniculata geniculata W. L. E. Schmidt, 1846: 128 **E**: AN AU BE BH BU BY CR CT CZ DE EN FI FR GB GE GG GR HU IT LA LT LU MD NL NR NT PL RO SK SL SP SV SZ UK YU **A**: ES HEB TM WS
cupreomicans Westhoff, 1881: 195
geniculata L. Heyden, 1877: 293 [HN]
nigricornis Westhoff, 1881: 195
saturata Westhoff, 1881: 195
viridis W. L. E. Schmidt, 1846: 128 [HN]
geniculata integricollis Heyden, 1886b: 275 **A**: FE NC SC
geniculata viatica Lewis, 1895a: 443 **A**: FE (Sakhalin, Kuriles) JA (Hokkaido, Honshu, Tsushima)
krali Švihla, 1982: 382
hamata Vázquez, 1989: 130 **E**: SP
himalaica Champion, 1920a: 74 **A**: HP NP PA UP
maroccana Håk. Lindberg, 1932: 34 **N**: MO
reitteri Seidlitz, 1899: 866 **E**: PT SP
rugicollis Champion, 1920a: 75 **A**: HP **ORR**
superba Reitter, 1872: 182 **E**: PT SP **N**: AG LB MO TU
oranensis Seidlitz, 1899: 867
valens Champion, 1920a: 75 **A**: HP
varipes planiceps Kiesenwetter, 1878: 256 **E**: AR GG ST (Kavkaz) **A**: TR
turcica Pic, 1899a: 15
varipes varipes Kiesenwetter, 1861a: 252 **E**: AL BH BU CR GR MC TR YU
viridissima Linnaeus, 1758: 403 (*Cantharis*) **E**: AL AN AU BE BH BU BY CR CT CZ EN FI FR GE GG GR HU IT LA LS LT MC NR NT PL RO SK SL SP SV SZ YU **A**: ES WS
cuprina Rey, 1892: 138
diversipes Pic, 1932h: 30
korbi Reitter, 1894c: 104
thalassina Fabricius, 1792b: 350 (*Necydalis*)
viridis DeGeer, 1775: 15 (*Cantharis*)

wittmeri obscurimembris Švihla, 2001a: 96 **A**: KA PA
wittmeri wittmeri Švihla, 1980: 56 **A**: KA

genus *Diasclera* Reitter, 1913d: 663 type species *Diasclera viridescens* Reitter, 1913
Eumecomera Arnett, 1951: 297 type species *Asclera obscura* LeConte, 1854
Nacerdasclera Munster, 1921: 8 type species *Oedemera sibirica* Gebler, 1830
sibirica laevithorax Pic, 1915h: 9 (*Asclera*) **A**: FE JIL
ussurica Pic, 1926e: 6 (*Asclera*)
sibirica sibirica Gebler, 1829a: 132 (*Oedemera*) **A**: ES MG WS "China"
viridescens oxiana Iablokoff-Khnzorian, 1974: 188 **A**: TD
viridescens viridescens Reitter, 1913d: 663 **E**: AB

genus *Diplectrus* Kirsch, 1866: 210 type species *Diplectrus ferrugineus* Kirsch, 1866
improvisus Švihla, 1999a: 77: 210 **A**: SCH
longipennis Fairmaire, 1896b: 126 (*Ananca*) **A**: SD
testaceipennis Pic, 1923c: 28 (*Patiala*)
piceipennis Pic, 1923c: 28 (*Patiala*) **A**: BT SD
variicollis Švihla, 1999a: 77 **A**: SCH

genus *Ditylus* Fischer von Waldheim, 1817: 469 type species *Ditylus helopioides* Fischer von Waldheim, 1817 (= *Helops laevis* Fabricius, 1787)
Mimetes Eschscholtz, 1818: 467 type species *Mimetes unicolor* Eschscholtz, 1818 (= *Helops laevis* Fabricius, 1792)
laevis laevis Fabricius, 1787a: 213 (*Helops*) **E**: AU BY CZ EN FI GE LA LT NT PL SK SV **A**: ES FE JA (Hokkaido, Honshu, Tsushima) NC NE
helopioides Fischer von Waldheim, 1817: pl. 5
unicolor Eschscholtz, 1818: 468 (*Mimetes*)
laevis oxianus Švihla, 1993a: 52 **A**: TM
praetermittus Švihla, 2005b: 72 **A**: SCH
sichuanus Švihla, 1996: 83 **A**: SCH
tibetanus Švihla, 2005b: 72 **A**: XIZ

tribe Nacerdini Mulsant, 1858

genus *Anogcodes* Dejean, 1834: 228 type species *Necydalis melanura* Fabricius, 1787
Anoncodes W. L. E. Schmidt, 1844: 46 type species *Necydalis melanura* Fabricius, 1787
Anoncodina Seidlitz, 1899: 786 type species *Necydalis ruficollis* Fabricius, 1781
Lethonymus Marseul, 1857: 130 type species *Lethonymus difformis* Marseul, 1857
Oedechira Motschulsky, 1872: 54 type species *Oedemera paradoxa* Faldermann, 1837 (= *Cantharis ustulatus* Scopoli, 1763)
Pachychirus L. Redtenbacher, 1845: 134 type species *Necydalis adusta* Panzer, 1795 (= *Cantharis ustulatus* Scopoli, 1763)
Peronocnemis Fairmaire, 1886c: 352 type species *Peronocnemis davidis* Fairmaire, 1886
caucasicus Pic, 1906b: 19 (*Nacerda*) **E**: "Caucasus"
latebasalis Pic, 1926e: 5 (*Nacerdes*)
coarctatus coarctatus Germar, 1824b: 165 (*Oedemera*) **E**: BY CT **A**: ES KZ MG WS
nigriventris Motschulsky, 1859d: 493 (*Anoncodes*)
coarctatus croceiventris Motschulsky, 1859d: 453 (*Anoncodes*) **A**: FE JA (Hokkaido, Honshu) NC
manciuricus Magistretti, 1939: 159 (*Nacerda*)
sambuceus Lewis, 1895a: 439 (*Anoncodes*)
davidis Fairmaire, 1886c: 352 (*Peronocnemis*) **A**: GAN QIN SCH XIZ YUN
femoratus Ganglbauer, 1889: 38 (*Anoncodes*)
difformis Marseul, 1857: 130 (*Lethonymus*) **E**: AB AR GR (Ródos) ST **A**: CY IN LE TR
bodemeyeri Pic, 1911m: 170 (*Anoncodes*)
signaticollis Pic, 1898a: 2 (*Anoncodes*)
tschitscherini Semenov, 1895b: 248 (*Nacerdes*)
versicolor Chevrolat, 1873: 205 (*Anoncodes*)
flaviceps Fairmaire, 1866a: 57 (*Anoncodes*) **N**: MO

fulvicollis Scopoli, 1763: 43 (*Cantharis*) **E**: AN AU BH BU CR CZ FR GE GR HU IT LS PL RO SK SL SP SZ UK YU

longicollis Roubal, 1936a: 257 (*Nacerda*)
medius Gredler, 1866: 295 (*Anoncodes*)
obscurus Gyllenhal, 1810: 630 (*Necydalis*)
therondi Méquignon, 1948: 78 (*Nacerda*)

geniculatus W. L. E. Schmidt, 1846: 99 (*Anoncodes*) **A**: TR

basalis W. L. E. Schmidt, 1846: 99 (*Anoncodes*)

melanurus Fabricius, 1787a: 170 (*Necydalis*) **E**: AU BH BU BY CR CT CZ EN FR GE GR HU IT LA LT MC NL PL RO SK SL SP ST SZ UK **A**: KZ TR WS XIN

fasciatus Villers, 1789: 283 (*Necydalis*)
rufus Fischer von Waldheim, 1820: pl. 5 (*Ditylus*)
scutellaris Waltl, 1839: 222 (*Anoncodes*)
ustulatus Fabricius, 1787a: 170 (*Necydalis*) [HN]

ruficollis Fabricius, 1781: 263 (*Necydalis*) **E**: AR AU BH BU CR CZ FR GE GG GR HU IT MC RO SK SP ST SZ UK YU **A**: IN IQ SY TR

coerulescens Rossi, 1790: 172 (*Necydalis*)

rufiventris Scopoli, 1763: 43 (*Cantharis*) **E**: AU BH BU BY CR CT CZ EN FR GE GR HU IT LA LT MC NT PL RO SK SL SP SV SZ UK YU

bipartitus Schrank von Paula, 1776: 68 (*Leptura*)
dorsalis A. G. Olivier, 1811: 444 (*Oedemera*)
melanocephalus Fabricius, 1794: 453 (*Necydalis*)

schatzmayri H. Wagner, 1928: 117 (*Nacerda*) **E**: FR SP

seladonius alpinus W. L. E. Schmidt, 1846: 121 (*Anoncodes*) **E**: AU FR GE IT SL SZ

azureus W. L. E. Schmidt, 1846: 123 (*Anoncodes*)

seladonius austriacus Ganglbauer, 1881: 103 (*Nacerdes*) **E**: AU HU RO SK

seladonius seladonius Fabricius, 1792b: 352 (*Necydalis*) **E**: FR IT PT SP SZ **N**: AG TU

amoenus W. L. E. Schmidt, 1846: 118 (*Anoncodes*)
aurosus Fairmaire, 1863: 645 (*Nacerdes*)
dispar Dufour, 1841: 8 (*Oedemera*)
meridionalis A. Costa, 1852: 10 (*Anoncodes*)
viridipes W. L. E. Schmidt, 1846: 116 (*Anoncodes*)

seladonius turcicus W. L. E. Schmidt, 1846: 120 (*Anoncodes*) **E**: AL BH BU CR GR MC TR YU

stenoderus Fairmaire, 1876: 50 (*Anoncodes*) **N**: AG

ustulatus Scopoli, 1763: 43 (*Cantharis*) **E**: AB AR AU BH BU BY CR CT CZ EN FR GE GG HU IT LT MC PL RO SK SL SP ST SV SZ UK YU **A**: IN TR WS

adustus Panzer, 1795: 279 (*Necydalis*)
axillaris Ménétriés, 1832: 212 (*Oedemera*)
collaris Panzer, 1795: 278 (*Necydalis*)
ferrugineus Schrank von Paula, 1776: 68 (*Leptura*)
flavipennis Motschulsky, 1872: 55 (*Oedechira*)
flaviventris Faldermann, 1837: 147
hybridus Rossi, 1794: 56 (*Necydalis*)
inlateralis Pic, 1930a: 1 (*Anoncodes*)
melanocephalus Fischer von Waldheim, 1829b: 370 (*Ditylus*) [HN]
paradoxus Faldermann, 1837: 147 (*Oedemera*)
zimmermanni Gistel, 1857a: 71 (*Anoncodes*)

wartmanni Pic, 1894m: 180 (*Anoncodes*) **E**: SP **N**: AG MO

genus *Nacerdes* Dejean, 1834: 228 type species *Necydalis notata* Fabricius, 1792 (= *Cantharis melanura* Linnaeus, 1758)

Nacerda Stephens, 1839: 337 type species *Cantharis melanura* Linnaeus, 1758
Nacerdoscuta Pic, 1915c: 6 type species *Nacerdoscuta semirufa* Pic, 1915

subgenus *Allagatha* Semenov & Ter-Minasjan, 1937: 227 type species *Allagatha decora* Semenov & Ter-Minasjan, 1937 (= *Xanthochroa crassifemur* Blair, 1935)

crassifemur Blair, 1935b: 302 (*Xanthochroa*) **A**: KA PA

decora Semenov & Ter-Minasjan, 1937: 228 (*Allagatha*)

subgenus *Nacerdes* Dejean, 1834: 228 type species *Necydalis notata* Fabricius, 1792 (= *Cantharis melanura* Linnaeus, 1758)

Nacerda Stephens, 1839: 337 type species *Cantharis melanura* Linnaeus, 1758

Nacerdoscuta Pic, 1915c: 6 type species *Nacerdoscuta semirufa* Pic, 1915

brancuccii Švihla, 1983b: 336 **A**: AF KA PA

melanura Linnaeus, 1758: 403 (*Cantharis*) **E**: AB AL AU BE BH BU CR CZ DE EN FI FR GB GE GG GR HU IR IT LA LT MA MR NL NR NT PL PT RO SK SP ST SV SZ UK YU **N**: AG CI MO TU **A**: AF JA NC TAI TR **AFR AUR NAR NTR**

acuta Marsham, 1802: 372 (*Cantharis*)

analis A. G. Olivier, 1811: 443 (*Oedemera*)

apicalis Say, 1835: 188 (*Oedemera*) [HN]

erminea Germar, 1817: 225 (*Necydalis*)

italica Chevrolat, 1877b: 121 (*Xanthochroa*)

lepturoides Thunberg, 1784: 18 (*Cantharis*)

nigripes Fabricius, 1781: 260 (*Cantharis*)

notata Fabricius, 1792b: 353 (*Necydalis*)

rufa Brullé, 1838: 70 (*Ditylus*)

sardea W. L. E. Schmidt, 1846: 34

testacea Geoffroy, 1785: 155 (*Cantharis*)

zoufali Reitter, 1907d: 161

semirufa Pic, 1915c: 6 (*Nacerdoscuta*) **A**: KA PA

subgenus *Xanthochroa* W. L. E. Schmidt, 1844: 46 type species *Oedemera carniolica* Gistel, 1832

Asiochroa Švihla, 1998a: 56 type species *Xanthochroa metallipennis* Fairmaire, 1889 (= *Xanthochroa waterhousei* Harold, 1875)

Axanthochroa Švihla, 1986a: 174 type species *Xanthochroa brachyptera* Švihla, 1980

Patiala Lewis, 1895a: 434 type species *Patiala antennata* Lewis, 1895

Patialomorpha Nakane, 1954: 174 type species *Xanthochroa ainu* Lewis, 1895

ainu Lewis, 1895a: 436 (*Xanthochroa*) **A**: FE (Sakhalin, Kuriles) JA

akiyamai Švihla, 1998a: 68 **A**: TAI

antennata Lewis, 1895a: 434 (*Patiala*) **A**: JA

apicipennis Švihla, 1998a: 50 **A**: TAI

asahinai Nakane, 1958: 91 (*Xanthochroa*) **A**: JA (Honshu, Shikoku)

atriceps atriceps Lewis, 1895a: 436 (*Xanthochroa*) **A**: FE JA "Korea"

atriceps shikokuana Iga & Nakane, 1954: 154 (*Xanthochroa*) **A**: JA

atripennis Pic, 1934b: 29 (*Xanthochroa*) **A**: SCH YUN

baibarana Kôno, 1932a: 139 (*Xanthochroa*) **A**: TAI

becvari Švihla, 1998a: 52 **A**: YUN

bicostata Lewis, 1895a: 436 (*Xanthochroa*) **A**: JA

brachyptera Švihla, 1980: 46 (*Xanthochroa*) **A**: BT

brendelli Švihla, 1987: 11 **A**: NP SD

carniolica atlantica Allemand, 1993: 8 **A**: FR SP

carniolica carniolica Gistel, 1834b: 150 (*Oedemera*) **E**: AL AU BH BU BY CR CZ FR GE GG GR HU IT LS MC RO SK SL SP SV SZ YU

blossevillei Guérin-Méneville, 1838: 40 (*Oedemera*)

carniolica foveata Fairmaire, 1892b: 157 (*Xanthochroa*) **A**: TR

carniolica peloponesiaca Švihla, 1991: 3 **E**: GR (Pelopponisos, Kríti)

caroli Švihla, 2001c: 226 **A**: BT

caudata Kôno, 1936d: 517 (*Xanthochroa*) **A**: JA

dedicata Švihla, 2001c: 228 **A**: AP **ORR**

deformis Lewis, 1895a: 435 (*Patiala*) **A**: JA

impressithorax Pic, 1937b: 24 (*Xanthochroa*)

fujiana Švihla, 1998a: 66 **A**: FUJ

fulvicrus Fairmaire, 1889b: 50 (*Xanthochroa*) **A**: GAN GUI HUB SHA SCH SHX

semenowi Ganglbauer, 1889: 37 (*Xanthochroa*)

fuscipennis Champion, 1927b: 247 (*Xanthochroa*) **A**: NP UP

gracilis bellieri Reiche, 1861c: 209 (*Xanthochroa*) **E**: FR SZ

testacea Pic, 1920g: 17 *Xanthochroa*) [HN]

tournieri Pic, 1894h: 72 (*Xanthochroa*)
gracilis gracilis W. L. E. Schmidt, 1846: 37 (*Xanthochroa*) **E**: AU BH CR FR GG GR IT RO SP ST SZ UK
grandis Pic, 1923c: 23 (*Xanthochroa*) **A**: BT NP "India"
guizhouensis Švihla, 2001c: 228 **A**: GUI
hesperica Magistretti, 1941b: 227 (*Xanthochroa*) **E**: SP
hilleri hilleri Harold, 1878a: 81 (*Xanthochroa*) **A**: FE (Kuriles) JA
hilleri kurosai Nakane, 1988a: 3 (*Xanthochroa*) **A**: JA (Hokkaido, Honshu)
himalaica himalaica Blair, 1930b: 181 (*Xanthochroa*) **A**: NP SD
hiromichii Švihla, 2004: 72 [RN] **A**: TAI
apicalis Kôno, 1932a: 141 (*Xanthochroa*) [HN]
holzschuhi Švihla, 1987: 10 **A**: NP XIZ
kantneri Švihla, 2001c: 230 **A**: TAI
katoi Kôno, 1932a: 140 (*Xanthochroa*) **A**: JA
konoi Nakane, 1954: 172 (*Xanthochroa*) **A**: JA
kubani Švihla, 1998a: 60 **A**: SCH YUN
ludmilae Švihla, 1998a: 52 **A**: HUB
luteipennis Marseul, 1877: 484 (*Xanthochroa*) **A**: FE JA NC NE
martensi Švihla, 1993a: 53 **A**: NP
ocularis Lewis, 1895a: 435 (*Patiala*) **A**: JA
osawai Nakane, 1954: 173 (*Xanthochroa*) **A**: JA
potanini Ganglbauer, 1889: 36 (*Xanthochroa*) **A**: SCH
raymondi canyellesi Vázquez, 2006: 40 **E**: SP (Mallorca)
raymondi ciliciensis Švihla, 1991: 6 **E**: GR (Ródos) **A**: TR
raymondi raymondi Mulsant & Godart, 1860: 158 (*Xanthochroa*) **E**: FR SP **N**: AG
barbara Peyerimhoff, 1918b: 143 (*Xanthochroa*)
schneideri Švihla, 2001c: 227 **A**: SCH
spinicoxis hirai Nakane, 1954: 175 (*Xanthochroa*) **A**: JA (Shikoku)
spinicoxis spinicoxis Nakane, 1954: 174 (*Xanthochroa*) **A**: JA (Honshu, Shikoku)
strandi Kôno, 1936d: 516 (*Xanthochroa*) **A**: FE (Kuriles) JA
subviolacea Pic, 1911f: 101 (*Xanthochroa*) **A**: YUN
yunnana Pic, 1923c: 23 (*Xanthochroa*)
taiwana Kôno, 1932a: 140 (*Xanthochroa*) **A**: TAI
transfretalis Švihla, 1998a: 66 **A**: TAI
umenoi okinawana M. Chûjô & Satô, 1972: 26 (*Xanthochroa*) **A**: JA (Okinawa)
umenoi umenoi Kôno, 1937a: 3 (*Xanthochroa*) **A**: JA (Amami-Ôshima, Okinawa)
violaceonotata Pic, 1922c: 22 (*Xanthochroa*) **A**: YUN **ORR**
wadai Nakane, 1954: 171 (*Xanthochroa*) **A**: JA
wardi Švihla, 1987: 5 **A**: NP XIZ
waterhousei Harold, 1875: 93 (*Xanthochroa*) **A**: FE (Sakhalin) GAN HUB JA SCH SHA "Korea"
cyanipennis Marseul, 1877: 483 (*Xanthochroa*) [HN]
metallipennis Fairmaire, 1889b: 49 (*Xanthochroa*)

genus *Opsimea* Miller, 1881: 224 type species *Opsimea ventralis* Miller, 1881
Ezonacerda Kôno, 1934b: 28 type species *Oedemera nigripennis* Matsumura, 1911
nigripennis Matsumura, 1911: 128 (*Oedemera*) **A**: FE (Sakhalin, Kuriles) JA
ventralis Miller, 1881: 225 **E**: AL BH BU CR GG GR ST (Kavkaz) YU **A**: TR
parallela Fairmaire, 1892b: 158 (*Stenaxis*)
quadrinervosa Reiche, 1862b: 296 (*Oedemera*) [HN]

tribe Oedemerini Latreille, 1810

genus *Dryopomera* Fairmaire, 1897j: 238 type species *Dryops indica* Fairmaire, 1896
subgenus *Dryopomera* Fairmaire, 1897j: 238 type species *Dryops indica* Fairmaire, 1896
Paroncomera Nakane, 1954: 179 type species *Paroncomera yatoi* Nakane, 1954
havai Švihla, 2001b: 128 **A**: SHA
kurosai H. Akiyama, 1998: 167 **A**: JA (Ryukyus) TAI

longissima Švihla, 1994: 244 **A**: NP SD
magna Švihla, 1999a: 84 **A**: NP
nepalensis Švihla, 1994: 246 **A**: NP
taiwana Švihla, 1997a: 89 **A**: TAI
yatoi tokarensis Nakane, 1954: 180 (*Paroncomera*) **A**: JA (Tokara Is.)
yatoi yatoi Nakane, 1954: 179 (*Paroncomera*) **A**: JA (Honshu, Shikoku, Kyushu)

genus *Oedemera* A. G. Olivier, 1789: 31 type species *Necydalis caerulea* Linnaeus, 1767 (= *Cantharis nobilis* Scopoli, 1763)
Oedemerastra Seidlitz, 1899: 923 type species *Necydalis podagrariae* Linnaeus, 1767
Oedemerella Seidlitz, 1899: 933 type species *Oedemera croceicollis* Gyllenhal, 1827
Oedemerina A. Costa, 1852: 31 type species *Necydalis lurida* Marsham, 1802
Oedemeronia Seidlitz, 1899: 919 type species *Necydalis flavipes* Fabricius, 1792
Stenolytra Dilwyn, 1829: 61 type species *Necydalis caerulea* Linnaeus, 1767 (= *Cantharis nobilis* Scopoli, 1763)
subgenus *Oedemera* A. G. Olivier, 1789: 31 type species *Necydalis caerulea* Linnaeus, 1767 (= *Cantharis nobilis* Scopoli, 1763)
Oedemerastra Seidlitz, 1899: 923 type species *Necydalis podagrariae* Linnaeus, 1767
Oedemerella Seidlitz, 1899: 933 type species *Oedemera croceicollis* Gyllenhal, 1827
Oedemerina A. Costa, 1852: 31 type species *Necydalis lurida* Marsham, 1802
Oedemeronia Seidlitz, 1899: 919 type species *Necydalis flavipes* Fabricius, 1792
Stenolytra Dilwyn, 1829: 61 type species *Necydalis caerulea* Linnaeus, 1767 (= *Cantharis nobilis* Scopoli, 1763)
alaiensis Semenov, 1894c: 544 **A**: KI KZ UZ
algerica Pic, 1899h: 116 **E**: IT (Lampedusa) **N**: AG LB MO TU
abdominalis Pic, 1897e: 42 [HN]
pici Schenkling, 1915: 58 [RN]
artvinensis Švihla, 1999b: 59 **A**: TR
atrata W. L. E. Schmidt, 1846: 77 **E**: AL CR FR GR IT SL SP YU **A**: TR
atriceps Abeille de Perrin, 1880: 261 **A**: IS JO
atrocyanea Švihla, 1999b: 12 **A**: KZ
barbara Fabricius, 1792b: 351 (*Necydalis*) **E**: CR FR GR IT MA PT SP TR **N**: AG MO MR (Salvages) TU **A**: TR
dalmatina Pic, 1898f: 73
subvirescens Pic, 1920c: 5
basalis Küster, 1849b: 70 **E**: SP **N**: TU
basipes Abeille de Perrin, 1892: 63 **A**: CY
brevipennis Ganglbauer, 1881: 108 **E**: GR (Sporades) TR **A**: SY TR
akbesiana Pic, 1901b: 11
pulchripes J. R. Sahlberg, 1908: 93
centrochinensis centrochinensis Švihla, 1999b: 15 **A**: SCH SHA
centrochinensis lixiana Švihla, 2003: 341 **A**: SCH
concolor Lewis, 1895a: 442 **A**: FE (Sakhalin, Kuriles) JA (Hokkaido, Honshu, Tsushima)
crassipes Ganglbauer, 1881: 114 **E**: AL BU CR FR GR IT MA PT SL SP **N**: AG LB MO TU TR **A**: CY IS JO LE SY TR
caudata Seidlitz, 1899: 919
cretica Švihla, 1999b: 33 **E**: GR (Kríti)
croceicollis Gyllenhal, 1827: 523 **E**: AB AR AU BE BU CR CT CZ DE EN FI FR GE GG GR HU IT NL PL RO SK SL ST SV SZ UK **A**: ES KZ
sarmatica F. F. Morawitz, 1861: 292
cuprata Reiche, 1864: 245 **N**: AG MO
atlasica Pic, 1932g: 28
danieli Magistretti, 1941b: 232 **E**: GG
femorata Scopoli, 1763: 45 (*Cantharis*) **E**: AB AL AN AR AU BE BU BY CR CT CZ DE EN FI FR GE GG GR HU IT LA LS MC MD NR NT PL RO SK SL SP ST SV SZ TR UK YU **A**: ES IN KZ TM TR WS
fallax Seidlitz, 1899: 913
flavescens Linnaeus, 1767: 642 (*Necydalis*)
praterrana Schrank von Paula, 1781: 168 (*Necydalis*)
similis W. L. E. Schmidt, 1846: 59

flavipennis W. L. E. Schmidt, 1846: 62 **E**: GR (Sporades) **A**: CY IS JO LE SY TR
coarcticollis Abeille de Perrin, 1880: 260
flavipes Fabricius, 1792b: 355 (*Necydalis*) **E**: AL AN AU BH BU CR CT CZ FR GE GR HU IT LA LS LT LU MC NL NR PL PT RO SK SL SP SV SZ TR UK YU **A**: IS JO LE SY TR
aenea A. G. Olivier, 1795: 14
clavipes Fabricius, 1801b: 373 (*Necydalis*)
delagrangei Pic, 1923i: 13
pusilla A. Costa, 1852: 29
graeca Švihla, 1978: 38 **E**: AL GR MC
hispanica Pic, 1904h: 91 **E**: SP
laufferi Reitter, 1905c: 206
inapicalis Pic, 1910o: 65 **A**: CY
jucunda Heyden, 1890c: 363 **A**: AF TD UZ
kantneri Švihla, 2004: 71 **A**: SY
kocheri Pic, 1934d: 21 **N**: MO
nigra Kocher, 1953: 51
lateralis Gebler, 1829a: 131 **E**: AB BU CR FR GG HU IT RO SP ST SZ UK YU **A**: TM WS
caucasica Kolenati, 1847: 142
impressithorax Pic, 1920c: 6
korbi Seidlitz, 1899: 918
lucidicollis flaviventris Fairmaire, 1891f: ccxix **A**: FUJ GUI HEB HEI HEN HUB HUN JIX NC SCH SHA SHN ZHE
coreana Pic, 1926e: 6
sieversi Seidlitz, 1899: 919
lucidicollis lucidicollis Motschulsky, 1866: 174 **A**: FE (Kuriles) JA
montana Marseul, 1877: 485
vitticollis Motschulsky, 1866: 174 (*Oncomera*)
lurida lurida Marsham, 1802: 360 (*Necydalis*) **E**: AB AL AN AR AU BE BU BY CR CZ DE EN FI FR GB GE GG GR HU IR IT LA LT LU MC NL NR NT PL PT RO RU SK SL SP ST SV SZ TR UK YU **A**: IN SY TM TR
viridula Seidlitz, 1899: 919
lurida sinica Švihla, 1999b: 41 **A**: SCH
manicata Lewis, 1895a: 442 **A**: JA
masatakai Švihla, 2003: 341 **A**: SCH
melanopyga W. L. E. Schmidt, 1846: 66 **E**: IT
maculiventris A. Costa, 1852: 24
merkli Seidlitz, 1899: 919 **A**: TR
mogadorica Escalera, 1913: 56 **N**: MO
basalis Chevrolat, 1877b: 121 [HN]
rufina Escalera, 1913: 56
theryi Pic, 1936e: 30
monticola Švihla, 1978: 35 **E**: AU CZ FR GE IT PL RO SK SL SZ UK
nasalis Reitter, 1896k: 290 **A**: KI KZ UZ XIN
atrofemoralis Pic, 1915a: 2
rostrifera Reitter, 1896k: 290
nigripes Ganglbauer, 1889: 44 (*Stenaxis*) **A**: GAN SCH
nigroapicata Švihla, 1982: 384 **A**: IN
nobilis Scopoli, 1763: 45 (*Cantharis*) **E**: AL AU BE CR DE FR GB GE GR IT LU MC NL PT SL SP SZ YU **N**: AG MO TU
caerulea Linnaeus, 1767: 642 (*Necydalis*)
ceramboides Forster, 1771: 47 (*Necydalis*)
gallica Pic, 1920c: 6
magica Dalla Torre, 1880: 23
sirgueyi Pic, 1927e: 5
subviolacea Pic, 1901b: 11
viridescens Pic, 1920c: 6
obscuripes Pic, 1912r: 220 **N**: MO

pallidicolor Pic, 1899h: 116 **N**: AG
pallidipes angustipennis Gressitt, 1939: 224 **A**: FUJ GUA
pallidipes pallidipes Pic, 1907j: 174 **A**: SCH
pallidipes shaanxiensis Švihla, 2003: 341 **A**: SHA
parallelipennis Seidlitz, 1899: 919 **A**: FE NC
penicillata W. L. E. Schmidt, 1846: 54 **E**: BU GG GR MC RO TR **A**: TR
petroschmidti L. Heyden, 1894b: 265 **A**: KI KZ UZ
confusa Reitter, 1895d: 158
uniformis Seidlitz, 1899: 915
podagrariae acutipalpis Abeille de Perrin, 1880: 259 **A**: IS TR
podagrariae podagrariae Linnaeus, 1767: 642 (*Necydalis*) **E**: AB AN AR AU BE BH BU BY CR CZ FR GE GG GR HU IT LA LS LT MA MC MD NL PL PT RO SK SL SP ST SV SZ TR UK YU **A**: IN TR
discoidalis Pic, 1921a: 5 [HN]
distinctipes Pic, 1921e: 10
diversipes Pic, 1921a: 5
femoralis Seidlitz, 1899: 925 [HN]
flavicrus Seidlitz, 1899: 925
fulva Geoffroy, 1785: 155 (*Cantharis*)
incerta Mulsant, 1858a: 192
inlateralis Pic, 1921e: 10
limbaticollis Pic, 1921a: 5
melanocephala Panzer, 1795: 277 (*Necydalis*) [HN]
reducta Pic, 1921a: 5
sebastiani Pic, 1901b: 10
sericans Mulsant, 1858a: 190
simplex Donovan, 1801: 90 (*Necydalis*) [HN]
testacea Fabricius, 1792b: 355 (*Necydalis*)
podagrariae ventralis Ménétriés, 1832: 212 **E**: AB **A**: IN TM
immunda Reitter, 1885d: 386
menetriesii Faldermann, 1837: 144
obscura Ganglbauer, 1881: 109
schmidti Gemminger, 1870a: 124 [RN]
ventralis W. L. E. Schmidt, 1846: 55 [HN]
pthysica Scopoli, 1763: 44 (*Cantharis*) **E**: AB AN AR AU BE BU CR CZ FR GE GG GR HU IT LU MC NL PL PT RO SK SL SP ST SV SZ TR UK **A**: IN TR
angusticollis A. Costa, 1852: 20
discoidalis Latreille, 1804b: 13
femorata Panzer, 1795: 279 (*Necydalis*) [HN]
marginata Fabricius, 1798: 155 (*Necydalis*)
stenoptera Faldermann, 1837: 146
subulata A. G. Olivier, 1795: 15
vittata J. Frivaldszky, 1877c: 325
qinlingensis Švihla, 1999b: 66 **A**: HUB SHA
reini L. Heyden, 1887d: 448 **N**: MO
robusta Lewis, 1895a: 443 **A**: FE HEI JA (Honshu) NC
laticollis Seidlitz, 1899: 919
rostralis deubeli Ganglbauer, 1897: 570 **E**: RO UK (Carpathes)
rostralis rostralis Reitter, 1885d: 386 **E**: AB AR GG ST (Kavkaz) UK
rufofemorata euboica Pic, 1902h: 64 **E**: GR **A**: TR
rufofemorata pruinosa Abeille de Perrin, 1880: 259 **A**: IS LE SY TR
rufofemorata rubricollis Reitter, 1885d: 386 **A**: TR
rufofemorata rufofemorata Germar, 1824b: 166 **E**: AL AU BH BU CR GR MC SL YU
brevicornis W. L. E. Schmidt, 1846: 78 [HN]
cyanescens W. L. E. Schmidt, 1846: 72
femorata Brullé, 1832: 227 [HN]
satoi Švihla, 2003: 339 **A**: SHA
sauteri luteipes Švihla, 1999b: 18 **A**: TAI
sauteri sauteri Pic, 1927i: 48 **A**: TAI

schrammi Pic, 1913k: 137 **E**: SP
sexualis ryukyuana Švihla, 1999b: 19 **A**: JA (Ryukyus)
sexualis sexualis Marseul, 1877: 486 **A**: JA (Honshu, Shikoku, Kyushu, Tsushima, Goto Is.)
sichuana Švihla, 1999b: 65 **A**: SCH SHA
simplex Linnaeus, 1767: 643 (*Necydalis*) **E**: FR IT MA PT SP **N**: AG LB MO TU
basipennis Pic, 1936e: 30
basitestacea Pic, 1936e: 30
brevicollis W. L. E. Schmidt, 1846: 56
flavimana W. L. E. Schmidt, 1846: 57
tibialis P. Lucas, 1846: 362
subrobusta Nakane, 1954: 180 (*Oedemerina*) **E**: AR CT CZ FR GG HU MC NT PL RO SK ST SV SZ **A**: ES FE GAN HUB JA (Honshu) KZ MG NC QIN SCH SHA TM TR WS
testaceithorax okinawana Kôno, 1937a: 11 **A**: JA (Amami-Ôshima, Okinawa)
testaceithorax testaceithorax Pic, 1927i: 49 **A**: FUJ GUA GUI JA (Ishigaki-jima, Iriomote-jima, Yonaguni-jima) JIX SCH TAI **ORR**
tristis W. L. E. Schmidt, 1846: 75 **E**: AU BE CR CZ FR GE HU IT LU PL RO SK SL SZ
unicolor W. L. E. Schmidt, 1846: 74 **E**: PT SP **N**: MO
olcesei Pic, 1898h: 121
tangeriana Pic, 1920c: 6
vilis pontica Švihla, 1999b: 44 **A**: TR
vilis siberoaltaica Švihla, 1999b: 45 **A**: ES WS
vilis vilis Roubal, 1917a: 44 **E**: AR GG
virescens chalybea Faldermann, 1837: 149 **E**: AR ST (Kavkaz)
virescens virescens Linnaeus, 1767: 650 (*Cantharis*) **E**: AN AU BE BH BU BY CR CT CZ DE EN FI FR GB GE GR HU IT LA LT LU MC NL NR NT PL RO SK SL SP ST SV SZ UK YU **A**: ES FE HEI JA (Honshu) KZ MG SHA SHX WS
abdominalis A. G. Olivier, 1795: 50: 14
cupreomicans Reitter, 1905a: 93
striata Herbst, 1784b: 109 (*Necydalis*)
subgenus *Oncomera* Stephens, 1829b: 251 type species *Dryops femorata* Fabricius, 1792 (= *Oedemera femoralis* A. G. Olivier, 1803)
Oncomerella Reitter, 1911a: 406 type species *Oedemera marmorata* Erichson, 1841
Oncomerina Seidlitz, 1899: 885 type species *Oedemera murinipennis* Kiesenwetter, 1859
antoinei Pic, 1932h: 30 (*Oncomera*) **N**: MO
mimeuri Pic, 1949c: 74 (*Oncomera*)
bomfordi Fairmaire, 1897j: 237 (*Dryops*) **A**: KA NP UP
diversecostata afghana Švihla, 1993b: 197 **A**: AF PA
diversecostata diversecostata Pic, 1915c: 6 (*Oncomera*) **A**: KA PA **ORR**
grandis Pic, 1915c: 6 (*Oncomera*)
farsica demissa Švihla, 1999a: 83 **A**: IN
farsica farsica Švihla, 1983a: 124 (*Oncomera*) **A**: IN
femoralis femoralis A. G. Olivier, 1803: 21 **E**: AN AU BE BH BU CR FR GB GE GR HU IT NL PT RO SL SP SZ UK **A**: TR
acutiventris Abeille de Perrin, 1896: 284 (*Oncomera*)
calopoides Germar, 1817: 225 (*Necydalis*)
femorata Fabricius, 1792b: 74 (*Dryops*) [HN]
femoralis purpureocoerulea Ganglbauer, 1882: 137 (*Oncomera*) **E**: BH CR GR YU
flavicans Fairmaire, 1860: 632 (*Dryops*) **E**: BU GR MC YU **A**: CY IS JO SY TR
parnassica Pic, 1920c: 5 (*Oncomera*)
stylifera Abeille de Perrin, 1896: 284 (*Oncomera*)
marmorata Erichson, 1841: 185 **E**: PT SP **N**: AG MO TU
murinipennis karpathosensis Švihla, 1999a: 83 **E**: GR (Karpathos)
murinipennis murinipennis Kiesenwetter, 1859: 192 **E**: GR (Kríti)
paganettii Stolz, 1915: 91
natolica Reiche, 1862c: 544 (*Oncomera*) **E**: GR **N**: EG **A**: IS LE SY TR
apicipennis Fairmaire, 1892b: 157

libanica Pic, 1920c: 6
pallida Fleischer, 1919a: 170
rhodosica Fleischer, 1919a: 171
nepalensis Švihla, 1993b: 194 **A**: NP
reitteri Švihla, 1993b: 199 [RN] **E**: GR
atriceps Reitter, 1894e: 303 (*Oncomera*) [HN]
venosa Lewis, 1895a: 441 (*Oncomera*) **A**: FE JA
subgenus *Stenaxis* W. L. E. Schmidt, 1844: 46 type species *Oedemera annulata* Germar, 1824
amurensis L. Heyden, 1884a: 295 **A**: FE HEI NC SHX
obscurior Pic, 1904d: 27
annulata Germar, 1824b: 166 **E**: AL AU BH BU CR IT MC RO SL YU

tribe Stenostomini Mulsant, 1858

genus ***Stenostoma*** **Latreille, 1810: 217** type species *Leptura rostrata* Fabricius, 1787
cossyrense Bologna, 1995b: 516 **E**: IT (Pantelleria)
lowei Wollaston, 1854: 524 (*Stenaxis*) **E**: MR
melitense Cameron, 1907: 226 **E**: MA
rostratum rostratum Fabricius, 1787a: 159 (*Leptura*) **E**: GR (Kós, Críti, Ródos) PT SP **N**: AG MO TU **A**: IS
schatzmayri Magistretti, 1941a: 77
rostratum septentrionale Švihla, 2005a: 72 [RN] **E**: AL CR FR GR (Kerkyra, Pelopponisos) IT PT SP YU
caeruleum V. Petagna, 1787: 14 (*Rhinomacer*) [HN]
ceramboides Rossi, 1794: 99 (*Necydalis*) [HN]

family STENOTRACHELIDAE C. G. Thomson, 1859

subfamily **Stenotrachelidae** C. G. Thomson, 1859

genus ***Scotodes*** **Eschscholtz, 1818: 454** type species *Scotodes annulatus* Eschscholtz, 1818
Pelmalopus Fischer von Waldheim, 1824: 163 type species *Pelmalopus hummeli* Fischer von Waldheim, 1824
annulatus Eschscholtz, 1818: 454 **E**: BY CT LA NT **A**: ES FE HEI JA WS
hummeli Fischer von Waldheim, 1824: 163 (*Pelmalopus*)
niponicus Lewis, 1895a: 275
uniformis Motschulsky, 1872: 46 **A**: FE (Kamchatka)

genus ***Stenocephaloon*** **Pic, 1932a: 2** type species *Stenocephaloon metallicum* Pic, 1932
metallicum Pic, 1932a: 2 **A**: JA

genus ***Stenotrachelus*** **Berthold, 1827: 371** type species *Lagria aenea* Fabricius, 1787
aeneus Fabricius, 1787a: 94 (*Lagria*) **E**: FI NT **A**: ES FE **NAR**
arctatus Say, 1824b: 283 (*Helops*)
fennicus Weber & Mohr, 1804: 182 (*Dryops*)
obscurus Mannerheim, 1852: 347
rouillieri Motschulsky, 1845a: 84

subfamily **Nematoplinae** LeConte, 1862

genus ***Nematoplus*** **LeConte, 1855: 275** type species *Nematoplus collaris* LeConte, 1855
konoi Nakane, 1963a: 30 (*Pedilocephaloon*) **A**: JA
semenovi Nikitsky, 1973a: 611 **A**: FE
yamato Nakane, 1987: 176 **A**: JA

subfamily **Cephaloinae** LeConte, 1862

genus ***Cephaloon*** **Newman, 1838: 376** type species *Cephaloon lepturides* Newman, 1838
pallens Motschulsky, 1860c: 140 (*Cephalaon*) **A**: JIL FE JA NC
sakurai Lewis, 1895a: 444 **A**: JA
variabile Motschulsky, 1860c: 141 (*Cephalaon*) **A**: CH FE NC

subfamily **Stolinae** Nikitsky, 1985

genus *Stolius* Lewis, 1895a: 433 type species *Stolius vagepictus* Lewis, 1895
Falsostolius Pic, 1932d: 27 type species *Falsostolius diversipes* Pic, 1932
diversipes Pic, 1932d: 27 (*Falsostolius*) **A**: JA
vagepictus Lewis, 1895a: 433 **A**: JA

family MELOIDAE Gyllenhal, 1810

subfamily **Eleticinae** Wellman, 1910

tribe Derideini Wellman, 1910

genus *Deridea* Westwood, 1875: 226 type species *Deridea curculionides* Westwood, 1875
notata minor Kaszab, 1960c: 153 **A**: YE
notata notata Thomas, 1897a: 389 **A**: SA YE

tribe Eleticini Wellman, 1910

genus *Eletica* Dejean, 1834: 224 type species *Lytta rufa* Fabricius, 1801 (= *Lytta bipustulata* Fabricius, 1801)
Iletica Gemminger & Harold, 1870: 2145 [unjustified emendation]
subgenus *Meteletica* Kaszab, 1955a: 18 type species *Cantharis testacea* A. G. Olivier, 1789
testacea A. G. Olivier, 1789: 277 (*Cantharis*) **A**: SD **ORR**
bicolor Schoenherr, 1817: 23 (*Lytta*)
gigantea Dohrn, 1873: 70
rufocincta Pic, 1915d: 8

subfamily **Meloinae** Gyllenhal, 1810

tribe Cerocomini Leach, 1815

genus *Anisarthrocera* Semenov, 1896e: 517 type species *Rhampholyssa batesi* Marseul, 1872
batesi batesi Marseul, 1872a: 416 (*Rhampholyssa*) **A**: IN IQ SA
batesi villiersi Kaszab, 1968f: 750 **A**: IN

genus *Cerocoma* Geoffroy, 1762: 357 type species *Meloe schaefferi* Linnaeus, 1758
subgenus *Cerocoma* Geoffroy, 1762: 357 type species *Meloe schaefferi* Linnaeus, 1758
Meloides Piller & Mitterpacher, 1783: 39 type species *Meloides adamovichiana* Piller & Mitterpacher, 1783
adamovichiana Piller & Mitterpacher, 1783: 39 (*Meloides*) **E**: AL BH CR GR HU MC SK TR YU **A**: TR
schraderi Kraatz, 1863: 111
albopilosa Dvořák, 1993a: 7 **A**: IN TR
azurea Reitter, 1914f: 192 (*Cerocoma*) **A**: IS JO LE SY
coeruleotincta Muche, 1963a: 13
humilis Muche, 1963a: 12
barthelemyi Baudi di Selve, 1878b: 358 **E**: AR GR **A**: SY TR
haifensis Kaszab, 1951d: 268
bernhaueri Pardo Alcaide, 1977: 64-65 **A**: IN IS TR
bodemeyeri Reitter, 1909c: 103 **A**: IN IQ
dahli Kraatz, 1863: 112 **E**: AR AU BU GG GR RO TR **A**: IN IS JO SY TR
aeneipes Reitter, 1914f: 193
nigrivestis Muche, 1963a: 13
obscuripes Reitter, 1885a: 14
gloriosa Mulsant, 1857a: 103 **E**: AR **A**: SY TR
pseudogloriosa Muche, 1963a: 14
vasvarii Kaszab, 1941f: 680
graeca Mařan, 1944c: 99 **E**: GR MC
kunzei Frivaldszky von Frivald, 1835: 265 **E**: AL AR BH BU CR GR MC RO TR YU "Caucasus" **A**: IN IS JO SY TR
iranica Muche, 1963a: 11
palaestinensis Muche, 1963a: 11
patelligera Baudi di Selve, 1878b: 358

macedonica Mařan, 1944c: 98 **E**: BH BU GR MC TR YU **A**: TR
malatyensis Kaszab, 1951d: 266 **E**: "Caucasus" **A**: TR
marginiventris Reitter, 1889a: 34 **A**: IN TR
pictiventris Reitter, 1890e: 174
muehlfeldi Gyllenhal, 1817: 13 **E**: AB AL AU BH BU CR CZ GE GR HU MC RO SK UK YU **A**: IN IQ IS JO SY TR
faldermanni Laporte, 1840: 267
gonocera Motschulsky, 1872: 49
micans Ménétriés, 1832: 206
prochaskana Reitter, 1896j: 267 **A**: IN IQ SY TR
rapillyi Pardo Alcaide, 1977: 61 **A**: IN
schaefferi Linnaeus, 1758: 249 (*Meloe*) **E**: AL AU BE BH BY CT CZ FR GE GR HU IT KZ LA MC NL PL PT RO SK SP ST SZ UK YU **A**: TR
affinis Baudi di Selve, 1878b: 358
orensis Reitter, 1914f: 193
viridis Geoffroy, 1785: 163
viridula Reitter, 1885a: 14
simplicicornis Reitter, 1914f: 193 **E**: AR "Caucasus" **A**: IN TR
syriaca Abeille de Perrin, 1880: 235 **E**: AL GR "Caucasus" **A**: IN SY TR
turcica Pardo Alcaide, 1977: 60 **A**: TR
subgenus *Cerocomina* Kaszab, 1951d: 259 type species *Cerocoma vahli* Fabricius, 1787
vahli taraudanti Escalera, 1913: 55 **N**: MO
vahli vahli Fabricius, 1787a: 217 **N**: AG MO TU
chalybaeiventris Chevrolat, 1840b: 268
wagneri Küster, 1844: 32
subgenus *Mesocerocoma* Kaszab, 1951d: 259 type species *Cerocoma scovitzii* Faldermann, 1837
latreillei latreillei Baudi di Selve, 1878b: 359 **A**: IN IQ
collaris Reitter, 1885a: 12
latreillei schah Kaszab, 1968f: 751 **A**: IN PA
scovitzi intermedia Mařan, 1944c: 94 **N**: EG **A**: IQ IN IS JO SA SY TR
jureceki Mařan, 1944c: 94
marani Kaszab, 1951d: 268
mirabilis Dvořák, 1993a: 7
sterbai Mařan, 1944c: 95
scovitzi scovitzi Faldermann, 1837: 117 **E**: AR "Caucasus" "Transcaucasia" **A**: IN
lateralis Reitter, 1890e: 174
picticornis Motschulsky, 1872: 49
rufiventris Reitter, 1890e: 173
subgenus *Metacerocoma* Kaszab, 1951d: 259 type species *Cerocoma schreberi* Fabricius, 1781
ephesica Reitter, 1885a: 12 **E**: BU GR MC TR **A**: IN TR
festiva Faldermann, 1837: 118 **E**: AB AR **A**: IN KZ TM TR
prevezaensis Dvořák, 1993a: 4 **E**: GR
schreberi Fabricius, 1781: 331 **E**: AB AL AU BH BU CR CT FR GR HU IT KZ MC PL PT SK SP ST UK YU **A**: KZ MG TD TM TR UZ XIN

genus *Diaphorocera* Heyden, 1863: 126 type species *Diaphorocera hemprichi* Heyden, 1863
carinicollis Chobaut, 1921: 298 **N**: AG TU
chrysoprasis Fairmaire, 1863: 644 **N**: AG LB MO TU
kerimii Fairmaire, 1875a: 530
hemprichi hemprichi Heyden, 1863: 127 **N**: AG EG MO TU **A**: IS JO
hemprichi saudita Kaszab, 1983a: 179 **A**: AE IN OM SA
johnsoni Kaszab, 1983a: 180 **A**: AE SA
obscuritarsis Fairmaire, 1885a: xxxviii **N**: AG EG MO TU **A**: IS JO
ruficornis Pic, 1923g: 10
peyerimhoffi Kocher, 1954: 284 **N**: MO
promelaena Fairmaire, 1876: 69 **N**: AG MO TU **A**: SI

sicardi Bedel, 1918: 364 **N**: MO

genus *Rhampholyssa* Kraatz, 1863: 110 type species *Cerocoma stevenii* Fischer von Waldheim, 1823
antennata Reitter, 1906b: 36 **A**: TM
stevenii Fischer von Waldheim, 1823: pl. 41 (*Cerocoma*) **E**: KZ **A**: IN KI KZ TD TM UZ
intermedia Semenov, 1896e: 518
komarowi Reitter, 1889b: 119
pallida Semenov, 1896e: 519
plagicollis Reitter, 1909c: 102
varentzowi Semenov, 1896e: 518

genus *Rhampholyssodes* Kaszab, 1983a: 182 type species *Rhampholyssodes pitcheri* Kaszab, 1983
pichteri Kaszab, 1983a: 182 **A**: AE SA

tribe Epicautini Denier, 1935

genus *Cyaneolytta* Péringuey, 1909: 249 type species *Lytta signifrons* Fåhraeus, 1870
coerulea Pfaff, 1834: 209 (*Lytta*) **A**: PA SD UP **ORR**
acteon Laporte, 1840: 273 (*Lytta*)
fryi Wollaston, 1861a: 252 (*Cantharis*) **A**: SA **AFR**
holocyanea Fairmaire, 1896c: 225 (*Cantharis*)
saphirina Mäklin, 1875: 609 (*Cantharis*)
maculifrons Mäklin, 1875a: 608 (*Cantharis*) **A**: SA YE **AFR**
frontalis Kolbe, 1883: 24 (*Lytta*)
frontosa Wellman, 1912: 34 (*Lytta*)
violacea Brandt & Ratzenburg, 1833: 123 (*Lytta*) **A**: UP **ORR**

genus *Denierella* Kaszab, 1952a: 81 type species *Cantharis incompleta* Fairmaire, 1896
minutiserra Tan, 1988: 291 **A**: XIZ
serrata Kaszab, 1952a: 88 **A**: FUJ HAI HUB JIX SCH ZHE

genus *Epicauta* Dejean, 1834: 224 type species *Meloe erythrocephalus* Pallas, 1771
subgenus *Epicauta* Dejean, 1834: 224 type species *Meloe erythrocephalus* Pallas, 1771
Anomalonyx Denier, 1935: 161 type species *Lytta fumosa* Germar, 1824
Anomalonychus Saylor, 1940: 46 [RN]
Apterospasta LeConte, 1862: 272 type species *Lytta segmenta* Say, 1823
Causima Dejean, 1834: 226 type species *Lytta vidua* Klug, 1825
Gnathospasta Horn, 1875: 154 type species *Gnathospasta mimetica* Horn, 1875
Henous Haldeman, 1852: 377 type species *Henous techanus* Haldeman, 1852 (= *Meloe conferta* Say, 1824)
Isopentra Mulsant & Rey, 1858a: 180 type species *Lytta megalocephala* Gebler, 1817
Maculicauta Dillon, 1952: 416 type species *Epicauta stuarti* LeConte, 1868
Nomaspis LeConte, 1866: 156 type species *Meloe parvus* Haldeman, 1852: 337 (=*Meloe parvulus* Haldeman, 1854)
Pleuropompha LeConte, 1862: 273 type species *Lytta costata* LeConte, 1854
aethiops Latreille, 1827: 286 (*Cantharis*) **N**: EG **AFR**
ambusta Pallas, 1781: 102 (*Meloe*) **A**: ES MG NMO
apicipennis Tan, 1958: 159, 165 **A**: SCH
aptera Kaszab, 1952e: 590 **A**: FUJ GUI GUX JIA SCH YUN ZHE
assamensis Waterhouse, 1871: 407 (*Cantharis*) **A**: FUJ NP **ORR**
atkinsoni Kraatz, 1880a: 16 **A**: IN UP "Himalaya"
niveolineata Haag-Rutenberg, 1880: 82, 85 (*Lytta*)
badeni badeni Haag-Rutenberg, 1880: 82 (*Lytta*) **A**: TAI XIN
taipina Kôno, 1936b: 90
badeni sinica Kaszab, 1960d: 256 **A**: YUN
brevitibialis Kaszab, 1952e: 589 **A**: YUN
cheni Tan, 1958: 164 **A**: JIX **ORR**
chinensis Laporte, 1840: 274 (*Lytta*) **A**: AHN GAN HEB HEI HEN HUB HUN JA JIA JIL NC NMO SC SCH SHN SHX TAI XIN "Korea"

cryptogramaca Y.-X. Yang & Ren, 2006: 272 **A**: YUN
curvispina Kaszab, 1952e: 587 **A**: FUJ GUI GUX SCH
dalihodi Dvořák, 1996: 39 **A**: TAI
dubia Fabricius, 1781: 329 (*Lytta*) **E**: ST **A**: ES GAN HEB HEI HUB JIA KZ MG NC NE NIN NMO QIN SC SCH SHX
sibirica LeConte, 1866: 159 [HN]
emmerichi emmerichi Pic, 1934f: 86 **A**: SCH
emmerichi yunnanensis Kaszab, 1960d: 256 **A**: SCH YUN
erythrocephala Pallas, 1771: 466 (*Meloe*) **E**: AL AR BU GR RO SK ST TR UK **A**: AF GUA HAI IN KZ TD WS TM TR UZ XIN
albivittis Pallas, 1781: 101 (*Meloe*)
latelineolata Mulsant & Rey, 1858a: 177
lineata Thunberg, 1791: 109 (*Meloe*)
sonchi Gmelin, 1790: 1896 (*Meloe*)
fasciceps Walker, 1871: 16 **N**: EG **A**: YE **AFR**
nigronotata Haag-Rutenberg, 1880: 67 (*Lytta*)
flabellicornis Germar, 1817: 208 (*Lytta*) **E**: BH CR IT (Venezia Giulia) SL YU
depressicornis Laporte, 1840: 274 (*Lytta*)
sibirica Reitter, 1905b: 194 [HN]
gorhami Marseul, 1873a: 227 (*Cantharis*) **A**: AHN FUJ GUA GUI GUX HAI HEN HKG HUB HUN JA JIA JIX SCH SHX SHN TAI ZHE "Korea"
taishoensis Lewis, 1879: 464
himalayica Kaszab, 1960g: 283 **A**: UP
hirticornis Haag-Rutenberg, 1880: 81 (*Lytta*) **A**: AP GUA HEN SCH SD TAI XIZ YUN **ORR**
afra Kaszab, 1960d: 257
hirtipes Waterhouse, 1871: 406 (*Cantharis*) **A**: IN KZ NP PA SD UP
impressicornis Pic, 1913q: 163 (*Lytta*) **A**: GUX YUN
indiana Kaszab, 1956e: 635 **A**: NP **ORR**
insularis Haag-Rutenberg, 1880: 82 (*Lytta*) **A**: SCH **ORR**
interrupta Fairmaire, 1889b: 48 (*Lytta*) **A**: SCH XIZ YUN
desgodinsi J. Frivaldszky, 1892: 115
kwangsiensis Tan, 1958: 166 **A**: SCH
laticornis Haag-Rutenberg, 1880: 82 (*Lytta*) **A**: "South China" **ORR**
mandibularis Anand, 1977: 82 **A**: HP NP
mannerheimi Mäklin, 1875a: 623 (*Cantharis*) **A**: BT FUJ GUA GUX NP UP YUN **ORR**
megalocephala Gebler, 1817: 318 (*Lytta*) **E**: ST **A**: BEI ES FE HEB HEI HEN JIL LIA KZ MG NIN NMO SCH SHX WS "Korea"
albinae Reitter, 1905b: 196
maura Faldermann, 1833: 61 (*Lytta*)
montana Anand, 1977: 81 **A**: HP NP
nepalensis Hope, 1831: 32 (*Lytta*) **A**: AP NP SD **ORR**
obscurocephala Reitter, 1905b: 195 **A**: ANH HEB HEN HUB JIA JIL JIX NIN NMO SHA SHN TIA ZHE
probsti Dvořák, 1996: 38 **A**: NP
promerotricha Dvořák, 1996: 37 **A**: UP **ORR**
quadraticollis Fairmaire, 1891d: ci **A**: HP KA
reini Kiesenwetter, 1879: 307 (*Cantharis*) **A**: JA
rubriceps L. Redtenbacher, 1844: 535 (*Epicauta*) **A**: KA PA XIN
haagi Bates, 1879: 483
reversa Gemminger, 1870a: 124 (*Cantharis*) [RN]
ruficeps Illiger, 1800: 140 (*Lytta*) **A**: ANH FUJ GUX HUB HUN JIX SCH YUN **ORR**
plumicornis Laporte, 1840: 274 (*Lytta*)
rufidorsum Goeze, 1777: 704 (*Meloe*) **E**: AL AU BH BU CR CZ FR GR HU IT MC PL RO SK SL SZ YU
algirica Sulzer, 1776: 66 (*Meloe*)
dubia A. G. Olivier, 1789: 279 (*Cantharis*)
erytrocephala Rossi, 1790: 259 (*Lytta*) [HN]
marginata Dorthes, 1787: 69 (*Lytta*)

rufa Gmelin, 1790: 2016 (*Lytta*)
verticalis Illiger, 1804: 172 (*Lytta*)
sanguiniceps Fairmaire, 1885a: xxxviii (*Cantharis*) **N**: AG
sharpi Marseul, 1875a: xxvii **A**: IN OM SA IS JO YE
suavis Haag-Rutenberg, 1880: 83 (*Lytta*)
sibirica Pallas, 1773: 720 (*Meloe*) **A**: ES GAN GUA HEB HEI HEN HUB HUN JA JIA JIL JIX KZ MG NIN NMO NO QIN WS XIN YUN ZHE **ORR**
dubia Fischer von Waldheim, 1823: pl. 42 (*Cantharis*)
pectinata Goeze, 1777: 701 (*Meloe*)
textilis Haag-Rutenberg, 1880: 82 (*Lytta*) **A**: IN IQ SA
tibialis Waterhouse, 1871: 406 (*Cantharis*) **A**: FUJ GUA GUX HEN HGK HUN JIA JIX NP SCH TAI YUN **ORR**
waterhousei Haag-Rutenberg, 1880: 81 (*Lytta*) **A**: ANH FUJ GUX HAI HEN HUB JA JIX TAI SHN YUN **ORR**
formosensis Wellman, 1912: 31
wellmani Kaszab, 1956e: 636 **A**: XIZ **ORR**
wittmeri Kaszab, 1978c: 331 **A**: SD
xantusi Kaszab, 1952e: 592 **A**: GUX JIA JIX NIN SCH SHG SHX

genus *Epicauta*, nomen dubium
punctata Pallas, 1781: 100 (*Meloe*) **E**: RU "Caucasus" **A**: "Siberia"

genus *Psalydolytta* Péringuey, 1909: 254 type species *Lytta lorigera* Gerstaecker, 1854
Gnathospastoides Fletcher, 1914: 302 type species *Lytta rouxii* Laporte, 1840
aegyptiaca Mäklin, 1875a: 612 (*Cantharis*) **N**: EG **AFR**
monardi Pic, 1951h: 219
antennata Saha, 1979: 34 **A**: UP **ORR**
atricollis Pic, 1920a: 18 (*Epicauta*) **A**: HP SD UP **ORR**
rouxi Laporte, 1840: 274 (*Lytta*) **A**: HP UP **ORR**

tribe Lyttini Solier, 1851

genus *Alosimus* Mulsant, 1857: 358 type species *Meloe syriacus* Linnaeus, 1758
Halosimus Gemminger & Harold, 1870: 2157 [unjustified emendation]
abeillei Escherich, 1896c: 215 (*Lydus*) **A**: TR
albovillosus Escalera, 1914: 391 (*Halosimus*) **N**: MO
araxis Reitter, 1892a: 63 (*Halosimus*) **E**: AB AR **A**: TR
armeniacus Faldermann, 1837: 135 (*Lytta*) **E**: AB AR RO BU **A**: IN IQ JO SY TR
babadjanidesi Kaszab, 1951b: 147
cupratus Abeille de Perrin, 1880: 251 (*Lydus*)
depilis Abeille de Perrin, 1880: 250 (*Lydus*)
elegantulus Mulsant & Rey, 1858a: 143
brevicornis Abeille de Perrin, 1880: 249 (*Lydus*) **A**: IQ IS JO LE SY
castaneus Escherich, 1896a: 27 (*Lydus*) **E**: AR "Caucasus" **A**: TR
chalybaeus Tauscher, 1812b: 153 (*Oenas*) **E**: AB AR BU CT GR KZ MC RO ST TR UK YU **A**: IQ IN SY TR
dichrous Fischer von Waldheim, 1824: 230 (*Lytta*)
obscurus Escherich, 1896c: 225 (*Lydus*)
cirtanus Lucas, 1847: 394 (*Cantharis*) **N**: AG TU
mendax Fairmaire, 1879c: 200 (*Cantharis*)
purpureicollis Chobaut, 1897a: 39 (*Halosimus*)
crabrerai Escalera, 1923e: 299 (*Halosimus*) **N**: MO
crassicornis Escalera, 1914: 391 (*Halosimus*) **N**: MO
decolor Abeille de Perrin, 1880: 249 (*Lydus*) **E**: BU GR TR **A**: SY TR
filitarsis Escalera, 1925f: 502 (*Halosimus*) **N**: MO
gracilis Abeille de Perrin, 1880: 251 (*Lydus*) **A**: IN IS JO SY TR
latitarsis Kaszab, 1951b: 145 **A**: TR
longicollis Escherich, 1896c: 213 (*Lydus*) **A**: IN

luteus Waltl, 1838: 467 (*Lytta*) **A**: IN IQ IS JO SY TR
maculicollis Mulsant & Wachanru, 1853: 172 (*Lydus*) **A**: TR
cinctus Schaufuss, 1862b: 310 (*Alosimus*)
noticollis Mulsant & Rey, 1858a: 137
marginicollis Haag-Rutenberg, 1880: 72 (*Lytta*) **A**: TR
pallidicollis Gyllenhal, 1806: 27 (*Lytta*) **A**: IQ IS JO SY TR
reitterianus Semenov, 1900b: 480 (*Lydus*) [RN] **E**: AB AR **A**: IN TR
armeniacus Reitter, 1889a: 36 (*Halosimus*) [HN]
robustus Escalera, 1914: 504 (*Halosimus*) **N**: MO
albopilosus J. R. Sahlberg, 1903e: 29
lateauratus Pic, 1949a: 9 (*Lydus*)
theryi Pic, 1949a: 9 (*Lydus*)
smyrnensis Mařan, 1942c: 91 (*Lydus*) **E**: GR MC **A**: IN TR
sulcicollis Abeille de Perrin, 1880: 248 (*Lydus*) **A**: IS JO SY
luteipennis Pic, 1899h: 115 (*Lydus*)
syriacus afghanicus Kaszab, 1973d: 272 **A**: AF
syriacus austriacus Schrank von Paula, 1781: 225 (*Meloe*) **E**: AL AU BU CR CT GR HU IT KZ LA MC MD PL RO ST SK SZ TR UK YU **A**: KZ TR
basinotatus Mařan, 1942c: 94 (*Lydus*)
binotatus Mařan, 1942c: 94 (*Lydus*)
crambes Pallas, 1782: 95 (*Meloe*)
cyanipennis Mařan, 1942c: 94 (*Lydus*)
danubianus Mařan, 1942c: 94 (*Lydus*)
nigricollis Escherich, 1896c: 219 (*Lydus*)
immaculifrons Mařan, 1942c: 94 (*Lydus*)
syriacus rauterbergi Reitter, 1907i: 485 (*Halosimus*) **N**: EG LB **A**: IS SI
syriacus syriacus Linnaeus, 1758: 419 (*Meloe*) **A**: IN IQ IS JO LE SA SY TR
albopilosus Kaszab, 1951b: 140
azureipennis J. R. Sahlberg, 1913b: 197 (*Lydus*)
myagri Fischer von Waldheim, 1823: pl. 42 (*Lytta*)
ruficollis Herbst, 1786: 179 (*Lytta*)
tenuicornis Escalera, 1914: 392 (*Halosimus*) **N**: MO
kocheri Pic, 1949a: 9
tyrrhenicus Bologna, 1989a: 191 **E**: IT
viridissimus Lucas, 1847: 393 (*Cantharis*) **N**: AG LB TU
bipartitus Pic, 1905d: 107
coeruleus Escherich, 1896c: 221 (*Lydus*)
intermedius Pic, 1897e: 42 (*Oelosimus*)
opacipennis Fairmaire, 1871a: 395 (*Cantharis*)
purpurascens Pic, 1939b: 17 (*Lydus*)

genus *Berberomeloe* Bologna, 1989b: 355 type species *Meloe majalis* Linnaeus, 1758
insignis Charpentier, 1818: 258 (*Meloe*) **E**: SP
majalis Linnaeus, 1758: 419 (*Meloe*) **E**: FR PT SP **N**: AG MO TU
fissicornis Gorriz y Muñoz, 1882: 38 (*Meloe*)
frontalis Baudi di Selve, 1878b: 355 (*Meloe*)
laevigatus A. G. Olivier, 1797: 650 (*Meloe*)
maculifrons Lucas, 1847: 399 (*Meloe*)
maculicollis Reitter, 1872: 176 (*Meloe*)

genus *Cabalia* Mulsant & Rey, 1858a: 154 type species *Lytta segetum* Fabricius, 1792
Paroenas Kolbe, 1894a: 190 type species *Paroenas limbata* Kolbe, 1894
arabica Kaszab, 1983a: 172 **A**: SA
longicollis Kaszab, 1948: 13 **N**: AG
rubriventris Fairmaire, 1860: 338 (*Cantharis*) **N**: AG LB MO TU
rufiventris Walker, 1871: 16 (*Cantharis*) **N**: AG LB MO TU **A**: SI
segetum Fabricius, 1792b: 84 (*Lytta*) **E**: IT (Sicilia) MA **N**: AG TU LB

bassii Laporte, 1840: 272 (*Cantharis*)
coerulea Escherich, 1895: 293
perroudi Mulsant & Rey, 1858a: 154 (*Cacalia*)

genus *Calydus* Reitter, 1896h: 192 [RN] type species *Caloenas pulcher* Reitter, 1889
Caloenas Reitter, 1889a: 34 [HM] type species *Caloenas pulcher* Reitter, 1889
ater Kaszab, 1960b: 129 **A**: IN IQ TR
escherichi Reitter, 1898f: 351 **A**: TR
anatolicus Muche, 1962a: 51 (*Lydus*)
ornaticollis Escherich, 1896b: 122 (*Caloenas*) **A**: IN SY
pulcher Reitter, 1889a: 35 (*Caloenas*) **E**: AB AR
semenovi Escherich, 1896b: 122 (*Caloenas*) **A**: IN SY TR
plagifrons Reitter, 1913d: 662 (*Zonabris*)
syriacus Kaszab, 1960b: 130 **A**: SY

genus *Eolydus* Denier, 1913: 231 type species *Eolydus atrocoeruleus* Denier, 1913 (= *Alosimus atripes* Pic, 1905)
afghanicus Kaszab, 1958c: 247 **A**: AF
atripes Pic, 1905f: 130 (*Alosimus*) **A**: IN IQ
atrocoeruleus Denier, 1913: 231
conspicuus Waterhouse, 1889: 129 (*Cantharis*) **A**: AF IN PA "Turkestan"
binotatithorax Kaszab, 1956e: 635
testaceicornis Pic, 1910j: 34 (*Lydus*)

genus *Lagorina* Mulsant & Rey, 1858a: 150 type species *Lytta sericea* Waltl, 1835
scutellata Laporte, 1840: 273 (*Lytta*) **N**: AG MO
janthina Fairmaire, 1860: 338 (*Cantharis*)
sericea Waltl, 1835: 76 (*Lytta*) **E**: SP **N**: AG MO

genus *Lydomorphus* Fairmaire, 1882a: 85 type species *Lydomorphus cinnamomeus* Fairmaire, 1882
subgenus *Lydomorphus* Fairmaire, 1882a: 85 type species *Lydomorphus cinnamomeus* Fairmaire, 1882
Cylindrothorax Escherich, 1896a: 28 [RN] type species *Lytta angusticollis* Haag-Rutenberg, 1880: 62
Mimovesperus Pic, 1923g: 10 type species *Mimovesperus pilosus* Pic, 1923 (= *Lytta verrucicollis* Karsch, 1881)
Sagitta Escherich, 1895: 295 [HM] type species *Lytta angusticollis* Haag-Rutenberg, 1880
angusticollis fairmairei Kaszab, 1955b: 250 (*Cylindrothorax*) **A**: SA YE **AFR**
angusticollis maculithorax Kaszab, 1955b: 251 (*Cylindrothorax*) **A**: YE
angusticollis suturellus Haag-Rutenberg, 1880: 63 (*Lytta*) **A**: AE IN OM PA SA YE **ORR**
insuturella Pic, 1914p: 106 (*Lytta*)
varicolor Wellman, 1910a: 24 (*Lytta*) [RN]
beccarii Haag-Rutenberg, 1880: 71 (*Lytta*) **A**: SA YE
arabicus Pic, 1912g: 34 (*Lytta*)
braeti Fairmaire, 1894a: 41 (*Cantharis*) **A**: SD **ORR**
brittoni Kaszab, 1953b: 853 (*Cylindrothorax*) **A**: AE IQ OM SA
buettikeri Kaszab, 1983a: 177 (*Cylindrothorax*) **A**: AE OM SA
femoralis Kocher, 1955a: 125 (*Cylindrothorax*) **N**: MO
janczyki Kaszab, 1959h: 437 (*Cylindrothorax*) **A**: IN IQ
kurdistanicus Pic, 1912g: 34 (*Lytta* (*Sagitta*)) **A**: "Kurdistan"
leucophthalmus bipallidus Kaszab, 1953b: 861 (*Cylindrothorax*) **A**: SA
leucophthalmus leucophthalmus Fairmaire, 1893d: 153 (*Cantharis*) **A**: SA YE **AFR**
bimaculatithorax Pic, 1913u: 457 (*Lytta*)
citernii Pic, 1914p: 105 (*Lytta*)

palaestinus Kirsch, 1871: 390 (*Lagorina*) **N**: AG EG LB MO TU **A**: IS JO SA SI
chanzyi Fairmaire, 1876: 38 (*Epicauta*)
picticollis picticollis Haag-Rutenberg, 1880: 70 (*Lytta*) **A**: SA YE **AFR**
breviceps Kaszab, 1955b: 247 (*Cylindrothorax*)

spurcaticollis Fairmaire, 1883c: 104 (*Cantharis*)
reymondi Selander, 1988: 34 (*Cylindrothorax*) [RN] **N**: AG
pici Reymond, 1956a: 41 (*Cylindrothorax*) [HN]
saharanus Kaszab, 1961g: 348 (*Cylindrothorax*) **N**: AG LB
semitestaceus Fairmaire, 1896a: 51 (*Cantharis*) **A**: NP
tenuicollis Pallas, 1798: 102 (*Meloe*) **A**: IN PA SA UP YE **AFR ORR**
ornatus Laporte, 1840: 275 (*Lytta*)
ruficollis A. G. Olivier, 1789: 278 (*Cantharis*) [HN]
verrucicollis Karsch, 1881: 49 (*Lytta*) **N**: AG EG LB MO TU
atriventris Pic, 1911d: 75 (*Sagitta*)
pilosus Pic, 1923g: 10 (*Mimovesperus*)
subgenus *Somalolydus* Bologna, 1992: 63 type species *Cantharis bifoveiceps* Fairmaire, 1897
szalaymarzsoi Kaszab, 1978c: 337 (*Cylindrothorax*) **A**: SA YE

genus *Lydulus* Semenov, 1893d: 265 type species *Lydulus albopilosus* Semenov, 1893
Lydoenas Reitter, 1894a: 51 type species *Lydus pygmaeus* Dokhtouroff, 1889
albopilosus Semenov, 1893d: 266 **A**: KZ TM UZ
tibialis Semenov, 1896b: 61
granulidorsis Semenov, 1896b: 59 **A**: KZ XIN
pygmaeus Dokhtouroff, 1889: 180 **A**: TD TM
discoloripes Semenov, 1896b: 58
semipurpureus Reitter, 1902c: 188 **A**: KZ TD UZ
uzbekistanus Kaszab, 1981a: 165 **A**: UZ

genus *Lydus* Dejean, 1821: 75 type species *Meloe algiricus* Linnaeus, 1758
algiricus Linnaeus, 1758: 420 (*Meloe*) **N**: AG MO TU
algericus Laporte, 1840: 271 (*Lytta*)
immaculatus Fabricius, 1775: 826 (*Mylabris*)
indicus Herbst, 1784c: 145 (*Lytta*)
europaeus Escherich, 1896c: 208 **E**: AL BH BU CR GR HU MC MD RO UK YU **A**: TR
gibbiger Escherich, 1896c: 208 **A**: SY TR
humeralis Gyllenhal, 1817: 16 (*Lytta*) **A**: IS JO SY TR
suturalis Abeille de Perrin, 1880: 250
marginatus Fabricius, 1792b: 88 (*Mylabris*) **N**: AG MO TU
codinai Pardo Alcaide, 1950b: 46
marginea Schoenherr, 1817: 27 (*Lytta*)
pilicollis Fairmaire, 1892b: 155 **A**: SY TR
praeustus L. Redtenbacher, 1850: 49 **A**: IN IQ SY
hauseri Mader, 1931: 136
stigmatifrons Marseul, 1872a: 419
quadrimaculatus Tauscher, 1812b: 141 (*Mylabris*) **E**: KZ ST UK **A**: KZ UK
quadrinotatus Wellman, 1910a: 25 [RN]
quadrisignatus Faldermann, 1835b: 415 [HN]
sanguinipennis Chevrolat, 1840b: 279 **N**: AG MO
taraudanti Escalera, 1913: 54
marrakensis Escalera, 1909b: 243
tarsalis Abeille de Perrin, 1880: 247 **A**: CY IN IS JO LE SY TR
cerastes Abeille de Perrin, 1880: 247
tenuicollis Reitter, 1905d: 246 **A**: IN
tenuitarsis Abeille de Perrin, 1880: 247 **E**: "Caucasus" **A**: IN IS JO LE SI SY TR
impressicollis Fairmaire, 1892b: 156 **A**: TR
trimaculatus italicus Kaszab, 1952a: 98 **E**: IT
trimaculatus trimaculatus Fabricius, 1775: 261 (*Mylabris*) **E**: AL BH BU CR GR IT HU KZ MC RO TR UK YU **A**: TR
communimaculatus Escherich, 1896c: 206
vexator Escherich, 1896c: 206
turcicus Kaszab, 1952a: 98 **A**: IN SY TR

rufus Kaszab, 1952a: 97
vasvarii Kaszab, 1952a: 97
unicolor Reitter, 1887e: 525 **E**: AB AR ST **A**: IN IS LE JO SY TM TR
bitlisiensis Pic, 1901c: 34
glabricollis Escherich, 1896c: 211
halbhuberi Escherich, 1896c: 210
signatithorax Pic, 1902b: 9

genus *Lytta* Fabricius, 1775: 260 type species *Meloe vesicatorius* Linnaeus, 1758
subgenus *Asiolytta* Kaszab, 1962b: 295 type species *Lytta badakschanica* Kaszab, 1958
badakschanica Kaszab, 1958c: 259 **A**: AF UZ
battonii Kaszab, 1962b: 291 **A**: QIN SCH
discipennis Fairmaire, 1891d: c (*Cantharis*) **A**: KA
limbata afghanica Kaszab, 1981a: 164 **A**: AF
limbata limbata L. Redtenbacher, 1844: 535 (*Epicauta*) **A**: AF KA NP UP
satiata Escherich, 1904: 31 **A**: TD XIN
selanderi Saha, 1979: 98 [DA] **A**: UP
subgenus *Indiolytta* Selander, 1960: 28 type species *Lytta apicalis* Haag-Rutenberg, 1880 (= *Lydus melanurus* Hope, 1831)
melanura Hope, 1831: 32 (*Lydus*) **A**: NP SD UP **ORR**
apicalis Haag-Rutenberg, 1880: 71
subgenus *Lytta* Fabricius, 1775: 260 type species *Meloe vesicatorius* Linnaeus, 1758
antennalis Marseul, 1873a: 230 (*Cantharis*) **A**: HP JA NP PA UP XIN
bieti Wellmana, 1912: 33 **A**: WP
caraganae Pallas, 1781: 97 (*Meloe*) **A**: ANH ES FE GAN HEB HEI HEN HUB HUN JA JIA JIL LIA MG NIN NMO NC QIN SHA SHN SHX XIN
pallasi Gebler, 1829a: 141 (*Lytta*)
flavipennis Motschulsky, 1860c: 145 (*Cantharis*) **A**: UP
flaviventris Ballion, 1878: 346 (*Cantharis*) [DA] **A**: XIN
kashmirensis Bologna, 1983: 125 **A**: KA
kryzhanovskyi Kaszab, 1962b: 292 **A**: QIN
menetriesi Faldermann, 1832: 210 **E**: AB AR **A**: IN KZ TD TM UZ
regiszahiri Kaszab, 1958c: 255 **A**: AF
roborowskyi Dokhtouroff, 1887a: 343 (*Cantharis*) **A**: SCH XIN XIZ
thibetana E. Olivier, 1888: lvi
suturella Motschulsky, 1860c: 144 (*Cantharis*) **A**: ES FE GUX HEB HEI JA JIL LIA NIN NMO QIN SHA SHX TD XIN "Korea"
vesicatoria freudei Kaszab, 1958c: 257 **E**: AR GG **A**: TR
vesicatoria heydeni Haag-Rutenberg, 1880: 73 **A**: TR
vesicatoria moreana Pic, 1941a: 1 **E**: GR
vittata Brullé, 1832: 232 (*Cantharis*) [HN]
vesicatoria togata Fischer von Waldheim, 1844: 135 **A**: AF ES KI KZ TD WS XIN
flavovittata Ballion, 1878: 346 (*Cantharis*)
vesicatoria vesicatoria Linnaeus, 1758: 419 (*Meloe*) **E**: AL AR AU BE BH BU BY CR CT CZ EN FR DE GB GE GG GR HU IR IT KZ LA LT LU MC MD NL PL PT RO SK SL SP ST SV SZ TR UK YU
armeniaca Reitter, 1886b: 99 [HN]
aurantica Escherich, 1895: 274
basipennis Pic, 1948b: 4
caeruleicollis Pic, 1948b: 3
coerulea Della Beffa, 1909: 200
costatella Reitter, 1916g: 294
crassicornis A. Costa, 1882: 3 (*Cantharis*)
dibapha Reitter, 1890a: 15
fauconneti Pic, 1948b: 4
lateopaca Pic, 1948b: 4
lebisi Pic, 1949b: 12
leodi Escherich, 1896a: 28

martialis Pic, 1948b: 4
paulocuprea Pic, 1948b: 3
semicuprea Pic, 1948b: 3
theresae Pic, 1948b: 3
viridicuprea Fleischer, 1923: 49
subgenus *Mesolytta* Selander, 1960: 27 type species *Oenas coccineus* Ménétriés, 1849
coccinea Ménétriés, 1849: 247 (*Oenas*) **A**: AF IN KZ TD TM UZ
fuscicrus Brancsik, 1899: 102
glabricollis Waterhouse, 1889: 129 (*Cantharis*)
minor Dokhtouroff, 1889: 179 (*Cantharis*)
deserti Semenov, 1891a: 377 **A**: KZ TD TM UZ
hauseri Heyden, 1894a: 51
laeta Waterhouse, 1889: 129 (*Cantharis*) **A**: AF
vrendenburgi Kaszab, 1962b: 295 **A**: AF UZ
subgenus *Poreospasta* Horn, 1868: 139 type species *Poreospasta polita* Horn, 1868 (= *Nomaspis sublaevis* Horn, 1868)
clematidis Pallas, 1782: 95 (*Meloe*) **A**: AF ES KZ MG TD TM UZ XIN WS
bivitta Schoenherr, 1817: 28 (*Cantharis*)
bivittis Pallas, 1782: 93 (*Meloe*)
fischeri Gebler, 1817: 317
korbi Pic, 1906b: 18
fissicollis Fairmaire, 1886c: 350 (*Cantharis*) **A**: SCH XIZ YUN
flavoangulata Fairmaire, 1891e: cxxxii (*Cantharis*) **A**: AF KA
grumi Semenov, 1893d: 268 **A**: XIN
kabakovi Kaszab, 1981a: 162 **A**: AF
luteovittata Kraatz, 1882d: 334 (*Cantharis*) **A**: KZ NW TD UZ WS
melancholica Wellman, 1910a: 24
tristis Escherich, 1895: 287 [HN]
manicata J. R. Sahlberg, 1903a: 38 (*Lytta*) **A**: TD UZ
poeciloptera Semenov, 1893d: 270 **A**: TD
rubra Hope, 1831: 32 **A**: AF NP
rubrinotata Tan, 1981: 409, 414 **A**: XIZ
sifanica Semenov, 1910a: 28 **A**: XIN XIZ
kukunoorensis Wellman, 1910a: 24 [RN]
tibetana Escherich, 1904: 30 [HN]
subgenus *Pseudolytta* Selander, 1960: 27 type species *Lytta aeneiventris* Haag-Rutenberg, 1880
aeneiventris Haag-Rutenberg, 1880: 75 **A**: GUA HKG HUB HUN JIA JIX
impressithorax Pic, 1924a: 21
kwanhsiensis Mařan, 1941b: 182 **A**: SCH
spissicornis Fairmaire, 1891c: xxiii (*Cantharis*) **A**: JIA
taliana Pic, 1915e: 9 **A**: SCH XIZ

genus *Lyttolydulus* Reitter, 1913d: 662 type species *Lyttolydulus suturalis* Reitter, 1913 (= *Lytta simplicicornis* Pic, 1899)
Lyttalydulus Reitter, 1913d: 662 [emended to *Lyttolydulus* by Reitter, 1915]
Lydopsis Escalera, 1914: 393 type species *Lydus susicus* Escalera, 1914
cinereovestitus Fairmaire, 1876: 49 (*Cantharis*) **N**: AG EG MO TU
mus Escherich, 1904: 32 (*Lagorina*)
deserticola Kaszab, 1952b: 94 **N**: LB MO
giganteus Kaszab, 1952b: 93 **N**: TU
mozabita Chobaut, 1897c: 171 (*Lytta*) **N**: AG
nigronotatus Pic, 1907l: 178 (*Lagorina*) **N**: AG
nubeculosus Kaszab, 1952b: 92 **N**: AG EG MO
rufulus Fairmaire, 1863: 645 (*Lydus*) **N**: AG LB MO TU
djerbensis Escherich, 1895: 285 (*Lytta*)
simplicicornis Pic, 1899n: 279 (*Lytta*) **A**: IS JO
suturalis Reitter, 1913d: 662

susicus Escalera, 1914: 393 (*Lydus*) **N**: MO
thiebaulti arabicus Kaszab, 1983a: 174 **A**: SA
thiebaulti thiebaulti Fairmaire, 1876: 49 (*Cantharis*) **N**: AG EG MO TU **A**: SI

genus *Lyttonyx* Marseul, 1876b: 36 type species *Cantharis bilateralis* Marseul, 1876 (= *Epicauta bicolor* Walker, 1871)
bicolor Walker, 1871: 17 (*Epicauta*) **N**: AG EG LB MO TU **A**: AE IN IS OM SA SI YE
beluchistanica Kaszab, 1959b: 2
bilateralis Marseul, 1876b: 35 (*Cantharis*)
myrmido Fairmaire, 1876: 93 (*Cantharis*)

genus *Muzimes* Aksentjev, 1988: 577 type species *Litta* [sic !] *collaris* Fabricius, 1787
Micromerus Mulsant & Rey, 1858a: 147 [HN]
bitlisiensis Kaszab, 1958a: 15 (*Micromerus*) **A**: TR
brodskyi Dvořák, 1983: 448 (*Micromerus*) **E**: AR
caucasicus Mařan, 1940a: 82 (*Lydus*) **E**: AB AR **A**: TR
unicoloriceps Kaszab, 1958a: 18
collaris Fabricius, 1787a: 215 (*Lytta*) **E**: BU GR HU KZ MC MD ST TR UK **A**: TR
erythrocynea Pallas, 1782: 96 (*Meloe*)
dersimensis Kaszab, 1968d: 447 (*Micromerus*) **A**: TR
erivanicus Mařan, 1941a: 62 (*Lydus*) **E**: AR **A**: TR
luteiceps Kaszab, 1958a: 17
nigrilabris Kaszab, 1958a: 17
nigriceps Kaszab, 1958a: 17
seminigriceps Kaszab, 1958a: 17
iranicus Mařan, 1940a: 86 (*Lydus*) **A**: IN
jureckovae Mařan, 1940a: 80 (*Lydus*) **E**: AR GG
luristanicus Mařan, 1940a: 83 (*Lydus*) **A**: IN
maceki Dvořák, 1983: 446 (*Micromerus*) **A**: IQ
marani Kaszab, 1958a: 16 (*Micromerus*) **E**: GG
nigricornis Escherich, 1896c: 232 (*Lydus*) **E**: AB AR
obenbergeri Kaszab, 1958a: 14 (*Micromerus*) **A**: TR
ressli Dvořák, 1993b: 2 (*Micromerus*) **A**: TR
semiobscurus Pic, 1898f: 73 (*Alosimus*) **A**: TR
sterbai Mařan, 1940a: 88 (*Lydus*) **A**: SY TR
tauricus Mařan, 1941a: 60 (*Lydus*) **A**: TR

genus *Oenas* Latreille, 1802: 186 type species *Lytta atrata* sensu Latreille (= *Meloe afer* Linnaeus, 1767)
afer Linnaeus, 1767: 680 (*Meloe*) **N**: AG MO TU
africanus Desmarest, 1825: 435
maculatus Normand, 1936a: 154
unicolor Laporte, 1840: 271 (*Aenas*)
brevicollis Abeille de Perrin, 1880: 244 **A**: IS JO SY
crassicornis Illiger, 1800: 142 (*Lytta*) **E**: AL AU BH BU CR CZ GR HU IT MC RO SK ST TR UK YU **A**: IN IQ SY TR
ruficollis A. G. Olivier, 1795: 47: 14 (*Mylabris*)
cribricollis Abeille de Perrin, 1880: 244 **A**: IS LE SY
fusicornis Abeille de Perrin, 1880: 242 **E**: PT SP
hispanus Abeille de Perrin, 1880: 243
graecus Bologna, 1991: 176 **E**: GR
laevicollis Abeille de Perrin, 1880: 245 **A**: IS JO
pseudafer Kaszab, 1951e: 278 **N**: EG
luteus Kaszab, 1951c: 276
rubriceps Dvořák, 2004: 103 **A**: IN
sericeus A. G. Olivier, 1795: 46: 18 **N**: AG MO TU
luctuosus Latreille, 1804a: 393

tarsensis Abeille de Perrin, 1880: 244 **E**: AR GG **A**: IN SY TR
tenuicornis Abeille de Perrin, 1880: 245
wilhelmsi Faldermann, 1832: 209 **E**: AB AR **A**: IN TR
armeniacus Kaszab, 1951e: 277

genus *Sybaris* Stephens, 1832: 70 type species *Sybaris immunis* Stephens, 1832
Prionotus L. Redtenbacher, 1844: 536 type species *Prionotus praeustus* Redtenbacher, 1844
flavus Thunberg, 1791: 108 (*Meloe*) **A**: "China"
garhwalensis Saha, 1979: 118 **A**: UP
praeustus L. Redtenbacher, 1844: 536 (*Prionotus*) **A**: KA UP
semivittatus L. Redtenbacher, 1844: 537 (*Prionotus*) **A**: HP KA
testaceus Fabricius, 1792b: 85 (*Lytta*) **A**: NP PA **ORR**
melophtalmos A. G. Olivier, 1795: 46: 10 (*Cantharis*)
tunicatus L. Redtenbacher, 1844: 537 (*Prionotus*) **A**: KA

genus *Syriolytta* Kaszab, 1962b: 295 type species *Lydus vulneratus* Fairmaire, 1892 (= *Lytta suturifera* Pic, 1899)
suturifera Pic, 1899n: 279 (*Lytta*) **A**: IQ IN SY TR
fairmairei Borchmann, 1917: 94 (*Lytta*)
vulnerata Fairmaire, 1892b: 155 (*Lydus*) [HN]

genus *Teratolytta* Semenov, 1894c: 533 type species *Cantharis dives* Brullé, 1832
carlae Bologna, 2006: 145 **A**: TR
dives Brullé, 1832: 232 (*Cantharis*) **E**: AB AL AR BU CR GR MC ME RO SB ST TR UK **A**: SY TR
excellens Escherich, 1895: 280 (*Lytta*)
phalerata Frivaldszky von Frivald, 1837: 182 (*Lytta*)
vittata Kiesenwetter, 1861a: 250 (*Lytta*) [HN]
dvoraki Bologna, 2006: 154 **A**: TR
eylandti Semenov, 1894c: 534 **A**: IN TM UZ
holzshuhi Dvořák, 1983: 443
flavipes Mulsant & Rey, 1858a: 163 (*Cantharis*) **A**: LE SY TR
saphirina Abeille de Perrin, 1895b: xxiv (*Lytta*)
semividua Abeille de Perrin, 1895b: xxiv (*Lytta*)
gentilis J. Frivaldszky, 1877b: 136 (*Cantharis*) **E**: GR (Kos) **A**: TR
coeruleipennis Escherich, 1895: 280 (*Lytta*)
cooensis J. Müller, 1936: 91
violaceipennis Beauregard, 1890: 493 (*Cantharis*)
kaszabi Kryzhanovskij, 1959: 857 **A**: TD TM UZ
klapperichi Kaszab, 1958c: 249 **A**: AF
kulzeri Kaszab, 1958c: 252 **A**: TR
monticola Bologna, 2006: 155 **A**: TR
optabilis Faldermann, 1832: 209 (*Lytta*) **E**: AB AR ST **A**: IN
pilosella Solsky, 1881a: 259 (*Cantharis*) **A**: KI KZ TD UZ
albovillosa Escherich, 1895: 282 (*Lytta*)
tadzhika Kryzhanovskij, 1959: 858
regina Kaszab, 1958c: 250 **A**: AF
senilis Abeille de Perrin, 1895b: xxiv (*Lytta*) **A**: IS SY TR
bytinskii Kaszab, 1957e: 230
jodina Kaszab, 1957e: 231
taurica Bologna, 2006: 147 **A**: TR
tricolor Haag-Rutenberg, 1880: 76 (*Lytta*) **A**: IN UZ
vanensis Kaszab, 1968d: 448 **A**: TR

genus *Trichomeloe* Reitter, 1911a: 391 type species *Meloe chrysocomus* Miller, 1861
chrysocomus Miller, 1861: 206 (*Meloe*) **A**: CY IS JO SY TR
conicicollis Reitter, 1907g: 215 (*Meloe*) **E**: GR (Castelorizon) **A**: TR

deflexus Reitter, 1889c: 106 (*Meloe*) **A**: IS JO SY TR
ottomanus Pliginskij, 1914: 254 (*Meloe*) **A**: SY TR
frivaldszkyi Kaszab, 1958c: 304 (*Meloe*)
ovatus Pripisnova, 1987a: 43 (*Meloe*) **A**: TD
sericellus Reiche, 1857: 273 (*Meloe*) **E**: IT (Sicilia) **A**: IS JO LE SY

tribe Mylabrini Laporte, 1840

genus *Actenodia* Laporte, 1840: 268 type species *Actenodia guttata* Laporte, 1840
Rusadiria Pardo Alcaide, 1952: 60 type species *Mylabris billbergi* Gyllenhal, 1817
billbergi Gyllenhal, 1817: 33 (*Mylabris*) **E**: FR IT (West Liguria) PT SP
baudii Magistretti, 1943: 36 (*Mylabris*)
circumfusa Baudi di Selve, 1878c: 1062 (*Coryna*)
clavicornis Duméril, 1825: 13 (*Mylabris*)
gallica Pic, 1916a: 7 (*Coryna*)
parcepunctata Pardo Alcaide, 1950a: 81 (*Coryna*)
confluens Reiche, 1866: 629 (*Coryna*) **E**: BU TR **A**: SI SY TR
caudanigra Abeille de Perrin, 1880: 237 (*Coryna*)
mediodisjuncta Pic, 1930a: 1 (*Coryna*)
mediopartita Mader, 1929: 109 (*Mylabris*)
ottomana Escherich, 1899: 109 (*Coryna*)
picta Baudi di Selve, 1878b: 361 (*Coryna*)
denticulata Marseul, 1872a: 418 (*Coryna*) **A**: AE SA
distincta Chevrolat, 1840b: 269 (*Hycleus*) **E**: IT (Sicilia, Lampedusa) **N**: AG MO TU
andalusiaca Pic, 1916a: 7 (*Coryna*)
anticefasciata Magistretti, 1943: 36 (*Mylabris*)
beccariana Pic, 1916a: 7 (*Coryna*)
gafsana Normand, 1950a: 81 (*Coryna*)
galloisi Pic, 1916a: 7 (*Coryna*)
hafidi Pic, 1916a: 7 (*Coryna*)
particularenotata Pic, 1949a: 10 (*Coryna*)
sexnotata Pic, 1949a: 10
sicula Baudi di Selve, 1878b: 361 (*Coryna*)
theresae Pic, 1949a: 10
lata Reiche, 1866: 628 (*Coryna*) **N**: EG
fasciata Baudi di Selve, 1878b: 360 (*Dices*)
perfuga Dvořák, 1993a: 3 (*Paractenodia*) [DA] **A**: SY
peyroni Reiche, 1866: 630 (*Coryna*) **A**: IS JO LE SI SY TR
septempunctata Baudi di Selve, 1878b: 361 (*Coryna*) **N**: AG MO TU
extera Dvořák, 1993a: 1
susica Escalera, 1913: 54 (*Hycleus*)
suturifera Pic, 1896d: 42 (*Coryna*) **N**: AG LB MO TU
alfredi Pic, 1913f: 112 (*Coryna*)
bleusei Chobaut, 1896b: 376 (*Coryna*)
indenticulata Pic, 1913f: 112 (*Coryna*)
vaulogeri Pic, 1913f: 112 (*Coryna*)

genus *Ceroctis* Marseul, 1870b: 168 type species *Mylabris corynoides* Reiche, 1866 (= *Mylabris trizonata* Reiche, 1866)
trizonata Reiche, 1866: 631 (*Mylabris*) **N**: AG EG LB MO TU **A**: IS JO SI
biinterrupta Pic, 1898l: 224
binotata Pic, 1897l: 126
breveinterrupta Pic, 1919a: 18 (*Zonabris*)
coronata Marseul, 1873b: 556 (*Mylabris*)
corynoides Reiche, 1866: 631 (*Mylabris*)
gamondana Pic, 1919a: 18 (*Zonabris*)
gafsana Pic, 1919a: 18 (*Zonabris*)
interrupta Pic, 1897l: 126
mediodivisa Pic, 1919a: 18 (*Zonabris*)

setafana Pic, 1919a: 18 (*Zonabris*)
subdivisa Pic, 1919a: 18 (*Zonabris*)
subhumeralis Pic, 1903g: 114
subhumeralis Normand, 1938b: 365 (*Mylabris*) [HN]
trinotata Pic, 1953d: 499
wittmeri Kaszab, 1983a: 152 **A**: SA
yerburyi Gahan, 1896b: 456 (*Mylabris*) **A**: SA YE
theresae Pic, 1908b: 33 (*Zonabris*)

genus *Croscherichia* Pardo Alcaide, 1950a: 63 type species *Mylabris circumflexa* Chevrolat, 1840 (= *Mylabris paykulli* Billberg, 1813)
Lybicisca Kuzin, 1954: 369 type species *Mylabris sanguinolenta* A. G. Olivier, 1811
albilaena Bedel, 1899: 82 (*Zonabris*) **N**: AG EG LB TU **A**: SA
bedeli Bleuse, 1899: 383 (*Zonabris*) **N**: AG LB MO
ainensis Pic, 1919f: 17 (*Zonabris*)
augusti Pic, 1919e: 13 (*Zonabris*)
crosi Pic, 1919f: 17 (*Zonabris*)
delarouzei Reiche, 1866: 639 (*Mylabris*) **A**: IS JO LB
sancta Pic, 1920e: 13 (*Zonabris*)
femorata Klug, 1845: nr. 8 (*Mylabris*) **A**: SA
fulgurita Reiche, 1866: 640 (*Mylabris*) **N**: AG EG LB MO **AFR**
bisbiinterrupta Pic, 1919a: 18
subundulata Pic, 1919f: 18
gilvipes Chevrolat, 1840b: 273 (*Mylabris*) **N**: AG EG LB MO TU **A**: IS JO SI
angulata Klug, 1845: nr. 18 (*Mylabris*)
diffinis Abeille de Perrin, 1880: 238 (*Mylabris*)
lameyi Marseul, 1885: cxlvii (*Mylabris*)
mediodestituita Pic, 1919f: 18 (*Zonabris*)
goryi Marseul, 1870b: 88 (*Mylabris*) **A**: AE IN IQ IS JO OM PA SA SY
litigiosa Chevrolat, 1840b: 271 (*Mylabris*) **N**: AG EG MO TU **A**: IN IS OM SA SI SY "Turkestan"
alepensis Pic, 1913s: 185 (*Zonabris*)
islamita Marseul, 1873b: 467 (*Mylabris*)
mozabita Pic, 1897a: 2 (*Zonabris*) **N**: AG LB MO TU
angulata Sumakov, 1915: 60 (*Mylabris*) [HN]
postbifasciata Pic, 1921f: 21 (*Zonabris*)
paykulli Billberg, 1813: 63 (*Mylabris*) **N**: AG LB MO TU
abbesensis Pic, 1921f: 22 (*Zonabris*)
alegensis Pic, 1921f: 22 (*Zonabris*)
antebifasciata Sumakov, 1934: 4 (*Mylabris*)
antebinotata Pic, 1921f: 22 (*Zonabris*)
antequadrinotata Pic, 1921f: 22 (*Zonabris*)
antetrinotata Pic, 1921f: 22 (*Zonabris*)
apicecincta Pic, 1921f: 22 (*Zonabris*)
atrojuncta Pic, 1921f: 21 (*Zonabris*)
aumalensis Pic, 1921f: 22 (*Zonabris*)
bisquinquenotata Pic, 1915k: 22 (*Zonabris*)
borealis Voigts, 1902: 178 (*Zonabris*)
circumflexa Chevrolat, 1840b: 273 (*Mylabris*)
comptei Pic, 1898b: 3 (*Zonabris*)
demoflyisi Battoni, 1961: 29
destituita Pic, 1921f: 23 (*Zonabris*)
discojuncta Pic, 1921f: 22 (*Zonabris*)
diversenotata Pic, 1921f: 22 (*Zonabris*)
distincta Baudi di Selve, 1878b: 369 (*Mylabris*)
edoughnensis Pic, 1921f: 21 (*Zonabris*)
fontainei Pic, 1921f: 23 (*Zonabris*)
goudoti Chevrolat, 1840b: 274 (*Mylabris*)
mediofasciata Pic, 1921f: 22 (*Zonabris*)
mediotrinotata Pic, 1921f: 22 (*Zonabris*)

octopunctata Sumakov, 1934: 4 (*Mylabris*)
oraniensis Pic, 1898b: 3 (*Zonabris*)
posticefasciata Pic, 1921f: 22 (*Zonabris*)
postrinotata Pic, 1921f: 22 (*Zonabris*)
quinquenotata Pic, 1921f: 22 (*Zonabris*)
reghaiensis Pic, 1921f: 22 (*Zonabris*)
rufoapicalis Pic, 1921f: 23 (*Zonabris*)
scapularis Chevrolat, 1840b: 278 (*Mylabris*)
septemnotata Pic, 1921f: 22 (*Zonabris*)
sexsignata Pic, 1921f: 22 (*Zonabris*)
subangulata Kocher, 1955b: 127 (*Mylabris*)
subapicalis Pic, 1921f: 21 (*Zonabris*)
subexternepunctata Pic, 1921b: 21 (*Zonabris*)
subfasciata Pic, 1921f: 21 (*Zonabris*)
tlemcensis Pic, 1915k: 22 (*Zonabris*)
trifasciata Chevrolat, 1840b: 275 (*Mylabris*)
unireducta Pic, 1921f: 21 (*Zonabris*)
quadrizonata Fairmaire, 1875a: 530 (*Mylabris*) **N**: AG LB TU
richteri Kaszab, 1957b: 50 **A**: AE IN OM SA YE
salavatiani Kaszab, 1968f: 756 **A**: AE AF IN PA SA YE
sanguinolenta arabica Bologna, 1991: 153 **A**: SA
sanguinolenta sanguinolenta A. G. Olivier, 1811: 95 (*Mylabris*) **N**: AG EG LB MO TU **A**: IN IS JO SI SY
andresi Pic, 1911d: 74 (*Zonabris*) [HN]
adolfi Pic, 1912o: 90 (*Zonabris*) [RN]
ghardaiensis Pic, 1897a: 2 (*Zonabris*)
incerta Klug, 1845: n. 17 (*Mylabris*)
latreillei Billberg, 1813: 36 (*Mylabris*)
margensis Pic, 1924: 30 (*Zonabris*)
nedjalensis Pic, 1920e: 13 (*Zonabris*)
notatipennis Pic, 1897l: 124 (*Zonabris*)
unijuncta Pic, 1897a: 2 (*Zonabris*)
usta Baudi di Selve, 1878c: 1143 (*Mylabris*)
sonyae Kaszab, 1983a: 165 (*Mylabris*) **A**: SA
tigrinipennis Latreille, 1827: 286 (*Mylabris*) **N**: AG EG LB MO TU **A**: IQ IS JO SA YE **AFR**
bisbiflexuosa Pic, 1911d: 73 (*Zonabris*)
diversithorax Pic, 1910e: 18 (*Zonabris*)
luxorensis Pic, 1911d: 74 (*Zonabris*)
menthae Klug, 1845: nr. 11 (*Mylabris*)
vigintipunctata A. G. Olivier, 1811: 97 (*Mylabris*) **N**: EG **A**: IN IS JO SI
wartmanni Pic, 1896f: 62 (*Zonabris*) **N**: AG MO
abrupta Wellman, 1910a: 22 (*Mylabris*) [RN]
interrupta Pic, 1896f: 62 [HN]

genus *Hycleus* Latreille, 1817a: 317 type species *Mylabris argentata* Fabricius, 1792
Arithmema Chevrolat, 1834: pl. 35 type species *Meloe decemguttata* Thunberg, 1791
Androfoveata Pardo Alcaide, 1954a: 81 type species *Mylabris duodecimpunctata* A. G. Olivier, 1811
Coryna Billberg, 1813: 73 [HN] type species *Mylabris argentata* Fabricius, 1792
Decatoma Dejean, 1821: 74 [HN] type species *Meloe lunata* Pallas, 1782
Decapotoma Voigts, 1902: 177 [RN] type species *Meloe lunata* Pallas, 1782
Dices Dejean, 1821: 74 type species *Cerocoma ocellata* A. G. Olivier, 1791
Euzonabris Kuzin, 1954: 357 type species *Meloe cichorii* Linnaeus, 1758
Gorrizia Pardo Alcaide, 1954a: 61 type species *Mylabris duodecimpunctata* A. G. Olivier, 1811
Megabris Des Gozis, 1881: cxiii type species *Meloe cichorii* Linnaeus, 1758
Mesogorbata Pardo Alcaide, 1954a: 78 type species *Mylabris apicipennis* Reiche, 1866
Mesoscutata Pardo Alcaide, 1954a: 81 type species *Mylabris duodecimpunctata* A. G. Olivier, 1811
Mesotaeniata Pardo Alcaide, 1955a: 4 type species *Meloe lunata* Pallas, 1782
Sphenabris Kuzin, 1954: 361 type species *Meloe balteata* Pallas, 1782
Tigrabris Kuzin, 1954: 364 type species *Meloe atrata* Pallas, 1773
Zonabris Harold, 1879: 134 type species *Meloe cichorii* Linnaeus, 1758

abiadensis Marseul, 1870b: 29 (*Mylabris*) **N**: EG **AFR**
aegyptiacus Marseul, 1870b: 142 (*Mylabris*) **N**: EG LB
 chakouri Pic, 1907h: 161 (*Zonabris*)
 conjugatus Baudi di Selve, 1878b: 369 (*Mylabris*)
aestuans Klug, 1845: nr. 31 (*Mylabris*) **A**: SA YE **AFR**
 rugosissimus Pic, 1909j: 130 (*Zonabris*)
afrotropicus Kaszab, 1983a: 171 (*Actenodia*) **A**: YE
allardi Marseul, 1870b: 193 (*Coryna*) **N**: AG MO TU
 punctofasciata Fairmaire, 1875a: 531 (*Mylabris*)
 sefrensis Pic, 1896b: 12 (*Coryna*)
apicipennis Reiche, 1866: 635 (*Mylabris*) **N**: EG LB **A**: IS JO SI
 fourtani Pic, 1921g: 45 (*Zonabris*)
 innesi Pic, 1912c: 1 (*Zonabris*)
arabicus Pallas, 1781: 89 (*Meloe*) **A**: SA YE
argentatus Fabricius, 1792b: 90 (*Mylabris*) **N**: EG
 caroli Pic, 1908b: 34 (*Coryna*)
 externenotatus Pic, 1908b: 34 (*Coryna*)
argentifer argentifer Pic, 1895e: 81 (*Zonabris*) **N**: AG LB MO TU
 auberti Chobaut, 1898a: 87 (*Decatoma*)
 henoni Fairmaire, 1897c: 121 (*Mylabris*)
argentifer bytinskii Kaszab, 1969a: 283 (*Decapotoma*) **A**: IS SA
argentifer pallidissimus Kaszab, 1983a: 168 (*Decapotoma*) **A**: SA
atratus Pallas, 1773: 722 (*Meloe*) **E**: AB AR CT GG ST UK **A**: AF IN KI KZ MG UZ TM WS XIN
 karelini Fischer von Waldheim, 1842: 25 (*Mylabris*)
 minutus Fabricius, 1798: 121 (*Mylabris*)
 mediomaculatus Sumakov, 1929a: 4
 metatarsalis Marseul, 1870b: 81 (*Mylabris*)
belli Borchmann, 1940b: 616 (*Mylabris*) **A**: PA
biguttatus Gebler, 1841a: 374 (*Mylabris*) **A**: AF IN KI KZ TD TM UZ XIN
 fasciculatus Dokhtouroff, 1889: 140 (*Zonabris*)
 karateginensis Sumakov, 1915: 73 (*Mylabris*)
 pullatus Heyden, 1883a: 65 (*Mylabris*)
 sexmaculatus Dokhtouroff, 1889: 141 (*Zonabris*)
 similis Voigts, 1902: 178 (*Zonabris*)
bipunctatus A. G. Olivier, 1811: 94 (*Mylabris*) **A**: IN IQ SA
birecurvus Marseul, 1870b: 184 (*Coryna*) **A**: IS JO SY TR
 caudaluteus Pic, 1930a: 1 (*Coryna*)
bistillatus Tan, 1981: 407, 409 (*Mylabris*) **A**: SCH XIZ
biundulatus Pallas, 1782: 78 (*Meloe*) **A**: FUJ YUN **ORR**
 cleroides Lichtenstein, 1796: 74 (*Mylabris*)
 pustulatus Thunberg, 1791: 113 (*Meloe*)
 undulatus Herbst, 1786: 179 (*Mylabris*)
borchmannianus Kaszab, 1983a: 164 (*Mylabris*) **A**: YE
brevetarsalis Kaszab, 1960d: 260 (*Mylabris*) **A**: FUJ GUX NP SD TAI YUN **ORR**
brevicollis Baudi di Selve, 1878b: 373 (*Mylabris*) **E**: SP (Almería) **N**: AG LB MO TU
 beauprei Pic, 1909q: 185 (*Zonabris*)
 mediobisbijunctus Pic, 1916a: 7
 multibijunctus Pic, 1921a: 5
 nigribasis Escalera, 1914: 399 (*Zonabris*)
 ouiouanensis Pic, 1936e: 30 (*Zonabris*)
 plagiellus Escalera, 1914: 398 (*Zonabris*)
brunnipes Klug, 1845: nr. 32 (*Mylabris*) **N**: AG EG LB MO TU **A**: AE IS OM SA YE
 adrarensis Pic, 1942d: 78 (*Zonabris*)
 humeronotatus Pic, 1939e: 29 (*Zonabris*)
 interruptus Pic, 1896f: 62 (*Zonabris*)
 letourneuxi Pic, 1896f: 62 (*Zonabris*)
 subruptus Wellman, 1910a: 23 (*Mylabris*)
chodschenticus Ballion, 1878: 337 (*Mylabris*) **E**: AB **A**: KI KZ TD TM UZ XIN

ataensis Pic, 1901h: 78 (*Zonabris*)
diabolicus Escherich, 1899: 108 (*Zonabris*)
mediojunctus Pic, 1901h: 78 (*Zonabris*)
mediopunctatus Pic, 1901h: 78 (*Zonabris*)
cichorii Linnaeus, 1758: 419 (*Meloe*) **A**: ANH AP FUJ GAN GUA HEB HEN HP HUN JA JIX NIN NP SD SHX TAI UP YUN **ORR**
formosensis Kaszab, 1959h: 438 (*Mylabris*)
postscutellomaculatus Kaszab, 1960d: 259 (*Mylabris*)
cingulatus Faldermann, 1837: 122 (*Mylabris*) **E**: "Transcaucasia" **A**: IN SA "Turkestan"
flavipennis Dokhtouroff, 1889: 157 (*Zonabris*) [HN]
flavis Mader, 1927: 196 [RN]
colligatus L. Redtenbacher, 1850: 49 (*Mylabris*) **A**: IN
succinctus Baudi di Selve, 1878b: 365 (*Mylabris*)
concinnus Marseul, 1870b: 191 (*Mylabris*) **A**: IS JO SY
contaminatus Abeille de Perrin, 1880: 236 (*Coryna*)
impressithorax Pic, 1930a: 2 (*Zonabris*)
cruentatus Klug, 1845: nr. 1 (*Mylabris*) **A**: SA
curticornis Pic, 1919f: 19 (*Zonabris*) [DA] **E**: AR
damohensis Saha, 1979: 60 (*Mylabris*) [DA] **A**: UP **ORR**
designatus Reiche, 1850: 377 (*Mylabris*) **N**: EG **AFR**
hacolyssa Rochebrun, 1883: 182 (*Mylabris*)
tristis Reiche, 1850: 379 (*Mylabris*)
diversesignatus Pic, 1919f: 18 (*Zonabris*) **N**: AG TU
diversipilis Sumakov, 1915: 42 (*Mylabris*) [DA] **A**: IS SY
dohrni Marseul, 1873b: 506 (*Mylabris*) **A**: UP **ORR**
dolens Marseul, 1872a: 417 (*Coryna*) **A**: SY "Palestine"
dubiosus Marseul, 1870b: 35 (*Mylabris*) **N**: EG **AFR**
antebiplagiatus Kaszab, 1960f: 278 (*Mylabris*)
bipartitus Marseul, 1873b: 427 (*Mylabris*)
cursorius Kaszab, 1960f: 278 (*Mylabris*)
dabbenei Pic, 1914p: 110 (*Zonabris*)
erythraeus Pic, 1908g: 60 (*Zonabris*)
katesheanus Kaszab, 1960f: 278 (*Mylabris*)
mimis Kaszab, 1960f: 278 (*Mylabris*)
peregrinus Kaszab, 1960f: 278 (*Mylabris*)
rollei Pic, 1907b: 119 (*Zonabris*)
taenia Kaszab, 1960f: 278 (*Mylabris*)
dufourii Graells, 1849: 621 (*Mylabris*) **E**: PT SP
duodecimmaculatus A. G. Olivier, 1811: 98 (*Mylabris*) **N**: MO
duodecimpunctatus A. G. Olivier, 1811: 98 (*Mylabris*) **E**: FR PT SP
cyanescens Rosenhauer, 1856: 231 (*Mylabris*)
fasciatus Fuente, 1910: 444 (*Zonabris*)
massiliensis Pic, 1902b: 9 (*Zonabris*)
postunipunctatus Mader, 1929: 108 (*Mylabris*)
duplicatus Klug, 1845: nr. 4 (*Mylabris*) **A**: SA YE **AFR**
zonatus Klug, 1845: nr. 5 (*Mylabris*)
esfandiarii Kaszab, 1969a: 279 (*Decapotoma*) **A**: IS
euphraticus Marseul, 1870b: 84 (*Mylabris*) **A**: IQ
fimbriatus Marseul, 1870b: 83 (*Mylabris*) **N**: EG **A**: SA YE
bafrechianus Pic, 1955e: 1047
bisbiseparatus Pic, 1955a: 1 (*Zonabris*)
fisselensis Pic, 1944b: 7 (*Zonabris*)
laurantianus Pic, 1955a: 1 (*Zonabris*)
posticeluteonotatus Pic, 1955e: 1047
villiersianus Pic, 1955e: 1047
flavohirtus Kaszab, 1958c: 298 (*Decapotoma*) **A**: AF
fuscus A. G. Olivier, 1811: 100 (*Mylabris*) **E**: AB AR CT ST **A**: IN IQ IS JO KZ SY TM TR
alienus J. R. Sahlberg, 1913b: 196 (*Zonabris*)

circascutellaris Pic, 1914o: 77 (*Zonabris*)
deleta Marseul, 1870b: 92 (*Mylabris*)
galatiensis Escherich, 1899: 39 (*Zonabris*)
ghorfii Pardo Alcaide, 1961b: 215 (*Gorrizia*) **N**: MO TU
gratiosus Marseul, 1870b: 135 (*Mylabris*) **N**: EG **A**: OM SA
hanguensis Kaszab, 1958c: 290 (*Mylabris*) **A**: PA
himalayaensis Saha, 1979: 66 (*Mylabris*) [DA] **A**: AP NP UP
hokumanensis Kôno, 1940: 58 (*Mylabris*) [DA] **A**: HEI
horai Saha, 1972: 20 (*Mylabris*) [DA] **A**: PA
humerosus Escherich, 1899: 97 (*Zonabris*) **E**: AB AR GG GR ST TR **A**: IN IQ JO KZ SY TM TR UZ
infasciculatus Pic, 1929e: 14 (*Zonabris*) **A**: SA
japonicus Sumakov, 1913b: 474 (*Mylabris*) **A**: JA
javeti javeti Marseul, 1870b: 63 (*Mylabris*) **A**: IN IQ TM "Turkestan"
javeti umbilicatus Kaszab, 1958c: 291 (*Mylabris*) **A**: AF
kirgisicus Aksentjev & Sadykova, 1990: 152 (*Mylabris*) **A**: KI
lacteus Marseul, 1870b: 140 (*Mylabris*) **N**: EG
ligatus Marseul, 1870b: 31 (*Mylabris*) **A**: SA **AFR**
lindbergi Kaszab, 1973d: 294 (*Mylabris*) **A**: AF
linnavuorii Pardo Alcaide, 1963: 295 (*Mylabris*) **A**: IS
maculiventris Klug, 1845: n. 2 (*Mylabris*) **N**: EG **A**: OM SA YE **AFR**
rubronotatithorax Pic, 1911j: 146 (*Zonabris*)
mannheimsi Kaszab, 1961g: 349 (*Mylabris*) **A**: CH
mediozigzagus Pic, 1924d: 25 (*Zonabris*) [DA] **A**: IN
mediobipunctatus Pic, 1919f: 19 (*Zonabris*) [DA] **A**: IN
mediofasciatellus Pic, 1908b: 33 (*Zonabris*) **N**: EG **A**: SI
heluanensis Pic, 1908b: 33 (*Zonabris*)
massariensis Pic, 1908b: 33 (*Zonabris*)
sesostrisi Pic, 1930a: 2 (*Zonabris*)
medioinsignatus Pic, 1909m: 166 (*Zonabris*) **A**: FUJ MG NP SCH TIA YUN
phaleratoides Kaszab, 1960d: 261
mylabroides Laporte, 1840: 268 **N**: EG **AFR**
mediofasciatus Pic, 1914p: 114 (*Coryna*)
nigriplantis Klug, 1845: n. 9 (*Mylabris*) **A**: IS SA YE **AFR**
nigrohirtus Kaszab, 1983: 167 (*Decapotoma*) **A**: SA
novemdecimpunctatus A. G. Olivier, 1811: 98 (*Mylabris*) **N**: AG EG MO TU **A**: IS SA SI
argenteopubescens Escalera, 1909b: 247 (*Zonabris*)
bicolorepilosus Sumakov, 1913a: 473 (*Mylabris*)
octodecimpunctatus Klug, 1845: no. 23 (*Mylabris*)
semifasciatus Pic, 1895e: 81 (*Zonabris*)
ocellaris A. G. Olivier, 1795: 48: 6 (*Cerocoma*) **N**: EG LB
multijunctus Pic, 1908b: 34 (*Coryna*)
octodecimmaculatus Marseul, 1870b: 147 (*Mylabris*) **N**: AG EG LB MO TU **A**: IS
melaini Escalera, 1909b: 249 (*Zonabris*)
ornatus Reiche, 1866: 630 (*Coryna*) **A**: IS JO LE SI SY
berytensis Pic, 1912q: 177 (*Coryna*)
diversesignatus Pic, 1939f: 147 (*Coryna*) [HN]
latebijunctus Pic, 1912q: 177 (*Coryna*)
parvulus J. Frivaldszky, 1892: 114 (*Zonabris*) **A**: "China"
lajoyei Pic, 1916i: 126 (*Zonabris*)
phaleratus choaicus Saha, 1972: 79 (*Mylabris*) [DA] **A**: PA
phaleratus phaleratus Pallas, 1781: 78 (*Meloe*) **A**: FUJ JIX HP KA NP PA UP XIZ YUN **ORR**
irigator Kaszab, 1960d: 259 (*Mylabris*)
moquinianus L. Ferrer, 1859: 540 (*Mylabris*)
patruelis J. Sturm, 1843: 172 (*Mylabris*)
sidae Fabricius, 1798: 120 (*Mylabris*)
pierrei Kaszab, 1968f: 765 (*Mylabris*) **A**: IN
pitcheri Kaszab, 1983a: 169 (*Coryna*) **A**: AE SA

polymorphus polymorphus Pallas, 1771: 465 (*Attelabus*) **E**: AB AN AR AU BH BU BY CR CT CZ FR GE GG GR HU IT KZ LS MC MD RO SK SL SP ST SZ TR UK YU **A**: IN FE KZ SY TD TM TR UZ WS XIN
agilis Baudi di Selve, 1878b: 367 (*Mylabris*)
alpestris Pic, 1925f: 13 (*Zonabris*)
australis Baudi di Selve, 1878b: 367 (*Mylabris*)
baregesius Pic, 1925f: 13 (*Zonabris*)
cichorii Schrank von Paula, 1781: 222 (*Meloe*) [HN]
fasciatus Fuessly, 1775: 20 (*Meloe*)
floralis Pallas, 1782: 82 (*Meloe*)
fuesslini Panzer, 1796a: 18 (*Mylabris*)
karateginensis Sumakov, 1915: 73
luteointerruptus Pic, 1919f: 18 (*Zonabris*)
monetierensis Pic, 1925f: 13 (*Zonabris*)
mulsanti Pic, 1925f: 13 (*Zonabris*)
multidisjunctus Pic, 1929a: 2 (*Zonabris*) [HN]
multijunctus Kaszab, 1958b: 190 (*Mylabris*) [RN]
nigritus Baudi di Selve, 1878b: 368 (*Mylabris*) [HN]
nigritulus Kaszab, 1958b: 190 (*Mylabris*) [RN]
octomaculatus Villers, 1789: 404 (*Meloe*)
pelissieri Pic, 1925f: 13 (*Zonabris*)
quadrifarius Marseul, 1870b: 79 (*Mylabris*)
spartii Germar, 1817: 210 (*Mylabris*)
subjunctus Pic, 1929a: 2 (*Zonabris*)
variabilis Olivier, 1795: 47: 10 (*Mylabris*) [HN]
viturati Pic, 1925f: 13 (*Zonabris*)
zoufali Reitter, 1908j: 272 (*Zonabris*)
polymorphus regeli Sumakov, 1929a: 4 (*Mylabris*) **A**: UZ
postbilunulatus Pic, 1919f: 21 (*Zonabris*) [DA] **A**: "Kurdistan"
kurdistanicus Pic, 1919f: 21 (*Zonabris*)
posticatus Fairmaire, 1892b: 154 (*Mylabris*) **A**: SY TR
praeustus Fabricius, 1792b: 88 (*Mylabris*) **N**: AG MO TU
apicalis Chevrolat, 1840b: 278 (*Mylabris*)
contextus Chevrolat, 1840b: 278 (*Mylabris*)
nigritus Marseul, 1870b: 57 (*Mylabris*)
semirufus Marseul, 1870b: 57 (*Mylabris*)
superfluus Marseul, 1870b: 57 (*Mylabris*)
pseudobrunnipes Kaszab, 1983a: 163 (*Mylabris*) **A**: AE SA
quatuordecimpunctatus Pallas, 1781: 80 (*Meloe*) **E**: AR CT ST UK **A**: ES IN KZ TR UZ WS
conbustus Tauscher, 1812b: 143 (*Mylabris*)
confusus Fischer von Waldheim, 1827: 6 (*Mylabris*)
meliloti A. G. Olivier, 1811: 99 (*Mylabris*)
setiger Waltl, 1838: 466 (*Mylabris*)
quatuordecimsignatus Marseul, 1870b: 133 (*Mylabris*) **N**: AG EG MO **A**: IS JO SA SI
peyerimhoffi Pic, 1929h: 126 (*Mylabris*)
pharius Pic, 1916a: 7 (*Zonabris*)
raphael Marseul, 1876a: 29 (*Mylabris*) **N**: EG **A**: IN IS SI
rotroui Pic, 1930a: 1 (*Coryna*) **N**: AG MO
rubricollis Marseul, 1875b: xxiv (*Coryna*) **A**: SA YE
ruficornis Fabricius, 1798: 121 (*Mylabris*) **N**: MO
deauratus Pardo Alcaide, 1954a: 84 (*Mylabris*)
igneopubescens Escalera, 1909b: 249 (*Zonabris*)
tangerianus Pic, 1903g: 114 (*Zonabris*)
rufipalpis Escalera, 1909b: 248 (*Zonabris*) **N**: CI MO
discrepans Pardo Alcaide, 1954a: 82 (*Mylabris*) [HN]
mamorensis Pardo Alcaide, 1954a: 82 (*Mylabris*)
pardoianus Kaszab, 1958b: 190 (*Mylabris*) [RN]
rungsi Peyerimhoff, 1935b: 20 (*Zonabris*) **N**: MO
saharicus Chobaut, 1901: 280 (*Myulabris*) **N**: AG MO TU
mayeti Chobaut, 1901: 2890 (*Mylabris*)

oasium Chobaut, 1901: 2890 (*Mylabris*)
vareillesi Chobaut, 1901: 2890 (*Mylabris*)
scabiosae djiroftanus Kaszab, 1959b: 3 (*Mylabris*) **A**: IN
scabiosae dressi Kaszab, 1958c: 296 (*Mylabris*) **A**: AF
scabiosae scabiosae A. G. Olivier, 1811: 99 (*Mylabris*) **E**: AB AR BU CT GG GR RO ST **A**: IQ IN IS JO KZ LE MG SY TD TM TR UZ XIN
alieniger Heyden, 1883a: 67 (*Mylabris*)
auliensis Pic, 1901l: 111 (*Zonabris*)
eichleri Mader, 1927: 194 (*Zonabris*)
euphraticus Marseul, 1870b: 84 (*Mylabris*)
ghorbandanus Kaszab, 1958c: 295 (*Mylabris*)
maracadensis Dokhtouroff, 1889: 130 (*Zonabris*)
marseuli Ballion, 1878: 336 (*Mylabris*)
sicheri Fleischer, 1921: 142 (*Zonabris*)
transcaspicensis Pic, 1928e: 5 (*Zonabris*)
scabratus Klug, 1845: nr. 33 (*Mylabris*) **A**: OM SA SI YE
scapularis Klug, 1845: nr. 6 (*Mylabris*) **A**: OM SA YE
schah shah Reiche, 1866: 632 (*Mylabris*) **A**: IN
postluteodisjunctus Kaszab, 1957b: 59 (*Mylabris*)
postluteoreductus Kaszab, 1957b: 59 (*Mylabris*)
schah tuxeni Kaszab, 1953a: 306 **A**: AF (*Mylabris*)
schauffelei Kaszab, 1957b: 57 (*Mylabris*) **A**: IN
anticebiconjuncta Kaszab, 1957b: 58 (*Mylabris*)
bisbisconjuncta Kaszab, 1957b: 58 (*Mylabris*)
postmediojuncta Kaszab, 1957b: 58 (*Mylabris*)
schoenherri pretiosus Kaszab, 1961g: 350 (*Mylabris*) **A**: FUJ
schoenherri schoenherri Billberg, 1813: 14 (*Mylabris*) **A**: FUJ GUA GUX HAI JIX TAI **ORR**
scutellatus Rosenhauer, 1856: 231 (*Mylabris*) **E**: PT SP
lichtensteini Gorriz y Muñoz, 1882: 124 (*Mylabris*)
sexmaculatus A. G. Olivier, 1811: 98 (*Mylabris*) **E**: AB AR GG ST UK **A**: IN IS JO KZ SA SY TR "Turkestan"
bimaculatus Klug, 1845: nr. 32 (*Mylabris*) [HN]
ledereri Marseul, 1870b: 57 (*Mylabris*)
oneraa Escherich, 1899: 109 (*Zonabris*)
superbus Faldermann, 1837: 123 (*Mylabris*)
sialanus Pic, 1929g: 110 (*Zonabris*) [DA] **N**: EG
silbermanni Chevrolat, 1840b: 277 (*Mylabris*) **N**: AG MO
affinis Lucas, 1847: 389 (*Mylabris*)
reducticornis Pic, 1939b: 22 (*Zonabris*)
saidensis Pic, 1897l: 124 (*Zonabris*)
vicinus Lucas, 1847: 389 (*Mylabris*)
solanensis Saha, 1979: 83 (*Mylabris*) [DA] **A**: HP **ORR**
solonicus Pallas, 1782: 87 (*Meloe*) **A**: ES FE HEI MG WS "Korea"
dianae J. R. Sahlberg, 1913b: 196 (*Zonabris*)
famelicus Ménétriés, 1854: 36 (*Mylabris*)
soumacovi Pic, 1930a: 2 (*Zonabris*) **A**: KZ TM
subparallelus Pic, 1919f: 19 (*Zonabris*) [DA] **A**: IN
talhouki Kaszab, 1983a: 160 (*Mylabris*) **A**: SA
talyshensis Sumakov, 1929a: 3 (*Mylabris*) [DA] **E**: AB
tekkensis marani Kaszab, 1958c: 297 (*Mylabris*) **A**: AF
antebistrinotatus Kaszab, 1958c: 297
tekkensis tekkensis Heyden, 1883e: 360 (*Zonabris*) **A**: AF IN KZ TD TM UZ
adnexus Mader, 1929: 109 (*Mylabris*)
atavus Mader, 1929: 109 (*Mylabris*)
imitatus Mader, 1929: 109 (*Mylabris*)
latebifasciatus Pic, 1919f: 18 (*Zonabris*)
tenerus Germar, 1836b: 5 (*Mylabris*) **E**: AU BU HU RO SK YU
tenuepictus Fairmaire, 1892b: 154 (*Mylabris*) **A**: IS JO SY TR

marsabensis Pic, 1898b: 3 (*Zonabris*)
terminatus Illiger, 1800: 143 (*Mylabris*) **N**: EG **AFR**
afzelii Billberg, 1813: 48 (*Mylabris*)
bertrandi Laporte, 1840: 270 (*Mylabris*)
ustulatus Reiche, 1866: 633 (*Mylabris*)
theryanus Pic, 1940a: 2 (*Zonabris*) **N**: MO
alluaudi Pic, 1936e: 30 (*Zonabris*) [HN]
lecerfi Pic, 1940a: 2 (*Zonabris*)
lehuensis Pic, 1940a: 2 (*Zonabris*)
tigrinus Klug, 1845: n. 24 (*Mylabris*) **N**: EG **A**: IS SA SI YE **AFR**
tigripennis Marseul, 1870b: 143 (*Mylabris*) **N**: LB EG **A**: IS
obscurior Pic, 1897l: 125 (*Zonabris*)
trianguliferus reitterioides Mader, 1929: 108 (*Mylabris*) **A**: IN
bushirensis Mader, 1929: 109 (*Mylabris*)
trianguliferus trianguliferus Heyden, 1883e: 359 (*Zonabris*) **A**: AF IN KZ TD TM UZ
varius A. G. Olivier, 1811: 96 (*Mylabris*) **N**: EG **A**: IS
wagneri Chevrolat, 1840b: 274 (*Mylabris*) **N**: AG TU
brevis Wellman, 1910a: 23 (*Mylabris*)
curtus Chevrolat, 1840b: 277 (*Mylabris*)
ernesti Pic, 1905i: 170 (*Zonabris*)
insignipennis Pic, 1901h: 77 (*Zonabris*)
mauricei Wellman, 1910a: 23 (*Mylabris*)
yemenicus Kaszab, 1983a: 171 (*Actenodia*) **A**: YE
zebraeus Marseul, 1870b: 79 (*Mylabris*) **E**: AB AL AR BH BU GG GR MC TR **A**: IN IQ IS JO SY TM TR UZ "Turkestan"
kervillei Pic, 1932e: 17 (*Zonabris*)
tenenbaumi Pic, 1928e: 5 (*Zonabris*)

genus *Lydoceras* Marseul, 1870b: 12 type species *Mylabris fasciata* Fabricius, 1775
Denierus Pic, 1913s: 175 type species *Zonabris stanleyana* Duvivier, 1890
fasciatus Fabricius, 1775: 261 (*Mylabris*) **A**: SA YE
unifasciatus A. G. Olivier, 1811: 92 (*Mylabris*)
lictor Gerstaecker, 1884: 57 (*Mylabris*) **A**: "Arabia" **AFR**
bilineatus Thomas, 1897b: 501 (*Mylabris*)

genus *Mylabris* Fabricius, 1775: 261 type species: see Comments in New Acts
subgenus *Ammabris* Kuzin, 1954: 359 type species *Mylabris elegans* A. G. Olivier, 1811
Mesorbata Pardo Alcaide, 1954a: 66 type species *Mylabris elegans* A. G. Olivier, 1811
amoenula Ménétriés, 1849: 247 **A**: KZ TM
andresi Pic, 1911l: 162 (*Zonabris*) [DA] **N**: EG
boghariensis Raffray, 1873: 380 **N**: AG MO
chudeaui Bedel, 1921: 199 (*Zonabris*) **N**: AG
elegans A. G. Olivier, 1811: 101 **N**: AG EG LB MO TU **A**: AE IS JO OM SA SY
nefischensis Pic, 1926k: 229
trijuncta Pic, 1899l: 208
elegantissima Zubkov, 1837: 70 **A**: AF IN KZ TD TM UZ XIN
kuzini Kryzhanovskij, 1956: 887 **A**: TM
laticollis Escherich, 1899: 99 (*Zonabris*) **E**: AB **A**: IN TM
alpha Sumakov, 1915: 50
myrmidon Marseul, 1870b: 154 **N**: AG MO TU **A**: IS SI
raml W. Schneider, 1991: 280 **A**: AE SA
sisymbrii Klug, 1845: nr. 12 **N**: EG MO
theryi Abeille de Perrin, 1894b: cxxxii (*Zonabris*) [DA] **N**: AG
zoltankaszabi Ruiz & García-París, 2007: 151 [RN] **A**: SA
excellens Kaszab, 1983a: 159 [HN]
subgenus *Argabris* Kuzin, 1954: 362 type species *Meloe ocellata* Pallas, 1773
audouini Marseul, 1870b: 141 **A**: IN KI KZ TD UZ XIN

bangi Heyden, 1886a: 181 (*Zonabris*)
impedita bogatshevi Kaszab, 1958c: 292 **A**: AF
impedita impedita Heyden, 1883a: 66 (*Zonabris*) **A**: AF KZ TD UZ
undulatella Kaszab, 1958b: 190 [RN]
undulata J. R. Sahlberg, 1908: 90 [HN]
impedita kryzhanovskii Pripisnova, 1987b: 563 **A**: TD
impedita lopatini Pripisnova, 1987b: 563 **A**: TD
klugi kabuliensis Kaszab, 1958c: 293 **A**: AF
klugi klugi L. Redtenbacher, 1850: 49 **A**: IN KZ TD TM XIN
graffi Pic, 1947b: 10 (*Zonabris*)
klugi lacustris Kaszab, 1958c: 294 **A**: AF
ocellata Pallas, 1773: 721 (*Meloe*) **A**: KZ TD TM UZ XIN
argus A. G. Olivier, 1811: 98
octonotata Fischer von Waldheim, 1844: 132
pupillata Schoenherr, 1817: 60
quatuordecimmaculata Dokhtouroff, 1889: 150 **A**: TD UZ
sedecimpunctata Gebler, 1825: 49 **A**: AF IN KI KZ MG TD TM UZ WS XIN
gilana Pic, 1924d: 26 (*Zonabris*)
quindecimpunctata Sumakov, 1909: 220
teuepunctata Dokhtouroff, 1889: 165 (*Zonabris*)
wilkinsi Dokhtouroff, 1889: 155 (*Zonabris*)
subgenus *Calydabris* Kaszab, 1960b: 133 type species *Calydus allousei* Kaszab, 1960
allousei Kaszab, 1960b: 131 (*Calydus*) **A**: IN IQ
cinctus Kaszab, 1960b: 133 (*Calydus*)
mirzayani Kaszab, 1968f: 769 (*Calydus*) **A**: IN
subgenus *Chalcabris* Kuzin, 1954: 346 type species *Meloe festiva* Pallas, 1773
bivulnera Pallas, 1781: 94 (*Meloe*) **E**: ST **A**: ES KI KZ MG WS XIN
bimaculatus Pallas, 1771: 466 (*Attelabus*)
caspica Gmelin, 1790: 1896 (*Cantharis*)
luteosexmaculata Pic, 1919f: 21
coeruleomaculata L. Redtenbacher, 1843: 987 **A**: IN IS JO LE SY
jarrahiana Kaszab, 1959b: 4 (*Decapotoma*)
shadegana Kaszab, 1959b: 4 (*Decapotoma*)
subhumeralis Pic, 1908k: 74
viridiflua Marseul, 1873b: 654
coerulescens Gebler, 1841a: 374 **E**: ST **A**: AF IN KZ TD TM UZ XIN
cyaneovaria Reitter, 1889b: 121 (*Zonabris*) **A**: AF KZ MG TM
speciosoides Sumakov, 1915: 7
festiva Pallas, 1773: 721 (*Meloe*) **E**: AR BU CT GR MC RO ST TR UK **A**: ES KZ MG TR WS XIN
sericea Pallas, 1782: 85 (*Meloe*)
formosa Wellman, 1910a: 23 [RN] **E**: BU GR MC TR **A**: TR
decora Frivaldszky von Frivald, 1835: 264 [HN]
koenigi Dokhtouroff, 1889: 143 (*Zonabris*) **A**: AF IN KZ TD TM UZ
ledebouri fortunata Kaszab, 1969c: 319 **A**: MG
ledebouri ledebouri Gebler, 1829b: 166 [DA] **E**: ST **A**: ES KZ
marginata Fischer von Waldheim, 1844: 136 **E**: AB AR **A**: ES IN XIN
mongolica Dokhtouroff, 1887b: 345 (*Zonabris*) **A**: HEN MG XHA XIN
alpha Sumakov, 1915: 7
aurora Escherich, 1904: 34 (*Zonabris*)
beta Sumakov, 1915: 7
chinensis J. Frivaldszky, 1892: 114 (*Zonabris*)
monozona Wellman, 1910a: 23 [RN] **E**: CT ST **A**: ES KI KZ MG WS XIN
unifasciata Ballion, 1878: 343 [HN]
munda Escherich, 1897a: 38 (*Zonabris*) **A**: TR
pallasi Gebler, 1829b: 170 **A**: KZ TM
vittata Kirsch, 1880: 77
pulchella Faldermann, 1833: 59 **A**: KI KZ MG TD UZ XIN

sairamensis Ballion, 1878: 342 [DA] **A**: XIN
perezi Pic, 1916a: 7 (*Zonabris*)
postinsignata Mader, 1929: 108
quaterna Mader, 1929: 108
terna Mader, 1929: 108
traizeti Pic, 1916a: 7 (*Zonabris*)
semivittata Pic, 1947: 7 (*Zonabris*) **A**: "Central Asia"
speciosa Pallas, 1781: 84 (*Meloe*) **A**: AF ES GAN HEI JIX LIA KZ MG NMO QIN SHA UZ WS
splendidula Pallas, 1781: 83 (*Meloe*) **A**: ES FE KI KZ MG WS XIN
apicetenuimarginata Sumakov, 1913a: 304
staudingeri Heyden, 1881: 328 (*Zonabris*)
uhagoni Martinez y Sáez, 1873a: 66 [DA] **E**: SP
subgenus *Chrysabris* Kuzin, 1954: 348 type species *Meloe trifascis* Pallas, 1773
doriai Marseul, 1870b: 164 **A**: IN IQ
boetcheri Pic, 1909n: 171
centropunctatella Kaszab, 1958b: 189 [RN]
centropunctata Bodemeyer, 1916: 112 [HN]
daligrandi Pic, 1947: 7 (*Zonabris*)
luristanica Pic, 1920d: 12
persicola Pic, 1909n: 171
magnificia Bodemeyer, 1916: 112
unifasciata Bodemeyer, 1916: 112 [HN]
viridibifasciata Pic, 1920e: 13 (*Zonabris*)
viridifasciata Pic, 1920e: 13 (*Zonabris*)
gemmula Dohrn, 1873: 73 **A**: IN TM
ghazniana Kaszab, 1973d: 278 **A**: AF
hemprichi Klug, 1845: n. 3 **N**: AG EG LB MO TU **A**: IN IQ IS JO SI SY
kocheri Pardo Alcaide, 1954a: 68
karakalensis Kryzhanovskij, 1956: 886 **A**: TM
kraatzi Heyden, 1881: 329 (*Decatoma*) **A**: IN
plurivulnera Dohrn, 1873: 73 **A**: IN TM
marseuli Kirsch, 1880: 77 [HN]
viridula Marseul, 1882c: 170 [RN]
smaragdina Gebler, 1841b: 597 **A**: KZ MG WS
trifascis Pallas, 1773: 721 (*Meloe*) **A**: AF IN KZ TM WS "Turkestan"
subgenus *Eumylabris* Kuzin, 1954: 355 type species *Mylabris decempunctata* Fabricius, 1792 (= *Mylabris fabricii* Sumakov, 1924)
Neabris Kaszab, 1958c: 284 type species *Mylabris klapperichi* Kaszab, 1958
aulica Ménétriés, 1832: 208 **E**: AB **A**: ES FE HEI KZ MG NMO NO WS
breveapicalis Pic, 1919f: 21 (*Zonabris*) **A**: TR
calida Pallas, 1782: 85 (*Meloe*) **E**: AB AR BU CT GG GR MC ST TR UK **N**: AG EG LB MO TU **A**: AF ES FE HEB HEI HUB IN IQ IS JIA JIL JO KI KZ LE LIA MG NC NE NMO SA SC SHX SI SY TD TM TR UZ XIN WS YE
baicalica Pic, 1919f: 22 (*Zonabris*)
bijuncta Pic, 1897a: 2 (*Zonabris*)
bimaculata A. G. Olivier, 1811: 93
bimaculaticeps Pic, 1920e: 13 (*Zonabris*)
decora A. G. Olivier, 1811: 94 [HN]
interrupta Eichler, 1924: 66 (*Zonabris*) [HN]
latifasciata Pic, 1896f: 62 (*Zonabris*)
maculata A. G. Olivier, 1795: 47: 7
maculata Eichler, 1923: 66 (*Zonabris*) [HN]
maroccana Escherich, 1899: 107 (*Zonabris*)
maura Chevrolat, 1840b: 273
niligena Reiche, 1866: 638
tlemceni Pic, 1930a: 3 (*Zonabris*)
transcaspica Escherich, 1899: 107 (*Zonabris*)

cincta A. G. Olivier, 1811: 93 **E**: AB AR CT GG GR MC ST TR UK **N**: AG EG LB MO TU **A**: AF HP IN IQ IS JO KZ PA SA SI SY TR UP
alfierii Pic, 1921g: 45 (*Zonabris*)
algerica Pic, 1898b: 3 (*Zonabris*)
bicolor Waltl, 1838: 465
diversepunctata Pic, 1921a: 5 (*Zonabris*)
gafsaensis Pic, 1915a: 2 (*Zonabris*)
jugatoria Reiche, 1866: 633
latecincta Kocher, 1955b: 128
lateripiuncta Kaszab, 1957b: 55
matthesi Faldermann, 1837: 120
mediocingulata Kaszab, 1957b: 55
pseudocincta Kaszab, 1957b: 55
rufobifasciata Pic, 1921a: 5 (*Zonabris*)
rufomaculata Pic, 1921a: 5 (*Zonabris*)
rufotrifasciata Pic, 1921a: 5 (*Zonabris*)
soubironi Pic, 1921a: 5 (*Zonabris*)
taeniata Waltl, 1838: 465
tiflensis Billberg, 1813: 13
tricingulata L. Redtenbacher, 1850: 49
cinctoides Kaszab, 1957b: 55 [DA] **A**: IN
crocata Pallas, 1781: 87 (*Meloe*) **E**: AB AR BU CT GE GG GR HU RO ST TR UK **A**: ES HEI IN IS JO KI KZ MG NO SY TD TR UZ WS XIN
daurica Baudi di Selve, 1878b: 372
duodecimpunctata Tauscher, 1812b: 139 [HN]
spectabilis Baudi di Selve, 1878b: 372
fabricii Sumakov, 1924: 27 **E**: AB AL AR BU CT GG GR IT MC MO ST TR UK YU **A**: IN IS JO KZ SI SY TD TM TR UZ
decimpunctata V. Petagna, 1787: 27 [HN]
forti Mulsant, 1857a: 133
sexpunctata Fuente, 1913: 308 (*Zonabris*) [HN]
sexpunctulata Kaszab, 1958b: 190
groschkei Kaszab, 1957b: 53 **A**: AF IN
impressa hoggarensis Pic, 1929h: 127 (*Zonabris*) **N**: AG
impressa impressa Chevrolat, 1840b: 275 **N**: AG MO TU
aini Pic, 1921a: 4 (*Zonabris*)
antejuncta Pardo Alcaide, 1955b: 54
bisbijuncta Pardo Alcaide, 1955b: 55
externobliterata Pardo Alcaide, 1955b: 55
oranensis Pic, 1909q: 185 (*Zonabris*)
vaucheri Escalera, 1914: 400 (*Zonabris*)
impressa stillata Baudi di Selve, 1878b: 375 **E**: IT (Sicilia)
ragusai Pic, 1897o: 217 (*Zonabris*)
klapperichi hirtipedes Kaszab, 1958c: 286 **A**: AF PA
klapperichi klapperichi Kaszab, 1958c: 285 **A**: AF PA
magnoguttata iranica Kaszab, 1957b: 56 **A**: IN
magnoguttata magnoguttata Heyden, 1881: 329 (*Zonabris*) **E**: ST **A**: AF IN KI KZ TD TM UZ XIN
angelicae Reitter, 1889b: 121
aschabadensis Voigts, 1902: 178 (*Zonabris*)
disjuncteguttata Pic, 1919f: 19 (*Zonabris*)
fasciculata Escherich, 1899: 108 (*Zonabris*)
heydeniana Dokhtouroff, 1889: 145 (*Zonabris*)
margarithae Reitter, 1889b: 120 (*Zonabris*)
magnoguttata pardoi Kaszab, 1958c: 278 **A**: AF PA
omocrates Wellman, 1910a: 23 [RN] **A**: ES MG
axillaris Motschulsky, 1872: 50 [HN]
posticalis Dokhtouroff, 1889: 136 (*Zonabris*) **A**: AF IN KI PA UZ XIN
pulchra Kaszab, 1973d: 287 **A**: AF

quinqueplagiata Kaszab, 1958c: 274 **A**: AF PA
schrenkii Gebler, 1841a: 374 **A**: AF IN KZ TD TM UZ XIN
reitteri Dokhtouroff, 1889: 157 (*Zonabris*)
sculptilis Kaszab, 1958c: 279 **A**: AF
subbrevicornis Kaszab, 1958c: 276 **A**: AF
tiflensis Billberg, 1813: 13 **A**: PA
adam Kaszab, 1958c: 287
antefracta Pardo Alcaide, 1958b: 2
subgenus *Lachnabris* Kuzin, 1953: 137 type species *Mylabris mannerheimii* Gebler, 1844
crux Escherich, 1899: 103 (*Zonabris*) **E**: ST **A**: TD TM UZ
opulenta Escherich, 1904: 33 (*Zonabris*)
mannerheimii Gebler, 1844: 103 **A**: AF KZ TD TM XIN
turkestanica Dokhtouroff, 1889: 168 (*Zonabris*) **A**: TD
undecimnotata Heyden, 1883a: 66 (*Zonabris*) **A**: IN KZ UZ
subgenus *Mauritabris* Pardo Alcaide, 1969: 375 type species *Mylabris tenebrosa* Laporte, 1840
abdelkaderi Escalera, 1909b: 246 (*Zonabris*) **N**: MO
damascena Reiche, 1866: 634 **A**: IN IQ IS JO LE SI SY
longipilis Pic, 1897l: 124 (*Zonabris*) **N**: EG
unifenestrata Pic, 1915a: 2 (*Zonabris*)
mateui Pardo Alcaide, 1954b: 338 **N**: MO
tenebrosa Laporte, 1840: 270 **N**: AG EG LB MO TU **A**: SA
diverseluteonotata Pic, 1953d: 499 (*Zonabris*)
exuta Normand, 1950a: 81
mohtari Escalera, 1910a: 284 (*Zonabris*)
oberthuri Pic, 1953d: 498 (*Zonabris*)
pauloluteobifasciata Pic, 1953d: 499 (*Zonabris*)
postluteofasciata Pic, 1921a: 6 (*Zonabris*)
satanas Mader, 1929: 108
soufensis Pic, 1921a: 6 (*Zonabris*)
septemnotata Pic, 1953d: 498 (*Zonabris*)
triluteofasciata Pic, 1921a: 6 (*Zonabris*)
uniluteofasciata Pic, 1921a: 6 (*Zonabris*)
subgenus *Mesosulcata* Pardo Alcaide, 1950a: 65 type species *Mylabris hieracii* Graells, 1849
Glaucabris Kuzin, 1954: 350 type species *Mylabris hieracii* Graells, 1849
hieracii Graells, 1849: 621 **E**: FR PT SP
subspiciosa Rosenhauer, 1856: 229
hirtipennis Raffray, 1873: 379 **N**: AG MO
ifranensis Kocher, 1955b: 128
kabylica Escherich, 1899: 101 (*Zonabris*)
thamii Kocher, 1963a: 36 **N**: MO
subgenus *Micrabris* Kuzin, 1954: 351 type species *Mylabris geminata* Fabricius, 1798
Mesolaevigata Pardo Alcaide, 1954a: 68 type species *Mylabris geminata* Fabricius, 1798
afghanica Kaszab, 1953a: 308 **A**: AF PA
alpina Ménétriés, 1832: 208 **E**: AB AR ST
apicefasciata Sumakov, 1929a: 3 [DA] **A**: TM
beckeri Escherich, 1890a: 24 **E**: ST
therondi Roubal, 1932a: 1
connata Rey, 1892: 138 (*Zonabris*) **E**: FR IT (Western Alps: Susa Valley) SZ
apicalifrons Puel, 1907: 116
camarguensis Puel, 1907: 116
eryngii Puel, 1907: 116
gracilior Puel, 1907: 116
fenestrata Puel, 1907: 118 [HN]
meridionalis Puel, 1907: 116
novickii Puel, 1907: 116
pseudobsoleta Puel, 1907: 116
prolongata Puel, 1907: 116

pueli Mader, 1927: 196 [RN]
pueli Sumakov, 1930a: 30 [HN]
subobliterata Pic, 1908j: 72
dashidorzsi Kaszab, 1964e: 357 **A**: MG
deferreri Ruiz & García-París, 2004: 186 **E**: SP
dejeani Gyllenhal, 1817: 35 **E**: PT SP
djebelina Pic, 1902i: 72 (*Zonabris*) **N**: MO
dokhtouroffi afghana Kaszab, 1958c: 269 **A**: AF
dokhtouroffi dokhtouroffi Escherich, 1899: 104 (*Zonabris*) **A**: IN TD UZ
coalescens Voigts, 1902: 178 (*Zonabris*)
excisofasciata Heyden, 1883a: 67 (*Zonabris*) **A**: AF IN KI TM UZ
externepunctata Faldermann, 1832: 208 **E**: AB AR ST **A**: IN TR
taurusiensis Pic, 1901h: 77
fenestrata Escherich, 1899: 108 (*Zonabris*) **A**: "Turkestan"
flexuosa A. G. Olivier, 1811: 101 **E**: AN FR IT SP SZ
abrutia Sumakov, 1915: 19 [RN]
aprutia Magistretti, 1943: 40 [HN]
cassetensis Pic, 1908j: 72
inflexuosa Pic, 1908j: 72
italica Magistretti, 1943: 34 [HN]
luteosubapicalis Pic, 1930a: 3 (*Zonabris*)
magistrettii Kaszab, 1958b: 190 [RN]
pyrenaea Pic, 1908j: 72
rarissima Pic, 1908j: 72
vittata Magistretti, 1943: 34 [HN]
vittatoides Kaszab, 1958b: 190 [RN]
frolovi aliensis Kuzin, 1927: 26 **A**: KZ
frolovi androssovi Sumakov, 1929a: 1 **A**: "Turkestan"
frolovi frolovi Fischer von Waldheim, 1823: pl. 40 **A**: AF ES IN KI KZ TD TM UZ WS XIN
confluens Fischer von Waldheim, 1823: 227
frolovi Germar, 1824b: 170 [HN]
obtecta Escherich, 1899: 108
sexguttata Pic, 1947: 7 (*Zonabris*)
sodalis Heyden, 1883a: 65 (*Zonabris*)
frolovi kopalensis Sumakov, 1930b: 13 [RN] **A**: KZ
schapinskii Sumakov, 1929a: 1 [HN]
frolovi kuziniana Kaszab, 1958c: 262 **A**: AF
frolovi shapinskyi Kuzin, 1927: 26 **A**: KZ
frolovi syrdajensis Sumakov, 1929a: 2 **A**: KZ
furcimacula Sumakov, 1915: 20 **E**: AB AR ST **A**: IN TR
geminata Fabricius, 1798: 120 **E**: AB AR CT GG ST UK **A**: IN KI KZ TR WS XIN
kushakiewitschi Dokhtouroff, 1889: 167 (*Zonabris*)
hauseri Escherich, 1899: 101 (*Zonabris*) **A**: KI KZ XIN
inculta Escherich, 1899: 98 (*Zonabris*) **E**: AB AR "Caucasus" **A**: SY TR
humeromaculata Mader, 1929: 109
intermedia Fischer von Waldheim, 1844: 132 **A**: ES KI KZ TD XIN
karacaevica Roubal, 1914: 145 (*Zonabris*) **E**: ST **A**: TM
laevicollis Marseul, 1870b: 109 **E**: AB AL AR BH GG GR MC TR **A**: IN KZ TM TR
liliputana Escherich, 1899: 106 (*Zonabris*) **A**: KI KZ TD
lucens Escherich, 1904: 32 **A**: AF KI KZ TD TM UZ XIN
alpinoides Sumakov, 1929a: 2
apicetenuimarginata Sumakov, 1913a: 304
macilenta Marseul, 1873b: 489 **A**: BT HP KA NP PA UP XIZ
maculosopunctata Graells, 1858: 75 **E**: PT SP
rosinae Escherich, 1899: 104 (*Zonabris*)
unicincta Voigts, 1902: 178 (*Zonabris*)
marakensis badakschanica Kaszab, 1958c: 266 **A**: AF

marakensis marakensis Kaszab, 1953a: 311 **A**: AF TD
modesta elbursiensis Aksentjev, 1985: 247 **A**: IN
modesta modesta Escherich, 1899: 99 (*Zonabris*) **E**: AB AR **A**: IN TR
neglecta Escherich, 1899: 102 (*Zonabris*) **E**: GR (Peloponnese)
nevadensis Escalera, 1915: 493 (*Zonabris*) **E**: SP
nuristanica nuristanica Kaszab, 1958c: 269 **A**: AF PA
nuristanica subalpestris Kaszab, 1958c: 271 **A**: AF
obsoleta Nowicki, 1872: 6 **E**: IT
italica Sumakov, 1929a: 6
pannonica Kaszab, 1956c: 333 **E**: AU HU RO SK
pauper Escherich, 1899 (*Zonabris*) **E**: PT SP
pentheri Ganglbauer, 1905: 279 **E**: AB or AR "Araxes valley" **A**: TR
platai Pardo Alcaide, 1975: 45 **E**: SP
pusilla bosnica Reitter, 1903f: 230 (*Zonabris*) **E**: AL BH CR YU
hercegovinensis Mader, 1927: 106 [RN]
pusilla latialis Magistretti, 1943: 34 **E**: IT (Central Apennines)
pusilla pusilla A. G. Olivier, 1811: 101 **E**: AB AL AR BU CT GG GR MO ST TR YU UK **A**: KI KZ TR UZ WS
quadrisignata Fischer von Waldheim, 1823: pl. 40 **A**: KI KZ UZ WS XIN
antenigra Mader, 1929: 108
longeundulata Pic, 1919f: 18 (*Zonabris*)
voigtsi Escherich, 1904: 33
sedilethorax Sumakov, 1929a: 3 **E**: AR **A**: IN
serena Escherich, 1899: 97: 91 (*Zonabris*) **E**: AB AR **A**: TR
sexnotata L. Redtenbacher, 1843: 987 **E**: AR **A**: SY TR
sibirica Fischer von Waldheim, 1823: pl. 40 **E**: CT ST UK **A**: ES KI KZ WS
besseri Baudi di Selve, 1878b: 374
kushakevitshi Dokhtouroff, 1889: 167
luteobiinterrupta Pic, 1919f: 19
sinuata Klug, 1845: n. 19 **A**: CY LE SY TR
sobrina Graells, 1851: 20 **E**: PT SP
steppensis Dokhtouroff, 1889: 137 (*Zonabris*) **E**: GG **A**: KI KZ TD UZ
tomentosa Escherich, 1899: 99 (*Zonabris*) **E**: AR **A**: IN TR
externeundulata Pic, 1919f: 19 (*Zonabris*)
kulpensis Pic, 1919f: 18 (*Zonabris*)
unicolor Faldermann, 1837: 127 **E**: AB AR **A**: IN TR
varians Gyllenhal, 1817: 34 **E**: PT SP
bipunctata Fuente, 1909: 307 [HN]
bipunctatafasciata Fuente, 1909: 307
decimspilota Chevrolat, 1865: 393
fuentana Kaszab, 1958b: 190 [RN]
fuentei Pic, 1909n: 171 (*Zonabris*)
inconstans Chevrolat, 1865: 393
luteipennis Marseul, 1870b: 106
pozuelensis Pic, 1909n: 171 (*Zonabris*)
quinquepunctata Fuente, 1909: 307
sexpunctata Fuente, 1909: 307 [HN]
sexpunctatella Kaszab, 1958b: 190 [RN]
tripunctatafasciata Fuente, 1909: 307
unipunctata Fuente, 1909: 307
waziristanica Kaszab, 1958c: 299 (*Coryna*) **A**: AF PA
subgenus *Monabris* Kuzin, 1953: 137 type species *Mylabris undecimpunctata* Fischer von Waldheim, 1842
undecimpunctata Fischer von Waldheim, 1844: 17 **A**: AF IN KZ TD TM UZ XIN
septempunctata Pic, 1901h: 78 (*Zonabris*)

subgenus *Mylabris* Fabricius, 1775: 261 type species: see Comments in New Acts

Mesopunctata Pardo Alcaide, 1954a: 74 type species *Meloe quadripunctata* Linnaeus, 1767
amori Graells, 1858: 76 **E**: PT SP
korbi Escherich, 1891b: 53 (*Zonabris*)
apicenigra Sumakov, 1915: 24 **A**: IN TR
batnensis Marseul, 1870b: 121 **N**: AG MO
ciliciensis Escherich, 1899: 106 (*Zonabris*) **A**: TR
concolor Marseul, 1870b: 112 **E**: AB AR **A**: IN TM TR
emiliae Escherich, 1899: 105 (*Zonabris*) **E**: AR "Caucasus" **A**: TR
guerini Chevrolat, 1840b: 271 **N**: AG MO TU
lambaesis Pic, 1953d: 499 (*Zonabris*)
normandi Pic, 1936f: 155 (*Mylabris*)
rubripennis Chevrolat, 1840b: 270
kodymi Mařan, 1944a: 37 **E**: GR (Kríti)
stepaneki Mařan, 1944a: 38
taborskyi Mařan, 1944a: 38
madani Escalera, 1909b: 247 (*Zonabris*) **N**: MO
tahari Escalera, 1909b: 246 (*Zonabris*)
transversa Kaszab, 1958b: 190 [RN]
transversalis Escalera, 1914: 396 (*Zonabris*) [HN]
madoni Marseul, 1883a: 178 **A**: CY
lividiana Pic, 1911n: 178
olivieri Billberg, 1813: 71 **E**: AB AR CT GG GR MC RO ST TR UK **A**: IN KZ TD TR
caspia Ménétriés, 1832: 206
grisescens Reiche, 1866: 638 [HN]
impar Baudi di Selve, 1878b: 375 [HN]
impexa Escherich, 1899: 107 (*Zonabris*)
kopetensis Pic, 1930a: 2 (*Zonabris*)
parumpicta Heyden, 1883d: 353 (*Zonabris*) **A**: IN IQ IS JO SY "Turkestan"
quadripunctata angustata Sumakov, 1915: 72 [DA] **A**: UZ
quadripunctata mesoasiatica Kuzin, 1953: 157 [DA] **A**: "Turkestan"
quadripunctata quadripunctata Linnaeus, 1767: 680 (*Meloe*) **E**: AB AL AR BU CT FR GG GR IT MC RO ST TR UK YU **A**: AF ES IN IQ IS JO KI KZ SY TD TM TR UZ WS XIN
adamsii Fischer von Waldheim, 1824: 224
antefasciata Sumakov, 1913a: 304 [RN]
armeniaca Faldermann, 1837: 125 [HN]
chobautiana Sumakov, 1930a: 37 [RN]
decimpunctata Fabricius, 1781: 86 (*Meloe*)
fasciatopunctata Adams, 1824: 224
husseini L. Redtenbacher, 1850: 49
melanura Pallas, 1781: 86 (*Meloe*)
octopunctata A. G. Olivier, 1811: 95
quadripunctata Billberg, 1813: 28 [HN]
tetraspilota Wellman, 1910a: 23 [RN]
vareillesi Chobaut, 1921: 280 [HN]
quadripunctata restricta Motschulsky, 1849: 132 [DA] **E**: PT SP
hispanica Motschulsky, 1849: 132
maldinesi Chevrolat, 1865: 393
beauregardi Gorriz y Muñoz, 1884: 1
schreibersi Reiche, 1866: 336 **E**: IT (Sicilia) **N**: AG MO TU
parumpunctata Magistretti, 1943: 35
terminata Chevrolat, 1840b: 276 [HN]
unifasciata Magistretti, 1943: 35 [HN]
unifasciatella Kaszab, 1958b: 190 [RN]
tauricola Marseul, 1870b: 73 **A**: TR
tricincta Chevrolat, 1840b: 270 **N**: AG MO
brevevittata Pic, 1935g: 159, [= 1935c: 4] (*Zonabris*)
nigrosuturata Escalera, 1914: 397 (*Zonabris*)

variabilis Pallas, 1781: 81 (*Meloe*) **E**: AB AL AR AU BU CR CT CZ FR GG GR HU IT MC MO PT RO SK SL SP ST SZ TR UK YU **A**: AF ES IN IS JO KZ TM TR UZ WS
disrupta Baudi di Selve, 1878c: 1160
erivanica Pic, 1901h: 77 (*Zonabris*)
guerini Magistretti, 1943: 34 [HN]
hypocrita Mulsant, 1857a: 125
italiana Kaszab, 1958b: 190 [RN]
lacera Fischer von Waldheim, 1827: 6
leonii Magistretti, 1943: 35
mutabilis Marseul, 1873b: 492
mutans Guérin-Méneville, 1837: 554
similaris Mulsant, 1857a: 125
sturmi Baudi di Selve, 1878c: 1160
subgenus *Zitunabris* Pardo Alcaide, 1969: 375 type species *Mylabris oleae* Chevrolat, 1840
atrofasciata Pic, 1921a: 4 (*Zonabris*) **A**: IN
baulnyi Marseul, 1870b: 49 **N**: AG MO
filicornis Marseul, 1870b: 64 **N**: EG **A**: IS JO SI
interrupta A. G. Olivier, 1801: 93 **N**: EG **A**: IN
excellens L. Redtenbacher, 1850: 49
mediounifasciata Pic, 1920e: 13 (*Zonabris*)
subunifasciata Pic, 1920e: 13 (*Zonabris*)
oleae Chevrolat, 1840b: 269 **N**: AG LB MO TU
bidivisa Pic, 1921f: 24 (*Zonabris*)
bireductenotata Pic, 1914a: 1 (*Zonabris*)
bisbireducta Pic, 1921f: 24 (*Zonabris*) [HN]
bisbireductella Kaszab, 1958b: 190 (*Mylabris*) [RN]
diversedivisa Pic, 1921f: 24 (*Zonabris*)
diversesuturalis Pic, 1933e: 13 (*Zonabris*)
drahamensis Pic, 1921f: 24 (*Zonabris*)
harterti Pic, 1921f: 23 (*Zonabris*)
hipponensis Pic, 1921f: 23 (*Zonabris*)
latenigra Pic, 1921f: 24 (*Zonabris*)
laghouatensis Pic, 1921f: 24 (*Zonabris*)
mediobliterata Pic, 1921f: 24 (*Zonabris*)
multinotata Pic, 1906c: 26 (*Zonabris*)
postdivisa Pic, 1921f: 24 (*Zonabris*)
rotroui Pic, 1933e: 13 (*Zonabris*) [HN]
separata Pic, 1897l: 123 (*Zonabris*)
tridivisa Pic, 1921f: 24 (*Zonabris*)
unidivisa Pic, 1921f: 24 (*Zonabris*)
rimosa Marseul, 1870b: 40 **N**: AG MO TU
suturalis Pic, 1898b: 2 (*Zonabris*) **A**: IN SY TR
morgani Pic, 1920d: 12 (*Zonabris*)
syriaca panjaoensis Kaszab, 1953a: 306 **A**: AF
syriaca syriaca Klug, 1845: nr. 13 **E**: AB AR GR **A**: IN IQ IS JO LE SY SI TR
antedivisa Mader, 1929: 108
intersecta Reiche, 1857: 274
leseleuci Pic, 1898b: 2 (*Zonabris*)
smyrnensis Escherich, 1899: 109 (*Zonabris*)

genus *Mylabris*, species incertae sedis
aurociliata Escherich, 1899: 98 (*Zonabris*) distribution unknown
axillaris Billberg, 1813: 24 **N**: EG
brevicornis Motschulsky, 1872: 50 **E**: UK **A**: "Siberia"
caroli Pic, 1899l: 200 (*Zonabris*) **N**: EG
delagrangei Pic, 1899l: 208 (*Zonabris*) **A**: SY
flavipennis Motschulsky, 1872: 50 **A**: ES
isis Marseul, 1876a: 28 **N**: EG

maceki Dvořák, 1985: 1 (*Croscherichia*) **A**: IQ
mimosae A. G. Olivier, 1811: 99 **A**: SA
paliji Kotschenov, 1970: 21 **A**: KI
quinquemaculata Okamoto, 1924: 1984 **A**: SC
sanguinosa Marseul, 1872a: 417 **A**: IN IQ
tarbagataiensis Pic, 1919f: 18 (*Zonabris*) **A**: KZ

genus *Mylabris*, nomina dubia
elongata Herbst, 1784c: 147 distr. unknown
gebleri Faldermann, 1837: 124 **A**: IN
trivittis Pallas, 1782: 92 (*Meloe*) "South Russia"

genus *Pseudabris* Fairmaire, 1894b: 222 type species *Pseudabris tigriodera* Fairmaire, 1894
hingstoni Blair, 1927: 254 (*Mylabris*) **A**: XIZ
　waltoni Blair, 1927: 255 (*Mylabris*)
longiventris Blair, 1927: 253 (*Mylabris*) **A**: XIZ
tigriodera Fairmaire, 1894b: 222 **A**: XIZ
przewalskyi Dokhtouroff, 1887a: 341 (*Zonabris*) **A**: SCH SD XIZ
　goutelli Fairmaire, 1889b: 48 (*Mylabris*)

genus *Semenovilia* Kuzin, 1954: 370 type species *Mylabris fischeri* Gebler, 1847
fischeri Gebler, 1847: 502 (*Mylabris*) **E**: ST **A**: AF ES IN KZ TM UZ WS XIN
　tauscheri Fischer von Waldheim, 1844: 130 (*Mylabris*) [HN]
komarowi Reitter, 1889b: 121 (*Zonabris*) **A**: "Turkestan"
　connecticornis Reitter, 1894a: 51 (*Zonabris*)
　pilimana Reitter, 1894a: 51 (*Zonabris*)

genus *Xanthabris* Kaszab, 1956e: 637 type species *Xanthabris baluchistana* Kaszab, 1956
baluchistana Kaszab, 1956e: 638 **A**: PA

tribe Mylabrini, species incertae sedis
fuscicornis Drapiez, 1821: 277 (*Mylabris*) **A**: "China"
heteroclitus Pic, 1919f: 21 (*Zonabris*) **E**: ST
jacquemontii Blanchard, 1844: 26 (*Mylabris*) **A**: KA
kuschkensis Pic, 1919f: 21 (*Zonabris*) **A**: IN TM UZ
pallidomaculatus L. Redtenbacher, 1850: 49 (*Mylabris*) **N**: EG **A**: IN

tribe Meloini Gyllenhal, 1810

genus *Meloe* Linnaeus, 1758: 419 type species *Meloe proscarabaeus* Linnaeus, 1758
subgenus *Afromeloe* K. Schmidt, 1913: 332 type species *Meloe angulatus* Leach, 1815
　Submeloegonius Pliginskij, 1935: 320 type species *Meloe adisabebae* Pliginskij, 1935
lefevrei Guérin-Méneville, 1849: 322 **A**: YE (Suqutra) **AFR**
sanaanus Borchmann, 1938b: 465 **A**: YE
trapeziderus Gahan, 1903: 282 **A**: YE (Suqutra)
subgenus *Alveomeloe* Pripisnova, 1987a: 41 type species *Meloe tadzhikistanicus* Pripisnova, 1987
babatagicus Pripisnova, 1987a: 42 **A**: TD
tadzikistanicus Pripisnova, 1987a: 41 **A**: TD
subgenus *Chiromeloe* Reitter, 1911a: 390 type species *Meloe nigropilosellus* Reitter, 1900
nigropilosellus Reitter, 1900b: 56 **A**: UZ
subgenus *Coelomeloe* Reitter, 1911a: 389 type species *Meloe tuccius* Rossi, 1792
tuccius corrosus Brandt & Erichson, 1832: 122 **E**: IT
tuccius tuccius Rossi, 1792: 283 **E**: AB AL AR AU BH BU CR FR GR HU IT MA PT RO SK SP ST UK YU **N**: AG CI EG LB MO MR TU **A**: AF CY IN IS JO KZ SI SY TD TM TR UZ XIN
subgenus *Desertimeloe* Kaszab, 1964d: 321 type species *Meloe centripubens* Reitter, 1897
alashana Kaszab, 1964d: 322 **A**: GAN

centripubens Reitter, 1897e: 223 **A**: MG XIN
auropictus Denier, 1921: 112
citoeni Reymond, 1933: 14
subgenus *Eurymeloe* Reitter, 1911a: 391 type species *Meloe brevicollis* Panzer, 1793
aeneus Tauscher, 1812b: 151 **E**: KZ ST **A**: KZ
affinis affinis Lucas, 1847: 398 **N**: AG MO TU
setosus Escherich, 1890b: 90
affinis apivorus Reitter, 1895a: 10 **N**: MO
affinis schatzmayri Bologna, 1988: 272 **N**: LB
affinis simillimus Escalera, 1914: 383 **N**: MO
apenninicus Bologna, 1988: 250 **E**: IT
asperatus Tan, 1981: 415 [DA] **A**: XIZ
austrinus Wollaston, 1854: 257 [DA] **N**: MR
baudii Leoni, 1907: 263 **E**: CR IT SL
scabriusculus Baudi di Selve, 1878b: 353 [HN]
baudueri Grenier, 1863: 92 **E**: FR IT (Sardegna) PT SP **N**: AG MO
brevicollis algiricus Escherich, 1890c: 283 **E**: IT (Sicilia) **N**: AG TU
brevicollis brevicollis Panzer, 1793d: 15 **E**: AB AL AR AU BE BH BU BY CR CT CZ DE EN FI FR GE GB GG GR HU IT KZ LA LT LU MC MO NL NR NT PL PT RO SK SL SP ST SV SZ TR UK **A**: AF ES FE HEB HEI IN JIX JO KI KZ MG NC NO TD TR WS
aestivus Baudi di Selve, 1878b: 353
cephalotes Curtis, 1829: 38
ibericus Reitter, 1895a: 7
laticollis Baudi di Selve, 1878b: 353
pallidotarsalis Kaszab, 1956b: 158
puncticollis Motschulsky, 1872: 46
splendens Escherich, 1889a: 105
brevicollis curticornis Escalera, 1914: 384 **N**: MO
brevicollis mistaniensis Aksentjev, 1985: 248: **E**: AB **A**: IN
corvinus Marseul, 1877: 482 **A**: FE JA NE "Korea"
crosi Peyerimhoff, 1926a: 94 **N**: MO
curticollis Kraatz, 1882c: 117 **A**: UZ
escherichi Reitter, 188b: 107 **A**: TR
fernandezi Pardo Alcaide, 1951: 252 **N**: CI
flavicomus Wollaston, 1854: 528 **N**: CI MR
frontalis Reitter, 1905d: 245 **A**: IN
gaberti Reitter, 1907g: 214 **A**: KZ
ganglbaueri Apfelbeck, 1905: 250 [= 1907: 643] **E**: AL BH BU CR IT FR (Corse) GR YU **A**: SY TR
glazunovi Pliginskij, 1910: 171 **E**: AR GR ST UK "Transcaucasia" **A**: AF IN KI KZ TD TM TR
rufotarsalis Pliginskij, 1911: 47
heptapotamicus Pliginskij, 1910: 170 **A**: KZ
kandaharicus Kaszab, 1958c: 302 **A**: AF IN
laevipennis Brandt & Erichson, 1832: 124 **A**: ES FE MG
lederi Reitter, 1895a: 7 **A**: MG NMO
lobicollis Fairmaire, 1891c: xxiii **A**: HUB HUN
luctuosus Brandt & Erichson, 1832: 122 **E**: IT (Sicilia) **N**: AG TU
mandli Borchmann, 1942b: 682 **A**: ES
marianii Kaszab, 1983a: 184 **N**: EG **A**: IS SA
mathiesseni Reitter, 1905d: 245 **A**: IN
mediterraneus J. Müller, 1925: 22 **E**: AL AR CR BH BU FR GB GR HU IT MA PT SK SP TR YU "Transcaucasia" **N**: AG CI EG LB MO MR TU **A**: IN IS TR
murinus Brandt & Erichson, 1832: 127 **E**: GR (Kríti) IT (Sicilia, Sardegna) SP **N**: AG LB MO TU
thalensis Pic, 1933e: 10
nanus Lucas, 1847: 400 **E**: PT SP **N**: AG EG LB MO TU **A**: IS SI
murinus Reitter, 1895a: 12 [HN]
omanicus Kaszab, 1983a: 186 **A**: OM

otini Peyerimhoff, 1949b: 275 **N**: MO
pallidicolor Escalera, 1909b: 242 **N**: MO
primaeveris Kaszab, 1958c: 301 **A**: AF
primulus Semenov, 1903c: 356 **A**: IN TM
pubifer Heyden, 1887c: 315 **A**: AF KI KZ TD TM
pusio Wellman, 1910a: 22 **A**: UZ
pygmaeus Kraatz, 1882d: 334 [HN]
rugosus Marsham, 1802: 483 **E**: AB AL AR AU BE BH BU CR CT CZ FR GB GE GG GR HU IT NL MC PL RO SK SL SP ST SZ UK YU **A**: AF IN KZ SY TD
abdominalis Escherich, 1890b: 89
autumnalis Leach, 1815b: 40 [HN]
saharensis Chobaut, 1898a: 86 **N**: AG
scabriusculus Brandt & Erichson, 1832: 125 **E**: AL AU BH BU CR CT CZ FR GE GR HU IT MC PL RO SL SK ST SZ TR UK YU "Caucasus" **A**: ES KZ TR UZ
nigra Pliginskij, 1911: 46 [HN]
pliginskyi Kaszab, 1958b: 191 [RN]
pygmaeus Baudi di Selve, 1878b: 353 [HN]
semipunctatus Krynicki, 1832: 140
tarsalis Depoli, 1912: 101 [HN]
schmidi Kaszab, 1978c: 341 **A**: HP SD UP
scutellatus Reitter, 1895a: 11 **E**: AR
servulus Bates, 1879: 483 **A**: AF KA NP XIZ
subsetosus Reitter, 1895a: 10 **A**: TM
transversicollis Fairmaire, 1891d: cii **A**: KA "Himalaya"
zolotarevi Pliginskij, 1914b: 264 **A**: KZ
subgenus *Lampromeloe* Reitter, 1911a: 388 type species *Meloe variegatus* Donovan, 1793
cavensis L. Petagna, 1819: 40 **E**: FR GR IT MA PT SP TR **N**: AG EG LB MO TU **A**: CY IN IQ IS JO LE SY TR
aeneus Laporte, 1840: 278
latreillei Reiche, 1867: 89
purpurascens Germar, 1836a: 12
sardous Gené, 1836b: iv [= 1836a: 198]
specularis Gredler, 1878: 518
stellatus Pliginskij, 1923b: 143
variegatus mandzhuricus Pliginskij, 1930: 110 **A**: HEI
variegatus variegatus Donovan, 1793: 81 **E**: AB AR AU BE BH BU BY CR CT CZ DE FR GB GE GG GR HU IT LT NL PL RO SK SL SP ST SV SZ UK YU **N**: AG MO TU **A**: AF ES HEI IN JO KA KI KZ LE TD TM TR WS
areolatus Reitter, 1895a: 8
cupreus Baudi di Selve, 1878b: 354
scabrosus Laporte, 1840: 278 [HN]
subgenus *Lasiomeloe* Reitter, 1911a: 389 type species *Meloe olivieri* Chevrolat, 1833
olivieri Chevrolat, 1833d: pl. 57 **E**: AB AR TR **N**: EG **A**: IN IQ JO SY TM TR
subgenus *Listromeloe* Reitter, 1911a: 390 type species *Meloe hungarus* Schrank von Paula, 1776
hungarus Schrank von Paula, 1776: 71 **E**: AR AU BU CR CZ FR (Alsace) GE GR HU MC RO SK ST TR UK YU "Caucasus" "Transcaucasia" **A**: IN IQ IS JO SY TR
limbatus Fabricius, 1801b: 588
marginatus Tauscher, 1812b: 152
subgenus *Meloe* Linnaeus, 1758: 419 type species *Meloe proscarabaeus* Linnaeus, 1758
Cnestocera C. G. Thomson, 1859: 124 type species *Meloe proscarabaeus* Linnaeus, 1758
Melittophagus Kirby, 1819: 164 type species *Pediculus melittae* Kirby, 1802 (= *Meloe violaceus* Marsham, 1802)
Proscarabaeus Schrank von Paula, 1781: 225 type species *Meloe proscarabaeus* Linnaeus, 1758
Triungulinus Dufour, 1828: 63 type species *Triungulinus andrenatarum* Dufour, 1828
aegyptius aegypytius Brandt & Erichson, 1832: 119 **E** IT (Sicilia) **N**: AG EG LB MO TU **A**: "Palestine"
plicatipennis Lucas, 1847: 400
siculus Baudi di Selve, 1878b: 352
aegyptius subcyaneus Wollaston, 1864: 514 **N**: CI

aruanachalae Saha, 1979: 128 **A**: AP BT NP UP
auriculatus Marseul, 1877: 480 **A**: JA NE XIZ "Northen, Central, East China"
bellus Jakovlev, 1897: 248 **A**: FE
bodemeyeri Ganglbauer, 1900: 160 **A**: TR
coarctatus Motschulsky, 1858a: 35 **A**: JA
formosensis Miwa, 1930: 12 **A**: TAI
gracilior Fairmaire, 1891c: xxii **A**: HUB
lobatus Gebler, 1832: 57 **A**: ES FE HEI JA NC "Korea"
granulifer Motschulsky, 1872: 47
menoko Kôno, 1936b: 92 **A**: JA
modestus Fairmaire, 1887b: 128 **A**: ANH FIJ JIA SCH SHX XIZ YUN
ovalicollis Reitter, 1908i: 246 **A**: KI
patellicornis Fairmaire, 1887b: 325 **A**: BEI FUJ
poteli Fairmaire, 1897a: 123 **A**: JIX
proscarabaeus afghanistanicus Kaszab, 1953a: 310 **A**: AF
proscarabeaus proscarabaeus Linnaeus, 1758: 419 **E**: AB AL AR AU BE BH BU BY CR CT CZ DE EN FI FR GB GE GG GR HU IR IT KZ LA LU LT MA MC MD NL NR NT PL PT RO SK SL SP ST SV SZ TR UK YU **N**: EG MO **A**: ANH ES GAN HEI HUB IN IS JO KI KZ LE LIA MG NC NE NMO NO SC SCH SY TD TM TR UZ WS XIZ
andrenatarum Dufour, 1828: 64
atratus Meyer, 1793: 15
brunsvicensis Meyer, 1793: 25
crispatus Fairmaire, 1884a: 173
cyanellus Brullé, 1832: 229
cyaneus Mulsant, 1857a: 47
exaratus Faldermann, 1832: 210
gallicus Baudi di Selve, 1878b: 351
incertus Tauscher, 1812b: 149
megacephalus Fischer von Waldheim, 1842: 27
pannonicus Baudi di Selve, 1878b: 351
punctatus Fabricius, 1792b: 518
rugicollis Stephens, 1832: 66
rugipennis Mannerheim, 1825: 31
tauricus Baudi di Selve, 1878b: 351
tectus Panzer, 1793d: 14
undulatus Baudi di Selve, 1878b: 351
volgensis Tauscher, 1812b: 149
vulgaris Stephens, 1832: 66
proscarabaeus sapporensis Kôno, 1936b: 91 **A**: JA
proscarabaeus sericeorugosus Aksentjev, 1987: 474 **A**: NP
rathjensi Borchmann, 1938b: 464 **A**: YE
semenowi Jakovlev, 1897: 251 **A**: CH ES FE JA "Korea"
semicoriaceus Fairmaire, 1891d: cii **A**: KA
simplicicornis Escherich, 1889a: 105 [DA] **E**: AR "Caucasus" **A**: TR
subcordicollis Fairmaire, 1887a: 129 **A**: GAN GUI JIX NMO XIZ YUN
tarsalis Jakovlev, 1897: 249 **A**: ES
tenuipes Jakovlev, 1897: 252 **A**: FE MG
violaceus Marsham, 1802: 482 **E**: AB AL AN AR AU BE BH BU BY CR CT CZ DE EN FI FR GB GE GG GR HU IR IT KZ LA LT MC NL NR NT PL PT RO SK SL SP ST SV SZ UK YU **N**: AG MO **A**: ES FE IN KA KI KZ TD TR UZ WS
angusticollis Rey, 1892: 137 [HN]
melittae Kirby, 1802: 168 (*Pediculus*)
montanuus Kaszab, 1958b: 190 [RN]
proscarabaeus Sulzer, 1761: 92 [HN]
similis Marsham, 1802: 482
strigosus Motschulsky, 1872: 48
tenuicollinus Kaszab, 1958b: 191 [RN]

subgenus *Meloegonius* Reitter, 1911a: 389 type species *Meloe cicatricosus* Leach, 1815
cicatricosus Leach, 1815b: 39 **E**: AB AL AR AU BH BU BY CR CT CZ FR GB GE GG GR HU IT MC MO PL RO SL SK SP ST SZ TR UK YU **A**: AF IN KI KZ TD TM TR UZ WS "Palestine"
iluronensis Salvaña, 1870: 37
ineditus Salvaña, 1870: 38
variolosus Fischer von Waldheim, 1844: 27
rufiventris himalaycus Kaszab, 1978c: 339 **A**: KA
rufiventris rufiventris Germar, 1832: 6 **E**: AB AU CR CT CZ FR GE GR HU IT MD PL RO SK ST SZ UK **N**: AG LB **A**: AF IN IQ JO KI KZ TD TR UZ
coriarius Brandt & Erichson, 1832: 131
hoffmannseggi Germar, 1832: 6
reticulatus Brandt & Ratzenburg, 1833: 108
subgenus *Meloenellus* Reitter, 1911a: 390 type species *Meloe griseopuberulus* Reitter, 1890
griseopuberulus Reitter, 1890c: 152 **A**: "Turkestan"
sulciceps Reitter, 1911b: 54 **A**: KZ
turkestanicus Escherich, 1890b: 88 **A**: KI
vlasovi Semenov & Arnoldi, 1937: 222 **A**: AF TM
klapperichi Kaszab, 1958c: 300
subgenus *Mesomeloe* Reitter, 1911a: 392 type species *Meloe sulcicollis* Kraatz, 1882 (= *Meloe xanthomelas* Solsky, 1881)
bytinskii Kaszab, 1969a: 279 **A**: IN
cinereovariegatus Heyden, 1885a: 294 **A**: KI UZ
coelatus Reiche, 1857: 271 **N**: AG CI MO TU **A**: IN IS JO SA SI
aegyptius Reitter, 1895a: 5 [HN]
nudus Wollaston, 1864: 514
xanthomelas Solsky, 1881a: 257 **A**: TD TM UZ
sulcicollis Kraatz, 1882d: 334
xanthomelaena Semenov, 1903d: 392
subgenus *Micromeloe* Reitter, 1911a: 390 type species *Meloe uralensis* Pallas, 1773
Trapezimeloe Pliginskij, 1935: 322 type species *Meloe conradti* Heyden, 1888
cavicornis Reitter, 1898a: 16 **A**: UZ
conradti Heyden, 1888: 42 **A**: KI TD
trapezicollis Dokhtouroff, 1889: 129
decorus Brandt & Erichson, 1832: 137 **E**: AU BU CR CZ FR GE HU IT KZ PL RO SK ST SZ UK YU **A**: KZ TD TR
pygmaeus L. Redtenbacher, 1849: 619
intermedius Escherich, 1904: 30 **A**: KZ TD UZ
kabuliensis Kaszab, 1981a: 164 **A**: AF
lopatini Pripisnova, 1987a: 44 **A**: TD
reitteri Escherich, 1889b: 112 **E**: KZ ST **A**: KI KZ
simulans Reitter, 1895a: 6 **A**: TM
terentjevi Kaszab, 1978c: 340 **A**: TD
uralensis Pallas, 1773: 25 **E**: AR AU CR CT CZ HU IT (South Tyrol) KZ MD RO SK ST UK **A**: ES KZ TM
marginalis Wellman, 1910a: 22
punctatus Meyer, 1793: 28

subgenus *Taphromeloe* Reitter, 1911a: 389 type species *Meloe erythrocnemus* Pallas, 1782
erythrocnemus Pallas, 1782: 76 **E**: AB AR BH BU CR CT GR IT KZ RO ST TR UK YU **N**: AG MO TU **A**: AF IN KI KZ TD TM TR UZ WS XIN
corallipes Gebler, 1847: 492
foveolatus Guérin-Méneville, 1842: 338 **E**: IT (Apulia) SP **N**: AG LB TU
roubali Mařan, 1942b: 56
subgenus *Treiodous* Dugès, 1869: 102 type species *Treiodous barranci* Dugès, 1869 (= *Meloe laevis* Leach, 1815)
Anchomeloe Iablokoff-Khnzorian, 1983: 88 type species *Meloe autumnalis* A. G. Olivier, 1797

autumnalis autumnalis A. G. Olivier, 1797: 650 **E**: AL AR AU BE BH BU CR CT CZ FR HU GE GB GG GR IT MC MO NL PL PT RO SK SL SP ST SZ TR UK YU **N**: AG MO TU **A**: CY IN IS JO SY TR

carnicus Katter, 1883: 41
cribripennis Baudi di Selve, 1878b: 352
cyaneus Fabricius, 1801b: 589
glabratus Leach, 1815b: 43
hiemalis Gredler, 1866: 289
impunctatus Wellman, 1910a: 22
laevis Gredler, 1866: 289
punctatus Marsham, 1802: 483
punctipennis Escherich, 1889a: 106

autumnalis heydeni Escherich, 1889a: 106 **E**: IT (Sicilia)
genus *Meloe*, species incertae sedis
elegantulus Semenov & Arnoldi, 1934: 215 **A**: XIZ
longipennis Fairmaire, 1891c: xxii **A**: HUB
medogensis Tan, 1988: 292 **A**: XIZ

genus *Meloe*, nomina dubia
Meloe aprilina Meyer, 1793: 21
Meloe bilineatus Aragona, 1830: 20
Meloe cinereus Brandt & Ratzenburg, 1833: 111
Meloe fascicularis Aragona, 1830: 20
Meloe marginatus Tauscher, 1812b: 152 [HN]
Meloe ovilis Mulsant, 1857a: 82
Meloe prolifericornis Motschulsky, 1872: 48
Meloe punctatoradius Latreille, 1804a: 391
Meloe rufipes Bremi-Wolf, 1856: 199
Meloe rugulosus Brullé, 1832: 230
Meloe sulcicollis Latreille, 1804a: 391
Meloe viennensis Schrank von Paula, 1776: 71

genus *Oreomeloe* Tan, 1981: 411 type species *Oreomeloe spinulus* Tan, 1981
spinulus Tan, 1981: 415 **A**: XIZ

genus *Physomeloe* Reitter, 1911a: 388 type species *Meloe corallifer* Germar, 1818
corallifer Germar, 1818: 259 (*Meloe*) **E**: PT SP

evae Flach, 1907: 19

subfamily **Nemognathinae** Laporte, 1840

tribe Horiini Latreille, 1802

genus *Horia* Fabricius, 1787a: 164 type species *Horia fabriciana* Betrem, 1929

Hoplozonitis Blackburn, 1892: 228 type species *Hoplozonitis mira* Blackburn, 1892

fabriciana Betrem, 1929: xxvii [RN] **N**: EG **A**: AE SA YE

africana Aurivillius, 1890: 203 (*Cissites*) [HN]

gahani Betrem, 1932: 95 **A**: SD

genus *Synhoria* Kolbe, 1897: **256** type species *Horia cephalotes* A. G. Olivier, 1793
maxillosa Fabricius, 1801b: 86 (*Horia*) **A**: HEN HKG JA SCH SD TAI **ORR**

anguliceps Fairmaire, 1885b: 111 (*Horia*)
sasakii Kôno, 1936b: 97 (*Cissites*)
tosana Kôno, 1936b: 96 (*Horia*)

tribe Nemognathini Laporte, 1840

genus *Allendesalazaria* Escalera, 1910b: **379** type species *Allendesalazaria nymphoides* Escalera, 1910
nymphoides Escalera, 1910b: 380 **N**: AG MO

genus *Apalus* Fabricius, 1775: 127 type species *Meloe bimaculatus* Linnaeus, 1760
Coriologiton Marseul, 1879b: 65 [DA] type species *Criolis hilaris* Marseul, 1879
Criolis Mulsant, 1858b: 240 type species *Criolis guerini* Mulsant, 1858
Deratus Motschulsky, 1872: 51 type species *Meloe necydalea* Pallas, 1773
Hapalus Illiger, 1801: 138 [unjustified emendation]
asiaticus Escherich, 1897b: 119 **A**: AF "Turkestan"
bimaculatus Linnaeus, 1760: 288 (*Meloe*) **E**: AU BH BU CR CZ DE EN FI FR GE HU IT KZ LA LT NR NT RO SK ST SP SV SZ UK **N**: AG LB MO TU **A**: JA KZ SY TM TR "Siberia"
caruanae Prochazka, 1892: 265 (*Hapalus*)
flava Escherich, 1897b: 120
lecomptei Pic, 1896j: 140 (*Hapalus*)
binaevus Prochazka, 1892: 270 (*Hapalus*) **A**: KI
bipunctatus Megerle von Mühlfeld, 1812: 18 **E**: AU CR CZ GR HU IT KZ RO SK SP ST UK YU "Caucasus" **A**: IS KZ TD TR "Palestine"
nigripes Kaszab, 1942b: 43 (*Hapalus*)
nigritarsis Ragusa, 1898: 222 (*Hapalus*)
melanipes Kaszab, 1942b: 43 (*Hapalus*)
rufus Ragusa, 1906: 6
cinctus Pic, 1896j: 140 (*Hapalus*) **N**: AG EG TU
creticus J. Frivaldszky, 1877a: 83 (*Hapalus*) **E**: GR (Kríti)
davidis Fairmaire, 1886c: 352 (*Hapalus*) **A**: HUB
guerini Mulsant, 1858b: 240 (*Criolis*) **E**: FR IT SP **N**: MO
haemapterus Fairmaire, 1889b: 49 **A**: SCH SHN
mongolicus Kaszab, 1965f: 558 **A**: MG
montanus Escherich, 1897b: 121 **A**: TR
mulsanti Reiche, 1878: lxxiii (*Criolis*) **A**: TR
necydaleus Pallas, 1782: 92 (*Meloe*) **E**: GR HU KZ MC RU UK "Caucasus" **A**: IN KI KZ TR
inornatus Prochazka, 1892: 266 (*Hapalus*)
moreanus Pic, 1922d: 19
phoebus Prochazka, 1892: 267 (*Hapalus*)
rufipennis Gebler, 1829a: 142
spectabilis Frivaldszky von Frivald, 1835: 265
spectabilis Schaum, 1859: 52 (*Hapalus*) [HN]
plagiatus Waterhouse, 1889: 130 **A**: AF
subfasciatus Pic, 1922d: 19 **A**: IN
sulcithorax Pic, 1901i: 79 (*Hapalus*) **A**: SY

genus *Cochliophorus* Escherich, 1891a: 16 type species *Cochliophorus reitteri* Escherich, 1891
reitteri Escherich, 1891a: 16 **E**: GR

genus *Ctenopus* Fischer von Waldheim, 1823: pl. 38 type species *Ctenopus melanogaster* Fischer von Waldheim, 1824
abdominalis Motschulsky, 1845a: 83 **E**: ST
aurantiacus Semenov, 1893a: 15 **A**: IN
carinifer Semenov, 1896d: 380 **A**: AF TM
eous Semenov, 1900b: 487 **A**: XIN
hauseri Reitter, 1894a: 50 **A**: KI TD TM
ahngeri Semenov, 1896d: 381
lama Escherich, 1904: 35 **A**: XIN
melanogaster Fischer von Waldheim, 1823: pl. 38 **E**: AB AR **A**: IN KI KZ TM
nudus Escherich, 1897b: 128 **A**: TD TM
oxianus Semenov, 1894c: 540 **A**: UZ
persicus Semenov, 1893a: 17 **A**: IN
pygmaeus Semenov, 1903c: 357 **A**: TM
reitteri Semenov, 1893a: 18 **E**: AR

roseni Escherich, 1897b: 128 **A**: TM
hauseri Semenov, 1896d: 388 [HN]
rufoscutellatus Reitter, 1889a: 38 **A**: TM
semenovi Reitter, 1895b: 157 **A**: XIN
sinuatipennis Fairmaire, 1892e: cliv (*Zonitis*) **A**: TM UZ XIN
testaceus Semenov, 1893a: 14
sogdianus Semenov, 1900b: 486 **A**: TM UZ
vitticollis Reitter, 1889b: 38 **E**: AB AR

genus *Euzonitis* Semenov, 1893d: 276 type species *Apalus sexmaculatus* A. G. Olivier, 1789
adustipennis Motschulsky, 1872: 53 (*Zonitis*) **E**: AR KZ ST "Caucasus"
inapicalis Pic, 1935d: 11
ahngeriana Semenov, 1896d: 379 (*Zonitis*) **A**: TM
alfierii Pic, 1929g: 110 (*Zonitis*) **N**: EG
fulvipennis Fabricius, 1792b: 49 (*Zonitis*) **E**: AB AR AU BU CR CT GG GR HU IT MC RO SK ST UK YU **A**: TR
funeraria Fairmaire, 1884g: cxlii (*Zonitis*)
nigra Tauscher, 1812b: 163 (*Zonitis*)
nigricollis Faldermann, 1832: 211 (*Zonitis*)
haroldi Heyden, 1870: 150 (*Zonitis*) **E**: SP **N**: MO
maculiceps Reitter, 1897c: 110 **A**: IN KI TD UZ
mzchetica Sumakov, 1929b: 113 (*Zonitis*) **E**: GG
paulinae Mulsant & Rey, 1858a: 193 (*Zonitis*) **A**: IN IS JO
quadrimaculata Pallas, 1782: 91 (*Meloe*) **E**: AB AR AU BH BU CR CT CZ FR GG HU IT KZ MC PT RO SK SP ST SZ UK YU **N**: AG CI MO MR TU **A**: CY IN IQ IS KZ MG SY TM TR
adanensis Pic, 1935d: 11
atra Schwarz, 1808: 340 (*Zonitis*)
auliensis Pic, 1913k: 138
bifasciata Schwarz, 1808: 310 (*Zonitis*)
bimaculata Tauscher, 1812b: 158 (*Zonitis*)
concolor Abeille de Perrin, 1880: 257 (*Zonitis*)
concolor Wellman, 1910a: 25 (*Zonitis*) [HN]
fasciata Tauscher, 1812b: 159 (*Zonitis*)
iconiensis Pic, 1901i: 79 (*Zonitis*)
immaculata Escherich, 1892: 247 (*Zonitis*) [HN]
imperialis Wollaston, 1861b: 106 (*Zonitis*)
korbi Escherich, 1890a: 24 (*Zonitis*)
latenotata Pic, 1907f: 154
obliquata Motschulsky, 1872: 53 (*Zonitis*)
palumboi Ragusa, 1898: 221 (*Zonitis*)
quadripunctata Fabricius, 1801b: 84 (*Mylabris*)
rufofasciata Fairmaire, 1884g: cxlii (*Zonitis*)
xanthoptera Fairmaire, 1876: 94 (*Zonitis*)
rubida Ménétriés, 1832: 211 (*Zonitis*) **E**: AB AR GR **A**: AF TM TR
pubescens Waltl, 1838: 467 (*Zonitis*)
sexmaculata A. G. Olivier, 1789: 90 (*Apalus*) **E**: AB AR BH BU CR CZ FR GG GR HU IT KZ RO SP ST UK YU **A**: CY IN IQ IS JO KZ LE SY TD TM TR
bipunctata Ragusa, 1881: 42 (*Zonitis*) [HN]
discolor Escherich, 1892: 243 (*Zonitis*) [HN]
distigma Wellman, 1910a: 25 (*Zonitis*) [RN]
fenestrata Pallas, 1782: 90 (*Zonitis*)
lunaris Wellman, 1910a: 25 (*Zonitis*) [RN]
lunata Tauscher, 1812b: 159 (*Zonitis*) [HN]
medioconjuncta Kaszab, 1942b: 43
obliterata Pic, 1922d: 18
octomaculata Ragusa, 1923: 7
pedemontana Pic, 1901i: 79 (*Zonitis*)
subconjuncta Pic, 1901i: 79 (*Zonitis*)

suramensis Pic, 1922d: 18
sogdiana Semenov, 1893d: 275 (*Zonitis*) **A**: TD TM
spectabilis Kraatz, 1881c: 326 (*Zonitis*) **A**: AF TD TM "Turkestan"
atricornis Reichardt, 1934: 237
terminata Abeille de Perrin, 1880: 255 (*Zonitis*) **E**: GR HU IT RO TR **N**: EG **A**: AF IN IS JO SY TM TR
armeniaca Pic, 1901i: 79 (*Zonitis*)
auricoma Escherich, 1891b: 54 (*Zonitis*)
bombycina Escherich, 1896a: 27 (*Zonitis*)
pallens Escherich, 1897a: 47
semideleta Pic, 1903m: 163 (*Zonitis*)
suprema Escherich, 1897b: 110
unimaculata Pic, 1933e: 14 **N**: MO
varentzowi Semenov, 1894c: 537 (*Zonitis*) **A**: IN TM
genus *Euzonitis*, species incertae sedis
jacobsoni Reichardt, 1934: 223 **A**: TD

genus *Glasunovia* Semenov, 1896c: 357 type species *Glasunovia caspica* Semenov, 1896
afghanica Semenov, 1896c: 358 **A**: AF IN TM
caspica Semenov, 1896c: 360 **A**: TM
sillemi Borchmann, 1935: 308 **A**: AF KA
sinica Pic, 1938e: 16 **A**: CH

genus *Leptopalpus* Guérin-Méneville, 1834a: pl. 35 type species *Leptopalpus chevrolatii* Guérin-Méneville, 1834 (= *Zonitis rostrata* Fabricius, 1792)
quadrimaculatus Gahan, 1896b: 457 **A**: YE **AFR**
rostratus Fabricius, 1792b: 50 (*Zonitis*) **E**: IT (Sicilia) PT SP **N**: AG MO TU
chevrolatii Guérin-Méneville, 1834a: pl. 35

genus *Megatrachelus* Motschulsky, 1845a: 83 type species *Zonitis polita* Gebler, 1832
Schroetteria Reitter, 1911a: 395 type species *Zonitis polita* Gebler, 1832
pallidipennis Motschulsky, 1845a: 83 (*Zonitis*) **A**: ES
politus Gebler, 1832: 58 (*Zonitis*) **A**: ES FE HEI JA MG NC SC
megacephalus Kôno, 1936a: 4
quadricollis Fairmaire, 1892e: cliii (*Zonitis*) **A**: "Turkestan"
sibiricus Tauscher, 1812b: 162 (*Zonitis*) **A**: KZ MG
subpolitus Reitter, 1911a: 395 (*Schroetteria*) **A**: NE

genus *Nemognatha* Illiger, 1807: 333 type species *Zonitis vittata* Fabricius, 1801 (= *Zonitis piazata* Fabricius, 1794)
Gnathonemula Aksentjev, 1981: 862 type species *Gnathonemula gracilis* Aksentjev, 1981
Nematognatha Gemminger & Harold, 1870: 2163 [unjustified emendation]
chrysomelina Fabricius, 1775: 126 (*Zonitis*) **E**: AB AR FR GR IT (Sicilia, Sardegna) PT SP ST UK **N**: AG EG MO TU **A**: AF CY IN IQ IS JO KZ LE SA SY TD TM TR UZ
akbesiana Pic, 1897d: 35
asiatica Motschulsky, 1872: 54
atripennis Chobaut, 1898b: 143
flavipes Ménétriés, 1849: 284
latenigra Pic, 1941a: 5
nigripes Suffrian, 1853: 236
maculicollis Fairmaire, 1875a: 523 (*Zonitis*)
unipunctata Beauregard, 1890: 466
flavicornis Stierlin, 1876: 478 **E**: AB AR **A**: IN
gemina Suffrian, 1853: 235 **N**: EG
gracilis Aksentjev, 1981: 863 (*Gnathonemula*) **A**: KZ UZ
transcaspica Kaszab, 1981a: 166
quinquemaculata Suffrian, 1853: 235 **N**: EG MO **A**: IN SA
sexmaculata Pic, 1911d: 74

sibirica Gebler, 1833: 290 (*Zonitis*) [DA] **A**: "Siberia"

genus *Nyadatus* Aksentjev, 1981: 860 type species *Nyadatus telejus* Aksentjev, 1981
sphalerus Aksentjev, 1981: 862 **A**: TM
telejus Aksentjev, 1981: 861 **A**: KZ TD UZ

genus *Sitaris* Latreille, 1802: 187 type species *Necydalis humeralis* Fabricius, 1775 (= *Necydalis muralis* Förster, 1771)
subgenus *Filalia* Escalera, 1906a: 162 type species *Sitaris cerambycina* Escalera, 1906
cerambycinus Escalera, 1906a: 161 **N**: LB MO
subgenus *Sitaris* Latreille, 1802: 187 type species *Necydalis humeralis* Fabricius, 1775 (= *Necydalis muralis* Förster, 1771)
balachowskyi Peyerimhoff, 1931b: 83 (*Apalus*) **N**: AG
emiliae Escherich, 1897b: 130 **E**: GR
ferrantei Pic, 1911d: 74 **N**: AG EG MO **A**: "Palestine"
grassei Reymond, 1956a: 41 (*Ctenopus*)
homonimus Kaszab, 1958b: 191 [RN] **N**: EG
atricornis Pic, 1927e: 6 [HN]
incantatus Peyerimhoff, 1929a: 124 (*Apalus*) **N**: AG MO
melanurus Küster, 1849a: 84 **E**: SP
mellicolor Biggs, 1945: 110 (*Apalus*) **A**: IN
minutus Pic, 1932e: 17 **A**: TR
muralis Förster, 1771: 48 (*Necydalis*) **E**: AU BE CR CZ FR GB GE GR HU IT NL PL PT RO SK SP ST SZ UK YU **N**: AG LB MO TU **A**: "Turkestan"
attenuatus Geoffroy, 1785: 154 (*Cantharis*)
flavus Hamm, 1909: 277
humeralis Fabricius, 1787a: 170 (*Necydalis*)
humuralis Fabricius, 1775: 209 (*Necydalis*)
mauritanicus Normand, 1950a: 82 (*Apalus*)
nitidicollis Abeille de Perrin, 1870: 42
splendidus Schaufuss, 1861: 49
pallens Escherich, 1904: 35 (*Hapalus*) **A**: "Turkestan"
pectoralis Bates, 1879: 484 (*Sitaris*) **A**: XIN
rufipennis Küster, 1849a: 83 **E**: SP
rufipes Gory, 1841: pl. 73 **N**: AG MO
ferdinandi Escalera, 1910b: 382
rufiventris Kraatz, 1884: 175 **E**: GR **A**: TR
solieri solieri Pecchioli, 1839: 529 **E**: CR FR GR IT PT SP **N**: AG CI MO TU **A**: TR
apicalis Escherich, 1897b: 130
corzoi Pardo Alcaide, 1950b: 45
lativentris Schaufuss, 1861: 49 (*Sitaris*)
longicornis Kraatz, 1858: 387, 388 (*Sitaris*)
tibialis Ragusa, 1898: 222
tauricus Motschulsky, 1872: 54 **E**: GR UK
tenuicornis Schaufuss, 1870: 47 **N**: EG

genus *Sitarobrachys* Reitter, 1883b: 309 type species *Sitarobrachys brevipennis* Reitter, 1883 (= *Stenoria thoracica* Kraatz, 1862)
thoracica Kraatz, 1862b: 126 (*Stenoria*) **E**: BU GR SP **N**: AG CI EG MO TU **A**: IS
alfierii Pic, 1913h: 122
brevipennis Reitter, 1883b: 309
diversipes Pic, 1924f: 29
buigasi Escalera, 1909b: 240
proxima Escalera, 1914: 385

genus *Sitaromorpha* Dokhtouroff, 1889: 190 type species *Sitaromorpha wilkinsi* Dokhtouroff, 1889
wilkinsi Dokhtouroff, 1889: 190 **A**: "Turkestan"

genus *Stenoria* Mulsant, 1857a: 186 type species *Sitaris apicalis* Latreille, 1804
subgenus *Gineremia* Pardo Alcaide, 1961a: 92 type species *Stenoria saharica* Pardo Alcaide, 1961
saharica Pardo Alcaide, 1961a: 92 **N**: MO
subgenus *Stenoria* Mulsant, 1857a: 186 type species *Sitaris apicalis* Latreille, 1804
Sitaropsis Iablokoff-Khnzoryan, 1958: 1896 type species *Hapalus erevanensis* Iablokoff-Khnzoryan, 1958
analis Schaum, 1859: 51 **E**: CR CZ FR GE IT PL PT SP ST SZ **A**: TR
adusta Schaum, 1859: 51 (*Sitaris*)
colletis Mayet, 1874: cxcviii (*Sitaris*)
acutipennis Fairmaire, 1881d: xliv (*Sitaris*)
sareptana Pic, 1912c: 9
antoinei Pardo Alcaide, 1953: 289 **E**: SP **N**: AG MO
gutticollis Pardo Alcaide, 1953: 291
apicalis apicalis Latreille, 1804a: 403 (*Sitaris*) **E**: CR FR GR HU IT PL PT RO SP ST UK YU "Caucasus" **N**: MO **A**: IN TM TR
basicollis Kaszab, 1956b: 158
bipunctata Kaszab, 1956b: 158
brunneicollis Pic, 1914m: 74 **A**: LE
catalonica Pardo Alcaide, 1958b: 18
communimacula Kaszab, 1956b: 158
iranica Kaszab, 1959h: 440
kaszabiana Pardo Alcaide, 1958b: 18
kraatzi Mulsant & Rey, 1861: 191
luteifrons Pardo Alcaide, 1958a: 18
nigroplagiata Kaszab, 1956b: 158
paucinigra Pardo Alcaide, 1958a: 17
picicollis Escherich, 1897b: 123 (*Apalus*)
sturmii Küster, 1846b: 72 (*Ctenopus*)
tristicula Kaszab, 1956b: 158
vitticollis Kaszab, 1956b: 158
canariensis Pic, 1902k: 80 (*Sitaris*) **N**: CI
erevanensis Iablokoff-Khnzorian, 1958: 1896 (*Hapalus*) **E**: AR
erythromelas Semenov, 1893d: 278 (*Hapalus*) **A**: UZ
falcipennis Semenov, 1893d: 277 (*Hapalus*) **A**: TM "Turkestan"
fasciata Faldermann, 1835b: 416 (*Hapalus*) **A**: HEI MG NMO
grandiceps Semenov, 1893d: 280 (*Hapalus*) **A**: IN
hauseri Escherich, 1904: 34 (*Hapalus*) **A**: QIN XIN
intermedia Dokhtouroff, 1889: 185 (*Hapalus*) **E**: AB AR **A**: IN TD TM
immaculaticeps Pic, 1911e: 97 (*Hapalus*)
klapperichi Kaszab, 1958c: 310 **A**: AF
komarowi Dokhtouroff, 1889: 90 (*Hapalus*) **A**: TM "Turkestan"
immaculatus Prochazka, 1892: 270 (*Hapalus*)
reitteri Prochazka, 1892: 270 (*Hapalus*)
kraatziana Semenov, 1900b: 490 (*Hapalus*) **A**: UZ
apicalis Kraatz, 1882d: 335 [HN]
kraatzi Semenov, 1893d: 277 (*Hapalus*)
laterimaculata Reitter, 1898e: 124 (*Hapalus*) **A**: MG NMO TD
longipennis Pic, 1933b: 7 (*Apalus*) **A**: XIN
muiri Kaszab, 1956e: 632 **A**: YE
arabica Kaszab, 1956e: 633
nigricollis Semenov, 1893d: 278 (*Hapalus*) **E**: AB AR
plagiata Escalera, 1914: 387 (*Sitaris*) **N**: MO
escalerai Pic, 1920d: 10 (*Apalus*)
richteri Kaszab, 1959b: 5 **A**: IN
sokolowi Semenov, 1900b: 491 (*Hapalus*) **A**: UZ

steppensis Kaszab, 1966a: 259 **A**: MG
testaceicornis Pic, 1939f: 147 (*Apalus*) **N**: EG **A**: SI
thakkola Schawaller, 1996b: 148 **A**: NP
tibetana Escherich, 1904: 35 (*Hapalus*) **A**: QIN
tibialis Motschulsky, 1872: 51 (*Deratus*) **A**: ES
tristis Escherich, 1897b: 124 (*Apalus*) **E**: CT KZ **A**: KZ

genus *Zonitis* Fabricius, 1775: **126** type species *Zonitis flava* Fabricius, 1775
subgenus *Zonitis* Fabricius, 1775: 126 type species *Zonitis flava* Fabricius, 1775
afghanica Kaszab, 1958c: 307 **A**: AF
ballionis Escherich, 1892: 242 **A**: AF KZ TD TM UZ XIN
scutellaris Fairmaire, 1892e: cliv
tristicula Semenov, 1900b: 481
bellieri Reiche, 1860: 731 **E**: IT **N**: AG LB MO TU **A** TR
turcica J. Frivaldszky, 1877b: 104
bomiensis Tan, 1988: 292 **A**: XIZ
bytinskii Kaszab, 1957e: 231 **A**: IS
escherichi Semenov, 1893d: 273 **A**: TD TM TR
ferganensis Pic, 1951c: 11 **A**: "Turkestan"
fernancastroi Pardo Alcaide, 1950b: 40 **E**: FR IT SP **N**: MO
flava Fabricius, 1775: 127 **E**: AB AL AR AU BU CR CZ FR GG GR HU IT KZ PT RO SK SP ST SZ UK YU **N**: AG LB MO TU **A**: CY IN IS JO KZ LE SY TM TR UZ XIN
afra Rossi, 1790: 240 (*Lytta*)
analis Abeille de Perrin, 1880: 256
ancoroides Escherich, 1892: 242
atriventris Pic, 1951a: 12
flaviventris J. Müller, 1902: 116
gaditana Pardo Alcaide, 1956: 18
hipponensis Pic, 1900e: 38
impressicollis Motschulsky, 1872: 53
inscutellaris Pic, 1922d: 18
latenigra Pic, 1951c: 11
melanopus Wellman, 1910a: 25 [RN]
metasternalis Csiki, 1953: 120 [HN]
metasternaloides Kaszab, 1958b: 191 [RN]
moltonii Schatzmayr, 1941a: 81
nigriceps Pardo Alcaide, 1956: 18
nigripennis Fabricius, 1794: 103
nigripes Motschulsky, 1872: 52
nigripes J. Müller, 1902: 116 [HN]
nigrithorax Motschulsky, 1872: 52
nigrithorax Pic, 1904h: 90 [HN]
obscuricepes Pic, 1907l: 178
praeusta Fabricius, 1792b: 48
ramirezi Pardo Alcaide, 1956: 18
reitteri Kaszab, 1958b: 191 [RN]
scutellaris Motschulsky, 1872: 52
semiobscura Schatzmayr, 1941a: 81
signatithorax Pic, 1900e: 38
sophiensis Nedelkov, 1905: 434
testacea Fabricius, 1781: 331 (*Mylabris*)
thoracica Laporte, 1840: 276
unicolor Ragusa, 1898: 220
fortuccii Fairmaire, 1887f: 326 **A**: BEI MG
testaceonotata Pic, 1914c: 3
fuscimembris Fairmaire, 1886c: 351 **A**: JIX YUN
glasunowi Semenov, 1893d: 271 **E**: KZ ST "Caucasus" **A**: KZ MG TD TM
atripennis Semenov, 1893d: 272 [HN]

semenovi Kaszab, 1958b: 191 [RN]
hauseri Escherich, 1897b: 127 **A**: TM UZ
immaculata A. G. Olivier, 1789: 166 (*Apalus*) **E**: AL AU BH BU CR GR HU IT FR MC PT RO SP ST SZ UK YU **N**: AG MO TU **A**: IS SY TR ZU
biimpressa Chevrolat, 1882: 4
binotaticollis Pic, 1913k: 137
fulva Rossi, 1792: 94 (*Mylabris*)
gibbicollis Abeille de Perrin, 1880: 254
gracilenta Fairmaire, 1892b: 156
ligurica Schellenberg, 1802: 11 (*Lymexylon*)
mahia Hübner, 1789: 44
mutica L. G. Scriba, 1790: 23
scutellaris Escherich, 1897a: 46 [HN]
scutellifera Wellman, 1910a: 25 [RN]
seminigra Reitter, 1884b: 93
silvestrii Schatzmayr, 1941b: 362
iphigeniae Pliginskij, 1914c: 429 **E**: UK
rufipes Pliginskij, 1914c: 429
rufoscutellata Pliginskij, 1914c: 429
kozlowi Semenov, 1900b: 482 **A**: XIN
microcephala Escherich, 1897a: 46 **A**: TR
nana Ragusa, 1881: 43 **E**: AL BH BU CR CZ FR GR HU IT SK ST SZ YU **A**: IS JO SY SZ TR
apicalis Motschulsky, 1872: 53 [HN]
laticollis Escherich, 1892: 238
motschulskyi Kaszab, 1958b: 191 [RN]
nigripectus Fairmaire, 1891e: cxxxiii **A**: KA
nigritarsis Stierlin, 1876: 477 (*Nemognatha*) **E**: AR "Caucasus" **A**: IN
nigriventris Motschulsky, 1872: 53 **E**: AR GG GR "Caucasus" **A**: TR
novercalis Escherich, 1891b: 236 **N**: AG
pseudopraeusta Kaszab, 1958c: 306 **A**: AF
ruficollis J. Frivaldszky, 1877b: 107 **E**: GR (Kríti) **N**: AG LB **A**: IS JO SY TR
abeillei Bedel, 1892: 248
rubricollis Abeille de Perrin, 1880: 256
thoracica Escherich, 1892: 237 [HN]
semirubra Pic, 1911f: 101 **A**: FUJ YUN
sikkimensis Pic, 1910t: 91 **A**: SD
sinensis Pic, 1935b: 24 **A**: SCH
turkestanica Semenov, 1900b: 480 **A**: TD UZ
genus *Zonitis*, species incertae sedis
scutellaris Fairmaire, 1892e: cliv **A**: UZ "Turkestan"
prescutellaris Pic, 1911i: 138

genus *Zonitomorpha* Péringuey, 1909: 272 type species *Zonitis sellata* Fåhraeus, 1870
angustithorax Pic, 1912f: 31 **A**: SHN
davidis Fairmaire, 1886c: 351 (*Zonitis*) **A**: BEI

genus *Zonitoschema* Péringuey, 1909: 274 type species *Lytta coccinea* Fabricius, 1801
Stenoderistella Reitter, 1911a: 395 type species *Stenodera pallidissima* Reitter, 1908
Zonitoides Fairmaire, 1883a: 31[HM] type species *Zonitoides megalops* Fairmaire, 1883
Zonitopsis Wellman, 1910b: 395 [RN] type species *Zonitoides megalops* Fairmaire, 1883
cothurnata Marseul, 1873a: 228 (*Zonitis*) **A**: JA NC TAI
elongatipennis Pic, 1915e: 16 (*Zonitis*) **A**: JIX
gibdoana Kaszab, 1956e: 631 (*Zonitopsis*) **A**: SA **AFR**
griseohirta Pic, 1914p: 107 **A**: SA **AFR**
iranica Kaszab, 1959b: 4 **A**: IN
japonica Pic, 1910t: 90 (*Zonitis*) **A**: JA NC TAI ZHE **ORR**
geniculata Borchmann, 1912a: 12 (*Zonitis*) [HN]

pallida Fabricius, 1794: 447 (*Zonitis*)
kimi Kôno, 1936b: 95 (*Zonitis*) **A**: "Korea"
klapperichi Borchmann, 1941: 23 **A**: FUJ
macroxantha yunnana Kaszab, 1960d: 262 **A**: YUN
miwai Kôno, 1936b: 94 (*Zonitis*.) **A**: TAI
oculatissima Peyerimhoff, 1929a: 125 [= 1931b: 84] **N**: AG
okinawensis Miwa, 1928: 76 (*Zonitis*) **A**: JA
pallidissima Reitter, 1908c: 50 (*Stenodera*) **N**: AG EG MO TU **A**: IS
rubricolor Pic, 1924g: 454 **A**: IS SA **AFR**

tribe Stenoderini Selander, 1991

genus *Stenodera* Eschscholtz, 1818: 469 type species *Stenodera sexpunctata* Eschscholtz, 1818
subgenus *Stenodera* Eschscholtz, 1818: 469 type species *Stendoera sexpunctata* Eschscholtz, 1818 (= *Meloe caucasica* Pallas, 1782)
Zonitides Abeille de Perrin, 1880: 253 type species *Zonitides oculifer* Abeille de Perrin, 1880
caucasica Pallas, 1781: 94 (*Meloe*) **E**: AB AL AR BU CR GG GR MC MD RO ST TR UK YU **A**: IN IS JO LE SY TM TR
bisbinotata Pic, 1909j: 130
crocata Abeille de Perrin, 1880: 253
inornata Escherich, 1897b: 101
sexmaculata Fabricius, 1794: 120 (*Mylabris*)
sexpunctata Eschscholtz, 1818: 469
djakonovi Aksentjev, 1978: 124 **A**: FE
foveicollis Fairmaire, 1897b: 231 (*Megatrachelus*) **A**: FUJ GUX
oculifera Abeille de Perrin, 1880: 253 (*Zonitides*) **A**: IS LE SY
bipunctata Chevrolat, 1882: 5 (*Zonitis*)
chevrolati Ragusa, 1882: 251 (*Zonitis*)
impressicollis Escherich, 1891b: 54
subgenus *Stenoderina* Aksentjev, 1988: 579 type species *Zonitis puncticollis* Chevrolat, 1834
anatolica J. Frivaldszky, 1884: 1 (*Zonitis*) **A**: SY TR
coeruleiceps Fairmaire, 1892b: 157 (*Zonitis*) **A**: TR
palestine Mařan, 1942a: 25 **A**: IS
puncticollis Chevrolat, 1834: pl. 35 (*Zonitis*) **A**: IS LE SY TR

family Meloidae, nomen dubium
Meloe lutea Pallas, 1773: 722

family MYCTERIDAE Blanchand, 1845
subfamily **Mycterinae** Blanchard, 1845

genus *Mycterus* Schellenberg, 1798: 124 type species *Rhinomacer curculioides* Fabricius, 1781
subgenus *Eutryptes* Gistel, 1856: 375 type species *Bruchus umbellatarum* Fabricius, 1792
Cyclopidius Seidlitz, 1890: 555 [RN] type species *Bruchus umbellatarum* Fabricius, 1792
Cyclops Mulsant, 1859: 18 [HN] type species *Bruchus umbellatarum* Fabricius, 1792
Triconatus Spilman, 1954: 92 [RN] type species *Bruchus umbellatarum* Fabricius, 1792
articulatus Reitter, 1911a: 419 **E**: AR
tibialis Küster, 1850a: 86 **E**: AU BH CR CZ GE GR HU IT SK ST SZ TR UK YU **A**: TR
ruficornis Mulsant & Godart, 1855d: 278
umbellatarum Fabricius, 1787a: 41 (*Bruchus*) **E**: FR (Corse) IT PT SP **N**: AG MO TU
algiricus Seidlitz, 1920: 1177
pulverulentus Chevrolat, 1834: pl. 33
pulverunculus Küster, 1850a: 85
siculus Baudi di Selve, 1883a: 6
subtruncatus Guillebeau, 1893a: 90
subgenus *Mycterus* Schellenberg, 1798: 124 type species *Rhinomacer curculioides* Fabricius, 1781
Rhinomacer Fabricius, 1781: 199 [HN] type species *Rhinomacer curculioides* Fabricius, 1781

curculioides Fabricius, 1781: 199 (*Rhinomacer*) **E**: AB AU CR CT CZ FR GE GR IT LA PL PT RO SK SP ST SZ UK **N**: MO TU **A**: CY LE TR
griseus Schellenberg, 1798: 124
gracilior Fairmaire, 1874a: 388

genus *Omineus* Lewis, 1895b: 119 type species *Omineus humeralis* Lewis, 1895
humeralis Lewis, 1895b: 119 **A**: JA

subfamily **Hemipeplinae** Lacordaire, 1854

***Hemipeplus* Berthold, 1827: 394** type species *Hemipeplus hemipterus* Lacordaire, 1854
Nemicelus LeConte, 1854: 79 type species *Nemicelus marginipennis* LeConte, 1854
Ochrosanis Pascoe, 1866: 444 type species *Ochrosanis dohrnii* Pascoe, 1866
miyamotoi H. Kamiya, 1961: 75 **A**: JA (Amami-Ôshima) TAI

family BORIDAE C. G. Thomson, 1859

subfamily **Borinae** C. G. Thomson, 1859

genus *Boros* Herbst, 1797: 318 type species *Boros elongatus* Herbst, 1797 (= *Helops schneideri* Panzer, 1796)
Borus Illiger, 1802a: 129 [unjustified emendation]
Borus Agassiz, 1846b: 49 [unjustified emendation]
schneideri Panzer, 1796b: 1 (*Helops*) **E**: BY CT CZ FI GE LA LT NT PL RO SK ST SV UK **A**: ES FE JA NE WS "Korea"
boros Fabricius, 1801b: 558 (*Hypophloeus*)
corticalis Paykull, 1798: 93 (*Trogosita*)
elongatus Herbst, 1797: 319
picipes Schoenherr, 1806: 155 (*Trogosita*)

family TRICTENOTOMIDAE Blanchard, 1845

genus *Autocrates* J. Thomson, 1860a: 28 type species *Trictenotoma aeneus* Parry, 1847
aeneus Westwood, 1846: 353 (*Trictenotoma*) **A**: BT NP SC SD UT
oberthueri Vuillet, 1910: 347 **A**: YUN
diversus Pic, 1920f: 15
vitalisi Vuillet, 1912: 297 **A**: YUN **ORR**
waldenfelsi De Lisle, 1973: 17

genus *Trictenotoma* Gray, 1832: 534 type species *Trictenotoma childreni* Gray, 1832
davidi H. Deyrolle, 1875: lx **A**: FUJ GUA GUX HAI HUN JIX SCH YUN
formosana Kriesche, 1919: 150 **A**: TAI
mniszechi H. Deyrolle, 1875: lx **A**: SD **ORR**
birmana Dohrn, 1882: 458

family PYTHIDAE Solier, 1834

genus *Pytho* Latreille, 1796: 23 type species *Cucujus coeruleus* Fabricius, 1787 (= *Tenebrio depressus* Linnaeus, 1767)
Enoptes Gistel, 1848: 10 [RN] type species *Cucujus coeruleus* Fabricius, 1792
Pytholus Rafinesque, 1815: 114 type species *Cucujus coeruleus* Fabricius, 1792
abieticola J. R. Sahlberg, 1875: 222 **E**: AU CZ FI GE NR NT PL SK SV **A**: ES FE
niger Reitter, 1911a: 414
depressus Linnaeus, 1767: 675 (*Tenebrio*) **E**: AU BH BY CT CZ DE EN FI FR GB GE GG HU IT KZ LA LI LT NL NR NT PL RO SK SP ST SV SZ UK **A**: ES FE WS
atriceps Pic, 1912j: 49
basipennis Pic, 1912k: 58
castaneus Fabricius, 1792b: 95 (*Cucujus*)

coeruleus Herbst, 1782: 6 (*Cucujus*)
coeruleus Fabricius, 1787a: 166 (*Cucujus*) [HN]
coeruleus Fabricius, 1792b: 94 (*Cucujus*) [HN]
festivus Fabricius, 1792b: 95 (*Cucujus*)
lignarius DeGeer, 1775: 38 (*Tenebrio*)
planus Herbst, 1799: 14 (*Cucujus*)
jezoensis Kôno, 1936c: 36 **A**: JA
kolwensis C. R. Sahlberg, 1833: 445 **E**: EN FI NT PL SV **A**: ES WS
sahlbergi Pic, 1912j: 49
nivalis Lewis, 1888: 109 **A**: FE JA

family Pythidae, genus incertae sedis
genus *Osphyoplesius* Winkler, 1915: 331 type species *Osphyoplesius anophthalmus* Winkler, 1915
anophthalmus Winkler, 1915: 333 **E**: UK (Krym)
loebli Español, 1975a: 107 **E**: GR (Epeiros)

family PYROCHROIDAE Latreille, 1807

subfamily **Tydessinae** Nikitsky, 1986

genus *Tydessa* Peacock, 1982: 362 type species *Dasytes constrictus* Lewis (= *Dasytes lewisi* Pic, 1937)
lewisi Pic, 1937c: 74 (*Dasytes*) [RN] **A**: FE JA
constrictus Lewis, 1895b: 118 (*Dasytes*) [HN]

subfamily **Pedilinae** Lacordaire, 1859

genus *Pedilus* Fischer von Waldheim, 1820: pl. 5 type species *Pedilus fuscus* Fischer von Waldheim, 1820
Corphyra Say, 1835: 189 type species *Anthicus lugubris* Say, 1827
Tosadendroides Kôno, 1935c: 158 type species *Tosadendroides okamotoi* Kôno, 1935
canescens Semenov, 1902a: 258 **E**: "Caucasus"
constricticollis Semenov, 1893e: 327 **A**: "Turkestan"
errans Faust, 1877b: 323 **E**: AR ST (Dagestan)
fuscus Fischer von Waldheim, 1820: pl. 5 **A**: "Altai"
dispar J. Sturm, 1843: 167
reitteri Pic, 1903b: 82
laevicollis Reitter, 1901d: 115 **E**: "South Ural" **A**: WS (Altayskiy Kray)
mongolicus Reitter, 1901d: 116 **A**: MG
okamotoi Kôno, 1935c: 157 (*Tosadendroides*) **A**: JA
polyxanthus Semenov, 1900e: 687 **A**: UZ
rubricollis Motschulsky, 1858d: 190 **E**: ST
rufipes Motschulsky, 1858d: 190 [RN] **A**: KI
fulvipes Motschulsky, 1845a: 82 [HN]
serricornis Semenov, 1902a: 260 **E**: ST (Dagestan) "Caucasus"
tibialis Semenov, 1893e: 327 **A**: KZ
pallidipes Pic, 1900a: 10
tristis Semenov, 1893e: 328 **A**: IN TM
lividipennis Reitter, 1901d: 114
pallidipennis Semenov, 1893e: 327
weberi Reitter, 1901d: 114 **E**: ST
willbergi Semenov, 1900e: 687 **A**: UZ
zubowskii Reitter, 1901d: 114
xanthopus Semenov, 1899b: 654 **A**: "Turkestan, Altai"
zubowskyi Semenov, 1900a: 319 **A**: UZ

subfamily **Pyrochroinae** Latreille, 1807

genus *Dendroides* Latreille, 1810: 212 type species *Dendroides canadensis* Latreille, 1810
lesnei Blair, 1914a: 314 **A**: JA
nakabusana Kôno, 1936b: 97

ussuriensis L. N. Medvedev, 1977: 48 **A**: FE NC NE SC

genus *Eupyrochroa* Blair, 1914a: 316 type species *Pyrochroa insignata* Fairmaire, 1894
insignata Fairmaire, 1894b: 220 (*Pyrochroa*) **A**: FUJ XIZ
rufoverticalis Pic, 1915c: 4
limbaticollis Pic, 1909r: 330 (*Pyrochroa*) **A**: JIX TAI
immaculata Pic, 1923b: 19 (*Pseudopyrochroa*)

genus *Frontodendroidopsis* Young, 2004b: 208 type species *Dendroides ocularis* Lewis, 1887
ocularis Lewis, 1887a: 174 (*Dendroides*) **A**: JA

genus *Hemidendroides* Ferrari, 1869: 198 type species *Dendroides ledereri* Ferrari, 1869
davidis Fairmaire, 1878a: 121 (*Pyrochhroa*[sic]) **A**: SCH

genus *Himalapyrochroa* Young, 2004a: **190** type species *Himalapyrochroa gibbosa* Young, 2004
gibbosa Young, 2004a: 191 **A**: SD

genus *Phyllocladus* Blair, 1914a: 315 type species *Dendroides magnificus* Blair, 1912
grandipennis Pic, 1906d: 29 (*Dendroides*) **A**: CH
kasantsevi Young, 2005b: 103 **A**: GAN SHA

genus *Pogonocerus* Fischer von Waldheim, 1812: 281 type species *Pogonocerus thoracicus* Fischer von Waldheim, 1812
thoracicus Fischer von Waldheim, 1812: 281 **E**: GG ST (Kavkaz)

genus *Pseudodendroides* Blair, 1914a: 314 type species *Dendroides niponensis* Lewis, 1887
amamiana Nakane, 1988a: 4 (*Pseudopyrochroa*) **A**: JA (Amami-Ôshima)
niponensis Lewis, 1887a: 174 (*Dendroides*) **A**: JA
nitidicollis Pic, 1955b: 12
ocularis Lewis, 1887a: 174 (*Dendroides*) **A**: JA
grandipennis Pic, 1906d: 29 (*Dendroides*) **A**: CH
umenoi Kôno, 1936a: 5 **A**: JA
uraiana Kôno, 1935c: 158 **A**: TAI

genus *Pyrochroa* Geoffroy, 1762: 338 type species *Cantharis coccinea* Linnaeus, 1760
coccinea Linnaeus, 1760: 202 (*Cantharis*) **E**: AU AR BE BH BY CR CT CZ DE EN FI FR GB GE GG GR HU IT LA LI LT RO SK SL ST SV NL NR NT PL PT RO SK SP ST SZ **N**: AG
coccinea Donovan, 1793: [51], pl. 56 [HN]
kabyliana Pic, 1898g: 98
purpurata Schrank von Paula, 1781: 173
rubra DeGeer, 1775: 20
daglariensis Young, 2004c: 933 [RN] **A**: SY
cyanipennis Pic, 1896j: 141 (*Dendroides*) [HN]
pubescens Pic, 1907g: 158 (*Pyrochroa*) **A**: SD
notaticeps Pic, 1912d: 18
rubricolor Pic, 1912d: 18
serraticornis kiesenwetteri Fairmaire, 1849: 424 **E**: IT (Sicilia, South Italy)
serraticornis serraticornis Scopoli, 1763: 42 (*Cantharis*) **E**: AU AR BE BH CZ DE EN FR HU GB GE IR IT LA LI LT NL PL SK SP ST SZ UK **A**: TR
purpurata O. F. Müller, 1788: 80
rubens Schaller, 1783: 301 (*Lampyris*)
ruberrima Geoffroy, 1785: 153
rutilans Voet, 1806: 72
satrapa Schrank von Paula, 1781: 174
tauricola Pic, 1912e: 26
subcostulata Fairmaire, 1891d: cii **A**: KA

genus *Pseudopyrochroa* Pic, 1906d: 28 type species *Pseudopyrochroa deplanata* Pic, 1906
Pyrochomima Pic, 1955b: 13 type species *Pyrochromima dentaticollis* Pic, 1955 (=*Pseudopyrochroa melanocephala* Blair, 1912)
acuminata Young, 2000: 249 **A**: TAI
atripennis Lewis, 1887a: 171 (*Pyrochroa*) **A**: JA
brevitarsis Lewis, 1887a: 170 (*Pyrochroa*) **A**: JA
cardoni Fairmaire, 1894a: 30 (*Pyrochroa*) **A**: IN **ORR**
carinifrons Kôno, 1929b: 68 **A**: TAI
choui Young, 2000: 250 **A**: TAI
costatipennis Pic, 1908l: 75 **A**: YUN
depressa Pic, 1914d: 17 **A**: TAI
taiwana Kôno, 1929b: 67
donckieri Pic, 1908l: 75 **A**: YUN
lyciformis Pic, 1909r: 330 (*Pyrochroa*)
episcopalis Lewis, 1895a: 432 (*Pyrochroa*) **A**: JA
facialis Fairmaire, 1891c: xx (*Pyrochroa*) **A**: HUB
fainanensis Pic, 1911k: 158 (*Pyrochroa*) **A**: TAI
flavilabris Blair, 1914a: 321 **A**: JA
gibbifrons Lewis, 1887a: 173 (*Schizotus*) **A**: JA
girardi Young, 2001: 274 **A**: SCH
harmandi Pic, 1912d: 18 (*Pyrochroa*) **A**: NP SD
nepalensis Paulus, 1972: 75
japonica Heyden, 1879: 88 (*Pyrochroa*) **A**: JA
higoniae Lewis, 1895a: 432 (*Pyrochroa*)
kazuoi kazuoi Ohbayashi, 1968: 35 **A**: JA
kazuoi teruhisai Saitô, 1990: 105 **A**: JA
klapperichi Pic, 1955c: 29 **A**: FUJ
lateraria Motschulsky, 1860c: 143 (*Pyrochroa*) **A**: FE NE "Korea"
latevittata Pic, 1939d: 135 **A**: ZHE
laticollis Lewis, 1887a: 169 (*Pyrochroa*) **A**: JA
aurita Lewis, 1887a: 173 (*Schizotus*)
lineaticollis Pic, 1914d: 17 **A**: NP XIZ
marginata Kôno, 1929b: 70 **A**: JA
melanocephala Blair, 1912: 527 **A**: SD
angustata Pic, 1955b: 12
dentaticollis Pic, 1955b: 13
obtusicristata Young, 2000: 252 **A**: FUJ TAI
peculiaris Lewis, 1887a: 170 (*Pyrochroa*) **A**: FE JA
punctifrons Young, 2000: 253 **A**: TAI
rufula Motschulsky, 1866: 173 (*Pyrochroa*) **A**: FE JA
aurita Lewis, 1887a: 173 (*Schizotus*)
etorofuensis Kôno, 1935a: 87
ogasawarai Kôno, 1929b: 69
rubricollis Lewis, 1887a: 172 (*Schizotus*)
satoi Young, 2003: 346 **A**: FUJ "Korea"
tumidifrons Young, 2000: 254 **A**: TAI
velutina Fairmaire, 1891c: xxi (*Pyrochroa*) **A**: HUB
vestiflua Lewis, 1887a: 169 (*Pyrochroa*) **A**: FE JA

genus *Schizotus* Newman, 1838: 374 type species *Schizotus cervicalis* Newman, 1838
Pyrochroella Reitter, 1911a: 385 type species *Pyrochroa pectinicornis* Linnaeus, 1758
cardinalis Mannerheim, 1853a: 302 (*Pyrochroa*) **A**: ES FE
innotaticeps Pic, 1904d: 26
fuscicollis Mannerheim, 1853a: 301 (*Pyrochroa*) **A**: ES FE
punctus Motschulsky, 1860c: 143

pectinicornis Linnaeus, 1758: 403 (*Cantharis*) **E**: AU BE BH BY CT CZ DE EN FI FR GE HU IR IT LA LI LT NL NR NT PL RO SK SP SV SZ
nigra DeGeer, 1775: 22 (*Pyrochroa*)
yamaguchii Kôno, 1936a: 5 **A**: TAI

genus *Sinodendroides* Young, 2005a: 94 type species *Sinodendroides chinensis* Young, 2005
chinensis Young, 2005a: 96 **A**: GAN SHA

family Pyrochoidae, subfamily incertae sedis

subfamily **Agnathinae** Lacordaire, 1859

genus *Agnathus* Germar, 1818: 232 type species *Notoxus decoratus* Germar, 1818
decoratus Germar, 1818: 229 (*Notoxus*) **E**: AU BH BY CR CZ FR GE GG IT PL RO SK ST SZ UK **A**: TR

family SALPINGIDAE Leach, 1815

subfamily **Othniinae** LeConte, 1861

genus *Elacatis* Pascoe, 1860a: 52 type species *Elacatis delusa* Pascoe, 1860
Othnius LeConte, 1861: 102 type species *Othnius umbrosus* LeConte, 1861
atrithorax Pic, 1920a: 9 (*Othnius*) **A**: JA TAI
bengalensis Sengupta, 1977: 537 **A**: AP SD
bhutanensis Sengupta, 1980: 28 **A**: BT
formosanus Borchmann, 1916: 106 (*Othnius*) **A**: TAI
kraatzii Reitter, 1879: 226 **A**: FE JA
ocularis Lewis, 1891b: 248 (*Othnius*) **A**: JA
rugicollis Borchmann, 1916: 108 (*Othnius*) **A**: TAI
similis Borchmann, 1916: 107 (*Othnius*) **A**: TAI

subfamily **Prostominiinae** Grouvelle, 1914

genus *Ocholissa* Pascoe, 1863b: 85 type species *Ocholissa laeta* Pascoe, 1863
harmandi Grouvelle, 1908b: 411 **A**: SD

genus *Prostominia* Reitter, 1889i: 315 type species *Prostominia lewisi* Reitter, 1889
lewisi Reitter, 1889i: 316 **A**: FE (Kunashir) JA

genus *Trogocryptoides* Champion, 1924a: 167 type species *Trogocryptoides brachyderus* Champion, 1924
brachyderus Champion, 1924a: 168 **A**: UP
shintaroi Sasaji, 1988a: 35 **A**: JA

subfamily **Agleninae** Horn, 1878

genus *Aglenus* Erichson, 1845: 285 type species *Hypophloeus brunneus* Gyllenhal, 1813
Monopis Dejean, 1835: 314 [HN]
brunneus Gyllenhal, 1813: 711 (*Hypophloeus*) **E**: AU AZ CR CZ DE FR GB GE GR HU IR IT NL PL PT SK SP ST SV SZ "Caucasus" **N**: EG CI LB MR **AFR NAR**
longior Pic, 1923g: 9
major Schaufuss, 1882a: 48
obsoletus Shuckard, 1839: 249
rosellae Ragusa, 1892a: 137 (*Anommatus*)
rugipennis Schaufuss, 1882a: 48

subfamily **Inopeplinae** Grouvelle, 1908

genus *Inopeplus* Smith, 1851: 4 [RN] type species *Ino picta* Laporte, 1835
Ino Laporte, 1835: 135 [HN] type species *Ino picta* Laporte, 1835
Euryplatus Motschulsky, 1860a: 95 type species *Euryplatus lateralis* Motschulsky, 1860
Pseudino Fairmaire, 1869b: 208 type species *Pseudino coquereli* Fairmaire, 1869

albonotatus Motschulsky, 1860a: 98 (*Euryplatus*) **A**: AP SD UP **ORR**
apatani Pal, 1992b: 347 **A**: AP
biocellatus Motschulsky, 1860a: 98 (*Euryplatus*) **A**: AP SD **ORR**
decisus Walker, 1858: 205 (*Prognatha*) **A**: AP SD UP **ORR**
distinctus Sengupta, Pal & Mukhopadhyay, 1977: 401 **A**: SD
jairajpurii Pal, 1992b: 344 **A**: AP
nitidus Sengupta, Pal & Mukhopadhyay, 1977: 400 **A**: AP SD
patkoicus Pal, 1992a: 238 **A**: AP
quadrinotatus Gorham, 1873: 258 (*Ino*) **A**: JA
sauteri Grouvelle, 1914: 58 **A**: TAI
syozoi Sasaji, 1986: 7 **A**: TAI
uenoi ohbayashii Hatta, 1988: 300 **A**: JA
uenoi uenoi Sasaji, 1984: 46 **A**: JA

genus *Uruminopeplus* Satô & Hatta, 1988: 85 type species *Uruminopeplus sakaii* Satô & Hatta, 1988
sakaii Satô & Hatta, 1988: 86 **A**: JA
monstrosus Sasaji, 1988: 41 (*Inopeplus*) **A**: JA

subfamily **Salpinginae** Leach, 1815

genus *Cariderus* Mulsant, 1859: 46 type species *Rhinosimus aeneus* A. G. Olivier, 1807
aeneus A. G. Olivier, 1807: 86: 455 (*Rhinosimus*) **E**: AU BY FR GE GR HU IT LA PL SZ **N**: AG
megricus Iablokoff-Khnzorian, 1956a: 135
numidicus Pic, 1904b: 10
ornithorrhynchus Abeille de Perrin, 1874b: 30

genus *Chilopeltis* Seidlitz, 1916: 424 type species *Chilopeltis insculpta* Seidlitz, 1916
Platylissodema Blair, 1919: 117 type species *Lanthanus rouyeri* Pic, 1914
laevipennis Marseul, 1876d: 331 (*Lissodema*) **A**: JA

genus *Colposis* Mulsant, 1859: 41 type species *Salpingus virescens* Mulsant, 1859
mutilatus Beck, 1817: 19 (*Salpingus*) **E**: AU BH CZ ES FR GE HU IT LA LT PL RO RU SK SZ YU **A**: FE "Siberia"
maritimus Pic, 1903k: 140 (*Salpingus*)
virescens Mulsant, 1859: 41 (*Salpingus*)

genus *Istrisia* Lewis, 1895a: 254 type species *Istrisia rufobrunnea* Lewis
rufobrunnea Lewis, 1895a: 254 **A**: FE JA

genus *Lissodema* Curtis, 1833: 187 type species *Lissodema heyana* Curtis, 1833 (= *Salpingus cursor* Gyllenhal, 1813)
Spinolissodema Pic, 1919b: 3 type species *Spinolissodema unifasciata* Pic, 1919
Stenolissodema Desbrochers des Loges, 1900: 6 type species *Salpingus lituratus* A. Costa, 1847
ainunum Lewis, 1895a: 255 **A**: JA
cursor Gyllenhal, 1813: 716 (*Salpingus*) **E**: AU CR CT CZ DE FI FR GB GE HU IT NL NR NT PL RO SK ST SV SZ "Caucasus" **A**: FE
heyanum Curtis, 1833: 187
kirkae Donisthorpe, 1925: 106
dentatum Lewis, 1895a: 257 **A**: FE JA
beatulum Lewis, 1895a: 257
denticolle Gyllenhal, 1813: 715 (*Salpingus*) **E**: AR AU BE BH BU CR CZ DE FR GB GE HU IR IT NL PL RO RU SK SP ST SV SZ "Caucasus"
gallicum Pic, 1933a: 2
humerale A. Villa & G. B. Villa, 1833: 35 (*Salpingus*)
quadriguttatum Le Peletier de Saint-Fargeau & Audinet-Serville, 1825: 288 (*Rhinosimus*)
quadrimaculatum Mulsant, 1830: 302 (*Rhinosimus*)
quadripustulatum Marsham, 1802: 297 (*Curculio*) [HN]

japonum Reitter, 1877g: 382 **A**: JA
lituratum A. Costa, 1847: 158 (*Salpingus*) **E**: FR IT **N**: AG TU
clouetii Guillebeau, 1896: 245
minutum Lewis, 1895a: 258 **A**: JA
morimotoi Sasaji, 1988a: 30 **A**: JA
munaguro Sasaji, 1988a: 28 **A**: JA
myrmido Marseul, 1876d: 332 **A**: JA
pictipenne Lewis, 1895a: 256 **A**: JA
plagiatum Lewis, 1895a: 255 **A**: FE JA
quadridentatum Champion, 1923: 79 **A**: UP
rosti Reitter, 1897d: 125 **E**: ST "Circassia"
teruhisai Sasaji, 1988: 27 **A**: JA
tomaroides Lewis, 1895a: 258 **A**: JA
uenoi Sasaji, 1988: 32 **A**: JA
unifasciatum Pic, 1919b: 3 **A**: JA
validicorne Lewis, 1895a: 258 **A**: JA

genus *Pseudorabocerus* Pic, 1903k: 140 type species *Salpingus lederi* Reitter, 1888
lederi Reitter, 1888b: 213 (*Salpingus*) **E**: ST "Circassia"

genus *Pseudosphaeriestes* Nikitsky, 1992c: 491 type species *Salpingus niponicus* Lewis, 1895
niponicus Lewis, 1895a: 255 (*Salpingus*) **A**: FE (Kunashir) JA

genus *Rabocerus* Mulsant, 1859: 43 type species *Salpingus foveolatus* Ljungh, 1823
foveolatus Ljungh, 1823: 269 (*Salpingus*) **E**: AU BE BY CT CZ DE EN ES FI FR GB GE HU IR IT LA LT NR NT PL RO SK SV SZ
bishopi Sharp, 1909: 245
championi Sharp, 1909: 246
impressithorax Pic, 1903k: 140 (*Salpingus*)
mutilatus Champion, 1886b: 160 (*Salpingus*)
gabrieli Gerhardt, 1901: 18 (*Salpingus*) **E**: AU BE BY CT CZ DE EN FI FR GB GE IR IT LA NR NT PL SK ST SV SZ

genus *Salpingus* Illiger, 1802a: 150 [= 1802b: 301] type species *Curculio roboris* Fabricius, 1787
Anthribus Schellenberg, 1798: 120 [HN] type species *Curculio ruficollis* Linnaeus, 1760
Rhinosimus Latreille, 1802: 192 type species *Curculio planirostris* Fabricius, 1787
caucasicus Reitter, 1905f: 312 **E**: AR ST "Caucasus" **A**: JA
depressifrons Nikitsky & Belov, 1983: 523 **A**: FE JA
morishimai Sasaji, 1987: 30 **A**: JA
planirostris Fabricius, 1787a: 119 (*Curculio*) **E**: BE BY CR CT DE EN FI FR GB GE HU IR IT LA LT NL NR NT PL PT RO SP ST SV "Caucasus" **N**: MO TU
fulvirostris Fabricius, 1787b: 381 (*Curculio*)
luteonitens Fairmaire, 1879c: 206
quadriimpressus Desbrochers des Loges, 1900: 22
spinolae A. Costa, 1847: 109 (*Rhinosimus*)
ruficollis Linnaeus, 1760: 177 (*Curculio*) **E**: AU BY CR CT CZ DE EN FI FR GE HU IR IT LA LT NL PL NR NT RO SK SL SP SV SZ
bicolor Stephens, 1831: 217
coeruleocephalus Latreille, 1804b: 28 (*Rhinosimus*)
roboris Fabricius, 1787a: 100 (*Curculio*)
rostratus DeGeer, 1775: 252 (*Curculio*)
viridicollis Pic, 1892c: 32 (*Rhinosimus*)
tapirus Abeille de Perrin, 1874b: 29 (*Rhinosimus*) **E**: IT **N**: AG TU
angustirostris Pic, 1898d: 35
elephas Chobaut, 1903a: 143
tapiroides Reitter, 1912b: 196

genus *Salpingoides* Nikitsky, 1987: 1851 type species *Salpingoides thoracicus* Nikitsky, 1987: 1852
thoracicus Nikitsky, 1987: 1852 **A**: FE

genus *Sphaeriestes* Stephens, 1831: 218 type species *Dermestes ater* Paykull, 1798 (= *Salpingus stockmanni* Biström, 1977)
Arabocerus Iablokoff-Khnzorian, 1985: 216 type species *Salpingus impressus* Wollaston, 1857
Salpingellus Reitter, 1911a: 415 type species *Sphaeriestes ater* Paykull, 1798
Salpingus Gyllenhal, 1810: 640 [RN, HN] type species *Dermestes ater* Paykull, 1798
Sphaeriesthes Schenkling, 1922: 194 [unjustified emendation]
Trichocolposinus Seidlitz, 1916: 489 type species *Vincenzellus hirtus* Broun, 1886
aeratus Mulsant, 1859: 34 (*Salpingus*) **E**: AU BH CT CZ GE GR HU IT PL SK ST SZ **N**: AG TU
cedri Pic, 1903k: 141 (*Salpingus*)
nitidus Chevrolat, 1860: 77 (*Salpingus*)
jacqueti Pic, 1903k: 140 (*Salpingus*)
bimaculatus Gyllenhal, 1810: 644 (*Salpingus*) **E**: CT CZ EN FI GE IT LA LT NR NT PL RU SV UK
castaneus Panzer, 1796a: 16: (*Notoxus*) **E**: AU BE BH BY CR CT CZ DE EN FI FR GB GE HU IR IT LA LT NL NR PL RU SK SL SP ST SV SZ "Caucasus"
achilleae Bonelli, 1812: 175 (*Salpingus*)
brunnescens Pic, 1892c: 32 (*Salpingus*)
immaculatus Stephens, 1831: 218
piceae Germar, 1824a: 9 (*Salpingus*)
exsanguis Abeille de Perrin, 1870: 89 (*Salpingus*) **E**: FR
impressus Wollaston, 1857a: 161 **N**: MR
reyi Abeille de Perrin, 1874b: 26 (*Salpingus*) **E**: DE FI FR GB GE GR (Kríti) IR IT NL NR PL SP SV **A**: CY SY
aeneus Stephens, 1835: 421
ater Stephens, 1831: 218 [HN]
fowleri Seidlitz, 1916: 467
palpalis Baudi di Selve, 1877a: 49
tetramerus Rey, 1892: 138
stockmanni Biström, 1977: 18 (*Salpingus*) **E**: AU BY CT CZ DE EN ES FI FR GB GE HU IR IT LA NL NR NT PL RU SK ST SV SZ UK **A**: ES FE WS **NAR**
ater Paykull, 1798: 298 (*Dermestes*) [HN]

genus *Vincenzellus* Reitter, 1911a: 418 type species *Rhinosimus viridipennis* Latreille, 1804 (= *Anthribus ruficollis* Panzer, 1794)
Colposinus Seidlitz, 1916: 390 type species *Rhinosimus viridipennis* Latreille, 1804
fascipennis Reitter, 1897d: 125 (*Rhinosimus*) **E**: ST "Caucasus"
ruficollis Panzer, 1794c: 19 (*Anthribus*) **E**: AU BE BH BU CR CZ DE FR GE HU IR IT NL PL RO SK SP SV UK YU
genei A. Costa, 1847: 89 (*Rhinosimus*)
roboris Fabricius, 1798: 161 (*Anthribus*)
ruficeps Bose, 1858: 96 (*Rhinosimus*)
viridipennis Latreille, 1804b: 26 (*Rhinosimus*)
vaulogeri Abeille de Perrin, 1901b: 234 (*Salpingus*) **N**: AG

subfamily **Aegialitinae** LeConte, 1862

genus *Aegialites* Mannerheim, 1853b: 178 type species *Aegialites debilis* Mannerheim, 1853
Eurystethes Seidlitz, 1916: 127 [RN] type species *Aegialites debilis* Mannerheim, 1853
Orygmus Gistel, 1848: xi [RN for *Aegialites* Dejean, nomen nudum]
antsiferovaensis Zerche, 2004: 47 **A**: FE (Kuril Is.)
beringensis Zerche, 2004: 40 **A**: FE (Bering Is.)
chirikotanensis Zerche, 2004: 75 **A**: FE (Kuril Is.: Chirinkotan)
ekarmaensis Zerche, 2004: 77 **A**: FE (Kuril Is.: Ekarma)
gracilis Zerche, 2004: 46 **A**: FE (Kuril Is.)
iturupensis Zerche, 2004: 88 **A**: FE (Kuril Is.: Iturup)

kharimkotanensis Zerche, 2004: 73 **A**: FE (Kuril Is.: Kharimkotan)
kunashorensis Zerche, 2004: 89 **A**: FE (Kuril Is.: Kunashir)
leleji Zerche, 2004: 71 **A**: FE (Kuril Is.: Onekotan)
matuaensis Zerche, 2004: 84 **A**: FE (Kuril Is.: Matua)
onekotanensis Zerche, 2004: 70 **A**: FE (Kuril Is.: Onekotan)
raikokensis Zerche, 2004: 82 **A**: FE (Kuril Is.: Raikoke)
shiashkotanensis Zerche, 2004: 80 **A**: FE (Kuril Is.: Shiashkotan)
shumshuensis Zerche, 2004: 44 **A**: FE (Kuril Is.)
stejnegeri Linell, 1898: 74 **A**: FE (Sachalin: Tjuleniy Ostrov)
sugiharai Kôno, 1938: 3 (*Aegialatis*) **A**: FE (Kuril Is.)
urupensis Zerche, 2004: 86 **A**: FE (Kuril Is.: Urup)

family ANTHICIDAE Latreille, 1819
subfamily **Anthicinae** Latreille, 1819

genus *Amblyderus* LaFerté-Sénectère, 1847: 368 [= 1849n: 62] type species *Anthicus scabricollis* LaFerté-Sénectère, 1847
Inamblyderus Pic, 1911i: 134 type species *Amblyderus obesus* Casey, 1895
brunneus Pic, 1893a: 15 **E**: IT
indicus Fairmaire, 1894a: 33 **A**: NP **ORR**
laxatus Bonadona, 1964: 228 **A**: SA **AFR**
longidentatus Pic, 1957: 436 **A**: YE
maculipennis Pic, 1898j: 181 **N**: EG **A**: SA YE **AFR**
sabulosus Pic, 1899i: 173 **N**: EG **A**: IS
scabricollis LaFerté-Sénectère, 1847: 368 (*Anthicus*) **E**: FR (Corse) IT PT SP **N**: AG MO TU
inhumeralis Pic, 1900j: 89
sulcithorax Pic, 1898e: 69 **A**: YE **AFR**
truncatus LaFerté-Sénectère, 1849e: 3[= 1849n: 64] **N**: EG **AFR**
tuberculatus Krekich-Strassoldo, 1928a: 88 **A**: UP **ORR**

genus *Anthelephila* Hope, 1833: 63 [= 1834b: 101] type species *Anthicus cyaneus* Hope, 1833
Formicoma Motschulsky, 1845a: 83 type species *Anthicus pedestris* Fabricius (= *Carabus pedestris* Rossi, 1790)
Formicomus LaFerté-Sénectère, 1849g: 1
Formicosoma Motschulsky, 1845c: [unnumbered page at end of 2nd issue] [RN]
Myrmecosoma Mannerheim, 1846: 227 [RN]
Orthauchen Krekich-Strassoldo, 1925: 516 type species *Formicomus aestimabilis* Krekich-Stassoldo, 1925
alfierii Pic, 1924c: 22 (*Formicomus*) **N**: EG
amaena LaFerté-Sénectère, 1849g: 12 [= 1849n: 81] (*Formicomus*) **N**: EG **A**: SA YE **AFR**
mellyi Pic, 1894f: 64 (*Formicomus*)
anastasei Pic, 1935c: 6 (*Formicomus*) [RN] **N**: EG
alfierii Pic, 1935c: 4 (*Formicomus*) [HN]
anceyi anceyi Pic, 1898d: 63 (*Formicomus*) **A**: YE **AFR**
anceyi rathjensi Pic, 1957: 440 (*Formicomus*) **A**: YE
ancorifera Telnov, 2003a: 130 **A**: SD
angustata Bonadona, 1964: 244 (*Formicomus*) **A**: AF IN
angustiformis Fairmaire, 1893e: 155 (*Formicomus*) **N**: EG **A**: IS SA SI **AFR**
filicollis Pic, 1898d: 63 (*Formicomus*)
animata Pic, 1903h: 119 (*Formicomus*) [RN] **A**: YUN **ORR**
annamita Pic, 1903f: 107 (*Formicomus*) [HN]
antiqua Krekich-Strassoldo, 1919a: 172 (*Formicomus*) **A**: KA NP PA UP **ORR**
iridipennis Krekich-Strassoldo, 1928a: 79 (*Formicomus*)
arabica Pic, 1899m: 264 (*Formicomus*) **A**: SA YE
armata Boheman, 1858: 103 (*Formicomus*) **A**: JA SE UP **ORR**
arunvallis Kejval, 2000a: 86 (*Formicomus*) **A**: NP
baeri Pic, 1903r: 645 (*Formicomus*) **A**: TAI **ORR**

formosana Pic, 1912a: 7 (*Formicomus*)
longispina Pic, 1912a: 7 (*Formicomus*)
spinosa Pic, 1929b: 14 (*Formicomus*)
vitalisi Pic, 1914d: 19 (*Formicomus*)

barbipes Krekich-Strassoldo, 1931: 9 (*Formicomus*) **A**: NP **ORR**
basalis Krekich-Strassoldo, 1931: 10 (*Formicomus*) **A**: NP UP **ORR**
basirufa Krekich-Strassoldo, 1928a: 76 (*Formicomus*) **A**: AP BT HP NP SD UP **ORR**
bejceki Kejval, 2002a: 110 **A**: YE (Suqutra)
bengalensis Wiedemann, 1823: 70 (*Anthicus*) **A**: JA **ORR**
bifurcata Kejval, 2000a: 94 (*Formicomus*) **A**: BT
bimaculatipennis Pic, 1939f: 146 (*Formicomus*) **A**: IS SA SI
biskrensis Fairmaire, 1883j: clvii (*Formicomus*) **N**: AG TU **AFR**
bispilifasciata bispilifasciata Pic, 1897n: 134 (*Formicomus*) **A**: SA YE **AFR**
bonadonai Kejval, 2000a: 95 (*Formicomus*) **A**: BT
bramina badia Krekich-Strassoldo, 1931: 8 (*Formicomus*) **A**: HP NP SD UP
bramina bramina LaFerté-Sénectère, 1849g: 10 [= 1849n: 79] (*Formicomus*) **A**: HUB JIA JIL PA TAI **ORR**
bramina candens Krekich-Strassoldo, 1931: 8 (*Formicomus*) **A**: JA SHA
bramina coiffaiti Bonadona, 1964: 243 (*Formicomus*) **A**: JA
bramina tokaraensis Nomura, 1962a: 3 (*Formicomus*) **A**: JA (Ryukyus)
brisouti Pic, 1892c: 33 (*Formicomus*) **N**: AG MO
bukharensis Kejval, 2000b: 408 (*Formicomus*) **A**: KZ UZ
caeruleipennis caeruleipennis LaFerté-Sénectère, 1847: 369 (*Anthicus*) **E**: SP **N**: AE AG CI EG LB MO TU **A**: IN IQ IS JO LE OM PA SA SI YE **AFR**

cyanoptera LaFerté-Sénectère, 1849g: 5 [= 1849n: 74] (*Formicomus*)

canaliculata LaFerté-Sénectère, 1849g: 22 [= 1849n: 90] (*Formicomus*) **E**: IT (Sicilia) **N**: AG EG MO TU **AFR**
cinchonae Krekich-Strassoldo, 1931: 2 (*Formicomus*) **A**: NP SD
coniceps Pic, 1913b: 13 (*Formicomus*) **A**: NP SD UP **ORR**
consul LaFerté-Sénectère, 1849g: 23 [= 1849n: 91] (*Formicomus*) **A**: GUA HAI NP **ORR**
corrusca Krekich-Strassoldo, 1931: 4 (*Formicomus*) **A**: UP SD
degener Kejval, 2006: 177 **A**: HUB SCH YUN **ORR**
fossicollis Kejval, 2002b: 270 **A**: YUN
fossulata Krekich-Strassoldo, 1928d: 7 (*Formicomus*) **A**: SD **ORR**
fuscipes Krekich-Strassoldo, 1931: 9 (*Formicomus*) **A**: TAI
gandaki Kejval, 2000a: 92 (*Formicomus*) **A**: NP
gladia Telnov, 2003a: 133 **A**: SD
gloriosa Krekich-Strassoldo, 1919a: 168 (*Formicomus*) **A**: GUI SCH YUN
gorkha Kejval, 2000a: 90 (*Formicomus*) **A**: NP
hauseri Pic, 1897g: 62 (*Formicomus*) **E**: AR ST **A**: AF IN TM UZ

sterbae Reitter, 1905c: 205 (*Formicomus*)

himalayana Krekich-Strassoldo, 1914b: 109 (*Formicomus*) **A**: BT HP NP SD UP **ORR**
imperatrix LaFerté-Sénectère, 1849f: 2 **A**: FE JA NP PA SE UP TAI **ORR**

cribriceps Marseul, 1876d: 459 (*Formicomus*)
formicaria Nietner, 1856a: 533 (*Anthicus*)
formicaria Nietner, 1857b: 383 (*Anthicus*) [HN]
imperator LaFerté-Sénectère, 1849n: 66 [HN]
myrmecodes Gemminger, 1870a: 123 (*Anthicus*) [RN]
punctaticeps Pic, 1916g: 13 (*Formicomus*)
quisquiliaria Nietner, 1857a: 22 (*Anthicus*) [RN]
ruficolor Pic, 1916a: 5 (*Formicomus*)

ionica LaFerté-Sénectère, 1849g: 12 [= 1849n: 81] (*Formicomus*) **E**: GR IT TR **N**: EG **A**: CY IN IS JO LE SA SY TR YE
jelineki Kejval, 2000b: 414 (*Formicomus*) **A**: IN
kaszabi Bonadona, 1964: 245 (*Formicomus*) **A**: AF UP
klapperichi Uhmann, 1988b: 399 (*Formicomus*) **A**: TAI
lacertosa Krekich-Strassoldo, 1931: 11 (*Formicomus*) **A**: NP SD UP **ORR**

lagenicollis Fairmaire, 1894a: 41 (*Formicomus*) **A**: BT NP SD
aestimabilis Krekich-Strassoldo, 1919a: 168 (*Formicomus*)
bhutanensis Pic, 1913t: 204 (*Formicomus*)
latipennis Pic, 1914i: 55 (*Formicomus*) **A**: NP SD
latro bicoloripes Pic, 1935c: 4 (*Formicomus*) **E**: SP
latro latro LaFerté-Sénectère, 1849g: 21 [= 1849n: 90] (*Formicomus*) **E**: CR IT (Sardegna, Sicilia) SP **N**: AG EG TU **A**: IN IS
lewisi Marseul, 1876d: 458 (*Formicomus*) **A**: JA TAI **ORR**
longiceps Pic, 1913t: 204 (*Formicomus*) **A**: BT HP NP SD UP
championi Krekich-Strassoldo, 1928a: 75 (*Formicomus*)
limaria Kejval, 2006: 181 **A**: NP UP
macilenta Bonadona, 1962: 349 (*Formicomus*) **A**: SA OM **AFR**
mandarina Boheman, 1858: 104 (*Formicomus*) **A**: HKG SCH **ORR**
multiformis Kejval, 2002a: 107 **A**: IN OM PA
mutillaria Saunders, 1834: 66 **A**: NP SD **ORR**
nadari Pic, 1893t: 175 (*Formicomus*) **N**: EG **A**: IS JO LE SI SY TR
sphinx Krekich-Strassoldo, 1919a: 171 (*Formicomus*)
nemrod LaFerté-Sénectère, 1849g: 6 [= 1849n: 75] (*Formicomus*) **A**: AF IN IQ TM UZ
phymopes Bonadona, 1964: 248 (*Formicomus*)
nepalensis Kejval, 2000a: 87 (*Formicomus*) **A**: NP SD
ninus LaFerté-Sénectère, 1849g: 11 [= 1849n: 79] (*Formicomus*) **E**: ST (Kavkaz) **A**: AF CY IN IQ IS LE SA SY TD TM TR UZ
nitescens Krekich-Strassoldo, 1931: 5 (*Formicomus*) **A**: HP KA SD UP
nuristanica Bonadona, 1964: 247 (*Formicomus*) **A**: AF KA PA
okinawana Nomura, 1962a: 2 (*Formicomus*) **A**: JA (Ryukyus)
panelii Hille, 1977: 199 (*Formicomus*) **A**: SA YE **AFR**
paolii Krekich-Strassoldo, 1928b: 384 (*Formicomus*) **A**: YE **AFR**
parallela Marseul, 1879a: 50 (*Formicomus*) **A**: IQ
pedestris Rossi, 1790: 224 (*Carabus*) **E**: AB AL AR AU BH BU CR CZ FR GE GG GR HU IT KZ LA MC PL PT RO SK SP ST SZ TR UK YU **N**: AG LB MO MR TU **A**: AF CY IN IS KI KZ TM TR UZ WS XIN
atratula Reitter, 1889a: 33 (*Formicomus*)
brevipilis Desbrochers des Loges, 1875: 42 (*Formicomus*)
cursor LaFerté-Sénectère, 1849g: 10 [= 1849n: 78] (*Formicomus*)
equestris Panzer, 1800: 8 (*Notoxus*)
fusca Geoffroy, 1799: 344 (*Cantharis*)
hispanica Motschulsky, 1849: 131 (*Formicoma*)
nobilis Faldermann, 1837: 107 (*Anthicus*)
rubida Reitter, 1878a: 249 (*Formicomus*)
sareptana Desbrochers des Loges, 1875: 42 (*Formicomus*)
thomsoni Pic, 1901j: 90 (*Formicomus*)
thoracica Panzer, 1794b: 6 (*Notoxus*)
tincta Reitter, 1889a: 33 (*Formicomus*)
pegasus Reitter, 1892a: 63 (*Formicomus*) **A**: KZ TD UZ
persica Kejval, 2000b: 415 (*Formicomus*) **A**: IN
pokharensis Kejval, 2000a: 91 (*Formicomus*) **A**: NP SD
praeses LaFerté-Sénectère, 1849g: 17 (*Formicomus*) **A**: SD **ORR**
nigrivestis Krekich-Strassoldo, 1931: 13 [= 1849n: 85] (*Formicomus*)
probsti Kejval, 2000a: 88 (*Formicomus*) **A**: NP
pseudocorrusca Kejval, 2000a: 85 (*Formicomus*) **A**: NP
psiloptera Krekich-Strassoldo, 1931: 1 (*Formicomus*) **A**: NP SD **ORR**
ruficollis Saunders, 1834: 65 **A**: CH FE JA NP SD **ORR**
cyanochroa Nomura, 1962a: 2
rufobasalis Pic, 1913a: 9 (*Formicomus*) **A**: HP KA NP PA SD UP **ORR**
montana Krekich-Strassoldo, 1914b: 111 (*Formicomus*)
sauteri Pic, 1912a: 6 (*Formicomus*) **A**: TAI
serdanga Marseul, 1884: 163 (*Formicomus*) **A**: YUN **ORR**

sikkimensis Pic, 1913a: 9 (*Formicomus*) **A**: AP BT HP NP SD UP **ORR**
montivaga Krekich-Strassoldo, 1928a: 80 (*Formicomus*)
simoni Pic, 1893u: xxxiii (*Formicomus*) **A**: YE
sinica Kejval, 2002b: 260 **A**: YUN
soror Kejval, 2002a: 95 **A**: IN PA
spiniventris Krekich-Strassoldo, 1914c: 226 (*Formicomus*) **A**: SD **ORR**
strandi Krekich-Strassoldo, 1931: 3 (*Formicomus*) **A**: PA SD UP
surata Baudi di Selve, 1879: 87 (*Formicomus*) **Ni**: MO
taliana Pic, 1913a: 9 (*Formicomus*) **A**: YUN **ORR**
kippenbergi Uhmann, 1978: 81 (*Formicomus*)
truncatipennis Pic, 1897g: 61 (*Formicomus*) **A**: TD
uhmanni Kejval, 2000a: 93 (*Formicomus*) **A**: NP
umbratilis Krekich-Strassoldo, 1928a: 73 (*Formicomus*) **A**: BT HP NP SD UP
uncinata Desbrochers des Loges, 1875: 42 (*Formicomus*) **A**: "Syria"
viridipennis lindbergi Bonadona, 1960a: 56 (*Formicomus*) **A**: AF IN
viridipennis viridipennis Krekich-Strassoldo, 1931: 7 (*Formicomus*) **A**: UP **ORR**
volkonskyi Pic, 1942d: 77 **N**: AG
walkeri Pic, 1895m: cxci (*Formicomus*) **A**: SA YE
subelongata Pic, 1899m: 264 (*Formicomus*)

genus *Anthicomorphus* Lewis, 1895a: 428 type species *Anthicomorphus suturalis* Lewis, 1895
subgenus *Anthicomorphus* Lewis, 1895a: 428 type species *Anthicomorphus suturalis* Lewis, 1895
cruralis Lewis, 1895a: 429 **A**: JA JIL SC TAI
niponicus amamiensis Nomura, 1962a: 3 **A**: JA (Ryukyus)
niponicus niponicus Lewis, 1895a: 429 **A**: JA JIL SC
puberulus Marseul, 1877: 467 (*Anthicus*) **A**: JA
suturalis Lewis, 1895a: 428 **A**: JA **ORR**

genus *Anthicus* Paykull, 1798: 253 type species *Meloe antherinus* Linnaeus, 1760
Birricollis Marseul, 1879a: 67 type species *Anthicus genei* LaFerté-Sénectère, 1849
Birricomus Pic, 1894c: 43 [RN] type species *Anthicus genei* LaFerté-Sénectère, 1849
Brevicollis Marseul, 1879a: 67 type species *Meloe antherinus* Linnaeus, 1760
Brevicomus Pic, 1894c: 46 [RN] type species *Meloe antherinus* Linnaeus, 1760
Cartolus Mulsant & Rey, 1866a: 104 [= 1866b: 192] type species *Notoxus sellatus* Panzer, 1796
Curticollis Pic, 1892d: 44 type species *Anthicus trotommidens* Pic, 1892
Curticomus Pic, 1894g: 76 [RN] type species *Anthicus trotommidens* Pic, 1892
Eonius C. G. Thomson, 1864: 366 type species *Notoxus bimaculatus* Illiger, 1801
Nathicus Casey, 1895: 655 type species *Nathicus virginiae* Casey, 1895
Nodolinus Mulsant & Rey, 1866a: 83 [= 1866b: 171] type species *Anthicus schmidtii* Rosenhauer, 1847
Platylorus Mulsant & Rey, 1866a: 84 [= 1866b: 172] type species *Notoxus bimaculatus* Illiger, 1801
Pubicollis Marseul, 1879a: 67 type species *Anthicus axillaris* W. L. E. Schmidt, 1842
Pubicomus Pic, 1894c: 48 [RN] type species *Anthicus axillaris* W. L. E. Schmidt, 1842
Recticollis Marseul, 1879a: 66 type species *Anthicus inderiensis* Marseul, 1879
abashirensis Nakane, 1988a: 5 **A**: JA
abyssinicus adenensis Pic, 1957: 442 **A**: YE
aemulus Krekich-Strassoldo, 1914a: 11 **A**: SHA ZHE
alaiensis Pic, 1906b: 19 **A**: AF TM
anceyi Pic, 1895c: 59 **A**: SA **AFR**
angustatus Curtis, 1838: 714 **E**: FR GB IR PT **N**: CI
antherinus antherinus Linnaeus, 1760: 228 (*Meloe*) **E**: AB AL AU AZ BE BH BU BY CR CT CZ DE EN FI FR GB GE GG GR HU IR IT KZ LA LT MC MD NL NR NT PL PT RO SK SL SP ST SV SZ TR UK YU **A**: AF CY ES FE IN IS KI KZ LE SA SY TM TR UZ WS
astrachanicus Csiki, 1901a: 113
cinctellus Rossi, 1792: 46 (*Notoxus*)
semitestaceus Pic, 1892e: 43
tripustulatus Fabricius, 1792a: 169 (*Cicindela*)

valens Pic, 1896m: 179
antherinus syriae Pic, 1892e: 43 **E**: FR **N**: "northern Africa" **A**: AF "Syria"
arabicus Pic, 1894s: cclxxvii [RN] **A**: IS SA TR
laticollis Marseul, 1879a: 186 [HN]
arisi Pic, 1905k: 186 **A**: MG TM UZ
armaticornis L. N. Medvedev, 1974: 152 **A**: MG
armatus Truqui, 1855: 359 **N**: EG **A**: CY IQ IS JO SA SI SY
heluanensis Pic, 1899i: 175
lessepsi Marseul, 1868a: 192
armiger LaFerté-Sénectère, 1849i: 12 [= 1849n: 113] **A**: UP **ORR**
ater Thunberg, 1787: 45 (*Notoxus*) **E**: AR AU BY CT CZ DE EN FI FR GE GR HU IT KZ LA LT NR NT PL PT SK SP ST SV SZ TR UK **A**: ES FE IQ IS KI KZ MG WS
ater Panzer, 1796a: 15 (*Notoxus*) [HN]
axillaris W. L. E. Schmidt, 1842c: 186 **E**: AB AL AU BH BU BY CR CT CZ DE EN FA FI GE GG GR HU IT LA LT MC NR NT PL RO SK ST SV SZ UK YU **A**: HEI KZ ES FE SHA TR WS
inflatus Kolenati, 1846: 36
varians Kolenati, 1846: 36
bajulus LaFerté-Sénectère, 1849i: 13 [= 1849n: 114] **A**: SD **ORR**
balachanus Pic, 1906h: 49 **E**: ST **A**: AF TM TR UZ
rosinae Pic, 1913n: 146
basithorax Pic, 1941b: 3 **E**: GR (Kríti)
bigoti Bonadona, 1988: 42 **N**: MO
biguttatus LaFerté-Sénectère, 1849k: 119 [= 1849n: 218] **E**: FR (Corse) GR IT SP
bilunulatus Pic, 1893j: 79 **A**: IS "Syria"
bimaculatus Illiger, 1801: 80 (*Notoxus*) **E**: AU BE BY CT CZ DE EN FI FR GB GE HU IT LA LT NL NR NT PL RO ST SV SZ UK **A**: ES FE KZ TR WS
dauricus Motschulsky, 1845a: 83
fasciatus Schilsky, 1888: 190 [HN]
pallens Schilsky, 1888: 190 [HN]
pallescens Pic, 1894g: 78 [RN]
sagitta Krynicki, 1829: 196
schilskyi Pic, 1894d: 56 [RN]
bruleriei Desbrochers des Loges, 1875: 43 **N**: EG **AFR**
honestus Marseul, 1878b: 50
brunneipennis Pic, 1896j: 141 **E**: FR (Corse) IT (Sardegna)
brunneus LaFerté-Sénectère, 1842: 249 **E**: FR IT PT SP **N**: AG CI MO MR TU
bousaadae Pic, 1893i: 74
lucasianus Desbrochers des Loges, 1899: 191 [RN]
quadrimaculatus Lucas, 1843: 146
buyssoni Pic, 1899f: 77 **A**: NP SD **ORR**
catalanus Bonadona, 1953: 264 **E**: FR IT
coiffaiti Bonadona, 1952: 236 **A**: LE
commaculatus Krekich-Strassoldo, 1931: 36 **A**: HP UP **ORR**
consimilis Krekich-Strassoldo, 1928a: 98 **A**: NP SD UP **ORR**
constellatus Krekich-Strassoldo, 1928c: 109 **E**: BH BU GR MC **A**: TR
crassicollis Pic, 1894c: 49 **N**: AG
cribripennis Desbrochers des Loges, 1875: 45 **E**: IT PT SP **N**: AG MO TU
maurus Marseul, 1879a: 160
crinitulus Pic, 1901k: 90 **A**: AF TM UZ
crinitus LaFerté-Sénectère, 1849k: 105 [= 1849n: 204] **E**: AB AL AR GR MA ST TR **N**: AG AZ CI EG LB MO MR TU **A**: AE AF AP CY HP IN IQ IS JA JO KU NP OM PA SA SD SI SY TAI TR UP UZ XIZ YE **AFR NARi NTRi ORR**
communimacula Fairmaire, 1896a: 47
flavisternus Marseul, 1879a: 149
laevaticeps Marseul, 1879a: 245 [RN]
laeviceps Marseul, 1879a: 105 [HN]
longipennis Desbrochers des Loges, 1875: 45 [HN]

manillanus Pic, 1903r: 646
uninotatus Pic, 1903s: 350
curvitarsis Krekich-Strassoldo, 1928a: 101 **A**: UP **ORR**
cylindricus Pic, 1899b: 20 **E**: RO
czernshorskyi Pic, 1912g: 33 **Ei**: IT **A**: SHN
demaisoni biuninotatus Pic, 1939f: 146 **N**: EG
demaisoni demaisoni Pic, 1893c: 27 **N**: EG
difformis Marseul, 1879a: 178 **E**: BU UK **A**: TR
dilatatus Krekich-Strassoldo, 1914b: 112 **A**: KI KZ TM
dimidiatipennis Desbrochers des Loges, 1875: 46 **N**: EG **A**: YE **AFR**
diversepunctatus Pic, 1913j: 132 **A**: TAI
diversicornis Pic, 1907o: 195 **A**: AF HP KA NP SD UP
diversus Marseul, 1879a: 174 **E**: AB AR ST **A**: ES FE IN KZ TR WS
drurei Pic, 1901d: 37 **A**: IQ
dzhungaricus L. N. Medvedev & Micheetchev, 1979: 142 **A**: MG
elegans Steven, 1806: 161 [= 1811: 124] [ND] **E**: ST (Kavkaz)
escorialensis Pic, 1893x: clxi [Pic, 1893q: 135] **E**: SP
euphraticus LaFerté-Sénectère, 1849j: 72 [= 1849n: 172] **A**: IN IQ IS JO SY
externus Pic, 1899f: 78 **A**: NP SD
fenestratus W. L. E. Schmidt, 1842c: 181 **E**: AL BU CR FR GR HU IT MA PT SP ST TR "Caucasus" **N**: AG CI EG LB MO TU **A**: IS LE SY TR
fauconneti Pic, 1928a: 2
nigricans Pic, 1898g: 98
submaculatus Pic, 1898g: 98
flavipes altaicus L. N. Medvedev, 1975: 92 **A**: FE MG
flavipes flavipes Panzer, 1796f: 22 (*Notoxus*) **E**: AL AU BE BH BU BY CR CT CZ DE EN FI FR GB GE GR HU IR IT LA LT MC MD NL NR NT PL RO SK SP ST SV SZ TR UK "Caucasus" **A**: ES FE TR WS
flavescens Pic, 1899k: 193
nigriceps Mannerheim, 1843b: 97
obscurus Küster, 1848a: 73
rufipes Paykull, 1800: 444
scoticus Rye, 1872: 10
fortepunctatus fortepunctatus Pic, 1892b: 18 **N**: AG **A**: IS **AFR**
fortiterpunctatus Pic, 1892b: 54 [RN]
fumosus Lucas, 1843: 146 **E**: BU FR (Corse) IT (Sardegna, Sicilia) **N**: AG CI MO TU **A**: TR
bicolor Lucas, 1843: 146
lethierryensis Pic, 1955a: 1 [RN]
lethierryi Pic, 1916a: 5 [HN]
sardiniensis Pic, 1928a: 2
fuscicornis LaFerté-Sénectère, 1849k: 116 [= 1849n: 215] **E**: FR PT SP **N**: AG EG LB MO TU
barbarus Pic, 1893h: 64
gebleri Mäklin, 1881: 46 **A**: ES WS
genei LaFerté-Sénectère, 1849k: 120 [= 1849n: 219] **E**: AL CR FR GR IT PT RO SP TR YU **N**: AG CI EG LB TU **A**: CY TR
gobiensis Pic, 1917i: 256 **A**: MG NO NW
gordeevae L. N. Medvedev, 1974: 154 **A**: MG
grandiceps Pic, 1902b: 10 **A**: KZ TM WS
gratiosus Pic, 1896m: 179 **E**: GR "Caucasus" **A**: KZ TR UZ
auliensis Pic, 1900i: 78
gravidulus Krekich-Strassoldo, *1931*: 22 **A**: UP **ORR**
guttifer Wollaston, 1864: 522 **N**: CI
hamai Nomura, 1964: 55 **A**: JA (Ryukyus)
hamicornis Marseul, 1880: 418 **E**: IT (Sardegna) PT SP **N**: AG LB TU **A**: JO SA **AFR**
bedeli Pic, 1892n: cxcvii
hissarensis Pic, 1912h: 41 **A**: TD UZ
hoggarensis Pic, 1929h: 126 **N**: AG

humeralis Gebler, 1841b: 596 **A**: KZ NC WS
impressicollis Chobaut, 1924a: 104 **N**: MO TU
inaequaliceps inaequaliceps Pic, 1911c: 61 **A**: YE **AFR**
incrassicornis Pic, 1913q: 164
inaequalis Marseul, 1879a: 173 **E**: AR GG RO ST UK **A**: KZ TD
inderiensis Marseul, 1879a: 78 **E**: ST **A**: KZ ES
inhamicornis Pic, 1942c: 10 **N**: MO
inimicus Bonadona, 1964: 230 **A**: MG TD TM XIN
insperatus Pic, 1906g: 46 **A**: KZ TM
invreai Koch, 1933: 151 **E**: BU CR FR GR IT SP YU **A**: LE TR
irregularis Pic, 1932e: 18 **N**: AG
jacobsoni Pic, 1903d: 89 **A**: IN KZ TM
karelinii Zubkov, 1833: 332 [ND] **A**: TM
korbi Pic, 1902b: 10 **E**: AB AR GG ST **A**: IN IQ SY TD TM TR UZ
laetus L. N. Medvedev, 1974: 153 **A**: MG
laeviceps Baudi di Selve, 1877d: 688 **E**: BH BU CR FR GR HU IT MA PT SP SZ YU "Caucasus" **N**: AG CI MO TU **A**: JO SY TR
cruciferus Ragusa, 1898: 231
lucidipes Pic, 1892h: 102
lucidithorax Pic, 1896c: 39
rummeli Pic, 1893s: 161
lapidosus Wollaston, 1864: 521 **N**: CI
latebasalis Pic, 1932f: 25 **A**: MG
latedecoratus Pic, 1892k: 140 **E**: AB **A**: IN TR
astrabadensis Pic, 1905e: 113
karsantiana Pic, 1908d: 43
latefasciatus Desbrochers des Loges, 1875: 46 **E**: ST
latens Bonadona, 1977b: 86 **N**: AG MO TU **A**: TR
lepaumi Pic, 1893c: 28 **N**: EG **A**: "Syria"
lepidus Marseul, 1876d: 462 **A**: "Inde boreale"
leveillei Pic, 1893u: xxxiii **E**: FR (Corse) IT (Sardegna, Sicilia)
libanicus Desbrochers des Loges, 1875: 44 **A**: LE SY
lii Uhmann, 1997: 21 (*Pseudoleptaleus*) **A**: HEB
lividipes Desbrochers des Loges, 1875: 47 **A**: LE SY
lubbockii Wollaston, 1857a: 166 **N**: CI MR

luteicornis W. L. E. Schmidt, 1842c: 187 **E**: AL AU CR CZ DE FR GE IT MC PL PT SK SL SP SZ UK YU
picicornis Rey, 1892: 101
luteipes Marseul, 1879a: 159 **A**: IS JO LE SY TR
luteobinotatus Pic, 1957: 442 **A**: YE
maderi Heberdey, 1938: 163 **A**: YUN
magnus Pic, 1914f: 35 **A**: PA **ORR**
mariaeantoniae Chobaut, 1894c: ccxxviii **N**: AG
mellyi Pic, 1894f: 65 **N**: EG **A**: OM
minutissimus Pic, 1894c: 43 **N**: MO
moka Marseul, 1878b: 51 **A**: SA
monstrator Telnov, 2005: 173 **A**: SCH
monstrosiceps Pic, 1912h: 41 **A**: TM
monstrosicornis Marseul, 1877: 465 **A**: AF HUB JA MG NP SD TAI **ORR**
inapicalis Pic, 1913j: 131
muehlei Uhmann, 1988a: 2 **A**: YE
niger A. G. Olivier, 1811: 397 (*Notoxus*) **E**: AL BH BU CR FR GR IT MC PT RO SP ST SZ TR UK YU **N**: AG TU **A**: CY IN IQ IS JO LE SY TR **AFR**
differens Pic, 1896m: 179
morio LaFerté-Sénectère, 1849k: 116 [= 1849n: 215]

paulobrunnescens Pic, 1951a: 10
romanus Pic, 1951a: 10
perileptoides Lewis, 1895a: 431 **A**: JA LIA
pexicollis Fairmaire, 1894a: 34 **A**: SD UP **ORR**
trabeatus Fairmaire, 1896a: 47
pilosus Marseul, 1877: 466 **A**: FE JA
plucheryi Pic, 1951c: 15 [RN] **N**: MO
testaceonitidus Pic, 1939b: 21 [HN]
protensus Marseul, 1877: 470 **A**: JA
proximus Marseul, 1879a: 179 **E**: BU GR IT TR **A**: CY IS OM TR
pseudoruficeps Nardi, 2004: 141 [RN] **A**: TM
uniformis Pic, 1938d: 13 [HN]
pullulus Bonadona, 1964a: 233 **A**: AF ES
punctiger Krekich-Strassoldo, 1931: 23 **A**: NP SD UP YUN
quadridecoratus Abeille de Perrin, 1885: 160 **E**: FR (Corse) IT (Sardegna)
quadrioculatus LaFerté-Sénectère, 1849k: 102 [= 1849n: 201] [RN] **E**: AU FR IT GR HU PT SK SP ST SZ YU **A**: AF KI KZ TR
pedemontanus Pic, 1901g: 66
quadriguttatus Waltl, 1835: 75 [HN]
rectus Bonadona, 1970: 379 **A**: IN
reymondi Pic, 1951d: 37 **N**: AG
rubens Krekich-Strassoldo, 1931: 20 **A**: AP HP NP PA UP YUN
rubripes Pic, 1894o: 206 **A**: AF
ruficeps Pic, 1913n: 146 **A**: TM
rufivestis Marseul, 1879a: 108 **E**: FR (Corse) IT (Sardegna) **N**: AG
grenieri Pic, 1910a: 325 [= 1910n: 58]
rufoaeneus Pic, 1915c: 10 **A**: JIA
sabulosus Marseul, 1879a: 172 **E**: AB GG MD RO **A**: ES IN UZ
schmidtii Rosenhauer, 1847: 35 **E**: AL AR AU BU CR CZ FR GG GR HU IT MD PL PT RO SK SP ST SZ TR UK **A**: AF KZ TR
subfasciatus LaFerté-Sénectère, 1849j: 79 [= 1849n: 179]
subobliteratus Pic, 1899b: 19
unipunctatus LaFerté-Sénectère, 1849j: 80 [= 1849n: 180]
scurrula Truqui, 1855: 362 **E**: GR **A**: CY TR "Syria"
brenskei Pic, 1893z: cclxx
sellatus Panzer, 1796f: 20 (*Notoxus*) **E**: AU BU BY CR CT CZ DE EN FI FR GE GR HU IT LA LT NL NR NT PL PT RO SK SP SV SZ UK
semicupreus Pic, 1893o: 126 **E**: ST
sexmaculatus Gebler, 1829a: 137 **A**: ES TM WS
sillemi Heberdey, 1935: 306 **A**: AF XIN
simplex Marseul, 1879a: 156 **A**: ES WS
sinaiticus Schatzmayr & Koch, 1934a: 235 **A**: SI
sivaschensis Blinstein, 1988: 76 **E**: UK
sollicitus LaFerté-Sénectère, 1849k: 98 [= 1849n: 197] **E**: AB **A**: AF IN IQ SY TD TR
spadiceus Bonadona, 1958: 287 **E**: SP **N**: MO
staudingeri Pic, 1893v: liv **A**: AF TM UZ
leonardi Pic, 1913n: 146
steppensis Marseul, 1879a: 151 **E**: ST **A**: AF IN KI KZ UZ
stygius LaFerté-Sénectère, 1849j: 61 [= 1849n: 161] **A**: YE **AFR**
subaeneus LaFerté-Sénectère, 1847: 378 [ND] **E**: SP
subcrucifer subcrucifer Pic, 1900i: 78 **A**: AF KZ UZ
subcrucifer kaszabi Bonadona, 1964: 231 **A**: AF
subtilis Gistel, 1831: 304 [ND] **E**: GR
suturadepressa Pic, 1892l: 219 **N**: AG
suturalis simillimus Krekich-Strassoldo, 1928a: 97 **A**: NP UP

suturalis suturalis LaFerté-Sénectère, 1849j: 81 [= 1849n: 181] **A**: NP PA UP **ORR**
takaosus Pic, 1913i: 130 **A**: TAI **ORR**
testaceicolor subsuturalis Pic, 1913j: 133 **A**: TAI
testaceicolor testaceicolor Pic, 1913j: 133 **A**: TAI
theryi Pic, 1892i: 115 [= 1893f: 53] **E**: IT (Sardegna) **N**: AG EG MO TU **A**: JO
batnensis Pic, 1892l: 217
mayeti Chobaut, 1898a: 84
thyreocephalus Solsky, 1866: 93 **E**: AR GG ST **A**: AF IN IQ KZ MG TM
tristis schaumii Wollaston, 1857b: 5721 **E**: FR GB PT SP **N**: MO
ater Stephens, 1832: 74 [HN]
tristis tristis W. L. E. Schmidt, 1842c: 172 **E**: AB AR BU CR FR GG GR IT MA RO SK SP ST SV TR UK **N**: AG CI EG LB MO TU **A**: AF CY IN IQ IS JO KZ LE SA SY TD TM TR UZ YE
discrepans Krekich-Strassoldo, 1928c: 106
fenestratus LaFerté-Sénectère, 1847: 375 [HN]
nigroapicalis Pic, 1913n: 145
sericeus LaFerté-Sénectère, 1847: 375
subhumeralis Pic, 1913n: 146
tristiculus Reitter, 1884b: 92
tristissimus Chobaut, 1923b: 130
truquii Desbrochers des Loges, 1875: 44
trotommidens Pic, 1892d: 44 **N**: AG
turgidipes Krekich-Strassoldo, 1928a: 99 **A**: NP PA **ORR**
turkestanicus Pic, 1893r: 155 **E**: ST **A**: AF KZ MG PA TM UZ
apicenotatus Pic, 1913o: 154
cinctipennis Pic, 1913n: 146
umbrinus LaFerté-Sénectère, 1849k: 115 [= 1849n: 214] **E**: CT FA FI LA NT SV SZ **A**: ES FE IN KZ MG NC SC TIA TM WS
setulosus Boheman, 1851: 220
urganensis Pic, 1896m: 181 **A**: AF BEI ES FE HEI KZ MG NMO WS
dachuricus Csiki, 1901a: 114
shishmareffi Csiki, 1901a: 114
vagesignatus Pic, 1940a: 1 **E**: "Caucasus"
vicarius Telnov, 2005: 177 **A**: SCH YUN
yemenitus Nardi, 2004: 138 [RN] **A**: YE
scotti Pic, 1957: 448 [HN]
yunnanus Pic, 1915c: 9 **A**: SCH YUN

subgenus *Odacanthicus* Sakai & Ohbayashi, 1994: 556 type species *Anthicus watarasensis* Sakai & Ohbayashi, 1994
watarasensis nazarovi Lafer, 1996: 40 **A**: FE
watarasensis watarasensis Sakai & Ohbayashi, 1994: 558 **A**: JA

genus *Aulacoderus* LaFerté-Sénectère, 1849l: 133 [= 1849n: 232] type species *Anthicus transversalis* LaFerté-Sénectère, 1849 (= *Anthicus mutatus* Gemminger, 1870)
Biscissicollis Marseul, 1879a: 68 type species *Anthicus friwaldszkyi* LaFerté-Sénectère, 1849
alitienensis Pic, 1897c: 27 (*Anthicus*) **A**: YE **AFR**
canariensis bucheti Pic, 1903a: 21 (*Anthicus*) **N**: CI
canariensis canariensis Wollaston, 1864: 523 (*Anthicus*) **N**: CI
coarcticollis Baudi di Selve, 1881: 295 (*Anthicus*) **A**: AF IN IS IQ JO LE SA SY TR
ferrantei Peyerimhoff, 1943: 16 (*Anthicus*) **A**: EG
friwaldszkyi LaFerté-Sénectère, 1849l: 168 [= 1849n: 267] (*Anthicus*) **E**: AU CR GR HU
funebris Reitter, 1884e: 257 (*Tomoderus*) **E**: AL CR GR
distinctipes Pic, 1901n: 180 (*Anthicus*)
sulcatus Pic, 1897c: 27 (*Anthicus*)
ghanensis Hille, 1984: 103 (*Anthicus*) **A**: YE **AFR**
maderae Bonadona, 1963: 105 **N**: MR

mecheriensis Chobaut, 1896b: 376 (*Anthicus*) **N**: AG MO
mohammedis Marseul, 1879a: 232 (*Anthicus*) **A**: OM SA YE **AFR**
peyroni Pic, 1900i: 78 (*Anthicus*) **A**: JO OM SA SI SY **AFR**
appli Pic, 1900i: 78 (*Anthicus*)
breveapicalis Pic, 1951g: 136
picheyrei Peyerimhoff, 1943: 16 (*Anthicus*) **N**: AG **A**: IS OM YE **AFR**
manzer Bonadona, 1964: 242
scydmaenoides Wollaston, 1864: 524 (*Anthicus*) **N**: CI
sefrensis Pic, 1894u: lxxviii (*Anthicus*) **N**: AG EG **A**: SI **AFR**
postnotatus Pic, 1910l: 39 (*Anthicus*)
zizyphi Pic, 1902h: 65 (*Anthicus*)
sulcicollis Pic, 1895j: 786 (*Anthicus*) **E**: AF AR **A**: IN IQ JO
sulcifer Pic, 1893v: liv (*Anthicus*) **N**: AG
miniaticollis Chobaut, 1898a: 85 (*Anthicus*)
sulcithorax melitensis Pic, 1903j: 139 (*Anthicus*) **E**: IT (Sicilia) MA
sulcithorax pallidior Pic, 1941a: 5 (*Anthicus*) **A**: IS
sulcithorax sulcithorax Desbrochers des Loges, 1875: 46 (*Anthicus*) **E**: GR (Kríti) **A**: JO IS LE
blechroides Baudi di Selve, 1883b: 150 (*Anthicus*)
nigrithorax Pic, 1897c: 27 (*Anthicus*)
vaulogeri Pic, 1897r: 297 (*Anthicus*) **N**: AG MO

genus *Clavicollis* Marseul, 1879a: 66 type species *Anthicus longiceps* LaFerté-Sénectère, 1849
Clavicomus Pic, 1894g: 70 [RN]
Pseudanthicus Desbrochers des Loges, 1868: 80 type species *Formicomus olivierii* Desbrochers des Loges, 1868 [precedence suppressed]
absconditus Telnov, 2000: 175 (*Clavicomus*) **A**: NP
adustus Krekich-Strassoldo, 1931: 34 (*Anthicus*) **A**: NP SD UP **ORR**
almorae almorae Krekich-Strassoldo, 1931: 33 (*Anthicus*) **A**: HP SD UP
almorae paradoxus Krekich-Strassoldo, 1931: 33 (*Anthicus*) **A**: HP UP
ambustus Krekich-Strassoldo, 1931: 33 (*Anthicus*) **A**: HP NP SD **ORR**
angulifer Pic, 1893z: cclxx (*Anthicus*) **A**: IS "Syria"
anomalus Telnov, 1998: 97 (*Clavicomus*) **A**: SCH
apicordiger Bonadona, 1954b: 273 (*Anthicus*) **N**: MO
aquatilis Krekich-Strassoldo, 1931: 30 (*Anthicus*) **A**: NP SD
assamensis Pic, 1907q: 338 (*Anthicus*) **A**: NP **ORR**
assequens Krekich-Strassoldo, 1931: 35 (*Anthicus*) **A**: HUB NC **ORR**
atratus Krekich-Strassoldo, 1931: 32 (*Anthicus*) **A**: BT NP SD UP
austriacus Pic, 1901g: 66 (*Anthicus*) **E**: AU
balazuci Bonadona, 1986: 35 (*Clavicomus*) **E**: HU
bicarinifrons Pic, 1892c: 33 (*Anthicus*) **N**: AG
pygidiolongus Pic, 1892k: 140 (*Anthicus*)
biguttatus Bonadona, 1964: 239 (*Clavicomus*) **A**: AF IN KA PA
brevipilis Pic, 1893y: ccxi (*Anthicus*) [RN] **N**: AG
pilosus Pic, 1892e: 43 (*Anthicus*) [HN]
bruckii Kiesenwetter, 1870: 147 (*Anthicus*) **E**: SP **N**: AG
brunneipes brunneipes Krekich-Strassoldo, 1931: 28 (*Anthicus*) **A**: HP NP SD UP **ORR**
brunneipes comparabilis Krekich-Strassoldo, 1931: 29 (*Anthicus*) **A**: UP
callimus Baudi di Selve, 1877d: 700 [= 1878a: 13] (*Anthicus*) **E**: SP
comes Krekich-Strassoldo, 1931: 28 (*Anthicus*) **A**: BT SD UP **ORR**
cordatus Krekich-Strassoldo, 1931: 27 (*Anthicus*) **A**: NP SD UP **ORR**
desertus Marseul, 1879a: 134 (*Anthicus*) **A**: ES
dichrous LaFerté-Sénectère, 1849l: 135 [= 1849n: 234] (*Anthicus*) **E**: IT **N**: AG
lacunosus Pic, 1901g: 66 (*Anthicus*)
doderoi Pic, 1902c: 19 (*Anthicus*) **E**: IT (Sardegna)
erythroderus Marseul, 1878c: 55 (*Anthicus*) **N**: AG TU
obscurans Pic, 1892c: 33 (*Anthicus*)

subcyaneus Pic, 1899b: 19 (*Anthicus*)
fugax LaFerté-Sénectère, 1849l: 136 [= 1849n: 235] (*Anthicus*) **A**: SD **ORR**
fugiens Marseul, 1876d: 463 (*Anthicus*) **A**: CH FE JA
gigas Pic, 1899g: 86 (*Anthicus*) **E**: GR TR **A**: TR
rhodius Pic, 1899g: 86 (*Anthicus*)
gravidus Krekich-Strassoldo, 1931: 25 (*Anthicus*) **A**: NP SD UP
harmandi Pic, 1899f: 78 (*Anthicus*) **A**: NP SD
hauseri Pic, 1906h: 49 (*Anthicus*) **A**: MG XIZ
henoni Pic, 1892h: 103 (*Anthicus*) **E**: SP **N**: AG MO
quinquenotatus Reitter, 1906a: 442 (*Anthicus*) [RN]
v-notatus Pic, 1899b: 20 (*Anthicus*)
heydeni Marseul, 1879a: 132 (*Anthicus*) **E**: PT SP
himalayanus Pic, 1909c: 57 (*Anthicus*) **A**: HP NP
hummeli Pic, 1933b: 10 (*Anthicus*) **A**: XIN
immaculipennis Krekich-Strassoldo, 1931: 29 (*Anthicus*) **A**: HP NP PA SD UP
inabsolutus Telnov, 2003b: 281 (*Clavicomus*) **A**: SD
indeprensus Telnov, 2000: 177 (*Clavicomus*) **A**: NP
kabylianus Pic, 1896l: 167 (*Anthicus*) **N**: AG TU
kocheri Pic, 1951f: 81 (*Anthicus*) **N**: MO
kuluensis Pic, 1914d: 20 (*Anthicus*) **A**: HP NP
laevipennis Marseul, 1877: 471 (*Anthicus*) **A**: JA LIA
lepidulus Marseul, 1876d: 461 (*Anthicus*) **A**: JA
longiceps LaFerté-Sénectère, 1849l: 134 [= 1849n: 233] (*Anthicus*) **E**: IT (Sicilia) SP **N**: AG TU
longicornis Uhmann, 1983: 203 (*Clavicomus*) **A**: SD **ORR**
manifestus Pic, 1907d: 135 (*Anthicus*) **A**: TAI YUN
martinezi Pic, 1932e: 18 (*Anthicus*) **E**: SP
mouzafferi Pic, 1910m: 42 (*Anthicus*) **A**: IN
muguensis Telnov, 2000: 178 (*Clavicomus*) **A**: NP
nigrocyanellus Marseul, 1877: 470 (*Anthicus*) **A**: FE FUJ GAN GUA JA LIA NMO SC XIN YUN ZHE **ORR**
nigrofuscus Telnov, 2000: 179 (*Clavicomus*) **A**: NP
nigroterminatus Pic, 1909d: 106 (*Anthicus*) **E**: SP
notatipennis Pic, 1909c: 56 (*Anthicus*) **A**: HP NP SD
obscuricolor Pic, 1909c: 57 (*Anthicus*)
olivierii Desbrochers des Loges, 1868: 79 (*Formicomus*) **N**: AG TU
optabilis LaFerté-Sénectère, 1849k: 88 [= 1849n: 187] (*Anthicus*) **E**: FR
paganettii Pic, 1909f: 114 (*Anthicus*) **E**: IT
piceus LaFerté-Sénectère, 1849l: 136 [= 1849n: 235] (*Anthicus*) **A**: NP SCH SD **ORR**
plagiostolus Bonadona, 1958: 288 (*Cordicomus*) **N**: MO
plicatipennis Pic, 1936e: 29 (*Anthicus*) **N**: MO
posthumus Krekich-Strassoldo, 1931: 32 (*Anthicus*) **A**: SD UP
postimpressus Pic, 1938b: 5 (*Anthicus*) **N**: MO
postluteofasciatus Pic, 1938b: 5 (*Anthicus*) **E**: SP
protervus Krekich-Strassoldo, 1931: 30 (*Anthicus*) **A**: SD UP
ragusae Pic, 1898m: 229 (*Anthicus*) **E**: IT (Sicilia)
schrammi Pic, 1913n: 146 (*Anthicus*) **E**: PT SP
semidepressus Pic, 1893h: 64 (*Anthicus*) **N**: AG
separatithorax Pic, 1914q: 184 (*Anthicus*) **A**: NP SD UP
shibatai Nomura, 1962a: 3 (*Anthicus*) **A**: JA (Ryukyus)
sikkimensis Pic, 1907o: 196 (*Anthicus*) **A**: BT PA SD
sinensis Pic, 1907d: 135 (*Anthicus*) **A**: SE TAI YUN
spinipennis Pic, 1898g: 99 (*Anthicus*) **N**: AG
sporadicus Krekich-Strassoldo, 1931: 26 (*Anthicus*) **A**: NP SD UP
striaticollis Krekich-Strassoldo, 1931: 26 (*Anthicus*) **A**: NP SD UP
subpiceus Pic, 1914f: 36 (*Anthicus*) **A**: "Himalaya"
tersus Krekich-Strassoldo, 1931: 31 (*Anthicus*) **A**: HP NP SD UP

turgidus Krekich-Strassoldo, 1928a: 100 (*Anthicus*) **A**: SD
uhagoni Pic, 1904g: 81 (*Anthicus*) **E**: SP
uniformis Krekich-Strassoldo, 1931: 27 (*Anthicus*) **A**: NP SD UP
variabilis Telnov, 2003b: 282 (*Clavicomus*) **A**: NP
versicolor Kiesenwetter, 1866: 383 (*Anthicus*) **E**: SP
weigeli Telnov, 2000: 180 (*Clavicomus*) **A**: NP
wuyishanensis Nardi, 2004: 137 [RN] **A**: FUJ
plicatipennis Pic, 1954b: 64 (*Anthicus*) [HN]

genus *Cordicollis* Marseul, 1879a: 66 type species *Anthicus instabilis* W. L. E. Schmidt, 1842
Bitumicollis Marseul, 1879a: 66 type species *Anthicus turca* Marseul, 1879
Cordicomus Pic, 1894c: 45 [RN]
Laticollis Marseul, 1879a: 67 type species *Anthicus baicalicus* Mulsant & Rey, 1866
Laticomus Pic, 1894g: 71 [RN]
baicalicus Mulsant & Rey, 1866a: 114 [= 1866b: 202] (*Anthicus*) **E**: RU **A**: BEI ES FE HEI JA JIL LIA NC MG WS
niponicus Lewis, 1895a: 431 (*Anthicus*)
usitatus Pic, 1896m: 179 (*Anthicus*)
collaciculus Marseul, 1879a: 113 (*Anthicus*) **N**: AG MO
cyclodiniformis Nardi, 2004: 140 [RN] **N**: MO
subopaculus Pic, 1936d: 25 (*Anthicus*) [HN]
galloisi Pic, 1910f: 20 (*Anthicus*) **A**: JA
gracilior Abeille de Perrin, 1885: 159 (*Anthicus*) **E**: AB BU CR FR GG GR HU IT MC RO SP ST TR **N**: EG **A**: KZ TM
auliatanus Pic, 1940a: 1 (*Anthicus*)
gracilis Panzer, 1796f: 21 (*Notoxus*) **E**: AB AU BU CR CZ FR GE GR HU IT KZ NR PL RO SK SL ST SV SZ UK **A**: TR WS **N**: AG
atricollis Abeille de Perrin, 1885: 160 (*Anthicus*)
lateripunctatus J. Sturm, 1826: 70 (*Anthicus*)
sinuatefasciatus Pic, 1940a: 1 (*Anthicus*)
stevenii LaFerté-Sénectère, 1849j: 69 [= 1849n: 169] (*Anthicus*)
haasi Pic, 1900j: 89 (*Anthicus*) **A**: AF TM UZ
infasciata Pic, 1908f: 58 (*Anthicus*)
inconspicuus alternans Krekich-Strassoldo, 1931: 21 (*Anthicus*) **A**: UP
inconspicuus inconspicuus Krekich-Strassoldo, 1931: 21 (*Anthicus*) **A**: HP SD UP YUN
infitialis Pic, 1894u: lxxvii (*Anthicus*) **A**: HEB MG SC
infuscatus LaFerté-Sénectère, 1849j: 82 [= 1849n: 182] (*Anthicus*) **N**: EG
insolubilis Pic, 1906b: 19 (*Anthicus*) **A**: TM UZ
instabilis franzi Bonadona, 1954b: 274 (*Cordicomus*) **E**: SP
instabilis geminipilis Desbrochers des Loges, 1875: 44 (*Anthicus*) **E**: SP
instabilis instabilis W. L. E. Schmidt, 1842c: 184 (*Anthicus*) **E**: AL BU CR DE FR GB GE GR IT MA NL NR PL PT SP ST SV SZ TR UK **N**: AG AZi CI EG LB MO MR TU **A**: IN IS JO TR
agilis Küster, 1849a: 75 (*Anthicus*)
deslogesi Pic, 1892e: 43 (*Anthicus*)
instabilis LaFerté-Sénectère, 1842: 259 (*Anthicus*) [HN]
mauritanicus Lucas, 1843: 146 (*Anthicus*)
olcesei Pic, 1894f: 65 (*Anthicus*)
puberulus Rey, 1892: 90 (*Anthicus*) [HN]
quittardi Pic, 1892o: 140 (*Anthicus*)
reyi Pic, 1893y: ccxii(*Anthicus*) [RN]
semiruber Pic, 1894c: 45 (*Anthicus*)
stabilis Pic, 1892k: 102 (*Anthicus*)
talaris Schiödte, 1893: 309 (*Anthicus*)
tibialis Curtis, 1838: 714 (*Anthicus*) [HN]
joannis Pic, 1901k: 90 (*Anthicus*) **A**: IQ
kuthyi Csiki, 1901a: 113 (*Anthicus*) **A**: ES MG NMO
laetipennis Krekich-Strassoldo, 1931: 20 (*Anthicus*) **A**: SD UP

lamprinulus Krekich-Strassoldo, 1931: 21 (*Anthicus*) **A**: PA SD UP **ORR**
litoralis Wollaston, 1854: 535 (*Anthicus*) **N**: CI MR
margelanicus Pic, 1893r: 155 (*Anthicus*) **E**: AB ST UK **A**: AF IN IQ KI KZ MG SY TM TR UZ
aralensis Pic, 1901c: 34 (*Anthicus*)
caucasicus Pic, 1893t: 175 (*Anthicus*)
tournieri Pic, 1894f: 65 (*Anthicus*)
medius Pic, 1905m: 493 (*Anthicus*) **A**: ES MG WS
muelleri Nardi, 2004: 138 [RN] **E**: CR RO UK
rufescens J. Müller, 1908: 316 (*Anthicus*) [HN]
nankineus Pic, 1913a: 12 (*Anthicus*) **A**: HEB JIA
nigronotatus Pic, 1894n: 182 (*Anthicus*) **A**: NP **ORR**
notoxoides Wollaston, 1864: 520 (*Anthicus*) **N**: CI
opaculus opaculus Wollaston, 1864: 519 (*Anthicus*) **E**: MA PT SP **N**: AG CI LB MA MO TU
mozabita Chobaut, 1898a: 84 (*Anthicus*)
semibrunneus Pic, 1892l: 216 (*Anthicus*)
opaculus wittmeri Schatzmayr & Koch, 1934a: 224 (*Anthicus*) **N**: EG LB **A**: IS
posticatus Pic, 1895i: 667 (*Anthicus*) **N**: EG LB **A**: IN SA YE **AFR**
posticus Motschulsky, 1849: 131 (*Anthicus*) **E**: SP
rubiginosus Krekich-Strassoldo, 1931: 21 (*Anthicus*) **A**: NP SD UP **ORR**
sabuleti LaFerté-Sénectère, 1849j: 64 [= 1849n: 164] (*Anthicus*) **N**: AG EG TU
harenosus Krekich-Strassoldo, 1913a: 223 (*Anthicus*)
saharensis Chobaut, 1896b: 375 (*Anthicus*) **N**: AG LB **AFR**
scapularis LaFerté-Sénectère, 1849k: 125 [= 1849n: 224] (*Anthicus*) **A**: ES FE HEI MG NC SC WS
solers LaFerté-Sénectère, 1849j: 80 [= 1849n: 180] (*Anthicus*) **A**: IQ UZ
subnitidus Pic, 1902b: 9 (*Anthicus*) **A**: KZ TM
tedjensis Pic, 1913i: 129 (*Anthicus*) **E**: ST **A**: IN TM UZ
trigonocephalus LaFerté-Sénectère, 1849i: 45 [= 1849n: 146] (*Anthicus*) **A**: NP **ORR**
turca Marseul, 1879a: 99 (*Anthicus*) **E**: TR
velatus Marseul, 1879a: 115 (*Anthicus*) **N**: AG EG LB **A**: IN PA SA YE **AFR**
mauritanicus Pic, 1952c: 492 (*Anthicus*)

genus *Cyclodinus* Mulsant & Rey, 1866a: 77 [= 1866b: 165] type species *Anthicus humilis* Germar, 1824
Lagenicollis Marseul, 1879a: 66 type species *Anthicus humilis* Germar, 1824
Spinicornus Krekich-Strassoldo, 1919b: 65 type species *Anthicus beckeri* Desbrochers des Loges, 1875 (= *Anthicus humilis* Germar, 1824)
Spiniferus Pic, 1911: 33 type species *Anthicus cerastes* Truqui, 1855
Thicanus Casey, 1895: 670 type species *Anthicus annectans* LeConte, 1851
angustulus Pic, 1892b: 18 (*Anthicus*) **E**: ST "Caucasus" **A**: AF
ataensis Pic, 1901c: 34 (*Anthicus*) **E**: ST **A**: KZ TM
basanicus Sahlberg, 1913a: 57 (*Anthicus*) **A**: SY
basanicus Krekich-Strassoldo, 1913a: 227 (*Anthicus*) [HN]
bicarinula Marseul, 1879a: 86 (*Anthicus*) **E**: GR **A**: CY IS JO TR
kraatzi Pic, 1893z: cclxviii (*Anthicus*)
biplicatulus Reitter, 1889b: 119 (*Anthicus*) **N**: EG **A**: AB IN MG SY TM UZ
latetestaceus Pic, 1917i: 256 (*Anthicus*)
biskrensis Pic, 1893b: 26 (*Anthicus*) **N**: AG TU
blandulus blandulus Baudi di Selve, 1877d: 677 (*Anthicus*) **E**: GR IT MA SP **N**: EG MO TU **AFR**
blandulus zinrami Koch, 1934b: 122 (*Anthicus*) **E**: GR SP
bremei LaFerté-Sénectère, 1842: 252 (*Anthicus*) **E**: AL AR FR GG GR IT PT RO SP ST TR YU **N**: AG EG LB MO TU **A**: AF CY IN IQ IS JO KI KU LE PA SI TD TM TR UZ **AFR**
impressithorax Pic, 1903q: 304 (*Anthicus*) [RN]
inapicalis Pic, 1919g: 67 (*Anthicus*)
latenotatus Pic, 1893o: 128 (*Anthicus*)
nigrolineatus Pic, 1892b: 26 (*Anthicus*)
riparius LaFerté-Sénectère, 1842: 254 (*Anthicus*)
stricticollis Fairmaire, 1903b: 217 (*Anthicus*) [HN]
subconiceps Pic, 1955e: 1043 (*Anthicus*)

subustulatus Pic, 1894a: 16 (*Anthicus*)
variabilis LaFerté-Sénectère, 1842: 253 (*Anthicus*)
brivioi Bucciarelli, 1962: 115 **E**: IT (Sardegna)
brunnescens Pic, 1893o: 137 (*Anthicus*) **A**: IN IQ IS SY TR
cerastes Truqui, 1855: 355 (*Anthicus*) **E**: AB GR ST TR **A**: AF CY IQ IN IS TM TR
corcyreus Pic, 1900i: 78 (*Anthicus*)
ciliatipes Pic, 1916a: 5 (*Anthicus*) **N**: TU
coniceps Marseul, 1879a: 83 (*Anthicus*) **E**: AL AU BU CR FR GR IT MA PT RO SL SP ST UK **N**: AG EG LB MO TU **A**: CY IS TR
femoralis Marseul, 1879a: 82 (*Anthicus*) [HN]
imitator Krekich-Strassoldo, 1919b: 61 (*Anthicus*)
lagunarum Krekich-Strassoldo, 1919b: 62 (*Anthicus*)
magistri Gozis, 1882: 203 (*Anthicus*) [RN]
marseulianus Desbrochers des Loges, 1899: 171 (*Anthicus*) [RN]
parvulus Krekich-Strassoldo, 1919b: 61 (*Anthicus*)
schatzmayri Koch, 1931: 68 (*Anthicus*)
sedjoumiensis Koch, 1931: 67 (*Anthicus*)
subopacicollis Pic, 1915l: 37 (*Anthicus*)
coniger Krekich-Strassoldo, 1927: 116 (*Anthicus*) **A**: SHN
constrictus constrictus Curtis, 1838: 714 (*Anthicus*) **E**: AB AL BU CR FR GB GE GR IR IT MA MC NR PT SL SP ST TR UK **N**: AG CI EG MO SI TU **A**: JO TR
albionis Krekich-Strassoldo, 1919b: 71 (*Anthicus*)
lameyi Marseul, 1879a: 88 (*Anthicus*)
marinus Krekich-Strassoldo, 1919b: 70 (*Anthicus*)
maroccanus Pic, 1893z: cclxix (*Anthicus*)
ruffoi Bucciarelli, 1959: 185
constrictus subconvexus Rey, 1892: 89 (*Anthicus*) **E**: FR
croissandeaui croissandeaui Pic, 1893z: cclxxi (*Anthicus*) **E**: FR IT MA **N**: AG MO TU **A**: TR **AFR**
vaucheri Chobaut, 1895a: 156 (*Anthicus*)
dagestanicus Pic, 1919d: 10 (*Anthicus*) **E**: ST (Kavkaz)
debilis LaFerté-Sénectère, 1849i: 28 [= 1849n: 129] (*Anthicus*) **E**: AB IT (Pantelleria) MA MC ST **N**: AG CI EG LB MO TU **A**: AE AF CY IN IQ IS JO LE KA KU KZ LB PA SA SI TD TM TR UZ YE **AFR ORR**
phoxus Marseul, 1879a: 87 (*Anthicus*)
wegeneri Pic, 1899i: 174 (*Anthicus*)
densifer Pic, 1900i: 78 (*Anthicus*) **A**: KZ TM
denticulatus Pic, 1900i: 78 (*Anthicus*)
dentaticornis Pic, 1898g: 99 (*Anthicus*) **A**: ST TM UZ
dentatus dentatus Pic, 1895b: 52 (*Anthicus*) **E**: AB ST UK **A**: KZ TM TR
dentatus transdanubianus Roubal, 1927: 116 (*Anthicus*) **E**: HU RO SK
deplanatus Desbrochers des Loges, 1875: 45 (*Anthicus*) **A**: SY
desbrochersi Pic, 1893m: 115 (*Anthicus*) [RN] **E**: PT SP **N**: AG MO
anguliceps Abeille de Perrin, 1885: 159 (*Anthicus*) [HN]
caroli Pic, 1893o: 128 (*Anthicus*)
dimidiatus Wollaston, 1864: 521 (*Anthicus*) **N**: CI **AFR**
doriae Pic, 1895j: 784 (*Anthicus*) **A**: IN IQ
episcopalis Pic, 1903n: 170 (*Anthicus*) **A**: IQ
insignatus Pic, 1904e: 57 (*Anthicus*)
erro Truqui, 1855: 348 (*Anthicus*) **E**: GR **N**: EG **A**: CY IS JO LE YE
lysholmi Pic, 1899i: 174 (*Anthicus*)
fatuus Truqui, 1855: 350 (*Anthicus*) **E**: BU GR MD ST UK **N**: EG **A**: AF CY IN IQ IS JO LE TM TR
dilutius Truqui, 1855: 350 (*Anthicus*)
forticornis Pic, 1893o: 128 (*Anthicus*) **E**: GR TR **A**: TR
hammami hammami Pic, 1892d: 43 (*Anthicus*) **N**: AG EG TU
scutelloniger Pic, 1892d: 43 (*Anthicus*)
hammami viberti Pic, 1906j: 284 (*Anthicus*) **N**: TU
humilis Germar, 1824a: 6 (*Anthicus*) **E**: AB AU AZ BU CR CZ DE FR GE GR HU IT MA MC MD NR PL RO SK ST SV UK **N**: AZi EG **A**: CY IS KZ LE SE TM TR

adriaticus Krekich-Strassoldo, 1919b: 68 (*Anthicus*)
beckeri Desbrochers des Loges, 1875: 43 (*Anthicus*)
depilis Rey, 1892: 89 (*Anthicus*)
detritus Rey, 1892: 89 (*Anthicus*)
fuscicrus Rey, 1892: 89 (*Anthicus*)
latinus Krekich-Strassoldo, 1919b: 69 (*Anthicus*)
leukoranus Pic, 1913n: 145 (*Anthicus*)
lucidulus LaFerté-Sénectère, 1849i: 29 [= 1849n: 130] (*Anthicus*)
nigrinus Zetterstedt, 1838: 159 (*Anthicus*)
nigrofasciatus Truqui, 1855: 354 (*Anthicus*)
orientalis Krekich-Strassoldo, 1919b: 69 (*Anthicus*)
peranxius Rey, 1892: 89 (*Anthicus*)
siciliae Pic, 1893z: cclxx (*Anthicus*)
incomptus Truqui, 1855: 350 (*Anthicus*) **E**: GR (Corfu) "Caucasus" **N**: EG **A**: CY IN IS LE TR
italicus Pic, 1901n: 178 (*Anthicus*) **E**: IT
kryzhanovskii Blinstein, 1988: 75 **E**: UK
lajoyei Pic, 1915c: 5 (*Anthicus*) **A**: HEB
larvipennis Marseul, 1879a: 85 (*Anthicus*) **E**: FR GR IT MA TR **N**: AG EG LB TU **A**: CY IQ IN SI TM
graecus Pic, 1896m: 178 (*Anthicus*)
motschulskyi Pic, 1894s: cclxxvii (*Anthicus*)
punicus Krekich-Strassoldo, 1919b: 73 (*Anthicus*)
testaceipes Pic, 1892l: 214 (*Anthicus*)
latior Pic, 1895l: cxxix (*Anthicus*) **N**: AG TU
latithorax Pic, 1893i: 74 (*Anthicus*) **A**: TR
lindbergi lindbergi Bonadona, 1964: 235 **A**: AF TM
lindbergi litatus Bonadona, 1964: 236 **A**: AF
longipilis C. Brisout de Barneville, 1863: 89 (*Anthicus*) **E**: CR FR GR RU SP YU **N**: AG MO TU
punctatithorax Pic, 1913o: 153 (*Anthicus*)
lotus Marseul, 1879a: 98 (*Anthicus*) **E**: ST **A**: IN KZ UZ
lucidicollis Marseul, 1879a: 89 (*Anthicus*) **E**: IT (Sardegna) SP **N**: AG TU
maltzevi Blinstein, 1988: 74 **E**: UK
mediobrunneus Pic, 1893o: 128 (*Anthicus*) **E**: ST
minutus minutus LaFerté-Sénectère, 1842: 255 (*Anthicus*) **E**: BU CR FR GR IT MA PT SP "Caucasus" **N**: AG TU **A**: TR
algiriensis Pic, 1893o: 126 (*Anthicus*)
lateralis Küster, 1849a: 77 (*Anthicus*)
sardous W. L. E. Schmidt, 1842c: 175 (*Anthicus*)
minutus truncaticeps Koch, 1934b: 121 (*Anthicus*) **E**: GR SP **N**: MO
misoloughii Pic, 1893o: 127 (*Anthicus*) **E**: GR
moltonii Bucciarelli, 1961: 272 **E**: CR (Lošinj)
mongolensis L. N. Medvedev, 1974: 153 (*Anthicus*) **A**: MG
montandoni Pic, 1909n: 171 (*Anthicus*) **E**: KZ MD RO ST UK
morawitzi Desbrochers des Loges, 1875: 43 (*Anthicus*) **E**: AB RO ST **A**: KZ MG TD UZ
detectus Pic, 1896m: 179 (*Anthicus*)
moricei Pic, 1899i: 174 (*Anthicus*) **A**: IS
nitidior Pic, 1893z: cclxix (*Anthicus*) **E**: AB **A**: AF IQ IN MG TM UZ WS
bagdatensis Pic, 1901k: 90 (*Anthicus*)
raffrayi Pic, 1894t: xi (*Anthicus*) **A**: NE NO
obscurior Pic, 1894t: xi (*Anthicus*)
reitteri Pic, 1892b: 18 (*Anthicus*) **E**: AB ST **A**: AF ES SY TM UZ WS
reductus Brancsik, 1899: 102 (*Anthicus*) [HM]
rhagis Marseul, 1879a: 98 (*Anthicus*) **A**: IQ SA SY
roberti Pic, 1892l: 213 (*Anthicus*) **N**: AG
salinus carthageniensis Pic, 1892n: cxcvii (*Anthicus*) **E**: IT (Sardegna) SP
rosai Bucciarelli, 1978: 158
subopaculus Pic, 1919g: 68 (*Anthicus*)
salinus salinus Crotch, 1867b: 439 (*Anthicus*) **E**: FR GB IT PT SP

chobauti Puel, 1919: 52 (*Anthicus*) [HN]
crotchi Pic, 1893m: 115 (*Anthicus*) [RN]
doctoris Pic, 1919g: 66 (*Anthicus*) [RN]
sociabilis Krekich-Strassoldo, 1919b: 67 (*Anthicus*)
sareptanus Pic, 1893p: 126 (*Anthicus*) **E**: GR ST "Caucasus"
sauteri Pic, 1913j: 131 (*Anthicus*) **A**: TAI
semiopacus Reitter, 1887e: 524 (*Anthicus*) **E**: ST **A**: AF IN KZ TD TM TR UZ
cerastoides Reitter, 1891a: 30 (*Anthicus*)
sibiricus Pic, 1893o: 128 (*Anthicus*) **A**: WS
subarmatus Pic, 1899c: 30 (*Anthicus*) **A**: AF IN KZ PA UZ
subbrunneus Pic, 1913n: 145 (*Anthicus*) **A**: IN
thessalius Marseul, 1879a: 97 (*Anthicus*) **E**: GR
ustulatus anatoliae Pic, 1893o: 126 (*Anthicus*) **A**: IN TR
persicus Pic, 1895j: 785 (*Anthicus*)
ustulatus ustulatus LaFerté-Sénectère, 1849i: 23 [= 1849n: 124] (*Anthicus*) **E**: CR GG GR RO **N**: EG **A**: CY IN IQ IS LE JO SI TR
lacustris Pic, 1901b: 11 (*Anthicus*)
verticalis Pic, 1893t: 175 (*Anthicus*) **E**: "Caucasus"
villosulus Truqui, 1855: 351 (*Anthicus*) **A**: AF JO TR

genus *Endomia* LaPorte, 1840: 259 type species *Anthicus elongatissimus* LaPorte, 1840 (= *Notoxus tenuicollis* Rossi, 1792)
Ochthenomus W. L. E. Schmidt, 1842d: 196 type species *Ochthenomus sinuatus* W. L. E. Schmidt, 1842 (= *Anthicus unifasciatus* Bonelli, 1812)
arabica Pic, 1913p: 162 **A**: SA YE
longicornis Pic, 1957: 448
castelsi Pic, 1929b: 14 **A**: IN OM PA **ORR**
concolor Desbrochers des Loges, 1874: 176 (*Ochthenomus*) **A**: TR
coniceps Reitter, 1901c: 79 (*Ochthenomus*) **A**: AF IN KI TD TM UZ
euphratica Krekich-Strassoldo, 1928a: 103 **A**: IN IQ TR
gratiosa Krekich-Strassoldo, 1928a: 103 **A**: PA UP **ORR**
lefebvrei LaFerté-Sénectère, 1849m: 9 [= 1849n: 289] (*Ochthenomus*) **N**: AG EG LB MO TU **A**: AE AF CY IN IQ IS JO OM QA SA SI TM TR YE **AFR**
bivittata Truqui, 1855: 369 (*Ochthenomus*)
leprieuri Pic, 1893d: 38 (*Ochthenomus*)
nigripennis Sahlberg, 1913b: 194
obscuriceps Pic, 1913p: 162
reducta Pic, 1919e: 13
susica Escalera, 1914: 380 (*Ochthenomus*)
lunulata Krekich-Strassoldo, 1928a: 104 **A**: NP UP **ORR**
occipitalis Dufour, 1843: 71 (*Anthicus*) **E**: AL BH BU CR FR GR IT PT SP ST (Kavkaz) SZ **N**: AG EG MO MR TU **A**: OM TR "Syria"
pallidicolor Pic, 1913p: 161
punctata LaFerté-Sénectère, 1847: 380 (*Ochthenomus*)
quadrinotata Pic, 1913p: 161
senilis Wollaston, 1864: 525 (*Ochthenomus*) **N**: CI
tenuicollis tenuicollis Rossi, 1792: 47 (*Notoxus*) **E**: AB AL AR AU BH BU CR CZ FR GG GR HU IT MA MC PT RO SK SP ST SZ TR UK YU **N**: AG EG LB MO TU **A**: CY IN IQ IS LE SA SY TR **AFR**
angustata LaFerté-Sénectère, 1847: 381 (*Ochthenomus*)
elongatissima LaPorte, 1840: 259 (*Anthicus*)
melanocephala Küster, 1847b: 57 (*Ochthenomus*)
obscuripennis Pic, 1913p: 162
unifasciata unifasciata Bonelli, 1812: 174 (*Anthicus*) **E**: AB AL AR BH BU CR FR GG GR IT MC PT SP ST SZ YU **N**: AG CI EG LB MO TU **A**: AE AF CY IN IS LE OM PA SA TM TR UZ YE **AFR**
caucasica Baudi di Selve, 1878a: 20 (*Ochthenomus*)
maculata Pic, 1919d: 11
sinuata W. L. E. Schmidt, 1842d: 199 (*Ochthenomus*)

genus *Hirticollis* Marseul, 1879a: 67 type species *Notoxus quadriguttatus* Rossi, 1792
Hirticomus Pic, 1894g: 69 [RN]
biplagiatus LaFerté-Sénectère, 1849k: 111 [= 1849n: 210] (*Anthicus*) **N**: EG **A**: SA **AFR**
balteatus LaFerté-Sénectère, 1849k: 112 [= 1849n: 211] (*Anthicus*)
basalis LaFerté-Sénectère, 1849k: 113 (*Anthicus*)
basalis LaFerté-Sénectère, 1849n: 212 (*Anthicus*) [HN]
bisuniguttatus Pic, 1951c: 15 (*Anthicus*)
chembanus Pic, 1932c: 18 (*Anthicus*)
densatus Fairmaire, 1898c: 241 (*Anthicus*)
floreus LaFerté-Sénectère, 1849k: 112 [= 1849n: 211] (*Anthicus*)
fuscobrunneus Bonadona, 1969: 267 (*Hirticomus*)
prebasalis Pic, 1955d: 130 (*Anthicus*)
hirsutus LaFerté-Sénectère, 1849k: 107 [= 1849n: 206] (*Anthicus*) **A**: NP **ORR**
hirsutulus Motschulsky, 1863: 492 (*Anthicus*)
subcruciatus Fairmaire, 1896a: 48 (*Anthicus*)
hispidus Rossi, 1792: 46 (*Notoxus*) **E**: AB AL AR AU BH BU CR CT CZ FR GB (Channel Is.) GE GG GR HU IT KZ LA MA MC MD NR PL PT RO SK SP ST SV SZ TR UK YU **N**: AG CI EG MO MR **A**: AF CY IN IQ IS JO KI KZ LE SA TD TM TR UZ WS YE XIZ **AFR ORR**
bicolor A. G. Olivier, 1795: 51: 5 (*Notoxus*) [HN]
hirtellus Creutzer, 1796b: 3 (*Notoxus*)
jacqueti Pic, 1893p: 126 (*Anthicus*) **E**: AB GG **A**: AF IN IS JO TD TM TR UZ
longus Pic, 1896n: 422 (*Anthicus*) **N**: AG
ocellatus LaFerté-Sénectère, 1849k: 106 [= 1849n: 205] (*Anthicus*) **A**: NP UP **ORR**
quadriguttatus Rossi, 1792: 48 (*Notoxus*) **E**: AL AU BH CR CZ FR GR HU IT MA PL PT RO SK SL SP SZ TR YU **N**: AG AZi CI LB MO TU **A**: TM TR
bifasciatus LaPorte, 1840: 259 (*Anthicus*)
guttatus LaFerté-Sénectère, 1842: 248 (*Anthicus*)
quadrinotatus Gyllenhal, 1810: 498 (*Anthicus*)
valettensis Pic, 1951b: 10 (*Anthicus*)

genus *Leptaleus* LaFerté-Sénectère, 1849i: 5 [= 1849n: 106] type species *Notoxus rodriguesi* Latreille, 1804
Ephippicollis Marseul, 1879a: 65 type species *Notoxus rodriguesi* Latreille, 1804
arabs Marseul, 1879a: 72 (*Anthicus*) **N**: AG LB TU **A**: IS OM SA SY TM YE **AFR**
atricolor Pic, 1957: 438 **A**: YE
brittoni Pic, 1957: 431 **A**: YE
chaudoiri Kolenati, 1846: 34 (*Formicoma*) **E**: AB AL BU GG GR YU **A**: AF CY IN LE TM TR
delicatulus LaFerté-Sénectère, 1849i: 9 [= 1849n: 110] (*Anthicus*) **A**: NP **ORR**
glabellus Truqui, 1855: 346 (*Anthicus*) **E**: GR **N**: AG EG **A**: IN IS JO LE SA SI SY YE **AFR**
glabricephalus Uhmann, 1983: 194 **A**: NP SD
klugii klugii LaFerté-Sénectère, 1849i: 6 [= 1849n: 107] (*Anthicus*) **N**: AG EG LB MO TU **AFR A**: AE AF IS SA SI YE
truncatulus Fairmaire, 1892a: 116
maximicollis Pic, 1893l: 111 **N**: AG EG MO **A**: IN IS OM SA SI **AFR**
mediaguttatus Uhmann, 1995: 38 **A**: OM
punctatissimus Fairmaire, 1893e: 157 **N**: EG **A**: IN LE OM SA YE **AFR**
retrofasciatus Motschulsky, 1863: 491 **A**: NP **ORR**
rodriguesi Latreille, 1804a: 357 (*Notoxus*) **E**: FR IT PT SP **N**: AG EG MO TU **A**: "Syria"
postreductus Pic, 1919d: 10
pulchellus LaFerté-Sénectère, 1842: 256 (*Anthicus*) [HN]
pulchellus W. L. E. Schmidt, 1842d: 195 (*Anthicus*)
rufescens Pic, 1891b: 117
subinfasciatus Pic, 1916a: 5
sasajii Sakai & Telnov, 2001: 328 **A**: JA
sennarensis Pic, 1907p: 322 (*Anthicus*) **A**: SA YE **AFR**
minutus Bonadona, 1969: 273
profanus Bonadona, 1969: 274

trisignatus Pic, 1955e: 1043
testaceonotatus Pic, 1922b: 20 **A**: NP **ORR**
triguttatus LaFerté-Sénectère, 1849n: 299 (*Anthicus*) **A**: IS LE SY
truncatipennis Desbrochers des Loges, 1875: 43
uralensis Pic, 1905m: 490 **E**: ST **A**: KZ MG TM

genus *Liparoderus* LaFerté-Sénectère, 1849k: 85 [= 1849n: 184] type species *Anthicus insignis* Lucas, 1843
Rotundicollis Marseul, 1879a: 67 type species *Anthicus insignis* Lucas, 1843
argenteovestitus Pic, 1893e: 51 (*Anthicus*) **N**: AG EG
subincisipes Pic, 1924c: 21 (*Anthicus*)
galeatus Bonadona, 1988: 46 **N**: AG
insignis insignis Lucas, 1843: 145 (*Anthicus*) **E**: FR SP **N**: AG LB MO TU
alutaceiceps Pic, 1916a: 5 (*Anthicus*)
insignior Pic, 1894g: 71 (*Anthicus*)
mutandus Desbrochers des Loges, 1899: 157 (*Anthicus*)
noualhieri Pic, 1896i: 132 (*Anthicus*)
obscuripes Pic, 1896i: 132 (*Anthicus*)
rubrithorax Pic, 1897f: 50 (*Anthicus*)
tripolitanus Pic, 1900m: 165 (*Anthicus*)
insignis leseleuci Pic, 1893j: 78 (*Anthicus*) **N**: MO
luteobisignatus Pic, 1942c: 11 (*Anthicus*)
panousei Pic, 1948a: 2 (*Anthicus*)
undatus Bonadona, 1986: 36 (*Microhoria*) **N**: MO
venator Dufour, 1849: 229 (*Anthicus*) **E**: SP **N**: AG MO TU
rubromaculatus Pic, 1896i: 132 (*Anthicus*)

genus *Microhoria* Chevrolat, 1877c: 168 type species *Anthicus oedipus* Chevrolat, 1860
Bifossicollis Marseul, 1879a: 67 type species *Anthicus ghilianiii* LaFerté-Sénectère, 1849
Immicrohoria Pic, 1894g: 72 type species *Notoxus fasciatus* Chevrolat, 1834
Monstrosipedes Marseul, 1879a: 67 type species *Anthicus varus* Marseul, 1875
Normalipedes Marseul, 1879a: 68 type species *Anthicus andalusiacus* LaFerté-Sénectère, 1849
Platyhoria Bonadona, 1952: 234 type species *Anthicus terminatus* W. L. E. Schmidt, 1842
Submicrohoria Bonadona, 1952: 234 type species *Anthicus plumbeus* LaFerté-Sénectère, 1842
abdeselami Escalera, 1914: 379 (*Anthicus*) **N**: MO
abeillei Pic, 1892b: 17 (*Anthicus*) **N**: AG
admirabilis Pic, 1894i: 87 (*Anthicus*) [= 1894l: 144] **N**: AG MO
advecta Krekich-Strassoldo, 1929: 173 (*Anthicus*) **N**: AG
adventicia Krekich-Strassoldo, 1929: 162 (*Anthicus*) **E**: CR
agtayi Bonadona, 1986: 42 **N**: MO
aguilari Bonadona, 1960a: 52 **E**: GR (Lesbos)
akbesiana Pic, 1896a: 5 (*Anthicus*) **A**: TR
albopilosa Krekich-Strassoldo, 1929: 179 (*Anthicus*) **E**: FR SP
amata Bonadona, 1958: 291 **N**: MO
amicitiae Dufour, 1849: 230 (*Anthicus*) **E**: SP
macularis Krekich-Strassoldo, 1929: 187 (*Anthicus*)
semicincta Desbrochers des Loges, 1875: 47 (*Anthicus*)
andalusiaca LaFerté-Sénectère, 1849l: 151 [= 1849n: 250] (*Anthicus*) **E**: SP
angulapex Koch, 1935c: 260 (*Anthicus*) **E**: GR **A**: TR
annulipes Pic, 1894i: 88 [= 1894l: 144] (*Anthicus*) **N**: AG
antoinei Chobaut, 1923a: 76 (*Anthicus*) **N**: MO
aphaenops Pic, 1902f: 41 (*Anthicus*) **A**: TM UZ
arcuatipes Krekich-Strassoldo, 1928a: 103 (*Anthicus*) **A**: HP NP UP
aspelia Truqui, 1855: 363 (*Anthicus*) **E**: GR **A**: CY
aspernata Chobaut, 1923b: 129 (*Anthicus*) **N**: MO
atlasica Pic, 1951f: 81 (*Anthicus*) **N**: MO
aubei LaFerté-Sénectère, 1849l: 152 [= 1849n: 251] (*Anthicus*) **E**: PT SP **N**: AG LB MO
calorificus Pic, 1900f: 56 (*Anthicus*)

balearica Pic, 1904g: 81 (*Anthicus*) **E**: SP (Baleares)
rubricula Krekich-Strassoldo, 1929: 182 (*Anthicus*)
barrosi Pic, 1938d: 13 (*Anthicus*) **E**: PT
baudii Pic, 1893w: lxxxvii (*Anthicus*) **N**: AG TU
obscurevestita Pic, 1894g: 73 (*Anthicus*)
baudueri Baudi di Selve, 1877d: 701 [= 1878a: 14] (*Anthicus*) **N**: AG
benigna Krekich-Strassoldo, 1929: 162 (*Anthicus*) **E**: PT SP
bergeri Bonadona, 1986: 44 **A**: CY
biargenteofasciata Pic, 1929d: 6 (*Anthicus*) **N**: MO
biauriculata Pic, 1920d: 10 (*Anthicus*) **N**: AG
bicoloriceps Pic, 1932b: 32 [= 1932g: 27] (*Anthicus*) **N**: MO
binotaticollis Pic, 1919d: 10 (*Anthicus*) **E**: FR (Corse)
bispilifasciata Marseul, 1878c: 55 (*Anthicus*) **E**: SP
bleusei Pic, 1892n: cxcviii (*Anthicus*) **E**: SP
boyadjeani Pic, 1904f: 73 (*Anthicus*) **A**: TR
bremondi Pic, 1936e: 29 (*Anthicus*) **N**: MO
breviuscula Desbrochers des Loges, 1875: 48 (*Anthicus*) **A**: LE
brisouti brisouti Desbrochers des Loges, 1875: 48 (*Anthicus*) **E**: FR PT SP
corrugata Krekich-Strassoldo, 1929: 188 (*Anthicus*)
incisa Baudi di Selve, 1877d: 697 (*Anthicus*)
nigrescens Krekich-Strassoldo, 1929: 188 (*Anthicus*)
brisouti peragalloi Pic, 1902h: 64 (*Anthicus*) **E**: SP
brisouti sternolobata Bonadona, 1954b: 280 **E**: PT SP
caliginosa caliginosa LaFerté-Sénectère, 1849l: 163 [= 1849n: 262] (*Anthicus*) **E**: AL BH BU CR GR IT SL TR YU
caliginosa epidaurosa Pic, 1936c: 22 (*Anthicus*) **E**: GR
caliginosa pannonica Kaszab, 1956c: 337 [= 1956b: 168] (*Anthicus*) **E**: HU
caliginosa sexualis Koch, 1933: 155 (*Anthicus*) **E**: CR
cantabrica cantabrica Marseul, 1879a: 212 (*Anthicus*) **E**: SP
manueli Pic, 1906a: 12 (*Anthicus*)
cantabrica relicta Krekich-Strassoldo, 1929: 172 (*Anthicus*) **E**: SP
capito LaFerté-Sénectère, 1849l: 160 (*Anthicus*) **E**: SP
capito LaFerté-Sénectère, 1849n: 259 (*Anthicus*) [HN]
caprai Bucciarelli, 1977: 179 **E**: IT
caspia Desbrochers des Loges, 1875: 45 (*Anthicus*) **E**: ST **A**: TM
cerrutii Bucciarelli, 1976: 139 **E**: GR (Kríti)
chakouri Pic, 1909l: 143 (*Anthicus*) **N**: EG
chardoni chardoni Pic, 1893s: 161 (*Anthicus*) **N**: AG
chardoni fezzarae Pic, 1893s: 162 (*Anthicus*) **N**: AG
chobauti Pic, 1892g: 75 (*Anthicus*) **E**: IT (Lampedusa) SP **N**: AG TU
lucidior Pic, 1892g: 76 (*Anthicus*)
cinctuta Marseul, 1878c: 54 (*Anthicus*) **N**: AG LB TU
rossii Pic, 1931f: 66 (*Anthicus*)
codinai Pic, 1919d: 11 (*Anthicus*) **E**: SP
constricticollis Desbrochers des Loges, 1870: 124 (*Anthicus*) **N**: AG TU
corallicollis Reitter, 1889e: 258 (*Anthicus*) **E**: GR **A**: IS
curticollis Pic, 1894k: 140 (*Anthicus*) **E**: PT SP
cyrtopyga Bonadona, 1952: 236 **E**: AB ST **A**: TM
decora Krekich-Strassoldo, 1929: 174 (*Anthicus*) **E**: SP
degener Baudi di Selve, 1881: 294 (*Anthicus*) **A**: LE
dejeanii corsica LaFerté-Sénectère, 1849l: 146 [= 1849n: 245] (*Anthicus*) **E**: FR (Corse)
valdoniellensis Pic, 1919d: 10 (*Anthicus*)
dejeanii dejeanii LaFerté-Sénectère, 1849l: 145 [= 1849n: 244] (*Anthicus*) **E**: CR FR (Corse) IT
dejeanii stocki Krekich-Strassoldo, 1929: 171 (*Anthicus*) **E**: FR (Corse)
delagrangei Pic, 1892e: 43 (*Anthicus*) **A**: SY

dentipalpis Bonadona, 1977a: 9 **N**: MO
digitalis Marseul, 1878c: 55 (*Anthicus*) **E**: SP **N**: AG MO
emaciata Pic, 1896m: 180 (*Anthicus*) **E**: GR (Kríti) **A**: LE
espunana Pic, 1930d: 10 (*Anthicus*) **E**: SP
faceta Bonadona, 1960a: 54 **A**: AF
fairmairei C. Brisout de Barneville, 1863: 90 (*Anthicus*) **E**: FR PT SP **N**: AG
decoloratipes Chobaut, 1908: 162 (*Anthicus*)
fasciata fasciata Chevrolat, 1834: pl. 34 (*Notoxus*) **E**: AL CR FR GE GR IT SL SP SZ YU
affinis LaFerté-Sénectère, 1842: 248 (*Anthicus*)
antoniae LaFerté-Sénectère, 1842: 249 (*Anthicus*)
basidensata Pic, 1936c: 21 (*Anthicus*)
bicincta LaFerté-Sénectère, 1849l: 148 [= 1849n: 247] (*Anthicus*) [HN]
francogallicus Koch, 1933: 153 (*Anthicus*) [RN]
fulvicollis Pic, 1892d: 43 (*Anthicus*) [HN]
genistae Rosenhauer, 1847: 36 (*Anthicus*)
histrio LaFerté-Sénectère, 1849l: 148 [= 1849n: 247] (*Anthicus*) [HN]
lata Pic, 1892c: 33 (*Anthicus*)
monogramma W. L. E. Schmidt, 1842c: 174 (*Anthicus*)
perobscura Krekich-Strassoldo, 1929: 170 (*Anthicus*)
unifasciata Krekich-Strassoldo, 1929: 169 (*Anthicus*) [HN]
fasciata fortis Krekich-Strassoldo, 1929: 170 (*Anthicus*) **N**: MO
rutila Krekich-Strassoldo, 1929: 170 (*Anthicus*)
fasciata helvetica Krekich-Strassoldo, 1929: 169 (*Anthicus*) **E**: SZ
fasciata opaca Rey, 1892: 101 (*Anthicus*) **E**: FR
ferianensis Pic, 1900f: 56 (*Anthicus*) **N**: AG TU
feroni Bonadona, 1960a: 51 **E**: GR
fornicata Krekich-Strassoldo, 1931: 39 (*Anthicus*) **A**: HP
fortissima Pic, 1894g: 78 [= 1895g: 196] (*Anthicus*) **N**: AG
franzi Bonadona, 1958: 289 **E**: SP **N**: MO
fumeoalata Krekich-Strassoldo, 1931: 39 (*Anthicus*) **A**: NP **ORR**
funeraria Marseul, 1879a: 225 (*Anthicus*) **E**: GR IT **A**: CY
fuscipes Marseul, 1879a: 199 (*Anthicus*) **E**: SP **N**: AG MO
ghilianiii LaFerté-Sénectère, 1849l: 150 [= 1849n: 249] (*Anthicus*) **E**: SP
globipennis globipennis Pic, 1897j: 120 (*Anthicus*) **A**: "Syria"
globipennis quercicola Sahlberg, 1913a: 59 [= 1913a: 227] (*Anthicus*) **A**: TR
gorgus Truqui, 1855: 366 (*Anthicus*) **E**: GR **A**: CY "Syria"
gourvesi Bonadona, 1986: 38 **N**: MO
hafidi Pic, 1923a: 3 (*Anthicus*) **N**: MO
hameti Escalera, 1914: 376 (*Anthicus*) **N**: MO
helenae Koch, 1933: 154 (*Anthicus*) **E**: IT (Apulia)
hingstoni Blair, 1927: 251 (*Anthicus*) **A**: XIZ
hipponensis Pic, 1893d: 38 (*Anthicus*) **N**: AG
hispanica Pic, 1899b: 20 (*Anthicus*) **E**: SP
humerifer Pic, 1902f: 41 (*Anthicus*) **E**: GR
imbasicornis Pic, 1931c: 6 (*Anthicus*) **E**: SP
informipes Krekich-Strassoldo, 1919a: 173 (*Anthicus*) **A**: HP UP
inoblita Krekich-Strassoldo, 1913a: 225 (*Aulacoderus*) **N**: AG **AFR**
ionica Pic, 1901e: 52 (*Anthicus*) **E**: BU GR TR **A**: TR
iscariotes LaFerté-Sénectère, 1849n: 301 (*Anthicus*) **E**: GR **N**: EG LB **A**: CY IS JO LE SI
kaifensis Pic, 1896m: 180 (*Anthicus*) **A**: IS JO LE SY TR
obscuriceps Pic, 1899i: 177 (*Anthicus*)
lafertei Truqui, 1855: 365 (*Anthicus*) **E**: BU GR **A**: CY IQ IS TR
lanata Krekich-Strassoldo, 1929: 177 (*Anthicus*) **E**: SP
latecincta Chobaut, 1923b: 128 (*Anthicus*) **N**: MO
latipennis Pic, 1892o: 313 (*Anthicus*) **A**: IS JO SI SY TR
lavocati Pic, 1951d: 37 (*Anthicus*) **N**: MO

lederi Marseul, 1879a: 252 (*Anthicus*) **E**: AR CR GR **A**: TM TR
leonhardi Krekich-Strassoldo, 1913a: 229 (*Anthicus*) **A**: IT
strassoldoi Pic, 1930d: 10 (*Anthicus*) [RN]
leprieuri Baudi di Selve, 1877d: 717 (*Anthicus*) **N**: AG TU
leptostemma Kolenati, 1846: 35 (*Anthicus*) **E**: AB AR GG KZ ST (Kavkaz) **A**: AF TM UZ
leuthneri Pic, 1897j: 120 (*Anthicus*) **E**: GR **A**: IS LE
creticus Pic, 1903n: 170 (*Anthicus*)
dissimilis Pic, 1897j: 120 (*Anthicus*)
lindbergi albocrispa Bonadona, 1954b: 277 **E**: MO
lindbergi flavostola Bonadona, 1958: 293 **E**: SP
lindbergi lindbergi Pic, 1932b: 33 [= 1932g: 27 (*Anthicus*) **N**: MO
lobata Pic, 1905c: 99 (*Anthicus*) **N**: AG
loebli Uhmann, 1989: 250 **A**: PA
ludovici Pic, 1893l: 111 (*Anthicus*) **N**: AG MO
mactae Pic, 1894g: 73 (*Anthicus*) **N**: AG
maculicollis Pic, 1893r: 156 (*Anthicus*) **N**: AG TU
madoni Pic, 1931d: 9 (*Anthicus*) **N**: AG
magnifica magnifica Pic, 1936e: 29 (*Anthicus*) **N**: MO
magnifica roulleaui Bonadona, 1954b: 277 **N**: MO
major major Pic, 1896l: 168 (*Anthicus*) **N**: AG MO TU
major rugata Krekich-Strassoldo, 1929: 186 (*Anthicus*) **E**: SP
mateui Bonadona, 1954b: 277 **E**: SP
mateui Bonadona, 1954a: 100 [HM]
medvedevi Telnov, 2004: 137 (*Anthicus*) [RN] **A**: MG
kaszabi L. N. Medvedev, 1975: 92 [HN]
melanocephala Bonelli, 1812: 173 (*Anthicus*) [ND] **E**: IT
miranda Krekich-Strassoldo, 1931: 37 (*Anthicus*) **A**: UP
mogadorica Escalera, 1914: 375 (*Anthicus*) **N**: MO
mollis Desbrochers des Loges, 1875: 46 (*Anthicus*) **E**: ST "Caucasus" **A**: IN
monodi Bonadona, 1977a: 5 **A**: IN
moroderi Pic, 1930d: 10 (*Anthicus*) **E**: SP
mylabrina Gené, 1839: 74 (*Anthicus*) **E**: FR (Corse) IT (Sardegna)
revelierei Pic, 1893b: 26 (*Anthicus*)
nectarina Panzer, 1794b: 7 (*Notoxus*) **E**: BU CR CZ GE GR HU PL SK UK ST **A**: AF TM WS UZ
atriceps Pic, 1904g: 82 (*Anthicus*)
bicincta Gebler, 1825: 49 (*Anthicus*)
maltae Pic, 1892f: 66 (*Anthicus*) [RN]
malvae Pic, 1892e: 43 (*Anthicus*)
testacea Pic, 1941a: 5 (*Anthicus*)
nepticula Bonadona, 1964: 240 **A**: AF PA TD TM
nicolasi Pic, 1919d: 10 (*Anthicus*) **A**: TM
normandi Pic, 1915l: 37 (*Anthicus*) **N**: TU
oberthuri Baudi di Selve, 1877d: 709 [= 1878a: 17] (*Anthicus*) **N**: AG MO
brevinotata Pic, 1896j: 141 (*Anthicus*)
juncea Pic, 1899b: 20 (*Anthicus*)
obuncata Normand, 1950a: 79 (*Anthicus*) **N**: TU
oedipa Chevrolat, 1860: 210 (*Anthicus*) **N**: AG MO TU
scaura Fairmaire, 1875b: 495 (*Anthicus*)
subsinuatipes Pic, 1920d: 10 (*Anthicus*)
tortiscelis Marseul, 1875a: 38 (*Anthicus*)
oertzeni oertzeni Pic, 1901n: 181 (*Anthicus*) **E**: AL BU GR MC
sporadensis Pic, 1901n: 181 (*Anthicus*)
oertzeni sicula Bucciarelli, 1978: 162 **E**: IT (Sicilia)
opipara Bonadona, 1977a: 7 **E**: PT SP
ovata Marseul, 1879a: 229 (*Anthicus*) **E**: GR
palicari LaPorte, 1840: 258 (*Anthicus*) **E**: GR TR

pallidioritarsis Pic, 1936g: 213 (*Anthicus*) **N**: MO
pallidula Pic, 1892e: 43 (*Anthicus*) **E**: BU CR GR KZ MC PL RO ST YU **A**: IS KZ TM TR WS XIN
mutata Pic, 1892e: 43 (*Anthicus*) [HN]
notata Pic, 1901m: 131 (*Anthicus*) [RN]
terminata LaFerté-Sénectère, 1849l: 144 [= 1849n: 243] (*Anthicus*) [HN]
pardoi Bonadona, 1952: 234 **N**: MO
paykullii Gyllenhal, 1808: 55 (*Anthicus*) **E**: PT SP **N**: CI AG TU
bijubata Motschulsky, 1849: 129 (*Anthicus*)
petraea Pic, 1902l: 187 (*Anthicus*) **A**: IS JO SA
piceodiscoidalis Pic, 1936d: 25 (*Anthicus*) **N**: MO
piciceps Desbrochers des Loges, 1875: 47 (*Anthicus*) **E**: ST "Macedonia" **A**: IS TR WS
melanocephala Marseul, 1879a: 216 (*Anthicus*) [HN]
pinicola Reitter, 1889e: 258 (*Anthicus*) **E**: GR
plagifer Krekich-Strassoldo, 1931: 38 (*Anthicus*) **A**: HP NP SD UP
planiceps Desbrochers des Loges, 1875: 49 (*Anthicus*) **N**: AG
plumbea alpina Pic, 1951b: 10 (*Anthicus*) **E**: FR
plumbea plumbea LaFerté-Sénectère, 1842: 250 (*Anthicus*) **E**: CR FR IT PT SP
brevis W. L. E. Schmidt, 1842c: 180 (*Anthicus*)
callosus W. L. E. Schmidt, 1842c: 181 (*Anthicus*)
melanarius W. L. E. Schmidt, 1842c: 178 (*Anthicus*)
subrubricollis Pic, 1951b: 10 (*Anthicus*)
postluteomaculata Pic, 1938b: 5 (*Anthicus*) **N**: MO
quadraticeps Desbrochers des Loges, 1875: 47 (*Anthicus*) **A**: JO SY TR
rabinovitchi Koch, 1935b: 141 (*Anthicus*) **N**: EG
raveli Pic, 1899b: 20 (*Anthicus*) **E**: BU GR IT **A**: TR
rectipennis LaFerté-Sénectère, 1849l: 144 (*Anthicus*) **E**: AB GG
rectipennis LaFerté-Sénectère, 1849n: 243 (*Anthicus*) [HN]
robustioriceps Pic, 1938c: 11 (*Anthicus*) **N**: MO
roseicollis Pic, 1892h: 103 (*Anthicus*) **E**: PT **N**: AG
portalegrea Pic, 1909p: 178 (*Anthicus*)
semirosea Pic, 1896c: 41 (*Anthicus*)
rosti Pic, 1906i: 79 (*Anthicus*) **A**: HP KA PA SD
rubriceps Pic, 1896m: 180 (*Anthicus*) **E**: GR **A**: LE TR
rubronotata Pic, 1896m: 180 (*Anthicus*)
rubrofasciata Pic, 1894g: 76 (*Anthicus*) **N**: AG
rufescens Pic, 1893i: 74 (*Anthicus*) **A**: TR
saidi Escalera, 1914: 378 (*Anthicus*) **N**: MO
schmiedeknechti Pic, 1899i: 177 (*Anthicus*) **E**: GR **A**: TR
scrobicollis LaFerté-Sénectère, 1849l: 166 [= 1849n: 265] (*Anthicus*) **E**: PT SP **N**: AG
selvei bicolorithorax Pic, 1930d: 10 (*Anthicus*) **E**: SP
selvei consona Krekich-Strassoldo, 1929: 178 (*Anthicus*) **E**: FR SP
selvei pullata Krekich-Strassoldo, 1929: 179 (*Anthicus*) **E**: SP
selvei selvei Pic, 1895g: 195 (*Anthicus*) **E**: SP
separanda Krekich-Strassoldo, 1929: 180 (*Anthicus*) **E**: SP
sicardi Pic, 1893n: 123 (*Anthicus*) **N**: TU
thalensis Pic, 1920d: 10 (*Anthicus*)
sidonia Truqui, 1855: 365 (*Anthicus*) **E**: "Eastern Europe" **A**: IS LE TR
simplicipes Pic, 1936d: 26 (*Anthicus*) **N**: MO
stettini Pic, 1892e: 43 (*Anthicus*) [RN] **E**: FR
ruficollis W. L. E. Schmidt, 1842c: 172 (*Anthicus*) [HN]
subgracilis Krekich-Strassoldo, 1929: 179 (*Anthicus*) **E**: SP
subtruncata Pic, 1920d: 10 (*Anthicus*) **N**: AG
succincta reymondi Bonadona, 1954b: 275 **N**: MO
succincta succincta Chevrolat, 1877c: 169 (*Anthicus*) **N**: AG MO
jurjurensis Pic, 1902d: 24 (*Anthicus*)
tumidipes Marseul, 1879a: 195 (*Anthicus*)

superba pueli Chobaut, 1903b: 170 (*Anthicus*) **N**: AG
superba superba Pic, 1896c: 40 (*Anthicus*) **N**: AG TU
sydowi olympiae Pic, 1941a: 5 (*Anthicus*) **E**: GR
sydowi reductior Pic, 1936c: 21 (*Anthicus*) **E**: GR
sydowi sydowi Pic, 1936c: 21 (*Anthicus*) **E**: GR
syrensis Pic, 1902f: 41 (*Anthicus*) **E**: GR **A**: CY
taeniata Baudi di Selve, 1877d: 712 [= 1878a: 18] (*Anthicus*) **N**: AG
taurica Pic, 1904f: 74 (*Anthicus*) **A**: TR
inobscura Pic, 1908i: 66 (*Anthicus*)
tenebricosa Pic, 1896a: 5 (*Anthicus*) **A**: LE
terminata W. L. E. Schmidt, 1842b: 128 (*Anthicus*) **E**: AL BH BU CR CT FR GR IT MC PT SP ST TR UK YU **N**: AG TU **A**: IN TM TR
chiosicola Reitter, 1889e: 259 (*Anthicus*)
dalmatina Pic, 1901m: 130 (*Anthicus*)
emgei Pic, 1897r: 296 (*Anthicus*)
latebrunnea Pic, 1901m: 131 (*Anthicus*)
latebrunnea Pic, 1903o: 178 (*Anthicus*) [HN]
sanguinicollis LaFerté-Sénectère, 1849l: 140 [= 1849n: 239] (*Anthicus*)
testaceofasciatus Pic, 1894g: 73 (*Anthicus*) **N**: AG
torretassoi Pic, 1931f: 66 (*Anthicus*) **N**: LB
tunisica Pic, 1893n: 123 (*Anthicus*) **N**: TU
unicolor calliger Marseul, 1879a: 225 (*Anthicus*) **E**: FR GR
unicolor unicolor W. L. E. Schmidt, 1842c: 179 (*Anthicus*) **E**: AL AU BU CR CZ FR GR HU IT PL SK **A**: TR
patagiata Kiesenwetter, 1861a: 247 (*Anthicus*)
validicornis LaFerté-Sénectère, 1849l: 165 (*Anthicus*)
validicornis LaFerté-Sénectère, 1849n: 264 (*Anthicus*) [HN]
valga Fairmaire, 1875b: 496 (*Anthicus*) **N**: AG MO
varus Marseul, 1875a: 38 (*Anthicus*)
valida Pic, 1896m: 179 (*Anthicus*) **E**: SP
dispar Pic, 1899b: 21 (*Anthicus*)
velutina LaFerté-Sénectère, 1849l: 162 (*Anthicus*) **E**: FR SP
velutinus LaFerté-Sénectère, 1849n: 261 (*Anthicus*) [HN]
venusta A. Villa & J. B. Villa, 1833: 35 (*Anthicus*) **E**: AU FR IT SZ
atrithorax Pic, 1912o: 90 (*Anthicus*)
birubronotata Pic, 1912o: 90 (*Anthicus*)
nigerrima Pic, 1892h: 103 (*Anthicus*)
nigrissima Pic, 1892e: 43 (*Anthicus*)
unifasciata W. L. E. Schmidt, 1842c: 173 (*Anthicus*) [HN]
veris Pic, 1893e: 52 (*Anthicus*) **N**: AG
vespertina Rosenhauer, 1856: 224 (*Anthicus*) **E**: PT SP TR
villiersi Bonadona, 1984: 71 **E**: PT
violaris Marseul, 1878b: 51 (*Anthicus*) **N**: AG
violaris seminiger Pic, 1899q: 412 (*Anthicus*) **N**: AG
volxemi Marseul, 1878a: xliii (*Anthicus*) **E**: PT SP
vosseleri melanoderus Chobaut, 1897a: 39 (*Anthicus*) **N**: AG
vosseleri vosseleri Pic, 1894i: 88 (*Anthicus*) **N**: AG MO
vosseleri Pic, 1894n: 145 (*Anthicus*) [HN]
winkleri Telnov, 2004: 137 [RN] **A**: "Syria"
longipygus Pic, 1919d: 10 (*Anthicus*) [HN]
zonata LaFerté-Sénectère, 1849l: 153 [= 1849n: 252] (*Anthicus*) **E**: IT (Sardegna)

genus *Nitorus* Telnov, 2007: 81 type species *Formicomus trigibber* Marseul, 1876
angustior Krekich-Strassoldo, 1931: 16 (*Pseudoleptaleus*) **A**: SD UP
apricans apricans Krekich-Strassoldo, 1931: 17 (*Pseudoleptaleus*) **A**: AF NP PA SD **ORR**
apricans lindbergi Bonadona, 1964: 238 (*Pseudoleptaleus*) **A**: AF
asiaticus Pic, 1907j: 173 (*Anthicus*) **A**: NP SD TAI **ORR**

asperatus Krekich-Strassoldo, 1928a: 90 (*Pseudoleptaleus*) **A**: HP NP SD UP **ORR**
aspericollis Krekich-Strassoldo, 1928a: 92 (*Pseudoleptaleus*) **A**: NP SD UP **ORR**
bigibbosus Pic, 1913l: 138 (*Leptaleus*) **A**: NP SD TAI YUN **ORR**
obscuriceps Pic, 1915c: 9 (*Pseudoleptaleus*)
carmichaeli Krekich-Strassoldo, 1931: 18 (*Pseudoleptaleus*) **A**: SD
fumatus Krekich-Strassoldo, 1931: 18 (*Pseudoleptaleus*) **A**: NP SD **ORR**
glabratus Uhmann, 1996: 30 (*Stricticomus*) **A**: NP SD **ORR**
guttulatus Krekich-Strassoldo, 1928a: 94 (*Pseudoleptaleus*) **A**: SD
lepidus Krekich-Strassoldo, 1913b: 137 (*Pseudoleptaleus*) **A**: HP NP SD
moestus Krekich-Strassoldo, 1928a: 94 (*Pseudoleptaleus*) **A**: HP NP SCH UP **ORR**
germanus Krekich-Strassoldo, 1928a: 95 (*Pseudoleptaleus*)
obfuscatus Krekich-Strassoldo, 1928a: 88 (*Pseudoleptaleus*) **A**: HP SD UP
pulchripennis Krekich-Strassoldo, 1928a: 93 (*Pseudoleptaleus*) **A**: NP SD UP **ORR**
sahlbergi Pic, 1898i: 124 (*Anthicus*) **A**: KZ TM
sculptus Krekich-Strassoldo, 1928a: 93 (*Pseudoleptaleus*) **A**: NP SD UP **ORR**
sensitivus Krekich-Strassoldo, 1928a: 91 (*Pseudoleptaleus*) **A**: HP NP SD UP **ORR**
sichuanus Telnov, 2005: 169 (*Pseudoleptaleus*) **A**: SCH
stevensi Krekich-Strassoldo, 1928a: 90 (*Pseudoleptaleus*) **A**: UP
subnudus Krekich-Strassoldo, 1928a: 90 (*Pseudoleptaleus*) **A**: HP NP UP
testaceicornis Uhmann, 1983: 197 (*Pseudoleptaleus*) **A**: SD **ORR**
trigibber Marseul, 1876d: 460 (*Formicomus*) **A**: FE JA NC NE NP SC SD TAI UP **ORR**
basalis Pic, 1915c: 9 (*Psedoleptaleus*)
mediomaculatus Pic, 1913j: 130 (*Anthicus*)

genus *Omonadus* Mulsant & Rey, 1866a: 104 [= 1866b: 192] type species *Meloe floralis* Linnaeus, 1758
Hemantus Casey, 1895: 682 type species *Hemantus rixator* Casey, 1895
Trapezicollis Marseul, 1879a: 66 type species *Meloe floralis* Linnaeus, 1758
Trapezicomus Pic, 1894c: 45 [RN]
Trapezonotus Sahlberg, 1913b: 191 type species *Anthicus phoenicius* Truqui, 1855
addendus Krekich-Strassoldo, 1928a: 102 (*Anthicus*) **A**: NP UP **ORR**
anticemaculatus Pic, 1900i: 78 (*Anthicus*) **E**: ST **A**: TM
beloni Pic, 1892d: 44 (*Anthicus*) **N**: AG
bifasciatus Rossi, 1792: 48 (*Notoxus*) **E**: AB AR AU BH BU CR CZ DE FR GB GE GG GR HU IT MA MC PL PT RO SK SP ST SV SZ TR UK **N**: AG MO TU **A**: AF IN IS KI KZ LE TD TM TR UZ
kolenatii Kolenati, 1846: 35 (*Anthicus*)
brevicornis Pic, 1894i: 86 (*Anthicus*) [= 1894l: 142] **N**: AG LB
cinctomus Marseul, 1879a: 106 (*Anthicus*) **A**: IS LE SA
confucii confucii Marseul, 1876d: 464 (*Anthicus*) **A**: FE JA HEB LIA NC SC SD SHG TAI **ORR**
curticeps Pic, 1923b: 18 (*Anthicus*)
obscuripennis Pic, 1913a: 11 (*Anthicus*)
quadraticollis Yawata, 1944: 1 (*Anthicus*)
confucii tokaraensis Nomura, 1962a: 5 (*Anthicus*) **A**: JA (Ryukyus)
floralis Linnaeus, 1758: 420 (*Meloe*) **E**: AB AL AR AU BE BH BU BY CM CR CT DE EN FI FR GB GE GG GR HU IR IT LA LS LT MA MC MD NL NR NT PL PT RO SK SP ST SV SZ TR UK YU **N**: AG AZi CI LB MO MR TU **A**: AF BT CY FE HUN IN IS JA JO LE MG NE NP OM PA SA SCH SD SI SY TM TR UP **COS**
basalis A. Villa & J. B. Villa, 1838: 63 (*Anthicus*)
basillaris Say, 1824a: 279 (*Anthicus*)
breviculus Philippi, 1864: 353 (*Anthicus*)
calycinus Panzer, 1792b: 3 (*Notoxus*)
fallax Broun, 1893b: 1168 (*Anthicus*)
formicarius A. G. Olivier, 1795: 51: 4 (*Notoxus*) [HN]
formicoides Geoffroy, 1785: 156 (*Cantharis*)
fuscus Marsham, 1802: 486 (*Lytta*)
massauensis Pic, 1900n: 575 (*Anthicus*)
myrmicocephalus Rossi, 1792: 46 (*Notoxus*)
pedicularius Schrank von Paula, 1781: 224 (*Meloe*)
reducteapicalis Pic, 1915k: 22 (*Anthicus*)

semirufus Fairmaire & Germain, 1860: 3 (*Anthicus*)
syriacus Baudi di Selve, 1881: 294 (*Anthicus*)
formicarius atropterus Nomura, 1962a: 5 (*Anthicus*) **A**: JA (Ryukyus)
formicarius formicarius Goeze, 1777: 706 (*Meloe*) **E**: AB AL AR AU BE BH BU BY CR CT DE EN FI FR GB GE GG GR HU IR IT LA LS LT MA MC MD NL NR NT PL PT RO SK SP ST SV SZ TR UK YU **N**: AG AZi EG MO TU **A**: AE AF AP BT CY ES FE FUJ HP IN IQ IS JA JO KA LE NP OM PA QA SA SD SY TAI TM TR UP UZ YE **COS**
enodis Casey, 1895: 684 (*Hemantus*)
picianus Koch, 1931: 75 (*Anthicus*)
quisquilius C. G. Thomson, 1864: 360 (*Anthicus*)
rixator Casey, 1895: 686 (*Hemantus*)
scenicus Casey, 1895: 685 (*Hemantus*)
semirufus Fairmaire, 1896a: 49 (*Anthicus*) [HN]
hadjaziensis Pic, 1899m: 264 (*Anthicus*) **A**: AE SA
ineditus LaFerté-Sénectère, 1849j: 54 [= 1849n: 154] (*Anthicus*) **A**: SY TR
lateriguttatus Marseul, 1879a: 107 (*Anthicus*) **E**: ST **A**: MG
inguttatus Pic, 1912o: 90 (*Anthicus*)
mongolicus Pic, 1896m: 181 (*Anthicus*)
longemaculatus Pic, 1938a: 18 (*Anthicus*) **A**: FUJ HEB JIA NP PA XIN UP **ORR**
bifenestratus Pic, 1938a: 19 (*Anthicus*)
bifenestratus Heberdey, 1938: 162 (*Anthicus*) [HN]
gardneri Heberdey, 1938: 163 (*Anthicus*)
patruelis LaFerté-Sénectère, 1849j: 54 [= 1849n: 154] (*Anthicus*) **A**: NP SD UP **ORR**
phoenicius Truqui, 1855: 356 (*Anthicus*) **E**: ST UK "Caucasus" **A**: CY IS KZ LE SA SY TR
dilutius Truqui, 1855: 357 (*Anthicus*)
obscurius Truqui, 1855: 357 (*Anthicus*)
quadrispilus Marseul, 1879a: 104 (*Anthicus*) **A**: AF IQ SY TM UZ WS
semicalvus Pic, 1895j: 785 (*Anthicus*) **A**: TR
signatellus Krekich-Strassoldo, 1928c: 111 (*Anthicus*) **E**: AB ST **A**: KZ UZ
spinosus Pic, 1912h: 42 (*Anthicus*) **A**: AF KI UZ

genus *Pseudoleptaleus* Pic, 1900o: 603 type species *Anthicus gibbipennis* Pic, 1900
unifasciatus unifasciatus Desbrochers des Loges, 1875: 43 (*Leptaleus*) **N**: EG **A**: IN **AFR**
sublatus Pic, 1914q: 181 (*Leptaleus*)

genus *Sapintus* Casey, 1895: 732 type species *Anthicus pubescens* LaFerté-Sénectère, 1849
subgenus *Sapintus* Casey, 1895: 732 type species *Anthicus pubescens* LaFerté-Sénectère, 1849
apicatus apicatus Fairmaire, 1896a: 48 (*Anthicus*) **A**: NP **ORR**
bataviensis Marseul, 1882a: 63 [= 1882b: 123] (*Anthicus*) **A**: NP **ORR**
breviceps LaFerté-Sénectère, 1849j: 70 [= 1849n: 170] (*Anthicus*) **A**: AF NP PA SD **ORR**
cohaeres Lewis, 1895a: 430 (*Anthicus*) **A**: FE JA
extus Lewis, 1895a: 430 (*Anthicus*) **A**: JA
fallax Bonadona, 1964: 234 **A**: AF PA
flavomaculatus LaFerté-Sénectère, 1849j: 72 [= 1849n: 172] (*Anthicus*) **A**: NP **ORR**
formosanus Pic, 1911b: 11 (*Anthicus*) **A**: TAI
immaturus LaFerté-Sénectère, 1849j: 76 [= 1849n: 176] (*Anthicus*) **A**: KA **ORR**
javanus Marseul, 1882a: 62 [= 1882b: 121] (*Anthicus*) **A**: SW **ORR**
litorosus Lewis, 1895a: 429 (*Anthicus*) **A**: FE JA
longepilosus Pic, 1942a: 13 (*Anthicus*) **A**: CH
marseuli Pic, 1893y: ccxi (*Anthicus*) [RN] **A**: FUJ JA NP **ORR**
scoticus Marseul, 1876d: 462 (*Anthicus*) [HN]
misellus LaFerté-Sénectère, 1849j: 74 [= 1849n: 174] (*Anthicus*) **A**: NP SCH **ORR**
nomurai Nardi, 2004: 136 [RN] **A**: JA (Ryukyus)
irregularis Nomura, 1962a: 4 (*Anthicus*) [HN]
oceanicus LaFerté-Sénectère, 1849j: 70 [= 1849n: 170] (*Anthicus*) **A**: JA (Ryukyus) **ORR**
guamensis Blair, 1942: 57 (*Anthicus*)

plectilis Pic, 1910p: 71 (*Anthicus*) **A**: SD SHG TAI **ORR**
sakishimanus Nomura, 1964: 56 (*Anthicus*) **A**: JA (Ryukyus)
serotinus Krekich-Strassoldo, 1931: 23 (*Anthicus*) **A**: PA **ORR**
siamensis Pic, 1914f: 35 (*Anthicus*) **A**: PA **ORR**
sodalis Pic, 1895d: 64 (*Anthicus*) **A**: JA TAI **ORR**
luteonotatus Pic, 1913j: 132 (*Anthicus*)

genus *Stenidius* LaFerté-Sénectère, 1847: 371 type species *Anthicus vittatus* Lucas, 1843
Stenicollis Marseul, 1879a: 65 type species *Anthicus vittatus* Lucas, 1843
Stenidius LaFerté-Sénectère, 1849k: 93 [HN] type species *Anthicus vittatus* Lucas, 1843
Stenidius LaFerté-Sénectère, 1849n: 192 [HN] type species *Anthicus vittatus* Lucas, 1843
aristidis Pic, 1893e: 51 (*Anthicus*) **N**: EG **A**: SA
aulicus LaFerté-Sénectère, 1849g: 13 [= 1849n: 82] (*Formicomus*) **A**: NP SD **ORR**
coloratus Motschulsky, 1863: 492 (*Leptaleus*)
contractipennis Pic, 1911o: 185 (*Anthicus*) **E**: GR MC **N**: TR
curtulus Krekich-Strassoldo, 1928a: 84 (*Anthelephilus*) **A**: HP KA NP PA SD UP
dolosus Kejval, 2002c: 193 **A**: NP UP
femoralis Desbrochers des Loges, 1875: 44 (*Anthicus*) **A**: IN IS JO LE SY TR
saulcyi Marseul, 1879a: 75 (*Anthicus*)
triangulum Desbrochers des Loges, 1875: 44 (*Anthicus*)
finicola Kejval, 2002c: 196 **A**: KA PA
glabratus Kejval, 2002c: 199 **A**: NP
lindbergi Bonadona, 1960a: 55 (*Clavicomus*) **A**: AF
luteonotatus Pic, 1909b: 38 (*Formicomus*) **A**: BT HP NP PA SD UP
pygidialis Pic, 1895j: 784 (*Anthicus*) **A**: IN IQ
schuhi Uhmann, 1994a: 409 (*Anthelephilus*) **A**: UP
tenuipes LaFerté-Sénectère, 1849k: 95 [= 1849n: 194] (*Anthicus*) **E**: AR CR GR ST (Kavkaz) **A**: AF CY IN IQ IS JO LE SA SY TM TR UZ
saturatus Pic, 1902l: 186 (*Anthicus*)
vittatus hartliebi Pic, 1899i: 174 (*Anthicus*) **N**: EG
vittatus vittatus Lucas, 1843: 145 (*Anthicus*) **E**: AB IT GR **N**: AG EG TU **A**: IN **AFR**

genus *Stricticollis* Marseul, 1879a: 66 type species *Anthicus longicollis* W. L. E. Schmidt, 1842
Stricticomus Pic, 1894g: 70 [RN]
Sulcicollus Marseul, 1879a: 66 type species *Anthicus giganteus* LaFerté-Sénectère, 1849
adleri Pic, 1911m: 170 (*Anthicus*) **E**: "Caucasus"
amaniensis amaniensis Pic, 1913q: 164 (*Anthicus*) **A**: SA **AFR**
ambulator LaFerté-Sénectère, 1849k: 91 [= 1849n: 190] (*Anthicus*) **A**: NP KA PA SD TD UP **ORR**
ampliatus Sahlberg, 1913a: 56 (*Anthicus*) **A**: SY
ampliatus Krekich-Strassoldo, 1913a: 228 (*Anthicus*) [HN]
araxicola Reitter, 1889a: 34 (*Anthicus*) **E**: AB AR
arcuaticeps Pic, 1900j: 90 (*Anthicus*) **A**: KZ TM UZ
argenteofasciatus Pic, 1894o: 205 (*Anthicus*) **A**: IN KI KZ TD TM
bistrimaculatus Pic, 1913g: 114 (*Anthicus*) **A**: TD TM
bodemeyeri Pic, 1913n: 146 (*Anthicus*) **A**: AF IN
capilliger Marseul, 1878b: 51 (*Anthicus*) **A**: SA
coreanus Pic, 1938d: 13 (*Anthicus*) **A**: ES FE JA NC SC
nipponicus Nomura, 1960: 47 (*Anthicus*)
giganteus LaFerté-Sénectère, 1849j: 60 [= 1849n: 160] (*Anthicus*) **E**: "Caucasus" **A**: AF IN IS IQ PA TD TM TR UZ
goebelii LaFerté-Sénectère, 1849k: 91 [= 1849n: 190] (*Anthicus*) **E**: AR GR MD TR **N**: AG EG LB TU **A**: AF IN IQ IS KU KZ PA SA TD TM TR UZ YE **AFR**
buchariensis Pic, 1897g: 62 (*Anthicus*)
herzi Pic, 1905m: 493 (*Anthicus*) **A**: TM
incrinitus Pic, 1929h: 126 (*Anthicus*) **N**: AG

longicollis W. L. E. Schmidt, 1842b: 130 (*Anthicus*) **E**: AB AL AR AU BH BU CR FR GG GR HU IT MC PT SK SP ST SZ **N**: AG MO **A**: CY IN IQ IS JO LE SY TR YE

pallidipes Pic, 1896g: 87 (*Anthicus*)
tenenbaumi Pic, 1925a: 1 (*Anthicus*)
uniguttatus Pic, 1896m: 179 (*Anthicus*)

longulus Pic, 1905m: 491 (*Anthicus*) **A**: ES TD

modestus LaFerté-Sénectère, 1849k: 92 [= 1849n: 191] (*Anthicus*) **N**: AG EG LB MO TU **A**: AE IN IQ IS JO KZ OM SA SI SY TR YE **AFR**

notabilis Krekich-Strassoldo, 1914c: 225 (*Anthicus*) **A**: NP UP

ophthalmicus Rottenberg, 1871: 259 (*Anthicus*) **E**: IT (Sicilia) **N**: AG EG MO TU **A**: IN IQ IS JO SA **AFR**

dromioides Pic, 1894u: lxxvii (*Anthicus*)
megalops Marseul, 1879a: 127 (*Anthicus*)

palustris Pic, 1905m: 490 (*Anthicus*) **A**: AF KZ NE NO TM

peplifer Marseul, 1879a: 126 (*Anthicus*) **A**: IN OM PA SA YE

adensis Pic, 1911m: 170 (*Anthicus*)

petri Pic, 1894o: 205 (*Anthicus*) **A**: TM

peyerimhoffi Pic, 1902l: 187 (*Anthicus*) **N**: EG **A**: IN OM SA SI

rarepunctatus Marseul, 1879a: 126 (*Anthicus*) **A**: AF IS SA

robustior Pic, 1908f: 57 (*Anthicus*) **A**: TD TM

rubricollis Pic, 1892j: 139 (*Anthicus*) **N**: MO

lecerfi Pic, 1936c: 23 (*Anthicus*)
obscuricollis Pic, 1892j: 139 (*Anthicus*)

rufithorax LaFerté-Sénectère, 1849k: 88 [= 1849n: 187] (*Anthicus*) **E**: GR **A**: IN TR "Syria"

sternbergsi Telnov, 2005: 171 **A**: YUN

tixieri Pic, 1908f: 57 (*Anthicus*) **A**: IN KZ TM UZ

tobias Marseul, 1879a: 125 (*Anthicus*) **E**: AB AL AR AU BE BH BU BY CR CT CZ DE FR GB GE GR HU IT LA MA MC MD NL NR PL PT RO SK SP ST SV SZ TR **N**: AG CI EG MO MR TU **A**: AF BEI CY IN IQ IS JA JO KA KZ LE MG NE NP PA SA SD SY TD TM TR UP UZ YE **COS**

binhanus Pic, 1927d: 30 (*Anthicus*)
corporaali Pic, 1923c: 20 (*Anthicus*)
mauritiensis Pic, 1898k: 182 (*Anthicus*)
mundulus Sharp, 1885: 168 (*Anthicus*)
parisiensis Saint-Albin, 1952: 21 (*Anthicus*)
postoculatus Fairmaire, 1896a: 49 (*Anthicus*)
tanakai Nomura, 1960: 48 (*Anthicus*)
turanicus Reitter, 1889b: 118 (*Anthicus*)

transcaspicus Pic, 1898i: 123 (*Anthicus*) **E**: AB **A**: TM

subnotatus Pic, 1910c: 10 (*Anthicus*)

transversalis meridionalis Pic, 1896c: 40 (*Anthicus*) **E**: IT (Lampedusa) MA **N**: AG LB MO TU

erichi Koch, 1931: 77 (*Anthicus*)

transversalis transversalis A. Villa & J. B. Villa, 1833: 35 (*Anthicus*) **E**: AB AL AR BU CR CZ FR GE GR IT MD PT SK SP ST (Kavkaz) UK **N**: AG EG LB MO TU **A**: AF CY IN IQ IS KZ LE SA SY TM TR

amoenus W. L. E. Schmidt, 1842c: 176 (*Anthicus*)
nigrovelutinus Pic, 1892h: 103 (*Anthicus*)
ornatus Truqui, 1855: 358 (*Anthicus*)
tenellus LaFerté-Sénectère, 1842: 251 (*Anthicus*)

truncatus Pic, 1895n: cclxxxiv (*Anthicus*) **E**: GR **A**: TR

decoloratus Pic, 1897j: 120 (*Anthicus*)
subcoloratus Pic, 1901h: 179 (*Anthicus*)

valgipes Marseul, 1875b: xxxviii (*Anthicus*) **A**: ES FE JA JIL NC SC SHA

walteri Uhmann, 1998b: 153 (*Stricticomus*) **A**: MG

genus *Tenuicollis* Marseul, 1879a: 67 type species *Anthicus olivaceus* LaFerté-Sénectère, 1849

Tenuicomus Pic, 1894g: 69 [RN]

agriliformis Pic, 1893j: 79 (Anthicus) **N**: MO

alfierii Pic, 1923a: 1 (*Anthicus*) **N**: EG

arenarius Bonadona, 1956: 119 (*Microhoria*) **N**: AG

armeniacus Pic, 1899m: 264 (*Anthicus*) **E**: AR
barnevillei Pic, 1892b: 26 (*Anthicus*) **E**: SP
bonnairii Fairmaire, 1883e: xlv (*Anthicus*) **N**: AG LB TU
nubilicollis Pic, 1906j: 283 (*Anthicus*)
cyanipennis Grilat, 1886: 187 (*Anthicus*) **N**: AG TU
grilati Pic, 1898g: 99 (*Anthicus*)
depressus LaFerté-Sénectère, 1849l: 156 [= 1849n: 255] (*Anthicus*) **E**: AB ST (Kavkaz) **A**: IN
dolichocephalus Baudi di Selve, 1877d: 702 [= 1878a: 15] (*Anthicus*) **N**: AG
duplex Nardi, 2004: 139 [RN] **E**: AR GG
subaeneus LaFerté-Sénectère, 1849l: 181 [= 1849n: 280] (*Anthicus*) [HN]
unicolor Kolenati, 1846: 37 (*Anthicus*) [HN]
edmondi Pic, 1893a: 16 (*Anthicus*) **A**: TM UZ
escalerai Pic, 1904g: 81 (*Anthicus*) **E**: SP
finalis Telnov, 2003b: 283 (*Tenuicomus*) **A**: AF
fuscomaculatus Pic, 1893v: liv (*Anthicus*) **N**: AG MO
tangerianus Pic, 1893j: 79 (*Anthicus*)
insignitus Pic, 1906j: 285 (*Anthicus*) **N**: TU
luristanicus Pic, 1911e: 97 (*Anthicus*) **A**: IN
meloiformis Reitter, 1890d: 152 (*Anthicus*) **N**: AG MO
merkli Pic, 1897r: 295 (*Anthicus*) **E**: GR TR **A**: TR
mesopotamicus Pic, 1912c: 9 (*Anthicus*) **A**: IQ
ocreatus algecirensis Koch, 1940b: 76 (*Anthicus*) **E**: SP
ocreatus ocreatus LaFerté-Sénectère, 1847: 379 (*Anthicus*) **E**: IT PT SP **N**: AG MO TU
algerinus Pic, 1892b: 25 (*Anthicus*)
basicollis Pic, 1916a: 5 (*Anthicus*)
posticus LaFerté-Sénectère, 1849l: 157 [= 1849n: 256] (*Anthicus*)
olivaceus LaFerté-Sénectère, 1849l: 154 [= 1849n: 253] (*Anthicus*) **E**: IT (Sicilia) PT SP **N**: AG MO TU
ottomanus LaFerté-Sénectère, 1849l: 155 [= 1849n: 254] (*Anthicus*) **E**: GR (Lesbos) **A**: TR
pallicrus Dufour, 1849: 230 (*Anthicus*) **E**: SP **N**: AG MO TU **A**: SA
paralleliceps Reitter, 1890d: 151 (*Anthicus*) **E**: AB
pierrei Bonadona, 1956: 118 (*Microhoria*) **N**: AG
pietschmi Pic, 1938c: 11 (*Anthicus*) **A**: IQ
platiai Degiovanni, 2000: 86 (*Tenuicomus*) **E**: IT (Calabria)
pauperculus LaFerté-Sénectère, 1847: 378 (*Anthicus*) **E**: IT (Sicilia) SP **N**: AG MO TU
lucasi Marseul, 1879a: 163 (*Anthicus*)
pumilus Baudi di Selve, 1877d: 702 [= 1878a: 16] (*Anthicus*)
siccensis Normand, 1950b: 107 (*Anthicus*) **N**: TU
subaereus pardoi Bonadona, 1977a: 4 (*Tenuicomus*) **N**: MO
subaereus quaesitus Bonadona, 1977a: 3 (*Tenuicomus*) **N**: MO
subaereus ramirezi Bonadona, 1977a: 5 (*Tenuicomus*) **E**: SP
subaereus subaereus Reitter, 1890d: 151 (*Anthicus*) **N**: AG MO
sericeus Pic, 1896c: 39 (*Anthicus*) [HN]
subcaeruleus lampros Krekich-Strassoldo, 1931: 36 (*Anthicus*) **A**: UP
subcaeruleus subcaeruleus Pic, 1906i: 78 (*Anthicus*) **A**: AF HP KA NP PA SD UP
subsericeus Pic, 1898m: 231 (*Anthicus*) **E**: IT (Sicilia)
tarifanus Pic, 1904a: 4 (*Anthicus*) **E**: SP
tibialis Waltl, 1835: 75 (*Anthicus*) **E**: PT SP
velox bucciarellii Nardi, 1995: 161 (*Tenuicomus*) [RN] **E**: IT (Venetian lagoon)
relictus Bucciarelli, 1959: 188 (*Anthicus*) [HN]
velox velox LaFerté-Sénectère, 1849l: 162 [= 1849n: 261] (*Anthicus*) **E**: IT MA
viridipennis metallescens Pic, 1899e: 75 (*Anthicus*) **A**: UP
viridipennis viridipennis Pic, 1899e: 75 (*Anthicus*) **A**: UP **ORR**
viturati Pic, 1893s: 160 (*Anthicus*) **N**: AG
obscurithorax Pic, 1896l: 168 (*Anthicus*)

genus *Yunnanomonticola* Telnov, 2002b: 47 type species *Yunnanomonticola nanzhao* Telnov, 2002
nanzhao Telnov, 2002b: 49 **A**: YUN

subfamily **Eurygeniinae** LeConte, 1862

genus *Abdullahia* Abdullah, 1968: 193 type species *Eurygenius africanus* Kolbe, 1897
africana Kolbe, 1897: 263 (*Eurygenius*) **N**: AG MO **AFR**

genus *Neosteropalpus* Abdullah, 1967: 265 type species *Eurygenius niponicus* Lewis, 1895
imasakai Nakane, 1983a: 5 **A**: JA
kyushuensis Nakane, 1975: 168 **A**: JA
niponicus Lewis, 1895a: 423 (*Eurygenius*) **A**: JA
oitaensis kidoi Nakane, 1988a: 6 **A**: JA
oitaensis oitaensis Nakane, 1983a: 5 **A**: JA

genus *Pergetus* Casey, 1895: 636 type species *Eurygenius campanulatus* LeConte, 1874
eous Semenov, 1896c: 361 (*Hypsogenia*) **A**: XIN
mandschuricus Kaszab, 1957a: 49 (*Hypsogenia*) **A**: HEI

genus *Pseudostereopalpus* Abdullah, 1964: 74 type species *Stereopalpus angusticollis* Pic, 1899
angusticollis Pic, 1899f: 76 (*Stereopalpus*) **A**: AP BT KZ NP SD UP XIZ YUN
  fulvopictus Champion, 1925: 109 (*Eurygenius*)

genus *Stereopalpus* LaFerté-Sénectère, 1849c: 1 [= 1849n: 4] type species *Stereopalpus mellyi* LaFerté-Sénectère, 1849
  Hypsogenia Semenov, 1893e: 332 type species *Hypsogenia centroasiatica* Semenov, 1893
  Stereopselaphus Gemminger & Harold, 1870: 2082 [unjustified emendation]
afghanicus Kaszab, 1957a: 47 (*Hypsogenia*) **A**: AF
asiaticus Pic, 1942b: 12 (*Eurygenius*) **A**: SCH
centroasiaticus Semenov, 1893e: 333 (*Hypsogenia*) **A**: KI KZ TD TM UZ
  obscuripes Pic, 1903e: 98 (*Hypsogenia*)
chinensis Abdullah, 1964: 38 **A**: FUJ
femoralis Lewis, 1895a: 424 **A**: JA
gigas Marseul, 1876d: 447 (*Macratria*) **A**: FE JA
minutus Pic, 1942b: 12 **A**: SCH
nagayamai Nakane, 1983a: 6 **A**: JA
tokioensis Nakane, 1983a: 6 **A**: JA

genus *Steriphodon* Abeille de Perrin, 1895a: cclxxi type species *Stereopalpus chobauti* Abeille de Perrin, 1894
bedeli Abeille de Perrin, 1894a: 94 (*Stereopalpus*) **A**: YE (Périm Is.)
chobauti Abeille de Perrin, 1894a: 94 (*Stereopalpus*) **N**: AG MO TU
indicum Pic, 1903h: 120 **A**: UP **ORR**

subfamily **Ischaliinae** Blair, 1920

genus *Ischalia* Pascoe, 1860a: 54 type species *Ischalia indigacea* Pascoe, 1860
  Eupleurida LeConte, 1862: 267 type species *Eupleurida costata* LeConte, 1862
arakii Saitô, 2003: 55 **A**: JA
arisana Kôno, 1935c: 158 **A**: TAI
brachyptera Nikitsky, 1994a: 35 **A**: FE
chinensis Young, 1976: 213 **A**: JIX ZHE
kunashirica Nikitsky, 1994a: 33 **A**: FE
latemarginata Ohbayashi & Tôyama, 1994: 146 **A**: TAI
luteolineata Pic, 1912p: 142 **A**: JA
martensi Paulus, 1972: 78 **A**: NP
nepalensis Paulus, 1972: 78 **A**: NP

patagiata Lewis, 1879: 463 **A**: FUJ JA LIA YUN "Korea" **ORR**
takane Saitô, 1994: 339 **A**: JA
uenoi Satô, 1990: 102 **A**: TAI

subfamily **Macratriinae** LeConte, 1862

genus *Macratria* Newman, 1838: 377 type species *Macratria linearis* Newman, 1838
Macrarthrius LaFerté-Sénectère, 1849d: 1 [= 1849n: 11] [unjustified emendation]
albofasciata Champion, 1916c: 209 **A**: BT **ORR**
antennalis Lewis, 1895a: 425 **A**: JA
apicalis Lewis, 1895a: 424 **A**: JA
arabica Uhmann, 1998a: 104 **A**: SA
atrata Nomura, 1962a: 6 **A**: JA (Ryukyus) TAI
basithorax Pic, 1905a: 48 **A**: SHG YUN
bicoloripes Pic, 1923b: 19 **A**: YUN
bipunctata Nomura, 1962a: 6 **A**: JA (Okinawa)
chitwana Telnov, 2003b: 284 **A**: NP
cingulifera Marseul, 1876d: 449 **A**: JA
concolor LaFerté-Sénectère, 1849d: 8 [= 1849n: 18] (*Macrarthrius*) **A**: NP **ORR**
dichroa Champion, 1916c: 211 **A**: NP UP **ORR**
flavicornis Champion, 1916c: 214 **A**: AP BT **ORR**
fluviatilis Lewis, 1895a: 425 **A**: JA
freyi Pic, 1938f: 71 **A**: JIA
fumosa Pascoe, 1860a: 56 **A**: UP **ORR**
rubroapicata Pic, 1903h: 120
griseosellata griseosellata Fairmaire, 1893b: 36 **A**: JA NP SCH TAI ZHE **ORR**
sauteri Pic, 1913j: 133
harmandi Pic, 1899f: 77 **A**: NP SD
hungarica Hampe, 1873: 165 (*Steropes*) **E**: BH HU
infima Fairmaire, 1896a: 44 **A**: NP **ORR**
japonica Harold, 1877: 359 **A**: FE JA SC
leprieuri Reiche, 1864: 243 **N**: AG EG TU **AFR**
brunnipennis Pic, 1899b: 18
gracilis Pic, 1907p: 322
minima Motschulsky, 1858b: 67 (*Macrarthrius*) **A**: UP **ORR**
nankinea Pic, 1909e: 109 **A**: JIA
nilgirica Champion, 1916c: 219 **A**: NP **ORR**
pallidicornis pallidicornis Pascoe, 1860a: 55 **A**: NP **ORR**
pelegrini Pic, 1923d: 28 **A**: JA
robustior Pic, 1923b: 19 **A**: SD **ORR**
rufescens Champion, 1916c: 218 **A**: NP SCH SD YUN
serialis Marseul, 1876d: 448 **A**: JA SC TAI
setigera Champion, 1916c: 225 **A**: NP SD **ORR**
sola Telnov, 2002a: 26 **A**: AF

subfamily **Notoxinae** Stephens, 1829

genus *Hypaspistes* Waterhouse, 1886: 39 type species *Hypaspistes armatus* Waterhouse, 1886
Pristocyphus Fairmaire, 1898b: 37 type species *Pristocyphus perrieri* Fairmaire, 1898
suturalis Pic, 1907a: 64 **A**: SA **AFR**

genus *Mecynotarsus* LaFerté-Sénectère, 1849b: 1 [= 1849n: 57] type species *Notoxus rhinoceros* Fabricius, 1798 (= *Notoxus serricornis* Panzer, 1796)
bicinctulus Marseul, 1879a: 42 **A**: AF IQ TR
bipustulatus Krekich-Strassoldo, 1928a: 70 **A**: HP NP PA UP **ORR**

bison A. G. Olivier, 1811: 394 (*Notoxus*) **E**: GR (Lesbos) **N**: AG CI EG LB MO TU **A**: AE CY IN IQ IS LE SA SI TR YE **AFR**
algericus Desbrochers des Loges, 1881: 150
beccarii Pic, 1894r: 583
bimaculatus Desbrochers des Loges, 1898b: 4
cornutus Pic, 1896m: 178
ferrantei Pic, 1910i: 26
latior Pic, 1897e: 6
macularis Baudi di Selve, 1877d: 661
mellyi Marseul, 1878b: 50
osiris Pic, 1893d: 38
sabulosus Pic, 1893s: 159
bordonii Uhmann, 2004: 1 **A**: PA
coronatus Chobaut, 1898a: 83 **N**: AG **A**: SA **AFR**
picheyrei Peyerimhoff, 1943: 15
faustii Seidlitz, 1891: 591 **E**: AL BU FR (Corse) GR IT RO ST TR UK "Caucasus" **A**: KI TR
flavipes Pic, 1913j: 129 **A**: TAI
fragilis LaFerté-Sénectère, 1849b: 5 [= 1849n: 61] **A**: NP **ORR**
karakumensis Semenov, 1891a: 375 **A**: SA TM UZ
ahngeri Semenov, 1900c: 598
minimus laticornis Nomura, 1962a: 1 **A**: JA (Ryukyus)
minimus minimus Marseul, 1876d: 461 **A**: FE JA NO TAI "Korea"
niponicus Lewis, 1895a: 427 **A**: JA
obliquemaculatus Marseul, 1882a: 59 [= 1882b: 117] **A**: AF UP **ORR**
quadrimaculatus Pic, 1913j: 129 **A**: TAI
semicinctus Wollaston, 1865: (appendix, p. 65) **N**: AG CI EG TU
sericellus Krekich-Strassoldo, 1931: 1 **A**: HKG NC NO
serricornis Panzer, 1796a: 17 (*Notoxus*) **E**: AL AU BH BU BY CR CZ FR GE GR HU IT PL PT RO SK SP ST SZ YU "Caucasus"
immaculatus Latreille, 1804a: 355 (*Notoxus*)
nigripennis Latreille, 1804a: 355 (*Notoxus*)
notatipennis Pic, 1915m: 41
rhinoceros Fabricius, 1798: 66 (*Notoxus*)
sinensis Heberdey, 1942: 472 **A**: FE FUJ NC SC SHA SHN **ORR**
tenuipes Champion, 1891b: 189 **A**: FE JA NO SC
obscurior Pic, 1943: 4
pallidior Pic, 1943: 4
truquii Marseul, 1879a: 41 **N**: EG **A**: CY IS JO LE SA SY TR YE **AFR**
alatus Koch, 1935b: 134
lysholmi Pic, 1899i: 172
vagepictus Fairmaire, 1893b: 37 **A**: UP **ORR**

genus *Notoxus* Geoffroy, 1762: 356 type species *Attelabus monoceros* Linnaeus, 1760
Ceratoderus Blanchard, 1845: 45 type species *Attelabus monoceros* Linnaeus, 1760
Monocerus A. Villa & J. B. Villa, 1833: 20 type species *Notoxus cornutus* Fabricius, 1792
abditus Nardi, 2004: 142 [RN] **A**: "Arabia"
unifasciatus Pic, 1947b: 15 [HN]
anatolicus Heberdey, 1936b: 162 **A**: IN TR
andrewesi Krekich-Strassoldo, 1913b: 131 **A**: NP UP **ORR**
angustulus Krekich-Strassoldo, 1919a: 166 **E**: AR GG GR **A**: IN TM TR
appendicinus Desbrochers des Loges, 1874: 165 **E**: AB AL AU BH BU CR FR GE GG GR HU IT MC MD PL RO SK ST SZ TR UK YU **A**: AF CY IN IS KI KZ LE TM TR UZ
moldaviensis Pic, 1909n: 170
reductus Pic, 1911i: 138
separatus Pic, 1900h: 67
assamensis Krekich-Strassoldo, 1913b: 129 **A**: NP SD UP **ORR**
bangi Pic, 1902a: 1 **A**: KZ TM

quadrimaculatus Heyden, 1887c: 315 [HN]
bellus Krekich-Strassoldo, 1928a: 67 **A**: UP
binotatus Gebler, 1829a: 136 (*Anthicus*) **A**: ES FE KI KZ MG NC NMO TIA TM WS
suturalis LaFerté-Sénectère, 1849a: 41
brachycerus Faldermann, 1837: 106 (*Monocerus*) **E**: AB AL AU BH BU BY CR CZ FR GE GG GR HU IT MD PL PT RO SK SP ST SZ UK **N**: AG MO TU **A**: KI KZ LE TR
hipponensis Pic, 1894b: 22
major W. L. E. Schmidt, 1842a: 83
notaticollis Pic, 1915j: 13
obliteratus Pic, 1930c: 6
caucasicus Pic, 1900h: 65 **E**: AB AR **A**: AF TR
cechovskyi Kejval, 1999: 20 **A**: AF TM UZ
centralasiae Kejval, 1999: 16 **A**: UZ
confluens Krekich-Strassoldo, 1928a: 70 **A**: UP
donckieri Pic, 1908n: 274 **A**: YUN
elongatus LaFerté-Sénectère, 1849a: 13 [= 1849n: 33] **E**: AB CT ST **A**: ES KI KZ MG WS
biconjunctus Pic, 1900k: 90
eurycerus Kiesenwetter, 1861a: 244 **E**: GG GR **A**: SY TR
histrix Pic, 1893t: 175
excisus Küster, 1848a: 68 **E**: FR PT SP **N**: AG MO TU
bicoronatus Bedel, 1869: 34
cavifrons LaFerté-Sénectère, 1849a: 11 [= 1849n: 31]
hispanicus Motschulsky, 1849: 128
maculatus Pic, 1916g: 13
meridionalis Pic, 1900h: 67
golestanus Kejval, 1999: 18 **A**: IN TM
haagi haagi Marseul, 1879a: 28 **A**: FE JA NC
galloisi Pic, 1915n: 152
haagi hokkaidensis Kôno, 1931: 2 **A**: JA
henoni Bedel, 1895: 193 **N**: AG
disjunctus Pic, 1915j: 13
sefrensis Pic, 1915j: 13
hirsutulus Uhmann, 1980b: 163 **A**: KA PA
hirtus LaFerté-Sénectère, 1849a: 8 [= 1849n: 28] **E**: ST **A**: AF KI KZ MG TD TM UZ WS XIZ
nigrosignatus Pic, 1900h: 65
semipunctatus Reitter, 1888d: 431
lancifer afghanus Heberdey, 1936a: 143 **A**: AF
lancifer hoggarensis Heberdey, 1936a: 142 **N**: AG
lancifer lancifer A. G. Olivier, 1811: 394 **A**: AF IN IQ IS JO SA SI UZ
chaldaeus LaFerté-Sénectère, 1849a: 29 [= 1849n: 49]
lancifer orientalis Heberdey, 1936a: 142 **N**: IN TM
lancifer sedilloti Pic, 1894g: 78 **N**: AG MO
bedeli Vauloger de Beaupré, 1895: 193
innotatus Pic, 1919e: 13
signatipennis Pic, 1919e: 13
latus Pic, 1928f: 9 **A**: UZ
lobicornis Reiche, 1864: 244 **E**: PT SP **N**: AG MO TU **A**: SY
impexus Kiesenwetter, 1870: 146
serridens Reitter, 1888d: 431
lonai Bucciarelli, 1973: 476 **E**: FR IT
martini Pic, 1903c: 83 **A**: YE **AFR**
mauritanicus latefasciatus Heberdey, 1936a: 153 **A**: IN TM
mauritanicus maroccanus Heberdey, 1936a: 153 **N**: MO
mauritanicus mauritanicus LaFerté-Sénectère, 1847: 365 **E**: GR PT SP **N**: AG TU
conjunctus Pic, 1900h: 67
lethierryi T. Pic, 1900: 32
semiobliteratus Pic, 1930c: 6

miles italicus Bucciarelli, 1978: 156 **E**: IT
miles miles W. L. E. Schmidt, 1842a: 86 **E**: AL BH BU CR GR HU RO TR YU
cinctus Pic, 1892a: 4
jacqueti Pic, 1892a: 4
miles rubrithorax Pic, 1900b: 160 **A**: TR
monoceros Linnaeus, 1760: 185 (*Attelabus*) **E**: AU BE BH BU BY CR CT CZ DE EN FI FR GB GE GR IT LA LT MC MD NL NR NT PL PT RO SK SL SP ST SV SZ UK YU **A**: ES FE KI TM UK UZ WS
biinterruptus Pic, 1900h: 66
cucullatus Geoffroy, 1785: 162
immaculatus Pic, 1899b: 19 [HN]
integer LaFerté-Sénectère, 1849a: 10 [= 1849n: 30]
latemaculatus Pic, 1892k: 139
obscuricollis Pic, 1916g: 13
platycerus LaFerté-Sénectère, 1849a: 12 [= 1849n: 32]
septemmaculata Roubal, 1933: 139
subjunctus Pic, 1911e: 97
unicornis Westhoff, 1882: 189
montivagus Kejval, 1999: 20 **A**: UZ
noctivagus Krekich-Strassoldo, 1913b: 133 **A**: HP NP **ORR**
numidicus Lucas, 1843: 145 (*Monocerus*) **N**: AG EG LB MO TU **A**: IS
aristidis Pic, 1893k: 88
cloueti Chobaut, 1895b: 347
mauritii T. Pic, 1900: 32
peregrinus Krekich-Strassoldo, 1914c: 215 **A**: HP NP UP **ORR**
pretiosus Hille, 1975: 6 **A**: SA **AFR**
tiganii Uhmann, 1980a: 138
raddei Pic, 1903d: 88 **A**: ES FE MG
raffrayi brittoni Pic, 1957: 436 **A**: OM YE
raffrayi erythreus Pic, 1914q: 183 **A**: YE **AFR**
raffrayi raffrayi Pic, 1894r: 582 **A**: YE **AFR**
rubetorum Truqui, 1855: 343 **A**: CY IS LE TR
dilutius Truqui, 1855: 343
leuthneri Pic, 1900k: 90
obscurius Truqui, 1855: 343
sahlbergi Pic, 1898i: 122 **A**: KI TM UZ
sareptanus Heberdey, 1936a: 160 **E**: ST
senegalensis LaFerté-Sénectère, 1849a: 28 [= 1849a: 48] **N**: AG **AFR**
siculus LaFerté-Sénectère, 1849a: 14 [= 1849n: 34] **E**: IT (Sicilia)
simulans coffini Bonadona, 1988: 40 [= 1989a: 11] **E**: AU FR IT SZ
simulans simulans Heberdey, 1935: 304 **E**: AB BH ST **A**: AF ES GAN KA KI KZ MG PA TD TM TR UZ WS XIN
sinensis Pic, 1907n: 190 **A**: YUN
strejceki Kejval, 1999: 18 **A**: TD
syriacus LaFerté-Sénectère, 1849n: 298 **N**: EG LB **A**: IS LE
transcaspicus Pic, 1899b: 19 **E**: "Caucasus" **A**: KZ TM
trifasciatus Rossi, 1792: 45 **E**: AB AL AR AU BH BU CR CT CZ DE FR GE GG GR HU IT MC MD NL PL PT RO SK SL SP ST SZ UK YU **N**: AG LB MO TU **A**: CY IN IQ IS JO KI KZ LE TM TR UZ
armatus W. L. E. Schmidt, 1842a: 86
cornutus Fabricius, 1792a: 211 [HN]
discolor Pic, 1897g: 61
inhumeralis Pic, 1900h: 68
pesruchesi Chobaut, 1897a: 38
rosti Pic, 1905e: 113
sexmaculatus Pic, 1893a: 16
trinotatus Pic, 1894t: x **A**: ES FE GAN HEI JA MG NC NMO SC SHN SHX TIA TM XIN
daimio Lewis, 1895a: 426
mongolicus Pic, 1914m: 74

subobliteratus Pic, 1903d: 88

genus *Pseudonotoxus* Pic, 1899j: 189 type species *Notoxus testaceus* LaFerté-Sénectère, 1849
championi Krekich-Strassoldo, 1928a: 71 (*Mecynotarsus*) **A**: HP UP **ORR**
testaceus ringens Peyerimhoff, 1943: 14 **N**: AG
testaceus testaceus LaFerté-Sénectère, 1849a: 34 [= 1849n: 54] (*Notoxus*) **E**: IT (Sardegna) **N**: AG EG TU **A**: SA **AFR**
feminus Buck, 1965: 192 (*Mecynotarsus*)
photophilus Hille, 1961: 218
viberti Bonadona, 1964: 226 [= 1965: 862] **N**: AG

subfamily **Steropinae** Jacquelin du Val, 1863

genus *Steropes* Steven, 1806: 166 type species *Steropes caspius* Steven, 1806
Blastanus Illiger, 1807: 334 type species *Blastanus colon* Illiger, 1807
caspius Steven, 1806: 166 [= 1811: 129] **E**: AB MD ST UK **A**: AF KI KZ MG TM UZ WS
colon Illiger, 1807: 334 (*Blastanus*)
colasi Abdullah, 1966: 209 **A**: NMO
latifrons Sumakov, 1908a: 133 **E**: ST (Kavkaz) **A**: GAN KI KZ MG TD TIA TM UZ
obscurans Pic, 1894j: 93 **E**: AB RO ST UK **A**: AF KZ MG PA TM TR UZ XIZ
pici Abdullah, 1966: 213 **E**: GR
popei Abdullah, 1966: 213 **E**: GR
vonhayekae Abdullah, 1966: 214 **A**: AF IQ

subfamily **Tomoderinae** Bonadona, 1961

genus *Macrotomoderus* Pic, 1901p: 741 type species *Macrotomoderus latipennis* Pic, 1901
Derarimus Bonadona, 1978: 655 type species *Derarimus carinatus* Bonadona, 1978
clavipes Champion, 1890b: 325 (*Tomoderus*) **A**: ANH JA
besucheti Uhmann, 1990a: 140
kurbatovi Telnov, 1998: 82 **A**: HUB
nigripennis Uhmann, 1994b: 663 **A**: TAI
satoi Saitô, 2003b: 321 (*Tomoderus*) **A**: JA (Ryukyus)
sichuanus Telnov, 1998: 84 **A**: SCH
uhmanni Telnov, 1998: 85 **A**: HUB SCH
yunnanus Telnov, 1998: 87 **A**: YUN
genus *Pseudotomoderus* Pic, 1892m: 240 type species *Anthicus compressicollis* Motschulsky, 1839
compressicollis compressicollis Motschulsky, 1839: 59 (*Anthicus*) **E**: AB AL BH BU CR FR GG GR IT PT RO SP ST SZ UK **N**: AG MO TU **A**: AF IN IQ MG NP TR UZ WS **AFR ORR**
melanophthalmus LaFerté-Sénectère, 1842: 255 (*Anthicus*)
harwoodi Blair, 1923a: 124 (*Tomoderus*) **A**: IQ
schawalleri Uhmann, 1990b: 335 **A**: NP
genus *Rimaderus* Bonadona, 1978: 649 type species *Rimaderus mussardi* Bonadona, 1978
meaticollis Bonadona, 1978: 651 **A**: NP SD **ORR**
montivagus Telnov, 2003b: 280 **A**: NP

genus *Tomoderus* LaFerté-Sénectère, 1849h: 1 [= 1849n: 94] type species *Tomoderus cruciatus* LaFerté-Sénectère, 1849
Curtotomoderus Pic, 1900j: 89 type species *Tomoderus brevipennis* Pic, 1900
Sirambeus Pic, 1901p: 802 type species *Tomoderus depressipennis* Pic, 1901
bosnicus Pic, 1892m: 242 **E**: BH
dalmatinus Reitter, 1881a: 224 **E**: AL BH CR HU GR MC YU
ehlersi Heyden, 1882: 47 **E**: SP
fasciatonitidus Telnov, 2001: 119 **A**: NP SD
fortepunctatus Uhmann, 1988c: 692 **A**: SD
italicus Marseul, 1879a: 61 **E**: IT
lineatopunctatus Uhmann, 1994b: 666 **A**: TAI

major Pic, 1895f: 174 **E**: GG ST (Kavkaz) TR
martensi Uhmann, 1982: 233 **A**: NP SD
nitidus Pic, 1894a: 16 **N**: EG
piochardi Heyden, 1870: 147 **E**: PT SP
praemontanus Telnov, 2001: 121 **A**: SD
reitteri Heberdey, 1938: 161 **A**: XIN
scydmaenoides Reitter, 1878a: 249 **E**: AB GG ST (Kavkaz)
smetanai Uhmann, 1989: 245 **A**: NP SD
sulcicollis LaFerté-Sénectère, 1849h: 3 [= 1849n: 96] **A**: NP SD UP **ORR**
syriacus Pic, 1907e: 138 **A**: IS SY TR
  innotatipennis Pic, 1919d: 10
tibialis Uhmann, 1988c: 696 **A**: SD
uhmanni Chandler, 2004: 142 [RN] **A**: TAI
  bimaculatus Uhmann, 1988c: 396 [HN]
unifasciatus Pic, 1907r: 341 **A**: SD UP YUN **ORR**
ventralis Marseul, 1878b: 50 **N**: AG TU

family ADERIDAE Winkler, 1927

genus *Aderus* Stephens, 1829b: 255 type species *Lytta boleti* Marsham, 1802 (= *Notoxus populneus* Creutzer, 1796)
  Hylophilus Berthold, 1827: 375 [HN] type species *Notoxus populneus* Creutzer, 1796
  Phomalus Casey, 1895: 785 type species *Xylophilus brunnipennis* LeConte, 1875
  Xylophilus Latreille, 1829: 73 [HN] type species *Notoxus populneus* Creutzer, 1796
diversimembris Pic, 1954a: 55 (*Hylophilus*) **A**: FUJ JA
flavus Fairmaire, 1863: 644 (*Xylophilus*) [DA] **N**: AG
gibbulus Marseul, 1876d: 454 (*Xylophilus*) **A**: JA
grouvellei Pic, 1910q: 75 (*Hylophilus*) **A**: JA
laticeps Pic, 1913j: 134 (*Hylophilus*) **A**: JA TAI
populneus Creutzer, 1796: 4 (*Notoxus*) **E**: AB AL AR AU BH BY CR CT CZ DE EN GB GE GG GR FI FR IT HU LA LT MA NL NR PL PT RO SK SP ST SV SZ TR UK **N**: AG EG MO TU **A**: CY IS KZ LE SY TM TR UZ **NAR**i
  biskrensis Pic, 1893s: 158 (*Xylophilus*)
  boleti Marsham, 1802: 486 (*Lytta*)
rufulus Marseul, 1876d: 451 (*Xylophilus*) **A**: JA
yaeyamanus Nomura, 1964: 54 **A**: JA

genus *Anidorus* Mulsant & Rey, 1866a: 22 type species *Xylophilus nigrinus* Germar, 1842
  Anidorus Mulsant & Rey, 1866b: 120 [HN]
caucasicus Pic, 1913g: 113 (*Hylophilus*) **E**: "Caucasus"
hipponensis Pic, 1893i: 75 (*Xylophilus*) **N**: AG
lateralis Gredler, 1866: 281 (*Xylophilus*) **E**: AU IT SZ
  tirolensis Gredler, 1866: 282 (*Xylophilus*)
nigrinus Germar, 1842: 7 (*Xylophilus*) **E**: AU BE BY CT CZ DE EN FR GE HU IT LA LT NL NT PL SK SP ST SV SZ UK
  alpestris Pic, 1899d: 51 (*Xylophilus*)
  nadeshdae Semenov, 1899a: 613 (*Euglenes*)
  patricius Abeille de Perrin, 1872: 37 (*Xylophilus*)
sanguinolentus Kiesenwetter, 1861a: 241 (*Xylophilus*) **E**: BH FR IT PT SP **N**: AG **A**: "Syria"
  tenietensis Pic, 1894e: 61 (*Xylophilus*)
sefrensis Pic, 1894i: 85 (*Xylophilus*) **N**: AG
  sefrensis Pic, 1894l: 141 (*Xylophilus*) [HN]
testaceipes Pic, 1892k: 139 (*Xylophilus*) **N**: AG

genus *Cnopus* Champion, 1893: 460 type species *Cnopus flohri* Champion, 1893
minor Baudi di Selve, 1877d: 640 (*Xilophilus* [sic!]) **E**: FR (Corse) IT
corsicus Pic, 1928a: 2 (*Hylophilus*)
minor Baudi di Selve, 1878a: 1 (*Xylophilus*) [HN]
pinicola Champion, 1919: 244 **A**: UP

genus *Cobososia* Collado & Alonso-Zarazaga, 1996: 98 type species *Hylophilus fulvicolor* Escalera, 1942
angulithorax Desbrochers des Loges, 1881: 146 (*Xylophilus*) **E**: IT (Sardegna) **N**: AG CI (Tenerife) EG TU
aristidis Pic, 1893i: 75 (*Xylophilus*) **N**: EG **A**: "Syria" **AFR**
benimorae Pic, 1893s: 157 (*Xylophilus*) **N**: AG
pallescens Wollaston, 1854: 538 (*Xylophilus*) **E**: FRi IT **N**: AG CI MO MR **A**: CY **AFR**
pallens Desbrochers des Loges, 1881: 147 (*Xylophilus*)
punctiger Mulsant & Rey, 1866a: 23 (*Xylophilus*)
punctiger Mulsant & Rey, 1866b: 111 (*Xylophilus*) [HN]

genus *Euglenes* Westwood, 1830: 59 type species *Anthicus oculatus* Paykull, 1798
Xylophilus Curtis, 1830: 299 [HN] type species *Anthicus oculatus* Paykull, 1798
kabylianus Pic, 1896c: 38 (*Xylophilus*) **N**: AG
klapperichi Pic, 1956: 6 **A**: AF
maronitus Pic, 1894q: 435 (*Xylophilus*) **A**: IS LE
nitidifrons C. G. Thomson, 1886: x (*Xylophilus*) **E**: AU CZ DE GE FI NR SV
oculatissimus Wollaston, 1864: 525 (*Xylophilus*) **N**: CI
oculatus Paykull, 1798: 256 (*Anthicus*) **E**: AU CT DE FI FR GE GB HU NR PL SP ST SV SZ
nigricollis Marsham, 1802: 487 (*Lytta*)
pygmaeus DeGeer, 1775: 80 (*Cerambyx*) **E**: AU BY CT CZ EN FI FR GE HU IT LA NR NT PL SK ST SV
fennicus Mannerheim, 1843b: 98 (*Euglenes*)
ferrugineus Paykull, 1798: 257 (*Anthicus*)
fulvus A. G. Olivier, 1795: 51: 5 (*Notoxus*)
lokvenci Roubal, 1938: 40 (*Anidorus*)
melanocephalus Creutzer, 1796: 5 (*Notoxus*)
serricornis Reitter, 1885d: 384 **E**: RO
trisignatus Pic, 1956: 6 **A**: AF
wollastoni Israelson, 1971: 117 **N**: CI

genus *Mixaderus* Collado & Alonso-Zarazaga, 1996: 98 type species *Hylophilus simplicis* Escalera, 1942
species incertae sedis
cooperi Báguena Corella, 1948: 418 ("*Mixaderus* Báguena Corella") **N**: "EG (Zeltoum)"
tamaii Nomura, 1964: 54 ("*Mixaderus* Báguena Corella") **A**: JA

genus *Otolelus* Klinger, 2000: 275 [RN] type species *Xylophilus pruinosus* Kiesenwetter, 1861
Olotelus Mulsant & Rey, 1866a: 22 [HN] type species *Xylophilus pruinosus* Kiesenwetter, 1861
Olotelus Mulsant & Rey, 1866b: 110 [HN] type species *Xylophilus pruinosus* Kiesenwetter, 1861
atlasicus Pic, 1932b: 28 (*Hylophilus*) [= 1932g: 28] **N**: MO
atomus A. Costa, 1884: 167 (*Euglenes*) **E**: IT (Sardegna)
brunneomaculatus Pic, 1907m: 182 (*Hylophilus*) **A**: HP PA UP
diversithorax Pic, 1915n: 153 (*Hylophilus*) **A**: JA
drusus Pic, 1894q: 435 (*Xylophilus*) **A**: "Antiliban" "Syria"
flaveolus Mulsant & Rey, 1866a: 26 (*Xylophilus*) [= 1866b: 114] **E**: CR FR GR IT MA SP SZ
galloisi Pic, 1910h: 23 (*Hylophilus*) **A**: JA
gratiosus alfierii Pic, 1924f: 30 (*Hylophilus*) **N**: EG
gratiosus gratiosus Pic, 1894q: 435 (*Xylophilus*) **N**: EG **A**: LE "Syria"
harmandi Pic, 1910h: 23 (*Hylophilus*) **A**: JA
longissimus Pic, 1909b: 39 (*Hylophilus*) [DA] **A**: HP "Himalaya"
neglectus Jacquelin du Val, 1863: pl. 85 (*Xylophilus*) **E**: FR IT PT SP SZ **N**: AG EG MO TU
algiricus Pic, 1897i: 81 (*Xylophilus*)
bicolor Schaufuss, 1882b: 536 (*Xylophilus*)

filicornis Schaufuss, 1882b: 536 (*Xylophilus*)
humeralis Favre, 1890: 249 (*Xylophilus*) [HN]
nigriceps Pic, 1899i: 172 (*Xylophilus*) **A**: LE TR "Syria"
obscurithorax Pic, 1899i: 172 (*Xylophilus*) **E**: GR (Ródos)
pruinosus pruinosus Kiesenwetter, 1861a: 241 (*Xylophilus*) **E**: AB AU BU CR CT CZ FR GG GR HU IT LA PL PT RO SK SP ST SZ UK "Caucasus" **N**: AG EG TU **A**: KI KZ TR UZ
montandoni Pic, 1925f: 13 (*Hylophilus*)
moreanus Pic, 1925f: 13 (*Hylophilus*)
obscurus Pic, 1892b: 27 (*Xylophilus*)
testaceus Baudi di Selve, 1877c: 415 (*Xylophilus*) [HN]
pruinosus uniformis Pic, 1925f: 13 (*Hylophilus*) **E**: "Caucasus" **N**: EG
reitteri Pic, 1897i: 81 (*Xylophilus*) **E**: AB (Araz Valley) "Caucasus"
atricornis Pic, 1900a: 12 (*Hylophilus*)
rosti Pic, 1909h: 118 (*Hylophilus*) **A**: HP "Himalaya"
ruficollis Rossi, 1794: 96 (*Notoxus*) **E**: CR GG GR IT "Caucasus" **N**: EG **A**: CY IS JO LE TR "Syria"
fulvipennis Pic, 1898l: 224 (*Xylophilus*) [RN]
nigripennis A. Villa & J. B. Villa, 1838: 63 (*Anthicus*)
nodicollis Reitter, 1885d: 384 (*Euglenes*)
subobscurus Pic, 1910b: 1 (*Hylophilus*)
testaceipennis Pic, 1897i: 81 (*Xylophilus*) [HN]
semiobscurus Pic, 1893b: 26 (*Euglenes*) **E**: GG ST "Caucasus" **A**: "Turkestan"
symphoniacus Klinger, 2000: 277 **E**: AU BU GE HU
testaceus Kolenati, 1846: 38 (*Xylophilus*) **E**: RU "Transcaucasus"

genus *Phytobaenus* R. F. Sahlberg, 1834: 276 type species *Phytobaenus amabilis* R. F. Sahlberg, 1834
amabilis amabilis R. F. Sahlberg, 1834: 277 **E**: BY CR CT CZ FI FR GE HU IT LA NT PL RO SK UK **A**: ES FE JA "NE China"
bisbimaculatus Hampe, 1850: 356 (*Xylophilus*)
amabilis amamiensis Nomura, 1964: 51 **A**: JA (Ryukyus)
amabilis scapularis Marseul, 1876d: 453 (*Xylophilus*) **A**: JA

genus *Picemelinus* Nakane, 1987: 175 type species *Hylophilus flabellicornis* Pic, 1910
flabellicornis Pic, 1910h: 22 (*Hylophilus*) **A**: JA

genus *Pseudanidorus* Pic, 1893s: 158 type species *Xylophilus monstrosipes* Pic, 1893
Pseudeuglenes Pic, 1897i: 81 type species *Xylophilus pentatomus* C. G. Thomson, 1864
cyprius Baudi di Selve, 1877d: 641 (*Xilophilus* [sic!]) **A**: CY
cyprius Baudi di Selve, 1877c: 414 (*Xylophilus*) [HN]
laesicollis Fairmaire, 1883h: cxiii (*Xylophilus*) **E**: FR (Corse) **N**: AG
monstrosipes Pic, 1893s: 157 (*Xylophilus*)
pentatomus C. G. Thomson, 1864: 371 (*Xylophilus*) **E**: CT CZ FI FR GE IT NR NT SV
quinquetomus Seidlitz, 1891: 546 (*Euglenes*)
rossicus Semenov, 1899a: 614 (*Euglenes*)
peyroni Pic, 1900a: 19 (*Hylophilus*) **A**: LE "Syria"
rubriventris Marseul, 1876d: 450 (*Xylophilus*) **A**: JA

genus *Pseudolotelus* Pic, 1901g: 67 type species *Euglenes punctatissimus* Reitter, 1885
cinctus Marseul, 1876d: 452 (*Xylophilus*) **A**: JA
curtipennis Pic, 1899i: 171 (*Xylophilus*) **A**: TR
distortus Champion, 1890a: 267 (*Xylophilus*) **A**: JA JIL
humeralis Nomura, 1964: 52 **A**: JA
japonicus Champion, 1890a: 268 (*Xylophilus*) **A**: JA
ogasawarensis Nakane, 1991: 8 **A**: JA
punctatissimus Reitter, 1885d: 384 (*Euglenes*) **E**: CR GR
kruperi Pic, 1901n: 178 (*Hylophilus*)
syriacus Pic, 1896e: 51 (*Xylophilus*) **A**: LE "Syria"

genus *Syzeton* Blackburn, 1891: 337 type species *Syzeton laetus* Blackburn, 1891 (= *Anthicus abnormis* King, 1873)
species incertae sedis
brunnidorsis Marseul, 1876d: 450 (*Xylophilus*) **A**: JA
quadrimaculatus Marseul, 1876d: 452 (*Xylophilus*) **A**: JA TAI
shibatai Nomura, 1964: 53 **A**: JA

genus *Tokiophilus* Pic, 1921h: 415 type species *Tokiophilus applicatus* Pic, 1921
applicatus Pic, 1921h: 415 **A**: JA

genus *Vanonus* Casey, 1895: 791 type species *Vanonus wickhami* Casey, 1895
Tanilotes Casey, 1895: 798 type species *Tanilotes lacustris* Casey, 1895 (= *Vanonus wickhami* Casey, 1895)
brevicornis brevicornis Perris, 1869: 20 (*Xylophilus*) **E**: CZ DE FR GB GE IT NL SV
omissus Pic, 1948c: 7 (*Hylophilus*)
brevicornis rotundaticollis Israelson, 1971: 112 **E**: CI (La Palma)

family Aderidae, species incertae sedis
Xylophilus basifasciatus Champion, 1924b: 489 **A**: UP
Xylophilus bifossulatus Champion, 1924b: 491 **A**: UP
Xylophilus breviscopa Champion, 1924b: 486 **A**: UP
Xylophilus bulbifer Champion, 1919: 243 **A**: UP
Xylophilus chinensis Champion, 1916a: 43 **A**: CH
Hylophilus formosanus Pic, 1913j: 133 **A**: TAI
Xylophilus haldwaniae Champion, 1924b: 486 **A**: UP
Hylophilus hedini Pic, 1933b: 9 **A**: XIN "NW China"
Xylophilus himalaicus Champion, 1920b: 405 **A**: UP
Xylophilus holosericeus Champion, 1916a: 28 **A**: HP
Xylophilus kumaonae Champion, 1924b: 488 **A**: UP
Xylophilus latidens Champion, 1924b: 489 **A**: UP
Xylophilus luteitarsis Champion, 1924b: 484 **A**: SD UP
Xylophilus melanostictus Champion, 1924b: 487 **A**: UP
Xylophilus parvidens Champion, 1916a: 42 **A**: ZHE
Xylophilus pectinatus Champion, 1916a: 40 **A**: HKG
Hylophilus praescutellaris Pic, 1914s: 449 **A**: "Korea"
Hylophilus quadratipennis klapperichi Pic, 1954a: 55 **A**: FUJ
Xylophilus quadratipennis quadratipennis Champion, 1916a: 44 **A**: CH
Hylophilus sauteri Pic, 1913j: 134 **A**: TAI
Xylophilus seminiger Champion, 1924b: 485 **A**: UP **ORR**
Hylophilus singularicornis Pic, 1910h: 22 **A**: JA
Xylophilus spinimanus Champion, 1916a: 42 **A**: HKG
Hylophilus subnitidus Pic, 1924e: 27 [RN] **A**: SD
harmandi Pic, 1921h: 417 (*Hylophilus*) [HN]
Xylophilus tetrastictus Champion, 1924b: 487 **A**: UP
Xylophilus trapeziderus Champion, 1924b: 485 **A**: UP

family SCRAPTIIDAE Mulsant, 1856 / Gistel, 1856
subfamily **Scraptiinae** Mulsant, 1856 / Gistel, 1856
tribe Scraptiini Mulsant, 1856 / Gistel, 1856

genus *Scraptia* Latreile, 1807: 199 type specie *Melyris dubius* A. G. Olivier, 1790
adenensis Franciscolo, 1956c: 453 **A**: YE
alutacea Reitter, 1889g: 268 **N**: MO
angustior Pic, 1944a: 9 **A**: JA

aureopubens Pic, 1903h: 119 **A**: SD
bifoveolata Küster, 1853: 90 **E**: GR **A**: TR
clairii Rey, 1892: 77
caucasicola Roubal, 1924: 182 **E**: ST
chinensis Pic, 1904c: 20 **A**: CH
cribriceps Champion, 1916d: 243 **A**: SE
curta Pic, 1944a: 9 **A**: JA
distincta Pic, 1944a: 10 **A**: "China"
dubia A. G. Olivier, 1790: 21: 12 (*Melyris*) **E**: CR FR IT SK SL SP SZ
fusca Latreille, 1804a: 342 (*Melandrya*)
fuscula Gyllenhal, 1810: 525 (*Dircaea*)
sericea Gyllenhal, 1817: 49 (*Dircaea*)
subdubia Reitter, 1889g: 267
ferruginea Kiesenwetter, 1861b: 387 **E**: FR IT SK SL SZ
forticornis Champion, 1916d: 236 **A**: JA
fuscula P. W. J. Müller, 1821: 202 **E**: AU BY CZ FI FR GB IT LA NL PL SP ST SV **N**: AG MO
algirica Pic, 1898h: 122
minuta Mulsant, 1856d: 142
nigricans Stephens, 1832: 39
nigriceps Heyden, 1870: 145
jakowleffi Reitter, 1889g: 268 **E**: ST
libanica Pic, 1917a: 2 **A**: IS LE
livens Marseul, 1876d: 456 **A**: JA
longicornis Kiesenwetter, 1861b: 243 **E**: GR
melina Champion, 1916d: 244 **A**: SD UP
nilgiriana Champion, 1916d: 246
mozabita Pic, 1898h: 123 (*Canifa*) **N**: AG
ophthalmica Mulsant, 1856d: 141 **E**: FR IT SP **N**: MO
pallescens Champion, 1916d: 245 **A**: SD
quadrisignata Champion, 1916d: 250 **A**: SD
roubali Winkler, 1927: 829 [RN] **E**: RO
forticornis Roubal, 1924: 181 [HN]
straminea Peyerimhoff, 1931b: 85 **N**: AG
testacea Allen, 1940: 56 **E**: GB FR SP SV
thoracica Baudi di Selve, 1877a: 83 **N**: MO
subfoveolata Reitter, 1889g: 267
trotommoides Pic, 1892a: 4 **A**: SY

genus *Trotomma* Kiesenwetter, 1851: 623 type specie *Trotomma pubescens* Kiesenwetter, 1851
subgenus *Macrotrotomma* Pic, 1917a: 3 type specie *Trotomma antennatum* Pic, 1902
antennatum Pic, 1902j: 80 **E**: IT
subgenus *Trotomma* Kiesenwetter, 1851: 623 type specie *Trotomma pubescens* Kiesenwetter, 1851
baborensis Pic, 1902e: 32 **N**: AG MO
impressicolle Pic, 1917a: 3 **E**: GR
latipenne Pic, 1905f: 130 **N**: TU
longipenne Pic, 1910k: 36 **N**: AG
pubescens Kiesenwetter, 1851: 624 **E**: FR GR IT **N**: AG
tuniseum Pic, 1917g: 5
vaulogeri Pic, 1917a: 3 **N**: TU

genus *Trotommidea* Reitter, 1883b: 307 type species *Trotommidea salonae* Reitter, 1883
revelierei elongata Pic, 1896e: 51 **E**: GR **N**: AG TU **A**: LE
oertzeni Schilsky, 1900: 28 (*Scraptia*)
revelierei revelierei Abeille de Perrin, 1885: 157 (*Scraptia*) **E**: FR IT
baudii Pic, 1901g: 67
ruficeps Pic, 1932b: 27 **N**: MO

salonae Reitter, 1883b: 307 **E**: CR GR FR
corcyrea Pic, 1901e: 52

subfamily **Anaspidinae** Mulsant, 1856
tribe Anaspidini Mulsant, 1856

genus *Anaspis* Geoffroy, 1762: 315 type specie *Mordella frontalis* Linnaeus, 1758
subgenus *Anaspis* Geoffroy, 1762: 315 type specie *Mordella frontalis* Linnaeus, 1758
Plesianaspis A. Costa, 1854: 28 type specie *Mordella thoracica* Linnaeus, 1758
acanthura Ermisch, 1962: 167 **A**: KI KZ UZ
apfelbecki Schilsky, 1894a: 98 **E**: BH SL YU
teratopyga Franciscolo, 1949c: 94
arctica Zetterstedt, 1828: 290 **E**: CT GE FI LA NT PL SK SV **A**: ES FE MG NE NO WS
imitata Schilsky, 1899: QQ
nigriceps Schilsky, 1895: 70
ruficeps Zetterstedt, 1840: 166
tenenbaumi Pic, 1928e: 5
atriventris Pic, 1932i: 32 **A**: GAN
auripes Ermisch, 1962: 165 **A**: KZ
basalis Emery, 1876: 118 **E**: AR
bohemica Schilsky, 1899: 88 **E**: AU CZ DE GB GE FI HU LA NL NR NT PL SV
forticornis Schilsky, 1899: 89
fortipes Reitter, 1911a: 379
norvegica Munster, 1924: 293
emarginata Schilsky, 1895: 64 **A**: MG
excellens Schilsky, 1908a: 100 **E**: HU
fasciata Forster, 1771: 65 (*Mordella*) **E**: AL BE CR FR GB GE IR IT NL PL PT SL SP SZ **N**: AG **A**: MG
bicolor Geoffroy, 1785: 142
biguttata Rossi, 1794: 103 (*Mordella*)
biinterrupta Pic, 1931c: 6
bipunctata Ragusa, 1898: 208 [HN]
bisbimaculata Pic, 1910b: 1 [RN]
cruciata A. Costa, 1854: 25
discicollis A. Costa, 1854: 25
fulvicollis Schilsky, 1895: 76
geoffroi P. W. J. Müller, 1821: 214
[*humeralis* Fabricius, 1775: 264 (*Mordella*)]
nigricollis Marsham, 1802: 492 (*Mordella*)
quadrimaculata A. Costa, 1854: 25 [HN]
quadrinotata Stephens, 1832: 46
quadripustulata Stephens, 1832: 46
scapularis Stephens, 1832: 47
sicula Pic, 1910b: 1 [RN]
subfasciata Stephens, 1832: 45
vulcanica A. Costa, 1854: 27
frontalis Linnaeus, 1758: 420 (*Mordella*) **E**: AU BE BE BU BY CT CZ FR GB.GE GR HU IR IT LA LU MC NL PL RO SK SL SP ST SZ UK **A**: FE LE MG SC TR WS
abollata Gozis, 1882: 201
assimilis Snellen van Vollenhoven, 1854: 70
flavifrons Eschscholtz, 1818: 471 (*Mordella*)
flavoatra Letzner, 1857: 119
hopffgarteni Schilsky, 1895: 61
lateralis Gyllenhal, 1810: 616 (*Mordella*) [HN]
maculicollis Motschulsky, 1860c: 110
nigra Rossi, 1792: 55 (*Mordella*)
punctata Eschscholtz, 1818: 472 (*Mordella*)
verticalis Faldermann, 1837: 112
funagata Kôno, 1928b: 150 **A**: JA
garneysi Fowler, 1889: 333 **E**: FR GB GE IR LA NL SP SZ

bilaciniata Pic, 1918c: 14
graeca Schilsky, 1895: 62 **E**: BU GR
helvetica Pic, 1918c: 14 **E**: SZ
hispanica Schilsky, 1899: 90 **E**: SP
incognita Schilsky, 1895: 74 **E**: FR GR IT SP
inderiensis Motschulsky, 1860c: 142 **A**: KZ
infuscata Motschulsky, 1860c: 142 **A**: ES FE SC
kiesenwetteri Emery, 1876: 15 **E**: AL AU BH BU CR CZ GE PL RO SK SZ
melanaria Schilsky, 1899: RR
picea Schilsky, 1895: 70
koenigi Schilsky, 1899: 87 **A**: "Kurdistan"
lucana Franciscolo, 1943b: 76 **E**: IT
lurida Stephens, 1832: 44 **E**: AL BU CR CT FR GB GE GR HU IR IT LA NL SZ **A**: IS SY TR
analis A. Costa, 1854: 30 (*Plesianaspis*)
flava A. Costa, 1854: 29 (*Plesianaspis*)
fusca Schrank von Paula, 1781: 299 (*Mordella*)
subtestacea Stephens, 1832: 45
testacea Marsham, 1802: 493 (*Mordella*) [HN]
unicolor A. Costa, 1854: 30 (*Plesianaspis*)
maculata Geoffroy, 1785: 142 (*Mordella*) [NP] **E**: BE CR FR GB GE HU IR IT NL PT SL SP SV SZ
bipunctata Bonelli, 1812: 173 (*Mordella*)
bimaculata Rossi, 1792: 245 (*Mordella*)
exoleta A. Costa, 1854: 27
florenceae Donisthorpe, 1928: 121
innotata Schilsky, 1892: 203
melanopa Forster, 1771: 64 (*Mordella*) [NO]
nigricollis Marsham, 1802: 492 (*Mordella*)
obscura Marsham, 1802: 492 (*Mordella*)
pallida Marsham, 1802: 492 (*Mordella*)
picta Hampe, 1871: 333 (*Silaria*)
testacea Snellen van Vollenhoven, 1854: 70
maroccana Fleischer, 1909: 241 **N**: MO
marseuli Csiki, 1915: 69 [RN] **A**: JA TAI
sericea Marseul, 1877: 477 [HN]
mongolica Ermisch, 1964b: 350 **A**: MG
nigripes C. Brisout de Barneville, 1866: 384 **E**: AB AR BU FR GG IT SP ST SZ
baudii Schilsky, 1894a: 97
curvicrus Schilsky, 1895: 65
nigritula Schilsky, 1894b: 331
orientalis Faldermann, 1837: 113 **E**: AB
oshimana Nomura, 1964: 42 **A**: JA
picimana Gebler, 1829a: 114 **A**: KZ XIN
poecila Faldermann, 1837: 114 **E**: AB
poggii Franciscolo, 1982: 198 **E**: IT
pulicaria A. Costa, 1854: 22 **E**: AB AL AU BH BU CR FR GB GE GG GR HU IT MC NL PL PT SL SP ST SZ YU **N**: AG MO TU **A**: IS
curtii Roubal, 1912c: 401
forcipata Mulsant, 1856f: 402
lateralis Stephens, 1832: 43
pyrenaea Fairmaire & C. Brisout de Barneville, 1859: 54 **E**: AL FR
regimbarti Schilsky, 1895: 69 **E**: BE FR GB GE IR IT NL PT SP SZ
fraudulenta Joy, 1912: 115
ruficollis Fabricius, 1792b: 115 (*Mordella*) **E**: AL AU BE BH BU DE FR GE IT LA NL PL SK SL SP ST SZ **A**: SY
alpicola Emery, 1876: 22
bickhardti Schilsky, 1906: 25
thoracica Herbst, 1784c: 149 (*Mordella*) [HN]

silvatica Gabriel, 1916: 183 **E**: PL
stussineri Fleischer, 1909: 242 **E**: AL IT MC
  albanica Csiki, 1940a: 260
subsulcata Motschulsky, 1845b: 368 **A**: ES
takeii M. Chûjô, 1957a: 8 **A**: JA
thoracica Linnaeus, 1758: 420 (*Mordella*) **E**: AU BE BH BY CR CT CZ FR GB GE IR IT LA NL PL SK SL SP SZ UK YU **A**: TR
  confusa Emery, 1876: 14
  gerhardti Schilsky, 1892: 201
  lateralis C. G. Thomson, 1864: 302 [HN]
  latipalpis Schilsky, 1895: 73
  marginicollis Håk. Lindberg, 1925: 22
  martialis Pic, 1918c: 14
  notatithorax Pic, 1918c: 14
  obscuricolor Stchegoleva-Barovskaya, 1931: 61
  occipitalis Stchegoleva-Barovskaya, 1931: 62
  schilskyana Hellén, 1935: 22
  scutellata Schilsky, 1899: OO
  septentrionalis Champion, 1891a: 104
vaulogeri Chobaut, 1896a: 200 **N**: AG
versicolor Baudi di Selve, 1878b: 341 **E**: GR IT
  ornata Schilsky, 1895: 98 (*Silaria*)
viennensis Schilsky, 1895: 66 **E**: AU BH BU CR CZ FR GR HU IT RO SK ST YU
subgenus *Nassipa* Emery, 1876: 13 type specie *Mordella flava* Linnaeus, 1758
adil Bogatchev, 1939b: 133 **E**: AB
costai Emery, 1876: 15 [NP] **E**: AU BE BH BU CR CZ DE FR GB GE HU IT MC NL PL SL SV SZ
  fulvipes Costa, 1854: 29
  fuscescens Stephens, 1832: 43 [NO]
  skalitzkyi Schilsky, 1895: 80
  thoracica A. Costa, 1854: 28 (*Plesianaspis*) [HN]
cypria Baudi di Selve, 1878b: 339 **E**: GR **A**: CY
flava Linnaeus, 1758: 420 (*Mordella*) **E**: AL AU BE BH BU BY CR CZ FR GE GR IT LA MC NL NT PL RO SK SL SP ST SZ UK YU **A**: LE
  fauconneti Pic, 1918c: 13
  pectoralis Schilsky, 1899: HH
  unicolor Schilsky, 1899: KK
  scutellaris Schilsky, 1899: II
himalayana Champion, 1927a: 54 **A**: UP
karafutona Kôno, 1933: 30 **A**: ES
melanostoma A. Costa, 1854: 23 **E**: AL AU BH CR CZ FR GE GR IT MC PL SL SV ST SZ **A**: CY TR
  monilicornis Mulsant, 1856f: 392
occipitalis Emery, 1876: 118 **E**: ST
rufilabris Gyllenhal, 1827: 521 (*Mordella*) **E**: AU BE BH BU BY CR CT CZ FR GB GE GG IR IT LA MC NL PL SK SL SP ST SV SZ UK **A**: LE
  atra Fabricius, 1792b: 115 (*Mordella*)
  clavicornis Forster, 1771: 66 (*Mordella*)
  hudsoni Donisthorpe, 1909: 800
  nigra Rossi, 1792: 213 (*Mordella*)
  obscuripes Schilsky, 1895: 81
  testaceomarginata Roubal, 1935b: 107
takeji M. Chûjô, 1957a: 8 **A**: JA
subgenus *Silaria* Mulsant, 1856f: 39 type specie *Silaria brunnipes* Mulsant, 1856
algerica Ermisch, 1963b: 20 **N**: AG
almorana Champion, 1927a: 78 **A**: UP
alticola Champion, 1927a: 55 **A**: UP
balthasari Roubal, 1935b: 106 **E**: ST
bernhaueri Roubal, 1910: 156 **E**: BU

binotata Champion, 1927a: 56 **A**: NP SD UP
brunnipes Mulsant, 1856f: 427 (*Silaria*) **E**: AL AU BH BU BY CR CZ FI FR GE GR IT LA MC NL NT PL SK SV SZ TR UK YU **A**: LE
brunnifrons Coucke, 1891: ccclxxxiv
fuscipes Mulsant, 1856f: 429
canariensis Ermisch, 1967c: 120 **E**: AZ **N**: CI
corcyrica Schilsky, 1899: 98 **E**: GR
distinguenda Ermisch, 1963b: 19 **E**: PT SP **N**: AG TU
diversipes Pic, 1933c: 28 **A**: YUN
major Pic, 1933c: 28
everestina Blair, 1927: 250 **A**: NP UP XIZ
ganglbaueri Schilsky, 1899: 100 **E**: CR SL
hayashii M. Chûjô & Nakane, 1953: 23 **A**: JA
humerosa Csiki, 1915: 57 **E**: AL GR IT **A**: CY JO LE SY
scapularis Emery, 1876: 46 [HN]
immaculata Pic, 1933c: 27 **A**: YUN
insularis Ermisch, 1963c: 67 **E**: GR **A**: CY
kochi Ermisch, 1944: 214 **E**: FR PT SP
kriegeri Ermisch, 1963a: 30 **E**: CZ GE IT FR MC RO SZ YU
latiuscula Mulsant, 1856f: 429 (*Silaria*) **E**: AU FR IT SZ
luteobrunnea Fleischer, 1909: 241 **E**: GR
luteola Marseul, 1877: 477 **A**: JA SC
mancinii Franciscolo, 1942b: 20 **E**: IT
minutula Champion, 1927a: 78 **A**: UP GAN
mitchyi M. Chûjô & Nakane, 1953: 22 **A**: JA
quadrimaculata Gyllenhal, 1817: 37 [NP] **E**: BE CR FR GE IT PT SP SZ **N**: AG MO
fulvicollis Schilsky, 1895: 100
fulvithorax Pic, 1910b: 1 [RN]
lateobscura Pic, 1909n: 162
bipustulata Bonelli, 1812: 172 (*Mordella*) [NO]
quadripustulata P. W. J. Müller,1821: 212 (*Silaria*)
subobliterara Pic, 1909n: 162
saharaensis Ermisch, 1966b: 43 **N**: MO
serbica Apfelbeck, 1931: 38 **E**: CR GR IT MC TR YU
shibatai Nakane, 1961: 87 **A**: JA
sibirica Schilsky, 1899: 97 **A**: MG
antennalis Sthegoleva-Barovskaya, 1930: 752
suturalis Emery, 1876: 16 **E**: FR IT
limbata Schilsky, 1895: 97
tenebrica Champion, 1927a: 56 **A**: UP
thoracoxantha Franciscolo, 1943b: 79 **E**: IT
touzalini Pic, 1933c: 27 **A**: CH
trifasciata Chevrolat, 1860: 455 **E**: FR IT PT SP **N**: AG MO
confluens Schilsky, 1895: 99
interrupta Schilsky, 1895: 99
quadripustulata Jacquelin du Val, 1863: 408
varians varians Mulsant, 1856f: 431 (*Silaria*) **E**: AL AU BE BH BU CZ FR GE GR IT NL PL RO SK SL SP ST SZ YU **A**: CY SY
collaris Mulsant, 1856f: 433
frontalis Costa, 1854: 21 (*Mordella*)
juvenilis Schilsky, 1899: Z
varians longicornis Roubal, 1910: 156 **E**: BU
subgenus *Larisia* Emery, 1876: 13 type specie *Anaspis rufitarsis* Lucas, 1854
akaira Franciscolo, 1991: 173 **E**: IT
bernikovi bernikovi Apfelbeck, 1931: 37 **E**: BH

bernikovi levisternalis Roubal, 1935b: 107 **E**: RO
caucasica Schilsky, 1894a: 100 **E**: AR
lurida Schilsky, 1894a: 100
cavipalpis Apfelbeck, 1931: 37 **E**: MC
chevrolati Mulsant, 1856f: 434 **E**: GE IT PT SP **N**: AG MO **A**: IS
aemula Schilsky, 1895: 92
ferruginea Schilsky, 1895: 92
nigripennis Schilsky, 1899: EE, GG
tristis Schilsky, 1899: GG
dichroa Emery, 1876: 15 **E**: FR GR IT TR **A**: CY IS SY TR
fausti Schilsky, 1899: 93 **A**: KI TM
ganglbaueri Schilsky, 1899: 100 **E**: CR IT
heydeni Csiki, 1915: 58 **A**: TD UZ
flavipennis Heyden, 1893: 329 [HN]
horni Schilsky, 1899: 99 **E**: FR PT
discolor Schilsky, 1899: 99
luridipennis Ermisch, 1962: 168 **A**: TM UZ
mariae Roubal, 1910: 154 **E**: BU
meridionalis Franciscolo, 1956c: 465 **A**: YE
mulsanti C. Brisout de Barneville, 1859: 234 **E**: FR GE IT SP **N**: MO
eppelsheimi Schilsky, 1895: 91
nigriventris Schilsky, 1899: 94 **A**: TR
palpalis Gerhardt, 1876: 381 (*Silaria*) **E**: AL AU BH BU CR CT CZ FR GE GR HU IT MC PL ST SZ TR UK YU
affinis Schilsky, 1895: 93
eucineticolor Roubal, 1935b: 106
impressa Guillebeau, 1893b: 328
rambouseki Roubal, 1910: 155 **E**: BU
reitteri Schilsky, 1895: 85 **A**: MG
revelieri Emery, 1876: 15 **E**: FR GR IT MC SP **A**: TR
rufitarsis Lucas, 1854: 35 **E**: GR MA **A**: CY IQ IS SY
salsolae cyrensis Bogatchev, 1939b: 132 **E**: AB
salsolae salsolae Bogatchev, 1939b: 131 **E**: AB
schneideri Emery, 1876: 119 **A**: ST
setulosa Schilsky, 1895: 90 **E**: GR
steppensis Motschulsky, 1860b: 142 **E**: BU ST
ochracea Stierlin, 1867: 223 (*Silaria*)
stierlini Emery, 1876: 15 **E**: BU GR ST **A**: LE SY TR
truquii Baudi di Selve, 1878b: 344 **E**: IT **A**: CY
turkmenica Schilsky, 1899: 92 **A**: TM
subgenus *Pinassa* Ernisch, 1950: 73 type specie *Anaspis angustata* Ermisch, 1950
eversi Ermisch, 1964a: 18 **N**: CI
imitator Ermisch, 1963d: 110 **N**: CI MR
lindbergi Franciscolo, 1956b: 48 **A**: CY
proteus Wollaston, 1854: 532 (*Silaria*) **E**: AZ **N**: CI MR
subgenus *Spanisa* Emery, 1876: 13 type specie *Anaspis labiata* A. Costa, 1854
aharonii Reitter, 1911a: 379 **A**: IS
labiata A. Costa, 1854: 22 **E**: FR GR SP IT **N**: MO
subtilis Hampe, 1871: 333 **E**: AU BH CR HU IT RO RU SL SK YU
genus *Anaspis*, species incertae sedis
lindbergianana Roubal, 1934a: 6 **N**: MO (Atlas)

genus *Anaspis*, nomina dubia
testacea Forster, 1771: 63 (*Mordella*) **E**: GB
oxyacanthae Gmelin, 1790: 2024 (*Mordella*)

genus *Cryptanaspis* Franciscolo, 1956c: 463 type specie *Cryptanaspis brittoni* Franciscolo, 1956
brittoni Franciscolo, 1956c: 464 **A**: YE

genus *Cyrtanaspis* Emery, 1876: 8 type specie *Anaspis phalerata* Germar, 1847
phalerata Germar, 1847: 8 (*Anaspis*) **E**: AU BY CT CZ FI FR GE IT NT PL SL SP SV SZ
interrupta Schilsky, 1908c:606
obscura Schilsky, 1895: 60
sauteri Pic, 1911p: 189 **A**: TAI
yunnana Pic, 1933c: 27 **A**: GAN YUN
diversitarsis Pic, 1933c: 27

tribe Pentariini Franciscolo, 1954

genus *Anaspella* Schilsky, 1895: 59 type specie *Anaspella clavicornis* Schilsky, 1895
antennata Franciscolo, 1953: 251 **N**: EG
clavicornis Schilsky, 1895: 59 **N**: AG MO TU
clavifera Marseul, 1876a: 25 (*Anaspis*) **N**: AG EG MO **A**: IS
desertifera Chobaut, 1898a: 86 (*Anaspis*) **N**: AG
inclavifera Pic, 1922e: 21 (*Anaspis*) **N**: EG
alfierii Pic, 1922e: 21 (*Anaspis*)
marioutensis Pic, 1922e: 21 (*Anaspis*)
lutea Marseul, 1876a: 26 (*Anaspis*) **N**: EG
subparallela Pankow, 1981: 273 **A**: SA

genus *Ectasiocnemis* Franciscolo, 1956a: 43 type specie *Pentaria platycnema* Champion, 1920
Konaspis M. Chûjô, 1957b: 12 type specie *Anthobates elongata* Kôno, 1929
Nipponanaspis M. Chûjô, 1957b: 11 type specie *Nipponanaspis shirozui* M. Chûjô, 1957
anchoralis Nomura, 1961b: 87 **A**: JA
elongata Kôno, 1929c: 84 (*Anthobates*) **A**: JA
caneparii Franciscolo, 1989: 108 **A**: KA
chloroptera Champion, 1920a: 76 (*Pentaria*) **A**: UP
contorta Batten, 1983b: 487 **A**: AF
hazariana Franciscolo, 1956a: 57 **A**: PA
horaki Batten, 1983b: 485 **A**: AF
inermis Franciscolo, 1956a: 51 **A**: AF
laciniata Franciscolo, 1961: 53 **A**: AF
luteicollis Champion, 1921: 206 (*Pentaria*) **A**: AF UP
maxima Franciscolo, 1956a: 53 **A**: PA UP
platycnema Champion, 1920a: 77 (*Pentaria*) **A**: UP
shirozui M. Chûjô, 1957b: 13 (*Nipponanaspis*) **A**: JA
straminea Champion, 1927a: 53 (*Pentaria*) **A**: UP
tibialis Schilsky, 1906: 24 (*Anaspis*) **A**: KA UP
kumaonensis Champion, 1920a: 76 (*Pentaria*)

genus *Pentaria* Mulsant, 1856d: 135 type specie *Anaspis badia* Rosenhauer, 1847
Anthobates LeConte, 1850: 231 [HN]
Anthobatula E. Strand, 1929: 20 [RN]
abderoides Chobaut, 1893: 97 (*Silaria*) **N**: AG MO TU
arabica Pankow, 1981: 275 **A**: SA
badia Rosenhauer, 1847: 39 (*Anaspis*) **E**: AU FR IT SK SP **N**: AG MO **A**: SY
sericaria Mulsant, 1856f: 439
bleusei Chobaut, 1897b: 133 **N**: AG
dimidiata Marseul, 1876a: 27 **A**: LE
gracilis Schilsky, 1894a: 99 (*Anaspis*) **E**: AB AR
reitteri Chobaut, 1894a: lxxv

mongolica Ermisch, 1969c: 176 **A**: MG
phoenix Franciscolo, 1956a: 459 **A**: YE
redemptus Peyerimhoff, 1931b: 86 **N**: AG
soderbomi Pic, 1933b: 10 (*Anthobates*) **A**: MG
spissicornis Peyerimhoff, 1946: 95 (*Anthobates*) **N**: MO

genus *Pseudopentaria* Franciscolo, 1956a: 460 type specie *Anaspis defarguesi* Abeille de Perrin, 1885
defarguesi Abeille de Perrin, 1885: 161 (*Anaspis*) **E**: FR IT SP **N**: AG MO TU **A**: IS
immaculata Pic, 1895l: cxxvii (*Pentaria*)
oberthueri Champion, 1892: 109 (*Pentaria*)
pictipennis Reitter, 1891a: 31 (*Anaspis*)
unifasciata Chobaut, 1897: 135 (*Pentaria*) **N**: AG
mesopotamica Blair, 1923a: 126 (*Pentaria*) **A**: IQ
scotti Franciscolo, 1956a: 461 **A**: YE

REFERENCES

Abdullah M. 1966: A revision of the genus Steropes Steven, with a proposed new subfamily Steropinae of the family Anthicidae (Coleoptera). *The Entomologist's Monthly Magazine* **101**: 206-216, 1 pl.

Abdullah M. 1967: A new genus (Neostereopalpus) of the Eurygeniini (Coleoptera: Anthicidae) from Japan. *Pakistan Journal of Scientific and Industrial Research* **10**: 265-267.

Abdullah M. 1968: New Old World genera of the Eurygeniini (Coleoptera: Anthicidae). *Pakistan Journal of Scientific and Industrial Research* **11**: 190-198.

Abeille de Perrin E. 1869: [Coléoptères nouveaux]. *Petites Nouvelles Entomologiques* **1** (11) [1869-1875]: 42.

Abeille de Perrin E. 1870: Salpingus exsanguis. Espèce nouvelle de la faune française. *L'Abeille, Mémoires d'Entomologie* **7**: 89-90.

Abeille de Perrin E. 1872: *Études sur les coléoptères cavernicoles suivies de la description de 27 coléoptères nouveaux français*. Marseille: Marius Olive, 40 + [1] pp.

Abeille de Perrin E. 1874a: *Essai monographique sur les cisides européens et circaméditerranéens*. Marseille: E. Camoin, [1] + 99 + [1] pp.

Abeille de Perrin E. 1874b: Étude sur les salpingiens européens. *Bulletin de la Société d'Histoire Naturelle de Toulouse* **8**: 24-31.

Abeille de Perrin E. 1874c: [Diagnoses de cisides nouveaux]. *Bulletin de la Société Entomologique de France* **1874**: lii-liii.

Abeille de Perrin E. 1876: Notes sur les cisides européens et circumméditerranéens. *Annales de la Société Entomologique de France* (5) **6**: 309-314.

Abeille de Perrin E. 1876: [new taxa]. *Bulletin de la Société Entomologique de France* **1876**: clxvi.

Abeille de Perrin E. 1880: Contribution à la faune coléoptérologique de la Méditerranée. *Bulletin de la Société d'Histoire Naturelle de Toulouse* **14**: 233-262.

Abeille de Perrin E. 1885: Coléoptères rares ou nouveaux de France. *Revue d'Entomologie* **4**: 153-161.

Abeille de Perrin E. 1892: Notices entomologiques. *Revue d'Entomologie* **11**: 62-67.

Abeille de Perrin E. 1894a: Diagnoses de coléoptères reputés nouveaux. *L'Échange, Revue Linnéenne* **10**: 91-94.

Abeille de Perrin E. 1894b: Description d'un coléoptère saharien nouveau. *Bulletin de la Société Entomologique de France* **1894**: cxxxii-cxxxiii.

Abeille de Perrin E. 1895a: [Description de deux genres et d'une espèce française de coléoptères]. *Bulletin de la Société Entomologique de France* **1894**: cclxxi- cclxxii.

Abeille de Perrin E. 1895b: [new taxa]. *Bulletin de la Société Entomologique de France* **1895**: xxiii-xxiv.

Abeille de Perrin E. 1896: Descriptions d´un serropalpide et de trois oedémerides nouveaux (Col.). *Bulletin de la Société Entomologique de France* **1896**: 283-285.

Abeille de Perrin E. 1899: Description d'une espèce nouvelle de colydien (Col.) des Alpes-Maritimes. *Bulletin de la Société Entomologique de France* **1899**: 408-409.

Abeille de Perrin E. 1901a: Nouvelles espèces de coléoptères français. *L'Échange, Revue Linnéenne* **17**: 59-62.

Abeille de Perrin E. 1901b: Descriptions de deux coléoptères nouveaux du nord de l'Afrique. *Bulletin de la Société Entomologique de France* **1901**: 234.

Abeille de Perrin E. 1909: Étude d'un Rhipidius nouveau de Provence, R. Boissyi Abeille (Présentation de l'insecte). *Comptes-Rendus de Séances de la Société de Biologie* **67**: 854-858.

Adams M. 1817: Descriptio insectorum novorum Imperii Rossici, inprimis Caucasi et Sibiriae. *Mémoires de la Société Impériale des Naturalistes de Moscou* **5**: 278-314, pl. X.

Agassiz J. L. R. 1846a: Nomina systematica generum Coleopterorum, tam viventium quam fossilium, secundum ordinem alphabeticum disposita, adjectis auctoribus, libris in quibus reperiuntur, anno editionis, etymologia et familis ad quas pertinent. Pp. xi + [1] + 1-170. In: *Nomenclator zoologicus, continens nomina systematica generum animalium, tam viventium quam fossilium, secundum ordinem alphabeticum disposita, adjectis auctoribus, libris, in quibus reperiuntur, anno editionis, etymologia et familiis, ad quas pertinen, in singulis classibus*. [1842-1846]. Soloduri: Jent & Gassmann.

Agassiz J. L. R. 1846b: *Nomenclatoris zoologici index universalis, continens nomina systematica classium, ordinum, familiarum et generum animalium omnium, tam viventium quam fossilium, secundum ordinem alphabeticum unicum disposita, adjectis homonymiis plantarum, nec non variis adnotationibus et emendationibus*. Soloduri: Jent & Gassmann, viii + 383 pp.

Ahrens A. 1814: *Fauna Insectorum Europae, Fasciculus II*. Halae: Car. Aug. Kümmelii, [25] pp., 25 pls.

Akita K. & Masumoto K. 2003: [new taxon]. In: Masumoto K. & Akita K.: New or little-known tenebrionid species from Japan (Part 4) (Coleoptera). *Entomological Review of Japan* **58**: 173-182.

Akiyama H. 1998: A new species of the subgenus Dryopomera (Coleoptera, Oedemeridae) from the Yaeyama Islands of the Ryukyus. *Elytra*, Tokyo **26**: 167-171.

Aksentjev S. I. 1978: Novyy vid naryvnikov roda Stenodera Esch. (Coleoptera, Meloidae) iz Primor'ya [A new species of meloid beetles of the genus Stenodera Esch. (Coleoptera, Meloidae) from Primorye Territory]. *Entomologicheskoe Obozrenie* **57**: 124-125.

Aksentjev S. I. 1981: Novye rody i vidy naryvnikov (Coleoptera, Meloidae) iz Stredney Azii [New genera and species of blister beetles (Coleoptera, Meloidae) from Middle Asia]. *Entomologicheskoe Obozrenie* **60**: 860-864.

Aksentjev S. I. 1985: Meloidae from Iran (Insecta: Coleoptera). *Senckenbergina Biologica* **65** [1984]: 245-250.

Aksentjev S. I. 1987: Meloidae from the Nepal Himalayas (Insecta: Coleoptera). *Courier Forschungs-Institut Senckenberg* **93**: 471-476.

Aksentjev S. I. 1988: Katalog taksonov rodovoy grupy zhestkokrylykh sem. Meloidae (Coleoptera) [A catalogue of the genus group taxa of the beetle family Meloidae (Coleoptera)]. *Entomologicheskoe Obozrenie* **67**: 569-582.

Aksentjev S. I. & Sadykova Kh. M. 1990: A new species of blister beetles (Coleoptera, Meloidae) from Kirghiziya. *Zoologiskiy Zhurnal* **69**: 152-153.

Aliquiò V. 1993: Dati nuovi e riassuntivi sui coleotteri tenebrionidi delle isole circum-siciliane (Coleoptera: Tenebrionidae). *Il Naturalista Siciliana* (Ser. IV) **17**: 111-125.

Allard E. 1865a: Quatre nouvelles espèces du genre Erodius. *Annales de la Société Entomologique de France* (4) **4** [1864]: 387-388.

Allard E. 1865b: Tableau synoptique des espèces du genre Erodius. *Annales de la Société Entomologique de France* (4) **4** [1864]: 389-398.

Allard E. 1869a: Révision du genre Asida (Latr.). *L'Abeille, Journal d'Entomologie* **6**: 159-304.

Allard E. 1869b: Description de quelques coléoptères nouveaux et notes diverses. *L'Abeille, Mémoires d'Entomologie* **5**: 465-478.

Allard E. 1870: [Coléoptères nouveaux]. *Petites Nouvelles Entomologiques* **1** [1869-1875]: 49-50.

Allard E. 1873: Monographie des espèces de coléoptères du genre Erodius, Fab. *Revue et Magasin de Zoologie Pure et Appliquée* (3) **1**: 122-234.

Allard E. 1874: Mémoire sur les coléoptères ténébrionides formant les genres Sepidium, Fabr. & Vieta, Cast. *Revue et Magasin de Zoologie Pure et Appliquée* (3) **2**: 120-151.

Allard E. 1876a: Révision des hélopines vrais de Lacordaire. *L'Abeille, Journal d'Entomologie* **14**: 1-80.

Allard E. 1876b: Descriptions d'espèces nouvelles. *L'Abeille, Journal d'Entomologie* **14**: 21-25.

Allard E. 1877a: Révision des helopides vrais. *Mitteilungen der Schweizerischen Entomologischen Gesellschaft* **5**: 13-268.

Allard E. 1877b: Description de quelques hélopides nouveaux. *Petites Nouvelles Entomologique* **9**: 97-98.

Allard E. 1880: Essai de classification des blapsides de l'ancien monde. 1e partie. *Annales de la Société Entomologique de France* (5) **10**: 269-320.

Allard E. 1881a: Essai de classification des blapsides de l'ancien monde. 2e partie. *Annales de la Société Entomologique de France* (6) **1**: 131-180.

Allard E. 1881b: [new taxa]. *Bulletin de la Société Entomologique de France* **1881**: ciii-civ.

Allard E. 1882a: Essai de classification des blapsides de l'ancien monde. 4e et dernière partie. *Annales de la Société Entomologique de France* (6) **2**: 77-140.

Allard E. 1882b: [new taxa]. *Bulletin de la Société Entomologique de France* **1882**: lxxxvi-lxxxvii.

Allard E. 1883: Mélanges Entomologiques. *Annales de la Société Entomologique de Belgique* **27**: 5-49.

Allard E. 1884: Diagnoses de quatre nouveaux coléoptères. *Wiener Entomologische Zeitung* **3**: 248.

Allard E. 1885a: Classification des adesmides et des mégagénides (Lacord.). *Annales de la Société Entomologique de France* (6) **5**: 154-208.

Allard E. 1885b: Coleoptera nova. *Wiener Entomologische Zeitung* **4**: 180.

Allard E. 1886: *Description de six nouvelles espèces de coléoptères hétéromères*. [7] pp.

Antoine M. 1945: Notes d'entomologie marocaine XXXVIII. Les Crypticus du Maroc (Col. Tenebr.). *Eos, Revista Española de Entomología* **20** [1944]: 257-276.

Antoine M. 1946a: Notes d'entomologie marocaine XLI. Sur les Arthrodeis du Maroc (Col. Tenebr.). *Bulletin de la Société des Sciences Naturelles du Maroc* **24**: 25-32.

Antoine M. 1946b: Notes d'entomologie marocaine XLII. Les Thalpobia Fairm. (Col. Tenebr.). *Bulletin de la Société des Sciences Naturelles du Maroc* **24**: 33-39.

Antoine M. 1946c: Notes d'entomologie marocaine. XLIII. – Notes complémentaires sur les Litoborinae et Opatrum du Maroc (Col. Tenebr.). *Bulletin de la Société d'Histoire Naturelle de l'Afrique du Nord* **37**: 51-59.

Antoine M. 1947: Notes d'entomologie marocaine XLVI. - Trois ténébrionides nouveaux de la montagne marocaine. *Bulletin de la Société Entomologique de France* **52**: 147-149.

Antoine M. 1948: Notes d'entomologie marocaine XLVII. Les Stenosis du groupe fulvipennis (Col. Tenebrionidae). *Revue Française d'Entomologie* **14** [1947]: 277-298.

Antoine M. 1949a: Notes d'entomologie marocaine XLVIII. - Tableaux de détermination des Pimelia du Maroc (Col. Tenebrionidae). *Annales de la Société Entomologique de France* **116** [1947]: 17-58.

Antoine M. 1949b: Notes d'entomologie marocaine XLIV. Matériaux pour l'étude des Helopinae du Maroc (Col. Tenebrionides). *Bulletin de la Société des Sciences Naturelles du Maroc* **25-27** [1945-1947]: 123-162.

Antoine M. 1949c: Notes d'entomologie marocaine LI. - Les Stenosis du Maroc (suite et fin). *Revue Française d'Entomologie* **16**: 209-233.

Antoine M. 1951a: Un ténébrionide nouveau du Sahara marocain. *Comptes Rendus de la Société des Sciences Naturelles du Maroc* **17**: 76-77.

Antoine M. 1951b: Notes d'entomologie marocaine LII. - Coléoptères nouveaux ou intéressants. *Bulletin de la Société des Sciences Naturelles du Maroc* **29** [1949]: 311-321.

Antoine M. 1951c: Notes d'entomologie marocaine LIV. - Captures et observations nouvelles (Coléoptères). *Bulletin de la Société des Sciences Naturelles du Maroc* **30** [1950]: 87-101.

Antoine M. 1952: Notes d'entomologie marocaine LV. - Additions et remarques taxonomiques. *Bulletin de la Société des Sciences Naturelles du Maroc* **31** [1951]: 103-119.

Antoine M. 1953a: Présentation de coléoptères troglobies. *Comptes Rendus de la Société des Sciences Naturelles du Maroc* **19**: 58-59.

Antoine M. 1953b: Notes d'entomologie marocaine. LVI. - Mélanges coléoptérologiques. *Revue Française d'Entomologie* **20**: 202-223.

Antoine M. 1954: Notes d'entomologie marocaine LX. - Sur quelques captures intéressantes (Coléopt. Carab. et Ténébr.). *Bulletin de la Société des Sciences Naturelles et Physiques du Maroc* **34**: 199-209.

Antoine M. 1955: Notes d'entomologie marocaine LXII. - Trois ténebrionides nouveaux pour le Maroc. *Bulletin de la Société des Sciences Naturelles et Physiques du Maroc* **34** [1954]: 362-365.

Antoine M. 1956: Notes d'entomologie marocaine LXIII. - Sur la systématique des Litoborini et révision des Melambiina marocains (Col. Ténéb.). *Bulletin de la Société des Sciences Naturelles et Physiques du Maroc* **36**: 341-366.

Antoine M. 1957: Notes d'entomologie marocaine LXIV. - Contribution à la connaissance des ténébrionides du Maroc. *Bulletin de la Société des Sciences Naturelles et Physiques du Maroc* **37**: 39-56.

Antoine M. 1962: Notes d'entomologie marocaine LXIX. – Coléoptères nouveaux. En particulier du Rif et du Maroc oriental. *Bulletin de la Société des Sciences Naturelles et Physiques du Maroc* **42**: 47-56.

Apfelbeck V. 1901: Bericht über eine entomologische Forschungsreise nach der Türkei und Griechenland im Jahre 1900. *Wissenschaftliche Mittheilungen aus Bosnien und der Herzegovina* (Wien) **8**: 447-469.

Apfelbeck V. 1905: Paeninsulae balcanicae coleopterorum species novae. *Glasnik Zemaljskog Muzeja u Bosni i Hercegovini* **17**: 230-252.

Apfelbeck V. 1907: Neue Koleopteren von der Balkanhalbinseln. *Wissenschaftliche Mitteilungen aus Bosnien und der Herzegowina* (Wien) **10**: 636-644.

Apfelbeck V. 1907: Neu Koleopteren, gesammelt während einer im Jahre 1905 mit Subvention der Kaiserlichen Akademie der Wissenschaften in Wien durchgeführten Forschungsreise nach Albanien und Montenegro. (I. Serie). *Sitzungsberichte der Kaiserlichen Akademie der Wissenschaften in Wien, Mathematish-Naturwissenschaftiche Klasse* **115** (Abt. I): 1663-1674.

Apfelbeck V. 1914: Revision der Palaearktischen Mordella-Arten aus der aculeata-Gruppe. *Annales Historico-Naturales Musei Nationalis Hungarici* **12**: 605-622.

Apfelbeck V. 1931: Neue Anaspis-Arten aus Jugoslavien. *Entomologischer Anzeiger* **11**: 37-39.

Aragona L. A. 1830: *De quisusquam insectis Italiae novis aut rarioribus.* Ticini Regii: Typographia Bizzomi, 31 pp.

Ardoin P. 1958a: Contribution à l'étude des Helopinae de France (Col. Tenebrionidae). *Annales de la Société Entomologique de France* **127**: 9-49.

Ardoin P. 1958b: Une nouvelle espèce d'Asida de la Faune de France (Coleopt. Tenebrionidae). *L'Entomologiste* **14**: 37-41.

Ardoin P. 1960a: Missions de MM. G. Colas et J. Mateu à l'archipel de Madère. Les ténébrionides (Col.). *Revue Française d'Entomologie* **27**: 103-143.

Ardoin P. 1960b: Une nouvelle espèce d'Uloma (Tenebrionidae) de l'Inde. *Entomologische Arbeiten aus dem Museum G. Frey* **11**: 273-275.

Ardoin P. 1964: Contribution à l'étude du genre Enicmosoma Geb. (Col. Tenebrionidae). *Entomologische Arbeiten aus dem Museum G. Frey* **15**: 686-690.

Ardoin P. 1965a: Contribution à l'étude des ténébrionides africains (Col.). *Bulletin de la Société Entomologique de France* **70**: 73-85, pl. II.

Ardoin P. 1965b: Note préliminaire sur les ténébrionides récoltés aux Monts Loma (Sierra-Leone). *Bulletin de l'Institut Français d'Afrique Noire* **27**: 1315-1320.

Ardoin P. 1969: Le Parc National du Niokolo-Koba, Fascicule III. XII. Coleoptera Tenebrionidae (note récapitulative) *Mémoires de l'Institut Fondamental d'Afrique Noire* **84**: 247-261.

Ardoin P. 1971a: Tenebrionidae (Coleoptera) récoltés par l'expédition Mares dans l'Archipel de la Galite, Tunisie. *Nouvelle Revue d'Entomologie* **1**: 45-52.

Ardoin P. 1971b: Contribution à l'étude des espèces africaines et malgaches du genre Anemia Laporte (Col. Tenebrionidae). *Annales de la Société Entomologique de France* (N. S.) **7**: 357-422.

Ardoin P. 1972: Deux nouvelles espèces du genre Isocerus Latr. (Col. Tenebrionidae). *Nouvelle Revue d'Entomologie* **2**: 239-241.

Ardoin P. 1973: Contribution à l'étude des Tenebrionidae (Coleoptera) de Sardaigne. *Annales de la Société Entomologique de France* (N. S.) **9**: 257-307.

Ardoin P. 1978a: Récoltes dans le Peloponnese oriental et description d'une sous-espèce nouvelle de Tenebrionidae (Coleoptera). *Entomops* **45**: 133-140.

Ardoin P. 1978b: Contribution à l'étude des Adesmia (Coleoptera: Tenebrionidae) du Sinaï, d'Israël et de Jordanie. *Nouvelle Revue d'Entomologie* **8**: 295-313.

Ardoin P. 1979: Une nouvelle espèce d'Akis (Coleoptera: Tenebrionidae) de l'île de Formentera, Baléares. *Entomologische Blätter* **74**: 155-157.

Arnett R. H. 1951: A revision of the Nearctic Oedemeridae (Coleoptera). *The American Midland Naturalist* **45**: 257-391.

Aubé C. 1842: Notes sur quelques coléoptères nouveaux. *Annales de la Société Entomologique de France* **11**: 225-237.

Aubé C. 1850: Description de quelques insectes coléoptères appartenant à l'Europe et à l'Algérie. *Annales de la Société Entomologique de France* (2) **8**: 299-346.

Aubé C. 1861: Description de six espèces nouvelles de coléoptères d'Europe dont deux appartenant a deux genres nouveaux et aveugles. *Annales de la Société Entomologique de France* (4) **1**: 195-199.

Audinet-Serville J. G. A. 1825: [new taxa]. In: *Encyclopédie méthodique ou par ordre de matières; par une Société de gens de lettres, de savants et d'artistes; Précédée d'un Vocabulaire universel, servant de Table pour tout l'Ouvrage, ornée des Portraits de MM. Diderot & d'Alembert, premiers Éditeurs de l'Encyclopédie. Histoire Naturelle. Entomologie, ou Histoire naturelle des Crustacés, des Arachnides et des Insectes. Tome dixième.* Paris: Agasse, (6) +832 + [1] pp.

Aurivillius C. 1890: Neue Käfer aus Afrika. *Entomologisk Tidskrift* **11**: 203-205.

Bach M. 1856: *Käferfauna für Nord- und Mitteldeutschland mit besonderer Rücksicht auf die preussischen Rheinlande. Dritter Band.* Coblenz: J. Bölscher, 292 pp.

Báguena Corella L. 1948: *Estudio sobre los Aderidae.* Instituto de Estudios Africanos, Madrid, xvi + 547 pp. + 1 p. [errata].

Báguena Corella L. 1962: *Exploration du Parc National de la Garamba - Mission H. De Saeger. Fascicule 26 Aderidae (Coleoptera, Heteromeroidea).* Bruxelles: Institut des Parcs Nationaux du Congo et du Ruanda-Urundi, 135 pp.

Ba Y.-B. & Ren G.-D. 2004: Taxonomic study on the genus Syachis Bates from China and its adjacent regions (Coleoptera, Tenebrionidae). *Acta Zootaxonomica Sinica* **29**: 761-765.

Bai M. & Ren G.-D. 2003: Four new species of the genus Cteniopinus Seidlitz, 1896 from China (Coleoptera, Tenebrionidae, Alleculinae). *Acta Zootaxonomica Sinica* **28**: 682-687.

Bakhuysen 1806: [new taxa]. In: Voet J. E.: *Catalogus systematicus Coleopterorum - Catalogue systématique des coléoptères — Systematische naamlyst van dat geslacht van Insecten dat men Torren noemt. Volume 2.* La Haye: Bakhuysen, 82 + 85 + 87 pp.

Ballion E. E. 1878: Verzeichniss der im Kreise von Kuldsha gesammelten Käfer. *Bulletin de la Société Impériale des Naturalistes de Moscou* **53** (1): 253-389.

Ballion E. E. 1888a: Kurze Notizen ueber einige russische Blaps-Arten. II. Artickel. *Bulletin de la Société Impériale des Naturalistes de Moscou* (N. S.) **2**: 269-276.

Ballion E. E. 1888b: Kurze Notizen ueber einige russische Blaps-Arten. III. Artickel. *Bulletin de la Société Impériale des Naturalistes de Moscou* (N. S.) **2**: 694-704.

Bates F. 1870: Descriptions of new genera and species of Heteromera. *The Entomologist's Monthly Magazine* **6** [1869-1870]: 268-275.

Bates F. 1872: Notes on Heteromera, and description of new genera and species (No. 1). *The Entomologist's Monthly Magazine* **9** [1872-1873]: 97-99.

Bates F. 1873a: Notes on Heteromera, and description of new genera and species (No. 8), (No. 9). *The Entomologist's Monthly Magazine* **9** [1873-1874]: 14-17, 45-52.

Bates F. 1873b: Notes on Heteromera, and description of new genera and species. *The Entomologist's Monthly Magazine* **9** [1872-1873]: 201-204.

Bates F. 1879: Characters of the new genera and species of Heteromera collected by Dr. Stoliczka during the Forsyth Expedition to Kashgar in 1873-74. *Cistula Entomologica* **2** [1875-1882]: 467-484.

Bates F. 1890: Heteromera. Pp. 55-79. In: Ball V. (ed.): *Scientific results of the second Yarkand Mission; based upon the collections and notes of the late Ferdinand Stoliczka, Ph. Dr.* Calcutta: Office of superintendent of Goverment Printing, India, 79 pp., 2 pls.

Batten R. 1977: Two new Mordellidae (Coleoptera) from Southern Europe, and a key to the Mordellistena micans group. *Entomologische Berichten* **37**: 167-176.

Batten R. 1978: New Mordellidae (Coleoptera) from Greece and Tunesia. *Entomologische Berichten* **38**: 78-80.

Batten R. 1980: Notes on the Mordellistena gemellata group: two new species and a case of synonymy. *Entomologische Berichten* **40**: 41-46.

Batten R. 1983a: A new Mordellistena from Asia Minor. *Entomologische Berichten* **43**: 88-89.

Batten R. 1983b: Two new Ectasiocnemis (Anaspidinae, Coleoptera Mordellidae) from Afganistan and a case of synonymy. *Bonner Zoologische Beiträge* **34**: 485-488.

Batten R. 1989: Two new Mordellidae (Coleoptera) from Kashmir and Sumatra. *Tijdschrift voor Entomologie* **132**: 105-107.

Baudi di Selve F. 1870: Coleopterorum messis in insula Cypro et Asia minore ab Eugenio Truqui congregate recensitio: de Europaeis notis quibusdam additis. Pars tertia. *Berliner Entomologische Zeitschrift* **14**: 49-90.

Baudi di Selve F. 1874a: Catalogo dei tenebrioniti della fauna europea e circummediterranea del Museo Civico di Genova. *Annali del Museo Civico di Storia Naturale di Genova* **6**: 89-115.

Baudi di Selve F. 1874b: Coleotteri tenebrioniti delle collezioni italiane. Parte prima. *Bullettino della Società Entomologica Italiana* **6**: 183-200.

Baudi a Selve F.1874c: Coleopterorum messis in insula Cypro et Asia Minore ab Eugenio Truqui congregatae recensitio: de Europaeis notis quibusdem additis. Pars quinta. *Berliner Entomologische Zeitschrift* **17** [1873]: 317-338.

Baudi di Selve F. 1875a: Coleotteri tenebrioniti delle collezioni italiane. *Bullettino della Società Entomologica Italiana* **6** [1874]: 275-301.

Baudi di Selve F. 1875b: Coleotteri tenebrioniti delle collezioni italiane. *Bullettino della Società Entomologica Italiana* **7**: 3-36.

Baudi di Selve F. 1875c: Europaeae et circummediterraneae faunae Tenebrionidum specierum, quae Comes Dejean in suo Catalogo, editio 3[a] consignavit, ex ejusdem collectione in R. Taurinensi Musaeo asservata, cum auctorum hodierne denominatione collatio. *Deutsche Entomologische Zeitschrift* **19**: 17-119.

Baudi di Selve F. 1875d: Catalogo dei tenebrioniti della fauna europea e circummediterranea appartenenti alle collezioni del Museo Civico di Genova. Parte seconda. *Annali del Museo Civico di Storia Naturale di Genova* **7**: 684-703.

Baudi di Selve F. 1876a: Europaeae et circummediterraneae faunae Tenebrionidum specierum, quae Comes Dejean in suo Catalogo, editio 3e consignavit, ex ejusdem collectione in R. Taurinensi Musaeo asservata, cum auctorum hodierne denominatione collatio. Pars altera. *Deutsche Entomologische Zeitschrift* **20**: 1-74.

Baudi di Selve F. 1876b: Coleotteri tenebrioniti delle collezioni italiane. *Bullettino della Società Entomologica Italiana* **7** [1875]: 91-106, 137-165, 209-237.

Baudi di Selve F. 1876c: Europaeae et circummediterraneae faunae Tenebrionidum specierum, quae Comes Dejean in suo Catalogo, editio 3e consignavit, ex ejusdem collectione in R. Taurinensi Musaeo asservata, cum auctorum hodierne denominatione collatio. Pars tertia. *Deutsche Entomologische Zeitschrift* **20**: 225-267.

Baudi di Selve F. 1876d: Generis Helops specierum methodicae dispositionis tentamen. *Deutsche Entomologische Zeitschrift* **20**: 268-281.

Baudi di Selve F. 1877a: *Coleotteri eteromeri esistenti nelle collezioni del R. Museo Zoologico di Torino ed in altre Italiane. Eteromeri delle famiglie susseguenti a quella dei tenebrioniti nei limiti della fauna europea e circummeditteranea. Volume 1.* Torino: Stamperia Reale, 163 pp. [note: this is a separate issue from the *Atti della Reale Accademia delle Scienze di Torino* **12**: 571-729].

Baudi di Selve F. 1877b: Coleotteri tenebrioniti delle collezioni italiane. *Bullettino della Società Entomologica Italiana* **9**: 25-54.

Baudi di Selve F. 1877c: Europae et circummediterraneae faunae Heteromerum specierum, quae Comes Dejean in suo Catalogo, editio 3[a], consignavit, ex ejusdem collectione in R. Taurinensi Musaeo asservata, cum auctorum hodiernae denominatione collatio. Pars quarta. *Deutsche Entomologische Zeitschrift* **21**: 385-416.

Baudi di Selve F. 1877d: Coleotteri eteromeri esistenti nelle collezioni del R. Museo Zoologico di Torino ed in altre Italiane. Eteromeri delle famiglie susseguenti a quella dei tenebrioniti nei limiti della fauna europea e circummeditteranea. *Atti della Reale Accademia delle Scienze di Torino* **12**: 571-729.

Baudi di Selve F. 1878a: Europaeae et circummediterraneae faunae Heteromerum specierum, quae Comes Dejean in suo Catalogo, edito 3a, consignavit, ex ejusdem collectione in R. Taurinensi Musaeo asservata, cum auctorum hodierne recepta denominatione collatio. Pars quinta. *Deutsche Entomologische Zeitschrift* **22**: 1-20.

Baudi di Selve F. 1878b: Europaeae et circummediterraneae faunae Heteromerum specierum, quae Comes Dejean in suo Catalogo, editio 3a, consignavit, ex ejusdem collectione in R. Taurinensi Musaeo asservata, cum auctorum hodierne recepta denominatione collatio. *Deutsche Entomologische Zeitschrift* **22**: 337-376.

Baudi di Selve F. 1878c: Coleotteri eteromeri esistenti nelle collezioni del R. Museo zoologico di Torino ed in altre italiane. *Atti della Reale Accademia delle Scienze di Torino* **13**: 765-866, 1027-1183.

Baudi di Selve F. 1879: [Une espèce nouvelle de *Formicomus*, de la collection du Musée royal.] *Annales de la Société Entomologique de Belgique* **22**: 87-88.

Baudi di Selve F. 1881: Heteromerum species ex Aegypto, Syria et Arabia. *Deutsche Entomologische Zeitschrift* **25**: 273-296.

Baudi di Selve F. 1883a: Note entomologiche. *Il Naturalista Siciliano* **3** [1883-1884]: 1-8.

Baudi di. Selve F. 1883b: Anthicus blechroides Baudi nov. spec. *Deutsche Entomologische Zeitschrift* **27**: 150.

Baudi di Selve F. 1883c: Descrizione d'una nuova specie di Zophosis. *Annali del Museo Civico di Storia Naturale di Genova* **18**: 760-761.

Baudi di Selve F. 1892: Di una nuova specie del genere Tritoma Geoffr. (Mycetophagus Hellw.). *Il Naturalista Siciliano* **11**: 153-154.

Baudi di Selve F. 1894: Viaggio del Dr. E. Festa in Palestina, nel Libano e regioni vicine. *Bollettino dei Musei di Zoologia e Anatomia Comparata della Reale Università di Torino* **9** (173): 1-13.

Beauregard H. 1890: *Les insectes vésicants*. Paris: Félix Alcan, xvi + 544 pp., 34 pls.

Beck L. von. 1817: *Beiträge zur beierischen Insektenfaune, oder Beschreibung und Abbildung neuentdeckter Käfer, mit angehängtem Namensverzeichniss der Eleuteraten des Landgerichtbezirks Zusmeshausen.* Augsburg: J. Wolff, 45pp., 7 pls.

Bedel E. M. L. 1869: [Une nouvelle espèce française du genre Notoxus.] *Bulletin de la Société Entomologique de France* **1869**: 34-35.

Bedel E. M. L. 1891a: Description de deux coléoptères hétéromères du nord de l'Afrique. *Annali del Museo Civico di Storia Naturale di Genova* **30**: 794-796.

Bedel E. M. L. 1891b: [Diagnose de trois coléoptères nouveaux du nord de l'Afrique.] *Bulletin de la Société Entomologique de France* **1890**: cxxxv-cxxxvi.

Bedel E. M. L. 1892: Reinseignements sur les moeurs et le développement des Meloidae. *L'Abeille, Journal d'Entomologie* **27**: 235-250.

Bedel E. M. L. 1893: Diagnoses de coléoptères nouveaux d'Algérie. *L'Abeille, Journal d'Entomologie* **28**: 102-104.

Bedel E. M. L. 1894: Révision des Omophlus et Heliotaurus de Barbarie. *L'Abeille, Journal d'Entomologie* **28**: 157-169.

Bedel E. M. L. 1895: [new taxon]. In: Bedel L. E. M. & Chobaut A.: Étude sur les Macrosiagon Hentz (Emenadia Lap.-Cast.) de la région méditerranéenne. *L'Abeille, Journal d'Entomologie* **28**: 181-192.

Bedel E. M. L. 1895: [new taxon]. In: Vauloger de Beaupré M.: Description de deux Notoxus nouveaux du nord de l'Afrique. *L'Abeille, Journal d'Entomologie* **28**: 193.

Bedel E. M. L. 1899: Diagnose d'un nouveau mylabre saharien. *Bulletin de la Société Entomologique de France* **1899**: 382-383.

Bedel E. M. L. 1903: Entomologie. Liste des principales espèces de coléoptères recueillis par le Mis de Segonzec, pp. 367-371. In: *Voyage au Maroc (1899-1901) par le Mis de Segonzec*. Paris: Armand Colin, xi + 408 + (1) pp., 11 pls, 1 map.

Bedel E. M. L. 1904: Communications. Liste de coléoptères récoltés à la Ferté-Alais (Seine-et-Oiso). *Bulletin de la Société Entomologique de France* **1904**: 210-212.

Bedel E. M. L. 1905: Communications. Description de deux coléoptères hétéromères nouveaux, de la côte occidentale du Maroc. *Bulletin de la Société Entomologique de France* **1905**: 272-273.

Bedel E. M. L. 1906a: Description de deux coléoptères hétéromères nouveaux de la côte occidentale du Maroc. *Bulletin de la Société Entomologique de France* **1905**: 272-273.

Bedel E. M. L. 1906b: Synonymies de coléoptères paléarctiques. *Bulletin de la Société Entomologique de France* **1906**: 91-93.

Bedel E. M. L. 1906c: Observations divers. *Bulletin de la Société Entomologique de France* **1906**: 177-178.

Bedel E. M. L. 1918: Diagnoses de coléoptères nouveaux du Maroc oriental. *Bulletin de la Société Entomologique de France* **1917**: 362-364.

Bedel E. M. L. 1921: Description d'un Zonabris de l'Ahaggar (Sahara central) (Col. Meloidae). *Bulletin de la Société Entomologique de France* **1921**: 199.

Belon R. P. M. J. 1882: [Des observations sur divers lathridiens]. *Bulletin de la Société Entomologique de France* **1882**: cxxiii-cxxv.

Berthold A. 1827: *Latreille's natürliche Familien des Thierreichs. Aus dem Französischen mit Anmerkungen und Zusätzen von Dr. Arnold Adolph Berthold.* Weimar: Landes-Industrie-Comptoir, x + 602 pp.

Bertolini S. 1868: Neue Käferarten des Trentinogebietes. *Verhandlungen der Kaiserlich-Königlichen Zoologisch-Botanischen Gesellschaft in Wien* **18**: 119-120.

Bertolini S. 1892: Contribuzione alla fauna Trentina dei Coleotteri. *Bullettino della Società Entomologica Italiana* **23**: 346-368.

Besser W. S. J. G. von. 1832: Additamenta et observatiunculae in Cel. Steven Tentyrias et Opatra collectionis Stevenianae nunc musei Universitatis Mosquensis. *Nouvelles Mémoires de la Société Impériale des Naturalistes de Moscou* **2**: 5-21.

Betrem G. 1929: Systematiek der Horiini. *Tijdschrift voor Entomologie* **72**: xxvi-xxix.

Betrem G. 1932: Beiträge zur Kenntnis des Tribus Horiini der Familie der Meloidae (Col.). 1. Die Systematik der Horiini. *Treubia* **14**: 85-101, 1 pl.

Betta V. 1847: *De quibusdam Coleopteris agri Ticinensis. Dissertatio.* Ticini regii: Fusi et Socii, 35

Biggs H. E. J. 1945: Two new species of Coleoptera (Scarabaeidae and Meloidae). *The Entomologist's Monthly Magazine* **81**: 110-111.

Billberg G. J. 1813: *Monographia Mylabridum.* Holmiae: Caroli Delén, 74 + 6 pp., 7 pls.

Billberg G. J. 1815: Insecta ex ordine Coleopterorum descripta. *Nova Acta Upsalae* **7**: 271-281, pl. IX.

Billberg G. J. 1820: *Enumeratio insectorum in Museo Gust. Joh. Billberg.* Stockholm: Typis Gadelianis, [4] + 138 pp.

Binaghi G. 1937: Studio sul genere Langelandia Aubé e descrizione di un nuovo genere di Coxelini (Col. Colydiidae). *Bollettino della Società Entomologica Italiana* **69**: 69-79.

Biström O. 1977: Nomenclatoric notes on Coleoptera. *Notulae Entomologicae* **57**: 17-18.

Blackburn T. 1888: Further notes on Australian Coleoptera, with description of new species. *Transactions of the Royal Society of South Australia* **10** [1886-1887]: 52-71, 177-287.

Blackburn T. 1891: Further notes on Australian Coleoptera, with descriptions of new genera and species. X. *Transactions of the Royal Society of South Australia* **14**: 292-345.

Blackburn T. 1907: Further notes on Australian Coleoptera, with description of new genera and species, xxxvii. *Transactions of the Royal Society of South Australia* **31**: 231-299.

Blair K. G. 1912: Descriptions of new species of Pyrochroidae. *The Annals and Magazine of Natural History* (8) **9**: 527-533.

Blair K. G. 1913: Some new species of Indian Tenebrionidae. *The Annals and Magazine of Natural History* (8) **12**: 56-58.

Blair K. G. 1914a: A revision of the family Pyrochroidae (Coleoptera). *The Annals and Magazine of Natural History* (8) **13**: 310-326.

Blair K. G. 1914b: On the Fabrician types of Tenebrionidae (Coleoptera) in the Banks collection. *The Annals and Magazine of Natural History* (8) **13**: 482-490.

Blair K. G. 1919: Notes on the Pythidae with descriptions of new species. *The Entomologist's Monthly Magazine* **55**: 112-124.

Blair K. G. 1921: Notes on the Indian species of Hypophloeus, Fabr. with description of new species. *The Entomologist's Monthly Magazine* (3) **7** [1920]: 1-7.

Blair K. G. 1922a: A new genus and some new species of Mordellidae. *The Entomologist's Monthly Magazine* **58**: 221-226.

Blair K. G. 1922b: The generic name Calycella Blair (Coleoptera). *The Entomologist's Monthly Magazine* **58**: 250.

Blair K. G. 1922c: Coleoptera of the Mt. Everest Expedition, 1921. *The Annals and Magazine of Natural History* (9) **9**: 558-562.

Blair K. G. 1923a: New species of heteromerous Coleoptera from Mesopotamia. *The Entomologist's Monthly Magazine* **59**: 118-126.

Blair K. G. 1923b: Coleoptera of the Second Mt. Everest Expedition, 1922. Part II. Heteromera. *The Annals and Magazine of Natural History* (9) **11**: 278-285.

Blair K. G. 1927: Heteromera of the Third Mt.Everest Expedition, 1924. *The Annals and Magazine of Natural History* (9) **19**: 241-255.

Blair K. G. 1928: Pythidae. Pars 99. In: Schenkling S. (ed.): *Coleopterorum Catalogus*. Berlin and s'Gravenhage: W. Junk, 41 pp.

Blair K. G. 1929: Some new species of myrmecophilous Tenebrionidae (Col.). *Zoologischer Anzeiger* **28** (*Wasmann-Festband*): 238-247.

Blair K. G. 1930a: The Indian species of Palorus, Muls. (Coleoptera: Tenebrionidae) and some associated beetles. *Indian Forest Records, Entomological Series* (Calcutta) **14**: 133-152, 1 pl.

Blair K. G. 1930b: Some new species of Indian Heteromera. *The Entomologist's Monthly Magazine* **66**: 177-181.

Blair K. G. 1931a: Some new species of Indian Heteromera (Col.). *The Entomologist's Monthly Magazine* **67**: 199-205.

Blair K. G. 1931b: A collection of Coleoptera from the Rub'Al Khali, southern Arabia, with descriptions of some new species and varieties. *The Entomologist's Monthly Magazine* **67**: 269-271.

Blair K. G. 1933: Tenebrionidae from Central Arabia collected by Major H. Sr. J. B. Philby. *The Entomologist's Monthly Magazine* **69**: 4-7.

Blair K. G. 1935a: Some synonymic notes in the family Tenebrionidae (Col.). *The Entomologist's Monthly Magazine* **71**: 102-104.

Blair K. G. 1935b: Oedemeridae. Pp. 302-303. In: Visser C. & Visser-Hooft J. (eds.): *Wissenschaftliche Ergebnisse der niederländischen Expeditionen in den Karakorum und die angrenzenden Gebiete in den Jahren 1922, 1925 und 1929/30*. Leipzig: F. A. Brockhaus, 501 + xviii pp., 8 pls.

Blair K. G. 1937: Some new species of Indian Heteromera (Col.) (3). *The Entomologist Monthly Magazine* **73**: 35-37.

Blair K. G. 1938: A new genus and species of tenebrionid beetles in bee-hives in India. *The Entomologist's Monthly Magazine* **74**: 222-223.

Blair K. G. 1942: Coleoptera Heteromera from Guam, pp.56-60. In: Insects of Guam I. *Bulletin of the Bernice P. Bishop Museum* **172**: v + 218 pp., 7 pls.

Blanchard É. 1843: Insectes de l'Amérique méridionale recueillis par Alcide d'Orbigny. In: Blanchard É. & Brullé A.: *Insectes du Voyage dans l'Amérique méridionale executé pendant les années 1828, 1829, 1830, 1831, 1832 et 1833 par Alcide d'Orbigny. Tome sixième 2e partie: Insectes.* [1837-1843]. Paris: P. Bertrand et Strasburg: V. Levrault, 222 pp., 32 pls.

Blanchard É. 1844: Zoologie. Insectes recueillis à l'Himalaya par Victor Jacquemont. Pp. 13-31. In: *Voyage dans l'Inde par Victor Jacquemont pendant les années 1828 à 1832. Publié sous les auspices de M. Guizot, Ministre de l'Instruction Publique. Descriptions des collections par MM. Isidore Geoffroy Saint-Hilaire.-- Milne Edwards.—Émile Blanchard. Valenciennes.-- Cambessèdes.-- J. Decaisne. Tome quatrième.* Paris: Typographie de Firmin Didot Frères [issued in parts].

Blanchard É. 1845: *Histoire des insectes, traitant de leur moeurs et de leur métamorphoses en générale et comprenant une nouvelle classification fondée sur leur rapports naturelles.* In: *Traité Complet d'Histoire Naturelle. Histoires des insectes. Tome second.* Paris: Firmin Didot Frères, 524 pp., pls. 11-20.

Blanchard É. 1853: Insectes. In: Hombrom J. B. & Jacquinot H. (eds): Atlas d'Histoire naturelle Zoologie. In: *Voyage au Pôle Sud et dans l'Océanie sur les corvettes l'Astrolabe et la Zélée, executé par l'ordre du Roi pendant les année 1837-1838-1839-1840 sous le commandement de M. J. Dumont-d'Urville, Capitaine de vaisseau; Zoologie. Tome Quatrième. Deuxième partie. Coléoptères et autres ordres.* Paris: J. Tastu, vii + 716 pp., 20 pls. [plates issued in 1847].

Bland J. H. B. 1863: Descriptions of a few supposed new species of North American Coleoptera. No. 2. *Proceedings of the Entomological Society of Philadelphia* **2**: 319.

Bleuse L. 1899: Description d'un mylabre du Sud-Oranais (Col.). *Bulletin de la Société Entomologique de France* **1899**: 383.

Blinstein S. J. 1988: Novye i maloizvestinye vidy bistryanok (Coleoptera, Anthicidae) s yuga Ukraini. Pp. 73-78. In: *Ekologiya i taksonomiya nasekomikh Ukrainskoi SSR. Sbornik Nauchnikh Trudov* [*Ecology and taxonomy of Ukrainian insects. Collection of Scientific Works*]. Kiev: Akademiya Nauk Ukrainskoy SSR.

Bodemeyer B. von. 1916: Neuheiten der paläarktischen Coleopterenfauna. *Deutsche Entomologische Zeitschrift* **61**: 111-112.

Bodemeyer B. von. 1927: *Ueber meine Entomologischen Reisen nach Kleinasien (1911), Ost-Sibirien, Schilka und Amur (1912), Tunis, Oasis Gafsa, Khroumerie (1913) und Iran, das Elbursgebirge (1914). Bd. IV. Iran, das Elbursgebirge.* Stuttgart: Albert Kerner, 96 pp., 1 pl.

Bogatchev A. V. 1934: Materialy k poznaniyu fauny zhukov Apsheronskogo poluostrova. *Trudy Azerbaydzhanskogo Otdeleniya Zakavkazkogo Filiala Akademii Nauk SSSR, Sektsiya Zoologii* **7**: 14-71.

Bogatchev A. V. 1939a: Novye vidy sem. Tenebrionidae. *Trudy Zoologicheskogo Instituta Akademii Nauk SSSR - Azerbaidschanskiy Filial* **10**: 123-130.

Bogatchev A. V. 1939b: Novye formy Mordellidae (Coleoptera) iz AzSSR. *Trudy Zoologicheskogo Instituta Akademiya Nauk SSSR - Azerbaidschanskiy Filial* **10**: 131-134.

Bogatchev A. V. 1939c: Fauna Binagadinskikh Kirovykh Plastov. Zhuki-Coleoptera. *Izvestiya Azerbaydzhanskogo Filiala Akademii Nauk SSSR* **2**: 137-160.

Bogatchev A. V. 1940: Kharakternye pribavki k faune SSSR po otryadu zhestkokrylykh (Coleoptera). I. *Byulleten Moskovskova Obshchestva Ispitalelov Prirody, Otdel Biologii* **49**: 201-209.

Bogatchev A. V. 1941a: Novye vidy roda Adesmia Fisch.-W. *Izvestiya Azerbaydzhanskogo Filiala Akademii Nauk SSSR* **3**: 96-100.

Bogatchev A. V. 1941b: Novye vidy triby Blaptini (Coleoptera). *Izvestiya Azerbaydzhanskogo Filiala Akademii Nauk SSSR* **4**: 109-111.

Bogatchev A. V. 1946a: Dve novykh Microdera Eschsch. (Coleoptera, Tenebrionidae). *Doklady Akademii Nauk AZSSR* (*Proceedings of the Academy of Sciences of the Azerbaijan SSR*) **2** (6): 210-211.

Bogatchev A. V. 1946b: Novyy Cylindronotus iz doliny Kury. *Doklady Akademii Nauk AZSSR* (*Proceedings of the Academy of Sciences of the Azerbaijan SSR*) **2** (6): 245-247.

Bogatchev A. V. 1946c: Novye chernotelki (Tenebrionidae) Palearctiki. *Doklady Akademii Nauk Azerbaydzhanskoy SSR* (*Proceedings of the Academy of Sciences of the Azerbaijan SSR*) **2** (9): 391-394.

Bogatchev A. V. 1946d: Novye vidy triby Platyscelini (Coleoptera, Tenebrionidae). *Doklady Akademii Nauk Azerbaydzhanskoy SSR* **2** (10): 435-437.

Bogatchev A. V. 1947a: Quelques espèces nouvelles des coléoptères d'Iran. *Entomologie et Phytopathologie Appliquées* (Teheran) **4**: 15-18.

Bogatchev A. V. 1947b: Novyy vid roda Cyphostethe Mars. iz doliny Araksa (Coleoptera, Tenebrionidae). *Doklady Akademii Nauk Armyanskoy SSR* (*Proceedings of the Academy of Sciences of the Armenian SSR*) **7** (2): 79-81.

Bogatchev A. V. 1947c: Novye vidy roda Platynoscelis Seidl. *Doklady Akademii Nauk Azerbaydzhanskoy SSR* (*Proceedings of the Academy of Sciences of the Azerbaijan SSR*) **3** (2): 84-88.

Bogatchev A. V. 1947d: Novye vidy Tentyriini. *Doklady Akademii Nauk Arerbaydzhanskoy SSR* **3** (10): 460-463.

Bogatchev A. V. 1947e: K systematike triby Blaptini (Tenebrionidae, Col.). *Doklady Akademii Nauk Arerbaydzhanskoy SSR* **3** (11): 513-515.

Bogatchev A. V. 1949a: Novye vidy Tenebrionidae Azerbaydzhana i drugikh oblastey Palearktiki. *Doklady Akademii Nauk Azerbaydzhanskoy SSR* **5** (1): 38-41.

Bogatchev A. V. 1949b: Novye palearcticheskie vidy Epitragini i Pimeliini (Tenebrionidae). *Doklady Akademii Nauk Azerbaydzhanskoy SSR* **5** (7): 277-280.

Bogatchev A. V. 1950: Novye maloizvestnye vidy triby Erodiini (Coleoptera, Tenebrionidae). *Entomologicheskoe Obozrenie* **31**: 231-236.

Bogatchev A. V. 1951: Novye formy roda Lobodera Muls. (Coleoptera, Tenebrionidae). *Trudi Krimskogo Filiala Academemii Nauk SSSR* **2**: 69-72.

Bogatchev A. V. 1952a: Novyy rod i vid zhukov-chernotelok iz severnych Kara-Kumov. *Doklady Akademii Nauk UZSSR* **7**: 44-45.

Bogatchev A. V. 1952b: Novye chernotelki (Tenebrionidae) iz SSSR i sopredelnykh stran. *Entomologicheskoe Obozrenie* **32**: 284-286.

Bogatchev A. V. 1953: Novye formy roda Pimelia F. (Coleoptera, Tenebrionidae). *Entomologicheskoe Obozrenie* **33**: 302-306.

Bogatchev A. V. 1957: Novye vidy chernotelok rodov Zophosis Latr., Erodius F. i Earophanta Sem. *Zoologicheskiy Zhurnal* **36** (4): 529-532.

Bogatchev A. V. 1959: Novye vidy zhukov-chernotelok (Tenebrionidae) iz Tadzhikistana. In: Zapovednik "Tigrovaya Balka". *Trudy Akademii Nauk Tadzhikskoy SSR* **115**: 63-67.

Bogatchev A. V. 1960a: Novyy rod zhukov-chernotelok iz Tadzhikistana Allotadzhikistania, gen. nov. (Tenebrionidae, Pimeliini). *Doklady Akademii Nauk Tadzhikskoy SSR* **3** (1): 43-46.

Bogatchev A. V. 1960b: Novyy svoeobraznyy rod zhukov-chernotelok, naydenykh v Tadzhikistane – Tadzhikistania, gen. nov. (Coleoptera, Tenebrionidae). *Doklady Akademii Nauk Tadzhikskoy SSR* **3** (3): 35-37.

Bogatchev A. V. 1960c: Novye vidy Prosodes (Coleoptera, Tenebrionidae) iz Tadzhikistana. *Izvestiya Otdeleniya Selskokhozyaystvennych i Biologicheskikh Nauk* **2**: 79-86.

Bogatchev A. V. 1960d: [new taxa]. In: Bogatchev A. V. & Kryzhanovskiy O. L.: Novye i maloizvestnye vidy chernotelok (Coleoptera, Tenebrionidae) iz Zapadnoy Turkmenii. *Trudy Zoologicheskogo Instituta Akademii Nauk SSSR* **27**: 264-275.

Bogatchev A. V. 1961a: Novye vidy zhukov-chernotelok (Tenebrionidae) iz Tadzhikskoy SSR, drugikh respublik Sredney Azii i sopredelnykh stran. *Trudy Instituta Zoologii i Parazitologii Akademii Nauk Tadzhikskoy SSR* **20**: 107-121.

Bogatchev A. V. 1961b: Novyy vid roda Diesia Fisch.-W. (Coleoptera, Tenebrionidae). *Entomologicheskoe Obozrenie* **40** (2): 369-370.

Bogatchev A. V. 1962: Novye vidy triby Akidini (Tenebrionidae) iz Srednej Azii. *Doklady Akademii Nauk Tadzhikskoy SSR* **5** (1): 44-46.

Bogatchev A. V. 1963a: Novyy rod zhukov-chernotelok iz pustiny Kyzylkum (Col., Tenebrionidae, Tentyriini). *Doklady Akademii Nauk Uzbekskoy SSR* **6**: 57-59.

Bogatchev A. V. 1963b: Novye vidy chernotelok (Tenebrionidae) s yuga Sredney Azii. *Trudy Instituta Zoologii i Parazitologii im. akad. E. N. Pavlovskogo Akademii Nauk Tadzhikskoy SSR* **24**: 94-109.

Bogatchev A. V. 1964a: Dve novye formy roda Pimelia Fabr. (Coleoptera, Tenebrionidae). *Uzbekskiy Biologichesdkiy Zhurnal* **2**: 42-45.

Bogatchev A. V. 1964b: Vnetrevidovye formy chernotelki Adesmia fischeri Fald. 1837 (Coleoptera, Tenebrionidae). *Entomologicheskoe Obozrenie* **43** (1): 175-179.

Bogatchev A. V. 1965: [new taxon, p. 372]. In: Bey-Bienko G. Ya. (ed.): *Opredilitel nasekomykh evropeyskoy chasti SSSR v pyati tomakh. II. Zhestkokrylye i veerokrylye.* Nauka, Moskva-Leningrad.

Bogatchev A. V. 1967a: Novyy rod i vid zhukov-chernotelok s Apsheronskogo poluostrova. *Trudy Instituta Zoologii Akademii Nauk Azerbaydzhanskoy SSR* **26**: 157-163.

Bogatchev A. V. 1967b: Chernotelki roda Epitrichia Gebl. (Coleoptera, Tenebrionidae). *Entomologicheskoe Obozrenie* **46**: 348-352.

Bogatchev A. V. 1967c: [new taxon]. In: Medvedev G. S.: Obzor chernotelok triby Leptodini (Coleoptera, Tenebrionidae). *Entomologicheskoe Obozrenie* **46**: 353-360.

Bogatchev A. V. 1972: Novye vidy zhukov-chernotelok triby Opatrini (Coleoptera, Tenebrionidae). *Entomologicheskoe Obozrenie* **51**: 625-632.

Bogatchev A. V. 1976: Novyy rod i vid zhukov-chernotelok (Coleoptera, Tenebrionidae) iz Tadzhikistana. *Entomologicheskoe Obozrenie* **55**: 98-100.

Bogatchev A. V. & Kryzhanovskiy O. L. 1955: Novyy vid chernotelok (Coleoptera, Tenebrionidae) iz zapadnoy Turkmenii. *Entomologicheskoe Obozrenie* **34**: 240-241.

Bogatchev A. V. & Kryzhanovskiy O. L. 1960: Novye i maloizvestnye vidy chernotelok (Coleoptera, Tenebrionidae) iz zapadnoy Turkmenii. *Trudy Zoologicheskogo Instituta Akademii Nauk SSSR* **27**: 264-275.

Bogatchev A. V. & Medvedev G. S. 1974: Novyy vid zhukov-chernotelok roda Blaps Fabr. (Coleoptera, Tenebrionidae) iz Turkmenii. *Entomologicheskoe Obozrenie* **53** (3): 645-646.

Bogatchev A. V. & Nepesova M. G. 1980: Zhuki-chernotelki roda Pachyscelis Sol. (Coleoptera, Tenebrionidae) fauny Turkmenii. *Entomologicheskoe Obozrenie* **59** (3): 607-611.

Bogatchev A. V. & Semenov A. P. 1947: Novyy rod triby Eurychorini (Col., Tenebrionidae). *Doklady Akademii Nauk Azerbayzhanskoy SSR* **3** (4): 175-177.

Bogdanov-Katjkov N. N. 1915: De speciebus novis vel parum cognitis Tentyriinorum (Coleoptera, Tenebrionidae). Novye i maloizvestnye vidy podsemeystva Tentyriinae (Coleoptera, Tenebrionidae). *Russkoe Entomologicheskoe Obozrenie* **15** (1): 1-7.

Bogdanov-Katjkov N. N. 1916a: Matériaux pour servir à l'étude des Gnathosia (Coleoptera, Tenebrionidae). *Revue Russe d'Entomologie* **16**: 68-71.

Bogdanov-Katjkov N. N. 1916b: Espèces nouvelles ou peu connues des tentyriines (Coleoptera, Tenebrionidae). *Ezhegodnik Zoologicheskago Muzeya Imperatorskoy Akademii Nauk* **20** [1915]: xxix-xxx.

Boheman C. H. 1851: Bidrag till Gottlands insekten-fauna. *Öfversigt af Kongliga Vetenskaps-Akademiens Förhandlingar* **1849**: 194-267.

Boheman C. H. 1858: Coleoptera. Species nova descripsit. Pp. 1-112. In: Virgin C. H. (ed.): *Kongliga Svenska Fregatten Eugenies Resa omkring jorden unter befäll af C. A. Virgin ären 1851-1853. Vetenskapliga Iakttagelser pa H. M. Konung Oscart den Försten befallning utgifna af K. Svenska Vetenskaps Akademien. II. Zoologi. 1. Insecta.* Uppsala et Stockholm: P. A. Norstedt & Söner, Almquist et Wiksells, 614 pp., 9 pls. [issued in parts, pp. 113-217 in 1859].

Böhm R. 1908: Les Adesmides des déserts égyptiens. *Bulletin de la Société Entomologique d'Egypte* **1** [1908-1909]: 151-165.

Böhm R. 1909: Les Zophosini et Erodiini de l'Egypte. *Bulletin de la Société Entomologique d'Egypte* **2**: 108-127.

Böhm R. 1916a: Sur les caractères spécifiques des Tentyria et Tentyrina d'Egypte. *Bulletin de la Société Entomologique d'Egypte* **4** [1914-1918]: 49-51.

Böhm R. 1916b: Sur les caractères spécifiques des Ocnera égyptiens. *Bulletin de la Société Entomologique d'Egypte* **4** [1914-1918]: 64-69.

Boieldieu A. 1865: Quelques coléoptères nouveaux des îles d'Eubée et Baléares. *Annales de Société Entomologique de France* (4) **5**: 5-11.

Boisduval J. B. A. D. de. 1835: [new taxa]. In: Hombrom J. B. & Jacquinot H. (eds): Atlas d'Histoire naturelle Zoologie. In: *Voyage de découvertes de l'Astrolabe exécuté par l'ordre du Roi, pendant les années 1826-1827-1828-1829, de M. Dumont d'Urville. Faune Entomologique de l'Océan pacifique, avec l'illustration des insectes nouveaux recueillis pendant le voyage. Deuxième partie. Coléoptères et autres ordres.* Paris: J. Tastu, vii + 716 pp., 20 pls. [plates issued in 1847].

Bolívar y Pieltain C. 1914: Asida (Alphasida) Merceti nov. sp. (Col. Tenebrionidae). *Boletín de la Real Sociedad*

Española de Historia Natural **14**: 237-238.

Bolívar y Pieltain C. 1917: Descripción de una especie española del género Rhipidius (Col. Rhipiphoridae). *Boletín de la Real Sociedad Española de Historia Natural* **17**: 249-252.

Bologna M. A. 1983: Una nuova specie di Lytta del Kashmir (Coleoptera, Meloidae). *Fragmenta Entomologica* **17**: 125-138.

Bologna M. A. 1988: Note su Eurymeloe e revisione delle specie euromediterrenee del gr. *rugosus* (Coleoptera, Meloidae). *Fragmenta Entomologica* **20**: 233-301.

Bologna M. A. 1989a: Un nuovo Alosimus del litorale tirrenico e note tassonomiche su altre specie congeneriche nord africane (Coleoptera, Meloidae). *Fragmenta Entomologica* **21**: 191-204.

Bologna M. A. 1989b: Berberomeloe, a new west Mediterranean genus of Lyttini for Meloe majalis Linné (Coleoptera, Meloidae). Systematics and bionomics. *Bollettino di Zoologia* **55** [1988]: 359-366.

Bologna M. A. 1991: Un nouvel Oenas de Grèce et note synonymique sur une autre espèce du Proche-Orient (Coleoptera, Meloidae). *Revue Française d'Entomologie* (N. S.) **13**: 175-179.

Bologna M. A. 1992: [new taxon]. In: Bologna M. A. & Aloisi G.: Systematics of Lydomorphus Fairmaire 1882, with description of the first instar larva of L. dusalti (Dufour, 1821) (Coleoptera Meloidae). *Tropical Zoology* **5**: 55-71.

Bologna M. A. 1995a: A new Eleticinae genus from Ethiopia and remarks on the tribe Morphozonitini (Coleoptera: Meloidae). *Annales de la Société Entomologique de France* (N. S.) **31**: 185-193.

Bologna M. A. 1995b: Coleoptera Meloidae ed Oedemeridae di Lampedusa, Pantelleria e delle isole Maltesi. *Il Naturalista Siciliano*, Suppl. **19**: 505-525.

Bologna M. A. 2006: [new taxa]. In: Bologna M. A. & Di Giulio A.: Revision of the genus Teratolytta (Coleoptera: Meloidae). *European Journal of Entomology* **103**: 137-161.

Bologna M. A. & Coco E. 1991: Revisione del genere Croscherichia Pardo Alcaide, 1950 (Coleoptera Meloidae). *Memorie della Società Entomologica Italiana* **69** [1990]: 97-180.

Bologna M. A. & Pinto J. D. 2001: Phylogenetic studies of Meloidae (Coleoptera), with emphasis on the evolution of phoresy. *Systematic Entomology* **26**: 33-72.

Bonadona P. 1952: Notes sur les Anthicidae paléarctiques. *Revue Française d'Entomologie* **19**: 233-237.

Bonadona P. 1953: Les Anthicus français du groupe du tristis Schmidt. *Revue Française d'Entomologie* **20**: 253-264.

Bonadona P. 1954a: In Mateu J.: Coleópteros de la Sierra Nevada. Familias: Dytiscidae, Hydrophilidae, Dryopidae, Catopidae, Silphidae, Pselaphidae, Histeridae, Cleridae, Cantharidae, Dasytidae, Aderidae, Anthicidae. *Archivos del Instituto de Aclimatación, Almería* **2**: 89-101, pl. 1.

Bonadona P. 1954b: Notes sur les anthicides paléarctiques. *Revue Française d'Entomologie* **21**: 272-280.

Bonadona P. 1956: Anthicides recueillis par. F. Pierre dans le Sahara algérien (Coleoptera). *Revue Française d'Entomologie* **23**: 116-120.

Bonadona P. 1958: Notes sur les Anthicidae paléarctiques (Coleoptera). *Revue Française d'Entomologie* **25**: 287-297.

Bonadona P. 1960a: Notes sur les Anthicidae paléarctiques (Coleoptera). *Revue Française d'Entomologie* **27**: 51-57.

Bonadona P. 1960b: Les Endomia Castelnau d'Afrique et de Madagascar (Coléoptères Anthicidae). *Revue de Zoologie et de Botanique Africaines* **62**: 293-328.

Bonadona P. 1962: Coléoptères Anthicidae d'Afrique Noire et de Madagascar. *Revue de Zoologie et de Botanique Africaines* **65**: 325-351.

Bonadona P. 1963: XII. Anthicidae de l'Archipel de Madère. *Commentationes Biologicae* **25**: 105-107.

Bonadona P. 1964: Note sur les Anthicidae paléarctiques. *Revue Française d'Entomologie* **31**: 225-249.

Bonadona P. 1965: Anthicidae, pp. 862-865. In: Thérond J. and Hollande A.: Contribution à l'étude des coléoptères de la région de Béni-Abbès et de la Vallée de la Saoura. *Annales de la Société Entomologique de France* **1**: 851-877.

Bonadona P. 1969: Le Parc National du Niokolo-Koba (Sénégal). XIII. Coleoptera Anthicidae. *Mémoires de l'Institut Fondamental d'Afrique Noire* **84**: 263-277.

Bonadona P. 1970: Contribution a la faune de l'Iran. 18. Coléoptères Anthicidae. *Annales de la Société Entomologique de France* **6**: 379-383.

Bonadona P. 1977a: Notes sur les Anthicidae paléarctiques (Col.). *L'Entomologiste* **33**: 2-11.

Bonadona P. 1977b: Anthicus genei et formes affines (Col. Anthicidae). *L'Entomologiste* **33**: 84-87.

Bonadona P. 1978: Les Tomoderini subendogés d'Afrique centrale et de l'Inde méridionale (Col. Anthicidae). *Revue Suisse de Zoologie* **85**: 645-656.

Bonadona P. 1979: Une Isomira nouvelle de France méditerranéenne (Col. Alleculidae). *L'Entomologiste* **35**: 2-5.

Bonadona P. 1984: Description d'une Microhoria nouvelle du Portugal (Coleoptera, Anthicidae). *Revue Française d'Entomologie* **6**: 71-72.

Bonadona P. 1986: Notes sur les Anthicidae paléarctiques. (Insecta, Coleoptera). XVI. Descriptions d'un Clavicomus Pic et de quatre Microhoria Chevrolat. *Biocosme Mésogéen* **3**: 35-47.

Bonadona P. 1988: Note sur les Anthicidae paléarctiques (suite). *Biocosme Mésogéen* **5**: 39-49.

Bonadona P. 1989a: Les espèces françaises du genre Notoxus Geoffroy (Coleoptera Anthicidae). *L'Entomologiste* **45**: 9-13.

Bonadona P. 1989b: Anthicidae (Coleoptera) nouveaux des collections du Muséum d'histoire naturelle de Genève. *Revue Suisse de Zoologie* **96**: 253-276.

Bonelli F. A. 1812: Specimen faunae subalpinae sistens Insecta Pedemontii hucusque inedita, aut rariora, aut ea quae commodi damnive gratia quod inferunt, prudentis agricolae magis interest cognoscere. Per genera naturalia et species digesta. Fasciculus I. Coleoptera plerumque inedita comprehendens. *Memorie della R. Società di Agricultura di Torino* **9**: 149-183, pls. 1-6.

Bonometto L. 1976: Le *Phaleria* di Sardegna. Note sistematiche ed ecologiche (Coleoptera Tenebrionidae). *Lavori, Socità Veneziana di Scienze Naturali* **1**: 28-31.

Borchmann F. 1909a: Neue asiatische und australische Lagriiden hauptsächlich aus dem Museum in Genua. *Bollettino della Società Entomologica Italiana* **41** [1911]: 201-234.

Borchmann F. 1909b: Systematische und synonymische Notizen über Lagriiden und Alleculiden. (Col.) *Deutsche Entomologische Zeitschrift* **53**: 712-714.

Borchmann F. 1912a: H. Sauter's Formosa-Ausbeute. Lagriidae, Alleculidae, Cantharidae (Col.). *Supplementa Entomologica* **1**: 6-12.

Borchmann F. 1912b: Neue afrikanische Lagriiden. (Material zur Monographie der afrikanischen Lagriiden.) *Coleopterologische Rundschau* **1**: 17-21.

Borchmann F. 1915a: Die Lagriinae (Unterfamilie der Lagriidae.). *Archiv für Naturgeschichte* **A 81** [1916]: 46-188.

Borchmann F. 1915b: Lagriidae und Alleculidae des „Indian Museum". *Records of the Indian Museum* **11**: 179-188.

Borchmann F. 1915c: Eine neue Gattung der Statirinae (Col.). *Entomologische Mitteilungen* **4**: 296-299.

Borchmann F. 1916: H. Sauter's Formosa-Ausbeute: Alleculidae und Othniidae (Col.). *Archiv für Naturgeschichte* **A 82** (5): 101-108.

Borchmann F. 1917: Meloidae, Cephaloidae. Pars 69. In S. Schenkling (ed.): *Coleopterorum Catalogus*. Berlin: W. Junk, 208 pp.

Borchmann F. 1925: Neue Heteromeren aus malayischen Gebiete. *Treubia* **6**: 329-354.

Borchmann F. 1929: Ueber die von Herrn J. B. Corporaal in Ost-Sumatra gesammelten Lagriiden, Alleculiden, Meloiden und Othniiiden. *Tijdschrift voor Entomologie* **71**: 1-39.

Borchmann F. 1930a: Eine neue Cistelina-Art aus der Ausbeute der Stötznerschen Szetschwan-Expedition. *Entomologische Blätter* **26**: 118-119.

Borchmann F. 1930b: Die Gattung Cteniopinus Seidlitz. *Koleopterologische Rundschau* **15**: 143-164.

Borchmann F. 1930c: Eine neue Alleculidenart aus Transbaikalien. *Koleopterologische Rundschau* **15**: 244-245.

Borchmann F. 1930d: Die Lagriiden-Fauna der Philippinen. *Philippine Journal of Science* **41**: 403-533.

Borchmann F. 1932: Fünf neue ostchinesische Heteromeren. (Aus der Kollektion Em. Reitter.) *Entomologisches Nachrichtenblatt* **6**: 91-96.

Borchmann F. 1935: Meloidae (Col.). Pp. 308-309. In: Visser C. & Visser-Hooft J. (eds.): *Wissenschaftliche Ergebnisse der Niederländischen Expeditionen in den Karakorum und die angrenzenden Gebiete in den Jahren 1922, 1925 und 1929/1930. Vol. 1.* Leipzig: F. A. Brockhaus, xviii + 499 pp., 1 map, 8 pls.

Borchmann F. 1936: Coleoptera Heteromera Fam. Lagriidae. In: Wytsman P. (ed.): *Genera Insectorum. Fasc. 204*. Brussels: Louis Desmet-Verteneuil, 561 pp., 9 pls.

Borchmann F. 1937: Neue Alleculiden aus dem Deutschen Entomologischen Institut, Berlin-Dahlem. (Coleoptera.). *Arbeiten über Morphologische und Taxonomische Entomologie aus Berlin Dahlem* **4**: 210-231.

Borchmann F. 1938a: Neue Alleculiden aus dem Museum der Stadt Stettin (Col.). *Stettiner Entomologische Zeitung* **99**: 292-298.

Borchmann F. 1938b: Über die von Herrn Dr. C. Rathjens im Jahre 1931 in Arabien gesammelten Meloidae. *Entomologische Blätter* **33**: 463-466.

Borchmann F. 1940a: Alleculidae aus dem Deutschen Entomologischen Institut. (Coleoptera.). *Arbeiten über Morphologische und Taxonomische Entomologie aus Berlin Dahlem* **7**: 154-158.

Borchmann F. 1940b: Neue Meloiden-Arten (Col.). *Mitteilungen der Münchener Entomologischen Gesellschaft* **30**: 592-621.

Borchmann F. 1941: Über die von Herrn J. Klapperich in China gesammelten Heteromeren. *Entomologische Blätter* **37**: 22-29.

Borchmann F. 1942a: Entomological results from the Swedish Expedition 1934 to Burma and British India. Coleoptera: Lagriidae und Alleculidae. Gesammelt von René Malaise. *Arkiv för Zoologi* **33 A** (9): 1-32.

Borchmann F. 1942b: Neue Meloiden-Arten (Col.) II. *Mitteilungen der Münchener Entomologischen Gesellschaft* **32**: 682-712.

Borges P. A. V. 1991: Two new species of Tarphius Erichson, 1848 (Coleoptera, Colydiidae) from the Azores. *Bocagiana* **143**: 1-11.

Bosc d'Antic L. A. G. 1791: Mémoire sur l'établissement d'un nouveau genre d'insectes coléoptères. Pp. 327-328. In: *La médecine éclairée par les sciences physiques, ou journal des découvertes relatives aux différentes parties de l'art de guérir. Vol. 1.*

Bosc d'Antic L. A. G. 1792: [untitled]. *Journal d'Histoire Naturelle* **2**: 293-296, pl. 40.

Bose F. C. 1858: Rhinosimus ruficeps nov. spec. *Stettiner Entomologische Zeitung* **19**: 96.

Bouchard P., Lawrence J. F., Davies A. E. & Newton A. F. 2005: Synoptic classification of the world Tenebrionidae (Insecta: Coleoptera) with a review of family-group names. *Annales Zoologici* **55**: 499-530.

Bouchard P., Löbl I. & Merkl O. 2007: Nomenclatural notes on tenebrionid beetles of the Palaearctic region (Insecta, Coleoptera). *Annales Zoologici* **57**: 385-394.

Bouskell F. 1899: Notes on some aberrations of Osphya bipunctata Fab. *The Entomologist`s Record and Journal of Variation* **11** (1): 265 – 267

Brahm N. J. 1790: *Insektenkalendar für Sammler und Oekonomen. Erster Theil.* Mainz: Kurfürstl. Privil. Universitätsbuchhandlung, lxvii + 248 pp.

Brancsik C. 1874: Neue Coleopteren aus Ungarn. *Berliner Entomologische Zeitschrift* **18**: 227-231.

Brancsik C. 1899: Aliquot Coleoptera novae Russiae asiaticae. *Évkönyve, A Trencsén Vármegyei Természettudományi Egylet* [*Jahresheft des Naturwissenschaftlichen Vereines des Trencsiner Comitates*] **21-22** [1898-1899]: 97-105 + [1], pl. 4.

Brancsik C. 1914: Coleoptera nova. *A Trencsén Megyei Múzeumi Egyesület Értesítője* **1914**: 58-69.

Brandt J. F. & Erichson W. F. 1832: Monographia generis Meloe. *Nova Acta Academiae Leopoldina* **16** (1): 101-142, 1 pl.

Brandt J. F. & Ratzenburg J. T. C. 1833: *Medizinische Zoologie, oder getreue Darstellung und Beschreibung der Thiere die in Arzeneimittellehre in Betracht kommen, in systematischer Folge herausgegeben. Tome II.* [1830-1833]. Berlin: Hirschwald, 4 + 364 pp., 39 pls. [Tome I, 1827-1829: 6 + 198 pp., 24 pls].

Brenske E. & Reitter E. 1884: Neuer Beitrag zur Käferfauna Griechenlands. *Deutsche Entomologische Zeitschrift* **28**: 17-100.

Breit J. 1908: Eine koleopterologische Sammelreise auf Mallorka (Balearen). *Verhandlungen der Kaiserlich-Königlichen Zoologisch-Botanischen Gesellschaft in Wien* **58**: 52-67.

Breit J. 1914: Neue Coleopterenformen aus Süd-Europa. *Coleopterologische Rundschau* **3**: 51-62.

Brême F. A. G. de. 1842: Note monographique sur le genre Misolampus. *Revue Zoologique* **1842**: 81-83.

Brême F. A. G. de. 1846: *Essai monographique et iconographique de la tribu des Cossyphides. Deuxième Partie. (Cossyphus proprement dit).* Paris: Victor Masson, 31 pp., 3 pls.

Bremer H. J. 1998: Revision der orientalischen Corticeus-Arten (Col., Tenebrionidae, Hypophloeini). I. Teil. *Acta Coleopterologica* **14**: 3-32.

Bremer H. J. 1999: Revision der orientalischen Corticeus-Arten (Col., Tenebrionidae, Hypophloeini). II. Teil. *Acta Coleopterologica* **15**: 3-64.

Bremer H. J. 2001: Revision der Gattung Amarygmus Dalman, 1823 und verwandten Gattungen. I. Allgemeine Bemerkungen, Status einiger Gattungen affine Amarygmus Dalman, 1823; neue Kombinationen von Arten der Gattung Amarygmus Dalman (Coleoptera: Tenebrionidae: Alleculinae: Amagygmini). *Coleoptera* **5**: 57-80.

Bremi-Wolf J. J. 1856: *Catalog der Schweizerischen Coleopteren, als Vorläufer der Beiträge für schweizerische Entomologie.* Zürich: Schulthess, [iii]-vi +78 pp.

Brenske E. 1884: [new taxa]. In: Brenske E. & Reitter E.: Neuer Beitrag zur Käferfauna Griechenlands. *Deutsche Entomologische Zeitschrift* **28**: 17-100, pls 1, 2.

Brisout de Barneville C. 1859: Descriptions de quelques espèces nouvelles de coléopterès français. *Bulletin de la Société Entomologique de France* **1859**: 231-235.

Brisout de Barneville C. 1862: Description de trois insectes coléoptères nouveaux qui se trouvent en France. *Revue et Magasin de Zoologie Pure et Appliquée* (2) **14**: 23-25.

Brisout de Barneville C. 1863: [new taxa]. In: Grenier A. (ed.): *Catalogue des coléoptères de France et matériaux pour servir à la faune des coléoptères français, par MM. E. Allard, D*[r]*. Ch. Aubé, Ch. Brisout de Barneville, A. Chevrolat, L. Fairmaire, Al. Fauvel, D*[r]*. A. Grenier, D*[r]*. Kraatz, J. Linder, L. Reiche et Félicien de Saulcy.* Paris: L. Toinon, iv + 3-79 + 135 pp.

Brisout de Barneville C. 1866: Coléoptères nouveaux trouvés en Espagne pendant l'excursion de la Société en 1865. *Annales de la Société Entomologique de France* (4) **6**: 355-426.

Brisout de Barneville C. 1867: [new taxa]. In: Grenier A. (ed.): *Matériaux pour servir à la faune des coléoptères de France, recueillis et publiés par le Dr. A. Grenier. 2*[e] *Cahier.* Paris: A. Grenier, [1] + iv + 131-194 pp.

Brisout de Barneville C. 1888: [new taxa]. *Bulletin de la Société Entomologique de France* **1887**: cxcii.

Brisout de Barneville C. & Brisout de Barneville H. 1877: [new taxon]. *Bulletin de la Société Entomologique de France* **1877**: cvii.

Broun T. 1880: *Manual of the New Zealand Coleoptera.* Wellington: J. Hughes, 652 pp.

Broun T. 1893a: Descriptions of new Coleoptera from New Zealand. *The Annals and Magazine of Natural History* (6) **12**: 161-195.

Broun T. 1893b: *Manual of the New Zealand Coleoptera. Parts V., VI., VII.* Wellington: Colonial Museum and Geological Survey Department, xvii + 975-1504 pp.

Brullé G. A. 1832: IVe Classe. Insectes. Pp. 1-240. In: Brullé G. A. & Guérin-Méneville F. M. (eds): *Expédition scientifique de Morée. Section des sciences physiques. Tome III. – 1.re partie. Zoologie. Deuxième Section.- Des animaux articulés.* Paris et Strasbourg: F. G. Levrault, [1] + 400 + [2] pp.

Brullé G. A. 1833: IVe Classe. Insectes. Pp. 241-400. In: Brullé G. A. & Guérin-Méneville F. M. (eds): *Expédition scientifique de Morée. Section des sciences physiques. Tome III. – 1.re partie. Zoologie. Deuxième Section.- Des animaux articulés.* Paris et Strasbourg: F. L. Levraut, 136 + 400 + [2 errata] pp., 27-52 pls. [note: plates issued in 1832-1836].

Brullé G. A. 1835: Coléoptères II. In: Audoin J. V. (ed.): *Histoire naturelle des insectes. Vol. 5.* Paris: Pillot, 436 pp.

Brullé G. A. 1838: Entomologie. Pp. 53-95. In: Barker-Webb P. & Berthelot S.: *Histoire naturelle des Iles Canaries.* (*Animaux articulés recueillis aux Iles Canaries.*). *Vol. II. Pars 2.* Paris: Mellier, 119 pp., 8 pls. [issued in parts, from 1836 to 1840].

Bucciarelli I. 1959: Coleotteri anticidi dell'Aspromonte (Appenino calabrese). *Memorie del Museo Civico di Storia Naturale, Verona* **7**: 183-189.

Bucciarelli I. 1961: Una nuova specie dalmata di Cyclodinus Muls. e Rey. *Atti della Società Italiana di Scienze Naturali e del Museo Civico di Storia Naturale in Milano* **100**: 269-274.

Bucciarelli I. 1962: Gli anticidi del Museo Entomologico del Pontificio Instituto Missioni estere di Monza. *Bollettino della Società Entomologica Italiana* **92**: 114-122.

Bucciarelli I. 1973: Note sui Notoxus mediterranei. *Atti del Museo Civico di Storia Naturale Trieste* **28**: 471-484.

Bucciarelli I. 1976: Su alcuni coleotteri anticidi raccolti da M. e. T. Cerruti nelle isole di Creta, Corfù e Thasos (Grecia), con descrizione di una nuova specie di Microhoria (Coleoptera, Anthicidae). *Fragmenta Entomologica* **12**: 133-142.

Bucciarelli I. 1977: Microhoria (Immicrohoria) caprai n. sp. della Liguria. *Memorie della Società Entomologica Italiana* **55**: 179-182.

Bucciarelli I. 1978: Anticidi nuovi, nuovi per l'Italia e sinonimie. *Bollettino della Società Entomologica Italiana* **110**: 154-163.

Buck F. D. 1965: Some Anthicidae (Col.) from southern Rhodesia including a species new to science. *The*

Entomologist's Monthly Magazine **101**: 192-194.

Buquet J. B. L. 1840: Note sur quelques coléoptères nouveaux d'Algérie et particulièrement de Constantine. *Revue Zoologique, par la Société Cuvierienne* **1840**: 240-243.

Bythinski-Salz H. 1955: Prionotheca coronata Ol. freyi ssp. n. *Entomologische Arbeiten aus dem Museum G. Frey* **6**: 313-316.

Caillol H. 1914: *Catalogue des coléoptères de Provence, d'après des documents recueillis et groupés. 3me Partie.* Marseille: Société Linnéenne de Provence, 594 pp.

Caillol H. 1925: Une espèce présumée nouvelle d'Atritomus Reitter (Col. Mycetophagidae). *Bulletin de la Société Entomologique de France* **1925**: 101.

Cameron M. 1907: Description of a new species of Stenostoma from Malta. *The Entomologist's Monthly Magazine* **43**: 226-227.

Campbell J. M. 1980: Insects of Saudi Arabia Coleoptera: Fam. Alleculidae. *Fauna of Saudi Arabia* **2**: 133-136.

Candèze E. C. A. 1861: Histoire de la métamorphose de quelques coléoptères exotiques. *Mémoires de la Société Royale des Scences de Liège* **16**: 325-410, 6 pls.

Canzoneri S. 1959a: Note sugli Xanthomus Muls. (Coleoptera, Tenebrionidae). *Bollettino della Società Entomologica Italiana* **89**: 145-148.

Canzoneri S. 1959b: Contributo ad una migliore conoscenza del genere Stenomax Allard (Col. Tenebrionidae). *Bollettino de la Società Entomologica Italiana* **89**: 148-149.

Canzoneri S. 1960a: Diagnosi preliminare di una nuova entità del genere Phaleria. (III. Contributo allo studio dei tenebrionidi). *Bollettino della Società Entomologica Italiana* **90**: 43-45.

Canzoneri S. 1960b: Note sistematiche e biogeografiche su alcuni tenebrionidi nuovi o poco noti. *Bollettino della Società Entomologica Italiana* **90**: 50-54.

Canzoneri S. 1961a: Note sul genere Oochrotus Luc. VII Contributo allo studio dei tenebrionidi. *Bollettino del Museo Civico di Storia Naturale di Venezia* **14**: 7-27.

Canzoneri S. 1961b: Sulla Phaleria bimaculata Lin. (IX Contributo allo studio dei tenebrionidi). *Bollettino della Società Entomologica Italiana* **91**: 24-27.

Canzoneri S. 1961c: Sulla Phaleria provincialis Fauv. (X Contributo allo studia dei tenebrionidi). *Bollettino della Società Entomologica Italiana* **91**: 57-61.

Canzoneri S. 1963a: Sulla Pimelia bipunctata Fabr. (XI Contributo allo studio di tenebrionidi). *Bollettino della Società Entomologica Italiana* **93**: 13-16.

Canzoneri S. 1963b: Sulla Alphasida grossa Sol., con descrizione di una nuova sottospecie. (XII Contributo allo studia dei tenebrionidi). *Bollettino della Società Entomologica Italiana* **93**: 35-36.

Canzoneri S. 1965: Una nuova specie del genere Falsaspila Koch (XV Contributo alla conoscenza dei Tenebrionidae). *Bollettino della Società Entomologica Italiana* **95**: 168-169.

Canzoneri S. 1968: Materiali per una monografia delle Phaleria del sottogenere Phaleria Latr. 20. Contributo alla conoscenza dei Coleoptera, Tenebrionidae. *Memorie della Società Entomologica Italiana* **47**: 117-167.

Canzoneri S. 1969: Blaps sulcata ssp. mercantii nov.: diagnosi preliminare (XIX Contributo allo studio dei Tenebrionidae). *Bollettino della Società Entomologica Italiana* **99-101**: 44.

Canzoneri S. 1970: I Tenebrionidae delle Isole Egadi. *Memorie del Museo Civico di Storia Naturale di Verona* **18**: 55-89.

Canzoneri S. 1971: Una nuova Akis di Persia (XXV Contrituto alla conoscenza dei Tenebrionidae). *Bollettino della Società Entomologica Italiana* **103**: 155.

Canzoneri S. 1972a: Annotazioni sui Colpophorus italiani. *Bollettino del Museo Civico di Storia Naturale di Venezia* **22/23** [1969/1970]: 187-191.

Canzoneri S. 1972b: Nuovi dati sui Tenebrionidae di "Piccole Isole" Italiane, con descrizione di Alphasida tirelli moltonii n.ssp. *Atti della Societã Italiana di Scienza Naturali e del Museo Civico di Storia Naturale di Milano* **113**: 288-296.

Canzoneri S. 1973: Una nuova specie del genere Phaleria Latr. *Bollettino della Società Entomologica Italiana* **105**: 41-42.

Canzoneri S. 1976: I Tenebrionidae delle isole Ponziane (Coleoptera). *Fragmenta Entomologica* **12**: 9-18.

Canzoneri S. 1977: Contributo alla conoscenza de tenebrionidi appenninici. (XXXII Contributo allo studio dei Tenebrionidi). *Bollettino del Museo Civico di Storia Naturale Verona* **4**: 227-285.

Canzoneri S. 1979: Primi appunti sui tenebrionidi dell'Arcipelago Maltese (XXXII Contributo allo studio dei Tenebrionidi). *Lavori - Società Veneziana di Sienza Naturali* **4**: 20-22.

Canzoneri S. 1984: Una nuova specie di Pedinus di Sicilia. *Lavori - Società Veneziana di Sienza Naturali* **9**: 3-4.

Carl M. 1989: Eine neue Tentyria-Art aus dem Irak. *Entomofauna* **10**: 25-30.

Carl M. 1990a: Übersicht über die irakischen Adesmia-Arten und Neubeschreibung von Adesmia micrograńulata sp. n. (Coleoptera, Tenebrionidae). *Mitteilungen der Münchener Entomologischen Gesellschaft* **80**: 71-83.

Carl M. 1990b: Zwei neue Zophosis-Arten aus dem Vorderen Orient (Coleoptera: Tenebrionidae). *Nachrichtenblatt der Bayerischen Entomologen* **39**: 79-81.

Carl M. 1990c: Eine neue Dichillus-Art aus dem Irak (Coleoptera, Tenebrionidae). *Entomofauna* **11**: 105-108.

Carl M. 1991a: Die Gattungen Adelostomoides gen. n. und Adelostoma Dup. aus Mesopotamien (Coleoptera: Tenebrionidae). *Nachrichtenblatt der Bayerischen Entomologen* **40**: 23-27.

Carl M. 1991b: Eine neue Tentyria-art aus Iraq (Coleoptera, Tenebrionidae). *Acta Zoologica Academiae Scientiarum Hungaricae* **37**: 171-175.

Carl M. 1992a: A revision of the genus Microtelus Solier, 1838 (Coleoptera: Tenebrionidae). *Zoology of the Middle East* **6**: 69-87.

Carl M. 1992b: Neue und bemerkenswerte Tenebrionidae (Coleoptera) aus dem Vorderen Orient. *Linzer Biologische Beiträge* **24**: 331-337.

Carl M. 1994: The taxonomy, biogeography and phylogeny of the genus Prochoma in the Middle East and South-East Asia (Coleoptera: Tenebrionidae). *Fauna of Saudi Arabia* **14**: 77-96.

Carl M. 2000: Beschreibung von Heinrichesia schaeferi gen. nov. et sp. nov. aus Sistan (Iran) (Coleoptera, Tenebrionidae). *Entomofauna* **21**: 257-261.

Carter H. J. 1911: Revision of Pterohelaeus (continued) and of Sargus; with descriptions of new species of Australian Tenebrionidae.*Proceedings of the Linnean Society of New South Wales* **36**: 179-223, pl. VIII.

Carter H. J. 1919: Notes on Australian Coleoptera, with description of new species. *Proceedings of the Linnean Society of New South Wales* **44**: 137-173, pls. IV, V.

Carter H. J. 1920: Notes on some Australian Tenebrionidae, with description of new species; – also of a new genus and species of Buprestidae. *Proceedings of the Linnean Society of New South Wales* **45**: 222-249.

Carter H. J. 1924a: Australian Coleoptera: notes and new species. No. III. *Proceedings of the Linnean Society of New South Wales* **49**: 19-45.

Carter H. J. 1924b: Australian Coleoptera: notes and new species. No. IV. *Proceedings of the Linnean Society of New South Wales* **49**: 521-544.

Carter H. J. 1928: Revision of the Australian species of the genera Curis, Neocuris and Trachys, together with notes and descriptions of new species of other Coleoptera. *Proceedings of the Linnean Society of New South Wales* **53**: 270-290.

Carter H. J. & Zeck E. H. 1937: A monograph of the Australian Colydiidae. *Proceedings of the Linnean Society of New South Wales* **62**: 181-208, pls. 8-9.

Casey T. L. 1898: Studies in the Ptinidae, Cioidae and Sphindidae of America. *Journal of the New York Entomological Society* **6**: 61-93.

Casey T. L. 1900: Review of the American Corylophidae, Cryptophagidae, Tritomidae and Dermestidae, with other studies. *Journal of the New York Entomological Society* **8**: 51-172.

Casey T. L. 1891: Coleopterological Notices. III. *Annals of the New York Academy of Sciences* **6**: 9-214.

Casey T. L. 1895: Coleopterological Notices. VI. *Annals of the New York Academy of Sciences* **8**: 435-838.

Casey T. L. 1924: Additions to the knowledge of the Coleoptera of North America. *Memoirs on the Coleoptera* **11**: 1-347 pp.

Champion G. C. 1886a: Tenebrionidae. Pp. 137-264. In: *Biolologia Centrali-Americana. Insecta. Coleoptera Heteromera (part). Vol. IV, part 1.* [1884-1893]. London: R. H. Porter, 572 pp., 23 pls.

Champion G. C. 1886b: Salpingus mutilatus, Beck, a British insect. *The Entomologist's Monthly Magazine* **23**: 160.

Champion G. C. 1889: Melandryidae. Pp. 75-103, 5 pls. In: *Biologia Centrali-Americana. Insecta, Coleoptera. Vol. IV, part 2.* [1889-1893]. London: R. H. Porter, x + 494 pp., 21 pls.

Champion G. C. 1890a: Some remarks on the genus Xylophilus, with description of two species from Japan. *The Entomologist's Monthly Magazine* **26**: 264-269.

Champion G. C. 1890b: On a new species of Tomoderus from Japan. *The Entomologist's Monthly Magazine* **26**: 325-326.

Champion G. C. 1891a: Description of a new species of Anaspis from Scotland, with some remarks on the black species occurring in Britain. *The Entomologist's Monthly Magazine* **27**: 104-105.

Champion G. C. 1891b: On two new species of Anthicidae. *The Entomologist's Monthly Magazine* **27**: 188-189.

Champion G. C. 1891c: Mordellidae. Pp. 257-350. In: *Biologia Centrali-Americana. Insecta, Coleoptera. Vol. IV, part 2.* [1889-1893]. London: R. H. Porter, x + 494 pp., 21 pls.

Champion G. C. 1891d: A list of the Coleoptera collected by Mr. J. J. Walker, R.N., F.L.S., in the region of the Straits of Gibraltar, with descriptions of four new species. *Transactions of the Entomological Society of London* **1891**: 375-401.

Champion G. C. 1892: Note on Pentaria badia Rosenh. (= sericaria, Muls.), with description of a second species from the Pyrénées-Orientales. *The Entomologist's Monthly Magazine* **28**: 108-109.

Champion G. C. 1893: Supplement (Fam. Xylophilidae). Pp. 455-461, pl. 21. In: *Biologia Centrali-Americana, Insecta, Coleoptera Heteromera (part). Vol. IV, part.2* [1889-1893]. London: R. H. Porter, x + 494 pp., 21 pls.

Champion G. C. 1894a: On the Tenebrionidae collected in Australia and Tasmania by Mr. James J. Walker, R.N., F.L.S., during the voyage of H.M.S. "Penguin" with descriptions of new genera and species. *Transactions of the Entomological Society of London* **1894**: 351-408, pl.viii.

Champion G. C. 1894b: On two species of tenebrionid Coleoptera from Namoa Island. *The Annals and Magazine of Natural History* (6) **14**: 448-449.

Champion G. C. 1895: A list of Tenebrionidae supplementary to the "Munich" Catalogue. *Mémoires de la Société Entomologique de Belgique* **1** (3): 1-264 pp.

Champion G. C. 1916a: On new or little-know Xylophilidae. *Transactions of the Entomological Society of London* **1916** (1): 1-64, I-II pls.

Champion G. C. 1916b: Notes on Melandryidae (2-3). *The Entomologist's Monthly Magazine* **52**: 1-10, 32-40, 52-59, 75-83, 99-108, 144-156.

Champion G. C. 1916c: On new and little-known Lagriidae and Pedilidae. *Transactions of the Entomological Society of London* **1916**: 181-242, pls. 63-64.

Champion G. C. 1916d: On new exotic Scraptiina. *The Entomologist's Monthly Magazine* **52**: 233-256, 265-274, pl. 7.

Champion G. C. 1917: Some Coleoptera from northen India. *The Entomologist's Monthly Magazine* **53**: 52-55.

Champion G. C. 1919: Some Indian Coleoptera (1). *The Entomologist's Monthly Magazine* **55**: 236-246.

Champion G. C. 1920a: Some Indian Coleoptera (2). *The Entomologist's Monthly Magazine* **56**: 68-77.

Champion G. C. 1920b: On some Eastern Xylophilids (Coleoptera). *The Annals and Magazine of Natural History* (9) **5**: 393-406.

Champion G. C. 1921: Some Indian Coleoptera (6). *The Entomologist's Monthly Magazine* **57**: 201-206.

Champion G. C. 1922a: Some Indian Coleoptera (7). *The Entomologist's Monthly Magazine* **58**: 31-34, 68-76.

Champion G. C. 1922b: Some Indian Coleoptera (8). *The Entomologist's Monthly Magazine* **58**: 126-131, 145-152.

Champion G. C. 1923: Some Indian Coleoptera (10). *The Entomologist's Monthly Magazine* **59**: 77-80.

Champion G. C. 1924a: Some Indian Coleoptera (14). *The Entomologist's Monthly Magazine* **60**: 159-172.

Champion G. C. 1924b: Additional species of Xylophilidae from the Philippines, India, Fiji and S. and W. Africa. *The Annals and Magazine of Natural History* (9) **12**: 481-498.

Champion G. C. 1924c: Some Indian Coleoptera. *The Annals and Magazine of Natural History* (9) **13**: 249-264.

Champion G. C. 1925: Some Indian (and Tibetan) Coleoptera (16). *The Entomologist's Monthly Magazine* (2) **36**: 101-112.

Champion G. C. 1927a: Some Indian Coleoptera (22). *The Entomologist's Monthly Magazine* **63**: 49-57, 78-85.

Champion G. C. 1927b: Some Indian Coleoptera (23). *The Entomologist's Monthly Magazine* **63**: 245-250.

Chandler D. S. 2004: [new name]. In: Chandler D. S., Nardi G. & Telnov D.: Nomenclatural notes on the Palaearctic Anthicidae (Coleoptera). *Mitteilungen des Internationalen Entomologischen Vereins e. V.* (Frankfurt am Main) **29**: 109-173.

Charpentier T. de. 1818: Meloe insignis, p. 258. In Germar E. F.: VIII. Vermischte Bemerkungen über einige Käferarten. *Germar's Magazin der Entomologie* **3**: 228-260.

Charpentier T. de. 1825: *Horae entomologicae, adjectis tabulis novem coloratis*. Wratislaviae: Apud A. Gosohorsky, xvi + 255 pp., 9 pls.

Chatanay J. 1914a: Sur trois ténébrionides faisant partie de la collection de M. Leboeuf (Coléoptères). *Bulletin de la Société d'Étude des Sciences Naturelles de la Haute-Marne* **1914**: 19-24.

Chatanay J. 1914b: Un nouveau Blaps nord-africain (Col. Tenebrionidae). *Bulletin de la Société Entomologique de France* **1914**: 75-77.

Chatanay J. 1914c: Description de deux Tentyriinae nouveaux. *Annales de la Société Entomologique de France* **83**: 215-224.

Chatanay J. 1914d: Diagnoses préliminaires des Zophosinae nouveaux (Col. Tenebrionidae). *Bulletin de la Société Entomologique de France* **1914**: 379-381.

Chatanay J. 1915: Description d'un genre nouveau d'Épitragides de Madagascar (Coléoptères, Tenebrionidae). *Bulletin du Muséum National d'Histoire Naturelle, Paris* **21**: 64-67.

Chatanay J. 1917a: Matériaux pour servir à l'étude de la faune entomologique de l'Indo-Chine française réunis par M. Vitalis de Salvaza. *Bulletin du Muséum National d'Histoire Naturelle, Paris* **23**: 229-255.

Chatanay J. 1917b: Révision des Zophosis de l'est de l'Afrique. *Annales de la Société Entomologique de France* **85** [1916]: 503-624.

Chatzimanolis S., Engel M. S. & Trichas A. 2002: Taxonomic changes for Aegean species of the Mediterranean darkling beetle genus Dendarus (Coleoptera: Tenebrionidae). *Journal of the Kansas Entomological Society* **75**: 259-267.

Chen B. 1997a: A study on the genus Donaciolagria Pic (Coleoptera: Lagriidae) from China. *Entomologia Sinica* **4**: 121-128.

Chen B. 1997b: Two new genera and two new species of Lagriidae (Coleoptera) from China. *Entomologia Sinica* **4**: 306-311.

Chen B. & Chou I. 1996: A new subgenus and two new species of the genus Cerogria Borchmann (Coleoptera: Lagriidae) from China. *Entomotaxonomia* **18**: 265-269.

Chen B. & Xia X. 2001: The genus Schevodera Borchmann: phylogeny and historical biogeography, with description of a new species (Coleoptera: Tenebrionidae: Lagriinae). *Oriental Insects* **35**: 3-27.

Chen B. & Yang X. 1997: Three new species and key to species of the genus Arthromacra Kirby from China (Coleoptera: Lagriidae). *Entomologia Sinica* **4**: 1-8.

Chen B. & Yuan F. 1996: A new genus and two new species of Statirinae (Coleoptera: Lagriidae) from China. *Entomotaxonomia* **18**: 183-287.

Chevrolat L. A. A. 1831: [new taxa, livraison 15]. In: Guérin-Méneville F. E.: [1829-1838]: *Iconographie du règne animal de G. Cuvier, ou représentation d'après la nature de l'une des espèces les plus remarquables et souvent non encore figurées, de chaque genre d'animaux. Avec un texte descriptif mis au courant de la science. Ouvrage pouvant servir d'atlas à tous les traités de zoologie. Iconographie du règne animal de G. Cuvier. II. Planches des animaux invertébrés. Insectes*. Paris: J. B. Bailière, 110 pls.

Chevrolat L. A. A. 1833a: [new taxa, livraison 28]. In: Guérin-Méneville F. E.: [1829-1838]: *Iconographie du règne animal de G. Cuvier, ou représentation d'après la nature de l'une des espèces les plus remarquables et souvent non encore figurées, de chaque genre d'animaux. Avec un texte descriptif mis au courant de la science. Ouvrage pouvant servir d'atlas à tous les traités de zoologie. Iconographie du règne animal de G. Cuvier. II. Planches des animaux invertébrés. Insectes*. Paris: J. B. Bailière, 110 pls.

Chevrolat L. A. A. 1833b: Mémoire sur un nouveau genre de coléoptères, de la famille de mélasomes. *Revue Entomologique* (G. Silbermann) **1**: 25-26.

Chevrolat L. A. A. 1833c: Notice sur un genre nouveau de la famille des diapérides. *Revue Entomologique* (G. Silbermann) **1**: 30-32.

Chevrolat L. A. A. 1833d: [new taxa]. *Magasin de Zoologie* **3**, cl. IX, pl. 57.

Chevrolat L. A. A. 1834: [new taxa, livraison 35]. In: Guérin-Méneville F. E.: [1829-1838]: *Iconographie du règne animal de G. Cuvier, ou représentation d'après la nature de l'une des espèces les plus remarquables et souvent non encore figurées, de chaque genre d'animaux. Avec un texte descriptif mis au courant de la science. Ouvrage pouvant servir d'atlas à tous les traités de zoologie. Iconographie du règne animal de G. Cuvier. II. Planches des animaux invertébrés. Insectes*. Paris: J. B. Bailière, 110 pls.

Chevrolat L. A. A. 1835a: [new taxa, livraison 38]. In: Guérin-Méneville F. E.: [1829-1838]: *Iconographie du règne animal de G. Cuvier, ou représentation d'après la nature de l'une des espèces les plus remarquables et souvent non encore figurées, de chaque genre d'animaux. Avec un texte descriptif mis au courant de la science. Ouvrage pouvant servir d'atlas à tous les traités de zoologie. Iconographie du règne animal de G. Cuvier. II. Planches des animaux invertébrés. Insectes*. Paris: J. B. Bailière, 110 pls.

Chevrolat L. A. A. 1835b: Mémoire sur un coléoptère tétramère de la famille des xylophages et observations sur plusiers espèces de cet ordre, rencontrées dans diverses fourmillières. *Revue Entomologique* (G. Silbermann) **3**: 263-269, pl. 34.

Chevrolat L. A. A. 1840a: Description de quelques coléoptères de la Galice et du Portugal provenant d'envoies de M. Deyrolles fils. *Revue Zoologique, par la Société Cuvierienne* **1840**: 8-18.

Chevrolat L. A. A. 1840b: Description de mylabrides de Barbarie. *Revue Entomologique* (G. Silbermann) **5** [1837]: 266-279, 1 pl.

Chevrolat L. A. A. 1844: [new taxa]. In: Guérin-Méneville F. E.: [1829-1844]: *Iconographie du règne animal de G. Cuvier, ou représentation d'après la nature de l'une des espèces les plus remarquables et souvent non encore figurées, de chaque genre d'animaux. Avec un texte descriptif mis au courant de la science. Ouvrage pouvant servir d'atlas à tous les traités de zoologie. III. Texte explicatif. Insectes*. Paris: J. B. Baillière, 576 pp. [issued in parts].

Chevrolat L. A. A. 1849: Trachyderme, p. 628. In: D'Orbigny C. V. D. (ed.): *Dictionnaire universel d'histoire naturelle résumant et complétant tous les faits présentés par les encyclopédies, les anciens dictionnaires scientifiques, les œuvres complètes de Buffon, et les meilleurs traités spéciaux sur les diverses branches des sciences naturelles; donnant la description des êtres et des divers phénomènes de la nature, l'étymologie et la définition des noms scientifiques, les principales applications des corps organiques et inorganiques, à l'agriculture, à la médecine, aux arts industriels, etc.; dirigé par M. Charles d'Orbigny, et enrichi d'un magnifique atlas de 288 planches gravées sur acier. Tome douzième*. Paris: C. Renard, Martinet et Cie.

Chevrolat L. A. A. 1860: Description de coléoptères nouveaux d'Algérie. *Revue et Magasin de Zoologie Pure et Appliquée* (2) **12**: 75-82, 208-212, 448-459.

Chevrolat L. A. A. 1861: Description des coléoptères nouveaux d'Algérie. *Revue et Magasin de Zoologie Pure et Appliquée* (2) **13**: 205-208.

Chevrolat L. A. A. 1865: Descriptions de coléoptères d'Espagne, nouveaux ou peu connus. *Revue et Magasin de Zoologie Pure et Appliquée* (2) **17**: 390-397.

Chevrolat L. A. A. 1873: Description de quelques espèces nouvelles de coléoptères de Syrie. *Annales de la Société Entomologique de France* (5) **3**: 203-218.

Chevrolat L. A. A. 1874: Coleopterorum species novarum descriptiones. *Anales de la Sociedad Española de Historia Natural* **3**: 157-159.

Chevrolat L. A. A. 1877a: Coléoptères nouveaux de la famille des ténébrionides. *Petites Nouvelles Entomologiques* **2** [1876-1879]: 113-114, 117-118.

Chevrolat L. A. A. 1877b: Coléoptères nouveaux. *Petites Nouvelles Entomologiques* **2** [1876-1879]: 121.

Chevrolat L. A. A. 1877c: Description d'un sous-genre nouveau de coléoptères hétéromères de la famille des Anthicides et indication des espèces que s'y rapportent. *Annales de la Société Entomologique de France* (5) **7**: 167-169.

Chevrolat L. A. A. 1877d: Diagnoses de diapérides nouveaux. *Petites Nouvelles Entomologiques* **2** [1876-1879]: 170, 173, 177-178.

Chevrolat L. A. A. 1878a: Diagnoses de diapérides. *Petites Nouvelles Entomologiques* **2** [1876-1879]: 221-222, 242-243.

Chevrolat L. A. A. 1878b: Diagnoses de diapérides nouveaux. *Comptes-Rendus des Séances de la Société Entomologique de Belgique*. **1878**: clii-cliii.

Chevrolat L. A. A. 1882: [Diagnoses de trois coléoptères nouveaux et des remarques sur une quatrième espèce]. *Bulletin de la Société Entomologique de France* **1882**: iv-v.

Chobaut A. 1891: Description d'une nouvelle espèce de Rhipidius avec quelques remarques sur ce genre. *Le Coléoptériste* **1**: 235-237, 1 pl.

Chobaut A. 1893: Coléoptères nouveaux du Nord de l'Afrique. *L'Abeille, Journal d'Entomologie* **28** [1892-1896]: 97-101.

Chobaut A. 1894a: [new taxa]. *Bulletin de la Société Entomologique de France* **1894**: lxxv-lxxvi.

Chobaut A. 1894b: [new taxa]. *Bulletin de la Société Entomologique de France* **1894**: ci-ciii.

Chobaut A. 1894c: [Description d'une espèce nouvelle d'Anthicus et des remarques sur Anthicus bedeli]. *Bulletin de la Société Entomologique de France* **1894**: ccxxviii-ccxxx.

Chobaut A. 1895a: Notes sur quelques Danacaea et description d'une espèce nouvelle d'Anthicus (Col.). *Bulletin de la Société Entomologique de France* **1895**: 155-157.

Chobaut A. 1895b: Description de deux variétés nouvelles de coléoptères circa-méditerranéens. *Bulletin de la*

Société Entomologique de France **1895**: 346-347.

Chobaut A. 1896a: Description d´une espèce nouvelle d´Anaspis d'Algérie. *Bulletin de la Société Entomologique de France* **1896**: 200.

Chobaut A. 1896b: Descriptions de trois nouveaux coléoptères héteromères d'Algérie. *Bulletin de la Société Entomologique de France* **1896**: 375-377.

Chobaut A. 1897a: Description de cinq variétés nouvelles de coléoptères de la faune circa-méditerranéenne. *Miscellanea Entomologica* **5**: 38-39.

Chobaut A. 1897b: Notes sur le genre Pentaria et description d´une espèce nouvelle. *Bulletin de la Société Entomologique de France* **1897**: 132-135.

Chobaut A. 1897c: Description de deux coléoptères nouveaux du Mzab (Sud-Algérien). *Bulletin de la Société Entomologique de France* **1897**: 170-171.

Chobaut A. 1897d: Description de deux espèces nouvelles de coléoptères du Mzab (Sud-Algérien). *Bulletin de la Société Entomologique de France* **1897**: 204-206.

Chobaut A. 1898a: Description de quelques espèces et variétés nouvelles de coléoptères algériens. *Revue d'Entomologie* **17**: 74-88.

Chobaut A. 1898b: Voyage chez les Beni-Mzab (Sud de la Province d'Alger). Contribution à l'étude de la faune entomologique du Sahara algérien. *Memoires de l'Académie de la Vaucluse* **17**: 131-235, 11 pls., 1 map.

Chobaut A. 1899: Description d'une espèces nouvelle de Zophosis (Col.) de la Tunisie mériodionale. *Bulletin de la Société Entomologique de France* **1899**: 409-411.

Chobaut A. 1900a: Description d'un Opatride nouveau (Col. Tenebrion.) de la Tunisie méridionale. *Bulletin de la Société Entomologique de France* **1900**: 31-32.

Chobaut A. 1900b: Description d'une Phaleria nouvelle d'Algérie. *Bulletin de la Société Entomologique de France* **1900**: 341-342.

Chobaut A. 1901: Description d'un Mylabris nouveau du Sahara algérien (Col.). *Bulletin de la Société Entomologique de France* **1901**: 280-281.

Chobaut A. 1902: Description d'un rhipiphoride nouveau (Col.) du nord de l'Afrique. *Bulletin de la Société Entomologique de France* **1902**: 89-90.

Chobaut A. 1903a: Description d'un salpingide nouveau du nord de la Tunisie. *Bulletin de la Société Entomologique de France* **1903**: 143.

Chobaut A. 1903b: Variabilité d'Anthicus superbus Pic (Col.). *Bulletin de la Société Entomologique de France* **1903**: 170-171.

Chobaut A. 1904a: Description d'un Rhipidius nouveau de la France méridionale avec tableau dichotomique des Rhipidiini (Col.). *Bulletin de la Société Entomologique de France* **1904**: 228-232.

Chobaut A. 1904b: Description de deux espèces nouvelles de coléoptères de l´Arabie. *Bulletin de la Société Entomologique de France* **1904**: 243-245.

Chobaut A. 1906: Note sur le Myiodes subdipterus Bosc (Col.). *Bulletin de la Société Entomologique de France* **1906**: 223-224.

Chobaut A. 1908: Variété nouvelle d'Anthicus fairmairei Bris. (Col.). *Bulletin de la Société Entomologique de France* **1908**: 162-163.

Chobaut A. 1918: Description de trois ténébrionides nouveaux de la Tunisie méridionale. *Bulletin de la Société Entomologique de France* **1918**: 74-76.

Chobaut A. 1919: Description des deux sexes, de l'oeuf et de la larve primaire d'un nouveau Rhipidius de Provence (Col. Rhipiphoridae). *Bulletin de la Société Entomologique de France* **1919**: 200-206.

Chobaut A. 1921: Description d'un nouveau Diaphorocera (Col., Meloidae) de l'Algérie méridionale. *Bulletin de la Société Entomologique de France* **1921**: 298-299.

Chobaut A. 1922: Description d'un Rhipidius nouveau du Maroc (Coléoptère Rhipiphoridae). *Bulletin de la Société des Sciences Naturelles du Maroc* **2**: 127-128.

Chobaut A. 1923a: Coléoptères nouveaux du Maroc. *Bulletin de la Société des Sciences Naturelles du Maroc* **3**: 75-77.

Chobaut A. 1923b: Description de trois Anthicus nouveaux du Maroc (Coléoptères Anthicides). *Bulletin de la Société des Sciences Naturelles du Maroc* **3**: 128-130.

Chobaut A. 1923c: Description d'un Oxycara nouveau (Col. Tenebrionidae) du Sud-Oranais. *Bulletin de la Société Entomologique de France* **1923**: 165-167.

Chobaut A. 1924a: Description d'un Anthicus nouveaux du Maroc (Coléoptères Anthicides). *Bulletin de la*

Société des Sciences Naturelles du Maroc **4**: 104.
Chobaut A. 1924b: Description d'un opatride (Col. Tenebrionidae) nouveau de l'Afrique du Nord. *Bulletin de la Société Entomologique de France* **1924**: 154-156.
Chobaut A. 1924c: Description d'un Cyphostethe (Col. Tenebrionidae) nouveau de l'Algérie méridionale. *Bulletin de la Société Entomologique de France* **1924**: 166-167.
Chobaut A. 1924d: Tableau synoptique des Stenalia paléarctiques à élytres maculés de jaune. *Bulletin de la Société Entomologique de France* **1924**: 179-183.
Chobaut A. 1925: Description d'un Allophylax (Col. Tenebrionidae). *Bulletin de la Société Entomologique de France* **1924**: 228-229.
Chûjô M. 1935a: Descriptions of three new mordellid-species from Formosa. *Sylvia, Taihoku* **6**: 72-75.
Chûjô M. 1935b: A systematic catalogue of Formosan Rhipiphoridae (Coleoptera). *Sylvia, Taihoku* **6**: 129-135.
Chûjô M. 1939: On the Japanese Ciidae (Coleoptera). *Mushi* **12**: 1-10.
Chûjô M. 1940a: Description of a new and a rare species of the Japanese ciid-beetles. *Insecta Matsumurana* **14**: 132-133.
Chûjô M. 1940b: Descriptions of new ciid-beetles from Japan proper and Manchuria. *Insecta Matsumurana* **14**: 134-137.
Chûjô M. 1941: Description of a new Ciid beetle from Japan. *Kontyû* 15 (2): 85-86.
Chûjô M. 1956: Descriptions of two new species of the Japanese mordellid-beetles with notices on three known species. *Kontyû* (Tokyo) **24**: 174-178.
Chûjô M. 1957a: Description of a new subgenus and five new species of the Japanese mordellid-beetles with notices on some known species. *Mushi* **30**: 1-10.
Chûjô M. 1957b: Descriptions of a new genus and six new species of the Japanese mordellid-beetles with notices on some known species. *Mushi* **30**: 11-22.
Chûjô M. 1959: Coleoptera of the Loo-choo Archipelago (I). *Memoirs of the Faculty of Liberal Arts & Education, Kagawa University, part II* **69**: 1-15.
Chûjô M. 1960: Description of a new Mordellid-species from the Island Naka-No-Shima. *Niponnius* **1**: 1-3.
Chûjô M. 1963a: A new species of the genus Basanus Lacordaire from Japan and Korea (Coleoptera, Tenebrionidae) *Niponius* **2**: 17-19.
Chûjô M. 1963b: A new species of Gonocephalum Solier from Okinawa and Takara Is. (Coleoptera, Tenebrionidae). *Kontyû* (Tokyo) **31**: 149-153.
Chûjô M. 1966: Taxonomic study on the Tenebrionidae (Coleoptera) of the Ryukyu Islands. *Journal of the Faculty of Agriculture, Kyushu University* **14**: 1-32, 2 pls.
Chûjô M. 1967: A check list of the Formosan Tenebrionidae (Coleoptera) I. *Kontyû* (Tokyo) **35**: 375-381.
Chûjô M. 1968a: A check list of the Formosan Tenebrionidae (Coleoptera) II. *Kontyû* (Tokyo) **36**: 14-17.
Chûjô M. 1968b: A check list of the Formosan Tenebrionidae (Coleoptera) III. *Kontyû* (Tokyo) **36**: 393-397.
Chûjô M. 1968c: Eine neue Basanus-Art aus Japan (Col., Tenebrionidae). *Kontyû* (Tokyo) **36**: 398-399.
Chûjô M. 1973: Five new species of the Tenebrionidae from Japan and her adjacent regions, with a new record of Gonocephalum coenosum Kaszab from the Senkaku island (Coleoptera). *Esakia* **8**: 13-23.
Chûjô M. 1975: Tenebrionidae of the Nansei Islands I (Coleoptera). *Esakia* **9**: 15-24, 2 pls.
Chûjô M. 1977: Tenebrionidae of the Nansei Islands II (Coleoptera). *Esakia* **10**: 1-19.
Chûjô M. 1978a: Tenebrionidae of the Nansei Islands III (Coleoptera). *Esakia* **11**: 63-80.
Chûjô M. 1978b: Tenebrionidae of the Nansei Islands IV (Coleoptera). *Esakia* **12**: 1-16.
Chûjô M. 1979: Tenebrionidae of the Nansei Islands V (Coleoptera). *Esakia* **14**: 31-56.
Chûjô M. 1980: Tenebrionidae of the Nansei Islands VI (Coleoptera). *Esakia* **15**: 1-10.
Chûjô M. 1981: Four new species of the Formosan Setenis (Coleoptera, Tenebrionidae). *Esakia* **17**: 131-140.
Chûjô M. 1983: Tenebrionidae of the Nansei Islands VIII (Coleoptera). *Esakia* **20**: 47-53.
Chûjô M. 1984: Tenebrionidae of the Nansei Islands IX (Coleoptera). *Esakia* **22**: 1-4.
Chûjô M. 1985: Notes on the Japanese Tenebrionidae (Coleoptera). *Esakia* **23**: 61-66.
Chûjô M. 1985: Lagriidae of the Nansei Islands. *Mushi* **50**. 87-90.
Chûjô M. & Imasaka S. 1982: Five new species of Misolampidius Solsky from Japan and Korea (Coleoptera, Tenebrionidae). *Esakia* **19**: 123-134.
Chûjô M. & Miyatake M. 1961: [new taxa]. In: Miyatake M.: The tribe Leiochrini of Japan (Coleoptera: Tenebrionidae). *Transactions of the Shikoku Entomological Society* **7**: 33-40.
Chûjô M. et Nakane T. 1953: Mordellidae of Shikoku, Japan. *Mushi* **25**: 17-24.

Chûjô M. & Nakane T. 1955: Description of a new mordellid-beetle from Japan. *Akitu* **4**: 27-29.

Chûjô M. & Satô M. 1972: 36. Family Oedemeridae. Coleoptera of the Loo-choo archipelago (IV). *Memoirs of the Faculty of Education Kagawa University* (2) **208**: 1-28.

Chûjô Mt. 1992: [new taxa]. In: Chûjô Mt. & Lee C. E.: Tenebrionidae from Chejudo Island, Korea (Insecta, Coleoptera). *Esakia* **32**: 31-46.

Chûjô Mt. 1995a: Gonocephalum ficifolium, a new species of Tenebrionidae from Japan (Coleoptera). *Esakia* **35**: 113-116.

Chûjô Mt. 1995b: Tenebrionid beetles of the genus Promethis Pascoe from the Nansei Islands (Coleoptera). *Esakia* **35**: 117-122.

Chûjô Mt. 1997a: Tenebrionidae of the Nansei Islands X (Coleoptera). *Esakia* **36**: 135-141.

Chûjô Mt. 1997b: Tenebrionidae from the Nansei Islands XI (Coleoptera). *Journal of the Faculty of Agriculture, Kyushu University* **41**: 197-204.

Chûjô Mt. & Lee C. E. 1993: Korean Tenebrionidae (Insecta, Coleoptera). *Esakia* **33**: 109-122.

Clermont J. 1925: Description d'une espèce nouvelle de Tenebrionidae (Col.). *Bulletin de la Société Entomologique de France* **1925**: 183-184.

Cobos A. 1950: Datos para el catálogo de los coleópteros de España. Especies de los alredadores de Málaga. *Boletín de la Real Sociedad Española de Historia Natural* **47** [1949]: 589-594.

Cobos A. 1953: Especies nuevas de coleópteros de la provincia de Almería, I. *Archivos del Instituto de Aclimatación* (Almería) **1**: 127-134.

Cobos A. 1954: Una breve campaña entomológica por las Sierras de Tejada y Almijara (Provincia de Málaga): Ins. Coleópteros. *Archivos del Instituto de Aclimatación* (Almería) **3**: 29-39.

Cobos A. 1961: Exploratión entomológica del central del Tizi-Ifri (macizo del Iguelmalet, Rif central, Marruecos). Coleópteros. *Archivos del Instituto de Aclimatación* (Almería) **10**: 63-98, 1 pl.

Cobos A. 1962: Sobre las Alphasida del subgénero Betasida Rtt. y descripción de une nueva especie (Coleoptera, Tenebrionidae). *Archivos del Instituto de Aclimatación* (Almería) **11**: 37-40, pl. 2.

Cobos A. 1963a: Ensayo sobre las Asida Latr. de Marruecos (Col. Tenebr.). *Archivos del Instituto de Aclimatación* (Almería) **12**: 9-37, 6 pls.

Cobos A. 1963b: El género "Coelometopus" Sol. (Col. Tenebrionidae). *Archivos del Instituto de Aclimatación* (Almería) **12**: 49-53, 1 pl.

Cobos A. 1965a: Adiciones y correcciones al "Ensayo sobre las Asida Latr., de Marruecos" (Col. Tenebrionidae). *Graellsia* **21**: 55-57.

Cobos A. 1965b: Una nueva especie de Helioctamenus de Canarias (Col. Colydiidae). *Graellsia* **21**: 59-63.

Cobos A. 1968: Dos nuevas Alphasida (Col. Tenebrionidae) de Marruecos. *Annales de la Société Entomologique de France* (N. S.) **4**: 911-913.

Cobos A. 1988a: La Coleopterofauna endémica almeriense. *Graëllsia* **43**: 3-17.

Cobos A. 1988b: Révision de las Alphasida Escalera 1905, del subgénero Betasida Reitter 1917 (Coleoptera, Tenebrionidae). *Eos, Revista Española de Entomología* **64**: 47-56.

Cockerell T. D. A. 1910: The classification of the Meloidae. *Entomological News* **21**: 307.

Codina A. 1918: Alguns coleópteros de les illes de Eivissa (Santa Eulalia i Illa Plana) i de Formentera (Saint Francesc Xavier). *Anuari Junta de Ciencies Naturals* (Barcelona) **1918**: 264-269.

Coiffait H. 1947: Un Rhipiphoride nouveau de la faune de France. *Revue Française d'Entomologie* **14**: 15-19.

Collado J. & Alonso-Zarazaga M. A. 1996: Báguena-Corella's genera of Aderidae (Coleoptera): nomenclatural corrections and validations. *The Coleopterist's Bullettin* **50**: 96-98.

Comolli A. 1837: *De Coleopteris novis ac rarioribus minusve cognitis provinciae Novocomi*. Ticini Regii: Fusi et Socii, 54 pp.

Compte A. 1970: Los coléopteros de las Islas Columbretes. *Eos, Revista Española de Entomología* **45** [1969]: 97-137.

Compte A. 1985: II. Congresso Ibérico de Entomologia: Mordellidae de las islas Baleares. *Boletim da Sociedade Portuguesa de Entomologia, Suplemento 1* **1985**: 57-66.

Costa A. 1847: Specie nuove e rare d'insetti delle montagne del Matese. *Annali dell'Accademia degli Aspiranti Naturalisti Napoli* (2) **1**: 89-131.

Costa A. 1847: Descrizione di alcuni coleotteri del Regno di Napoli. *Annali dell'Accademia degli Aspiranti Naturalisti Napoli* (2) **1**: 134-162.

Costa A. 1852: *Parte Ia. Coleotteri Eteromeri. Famiglia degli Edemeridei - Oedemeridea.* In: *Fauna del regno di Napoli ossia enumerazione de tutti gli animali che abitano le diverse regioni di questo regno e la acque che la bagnano contenente la descrizione de'nuovi o poco esattamente conosciuti noc figure ricavete da originali vivente e dipinte al naturale.* Napoli: Gaetano Sautto, 36 pp., pls. ix-xi.

Costa A. 1853: *Parte Ia. Coleotteri, Eteromeri. Famiglia de'Mordellidei – Mordellidea.* In: *Fauna del regno di Napoli ossia enumerazione de tutti gli animali che abitano le diverse regioni di questo regno e la acque che la bagnano contenente la descrizione de'nuovi o poco esattamente conosciuti noc figure ricavete da originali vivente e dipinte al naturale.* Napoli: Gaetano Sautto, 1-8 pp.

Costa A. 1854: *Parte Ia. Coleotteri, Eteromeri. Famiglia de'Mordellidei – Mordellidea.* In: *Fauna del regno di Napoli ossia enumerazione de tutti gli animali che abitano le diverse regioni di questo regno e la acque che la bagnano contenente la descrizione de'nuovi o poco esattamente conosciuti noc figure ricavete da originali vivente e dipinte al naturale.* Napoli: Gaetano Sautto, 9-32 pp.

Costa A. 1859: *Parte IIa. Supplementi alle famiglie di coleotteri.* In: *Fauna del regno di Napoli ossia enumerazione de tutti gli animali che abitano le diverse regioni di questo regno e la acque che la bagnano contenente la descrizione de'nuovi o poco esattamente conosciuti noc figure ricavete da originali vivente e dipinte al naturale.* Naples: Gaetano Sautto, 24 pp.

Costa A. 1878: Relazione di un viaggio per l'Egitto, la Palestina e le coste della Turchia asiatica per richerche zoologiche. *Atti della Realle Accademia delle Scienze Fisiche e Matematiche* **7** (2): 1-40.

Costa A. 1882: Relazione di un viaggio nelle calabrie per ricerche zoologiche fatto nella state del 1876. *Atti della reale Accademia delle Scienze Fisiche e Matematiche di Napoli* **9**: 1-62, 1 pl.

Costa A. 1884: Notizie ed osservazioni sulla geo-fauna sarda. Memoria quarta. *Atti della Reale Accademia delle Scienze Fisiche e Matematiche di Napoli* (2) **1** (13): 1-31.

Costa A. 1884: Notizie ed osservazioni sulla geo-fauna sarda; memoria quarta. *Rendiconto della Reale Accademia delle Scienze Fisiche e Matematiche* **23** (12): 167-174.

Coucke L. 1891: Tables dichotomiques pour servir à la détermination des espèces Belges de coléoptères hétèromères. *Comptes-Rendus des Séances de la Société Entomologique de Belgique* **1891**: ccclcccii-ccclxxxviii.

Creutzer C. 1796: [new taxa]. In: Panzer G. W. F. 1796b: *Faunae Insectorum Germanicae initia oder Deutschlands Insecten. Ed. 2. XXXV.* Nürnberg: Felsecker, 24 pp. + 24 pls.

Creutzer C. 1796: [new taxa]. In: Panzer G. W. F. 1796b: *Faunae Insectorum Germanicae initia oder Deutschlands Insecten. Ed. 2. XXXVI.* Nürnberg: Felsecker, 24 pp. + 24 pls.

Creutzer C. 1799: *Entomologische Versuche.* Wien: Karl Schaumburg und Comp., 140 + [8] pp., 3 pls.

Crotch G. R. 1867a: On the Coleoptera of the Azores. *Proceedings of the Scientific Meetings of the Zoological Society of London* **1867**: 359-391, pl. 23.

Crotch G. R. 1867b: [new taxa]. In: Crotch G. R. & Sharp D.: Additions to the Catalogue of British Coleoptera, with descriptions of new species. *Transactions of the Entomological Society of London* **5**: 435-451.

Crotch G. R. 1872: List of the Coleoptera found during the Progress of the Survey. Pp. 263-268. In: Wilson C. W. & Palmer H. S. (eds): *Ordonance Survey of the Peninsula of Sinaï.* South Hampton.

Crotch G. R. 1873: *Check list of the Coleoptera of America, north of Mexico.* Salem: Mass: Naturalists' Agency, 136 pp.

Csiki E. 1901a: Bogarak. Coleopteren. Pp. 77-120. In: Horváth G. (ed.): *Zichy Jenõ gróf harmadik ázsiai utazása. II. kötet. Zichy Jenõ gróf harmadik ázsiai utazásának állattani eredményei.* [*Dritte asiatische Forschungsreise des des Grafen Eugen Zichy. Band II. Zoologische Ergebnisse der dritten asiatischen Forschungsreise des Grafen Eugen Zichy*]. Budapest: Victor Hornyánszky and Leipzig: Karl w. Hiersemann, 470 pp., 28 pls.

Csiki E. 1901b: Coleoptera nova ex Hungaria. *Természetrajzi Füzetek.* **24**: 486-490.

Csiki E. 1913: Rhipiphoridae. Pars 54. In: S. Schenkling (ed.): *Coleopterorum Catalogus.* Berlin: W. Junk, 29 pp.

Csiki E. 1915: Mordellidae. Pars 63. In: S. Schenkling (ed.): *Coleopterorum Catalogus.* Berlin. W. Junk, 84 pp.

Csiki E. 1924: Serropalpidae. Pars 77. In: S. Schenkling (ed.): *Coleopterorum Catalogus.* Berlin: W. Junk, 62 pp.

Csiki E. 1940a: Bogarak. Coleopteren. In: Csiki E.: Allattani Kutatásai Albániában (Explorationes zoologicae ab E. Csiki in Albania paractae). *A Magyar Tudomanyos Academia Balkán-Kutatásainak Tudományos Eredményei, Budapest,* **1** (2): 208-288.

Csiki E. 1940b: Magyarországi új bogarak. (Coleoptera nova ex Hungaria.). *Matematikai Természettudományi Értesito* **59**: 916-920.

Csiki E. 1949: Über eine neue Mordellistena aus der Krim. *Folia Entomologica Hungarica* (S. N.) **3**: 45-46.

Csiki E. 1953: Ueber neue und bekannte Coleoptera aus Ungarn und der angrenzendern Ländern. *Annales Historico-Naturales Musei Nationalis Hungarici* (N. S.) **3**: 115-135.

Curtis J. 1826: *British Entomology; being illustrations and descriptions of the genera of insects found in Great Britain and Ireland: containing coloured figures from nature of the most rare and beautiful species, and in many instances of the plants upon which they are found. Vol. III.* London: J. Curtis, [8] pp + 99-146 pls. and sheets.

Curtis J. 1827: *British Entomology; being illustrations and descriptions of the genera of insects found in Great Britain and Ireland: containing coloured figures from nature of the most rare and beautiful species, and in many instances of the plants upon which they are found. Vol. IV.* London: J. Curtis, [8] pp + 147-194 pls. and sheets.

Curtis J. 1829: *A guide to an arrangement of British Insects*. London: Westely, vi + 248 pp.

Curtis J. 1829: *British entomology; being illustrations and descriptions of the genera of insects found in Great Britain and Ireland: containing coloured figures from nature of the most rare and beautiful species, and in many instances of the plants upon which they are found. Vol. VI.* London: J. Curtis, [8] pp. + 242-289 pls and sheets.

Curtis J. 1830: *British entomology; being illustrations and descriptions of the genera of insects found in Great Britain and Ireland: containing coloured figures from nature of the most rare and beautiful species, and in many instances of the plants upon which they are found. Vol. VII.* London: J. Curtis, [8] pp. + 290-337 pls and sheets.

Curtis J. 1833: Characters of some undescribed genera and species indicated in the guide to an arrangement of British Insects. *The Entomological Magazine* **1**: 186-199.

Curtis J. 1836: *British Entomology being illustrations and descriptions of the genera of insects found in Great britain and Ireland: containing coloured figures from nature of the most rare and beautiful species, and in many instances of the plants upon which they are found. Vol. XIII.* London: J. Curtis, 578-625 pls and sheets.

Curtis J. 1838: *British Entomology; being illustrations and descriptions of the genera of insects found in Great Britain and Ireland: containing coloured figures from nature of the most rare and beautiful species, and in many instances of the plants upon which they are found. Vol. XV.* London: J. Curtis, [4] pp. + 674-721 pls and sheets.

Czetó Z. 1990: New Mordellidae (Coleoptera) from the Mediterranean Region, and a key to the genus Mordellistenula Sthegolewa-Barowskaya. *Folia Entomologica Hungarica* (N. S.) **51**: 25-30.

Dadabaev H. R. 1986: Novyy vid roda Prosodes Eschsch. (Coleoptera, Tenebrionidae) s Ghissarskogo Khrebta. *Doklady Akademii Nauk Tadzhikskoy SSR* **29**: 127-128.

Dadabaev Kh. 1986: Dva novykh vida roda Prosodes Eschsch. (Coleoptera, Tenebrionidae) iz Tadzhikistana. *Vestnik Zoologii* **4**: 78-81.

Daffner H. 1983: Eine neue Art der Gattung Langelandia Aubé von den Kanarischen Inseln (Coleoptera, Colydiidae). *Nachrichtenblatt der Bayerischen Entomologen* **32**: 20-22.

Daffner H. 1985: Eine neue Art der Gattung Langelandia Aubé aus Griechenland. *Nachrichtenblatt der Bayerischen Entomologen* **34**: 36-38.

Dahlgren G. 1976: Eine neue Asclera-Art aus Griechenland (Col., Oedemeridae). *Entomologisk Tidskrift* **97**: 21-22.

Dajoz R. 1960: Les coléoptères mycetophiles de la forêt de la Massane (Pyrénées-Orientales). Note préliminaire. *Vie et Milieu* **11**: 195-208.

Dajoz R. 1964: Contribution à l'étude des coléoptères Mycetophagidae I. Revision du genre Esarcus et description de trois espèces nouvelles d'Espagne. *L'Entomologiste* **20**: 7-15.

Dajoz R. 1965: Catalogue des coléoptères de la forêt de la Massane. *Vie et Milieu, Supplement* **15** (4): 1-207 + [2] pp.

Dajoz R. 1968: Révision des Colydiidae anophtalmes de la faune paléarctique (Col.) IV. Études sur les genres Anommatus et Langelandia. *Annales de la Société Entomologique de France* (N. S) **4**: 975-988.

Dajoz R. 1969: Révision des Colydiidae anophtalmes de la faune paléarctique. 3. Étude morphologique et biometrique de Langelandia anophtalma Aubé et de Langelandia reitteri Bellon. *Bulletin du Muséum National d'Histoire Naturelle, Paris* **40**: 710-717.

Dajoz R. 1971: Coléoptères Colydiidae nouveaux ou peu connus de la région paléarctique. *L'Entomologiste* **27**: 83-101.

Dajoz R. 1973: Coléoptères Colydiidae paléarctiques nouveaux ou peu connus. *Cahiers des Naturalistes* **29**: 79-84.

Dajoz R. 1975: Ergebnisse der Bhutan-Expedition 1972 des Naturhistorischen Museums in Basel. Coleoptera: Fam. Colydiidae & Cerylonidae. *Entomologica Basiliensia* **1**: 293-311.

Dajoz R. 1977: Coléoptères Colydiidae et Anommatidae paléarctiques. *Faune de l'Europe et du Bassin Méditerranéen* 8. Paris: Masson, 275 pp.

Dajoz R. 1983: Les peuplements de coléoptères terricoles de Tunisie. Études écologique et biogéographique des Tenebrionidae (incl. 2 taxa nouveaux) et Carabidae. *Cahiers des Naturalistes* **38** [1982]: 33-67.

Dajoz R. 1984: Ténébrionides européens nouveaux ou peu connus (Coleoptera). *Cahiers des Naturalistes* (N. S.) **39** [1983]: 89-92.

Dajoz R. 1985: Description de deux Tenebrionidae nouveaux de Tunisie (Coleoptera). *Nouvelle Revue d'Entomologie* (N. S.) **2**: 193-196.

Dalla Torre K. W. von. 1879: Die Käferfauna von Oberösterreich. Systematisches Verzeichnis der in Oberösterreich bisher beobachteten Käfer. *Jahres-Bericht des Vereines für Naturkunde in Oesterrreich ob der Enns zu Linz* **10**: 1-125.

Dalla Torre K. W. von. 1880: Die Käferfauna von Oberösterreich. Systematisches Verzeichnis der in Oberösterreich bisher beobachteten Käfer. *Jahres-Bericht des Vereines für Naturkunde in Oesterreich ob der Enns zu Linz* **11**: 1-81.

Dalman J. W. 1823: *Analecta Entomologica*. Holmiae: Typis Lindhianis, vii + 104 + [4] pp., 4 pls.

D'Amore-Fracassi A. 1907: Gen. Colpotus Muls. et Rey. Tavola sinottica e descrizione delle specie appartenenti alla fauna europea, caucasica e circummediterranea. *Il Naturalista Siciliano* **19**: 177-199.

DeGeer C. 1775: *Mémoires pour servir à l'histoire des Insectes. Tome cinquième.* Stockholm: L'imprimerie Pierre Hesselberg, vii + 448 pp., 16 pls.

Degiovanni A. 2000: Contributo alla conoscenza dei coleotteri anticidi italiani con descrizione di Tenuicomus platiai n. sp. (Insecta Coleoptera Anthicidae). *Quaderno di Studi e Notizie di Storia Naturale della Romagna* **13** (*Supplement*): 85-89.

Dejean P. F. M. A. 1821: *Catalogue de la collection de coléoptères de M. le Baron Dejean.* Paris: Crevot, viii + 136 + 2 (errata) pp.

Dejean P. F. M. A. 1834: *Catalogue des coléoptères de la collection de M. le Comte Dejean. Deuxième édition. 3e Livraison.* Paris: Méquignon-Marvis Pères et Fils, pp. 177-256.

Dejean P. F. M. A. 1835: *Catalogue des coléoptères de la collection de M. le Comte Dejean. Deuxième édition. 4e Livraison.* Paris: Méquignon-Marvis Pères et Fils, pp. 257-360.

Dejean P. F. M. A. 1836: *Catalogue des coléoptères de la collection de M. le Comte Dejean. Troisième édition revue, corrigée et augmentée. Livraisons 1-4.* Paris: Méquignon-Marvis Père et Fils, 1-384 pp.

De Lisle M. O. 1973: Description d'un nouveau coléoptère Trictenotomidae de Malaisie. *Nouvelle Revue d'Entomologie* **3**: 17-18.

Della Beffa G. 1909: Note coleotterologiche. *Rivista Coleotterologica Italiana* **7**: 192-201.

Della Beffa G. 1931: Himalaia cashmiriano. Appendici scientifiche. Insetti del Kashmir raccolti dal dott. Cesare Calciati nelle spedizione Mario Piacenza (1913). Milano: Rizzoli & C., 173-187.

Demaison C. 1905: Description d'une variété et d'une espèce nouvelle d'Asie mineure (Col.). *Bulletin de la Société Entomologique de France* **1904**: 286-287.

Denier P. 1913: Description d'un genre et d'une espèce nouvelle de Meloidae (Col.). *Bulletin de la Société Entomologique de France* **1913**: 231-232.

Denier P. 1921: Description d'une espèce nouvelle de Meloe paléarctique. *Bulletin de la Société Entomologique de France* **1921**: 112-113.

Denier P. 1935: Coleopterorum americanorum familiae Meloidarum enumeratio synonymica. *Revista de la Sociedad Entomólogica Argentina* **7**: 139-176.

Denisova L. A. 1940: Obzor palearkticheskikh vidov podsemeystva Tenebrioninae (Coleoptera, Tenebrionidae). *Trudy Zoologicheskogo Instituta Akademii Nauk* SSSR **6**: 222-251.

Desbrochers des Loges J. 1868: Description de deux coléoptères nouveaux des environs de Bone. *Bulletin de l'Académie d'Hippone* **4**: 77-80.

Desbrochers des Loges J. 1870: Descriptions de coléoptères nouveaux d'Europe et confins. *L'Abeille, Mémoires d'Entomologie* **7**: 97-135.

Desbrochers des Loges J. 1874: [new taxa]. In: Marseul S.-A.: Répertoire des coléoptères d'Europe décrits isolément depuis 1863. Troisième partie. *L'Abeille, Journal d'Entomologie* **12** [1874-1875]: 1-456.

Desbrochers des Loges J. 1875: Anthicides nouveaux, pp.42-49. In: *Opuscules Entomologiques (Coléoptères)*. **1**er Cahier [1874-1875]: 1-56. Paris: A. Gaudon.

Desbrochers des Loges J. 1881: Insectes coléoptères du nord de l'Afrique nouveaux ou peu connus. 1er Mémoire. Ténébrionides. *Bulletin de l'Académie d'Hippone* **16**: 51-168.

Desbrochers des Loges J. 1884: Diagnoses de coléoptères nouveaux algériens. *Bulletin de l'Academie d'Hippone* **19**: 169.

Desbrochers des Loges J. 1890: [Diagnoses des coléoptères nouveaux d'Algérie]. *Bulletin de la Société Entomologique de France* **1894**: clxxxv-clxxxvi.

Desbrochers des Loges J. 1894: Révision monographique des ténébrionides appartenant à la tribu des cossyphides d'Europe et des contrées méditerranéennes en Afrique et en Asie. *Le Frelon* **4**: 2-21.

Desbrochers des Loges J. 1898a: Quelques matériaux pour la faune des coléoptères de Barbarie. *Le Frelon* **6** [1897-1898]: 65-82.

Desbrochers des Loges J. 1898b: Quelques matériaux pour la faune des coléoptères de Barbarie. *Le Frelon* **7**: 1-51.

Desbrochers des Loges J. 1899: Faunule des coléoptères de la France et de la Corse. Famille des Anthicidae. *Le Frelon* **7**: 151-192.

Desbrochers des Loges J. 1900: Faunule des coléoptères de la France et de la Corse. *Le Frelon* **8**: 1-56.

Des Gozis M. 1886: *Recherche de l'espèce typique de quelques anciens genres. Rectifications synonymiques et notes diverses.* Motluçon: Imprimerie Herbin, 36 pp.

Desmarest A. G. 1825: Oenas africana. P. 435. In: Cuvier F. G. (ed.) *Dictionnaire de Sciences Naturelles*. Paris: Levrault, 1816-1830.

Deyrolle A. 1864: [new taxa]. In: Wollaston T. W.: *A catalogue of the coleopterous insects of the Canaries in the collection of the British Museum*. London: Trustees, xiii + 648 pp.

Deyrolle A. 1867: Monographie de la tribu des zophosites. *Annales de la Société Entomologique de France* (4) **7**: 73-248, 4 pls.

Deyrolle H. 1875: Tableau synoptique des espèces de tricténotomides. *Bulletin de la Société Entomologique de France* **1875**: lix-lxi.

Dillon L. S. 1952: The Meloidae (Coleoptera) of Texas. *The American Midland Naturalist* **48**: 330-420.

Dilwyn L. W. 1829: *Memoranda relating to Coleopterous insects, found in the neighbourhood of Swansea.* Swansea: W.C. Murray and D. Rees, Cambrian-office, 75 pp.

Dodero A. 1916: Appunti Coleotterologici II. *Annali del Museo Civico di Storia Naturale Giacomo Doria* 47: 337-354.

Dohrn C. A. 1873: Exotisches. *Entomologische Zeitung* (Stettin) **34**: 70-83.

Dohrn C. A. 1880: Exotisches. *Entomologische Zeitung* (Stettin) **41**: 367-382.

Dohrn C. A. 1882: Exotisches. *Entomologische Zeitung* (Stettin) **43**: 245-259, 457-470.

Dohrn C. A. 1884: Exotisches. *Entomologische Zeitung* (Stettin) **45**: 39-46.

Dokhtouroff W. 1887b: Description de deux coléoptères nouveaux de la faune aralo-caspienne. *Horae Societatis Entomologicae Rossicae* **21**: 344-345.

Dokhtouroff W. 1889: Faune coléoptérologique aralo-caspienne. XIII partie. Meloïdes et cantharides (Mylabrides). *Horae Societatis Entomologicae Rossicae* **24** [1889-1890]: 121-192, 1 pl.

Donisthorpe H. S. K. 1909: Anaspis hudsoni, Donisthorpe, fem. bred. *The Entomologist's Record and Journal of Variation* **23**: 800.

Donisthorpe H. S. K. 1916: Cis dentatus, Mellie, not a British insect. *The Entomologist's Record and Journal of Variation* **28**: 155-156.

Donisthorpe H. S. K. 1925: Lissodema kirkae n.sp. A species of Coleoptera new to science. *The Entomologist's Record and Journal of Variation* **37**: 106-107.

Donisthorpe H. S. K. 1928: Anaspis florenceae nov. sp.; a species of Coleoptera new to science; with a few notes on some other species of the genus. *The Entomologist's Record and Journal of Variation* **40**: 121-122.

Donisthorpe H. S. K. 1936: Cis savilli n. sp., a species of Coleoptera new to science. *The Entomologist's Monthly Magazine* **72**: 220.

Donovan E. 1793: *The natural history of British Insects; explaining them in their several states with the period of their transformations, their food, oeconomy, &c. Together with the history of such minute insects as require investigations by the microscope. The whole illustrated by coloured figures designed and executed from living specimens. Vol. II.* London: printed for the author, and for F. and C. Rivington, 96 + [7] pp., pls. 37-72.

Donovan E. 1801: *The natural history of British Insects; explaining them in their several states, with the period of their transformations, their food, oeconomy, &c. together with the history of such minute Insects as require investigation by the microscope, the whole illustrated by coloured figures, designed and executed from living specimens. Vol. X.* London: F. C. and J. Rivington, 95+vii pp., 35 pls.

Dorthes J. A. 1787: Observations sur quelques insectes nuisibles aux blés et à la luzerne. *Mémoires de la Société Agricole de Paris* **1787** (trim. de printemps): 61-71.

Drapiez P. A. J. 1820a: Description de six espèces d'insectes nouveaux. *Annales Générales des Sciences Physiques* (Bruxelles) **4**: 349-354.

Drapiez P. A. J. 1820b: Description de huit espèces d'insectes nouveaux. *Annales Générales des Sciences Physiques* (Bruxelles) **5**: 117-123, pl. 82.

Drapiez P. A. J. 1821: Description de cinq espèces d'insectes nouveaux. *Annales Générales des Sciences Physiques* (Bruxelles) **8**: 275-282, 1 pl.

Drogvalenko A. N. 2002: Novye i redkie dlya fauny Ukrainy vidy zhestkokrylykh nasekomykh (Insecta: Coleoptera). Soobshcheniye 2 [New and rare species of beetles (Insecta: Coleoptera) for fauna of Ukraine. Information 2]. *Izvestiya Kharkovskogo Entomologicheskogo Obshchestva* **9**: 9-19.

Dubrovin N. N. 1992a: [new taxa]. In: Dubrovin N. N. & Kompantseva T. V.: Morpho-ecological review of larvae of Isomira (Coleoptera, Alleculidae) with the description of a new species from central Asia. *Zoologicheskiy Zhurnal* **71** (7): 14-22.

Dubrovin N. N. 1992b: Sem. Alleculidae – pyl'ceedy. Pp. 510-517. In: Ler P. A.: *Opredelitel' nasekomykh dalnego vostoka SSSR v shesti tomakh. Tom III, zhestkokrylye, ili zhuky, Chast' 2.* Sankt-Peterburg: Nauka, 704 pp.

Dubrovina M. I. 1973a: Novyy vid roda Stenerophlina (Coleoptera, Alleculidae) iz Uzbekistana [A new species of the genus Stenerophlina (Coleoptera, Alleculidae) from Uzbekistan]. *Zoologicheskiy Zhurnal* **52**: 284-286.

Dubrovina M. I. 1973b: Novyy podrod i novye vidy zhukov-pyl'ceedov roda Isomira Muls. (Coleoptera, Alleculidae) iz Stredney Azii [New subgenus and new species of pollen beetles of the genus Isomira Muls. (Coleoptera, Alleculidae) from Central Asia]. *Entomologichekoe Obozrenie* **52**: 367-376.

Dubrovina M. I. 1975: Novyy podrod i novye vidy zhukov-pyl'ceedov roda Hymenalia Muls. (Coleoptera, Alleculidae) iz Mongolii i sopredel'nykh rayanov Kitaya [A new subgenus and new species of pollen beetles of the genus Hymenalia Muls. (Coleoptera, Alleculidae) from Mongolia and from adjacent regions of China]. *Nasekomye Mongolii* **3**: 165-172.

Dubrovina M. I. 1976: Taksonomya podroda Pterna Siedl. roda Mycetochara Berth. (Coleoptera, Alleculidae). *Zoologicheskiy Zhurnal* **55**: 1407-1409.

Dubrovina M. I. 1978: Novye dannye o vidakh triby Alleculini (Coleoptera, Alleculidae) evropeyskoy chasti SSSR. *Nauchnye Doklady Vysshey Shkoly Biologicheskie Nauki* **7**: 53-55.

Dubrovina M. I. 1982: Obzor zhukov-pyl'ceedov roda Isomira Muls. (Coleoptera, Alleculidae) fauny SSSR [A review of pollen-beetles of the genus Isomira Muls. (Coleoptera, Alleculidae) of the USSR]. *Entomologichekoe Obozrenie* **61**: 131-144.

Dufour L. J. M. 1820: Description de dix espèces nouvelles ou peu connues d'insectes recueillis en Espagne. *Annales Générales des Sciences Physiques* (Bruxelles) **6**: 307-317, pl. XCVI.

Dufour L. J. M. 1828: Description d'un genre nouveau d'Insectes de l'ordre des Parasites. *Annales des Sciences Naturelles* (Paris) (1) **13**: 62-66, 1 pl.

Dufour L. J. M. 1841: Histoire de metamorphoses d'une Oedémère. *Annales de la Société Entomologique de France* **10**: 5-10.

Dufour L. J. M. 1843: Excursion entomologique dans les montagnes de la Vallée d'Ossau. *Bulletin des Sciences, Lettres et Arts de Pau* **3**: 1-118.

Dufour L. J. M. 1849: Note sur trois espèces du genre Anthicus. *Annales des Sciences Naturelles. Zoologie* (3) **11**: 229-230, pl. 5.

Dufour L. J. M. 1851: Des zones entomologiques dans nos Pyrénées occidentales, et designation des insectes qui les habitent (2). *Actes de la Société Linnéenne de Bordeaux* **17**: 304-364.

Duftschmid C. E. 1812: *Fauna Austriaca oder Beschreibung der österreichischen Insekten für angehende Freunde der Entomologie. Zweiter Theil.* Linz und Leipzig: K. K. priv. akademischen Kunst-, Mussik- und Buchhandlung, viii + 311+ [4] pp.

Duftschmid C. E. 1825: *Fauna Austriaca, oder Beschreibung der österreichischen Insekten, für angehende Freude der Entomologie. Dritter Theil.* Linz: Priv. akademischen Kunst-, Musik- und Buchhandlung, 289 pp.

Dugès E. 1869: Descripción de algunos melóideos indigenas. *Naturaleza Mexicana* **1** [1869]: 100-113.

Dugès E. 1870: Descripción de algunos melóideos indigenas. *Naturaleza Mexicana* **1** [1869]: 125-128, 157-168, 1 pl.

Duméril A. M. C. 1825: Mylabre, Mylabris. Pp. 11-13. In: Cuvier F. G. (ed.): *Dictionnaire des Sciences naturelles, dans lequel on traite méthodiquement des différents êtres de la nature, considérés soit en eux-mêmes, d'après l'état actuel de nos connaissances, soit relativement a l'utilité qu'en peuvent retirer la médecin, l'agriculture, le commerce et les arts. Suivi d'une biographie des plus célébres naturalistes. Tome 34.* Paris: F. G. Levrault, 560 pp.

Duponchel P.-A.-J. 1827: Mémoires sur quelques insectes nouveaux ou peu connus. *Annales de la Société Linnéenne de Paris* **6**: 338-343, 2 pls.

Duponchel P.-A.-J. 1845: [new taxa]. In: D'Orbigny C. V. D. (ed.): *Dictionnaire universel d'histoire naturelle résumant et complétant tous les faits présentés par les encyclopédies, les anciens dictionnaires scientifiques, les œuvres complètes de Buffon, et les meilleurs traités spéciaux sur les divers branches des sciences natrurelle; donnant la description des êtres et des divers phénomènes de la nature, l'étymologie et la définition des noms scientifiques, les principales applications des corps organiques et inorganiques, à l'agriculture, à la médicine, aux arts industriels, etc.; gravé sur acier. Tome 6.* Paris: C. Renard, 792 pp.

Dury C. A. 1902: A revised list of Coleoptera observed near Cincinnati, Ohio, with notes on localities, bibliographical references, and descriptions of new species. *Journal of the Cincinnati Society of Natural History* **20**: 107-196.

Dury C. A. 1917: Synopsis of the coleopterous family Cisidae (Cioidae) of America north of Mexico. *The Journal of the Cincinnati Society of Natural History* **22** (2): 2-28.

Dury C. A. 1919: A new ciside genus with new species from Manitoba. *Canadian Entomologist* **51**: 158.

Dvořák M. 1983: Drei neue Arten und Bemerkungen zur Familie Meloidae (Coleoptera). *Acta Entomologica Bohemoslovaca* **80**: 441-450.

Dvořák M. 1985: Croscherichia maceki sp. n. aus Irak (Coleoptera, Meloidae). *Annotationes Zoologicae et Botanicae* (Bratislava) **169**: 1-4.

Dvořák M. 1990: Taxonomische Bemerkungen zur Gattung Cerocoma Geoffroy (Coleoptera, Meloidae, Cerocomini). *Annotationes Zoologicae et Botanicae* (Bratislava) **199**: 1-12.

Dvořák M. 1993a: Einige neue paläarktische Meloiden-Arten (Coleoptera). *Annotationes Zoologicae et Botanicae* (Bratislava) **212**: 1-10.

Dvořák M. 1993b: Revision der Gattung Micromerus Mulsant et Rey, 1858 (Coleoptera, Meloidae). *Annotationes Zoologicae et Botanicae* (Bratislava) **214**: 1-9.

Dvořák M. 1996: Neue asiatische Epicauta-Arten (Coleoptera: Meloidae). *Folia Heyrovskiana* **4**: 35-42.

Dvořák M. 2004: Neue Oenas Latreille, 1802 Art aus Iran (Coleoptera, Meloidae). *Acta Musei Moraviae, Scientiae Biologicae* **89**: 103-104.

Egorov L. V. 1987: Novye vidy zhukov-chernotelok roda Oodescelis Motsch. (Coleoptera, Tenebrionidae). *Entomologicheskoe Obozrenie* **66**: 800-802.

Egorov L. V. 1989a: Novye vidy i podvid roda Oodescelis Motsch. (Coleoptera, Tenebrionidae) fauny SSSR. *Trudy Zoologicheskogo Instituta Akademii Nauk SSSR* **208**: 102-106.

Egorov L. V. 1989b: Obzor zhukov-chernotelok roda Platyscelis Latr. (Coleoptera, Tenebrionidae) fauny SSSR. *Entomologicheskoe Obozrenie* **68**: 336-352.

Egorov L. V. 1990a: Novye vidy chernotelok roda Bioramix Bat. (Coleoptera, Tenebrionidae). *Trudy Zoologicheskogo Instituta Akademii Nauk SSSR* **211**: 117-134.

Egorov L. V. 1990b: K sinonimike dvukh vidov roda Platyscelis Latr. i opisanie novogo vida roda Bioramix Bat. (Coleoptera, Tenebrionidae). *Entomologicheskoe Obozrenie* **69**: 890-892.

Egorov L. V. 1991: Die Platyscelidini des Staatlichen Museums für Tierkunde in Dresden (Insecta, Coleoptera: Tenebrionidae). *Reichenbachia* **28**: 145-148.

Egorov L. V. 1992: Novye vidy chernotelok triby Platyscelidini (Coleoptera, Tenebrionidae). *Entomologicheskoe Obozrenie* **71**: 796-799.

Egorov L. V. 1998: Obzor zhukov-chernotelok roda Somocoelia Kraatz, 1882 (Coleoptera, Tenebrionidae). *Entomologicheskoe Obozrenie* **77**: 209-215.

Egorov L. V. 2004: O sisteme zhukov-chernotelok Platyscelidini (Coleoptera, Tenebrionidae) mirovoy fauny. *Entomologicheskoe Obozrenie* **83**: 581-613.

Egorov L. V. 2006: Novye vidy zhukov-chernotelok podroda Cardiobioramix Kasz. roda Bioramix Bat. (Coleoptera, Tenebrionidae, Platyscelidini) iz provintsiy Gan"su i Sychuan" Kitaya. *Entomologicheskoe Obozrenie* **85**: 789-797.

Eichler W. 1924: Nouvelles espèces et variétés de coléoptères du Caucase et de l'Asie mineure. *Polskie Pismo Entomologiczne* **3**: 61-68.

Emery M. C. 1876: Essai monographique sur les mordellides de l'Europe et des contrées limitrophes. *L'Abeille, Journal d'Entomologie* **14** [1874-1876]: 1-128.

Emery M. C. 1891: [new taxa]. *Bulletin de la Société Entomologique de France* **1891**: xxxix-xl.

Erichson W. F. 1841: Über die Insecten von Algier mit besonderer Rücksicht auf ihre geographische Verbreitung. Pp. 140-194. In: Wagner M. F. (ed.): *Reisen in der Regentschaft Algier 1836, 1837 und 1838 nebst einem naturhistorischen Anhang und einem Kupferatlas. Dritter Band.* Leipzig: L. Voss, xviii + 298 pp., 17 pls.

Erichson W. F. 1842: Beitrag zur Insecten-Fauna von Vandiemensland, mit besonderer Berücksichtigung der geographischen Verbreitung der Insecten. *Archiv für Naturgeschichte* **8** (1): 83-287, pls 4, 5.

Erichson W. F. 1843: Beitrag zur Insecten-Fauna von Angola in besonderer Beziehung zur geographischen Verbreitung der Insecten in Afrika. *Archiv für Naturgeschichte* **9** (1): 199-267.

Erichson W. F. 1845: Lieferung II. Pp 161-320. In: *Naturgeschichte der Insecten Deutschlands. Erste Abtheilung. Coleoptera. Dritter Band.* Berlin: Nicolaische Buchhandlung, 320 pp. [issued in parts, pp. 1-320 (Lieferung I, II) in 1845].

Erichson W. F. 1846: Lieferung III. Pp. 321-480. In: *Naturgeschichte der Insecten Deutschlands. Erste Abtheilung. Coleoptera. Dritter Band.* Berlin: Nicolaischen Buchhandlung, vii + [2] + 968 pp., 1 pl. [issued in parts, pp. 321-480 in 1846, pp. 801-968 in 1848].

Ermisch K. 1940: Revision der ostasiatischen Glipa-Arten. *Entomologische Blätter* **36**: 161-173.

Ermisch K. 1941a: Stenomordella, eine neue Mordellidengattung aus der chinensischen Provinz Fukien. *Entomologische Blätter* **37**: 115-117.

Ermisch K. 1941b: Mordellistenoda, eine neue Mordellidengattung aus der chinesischen Provinz Fukien. *Mitteilungen Münchener Entomologische Gesellschaft* **37**: 589-593.

Ermisch K. 1941c: Tribus Mordellistenini (Col. Mordell.). *Mitteilungen der Münchener Entomologischen Gesellschaft* **31**: 710-726.

Ermisch K. 1942a: Eine neue paläarktische Mordellide. *Arbeiten über Morphologische und Taxonomische Entomologie aus Berlin-Dahlem* **9**: 85-88.

Ermisch K. 1942b: Tolidostena, ein neues Mordellidengenus aus der chinesischen Provinz Fukien. *Mitteilungen der Münchener Entomologischen Gesellschaft* **32**: 674-678.

Ermisch K. 1943: Tolida Tournieri Em. und ihre Synonyme Schusteri Schilsky und similis Stchegoleva. *Entomologische Blätter* **39**: 18-21.

Ermisch K. 1944: Mordellidae. In: G. Frey: Tagebuch meiner Spanienreise und deren wissenschaftliche Ergebnisse. I. Reisetagebuch. *Mitteilungen der Münchener Entomologischen Gesellschaft* **34**: 210-216.

Ermisch K. 1948: Neue Arten der Gattung Tomoxia und verwandter Gattungen aus Afrika. *Entomologische Blätter* **41-44**: 98-110.

Ermisch K. 1950a: Mordellidae. In: *Missiòn G. F. de Witte, Exploration du Parc National Albert 1933-1935.* Brussels: Institut des Parcs Nationaux du Congo Belge **71**: 1-95.

Ermisch K. 1950b: Die Gattungen der Mordelliden der Welt. *Entomologische Blätter* **45-46**: 34-92.

Ermisch K. 1952a: Mordellidae des Belgischen Congogebietes des Musée Royal du Congo Belge in Tervuren. *Annales du Musée Royal du Congo Belge, Tervuren* **22**: 7-105.

Ermisch K. 1952b: Neue Mordelliden aus der chinesischen Provinz Fukien. *Entomologische Blätter* **47-48**: 143-157.

Ermisch K. 1954a: Über Typen und neue Arten der afrikanischen Mordelliden. *Memorie della Società Entomologica Italiana* **33**: 167-200.

Ermisch K. 1954b: Mordellistena-Studien I. Mordellistena micans Germar und die micans der Autoren. *Beiträge zur Entomologie* **4**: 173-180.

Ermisch K. 1956: Mordellidae. Pp. 269-328. In: Horion A.: *Faunistik der mitteleuropäischen Käfer. Band V: Heteromera.* Tutzing: A. Frey, 336 pp., 6 pls.

Ermisch K. 1957: Revision der Gattung Mordellistenula Stshegol.-Bar. *Entomologische Blätter* **53**: 45-51.

Ermisch K. 1962: Drei neue Anaspis-Arten aus Turkestan. *Senckenbergiana Biologica* **43**: 165-170.

Ermisch K. 1963a: Neue Mordelliden (Heteromera, Mordellidae) aus Deutschland und Nachträge zur Faunistik der mitteleuropäischen Mordelliden. *Entomologische Blätter* **59**: 1-36.

Ermisch K. 1963b: Beitrag zur Mordelliden-Fauna Portugals. *Notulae Entomologicae* **43**: 14-21.

Ermisch K. 1963c: Die Mordelliden der Insel Cypern. *Notulae Entomologicae* **43**: 49-67.

Ermisch K. 1963d: XIII. Anaspis (Pinassa) proteus Woll. und imitator n. sp. In: Lindberg Har.: A contribution to the study of beetles in the Madeira Islands. *Commentationes Biologicae* **25**: 108-111.

Ermisch K. 1964a: Anaspis (Pinassa) eversi n. sp. von der Kanareninsel Fuerteventura. *Entomologische Blätter* **60**: 18-19.

Ermisch K. 1964b: Ergebnisse der zoologischen Forschungen von Dr. Z. Kaszab in der Mongolei. *Annales Historico-Naturales Musei Nationalis Hungarici* **56**: 341-351.

Ermisch K. 1965a: 49. Mordellidae II. Ergebnisse der zoologischen Forschungen von Dr. Z. Kaszab in der Mongolei. *Reichenbachia* **7**: 63-80.

Ermisch K. 1965b: Mordellistena-Arten von den Kanarischen Inseln. *Entomologische Blätter* **61**: 67-73.

Ermisch K. 1965c: Synonymische und nomenklatorische Feststellungen in der Familie Mordellidae. *Reichenbachia* **5**: 197-201.

Ermisch K. 1965d: Neue Mordelliden von der Balkanhalbinsel. *Reichenbachia* **5**: 251-272.

Ermisch K. 1966a: Neue westpaläarktische Mordellistena-Arten. *Entomologische Blätter* **62**: 30-39.

Ermisch K. 1966b: Neue Mordelliden aus Nordafrika. *Entomologische Blätter* **62**: 40-43.

Ermisch K. 1967a: 90. Mordellidae III. Ergebnisse der zoologischen Forschungen von Dr. Z. Kaszab in der Mongolei. *Reichenbachia* **9**: 45-52.

Ermisch K. 1967b: Neue Mordellistena-Arten aus Mitteleuropa und der Balkanhalbinsel. *Entomologische Blätter* **63**: 110-119.

Ermisch K. 1967c: Anaspis (Silaria) canariensis n. sp. von den Kanaren-Insel Gomera und Hierro. *Entomologische Blätter* **63**: 120.

Ermisch K. 1967d: Ergebnisse der 1. Mongolisch-tschechoslowakischen entomologisch-botanischen Expedition in der Mongolei. *Acta Faunistica Entomologica Musei Nationalis Pragae* **12**: 125-129.

Ermisch K. 1968a: Neue Mordellini aus der chinensischen Provinz Fukien. *Reichenbachia* **10**: 279-292.

Ermisch K. 1968b: 132. Mordellidae IV. Ergebnisse der zoologischen Forschungen von Dr. Z. Kaszab in der Mongolei. *Reichenbachia* **11**: 17-24.

Ermisch K. 1969a: Eine neue Hoshihananomia-Art aus Ussuri. *Entomologische Blätter* **65**: 102-103.

Ermisch K. 1969b: Neue Mordelliden aus Europa, Nordafrika und Nahen Osten. *Entomologische Blätter* **65**: 104-115.

Ermisch K. 1969c: 183. Mordellidae V. Ergebnisse der zoologischen Forschungen vin Dr. Z. Kaszab in der Mongolei. *Reichenbachia* **12**: 171-177.

Ermisch K. 1970: 204. Mordellidae VI. Ergebnisse der zoologischen Forschungen von Dr. Z. Kaszab in der Mongolei. *Reichenbachia* **13**: 169-187.

Ermisch K. 1977: Die Mordellistena-Arten Ungarns und benachbarter Gebiete sowie Beschreibung einer neuen Hoshihananomia-Art aus Siebenbürgen. *Folia Entomologica Hungarica* (N. S.) **30**: 151-177.

Escalera M. M. de la. 1901: Materiales para una revisión del género Asida. *Boletín de la Sociedad Española de Historia Natural* **1**: 172-175.

Escalera M. M. de la. 1903: Otra Asida nueva de España del grupo de las aterciopeladas. *Boletín de la Sociedad Española de Historia Natural* **3**: 75-76.

Escalera M. M. de la. 1905a: Sistema de las especies ibéricos del gen. Asida Latr. *Boletín de la Real Sociedad Española de Historia Natural* **5**: 377-402, 430-451.

Escalera M. M. de la. 1905b: Una nueva especie de «Eulipus» Woll. (Tentyriini) de Río de Oro. *Boletín de la Real Sociedad Española de Historia Natural* **5**: 467-468.

Escalera M. M. de la. 1906a: Especies nuevas de coleópteros de Marruecos. *Boletín de la Real Sociedad Española de Historia Natural* **6**: 161-162.

Escalera M. M. de la. 1906b: Adiciones al sistema de las especies ibéricas del género Asida. Una Globasida nueva de la provincia de Murcía. *Boletín de la Real Sociedad Española de Historia Natural* **6**: 193.

Escalera M. M. de la. 1906c: Una nuova Alphasida de Motril. *Boletín de la Real Sociedad Española de Historia Natural* **1906**: 235.

Escalera M. M. de la. 1906d: Sistema de las especies ibéricas del gén. Asida Latr. *Boletín de la Real Sociedad Española de Historia Natural* **6**: 306-317.

Escalera M. M. de la. 1906e: Sobre la variabilidad de las especies de Alphasida de las ramas orientales y descrición de especies nuevas. *Boletín de la Real Sociedad Española de Historia Natural* **1906**: 381-385.

Escalera M. M. de la. 1907: Especies nuevas de Marruecos. *Boletín de la Real Sociedad Española de Historia Natural* **6**: 337-340.

Escalera M. M. de la. 1909a: Especies nuevas de Marruecos. *Boletín de la Real Sociedad Española de Historia Natural* **9**: 135-136.

Escalera M. M. de la. 1909b: Especies nuevas de meloidos del SW. de Marruecos. *Boletín de la Real Sociedad Española de Historia Natural* **9**: 240-249.

Escalera M. M. de la. 1910a: Nuevos coleópteros de Marruecos. *Boletín de la Real Sociedad Española de Historia Natural* **1910**: 283-285.

Escalera M. M. de la. 1910b: Coleópteros nuevos de Marruecos. *Boletín de la Real Sociedad Española de Historia Natural* **10**: 379-382.

Escalera M. M. de la. 1910c: Especies nuevas de Marruecos. *Boletín de la Real Sociedad Española de Historia Natural* **10**: 409-416.

Escalera M. M. de la. 1911: Coleópteros nuevos del S.W. de Marruecos. *Boletín de la Real Sociedad Española de Historia Natural* **11**: 299-304.

Escalera M. M. de la. 1912: Dos nuevas especies españolas del género Asida. *Boletín de la Real Sociedad Española de Historia Natural* **12**: 166-168.

Escalera M. M. de la. 1913: Coleópteros nuevos del Sus (Marruecos). *Trabajos del Museo de Ciencias Naturales, Serie Zoológica* (Madrid) **8**: 3-56.

Escalera M. M. de la. 1914: Los coleópteros de Marruecos. *Trabajos del Museo Nacional de Ciencias Naturales Serie Zoológica* (Madrid) **11**: 1-553.

Escalera M. M. de la. 1915: Una specie nueva de Zonabris de España. *Boletín de la Real Sociedad Española de Historia Natural* **15**: 493-494.

Escalera M. M. de la. 1921a: Descripciones de especies nuevas de Asida de la Península Ibérica. *Boletín de la Real Sociedad Española de Historia Natural* **21**: 117-120.

Escalera M. M. de la. 1921b: Especies nuevas de Asida de la Península Ibérica. *Boletín de la Real Sociedad Española de Historia Natural* **21**: 350-363.

Escalera M. M. de la. 1921c: Especies nuevas de Asida de la Península Ibérica. *Boletín de la Real Sociedad Española de Historia Natural* **21**: 427-437.

Escalera M. M. de la. 1922a: Especies ibéricos del género Asida (Col. Tenebriónidos). *Boletín de la Real Sociedad Española de Historia Natural* **22**: 64-71.

Escalera M. M. de la. 1922b: Especies nuevas de Asida de Marruecos. *Boletín de la Real Sociedad Española de Historia Natural* **22**: 170-175.

Escalera M. M. de la 1922c: Los Heliotaurus (Col. Cistelidae) de Marruecos de protórax rojo. *Boletín de la Real Sociedad Española de Historia Natural* **22**: 358-362.

Escalera M. M. de la. 1922d: Especies nuevas de Arthrodeis de Marruecos (Col. Tenebriónidos). *Boletín de la Real Sociedad Española de Historia Natural* **22**: 373-377.

Escalera M. M. de la. 1922e: Sistema de las especies del género Asida de la Península Ibérica. Subgén. Glabrasita Esc. 1. nota. *Boletín de la Real Sociedad Española de Historia Natural* **22**: 388-391.

Escalera M. M. de la. 1922f: Sistema de las especies españolas del género Asida Latr. *Boletín de la Real Sociedad Española de Historia Natural* **22**: 431-451.

Escalera M. M. de la. 1923a: Sistema de las especies del género Asida de la Península Ibérica. Subgén. Glabrasida Esc. 2.a nota (1). *Boletín de la Real Sociedad Española de Historia Natural* **22** [1922]: 463-473.

Escalera M. M. de la. 1923b: Sistema de las especies del género Asida de la Península Ibérica. Subgén. Glabrasida Esc. 3.a nota. *Boletín de la Real Sociedad Española de Historia Natural* **23**: 37-42.

Escalera M. M. de la. 1923c: Sistema de las especies del género Asida de la Península Ibérica. Subgén. Glabrasida Esc. 4.a nota. *Boletín de la Real Sociedad Española de Historia Natural* **23**: 79-87.

Escalera M. M. de la. 1923d: Tenebriónidos nuevos de Marruecos y Rio de Oro. *Boletín de la Real Sociedad Española de Historia Natural* **23**: 128-132.

Escalera M. M. de la. 1923e: Una specie nueva de meloido de Marueccos. *Boletín de la Real Sociedad Española de Historia Natural* **23**: 299-300.

Escalera M. M. de la. 1925a: Varios tenebriónidos y un curculiónido nuevos de Marruecos. *Boletín de la Real Sociedad Española de Historia Natural* **24** [1924]: 324-337.

Escalera M. M. de la 1925b: Una especie nueva de Glabrasida Esc. (Col. Tenebrionidae) de España. *Boletín de la Real Sociedad Española de Historia Natural* **25**: 153-154.

Escalera M. M. de la. 1925c: Especies nuevas de coleópteros del Mediano Atlas marroquí. *Boletín de la Real Sociedad Española de Historia Natural* **25**: 311-316.

Escalera M. M. de la. 1925d: Especies de Pachychila y otros tenebriónidos nuevos de Marruecos. *Boletín de la Real Sociedad Española de Historia Natural* **25**: 372-385.

Escalera M. M. de la. 1925e: Especies nuevas y variedades de Asida de Baleares (Col. Tenebr.). *Boletín de la Real Sociedad Española de Historia Natural* **25**: 404-410.

Escalera M. M. de la. 1925f: Tenebriónidos y meloidos del Mediano Atlas. *Boletín de la Real Sociedad Española de Historia Natural* **25**: 498-505.

Escalera M. M. de la 1926: Una especie más de Heliotaurus de Marruecos (Col. Cistelidae). *Boletín de la Real Sociedad Española de Historia Natural* **26**: 184-185.

Escalera M. M. de la. 1927a: Las Machasida Esc. (Col. Tenebrionidae) de Marruecos. *Bulletin de la Société des Sciences Naturelles du Maroc* **7**: 135-149.

Escalera M. M. de la 1927b: Una nueva especie española de Crypticus Latr. *Eos, Revista Española de Entomología* **3**: 501-504.

Escalera M. M. de la. 1929: Las Glabrasida fortitertricosttae de Marruecos (Col. Tenebr.). *Memorias de la Real Sociedad Española de Historia Natural* **15**: 105-108.

Escalera M. M. de la 1933: Una especie y tres variedades nuevas de Heliotaurus de protórax rojo de Marruecos (Col. Allecul.). *Boletín de la Real Sociedad Española de Historia Natural* **33**: 129-131.

Escalera M. M. de la. 1934a: Especies nuevas de Crypticus Latr. de Sierras de Segura (Col. Tenebr.). *Boletín de la Real Sociedad Española de Historia Natural* **34**: 211-212.

Escalera M. M. de la. 1934b: Especies nuevas de Akis de Marruecos y del Sahara occidental (Col. Ten.). *Eos, Revista Española de Entomología* **9**: 301-309, pl. XVII.

Escalera M. M. de la. 1935a: Especies de Pachychila Esch. y otros tenebriónidos de Ifni y del Sáhara español. *Boletín de la Sociedad Española de Historia Natural* **35**: 295-309.

Escalera M. M. de la. 1935b: Especies de Pachychila Esch. y otros tenebriónidos nuevos de Marruecos. *Boletín de la Sociedad Española de Historia Natural* **35**: 372-380.

Escalera M. M. de la. 1940: Especies de Sepidium F. de Ifni (Col. Tenebrionidae). *Eos, Revista Española de Entomología* **13** [1937]: 5-11, 1 pl.

Escalera M. M. de la. 1943a: Especies nuevas de Asida (Col. Tenebrionidae) del Reino de Valencia y zonas limitrofes existentes en las colecciones de D. Emilio Moroder y Dr. Baguena en 1936. *Boletín de la Real Sociedad Española de Historia Natural* **43**: 199-207.

Escalera M. M. de la. 1943b: Datos sobre coléopteros de Ifni del Sahara español. 1. Una nueva especie de Eurycaulus Fairm. del litoral entre C. Juby y C. Bojador. *Eos, Revista Española de Entomología* **19**: 215-221.

Escalera M. M. de la. 1944: De la disparidad especifica de Dendarus zariquieyi Esp. (Col. Tenebrionidae) de sus similares ibéricios y africanos. *Eos, Revista Española de Entomología* **20**: 83-92.

Escherich K. 1889a: Beschreibung einer neuen deutschen Meloë-Art und mehrerer Varietäten. *Wiener Entomologische Zeitung* **8**: 105-106.

Escherich K. L. 1889b: Meloë Reitteri, eine neue russische Meloë-Art. *Wiener Entomologische Zeitung* **8**: 112.

Escherich K. L. 1890a: Meloiden-Studien. *Wiener Entomologische Zeitung* **9**: 23-24.

Escherich K. L. 1890b: Revision der behaarten Meloë-Arten der Alten Welt. *Wiener Entomologische Zeitung* **9**: 87-96.

Escherich K. L. 1890c: Bemerkungen über Meloë algirus nov. spec. aus Algier. *Deutsche Entomologische Zeitschrift* **1890**: 283.

Escherich K. L. 1891a: Cochliophorus nov. gen. Meloidarum. *Deutsche Entomologische Zeitschrift* **1891**: 16.

Escherich K. L. 1891b: Meloiden-Studien. II. Theil. *Wiener Entomologische Zeitung* **10**: 53-55.

Escherich K. L. 1892: Die palaärktischen Vertreter der Coleopteren-Gattung Zonitis Fab. Versuch einer Monographie. I. *Deutsche Entomologische Zeitschrift* **1891**: 225-250, 1 pl.

Escherich K. L. 1895: Beiträge zur Naturgeschichte der Meloidengattung Lytta Fab. *Verhandlungen der Kaiserlich-Königlichen Zoologisch-Botanischen Gesellschaft in Wien* **44** [1894]: 251-298, 4 pls.

Escherich K. L. 1896a: Meloiden-Studien. IV. Theil. *Wiener Entomologische Zeitung* **15**: 27-30.

Escherich K. L. 1896b: Zur Kenntnis der Gattung Caloenas Reitt. *Wiener Entomologische Zeitung* **15**: 121-123.

Escherich K. L. 1896c: Revision der Meloiden-Gattung Lydus Latr. *Deutsche Entomologische Zeitschrift* **1896**: 193-235.

Escherich K. L. 1896d: Beitrag zur Fauna der tunisischen Insel Djerba. *Verhandlungen der Kaiserlich-Königlichen Zoologisch-Botanischen Gesellschaft in Wien* **46**: 268-279.

Escherich K. L. 1897a: Zoologische Ergebnisse einer von Dr. K. Escherich und L. Kathariner nach Central-Kleinasien unternommen Reise. IV. Theil. Coleopteren. Under the title of the paper the following text is read: „Unter Mitwirkung von E. Brenske, L. Ganglbauer, Dr. L. von Heyden, E. Wasmann und J. Weise". *Entomologische Zeitung* (Stettin) **58**: 3-69.

Escherich K. L. 1897b: Revision der palaearktischen Zonitiden, einer Unterfamilie der Meloiden. *Verhandlungen des Naturforschenden Vereines in Brünn* **35** [1896]: 96-133.

Escherich K. L. 1899: Zur Kenntniss der Coleopteren-Gattung Zonabris Harold. *Wiener Entomologische Zeitung* **18**: 84-92, 97-109.

Escherich K. L. 1904: Neue paläarktische Meloiden aus der F. Hauser'schen Sammlung. *Münchener Entomologische Zeitschrift* **2**: 30-36.

Eschscholtz J. F. 1818: Decades tres Eleutheratorum novorum. *Mémoires de l'Académie Impériale des Sciences de St. Peterburg* **6**: 451-484.

Eschscholtz J. F. 1829: *Zoologischer Atlas, enthaltend Abbildungen und Beschreibungen neuer Tierarten, während des Flottscapitains v. Kotzebue zweiter Reise um die Welt, auf der Russisch-Kaiserlichen Kriegsschlupp Predpriaetië in den Jahren 1823-1826 beobachtet. Drittes Heft*. Berlin: G. Reimer, 18 pp., pls xii-xv.

Eschscholtz J. F. 1831: *Zoologischer Atlas, enthaltend Abbildungen und Beschreibungen neuer Tierarten, während des Flottscapitains V. Kotzebue zweiter Reise um die Welt, auf der Russisch-Kaiserlichen Kriegsschlupp Predpriaetië in den Jahren 1823-1826 beobachtet. Viertes Heft*. Berlin: G. Reimer, 19 pp., pls xvi-xx.

Español F. 1937: Assaig monogràfic del subgènere Dendarus s.str. (Col. Tenebrionidae). *Treballs del Museu de Ciències Naturals de Barcelona* (*Sèrie Entomolõgica* 12) **11**: 1-89, 13 pls.

Español F. 1940: Contribución al conocimiento de los coleópteros de Ibiza y Formentera. *Las Ciencias* **8**: 9.

Español F. 1942: Los representantes catalanes de la familia "Rhipiphoridae" (Col.). *Anales de la Escuela de Peritos Agrícolas y Superior de Agricultura y de los Servicios Técnicos de Agricultura* **2**: 335-346.

Español F. 1943a: Datos para el conocimento de los Tenebrionidos del Mediterráneo occidental. *Graellsia* **1**: 23-28.

Español F. 1943b: Contribución al conocimiento de los coleópteros de Ibiza y Formentera y un nuevo Heliophilus ibérico (Col. Tenebrionidae). *Anales de la Asociación Española para el Progreso de las Ciencias* **8**: 93-108.

Español F. 1943c: Misión cientifica E. Morales Agacino, Ch. Rungs y B. Zolotarevsky a Ifni y Sáhara Español. Tenebrionidae (Col.). 1.a parte. *Eos, Revista Española de Entomologia* **19**: 119-148.

Español F. 1944: Nuevos datos para el conocimiento de los tenebriónidos (Col.) del Sáhara Español. *Eos, Revista Española de Entomología* **20**: 7-30.

Español F. 1945a: Nuevos comentarios sistematicos sobre la subfamilia Opatrinae Reitt. con la descripción de un nuevo representante del Sáhara Español (Col. Tenebrionidae). *Eos, Revista Española de Entomologia* **20** [1944]: 213-232, pls XII-XVII.

Español F. 1945b: Revisión de los Phylan ibéricos (Col. Tenebrionidae). *Eos, Revista Española de Entomologia* **21**: 297-357, pl. xiii.

Español F. 1946: Tenebriónidos (Col.) nuevos o interesantes recogidos por D. Eugenio Mordes y D. Joaquín Mateu en el Sáhara Español. *Eos, Revista Española de Entomologia* **22**: 107-122, pls. VIII-X.

Español F. 1947a: Revisión del género Micrositus (Col. Tenebrionidae). *Trabajos del Museo de Ciencias Naturales de Barcelona* (*Nueva Seria Zoológica*) **1**: 5-60.

Español F. 1947b: Algunos Stenosinae y Opatrinae recogidos durante una breve campaña entomologica por la provincia de Castellon (Col. Tenebrionidae). *Eos, Revista Española de Entomología* **23**: 185-200.

Español F. 1948a: Notas sombre coleópteros africanos. *Eos, Revista Española de Entomología* **24**: 311-318.

Español F. 1948b: Contribución al estudio de los Crypticus paléarticos: El subgénero Lamprocrypticus. *Eos, Revista Española de Entomología* **24**: 475-491.

Español F. 1949: Contribución al estudio de los Crypticus del grupo del pruinosus: el subgénero Pseudoseriscus Españ. (Col. Tenebrionidae). *Eos, Revista Española de Entomología* **25**: 199-239.

Español F. 1950a: Los Crypticus s.str. de la Península Ibérica y del norte de Africa (Col. Tenebrionidae). *Eos, Revista Española de Entomología* **26**: 115-156.

Español F. 1950b: Contribución al estudio de los Crypticus paléarticos (Col. Tenebrionidae). Pp. 124-129. In: *Transactions of the Eighth International Congress of Entomology, Stockholm 1948*. Stockholm: Ales R. Elfström, 1030 pp.

Español F. 1951a: Tenebriónidos de Las Pitiusas (Baleares occidentales). *Eos, Revista Española de Entomología* **27**: 4-41.

Español F. 1951b: Un nuevo Arthrodeis de la region del Draa: A. (Apentanes) bolivari n.sp. *Eos, Revista Española de Entomología* (Tomo extraordinario) [1950]: 363-366.

Español F. 1952a: Mas Tenebriónidos del Sahara Español (Col.). *Eos, Revista Española de Entomología* **28**: 71-76.

Español F. 1952b: Note sur les Crypticus appartenant au groupe de C. viaticus Fairm. (Col. Tenebrionidae). *Transactions of the IXth International Congress of Entomology, Amsterdam* **1**: 117-120.

Español F. 1953a: Helopinae de la zona mediterránea de Marruecos (Col. Tenebrionidae). *Eos, Revista Española de Entomología* **29**: 71-83.

Español F. 1953b: Los Colpotus del Mediterráneo Occidental (Col. Tenebrionidae). *Eos, Revista Española de Entomología* **29**: 123-142.

Español F. 1953c: Misiones saharianas de l'Institut scientifique Chérifien (1950-1951) Col. Tenebrionidae. *Bulletin de la Société des Sciences Naturelles du Maroc* **31** [1951]: 287-312.

Español F. 1954a: Los tenebriónidos (Col.) de Baleares. *Trabajos del Museo de Ciencias Naturales de Barcelona* (*Nueva Seria Zoológia*) **1**: 3-96.

Español F. 1954b: Los Crypticus de Canarias (Col. Tenebrionidae). *Eos, Revista Española de Entomología* **30**: 99-117.

Español F. 1954c: Coleópteros de Sierra Nevada "Tenebrionidae". *Archivos des Instituto de Aclimatación* **2**: 111-122, 1 pl.

Español F. 1954d: Datos para el conocimiento de las tenebrionidos del mediterranea occidental. *Eos, Revista Española de Entomología* **30**: 161-162.

Español F. 1954e: Los Opatrum andaluces del subgénero Colpophorus (Col. Tenebrionidae). *Eos, Revista Española de Entomología* **30**: 315-330.

Español F. 1955a: Los Crypticini paléarticos (Col. Tenebrionidae). *Eos, Revista Española de Entomología* **31**: 7-38.

Español F. 1955b: Nuevos datos sobre los Tenebrionidos de las Pitiusas (Baleares occidentales). *Boletin de la Sociedad de Historia Natural de Baleares* **1**: 25-29.

Español F. 1955c: Soble algunos tenebriónidos gallegos recognidos por el profesor H. Franz. *Publicationes del Instituto de Biología Aplicada* **19**: 95-103.

Español F. 1955d: Colydiidos endogeos del Levante español. *Publicaciones del Instituto de Biología Aplicada* **21**: 101-106.

Español F. 1956a: Los Probaticus de España (Col. Tenebrionidae). *Eos, Revista Española de Entomología* **32**: 83-123.

Español F. 1956b: Nuevos Opatrini de Marruecos y del Sahara atlántico (Col. Tenebrionidae). *Publicationes de la Facultad de Ciencias de Barcelona, Universidad de Barcelona* **1956**: 277-286.

Español F. 1957a: Los Hegeter de las Canarias Orientales (Col. Tenebrionidae). *Trabajos del Museo Ciencas Naturales de Barcelona* (*Nueva Seria Zoológica*) **2**: 3-16.

Español F. 1957b: Contribución al estudio de los Stenohelops Ibericos (Col. Tenebrionidae). *Eos, Revista Española de Entomología* **33**: 19-38.

Español F. 1957c: Un nuevo Helioctamenus de Marruecos. *Publicaciones del Instituto de Biología Aplicada* 25: 147-150.

Español F. 1957d: Contributión al conocimiento de los Tentyriini de las Canarias orientales: sobre el pretendido gen. Pseudotalpophila Reitt. (Col. Tenebrionidae). *Eos, Revista Española de Entomología* **33**: 157-176.

Español F. 1958a: Contribución al conocimiento de los artrópodos y moluscos terrestres de las Islas Columbretes. *Miscelánea Zoológica* **1**: 3-37.

Español F. 1958b: Avance al estudio de los Stenosis ibéricos (Col. Tenebrionidae). *Eos, Revista Española de Entomología* **34**: 33-54.

Español F. 1958c: Datos para el conocimiento de los tenebriónidos de Mediterráneo occidental (Coleoptera). *Graellsia* **96**: 107-116.

Español F. 1959: Opatrinae del Rif, Marruecos (Col. Tenebrionidae). *Eos, Revista Española de Entomología* **35**: 243-255.

Español F. 1960a: Datos para el conocimiento de los tenebriónidos de Mediterráneo occidental. *Miscellánea Zoológica* **1** (3): 109-118, 1 pl.

Español F. 1960b: Un nuevo tipo de tenebriónido sabulicola de las costas del Perú. *Publicationes del Instituto de Biología Aplicada* **31**: 113-117.

Español F. 1961a: Revisión de los Dendarus s.str. (Col. Tenebrionidae). *Eos, Revista Española de Entomología* **37**: 41-70.

Español F. 1961b: El género Tarphius, nuevo par la fauna española (Col. Colydiidae). *Graellsia* **19**: 81-85.

Español F. 1961c: Sobre alcunas Solskyia del Paquistán (Co. Tenebrionidae). *Miscellánea Zoológica* **1** (4): 123-131.

Español F. 1961d: Los Cylindronotini de la Península Ibérica (Col. Tenebrionidae). *Eos, Revista Española de Entomología* **37**: 135-160.

Español F. 1961e: Los Nesotes de España (Col. Tenebrionidae). *Eos, Revista Española de Entomología* **37**: 289-308.

Español F. 1961f: Los Blaps de la Península Ibérica (Col. Tenebrionidae). *Eos, Revista Española de Entomología* **37**: 399-414.

Español F. 1962a: Los Opatrinae de las Islas Canarias (Col. Tenebrionidae). *Eos, Revista Española de Entomología* **38**: 203-221.

Español F. 1962b: Los Nesotes de Gran Canaria (Col. Tenebrionidae). *Eos, Revista Española de Entomología* **38**: 339-346.

Español F. 1963a: El género "Coelometopus" Sol. (Col. "Tenebrionidae"). *Archivos del Instituto de Aclimatación* **12**: 49-53.

Español F. 1963b: Nuevos tenebriónidos africanos. *Miscellánea Zoológica* **1** (5): 73-79.

Español F. 1963c: Los Opatrum ibéricos (Col. Tenebrionidae). *Eos, Revista Española de Entomología* **38**: 471-483.

Español F. 1963d: Datos para el conocimiento de los tenebriónidos del Mediterráneo occidental (Coleoptera). *Eos, Revista Española de Entomología* **39**: 185-209.

Español F. 1963e: Datos para el conocimiento de los tenebriónidos del Mediterráneo occidental (Coleoptera). *Eos, Revista Española de Entomología* **39**: 379-388.

Español F. 1963f: Más datos sobre los Helopinae de Marruecos (Col. Tenebrionidae). *Eos, Revista Española de Entomología* **39**: 389-396.

Español F. 1966: Datos para el conocimiento de los tenebriónidos del Mediterráneo occidental (Coleoptera). *Graellsia* **22**: 167-175.

Español F. 1967a: Misión entomológica Hakan Lindberg y M. Meinander a Marruecos. Col. Tenebrionidos. *Publicaciones del Instituto de Biología Aplicata* **42**: 17-47.

Español F. 1967b: Blaps propheta fiorii subspecie nueva de la Tripolitana. *Studi Sassar* (*Sez. 3*) **14** [1966]: 197-198.

Español F. 1969: Tenebrionidos de la región de Sax-Salinas, provincia de Alicante (Coleoptera). *Graellsia* **24**: [1968]: 79-86.

Español F. 1974: Nuevas localizaciones de Nalassus ibéricos (Coleoptera, Tenebrionidae). *Boletín de la Real Sociedad Española de de Historia Natural* (*Biología*) **72**: 213-217.

Español F. 1975a: Sobre un nuevo representante europeo del género Osphyoplesius Winkler (Col. Boridae). *Eos, Revista Española de Entomología* **49** [1973]: 105-111.

Español F. 1975b: Un nuevo Litoborini de las Islas Canarias (Col. Tenebrionidae, Opatrinae). *Vieraea* **4** [1974]: 237-244.

Español F. 1980: Sobre la presencia del género Probaticus Seidl., 1896, en las Islas Baleares (Col. Tenebrionidae, Helopinae). *Eos, Revista Española de Entomología* **56**: 107-110.

Español F. 1981: Sobre un nuevo Stenosis ibérico (Col. Tenebrionidae). *Publicaciones del Departamento de Zoología, Barcelona* **7**: 57-60.

Español F. & Viñolas A. 1981a: Eine neue Eurycaulus-Art aus der algerischen Sahara (Coleoptera, Tenebrionidae, Opatrinae). *Stuttgarter Beiträge zur Naturkunde Serie A* (*Biologie*) **352**: 1-7.

Español F. & Viñolas A. 1981b: Una puesta al dia de los Phylan Steph. ibéricos (Col. Tenebrionidae). *Eos, Revista Española de Entomología* **57**: 97-112.

Español F. & Viñolas A. 1983a: Revisión de los Gonocephalum del grupo prolixum (Col., Opatrinae). *Eos, Revista Española de Entomología* **59**: 31-39.

Español F. & Viñolas A. 1983b: Revisión del género Heliopathes Muls. 1854 (Col. Tenebrionidae). 1: Grupo del strigicollis. *Nouvelle Revue d'Entomologie* **13**: 213-218.

Español F. & Viñolas A. 1987: Los Erodius ibéricos (Col. Tenebrionidae). *Eos, Revista Española de Entomología* **63**: 21-29.

Eversmann E. F. von. 1836: [new taxon]. In: Faldermann F. 1836: Bereicherung zur Käfer-Kunde des russischen Reiches. *Bulletin de la Société des Naturalistes de Moscou* **9**: 351-398, 3 pls.

Fabre J. H. 1886: *Souvenirs entomologiques* (*troisième série*). *Études sur l'instinct et les moeurs des insectes.* (*troisième édition*). Paris. Librairie Ch. Delagrave, 433 pp.

Fabricius J. C. 1775: *Systema entomologicae, systens insectorum classes, ordines, genera, species, adiectis synonymis, locis, descriptionibus, observationibus.* Flensburgi et Lipsiae: Libraria Kortii, [32] + 832 pp.

Fabricius J. C. 1777: *Genera insectorum eorumque characteres naturales secundum numerum, figuram, situm et proportionem omnium partium oris adiecta mantissa specierum nuper detectarum.* Chilonii: Mich. Friedr. Bartschii: 14 + 310 pp.

Fabricius J. C. 1781: *Species insectorum, exhibens eorum differentias specificas, synonyma auctorum, loca natalia, metamorphosis, adiecitis observationibus, descriptionibus. Tom I.* Hamburgi et Kilonii: Carol Ernest Bohnii, viii +552 pp.

Fabricius J. C. 1787a: *Mantissa insectorum sistens eorum species nuper detectas adiectis characteribus generis, differentiis specificis, emendationibus observationibus. Tom I.* Hafniae: Christ. Gottl. Proft, xx + 348 pp.

Fabricius J. C. 1787b: *Mantissa insectorum sistens eorum species nuper detectas adiectis characteribus generis, differentiis specificis, emendationibus observationibus. Tom II.* Hafniae: Christ. Gottl. Proft, 382 pp.

Fabricius J. C. 1790: Nova insectorum genera. *Skrivter af Naturhistorie Selskabet* **1**: 213-228.

Fabricius J. C. 1792a: *Entomologica systematica emendata et aucta. Secundum classes, ordines, genera, species adjectis synonimis, locis, observationibus, descriptionibus. Tom I. Pars 1.* Hafniae: Christ. Gottl. Proft., xx + 330 pp.

Fabricius J. C. 1792b: *Entomologica systematica emendata et aucta. Secundum classes, ordines, genera, species adjectis synonimis, locis, observationibus, descriptionibus. Tom I. Pars II.* Hafniae: Christ. Gottl. Proft., 538 pp.

Fabricius J. C. 1794: *Entomologia systematica emendata et aucta. Sécundum classes, ordines, genera, species adjectis synonimis, locis, observationibus, descriptionibus. Tom. IV.* [Appendix specierum nuper detectarum: pp. 435-462]. Hafniae: C. G. Proft, Fil. et Soc., [6] + 472 + [5] pp.

Fabricius J. C. 1798: *Supplementum entomologiae systematicae.* Hafniae: Proft et Storch, [2] + 572 pp.

Fabricius J. C. 1801a: *Systema Eleutheratorum secundum ordines, genera, species adiectis synonymis, locis, observationibus, descriptionibus. Tomus I.* Kiliae: Bibliopolii Academici Novi, xxiv + 506 pp.

Fabricius J. C. 1801b: *Systema Eleutheratorum secundum ordines, genera, species adiectis synonymis, locis, observationibus, descriptionibus. Tomus II.* Kiliae: Bibliopolii Academici Novi, 687 pp.

Fåhraeus O. I. von. 1870: Coleoptera Caffrariae, annis 1838-1854 a J. A. Wahlberg collecta. Heteromera. *Öfversigt af Kongliga Svenska Vetenskaps-Akademiens Förhandlingar* **27**: 234-358.

Failla-Tebaldi L. 1887: Excursione entomologica all'Isola di Lampedusa. *Il Naturalista Siciliana* **6** [1886-1887]: 102-104.

Fairmaire L. 1848: Description de quelques coléoptères nouveaux. *Annales de la Société Entomologique de France* (2) **6**: 167-176.

Fairmaire L. 1849: Description de quelques coléoptères nouveaux d'Europe et de France. *Annales de la Société Entomologigue de France* (2) **7**: 419-427.

Fairmaire L. 1849: Essai sur les coléoptères de la Polynésie (suite). *Revue et Magazin de Zoologie Pure et Appliquée* (2) **1**: 445-460.

Fairmaire L. 1850a: Essai sur les coléoptères de la Polynésie. *Revue et Magasin de Zoologie* (2) **2**: 50-64, 115-122, 181-185.

Fairmaire L. 1850b: Description de quelque coléoptères nouveaux d'Europe et de France. *Annales de la Société Entomologique de France* (2) **7** (1849): 419-427.

Fairmaire L. 1851: Description de sept coléoptères nouveaux provenant du Midi de France et du Maroc. *Revue et Magazin de Zoologie Pure et Appliquée* (2) **3**: 527-528.

Fairmaire L. 1852: Quelques coléoptères nouveaux du midi de l'Europe et du nord de l'Afrique. *Annales de la Société Entomologique de France* (2) **10**: 69-93.

Fairmaire L. 1856: Miscellanea entomologica. *Annales de la Société Entomologique de France* (3) **4**: 517-548, pl. 16.

Fairmaire L. 1860: Diagnoses de nouvelles espèces de coléoptères. *Annales de la Société Entomologique de France* (3) **8**: 338, 629-632.

Fairmaire L. 1861: [new taxon]. In: Jacquelin du Val C.: *Manuel Entomologique. Genera des coléoptères d'Europe comprenant leur classification en familles naturelles, la description de tous les genres, des tableaux synoptiques destinés à faciliter l'étude, le Catalogue de toutes les espèces de nombreux dessins au trait de charactères et plus de treize cents types représentant un ou plusieurs insectes de chaque genre dessinés et peints d'apres nature avec le plus grand soin par M. Jules Migneaux. Tome troisième.* Paris: A. Deyrolle Deyrolle [1859-1863, 464 + 200 pp., 100 pls].

Fairmaire L. 1862: Miscellanea entomologica. Quatrième partie. *Annales de la Société Entomologique de France* (4) **1** [1861]: 577-596.

Fairmaire L. 1863: Coléoptères d'Algérie rapportés par M. Ogier de Baulny. *Annales de la Société Entomologique de France* (4) **3**: 637-648.

Fairmaire L. 1864: Espèces nouvelles de coléoptères d'Algérie. *Annales de la Société Entomologique de France* (4) **4**: 233-246.

Fairmaire L. 1866a: [new taxa]. In: Fairmaire L. & Coquerel C.: Essai sur les coléoptères de Barbarie. Quatrième partie. *Annales de la Société Entomologique de France* (4) **6**: 17-74.

Fairmaire L. 1866b: Notice sur les coléoptères récoltés par M. J. Lédérer sur le Bosz-Dagh (Asie Mineure). *Annales de la Société Entomologique de France* (4) **6**: 249-280.

Fairmaire L. 1867: Essai sur les coléoptères de Barbarie. Cinqième partie. *Annales de la Société Entomologique de France* (4) **7**: 387-416.

Fairmaire L. 1868: Essai sur les coléoptères de Barbarie. Sixième partie. *Annales de la Société Entomologique de France* (4) **8**: 471-502.

Fairmaire L. 1869a: Notes sur les coléoptères recueillis par Charles Coquerel à Madagascar et sur les côtes d'Afrique. *Annales de la Société Entomologique de France* (4) **8** [1868]: 753-820.

Fairmaire L. 1869b: Notes sur les coléoptères recueillis par Charles Coquerel à Madagascar et sur les cotes d'afrique. *Annales de la Société Entomologique de France* (4) **9**: 179-260.

Fairmaire L. 1871a: Essai sur les coléoptères de Barbarie. Septième partie. *Annales de la Société Entomologique de France* (4) **10** [1870]: 369-404.

Fairmaire L. 1871b: Notes sur les coléoptères recueillis par Charles Coquerel à Madagascar et sur les côtes d'Afrique, 3e partie. *Annales de la Société Entomologique de France* (5) **1**: 29-78.

Fairmaire L. 1873a: [new taxa]. In: Fairmaire L. & Raffray A.: Coléoptères du nord de l'Afrique. *Revue et Magasin de Zoologie Pure et Appliquée* (3) **1**: 331-385.

Fairmaire L. 1873b: Diagnoses de divers coléoptères nouveaux découverts en Algérie. *Annales de la Société Entomologique de France* (5) **3**: 391-392.

Fairmaire L. 1874a: Diagnoses de coléoptères nouveaux du nord de l'Afrique. *Petites Nouvelles Entomologiques* **1** [1869-1875]: 388-389.

Fairmaire L. 1874b: Note sur quelques coléoptères des environs de Tuggurth. *Petites Nouvelles Entomologiques* **1** [1869-1875]: 407-408.

Fairmaire L. 1875a: Coléoptères de la Tunisie récoltés par Mr Abdul Kerim. *Annali del Museo Civico di Storia Naturale di Genova* **7**: 475-540.

Fairmaire L. 1875b: Diagnose de quelques coléoptères nouveaux d'Algérie. *Petites Nouvelles Entomologiques* **1** [1869-1875]: 495-496.

Fairmaire L. 1875c: Descriptions de quelques coléoptères du Maroc (1). *Petites Nouvelles Entomologiques* **1** [1869-1875]: 543-544.

Fairmaire L. 1875d: Diagnoses d'hétéromères de Madagascar. *Bulletin de la Société Entomologique de France* **1875**: xxxiii-xxxiv.

Fairmaire L. 1876: Diagnoses de coléoptères du nord de l'Afrique. *Petites Nouvelles Entomologiques* **2**: 37-38, 49-50, 93-94.

Fairmaire L. 1877: Coléoptères nouveaux du nord de l'Afrique. *Petites Nouvelles Entomologiques* **3**: 141.

Fairmaire L. 1878a: [new taxa]. In: Deyrolle H. & Fairmaire L.: Descriptons de coléoptères recueillis par M. l'abbé David dans la Chine centrale. *Annales de la Société Entomologique de France* (5) **8**: 87-140.

Fairmaire L. 1878b: [new taxa]. *Bulletin de la Société Entomologique de France* **1878**: lv-lvi.

Fairmaire L. 1879a: Diagnoses de coléoptères du nord de l'Afrique. *Le Naturaliste* **1**: 3-4.

Fairmaire L. 1879b: Descriptions de coléoptères nouveaux du nord de l'Afrique. *Annales de la Société Entomologique de France* (5) **9**: 155-172.

Fairmaire L. 1879c: Coléoptères du nord de l'Afrique. *Revue et Magasin de Zoologie Pure et Appliquée* (3) **7**: 178-218.

Fairmaire L. 1879d: Description de quelques coléoptères nouveaux recueillis en Espagne par D. Manuel Martorell y Peña. *Annales de la Société Entomologique de France* (5) **9**: 241-244.

Fairmaire L. 1880a: Descriptions des coléoptères nouveaux du nord de l'Afrique. 3e partie. *Annales de la Société Entomologique de France* (5) **10**: 5-32.

Fairmaire L. 1880b: Diagnoses de coléoptères du Maroc. *Le Naturaliste* **2**: 190.

Fairmaire L. 1880c: Descriptions de coléoptères d'Espagne et de Turquie. *Annales de la Société Entomologique de France* (5) **10**: 237-244.

Fairmaire L. 1880d: Descriptions des coléoptères nouveaux du nord de l'Afrique. 4e partie. *Annales de la Société Entomologique de France* (5) **10**: 245-254.

Fairmaire L. 1881a: Essai sur les coléoptères des îles Viti (Fidgi). *Annales de la Société Entomologique de France* (6) **1**: 243-318.

Fairmaire L. 1881b: [Crypticus pubens]. *Bulletin de la Société Entomologique de France* **1880**: cxxxii.

Fairmaire L. 1881c: [new taxa]. *Bulletin de la Société Entomologique de France* **1881**: i-xii.

Fairmaire L. 1881d: [Descriptions de deux espèces nouvelles de coléoptères trouvées en Catalogne: Abax, Sitaris]. *Bulletin de la Société Entomologique de France* **1881**: xliii-xliv.

Fairmaire L. 1882a: Coléoptères recueillis par M. G. Révoil chez les çomalis. In: Révoil G.: *Faune et flore des pays çomalis*. Paris: Challames Ainé, iv + 104 pp., 1 pl.

Fairmaire L. 1882b: Notes sur quelques coléoptères du Soudan et de l'Inde boréale. *Annales de la Société Entomologique de France* (6) **2**: 65-68.

Fairmaire L. 1882c: Coléoptères hétéromères de Sumatra. *Notes from the Leyden Museum* **4**: 219-265.

Fairmaire L. 1883a: Essai sur les coléoptères de l'archipel de la Nouvelle Bretagne. *Annales de la Société Entomologique du Belgique* **27**: 1-58.

Fairmaire L. 1883b: Description de coléoptères hétéromères de l'île de Saleyer. *Notes from the Leyden Museum* **5**: 31-40.

Fairmaire L. 1883c: Coléoptères d'Abyssinie. *Annales de la Société Entomologique de France* (6) **3**: 89-112.

Fairmaire L. 1883d: Coléoptères de Tripoli. *Entomologische Zeitung* (Stettin) **44**: 459-460.

Fairmaire L. 1883e: Description de coléoptères recueillis par M. le Baron Bonnaire en Algérie. *Comptes-Rendus des Séances de la Société Entomologique de Belgique* **1883**: xli-xlviii.

Fairmaire L. 1883f: [Descriptions suivantes de deux coléoptères de l'Hindoustan]. *Bulletin de la Société Entomologique de France* **1883**: lv-lvi.

Fairmaire L. 1883g: Description de quelques coléoptères marocains. *Comptes-Rendus des Séances de la Société Entomologique de Belgique* **1883**: cviii-cxi.

Fairmaire L. 1883h: Description de coléoptères recueillis par M. le Baron Bonnaire en Algérie. *Comptes-Rendus des Séances de la Société Entomologique de Belgique* **1883**: cxi-cxiv.

Fairmaire L. 1883i: [Communications]. *Bulletin de la Société Entomologique de France* **1883**: cxxiv-cxxvi.

Fairmaire L. 1883j: Descriptions de coléoptères recueillis par le Baron Bonnaire en Algérie. *Comptes-Rendus des Séances de la Société Entomologique de Belgique* **27**: clvi-clix.

Fairmaire L. 1883k: Descriptions de deux Helops nouveaux. *Bulletin de la Société Entomologique de France* **1882**: clxix-clxx.

Fairmaire L. 1883l: [new taxa]. *Bulletin de la Société Entomologique de France* **1882**: clxxvii-clxxviii.

Fairmaire L. 1884a: Liste des coléoptères reueillis par M. l'abbé David à Akbès (Asie-Mineure). *Annales de la Société Entomologique de France* (6) **4**: 165-180.

Fairmaire L. 1884b: Note sur les coléoptères recueillis par M. Ach. Raffray à Madagascar et descriptions des espèces nouvelles. 1. partie. *Annales de la Société Entomologique de France* (6) **4**: 225-242.

Fairmaire L. 1884c: Description de coléoptères marocains. *Le Naturaliste* **6**: 446-447.

Fairmaire L. 1884d: Descriptions de coléoptères recueillis par le Baron Bonnaire en Algérie. *Comptes-Rendus des Séances de la Société Entomologique de Belgique* **1884**: lix-lxx.

Fairmaire L. 1884e: Diagnoses de coléoptères del'Afrique orientale. *Comptes-Rendus des Séances de la Société Entomologique de Belgique* **1884**: lxx-lxxviii.

Fairmaire L. 1884f: Desciptions de coléoptères nouveaux du Maroc. *Comptes-Rendus des Séances de la Société Entomologique de Belgique* **1884**: cxvii-cxxi.

Fairmaire L. 1884g: [Description de trois coléoptères nouveaux, trouvés en Bulgarie par notre collègue M. Alléon]. *Bulletin de la Société Entomologique de France* **1883**: cxli-cxliii.

Fairmaire L. 1884h: Diagnoses de coléoptères del'Afrique orientale. *Comptes-Rendus des Séances de la Société Entomologique de Belgique* **1884**: cxlii-cxlix.

Fairmaire L. 1884i: Diagnoses de coléoptères del'Afrique orientale. *Comptes-Rendus des Séances de la Société Entomologique de Belgique* **1884**: cccxlvi-cccxlvii.

Fairmaire L. 1885a: [Descriptions de deux coléoptères algériens]. *Bulletin de la Société Entomologique de France* **1885**: xxxviii-xxxix.

Fairmaire L. 1885b: Malacodermes, lymexylonide et hétéromères nouveaux recueillis par M. Deby à Sumatra et à Bornéo. *Comptes-Rendus des Séances de la Société Entomologique de Belgique* **1885**: cv-cxii.

Fairmaire L. 1886a: Notes sur les coléoptères recueillis par M. Laligant à Obock. *Annales de la Société Entomologique de France* (6) **5** [1885]: 435-462.

Fairmaire L. 1886b: Notes sur les coléoptères recueillis par M. Raffray, à Madagascar et descriptions des espèces nouvelles. 2e partie. *Annales de la Société Entomologique de France* (6) **6**: 31-96, pl. 2.

Fairmaire L. 1886c: Descriptions de coléoptères de l'intérieure de la Chine. *Annales de la Société Entomologique de France* (6) **6**: 303-356.

Fairmaire L. 1887a: Coléoptères des voyages de M. G. Revoil chez les Somâlis et dans l'intérieur du Zanguebar. *Annales de la Société Entomologique de France* (6) **7**: 69-186.

Fairmaire L. 1887b: Coléoptères de l'intérieur de la Chine. *Annales de la Société Entomologique de Belgique* **31**: 87-136.

Fairmaire L. 1887c: Coléoptères nouveaux ou peu connus du Musée de Leyden. *Notes from the Leyden Museum* **9**: 145-162.

Fairmaire L. 1887d: Description de cinq espèces nouvelles de la famille des Cantharides. *Notes from the Leyden Museum* **9**: 193-196.

Fairmaire L. 1887e: Coléoptères des voyages de M. G. Révoil chez les Somâlis et dans l'intérieur du Zanguebar. Heteromera (2e Parie). *Annales de la Société Entomologique de France* (6) **7**: 277-368, pls 1-3.

Fairmaire L. 1887f: Notes sur les coléoptères des environs de Pékin (1e partie). *Revue d'Entomologie* **6**: 312-336.

Fairmaire L. 1887g: [new taxa]. *Bulletin de la Société Entomologique de France* **1886**: clxxviii-clxxix.

Fairmaire L. 1888a: Coléoptères de l'intérieur de la Chine. *Annales de la Société Entomologique de Belgique* **32**: 7-46.

Fairmaire L. 1888b: Notes sur les coléoptères des environs de Pékin (2e partie). *Revue d'Entomologie* **7**: 111-160.

Fairmaire L. 1888c: Énumération des coléoptères recueillis par M. le Dr Hans Schinz dans le sud de l'Afrique. *Annales de la Société Entomologique de France* (6) **8**: 173-202.

Fairmaire L. 1889a: Descriptions de coléoptères de l'Indo-Chine. *Annales de la Société Entomologique de France* (6) **8** [1888]: 333-378.

Fairmaire L. 1889b: Coléoptères de l'intérieur de la Chine 5e partie. *Annales de la Société Entomologique de France* (6) **9**: 5-81.

Fairmaire L. 1889c: [Diagnoses de quelques coléoptères recueillis à Pékin par M. Provost]. *Bulletin de la Société Entomologique de France* **1888**: cc-cci.

Fairmaire L. 1891a: Note supplémentaire sur les coléoptères d'Obock. *Annales de la Société Entomologique de France* (6) **10** [1890]: 547-554.

Fairmaire L. 1891b: Notes sur quelques coléoptères de l'Afrique intertropicale et descriptions d'espèces nouvelles. *Annales de la Société Entomologique de France* **60**: 231-274, pl. 5.

Fairmaire L. 1891c: Description de coléoptères de l'intérieur de la Chine (Suite, 6e partie). *Comptes-Rendus des Séances de la Société Entomologique de Belgique* **1891**: vi-xxiv.

Fairmaire L. 1891d: Descriptions de coléoptères des Montagnes de Kashmir. *Comptes-Rendus des Séances de la Société Entomologique de Belgique* **1891**: lxxxviii-ciii.

Fairmaire L. 1891e: Descriptions de coléoptères des Montagnes de Kashmir (Suite). *Comptes-Rendus des Séances de la Société Entomologique de Belgique* **1891**: cxxi-cxxxiv.

Fairmaire L. 1891f: Coléoptères de l'intérieur de la Chine (Suite, 7e partie). *Comptes-Rendus des Séances de la Société Entomologique de Belgique* **1891**: clxxxvii-ccxix.

Fairmaire L. 1891g: Coléoptères de l'Afrique autrale. *Comptes-Rendus des Séances de la Société Entomologique de Belgique* **1891**: cckxxix-cccvii.

Fairmaire L. 1892a: Coléoptères d'Obock. Troisième partie. *Revue d'Entomologie* **11**: 77-127.

Fairmaire L. 1892b: Descriptions de coléoptères des environs d'Akbès (Syrie). *Annales de la Société Entomologique du Belgique* **36**: 144-159.

Fairmaire L. 1892c: [new taxa]. *Bulletin de la Société Entomologique de France* **1892**: vii-viii.

Fairmaire L. 1892d: [new taxa]. *Bulletin de la Société Entomologique de France* **1892**: cix-cxi.

Fairmaire L. 1892e: Descriptions de quelques coléoptères nouveaux du Turkestan. *Bulletin de la Société Entomologique de France* 1892: clii-clv.

Fairmaire L. 1893a: Notes sur les coléoptères du Chua. *Annales de la Société Entomologique de Belgique* **37**: 9-50.

Fairmaire L. 1893b: Contributions à la faune Indo-Chinoise. 11e Mémoire (1). Coléoptères hétéromères. *Annales de la Société Entomologique de France* **62**: 19-38.

Fairmaire L. 1893c: Coléoptères nouveaux des Indes Orientales, de la famille des Scarabaeidae, Rhipidoceridae, Tenebrionidae et Oedemeridae. *Notes from the Leyden Museum* **15**: 17-64.

Fairmaire L. 1893d: Note sur quelques coléoptères des pays Somalis. *Annales de la Société Entomologique du Belgique* **37**: 144-156.

Fairmaire L. 1893e: Matériaux pour la faune coléoptérique du Sénégal. *Annales de la Société Entomologique de France* **62**: 147-158.

Fairmaire L. 1893f: Notes sur quelques coléoptères des environs de Lang-Song. *Annales de la Société Entomologique de Belgique* **37**: 287-302.

Fairmaire L. 1893g: Coléoptères du Haute Tonkin. *Annales de la Société Entomologique de Belgique* **37**: 303-325.

Fairmaire L. 1893h: Coléoptères des îles Comores. *Annales de la Société Entomologique de Belgique* **37**: 521-555.

Fairmaire L. 1893i: Descriptions de quelques coléoptères de l'Afrique septentrionale. *Bulletin de la Société Entomologique de France* **1893**: cxlvi-cxlix.

Fairmaire L. 1894a: Hétèromères du Bengale. *Annales de la Société Entomologique de Belgique* **38**: 16-43.

Fairmaire L. 1894b: Quelques coléoptères du Thibet. *Annales de la Société Entomologique de Belgique* **38**: 216-225.

Fairmaire L. 1894c: Descriptions de coléoptères d'Algérie. *Annales de la Société Entomologique de Belgique* **38**: 310-313.

Fairmaire L. 1894d: Coléoptères de l'Afrique intertropicale et australe. Deuxième note. *Annales de la Société Entomologique de Belgique* **38**: 314-335.

Fairmaire L. 1894e: Coléoptères du Kilimandjaro et des environs. *Annales de la Société Entomologique de Belgique* **38**: 386-395.

Fairmaire L. 1894f: Coléoptères de l'Afrique intertropicale et australe. Troisième note. *Annales de la Société Entomologique de Belgique* **38**: 651-679.

Fairmaire L. 1895a: Description de coléoptères de Madagascar et îles voisines. *Annales de la Société Entomologique de Belgique* **39**: 443-454.

Fairmaire L. 1895b: Coléoptères nouveaux d´Akbès, des dernières chasses de M. Charles Delagrange. *Bulletin de la Société Entomologique de France* **66**: cix-cxi.

Fairmaire L. 1896a: Hétéromères de l'Inde recueillis par M. Andrewes. *Annales de la Société Entomologique de Belgique* **40**: 6-62.

Fairmaire L. 1896b: Coléopterès de l'Inde boréale, Chine et Malaisie. *Notes from the Leyden Museum* **18**: 81-129.

Fairmaire L. 1896c: Description de coléoptères recueillis pa le Dr. Ch. Matin à Obock, Djibouti et Aden. *Bulletin de la Société Entomologique de France* **1896**: 223-225.

Fairmaire L. 1896d: Coléopterès de l'Inde et de Malaisie. *Notes from the Leyden Museum* **18**: 225-240.

Fairmaire L. 1896e: Description de quatre coléoptères nouveaux de la région méditerranéenne. *Bulletin de la Société Entomologique de France* **1896**: 344-345.

Fairmaire L. 1897a: Pseudadrus, nouveau genre d'opatrides. *Revue d'Entomologie* **16**: 39.

Fairmaire L. 1897b: Description de six coléoptères de l'île Ishigaki-Sima. *Bulletin de la Société Entomologique de France* **1897**: 68-72.

Fairmaire L. 1897c: Coléoptères de l'Inde boréale, Chine et Malaisie. *Notes from the Leyden Museum* **18**: 81-129.

Fairmaire L. 1897d: Matériaux pour la faune coléoptérique de la région malgache. *Annales de la Société Entomologique de Belgique* **41**: 92-119.

Fairmaire L. 1897e: Coléoptères nouveaux de l'Afrique intertropicale et australe. *Annales de la Société Entomologique de France* **66**: 109-155.

Fairmaire L. 1897f: Description de trois coléoptères hétéromères d'Algérie et d'Egypte. *Bulletin de la Société Entomologique de France* **1897**: 120-121.

Fairmaire L. 1897g: Descriptions de deux coléoptères nouveaux de Japon. *Bulletin de la Société Entomologique de France* **1897**: 166-167.

Fairmaire L. 1897h: Description d'une espèce nouvelle du genre Hionthis (Col.). *Bulletin de la Société Entomologique de France* **1897**: 180-181.

Fairmaire L. 1897i: Description de coléoptères nouveaux de la Malaisie, de l'Inde et de la Chine. *Notes from the Leyden Museum* **19**: 209-223.

Fairmaire L. 1897j: Coléoptères de l'Inde et de la Malaisie. *Notes from the Leyden Museum* **18**: 225-240.

Fairmaire L. 1897k: Coléoptères de Szé-Tschouen et de Koui-Tchéou (Chine). *Notes from the Leyden Museum* **19**: 241-255.

Fairmaire L. 1898a: Description de trois coléoptères des îles Ishigaki (Japon méridional). *Bulletin de la Société Entomologique de France* **1898**: 14-15.

Fairmaire L. 1898b: Descriptions de quatre coléoptères de Madagascar. *Bulletin de la Société Entomologique de France* **1898**: 37-39.

Fairmaire L. 1898c: Matériaux pour la faune coléoptérique de la région malgache. 5e note. *Annales de la Société Entomologique de Belgique* **42**: 222-260.

Fairmaire L. 1898d: Descriptions de coléoptères d'Asie et de Malaisie. *Annales de la Société Entomologique de France* **67**: 383-400.

Fairmaire L. 1899: Description de coléoptères nouveaux recueillis en Chine par M. de Latouche. *Annales de la Société Entomologique de France* **68**: 616-643.

Fairmaire L. 1901a: Matériaux pour la faune coléoptèrique de la région Malgache (11e Note). *Revue d'Entomologie* **20**: 101-248.

Fairmaire L. 1901b: Description de coléoptères des montagnes de Sikkim. *Bulletin de la Société Entomologique de France* **1901**: 255-268.

Fairmaire L. 1903a: Descriptions de quelques hétéromères recueillis par M. Fruhstorfer dans le Haut-Tonkin. *Annales de la Société Entomologique de Belgique* **47**: 13-20.

Fairmaire L. 1903b: Matériaux pour la faune coléoptérologique de la région malgache. 16e note. *Annales de la Société Entomologique de France* **72**: 181-259.

Fairmaire L. 1903c: Description de coléoptères recueillis en Chine par M. de Latouche. *Bulletin de la Société Entomologique de France* **1902**: 316-318.

Fairmaire L. & Brisout de Barneville C. N. F. 1859: [new taxon]. In: Fairmaire L.: Miscellanea entomologica. Troisième partie. *Annales de la Société Entomologique de France* (3) **7**: 21-64.

Fairmaire L. & Coquerel C. 1866: Essai sur les coléoptères de Babarie. Quatrième partie. *Annales de la Société Entomologique de France* (4) **6**: 17-74.

Fairmaire L. & Germain P. 1860: Coleoptera Chilensia. *Extrait des Annales de la Société Entomologique de*

France. Paris: F. Malteste et Cie, 8 pp.

Fairmaire L. & Raffray A. 1873: Coléoptères du nord de l'Afrique. *Revue et Magazin de Zoologie Pure et Appliquée* **3**: 331-385.

Faldermann F. 1832: [new taxa]. In: Ménétriés E.: *Catalogue raisonné des objets de zoologie recueillis dans un voyage au Caucase et jusqu'aux frontières actuelles de la Perse entrepris par l'ordre de S. M. l'Empereur*. Académie des Sciences St.-Pétersbourg, xxxiii + 272 + iv + [1].

Faldermann F. 1833: Species novae Coleopterorum Mongoliae et Sibiriae. *Bulletin de la Société des Naturalistes de Moscou* **6**: 46-72.

Faldermann F. 1835a: Lettre sur deux nouveaux melasomes. *Bulletin de la Société des Naturalistes de Moscou* **8**: 165-170.

Faldermann F. 1835b: Coleopterorum ab illustrissimo Bungio in China boreali, Mongolia, et Montibus Altaicis collectorum, nec non ab ill. Turczaninoffio et Stchukino e provincia Irkutsk missorum illustrationes. *Mémoires de l'Académie Impériale des Sciences de St. Pétersbourg. Sixième Série. Sciences Mathematiques, Physiques et Naturelles* **3** (1): 337-464, pls I-V.

Faldermann F. 1836: Bereicherung zur Käfer-Kunde des Russischen Reiches. *Bulletin de la Société des Naturalistes de Moscou* **9**: 351-398, 3 pls.

Faldermann F. 1837: Fauna entomologica trans-caucasica. Pars II. Coleoptera trans-caucasica. *Nouveaux Mémoires de la Société Impériale des Naturalistes de Moscou* **5**: 1-433, 15 pls.

Fan X. 1995a: A new species and a new record species of genus Mordellistena. *Entomotaxonomia* **17**: 23-25. (in Chinese with English summary).

Fan X. 1995b: Mordellidae. Pp.: 224-224. In: Wu H. (ed.): *Insects of Baishanzu Mountain, eastern China*. Beijing: China Forestry Publishing House, xiii, + 586 pp. (in Chinese with English summary).

Fan X. & Yang J. 1993: Revision of the genus Glipa of China. *Memoirs of Beijing Natural History Museum* **53**: 45-68. (in Chinese with English summary).

Fan X. & Yang J. 1995: Mordellidae. Pp. 95-101. In: Zhu T. (ed.): *Insects and macrofungi of Gutianshan, Zhejiang*. Hangzhou: Zhejiang Science and Technology Publishing House, 318 pp. (in Chinese with English summary).

Fan Z.-T. & Huang R.-X. 2005: [new taxon]. In: Wu W., Fan Z.-T. & Huang R.-X.: Three new species of Platyope (Coleoptera: Tenebrionidae) from Xinjiang, China. *Entomotaxonomia* **27**: 283-288. (in Chinese with English summary).

Fancello L. & Leo P. 1991: Lyreus septemstriatus, nuova specie di Colydiidae anoftalma dello Sardegna, appartenente ad un genere nuovo per la fauna italiana (Coleoptera). *Bollettino della Società Entomologica Italiana* **123**: 129-31.

Faust J. 1875: Beiträge zur Kenntnis der Käfer des Europäischen und Asiatischen Russlands mit Einschluss der Küsten des Kaspischen Meeres. *Horae Societatis Entomologicae Rossicae* **11**: 163-252.

Faust J. 1877: Aeltere und einige neue Käfer der russischen Fauna. *Bulletin de la Société Impériale des Naturalistes de Moscou* **52** (2): 34-45.

Faust J. 1877: Beiträge zur Kenntnis der Käfer des europäischen und asiatischen Russlands mit Einschluss der Küsten des Kaspischen Meeres. *Horae Societatis Entomologicae Rossicae* **12** [1876-1877]: 300-330.

Fauvel A. 1867: Catalogue des coléoptères de la Nouvelle Calédonie et dépendances, avec descriptions, notes et synonymies nouvelles. *Bulletin de la Société Linnéenne de Normandie* (2) **1** [1865-1866]: 172-215, pl. 1.

Fauvel A. 1899: Tableau des Phaleria françaises. *Revue d'Entomologie* **18**: 77-79.

Fauvel A. 1901: Phaleria nouvelle de France. *Revue d'Entomologie* **20**: 253-255.

Fauvel A. 1904: Faune analytique des coléoptères de la Nouvelle Calédonie. 2e Partie. *Revue d'Entomologie* **23**: 113-208.

Favre E. 1890: *Faune des coléoptères du Valais et des régions limitrophes*. Zürich: Zurcher & Furrer, xliv + 448 pp.

Ferrari J. A. 1869: Drei neue westasiatische Käfer. *Verhandlungen der Kaiserlich-Königlichen Zoologisch-Botanischen Gesellschaft in Wien* **19**: 193-200.

Ferrer J. 1983: Seven new species of Tenebrionidae from the Old World (Coleoptera). *Entomologica Scandinavica* **14**: 137-144.

Ferrer J. 1992: Dos nuevas species de Tenebrionidae (Coleoptera). *Nouvelle Revue d'Entomologie* (N. S.) **9**: 83-89.

Ferrer J. 1993a: Essai de révision des espèces africaines et européennes appartenant au genre Gonocephalum

Solier (Coleoptera, Tenebrionidae). *Atti del Museo Civico di Storia Naturale di Trieste* **45**: 59-150.

Ferrer J. 1993b: Description of new genus and species of Opatrini from the Canary Island (Coleoptera, Tenebrionidae). *Nouvelle Revue d'Entomologie* (N. S.) **10**: 121-125.

Ferrer J. 1997a: Deux nouveaux Mesomorphus de la région orientale (Coleoptera, Tenebrionidae, Opatrini). *Entomofauna* **18**: 73-80.

Ferrer J. 1997b: Nouvelle espèce de Gonocephalum (Solier) de l'Afrique et l'Oman. *Nouvelle Revue d'Entomologie* (N. S.) **14**: 211-216.

Ferrer J. 2000a: Contributions to the knowledge of the genera Scleropatrum Reitter, 1890 and Polycoelogastridium Reichardt, 1936 (Coleoptera, Tenebrionidae, Opatrini). *Entomofauna* **21**: 33-48.

Ferrer J. 2000b: Révision des espèces africaines et européennes appartennantes au genre Gonocephalum Solier (Coleoptera, Tenebrionidae, Opatriini). Troisième partie. *Atti del Museo Civico di Storia Naturale di Trieste* **48**: 69-153.

Ferrer J. 2003: Les Caedius Mulsant et Cyptus Gerstaecker de la région afrotropicale (Coleoptera, Tenebrionidae, Opatrini). *Nouvelle Revue d'Entomologie* **19** [2002]: 295-349.

Ferrer J. & Picka J. 1990: The Blaps species of Sweden, with review of the B. lethifera group (Coleoptera, Tenebrionidae). *Entomologisk Tidskrift* **111**: 25-32.

Ferrer J. & Soldati L. 1999: Contribution à l'étude des Tenebrionidae de Turquie (Insecta, Coleoptera). *Entomofauna* **20**: 53-92.

Ferrer J. & Whitehead P. F. 2002: The genus Xanthomus Mulsant, 1854 (Coleoptera: Tenebrionidae), its evolutionary history and conservation significance. *Annales Zoologici* **52**: 383-401.

Ferrer J. & Yvinec J. H. 2004: Révision de la tribu des Lychnogyini Reitter, 1904 sensu nov. et description d'un nouveau genre et d'une espèce nouvelle de Taklamakan, Chine (Coleoptera: Tenebrionidae: Pimeliinae). *Annales de la Société Entomologique de France* (N. S.) **40**: 41-49.

Ferrer J. & Yvinec J. H. 2005: Changement du nom homonymique d'un genre de coléoptères Tenebrionidae du désert froid de Taklamakan, Chine (Coleoptera). *Nouvelle Revue d'Entomologie* **22**: 121-122.

Ferrer L. 1859: Note sur la Mylabris Moquinia n. sp. *Revue et Magasin de ZoologiePure et Appliquée* (2) **11**: 539-540.

Fiori A. 1903: Studio critico dei Tenebrio Lin. italiani. *Rivista Coleopterologica Italiana* **1**: 221-227.

Fischer J. B. 1827: *Tentamen conspectus Cantharidiarum. Dissertatio inauguralis quam pro summis in medicina et chirurgia honoribus legitime obtinendis eruditorum examini.* Monachi: Lindauer, 26 pp.

Fischer von Waldheim G. 1809: Sur deux genres nouveaux du coléoptères. *Mémoires de la Société Impériale des Naturalistes de Moscou* **2**: 293-304, pl. 18.

Fischer von Waldheim G. 1812: Pogonocerus, novum genus insectorum Caucasi meridionalis. *Mémoires de la Société Impériale des Naturalistes de Moscou* **3**: 281-283.

Fischer von Waldheim G. 1817: Adversaria Zoologica. Fasciculus secundus. Insecta. De coleopteris quibusdam novis. *Mémoires de la Société Impériale des Naturalistes de Moscou* **5**: 463-471.

Fischer von Waldheim G. 1820: *Entomographie de la Russie* [*Entomographia Imperii Rossici*]. *Auctoritate Societatis Caesareae Mosquensis naturae scrutatorum collecta et in lucem edita. Volume I.* Mosquae: Auguste Semen, 17 pls [Coleoptera] + 9 pls [other insects].

Fischer von Waldheim G. 1821: *Lettre adressée au nom de la Société Impériale des Naturalistes de Moscou, à l'un de ses membres M. le docteur Chrétien-Henri Pander, par Gotthelf Fischer de Waldheim, directeur de la Société; Contenant une notice sur un nouveau genre d'oiseau et sur plusiers nouveaux insectes.* Moscou: Auguste Semen, 15 pp.

Fischer von Waldheim G. 1822: *Entomographie de la Russie* [*Entomographia Imperii Rossici*]. *Auctoritate Societatis Caesareae Mosquensis naturae scrutatorum collecta et in lucem edita. Volumen I.* Mosquae: Auguste Semen, viii + 210 pp.

Fischer von Waldheim G. 1823: *Entomographie de la Russie, et genres des Insectes. Entomographia Imperii Rossici, sue Caesareae Majestati Alexandro I dicata. Volume II.* Plates XVIII-L [Coleoptera] + VI-XI [Lepidoptera].

Fischer von Waldheim G. 1824: *Entomographie de la Russie, et genres des Insectes. Entomographia Imperii Rossici, sue Caesareae Majestati Alexandro I dicata. Volume II.* Mosquae: Augusti Semen Typographi, xx + 264 pp.

Fischer von Waldheim G. 1829a: *Museum Historiae Naturalis Universitatis Caesareae Mosquensis. Pars 2. Insecta.* Mosquae: Typis Universitatis Caesareae, 147 pp.

Fischer von Waldheim G. 1829b: Notice sur quelques insectes de la collection de C. Steven. *Bulletin de la Société Impériale des Naturalistes de Moscou* **1** (1-2): 368-370, 1 pl.

Fischer von Waldheim G. 1830: Note sur quelques nouvelles espèces d'insectes; dans une lettre à Mr. de Steven. *Bulletin de la Société Impériale des Naturalistes de Moscou* **2**: 183-188.

Fischer von Waldheim G. 1832a: [new taxa]. In: Ménétriés E.: *Catalogue raisonné des objets de zoologie recueillis dans un voyage au Caucase et jusqu'aux frontières actuelles de la Perse entrepris par ordre de S. M. l'Empereur*. St.-Pétersbourg: Imprimerie Impériale des Sciences, xxxiii + 271 + iv + (1) pp.

Fischer von Waldheim G. 1832b: Analecta ad faunam insectorum rossicam. *Bulletin de la Société Impériale des Naturalistes de Moscou* **4**: 423-440, pls. 5-6.

Fischer von Waldheim G. 1835: Adesmiae genus speciebus quibusdam novis illustratum. *Bulletin de la Société Impériale des Naturalistes de Moscou* **8**: 310-316, pl. VIII.

Fischer von Waldheim G. 1836: Analecta ad faunam insectorum rossicam. *Bulletin de la Société Impériale des Naturalistes de Moscou* **9**: 154-166, pl. 5.

Fischer von Waldheim G. 1837a: [new taxa]. In: Faldermann F.: Fauna entomologica trans-caucasica. Pars II. Coleoptera trans-caucasica. *Nouveaux Mémoires de la Société Impériale des Naturalistes de Moscou* **5**: 1-433, 15 pls.

Fischer von Waldheim G. 1837b: Notice sur les mélasomes. *Bulletin de la Société Impériale des Naturalistes de Moscou* **10** (4): 3-18, pl. II.

Fischer von Waldheim G. 1838: Nouvelle espèce d'Adesmia, genre de coléoptères. *Revue Zoologique, par la Société Cuvierienne* **1838**: 316.

Fischer von Waldheim G. 1842: *Catalogus Coleopterorum in Siberia orientali a Cel. Gregorio Silide Karelin collectorum*. Moscou, 28 pp.

Fischer von Waldheim G. 1844: Spicilegum Entomographiae Rossicae. II. Heteromera. *Bulletin de la Société Impériale des Naturalistes de Moscou* **17**: 3-144, pls 1-3.

Flach C. 1882: Zwei neue deutsche Cis-Arten. *Deutsche Entomologische Zeitschrift* **26**: 249-251.

Flach K. 1907: Beiträge zur Käferfauna der Iberischen Halbinsel. *Wiener Entomologiche Zeitung* **26**: 17-19.

Fleischer A. 1900: Uebersichtstabelle der Arten der Coleopteren-Gattung Palorus Duv. *Wiener Entomologische Zeitung* **19**: 236-237.

Fleischer A. 1909: Neue Mordellidae. *Wiener Entomologische Zeitung* **28**: 241-242.

Fleischer A. 1916: Neue Varietät der Pelecotoma fennica Payk. aus Bosnien. *Wiener Entomologische Zeitung* **35**: 120.

Fleischer A. 1919a: Neue palaärktische Oedemeriden. *Entomologische Blätter* **15**: 168-171.

Fleischer A. 1919b: Eine neue Prostomis-Art aus dem Elbrusgebirge. *Entomologische Blätter* **15**: 211.

Fleischer A. 1921: Bemerkenswerte Käfer-Aberrationen. *Wiener Entomologiche Zeitung* **38**: 142.

Fleischer A. 1923: Lytta vesicatoria Linn var. dibapha Reitt. *Časopis České společnosti Entomologické* **20**: 89.

Fletcher T. B. 1914: *Some South Indian Insects*. Madras: Superintendent Goverment Press, vii + 565 pp.

Fonscolombe E. L. J. H. de. 1845: Calendrier de faune et de flore pour les environs de Aix ou première apparition des principaux insectes et première floraison des végétaux qui s'y trouvent. *Mémoires de l'Académie des Sciences, Agriculture, Arts et Belles-Lettres* **5**: 357-682.

Forel A. 1893: Sur la classification de la famille des formicides, avec remarques synonymiques. *Annales de la Société Entomologique de Belgique* **37**: 161-167.

Forskål P. 1775: *Descriptiones animalium avius, amphibiorum, piscium, insectorum, vermium quae in itinere orientali observavit*. Hauniae: Mölleri, aulae Typographi, 164 pp., 1 pl.

Forster J. R. 1771: *Novae species insectorum, Centuria I.* Londoni: T. Davies et B. White, viii + 100 pp.

Fowler M. A. 1889: On the British species of the genus Anaspis, Geoffroy, with description of a new species. *The Entomologist's Monthly Magazine* **25**: 331-336.

Franciscolo M. 1942a: Mordellidi raccolti dal marchese G. Doria all'isola del Giglio negli anni 1900-1902. *Bollettino della Società Entomologica Italiana* **74**: 6-10.

Franciscolo M. 1942b: Raccolte entomologiche nell'isola di Capraia fatte da C. Mancini e da F. Capra (1927-1931). *Bollettino della Società Entomologica Italiana* **74**: 18-23.

Franciscolo M. 1942c: Diagnosi di nuovi mordellidi Italiani. *Bollettino della Società Entomologica Italiana* **74**: 76-80.

Franciscolo M. 1942d: Mordellokoides grandii nuovo sottogenere e nuova specie di Mordellistena dell'Italia meridionale. *Bollettino dell'Instituto di Entomologia della R. Universita di Bologna* **13**: 133-136.

Franciscolo M. 1943a: Machairophora, nuovo genere della tribú dei Mordellini ed alcune note sistematiche e sinonimiche sui Mordellidae. *Bollettino della Società Entomologica Italiana* **75**: 33-40.

Franciscolo M. 1943b: Su due nuove Anaspis Geoffroy. *Bollettino della Società Entomologica Italiana* **75**: 76-80.

Franciscolo M. 1949a: Un nuovo mordellino di Birmania. *Doriana* **1** (1): 1-4.

Franciscolo M. 1949b: Note su alcuni mordellidi. *Bollettino della Società Entomologica Italiana* **79**: 52-58.

Franciscolo M. 1949c: XIII° Contributo alla conoscenza dei mordellidi. *Memorie della Società Entomologica Italiana* **28**: 81-95.

Franciscolo M. 1950: Ermischiella papuana n. ge. n. sp. di mordellide della Nuova Guinea e diagnosi preliminari di tre nuovi generi africani. *Doriana* **1** (5): 1-8.

Franciscolo M. 1951: Una nuova Mordella dell'Albania. *Bollettino della Società Entomologica Italiana* **81**: 7-9.

Franciscolo M. 1952: La Neoglipa e generi vicini della Nuova Guinea. *Annali del Museo Civico di Storia Naturale di Genova* **65**: 325-357.

Franciscolo M. 1953: Results of the Armstrong College Expedition to Siwa Oasis (Lybian desert), 1935, under the leadership of Prof. J. Omer-Cooper. *Bulletin de la Société Fouad Ier d'Entomologie* **37**: 251-253.

Franciscolo M. 1954: On two species of Anaspidinae (Col. Scraptiidae) taken on flowers of Protea abyssina in Natal. *Proceedings of the Royal Entomological Society of London* (B) **23 (3-4)**: 63-73.

Franciscolo M. 1956a: Su un nuovo genere di Scraptiidae dell'India settentr. *Memorie della Società Entomologica Italiana* **35**: 43-70.

Franciscolo M. 1956b: Su due interessanti specie del genere Anaspis L. del bacino del Mediterráneo. *Bollettino della Società Entomologica Italiana* **86**: 47-56.

Franciscolo M. 1956c: 30. Coleoptera: Mordellidae and Scraptiidae. Pp. 451-469. In: *British Museum Expedition to South-West Arabia, 1937-38.* [1941-1957] *Volume 1.* London: Trustees of the British Museum, 504 pp., 1 pl., 1 map.

Franciscolo M. 1961: On a new scraptiid beetle from Afganistan. *Proceedings of the Royal Entomological Society of London* **26**: 53-57.

Franciscolo M. 1974: New and little-known Mordellistena Costa from Pakistan and India. *Oriental Insects* **8**: 71-84.

Franciscolo M. 1982: Le Anaspis (s. str.) del gruppo maculata Fourcr., con descrizione di una nuova specie sarda. *Memorie della Società Entomologica Italiana* **60** [1981]: 197-207.

Franciscolo M. 1989: About Ectasiocnemis Franciscolo, 1956. *Bollettino della Società Entomologica Italiana* **121**: 108-123.

Franciscolo M. 1991: Su alcuni mordellidi e scraptiidi (Coleoptera-Heteromera) delle isole Pelagie. *Il Naturalista Siciliano* (4) **15**: 167-178.

Franz H. 1967: Revision der Tarphius-Arten Europas, Nordwestafrikas und der Kanarischen Inseln (Coleopt. Colydiidae). *Eos, Revista Española de Entomología* **43**: 61-91.

Franz H. 1970: Langelandia-Arten von Madeira-Gruppe und von den Kanaren (Coleoptera, Colydiidae). *Bocagiana* **26**: 1-5.

Franz H. 1980: Neue und ungenügend bekannte Coleoptera (Scydmaenidae et Colydiidae) aus Mazedonien. *Zeitschrift der Arbeitsgemeinschaft Österreichischer Entomologen* **31** [1979]: 107-110.

Franz H. 1982: Beiträg zur Kenntnis der Coleopterenfauna der Insel Galitia (mit einem Beitrag von Cl. Besuchet). *Sitzungsberichte der Österreichischen Akademie der Wissenschaften in Wien (Abteilung 1)* **191**: 231-240.

Franz H. 1984a: Beitrag zur Kenntnis der Coleopterenfauna Mazedoniens. *Zeitschrift der Arbeitsgemeinschaft Österreichischer Entomologen* **36**: 57-62.

Franz H. 1984b: Weitere neue Coleopterenarten von den Kanarischen Inseln. *Vieraea* **13** [1983]: 75-81.

Franz H. 1985: Revision der Genera Holoxantha Semenow und Alloxantha Seidlitz (Coleoptera, Oedemeridae). *Eos, Revista Española de Entomología* **61**: 95-105.

Franz H. 1990a: Weitere Beiträge zur Kenntnis der Coleopterenfauna der Canarishen Inseln (Colydiidae und Ptinidae). *Zeitschrift der Arbeitsgemeinschaft Österreichischer Entomologen* **42**: 27-31.

Franz H. 1990b: Stenosis canariensis n. sp. (Coleoptera, Tenebrionidae). *Zeitschrift der Arbeitsgemeinschaft der Österreichischer Entomologen* **42**: 93-94.

Franz H. 1996: Die Ergebnisse meiner langjährigen Aufsammlungen der Coleopteenfauna [sic] auf der Insel hierro (Kanarische Inseln). *Sitzungsberichte Mathematisch-naturwissenschaftliche Klasse Abt. 1* **202** [1995]: 71-138.

Freude H. 1955: Die Monommiden der Welt. I. Teil: Die Monommiden der indo-australischen Region. *Entomologische Arbeiten aus dem Museum G. Frey* **6**: 1-73, pls 1-11.

Freude H. 1958: Die Monommidae der Afrikanischen Region (Coleoptera). (IV Teil der Monommiden der Welt mit Zusammenfassung der Ergebnisse). *Annales du Musée Royal du Congo Belge, Série in 8o, Sciences Zoologiques* **61**: 1-115.

Freude H. 1978: Monommidae des Bishop Museums in Honolulu (Coleoptera). *Pacific Insects* **18**: 217-221.

Friedenreich C. W. 1883: Pilzbewohnende Käfer in der Provinz Santa Catharina (Südbrasilien). *Entomologische Zeitung* (Stettin) **44**: 375-380.

Frivaldszky J. 1865: Adatok a magyarhoni barlangok faunájához. *Mathematikai és Természettudományi Közlemények Vonatkozólag a Hazai Viszonyokra* **3**: 17-53.

Frivaldszky J. 1877a: Coleoptera nova ex ins. Creta et Asia Minore. *Természetrajzi Füzetek* **1**: 83-87.

Frivaldszky J. 1877b: Adnotationes coleopterologicae. *Természetrajzi Füzetek* **1**: 89-136.

Frivaldszky J. 1877c: Adatók Temes és Krassó megyek faunájához. *Mathematikai és Természettudományi Közlemenyek* **13**: 285-378.

Frivaldszky J. 1880: Coleoptera nova in Europa orientali et Asia Minor a D. Eduardo Merkl dedecta [sic.]. Ujtéhélyröpüek Európa keleti részében es Kis-Azsiában Merkl Ede urtól gyüjtve. *Természetrajzi Füzetek* **4**: 260-266.

Frivaldszky J. 1884: Coleoptera nova ex Asia Minore. *Természetrajzi Füzetek* **8**: 1-8.

Frivaldszky J. 1889: Coleoptera in expeditione D. Comitis Belae Széchenyi in China, praecipue borealis, a Dominis Gustavo Kreitner et Ludovico Lóczy anno 1879 collecta. *Természetrajzi Füzetek* **12**: 197-210.

Frivaldszky J. 1892: Coleoptera in expeditione D. Comitis Bolae Széchenyi in China, praecipuae boreali, a Dominus Gustavo Kreitner et Ludovico Loczy, anno 1879 collecta. Pars seconda. *Természetrajzi Füzetek* **15**: 114-125.

Frivaldszky J. 1894: Coleoptera nova. *Természetrajzi Füzetek* **16** [1893]: 85-86.

Frivaldszky von Frivald I. (= E.) 1835: Közlések a' Balkány vidékén tett természettudományi utazásról. *A Magyar Tudós Társaság Évkönyvei* **2**: 235-276, 7 pls.

Frivaldszky von Frivald I. (= E.) 1837: Balkány vidéki természettudományi utazás. Második közlés. *A Magyar Tudós Társaság Évkönyvei* **3**: 156-184, 8 pls.

Frivaldszky von Frivald I. (= E.) 1845: Rövid áttekintése egy természetrajzi utazásnak, az európai Törökbirodalomban, egyszersmind nehány a közben újonnat fölfedezett állatnak leirása. *A Királyi Magyar Természettományi Társulat Évkönyvei* **1**: 163-187, 3 pls.

Frivaldszky von Frivald I. (= E.) 1865: Jellemzö adatok Magyarország faunájához. *A Magyarország Tudományi Akadémia Évkönyvei* 11 (4). Pest: Emich Gustáv, 274 pp., xiii pls.

Frölich J. A. von. 1792: Bemerkungen über einige seltene Käfer aus der Insectensammlung des Herrn Hofr. und Prof. Rudolph in Erlangen. *Naturforscher* **26**: 68-165.

Fuente J. M. de la. 1907: Datos para la fauna de la provincia de Ciudad Real. XIX. Coleópteros. *Boletín de la Real Sociedad Española de Historia Natural* **7**: 317-323.

Fuente J. M. de la. 1909: Datos para la fauna de la provincia de Cuiudad Real. *Boletín de la Real Sociedad Española de Historia Natural* **9**: 306-310.

Fuente J. M. de la. 1910: Datos para la fauna de la provincia de Cuiudad Real. *Boletín de la Real Sociedad Española de Historia Natural* **10**: 442-446.

Fuente J. M. de la. 1913: Descripción de algunos Coleópteros nuevos de España. *Boletín de la Real Sociedad Española de Historia Natural* **13**: 475-476.

Fuessly G. 1775: *Verzeichniss der ihm bekannten Schweizerischen Inseckten mit einer ausgemahlten Kupfertafel nebst der Ankündigung eines neuen Insecten Werks.* Zürich und Winterthur: Heinrich Steiner und Compagnie, xii + 62 pp., 1 pl.

Fursov N. I. 1935: Drei neue kaukasische Arten aus Familie Cucujidae nützlich in der Forstwirtschaft. *Bulletin de la Société des Naturalistes de Moscou, Section Biologique* (N. S.) **44**: 90-91.

Fursov N. I. 1939: Eine neue Gattung und 6 neue Coleopterenarten von Mittelasien und Caucasus. *Bulletin de la Société des Naturalistes de Moscou, Section Biologique* (N. S.) **48**: 88-90.

Gabriel J. J. C. 1916: Eine neue Anaspis-Art. *Entomologische Mitteilungen* **5**: 183-184.

Gahan C. J. 1896a: On the Coleoptera obtained by Dr. Anderson's collector during Mr. T. Bent's expedition to the Hadramaut, South Arabia. *Journal of the Linnean Society* **25**: 285-291.

Gahan C. J. 1896b: On Coleoptera from Aden and Somaliland. *The Annals and Magazine of Natural History* (6) **18**: 448-461.

Gahan C. J. 1900: Description of new genera and species of Coleoptera collected by Mr. Ogilvie-Grant and Dr. Forbes in the Islands of Sokotra and Abd-el-Kuri. *Bulletin of the Liverpool Museum* **3**: 8-13.

Gahan C. J. 1903: Coleoptera. In: Forbes H. O.: Natural history of Sokotra and Abd-el-Kuri: Being the report upon the results of the conjoint expedition to these islands in 1898-9, by Mr. W. R. Ogilvie-Grant, of the British Museum, and Dr. H. O. Forbes, of the Liverpool Museums, together with information from other available sources. Forming a monogragh of the islands. *Bulletin of the Liverpool Museums* **1903**: 263-292, pl. 17.

Ganglbauer L. 1881: Bestimmungs-Tabellen der europäischen Coleopteren. IVa. Oedemeridae. *Verhandlungen der Kaiserlich-Königlichen Zoologisch-Botanischen Gesellschaft in Wien* **31**: 97-116.

Ganglbauer L. 1882: Coleopterologische Mittheilungen. *Wiener Entomologische Zeitung* **1**: 135-140.

Ganglbauer L. 1889: Insecta a Cl. G. N. Potanin in China et in Mongolia novissime lecta. VII. Buprestidae, Oedemeridae, Cerambycidae. *Horae Societatis Entomologicae Rossicae* **24** [1889-1890]: 21-85.

Ganglbauer L. 1891: Eine neue Art der Melandryiden-Gatung Zilora. *Wiener Entomologische Zeitung* **10**: 132-133.

Ganglbauer L. 1897: Einige neue Coleopteren des mitteleuropäischen Faunengebietes. *Verhandlungen der Kaiserlich-Königlichen Zoologisch-Botanischen Gesellschaft in Wien* **47**: 565-573.

Ganglbauer L. 1899: *Die Käfer von Mitteleuropa. Die Käfer der österreichisch-ungarischen Monarchie, Deutschlands, der Schweiz, sowie des französischen und italienischen Alpengebietes. Dritter Band. Familienreihe Staphylinoidea. II. Theil. Scydmaenidae, Silphidae, Clambidae, Leptinidae, Platypsyllidae, Corylophidae, Sphaeriidae, Trichopterygidae, Hydroscaphidae, Scaphidiidae, Histeridae. Familienreihe Clavicornia. Sphaeritidae, Ostomidae, Byturidae, Nitidulidae, Cucujidae, Erotylidae, Phalacridae, Thorictidae, Lathridiidae, Mycetophagidae, Colydiidae, Endomychidae, Coccinellidae.* Wien: Carl Gerold's Sohn, III +1046 pp.

Ganglbauer L. 1900: [new taxa]. In: Bodemeyer E. von: *Quer durch Klein-Asien in den Bulghar-Dagh. Eine naturwissenschatliche Studien-Reise.* Emmendingen: Verlangs-Aktiengesellschaft vormals Dölter, v + 169 pp.

Ganglbauer L. 1905: Coleoptera. Ergebnisse einer naturwissenschaftlichen Reise zum Erdschias-Dagh (Kleinasien). Ausgeführt von Dr. Arnold Penther und Dr. Emerich Zederbauer auf Kosten der "Gesellschaft zur Förderung des naturhistorischen Erforschung des Orients in Wien im Jahre 1902". *Annalen des K. K. Naturhistorischen Hofmuseums* **20**: 246-290.

Garciá R. & Oromí P. 2001: Melansis reyesi n. sp., un nuevo Litoborini de las islas Canarias (Coleoptera, Tenebrionidae). *Vieraea* **28**: 51-59.

Gardini G. 1975: Materiali per lo studio dei Tenebrionidi dell'arcipelago Toscano (Col. Heteromera). *Lavori della Società Italiana di Biogeografia* (Nuova Seria) **5** [1974]: 1-87.

Gardini G. 1979: Nuovi dati su Tenebrionidi (Col.) dell'Archipelago Toscano. *Bollettino di Museo Civico di Storia Naturale Verona* **6**: 73-77.

Gebien H. 1906a: Ueber die von Fabricius beschriebenen Typen von Tenebrioniden in den Museen von Kopenhagen und Kiel. *Deutsche Entomologische Zeitschrift* **1906**: 209-237.

Gebien H. 1906b: Notiz über Asida convexicollis. *Wiener Entomologische Zeitung* **25**: 248.

Gebien H. 1910a: Tenebrionidae I, II. Pars 15, 22. In: Schenkling S. (ed.): *Coleopterorum Catalogus. Vol. 18.* Berlin: W. Junk, 1-354 pp.

Gebien H. 1910b: 7. Coleoptera. 19. Tenebrionidae. Pp. 363-397. In: Sjöstedt Y. (ed.): *Wissenschaftliche Ergebnisse der Schwedischen Zoologischen Expedition nach dem Kilimandjaro, dem Meru und den umgebenden Massaisteppen Deutsch-Ostafrikas 1905-1906 unter der Leitung von Prof. Dr. Yngve Sjörstedt. 1. Band.* Stockholm: Königl. Schwedisches Akademie der Wissenschaften, 442 pp.

Gebien H. 1911: Tenebrionidae III, IV. Pars 28, 37. In: Schenkling S. (ed.): *Coleopterorum Catalogus. Vol. 18.* Berlin: W. Junk, 355-740 pp.

Gebien H. 1913: Neue Tenebrioniden der palaearktischen Fauna (Col. Heter.) nebst einer Übersicht über die Gattung Iphthimus. *Coleopterologische Rundschau* **2**: 1-8.

Gebien H. 1914a: Die Tenebrionidenfauna Borneos. Erster Teil. *Sarawak Museum Journal* **2** (5): 1-58, pl. 1.

Gebien H. 1914b: H. Sauter's Formosa-Ausbeute. Tenebrionidae (Coleopt.). *Archiv für Naturgeschichte* **A 79** (9) [1913]: 1-60, 1 pl.

Gebien H. 1914c: Fauna Simularensis. Coleoptera, Fam. Tenebrionidae. *Notes from the Leyden Museum* **36**: 61-80.

Gebien H. 1914d: Die Tenebrioniden der Philippinen. *The Philippine Journal of Science* **8** (D) [1913]: 373-433.

Gebien H. 1918: Beitrag zur Kenntnis der Gattung Setenis (Col., Heterom.). *Entomologische Mitteilungen* **7**: 121-130, pl. 2.

Gebien H. 1919a: Beitrag zur Kenntnis der Gattung Setenis (Col., Heterom.). *Entomologische Mitteilungen* **8**: 1-14.

Gebien H. 1919b: Monographie der südamerikanischen Camarien (Coleopt. Heterom.) nebst einer Übersicht über die indischen Gattungen der Camariinen. *Archiv für Naturgeschichte* A **83** (3) [1917]: 25-167, 2 pls.

Gebien H. 1920: *Käfer aus der Familie Tenebrionidae gesammelt auf der "Hamburger deutsch-südwestafrikanischen Studienreise 1911". Hamburgische Universität, Abhandlungen aus der Auslandskunde. Band 5. Reihe C Naturwissenschaften Band 2.* Hamburg: L. Friederichsen & Co., 168 pp., 2 pls.

Gebien H. 1920: Coleoptera Tenebrionidae. Pp. 213-500, pls IX-XI. In: *Nova Guinea. Résultats de l'Expédition scientifique néerlandaise la Nouvelle-Guinée.* Vol. XIII (3), Zoologie. Leide: E. J. Brill.

Gebien H. 1921: Die Tenebrioniden Westafrikas. *Archiv für Naturgeschichte* A **86** (6) [1920]: 1-256.

Gebien H. 1922a: The Percy Sladen trust Expedition to the Indian Ocean in 1905. No. V. – Coleoptera, Heteromera: Tenebrionidae. *Transactions of the Linnean Society of London* (Series 2, Zoology) **18** [1922-1925]: 261-323, pl. 23.

Gebien H. 1922b: Philippine Tenebrionidae, II. *The Philippine Journal of Science* **19** [1921]: 439-515, 2 pls.

Gebien H. 1925a: Die Tenebrioniden (Coleoptera) des Indomalayischen Gebietes, unter Beruecksichtigung der benachbarten Faunen. IV. Die Gattungen Phloeopsidus, Dysantes, Basanus, und Diaperis. *The Philippine Journal of Science* **27** (1): 131-157, 257-288, 1 pl.

Gebien H. 1925b: Die Tenebrioniden (Coleoptera) des Indomalayischen Gebietes, unter Beruecksichtigung der benachbarten Faunen. VI. Die Gattung Platydema Castenau und Brullé. *The Philippine Journal of Science* **27** (3): 423-453, 2 pls.

Gebien H. 1925c: Die Tenebrioniden (Coleoptera) des Indomalayischen Gebietes, unter Beruecksichtigung der benachbarten Faunen. VII. Die Gattungen Ischnodactylus, Hoplocephala, und Martianus. *The Philippine Journal of Science* **27** (4): 539-595, 1 pl.

Gebien H. 1925d: Die Tenebrioniden (Coleoptera) des Indomalayischen Gebietes, unter Beruecksichtigung der benachbarten Faunen. VIII. *The Philippine Journal of Science* **28**: 101-128, 1 pl.

Gebien H. 1927: Fauna Sumatrensis (Beitrag Nr. 31) Tenebrionidae (Col.). *Supplementa Entomologica* **15**: 22-58.

Gebien H. 1935: Tenebrionidae. Pp. 53-77, 1 pl. In: Van Stralen V. (ed.): Résultats Scientifiques du Voyage aux Indes Orientales Néerlandaises de LL. AA. RR. le Prince et la Princesse Léopold de Belgique. *Mémoires du Musée Royal d'Histoire Naturelle de Belgique. Hors Série.*Volume IV. Invertébrés: Hexapodes.

Gebien H. 1937: Katalog der Tenebrioniden (Col. Heteromera) Teil I. *Pubblicazioni del Museo Entomologico "Pietro Rossi"* **2**: 505-883.

Gebien H. 1938: [new taxa]. In: Schuster A: & Gebien H.: Tenebrioniden (Col.) aus Arabien. *Entomologische Blätter* **34**: 49-62.

Gebien H. 1939: Katalog der Tenebrioniden. *Mitteilungen der Münchener Entomologischen Gesellschaft* **29**: 443-474.

Gebien H. 1940a: Die von Dr. Trinkler in Tibet gesammelten Tenebrioniden (Col. Tenebr.). *Mitteilungen der Münchener Entomologischen Gesellschaft* **30**: 1-3.

Gebien H. 1940b: Katalog der Tenebrioniden, Teil II. *Mitteilungen der Münchener Entomologischen Gesellschaft* **30**: 755-786.

Gebien H. 1943: Katalog der Tenebrioniden, Teil III. *Mitteilungen der Münchener Entomologischen Gesellschaft* **33**: 895-926.

Gebler F. A. von. 1817: Insecta Siberia rariora, decas prima. *Mémoires de la Société Impériale des Naturalistes de Moscou* **5**: 315-333.

Gebler F. A. von. 1825: Coleopterorum Sibiriae species novae. Pp. 42-57. In: Hummel A. D.: *Essais Entomologiques, Insectes de 1824. Novae species. Vol. 1, Nr. 4.* St. Petersbourg: Chancellerie privée du Ministère de l'Intérieur, 71 pp. + index.

Gebler F. A. von. 1829a: Bemerkungen über die Insekten Sibiriens, vorzüglich des Altai. [Part 3]. Pp. 1-228. In: Lederbour C. F. von. (ed.): *Reise durch das Altai-Gebirge und die soongorische Kirgisen-Steppe. Auf Kosten der Kaiserlichen Universität Dorpat unternommen im Jahre 1826 in Begleitung der Herren D. Carl Anton Miecherund D. Alexander von Bunge K. K. Collegien Assessors. Zweiter Theil.* Berlin: G. Reimer, 427 pp.

Gebler F. A. von. 1829b: Des mylabrides de la Sibérie occidentale et des confins de la Tartarie. *Nouvelles Mémoires de la Société Impériale des Naturalistes de Moscou* **1**: 145-171.

Gebler F. A. von. 1830: Bemerkungen über die Insekten Sibiriens, vorzüglich des Altai. [Part 3]. Pp. 1-228. In: Lederbour C. F. von. (ed.): *Reise durch das Altai-Gebirge und die soongorische Kirgisen-Steppe. Auf Kosten der Kaiserlichen Universität Dorpat unternommen im Jahre 1826 in Begleitung der Herren D. Carl Anton Miecherund D. Alexander von Bunge K. K. Collegien Assessors. Zweiter Theil.* Berlin: G. Reimer, 427 pp. [Note: this seems to be a second edition or reprint of the 1829a work, dated from 1830].

Gebler F. A. von. 1832: Notice sur les coléoptères qui se trouvent dans le district des mines de Nertschinsk, dans la Sibérie orientale, avec la description de quelques espèces nouvelles. *Nouveaux Mémoires de la Société des Naturalistes de Moscou* **2** (8): 23-78.

Gebler F. A. von. 1833: Notae et addidamenta ad catalogum Coleopterorum Sibiriae occidentalis et confinis Tatariae operis, C. F. von Ledebours Reise in das Altaigebirge und die soongarische Kirgisensteppe (Zwyter Theil. Berlin 1830). *Bulletin de la Société Impériale des Naturalistes de Moscou* **6**: 262-309.

Gebler F. A. von. 1836: Lettre de M. le Conseiller d'Etat et Chevalier de Gebler, contenant un rapport d'un voyage dans les hautes montagnes catouniennes jusqu'à la frontière de la Chine; et description de trois nouvelles espèces de coléoptères. *Bulletin de la Société Impériale des Naturalistes de Moscou* **9**: 329-345.

Gebler F. A. von. 1841a: Characteristik mehrerer neuer sibirischer Coleopteren. *Bulletin Scientifique de l'Académie Impériale des Sciences de Staint-Pétersbourg* **8** (24): 369-376 [columns].

Gebler F. A. von. 1841b: Notae et addimenta ad catalogum Coleopterorum Sibiriae occidentalis et confinis Tatariae operis: C. F. Ledebour's Reise in das Altaigebirge und die soongarische Kirgisensteppe. Zweyter Theil. Berlin 1830. Fasciculus secundus. *Bulletin de la Société Impériale des Naturalistes de Moscou* **14** (4): 577-625.

Gebler F. A. von. 1843: Charakteristik der vom Hn. Dr. Schrenk im Jahre 1841 in den Steppen und Gebirgen der Songarei gefundenen neuen Coleopteren-Arten. *Bulletin de la Classe Physico-Mathématique de l'Académie Impériale des Sciences de St.-Pétersbourg* **1**: 36-42 [columns].

Gebler F. A. von. 1844: Charakteristik der von Hn. Dr. Schrenk in den Jahren 1842 und 1843 in den Steppen der Dsungarei gefundenen neuen Coleopterenarten. *Bulletin de la Classe Physico-Mathématique de l'Académie Impériale des Sciences de St.-Pétersbourg* (2) **3** [1844-1845]: 97-106 [columns].

Gebler F. A. von. 1847: Verzeichniss der im Kolywano-Woskresenskischen Hüttenbezirke Süd-West Sibiriens beobachteten Kaefer mit Bemerkungen und Beschreibungen. *Bulletin de la Société Impériale des Naturalistes de Moscou* **20** (3): 391-512.

Gebler F. A. von. 1859: Verzeichniss der von Herrn Dr. Schrenk in den Kreisen Ajagus und Karakaly in der östlichen Kirkisiensteppe und in der Songarey in den Jahren 1840 bis 1843 gefundenen Kaeferarten. *Bulletin de la Société Impériale des Naturalistes de Moscou* **32** (1): 426-519, (4): 315-356.

Gebler F. A. von. 1860: Coleopterorum species novae, a Dr. Schrenk in deserto kirgiso-songorico anno 1843 detectae. *Bulletin de la Société Impériale des Naturalistes de Moscou* **33** (2): 1-39.

Gemminger M. 1851: *Systematische Uebersicht der Käfer um München.* Jena: Mauke, 65 pp., 1 pl.

Gemminger M. 1870a: [new names]. In: Harold E. von.: Geänderte Namen. *Coleopterologische Hefte* **6**: 119-124.

Gemminger M. 1870b: [new names]. In: Gemminger M. & Harold E. von.: *Catalogus coleopterorum hucusque descriptorum synonymicus et systematicus. Tom. VII. Tenebrionidae, Nilionidae, Pythidae, Melandryidae, Lagriidae, Pedilidae, Anthicidae, Pyrochroidae, Mordellidae, Rhipidophoridae, Cantharidae, Oedemeridae* Monachii: E. H. Gummi, pp. 1801-2180 + [9].

Gené J. 1836a: De quisbusdam insectis Sardiniae novis aut minus cognitis. Fasciculus I. *Memoralia Regiae Scientiarum Academiae Taurinensis* **39**: 161-199, 1 pl.

Gené J. 1836b: [Dix-sept coléoptères nouveaux de Sardeigne]. *Bulletin de la Société Entomologique de France* **1836**: ii-iv.

Gené J. 1839: De quisbusdam insectis Sardiniae novis aut minus cognitis. Fasciculus II. *Memoralia Regiae Scientiarum Academiae Taurinensis* (2) **1**: 43-84, 2 pls.

Geoffroy E. L. 1762: *Histoire abrégée des insectes qui se trouvent aux environs de Paris; dans laquelle ces animaux sout rangés suivant un ordre méthodique. [1762-1763] Tome première.* Paris: Durand, xxviii + 523 pp. + 22 pls.

Geoffroy E. L. 1785: [new taxa]. In: Fourcroy A. F. de.: *Entomologia parisiensis; sive Catalogus Insectorum quae in Agro Parisiensi reperiuntur; secundum methodum Geoffraeanam in sectiones, genera & species distributus: cui addita sunt nomina trivalia & fere trecentae novae species. Vol. 1.* Parisiis: Serpentineis, vii + 231 pp.

Geoffroy E. L. 1799: *Histoire abrégée des insectes, dans laquelle ces animaux sont rangés suivant un ordre méthodique. Nouvelle édition, revue, corrigée et augmentée d'un supplément considérable. Vol. 1.* Paris: Delalain Fils, xxviii + 556 pp., 10 pls.

Gerhardt J. 1876: Anaspis (Silaria) palpalis n. sp. *Deutsche Entomologische Zeitschrift* **1876**: 381-383.

Gerhardt J. 1901: Eine neue Käferart. *Zeitschrift für Entomologie* (N. F.) (Breslau) **26**: 18-21.

Gerhardt J. 1904: Eine verkannte deutsche Käferart. *Zeitschrift für Entomologie* (Breslau) **29**: 79-82.

Germar E. F. 1813: Neue Insekten. *Magazin der Entomologie* **1** (1): 114-133.

Germar E. F. 1814: [new taxon]. In: Ahrens A.: *Fauna Insectorum Europae. Fasciculus 2.* Halae: Car. Aug. Kümmelii, [25] pp., 25 pls.

Germar E. F. 1817: *Reise nach Dalmatien und in das Gebiet von Ragusa. Band 2.* Leipzig und Altenburg: F. A. Brockhaus, xii + 323 pp., 8 pls, 2 maps.

Germar E. F. 1818: Vermischte Bemerkungen über einige Käferarten. *Magazin der Entomologie* **3**: 228-260, pl. 3.

Germar E. F. 1821: *Fauna Insectorum Europae. Fasciculus 4.* Halae: Car. Aug. Kümmelii, [25] pp., 25 pls.

Germar E. F. 1824a: *Fauna Insectorum Europae. Fasciculus 10.* Halae: Car. Aug. Kümmelii, [25] pp., 25 pls.

Germar E. F. 1824b: *Species insectorum novae aut minus cognitae, descriptionibus illustratae. Volumen Primum. Coleoptera.* Halae: J. C. Hendelii et filii, xxiv + 624 pp., 2 pls.

Germar E. F. 1827: *Fauna Insectorum Europae. Fasciculus 13.* Halae: Car. Aug. Kümmelii, [25] pp., 25 pls.

Germar E. F. 1831: *Fauna Insectorum Europae. Fasciculus 14.* Halae: Car. Aug. Kümmelii, [25] pp., 25 pls.

Germar E. F. 1832: *Fauna Insectorum Europae. Fasciculus 15.* Halae: Car. Aug. Kümmelii, [25] pp., 25 pls.

Germar E. F. 1836a: *Fauna Insectorum Europae. Fasciculus 16.* Halae: Car. Aug. Kümmelii, [25] pp., 25 pls.

Germar E. F. 1836b: *Fauna Insectorum Europae. Fasciculus 17.* Halae: Car. Aug. Kümmelii, [25] pp., 25 pls.

Germar E. F. 1836c: *Fauna Insectorum Europae. Fasciculus 18.* Halae: Car. Aug. Kümmelii, [25] pp., 25 pls.

Germar E. F. 1842: *Fauna Insectorum Europae. Fasciculus 22.* Halae: Car. Aug. Kümmelii, [25] pp., 25 pls.

Germar E. F. 1847: *Fauna Insectorum Europae. Fasciculus 24.* Halae: [2 + 25] pp; 25 pls.

Gerstaecker C. E. A. 1854: Diagnosen der von Peters in Mossambique gesammelten und von Dr. Gestaecker bearbeiteten Kaefer aus der Familien der Vesicantia. *Monatsberichte der Königlichen Preussischen Akademie der Wissenschaften zu Berlin* **1854**: 694-695.

Gerstaecker C. E. A. 1855a: *Rhipiphoridum Coleopterorum familiae dispositio systematica. Dissertatio Inauguralis Zoologica quam consensu et auctoritate amplissimi philosophorum ordinis in alma litterarum universitate Friderica Guilelma pro summis in philosophia honoribus rite capessendis die IV. M. Julii a MDCCCLV. Hora XII. Publice defendet auctor A. Gerstaecker, med. Dr. Berolinensis.* Berolini: Schlesinger, 36 pp.

Gerstaecker C. E. A. 1855b: *Rhipiphoridum Coleopterorum familiae dispositio systematica.* Berolini: F. Nicolai, 36 pp., 1 pl.

Gerstaecker C. E. A. 1871: Beitrag zur Insektenfauna von Zanzibar, nach dem während der Expedition des Baron v. d. Decken gesammelten Material zusammengestellt. *Archiv für Naturgeschichte* **37** (1): 42-86.

Gerstaecker C. E. A. 1884: Bestimmung der von Herrn Dr. G. A. Fischer während seiner Reise nach dem Massai-Land gesammelten Coleopteren. *Jahrbuch der Hamburgischen Wissenschaflichen Anstalten* **1**: 41-63.

Gestro R. 1880: Appunti sull´entomologia tunisina. *Annali del Museo Civico di Storia Naturale di Genova* **15** [1879-1880]: 405-424.

Gestro R. 1881: Aliquot Coleopterorum Musei Civici Januensis diagnoses. *Annali del Museo Civico di Storia Naturale di Genova* **16** [1880-1881]: 658-664.

Gestro R. 1892: Di alcuni coleotteri raccolti nel paese dei Somali dall'Ing. L. Bricchetti Robecchi. *Annali di Museo Civico di Storia Naturale di Genova* **32**: 747-790.

Gestro R. 1895: Esplorazione del Giuba e dei suoi affluenti compiuta dal cap. V. Bottego durante gli anni 1892-93 sotto gli auspicii della Società geografica italiana. XVI. Coleotteri. *Annali di Museo Civico di Storia Naturale di Genova* **35**: 247-478.

Gillerfors G. 1985: Two new species of the genus Tarphius Erichson from the Azores and redescription of Tarphius wollastoni Crotch (Coleoptera: Colydiidae). *Bocagiana* **85**: 1-10.

Gillerfors G. 1986a: Two new species of Tarphius Erichson (Coleoptera, Colydiidae) and a new species of Mniophlosoma Wollaston (Coleoptera, Chrysomelidae) from the Azores. *Bocagiana* **98**: 1-10.

Gillerfors G. 1986b: Two new species of the genus Tarphius Erichson from La Gomera (Canary Islands) and desigantion of lectotypes of Tarphius humerosus Wollaston and T. gomerae Franz (Coleoptera, Colydiidae). *Vieraea* **16**: 3-10.

Gillerfors G. 1990: Tarphius palmensis sp. n., a new species of the genus Tarphius Erichson from La Palma, Canary Islands (Col., Colydiidae). *Vieraea* **18**: 149-151.

Gillerfors G. 1991: Three new species of Coleoptera from the Canary Islands. *Vieraea* **20**: 21-26.

Gillerfors G. 1997: A new Tarphius (Coleoptera, Colydiidae) from Madeira Island. *Bocagiana* **187**: 1-4.

Gillerfors G. & Oromi P. 1991: Tarphius stagosus sp. n., a new species from La Gomera, Canary Islands (Coleoptera, Colydiidae). *Nouvelle Revue d'Entomologie* (N. S.) **8**: 85-87.

Giorna M. S. 1804: Mémoire d'entomologie. *Mémoires de l'Académie des Sciences, Littérature et Beaux-Arts de Turin. Sciences Physiques et Mathématiques* **1**: 215-223, pls. 1, 2.

Girard C. 1965: Contribution à l'étude du genre Tentyria Latreille (Col. Tenebrionidae). *Bulletin de la Société Entomologique de France* **70**: 138-139.

Girard C. 1966a: Contribution à l'étude du genre Tentyria Latreille (Col. Tenebrionidae). (2e note). *Bulletin de la Société Entomologique de France* **71**: 84-86.

Girard C. 1966b: Note sur quelques Tentyriini du Moyen-Orient. *Bulletin de la Société Entomologique de France* **71**: 203-206.

Girard C. 1968: Contribution à la faune de l'Iran 9. Coléoptères Tenebrionidae, genre Tentyria. *Annales de la Société Entomologique de France* (N. S.) **4**: 867-874.

Girard C. & Pierre F. 1965: Sur quelques Tentyriini du Sudan (Republic of Sudan) (Col. Tenebrionidae). *Bulletin de la Société Entomologique de France* **70**: 134-137.

Gistel J. N. F. X. 1831: Entomologische Fragmente. *Isis von Oken* **1831**: 301-310.

Gistel J. N. F. X. 1834a: *Die Insecten-Doubletten aus der Sammlung des Herrn Grafen Rudolph von Jenison Walworth zu Regensburg, welche sowohl im Kauf als im Tausche abgegeben werden. Nro. I. Käfer.* München: Jaquet, 36 pp.

Gistel J. N. F. X. 1834b: XIV. Entomologische Notizen. *Faunus* **3**: 128-151.

Gistel J. N. F. X. 1848: *Naturgeschichte des Thierreichs, für höhere Schulen.* Stuttgart: R. Hoffmann'sche Verlags-Buchhandlung, xvi + 216 + [4] pp., 32 pls.

Gistel J. N. F. X. 1856: *Die Mysterien der europäischen Insectenwelt. Ein geheimer Schlüssel für Sammler aller Insecten-Ordnungen und Stände, behufs des Fangs, des Aufenthalts-Orts, der Wohnung, Tag- und Jahreszeit u. s. w., oder autoptische Darstellung des Insectenstaats in seinem zusammenhange zum Bestehen des Naturhaushaltes überhaupt und insbesondere in seinem Einflusse auf die phanerogamische und cryptogamische Pflanzenbevölkerung Europa's. Zum ersten Male nach fünfundzwanzigjährigen eigenen Erfahrungen zusammengestellt und herausgegeben.* Kempten: Tobias Dannheimer. xii + 532 pp.

Gistel J. N. F. X. 1857a: *Achthundert und zwanzig neue oder unbeschriebene wirbellose Thiere.* Straubing: Verlag der Schorner'schen Buchhandlung, 92 pp. [note: separate issue from Vacuna].

Gistel J. N. F. X. 1857b: Achthundert und zwanzig neue oder unbeschriebene wirbellose Thiere. Pp. 513-606. In: *Vacuna oder die Geheimnisse aus der organischen und leblosen Welt. Unterdruckte Originalien-Sammlung von grösstentheils noch lebenden und verstorbenen Gelehrten aus dem Gebiete sämmtlicher Naturwissenschaften, der Medizin, Litteraturgeschichte, des Forst- und Jagtwesens, der Oekonomie, Geschichte, Biographie, und der freien schönen Künste, herausgegeben von Professor Dr. Johannes Gistel. Zweiter Band.* Straubing: Schorner, 1031 pp.

Gmelin J. F. 1790: *Systema Naturae per regna tria naturae, secundum classes, ordines, genera, species, cum characteribus, differentiis, synonymis, locis. Editio decima tertia, aucta, reformata. Tom. I. Pars IV. Classis V. Insecta.* Lipsiae: Georg Emanuel Beer, 1517-2224 pp.

Goeze J. A. F. 1777: *Entomologische Bayträge zu des Ritter Linné zwölften Ausgabe des Natursystems. Erster Theil.* Leipzig: Weidemanns Erben und Reich, xvi +736 pp.

Gorham H. S. 1873: Descriptions of a new genus and two new species of Coleoptera from Japan. *The Entomologist's Monthly Magazine* **9**: 257-258.

Gorriz y Muñoz R. J. 1882: *Ensayo para la monografía de los coléopteros melóidos indigenas con aplicación a las ciencias médicas*. Zaragoza: Sanz y Navarro, 201 pp., 2 pls.

Gory H. L. 1841: [new taxa, no. 71-74]. *Magasin de Zoologie* (2) **3**: 9 pp., 4 pls.

Gozis M. des. 1881: [Quelques rectifications synonymiques touchant différents genres et espèces de coléoptères français (1re partie)]. *Bulletin de la Société Entomologique de France* **1881**: cxii-cxiii.

Gozis M. des. 1882: Notes et remarques pour le futur catalogue des coléoptères gallo-rhénane. *Revue d'Entomologie* **1**: 193-207.

Gozis M. des. 1886: *Recherche de l'espèce typique de quelques anciens genres. Rectifications synonymiques et notes diverses*. Montluçon: Herbin, 36 pp.

Gradl J. 1882: Aus der Fauna des Egerlandes. *Entomologische Nachrichten* **8**: 323-327.

Graells M. de la Paz. 1849: [new taxa]. *Revue de Zoologie Pure et Appliquée* (2) **1**: 620-622.

Graells M. de la Paz. 1851: Description de quelques coléoptères nouveaux de la faune centrale d'Espagne. *Annales de la Société Entomologique de France* (2) **9**: 5-27, 1 pl.

Graells M. de la Paz. 1858: Insectos nuevos de España. *Memorias de la Comision del Mapa Geológico de España, año de 1855. Parte zoológica.* Madrid: Imprenta Nacional, 111 + [1] pp., 7 pls.

Gravely F. H. 1915: XLII. Coleoptera, IX: Tenebrionidae. In: Zoological results of the Abor Expedition, 1911-12. XLII. Coleoptera IX: Tenebrionidae. *Records of the Indian Museum* **8** [1912-1922]: 519-536, pls XLIII, XLIV.

Gray G. 1832: *The class insecta arranged by the Baron Cuvier, with supplementary additions to each order by Edward Griffith, F.L.S., A.S., &c. and Edward Pidgeon, Esq. and notices of new genera and species by George Gray, Esq.* London: Whittaker, Treacher, and Co., 570 pp.

Gredler I. V. M. 1854: Die Käfer von Passier. *Zeitschrift des Ferdinandeums für Tirol und Vorarlberg* **3**: 1-20. [issued also as *Die Käfer von Passier. 1.Heft*. Innsbruck: Wagner'sche Buchdruckerei, 20 pp.].

Gredler I. V. M. 1866: *Die Käfer von Tirol nach ihrer horizontalen und vertikalen Verbreitung. II. Hälfte: Dascillidae - Schluss*. Bozen: Druk u. Verlag von G. Ferrari, vormals Eberle, 235-491 pp.

Gredler I. V. M. 1878: Zur Käfer-Fauna Central-Afrikas. *Verhandlungen der Kaiserlich-Königlichen Zoologisch-Botanischen Gesellschaft in Wien* **27**: 501-522.

Grenier A. 1863: *Catalogue des coléoptères de France et matériaux pour servir à la faune des coléoptères français, par MM. E. Allard, Dr. Ch. Aubé, Ch. Brisout de Barneville, A. Chevrolat, L. Fairmaire, Al. Fauvel, Dr. A. Grenier, Dr. Kraatz, J. Linder, L. Reiche et Félicien de Saulcy*. Paris: L. Toinon, iv + 3-79 + 135 pp.

Gressitt J. L. 1939: Some Oedemeridae from South China, Hainan I., Formosa and the Riukiu Islands (Coleoptera). *Lingnan Science Journal* **18**: 217-231.

Gressitt J. L. 1941: Rhipiphoridae from south China (Coleoptera). *Annals of the Entomological Society of America* **34**: 525-536.

Gridelli E. 1928: Appunti su alcuni Tenebrionidae dell'Africa settentrionale. *Bollettino della Società Entomologica Italiana* **60**: 15-18.

Gridelli E. 1929a: Appunti su alcuni Tenebrionidae dell'Africa settentrionale. III. - Revisione del genere Doderoella. *Bollettino della Società Entomologica Italiana* **61**: 1-10.

Gridelli E. 1929b: Appunti su alcuni Tenebrionidae dell'Africa settentrionale. IV. - Revisione del genere Tentyriina Reitt. *Bollettino della Società Entomologica Italiana* **61**: 49-63.

Gridelli E. 1930: Risultati zoologici della Missione inviata dalla R. Società Geografica Italiana per l'esplorazione dell' oasi di Giarabub (1926-1927). Coleotteri. *Annali del Museo Civico di Storia Naturale di Genova* **54**: 1-485 + [1], 1 map.

Gridelli E. 1931: Spedizione des Barone Raimondo Franchetti in Dancalia. Coleotteri – Tenebrionidae. *Annali del Museo Civico di Storia Naturale di Genova* **55**: 220-233.

Gridelli E. 1933: Spedizione scientifica all'Oasi di Cufra (Marzo-Guglio 1931). Coleotteri. *Annali del Museo Civico di Storia Naturale di Genova* **56**: 155-258, pl.V.

Gridelli E. 1934: Materiali zoologici raccolti dalla spedizione italiana al Karakoram (1929 - Anno VII). Coleoptera - Tenebrionidae. *Atti del Museo Civico di Storia Naturale Trieste* **12**: 37-68, pls. IX, X.

Gridelli E. 1938: Coleotteri dell'Africa orientale italiana. 5° Contributo. Specie a me note appartenenti ai generi Mesostena e Mesostenopa auct. (Tenebrionidae). *Bollettino della Società Entomologica Italiana* **70**: 103-121.

Gridelli E. 1939a: Coleotteri dell'Africa orientale italiana. 6e Contributo. *Bollettino della Società Entomologica Italiana* **71**: 75-79.

Gridelli E. 1939b: Coleotteri dell'Africa orientale italiana 8. contributo. Revisione delle species riferite dagli autori ai generi Helopinus, Drosochrus, Emyon. (Coleopt. Tenebrionidae). *Atti del Museo Civico di Storia Naturale Trieste* **14**: 187-286.

Gridelli E. 1939c: Coleotteri dell'Africa orientale italiana 10. contributo. Revisione delle species del genere Leichenum Blch. (Coleopt. Tenebrionidae). *Atti del Museo Civico di Storia Naturale Trieste* **14**: 207-242.

Gridelli E. 1939d: Coleotteri del Fezzan e dei Tassili d'Agger (Missione Scortecci 1936). *Atti della Società Italiana di Scienze Naturali* **78**: 385-456, pl. XXIII.

Gridelli E. 1940: Coleotteri dell'Africa orientale italiana 11° Contributo. Materiali per lo studio della fauna Eritrea raccolti nel 1901-03 dal Dott. Alfredo Andreini. *Memorie della Società Entomologica Italiana* **18** [1939]: 219-258.

Gridelli E. 1945: Coleotteri dell'Africa orientale (XIV. contributo). Appunti per una monografia delle specie etiopiche del genere Gonocephalum Sol. (Coleopt. Tenebr.). *Atti del Museo Civico di Storia Naturale Trieste* **16**: 1-28.

Gridelli E. 1948: Coleotteri dell'Africa tropicale XVII contributo. Ulteriori appunti per una monografia delle specie africaine del genere Gonocephalum Sol. (Coleopt. Tenebr.). *Atti del Museo Civico di Storia Naturale Trieste* **17**: 1-56.

Gridelli E. 1949: [new taxon]. In: Porta A.: *Fauna Coleopterorum italica. Supplementum II.* Sanremo: Soc. An. G. Gandolfi, 386 pp.

Gridelli E. 1950: Il problema della specie a diffusione transadritica, con particolare riguardo ai coleotteri. *Memorie di Biogeografia Adriatica* (Venezia) **1**: 1-233.

Gridelli E. 1952: Contribution à l'étude de peuplement de la Mauritanie. Coléoptères ténébrionides. *Bulletin de l'Institut Français d'Afrique Noire* **14**: 60-96.

Gridelli E. 1953: Catalogo regionato delle specie di coleotteri tenebrionidi dell'Arabia. *Atti del Museo Civico di Storia Naturale Trieste* **19**: 1-70, pl. 1.

Gridelli E. 1954: Catalogo regionato delle species della famiglia Tenebrionidae a me nota dell'Afghanistan (Coleoptera). *Atti del Museo Civico di Storia Naturale Trieste* **19**: 169-292.

Gridelli E. 1955: The 3rd Danish Expedition to Central Asia. Zoological Results 15. Tenebrionidae (Insecta) from Afghanistan. *Videnskabelige Meddelelser fra Dansk Naturhistorisk Forening i København* **117**: 23-74.

Gridelli E. 1957: Una nuova specie del genera Syachis F. Bat. ed osservazioni su alcune specie note (Coleoptera Tenebrionidae). *Atti del Museo Civico di Storia Naturale Trieste* **21**: 49-61.

Gridelli E. 1972: Materiali per lo studio dei tenebrionidi italiani. Seconda nota postuma. *Atti del Museo Civico di Storia Naturale Trieste* **27**: 187-256.

Grilat R. 1886: [Une nouvelle espèce de coléoptère]. *Bulletin de la Société Entomologique de France* **1886**: 187-188.

Grimm R. 1981: Die Fauna der Ägäis-Insel Santorin. Teil 2 Tenebrionidae (Coleoptera). *Stuttgarter Beiträge zur Naturkunde Serie A (Biologie)* **348**: 1-14.

Grimm R. 1985: Zur Kenntnis der Tenebrioniden aus Süditalien (Insecta: Coleoptera). *Stuttgarter Beiträge zur Naturkunde Serie A (Biologie)* **379**: 1-32.

Grimm R. 1987: Pachychila denticollis sp.n. aus Südspanien (Coleoptera, Tenebrionidae). *Nachrichtenblatt der Bayerischen Entomologen* **36**: 21-24.

Grimm R. 1991: Tenebrioniden von der Insel Zypern (Insecta: Coleoptera). *Biocosme Mésogéen* **8**: 15-49.

Grimm R. 2000: Die Gattung Eurycaulus Fairmaire, 1868, in Algerien und Tunesien. *Entomologische Zeitschrift* (Stuttgart) **110**: 89-91.

Grimm R. 2003: Die Arten der Gattung Palorinus Blair, mit Beschreibungen von vier neuen Arten (Coleoptera: Tenebrionidae). *Stuttgarter Beiträge zur Naturkunde Serie A (Biologie)* **646**: 1-9.

Grimm R. & Lillig M. 2000: Erodius graniventris in der westlichen Sahara: eine Artengruppe (Coleoptera, Tenebrionidae). *Bulletin de la Société Entomologique de France* **105**: 31-35.

Grimm R. & Schawaller W. 2000: The genus Colpotus Mulsant & Rey (Coleoptera: Tenebrionidae) in the Eastern Mediterranean region with descriptions of two new species. *Stuttgarter Beiträge zur Naturkunde*

Serie A (*Biologie*) **615**: 1-15.

Grouvelle A. 1903: Clavicornes de l'Inde septentrionale récoltés par M. Harmand. Nitidulidae.-Colydiidae.-Cucujidae.-Monotomidae.-Dryopidae. *Annales de la Société Entomologique de France* **72**: 108-124.

Grouvelle A. 1908a: Etude sur les Colbicus vrais de l'Europe, de l'Asie et de l'Australie. *Notes from the Leyden Museum* **30**: 113-122.

Grouvelle A. 1908b: Coléoptères de la région indienne. Rhysodidae, Trogositidae, Nitidulidae, Colydiidae, Cucujidae (1er mémoire). *Annales de la Société Entomologique de France* **77**: 315-495, pls 6-9.

Grouvelle A. 1911: Descriptions of five new Indian species of clavicorn Coleoptera. *Records of the Indian Museum* **6**: 315-317.

Grouvelle A. 1913: Zoological results of the Abor Expedition, 1912-1912. Coleoptera I. Rhysodidae, Nitidulidae, Colydiidae, Cucujidae, Passandridae, Discolomidae, Cryptophagidae, Mycetophagidae, Dryopidae. *Records of the Indian Museum* **8**: 99-117.

Grouvelle A. 1914: H. Sauters Formosa-Ausbeute. Rhysodidae, Nitidulidae, Ostomidae, Colydiidae, Passandridae, Cucujidae, Cryptophagidae, Diphyllidae, Lathridiidae, Mycetophagidae, Dermestidae. *Archiv für Naturgeschichte* A **79** 11 [1913]: 33-76.

Grouvelle A. 1914: Coleoptera: Cucujidae, Cryptophagidae [of the Seychelles]. *Transactions of the Linnean Society of London* (*Zoology*) (2) **17**: 141-156.

Grouvelle A. 1918: Coleoptera of the families Ostomidae, Monotomidae, Colydiidae und Notiophygidae from the Seychelles and Aldabra Islands. *Transactions of the Entomological Society of London* **1918**: 1-57, pls. 1, 2.

Guérin-Méneville F. E. 1825a: [new taxon]. In: Latreille P. A., Le Peletier de Saint-Fargeau A., Audinet-Serville A. J. G. de. & Guérin-Méneville F.: *Encyclopédie méthodique ou par ordre de matières; par une société de gens de lettres, de savants et d'artistes; Précédée d'un vocabulaire universel, servant de table pour tout l'ouvrage, ornée des portraits de MM. Diderot & d'Alembert, premiers Éditeurs de l'Encyclopédie. Histoire naturelle. Entomologie, ou histoire naturelle des crustacées, des arachnides et des insectes. Tome Dixième.* Paris: H. Agasse, [7] + 344 pp. [note: Pars 2, pp. 345-832, issued in 1828].

Guérin-Méneville F. E. 1825b: Helops. Pp. 111-112. In: Bory de Saint-Vincent J. B. G. M. (ed.): *Dictionnaire classique d'histoire naturelle, par Messieurs Audoin, Isid. Bordon, Ad. Brongniart, De Candolle, Daudebard de Férussac, A. Desmoulins, Drapiez, Edwards, Flourens, Geoffroy de Saint-Hilaire, A. de Jussieu, Kunth, G. de Lafosse, Lamouroux, Latreille, Lucas fils, Presle-Duplessis, C. Prévost, A Richard, Thiébaut de Berneaud, et Bory de Saint-Vincent. Ouvrage dirigé par ce dernier collaborateur, et dans lequel on a ajouté, pour le porter au niveau de la science, un grand nombre de mots qui n'avaient pu faire partie de la plupart des dictionnaires antérieurs. Tome huitième.* Paris: Ray et Gravier [et] Beaudoin Frères.

Guérin-Méneville F. E. 1831 [livraison 12]: In: *Iconographie du règne animal de G. Cuvier, ou représentation d'après la nature de l'une des espèces les plus remarquables et souvent non encore figurées, de chaque genre d'animaux. Avec un texte descriptif mis au courant de la science. Ouvrage pouvant servir d'atlas à tous les traités de zoologie. II. Planches des animaux invertébrés. Insectes.* [1829-1838]. Paris: J. B. Baillière, 110 pls.

Guérin-Méneville F. E. 1834a [livraison 35]: In: *Iconographie du règne animal de G. Cuvier, ou représentation d'après la nature de l'une des espèces les plus remarquables et souvent non encore figurées, de chaque genre d'animaux. Avec un texte descriptif mis au courant de la science. Ouvrage pouvant servir d'atlas à tous les traités de zoologie. II. Planches des animaux invertébrés. Insectes.* [1829-1838]. Paris: J. B. Baillière, 110 pls.

Guérin-Méneville F. E. 1834b: Matériaux pour une classification des mélasomes. (Extraits d'une monographie de cette famille). *Magasin de Zoologie* **1**: 1-39, pls. 101-118.

Guérin-Méneville F. E. 1835: [new taxon]. In: Guérin-Méneville F. E. & Percheron A. R.: *Genera des insectes ou exposition detaillée de tous les caractères propres à chacun des genres de cette classe d'animaux. Première livraison.* Paris: Mécquignon-Marvis, 10 pls + [unnumb. pp.]. (1) no. 2.

Guérin-Méneville F. E. 1837: *Dictionnaire pittoresque d'histoire naturelle et des phénomènes de la nature par une société de naturalistes sous la direction de Guérin. Tome 5.* Paris, 640 pp., 1 pl.

Guérin-Méneville F. E. 1838: Genre oedémère, Oedemera. *Revue Zoologique par la Société Cuvierienne* **1**: 39-40.

Guérin-Méneville F. E. 1842: Description abrégée de trois espèces nouvelles du genre Meloe. *Revue de Zoologique par la Société Cuvierienne* **5**: 338-339.

Guérin-Méneville F. E. 1843a: Description d'une nouvelle espèce de coléoptère du genre Myrmechixenus, découverte dans les serres aux ananas de M. Panckoucke; à sa campagne de Fleury, près Paris. *Annales de la Société Entomologique de France* (2) **1**: 69-71, pl. 2.

Guérin-Méneville F. E. 1843b: [new taxa]. *Bulletin de la Société Entomologique de France* **1859**: clxxxvi-cxciii.

Guérin-Méneville F. E. 1844: *Iconographie du Règne Animal de G. Cuvier, ou représentation d'après nature de l'une des espèces les plus remarquables, et souvent non encore figurées, de chaque genere d'animaux. Avec un texte descriptif mis au courant de la science. Ouvrage pouvant servir d'atlas à tous les traités de zoologie. III. Texte explicatif. Insectes.* Paris: J. B. Baillière, 576 pp.

Guérin-Méneville F. E. 1849: Insectes. Pp. 239-390, 12 pls. In: Lefebvre C. T. (ed.): *Voyage en Abyssinie exécuté pendant les années 1839-43 par une commission scientifique composée de Th. Lefebvre, A. Petit et R. Quartin-Dillon naturalistes, et Vignaud dessinateur. Histoire naturelle. Zoologie, par MM. O. des Murs, F. Prévost, Guicherot et Guérin de Méneville. Zoologie 6 (4).* Paris: Bertrand.

Guérin-Méneville F. E. 1857: Matériaux pour une monographie des coléoptères du group des eumorphides et plus spécialement du genre Eumorphus. Pp. 237-280. In: Thompson J. (ed.): *Archives Entomologiques ou recueil contenant des illustrations d'insectes nouveaux ou rares. Tome premier.* Paris: Société entomologique de France, 512 + (1) pp., XXI pls.

Guérin-Méneville F. E. 1858: Description de deux coléoptères du genre Sepidium, dont l'un est pentamère et paraît être le mâle, et l'autre est hétéromère. *Revue de Zoologie Pure et Appliquée* (2) **10**: 127-129, pl. IV.

Guérin-Méneville F. E. 1859: [new taxa]. *Bulletin de la Société Entomologique de France* **1859**: clxxxvi-cxciii.

Guérin-Méneville F. E. 1862: Description d'une nouvelle espèce de coléoptères du genre Melancrus et rectification relative à une note publiée dans le Bulletin entomologique de 1859. *Annales de la Société Entomologique de France* (4) **1** [1861]: 375-376.

Guillebeau F. 1887: Révision du genre Orchesia Latreille. *L'Échange, Revue Linnéenne* **3** (29-35): 1-8.

Guillebeau F. 1888: Révision du genre Orchesia Latreille. *L'Échange, Revue Linnéenne* **4** (37-41): 9-15.

Guillebeau F. 1893a: Synopsis des espèces connues du genre Mycterus d'Europe et du Bassin de la Méditerranée. *L'Échange, Revue Linnéenne* **9**: 89-90.

Guillebeau F. 1893b: Description d'une espèce inédite du genre Anaspis Geoffroy. *Mittheilungen der Schweizerischen Entomologischen Gesellschaft* **8**: 328.

Guillebeau F. 1896: Description de quelques espèces de coléoptères inédites (suite). *Bulletin de la Société Entomologique de France* **1896**: 239-245.

Günther K. 1937: Ueber einige von M. Beick am Tien-Shan und in Nordkansu gesammelte Käfer. *Koleopterologische Rundschau* **23**: 130-135.

Gusakov A. A. 2005: New taxa of false darkling beetles (Coleoptera: Melandryidae) from China. *Russian Entomological Journal* **14**: 117-118.

Gyllenhal L. 1806: [new taxa]. In Schoenherr C. J.: *Synonymia Insectorum, oder: Versuch einer Synonymie Aller bisher bekannten Insecten; nach Fabricii Systema Eleutheratorum geordnet. Mit Berichtigungen und Anmerkungen, wie auch Beschreibungen neuer Arten und illuminierten Kupfern. Erster Band. Eleutherata oder Käfer. Erster Theil. Lethrus - Scolytes.* Stockholm: A. Nordström, xxii + 294 pp., 3 pls.

Gyllenhal L. 1808: [new taxa]. In Schoenherr C. J.: *Synonymia Insectorum, oder: Versuch einer Synonymie aller bisher bekannten Insecten; nach Fabricii Systema Eleutheratorum &c. geordnet. Band 1. Eleutherata oder Käfer. Zweiter Theil. Spercheus - Cryptocephalus.* Stockholm: C. F. Marquard, ix + 1 + 424 pp., 4 pls.

Gyllenhal L. 1810: *Insecta Suecica descripta. Classis I, Coleoptera sive Eleuterata. Tom I. Pars II.* Scaris: F. J. Leverentz, xx + 660 pp.

Gyllenhal L. 1813: *Insecta Suecica descripta. Classis I. Coleoptera sive Eleutherata. Volume 1 Pars III.* Scaris: Leverentz, iv + 730 pp.

Gyllenhal L. 1817: [new taxa]. In: Schoenherr C. J.: *Synonymia Insectorum oder Versuch einer Synonymie aller bisher bekannten Insecten; nach Fabricii Systema Eleutheratorum etc. geordnet. Erster Band, Eleutherata oder Käfer, Dritter Theil, Hispa---Molorchus.* Upsala, E. Bruzelius, xi + 506 pp.

Gyllenhal L. 1827: *Insecta Svecica descripta. Classis I. Coleoptera sive Eleuterata. Tom I. Pars IV.* Lipsiae: Friedericum Fleischer, viii + [2] + 762 pp.

Haag-Rutenberg G. J. 1870a: [new taxa]. In: Heyden L. F. J. D. von. 1870: Zweiter Theil. Beschreibungen der neuen Arten. Pp. 58-176. In: Heyden L. F. J. D. von: Entomologische Reise nach dem südlichen Spanien, der Sierra Guadarrama und Sierra Morena, Portugal und den Cantabrischen Gebirgen. *Berliner Entomologische Zeitschrift* **14** *Beiheft*, 218 pp., 2 pls.

Haag-Rutenberg G. J. 1870b: Beiträge zur Familie der Tenebrioniden. *Coleopterologische Hefte* **6**: 84-93.

Haag-Rutenberg G. J. 1875a: *Beiträge zur näheren Kenntniss einiger Gruppen aus der Familie der Tenebrioniden*. Berlin: Berliner Entomologischer Verein, 56 pp.

Haag-Rutenberg G. J. 1875b: Beiträge zur Familie der Tenebrioniden. (5. Stück). *Coleopterologische Hefte* **14**: 67-92.

Haag-Rutenberg G. J. 1875c: Hidrosis, eine neue Adelostomiden-Gattung. *Deutsche Entomologische Zeitschrift* **19** (1-2): 120.

Haag-Rutenberg G. J. 1875d: Monographie der Eurychoriden (Adelostomides Lacord.). *Berliner Entomologische Zeitschrift* **19** (4): 359-428. [Issued also as separat publication in Berlin: Nicolai'sche Verlags-Buchhandlung, 70 pp.

Haag-Rutenberg G. J. 1876a: Eine neue Melasomen-Art aus Marocco. *Entomologische Monatsblätter* **1**: 10.

Haag-Rutenberg G. J. 1876b: Neue Pimelien. *Entomologische Monatsblätter* **1**: 75-76.

Haag-Rutenberg G. J. 1876c: Akis Heydeni n. sp. *Entomologische Monatsblätter* **1**: 76.

Haag-Rutenberg G. J. 1877a: Revision der Gattung Himatismus Er. *Deutsche Entomologische Zeitschrift* **21**: 273-283.

Haag-Rutenberg G. J. 1877b: Zwei neue Adesmiiden-Arten. *Deutsche Entomologische Zeitschrift* **21**: 283-284.

Haag-Rutenberg G. J. 1878: Beschreibung neuer Arten von Heteromeren als Nachtrag zur Monographie. *Mittheilungen des Münchener Entomologischen Vereins* **2**: 77-95.

Haag-Rutenberg G. J. 1879a: Fernere Nachträge zu den Heteromeren-Monographien der Moluriden, Eurychoriden und Adesmiiden. *Deutsche Entomologische Zeitschrift* **23**: 289-296.

Haag-Rutenberg G. J. 1879b: Leptodopsis, eine neue Heteromeren-Gattung. *Deutsche Entomologische Zeitschrift* **24**: 408-410.

Haag-Rutenberg G. J. 1880: Beiträge zur Kenntniss der Canthariden. *Deutsche Entomologische Zeitschrift* **24**: 17-90.

Haldeman S. S. 1848: Description of North American Coleoptera, chiefly in the cabinet of J. L. Leconte, with references to described species. *Journal of the Academy of Natural Sciences of Philadelphia* (2) **1**: 95-100.

Haldeman S. S. 1852: Appendix C, Insects. Pp. 366-378, 2 pls. In: Stansbury H.: *Exploration and survey of the Valley of the Great Salt Lake of Utah, including a reconnaissance of a new route through the Rocky Mountains*. Philadelphia: Lippincott, Grambo & Co., 487 pp.

Hamm A. H. 1909: Sitaris muralis, Forst., n. var. flava. *The Entomologist's Monthly Magazine* **45**: 277.

Hampe C. 1850: Beschreibung einiger neuen Käfer-Arten. *Entomologische Zeitung* (Stettin) **11**: 346-360.

Hampe C. 1852: Verzeichnis der vom M. Wagner im Kaukasus, in Trans-Kaukasien, Armenien u. West-Persien gesammelten Koleopteren, mit Beschreibung von 16 neuen Arten. Pp. 302-315. In: Wagner M. F.: *Reise nach Persien und dem Lande der Kurden. Band 2*. Leipzig: Arnoldi, 4 + 316 pp.

Hampe C. 1871: Beschreibungen einiger neuen Käfer. *Berliner Entomologische Zeitschrift* **14** [1870]: 331-336.

Hampe C. 1873: Zwei neue Anthicinen. *Verhandlungen der Kaiserlich-Königlichen Zoologisch-Botanischen Gesellschaft in Wien* **23**: 165-166, pl. 2.

Hanatsuka M., Masumoto K. & Kon M. 2005: A new Isomira (Coleoptera, Tenebrionidae, Alleculinae) from Ishigaki Island of the Ryukyus. *Elytra* **33**: 413-416.

Hanatsuka M., Masumoto K. & Kon M. 2006: Two new alleculine species (Coleoptera, Tenebrionidae, Alleculinae) from Japan. *Elytra* **34**: 179-183.

Harold E. von. 1875a: V. Beschreibung eines neuen Oedemeriden aus Japan. *Coleopterologische Hefte* **14**: 93.

Harold E. von. 1875b: Verzeichniss der von Herrn T. Lenz in Japan gesammelten Coleopteren. *Abhandlungen vom Naturwissenschaftlichen Vereine zu Bremen* **4**: 283-296.

Harold E. von. 1877: Beiträge zur Käferfauna von Japan. (Zweites Stück.) Japanische Käfer des Berliner Königl. Museums. *Deutsche Entomologische Zeitschrift* **21**: 337-367.

Harold E. von. 1878a: Beiträge zur Käferfauna von Japan. (Viertes Stück). *Deutsche Entomologische Zeitschrift* **22**: 65-88.

Harold E. von. 1878b: Bericht über eine Sendung Coleopteren aus Hiogo. *Abhandlungen vom Naturwissenschaftlichen Vereine zu Bremen* **5** [1886]: 115-135.

Harold E. von. 1878c: Zur Kenntniss der Gattung Ceropria (Coleoptera Heteromera). *Entomologische Zeitung* (Stettin) **39**: *345-355*.

Harold E. von. 1879: Bericht über die von den Herren A. von Homeyer und P. Pogge in Angola und im Luanda-Reiche gesammelten Coleopteren. *Coleopterologische Helfte* **16**: 1-224, 3 pls.

Harold E. von. 1880: Einige neue Coleopteren. *Mittheilungen des Münchener Entomologischen Vereins* **4**: 148-171.

Hatayama T. 1985: Mordellidae. Pp. 376-397. In: Kurosawa Y., Hisamatsu S. & Sasaji H.: *Colored Illustrations of the Coleoptera of Japan. Vol. 3.* Osaka: Hoikusha, x +500 pp., 72 pls. (in Japanese).

Hatch M. H. 1965: *The beetles of the Pacific North-West. Part. IV. Macrodactyles, Palpicornes, and Heteromera with the collaboration of David V. Miller, David V. McCorkle, Floyd Werner and Dennis W. Boddy.* Seattle: University of Washington Press, 268 pp., 28 pls.

Hatta K. 1988: The inopeplid beetles of Japan. *Kontyû* **56**: 299-302.

Hayashi M. 1960: Studies on Melandryidae from Japan 1. (Col.). *Entomological Review of Japan* **11**: 42-46.

Hayashi M. and Katô A. 1956: A new species of Melandryidae from Japan (Col.). *Entomological Review of Japan* **7** (2): 37-38.

Heberdey R. F. 1935: Anthicidae (Col.). Pp. 304-307. In: Visser P. C. & Visser-Hooft J. (eds.): *Wissenschaftliche Ergebnisse der niederländischen Expeditionen in den Karakorum und die angrenzenden Gebiete 1922, 1925 und 1929/30. Zoologie. Band I.* Leipzig: F. A. Brockhaus, xviii + 499 pp., 8 pls.

Heberdey R. F. 1936a: Revision der paläarktischen Arten der Gattung Notoxus Geoffr. I. Die Verwandten des Notoxus monoceros L. *Koleopterologische Rundschau* **22** (2/3): 125-160.

Heberdey R. F. 1936b: Revision der paläarktischen Arten der Gattung Notoxus Geoffr. I. Die Verwandten des Notoxus monoceros L. *Koleopterologische Rundschau* **22** (4): 161-180.

Heberdey R. F. 1938: Neue Anthiciden aus Ostasien. *Entomologisches Nachrichtenblatt* **12**: 161-164.

Heberdey R. F. 1942: Revision der paläarktischen Arten der Gattung Mecynotarsus Laf. (Coleopt., Anthicidae). *Mitteilungen der Münchener Entomologischen Gesellschaft* **32**: 445-485.

Heineken C. 1835: Description of a new species of Hemiptera, and of a species of Hegeter. *Zoological Journal* **5** [1832-1834]: 35-41.

Heinze E. 1943b: Langelandia (Normandella) zacynthia sp. nov., eine neue Langelandia aus Griechenland. (Colydiidae). *Entomologische Blätter* **39**: 127-129.

Hellén W. 1935: Koleopterologische Mitteilungen aus Finnland. XIII. *Notulae Entomologicae* **15**: 89-93.

Hellenius C. N. 1787: Försök, til beskrifning paa et nytt genus blandinsecterna, som kunde kallas Serropalpus. *Kongliga Vetenskaps Academiens Nya Handlingar* **7**: 310-319, pl. VIII.

Heller K. M. 1923: Die Coleopterenausbeute der Stötznerschen Sze-Tschwan-Expedition (1913-1915). *Entomologische Blätter* **19**: 61-79.

Hellwig J. C. L. 1792: Dritte Nachricht von neuen Gattungen im entomologischen System. *Neuestes Magazin für die Liebhaber der Entomologie* **1** (3): 385-408.

Hentz N. M. 1830: Remarks on the use of the maxillae in coleopterous insects, with an account of two species of the family Telephoridae, and of three of the family Mordellidae, which ought to be type of two distinct genera. *Transactions of the American Philosophical Society, Philadelphia* **2**: 458-463, pl. 15.

Herbst J. F. W. 1782: Nähere Bestimmung des Käfergeschlechts. Platt= oder Rindkäfer. *Archiv der Insectengeschichte herausgegeben von J. C. Fuessly* **2**: 1-8, 1 pl.

Herbst J. F. W. 1783: Kritisches Verzeichniss meiner Insektensammlung. Erste Klasse. Erste Abteilung. Scarabaeus-Lagria. *Archiv der Insectengeschichte, herausgegeben von J. C. Fuessly* **4**: 1-68, pls. 19-23.

Herbst J. F. W. 1784a: Kurze Einleitung zur Kenntniss der Insecten für Ungeübte und Anfänger. In: Borowski G. H.: *Gemeinzzige* [sic] *Naturgeschichte des Thierreichs, darin die merkwürdigsten und nüzlichsten Thiere in systematischer Ordnung beschrieben, und die Geschlechter in Abbildungen nach der Natur vorgestellt werden. Band 6.* Berlin und Straslund: Lange, 278 pp., 48 pls.

Herbst J. F. W. 1784b: Kritisches Verzeichniss meiner Insektensammlung. Fortsetzung. Curculio-Dytiscus. *Archiv der Insectengeschichte, herausgegeben von J. C. Fuessly* **5** (1): 73-128, pls. 24-28.

Herbst J. F. W. 1784c: Kritisches Verzeichniss meiner Insektensammlung. Fortsetzung. Gyrinus-Paederus. *Archiv der Insectenschichte, herausgegeben von J. C. Fuessly* **5** (2): 129-151, pls. 29-30.

Herbst J. F. W. 1786: Erste Mantisse zum Verzeichniss der ersten Klasse meiner Insektensammlung. *Archiv der Insectenschichte, herausgegeben von J. C. Fuessly* **7**: 153-182, pls. 43-48.

Herbst J. F. W. 1792: *Natursystem aller bekannten in- und ausländischen Insekten, als eine Fortsetzung der von Buffonschen Naturgeschichte. Der Käfer vierter Theil. Mit zwölf illuminierten Kupferstafeln.* Berlin: Joachim Pauli, viii + 197 pp., 12 pls.

Herbst J. F. W. 1793: *Natursystem aller bekannten in- und ausländischen Insekten, als eine Fortsetzung der von Büffonschen Naturgeschichte. Der Käfer fünfter Theil.* Berlin: Pauli, 392 pp., pls. 44-59 + J-N.

Herbst J. F. W. 1797: *Natursystem aller bekannten in- und ausländischen Insekten, als eine Fortsetzung der von Buffonschen Naturgeschichte. Der Käfer siebenter Theil.* Berlin: Geh. Commerzien-Raths Pauli, xi + 346 pp.

Herbst J. F. W. 1799: *Natursystem aller bekannten in- und ausländischen Insekten, als eine Fortsetzung der von Buffonschen Naturgeschichte. Der Käfer achter Theil.* Berlin: Geh. Commerzien-Raths Pauli, xiv + 420 pp., 121-137 pls.

Heyden L. F. J. D. von. 1863: Zwei neue Coleopteren-Gattungen aus dem Mittelmeergebiet. *Berliner Entomologische Zeitschrift* **7**: 126-130.

Heyden L. F. J. D. von. 1870: Entomologische Reise nach dem südlichen Spanien, der Sierra Guadarrama und Sierra Morena, Portugal und den Cantabrischen Gebirgen. *Berliner Entomologische Zeitschrift* **14** *Beiheft*, 218 pp., 2 pls.

Heyden L. F. J. D. von. 1875: Bericht über die von Herrn Prof. Dr. Freiherr von Frisch und Dr. J. J. Ren auf den Canarischen Inseln gesammelten Käfer. *Bericht über die Senckenbergische Naturforschende Gesellschaft in Frankfurt am Main* **1874-1875**: 135-145.

Heyden L. F. J. D. von. 1877: Die Käfer von Nassau und Frankfurt. *Jahrbücher des Nassauischen Vereins für Naturkunde* **29-30**: 55-412.

Heyden L. F. J. D. von. 1879: Diagnoses Coleopterorum aliquot novorum in Japonia a Dom. Prof. J. J. Rein, Doct. phil., collectorum. *Bericht über die Senckenbergische Naturforschende Gesellschaft in Frankfurt am Main* **1878-1879**: 87-88.

Heyden L. F. J. D. von. 1881: Beiträge zur Käferfauna von Turkestan. Zonabris-Arten von Margelan. *Deutsche Entomologische Zeitschrift* **25**: 324-330.

Heyden L. F. J. D. von. 1882: Beitrag zur Coleopteren-Fauna der Sierra de Cordoba. *Deutsche Entomologische Zeitschrift* **26**: 43-48.

Heyden L. F. J. D. von. 1883a: Neue Zonabris (Mylabris Ol.) aus Turkestan. *Deutsche Entomologische Zeitschrift* **27**: 65-68.

Heyden L. F. J. D. von. 1883b: Verzeichniss der von Herrn Dr. Med. W. Kobelt in Nordafrika und Spanien gesammelten Coleopteren. *Bericht über die Senckenbergische Naturforschende Gesellschaft* **1882-1883**: 217-237.

Heyden L. F. J. D. von. 1883c: [new taxon]. In: Heyden L. F. J. D. von & Weise J.: Neue südost-europäische und klein-asiatische Käfer. *Deutsche Entomologische Zeitschrift* **27**: 310-314.

Heyden L. F. J. D. von. 1883d: Käfer aus Ochs (in Turkestan). *Deutsche Entomologische Zeitschrift* **27**: 337-353.

Heyden L. F. J. D. von. 1883e: Ueber Käfer aus Tekke-Turkmenien. *Deutsche Entomologische Zeitschrift* **27**: 354-360.

Heyden L. F. J. D. von. 1884a: Beitrag zur Coleopterenfauna der Insel Askold und andere Theile des Amurgebietes. *Deutsche Entomologische Zeitschrift* **28**: 273-300.

Heyden L. F. J. D. von. 1884b: Coleopterologische Ausbeute einer Excursion nach der Sierra d'Espuñas. *Deutsche Entomologische Zeitschrift* **28**: 355-361.

Heyden L. F. J. D. von. 1885a: [new taxon]. In: Heyden L. F. J. D. von. & Kraatz G.: Beiträge zur turkestanischen Coleopteren-Fauna. *Deutsche Entomologische Zeitschrift* **29**: 273-298.

Heyden L. F. J. D. von. 1885b: Weitere Beiträge zur Coleopteren-Fauna des Amur-Gebietes. *Deutsche Entomologische Zeitschrift* **29**: 298-307, pl. iv.

Heyden L. F. J. D. von. 1886a: [new taxa]. In: Heyden L. F. J. D. von. & Kraatz G.: Beiträge zur Coleopteren-fauna von Turkestan, namentlich des Alai-Gebirges, unter Beihilfe der Herren Dr. Candèze, Ganglbauer, Dr. Stierlin, und Weise. *Deutsche Entomologische Zeitschrift* **30**: 177-194.

Heyden L. F. J. D. von. 1886b: Die Coleopteren-Fauna des Suyush-Flusses (Amur). *Deutsche Entomologische Zeitschrift* **30**: 269-277.

Heyden L. F. J. D. von. 1887a: Verzeichniss der von Herrn Otto Herz auf der chinesischen Halbinsel Korea gesammelten Coleopteren. *Horae Societatis Entomologicae Rossicae* **21**: 243-273.

Heyden L. F. J. D. von. 1887b: Vierter Beitrag zur Kenntniss der Coleopteren-Fauna der Amurländer. *Deutsche Entomologische Zeitschrift* **31**: 297-304.

Heyden L. F. J. D. von. 1887c: XI. Beitrag zur Coleopteren-Fauna von Turkestan. *Deutsche Entomologische Zeitschrift* **31**: 305-336.

Heyden L. F. J. D. von. 1887d: Bericht über von den Herren Prof. Dr. Freiherrn von Fritsch und Prof. Dr. Rein in Marocco und dem Atlas-Gebirge gesammelten Käfer. *Deutsche Entomologische Zeitschrift* **31**: 433-448.

Heyden L. F. J. D. von. 1888: Ueber die von L. Conradt aus den Alai-Gebirge mitgebrachten Coleopteren. - Beitrag XII. Mit Beiträgen von J. Faust. *Deutsche Entomologische Zeitschrift* **32**: 39-44.

Heyden L. F. J. D. von. 1889a: Descriptiones Coleopterorum novorum Regionis Palaearcticae. *Deutsche Entomologische Zeitschrift* **1889**: 325-330.

Heyden L. F. J. D. von. 1889b: Insecta, a Cl. G. A. Potanin in China et in Mongolia novissime lecta. XII. Scarabaeidae, Cantharidae, Cleridae, Lagriidae, Melandryidae, Pedilidae, Anthicidae. *Horae Societatis Entomologicae Rossicae* **23**: 654-677.

Heyden L. F. J. D. von. 1890a: Aufzählung der Käfer-Arten aus Tunis und Tripolis aus Loosen von M. Quedenfeldt. *Deutsche Entomologische Zeitschrift* **1890**: 65-75.

Heyden L. F. J. D. von. 1890b: Die Varietäten der Diaperis boleti L. *Deutsche Entomologische Zeitschrift* **1890**: 78.

Heyden L. F. J. D. von. 1890c: XIII. Beitrag zur Coleopteren-Fauna von Turkestan. *Deutsche Entomologische Zeitschrift* **1890**: 353-367.

Heyden L. F. J. D. von. 1890d: Beiträge zur Coleopteren-Fauna von Amasia und Samsoun in Nord-Kleinasien. *Deutsche Entomologische Zeitschrift* **1890**: 369-374.

Heyden L. F. J. D. von. 1892a: XIV. Beitrag zur Coleopteren-Fauna von Turkestan. Unter Mitwirkung der Herren J. Faust, A. Kuwert und J. Weise. *Deutsche Entomologische Zeitschrift* **1892**: 99-104.

Heyden L. F. J. D. von. 1892b: XV. Beitrag zur Coleopteren-Fauna von Turkestan, Turkmenien und Süd-West-Sibirien. Aus Sendungen des Herrn Dr. Staudinger. *Deutsche Entomologische Zeitschrift* **1892**: 105-110.

Heyden L. F. J. D. von. 1894a: [new taxa]. In: Hauser F.: Beitrag zur Coleopteren-Fauna von Transcaspien und Turkestan. Zusammengestellt unter Mitwirkung der Herren Dr. Eppelsheim, Escherich, Faust, Dr. von Heyden, Kuwert, Reitter und Weise [u. Pastor Schmidt]. *Deutsche Entomologische Zeitschrift* **1894**: 17-74.

Heyden L. F. J. D. von. 1894b: Beiträge zur Coleopteren-Fauna von Turkestan. *Horae Societatis Entomologicae Rossicae* **28**: 262-268.

Heyden L. F. J. D. von. 1897: [new taxon]. In: Escherich K.: Zoologische Ergebnisse einer von Dr. K. Escherich und Dr. L. Kathariner nach Central-Kleinasien unternommenen Reise. IV. Theil. Coleopteren. Unter Mitwirkung von E. Brenske, L. Ganglbauer, Dr. L. von Heyden, E. Wasmann und J. Weise. *Entomologische Zeitung* (Stettin): **58**: 3-69.

Heyden L. F. J. D. von. 1908: Über die Coleopteren-Gattungen Parablops. *Deutsche Entomologische Zeitschrift* **1908**: 132.

Heyden L. F. J. D. von. 1909: Die kanarische Phaleria ornata Woll. und ihre Varietäten. *Deutsche Entomologische Zeitschrift* **1909**: 486.

Heyden L. F. J. D. von. & Kraatz G.: Käfer um Samarkand, gesammelt von Haberhauer. *Deutsche Entomologische Zeitschrift* **26**: 297-338.

Heyden L. F. J. D. von., Reitter E. & Weise J. 1891: *Catalogus Coleopterorum Europae, Caucasi et Armeniae rossicae*. Berlin: Friedländer & Sohn; Mödling: E. Reitter; Caen: Revue d'Entomologie, 812 pp.

Hille J. C. van. 1961: Coleoptera: Anthicidae, pp. 217-258. In: *South African Animal Life, results of the Lund University Expedition in 1950-1951. Vol. VIII*. Uppsala: Almqvist & Wiksells Boktryckeri Ab., 557 pp.

Hille J. C. van. 1975: Ten African species of Notoxus (Col.Anthicidae). *Koleopterologische Rundschau* **52**: 3-13.

Hille J. C. van. 1977: Species of the genus Formicomus Laf. from North-East Africa and adjacent parts of the Arabian peninsula (Coleoptera: Anthicidae). *Entomologica Scandinavica* **9**: 197-204.

Hille J. C. van. 1984: Anthiciden (Coleoptera) aus Afrika. *Folia Entomologica Hungarica* (N. S.) **45**: 103-104.

Hinton H. E. 1948: A synopsis of the genus Tribolium Macleay, with some remarks on the evolution of its species-groups (Coleoptera, Tenebrionidae). *Bulletin of Entomological Research, London* **39**: 13-55.

Holdhaus K. 1919: Ergebnisse der wissenschaflichen Expedition nach Mesopotanien, 1910. Koleopteren aus Mesopotamien. *Annalen des Naturhistorischen Museums in Wien* **33**: 39-58.

Hölzel E. 1958: Die mitteleuropäischen Arten der Gattung Isomira Muls. (Col. Alleculidae) mit Beschreibung der Untergattung Heteromira subgen. nov. und Art moroi spec. nov. aus den Kärntner Karawanken. *Nachrichtenblatt der Bayerischen Entomologen* **7**: 17-25.

Hope F. W. 1831: Synopsis of the new species of Nepaul insects in the collection of Major General Hardwicke. Pp. 21-31. In: Gray J. E. (ed.): *The Zoological Miscellany. Vol. 1*. London: Treuttel, Wurtz and Co., 40 pp., 4 pls.

Hope F. W. 1833: On the characters of several new genera and species of coleopterous insects. *Proceedings of the Zoological Society of London* **1**: 61-64.

Hope F. W. 1834a: Descriptions of some hitherto uncharacterized exotic Coleoptera, chiefly from New Holland. *Transactions of the Entomological Society of London* **1**: 11-20, pls. 1, 2.

Hope F. W. 1834b: Characters and descriptions of several new genera and species of coleopterous insects. *Transactions of the Zoological Society of London* **1**: 91-112, pls. 14-16.

Hope F. W. 1840: *The Coleopterist's Manual, part the third, containing various families, genera, and species, of beetles, recorded by Linneus and Fabricius. Also, descriptions of newly discovered and unpublished insects.* London: J. G. Bridgewater and Bowdery & Keby, [6] + 191 pp., 3 pls.

Hope F. W. 1842: Description of the coleopterous insects sent to England by Dr. Cantor from Chusan and Canton, with observation on the entomology of China. *Journal of Proceedings of the Entomological Society of London* **1842**: 49-52 [note: pp. 49 and 50 are lacking in original issue].

Hope F. W. 1843: Description of the coleopterous insects sent to England by Dr. Cantor from Chusan and Canton, with observation on the entomology of China. *The Annals and Magazine of Natural History* **(6) 11**: 62-66.

Hope F. W. 1845: Descriptions of the coleopterous insects sent to England by Dr. Cantor from Chusan and Canton, with observations on the Entomology of China. *Journal of the Proceedings of the Entomological Society of London*, pp. 49-52. In: *The Transactions of the Entomological Society of London* **4**.

Horák J. 1978: Stenalia bilyi sp.n. aus Tadschikistan. *Acta Entomologica Bohemoslovaca* **75**: 400-403.

Horák J. 1979: Mordella hofferi sp. n. von der Balkanhalbinsel. *Acta Entomologica Bohemoslovaca* **76**: 104-107.

Horák J. 1980: Drei neue Arten der Gattung Mordellistena aus Tadschikistan. *Acta Entomologica Bohemoslovaca* **77**: 280-286.

Horák J. 1981: Zwei neue Arten der Familie Mordellidae aus der paläarktischen Region. *Acta Entomologica Bohemoslovaca* **78**: 335-339.

Horák J. 1982: Mordellistenochroa gen.n. und Beschreibung vier neuer ostpaläarktischer Arten (Col. Mordellidae). *Acta Entomologica Bohemoslovaca* **79**: 46-55.

Horák J. 1983: Revision der Mordellistena-Arten aus der pentas-Gruppe. *Entomologische Abhandlungen* **47**: 1-13.

Horák J. 1985: Ergebnisse der tschechoslowakisch-iranischen entomologischen Expeditionen nach Iran 1970, 1973 und 1977. Coleoptera: Mordellidae 1 (Stenaliini, Mordellini). *Entomologische Abhandlungen* **49**: 1-25.

Horák J. 1986: Beitrag zur Systematik der Gattung Hoshihananomia KONO. *Reichenbachia* **23**: 161-164.

Horák J. 1989: Brodskyella gen.n. des Tribus Stenaliini aus der orientalischen Region (Col. Mordellidae). *Entomologische Abhandlungen* **53**: 37-44.

Horák J. 1990: Typenrevision einiger wenig bekannter Arten aus der Gattung Mordellistena Costa. *Entomologische Abhandlungen* **53**: 125-142.

Horák J. 1991: Zwei neue Arten der Gattung Mordellistena Costa aus Marokko. *Reichenbachia* **28**: 135-140.

Horák J. 1994: Description of Glipa (Macroglipa) thoracica sp.n. and revision of some species of the genus Glipa Leconte from the Oriental region. *Folia Heyrovskyana* **2**: 101-110.

Horák J. 1995a: The genus Stenalia from Oriental region. *Acta Societatis Zoologicae Bohemicae* **59**: 89-99.

Horák J. 1995b: Revision of species of the tribus Mordellistenini from Oriental region. Part 1. *Acta Societatis Zoologicae Bohemicae* **59**: 165-176.

Horák J. 1995c: Revision of some Oriental Mordellini with description of two new species. *Entomological Problems* **26**: 165-184.

Horák J. 1996: Revision of some little known species of genus Mordellistena with description of two new species. *Klapalekiana* **32**: 171-184.

Horn G. H. 1868: New species of Coleoptera from the Pacific district of the United States. *Transactions of the American Entomological Society* **2**: 129-140.

Horn G. H. 1870: Revision of the Tenebrionidae of America, north of Mexico. *Transactions of the Americal Philosophical Society, New Series* **14**: 253-404.

Horn G. H. 1875: Synonymical notes and descriptions of new species of North American Coleoptera. *Transactions of the American Entomological Society* **5**: 125-156.

Horn G. H. 1888: Miscellaneous coleopterous studies. *Transactions of the American Entomological Society* **15**: 26-48.

Houlbert C. 1934: [new taxa]. In: Barthe E.: Tableaux analytiques des coléoptères de la faune franco-rhénane (France, Hollande, Belgique, région rhénane, Valais). Famille LXX Melandryidae. *Miscellanea Entomologica* **35** (12): 1-34.

Hozawa S. 1914: Note on a new termitophilous Coleoptera found in Formosa (Ziaelas formosanus). *Annotationes Zoologicae Japonensis* **8**: 483-488, pl. VI.

Hubenthal W. 1905: Phryganophilus ruficollis F. var. nov. rosti aus Japan. *Deutsche Entomologische Zeitschrift* **1905**: 57-58.

Hubenthal W. 1911: Beschreibungen neuer europäischer Coleopteren und Bemerkungen zu bekannten Arten. *Entomologische Blätter* **7**: 185-193.

Hübner J. 1789: Beiträge zur Naturgeschichte der Insecten. *Der Naturforscher* **24**: 36-59, 1 pl.

Hummel A.-D. 1827: *Essais Entomologiques. No. VI. Insectes de 1826. Novae species. Catalogus Insectorum quaeitinere Petropoli in Chersonesum Tauricum et Iberiam anno 1825 collegit Benedictus Jäger.* St. Pétersburg: Chancellerie privée du Ministère de l'Intérieur, 48 + (1) pp.

Iablokoff-Khnzorian S. M. 1956a: Chetyre novykh vida zhestkokrylykh iz Armyanskoy SSR (Coleoptera, Insecta) [Four new species of Coleoptera from the Armenian SSR]. *Doklady Akademii Nauk Armyanskoy SSR* **22** (3): 135-139.

Iablokoff-Khnzorian S. M. 1956b: Novyy vid chernotelok iz Armyanskoy SSR. *Doklady Akademii Nauk Armyanskoy SSR* **23**: 41-43.

Iablokoff-Khnzorian S. M. 1957a: Novye vidy zhestkokrylykh iz Armyanskoy SSR i Nakh. ASSR. *Zoologicheskiy Sborkik Akademii Nauk Armyanskoy SSR* **3**: 153-183.

Iablokoff-Khnzorian S. M. 1957b: Novyy veeronosets iz Armyanskoi SSR (Coleoptera, Rhipiphoridae). *Doklady Akademii Nauk Armyanskoi SSR* **24**: 231-233.

Iablokoff-Khnzorian S. M. 1958: Dva novykh vida zhestkokrylykh – Heteromera iz Armenskoy SSR (Insecta, Coleoptera) [Deux nouveaux coléoptères hétéromères de l'Arménie de SSR (Insecta, Coleoptera)]. *Zoologiskiy Zhournal Akademii Nauk SSSR* **37**: 1896-1898.

Iablokoff-Khnzorian S. M. 1959a: Novye vidy zhestkokrylykh iz Talysha. *Izvestiya Akademii Nauk Azerbaydzhanskoy SSR* **3**: 57-62.

Iablokoff-Khnzorian S. M. 1959b: Novyy zhuk pylceed iz Armyanskoy SSR (Coleoptera, Alleculidae) Heliotaurus emmae Khnzorian, sp. nov. *Doklady Akademii Nauk Armyanskoy SSR* **28**: 191-192.

Iablokoff-Khnzorian S. M. 1959c: Novyy pylceed iz doliny Araksa (Coleoptera, Alleculidae) Mycetocharina riabovi Khnzorian sp. nov. *Doklady Akademii Nauk Armyanskoy SSR* **29**: 141-143.

Iablokoff-Khnzorian S. M. 1964: Bemerkungen über einige Reitter'sche Typen aus dem Ungarischen Naturwissenschaftlichen Museum (Coleoptera). *Folia Entomologica Hungarica* (N. S.) **17**: 293-315.

Iablokoff-Khnzorian S.M. 1970: Novye vidy zhestkokrylykh iz Armenii i drugikh chastey SSSR (Insecta, Coleoptera). *Sbornik Zoologicheskogo Instituta Akademii Nauk Armyanskoy SSR* **15**: 49-80.

Iablokoff-Khnzorian S. M. 1973: Dva novych vida veeronostsev iz SSSR (Coleoptera, Rhipiphoridae). *Doklady Akademii Nauk Armyanskoi SSR* **57**: 307-312.

Iablokoff-Khnzorian S. M. 1974: Dva novykh vida zhestkokrylykh - uzkotelok iz Tadzhikistana. *Doklady Akademii Nauk Armyanskoy SSR* **59**: 188-192.

Iablokoff-Khnzorian S. M. 1976: Les Isomira Muls. du Caucase (Coleoptera, Alleculidae). *Acta Zoologica Cracoviensia* **21**: 315-329.

Iablokoff-Khnzorian S. M. 1978: Novyy vid zhestokrylykh-uzkotelok iz Primorya. *Doklady Akademii Nauk Armyanskoy SSR, Entomologia* **46**: 252-255.

Iablokoff-Khnzorian S. M. 1983: *Fauna Armenia. Vol. 5. Coleoptera, Meloidae & Alleculidae.* Erevan: Akademia Nauk Armyanskoy SSR, 155 pp.

Iablokoff-Khnzorian S. M. 1984: Novyy vid zhukov-chernotelok iz Armyanskoy SSR. *Doklady Akademii Nauk Armyanskoy SSR* **78**: 43-44.

Iablokoff-Khnzorian S. M. 1985: Les Pythidae paléarctiques (Coleoptera). *Deutsche Entomologische Zeitschrift* (N. F.) **32**: 193-229.

Iablokoff-Khnzorian S. M. 1986: Novyy rod i vid zhukov-veeronostsev iz Tadzhikistana (Coleoptera, Rhipiphoridae). *Doklady Akademii Nauk Armyanskoi SSR* **82**: 89-92.

Iga M. & Nakane T. 1954: Oedemeridae. Pp. 154-155. In: Nakane T. (ed.): *Coloured illustrations of the insects of Japan. Coleoptera.* Osaka: Hoikusha Co. Ltd., 196 pp., 64 pls. (in Japanese, with English descriptions).

Illiger J. C. W. 1794: Beschreibung einiger neuen Kaferarten aus der Sammlung des Herrn Professore Hellwig in Braunschweig. *Neuestes Magazin für die Liebhaber der Entomologie* (5) **1**: 593-620.

Illiger J. C. W. 1798: *Verzeichniss der Käfer Preussens. Entworfen von Johann Gottlieb Kugelann Apotheker in Osterode. Mit einer Vorrede des Professors und Pagenhofmeisters Hellwig in Braunschweig, und dem angehängten Versuche einer natürlichen Ordnungs- und Gattungsfolge der Insekten.* Halle: Johann Jacob Gebauer, xlii + [1] + 510 + [1] pp.

Illiger J. C. W. 1800: Vierzig neue Insekten aus der Hellwigschen Sammlung in Braunschweig. *Archiv für Zoologie und Zootomie* (Wiedemann) **1** (2): 103-150.

Illiger J. C. W. 1801: Nachtrag und Berichtigungen zum Verzeichnisse der Käfer-Preussens. *Magazin für Insektenkunde* **1**: 1-94.

Illiger J. C. W. 1802a: Namen der Insekten-Gattungen, ihr Genitiv, ihr grammatisches Geschlecht, ihr Silbenmass, ihre Herleitung; zugleich mit den Deutschen Benennungen. *Magazin für Insektenkunde* **1**: 125-155.

Illiger J. C. W. 1802b: Aufzählung der Käfergattungen nach der Zahl der Fussglieder. *Magazin für Insektenkunde* **1**: 285-305.

Illiger J. C. W. 1804: Bemerkungen zum zweiten Theile von Fabricius Systema Eleutheratorum. *Magazin für Insektenkunde* **3**: 160-180.

Illiger J. C. W. 1805: [new taxon]. In: Panzer G. W. F.: *Kritische Revision der Insektenfauna Deutschlands nach dem System bearbeitet. Zum I-XCVI. Heft. I. Baendchen.* Nürnberg: Felseckerschen Buchhandlung, 12 + 144 pp.

Illiger J. C. W. 1807: Vorschlag zur Aufnahme im Fabricischen Systeme fehlende Käfergattungen. *Magazin für Insektenkunde* **6**: 318-349.

Illiger J. C. W. 1807: Vermischte Nachrichten und Bemerkungen. Literatur. *Magazin für Insektenkunde* **6**: 350-370.

Illiger J. C. W. 1810: [new taxon]. In: Gyllenhal L.: *Insecta Suecica descripta. Classis I, Coleoptera sive Eleuterata. Tom I. Pars II.* Scaris: F. J. Leverentz, xx + 660 pp.

Imasaka S. 2005: Review of the genus Arthromacra (Coleoptera, Tenebrionidae, Lagriinae) in Japan. *Miscellaneous Reports of the Hiwa Museum for Natural History* 44: 61-163, pls 1-28.

Ishikawa H. & Sakai M. 2001: A new species of the genus *Lederina* (Coleoptera, Melandryidae) from Japan. *Japanese Journal of Systematic Entomology* **7**: 333-336.

Ishikawa H. & Sakai M. 2004: The genus Microscapha (Coleoptera, Melandryidae) of Japan. *Japanese Journal of Systematic Entomology* **10**: 127-132.

Israelson G. 1971: On the Aderidae of the Canary Islands (Coleoptera). *Eos, Revista Española de Entomología* **46** [1970]: 109-122 (in Japanese and English).

Israelson G. 1980: Taxonomical and nomenclatoral notes on some Canarian Coleoptera. *Vieraea* **9** [1979]: 183-209.

Israelson G. 1984: Coleoptera from the Azores. *Boletim do Museu Municipal do Funchal* **36**: 142-161.

Israelson G. 1985: A new Macaronesian genus of Cisidae (Coleoptera). *Entomologische Blätter* **81**: 80-84.

Iwan D. 1997: Revision of the Asian genera of the tribe Platynotini (Coleoptera, Tenebrionidae). *Annales Zoologici* **47**: 243-272.

Iwan D. 2001: Two new species of Pseudoblaps Guérin, 1834 (Coleoptera: Tenebrionidae: Platynotini) from Asia. *Annales Zoologici* **51**: 147-152.

Iwan D. 2003: A review of the genus Pseudoblaps Guérin, 1834 (Coleoptera: Tenebrionidae: Platynotini). *Annales Zoologici* **53**: 489-507.

Iwan D. & Löbl I. 2007: Nomenclatural notes on tenebrionid beetles of the Palaearctic region (Insecta: Coleoptera). *Annales Zoologici* **57**: 733-739.

Jacquelin du Val C. 1858: Pp. 129-232. In: *Manuel Entomologique. Genera des coléoptères d'Europe comprenant leur classification en familles naturelles, la description de tous les generes, des tableaux synoptiques destinés à faciliter l'étude, le catalogue de toutes les espèces de nombreux dessins au trait de charactères et plus de treize cents types représentant un ou plusieurs insectes de chaque genre dessinés et peints d'apres nature avec le plus grand soin par M. Jules Migneaux. Tome deuxième.* Paris: A. Deyrolle [1857-1859, 285 + [3] +122 + [2] pp., 67 pls].

Jacquelin du Val C. 1860: Pp. 137-272. In: *Manuel Entomologique. Genera des coléoptères d'Europe comprenant leur classification en familles naturelles, la description de tous les genres, des tableaux synoptiques destinés à faciliter l'étude, le Catalogue de toutes les espèces de nombreux dessins au trait de charactères et plus de treize cents types représentant un ou plusieurs insectes de chaque genre dessinés et peints d'après nature avec le plus grand soin par M. Jules Migneaux. Tome troisième.* Paris: A. Deyrolle Deyrolle [1859-1863, 464 + 200 pp., 100 pls].

Jacquelin du Val C. 1861: Pp. 273-352. In: *Manuel Entomologique. Genera des coléoptères d'Europe comprenant leur classification en familles naturelles, la description de tous les genres, des tableaux synoptiques destinés à faciliter l'étude, le Catalogue de toutes les espèces de nombreux dessins au trait de charactères et plus de treize cents types représentant un ou plusieurs insectes de chaque genre dessinés et peints d'après nature avec le plus grand soin par M. Jules Migneaux. Tome troisième.* Paris: A. Deyrolle Deyrolle [1859-1863, 464 + 200 pp., 100 pls].

Jacquelin du Val C. 1863: Pp. 353-464. In: *Manuel Entomologique. Genera des coléoptères d'Europe comprenant leur classification en familles naturelles, la description de tous les genres, des tableaux synoptiques destinés à faciliter l'étude, le Catalogue de toutes les espèces de nombreux dessins au trait de charactères et plus de treize cents types représentant un ou plusieurs insectes de chaque genre dessinés et peints d'après nature avec le plus grand soin par M. Jules Migneaux. Tome troisième.* Paris: A. Deyrolle [1859-1863, 464 pp., 100 pls].

Jakobson G. 1896: Tria Coleoptera nova e Rossia europaea. *Horae Societatis Entomologicae Rossicae* **29** [1894-1895]: 520-524.

Jakowlew B. E. 1897: Espèces nouvelles du genre Meloë (subg. Proscarabaeus Steph.) de la Sibérie orientale. *Horae Societatis Entomologicae Rossicae* **31**: 248-252.

Jedlička A. 1965: Neue Carabiden aus Nepal (Coleoptera). *Khumbu Himal* **1**: 98-107.

Johnson C. 1989: Tenebrionidae (Coleoptera) collected in the Eastern Province of Saudi Arabia. *Fauna of Saudi Arabia* **10**: 123-133.

Joy N. H. 1912: A note on Anaspis ruficollis, Brit. Colls. *The Entomologist's Monthly Magazine* **48**: 115.

Kamiya H. 1961: Discovery of the family Hemipeplidae in Japan (Coleoptera: Heteromera). *Mushi* **35**: 75-77.

Kamiya H. 1963: On the systematic position of the genus Usechus Motschulsky, with a description of a new species from Japan (Coleoptera). *Mushi* **37**: 19-26.

Kangas E. 1986: Eine neue Mordellistena-Art. *Annales Entomologici Fennici* **52**: 139-141.

Kangas E. 1988: Mordellistena dalmatica Ermisch sensu Kangas (1976) (Coleoptera, Mordellidae), eine neue unbeschriebene Art. *Annales Entomologici Fennici* **54**: 89-92.

Karsch F. A. F. 1881: Die Käfer der Rohlfs'schen Afrikanischen Expedition 1878-79. *Berliner Entomologische Zeitschrift* **25**: 41-50, pl. 2.

Kaszab Z. 1938a: Species nova generis Platyscelis Latr. (Col., Tenebrionidae) Pl. székessyi spec. nov. *Folia Entomologica Hungarica* **4**: 1-3.

Kaszab Z. 1938b: I. Gruppe des Subgenus Oodescelis Motsch. (Col. Tenebr.). *Stettiner Entomologische Zeitung* **99**: 49-56.

Kaszab Z. 1938c: Die Arten der Gattung Cnemeplatia Costa und Psilachnopus Reitter (Col. Tenebr. Opatrinae). *Entomologisk Tidskrift* **59**: 77-83.

Kaszab Z. 1938d: Zwei neue Platyscelis-Arten aus Turkestan (Col. Tenebr.). *Koleopterologische Rundschau* **24**: 91-94.

Kaszab Z. 1938e: Neue Tenebrioniden-Arten aus Indien. *Entomologische Blätter* **34**: 226-231.

Kaszab Z. 1939a: Neue indomalayische Tenebrioniden. *Arbeiten über Morphologische und Taxonomische Entomologie aus Berlin Dahlem* **6**: 95-111.

Kaszab Z. 1939b: Tenebrioniden aus Neu-Guinea. *Nova Guinea (N. S.)* **3**: 185-267.

Kaszab Z. 1940a: Beiträge zur Kenntnis der Tenebrioniden-Fauna Kleinasiens. *Folia Entomologica Hungarica* **5**: 2-7.

Kaszab Z. 1940b: Revision der Tenebrioniden-Tribus Platyscelini (Col. Teneb.). *Mitteilungen der Münchener Entomologischen Gesellschaft* **30**: 119-235.

Kaszab Z. 1940c: Neue Heterotarsinen (Coleopt. Tenebr.) aus der Sammlung des Ungarischen Nationalmuseums. *Annales Musei Nationalis Hungarici* **33**: 153-160.

Kaszab Z. 1940d: Revision der Tenebrioniden-Tribus Platyscelini (Col. Teneb.). *Mitteilungen der Münchener Entomologischen Gesellschaft* **30**: 896-1003.

Kaszab Z. 1941a: Die indomalayischen Misolampinen (Coleopt., Tenebr.). *Annales Musei Nationalis Hungarici* **34**: 1-44, pl. 1.

Kaszab Z. 1941b: Die indischen Arten der Gattung Pachypterus Luc. *Folia Entomologica Hungarica* **6**: 16-20.

Kaszab Z. 1941c: Tenebrioniden aus Formosa (Col.). *Stettiner Entomologische Zeitung* **102**: 51-72.

Kaszab Z. 1941d: Neue orientalische Tenebrioniden (Coleoptera). *Arbeiten über Morphologische und Taxonomische Entomologie aus Berlin Dahlem* **8**: 118-127.

Kaszab Z. 1941e: Zwei neue Pedinus-Arten (Coleopt., Tenebr.) von der Balkanhalbinsel. *Annales Musei Nationalis Hungarici* **34**: 161-164.

Kaszab Z. 1941f: Zoologische Ergebnisse der ersten (VI.-X.1936.) und zweiten (V.-VIII.1937) Forschungsreisen Vasvári's in Kleinasien. IV. Meloidae. *Mathematischer und Naturwissenschaftlicher Anzeiger der Ungarischen Akademie der Wissenschaften* **60**: 674-681.

Kaszab Z. 1942a: Beiträge zur Kenntnis der orientalischen Opatrinen (Col. Teneb.). *Mitteilungen der Münchener Entomologischen Gesellschaft* **32**: 1-43.

Kaszab Z. 1942b: Die Meloiden Ungarns (Coleop.). *Fragmenta Faunistica Hungarica* **5**: 33-45.

Kaszab Z. 1942c: Über die von J. Klapperich in China gesammelten Leiochroinen (Col., Tenebrionidae). *Entomologische Blätter* **38**: 214-217.

Kaszab Z. 1943: Über die Gattungen Tearchus Kr. und Tonkinius Fairm. (Col., Tenebr.). *Folia Entomologica Hungarica* **8**: 71-74.

Kaszab Z. 1944a: [new taxon]. In: Español F.: Nuevos datos para el conocimiento de los Tenebriónidos (Col.) del Sáhara Español. *Eos, Revista Española de Entomología* **20**: 7-30.

Kaszab Z. 1944b: Über die Arten der Gattung Tetraphyllus Lap. & Brill. (Coleopt.). *Annales Musei Nationalis Hungarici* **37**: 25-71.

Kaszab Z. 1946: *Monographie der Leiochrinen*. Budapest: Ungarisches Naturwissenschaftliches Museum, 221 pp., 1 pl.

Kaszab Z. 1948: Die Arten der Gattung Cabalia Muls. (Col., Meloidae). *Folia Entomologica Hungarica* (S. N.) **3**: 12-14.

Kaszab Z. 1951a: Eine neue Subspezies von Trigonoscelis fasciculitarsis Reitt. aus Turkmenien (Coleopt., Tenebr.). *Folia Entomologica Hungarica* (N. S.) **4**: 1-4, 1 pl.

Kaszab Z. 1951b: Neue Revision der Gattung Alosimus Muls. (Col., Meloidae). *Annales Historico-Naturales Musei Nationalis Hungarici* **1**: 138-150.

Kaszab Z. 1951c: A new species of Gonocephalum (Col., Tenebrionidae) from the Philippine Islands. *The Annals and Magazine of Natural History* (12) **4**: 182-185.

Kaszab Z. 1951d: Revision der Cerocominen. *Acta Biologica Academiae Scientiarum Hungaricae* **2**: 255-274.

Kaszab Z. 1951e: Über die Arten der Gattung Oenas. *Acta Biologica Academiae Scientiarum Hungaricae* **2**: 275-279.

Kaszab Z. 1952a: Neue Epicautinen (Col. Meloidae) aus der Orientalischen Region. *Entomologische Arbeiten aus dem Museum G. Frey* **3**: 79-89.

Kaszab Z. 1952b: Über Lydus susicus Esc. und die neuen Arten der Gattung Lyttolydulus Reit. (Col. Meloidae). *Entomologische Arbeiten aus dem Museum G. Frey* **3**: 89-95.

Kaszab Z. 1952c: Eine neue Lydus-Art aus Kleinasien, nebst einer Bestimmungstabelle der bisher bekannten Arten der Gattung Lydus sensu stricto (Col., Meloidae). *Annales Historico-Naturales Musei Nationalis Hungarici* (S. N.) **2**: 95-99.

Kaszab Z. 1952d: Die indomalaischen und ostasiatischen Arten der Gattung Gonocephalum Solier (Coleoptera Tenebrionidae). *Entomologische Arbeiten aus dem Museum G. Frey* **3**: 416-688.

Kaszab Z. 1952e: Die paläarktischen und orientalischen Arten der Meloiden-Gattung Epicauta Redtb. *Acta Biologica Academiae Scientiarum Hungaricae* **3**: 573-599.

Kaszab Z. 1953a: The 3rd Danish expedition to central Asia. Zoological results 11. Meloiden (Insecta) aus Afghanistan. *Videnskabernas Meddelelser fra Dansk Naturhistorisk Forening i København* **115**: 305-311.

Kaszab Z. 1953b: Einige neue Cylindrothorax-Arten (Col., Meloidae). *The Annals and Magazine of Natural History* (12) **6**: 857-861.

Kaszab Z. 1954: Über die von Herrn J. Klapperich in der chinesischen Provinz Fukien gesammelten Tenebrioniden (Coleoptera). *Annales Historico-Naturales Musei Nationalis Hungarici* (S. N.) **5**: 248-264.

Kaszab Z. 1955a: Revision der Meloiden-Gattung Eletica Lac. (Coleoptera-Heteromera). *Annales du Musée Royal du Congo Belge Tervuren, Sciences Zoologiques* **41**: 1-121, 11 pls.

Kaszab Z. 1955b: Die Arten der Meloiden-Gattung Cylindrothorax Escher. (Coleoptera). *Annales Historico-Naturales Musei Nationalis Hungarici* (S. N.) **6**: 225-258.

Kaszab Z. 1955c: Tenebrioniden der Fiji-Inseln. *Proceedings of the Hawaiian Entomological Society* **15**: 657-671.

Kaszab Z. 1956a: Neue Tenebrioniden (Coleoptera) aus der papuanischern und indomalayischen Region. *Annales Historico-Naturales Musei Nationalis Hungarici* (S. N.) **7**: 93-108.

Kaszab Z. 1956b: Neue Meloiden-, Alleculiden-, Serropalpiden-, Anthiciden- und Oedemeriden-Formen aus Ungarn und den angrenzenden Gebieten (Coleoptera). *Folia Entomologica Hungarica* (N. S.) **9**: 142-172.

Kaszab Z. 1956c: Neue Heteromera (Coleoptera) aus Ungarn. *Acta Zoologica Academiae Scientiarum Hungaricae* **2**: 333-338.

Kaszab Z. 1956d: Zwei neue Prosodes-Arten aus der Untergattung Oligoprosodes Reitt. (Coleoptera, Tenebrionidae). *The Annals and Magazine of Natural History* (12) **9**: 418-422.

Kaszab Z. 1956e: Neue Meloiden aus der Sammlung des British Museum (Coleoptera). *The Annals and Magazine of Natural History* (12) **9**: 631-640.

Kaszab Z. 1957a: Einige neue Heteromeren aus Asien (Coleoptera). *Opuscula Zoologica* (Budapest) **2**: 47-53.

Kaszab Z. 1957b: Neue Meloiden aus Iran 1954 (Coleopt.) (Ergebnisse der Entomologischen Reisen Willi Richter, Stuttgart, in Iran 1954 und 1956 – Nr. 6). *Jahreshefte des Vereins für Vaterländische Naturkunde in Württemberg* **112**: 50-59.

Kaszab Z. 1957c: Weitere neue Tenebrioniden (Coleoptera) aus Fukien. *Bonner Zoologische Beiträge* **8**: 56-63.

Kaszab Z. 1957d: Neue Tenebrioniden aus Iran 1954 (Coleopt.) (Ergebnisse der Entomologischen Reisen Willi Richter, Stuttgart, im Iran 1954 und 1956 – Nr. 7). *Jahreshefte des Vereins für Vaterländische Naturkunde in Württemberg* **112**: 60-65.

Kaszab Z. 1957e: New Meloidae (Coleoptera) from Israel. *Bulletin of the Research Council of Israel, Section Biology and Geology* **6 b**: 229-232.

Kaszab Z. 1957f: Zehn neue Tenebrioniden aus Asien (Coleoptera). *Annales Historico-Naturales Musei Nationalis Hungarici* (S. N.) **8**: 289-299.

Kaszab Z. 1958a: Die Arten der Meloiden-Gattung Micromerus Muls. und Rey (Coleoptera). *Acta Entomologica Musei Nationalis Pragae* **32** (488): 5-18.

Kaszab Z. 1958b: Neue Meloiden-Namen (Coleoptera). *Annales Historico-Naturalis Musei Nationalis Hungarici* **50**: 189-191.

Kaszab Z. 1958c: Die Meloiden Afghanistans (Coleoptera). *Acta Zoologica Academiae Scientiarum Hungaricae* **3**: 245-312.

Kaszab Z. 1959a: Beiträge zur Kenntnis der Tenebrioniden-Fauna von Iran (Coleoptera). *Stuttgarter Beiträge zur Naturkunde* **19**: 1-3.

Kaszab Z. 1959b: Neue Meloiden aus Iran 1954, 1956 (Coleoptera). *Stuttgarter Beiträge zur Naturkunde* **20**: 1-5.

Kaszab Z. 1959c: Drei neue Blaps- und eine neue Laena-Art aus der Türkei und den angrenzenden Gebieten (Coleoptera Tenebrionidae). *Kungliga Fysiografiska Sällskapets i Lund Förhandlingar* **29**: 51-56.

Kaszab Z. 1959d: Wissensschaftliche Ergebnisse der zoologischen Expedition des National-Museums in Prag nach der Türkei. 24. Coleoptera Tenebrionidae. *Acta Entomologica Musei Nationalis Pragae* **33**: 69-82.

Kaszab Z. 1959e: The 3rd Danish Expedition to Central Asia. Zoological Results 24. Ergänzungen zur Tenebrioniden (Insecta). Aufarbeitung von Dr. E. Gridelli der 3. Dänischen Zentralasiatischen Expedition. *Videnskabelige Meddelelser fra Dansk Naturhistorisk Forening i København* **120** [1958]: 237-255.

Kaszab Z. 1959f: Die Tenebrioniden Afghanistans, auf Grund der Ergebnisse der Sammelreise des Herrn J. Klapperich in den Jahren 1952/53 (Col.). *Entomologische Arbeiten aus dem Museum G. Frey* **10**: 321-404, 10 pls.

Kaszab Z. 1959g: Neue Leptodes-Arten aus Asien, nebst einer Revision der Leptodini (Coleoptera, Tenebrionidae). *Acta Zoologica Academiae Scientiarum Hungaricae* **4**: 349-368.

Kaszab Z. 1959h: Eine neue Meloiden-Art und einige neue Formen der Koleopteren-Sammlung des Naturhistorischen Museums in Wien. *Annalen des Naturhistorischen Museums in Wien* **63**: 436-441.

Kaszab Z. 1960a: Die Tenebrioniden Afghanistans, auf Grund der Ergebnisse der Sammelreise des Herrn J. Klapperich in den Jahren 1952/53 (Col.). 1. Fortsetzung und Schluss. *Entomologische Arbeiten aus dem Museum G. Frey* **11**: 1-179.

Kaszab Z. 1960b: Die Arten und systematische Stellung der Meloiden-Gattung Calydus Reitter (Coleoptera). *Acta Zoologica Academiae Scientiarum Hungaricae* **6**: 125-134.

Kaszab Z. 1960c: Einige neue Meloiden (Coleoptera) aus der Sammlung des British Museum. *The Annals and Magazine of Natural History* (13) **3**: 153-159.

Kaszab Z. 1960d: Wissenschaftliche Ergebnisse der chinesisch-sowjetischen zoologischen Expedition nach SW. China. Meloidae (Coleoptera). *Annales Historico-Naturalis Musei Nationalis Hungarici* **52**: 255-263.

Kaszab Z. 1960e: Neue orientalische Misolampinen (Coleoptera, Tenebrionidae). *Annales Historico-Naturales Musei Nationalis Hungarici* **52**: 265-294.

Kaszab Z. 1960f: Mission zoologique de l'I.R.S.A.C. en Afrique orientale (P. Basilewsky et N. Leleup, 1957). XLV. Coleoptera Meloidae. *Annales du Musée Royal du Congo Belge Tervuren, Sciences Zoologiques* **88**: 269-286.

Kaszab Z. 1960g: Einige neue Meloiden (Coleoptera) des Museum G. Frey. *Entomologische Arbeiten aus dem Museum G. Frey* **11**: 283-290.

Kaszab Z. 1961a: Wissenschaftliche Ergebnisse der von Dr. F. Schmid in Pakistan und Iran gesammelte Tenebrioniden (Coleoptera). *Miscelánea Zoológica* **1**: 115-121.

Kaszab Z. 1961b: Revision der Tenebrioniden-Gattung Derispia Lewis (Coleoptera). *Acta Zoologica Academiae Scientiarum Hungarica* **7**: 139-184.

Kaszab Z. 1961c: Zwei neue Tenebrioniden (Coleoptera) aus Arabien und Kleinasien. *Abhandlungen und Berichte aus dem Staatlichen Museum für Tierkunde in Dresden* **26**: 169-175.

Kaszab Z. 1961d: Neue Tenebrioniden (Coleoptera) aus der Zoologischen Staatssammlung in Münnchen. *Mitteilungen der Münchener Entomologischen Gesellschaft* **51**: 213-230.

Kaszab Z. 1961e: Ergebnisse der Deutschen Afghanistan-Expedition 1956 der Landessamlungen für Naturkunde Karlsruhe. Tenebrionidae (Coleoptera). *Beiträge zur Naturkundlichen Forschung in Südwestdeutschland* **19**: 235-238.

Kaszab Z. 1961f: Ergebnisse der österreichischen Karakorum-Expedition 1958 Coleoptera – Tenebrionidae. *Annales Historico-Naturales Musei Nationalis Hungarici* **53**: 341-355.

Kaszab Z. 1961g: Neue Meloiden (Coleoptera) aus dem Museum A. Koenig. *Bonner Zoologische Beiträge* **12**: 343-353.

Kaszab Z. 1961h: Beiträge zur Kenntnis der Tenebrioniden-Tribus Leiochrini (Coleoptera). *Annales Historico-Naturales Musei Nationalis Hungarici* **53**: 357-380.

Kaszab Z. 1961i: Neue Arten der Gattung Leiochrodes Westwood (Coleoptera: Tenebrionidae). *Acta Zoologica Academiae Scientiarum Hungarica* **7**: 433-466.

Kaszab Z. 1962a: Beiträge zur Kenntnis der chinesischen Tenebrioniden-Fauna (Coleoptera). *Acta Zoologica Academiae Scientiarum Hungaricae* **8**: 75-86.

Kaszab Z. 1962b: Über das System der asiatischen Lytta-Arten, nebst Beschreibung drei neuer Arten (Coleoptera, Meloidae). *Annales Historico-Naturales Musei Nationalis Hungarici* **54**: 289-298.

Kaszab Z. 1962c: Beiträge zur Kenntnis einiger asiatischen Tenebrioniden-Gattungen und Arten (Coleoptera). *Annales Historico-Naturales Musei Nationalis Hungarici* **54**: 299-317.

Kaszab Z. 1963a: Ergebnisse der österreichischen Iran-Expedition 1949/50. X. Fünf neue Tenebrioniden aus Iran. *Anzeiger der Österreichischen Akademie der Wissenschaften Mathematisch-Naturwissenschaftliche Klasse* **100** (5): 69-77.

Kaszab Z. 1963b: Die paläarktischen und orientalischen Arten der Gattung Mesomorphus Seidl. (Coleoptera, Tenebrionidae). *Acta Zoologica Academiae Scientiarum Hungaricae* **9**: 333-354.

Kaszab Z. 1963c: Angaben zur Kenntnis der Tenebrioniden des Tschadsee-Gebietes, nebst einer Revision der afrikanischen Mesomorphus-Arten (Coleoptera). *Revue de Zoologie et de Botanique Africaines* **68**: 341-385.

Kaszab Z. 1964a: Über die Tenebrioniden einiger japanischen Inseln (II). *Entomological Review of Japan* **17**: 1-10.

Kaszab Z. 1964b: Tenebrionidae der mongolisch-deutschen biologischen Expedition 1962 (4. Coleoptera, Tenebrionidae). *Entomologische Abhandlungen* **32**: 1-26.

Kaszab Z. 1964c: Über die Tenebrioniden einiger japanischen Inseln (I). *Entomological Review of Japan* **16**: 39-49.

Kaszab Z. 1964d: Ergebnisse der mongolisch-deutschen biologischen Expedition 1962. Nr. 3. Meloidae (Coleoptera). *Folia Entomologica Hungarica* (N. S.) **17**: 317-324.

Kaszab Z. 1964e: Ergebnisse der zoologischen Forschungen von Dr. Z. Kaszab in der Mongolei. 15. Meloidae (Coleoptera). *Folia Entomologica Hungarica* (N. S.) **17**: 345-363.

Kaszab Z. 1964f: The Zoological Results of the Gy. Topál's collectings in South Argentina. *Annales Historico-Naturales Musei Nationalis Hungarici* **56**: 353-387.

Kaszab Z. 1964g: Beiträge zur Kenntnis der Tenebrioniden-Fauna des mittleren Teiles der Mongolischen Volksrepublik (Coleoptera). *Acta Zoologica Academiae Scientiarum Hungaricae* **10**: 363-404.

Kaszab Z. 1965a: Wissenschaftliche Ergebnisse der von Dr. F. Schmid in Indien gesammelten Tenebrioniden (Coleoptera). *Miscelánea Zoológica* **2**: 107-130.

Kaszab Z. 1965b: Meloiden und Tenebrioniden aus der Mongolei 1964 von H. Muche gesammelt (Coleoptera). *Reichenbachia* **5**: 131-134.

Kaszab Z. 1965c: Neue Tenebrioniden (Coleoptera) aus China. *Annales Historico-Naturales Musei Nationalis Hungarici* **57**: 279-285.

Kaszab Z. 1965d: Zoologische Ergebnisse der Forschungen von Dr. T. Pócs in der Volksrepublik Vietnam. Tenebrionidae (Coleoptera). *Annales Historico-Naturales Musei Nationalis Hungarici* **57**: 287-296.

Kaszab Z. 1965e: Angaben zur Kenntnis der Tenebrioniden-Fauna der Mongolischen Volksrepublik (Coleoptera). *Acta Zoologica Academiae Scientiarum Hungaricae* **11**: 295-346.

Kaszab Z. 1965f: Ergebnisse der zoologischen Forschungen von Dr. Z. Kaszab in der Mongolei. 38. Meloidae (Coleoptera). *Folia Entomologica Hungarica* (N. S.) **18**: 551-561.

Kaszab Z. 1966a: 74. Meloidae der III. Expedition. Ergebnisse der zoologischen Forschungen von Dr. Z. Kaszab in der Mongolei (Coleoptera). *Reichenbachia* **7**: 259-264.

Kaszab Z. 1966b: Revision der Tenebrioniden-Gattung Microdera Eschsch. (Coleoptera). *Acta Zoologica Academiae Scientarum Hungaricae* **12**: 279-305.

Kaszab Z. 1967a: Coleoptera: Tenebrionidae der mongolisch-deutschen biologischen Expedition 1964. *Mitteilungen aus dem Zoologischen Museum in Berlin* **43**: 3-33.

Kaszab Z. 1967b: Die Tenebrioniden der Westmongolei (Coleoptera). *Acta Zoologica Academiae Scientarum Hungaricae* **13**: 279-351.

Kaszab Z. 1968a: Tenebrionidae und Meloidae (Coleoptera) aus Nord-Korea, gesammelt von M. Mroczkowski und A. Riedel im Jahre 1965. *Annales Zoologici* **26**: 7-14.

Kaszab Z. 1968b: Beiträge zur Kenntnis der Fauna Afghanistans (Sammelergebnisse von O. Jakeš 1963-64, D. Povolný & Fr. Tenora 1966, J. Šimek 1965-66, D. Povolný, J. Geiser, Z. Šebek & Fr. Tenora 1967) Tenebrionidae, Col. *Časopis Moravského Musea, Vědy přírodní* **53**, *Supplementum*, 7-124 pp., 4 pls.

Kaszab Z. 1968c: Ergebnisse der zoologischen Forschungen von Dr. Z. Kaszab in der Mongolei. 168. Tenebrionidae (Coleoptera). *Acta Zoologica Academiae Scientiarum Hungaricae* **14**: 339-397.

Kaszab Z. 1968d: Ergebnisse zoologischer Sammelreisen in der Türkey. Coleoptera: Meloidae. *Annalen des Naturhistorischen Museums in Wien* **72**: 443-450.

Kaszab Z. 1968e: Ergebnisse zoologischer Sammelreisen in der Türkei. Coleoptera: Tenebrionidae. *Annalen des Naturhistorischen Museums in Wien* **72**: 451-463.

Kaszab Z. 1968f: Contribution à la faune de l'Iran, 8. Coléoptères Meloidae. *Annales de la Société Entomologique de France* (N. S.) **4**: 749-776.

Kaszab Z. 1969a: Three new East Mediterranean meloids (Coleoptera, Meloidae). *Israel Journal of Entomology* **4**: 279-284.

Kaszab Z. 1969b: Ergebnisse der zologischen Forschungen von Dr. Z. Kaszab in der Mongolei. 168. Tenebrionidae (Coleoptera) der VI. Expedition. *Acta Zoologica Academiae Scientarum Hungaricae* **15**: 299-332.

Kaszab Z. 1969c: 177. Meloidae der V. Expedition. Ergebnisse der zoologischen Forschungen von Dr. Z. Kaszab in der Mongolei (Coleoptera). *Faunistische Abhandlungen, Staatalisches Museum für Tierkunde in Dresden* **2**: 317-322.

Kaszab Z. 1969d: Contribution à la faune de l'Iran 15. Coléoptères Tenebrionidae, genre Lobodera. *Annales de la Société Entomologique de France* (N. S.) **5**: 451-456.

Kaszab Z. 1970a: Beiträge zur Kenntnis der Fauna Afghanistans (Sammelergebniss von O. Jakeš 1963-64, D. Povolný & Fr. Tenora 1966, J. Šimek 1965-66, D, Povolný, J. Geiser, Z. Šebek & Fr. Tenora 1967). Tenebrionidae, Col. *Časopis Moravského Musea, Vědy přírodní* **54**, *Supplementum*: 5-182 pp., 23 pls.

Kaszab Z. 1970b: Fünf neue Tenebrioniden aus Asien (Coleoptera). *Entomologische Arbeiten aus dem Museum G. Frey* **21**: 112-122.

Kaszab Z. 1970c: Beiträge zur Kenntnis der Tenebrioniden-Fauna von Nepal (Coleoptera). *Khumbu Himal* **3**: 422-434.

Kaszab Z. 1972: Missione 1965 del Prof. Giuseppe Scortecci nello Yemen (Arabia meridionale). *Atti della Società Italiana di Scienze Naturali e del Museo Civico di Storia Naturale di Milano* **113** (4): 366-384, pl. 17.

Kaszab Z. 1973a: Tenebrionidae (Coleoptera) aus Nepal. *Acta Zoologica Academiae Scientiarum Hungaricae* **19** (1972): 23-74, pls I-V.

Kaszab Z. 1973b: Neue Beiträge zur Kenntnis der Tenebrioniden-Fauna van Afghanistan (Coleoptera). *Entomologia Scandinavica* **4**: 35-58.

Kaszab Z. 1973c: [new taxon]. In: Medvedev G. S. & Kaszab Z. 1973: Ergebnisse der Mongolisch-sowietischen biologischen Expeditionen in der Mongolischen Volksrepublik Tenebrionidae, excl. Tentyriini (Coleoptera). *Folia Entomologica Hungarica* (N. S.) **26**: 79-111.

Kaszab Z. 1973d: Beiträge zur Kenntnis der Fauna Afghanistans (Sammelergebnisse von O. Jakeš 1963-64, D. Povolný 1965, D. Povolný & Fr. Tenora 1966, J. Šimek 1965-66, D. Povolný, J. Gaisler, Z. Šebenek & Fr. Tenora 1967). Meloidae. Col. *Časopis Moravského Musea, Vědy přírodní* **56-57**: 267-308.

Kaszab Z. 1974: Neue Angaben zur Kenntniss der Tenebrioniden (Coleoptera) von Afghanistan. *Annales Historico-Naturales Musei Nationalis Hungarici* **66**: 187-203.

Kaszab Z. 1975a: Tenebrioniden aus Indien (Coleoptera). *Acta Zoologica Academiae Scientarum Hungaricae* **21**: 1-38, 7 pls.

Kaszab Z. 1975b: Tenebrioniden (Coleoptera) aus Nepal, II. *Annales Historico-Naturales Musei Nationalis Hungarici* **67**: 119-126.

Kaszab Z. 1975c: Die Arten der Gattung Scaphidema Redt. (Coleoptera, Tenebrionidae). *Annales Historico-Naturales Musei Nationalis Hungarici* **67**: 127-130.

Kaszab Z. 1975d: Revision der asiatischen Platynotinen (Coleoptera: Tenebrionidae). *Acta Zoologica Academiae Scientiarum Hungaricae* **21**: 277-367, 7 pls.

Kaszab Z. 1975e: Ergebnisse der Bhutan-Expedition 1972 des Naturhistorischen Museums in Basel. Coleoptera: Fam. Tenebrionidae. *Entomologia Basiliensia* **1**: 313-333.

Kaszab Z. 1976a: Revision der Arten der Gattung Heterotarsus Latreille, 1829 (Coleoptera: Tenebrionidae). *Acta Zoologica Academiae Scientiarum Hungaricae* **22**: 33-63.

Kaszab Z. 1976b: Zwei neue Arten der Gattung Epitrichia Mäklin, 1872 (Coleoptera, Tenebrionidae). *Annales Historico-Naturales Musei Nationalis Hungarici* **68**: 99-104.

Kaszab Z. 1976c: Ergebnisse der mongolisch-sowjetischen biologischen Expedition in der Mongolischen Volksrepublik seit 1967. Tenebrionidae: Tentyriini (Coleoptera), II. *Annales Historico-Naturales Musei Nationalis Hungarici* **68**: 105-115.

Kaszab Z. 1976d: [new taxon]. In: Kaszab Z. & Knor I.: Zur Tenebrionidenfauna der Tuvinischen ASSR in Südsibirien (Coleoptera). *Acta Zoologica Academiae Scientiarum Hungaricae* **22**: 293-297.

Kaszab Z. 1977a: Neue Tenebrioniden (Coleoptera) von den Galapagos und Antillen, sowie aus Südamerika und Ostasien. *Annales Historico-Naturales Musei Nationalis Hungarici* **69**: 117-131.

Kaszab Z. 1977b: Tenebrionidae der Nepal-Expeditionen von Dr. J. Martens (1969-1974) (Insecta: Coleoptera). *Senckenbergiana Biologica* **57** [1976]: 241-283.

Kaszab Z. 1977c: Ergänzungen zur Tenebrionidenfauna der Westmongolei (Coleoptera). Ergebnisse der Mongolisch-Deutschen biologischen Expeditionen seit 1962, Nr. 76. *Mitteilungen aus dem Zoologischen Museum in Berlin* **53**: 277-279.

Kaszab Z. 1978a: Vier neue Tenebrioniden aus Nordbengal (Coleoptera). *Folia Entomologica Hungarica* (N. S.) **31** (2): 173-176.

Kaszab Z. 1978b: [new taxa]. In: Kaszab Z., Schawaller W. & Skopin N. G.: Systematik und Ökologie einiger Tenebrionidae aus Kashmir und Ladakh. *Senckenbergiana Biologica* **59**: 215-234.

Kaszab Z. 1978c: Neue Meloiden aus Asien, Afrika und Südamerika (Coleoptera). *Acta Zoologica Academiae Scientiarum Hungaricae* **24**: 331-342.

Kaszab Z. 1979a: Revision der asiatischen Arthrodosis-Artigen Erodiini (Coleoptera: Tenebrionidae). *Acta Zoologica Academiae Scientiarum Hungaricae* **25**: 69-119, 1 pl.

Kaszab Z. 1979b: Die Arten der Gattung Sivacrypticus Kaszab, 1964 (Coleoptera, Tenebrionidae). *Annales Historico-Naturales Musei Nationalis Hungarici* **71**: 185-204.

Kaszab Z. 1979c: Insects of Saudi Arabia. Coleoptera: Fam. Tenebrionidae. *Fauna of Saudi Arabia* **1**: 257-288.

Kaszab Z. 1980a: Neue Tenebrioniden (Coleoptera) aus Sri Lanka I. *Acta Zoologica Academiae Scientiarum Hungaricae* **26**: 123-196.

Kaszab Z. 1980b: Angaben zur Kenntnis der Tenebrioniden Nordvietnams (Coleoptera). *Annales Historico-Naturales Musei Nationalis Hungarici* **72**: 169-221.

Kaszab Z. 1981a: Faunistische und taxonomische Studien über Meloiden (Coleoptera). *Annales Historico-Naturales Musei Nationalis Hungarici* **73**: 159-181.

Kaszab Z. 1981b: Neue orientalische Stenosini (Coleoptera: Tenebrionidae). *Acta Zoologica Academiae Scientiarum Hungaricae* **27**: 273-313.

Kaszab Z. 1981c: Insects of Saudi Arabia. Coleoptera: Fam. Tenebrionidae (Part 2) Zusammenfassung der bis jetzt bekannt gewordenen Tenebrioniden (First part). *Fauna of Saudi Arabia* **3**: 276-401.

Kaszab Z. 1982a: Neue orientalische Tenebrioniden. *Acta Zoologica Academiae Scientiarum Hungaricae* **28**: 57-80.

Kaszab Z. 1982b: Insects of Saudi Arabia. Coleoptera: Fam. Tenebrionidae (Part 2) Zusammenfassung der bis jetzt bekanntgewordenen Tenebrioniden (Second part). *Fauna of Saudi Arabia* **4**: 124-243.

Kaszab Z. 1983a: Insects of Saudi Arabia. Coleoptera: Fam. Meloidae. A synopsis of the Arabian Meloidae. *Fauna of Saudi Arabia* **5**: 144-204.

Kaszab Z. 1983b: Über die Gattung Leprocaulus Fairmaire, 1896 (Coleoptera, Tenebrionidae). *Annales Historico-Naturales Musei Nationalis Hungarici* **75**: 177-183.

Kaszab Z. 1985: Drei neue Tenebrioniden (Coleoptera) aus Asien. *Bollettino di Museo Civico di Storia Naturale di Verona* **12**: 449-459.

Kaszab Z. 1986: Beiträge zur Kenntnis der Tenebrioniden (Coleoptera) aus Arabien. *Annales Historico-Naturales Musei Nationalis Hungarici* **78**: 147-150.

Kaszab Z. 1988: Katalog und Bestimmungstabelle der Gattung Promethis Pascoe, 1869 (Coleoptera, Tenebrionidae). *Acta Zoologica Hungarica* **34**: 67-170, pls I-XV.

Kaszab Z. & Knor I. 1976: Zur Tenebrionidenfauna der Tuvischen ASSR in Südsibirien (Coleoptera). *Acta Zoologica Academiae Scientiarum Hungaricae* **22**: 293-297.

Kaszab Z. & Medvedev G. S. 1972: Ergebnisse der mongolisch-sowjetischen biologischen Expedition in der Mongolischen Volksrepublik seit 1967. Tenebrionidae: Tentyriini (Coleoptera). *Annales Historico-Naturales Musei Nationalis Hungarici* **64**: 203-218.

Kaszab Z. & Medvedev G. S. 1976: Ergebnisse der mongolisch-sowjetischen biologischen Expeditionen in der Mongolischen Volksrepublik seit 1967. Tenebrionidae: Tentyriini (Coleoptera), II. *Annales Historico-Naturales Musei Nationalis Hungarici* **68**: 105-115.

Kaszab Z. & Medvedev G. S. 1977: Tentyriini (Coleoptera, Tenebrionidae) aus der Mongolei, III. *Annales Historico-Naturales Musei Nationalis Hungarici* **69**: 133-143.

Kaszab Z. & Medvedev G. S. 1984: Drei neue asiatische Tenebrioniden (Coleoptera). *Acta Zoologica Hungarica* **30**: 79-85.

Kaszab Z. & Pinheiro M. F. 1972: Una nova espécie de Sepidium (Coleoptera, Tenebrionidae) em Portugal. *Estudos e Divulgaçao Téchnica (C) Entomologia Forestal* **1972**: 5-17.

Katter F. 1883: *Monographiae der europäischen Arten der Gattung Meloë mit besonderer Berücksichtigung der Biologie dieser Insekten.* Putbus: Dose, 61 pp.

Kawanabe M. 1993: A new genus of the family Ciidae (Coleoptera) from Yakushima Island, Southwest Japan. *Japanese Journal of Entomology* **61**: 763-76.

Kawanabe M. 1994a: A new genus and species of the family Ciidae (Coleoptera) from Southwest Japan. *Japanese Journal of Entomology* **62**: 186-192.

Kawanabe M. 1994b: A new species of the genus Orthocis Casey from Japan (Coleoptera: Ciidae). *Transactions of the Shikoku Entomological Society* **20**: 187-190.

Kawanabe M. 1995: Taxonomic notes on the genus Nipponapterocis (Coleoptera, Ciidae), with descriptions of two new species from Japan. *Elytra* **23**: 167-175.

Kawanabe M. 1996a: A new species of the genus Ennearthron (Coleoptera, Ciidae) from Hokkaido, Japan. *Japanese Journal of Systematic Entomology* **2**: 105-108.

Kawanabe M. 1996b: A new genus of the family Ciidae (Coleoptera), with description of a new species from the Ryukyu Islands, southwest Japan. *Elytra* **24**: 331-336.

Kawanabe M. 1997a: A taxonomic study on the Japanese species of the genus Sulcacis (Coleoptera, Ciidae). *Elytra* **25**: 67-78.

Kawanabe M. 1997b: The Japanese species related to the group of Cis nitidus (Coleoptera, Ciidae). *Elytra* **25**: 311-331.

Kawanabe M. 2001: A new ciid beetle of the genus Cis from Hokkaido, Japan (Coleoptera: Ciidae). *Sukunahikona. Special Publication of the Japan Coleopterological Society* (Osaka) **1**: 267-270.

Kawanabe M. 2002: A revision of the genus Octotemnus (Coleoptera, Ciidae) from Japan. *Japanese Journal of Systematic Entomology* **8**: 243-267.

Kawanabe M. 2005: A new species of the genus Octotemnus(Coleoptera, Ciidae) from Nantou County, central Taiwan. *Elytra* **33**: 47-50.

Kejval Z. 1998: Review of oriental and some Palaearctic species of the genus Endomia (Coleoptera: Anthicidae). *European Journal of Entomology* **95**: 99-131.

Kejval Z. 1999: Five new species of Notoxus (Coleoptera: Anthicidae) from Palaearctic Asia, with redescription of N. quadrimaculatus. *Klapalekiana* **35**: 13-23.

Kejval Z. 2000a: Revisional notes on Oriental Formicomus LaFerté-Sénectére (Coleoptera, Anthicidae) 2. Himalayan species related to F. lagenicollis Fairmaire, 1894. *Acta Musei Moraviae, Scientiae Biologicae* (Brno) **85**: 73-111.

Kejval Z. 2000b: Revisional notes on the Palaearctic Formicomus (Coleoptera: Anthicidae). *Acta Societatis Zoologicae Bohemicae* **64**: 403-417.

Kejval Z. 2002a: The species of Anthelephila (Coleoptera: Anthicidae) related to A. angustiformis and A. walkeri. *Folia Heyrovskyana* **10**: 83-114.

Kejval Z. 2002b: New species, synonymy and records of Anthelephila (Coleoptera: Anthicidae). *Acta Societatis Zoologicae Bohemicae* **66**: 243-278.

Kejval Z. 2002c: Contribution to the knowledge of the genus Stenidius (Coleoptera: Anthicidae). *Klapalekiana* **38**: 185-212.

Kejval Z. 2006: Studies of the genus Anthelephila Hope, 1833 (Coleoptera, Anthicidae). 9. Descriptions of five new species from the Oriental region. *Entomologica Basiliensia et Collectionis Frey* **28**: 175-191.

Kelejnikova S. I. 1963: Novyy rod i vid triby Tentyriini (Coleoptera, Tenebrionidae) iz Kirgizii. *Zoologicheskiy Zhurnal* **42** (4): 622-623.

Kelejnikova S. I. 1970: Novyy vid roda Blaps (Col., Tenebr.) iz Altaya. Pp. 27-29. In: Tscherepanov A. I. (ed.): *Novye i maloizvetsnye vidy sibirskoy fauny*. 3. Institut Biologii, Sibirskoye Otdelenie Akademiy Nauk SSSR. Novosibirsk: Nauka, 112 pp.

Kelejnikova S. I. 1977: Novye chernotelki (Coleoptera, Tenebrionidae) iz Turkmenii. *Entomologicheskoe Obozrenie* **56** (3): 653-655.

Kiesenwetter E. A. H. von. 1851: Énumération des coléoptères. *Annales de la Société Entomologique France* (2) **9**: 577-656.

Kiesenwetter E. A. H. von. 1859: Beitrag zur Käferfauna Griechenlands. Sechstes Stück: Malacodermata, Cleridae, Ptinidae, Anobiadae. *Berliner Entomologische Zeitschrift* **3**: 158-192.

Kiesenwetter E. A. H. von. 1861a: Beitrag zur Käferfauna Griechenlands. Siebentes Stück: Tenebrionidae, Cistelidae, Lagriariae, Pedilidae, Anthicidae, Mordellonae, Meloidae, Oedemeridae. *Berliner Entomologische Zeitschrift* **5**: 221-252.

Kiesenwetter E. A. H. von. 1861b: Eine entomologische-Excursion in das Wallis und nach dem Monte Rosa. *Berliner Entomologische Zeitschrift* **5**: 360-395.

Kiesenwetter E. A. H. von. 1863: Beiträge zur Kenntniss der zur Untergattung Isomira Muls. gehörenden Cistela-Arten. *Berliner Entomologische Zeitschrift* **7**: 423-427.

Kiesenwetter E. A. H. von. 1866: Eine entomologische Excursion nach Spanien im Sommer 1865. *Berliner Entomologische Zeitschrift* **9** [1865]: 359-396.

Kiesenwetter E. A. H. von. 1870: [new taxa]. In: Heyden L. F. J. D. von: Entomologische Reise nach dem südlichen Spanien, der Sierra Guadarrama und Sierra Morena, Portugal und den Cantabrischen Gebirgen. *Berliner Entomologische Zeitschrift* **14** *Beiheft*, 218 pp., 2 pls.

Kiesenwetter E. A. H. von. 1873: Revision der Cisteliden-Gattung Podonta Muls. *Berliner Entomologische Zeitschrift* **17**: 9-22.

Kiesenwetter E. A. H. von. 1878: [new taxa]. In: Schneider O. & Leder H.: Beiträge zur Kenntniss der kaukasischen Käferfauna. *Verhandlungen des Naturforschenden Vereines in Brünn* **16** [1877]: 3-258, pls 1-4.

Kiesenwetter E. A. H. von. 1879: Coleoptera Japoniae collecta a Domino Lewis et aliis. *Deutsche Entomologische Zeitschrift* **23**: 305-320.

Kiesenwetter E. A. H. von. 1880: Beschreibung neuer Pedinus-Arten. *Entomologische Monatsblätter* **2** (15): 65-70.

Kirby W. 1802: *Monographia Apum Angliae*. Vol. 2. Ipswich: White, 1-388 pp., 4 pls.

Kirby W. 1819: A century of insects, including several new genera described from his cabinet. *Transactions of the Linnean Society of London* **12**: 375-453, pls 21-23.

Kirby W. 1837: The insects. In: Richardson J. (ed.): *Fauna Boreali-Americana; or the zoology of the northern parts of British America: containing descriptions of the objects of natural history collected on the late northern land expedition, under the command of Captain Sir John Franklin, R. N. Part the fourth and last.* Norwich: Josian Flechter, xxxix + 325 + [2] pp., 8 pls.

Kirsch T. F. W. 1866: Beiträge zur Käferfauna von Bogotà. *Berliner Entomologische Zeitschrift* **10**: 173-216.

Kirsch T. F. W. 1869: Beitrag zur Kenntniss der Gattung Omophlus Sol. *Berliner Entomologische Zeitschrift* **13**: 97-128.

Kirsch T. F. W. 1871: Neue Käfer-Arten aus Egypten gesammelt von Dr. Schneider. *Berliner Entomologische Zeitschrift* **14** [1870]: 389-396.

Kirsch T. F. W. 1878: [new taxa]. In: Schneider O. & Leder H.: Beiträge zur Kenntniss der kaukasischen Käferfauna. *Verhandlungen des Naturforschenden Vereines in Brünn* **16** [**1877**]: 3-262.

Kirsch T. F. W. 1880: Zwei neue Mylabris. *Entomologische Monatsblätter* **2**: 77.

Kirchsberg O. von. 1876: Pimelia balcanica Kirchsberg var. nov. *Entomologische Monatsblätter* **1**: 30-31.

Kirchsberg O. von. 1877a: Vieta Millingenii nov. spec. *Deutsche Entomologische Zeitschrift* **21**: 203.

Kirchsberg O. von. 1877b: Anthrodeis arabicus nov. spec. *Deutsche Entomologische Zeitschrift* **21**: 204.

Kiyoyama Y. 1975: A new species of genus Stenomordella Ermisch from Is. Amami-Ôshima with new record. *The Entomological Review of Japan* **28**: 46-48.

Kiyoyama Y. 1987: Two new species of the genus Mordellochroa. *The Entomological Review of Japan* **42**: 31-36.

Kiyoyama Y. 1991: A study on the genus Tolidostena Ermisch from Taiwan. *The Entomological Review of Japan* **46**: 53-61.

Kiyoyama Y. 1993: A study on the genus Hoshohananomia Kôno from Japan and Taiwan. *Transaction of the Shikoku Entomological Society* **20**: 71-86.

Kiyoyama Y. 1994: A new species of the genus Variimorda Méquignon from Taiwan. *Transaction of the Shikoku Entomological Society* **20**: 193-196.

Klinger R. 2000: Revision des Otolelus pruinosus (Kiesenwetter) mit Beschreibung einer neuen Art (Col., Aderidae). *Entomologische Nachrichten und Berichte* **44**: 275-278.

Klug J. C. F. 1830: *Symbolae Physicae, seu icones et descriptiones Insectorum quae ex itinere per Africam borealem et Asiam occidentalem Friderici Guilelmi Heimprich et Cristiani Godofredi Ehrenberg medicinae te chirurgiae doctorum studio novae aut illustratae redierunt. Decas secunda.* Berolini: Oficina Academica, [20] pp., pls xi-xx.

Klug J. C. F. 1833: Bericht über eine auf Madagarscar veranstaltete Sammlung von Insecten aus der Ordnung Coleoptera. *Abhandlungen der Königlichen Akademie der Wissenschaften* (Berlin) **1832-1833**: 91-223, 5 pls.

Klug J. C. F. 1845: *Symbolae physicae seu icones et dscriptiones Insectorum que ex itinere per Africam borealem et Asiam occidentalem Friderici Guilelmi Hemprich et Christiani Godofredi Ehrenberg medicinae et chirurgiae doctorum studio nova autillustrate redierunt. Insecta.* Berolini: Reumen, 1-24, pls. 31-32.

Knoch A. W. 1801: *Neue Beyträge zur Insectenkunde. Erster Theil.* Leipzig: Schwickert 1: xii + 208 pp., pls i-ix.

Knor I. B. 1970: [new taxon]. In: O faune chernotelok (Coleoptera, Tenebrionidae) altaysko-sayanskoy gornoy strany. Pp. 82-87: In: *Fauna Sibiri.* Novosibirsk: Nauka.

Knor I. B. 1976: Novye dl'a fauny Sovietskogo Soyuza chernotelki (Coleoptera, Tenebrionidae) iz Tuvy. Pp. 102-109. In: Tsherepanov A. I. (ed.): *Novosti fauny Sibiri. Novye i maloizvestnye vidy fauny Sibiri. Vypusk 10.* Novosibirsk: Nauka, 208 pp.

Koch C. 1931: Risultati scientifici delle spedizioni entomologiche di S. A. S. il Principe Alessandro della Torre e Tasso nell'Africa settentrionale e in Sicilia. II. Anthicus Payk. *Memorie della Società Entomologica Italiana* **10**: 66-79.

Koch C. 1933: Risultati scientifici delle caccie entomologiche di S. A. S. il Principe Alessandro della Torre e Tasso in Italia. II Anthicus. *Bollettino della Società Entomologica Italiana* **65**: 149-159.

Koch C. 1934a: Beitrag zur Kenntnis der Tenebrioniden vom Gebel Elba und Mersa Halaib (Coleoptera). *Bulletin de la Société Royale Entomologique d'Egypte* **18**: 92-103.

Koch C. 1934b: Studienergebnisse der entomologischen Expeditionen Seiner Durchlaucht des Fürsten della Torre e Tasso in Nordafrika und Sizilien. IV. Anthicus (Col. Anthic.). *Deutsche Entomologische Zeitschrift* **1934**: 117-127.

Koch C. 1935a: Wissenschaftliche Ergebnisse der entomologischen Expedition seiner Durchlaucht des Fürsten A. della Torre e Tasso nach Aegypten und auf der Halbinsel Sinai. VII. Tenebrionidae (Coleoptera). *Bulletin de la Société Royale Entomologique d'Egypte* **19**: 2-111.

Koch C. 1935b: Wissenschaftliche Ergebnisse der entomologischen Expeditionen seiner Durchlaucht des Fürsten Alessandro C. della Torre e Tasso nach Aegypten und auf die Halbinsel Sinai. VIII. Anthicidae (Coleoptera). *Bulletin de la Société Royale Entomologique d'Égypte* **19**: 132-144.

Koch C. 1935c: Risultati scientifici delle cacce entomologiche di S. A. S. il Principe della Torre e Tasso nelle Isole dell' Egeo. III. Anthicidae. *Bollettino del Laboratorio di Zoologia Generale e Agraria del R. Instituto Superiore Agrario in Portici* **28**: 257-281.

Koch C. 1935d: Wissenschaftliche Ergebnisse der entomologischen Expedition Seiner Durchlaucht des Fürsten A. Della Torre e Tasso in Palaestina. I. Tenebrionidae (Coleoptera). *Bulletin de la Société Royal Entomologique d'Egypte* **19**: 262-288.

Koch C. 1935e: Risultati scientifici delle cacce entomologiche di S. A. S. il Principe Alessandro della Torre e Tasso nelle isole dell'Egeo. V. Tenebrionidae. *Bollettino del Laboratorio di Zoologia Generale e Agraria del R. Instituto Superiore Agrario di Portici* **28**: 309-320.

Koch C. 1936: Neue Tenebrionidae aus dem östlichen Mediterrangebiet. *Pubblicazioni del Museo Entomologico "Pietro Rossi" Duino* **1**: 291-301.

Koch C. 1937: Wissenschaftliche Ergebnisse über die während der Expedition Seiner Durchlaucht des Fürsten Alessandro C. della Torre e Tasso in Lybien aufgefundenen Tenebrioniden. *Pubblicazioni del Museo Entomologico "Pietro Rossi" Duino* **2**: 285-500.

Koch C. 1939: Die Käfer der libyschen Ausbeute des Herrn Georg Frey. *Mitteilungen der Münchener Entomologischen Gesellschaft* **29**: 216-293, pls. viii, ix.

Koch C. 1940a: Der saharo-sindhische Verbreitungs-Typus bei der ungeflügelten Tenebrioniden-Gattung Mesostena unter Berücksichtigung ähnlicher biogeographischer Verhältnisse der Tentyria-Stamm-Gruppe Tentyrina. *Rivista di Biologia Coloniale* **3**: 1-137.

Koch C. 1940b: Resultados cientificos de los viajes entomológicos en España patrocinados por Su Alteza el Principe Alessandro C. della Torre e Tasso. Pp. 73-80. *VI Congreso Internacional de Entomología. Madrid, 6-12 de Septiembre de 1935. Vol. 1.* Madrid: Laboratorio de Entomología del Museo Nacional de Ciencias Naturales, 406 pp.

Koch C. 1940c: Über einige neue und interessante, von Johannes Roth 1857 im Wadi Arabad gesammelte Tenebrioniden (Col.). *Mitteilungen der Münchener Entomologischen Gesellschaft* **30**: 84-107.

Koch C. 1940d: Phylogenetische, biogeographische und systematische Studien über ungeflügelte Tenebrioniden (Col. Tenebr.). *Mitteilungen der Münchener Entomologischen Gesellschaft* **30**: 254-337.

Koch C. 1940e: Phylogenetische, biogegraphische und systematische Studien über ungeflügelte Tenebrioniden (Col. Tenebr.). II. *Mitteilungen der Münchener Entomologischen Gesellschaft* **30**: 683-750, pls 18-20.

Koch C. 1941a: Die Verbreitung und Rassenbildung der marokkanischen Pimelien (Col. Tenebr.) (Eine biogeographische Studie). *Eos, Revista Española de Entomología* **16** [1940]: 7-123, pls 1-3.

Koch C. 1941b: Contributo alla conoscenza del genere Mitotagenia Rtt. (Col. Tenebr.). *Memorie della Società Entomologica Italiana* **20**: 37-42.

Koch C. 1941c: Phylogenetische, biogegraphische und systematische Studien über ungeflügelte Tenebrioniden (Col. Tenebr.). III. *Mitteilungen der Münchener Entomologischen Gesellschaft* **31**: 252-314, pls 12-13.

Koch C. 1943a: Phylogenetische, biogeographische und systematische Studien über ungeflügelte Tenebrioniden (Col. Tenebr.). IV. *Mitteilungen der Münchener Entomologischen Gesellschaft* **33**: 479-597.

Koch C. 1943b: Revision der Tenebrionidengattungen Thalpophila und Rhytinota (Col. Tenebr.). *Mitteilungen der Münchener Entomologischen Gesellschaft* **33**: 759-889, pls 28-37.

Koch C. 1944a: Die Adesmiini der tropischen und subtropischen Savannen Afrikas. *Revue de Zoologie et Botanique Africaine* **38**: 139-191.

Koch C. 1944b: Tenebrionidae I (I. Beitrag zur Kenntnis der iberischen Fauna). *Mitteilungen der Münchener Entomologischen Gesellschaft* **34**: 216-254.

Koch C. 1945: Über die Tenebrioniden der Gipfelfauna des Hohen und Mittleren Atlas Marokkos. Ergebnisse der Forschungsreise R. Paulian und A. Villiers nach dem Hohen Atlas Marokkos 1938. XIX. Beitrag. *Eos,*

Revista Española de Entomología **20** [1944]: 387-433.

Koch C. 1948a: Die Tenebrioniden Kretas (Col.). *Mitteilungen der Münchener Entomologischen Gesellschaft* **34** [1944]: 255-363, pls. XVIII-XXIX.

Koch C. 1948b: Beitrag zur Kenntnis der Tribus Litoborini der Tenebrioniden-Unterfamilie der Opatrinae (Col. Ten.). *Eos, Revista Española de Entomología* **24**: 403-433.

Koch C. 1949: Beitrag zur Kenntnis der asiatischen Tentyriini (Col., Tenebr.). *Entomologische Blätter* **41-44** [1945-48]: 76-81.

Koch C. 1950a: The Tenebrionidae (Col.) of Southern Africa. V. – Contribution to the knowledge of the Caedius-group of Opatrini and the Crypticini from south of the Sahara. *Memórias do Museu Dr. Álvaro de Castro* **1**: 35-90, 6 pls.

Koch C. 1950b: Die Tenebrioniden des südlichen Afrikas. IV. – Über vier neue afrikanische Himatismus-Arten. *Entomologische Arbeiten aus dem Museum G. Frey* **1**: 186-197.

Koch C. 1951: Beitrag zur Kenntnis der Gattung Adesmia (Col. Tenebrionid.). *Entomologische Blätter* **45-46** [1949-50]: 121-130.

Koch C. 1952: The Tenebrionidae of Southern Africa. VIII. Materials for a monographic study on Euchychorini (Coleoptera). *Bulletin de la Société Fouad Ier d'Entomologie* **45**: 1-125, pls I-XIII.

Koch C. 1953: Tenebrioniden aus südlichen Afrikas XIV. Über einige neue Molurini aus dem Ungarischen Naturwissenschaftlichen Museum zu Budapest (Vorarbeiten zu einer Monographie der Molurini 4.). *Annales Historico-Naturales Musei Nationalis Hungarici, Series nova* **3** [1952]: 137-181, pls. V, VI.

Koch C. 1956: II. Tenebrionidae (Coleoptera Polyphaga), Opatrinae, First part: Platynotini, Litoborini and Loensini. *Exploration du Parc National de l'Upemba* **40**: 1-472, 35 pls.

Koch C. 1960: Zweiter taxonomischer Beitrag zur Kenntnis der Tenebrioniden Somalias. *Entomologische Arbeiten aus dem Museum G. Frey* **11**: 325-415, 2 pls.

Koch C. 1965a: Missione 1962 del Prof. Giuseppe Scortecci nell'Arabie meridionale. Coleoptera Tenebrionidae. Includendo materiale di viaggi in Arabia del Sig. G. Popov (1962) e del Dr. G. Benardelli (1962-63). *Atti della Società Italiana di Scienze Naturali e del Museo Civico di Storia Naturale di Milano* **104**: 99-154, pls V-VII.

Koch C. 1965b: Sur les types de Fairmaire des tribus Blaptini et Platyscelini conservés au Muséum de Paris. *Annales de la Société Entomologique de France* (N. S.) **1**: 125-135.

Koch C. 1969: Storthocnemis kulzeri, ein bisher unbekanntes Bindeglied zwischen den Tribus Pimellini/Platyopini und Leucolaepini (Col., Tenebrionidae). *Entomologische Arbeiten aus dem Museum G. Frey* **20**: 484-493.

Koch C. 1970: Die Tenebrioniden (Coleoptera) des Archipels von Socotra. *Monitore Zoologico Italiano* N. S. *Supplemento* **3**: 69-132.

Kocher L. 1951a: Notes sur les Erodius marocains (Contribution à l'étude systématique du genre). *Bulletin de la Société des Sciences Naturelles et Physiques du Maroc* **30** [1950]: 195-230.

Kocher L. 1951b: Sur le genre Cimipsa Peyerh. (Col. Tenebr.) avec description d'une espèce et d'une variété nouvelles. *Bulletin de la Société des Sciences Naturelles et Physiques du Maroc* **31**: 125-130.

Kocher L. 1953: Localisations nouvelles ou intéressantes de coléoptères marocains. *Travaux de l'Institut Scientifique Chérifien* **7**: 1-142 pp., 1 map.

Kocher L. 1954: Prospection entomologique (Coléoptères) de la moyenne Moulouya. *Bulletin de la Société de Sciences Naturelles et Physiques du Maroc* **34**: 263-286.

Kocher L. 1955a: Observations sur certains meloides marocains avec descriptions d'une nouvelle espèce. *Comptes Rendus des Séances Mensuelles de la Société des Sciences Naturelles et Physiques du Maroc* **7**: 125-126.

Kocher L. 1955b: Variation chez certains Mylabris marocains. *Comptes Rendus des Séances Mensuelles de la Société des Sciences Naturelles et Physiques du Maroc* **7**: 127-128.

Kocher L. 1956a: Contribution à l'étude des Heliomophlus de l'Atlas marocain. *Bulletin Société des Sciences Naturelles et Physiques du Maroc* **35** [1955]: 235-242.

Kocher L. 1956b: Note complémentaire sur les Erodius marocains. *Bulletin de la Société des Sciences Naturelles et Physiques du Maroc* **35** [1955]: 243-249.

Kocher L. 1957: Nouvelles espèces ou variétés marocaines de coléoptères. *Bulletin de la Société des Sciences Naturelles et Physiques du Maroc* **36** [1956]: 367-373.

Kocher L. 1958a: Observations sur le genre Adelphinus Fairm. (Col. Ténébr.), avec description d'une nouvelle espèce du Maroc. *Comptes Rendus des Séances Mensuelles de la Société des Sciences Naturelles et Physiques du Maroc* **24**: 123-125.

Kocher L. 1958b: Un nouveau Scaurus cavernicole du Maroc (Col. Ténébr.): Scaurus camelus sp. nov. *Comptes Rendus des Séances Mensuelles de la Société des Sciences Naturelles et Physiques du Maroc* **24**: 195-196.

Kocher L. 1960: Nouvelles espèces de coléoptères du Maroc. *Bulletin de la Société des Sciences Naturelles et Physiques du Maroc* **40**: 233-241.

Kocher L. 1962a: Coléoptères nouveaux ou intéressants du Maroc présaharien. *Bulletin de la Société des Sciences Naturelles et Physiques du Maroc* **42**: 195-214.

Kocher L. 1962b: Contribution à l'étude du genre Diplocyrtus Quedf. (Col. Tenebr.). *Bulletin de la Société des Sciences Naturelles et Physiques du Maroc* **42**: 301-310.

Kocher L. 1963a: Nouvelle recolte des coléoptères dans le Bas Drâ (Sahara occidental marocain). *Bulletin de la Société de Sciences Naturelles et Physiques du Maroc* **43**: 35-37.

Kocher L. 1963b: Les Diplocyrtus de la collection Escalera (Col. Tenebrionidae). *Bulletin de la Société des Sciences Naturelles et Physiques du Maroc* **43**: 113-116.

Kocher L. 1963c: Coléoptères nouveaux ou intéressants de la région rifaine. *Bulletin de la Société des Sciences Naturelles et Physiques du Maroc* **43**: 193-199.

Kocher L. 1964: Présentation de coléoptères rares recueillis récemment au Maroc et de quelques espèces nouvelles pour la sciences. *Comptes Rendus des Séances Mensuelles de la Société des Sciences Naturelles et Physiques du Maroc* **30**: 43-45.

Kocher L. 1965a: Contribution à l'étude des Stenosis du groupe elvirae Ant. (Col. Tenebr.). *Bulletin de la Société des Sciences Naturelles et Physiques du Maroc* **44** (3e et 4e trimestre) [1964]: 25-27.

Kocher L. 1965b: Scaurus cavernicoles de la collection Antoine (Col. Ténébr.). *Bulletin de la Société des Sciences Naturelles et Physiques du Maroc* **44** (3e et 4e trimestre) [1964]: 33-36.

Kocher L. 1966a: Alphasida nouvelles ou intéressants de la région rifaine (Col. Tenebr.). *Bulletin de la Société des Sciences Naturelles et Physiques du Maroc* **45** [1965]: 43-48.

Kocher L. 1966b: Coléoptères marocains nouveaux ou intéressants. *Bulletin de la Société des Sciences Naturelles et Physiques du Maroc* **45** [1965]: 133-136.

Kocher L. 1968: Description de nouveaux coléoptères du Maroc. *Bulletin de la Société des Sciences Naturelles et Physiques du Maroc* **48**: 107-113.

Kolbe H. J. 1883: Neue Coleoptera von Westafrika. *Berliner Entomologische Zeitschrift* **27**: 15-36.

Kolbe H. J. 1886: Beiträge zur Kenntnis der Coleopteren-Fauna Koreas, bearbeitet auf Grund der von Herrn C. Gottsche veranstalteten Sammlung; nebst Bemerkungen über die zoogeographische Verhältnisse dieses Faunengebietes und Untersuchungen über einen Sinnensapparat im Gaumen von Microlampidius morio. *Archiv für Naturgeschichte* **52** (1): 139-240, pls X, XI.

Kolbe H. J. 1894a: Über Coleopteren aus Afrika. *Entomologische Zeitung* (Stettin) **55**: 178-190.

Kolbe H. J. 1894b: Über Coleopteren aus Afrika. II. *Entomologische Zeitung* (Stettin) **55**: 361-397.

Kolbe H. J. 1897: Coleopteren. Die Käfer Deutsch-Ost-Afrikas, 367 pp., 4 pls. In: Kolbe H. J.: Käfer und Netzflügler Ost-Africas. In: Möbius K. (ed.): *Deutsch-Ost-Afrika* **4** (4). Berlin: D. Reimer, 367 + 42 pp., 5 pls.

Kolbe H. J. 1900: Ein vergessener Nyctobates. *Entomologische Nachrichten* **26**: 72-74.

Kolbe H. J. 1901: Ein Schädling des Affenbrotbaumes, Adansonius fructum n. sp. aus der Familie der Curculioniden. *Allgemeine Zeitschrift für Entomologie* **6**: 341-343.

Kolbe H. J. 1908: [new taxa]. In: Kolbe H. J., Obst P. & Weise J.: Coleoptera. Pp. 82-96, pl. III. In: *Expedition Filchner nach China und Tibet 1903-1905. Wissenschaftliche Ergebnisse X. Band - 1. Teil.* Berlin: Siegfried Mittler und Sohn.

Kolbe H. J. 1915: Eine neue isolierte Tenebrionidengattung von den Philippineninseln. *Deutsche Entomologische Zeitschrift* **1915**: 261-264.

Kolenati F. A. 1846: *Meletemata Entomologica. Fasc. III. - Brachelytra Caucasi cum distributione geographica adnexis Pselaphinis, Scydmaenis, Notoxidibus et Xylophagis.* Petropoli: Typis Imperialis Academiae Scientiarum, 44 pp., pls 12-14.

Kolenati F. A. 1847: Über einige russische Oedemeriden. *Bulletin de la Société Impériale des Naturalistes de Moscou* **20** *(1-2): 132-147.*

Kollar V. & Redtenbacher L. 1842: Aufzählung und Beschreibung der von Freiherr Carl v. Hügel auf seiner Reise durch Kaschmir und das Himalayagebirge gesammelten Insecten. Pp. 395-564. In: Hügel C. von:

Kashmir und das Reich der Siek. Vierter Band. Zweiter Abtheilung. Stuttgart: Hallbergerische Verlag, [6] + 247-865 pp.

Kompantsev A. V. 1996: A new species of the genus Cis Latr. (Coleoptera, Ciidae) from Middle Asia and notes on Palaearctic species of the comptus group. *Entomologicheskoe Obozreniye* **75**: 587-595.

König E. 1906: Dritter Beitrag zur Coleopteren-Fauna des Kaukasus. *Wiener Entomologische Zeitung* **25**: 23-27.

Kôno H. 1927: Beitrag zu den Rhipiphoriden-Arten von Japan. *Insecta Matsumurana* **1**: 177-182.

Kôno H. 1928a: Die Mordelliden Japans. *Transaction of Sapporo Natural History Society* **10**: 31-46.

Kôno H. 1928b: Einige Mordelliden-Arten Japans. *Insecta Matsumurana* **2**: 146-150.

Kôno H. 1929a: Die Lagriiden Japans (Col.). *Insecta Matsumurana* **4**: 25-35.

Kôno H. 1929b: Die Pyrochroiden Japans. *Insecta Matsumurana* **3**: 62-72.

Kôno H. 1929c: Die Mordelliden Japans. *Insecta Matsumurana* **3**: 83-86.

Kôno H. 1930a: Cerogria miwai nova spec. (Col. Lagriidae). *Transactions of the Natural History Society of Formosa* **20**: 18-19.

Kôno H. 1930b: Einige Alleculiden-Arten aus Japan und Formosa (Col.). *Insecta Matsumurana* **4**: 95-99.

Kôno H. 1930c: Die Mordelliden Japans. *Insecta Matsumurana* **4**: 104.

Kôno H. 1930d: Die Serropalpiden aus Sachalin (Col.). *Insecta Matsumurana* **5**: 48-51.

Kôno H. 1931: Die Notoxus-Arten aus Japan (Col. Anthicidae). *Insecta Matsumurana* **6**: 1-2.

Kôno H. 1932a: Die Nacerdinen Japans (Col.). *Insecta Matsumurana* **6**: 138-144.

Kôno H. 1932b: Die Mordelliden Japans. *Transaction of Sapporo Natural History Society* **12**: 152-160, 1 pl.

Kôno H. 1933: Die Mordelliden Japans. *Transaction of Sapporo Natural History Society* **13**: 29-31.

Kôno H. 1934a: H. Sauter's Formosa Ausbeute: Mordellidae. *Insecta Matsumurana* **8**: 116-118.

Kôno H. 1934b: Die Heteromeren von Sachalin. *Insecta Matsumurana* **9**: 27-40.

Kôno H. 1935a: Die Heteromeren aus den Kurilen (Col.). (Erster Beitrag zur Kenntnis der Käferfauna der Kurilen). *Insecta Matsumurana* **9** [1934]: 85-89.

Kôno H. 1935b: Die Mordelliden Japans. *Transaction of Sapporo Natural History Society* **14**: 123-130.

Kôno H. 1935c: Drei neue Heteromeren aus Japan und Formosa, mit einer Liste der geographischen Verbreitung der japanischen Pyrochroiden und Pediliden. *Insecta Matsumurana* **9**: 157-161.

Kôno H. 1936a: Über die Käfersammlung des Museums "Umeno Konchû Kenkiujo" I. Pyrochroidae, Meloidae, Rhipiphoridae. *Bulletin of the Umeno Entomological Laboratory, Kurume* **3**: 1-6.

Kôno H. 1936b: Neue und wenig bekannte Käfer Japans. I. *Insecta Matsumurana* **10**: 87-98.

Kôno H. 1936c: Beitrag zu den Pytho-Arten Japans (Col.). *Insecta Matsumurana* **11**: 35-37.

Kôno H. 1936d: Zwei neue Xanthochroa-Arten (Coleoptera, Oedemeridae). *Festschrift zum 60. Geburtstage von Professor Dr. Embrik Strand* (Riga) **1**: 516-518.

Kôno H. 1937a: Über die Käfersammlung des Museums "Umeno Konchû Kenkiujo". III Oedemeridae. Cephaloidae. *Bulletin of the Umeno Entomological Laboratory, Kurume* **5**: 1-15. (in Japanese, German title and descriptions).

Kôno H. 1937b: Neue und wenig bekannte Käfer Japans. II Oedemeridae. *Insecta Matsumurana* **11**: 135-146.

Kôno H. 1938: An inter-tidal rock-dwelling beetle, *Aegialatis* [sic] *stejnegeri*, and its one new subspecies. *The Entomological World* **6** (46): 1-5 (in Japanese).

Kôno H. 1939: Die Tenebrioniden in Tannen und Fichtenwald in Japan. *Insecta Matsumurana* **13**: 85-87.

Kôno H. 1940: Die Meloiden von Mandschukuo. *Insecta Matsumurana* **15**: 57-62.

Kontkanen P. 1956: Zur Kenntnis der paläarktischen Käferfauna. I. *Annales Entomologici Fennici* **22**: 55-63.

Kotschenov V. G. 1970: Novyy vid naryvnika (Coleoptera, Meloidae) iz Kirghizii. Pp. 21-22. In: Novye i maloizvestnyye vidy bespozvonochnykh zhivotnykh. *Entomologicheskiya Issledovaniya v Kirgizii* [Vypusk 6]. Frunze: Ilim, 87 pp.

Korge H. 1971: Beiträge zur Kenntnis der Koleopterenfauna Kleinasiens. *Annotationes Zoologicae et Botanicae* (Bratislava) **67**: 1-68.

Kraatz G. 1853: Wissenschaftliche Mittheilungen. *Entomologische Zeitung* (*Stettin*) **14**: 255-257.

Kraatz G. 1855: Synonymische Bemerkungen. *Entomologische Zeitung* (*Stettin*) **16**: 372-374.

Kraatz G. 1858: Neue Literatur – Histoire naturelle des coléoptères de France par M. E. Mulsant. Vésicants. Paris, 1857, 8, 198 S., 1 Tafel. *Berliner Entomologische Zeitschrift* **2**: 387-392.

Kraatz G. 1862a: Ueber die Arten der Gattungen Hyperops, Microtelus und Dichillus. *Berliner Entomologische Zeitschrift* **6**: 91-99.

Kraatz G. 1862b: Beiträge zur Käferfauna Griechenlands.Von Dr. Krüper neu aufgefundene Arten. *Berliner Entomologische Zeitschrift* **6**: 121-126.

Kraatz G. 1863: Revision der Cerocomiden-Gruppe. *Berliner Entomologische Zeitschrift* **7**: 109-115.

Kraatz G. 1865a: *Revision der Tenebrinoiden der alten Welt aus Lacordaire's Gruppen der Erodiides, Tentyriides, Akisides, Piméliides, und der europäischen Zophosis-Arten*. Berlin, 393 pp.

Kraatz G. 1865b: Synonymische Bemerkungen. *Berliner Entomologische Zeitschrift* **9**: 414.

Kraatz G. 1868a: Beiträge zur Kenntniss der deutschen Käferfauna. *Berliner Entomologische Zeitschrift* **12**: 283-304.

Kraatz G. 1868b: Ueber deutsche Käferarten. *Berliner Entomologische Zeitschrift* **12**: 331-338.

Kraatz G. 1870: [new taxa]. Heyden L. F. J. D. von. 1870: Zweiter Theil. Beschreibungen der neuen Arten. Pp. 58-176. In: Heyden L. F. J. D. von: Entomologische Reise nach dem südlichen Spanien, der Sierra Guadarrama und Sierra Morena, Portugal und den Cantabrischen Gebirgen. *Berliner Entomologische Zeitschrift* **14** *Beiheft*, 218 pp., 2 pls.

Kraatz G. 1872: [new taxa]. In: Reitter E.: Neue Käferarten von Oran gesammelt von Hans Leder. (Erstes Stück). *Berliner Entomologische Zeitschrift* **16**: 167-186.

Kraatz G. 1873: Ueber Uloma Perroudi Muls. aus Deutschland und eine neue Art von Cypern. *Berliner Entomologische Zeitschrift* **17**: 196-197.

Kraatz G. 1874: Asida pusillima, eine neue Art aus der Sierra Nevada. *Berliner Entomologische Zeitschrift* **18**: 112-114.

Kraatz G. 1877: Apolites graecus Kraatz n. sp. *Deutsche Entomologische Zeitschrift* **21**: 304.

Kraatz G. 1880a: Uebersicht der in den folgenden "Beiträgen zur Kenntniss der Canthariden von Dr. Haag-Rutenberg" beschriebenen und in der Uebersichtstabellen erwähnten Lytta-Arten. *Deutsche Entomologische Zeitschrift* **24**: 15-16.

Kraatz G. 1880b: Ueber die Heteromeren-Gattung Rhytinota Eschsch. und Verwandte. *Deutsche Entomologische Zeitschrift* **24**: 91-95.

Kraatz G. 1880c: Beitrag zur Kenntnis der asiatischen Cnodaloniden (Tenebrioniden Coh. II. Trib. XL. Lacord.). *Deutsche Entomologische Zeitschrift* **24**: 97-120.

Kraatz G. 1880d: [new taxa]. In: Heyden L. F. J. D. von. Verzeichnis von Coleopteren aus Asturien, mit Beschreibungen neuer Arten von Candèze, v. Heyden, Kirsch, Kraatz, Stierling. *Deutsche Entomologische Zeitschrift* **24**: 281-303.

Kraatz G. 1881a: Bemerkungen über typische Exemplare von Arten der Heteromeren-Gattung Blaps aus Fischer von Waldheim's Sammlung. *Deutsche Entomologische Zeitschrift* **25**: 53-60.

Kraatz G. 1881b: Ueber Veränderlichkeit der Prosodes-Arten. *Deutsche Entomologische Zeitschrift* **25**: 61-62.

Kraatz G. 1881c: Zonitis spectabilis nov. sp. *Deutsche Entomologische Zeitschrift* **25**: 326.

Kraatz G. 1881d: Einige neue Pimeliinen von Margelan. *Deutsche Entomologische Zeitschrift* **25**: 331-333.

Kraatz G. 1882a: In: Heyden L. F. J. D. von.: Beitrag zur Coleopteren-Fauna der Sierra de Cordoba. *Deutsche Entomologische Zeitschrift* **26**: 42-48.

Kraatz G. 1882b: Beiträge zur Käferfauna von Turkestan. II. Neue Tenebrioniden von Margelan. *Deutsche Entomologische Zeitschrift* **26**: 81-95.

Kraatz G. 1882c: [new taxa]. In: Heyden L. F. J. D. von. & Kraatz G.: Käfer um Margelan, gesammelt von Haberhauer. *Deutsche Entomologische Zeitschrift* **26**: 99-118.

Kraatz G. 1882d: [new taxa]. In: Heyden L. F. J. D. von. & Kraatz G.: Käfer um Samarkand, gesammelt von Haberhauer. *Deutsche Entomologische Zeitschrift* **26**: 297-338.

Kraatz G. 1883a: [new taxa]. In: Heyden L. F. J. D. von. & Kraatz G.: Käfer aus Osch (in Turkestan). *Deutsche Entomologische Zeitschrift* **27**: 337-353.

Kraatz G. 1883b: [new taxa]. In: Heyden L. F. J. D. von. & Kraatz G.: Käfer aus Tekke-Turcmenien. *Deutsche Entomologische Zeitschrift* **27**: 354-360.

Kraatz G. 1883c: Zwei neue Nalassus- (Helops-) Arten. *Deutsche Entomologische Zeitschrift* **27**: 395.

Kraatz G. 1884: Sitaris rufiventris nov. sp. *Deutsche Entomologische Zeitschrift* **28**: 175.

Kraatz G. 1885: [new taxa]. In: Heyden L. F. J. D. von. & Kraatz G.: Beiträge zur turkestanischen Coleopteren-Fauna. *Deutsche Entomologische Zeitschrift* **29**: 273-298.

Kraatz G. 1886: [new taxa]: In: Heyden L. F. J. D. von. & Kraatz G.: Beiträge zur Coleopteren-fauna von Turkestan, namentlich des Alai-Gebirges, unter Beihilfe der Herren Dr. Candèze, Ganglbauer, Dr. Stierlin, und Weise. *Deutsche Entomologische Zeitschrift* **30**: 177-194.

Kraatz G. 1887: Pimelia laevisulcata Kraatz aus Morocco, eine auffällige Varietät der P. Fairmairei Kraatz. *Deutsche Entomologische Zeitschrift* **31**: 361.

Kraatz G. 1888a: Zwei neue Blaps von Alai (Turkestan). *Deutsche Entomologische Zeitschrift* **32**: 195-197.

Kraatz G. 1888b: Ueber die Tenebrioniden-Gattungen Microdera und Dordanea. *Deutsche Entomologische Zeitschrift* **32**: 294.

Kraatz G. 1891: Rhipidius apicipennis nov. spec. aus Thüringen. *Deutsche Entomologische Zeitschrift* **35**: 358-360, pl. 5.

Kraatz G. 1896: [new taxa]. In: Ragusa E.: Catalogo ragionato dei coleotteri di Sicilia. *Naturalista Siciliano* (N. S.) **1**: 69-106.

Krauss H. 1911: Typhaea fumata var. n. obscura. *Rivista Coleopterologica Italiana* **9**: 109.

Krekich-Strassoldo H. von. 1913a: Neue Anthiciden und Mitteilungen über die Verbreitung bekannter Anthiciden. *Wiener Entomologische Zeitung* **32**: 223-232.

Krekich-Strassoldo H. von. 1913b: Beschreibungen neuer Anthiciden. *Verhandlungen der Kaiserlich-Königlichen Zoologisch-Botanischen Gesellschaft in Wien, Bericht der Sektion für Koleopterologie* **63**: 129-140.

Krekich-Strassoldo H. von. 1914a: Neue Anthiciden und Mitteilungen über die Verbreitung bekannter Anthiciden. *Wiener Entomologische Zeitung* **33**: 1-14.

Krekich-Strassoldo H. von. 1914b: Neue Anthiciden. *Verhandlungen der Kaiserlich-Königlichen Zoologisch-Botanischen Gesellschaft in Wien, Bericht der Sektion für Koleopterologie* **64**: 108-114.

Krekich-Strassoldo H. von. 1914c: Beiträge zur Kenntnis der Anthiciden. *Verhandlungen der Kaiserlich-Königlichen Zoologisch-Botanischen Gesellschaft in Wien, Bericht der Sektion für Koleopterologie* **64**: 215-228.

Krekich-Strassoldo H. von. 1919a: Beiträge zur Kenntnis der Anthiciden und Hylophiliden (Col.). *Entomologische Mitteilungen* **8**: 166-175.

Krekich-Strassoldo H. von. 1919b: Über Anthicus humilis Germ. und verwandte Arten. *Koleopterologische Rundschau* **8**: 60-76.

Krekich-Strassoldo H. von. 1925: Anthicidae of the Philippines I. *The Philippine Journal of Science* **27**: 515-535, pls. 1-3.

Krekich-Strassoldo H. von. 1927: Ein neuer Anthicus von China. *Entomologische Blätter* **22** [1926]: 116-117.

Krekich-Strassoldo H. von. 1928a: Beiträge zur Kenntnis indischer Anthiciden. *Archiv für Naturgeschichte* (*Abteilung A*) **5** [1926]: 67-105.

Krekich-Strassoldo H. von. 1928b: Il Formicomus canaliculatus LaF. e specie affini d'Africa. *Annali del Museo Civico di Storia Naturale Giacomo Doria, Genova* **52** [1925-1928]: 383-386.

Krekich-Strassoldo H. von. 1928c: Ueber Anthicus tristis Schm. und einige verwandte Arten. *Coleopterologisches Centralblatt* **3**: 104-111.

Krekich-Strassoldo H. von. 1928d: Beiträge zur Kenntnis malayischer Anthiciden. *Tijdschrift voor Entomologie* **71**: 1-10.

Krekich-Strassoldo H. von. 1929: Revision der paläarktischen Anthicus-Arten der Gruppe "Bifossicolles" (Gruppe XVIII in Marseuls Monographie). *Koleopterologische Rundschau* **15**: 151-189.

Krekich-Strassoldo H. von. 1931: Beiträge zur Kenntnis indischer Anthiciden II. *Folia Zoologia et Hydrobiologia* **3**: 1-41, pls. 1-3.

Kriechbaumer J. 1848: Wissenschaftliche Mittheilungen. Osphya ?aeneipennis. *Entomologische Zeitung* (Stettin) **9**: 163-165.

Kriesche R. 1919: Bemerkungen über einige Trictenotomiden. (Coleopt.). *Archiv für Naturgeschichte* A **83** (11): 150.

Kriesche R. 1921: Zweiter Beitrag zur Kenntnis der Trictenotomiden. *Deutsche Entomologische Zeitschrift* 1921: 278-279.

Królik R. 2002: Cis tauriensis n. sp. from Turkey (Coleoptera: Ciidae). *Genus* **13**: 197-202.

Krynicki I. 1829: [Corréspondence "Literae Cel. Professoris J. Krynicki, Directori datae, Charkoviae 28 Julii. 1829"]. *Bulletin de la Société Impériale des Naturalistes de Moscou* **1**: 187-199.

Krynicki I. 1832: Enumeratio Coleopterorum Rossiae meridionalis et praecipue in Universitatis Caesareae Charkoviensis circulo obvenientium, quae annorum 1827 - 1831 spatio observavit. *Bulletin de la Société Impériale des Naturalistes de Moscou* **5**: 65-179, 2 pls.

Krynicki I. 1834: Adenda et nonnulla synonyma Rosiae meridionalis Coleopterorum. *Bulletin de la Société Impériale des Naturalistes de Moscou* 7: 166-173.

Kryzhanovskij O. L. 1956: Two new species of genus Mylabris (Col. Meloidae) from Turkmenian SSR. *Entomologicheskoe Obozrenie* **35**: 886-889.

Kryzhanovskij O. L. 1959: Obzor vidov roda Teratolytta (Coleoptera, Meloidae) vstrechayushchikhsya v SSSR, s opisaniem dvukh novykh form [The key of the species Teratolytta (Coleoptera, Meloidae), inhabiting the U.S.S.R., with the description of two new forms from Tadjikistan]. *Entomologicheskoe Obozrenie* **38**: 855-859.

Kugelann J. G. 1794: Verzeichniss der in einigen Gegenden Preussens bis jetzt entdeckten Käferarten, nebst kurzen Nachrichten von denselben. *Neuestes Magazin für die Liebhaber der Entomologie* **1** (5): 513-582.

Kühnelt W. 1951: Zwei bemerkenswerte Tenebrioniden aus Iran. *Verhandlungen der Naturforschenden Gesellschaft in Basel* **62**: 245-247.

Kühnelt W. 1957: Ergebnisse der österreichischen Iran-Expedition 1949/50. Die Tenebrioniden Irans. *Sitzungsberichte der Österreichischen Akademie de Wissenschaften, Mathematisch-Naturwissenschaftliche Klasse* **166**, 2: 65-102.

Kulzer H. 1950: Drei neue Arten der Gattung Toxicum und Cryphaeus aus dem palaearktischen Faunengebiet (Tenebr.). *Entomologische Arbeiten aus dem Museum G. Frey* **1**: 37-42.

Kulzer H. 1952: Siebenter Beitrag zur Kenntnis der Tenebrioniden (Col.). Einige neue Gattungen und Arten der Tribus Cnodalonini aus dem Nachlass von H. Gebien in col. G. Frey. *Entomologische Arbeiten aus dem Museum G. Frey* **3**: 19-764, 2 pls.

Kulzer H. 1954: Achter Beitrag zur Kenntnis der Tenebrioniden (Col.). *Entomologische Arbeiten aus dem Museum G. Frey* **5**: 20-73, pls V-VII.

Kulzer H. 1956: Bemerkenswerte Tenebrioniden aus der Thar-Wüste. 14. Beitrag zur Kenntnis der Tenebrioniden (Col.). *Entomologische Arbeiten aus dem Museum G. Frey* **7**: 635-653.

Kulzer H. 1957: Insect of Micronesia Coleoptera: Tenebrionidae. *Insect of Micronesia* **17**: 185-256.

Kulzer H. 1959: Neue Tenebrioniden aus Süd-America (Col.). (18. Beitrag zur Kenntnis der Tenebrioniden). *Entomologische Arbeiten aus dem Museum G. Frey* **10**: 523-567.

Kulzer H. 1960: Einige neue Tenebrioniden (Col.) (20. Beitrag zur Kenntnis der Tenebrioniden). *Entomologische Arbeiten aus dem Museum G. Frey* **11**: 304-317.

Kulzer H. 1964: Über neue Tenebrionidenarten. *Entomologische Arbeiten aus dem Museum G. Frey* **15**: 221-276.

Küster H. C. 1844: *Die Käfer Europa's. Nach der Natur Beschrieben.* Nürnberg: Bauer & Raspe. Heft **1**: [4] + 100 sheets, 2 pls.

Küster H. C. 1845a: *Die Käfer Europa's. Nach der Natur Beschrieben.* Nürnberg: Bauer & Raspe. Heft **2**: **[4]** + 100 sheets, 2 pls.

Küster H. C. 1845b: *Die Käfer Europa's. Nach der Natur Beschrieben.* Nürnberg: Bauer & Raspe. Heft **3**: [4] + 100 sheets, 2 pls.

Küster H. C. 1846a: *Die Käfer Europa's. Nach der Natur Beschrieben.* Nürnberg: Bauer & Raspe. Heft **4**: [6] + 100 sheets., 3 pls.

Küster H. C. 1846b: *Die Käfer Europa's. Nach der Natur Beschrieben.* Nürnberg: Bauer & Raspe. Heft **5**: [4] + 100 sheets, 3 pls.

Küster H. C. 1846c: *Die Käfer Europa's. Nach der Natur Beschrieben.* Nürnberg: Bauer & Raspe. Heft **7**: [4] + 100 sheets, 2 pls.

Küster H. C. 1847a: *Die Käfer Europa's. Nach der Natur beschrieben.* Nürnberg: Bauer & Raspe. Heft **8**: [4] + 100 sheets, 2 pls.

Küster H. C. 1847b: *Die Käfer Europa's. Nach der Natur Beschrieben.* Nürnberg: Bauer & Raspe. Heft **9**: [4] + 100 sheets, 3 pls.

Küster H. C. 1847c: *Die Käfer Europa's. Nach der Natur Beschrieben.* Nürnberg: Bauer & Raspe. Heft **12**: [4] + 100 sheets, 2 pls.

Küster H. C. 1848a: *Die Käfer Europa's. Nach der Natur Beschrieben.* Nürnberg: Bauer & Raspe. Heft **13**: [4] + 100 sheets, 3 pls.

Küster H. C. 1848b: *Die Käfer Europa's. Nach der Natur Beschrieben.* Nürnberg: Bauer & Raspe. Heft **14**: [4] + 100 sheets, 2 pls.

Küster H. C. 1849a: *Die Käfer Europa's. Nach der Natur Beschrieben.* Nürnberg: Bauer & Raspe. Heft **16**: [4] + 100 sheets, 3 pls.

Küster H. C. 1849b: *Die Käfer Europa's. Nach der Natur beschrieben.* Nürnberg: Bauer & Raspe. Heft **18**: [4] + 100 sheets pls.

Küster H. C. 1849c: *Die Käfer Europa's. Nach der Natur Beschrieben.* Nürnberg: Bauer & Raspe. Heft **19**: [4] + 100 sheets, 3 pls.

Küster H. C. 1850a: *Die Käfer Europa's. Nach der Natur Beschrieben.* Nürnberg: Bauer & Raspe. Heft **20**: [4] + 100 sheets, 2 pls.

Küster H. C. 1850b: *Die Käfer Europa's. Nach der Natur Beschrieben.* Nürnberg: Bauer & Raspe. Heft **21**: [4] + 100 sheets, 2 pls.

Küster H. C. 1851: *Die Käfer Europa's. Nach der Natur Beschrieben.* Nürnberg: Bauer & Raspe. Heft **22**: [4] + 100 sheets, 3 pls.

Küster H. C. 1852a: *Die Käfer Europa's. Nach der Natur Beschrieben.* Nürnberg: Bauer & Raspe. Heft **24**: [6] + 100 sheets, 2 pls.

Küster H. C. 1852b: *Die Käfer Europa's. Nach der Natur Beschrieben.* Nürnberg: Bauer & Raspe. Heft **25**: [4] + 100 sheets, 2 pls.

Küster H. C. 1853: *Die Käfer Europa's. Nach der Natur Beschrieben.* Nürnberg: Bauer & Raspe. Heft **25**: [8] + 100 sheets, 2 pls.

Kuzin V. S. 1927: O geographytcheskoy izmentchivosti v rode Mylabris Fabr. (Coleoptera Meloidae) [Zur geographischen Variabilität in der Gattung Mylabris Fabr. (Coleoptera Meloidae)]. *Revue Russe d'Entomologie* **21**: 17-26.

Kuzin V. S. 1934: Novyy vid roda Psammocryptus (Coleoptera, Tenebrionidae) iz Srednei Azii. *Sbornik Trudov Gosudarstvenogo Zoologicheskogo Muzeya (pri Moskovskom Gosudarstvennom Universitete)* **1**: 160.

Kuzin V. S. 1953: Naryvniki Kazakhstana [The blister beetles of Kazakhstan]. *Trudy Respublikanskoy Stantcii Zashchity Rasteniy Kazakhstanskiy Filial Vaskhnil* (*Alma-Ata*) **1**: 72-154.

Kuzin V. S. 1954: K poznanyyu systemy narybnikov (Coleoptera, Meloidae, Mylabrini). *Trudy Vsesoyznogo Entomologicheskogo Obshchestva* **44**: 336-379.

Kuznetzova E. A. 1935: Zwei neue zentralasiatische Oedemeriden (Coleoptera). *Entomologicheskoe Obozrenie* **25**: 298-303.

Kwieton E. 1976: Une espèce nouvelle du genre Pimelia F. (Coleoptera, Tenebrionidae). *Annotationes Zoologicae et Botanicae* (Bratislava) **113**: 1-3.

Kwieton E. 1977a: Révision phylogénétique du groupe de Pimelia obsoleta (Col., Tenebrionidae). *Bulletin de la Société Entomologique de Mulhouse* **1977**: 17-24.

Kwieton E. 1977b: Esquisse phylogénétique du genre Pimelia F. *Acta Entomologica Musei Nationalis Pragae* **39**: 559-589.

Kwieton E. 1978a: Espèces nouvelles des genres Adesmia Fisch., Pimelia Sol. et Vieta (Col. Tenebrionidae). *Bulletin de la Société Entomologique de Mulhouse* **1978**: 8-12.

Kwieton E. 1978b: Sur la division de l'ancien genre Pachyscelis Sol. (Col., Tenebrionidae). *Bulletin de la Société Entomologique de Mulhouse* **1978**: 28-34.

Kwieton E. 1980a: Synopsis des espèces du genre Sepidium F. d'Algérie et de Tunisie (Col., Tenebrionidae). *Annotationes Zoologicae et Botanicae* (Bratislava) **138**: 1-19.

Kwieton E. 1980b: Contribution à la connaissance des genres Erodius Fab. et Morica Sol. (Col. Tenebrionidae) *Bulletin de la Société Entomologique de Mulhouse* **1980**: 25-28.

Kwieton E. 1981a: Coléoptères Tenebrionidae constatés par l'expédition de Tis au Pakistan. *Annotationes Zoologicae et Botanicae* (Bratislava) **143**: 1-20.

Kwieton E. 1981b: Insects of Saudi Arabia Coleoptera: Fam. Tenebrionidae, Tribe Pimeliini. *Fauna of Saudi Arabia* **3**: 402-407.

Kwieton E. 1982a: Contributions ultérieures à la connaissance du genre Pimelia F. (Col., Tenebrionidae). *Annotationes Zoologicae et Botanicae* (Bratislava) **145**: 1-38.

Kwieton E. 1982b: Notes sur les Tenebrionidae Pimeliinae du Pakistan. *Bulletin de la Société Entomologique de Mulhouse* **1982**: 21-26.

Kwieton E. 1986a: Synopsis des Asidini d'Algérie et de Tunisie I (Coleoptera, Tenebrionidae). *Annotationes Zoologicae et Botanicae* (Bratislava) **175**: 1-29.

Kwieton E. 1986b: Synopsis des Asidini d'Algérie et de Tunisie II (Coleoptera, Tenebrionidae). *Annotationes*

Zoologicae et Botanicae (Bratislava) **176**: 1-14.

Kwieton E. 1987: Mise au point sur la phylogénie et la paléogéographie du g. Pimelia F. (Col. Tenebrionidae). *Bulletin de la Société Entomologique de Mulhouse* **1987**: 9-15.

Labrique H. 1992: Description d'une nouvelle espèce de Scaurus du Maroc (Coleoptera, Tenebrionidae). *Revue Française d'Entomologie* (N. S.) **14**: 123-125.

Labrique H. 1995: Les Scaurus du groupe tingitanus Peyerimhoff (Coleoptera, Tenebrionidae). *Bulletin de la Société Entomologique de France* **100**: 119-126.

Labrique H. & Chavanon G. 2005: Une nouvelle sous-espèce de Scaurus F. (Coleoptera Tenebrionidae). *Bulletin Mensuel de la Société Linnéenne de Lyon* **74**: 41-43.

Lacordaire J. T. 1859: *Histoire naturelle des insectes. Genera des coléoptères ou exposé méthodique et critique de tous les genres proposés jusqu'ici dans ce ordre d'Insectes. Tome Cinquième.* Paris: Librairie Encyclopédique de Roret. Première partie, 1-400 pp., seconde partie, 401-750 pp.

Lacordaire J. T. 1866: *Histoire naturelle des insectes. Genera des coléoptères ou exposé méthodique et critique de tous les genres proposés jusqu'ici dans ce ordre d'Insectes. Tome Septième.* Paris: Librairie Encyclopédeique de Roret, 620 pp.

Lacordaire J. T. 1876: *Histoire naturelle des insectes. Genera des coléoptères ou exposé méthodique et critique de tous les genres proposés jusqu'ici dans ce ordre d'Insectes. Atlas.* Paris: Librairie Encyclopédique de Roret, 47 pp., 134 pls.

Lafer L. S. 1996: 100a, 101. Sem. Anthicidae (Pedilidae) – bistryanki. Pp. 26-45. In: Ler P. A. (ed.): *Opredelitel' nasekomykh Dal'nego Vostoka Rossii* [*Key to the insects of the Russian Far East*]. *Tom III. Zhestkokrylye, ili zhuki. Chast' 1.* Vladivostok: Nauka, 555 pp.

LaFerté-Sénectère F. T. de. 1842: Sur les Anthicus recueillis aux environs de Perpignan, et description de quelques espèces nouvelles. *Annales de la Société Entomologique de France* **11**: 247-260, pl. 10.

LaFerté-Sénectère F. T. de. 1847: [new taxa of Anthicidae, pp. 365-381, pl. 32]. In: Lucas P. H.: *Exploration scientifique de l'Algérie pendant les années 1840, 1841, 1842 publiée par ordre du gouvernement et avec le concours d'une Commission Académique. Sciences physiques Zoologie. Vol. II. Histoire naturelle des animaux articulés. Cinquième classe. Insectes. Premier ordre. Les coléoptères.* Paris: Imprimerie Nationale [1849], 590 pp., 47 pls.

LaFerté-Sénectère F. T. de. 1849: [new taxa]. In: Guérin-Méneville F.-E. (ed.): *Species et iconographie générique des animaux articulés ou representation des genres, avec leur description et celle de toutes les espèces de cette grande division du règne animal. Première Partie: Insectes Coléoptères (1846-1847).* Paris: de Fain et Thunot.

1849a. Notoxus. Livraison 4, No. 21, 36 pp., 1 pl.
1849b. Mecynotarsus, Livraison 4, No. 22, 5 pp., 1 pl.
1849c. Stereopalpus, Livraison 5, No. 18, 3 pp., 1 pl.
1849d. Macrarthrius, Livraison 5, No. 20, 10 pp., 1 pl.
1849e. Amblyderus, Livraison 6, No. 23, 3 pp., 1 pl.
1849f. Anthelephilus, Livraison 6, No. 24, 5 pp., 1 pl.
1849g. Formicomus, Livraison 7, No. 25, 25 pp., erratum p., 1 pl.
1849h. Tomoderus, Livraison 7, No. 26, 8 pp., 1 pl.
1849i. Anthicus (Première division), Livraison 7, No. 27, pp. 1-45, 1 pl.
1849j. Anthicus (Seconde Division), Livraison 8, No. 28, pp. 47-83, 1 pl.
1849k. Anthicus (Troisième Division), Livraison 8, No. 29, pp. 85-132, 1 pl.
1849l. Anthicus (Quatrième Division), Livraison 9, No. 30, pp. 133-181, 1 pl.
1849m. Ochthenomus, Livraison 9, No. 31, 9 pp., 1 pl.

LaFerté-Sénectére F. T. de. 1849n: *Monographie des Anthicus et genres voisins, coléoptères hétéromères de la tribu des Trachélides (1848).* Paris: De Sapia, xxii + 340 pp., pls. 17-32.

Laporte F. L. N. de Caumont de Castelnau. 1832: Mémoire sur cinquante espèces nouvelles ou peu connues d'insectes. *Annales de la Société Entomologique de France* **1**: 386-415.

Laporte F. L. N. de Caumont de Castelnau. 1833: Coléoptères et hémiptères nouveaux. *Revue Entomologique* (G. Silbermann) **1**: 32-36.

Laporte F. L. N. de Caumont de Castelnau. 1835: *Études entomologiques, ou description d'insectes nouveaux, et observations sur la synonymie.* Paris: Méquignon-Marvis, 159 pp.

Laporte F. L. N. de Caumont de Castelnau. 1840: *Histoire naturelle des insectes coléoptères; avec une introduction renfermant l'anatomie et la physiologie des animaux articulés, par M. Brullé. Tome deuxième.* Paris: P. Duménil, 563 + [1] pp., pls 20-37.

Laporte F. L. N. de Caumont de Castelnau & Brullé G. A. 1831: Monographie du genre Diaperis. *Annales des Sciences Naturalles* **23**: 325-410, pl. 10.

Latreille P. A. 1796: *Précis des caractères génériques des insectes, disposés dans un ordre naturel.* Paris: F. Bourdeaux, xiii+201 + [7] pp.

Latreille P. A. 1802: *Histoire naturelle, générale et particulière des crustacés et des insectes. Ouvrage faisant suite à l'histoire naturelle générale et particulière, composée par Leclerc de Buffon, et rédigée par C. S. Sonnini, membre de plusieurs sociétés savantes. Familles naturelles des genres. Tome troixième.* Paris: F. Dufart, xii +13-467 + [1] pp.

Latreille P. A. 1804a: *Histoire naturelle, genérale et particuliere, des crustacés et des insectes. Ouvrage faisant suite aux oeuvres de Leclerc de Buffon, et partie du cours complet d'histoire naturelle rédigé par C. S. Sonnini, membre de plusieurs Sociétés savantes. Tome dixième.* Paris: F. Dufart, 445 pp., pls 81-90.

Latreille P. A. 1804b: *Histoire naturelle, genérale et particuliere, des crustacés et des insectes. Ouvrage faisant suite aux oeuvres de Leclerc de Buffon, et partie du cours complet d'histoire naturelle rédigé par C. S. Sonnini, membre de plusieurs Sociétés savantes. Tome onzième.* Paris: F. Dufart, 422 pp.

Latreille P. A. 1807: *Genera crustaceorum et insectorum secundum ordinem naturalem in familias disposita, iconibus exemplisque plurimis explicita. Tomus secundus.* Parisiis et Argentorati: Amand Koenig, 280 pp.

Latreille P. A. 1809: *Genera crustaceorum et insectorum secundum ordinem naturalem in familias disposita, iconibus exemplisque plurimis explicita. Tomus quartus et ultimus.* Parisiis et Argentorati: Amand Koenig, 399 pp.

Latreille P. A. 1810: *Considérations générales sur l'ordre naturel des animaux composant les classes des crustacés, des arachnides, et des insectes; avec un tableau méthodique de leurs genres, disposés en familles.* Paris: F. Schoell, 444 pp.

Latreille P. A. 1816: [new taxa]. In: *Nouveau Dictionnaire d'Histoire Naturelle, appliquée aux arts, à l'agriculture, à l'économie rurale et domestique, à la médecine, etc. Par une société de naturalistes et d'agriculteurs. Nouvelle édition presque entièrement refondue et considérablement augmenté; avec des figures tirées des trois règnes de la nature. Tome I.* Paris: Deterville, [4] + lxxix + 552 pp.

Latreille P. A. 1817a: [new taxa]. In: Cuvier G.: *Le règne animal distribué d'après son organisation, pour servir de base à l'histoire naturelle des animaux et d'introduction à l'anatomie comparée. Tom III, contenant les crustacés, les arachnides et les insectes.* Paris: Deterville, xxiv + 660 pp., 2 pls.

Latreille P. A. 1817b: [new taxa]. In: *Nouveau Dictionnaire d'Histoire Naturelle, appliquée aux arts, à l'agriculture, à l'économie rurale et domestique, à la médecine, etc. Nouvelle édition presque entièrement refondue et considérablement augmentée, avec des figures tirées des trois règnes de la nature. Cor-Cun. Tome VIII.* Paris: Deterville, [4] + 602 pp., 10 pls.

Latreille P. A. 1818a: [new taxa]. In: *Nouveau Dictionnaire d'Histoire Naturelle, appliquée aux arts, à l'agriculture, à l'économie rurale et domestique, à la Médecine, etc. Nouvelle édition presque entièrement refondue et considérablement augmentée, avec des figures tirées des trois règnes de la nature. Med-Min. Tome XX.* Paris: Deterville, [2] + 586 pp.

Latreille P. A. 1818b: [new taxa]. In: *Nouveau Dictionnaire d'Histoire Naturelle, appliquée aux arts, à l'agriculture, à l'économie rurale et domestique, à la Médecine, etc. Nouvelle édition presque entièrement refondue et considérablement augmentée, avec des figures tirées des trois règnes de la nature. Mu-Nil. Tome XXII.* Paris: Deterville, [2] + 618 pp.

Latreille P. A. 1818c: [new taxa]. In: *Nouveau Dictionnaire d'Histoire Naturelle, appliquée aux arts, à l'agriculture, à l'économie rurale et domestique, à la médecine, etc. Nouvelle édition presque entièrement refondue et considérablement augmentée, avec des figures tirées des trois règnes de la nature. Pas-Pho. Tome XXV.* Paris: Deterville, [4] + 610 pp., 7 pls.

Latreille P. A. 1818d: [new taxa]. In: *Nouveau Dictionnaire d'Histoire Naturelle, appliquée aux arts, à l'agriculture, à l'économie rurale et domestique, à la médecine, etc. Nouvelle édition presque entièrement refondue et considérablement augmentée, avec des figures tirées des trois règnes de la nature. Pla-Por. Tome XXVII.* Paris: Deterville, [4] + 586 pp., 4 pls.

Latreille P. A. 1819a: [new taxa]. In: *Nouveau Dictionnaire d'Histoire Naturelle appliquée aux arts, à l'agriculture, à l'économie rurale et domestique, à la médecine, etc. par une société de naturalistes et d'agriculteurs. Nouvelle édition presqu'entièrement refondue et considérablement augmentée; avec des figures des trois règnes de la nature. Por-Ral. Tome XXVIII.* Paris: Deterville, [4] + 570 pp.

Latreille P. A. 1819b: [new taxa]. In: *Nouveau Dictionnaire d'Histoire Naturelle appliquée aux arts, à l'agriculture, à l'économie rurale et domestique, à la médecine, etc. par une société de naturalistes et d'agriculteurs. Nouvelle édition presqu'entièrement refondue et considérablement augmentée; avec des figures des trois règnes de la nature. Ram-Ryz. Tome XXIX.* Paris: Deterville, [4] + 576 pp.

Latreille P. A. 1827: Description d'insectes d'Afrique recueillis par M. Caillaud. Pp. 271-292. In: Cailliaud F.: *Voyage à Méroé, au fleuve Blanc, au-delà de Fazoql dans le midi du royaume de Sennâr, à Syouah et dans cinq autres oasis; fait dans les années 1819, 1820, 1821 et 1822. Tome IV.* Paris: Imprimerie Royale, 416 pp.

Latreille P. A. 1828: Piméliaires. Pp. 573-583. In: Bory de Saint-Vincent J. B. G. M. (ed.): *Dictionnaire classique d'histoire naturelle, par Messieurs Audoin, Isid. Bordon, Ad. Brongniart, De Candolle, Daudebard de Férussac, A. Desmoulins, Drapiez, Edwards, Flourens, Geoffroy de Saint-Hilaire, A. de Jussieu, Kunth, G. de Lafosse, Lamouroux, Latreille, Lucas fils, Presle-Duplessis, C. Prévost, A. Richard, Thiébaut de Berneaud, et Bory de Saint-Vincent. Ouvrage dirigé par ce dernier collaborateur, et dans lequel on a ajouté, pour le porter au niveau de la science, un grand nombre de mots qui n'avaient pu faire partie de la plupart des dictionnaires antérieurs.* Paris: Ray et Gravier [et] Beaudoin Frères.

Latreille P. A. 1829: Suite et fin des insectes. In: Cuvier G.: *Le règne animal distribué d'après son organisation, pour servir de base à l'histoire naturelle des animaux et d'introduction à l'anatomie comparée. Nouvelle édition, revue et augmentée. Tome V.* Paris: Déterville, xxii + 556 pp.

Lauffer G. 1905: Neue Arten und Varietäten von Coleopteren der pyrenäischen Halbinsel. *Boletín de la Real Sociedad Española de Historia Natural* **5**: 403-407.

Lawrence J. F. 1980: A new genus of Indo-Australian Gempylodini with notes on the constitution of the Colydiidae (Coleoptera). *Journal of the Australian Entomological Society* **19**: 293-310.

Lawrence J. F. 1991: Three new Asiatic Ciidae (Coleoptera: Tenebrionoidea) associated with commercial, dried fungi. *The Coleopterists Bulletin* **45**: 286-292.

Lazorko W. 1974: Zwei neue Stenalia-Arten aus der Ukraine. *Reichenbachia* **15**: 109-115.

Leach W. E. 1815a: Entomology. Pp. 57-172. In: Brewster D. (ed.): *The Edinburg Encyclopedia* **9** (1). Edinburg: Baldwin, 384 pp.

Leach W. E. 1815b: III. An essay on the British species of the genus Meloë, with descriptions of two exotic species. *Transactions of the Linnean Society of London* **11** [1810]: 35-49, 2 pls.

Lebedev A. G. 1931: Zwei neue Coleopteren aus Central-Asien. *Entomologisches Nachrichtenblatt* **5**: 78-79.

Lebedev A. G. 1932: Caediexis arenicola gen. et sp. n. (Tenebrionidae). *Entomologische Blätter* **28**: 125-126.

Leblanc P. 2002: Description de deux nouvelles espèces ouest-européennes du genre Mediimorda Méquignon, 1946. *Bulletin Mensuel de la Société Linnéenne de Lyon* **71**: 317-324.

LeConte J. L. 1850: *General remarks upon the Coleoptera of Lake Superior.* Pp. 201-242. In: Agassiz J. L. R. & Cabot J. E. (eds): *Lake Superior, its physical character, vegetation and animals, compared with those of other and similar regions. With a narrative of the tour, by J. Eliott Cabot. And contributions by other scientific gentlemen.* Boston: Gould, Kendall and Lincoln, x + 428 pp., 8 pls.

LeConte J. L. 1851: Descriptions of new species of Coleoptera, from California. *Annals of the Lyceum of Natural History, New York* **5**: 125-216.

LeConte J. L. 1854: Notice of some Coleopterous Insects, from the Collections of the Mexican Boundary Commission. *Proceedings of the Academy of Natural Sciences of Philadelphia* **7**: 79-85.

LeConte J. L. 1855: Synopsis of the Pyrochroidae of the United States. *Proceedings of the Academy of Natural Sciences of Philadelphia* **7**: 270-277.

LeConte J. L. 1856: Synopsis of the Mycetophagidae of the United States. *Proceedings of the Academy of Natural Sciences of Philadelphia* **8**: 12-15

LeConte J. L. 1857: The Coleoptera of Kansas and eastern New Mexico. *Smithsonian Contribution (to Zoology)* **11**: 1-58.

LeConte J. L. 1861: Classification of the Coleoptera of North America. Part 1. Prepared for the Smithonian Institution. *Smithsonian Miscellaneous Collections* **3**: i-xxv + 1-208.

LeConte J. L. 1862: Classification of the Coleoptera of North America. Part 1. (cont.). Prepared for the Smithonian Institution. *Smithsonian Miscellaneous Collections* **3**: 209-286.

LeConte J. L. 1863: New species of North American Coleoptera. *Smithsonian Miscellaneous Collections* **167** [1863-1866]: 1-92.

LeConte J. L. 1866: New species of North American Coleoptera. *Smithsonian Miscellaneous Collections* **6** (167): 87-168 [index 169-177].

LeConte J. L. 1874a: Descriptions of new Coleoptera, chiefly from the Pacific Slope of North America. *Transactions of the American Entomological Society* **5**: 43-72.

LeConte J. L. 1874b: [new taxa]. In: Austin E. P.: Catalogue of the Coleoptera of Mt. Washington, N. H. by E. P. Austin; with descriptions of new species, by J. L. LeConte, M. D. *Proceedings of the Boston Society of Natural History* **16**: 265-276.

LeConte J. L. 1879: New Coleoptera. *The North American Entomologist* **1** (1): 1-5, 1 pl.

Leo P. 1980a: Psammoardoinellus, nuovo genere di Opatrinae della Sardegna (Coleoptera Tenebrionidae). *Memorie della Società Entomologica Italiana* **59**: 34-36.

Leo P. 1980b: Un nuovo Stenohelops Reitt. della Sardegna (Coleoptera Tenebrionidae Helopinae). *Bollettino dela Società Entomologica Italiana* **112**: 36-39.

Leo P. 1980c: Asida argentierae n.sp. di Sardegna (Coleoptera Tenebrionidae). *Bollettino della Società Sarda di Scienzie Naturale* **19**: 189-192.

Leo P. 1982a: Nalassus genei melonii ssp. nov., dell'isola di San Pietro (Sardegna sud-occidentale) (Coleoptera Tenebrionidae). *Doriana* **5 (246)**: 1-4.

Leo P. 1982b: Gli Pseudoseriscius italiani del gruppo helvolus (Küster) (Coleoptera Tenebrionidae). *Natura* **73**: 97-107.

Leo P.1984: Belopus (Centorus) proceroides n. sp. della Sardegna (Coleoptera Tenebrionidae). *Bollettino della Società Entomologica Italiana* **116**: 165-167.

Leo P. 1985a: Due nuove Asida Latr. di Sardegna (Coleoptera Tenebrionidae). *Annali del Museo Civico di Storia Naturale «Giacomo Doria», Genova* **85**: 73-77.

Leo P. 1985b: Le popolazioni italiane di Nalassus aemulus (Küster) e descrizione di una nuova razza della Sardegna (Coleoptera Tenebrionidae Helopinae). *Bollettino del Museo Civico di Storia Naturale di Verona* **12**: 141-148.

Leo P. 1998: Nuovi dati sui tenebrionidi delle isole Toscane e descrizione di Asida (s. str.) gestroi Leoni lanzai n. ssp. (Coleoptera, Heteromera). *Atti del Museo di Storia Naturale della Maremma* **17**: 73-77.

Leo P. & Liberto A. 2003: Un nuovo genere di Helopini della Grecia (Coleoptera, Tenebrionidae). *Fragmenta Entomologica* **34** [2002]: 299-309.

Leo P. & Pisano P. 1978: Tassonomia, geonemia ed ecologia delle razze di Pimelia angusticollis Sol. e descrizione di una nuova sottospecie (Coleoptera Tenebrionidae). *Atti XI Congresso Nazionale Italiano di Entomologia Portici-Sorrento, 10-15 maggio 1976*: 79-83.

Leoni G. 1909: Le Asida italiane. *Rivista Coleopterologica Italiana* **7**: 142-189, 211-226.

Leoni G. 1910: Le Asida italiane. *Rivista Coleopterologica Italiana* **8**: 9-16, 40-55, 69-95.

Leoni G. 1911: Specie e varietã nuove o rare di coleotteri italiani. *Rivista Coleopterologica Italiana* **9**: 1-10.

Leoni G. 1929: Apunti sulle Asida (Coleoptera-Tenebrionidae) italiane e descrizione di nuove forme. *Bollettino dal Laboratorio di Entomologia del R. Instituto Superiore Agraria di Bologna* **2**: 148-154.

Le Peletier de Saint-Fargeau A. L. M. & Audinet-Serville A. J. G. 1825: [new taxa]. In: Latreille P. A., Le Peletier de Saint-Fargeau A., Audinet-Serville A. J. G. de. & Guérin-Méneville F.: *Encyclopédie méthodique. Histoire naturelle. Entomologie, ou histoire naturelle des crustacés, des arachnides et des insectes. Tome dixième.* Paris: Agasse, [4] + 832 + [1] pp.

Lepesme P. 1943: Un Tribolium inédit du Sénegal (Col. Tenebrionidae). *Revue Française d'Entomologie* **10**: 45-46.

Leske N. G. 1785: *Reise durch Sachsen in Rücksicht der Naturgeschichte und Őkonomie unternommen und beschrieben. Erstes Heft.* Leipzig: J. G. Müllersche Buchhandlung, xxx + 549 pp., 38 pls.

Lesne P. 1902a: Diagnose d'un Rhipidius nouveau des environs de Paris (Col.). *Bulletin de la Société Entomologique de France* **1902**:181-182.

Lesne P. 1902b: Notes sur deux espèces françaises de coléoptères du genre Rhipidius. *Bulletin du Muséum d'Histoire Naturelle* **8**: 420-423.

Lesne P. 1915: Les érodiens de l'Afrique orientale (coléoptères ténébrionides). *Bulletin du Muséum National d'Histoire Naturelle, Paris* **21**: 225-240.

Lesne P. 1917: Notes sur divers cisides (Col. Ciidae). *Bulletin de la Société Entomologique de France* **1917**: 190-192.

Lesne P. 1920: Notes sur divers zophosites. *Bulletin du Muséum National d'Histoire Naturelle, Paris* **26**: 14-19.

Lesne P. 1922: Bostrychides, clérides, sphindides et ténébrionides. Pp. 649-704. In: *Voyage M. le Baron Maurice de Rothschild en Éthiopie et en Afrique orientale anglaise (1904-1905). Résultats scientifiques. Animaux articulés. Deuxième partie.* Paris, Imprimerie Nationale, 1041 pp., 3 maps, 100 pls.

Letzner K. 1857: Ueber Anaspis flavo-atra, Mordella pumila, die Puppe von Opilus domesticus, die früheren Zustgände von Chrysom. cuprea F. u. 20-punctata Scop., litura F., Helodes phellandrii L., beccabungae Ill., Puppe u. Larvensack von Cryptocephalus sericeus L., Crioceris asparagi L., Rhagium bifasciatum F., Puppe von Ampedus nigrinus Payk. *Arbeiten und Veränderungen der Schlesischen Gesellschaft für Vaterländische Kultur* (Breslau) **1857**: 119-138.

Levrat J. N. B. G. 1853: Description d'une espèce nouvelle du genre Pimelia. *Annales de la Société Lynnéenne* (2) **1**: 1, 1 pl.

Levrat J. N. B. G. 1859a: Description de quelques coléoptères nouveaux des environs de Tunis. *Études Entomologiques* (Lyon) **1**: 33-36.

Levrat J. N. B. G. 1859b: Description d'une nouvelle espèce du genre Pimelia. *Études Entomologiques* (Lyon) **1**: 41-42.

Lewis G. 1879: On certain new species of Coleoptera from Japan. *The Annals and Magazine of Natural History* (5) **4**: 459-467.

Lewis G. 1887a: On the Pyrochroidae of Japan. *The Annals and Magazine of Natural History* (5) **20**: 165-175.

Lewis G. 1887b: On a new species of Diaperis from Japan. *The Entomologist* **20**: 217-218.

Lewis G. 1887c: On a new species of Phellopsis found in Japan and Siberia. *The Entomologist* **20**: 218-220.

Lewis G. 1888: On the capture of a new species of Pytho in Japan. *The Entomologist* **21**: 108-109.

Lewis G. 1891a: On two new species of Heteromera from Japan. *The Entomologist's Monthly Magazine* **27**: 70-71.

Lewis G. 1891b: On two species of Elacatis (Othnius) found in Japan. *The Entomologist's Monthly Magazine* **27**: 247-248.

Lewis G. 1894: On the Tenebrionidae of Japan. *The Annals and Magazine of Natural History* (6) **13**: 377-400, 465-485, pl. XIII.

Lewis G. 1895a: On the Cistelidae and other heteromerous species of Japan. *The Annals and Magazine of Natural History* (6) **15**: 250-278, 422-448, pl. 8.

Lewis G. 1895b: On the Dascillidae and malacoderm Coleoptera of Japan. *The Annals and Magazine of Natural History* (6) **16**: 98-122, pl. 6.

Lewis G. 1896: On new species of Coleoptera from Japan, and notice of others. *The Annals and Magazine of Natural History* (6) **17**: 329-343.

Li J.-K. 1992: *The Coleoptera Fauna of Northeast China.* Jilin: Education Publishing House, [8] + 205 pp. (in Chinese with English title).

Li J.-K. & Chen P. 1993: *Studies on fauna and ecogeography of soil animal.* Chanchun: Northeast Normal University Press, [8] + 265 pp. (in Chinese with English title).

Li Z. & Ren G.-D. 2004: A systematic study on Gnaptorina Reitter (Coleoptera: Tenebrionidae) from China. *Oriental Insects* **38**: 251-275.

Liberto A. & Leo P. 2002: Descrizione di una nuova specie di Melansis Wollaston, 1864 di Gran Canaria, Isole Canarie (Coleoptera, Tenebrionidae). *Vieraea* **30**: 19-29.

Lichtenstein A. A. 1796: *Catalogus musei Zoologici ditissimi Hamburgi, d III. Februar 1796. Auctionis lege distrahendi. Sectio Tertia Continens Insecta. Verzeichniss von höchstseltenen, aus allen Welttheilen mit vieler Mühe und Kosten zusammen gebrachten, auch aus unterschiedlichen Cabinettern, Sammlungen und Auctionen ausgehobenen Naturalien welche von einem Liebhaber, als Mitglied der Batavischen und verschiedener anderer Naturforschenden Gesellschaften gesammlet worden. Dritter Abschnitt, bestehend in wohlerhaltenen, mehrentheils ausländischen und höchstseltenen Insecten, die theils einzeln, theils mehrere zusammen in Schachteln festgesteckt sind, und welche am Mittwochen, den 3ten Februar 1796 und den folgenden Tagen auf dem Eimbeckschen Hause öffentlich verkauft werden sollen durch den Mackler Peter Hinrich Pachischefsky.* Hamburg: Schniebes, 13 + 222+ 2 pp.

Liljeblad E. 1945: *Monograph of the family Mordellidae (Coleoptera) of North America, north of Mexico.* Miscellaneous publications Museum of Zoology, University of Michigan, no. 62. 226 pp., 7 pls.

Lillig M. 1994: The subgenus Eodirosis Kwieton of the genus Erodius Fabricius, with two new species and a key to the species (Coleoptera: Tenebrionidae: Pimeliinae: Erodiini). *Israel Journal of Entomology* **28**: 151-158.

Lillig M. 1997: A new species of the genus Erodius from Israel and Egypt (Coleoptera: Tenebrionidae:

Pimeliinae). *Israel Journal of Entomology* **31**: 55-58.

Lillig M. 2001: Revision der Untergattung *Symphoxycara* Koch, 1943 der Gattung *Oxycara* Solier, 1835 (Coleoptera: Tenebrionidae, Tentyriini) *Coleoptera* **5**: 363-387.

Lillig M. 2006: Cryphaeus laticeps sp. n. from the Middle East (Coleoptera, Tenebrionidae, Toxidini). *Mittleilungen der Münchener Entomologischen Gesellschaft* **95**: 49-53.

Lillig M. & Pavlíček T. 2002: Astorthocnemis becvarorum, a new genus and new species from Middle East (Coleoptera: Tenebrionidae: Pimeliinae: Platyopini). *Mitteilungen des International Entomologischen Vereins* **27**: 97-104.

Lindberg Håk. 1925: Bidrag till kännedomen om åländska ögruppens skalbaggsfauna. *Meddelanden af Societas pro Fauna et Florà Fennica* **48**: 29-60.

Lindberg Håk. 1932: [new taxa]. In: Pic M. & Lindberg Håk.: Inventa entomologica, itineris Hispanici et Maroccani, quod a. 1926 fecerunt Harald et Håkan Lindberg. XII. Anobiidae, Cleridae, Malacodermata, Heteromera (ex parte). *Commentationes Biologicae* **3** (18): 1-37.

Lindberg Håk. 1962: Coleoptera Insularum Canariensium. III. Tenebrionidae. *Commentationes Biologicae* **25** (1): 1-85, 11 plates.

Lindberg Har. 1950: Beitrag zur Kenntnis der Käferfauna der Kanarischen Inseln. *Commentationes Biologicae* **10** (18) 1-20.

Lindberg Har. 1953: Zweiter Beitrag zur Kenntnis der Käferfauna der Kanarischen Inseln. *Commentationes Biologicae* **13** (12): 1-18.

Linder M. J. 1864: Description de trois coléoptères européens nouveaux. *Annales de la Société Entomologique de France* (4) **4**: 250-252.

Linell M. L. 1898: A new species of Aegialites. *The Canadian Entomologist* **30**: 74-75.

Linnaeus C. 1758: *Systema Naturae per regna tria naturae, secundum classes, ordines, genera, species, cum characteribus, differentiis, synonymis, locis. Tomus I. Ed. Decima, Reformata*. Holmiae: Laurentii Salvii, [5] + 6-823 + [1] pp.

Linnaeus C. 1760: *Fauna Suecica sistens animalia Sveciae regni: Mammalia, Aves, Amphibia, Pisces, Insecta, Vermes, distributa per classes & ordines, genera & species, cum differentiis specierum, synonymis auctorum, locis natalium, descriptionibus insectorum. Editio altera, auctiora.* "1761". Stockholmiae: Laurentii Salvii, [48] + 578 pp., 2 pls.

Linnaeus C. 1764: *Museum S:ae R:ae Ludovicae Ulricae Reginae. Svecorum, Gothorum, Vandalorumque &c. &c. &c. In quo Animalis rariora, exotica, imprimis Insecta & conchilia describuntur & determinantur. Prodromus instar editum*. Holmiae: L. Salvii, vi + 720 + (2).

Linnaeus C. 1767: *Systema naturae per regna tria naturae, secundum classes, ordines, genera, species, cum characteribus, differentiis, synonymis, locis. Tom I. Pars II. Editio duodecima reformata*. Holmiae: Laurentii Salvii, 533-1327 + [37] pp.

Linnaeus C. 1771: *Mantissa plantarum altera generum editionis VI & specierum editionis II*. Holmiae: Impensis Direct. Laurentii Salvii, [4] + 588 pp.

Ljungh S. I. 1823: Nya insecter, ur egan samling. *Svenska Vetenskaps Akademien Nya Handlingar* **1823** (2): 265-273.

Löbl I. 2005: Laenagenius apterus gen. et sp. n. from China (Coleoptera: Pterogeniidae). *Entomological Problems* **35**: 153-155.

Löbl I. & Merkl O. 2003: On the type species of several tenebrionid genera and subgenera (Coleoptera, Tenebrionidae). *Acta Zoologica Academiae Scientorum Hungaricae* **49**: 243-253.

Lo Cascio P. 2000: [new taxon]. In: Lo Cascio P., Bartolozzi L., Cecchi B., Dapporto L. & Sforzi A.: Contributi alla conoszenza dell'arthropodofauna dell'Isola di Pianoso (Arcipelago Toscano). 3. Coleoptera Tenebrionidae. *Bollettino della Società Entomologica Italiano* **132**: 157-174.

Lohse G. A. 1964: Die in Mitteleuropa vertretene Gattungen der Cisidae. (1. Beitrag zur Kenntnis der mitteleuropäischen Cisidae). *Entomologische Blätter* **60**: 116-122.

Lohse G. A. 1965: Über einige umstrittene Cis-Arten (2. Beitrag zur Kenntnis europäischer Cisidae). *Entomologische Blätter* **60** [1964]: 174-181.

Lohse G. A. 1966: Cisiden-Studien (3. Beitrag zur Kenntnis der Mitteleuropäischen Cisiden). *Entomologische Blätter* **62**: 25-29.

Lohse G. A. 1969: Cisiden Studien IV. *Entomologische Blätter* **65**: 48-52.

Lohse G. A. 1989: Typhaea stercorea (L.), T. crenata (Melsheimer) und T. decipiens sp. n. *Entomologische Blätter* **85**: 144-146.

Lokay E. 1907: Nový druh myrmekofilního rodu Oochotus Lucas. *Časopis České Společnosti Entomologické* **4**: 91-92.

Lopes-Andrade C. 2002: Two cases of homonymy in the genus Cis Latreille (Coleoptera: Tenebrionoidea: Ciidae). *Dugesiana* **9**: 54.

Lucas P. H. 1843: Note sur quelques nouvelles espèces d'insectes de la famille des Trachélides qui habitent les possessions françaises du nord de l'Afrique. *Revue Zoologique, par la Société Cuvierienne* (2) **6**: 145-147.

Lucas P. H. 1844: Note sur quelques nouvelles espèces d'insectes qui habitent les possessions françaises du nord de l'Afrique. *Revue Zoologique, par la Société Cuvierienne* (2) **7**: 262-267.

Lucas P. H. 1846: Pp. 1-360. In: *Exploration scientifique de l'Algérie pendant les années 1840, 1841, 1842 publiée par ordre du gouvernement et avec le concours d'une Commission Académique. Sciences physiques Zoologie. Vol. II. Histoire naturelle des animaux articulés. Cinquième classe. Insectes. Premier ordre. Les coléoptères.* Paris: Imprimerie Nationale [1849], 590 pp., 47 pls.

Lucas P. H. 1847: Pp. 361-448. In: *Exploration scientifique de l'Algérie pendant les années 1840, 1841, 1842 publiée par ordre du gouvernement et avec le concours d'une Commission Académique. Sciences physiques Zoologie. Vol. II. Histoire naturelle des animaux articulés. Cinquième classe. Insectes. Premier ordre. Les coléoptères.* Paris: Imprimerie Nationale [1849], 590 pp., 47 pls.

Lucas P. H. 1849: Pp. 449-590. In: *Exploration scientifique de l'Algérie pendant les années 1840, 1841, 1842 publiée par ordre du gouvernement et avec le concours d'une Commission Académique. Sciences physiques Zoologie. Vol. II. Histoire naturelle des animaux articulés. Cinquième classe. Insectes. Premier ordre. Les coléoptères.* Paris: Imprimerie Nationale, 590 pp., 47 pls.

Lucas P. H. 1850a: [new taxa]. *Bulletin de la Société Entomologique de France* **1850**: iii-iv.

Lucas P. H. 1850b: [new taxa]. *Bulletin de la Société Entomologique de France* **1850**: vi-vii.

Lucas P. H. 1850c: [new taxa]. *Bulletin de la Société Entomologique de France* **1850**: xi-xiii.

Lucas P. H. 1852: [new taxa]. *Bulletin de la Société Entomologique de France* **1852**: xxviii-xxix.

Lucas P. H. 1853: Essai sur les animaux articulés qui habitent l'île de Crête. *Revue et Magasin de Zoologie Pure et Appliquée* (2) **5**: 565-576.

Lucas P. H. 1854: Essai sur les animaux articulés qui habitent l'île de Crête. *Revue et Magasin de Zoologie Pure et Appliquée* (2) **6**: 28-44.

Lucas P. H. 1855a: Note sur quelques nouvelles espèces de coléoptères qui habitent les possessions français de nord de l'Afrique. *Revue et Magasin de Zoologie Pure et Appliquée* (2) **7**: 290-292.

Lucas P. H. 1855b: [new taxa]. *Bulletin de la Société Entomologique de France* **1855**: xxxiv.

Lucas P. H. 1856: [new taxa]. *Bulletin de la Société Entomologique de France* **1856**: lv-lvii.

Lucas P. H. 1857: [new taxa]. *Bulletin de la Société Entomologique de France* **1857**: lv-lvii.

Lucas P. H. 1858a: [new taxa]. *Bulletin de la Société Entomologique de France* **1858**: clxxvii-clxxxi.

Lucas P. H. 1858b: [new taxon]. *Bulletin de la Société Entomologique de France* **1858**: clxxvii-clxxxi.

Lucas P. H. 1858c: [new taxa]. *Bulletin de la Société Entomologique de France* **1858**: ccxx-ccxxiii.

Lucas P. H. 1859a: [new taxa]. *Bulletin de la Société Entomologique de France* **1859**: xxii-xxiv.

Lucas P. H. 1859b: [new taxa]. *Bulletin de la Société Entomologique de France* **1859**: cxiii-cxv.

Luigioni P. 1921: Una varietà nuova dell'Akis italica Sol. (Coleoptera – Fam. Tenebrionidae). *Atti della Pontificia Academia Romana dei Nuovi Lincei* **74**: 61.

Machado A. 1979: Consideraciones sobre el género Pimelia (Col. Tenebrionidae) en las Islas Canarias y descriptión de una nueva especie. *Boletín de la Asociación Española de Entomología* **3**: 119-127.

MacLeay W. J. 1873a: Notes on a collection of insects from Gayndah. *Transactions of the Entomological Society of New South Wales* **2** [1871]: 79-205.

MacLeay W. J. 1873b: Notes on a collection of insects from Gayndah. *Transactions of the Entomological Society of New South Wales* **2** [1872]: 239-318.

MacLeay W. J. 1887: The insects of the Cairns District, Northern Queensland. Part II. *Proceedings of the Linnean Society of New South Wales* **2**: 307-328.

MacLeay W. S. 1825: *Annulosa Javanica, or an attempt to illustrate the natural affinities and analogies of the insects collected in Java by Thomas Horsfield, M.D., F.L. & G.S. and deposited by him in the honourable East-India Company.* London: Kingsbury, Parbury, and Allen, xii + 9-50 pp., 1 pl.

Mader L. 1927: Coleopterologische Notizen. *Entomologischer Anzeiger* **7**: 193-197.

Mader L. 1929: Zur Variabilität der Meloiden und Coccinelliden. *Entomologischer Anzeiger* **9**: 108-114.

Mader L. 1931: Über paläarktische Coleopteren II. *Entomologischer Anzeiger* **11**: 134-136.

Mader L. 1943: Beitrag zur Kenntnis der Koleopteren. *Entomologische Blätter* **39**: 41-46.

Maeda M. & Nakane T. 1988: New or little-known Coleoptera from Japan and its adjacent regions, XL. Family Alleculidae. *Review of Miyazaki Sangyo-Keiei University* **1**: 1-10.

Maeda M. & Nakane T. 1991: [new taxa]. In: Nakane T.: Notes on some little known beetles (Coleoptera) in Japan. 7. *Kita-Kyūshū no Konchū* **38**: 1-9, pl. 1 (in Japanese, with title and descriptions in English).

Magistretti M. 1939: Contributo alla conoscenza degli Oedemeridi paleartici II. *Bollettino della Società Entomologica Italiana* **71**: 159-160.

Magistretti M. 1941a: Contributo alla conoscenza degli Oedemeridi paleartici IV. *Bollettino della Società Entomologica Italiana* **73**: 76-77.

Magistretti M. 1941b: Contributo alla conoscenza degli Oedemeridi paleartici III. *Mitteilungen der Münchener Entomologischen Gesselschaft* **31**: 227-234.

Magistretti M. 1943: Le specie italiane del genere Mylabris (Col. Meloidae). *Annali del Museo Civico di Storia Naturale di Genova* **62**: 30-54.

Mäklin F. W. 1863: Die Gattung Praogena und deren Repräsentanten. *Acta Societatis Scientiarum Fennicae* **7**: 555-583.

Mäklin F. W. 1864: *Monographie der Gattung Strongylium Kirby, Lacordaire und der damit zunächst verwandten Formen.* 410 pp., 3 pls.

Mäklin F. W. 1872: Synonymische und systematische Bemerkungen. *Entomologische Zeitschrift* (Stettin) **33**: 242-249.

Mäklin F. W. 1875a: Neue Canthariden. *Acta Societatis Scientificae Fennicae* **10**: 597-632.

Mäklin F. W. 1875b: Neue Cisteliden. *Acta Societatis Scientiarum Fennica* **10**: 661-682.

Mäklin F. W. 1881: Coleoptera insamlade under den Nordenskiöld'ska Expeditionen 1875 på några öar vid Norges Nordvestkust. på Novaja Semlja och ön Waigatsch samt vid Jenissej i Sibirien. *Kongliga Svenska Vetenskaps-Akademiens Handlingar* **18** (4): 1-48.

Mal N. 1984: Les espèces de Sepidium affines à S. bidentatum Solier, et description d'une espèce nouvelle du Portugal (Col. Tenebrionidae). *L'Entomologiste* **40**: 193-204.

Mamaev B. M. 1977: Redkie vidy zhestkokrylykh iz semeystva Colydiidae (Coleoptera) v faune sredney polosy evropeyskoy chasti SSSR. Pp. 27-30. In: Skarlato O. (ed.): *Novye i maloizvestnye vidy nasekomykh evropeyskoy chasti SSSR.* Leningrad: Nauka, 116 pp.

Mamaev B. M. 1978: [new taxon]. In: Mamaev B. M. & Kompantsev A. V.: Novye dannye o mycetophilnykh tenelubakh, vkluchaya rod Euryzilora Lew. (Coleoptera, Melandryidae). *Nauchnye Doklady Vyshey Shkoly. Biologicheskye Nauki* **3**: 46-49.

Mannerheim C. G. von. 1825: Novae coleopterorum species Imperii Rossici incolae descriptae. Pp. 19-41. In: Hummel A. D.: *Essais entomologiques. Bd. 1. Nr. 4.* St. Pétersbourg: Chancellerie privée du Ministère de l'Intérieur, 72 pp.

Mannerheim C. G. von. 1843a: Beitrag zur Kaefer-fauna der Aleutischen Inseln, Insel Sitkha und Neu-Californiens. *Bulletin de la Société des Naturalistes de Moscou* **16** (2): 175-314.

Mannerheim C. G. von. 1843b: Description de quelques autres nouvelles espèces de coléoptères de Finlande. *Bulletin de la Société Impériale des Naturalistes de Moscou* **16** (3): 88-99.

Mannerheim C. G. von. 1844a: Description de quelques nouvelles espèces de coléoptères de Finlande. *Bulletin de la Société Impériale des Naturalistes de Moscou* **17** (1): 189-202.

Mannerheim C. G. von. 1844b: Lettre à S. E. Mr. Fischer de Waldheim ou relation d'un voyage fait en 1844, en Suède, en Danemarck et dans le Nord de l'Allemagne. *Bulletin de la Société Impériale des Naturalistes de Moscou* **17**: 844-872.

Mannerheim C. G. von. 1846: Revue critique de quelques ouvrages récents de M. Victor de Motschulsky. *Bulletin de la Société Impériale des Naturalistes de Moscou* **19** (1): 194-254.

Mannerheim C. G. von. 1849: Insectes coléoptères de la Sibérie orientale nouveaux ou peu connus. No. 1. *Bulletin de la Société Impériale des Naturalistes de Moscou* **22** (1): 220-249.

Mannerheim C. G. von. 1852: Zweiter Nachtrag zur Kaefer-Fauna der nord-amerikanischen Laender des russischen Reiches. *Bulletin de la Société Impériale des Naturalistes de Mouscou* **25** (2) [1852-1853]: 283-372.

Mannerheim C. G. von. 1853a: Insectes coléoptères de la Sibérie orientale, nouveaux ou peu connus. *Bulletin de la Société Impériale des Naturalistes de Moscou* **25** (4) [1852-1853]: 273-309.
Mannerheim C. G. von. 1853b: Dritter Nachtrag zur Käferfauna der Nord-Amerikanischen Länder des russischen Reiches. *Bulletin de la Société Impériale des Naturalistes de Moscou* **26** (3) [1853-1854]: 95-273.
Mařan J. 1933: Beitrag zur Kenntnis der Gattung Marolia Muls. *Entomologisches Nachrichtenblatt* **7**: 110-113.
Mařan J. 1935a: De novis Alleculidarum speciebus formisque regionis palaearcticae. *Sborník Entomologického Oddělení Národního Musea v Praze* **13** (118): 141-146.
Mařan J. 1935b: Specierum subgeneris Dichillocerus Rttr. revisio (Col. Tenebrionidae Gen. Dichillus Duv.). *Sborník Entomologického Oddělení Národního Musea v Praze* **13**: 147-148.
Mařan J. 1936: De novis Alleculidarum speciebus formisque regionis palaearcticae Part II. *Sborník Entomologického Oddělení Národního Musea v Praze* **14** (137): 205-207.
Mařan J. 1940a: Conspectus specierum generis Lydus Latr. (subg. Alosimus Muls.) ex affinitate speciei Lydus collaris Fabr. *Věstník Československé Zoologické Společnosti v Praze* **8**: 73-90.
Mařan J. 1940b: Nový druh rodu Penthe Newm. z Číny. Novae speciei generis Penthe Newm. descriptio (Coleoptera: Melandryidae). *Časopis Československé Společnosti Entomologické* **37**: 87-88.
Mařan J. 1940c: Generis Allecula Fabr. regionis palaearcticae orientalis species novae. *Sborník Entomologického Oddělení Národního Musea v Praze* **18** (191): 168-171.
Mařan J. 1941a: Duae novae species Lydus Latr. (subg. Alosimus) ex affinitate speciei Lydus collaris Fabr. *Sborník Entomologického Oddělení Národního Musea v Praze* **19** (202): 60-63.
Mařan J. 1941b: De novis parumque cognitis chinensibus generis Lytta F. speciebus ex affinitate L. aeneiventris Haag. Coleoptera Meloidae. *Sborník Entomologického Oddělení Národního Musea v Praze* **19** (**223**): 180-185.
Mařan J. 1942a: Revise rodu Stenodera Eschschz. [Generis Stenodera Eschschz. revisio]. Coleoptera. Meloidae. *Sborník Entomologického Oddělení Národního Musea v Praze* **20**: 14-29.
Mařan J. 1942b: Revise podrodu Taphromeloë Rttr. [Subgeneris Taphromeloë Rttr. revisio]. Coleoptera Meloidae. Gen. Meloë L. *Sborník Entomologického Oddělení Národního Musea v Praze* **20**: 50-58.
Mařan J. 1942c: Přehled druhů rodu Lydus (subg. Alosimus Muls.) ze skupiny Lydus syriacus L. [Specierum generis Lydus (subg. Alosimus Muls.) ex affinitate speciei Lydus syriacus L. revisio]. Coleoptera Meloidae. *Sborník Entomologického Oddělení Národního Musea v Praze* **20** (234): 78-98.
Mařan J. 1944a: Novae speciei generis Mylabris F. ex insula Creta descriptio. Coleoptera. Meloidae. *Časopis České Společnosti Entomologické* **41**: 35-39.
Mařan J. 1944b: De novis Alleculidarum speciebus chinensibus. *Sborník Entomologického Oddělení Národního Musea v Praze* **21-22**: 76-77.
Mařan J. 1944c: Vorarbeiten für eine monographische Bearbeitung der Gattung Cerocoma Geoffr. *Věstník Československé Zoologické Společnosti v Praze* **9** [1941-1944]: 78-101.
Mařan J. 1944d: Generis Gonodera Muls. revisio. Coleoptera Alleculidae. *Sborník Entomologického Oddělení Národního Musea v Praze* **21-22**: 184-196.
Mařan J. 1954a: Nový palearktický druh rodu Bitoma Herbst z entomologických sbírek Národního musea v Praze. Bitoma iranica m. nov. spec. (Col., Colydiidae). *Časopis Československé Společnosti Entomologické* **50** [1953]: 132-133.
Mařan J. 1954b: Descriptions de deux espèces nouvelles du genre Mycetochara Berth. (Coleoptera, Alleculidae). *Sborník Entomologického Oddělení Národního Musea v Praze* **29**: 169-170.
Marquet C. 1897: Catalogue des coléoptères du Languedoc. Espèces observées dans quelques régions de cette province, notamment à Toulouse, Bériers, Cette, etc. *Bulletin de la Société d'Histoire Naturelle de Touluse* **31**: 5-240.
Marcuzzi G. 2001: Description of a new species of tenebrionid [sic] from the Isle of Djerba (Gulf of Gabès), Tunisia) (Coleoptera, Heteromera). *Doriana* **7** (328): 1-3.
Marcuzzi G. & Turchetto, M. 1982: Eutagenia elvirae, nuova species di coleotteri tenebrionide dell'Italia meridionale (Heteromera). *Memorie dela Società Entomologica Italiana* **60** [1981]: 235-238.
Marseul S. A. de. 1857: *Catalogue des coléoptères d'Europe.* Paris: S.-A. de Marseul, xvi + 200 pp.
Marseul S. A. de. 1867: Descriptions d'espèces nouvelles. *L'Abeille, Mémoires d'Entomologie* **4**: xxxiii-lx.
Marseul S. A. de. 1868a: Descriptions des espèces nouvelles. *L'Abeille, Journal d'Entomologie* **5**: 171-218.
Marseul S. A. de. 1868b: Archéologie entomologique. *L'Abeille, Mémoires d'Entomologie* **6** [1867-1869]: 1-82.
Marseul S. A. de. 1869: Genre Iphthimus par E. Truqui. *L'Abeille, Journal d'Entomologie* **5**: 271-275.

Marseul S. A. de. 1870a: Descriptions de coléoptères nouveaux. *L'Abeille, Mémoires d'Entomologie* **6** [1867-1869]: 368-386.

Marseul S. A. de. 1870b: Monographie des mylabrides d'Europe et des contrées limitrophes en Afrique et en Asie. *L'Abeille, Mémoires d'Entomologie* **7** (2): 1-204, 2 pls.

Marseul S. A. de. 1872a: Description d'espèces nouvelles. *L'Abeille, Mémoires d'Entomologie* **8** [1871-1872]: 413-420.

Marseul S. A. de. 1872b: Pp. 1-160. In: Repertoire des coléoptères d'Europe décrits isolément depuis 1863 (2e partie). *L'Abeille, Mémoire d'Entomologie* **9**: 1-448. [pp. 161-448 issued in 1873].

Marseul S. A. de. 1873a: Coléoptères du Japon recueillis par M. Georges Lewis. Énumération des histerides et des hétéromères avec la description des espèces nouvelles. *Annales de la Société Entomologique de France* (5) **3**: 219-230.

Marseul S. A. de. 1873b: Monographie des mylabrides. *Mémoires de la Société Royale des Sciences de Liège* (2) **3**: 363-662, 6 pls.

Marseul S. A. de. 1875a: Nouvelles et faits divers de l'Abeille No. **7**: *L'Abeille, Journal d'Entomologie* **14** [1874-1876]: xxv-xxviii.

Marseul S. A. de. 1875b: Nouvelles et faits divers de L'Abeille No. 10. Mélanges (suite). *L'Abeille, Journal d'Entomologie* **14** [1874-1876]: xxxviii-xxxix.

Marseul S. A. de. 1876a: Descriptions d'espèces nouvelles. *L'Abeille, Journal d'Entomologie* **14** [1874-1876]: 25-30.

Marseul S. A. de. 1876b: Nouvelles et faits divers de l'Abeille. Mélanges (suite). *L'Abeille, Journal d'Entomologie* **16** [1876-1878]: 35-36, 38-40.

Marseul S. A. de. 1876c: Coléoptères du Japon recueillis par M. Georges Lewis. Énumération des Hétéromères avec la description des espèces nouvelles. *Annales de la Société Entomologique de France* (5) **6**: 93-142.

Marseul S. A. de. 1876d: Coléoptères du Japon recueillis par M. Georges Lewis. 2. mémoire. Énumération des hétéromères avec la description des espèces nouvelles. 2. Partie. *Annales de la Société Entomologique de France* (5) **6**: 315-349, 447-464.

Marseul S. A. de. 1877: Coléoptères du Japon recueillis par M. Georges Lewis. 2. mémoire. Énumération des hétéromères avec la description des espèces nouvelles. 2. Partie. *Annales de la Société Entomologique de France* (5) **6** [1876]: 465-486.

Marseul S. A. de. 1878a: Anthicides recueillis par Camille van Volxem dans son voyage en Portugal, Andalousie et partie boréale du Maroc, en 1871. *Bulletin de la Société Entomologique de Belgique* **21**: xlii-xliii.

Marseul S. A. de. 1878b: Nouvelles et faits divers de l'Abeille. Deuxième Série, No. 13. Mélanges. *L'Abeille, Journal d'Entomologie* **16** [1876-1878]: 50-51.

Marseul S. A. de. 1878c: Nouvelles et faits divers de l'Abeille, Deuxième Série. - Nos. 14 et 15. Mélanges (suite). *L'Abeille, Journal d'Entomologie* **16** [1876-1878]: 54-56.

Marseul S. A. de. 1879a: Monographie des Anthicides de l'Ancien-Monde. *L'Abeille, Journal d'Entomologie* **17**: 1-268, pls. 1-2.

Marseul S. A. de. 1879b: Études sur les insectes d'Angola qui se trouvent au Muséum National de Lisbonne. *Journal de Sciences Mathématiques, Physiques et Naturelles* **25**: 43-67.

Marseul S. A. de. 1879c: Nouvelles et faits divers de l'Abeille. Mélanges (suite). *L'Abeille, Journal d'Entomologie* **17**: 74-76, 99-100.

Marseul S. A. de. 1880: [new taxa]. In: Gestro R.: Appunti sull'entomologia Tunisina. *Annali del Museo Civico di Storia Naturale di Genova* **15** [1879-1880]: 405-424.

Marseul S. A. de. 1882a: Espèces nouvelles de coléoptères de la famille des Pédilides et Anthicides du Musée Royal d'hist. nat. à Leyde. *Tijdschrift voor Entomologie* **25** [1881-1882]: 54-64.

Marseul S. A. de. 1882b: New species of Coleoptera belonging to the families Pedilidae and Anthicidae. *Notes from the Leyden Museum* **4**: 112-124.

Marseul S. A. de. 1882c: Nouvelles et faits divers de l'Abeille. Deuxième série – No 43. *L'Abeille, Journal d'Entomologie* **20**: 169-172.

Marseul S. A. de. 1883a: Nouvelles et faits divers de l'Abeille. Deuxième série – No 45. *L'Abeille, Journal d'Entomologie* **21**: 178.

Marseul S. A. de. 1883b: Nouvelles et faits divers de l'Abeille. Deuxième série – No 46. *L'Abeille, Journal d'Entomologie* **21**: 181-184.

Marseul S. A. de. 1885: Nouvelles et faits divers de l'Abeille. Première sàrie No. 36. Mélanges. *L'Abeille, Journal d'Entomologie* **22** 1886-1887]: cxlvii.

Marseul S. A. de. 1887: Catalogue des coléoptères de l'Ancien Monde Europe et contrées limitrophe en Afrique et en Asie. *L'Abeille, Journal d'Entomologie* **24** [1886-1887]: 193-360.

Marsham T. 1802: *Entomologia britannica, sistens Insecta Britanniae indigena, secundum methodum Linnaeanum disposita. Tomus I. Coleoptera.* London: Wilks & Tailor, xxxi + 547 + [1] pp.

Martinez y Sáez A. 1873a: Datos sobre algunos coleópteros de los alredadores de Cuenca. *Anales de la Sociedad Española de Historia Natural* **2**: 53-75, 1 pl.

Martinez y Sáez F. 1873b: Descripciones de coleópteros de España. *Anales de la Sociedad Española de Historia Natural* **2**: 407-417.

Masumoto K. 1981a: Tenebrionidae of Formosa (1). *Elytra* **8**: 37-53.

Masumoto K. 1981b: Notes and descriptions of Japanese Tenebrionidae (I). *Entomological Review of Japan* **35**: 29-32.

Masumoto K. 1981c: Tenebrionidae of Formosa (2). *Elytra* **9**: 15-52.

Masumoto K. 1981d: New or little known Tenebrionidae from Formosa (I). *Entomological Review of Japan* **36**: 15-26.

Masumoto K. 1981e: Tenebrionidae of Formosa (3). *Elytra* **9**: 79-99.

Masumoto K. 1982a: New or little known Tenebrionidae from Formosa (II). *Entomological Review of Japan* **36**: 143-152.

Masumoto K. 1982b: New or little known Tenebrionidae from Formosa (III). *Entomological Review of Japan* **37**: 49-53.

Masumoto K. 1982c: Tenebrionidae of Formosa (4). *Elytra* **10**: 17-32.

Masumoto K. 1982d: Tenebrionidae of Formosa (5). *Elytra* **10**: 53-72.

Masumoto K. 1983a: Notes and descriptions of Japanese Tenebrionidae (II). *Entomological Review of Japan* **38**: 81-92.

Masumoto K. 1983b: Notes and descriptions of Japanese Tenebrionidae (III). *Entomological Review of Japan* **38**: 173-176.

Masumoto K. 1984: Tenebrionidae of Formosa (6). *Elytra* **11**: 16-24.

Masumoto K. 1985a: Notes and descriptions of Japanese Tenebrionidae (IV). *Entomological Review of Japan* **40**: 21-27.

Masumoto K. 1985b: The Japanese species of Gonocephalum (Coleoptera, Tenebrionidae). *Elytra* **12**: 27-37.

Masumoto K. 1985c: Two new species of tenebrionid beetles from China. *Entomological Review of Japan* **40**: 121-124.

Masumoto K. 1986a: Two new Augolesthus (Coleoptera, Tenebrionidae) from East Asia, with notes on the known species of the genus, 250-256 pp. Entomological papers presented to Yoshihiko Kurosawa on the Occasion of his retirement. Tokyo: Coleopterist's Association of Japan.

Masumoto K. 1986b: Tenebrionidae of East Asia (III). A new genus and three new species from Taiwan. *Elytra* **14**: 61-68.

Masumoto K. 1987a: A study of the Japanese Lagriidae. *Entomological Review of Japan* **42**: 37-60.

Masumoto K. 1987b: Notes and descriptions of Japanese Tenebrionidae (V). *Entomological Review of Japan* **42**: 87-94.

Masumoto K. 1988: A study of the Taiwanese Lagriidae. *Entomological Review of Japan* **43**: 33-52 + 6 pls.

Masumoto K. 1989a: Plesiophthalmus and its allied genera (Coleoptera, Tenebrionidae, Amarygnini) (Part 3). *Japanese Journal of Entomology* **57**: 96-121.

Masumoto K. 1989b: Plesiophthalmus and its allied genera (Coleoptera, Tenebrionidae, Amarygnini) (Part 4). *Japanese Journal of Entomology* **57**: 295-317.

Masumoto K. 1989c: Plesiophthalmus and its allied genera (Coleoptera, Tenebrionidae, Amarygnini) (Part 5). *Japanese Journal of Entomology* **57**: 536-564.

Masumoto K. 1989d: Plesiophthalmus and its allied genera (Coleoptera, Tenebrionidae, Amarygnini) (Part 6). *Japanese Journal of Entomology* **57**: 742-767.

Masumoto K. 1989e: Four new species of flat large-sized camariine genera from Southeast Asia (Coleoptera, Tenebrionidae, Cnodalonini). *Kanagawa-Chûjô* (Yokohama) **90**: 229-236.

Masumoto K. 1990a: Plesiophthalmus and its allied genera (Coleoptera, Tenebrionidae, Amarygnini) (Part 7). *Japanese Journal of Entomology* **58**: 35-64.

Masumoto K. 1990b: A new Phellopsis species from South Korea (Zopheridae, Coleoptera). *Entomological Review of Japan* **45**: 87-91.

Masumoto K. 1990c: Tenebrionidae of East Asia (VI) On the genus Falsocamaria Pic (Cnodalonini). *Elytra* **18**: 93-100.

Masumoto K. 1990d: Plesiophthalmus and its allied genera (Coleoptera, Tenebrionidae, Amarygnini) (Part 8). *Japanese Journal of Entomology* **58**: 243-274.

Masumoto K. 1990e: Plesiophthalmus and its allied genera (Coleoptera, Tenebrionidae, Amarygnini) (Part 9). *Japanese Journal of Entomology* **58**: 475-505.

Masumoto K. 1990f: New Himalayan species of Laena (Coleoptera, Tenebrionidae) preserved in the collection of the National Science Museum, Tokyo. *Bulletin of the National Science Museum, Tokyo* (Ser. A) **16**: 175-196.

Masumoto K. 1990g: Plesiophthalmus and its allied genera (Coleoptera, Tenebrionidae, Amarygnini) (Part 10). *Japanese Journal of Entomology* **58**: 693-724.

Masumoto K. 1991a: Plesiophthalmus and its allied genera (Coleoptera, Tenebrionidae, Amarygnini) (Part 11). *Japanese Journal of Entomology* **59**: 1-36.

Masumoto K. 1991b: Two new tenebrionid species (Coleoptera, Tenebrionidae) from Japan. *Elytra* **19**: 25-29.

Masumoto K. 1992: Tenebrionidae of East Asia (VIII). Three new species of Campsiomorpha and Strongylium from Taiwan. *Elytra* **20**: 89-94.

Masumoto K. 1993a: Larger flattened species of Camariine genera from Asia (Part 1). *Japanese Journal of Entomology* **61**: 137-148.

Masumoto K. 1993b: Larger flattened species of Camariine genera from Asia (Part 2). *Japanese Journal of Entomology* **61**: 217-234.

Masumoto K. 1993c: A study of the genus Augolesthus Motschulsky, 1872 (Coleoptera, Tenebrionidae, Cnodalini). *Entomological Review of Japan* **48**: 37-43.

Masumoto K. 1993d: Notes and descriptions of Japanese Tenebrionidae, VI. *Entomological Review of Japan* **48**: 127-136.

Masumoto K. 1994a: Notes and descriptions of Japanese Tenebrionidae, VII. *Entomological Review of Japan* **49**: 41-46.

Masumoto K. 1994b: A study of the Asian species of the genus Ceropria (Coleoptera, Tenebrionidae, Diaperini) (part 1). *Japanese Journal of Entomology* **62**: 763-774.

Masumoto K. 1995a: A study of the Asian species of the genus Ceropria (Coleoptera, Tenebrionidae, Diaperini) (Part 2). *Japanese Journal of Entomology* **63**: 1-13.

Masumoto K. 1995b: A study of the Asian species of the genus Ceropria (Coleoptera, Tenebrionidae, Diaperini) (Part 3). *Japanese Journal of Entomology* **63**: 723-734.

Masumoto K. 1996a: Two new tenebrionid species (Coleoptera) from the Ryukyu Islands. *Japanese Journal of Entomology* **64**: 211-214.

Masumoto K. 1996b: Study of Asian Strongyliini (Coleoptera, Tenebrionidae) I. Six new species from Thailand, Laos and Taiwan, together with a new replacement name. *Elytra* **24**: 131-140.

Masumoto K. 1996c: A new toxicine genus from Taiwan (Tenebrionidae, Coleoptera). *Entomological Review of Japan* **51**: 67-69.

Masumoto K. 1996d: Fourteen new Laena (Coleoptera, Tenebrionidae) from China, Vietnam and Thailand. *Bulletin of the National Science Museum, Tokyo* (Ser. A) **22**: 165-187.

Masumoto K. 1998a: Study on Asian Strongyliini (Coleoptera, Tenebrionidae) V. Twenty new species of the genus Strongylium from East Asia. *Elytra* **26**: 173-200.

Masumoto K. 1998b: New tenebrionid beetles from East Asia (Coleoptera, Tenebrionidae). *Japanese Journal of Systematic Entomology* **4**: 305-319.

Masumoto K. 1999a: Study on Asian Strongyliini (Coleoptera, Tenebrionidae) VII. Brachypterous Strongyliines. *Elytra* **27**: 113-125.

Masumoto K. 1999b: Study on Asian Strongyliini (Coleoptera, Tenebrionidae) VIII. Ten new species of the genus Strongylium from East Asia. *Elytra* **27**: 335-352.

Masumoto K. 1999c: Additions to Plesiophthalmus and its allied genera (Coleoptera, Tenebrionidae, Amarygmini) from East Asia, Part I. *Elytra* **27**: 353-370.

Masumoto K. 2000a: Additions to Plesiophthalmus and its allied genera (Coleoptera, Tenebrionidae, Amarygmini) from East Asia, Part 2. *Elytra* **28**: 145-161.

Masumoto K. 2000b: Study of Asian Strongyliini (Coleoptera, Tenebrionidae) IX. Hairy Strongylium species from Southeast Asia. (Part 1. Species group of Strongylium gibbosulum). *Elytra* **28**: 163-172.

Masumoto K. 2000c: Study of Asian Strongyliini (Coleoptera, Tenebrionidae) X. Ten new species of the genus Strongylium from Southeast Asia. *Elytra* **28**: 391-407.

Matsumoto K. 2001a: A new genus and species of the tribe Adeliini (Coleoptera, Tenebrionidae) from the upper hypogean zone of Southwest China. *Journal of the Speleological Society of Japan* **26**: 44-49.

Masumoto K. 2001b: Additions to Plesiophthalmus and its allied genera (Coleoptera, Tenebrionidae, Amarygmini) from East Asia, Part 3. *Elytra* **29**: 57-73.

Masumoto K. 2002: Two new species of the genus Allopezus (Coleoptera, Tenebrionidae, Cnodalonini). *Special Bulletin of the Japanese Coleopterological Society, Tokyo* **5**: 371-376.

Masumoto K. 2004: Study on Asian Strongyliini (Coleoptera, Tenebrionidae). XV. Ten new Strongylium species from Southeast Asia. *Elytra* **32**: 371-388.

Masumoto K. 2005a: A new Plesiophthalmus species (Coleoptera, Tenebrionidae) from Taiwan. *Elytra* **33**: 76-78.

Masumoto K. 2005b: Study on Asian Strongyliini (Coleoptera, Tenebrionidae). XVI. Eight new Strongylium species from Taiwan. *Elytra* **33**: 193-209.

Masumoto K. & Akita K. 2001a: New or little-known tenebrionid species from Japan (Part 1). *Entomological Review of Japan* **56**: 15-22.

Masumoto K. & Akita K. 2001b: New or little-known tenebrionid species from Japan (Part 2). *Entomological Review of Japan* **56**: 75-79.

Masumoto K. & Akita K. 2001c: Two new tenebrionid species from Japan. *Special Publication of the Japan Coleopterological Society* (Osaka) **1**: 247-250.

Masumoto K. & Akita K. 2002: New or little-known tenebrionid species from Japan (Part 3). *Entomological Review of Japan* **57**: 147-153.

Masumoto K. & Akita K. 2003: New or little-known tenebrionid species from Japan (Part 4) (Coleoptera). *Entomological Review of Japan* **58**: 173-182.

Masumoto K., Akita K. & Lee C.-F. 2005: New tenebrionid beetles from Taiwan (1). *Entomological Review of Japan* **60**: 247-254.

Masumoto K., Akita K. & Utsunomiya Y. 2003: A new Misolampidius (Coleoptera, Tenebrionidae) from Sichuan, Southwest China. *Special Bulletin of the Japanese Society of Coleopterology* **6**: 327-330.

Masumoto K., Ando K. & Akita K. 2006: New or little known tenebrionid species from Japan (Part 5). A new species belonging to a new genus (Coleoptera: Tenebrionidae: Helopini). *Entomological Revue of Japan* **61**: 33-38.

Masumoto K. & Grimm R. 2004: A new genus and a species of the Palorinae (Coleoptera: Tenebrionidae) from Japan. *Entomological Review of Japan* **59**: 127-130.

Masumoto K. & Makihara H. 1997: Study on the tenebrionid beetles in South Sumatra. *Bulletin of the Forestry and Forest Products Research Institute Ibaraki* **374**: 115-153.

Masumoto K. & Merkl O. 2003: A new Spiloscapha and a new Basanus from Taiwan (Coleoptera: Tenebrionidae: Scaphidemini). *Entomological Review of Japan* **58**: 165-171.

Masumoto K. & Nishikawa N. 1986: A revisional study of the species of the genus Uloma from Japan, Korea and Taiwan (Tenebrionidae, Coleoptera). *Insecta Matsumurana* (N. S.) **35**: 17-43.

Masumoto K. & Yin Weng-y. 1993: Two new Yunnanese beetles of the genus Laena (Tenebrionidae) and Sivacrypticus (Archeocrypticidae) (Coleoptera). *Elytra* **21**: 239-243.

Masumoto K. & Yin Weng-Y. 1994: Four new Lena species (Coleoptera, Tenebrionidae) from Yunnan, Southwest China. *Elytra* **22**: 165-170.

Matsumura S. 1911: Erster Beitrag zur Insekten-Fauna von Sachalin. *The Journal of the College of Agriculture, Tohoku Imperial University, Sapporo, Japan* **4**: 1-145.

Matsumura S. 1906: *Thousand Insects of Japan* **3**. Tokyo: Keisesha Shoten, p.1-8, no. 450-737, pl. 36-55.

Matsumura M. 1915: *Taxonomy of Insects, 2.* Tokyo: Keiseisha-shotem, 316 + 10 + 10 pp., 5 pls. (in Japanese).

Mayet V. 1874: [Note sur Sitaris colletis]. *Bulletin de la Société Entomologique de France* **1873**: cxcviii-cxcix.

Mayet V. 1890: [Description d'un Opatrum nouveau de Tunisie]. *Bulletin de la Société Entomologique de France* **1890**: civ-cv.

Méhely L. 1890: Ujabb adatok Erdély s különösen a Barczaság bogárvilágának ismeretéhez [Supplementa ad faunam Coleopterorum Transsylvaniae]. *Orvos-természettudományi Értesítő* **2**: 257-360.

Medvedev G. S. 1961: Novye vidy roda Sarathropus (Coleoptera, Tenebrionidae) iz Sredney Azii. *Entomologicheskoe Obozrenie* **40**: 365-368.

Medvedev G. S. 1962: Novoe podsemeystvo chernotelok (Coleoptera, Tenebrionidae) iz Turkmenii. *Zoologicheskiy Zhurnal* **51**: 1184-1189.

Medvedev G. S. 1964a: Zoogeograficheskaya kharakteristika chernotelok (Coleoptera, Tenebrionidae) Kugitanga i opisanie novykh form. *Zoologicheskiy Zhurnal* **43**: 54-64.

Medvedev G. S. 1964b: Novye formy zhukov-chernotelok (Coleoptera, Tenebrionidae) iz Kazakhstana. *Trudy Zoologicheskogo Instituta* **34**: 159-163.

Medvedev G. S. 1964c: Novye formy chernotelok (Coleoptera, Tenebrionidae) iz Stredney Azii. *Entomologicheskoe Obozrenie* **43**: 651-661.

Medvedev G. S. 1966: Novyy rod i vid zhukov-chernotelok (Coleoptera, Tenebrionidae) iz Turkmenii. *Trudy Zoologicheskogo Instituta* **37**: 98-99.

Medvedev G. S. 1967: Obzor chernotelok triby Leptodini (Coleoptera, Tenebrionidae). *Entomologicheskoe Obozrenie* **46**: 353-360.

Medvedev G. S. 1968a: *Zhuki-chernotelki (Tenebrionidae) podsemeystvo Opatrinae Triby Platynotini, Dendarini, Pedinini, Dissonomini, Pachypterinmi, Opatrini (chast) i Heterotarsini.* Fauna SSSR Zhestkokrylye Tom XIX vypusk 2. Leningrad: Nauka, 285 pp

Medvedev G. S. 1968b: Novyy vid roda Pseudoblaps Guérin (Coleoptera, Tenebrionidae) iz Irana. *Entomologicheskoe Obozrenie* **47**: 179-180.

Medvedev G. S. 1968c: Novye chernotelki triby Akidini (Coleoptera, Tenebrionidae). *Entomologicheskoe Obozrenie* **47**: 892-898.

Medvedev G. S. 1970: Novye vidy zhukov-chernotelok (Coleoptera, Tenebrionidae) iz aridnych rayonov SSSR i Afganistana. *Entomologicheskoe Obozrenie* **49**: 393-408.

Medvedev G. S. 1972a: Novyy vid roda Leptodes Sol. (Coleoptera, Tenebrionidae) iz Tadzhikistana. *Trudy Zoologicheskogo Instituta Akademia Nauk SSSR* **52**: 309-310.

Medvedev G. S. 1972b: Novyy vid roda Dilamus Jacquelin du Val (Coleoptera, Tenebrionidae) iz Mongolskoy Narodnoy Respubliki. *Nasekomye Mongolii* **1**: 491-494.

Medvedev G. S. 1973: Polozhenie rodov Leichenum Dej. i Idisia Pask. (Coleoptera, Tenebrionidae) v sisteme i opisanie novogo roda iz severnykh Karakumov. *Entomologicheskoe Obozrenie* **52**: 644-650, 2 pls.

Medvedev G. S. 1974: Novyy vid roda Eurycaulus (Coleoptera, Tenebrionidae) v sredney Azii i Afganistane. *Zoologicheskiy Zhurnal* **53** (6): 946-948.

Medvedev G. S. 1975: Obzor zhukov-chernotelok roda Dichillus Jacquelin du Val (Coleoptera, Tenebrionidae) fauny SSSR. *Entomologicheskoe Obozrenie* **54**: 591-604, 2 pls.

Medvedev G. S. 1976: Novye vidy zhukov-chernotelok (Coleoptera, Tenebrionidae) iz Irana. *Entomologicheskoe Obozrenie* **55**: 889-902.

Medvedev G. S. 1978a: Novye vidy zhukov-chernotelok roda Dilamus Jacquelin du Val (Coleoptera, Tenebrionidae) iz Sredney Azii. *Trudy Zoologicheskogo Institute Akademia Nauk SSSR* **59**: 147-152.

Medvedev G. S. 1978b: Novye vidy zhukov-chernotelok triby Helopini (Coleoptera, Tenebrionidae) iz stredney Azii. *Trudy Zoologicheskogo Instituta Akademia Nauk SSSR* **71** [1977]: 50-52.

Medvedev G. S. 1978c: Dva novye vida zhukov-chernotelok (Coleoptera, Tenebrionidae) iz Irana. *Entomologicheskoe Obozrenie* **57**: 855-857.

Medvedev G. S. 1979a: Novyy vid zhukov-chernotelok roda Sternoplax Friv. (Coleoptera, Tenebrionidae) iz Kugitanga. *Trudy Zoologicheskogo Instituta Akademia Nauk SSSR* **82**: 40-42.

Medvedev G. S. 1979b: Zhuky-chernotelki roda Philhammus Farim. (Coleoptera, Tenebrionidae) fauny SSSR. *Trudy Zoologicheskogo Instituta Akademii Nauk SSSR* **83**: 42-45.

Medvedev G. S. 1979c: Novye vidy roda Leptodes Sol. (Coleoptera, Tenebrionidae) iz Yugo-Zapadnogo Tadzhikistana. *Trudy Zoologicheskogo Instituta Akademia Nauk SSSR* **88**: 69-73.

Medvedev G. S. 1979d: Novyy vid zhukov-chernotelok roda Gnathosia Fischer-Waldheim (Coleoptera, Tenebrionidae) iz Tadzhikistana. *Trudy Vsesoyuznogo Entomologicheskogo Obshchestva* **61**: 78-79.

Medvedev G. S. 1981: In: Angelov P. A. & Medvedev G. S.: Zhuki chernotelki (Coleoptera, Tenebrionidae) Bolgarii. *Entomologicheskoe Obozrenie* **60**: 302-315.

Medvedev G. S. 1982: Novyy vid zhukov-chernotelok roda Cnemeplatia Costa (Coleoptera, Tenebrionidae) iz Karakalpakskoy ASSR. *Trudy Zoologicheskogo Instituta Akademii Nauk SSSR* **110**: 40-41.

Medvedev G. S. 1984: Novyy vid zhukov-chernotelok roda Penthicus Faldermann (Coleoptera, Tenebrionidae)

iz Tuvy. *Nasekomye Mongolii* **9**: 298-300.

Medvedev G. S. 1987a: Zhuki-chernotelki roda Zophohelops Rtt. i blizkikh rodov (Coleoptera, Tenebrionidae) sredney Azii i Kazakhstana. *Trudy Zoologicheskogo Instituta Akademii Nauk SSSR* **164**: 95-129.

Medvedev G. S. 1987b: Obzor zhukov-chernotelkok roda Cylindrinotus Fald. (Coleoptera, Tenebrionidae) Kazakhastana i Sredney Azii. *Trudy Zoologicheskogo Instituta Akademii Nauk SSSR* **170**: 99-104.

Medvedev G. S. 1988a: Obzor chernotelok roda Penthicinus Rtt. (Coleoptera, Tenebrionidae). Pp. 55-65. In: Tobias V. I. (ed.): Sistematika nasekomykh is kleschey. *Trudy Vsesoyuznogo Entomologicheskogo Obschestva* **70**. Leningrad: Nauka, 226 pp.

Medvedev G. S. 1988b: Obzor zhukov-chernotelok roda Alcinoeta Strand (Coleoptera, Tenebrionidae). *Entomologicheskoe Obozrenie* **67** (4): 798-802.

Medvedev G. S. 1989a: Novyy vid zhukov-chernotelok roda Remipedella Sem. i polozhenie triby Remipedellini (Coleoptera, Tenebrionidae) v sisteme. *Trudy Zoologicheskogo Instituta Akademii Nauk SSSR* **208**: 90-97.

Medvedev G. S. 1989b: K nomeklature i systematike zhukov-chernotelok (Coleoptera, Tenebrionidae) Mongolii. *Nasekomye Mongolii* **10**: 371-388.

Medvedev G. S. 1990a: Novye vidy zhukov-chernotelok roda Leptodes Solier (Coleoptera, Tenebrionidae). *Trudy Zoologicheskogo Instituta Akademii Nauk SSSR* **211**: 104-116.

Medvedev G. S. 1990b: Novye vidy zhukov-chernotelok (Coleoptera, Tenebrionidae) iz Mongolii i Kitaya, pp. 132-138. In: *Nasekomye Mongolii.* Vol. 11. Leningrad, Nauka, 624 pp.

Medvedev G. S. 1991: Novye chernotelki trib Stenosini i Cnemeplatiini (Coleoptera, Tenebrionidae) mirovoy fauny. *Entomologicheskoe Obozrenie* **70**: 557-570.

Medvedev G. S. 1992: Novyy vid chernotelok roda Pachyscelis Sol. (Coleoptera, Tenebrionidae) iz yuzhnogo Kyrgyzstana. *Entomologicheskoe Obozrenie* **71**: 793-795.

Medvedev G. S. 1993: K poznaniyu zhukov-chernotelok roda Lasiostola Sol. (Coleoptera, Tenebrionidae). *Entomologicheskoe Obozrenie* **73**: 106-118.

Medvedev G. S. 1994a: Tenebrionid beetles of the tribe Stenosini from China (Coleoptera: Tenebrionidae). *Zoosystematica Rossica* **2** (2) [1993]: 307-309.

Medvedev G. S. 1994b: Novye dannye po sistematike zhukov-chernotelok triby Stenosini (Coleoptera, Tenebrionidae). *Entomologicheskoe Obozrenie* **73**: 844-867.

Medvedev G. S. 1995: Novye chernotelki roda Prosodes Eschsch. (Coleoptera, Tenebrionidae) iz Sredney Azii. I. Podrody Prosodes s. str., Mesoprosodes subgen. n., Peltarium F.-W. i Uroprosodes Rtt. *Entomologicheskoe Obozrenie* **74**: 811-854.

Medvedev G. S. 1996: Novye chernotelki roda Prosodes Eschsch. (Cvoleoptera, Tenebrionidae) iz Sredney Azii. II. Podrody Mesopersina Rtt., Dilopersina Rtt., Iranosodes subgen. n., Prosodura Rtt., Megaprosodes Rtt. i Prosodella Rtt. *Entomologicheskoe Obozrenie* **75**: 596-636.

Medvedev G. S. 1997a: Novye vidy zhukov-chernotelok roda Leptodes Sol. (Coleoptera, Tenebrionidae) iz gornykh sistem Tyan-Shanya i Paropamiza. *Entomologicheskoe Obozrenie* **76**: 176-182.

Medvedev G. S. 1997b: Novye chernotelki roda Prosodes Eschsch. (Coleoptera, Tenebrionidae) iz Sredney Azii. III. Podrody Prosodinia Rtt., Diprosodes Rtt. i Ferganoprosodes subgen. n. *Entomologicheskoe Obozrenie* **76**: 563-605.

Medvedev G. S. 1998a: A new species of tenebrionid beetle of the genus Cylindronotus from Tukmenistan (Coleoptera: Tenebrionidae). *Zoosystema Rossica* **7** (1): 138.

Medvedev G. S. 1998b: K poznaniyu zhukov-chernotelok triby Blaptini (Coleoptera, Tenebrionidae) vostochnoy chasti Tibetskogo nagorya. *Entomologicheskoe Obozrenie* **77**: 171-208.

Medvedev G. S. 1998c: Novye vidy zhukov-chernotelok triby Blaptini (Coleoptera, Tenebrionidae) iz Gissaro-Darvaza i s Tibetskogo Nagorya. *Entomologicheskoe Obozrenie* **77**: 555-586.

Medvedev G. S. 1999a: Taksonomicheskoe znachenie stroenia polovykh putey samok zhukov-chernotelok triby Blaptini (Coleoptera, Tenebrionidae) i opisanie novogo podroda i novogo vida. *Entomologicheskoe Obozrenie* **78**: 391-406.

Medvedev G. S. 1999b: Sistema zhukov-chernotelok roda Prosodes Eschsch. (Coleoptera, Tenebrionidae). I. Podrody Oliprosodes Rtt. i Prosodina Rtt. *Entomologicheskoe Obozrenie* **78**: 849-886.

Medvedev G. S. 2001: Novye vidy zhukov-chernotelok rodov Asidoblaps Fairm. i Prosodes Eschsch. (Coleoptera, Tenebrionidae) iz Kitaya i Mesopotamii. *Entomologicheskoe Obozrenie* **80**: 81-89.

Medvedev G. S. 2002: Obzor zhukov-chernotelok roda Asidoblaps Fairm. (Coleoptera, Tenebrionidae). *Entomologicheskoe Obozrenie* **81**: 64-110.

Medvedev G. S. 2003: Sistema zhukov-chernotelok roda Prosodes Eschsch, (Coleoptera, Tenebrionidae). II. Podrod Prosodinia Rtt. (Dopolnenie) *Entomologicheskoe Obozrenie* **82**: 362-415.

Medvedev G. S. 2004a: Novye vidy zhukov-chernotelok triby Blaptini (Coleoptera, Tenebrionidae) iz Indii, Nepala i Kitaya. *Entomologicheskoe Obozrenie* **83**: 163-189.

Medvedev G. S. 2004b: Novye vidy zhukov-chernotelok (Coleoptera, Tenebrionidae) iz pustyn Paleraktiki. *Entomologicheskoe Obozrenie* **83**: 553-580.

Medvedev G. S. 2005a: Sistema zhukov-chernotelok roda Prosodes Eschsch. (Coleoptera, Tenebrionidae). IV. Podrody Prosodestes Rtt. i Gebleria Motsch. *Entomologicheskoe Obozrenie* **84**: 62-107.

Medvedev G. S. 2005b: On collections of the sand desert faunas of Tenebrionidae (Coleoptera) of Middle Asia, Iran, and Afghanistan. Pp. 299-314. In: *Contribution to Systematics and Biology of Beetles. Papers Celebrating the 80th Birstday of Igor Konstantinovich Lopatin.* Sofia, Moscow: Pensoft.

Medvedev G. S. 2005c: Novye vidy zhukov-chernotelok roda Asidoblaps Fairm. (Coleoptera, Tenebrionidae) iz provinciy Gansu i Sychuanu Kitaya. *Entomologicheskoe Obozrenie* **84**: 531-568.

Medvedev G. S. 2006: New data on the tenebrionid genus Agnaptorina Rtt. (Coleoptera, Tenebrionidae) of south-western China. *Entomologicheskoe Obozrenie* **85**: 176-205, 261.

Medvedev G. S. & Abdurakhmanov G. M. 1984: Novyy vid zhukov-chernotelok roda Dila (Coleoptera, Tenebrionidae) iz Dagestana. *Zoologicheskiy Zhurnal* **63** (2): 291-293.

Medvedev G. S. & Iwan D. 2006: Notes on the taxonomy of tenebrionid beetles (Coleoptera: Tenebrionidae). *Annales Zoologici* **56**: 613-621.

Medvedev G. S. & Kaltaev T. 1979: Novye vidy zhukov-chernotelok (Coleoptera, Tenebrionidae) iz Kirgizii. *Izvestya Akademii Nauk Kirgizskoy SSR* **4**: 57-60.

Medvedev G. S. & Kaszab Z. 1973: Ergebnisse der Mongolisch-sowietischen biologischen Expeditionen in der Mongolischen Volksrepublik. Tenebrionidae, excl. Tentyriini (Coleoptera). *Folia Entomologica Hungarica* (N. S.) **26**: 79-111.

Medvedev G. S. & Kompantseva T. V. 1989: Novye vostochnopalearkticheskie vidy zhukov-chernotelok (Coleoptera, Tenebrionidae). *Trudy Zoologicheskogo Instituta Akademii Nauk SSR* **208**: 98-101.

Medvedev G. S. & Merkl O. 2001: Novye vidy zhukov-chernotelok triby Blaptini (Coleoptera, Tenebrionidae) iz yugo-zapadnogo Kitaya. *Entomologicheskoe Obozrenie* **80**: 620-626.

Medvedev G. S. & Merkl O. 2005: Two new species of Prosodes Eschscholtz, 1829 (Coleoptera, Tenebrionidae: Blaptini) from Iran. *Acta Zoologica Academiae Scientarum Hungaricae* **51**: 171-180.

Medvedev G. S. & Mihkailov V. A. 1978: Novyy vid roda Prosodes Eschsch. (Coleoptera, Tenebrionidae) iz Darvazskogo Khrebta. *Doklady Akademii Nauk Tadzikskoy* SSR **21**: 61-63.

Medvedev G. S. & Mordkovich V. G. 1970: Novyy vid zhukov chernotelok (Coleoptera, Tenebrionidae) iz Tuvy. Pp. 30-32. In: Tsherepanov A. I. (ed.): *Novye i maloizvestnye vidy sibirskoy fauny*. 3. Institut Biologii, Sibirskoye Otdelenie Akademiy Nauk SSSR. Novosibirsk: Nauka, 112 pp.

Medvedev G. S. & Nepesova M. G. 1985a: *Opredelitel zhukov-chernotelok Tukmenistana*. Ashkhabad: Ylym, 177 pp.

Medvedev G. S. & Nepesova M. G. 1985b: Novye vidy zhukov-chernotelok (Coleoptera, Tenebrionidae) iz Turkmenii i Karakalpakii. *Entomologicheskoe Obozrenie* **64**: 358-362.

Medvedev G. S. & Pirnazarov B. P. 1972: Novyy vid zhuka-chernotelki roda Leptodes Sol. (Coleoptera, Tenebrionidae) s poberezhya Aralskogo Morya. *Entomologicheskoe Obozrenie* **51**: 125-126.

Medvedev L. N. 1974: On the fauna of Anthicidae (Coleoptera) of Mongolia. Pp. 149-155. In: *Nasyekomiye Mongolii [Insects of Mongolia]. Vol. 4, Nr. 2.* Leningrad: Nauka.

Medvedev L. N. 1975: Anthicidae (Coleoptera) from Mongolia. *Acta Zoologica Academiae Scientiarum Hungaricae* **21**: 87-95.

Medvedev L. N. & Micheetchev A. N. 1979: New data of Anthicidae from Mongolia (Coleoptera). *Faunistische Abhandlungen Staatliches Museum für Tierkunde in Dresden* **7**: 141-143.

Megerle von Mühlfeld J. K. 1812: *Bemerkungen, Berichtigungen und Zusaetze zu Illiger's Zusaetzen, Berichtigungen und Bemerkungen zu Fabricii Systema Eleytheratorum.* Linz: Akad. Buchhandlung, 44 pp.

Melichar L. 1912: Neue palaearktische Coleopteren. *Coleopterologische Rundschau* **1**: 33-36.

Mellié J. 1847: [Mélanges et nouvelles]. *Revue Zoologique, par la Société Cuvierienne* **10**: 108-110.

Mellié J. 1848: Monographie de l´ancien genre Cis des auteurs. *Annales de la Société Entomologique de France* (2) **6**: 205-274, 313-396, pls 9-12.

Mellié J.1849: [Communication]. *Bulletin de la Société Entomologique de France* (2) **7**: lxxxvi-lxxxvii.

Melsheimer F. E. 1844: Descriptions of new species of Coleoptera of the United States. *Proceedings of the Academy of Natural Sciences of Philadelphia* **2** [1844-1845]: 98-118.

Melsheimer F. E. 1846: Descriptions of new species of Coleoptera of the United States. *Proceedings of the Academy of Natural Sciences of Philadelphia* **3**: 53-66.

Melsheimer F. E. 1847: Descriptions of new species of Coleoptera of the United States. *Proceedings of the Academy of Natural Sciences of Philadelphia* **3**: 158-181.

Ménétriés E. 1832: *Catalogue raisonné des objets de zoologie recueillis dans un voyage au Caucase et jusqu'aux frontières actuelles de la Perse entrepris par l'ordre de S. M. l'Empereur*. St. Petersburg: Académie des Sciences, xxxiii + 272 + iv + [1].

Ménétriés E. 1836a: Insectes nouveaux de la Turquie. *Bulletin Scientifique publié par l'Académie Impériale des Sciences de Saint-Pétersbourg* **1**: 149-151.

Ménétriés E. 1836b: Sur quelques insectes de la Russie. *Bulletin Scientifique publié par l'Académie Impériale des Sciences de Saint-Pétersbourg* **1**: 180-181.

Ménétriés E. 1838: Catalogue des insectes recueillis entre Constantinople et le Balkan. *Mémoires de l'Académie Impériale des Sciences de St.-Pétersbourg* (6) **5**: 1-52, 2 pls.

Ménétriés E. 1848: Descriptions des insectes recueillis par feu M. Lehmann. *Mémoires de l'Académie Impériale des Sciences de St.-Pétersbourg* **8** (2) (6: *Sciences Mathématiques, Physiques et Naturelles* **6** livr. 1/2): 17–66.

Ménétriés E. 1849: Catalogue des insectes recueillis par feu M. Lehmann avec les descriptions des nouvelles espèces. (Seconde et dernière partie) Coléoptères hétéromères. *Mémoires de l'Académie Impériale des Sciences de St.-Pètersbourg* **8** (2) (6: *Sciences Mathématiques, Physiques et Naturelles* **6** livr. 4): 217-328, pls 3-5.

Ménétriés E. 1854: Coléoptères recueillis dans la Mongolie chinoise et aux environs de Pékin. *Études Entomologiques* **3**: 26-41.

Meng L. & Ren G.-D. 2005: A systematic study of the genus Myatis Bates from China and adjacent areas (Coleoptera, Tenebrionidae). *Acta Zootaxonomica Sinica* **30**: 104-110.

Méquignon A. E. 1937: Note synonymique. *Bulletin de la Société Entomologique de France* **42**: 194.

Méquignon A. E. 1941: Description d'un Mordellidae (Col.) nouveau paléartique. *Bulletin de la Société Entomologique de France* **46**: 147.

Méquignon A. E. 1946: Contribution à l'étude des mordellides paléarctiques. *Revue Française d'Entomologie* **13**: 52-76.

Méquignon A. E. 1947: Notes diverses sur lers coléoptères de France (5ième note). *Bulletin de la Société Entomologique de France* **52**: 58-61.

Méquignon A. E. 1948: Notes diverses sur coléoptères de France. *Bulletin de la Société Entomologique de France* **53**: 75-80.

Merkl O. 1990: A review of Bothynogria Borchmann (Coleoptera, Tenebrionidae: Lagriini). *Acta Zoologica Hungarica* **36**: 279-294.

Merkl O. 1991a: Lagriini of the Nepal-Himalayas (Coleoptera: Tenebrionidae). *Stuttgarter Beiträge zur Naturkunde, Series A* (*Biologie*) 470: 1-18.

Merkl O. 1991b: Epitragini from Iran (Coleoptera, Tenebrionidae). *Acta Zoologica Hungarica* **37**: 59-66.

Merkl O. 2004: On taxonomy, nomenclature, and distribution of some Palaearctic Lagriini, with description of a new species from Taiwan (Coleoptera: Tenebrionidae). *Acta Zoologica Academiae Scientiarum Hungaricae* **50**: 283-305.

Merkl O. 2007: Notes on Asian Lagriini, with description of Cerogria gozmanyi sp. n. (Coleoptera: Tenebrionidae). *Acta Zoologica Academiae Scientiarum Hungaricae* **53** (Suppl. 1): 255-272.

Merkl O. & Chen B. 1997: A review of Mimoborchmania Pic (Coleoptera, Tenebrionidae: Lagriini). *Acta Zoologica Academiae Scientiarum Hungaricae* **43**: 111-119.

Merkl O. & Masumoto K. 2007: A review of Taiwanese Paramisolampidius (Coleoptera, Tenebrionidae: Cnodalonini). *Acta Zoologica Academiae Scientiarum Hungaricae* **53**: 127-137.

Meyer F. A. A. 1793: *Tentamen monographiae generis Meloës*. Gottingae: Vanderhoek et Ruprecht, 32 pp.

Miedel J. 1880: Observations sur les Opatrum. *Deutsche Entomologische Zeitschrift* **24**: 136-140.

Mikhaylov G. S. & Chikatunov M. G. 1987: *Zhetskokrylye yugo-vostochnoy chasti Pamiro-Alaya (v predelach Gorno-Badakhshanckoy avtonomnoy oblasti Tadzhikistoy SSR)*. Dushambe, 59 pp.

Miller L. 1851: [new taxa, in:Versammlung am 3. Dezember 1851]. *Verhandlungen des Zoologisch-Botanischen Vereins in Wien* **2**: 83-113.

Miller L. 1858: Zwei neue Erodiinen-Gattungen. *Wiener Entomologische Monatschrift* **2**: 115-124.

Miller L. 1861: Neue Käfer aus Kindermann's Vorräthen. *Wiener Entomologische Monatsschrift* **5**: 169-182, 201-209, 2 pls.

Miller L. 1881: [new taxa]. In: Reitter E. 1881: Coleopterologische Ergebnisse einer Reise nach Croatien, Dalmatien und der Herzegowina im Jahre 1879. *Verhandlungen der Kaiserlich-Königlichen Zoologisch-Botanischen Gesellschaft in Wien* **30** [1880]: 201-228.

Miller L. 1883: Neue Coleopteren aus Griechenland, gesammelt von E. v. Oertzen. *Verhandlungen der Kaiserlich-Königlichen Zoologisch-Botanischen Gesselschaft in Wien* **33**: 263-266.

Miwa Y. 1928: A study on the species of Meloidae in the Japanese Empire. *Transactions of the Sapporo Natural History Society* **10**: 63-78.

Miwa Y. 1930: A new Formosan blister-beetle of the genus Meloe. *Transactions of Natural History Society of Formosa* **20** (106): 12-13.

Miwa Y. 1931: A systematic catalogue of Formosan Coleoptera. *Reports of the Department of Agriculture of the Governement Research Institute, Taihoku* **55**: 1-359.

Miwa Y. 1933: An enumeration of Coleoptera from the Island Irimote in Loochoo, with descriptions of new species. *Transaction of the Natural History Society of Formosa* **23**: 4-11.

Miwa Y. 1935: Coleoptera from Amami-Islands in Loo-Choo Archipelago. *Transactions of the Kansai Entomological Society* **6**: 11-30, 2 pls. (in Japanese, English titel).

Miwa Y. 1939: Descriptions of four new species belonging to the family Tenebrionidae from Formosa. *Zoological Magazine* (Tokyo) **51**: 412-415.

Miyatake M. 1954: Studies on the Japanese Ciidae, I. *Scientific Reports of the Matsuyama Agricultural College* no.**14**: 40-67, pls. 1-11.

Miyatake M. 1955a: Descriptions of a new species of the Tetratomidae from Japan (Coleoptera). *Transactions of the Shikoku Entomological Society* **4**: 93-94.

Miyatake M. 1955b: Notes on some Japanese Ciidae (Coleoptera). *Transactions of the Shikoku Entomological Society* **4**: 94-95.

Miyatake M. 1957: Two genera of Mycetophagidae Litargus Erichson and Litargops Reitter of Japan (Coleoptera). *Transactions of the Shikoku Entomological Society* **5**: 33-45, 1 pl.

Miyatake M. 1959a: A revision of the Japanese species related to the genera Triphyllus and Pseudotriphyllus (Coleoptera, Mycetophagidae). *Entomological Review of Japan* **5**: 24-30.

Miyatake M. 1959b: The Ciidae of Amami-Oshima. (Studies on Japanese Ciidae, II). *Memoirs of the Ehime University* (6) **4**: 119-124.

Miyatake M. 1960: The genus Pisenus Casey and some notes on the family Tetratomidae (Coleoptera). *Transactions of the Shikoku Entomological Society* **6**: 121-135.

Miyatake M. 1961: The tribe Leiochrini of Japan (Coleoptera: Tenebrionidae). *Transactions of the Shikoku Entomological Society* **7**: 33-40.

Miyatake M. 1964: Notes on the tribe Bolitophagini of Japan, with the descriptions of four new genera and two new species Japan (Coleoptera: Tenebrionidae). *Transactions of the Shikoku Entomological Society*. **8**: 59-84.

Miyatake M. 1968: Two new species of Mycetophagidae from Japan (Coleoptera). *Entomological Review of Japan* **20**: 29-32.

Miyatake M. 1970: A revision of the genus Byrsax Pascoe of Japan, with some notes on the Japanese Bolitophagini (Coleoptera: Tenebrionidae). *Transactions of the Shikoku Entomological Society* **10**: 116-124, pls 4, 5.

Miyatake M. 1974: A new species of the genus Pisenus Casey from Formosa (Coleoptera, Tetratomidae). *Transactions of the Shikoku Entomological Society* **12**: 60-62.

Miyatake M. 1982a: A new genus and two new species of the tribe Hypulini from Japan (Coleoptera, Melandryidae). *Transactions of the Shikoku Entomological Society* **16**: 29-35.

Miyatake M. 1982b: Three new species of the Ciidae from Taiwan (Coleoptera). Pp.73-79. In: Satô M., Hori Y., Arita Y. & Okadome T. (eds): *Special Issue to the Memory of Retirement of Emeritus Professor Michio Chûjô*. Nagoya: Biological Laboratory, Nagoya Women's University, [2] + 185 pp., 1 pl.

Miyatake M. 1985: [new taxa]. In: Kurosawa Y., Hisamatsu S. & Sasaji H. (eds): *The Coleoptera of Japan in Color, Vol. III*. Osaka: Hoikusha Publishing Co., x + 500 pp., 192 pls. (in Japanese).

Mizota K. 1999a: A revision of the genus Indasclera in Japan (Coleoptera, Oedemeridae). *Insecta Matsumurana* **56**: 69-94.

Mizota K. 1999b: A new genus and species of Asclerini (Coleoptera: Oedemeridae) from the Yaeyama Islands (Japan) and Taiwan. *The Coleopterist's Bulletin* **53**: 287-292.

Mollandin de Boissy R. 1940: Un Rhipidius nouveau de Provence. *Revue Française d'Entomologie* **7**: 94-95.

Montrouzier X. 1860: Essai sur la faune entomologique de la Nouvelle-Calédonie (Balade). *Annales de la Société Entomologique de France* (3) **8**: 229-308.

Morawitz A. F. 1865: Ueber die in Russland und den angrenzenden Ländern vorkommenden Akis-Arten. *Horae Societatis Entomologicae Rossicae* **3** [1865-1866]: 3-48.

Morawitz F. F. 1861: Einige für die russisch-europäische Fauna neue Käfer. *Bulletin de la Société Impériale des Naturalistes de Moscou* **34** (1) [1861-1862]: 284-294.

Morishima N. 1988: A new genus and new species of the tribe Melandryini (Coleoptera, Melandryidae). *Elytra* **16**: 41-44.

Motschulsky V. de. 1839: Insectes du Caucase et des provinces transcaucasiennes. *Bulletin de la Société Impériale des Naturalistes de Moscou* **12**: 44-93, pls I, II, V, VI.

Motschulsky V. de. 1845a: Remarques sur la collection de coléoptères russes de Victor de Motschulsky. *Bulletin de la Société Impériale des Naturalistes de Moscou* **18** (1): 3-127, pls. 1-3.

Motschulsky V. de. 1845b: Observations sur le Musée entomologique de l'Université Impériale de Moscou. *Bulletin de la Société des Naturalistes de Moscou* **18** (4): 332-388.

Motschulsky V. de. 1845c: Corrections à faire dans l'article intitulé: "Remarques sur la Collection de coléoptères russes de Victor de Motchoulsky." Bulletin N° 1. 1845. *Bulletin de la Société Impériale des Naturalistes de Moscou* **18** (4): [unnumbered page at end of issue].

Motschulsky V. de. 1849: Coléoptères reçus d'un voyage de M. Handschuh dans le midi de l'Espagne, énumerés et suivis de notes. *Bulletin de la Société Impériale des Naturalistes de Moscou* **22** (3-4): 52-163.

Motschulsky V. de. 1851a: Lettres adressées à Monsieur le premier secrétaire de la Société impériale des naturalistes de Moscou. *Bulletin de la Société Impériale des Naturalistes de Moscou* **24** (2): 592-616.

Motschulsky V. de. 1851b: Lettre à Monsieur le Dr. Renard, sécretaire de la Société impériale des naturalistes de Moscou. *Bulletin de la Société Impériale des Naturalistes de Moscou* **24** (2): 649-657.

Motschulsky V. de. 1853: Nouveautés. *Études Entomologiques* **1** [1852-1853]: 77-80.

Motschulsky V. de. 1854: Diagnoses de coléoptères nouveaux, trouvés par MM. Tatarinoff et Gaschkéwitsch aux environs de Pékin. *Études Entomologiques* **2**: 44-51.

Motschulsky V. de. 1856: Voyages. Lettres de M. de Motschulsky à M. Ménétriés. N. 4. St. Pétersburg le 25 oct. 1855. *Études Entomologiques* **5**: 3-38, 1 pl.

Motschulsky V. de. 1858a: Insectes du Japon. *Études Entomologiques* **6** [1857]: 25-41.

Motschulsky V. de. 1858b: Insectes des Indes orientales. 1:ière Série. *Études Entomologiques* **7**: 20-122, 1 pl.

Motschulsky V. de. 1858c: Sur les collections coléoptèrologiques de Linné et de Fabricius. *Études Entomologiques* **7**: 123-152.

Motschulsky V. de. 1858d: Nouveautés. *Études Entomologiques* **7**: 187-190.

Motschulsky V. de. 1859a: Coléoptères rapportés de la Songarie par M. Séménof. *Mélanges Biologiques tirés du Bulletin de l'Académie Impériale des Sciences de St.-Pétersbourg* **3**: 290-309.

Motschulsky V. de. 1859b: Coléoptères rapportés de la Songarie par M. Séménov. *Bulletin de l'Académie Impériale des Sciences de Staint-Pétersburg* **1**: 301-314 [columns].

Motschulsky V. de. 1859c: [new taxa]. In: Gebler F. A.: Verzeichniss der von Herrn Schrenk in den Kreisen Ajagur und Kakaraly in den östlichen Kirgisensteppen und in der Songarey in den Jahren 1840 bis 1843 gefundeten Kaeferarten. *Bulletin de la Société Inpériale des Naturalistes de Moscou* **32** (1): 426-519.

Motschulsky V. de. 1859d: Catalogue des Insectes rapportés de environs du fl. Amour, dépuis la Schilka jusqu'à Nikolaëvsk. *Bulletin de la Société Impériale des Naturalistes de Moscou* **32** (3-4): 487-507.

Motschulsky V. de. 1860a: Insectes des Indes orientales, et de contrées analogue. *Études Entomologiques* **8** [1859]: 25-118.

Motschulsky V. de. 1860b: Insects nouveaux ou peu connus des bassins de la Méditerranée et de la mer Noire jusqu'à la mer Caspienne. *Études Entomologiques* **8** [1859]: 119-144.

Motschulsky V. de. 1860c: Coléoptères rapportés de la Sibérie orientale et notamment des pays situés sur les bords du fleuve Amour par MM. Schrenck, Maack, Ditmar, Voznessenski etc. Pp. 77-257 + [1] pp., pls 6-11, 1 map. In: Schrenck P. L.: *Reisen und Forschungen im Amur-Lande in den Jahren 1854-1856 im Auftrage der Kaiserl. Akademie der Wissenschaften zu St. Peterburg ausgeführt und in Verbindung mit mehreren Gelehrten herausgegeben von Dr.Leopold Schrenck. Band II. Zweite Lieferung. Coleopteren.* St. Peterburg: Kaiserliche Akademie der Wissenschaften, 976 pp.

Motschulsky V. de. 1860d: Coléoptères rapportés en 1859 par M. Sévertsef des steppes méridionales des Kirghises, et énumérés. *Mélanges Biologiques tirés du Bulletin de l'Académie Impériale des Sciences de St.-Pétersbourg* **3**: 408-452.

Motschulsky V. de. 1860e: Coléoptères rapportés en 1859 par M. Sévertsef des steppes méridionales des Kirghises, et énumérés. *Bulletin de l'Académie Impériale des Sciences de St.-Pétersbourg* **2**: 513-544 [columns].

Motschulsky V. de. 1861: Insectes du Japon. *Études Entomologiques* **9** [1860]: 4-39.

Motschulsky V. de. 1862: Insectes du Japon. *Études Entomologiques* **10** [1861]: 3-24.

Motschulsky V. de. 1863: Essai d'un catalogue des insectes de l'île de Ceylan. *Bulletin de la Société Impériale des Naturalistes de Moscou* **36** (2): 421-532.

Motschulsky V. de. 1866: Catalogue des insectes reçus du Japon. *Bulletin de la Société Impériale des Naturalistes de Moscou* **39** (1-2): 163-200.

Motschulsky V. 1868: Énumération des nouvelles espèces de coléoptères rapportés de ces voyages. 6-ième Article. *Bulletin de la Société Impériale des Naturalistes de Moscou* **41** (2) [1868-1869]: 170-201, pl. VIII.

Motschulsky V. de. 1870: Énumération des nouvelles espèces de coléoptères rapportés de ses voyages *Bulletin de la Société Impériale des Naturalistes de Moscou* **43** (1): 379-407, pls III, IV.

Motschulsky V. de. 1872: Énumération des nouvelles espèces de coléoptères rapportés de ses voyages. *Bulletin de la Société Impériale des Naturalistes de Moscou* **45** (3-4): 23-55.

Motschulsky V. de. 1873: Énumération des nouvelles espèces de coléoptères rapportés de ses voyages. *Bulletin de la Société Impériale des Naturalistes de Moscou* **46** (1): 466-482.

Muche W. H. 1962a: Lydus anatolicus spec. nov. (Col., Meloidae). *Reichenbachia* **1**: 51-53.

Muche W. H. 1962b: Die Podonta meiner Anatolienausbeute (Coll., Alleculidae, Omophlini). *Reichenbachia* **1**: 145-147.

Muche W. H. 1963a: Beitrag zur Kenntnis der Genus Cerocoma Geoffr. (Coleoptera, Meloidae, Cerocomini). *Reichenbachia* **2**: 11-15.

Muche W. H. 1963b: Über einige von Herrn Dr. Eckerlein in Kleinasien gesammelte Omophlini (Coleoptera, Alleculidae). *Reichenbachia* **2**: 61-65.

Muche W. H. 1964a: Neue Omophlus-Arten (Coleoptera, Alleculidae, Omophlinae). *Reichenbachia* **4**: 235-240.

Muche W. H. 1964b: Revision des Genus Omophlus Sol. (Coleoptera, Alleculidae, Omophlinae). *Entomologische Abhandlungen* **29**: 591-624.

Muche W. H. 1965a: Eine neue Unterart von Cteniopinus altaicus Gebl. (Alleculidae, Omophlinae). *Reichenbachia* 5: 207-210.

Muche W. H. 1965b: Neue Omophlus des Naturhistorischen Museums, Budapest (Coll. Alleculidae, Omophlini). *Entomologische Blätter* **61**: 85-99.

Muche W. H. 1965c: Neue Podonta-Arten (Coleoptera, Alleculidae, Omophlinae). *Reichenbachia* **6**: 83-87.

Muche W. H. 1967: Revision des Genus Heliotaurus Mulsant (Col., Alleculidae, Omophlinae). *Entomologische Zeitschrift* **77**: 17-26.

Muche W. H. 1972a: 2. Beitrag zur Alleculidenfauna der Kaukasusländer. *Entomologische Nachrichten* **10**: 127-131.

Muche W. H. 1972b: Ergebnisse der zoologischen Forschungen von Dr. Z. Kaszab in der Mongolei 288. Coleoptera, Alleculidae. *Annales Historico-Naturales Musei Nationalis Hungarici* **64**: 223-225.

Muche W. H. 1973: Die Omophlinen meiner Mittelasienausbeute 1972 (Col. Alleculidae). *Mitteilungen der Münchener Entomologischen Gesellschaft* **63**: 36-41.

Muche W. H. 1974: Eine neue Allecula aus Anatolien (Coleoptera, Alleculidae). *Fragmenta Entomologica* **10**: 219-222.

Muche W. H. 1978: Beitrag zur Kenntnis der Insektenfauna Mittelasiens (Hymenoptera, Tenthredinidae; Coleoptera, Alleculidae et Rhipiphoridae). *Entomologische Abhandlungen* **42**: 355-361.

Muche W. H. 1979a: Zwei neue Omophlus-Arten aus dem Naturhistorischen Museum Basel. (Coleoptera, Alleculidae). *Reichenbachia* **17**: 171-174.

Muche W. H. 1979b: Eine neue Allecula aus dem Zoologischen Museum Berlin (Coleoptera, Alleculinae). *Mitteilungen aus dem Zoologischen Museum in Berlin* **55**: 269-270.

Muche W. H. 1980: Omophlina muminovi sp. n. aus Tadshikistan, mit einem Beitrag zur Verbreitung tadshikischer Omophlina-Arten. *Faunistische Abhandlungen* **7**: 291-293.

Muche W. H. 1981: Eine neue Isomira-Art (Untergattung Asiomira) aus China (Coleoptera, Alleculidae). *Reichenbachia* **19**: 157-158.

Muche W. H. 1982: Insects of Saudi Arabia Coleoptera: Fam. Alleculidae. *Fauna of Saudi Arabia* **4**: 116-123.

Muche W. H. 1984: Eine neue Isomira-Art von Mallorca (Spanien). (Coleoptera, Alleculidae). *Reichenbachia* **22**: 207-208.

Muche W. H. 1986: Eine neue Mycetocharina-Art aus Afghanistan (Coleoptera, Alleculidae). *Reichenbachia* **23** (32): 181-183.

Muche W. H. 1987: Eine neue Art der Gattung Megischia SOLIER aus Griechenland (Coleoptera, Alleculidae, Omophlinae). *Reichenbachia* **24**: 135-137.

Müller C. 1887: Vierzehn neue Heteromeren, von Bradshaw im Zambesi-Gebiete aufgefunden und im Museum der Königlichen Zoologischen Gesellschaft „Natura Artis Magistra" zu Amsterdam befindlich. *Tijdschrift voor Entomologie* **30**: 297-308.

Müller J. [G.] 1902: Coleopterologische Notizen. III. *Wiener Entomologische Zeitung* **21**: 115-116.

Müller J. [G.] 1904: Beschreibung neuer dalmatinischer Koleopteren. *Münchener Koleopterologische Zeitschrift* **2** [1904-1906]: 208-210.

Müller J. 1908: Beschreibungen neuer dalmatinischer Koleopteren. *Münchner Koleopterologische Zeitschrift* **3**: 314-316.

Müller J. [G.] 1912: Ueber zwei neue Dunkelkäfer (Tenebrionidae) von der Insel Pelagosa. *Entomologische Blätter* **8**: 289-291.

Müller J. [G.] 1916: Coleopterologische Beiträge zur Fauna der österreichischen Karstprovinzen und ihrer Grenzgebiete. *Entomologische Blätter* **12**: 73-109.

Müller J. [G.] 1917: Die ostadriatischen Asida-Arten. *Wiener Entomologische Zeitung* **36**: 1-17.

Müller J. [G.] 1925: Sulle Meloë europee del gruppo rugosus. *Studi Entomologici* **1**: 21-25.

Müller J. [G.] 1929: Un nuovo Calopus (Coleopt. Oedemeridae). *Bollettino della Società Entomologica Italiana* **61**: 63-64.

Müller J. [G.] 1936: Sopra alcuni Meloidi dell'Isola di Coo (Egeo). *Bollettino della Società Entomologica Italiana* **68**: 91-92.

Müller J. [G.] 1946: Nuovi coleotteri della regione balcanica occidentale. *Redia* **31**: 107-122.

Müller J. [G.] 1950: Appunti sulle Asida di Pelagosa, Pomo ed altri scogli dell'Adriatico meridionale (Coleoptera, Tenebrionidae). *Bollettino della Società Entomologica Italiana* **80**: 12-13.

Müller O. F. 1764: *Fauna insectorum Fridrichsdalina sive methodica descriptio insectorum agri fridrichsdalensis cum characteribus genericis et specificis, nominibus trivialibus, locis natalibus, iconibus allegatis, novisve pluribus speciebus additis*. Hafniae et Lipsiae: Officina libraria Io. Frid. Gleditschii, xxiv + 96 pp.

Müller O. F. 1788: *Zoologiae Danicae prodromus, seu animalium Daniae et Norvegiae indigenarum characteres, nomina, et synonyma imprimis popularium*. Hafniae: Hallager, 32 + 282 pp.

Müller P. W. J. 1805: Bemerkungen über die Fussgliederzahl einiger Käfergattungen, in Beziehung auf Illiger's Abhandlungen über diesen Gegenstanden in dem Magazin für Insektenk. 1.B. p. 285 und folg. *Magazin der Insektenkunde* **4**: 197-219.

Müller P. W. J. 1821: Neue Insecten. *Magazin der Entomologie* (Halle) **4**: 184-230.

Mulsant E. 1830: *Lettres à Julie sur l'entomologie (en prose et en vers), suivies d'une description méthodique de la plus grande partie des insectes de France. Volume 2*. Lyon: Babeuf, 402 pp.

Mulsant E. 1851: Description d'un genre et d'un coléoptère nouveaux dans la familie des cistéliens. *Mémoires del'Academie Société de Lyon* **1**: 201-204.

Mulsant E. 1852: Description d'un genre et d'un coléoptère nouveaux dans la familie des cistéliens. *Opuscules Entomologiques* **1**: 68-71.

Mulsant E. 1854: *Histoire naturelle des coléoptères de France. Latigènes*. Paris: L. Maison, x + 396 + 2 pp.

Mulsant E. 1856a: *Histoire naturelle des coléoptères de France. Latipennes*. Paris: L. Maison, [4] + 44 + [5] pp.

Mulsant E. 1856b: *Histoire naturelle des coléoptères de France. Pectinipèdes*. Paris: L. Maison, [6] + 96 pp.

Mulsant E. 1856c: *Histoire naturelle des coléoptères de France. Barbipalpes*. Paris: L. Maison, [8] + 116 pp.

Mulsant E. 1856d: *Histoire naturelle des coléoptères de France. Longipèdes*. Paris: L. Maison, [8] + 172 pp., 2 pls.

Mulsant E. 1856e: Notes relatives à quelques insectes coléoptères de la tribu des pectinipèdes. *Opuscules Entomologiques* **7**: 17-59.

Mulsant E. 1856f: Histoire des coléoptères de France. Barbipalpes, longipèdes, latipennes. *Annales de la Société Linnéenne de Lyon* (N. S.) **3**: 193-544.

Mulsant E. 1857a: *Histoire naturelle de coléoptères de France. Vésicants*. Paris: Magnin & Blanchard et Cie., 201 pp., 1 pl. + [9pp. suppl.].

Mulsant E. 1857b: [new taxon]. In: Mulsant E. & Rey C.: Description de quelques coléoptères de la tribu des longipèdes. *Annales de la Société Linnéenne de Lyon* (N. S.) **4**: 410-413.

Mulsant E. 1858a: Tableau méthodique des coléoptères angustipennes de France. *Annales de la Société Linnéenne de Lyon* (N. S.) **5**: 65-238.

Mulsant E. 1858b: Description d'un coléoptère nouveau de la tribu des vésicants. *Annales de la Société Linnéenne de Lyon* (N. S.) **8**: 239-242.

Mulsant E. 1859: *Histoire naturelle des coléoptères de France. Rostrifères*. Paris: Magnin et Blanchard et Cie, [8] + 56 pp.

Mulsant E. & Godart A. 1855a: Description d'une espèce nouvelle d'Helops. *Opuscules Entomologiques* **6**: 83-86.

Mulsant E. & Godart A. 1855b: Description d'une espèce nouvelle d'Helops. *Annales de la Société Linnéenne de Lyon* (N. S.) **2**: 153-156.

Mulsant E. & Godart A. 1855c: Description de quelques coléoptères nouveaux ou peu connus. *Opuscules Entomologiques* **6**: 161-183.

Mulsant E. & Godart A. 1855d: Description de quelques coléoptères nouveaux ou peu connus. *Annales de la Société Linnéenne de Lyon* (N. S.) **2**: 261-283.

Mulsant E. & Godart A. 1856: Espèce nouvelle de coléoptère du genre Orchesia. *Annales de la la Société Linnéenne de Lyon* (N. S.) **3**: 108-110.

Mulsant E. & Godart A. 1859a: Description de trois espèces nouvelles de coléoptères. *Opuscules Entomologiques* **9**: 177-183.

Mulsant E. & Godart A. 1859b: Description de trois espèces nouvelles de coléoptères. *Annales des Sciences Physiques et Naturelles, d'Agriculture et d'Industrie publiées par la Société Impériale d'Agriculture, etc. de Lyon* (3) **3**: 241-252.

Mulsant E. & Godart A. 1860: Description d'une espèce nouvelle de coléoptère angustipenne. *Annales de la Société Linnéenne de Lyon* (N. S.) **7**: 158-161.

Mulsant E. & Godart A. 1865: Description de quelques coléoptères nouveaux ou peu connus. *Annales de la Société Linnéenne de Lyon* (N. S.) **12**: 447-456.

Mulsant E. & Godart A. 1868: Description de deux espèces nouvelles d'Alphitobius (Coléoptères de la tribu des latigènes, famille des ulomiens). *Annales de la Société Linnéenne de Lyon* (N. S.) **16**: 288-291.

Mulsant E. & Godart A. 1870: Description de trois espèces nouvelles de coléoptères. *Opuscules Entomologiques* **14**: 49-53.

Mulsant E. & Godart A. 1876a: Description d'une espèce nouvelle de coléoptères latigènes servant à former un genre nouveau. *Annales de la Société Linnéenne de Lyon* (N. S.) **22** [1875]: 181-183.

Mulsant E. & Guillebeau F. 1855a: Notes pour servir à l'histoire des lagries (Insectes coléoptères hétéromères). *Annales de la Société Linnéenne de Lyon* (N. S.) **2**: 65-74.

Mulsant E. & Guillebeau F. 1855b: Description d'une espèce nouvelle d'Uloma (Coléoptère de la tribu des latigènes). *Annales de la Société Linnéenne de Lyon* (N. S.) **2**: 421-424.

Mulsant E. & Guillebeau F. 1857: Description d'une espèces nouvelle du genre Orchesia de la tribu des coléoptères brevipalpes. *Annales de la la Société Linnéenne de Lyon* (N. S.) **4**: 414-416.

Mulsant E. & Mulsant V. 1855: Description d'une espèce nouvelle de genre mycétochare (Coléoptère de la tribu des pectinipèdes). *Opuscules Entomologiques* **6**: 155-157.

Mulsant E. & Revelière E. 1860: Description d'un coléoptère nouveau constituant un genre nouveau dans la tribu des opatates. *Annales de la Société Linnéenne de Lyon* (N. S.) **7**: 153-157.

Mulsant E. & Rey C. 1853a: Description de quelques coléoptères nouveaux ou peu connus. *Annales de la Société Linnéenne de Lyon* (N. S.) **1** [1852-1853]: 2-9.
Mulsant E. & Rey C. 1853b: Description de quelques coléoptères nouveaux ou peu connus. *Opuscules Entomologiques* **2**: 4-11.
Mulsant E. & Rey C. 1853c: Essai d'une division des derniers mélasomes. *Opuscules Entomologiques* **4**: 1-235 + [4] pp., pls I-IV.
Mulsant E. & Rey C. 1853d: Essai d'une division des derniers mélasomes. *Mémoires de l'Académie des Sciences, Belles-Lettres et Arts de Lyon, Classe des Sciences* (2) **2** [1852]: 226-329.
Mulsant E. & Rey C. 1853e: Essai d'une division des derniers mélasomes. *Mémoires de l'Académie des Sciences, Belles-Lettres et Arts de Lyon, Classe des Sciences* (2) **3**: 20-151 + [4], pls I-IV.
Mulsant E. & Rey C. 1853f: Description d'un coléoptère constituant un genre nouveau parmi les taxicornes. *Opuscules Entomologiques* **2**: 185-188.
Mulsant E. & Rey C. 1854a: Essai d'une division des derniers mélasomes. *Opuscules Entomologiques* **5**: 9-255.
Mulsant E. & Rey C. 1854b: Essai d'une division des derniers mélasomes. *Mémoires de l'Académie Impériale des Sciences, Belles-Lettres et Arts de Lyon. Classe des Sciences* (N. S.) **4**: 155-332.
Mulsant E. & Rey C. 1855a: Description des omocrates et des héliopathes derniers insectes de la tribu des pandarites. *Annales de la Société Linnéenne de Lyon* (N. S.) **2**: 77-142.
Mulsant E. & Rey C. 1855b: Description d'une nouvelle espèce de Pandarinus. *Opuscules Entomologiques* **6**: 94-96.
Mulsant E. & Rey C. 1855c: Description d'une nouvelle espèce de Pandarinus (Coléoptère de la tribu des pandarites). *Annales de la Société Linnéenne de Lyon* (N. S.) **2**: 194-197.
Mulsant E. & Rey C. 1856: Description d'une espèce nouvelle de coléoptères de la tribu des latigènes. *Opuscules Entomologiques* **7**: 15-16.
Mulsant E. & Rey C. 1858a: Coup d'oeil sur les insectes de la famille des cantharidiens accompagné de la description de diverses espèces nouvelles ou peu connues. *Mémoires de l'Académie Impériale des Sciences, Belles-Lettres et Arts de Lyon. Classe des Sciences* (N. S.) **8**: 122-220, 3 pls.
Mulsant E. & Rey C. 1858b: Description d'une espèce nouvelle de Phaleria (Coléoptères latigène). *Annales des Sciences Physiques et Naturelles, d'Agriculture et d'Industrie publiées par la Société Impériale d'Agriculture, etc. de Lyon* (3) **2**: 310-312.
Mulsant E. & Rey C. 1858c: Description d'une espèce constituant un genre nouveau dans la famille des mordelliens. *Annales des Sciences Physiques et Naturelles, d'Agriculture et d'Industrie publiées par la Société Impériale d'Agriculture, etc. de Lyon* (3) **2**: 313-315.
Mulsant E. & Rey C. 1859a: Description d'une nouvelle espèce de Phaleria (Coléoptères latigènes). *Opuscules Entomologiques* **9**: 52-54.
Mulsant E. & Rey C. 1859b: Essai d'une division des derniers mélasomes. *Mémoires de l'Académie des Sciences, Belles-Lettres et Arts de Lyon. Classe des Sciences* (N. S.) **9**: 62-155.
Mulsant E. & Rey C. 1859c: Essai d'une division des derniers mélasomes. *Opuscules Entomologiques* **9**: 65-137.
Mulsant E. & Rey C. 1859d: Description de quelques coléoptères nouveaux. *Opuscules Entomologiques* **9**: 159-176.
Mulsant E. & Rey C. 1859e: Essai d'une division des derniers mélasomes. Familles des parvilabres quatrième tribu opatrites. *Opuscules Entomologiques* **10**: 1-160.
Mulsant E. & Rey C. 1859f: Description de quelques coléoptères nouveaux. *Annales des Sciences Physiques et Naturelles, d'Agriculture et d'Industrie, publiées par du Société Impériale d'Agriculture, etc., de Lyon* (3) **3**: 222-240.
Mulsant E. & Rey C. 1860: Essai d'une division des derniers mélasomes. Famille des parvilabres quatrième tribu opatrites. *Mémoires de l'Académie des Sciences, Belles-Lettres et Arts de Lyon. Classe des Sciences* (N. S.) **10**: 1-68.
Mulsant E. & Rey C. 1861: Description de deux coléoptères nouveaux ou peu connus. *Opuscules Entomologiques* **12**: 189-192.
Mulsant E. & Rey C. 1866a: *Histoire naturelle des coléoptères de France. Colligères. Volume 17*. Paris: F. Savy, 4 + 188 pp., 3 pls.
Mulsant E. & Rey C. 1866b: Histoire naturelle des coléoptères de France. Tribu des colligères. *Annales de la Société Linnéenne de Lyon* (N. S.) **3**: 89-282.

Mulsant E. & Wachanru A. 1853: Première série de coléoptères nouveaux ou peu connus. *Mémoires de l'Académie des Sciences, Belles-Lettres et Arts de Lyon, Classe des Sciences* **2** [1852]: 1-17.

Munster T. G. 1921: Insecta, ex Sibiria meridionali et Mongolia, in itinere Ørjan Olsen 1914 collecta. A. Coleoptera, a Fritz Jensen lecta. I. Oedemeridae. *Norsk Entomologisk Tidsskrift* **1**: 8-10.

Munster T. G. 1924: Nova etc. ex Norvegia. *Norsk Entomologisk Tidsskrift* **1**: 288-294.

Munster T. G. 1926: Tillaeg og bemaerkninger til norges koleopterfauna. *Norsk Entomologisk Tidsskrift* **2** (3): 158-168 [cont. **2**(4) (1927): 169-200, **2**(5) (1928): 262-298].

Munster T. G. 1928: Tillaeg og bemaerkningen til norges koleopterfauna. *Norsk Entomologisk Tidsskrift* **2**: 262-298.

Nabozhenko M. V. 2000: Novye vidy zhukov-chernotelok roda Cylindronotus Fald. (Coleoptera, Tenebrionidae) s Severnogo Kavkaza. *Entomologicheskoe Obozrenie* **79**: 107–111.

Nabozhenko M. V. 2001a: Taxonomic notes on the genus Zophohelops Reitter, 1901with description of a new species from Tadzhikistan and a new genus Pseudoprobaticus gen. nov. *Annales Zoologici* **51**: 511-515.

Nabozhenko M. V. 2001b: O sisteme zhukov-chernotelok tribi Helopini i obzor rodov Nalassus Mulsant i Odocnemis Allard (Coleoptera, Tenebrionidae) evropeyskoy chasti SNG i Kavkaza. *Entomologicheskoe Obozrenie* **80**: 627-668.

Nabozhenko M. V. 2002a: Novyy rod zhukov-chernotelok triby Helopini (Coleoptera, Tenebrionidae). *Vestnik Zoologii* **36**: 41-46.

Nabozhenko M. V. 2002b: Zhuki-chernotelki rodov Hedyphanes Fischer i Entomogonus Solier (Coleoptera, Tenebrionidae, Helopini) Kavkaza. *Entomologicheskoe Obozrenie* **81**: 684-692.

Nabozhenko M. V. 2005: New synonymies and new species of the genus Hedyphanes Fischen von Waldheim, 1822 (Coleoptera, Tenebrionidae, Helopini). *Acta Zoologica Academiae Scientarum Hungaricae* **51**: 349-355.

Nabozhenko M. V. 2006a: Obzor iranskikh vidov podroda Helopocerodes Reitter, 1922 roda Nalassus Mulsant, 1854 (Coleoptera: Tenebrionidae). *Trudy Russkogo Entomologicheskogo Obshestva* **77**: 245-249.

Nabozhenko M. V. 2006b: Revisiya roda Catomus Allard, 1876 i sblizhayemykh s nim rodov (Coleoptera, Tenebrionidae) fauny Kavkaza, Sredney Azii i Kitaya. *Entomologicheskoe Obozrenie* **85**: 798-857.

Nabozhenko M. V. & Dzambazishvili M. Ya. 2001: Novye vidy zhukov-chernotelok roda Nalassus Muls. (Coleoptera, Tenebrionidae) iz Gruzii. *Entomologicheskoe Obozrenie* **80**: 90-95.

Nabozhenko M. V. & Gurgenidze L. N. 2006: Description of the larva of Cylindrinotus gibbicollis Faldermann, 1837 and notes on the classification of the subtribe Cylindrinotina Español, 1956 (Coleoptera, Tenebrionidae: Helopini). *Kavkazskiy Entomologichesky Bulleten* **2** (1): C. 79-82.

Nabozhenko M. V. & Medvedev G. S. 2006: [new taxon]. In: Nabozhenko M. V.: Revisiya roda Catomus Allard, 1876 i sblizhayemykh s nim rodov (Coleoptera, Tenebrionidae) fauny Kavkaza, Sredney Azii i Kitaya. *Entomologicheskoe Obozrenie* **85**: 798-857.

Nabozhenko M. V. & Tichý V. 2006: A new species of the genus Odocnemis Allard, 1876 (Coleoptera, Tenebrionidae) from Turkey. *Kavkazskiy Entomologichesky Bulleten* **2** (2): 183-185.

Naezen D. E. 1794: Beskrifning pa nagrsa vid Umea funde Insekter dels okande, dels forut otydeligen bemarkte, och i Fauna Suecica ej uptagne. *Kongliga Vetenskaps Academiens nya Handlingar* **15**: 256-274, pl. 1.

Nakakita T. 1987: Two new species of the genus Addia from Taiwan (Coleoptera, Tenebrionidae). *Entomological Review of Japan* **42** *Supplement*: 61-64.

Nakane T. 1949: Descriptions of two new species of mordellid-beetles from Japan. *Mushi* **20**: 39-41.

Nakane T. 1950: New or little known Coleoptera from Japan and its adjacent regions V. *Transactions of Kansai Entomological Society* **15**: 20-24.

Nakane T. 1954: New or little-known Coleoptera from Japan and its adjacent regions, XI. Oedemeridae. *The Scientific Reports of the Saikyo University* (A) **1**: 171-188.

Nakane T. 1955: New or little-known Coleoptera from Japan and its adjacent regions. XII. *The Scientific Reports of the Saikyo University* (A) **2**: 24-42.

Nakane T. 1956a: Entomological results from the scientific survey of the Tokara islands. *Publication of the Osaka Municipal Museum of Natural History* **9**: 1-10.

Nakane T. 1956b: New and little-known Coleoptera from Japan and its adjacent regions. XIII. *The Scientific Reports of the Saikyo University* (A) **2** (3): 164-174.

Nakane T. 1957: Einige Mordelliden aus Japan. *Insecta Matsumurana* **21**: 48-52.

Nakane T. 1958: On the Coleoptera of Shimokita Peninsula, northern end of Honshu, Japan (Insecta) I. *Miscellaneous Reports of the Research Institute for Natural Resources* **46-47**: 83-92 (in Japanese, with English title and summary).

Nakane T. 1960: Einige Bemerkungen über die japanischen Mordelliden. *Entomological Review of Japan* **12**: 17-19.

Nakane T. 1963a: New or little known Coleoptera from Japan and its adjacent regions, XIX. *Fragmenta Coleopterologica* **6-7**: 26-30.

Nakane T. 1963b: New or little-known Coleoptera from Japan and its adjacent regions. *The Scientific Reports of Kyoto prefectural University (Natural Science and Living Science)* **3** (5): A 221-A 226.

Nakane T. 1966: Coleoptera from Afghanistan. *Results of the Kyoto University Scientific Expedition to the Karakorum and Hindukush, 1955*. Vol. **8**: 231-248.

Nakane T. 1968a: New or little known Coleoptera from Japan and its adjacent regions, XXVI. *Fragmenta Coleopterologica* **19**: 75-76.

Nakane T. 1968b: New or little known Coleoptera from Japan and its adjacent regions, XXVII. *Fragmenta Coleopterologica* **19-21**: 76-85.

Nakane T. 1968c: New or little-known Coleoptera from Japan and its adjacent regions. XXVIII. *Fragmenta Coleopterologica* **21**: 85-86.

Nakane T. 1973a: Notes on the synonymy and on some types of Japanese Coleoptera in certain European collections II: Heteromera (1) (Insecta). *Bulletin of the National Science Museum* (Tokyo) **16**: 1-8.

Nakane T. 1973b: Notes on some species of Tenebrionidae from the Yaeyama Islands (Insecta: Coleoptera). *Memoirs of the National Science Museum* (Tokyo) **6**: 103-108, pl. 11.

Nakane T. 1975: On the heteromerous Coleoptera occurring in Kyushu including Yakushima and Tanegashima. *Memoirs National Science Museum* (Tokyo) **8**: 161-172, pl. 16.

Nakane T. 1977: Notes on some species of Coleoptera (Insecta) occurring in the Ogasawara and the Volcano Islands, with descriptions of new species. *Memoirs of the National Science Museum* (Tokyo) **10**: 147-162 (in Japanese, with title and descriptions in English).

Nakane T. 1978: On some interesting species of Coleoptera from Izu and Agasawara Islands, with descriptions of new species. *Memoirs of the National Science Museum* (Tokyo) **11**: 155-161 (in Japanese, with title and descriptions in English).

Nakane T. 1979: New or little-known Coleoptera from Japan and its adjacent regions. XXXI. *Fragmenta Coleopterologica* **25/28**: 112-114.

Nakane T. 1983a: On the Eurygeniini in Japan (Anthicidae). *Kita-Kyūshū no Konchū* **30** (1): 1-6, pl.

Nakane T. 1983b: New or little-known Coleoptera from Japan and its adjacent regions. XXXI. *Fragmenta Coleopterologica* **35/37**: 139-150.

Nakane T. 1984: On the Coleoptera occurring in Yaku-shima Island. Pp. 587-631. Conservation Reports of the Yaku-shima Wilderness Area, Kyushu, Japan.

Nakane T. 1987: Notes on some little known beetles (Coleoptera) in Japan. *Kita-Kyūshū no Konchū* **34**: 171-176, 1 pl.

Nakane T. 1988a: Notes on some little-known beetles (Coleoptera) in Japan, 2. *Kita-Kyūshū no Konchū* **35**: 1-6, 1 pl. (in Japanese, with English title and descriptions).

Nakane T. 1988b: Notes on some little-known beetles (Coleoptera) in Japan. *Kita-Kyūshū no Konchū* **35**: 77-83, pl. 9 (in Japanese, with English title and descriptions).

Nakane T. 1988c: New or little-known Coleoptera from Japan and its adjacent regions. XXXVIII. *Fragmenta Coleopterologica* **38-40**: 153-164.

Nakane T. 1989: Notes on some little-known beetles (Coleoptera) in Japan. *Kita-Kyūshū no Konchū* **36**: 1-10.

Nakane T. 1990: Notes on some little-known beetles (Coleoptera) in Japan. 6. *Kita-Kyūshū no Konchū* **37**: 61-68, 1 pl.

Nakane T. 1991: Notes on some little known bettles (Coleoptera) in Japan. 7. *Kita-Kyūshū no Konchū* **38**: 1-9, pl. 1 (in Japanese, with English title and descriptions).

Nakane T. 1994a: Notes on some little-known beetles (Coleoptera) in Japan. 12. *Kita-Kyūshū no Konchū* **41**: 81-86 (in Japanese, with English title and descriptions).

Nakane T. 1994b: On the species of the genus Lagria Fabricius occurring in the main islands of Japan (Coleoptera, Lagriidae). *Essa-Konchū dokokai Tokubetsu hokoku* (2): 241-246 (in Japanese, with English title and descriptions).

Nakane T. 1997: Notes on some little-known beetles (Coleoptera) in Japan. 14. *Kita-Kyūshū no Konchū* **44**: 1-6 (in Japanese, with English title and descriptions).

Nakane T. & Hayashi N. 1955: Melandryidae. Pp. 77-82. In: *Coloured illustrations of the insects of Japan. Coleoptera*. The Kinki Coleopterologigal Society, [xii] + 196 + [19] pp., 64 pls.

Nakane T. & Masumoto K. 1979: [new taxon]. In: New or little-known Coleoptera from Japan and its adjacent regions. XXXI. *Fragmenta Coleopterologica* **25/28**: 112-114.

Nakane T. & Nobuchi A. 1955: On a new genus and six new species of ciid-beetles from Japan (Ciidae, Coleoptera). *Scientific Report of Saikyo University. Natural Science and Living Science*(A) **2** (1): 47-52.

Nakane T. & Nobuchi A. 1956: Entomological results from the scientific survey of the Tokara Islands. II. Coleoptera: Ciidae. *Scientific Report of Saikyo University. Mathematics and Natural Science*(A) **2** (2): 107-108.

Nakane T. & Nomura S. 1950: New or little known Coleoptera from Japan and its adjacent regions. I. *Transactions of the Kansai Entomological Society* **15**: 1-17.

Nakane T. & Nomura S. 1957: Eine neue Hoshihananomia-Art aus Japan. *Entomological Review of Japan* **8**: 3-4.

Nardi G. 1995: Tenuicomus velox bucciarellii nomen novum per Tenuicomus velox relictus (Coleoptera, Anthicidae). *Bollettino dell'Associazione Romana di Entomologia* **49** [1994]: 161-163.

Nardi G. 2004: [new names]. In: Chandler D. S., Nardi G. & Telnov D.: Nomenclatural notes on the Palaearctic Anthicidae (Coleoptera). *Mitteilungen des Internationalen Entomologischen Vereins e. V.* **29**: 109-173.

Nedelkov N. 1905: Prinos kam entomologichnata fauna na Bulgariya [A contribution to the entomological fauna of Bulgaria]. *Priodichesko Spisanije ma Bulgaskoto Knizhovno Druzhestvo v Sofia* (17) **66**: 404-439.

Neervoort van de Poll J.R.H. 1886: A new species of the Heteromerous genus Leiochrinus, Westwood. *Notes from the Leyden Museum* **8**: 34.

Nevinson B. G. 1889: On a new Pimelia brought by Mr. Joseph Thomson from Morocco. *The Entomologist's Monthly Magazine* **25**: 255.

Newman E. 1838: Entomological notes. *The Entomological Magazine* **5**: 168-181, 202-205, 372-402, 483-500.

Nicolas J.-P. 1977: Description de Marolia leseigneuri n. sp. (Col. Melandryidae). *Bulletin de la Société Linnéenne de Lyon* **46**: 281-283.

Nietner J. 1856: Entomological papers - being descriptions of new Ceylon Coleoptera, with such observations on their habits as appear any way interesting. *Journal of the Asiatic Society of Bengal* **25**: 381-394, 523-539.

Nietner J. 1857a: Entomological papers; being chiefly descriptions of new Ceylon Coleoptera, with such observations on their habits, &c., as appear in any way interesting. *Journal of the Ceylon Branch of the Royal Asiatic Society* **3** (9): 1-75.

Nietner J. 1857b: Descriptions of new Ceylon Coleoptera. *The Annals and Magazine of Natural History* (2) **9**: 374-388.

Nikitsky N. B. 1972: Novye i maloizvestnye vidy zhukov-tenelubov (Coleoptera, Melandryidae) iz yuzhnogo Primorya razvivauschsiesia v drevesine. *Entomologicheskoe Obozrenie* **51**: 203-207.

Nikitsky N. B. 1973a: Novyy vid zhukov-golovachey roda Nematoplus LeC. (Coleoptera, Cephaloidae) iz yuzhnogo Primorya [A new species of the genus Nematoplus LeC. (Coleoptera, Cephaloidae) from the South Maritim Territory]. *Zoologicheskiy Zhurnal* **52**: 609–612.

Nikitsky N. B. 1973b: Novyy vid zhukov-tenelubov roda Enchodes (Coleoptera, Melandryidae) iz yuzhnogo Primorya. *Zoologicheskiy Zhurnal* **52**: 1728-1730.

Nikitsky N. B. 1974: Melandrya fasciata n. sp. du sud de l'Extrême-Orient (Coleoptera, Melandryidae). *Bulletin de la Société d'Entomologique Mulhouse* **1974**: 61-62.

Nikitsky N. B. 1983: Un nouveau genre de la famille de Melandryidae (Coleoptera) de Chine. *Bulletin de la Société d'Entomologique Mulhouse* **1983**: 37-40.

Nikitsky N. B. 1984: Vidy rodov Wanachia i Abdera (Coleoptera, Melandryidae) vostochnoy Palearktiki. *Zoologicheskiy Zhurnal* **63** (5): 677-681.

Nikitsky N. B. 1985a: Zhuki semeyst Tetratomidae i Cephaloidae vostochnoy palearktiki. *Sbornik Trudov Zoologicheskogo Muzea MGU* **23**: 26-37.

Nikitsky N. B. 1985b: Vidy roda Dircaea F. (Coleoptera, Melandryidae) vostochnoy Palearktiki. *Byulleten' Moskovskogo Obshchestva Ispytateley Prirody, Otdel Biologicheskiy* (N. S.) **90** (4): 53-64.

Nikitsky N. B. 1985c: The genera Mycetoma Mulsant, 1856 and Orchesia Latreille, 1807 in the Eastern Palaearctic (Insecta: Coleoptera: Tetratomidae, Melandryidae). *Senckenbergiana Biologica* **65**: 265-277.

Nikitsky N. B. 1985d: Zhuki-teneluby roda Melandrya F. (Coleoptera, Melandryidae) vostochnoy Palearktiki. *Entomologicheskoe Obozrenie* **64**: 748-759.

Nikitsky N. B. 1986: Semeystvo Pilipalpidae stat. nov. (Coleoptera, Heteromera), ego sostav i sistematicheskoie Svyazi [The family Pilipalpidae stat. n. (Coleoptera, Heteromera), its composition and taxonomic relationships]. *Zoologicheskiy Zhurnal* **65**: 1178-1189.

Nikitsky N. B. 1988a: Vidy roda Phryganophilus (Coleoptera, Melandryidae) Palearktiki. *Zoologicheskiy Zhurnal* **67**: 1426-1430.

Nikitsky N. B. 1988b: Novye taksony zhukov-griboedov (Col. Mycetophagidae) i Salpingidae Dalnego Vostoka SSSR [New taxa of Coleoptera of the families Mycetophagidae and Salpingidae from the Far East of the USSR]. *Zoologicheskiy Zhurnal* **67**: 1845-1853.

Nikitsky N. B. 1989a: Zhestkokrylye semeystv Tetratomidae i Melandryidae Dalnego Vostoka SSSR. Pp. 3-87. In: Shatalkin A. I. (ed.): *Nasekomye v ecosistemakh Sibiri i Dalnego Vostoka*. Sbornik trudov Zoologicheskogo muzeja MGU. Vol. 27. Izdatelstvo Moskovskogo Universiteta.

Nikitsky N. B. 1989b: Novye taksony zhukov-griboedov (Col. Mycetophagidae) i Tetratomidae Dalnego Vostoka SSSR. *Zoologicheskiy Zhurnal* **68**: 56-64.

Nikitsky N. B. 1992a: A new Microscapha from the Nepal Himalayas (Coleoptera, Melandryidae). *Mitteilungen der Münchener Entomologischen Gesellschaft* **82**: 83-85.

Nikitsky N. B. 1992b: Sem. Mycetophagidae. Pp. 406-423. Sem. Tetratomidae. Pp. 426-435. In: Ler P. A. (ed.): *Opredelitel nasekomykh Dalnego Vostoka SSSR. Tom III. Zhestkokrylye, ili zhuki. Chast 2* [*Keys to the identification of insects of the Soviet Far East. 3 - Coleoptera, or beetles. Part 2.*]. St. Petersburg: Nauka 704 pp.

Nikitsky N. B. 1992c: Sem. Salpingidae. Pp. 482-493. In: Ler P. A. (ed.): *Opredelitel' nasekomykh Dal'nego Vostoka SSSR. Tom 3 - Zhestkokrylye, ili zhuki. Chast 2* [*Keys to the identification of insects of the Soviet Far East. 3 - Coleoptera, or beetles. Part 2.*]. St. Petersburg: Nauka, 704 pp.

Nikitsky N. B. 1992d: Sem. Melandryidae. P. 659. In: Ler P. A. (ed.): *Opredelitel' nasekomykh Dal'nego Vostoka SSSR. Tom 3 - Zhestkokrylye, ili zhuki. Chast 2* [*Keys to the identification of insects of the Soviet Far East. 3 - Coleoptera, or beetles. Part 2.*]. St. Petersburg: Nauka, 704 pp.

Nikitsky N. B. 1994a: Vidy roda Ischalia (Coleoptera, Ischaliidae) dal'nego Vostoka Rossii. *Zoologicheskiy Zhurnal* **73**: 33-38.

Nikitsky N. B. 1994b: New species of the genus Lederina Nikitsky et Belov, 1982, stat. nov. from the Nepal Himalayas (Coleoptera, Melandryidae). *Entomofauna* **15**: 345-352.

Nikitsky N. B. 1996a: Novye zhestkokrylye (Coleoptera) iz Kitaya. *Zoologicheskiy Zhurnal* **75**: 1366-1373.

Nikitsky N. B. 1996b: Novyy vid roda Langelandia (Coleoptera, Colydiidae) s severo-zapadnogo Kavkaza. *Zoologicheskiy Zhurnal* **75**: 1434-1435.

Nikitsky N. B. 1998a: *Generic classification of the beetle family Tetratomidae* (*Coleoptera, Tenebrionoidea*) *of the world, with description of new taxa*. Sofia-Moscow: Pensoft, Series Faunistika (9), 80 pp.

Nikitsky N. B. 1998b: [new taxa]. In: Nikitsky N. B., Semenov V. B. & Dolgin M. M. (eds): *Zhestkokrylye - xylobionty, mycetobionty i plastinchatoyusye Prioksko-Terrasnogo Biosphernogo zapovednika* (*s obzorom fauny etikh grupp Moskovskoy oblasti*). *Dopolnenie 1* (*s zamechaniami po nomenklature i systematike nekotorykh zhukov Melandryidae mirovoy fauny*). Sbornik trudov Zoologicheskogo muzeja MGU. Vol. 36 (1). Moskva: Izdatelstvo Moskovskogo Universiteta, 60 pp.

Nikitsky N. B. 2000: K poznaniyu zhestkokrylykh semeystva Synchroidae (Coleoptera, Tenebrionoidea) mirovoy fauny. *Zoologicheskiy Zhurnal* **78** (1): 42-48.

Nikitsky N. B. 2001: Novye vidy zhukov-tenelubov (Coleoptera, Melandryidae). Pp. 19-25. In: *Ekologia, monitoring i ratsionalnoye prirodopolzovanie*. Vol. 307 (1). Moskva: MGUL.

Nikitsky N. B. 2002: Lichinki zhukov-tenelubov (Coleoptera, Melandryidae) Rossii, s zametkami po taksonomii. *Byulleten' Moskovskogo Obshchestva Ispytateley Prirody, Otdel Biologicheskiy* (N. S.) **107** (1): 8-30.

Nikitsky N. B. 2003: Novyy vid roda Pseudotryphillus Reitter, 1880 (Coleoptera, Mycetophagidae) iz Nepala. *Byulleten' Moskovskogo Obshchestva Ispytateley Prirody, Otdel Biologicheskiy* (N. S.) **108** (3): 49-50.

Nikitsky N. B. 2005: Zhestkokrylye podsemeystva Penthinae Lacordaire, 1859 (Coleoptera, Tenebrionoidea, Tetratomidae) mirovoi fauny. *Byulleten' Moskovskogo Obshchestva Ispytateley Prirody, Otdel Biologicheskiy* (N. S.) **110** (5): 16-26.

Nikitsky N. B. 2007: Vidy roda Holostrophus Horn, 1888 (Coleoptera, Tetratomidae, Eustrophinae, Holostrophini) mirovoy fauny. *Byulleten' Moskovskogo Obshchestva Ispytateley Prirody, Otdel Biologicheskiy* (N. S.) **112**: 13-30.

Nikitsky N. B. 2007: Novye taksony zhukov-gryboedov i zhukov-tenelubov (Coleoptera, Mycetophagidae, Melandryidae). *Bulleten' Moscovskogo Obschestva Ispytateley Prirody, Otdel Biologicheskiy* (N. S.) **112**: 58-62.

Nikitsky N. B. & Belov V. V. 1982: The false darkling beetle genus Lederia Rtt. (Coleoptera, Melandryidae). *Folia Entomologica Hungarica* (N. S.) **43**: 111-123.

Nikitsky N. B. & Belov V. V. 1983: Palearkticheskie vidy roda Salpingus Ill. (Col. Salpingidae) [Palaearctic species of the genus Salpingus (Coleoptera, Salpingidae)]. *Zoologicheskiy Zhurnal* **62** (4): 516-529.

Nobuchi A. 1955a: Studies on the ciid-beetles from Japan (I) With the descriptions of a new genus and some new species. *Entomological Review of Japan* **6**: 53-58.

Nobuchi A. 1955b: Studies on the ciid-beetles from Japan (II). *Akitu* **4**: 105-108.

Nobuchi A. 1959: Two new species of ciid-beetles. *Entomological Review of Japan* **10**: 51-52.

Nobuchi A. 1960a: Ciidae from Hokkaido (Coleoptera). *Entomological Review of Japan* **11**: 37-41, pl. 5.

Nobuchi A. 1960b: On a new species of Cis Latreille from Japan (Coleoptera, Ciidae). *Akitu* **9** (3-4): 65-66.

Nobuchi A. & Wada Y. 1955a: [new taxon]. In: Nobuchi A. 1955a: Studies on the ciid-beetles from Japan (I) with the descriptions of a new genus and some new species. *Entomological Review of Japan* **6**: 52-58.

Nobuchi A. & Wada Y. 1955b: [Cis mikagensis sp. n.]. P. 106. In: Nobuchi A.: Studies on the ciid-beetles from Japan (II). *Akitu* **4**: 105-108.

Nobuchi A. & Wada Y. 1956: A new species of Japanese Xylographus (Ciidae, Coleoptera). *Entomological Review of Japan* **7**: 35-36.

Nobuchi A. & Wada Y. 1959: [new taxon]. In: Nobuchi A.: Two new species of ciid-beetles. *Entomological Review of Japan* **10**: 51-52.

Nomura S. 1951: Zur Kenntnis der Mordellistenini (Col. Mordellidae) aus Japan, Korea und Formosa. *Tôhô-Gakuhô* **1**: 41-70.

Nomura S. 1957: Mordellid- and elmid-beetles of Yakushima. *Entomological Review of Japan* **8**: 40-44.

Nomura S. 1958: Zur Kenntnis der Gattung Mordella aus Japan und dessen Umgebung. *Tôhô-Gakuhô* **7**: 35-58.

Nomura S. 1959: Notes on the Japanese Melandryidae (Coleoptera). I. *Entomological Review of Japan* **10**: 43-45.

Nomura S. 1960: Two new species of the genus Anthicus from Japan (Coleoptera: Anthicidae). *The Entomological Review of Japan* **11**: 47-48.

Nomura S. 1961a: Two new Cteniopinus-species from Japan and Formosa (Alleculidae). *Entomological Review of Japan* **12**: 38-40 + pl. 6.

Nomura S. 1961b: Some new species of Coleoptera from Japan. *Tôhô-Gakuhô* **11**: 70-89.

Nomura S. 1962a: The anthicid-beetles from the Loochoo Islands. *Entomological Review of Japan* **14**: 1-7, pl. 1.

Nomura S. 1962b: Some new and remarkable species of Coleoptera from Japan and its adjacent regions. *Tôhô-Gakuhô* **12**: 35-51.

Nomura S. 1963a: [new taxa]. In: Nakane T., Obhayashi K., Nomura S. & Kurosawa Y. (eds): *Iconographia Insectorum Japonicum. Vol. 2* (*Coleoptera*). Tokyo: Hokuryukan, 443 pp., 192 pls (in Japanese).

Nomura S. 1963b: Some new species of the Coleoptera from Loochoo Is. and its adjacent regions. *Entomological Review of Japan* **15**: 37-42, pl. 5.

Nomura S. 1964: Some new species of the Coleoptera from Loochoo Is. and its adjacent regions, II. *Entomological Review of Japan* **17**: 47-57, pl. 3.

Nomura S. 1966: Mordellid-Fauna of the Loochoo Islands, with descriptions of some new forms. *Entomological Review of Japan* **18**: 41-53.

Nomura S. 1967a: The Mordellidae from Formosa. *Entomological Review of Japan* **19**: 5-34.

Nomura S. 1967b: Some new and remarkable species of the Coleoptera from Japan and its adjacent regions, II. *Entomological Review of Japan* **19**: 52-62.

Nomura S. 1975: Mordellidae of the Bonin Islands. *Entomological Review of Japan* **28**: 29-45.

Nomura S. & Katô A. 1957: On the subgenus Pseudomordellistena Ermisch, from Japan, Formosa and south China. *Transactions of the Shikoku Entomological Society* **5**: 77-90.

Nomura S. & Katô A. 1958: A new subgenus and two new species of the family Melandryidae from Japan (Coleoptera). *Entomological Review of Japan* **9**: 1-3.

Nomura S. & Katô A. 1959: Descriptions of some new mordellid- and melandriid-species and notes on others. *Entomological Review of Japan* **10**: 5-9.

Nomura S. & Katô A. 1968: A new species of the genus Melandrya from Japan (Coleoptera, Melandryidae). *Entomological Review of Japan* **21**: 1-2

Nomura S. & Nakane T. 1959: A new species of the genus Pelecotomoides from Japan (Coleoptera, Rhipiphoridae). *Insecta Matsumurana* **22**: 115-118.

Nomura S. & Yamazaki H. 1960: A new species and two new subspecies of the genus Strongylium from Japan (Tenebrionidae). *Entomological Review of Japan* **12**: 14-16.

Normand H. 1916: Nouveaux coléoptères de la faune tunisienne. *Bulletin de la Société Entomologique de France* **1916**: 283-287.

Normand H. 1936a: Contribution au catalogue des coléoptères de la Tunisie (9e fascicule). *Bulletin de la Société d'Histoire Naturelle de l'Afrique du Nord* **27**: 81-100, 144–164.

Normand H. 1936b: Contribution au catalogue des coléoptères de la Tunisie, 10e fascicule. *Bulletin de la Société d'Histoire Naturelle de l'Afrique du Nord* **27**: 355-383.

Normand H. 1938a: Contribution au catalogue des coléoptères de la Tunisie. Premier supplément. *Bulletin de la Société d'Histoire Naturelle de l'Afrique du Nord* **29**: 274-308.

Normand H. 1938b: Contribution au catalogue des coléoptères de la Tunisie. 13e fascicule. *Bulletin de la Société d'Histoire Naturelle de l'Afrique du Nord* **29**: 337-371.

Normand H. 1939: Contribution au catalogue des coléoptères de la Tunisie, 14e fascicule. Deuxième supplément. *Bulletin de la Société d'Histoire Naturelle de l'Afrique du Nord* **29** [1938]: 481-515, pl. 20.

Normand H. 1950a: Contribution au catalogue des coléoptères de la Tunisie. Troisième Supplement (Fascicule 4). *Bulletin de la Société des Sciences Naturelles de Tunisie* **2** [1949]: 79-104, pls. 6-7.

Normand H. 1950b: Nouveaux coléoptères du Nord de l'Afrique. *Revue Française d'Entomologie* **17**: 105-108.

Normand H. 1955: Heliotaurus nouveaux de Tunisie. *Bulletin de la Société des Sciences Naturelles de Tunisie* **8**: 85-88.

Novák V. 2005: Three new species of the genus Borboresthes Fairmaire, 1897 (Coleoptera: Tenebrionidae: Alleculinae). *Studies and Reports of District Museum Prague-East, Taxonomical Series* (1-2) **1**: 121-132.

Novák V. 2006a: New species of the genus Mycetocharina Seidlitz, 1891 (Coleoptera: Tenebrionidae: Alleculinae) from Iran. *Studies and reports of District Museum Prague-East, Taxonomical Series* **2**: 85-98.

Novák V. 2006b: Megischina bozdaglariensis sp. n. (Coleoptera: Tenebrionidae: Alleculinae) from Turkey. *Studies and reports of District Museum Prague-East, Taxonomical Series* **2**: 99-104.

Novák V. 2006c: Hymenophorus evae sp. nov. and H. gerdae sp. nov. (Coleoptera: Tenebrionidae: Alleculinae) from Iran. *Acta Societatis Zoologicae Bohemicae* **69**: 317-325.

Nowicki M. 1873: *Beschreibung neuer Käferarten*. Krakau: Jagellonische Universitäts Buchdrückerei, 6 pp.

Nyholm T. 1953: Bemerkungen über einige schwedische Cis-Arten (Col., Ciidae). *Entomologisk Tidskrift* **74**: 187-196.

Obenberger J. 1914: Beitrag zur Kenntnis der palaearktischen Käferfauna. (Fortsetzung und Schluss). *Coleopterologische Rundschau* **3**: 97-115, 129-144.

Obenberger J. 1917: II. Beitrag zur Kenntnis der palaearktischen Käferfauna. *Archiv für Naturgeschichte* A **82** (4) [1916]: 9-45, 2 pls.

Obenberger J. 1918: Vier neue palearktische Coleopteren. *Entomologische Blätter* **14**: 58-61.

Obenberger J. 1922: II. Beitrag zur Kenntnis der palaearktischen Käferfauna. *Archiv für Naturgeschichte* A **82** (4) **[1917]**: 9-45, pls. 2.

Oberthür R. 1903: Une nouvelle Asida appartenant au groupe des espèces espagnoles aterciopeladas. *Boletín de la Sociedad Española de Historia Natural* **1903**: 74-75.

Odnosum V. K. 1990: Novyy vid zhuka-gorbatki roda Mordellistena. *Novosti faunistiki i systematiky* **1990**: 54-55.

Odnosum V. K. 2000: Mordellid beetles of the genus Stenalia (Coleoptera, Mordellidae) of Central and Eastern Palaearctics. Communication 1. *Vestnik Zoologii* **34**: 37-50.

Odnosum V. K. 2001: Mordellid beetles of the genus Stenalia (Coleoptera, Mordellidae) of Central and Eastern Palaearctics. Communication 2. *Vestnik Zoologii* **35**: 21-26.

Odnosum V. K. 2002: Mordellid beetles (Coleoptera, Mordellidae) in the fauna of Kazakhstan and Middle Asia. Communication 1. *Vestnik Zoologii* **36**: 13-23.

Odnosum V. K. 2003: Mordellid beetles (Coleoptera, Mordellidae) in the fauna of Kazakhstan and Middle Asia. Communication 2. *Vestnik Zoologii* **37**: 33-49.

Odnosum V. K. 2005: Mordellistena dolini sp. n., a new mordellid beetle of the gemellata group. *Vestnik Zoologii* **39**: 53-54.

Ogloblin D. A. 1950: [new taxa]. In: Ogloblin D. A. & Znojko D. V.: *Fauna SSSR, Zhestkokrylye, Tom 18, vyp. 8. Pylceedy (Cem. Alleculidae), ch. 2, podsem. Omophlinae*. Moskva, Leningrad: Akademiya nauk SSSR, 133 pp.

Ohbayashi N. & Tôyama M. 1994: Two new pyrochroid species (Coleoptera, Pyrochroidae) from Taiwan and Thailand. *Japanese Journal of Entomology* **62**: 146-150.

Okada K. 2005: Occurrence of the genus Colobicones Grouvelle (Coleoptera, Zopheridae, Colydiinae) in Japan, with descriptions of two new species. *Elytra* **33**: 425-431.

Okamoto H. 1924: The insect-fauna of Quelpart Island (Saishuto). *Bulletin of the Agricultural Experimental Station Chosen, Korea* **1**: 47-233, 4 pls., 1 map.

Olexa A. 1955: Dircaea australis Fairm. var. imitata var. nov. *Bulletin de la Société d'Entomologique Mulhouse* **1955**: 38-39.

Olivier A. G. 1789: *Encyclopédie méthodique, ou par ordre de matières; par une société de gens de lettres, de savans et d'artistes; précédée d'un vocabulaire universel, servant de table pour tout l'ouvrage, ornée des portraits de Mm. Diderot & d'Alembert, premiers éditeurs de l'Encyclopédie. Histoire naturelle. Insectes. Tome quatrième*. Paris: Panckoucke, 1-331 pp.

Olivier A. G. 1790: *Entomologie, ou histoire naturelle des insectes, avec leurs caractères génériques et spécifiques, leur description, leur synonymie, et leur figure enluminée. Coléoptères. Tome second*. Paris: de Baudoin, 485 + xix + [3] pp., 63 pls. [Note: genera 9-34, each genus with separate pagination].

Olivier A. G. 1791: *Encyclopédie méthodique, ou par ordre de matières; par une société de gens de lettres, de savans et d'artistes; précédée d'un vocabulaire universel, servant de table pour tout l'ouvrage, ornée des portraits de Mm. Diderot & d'Alembert, premiers éditeurs de l'Encyclopédie. Histoire naturelle. Insectes. Tome sixième. Pars I*. Paris: C. J. Panckoucke, 1-368 pp.

Olivier A. G. 1792: *Encyclopédie méthodique, ou par ordre de matières; par une société de gens de lettres, de savans et d'artistes; précédée d'un vocabulaire universel, servant de table pour tout l'ouvrage, ornée des portraits de Mm. Diderot & d'Alembert, premiers éditeurs de l'Encyclopédie. Histoire naturelle. Insectes. Tome sixième. Pars II*. Paris: C. J. Panckoucke, 369-704 pp.

Olivier A. G. 1793: *Encyclopédie méthodique, ou par ordre de matières; par une société de gens de lettres, de savans et d'artistes; précédée d'un vocabulaire universel, servant de table pour tout l'ouvrage, ornée des portraits de Mm. Diderot & d'Alembert, premiers éditeurs de l'Encyclopédie. Histoire naturelle. Insectes. Tome septième*. (1). Paris: Panckoucke, 1-368 pp.

Olivier A. G. 1795: *Entomologie, ou histoire naturelle des insectes, avec leurs caractères génériques et spécifiques, leur description, leur synonymie, et leur figure enluminée. Coléoptères. Tome troisième*. Paris: de Lanneau, 557 + xxviii pp., 65 pls. [Note: genera 35-65, each genus with separate pagination].

Olivier A. G. 1797: *Encyclopédie méthodique, ou par ordre de matières; par une société de gens de lettres, de savans et d'artistes; précédée d'un vocabulaire universel, servant de table pour tout l'ouvrage, ornée des portraits de Mm. Diderot & d'Alembert, premiers éditeurs de l'Encyclopédie. Histoire naturelle. Insectes. Tome septième*. (2). Paris: Panckoucke, 369-827 pp.

Olivier A. G. 1807: *Entomologie, ou histoire naturelle des insectes, avec leurs caractères génériques et spécifiques, leur description, leur synonymie et leur figure enluminée. Coléoptères. Tome cinqième*. Paris: Baudonin, 612 pp. [Note: each genus with separate pagination].

Olivier A. G. 1803: L´Oedemère fémorale (Oedemera) femoralis. P. 21. In: *Nouveau dictionaire d´histoire naturelle, appliqée aux arts, principalement à l´agriculture et à l´economie rurale et domestique. Tome XVI*. Paris: Société de naturalistes et d´agriculteurs, 575 pp.

Olivier A. G. 1811: *Encyclopédie méthodique, ou par ordre de matières; par une société de gens de lettres, de savans et d'artistes; précédée d'un vocubulaire universel, servant de table pour tout l'ouvrage, ornée des portraits de Mm. Diderot & d'Alembert, premiers éditeurs de l'Encyclopédie. Histoire naturelle. Insectes. Tome huitième*. Paris: H. Agasse, 722 pp.

Olivier E. 1888: Description d'une nouvelle espèce de vésicants. *Bulletin de la Société Entomologique de France* **1888**: lxi-lxii.

Olliff A. S. 1883a: Description of two larvae and new genera and species of clavicorn Coleoptera, and a synopsis of Helota, MacLeay. *Cistula Entomologica* **3**: 49-61.

Olliff A. S. 1883b: Remarks on a small collection of Clavicorn Coleoptera from Borneo, with descriptions of new species. *Transactions of the Entomological Society of London* **1883**: 173-186.

Oromí P. 1977: Los Arthrodeis Sol. de las Islas Canarias (Col. Tenebrionidae, Erodiini). *Vieraea* **7**: 3-22.

Oromí P. 1978: Sobre los Crypticus (Col. Tenebrionidae) (Coleoptera) de Canarias: Descripción de C. gomerensis sp. n. *Boletín de la Asociación Española de Entomología* **7** (*Suppl. A*): 215-231.

Oromí P. 1982: Los Tenebrioniodos de las Islas Canarias. - Inst. Estud. Canarias, 50 Aniversario 1932-1982: 267-292.

Oromí P. 1984a: Helioctamenus occidentalis, un nuevo Colydiidae de Canarias (Coleoptera). *Vieraea* **13** [1983]: 225-232.

Oromí P. 1984b: Nuevos aportiacones al conocimiento de la distribución de los coléopteros de Canarias. *Vieraea* **13** [1983]: 283-240.

Oromí P. 1990: Una nueva especie del género Pimelia de la Isla de Grand Canaria (Coleoptera, Tenebrionidae). *Vieraea* **19**: 245-249.

Oromí P. & Garcia R. 1986: Canaricerus aeonii, un nuevo Orthocerini de Tenerife, Islas Canarias (Coleoptera, Colydiidae). *Nouvelle Revue d'Entomologie* (N. S.) **3**: 303-307.

Osculati G. 1844: *Note d'un viaggio nella Persia e nelle India orientale negli anni 1841, 1842.* Monza: Luca Corbetta, 73 pp., 1 pl.

Pal T. K. 1984: Two new species of Pseudendestes Lawrence (Coleoptera: Colydiidae) from India. *Bulletin of the Zoological Survey of India* **6**: 31-35.

Pal T. K. 1992a: A new inopeplid species (Coleoptera: Cucujoidea) from Patkoi Hill Range of Arunachal Pradesh. *Journal of the Bombay Natural History Society* **89**: 238-241.

Pal T. K. 1992b: Inopeplidae and elacatidae (Coleoptera: Heteromera) from Arunachal Pradesh, India. *Records of the Zoological Survey of India* **91**: 343-351.

Pal T. K. 2003: Insecta: Coleoptera: Colydiidae (sensu lato). Pp. 163-176. In: *State Fauna Series* **9**, *Fauna of Sikkim* (*Part 3*).

Pal T. K. & Ślipiński S. A. 1984: Notes on the Nematidium Erichson (Coleoptera, Colydiidae) with description of new species. *Polskie Pismo Entomologiczne* **53**: 531-43.

Palisot de Beauvois A. M. F. J. 1805: *Insectes recueillis en Afrique et en Amérique, dans le royaumes d'Owareet et de Benin, à Saint-Domingue et dans les États-Unis, pendant les années 1786-1797.* Paris: Levrault, Schoell et Cie, xvi + 224 pp., 34 + 20 + 9 +23 +3 +7+ 14 pls.

Pallas P. S. 1771: *Reise durch verschiedene Provinzen des russischen Reichs. Erster Theil.* St. Petersburg: Kayserliche Akademie der Wissenschaften, [10] + 3-504 pp., 11 pls.

Pallas P. S. 1773: *Reise durch verschiedene Provinzen des russischen Reichs. Zweyter Theil. Zweytes Buch vom Jahr 1771.* St. Petersburg: Kayserliche Akademie der Wissenschaften, pp. 371-744.

Pallas P. S. 1781: *Icones Insectorum praesertim Rossiae Sibiriaeque peculiarum quae collegit et descritionibus illustravit. Fasciculus primus.* Erlangae: W. Walther, 1-56 pp.

Pallas P. S. 1782: *Icones Insectorum praesertim Rossiae Sibiriaeque peculiarum quae collegit et descritionibus illustravit. Fasciculus secundus.* Erlangae: W. Waltheri, 57-96 pp., pls A-F.

Pallas P. S. 1798: *Icones Insectorum praesertim Rossiae Sibiriaeque peculiarum quae collegit et descritionibus illustravit. Fasciculus tertius.* Erlangae: W. Waltheri, 97-104 pp., pls. G-H.

Palm T. 1976: Eine neue Mordellistena-Art aus Tenerife. *Entomologisk Tidskrift* **97**: 31-33.

Pandellé L. 1867: [new taxon]. In: Grenier A. (ed.): *Matériaux pour servir à la faune des coléoptères de France, recueillis et publiés par le Dr. A. Grenier. 2e Cahier.* Paris: A. Grenier, [1] + iv + 131-194 pp.

Panin S. 1943: Supplément à la connaissance des coléoptères roumains (2-ème supplément). *Bulletin de la Section Scientifique de l'Académie Roumaine* **24** [1942]: 178-182.

Pankow W. 1981: Insect of Saudi Arabia. Coleoptera: Fam. Mordellidae and Scraptiidae, Subfam. Anaspidinae. *Fauna of Saudi Arabia* **3**: 273-275.

Panzer G. W. F. 1792a: *Faunae insectorum germanicae initia oder Deutschlands Insecten.* Heft 2. Norinbergae: Felsecker, 24 pp. + 24 pls.

Panzer G. W. F. 1792b: *Faunae insectorum germanicae initia oder Deutschlands Insecten.* Heft 3. Norinbergae: Felsecker, 24 pp. + 24 pls.

Panzer G. W. F. 1793a: *Faunae insectorum germanicae initia oder Deutschlands Insecten*. Heft 6. Norinbergae: Felsecker, 24 pp. + 24 pls.

Panzer G. W. F. 1793b: *Faunae insectorum germanicae initia oder Deutschlands Insecten*. Heft 8. Norinbergae: Felsecker, 24 pp. + 24 pls.

Panzer G. W. F. 1793c: *Faunae insectorum germanicae initia oder Deutschlands Insecten*. Heft 9. Norinbergae: Felsecker, 24 pp. + 24 pls.

Panzer G. W. F. 1793d: *Faunae insectorum germanicae initia oder Deutschlands Insecten*. Heft 10. Norinbergae: Felsecker, 24 pp. + 24 pls.

Panzer G. W. F. 1793e: *Faunae insectorum germanicae initia oder Deutschlands Insecten*. Heft 16. Norinbergae: Felsecker, 24 pp. + 24 pls.

Panzer G. W. F. 1794a: *Faunae insectorum germanicae initia oder Deutschlands Insecten*. Heft 22. Norinbergae: Felsecker, 24 pp. + 24 pls.

Panzer G. W. F. 1794b: *Faunae insectorum germanicae initia oder Deutschlands Insecten*. Heft 23. Norinbergae: Felsecker, 24 pp. + 24 pls.

Panzer G. W. F. 1794c: *Faunae insectorum germanicae initia oder Deutschlands Insecten*. Heft 24. Norinbergae: Felsecker, 24 pp. + 24 pls.

Panzer G. W. F. 1795: *Deutschlands Insectenfaune oder entomologisches Taschenbuch für das Iahr 1795* [*Entomologia Germanica exhibens insecta per Germania indigena*]. Nürnberg: Felsecker, iv + 372 pp., 12 pls.

Panzer G. W. F. 1796a: *Faunae insectorum germanicae initia oder Deutschlands Insecten*. Heft 31. Norinbergae: Felsecker, 24 pp. + 24 pls.

Panzer G. W. F. 1796b: *Faunae insectorum germanicae initia oder Deutschlands Insecten*. Heft 34. Norinbergae: Felsecker, 24 pp. + 24 pls.

Panzer G. W. F. 1796c: *Faunae insectorum germanicae initia oder Deutschlands Insecten*. Heft 35. Norinbergae: Felsecker, 24 pp. + 24 pls.

Panzer G. W. F. 1796d: *Faunae insectorum germanicae initia oder Deutschlands Insecten*. Heft 36. Norinbergae: Felsecker, 24 pp. + 24 pls.

Panzer G. W. F. 1796e: *Faunae insectorum germanicae initia oder Deutschlands Insecten*. Heft 37. Norinbergae: Felsecker, 24 pp. + 24 pls.

Panzer G. W. F. 1796f: *Fauna insectorum germanicae initia oder Deutschlands Insecten*. Heft 38. Norinbergae: Felsecker, 24 pp. + 24 pls.

Panzer G. W. F. 1797: *Faunae insectorum germanicae initia oder Deutschlands Insecten*. Heft 45. Norinbergae: Felsecker, 24 pp. + 24 pls.

Panzer G. W. F. 1798: *Faunae insectorum germanicae initia oder Deutschlands Insecten*. Heft 57. Norinbergae: Felsecker, 24 pp. + 24 pls.

Panzer G. W. F. 1799a: *Faunae insectorum germanicae initia oder Deutschlands Insecten*. Heft 61. Norinbergae: Felsecker, 24 pp. + 24 pls.

Panzer G. W. F. 1799b: *Faunea insectorum germanicae initia oder Deutschlands Insecten*. Heft 67. Norinbergae: Felsecker, 24 pp. + 24 pls.

Panzer G. W. F. 1799c: *Faunae insectorum germanicae initia oder Deutschlands Insecten*. Heft 68. Norinbergae: Felsecker, 24 pp. + 24 pls.

Panzer G. W. F. 1800: *Faunae insectorum germanicae initia oder Deutschlands Insecten*. Heft 74. Norinbergae: Felsecker, 24 pp. + 24 pls.

Panzer G. W. F. 1803a: *Faunae insectorum germanicae initia oder Deutschlands Insecten*. Heft 90. Norinbergae: Felsecker, 24 pp. + 24 pls.

Panzer G. W. F. 1803b: *Faunae insectorum germanicae initia oder Deutschlands Insecten*. Heft 94. Norinbergae: Felsecker, 24 pp. + 24 pls.

Panzer G. W. F. 1805a: *Kritische Revision der Insektenfaune Deutschlands nach dem System bearbeitet. Zum I-XCVI. Heft. I. Baendchen*. Nürnberg: Felfseckerschen Buchhandlung, 12 + 144 pp.

Panzer G. W. F. 1805b: *Faunae insectorum germanicae initia oder Deutschlands Insecten*. Heft 98. Norinbergae: Felsecker, 24 pp. + 24 pls.

Papp C. H. 1946: Verzeichnis der bei Madarasz-See (Ungarn) gesammelten Käferarten im Jahre 1939 – 1943. (II. Stück Hygrobiidae bis Ptinidae). *Additamenta Faunistica Coleopterorum* **6** (2): 34-38.

Pardo Alcaide A. 1948: Estudios sobre Meloidae. I. Acerca de la validez especifica de Mylabris maculoso-punctata Grll., rosinae Escher. y pauper Escher. *Eos, Revista Española de Entomología* **24**: 493-502.

Pardo Alcaide A. 1950a: Estudios sobre Meloidae. II. Los Mylabrini de la Península Ibérica. *Boletín de Patólogia Vegetal y de Entomología Agricola* **17**: 61-82.

Pardo Alcaide A. 1950b: Contribución al conocimiento de la fauna entomológica marroquí III. *Instituto General Franco de Estudios e investigación Hispano-Arabe. Publicaciones fuera de series*, 70 pp.

Pardo Alcaide A. 1951: Estudios sobre Meloidae III. Una nueva especie de Meloe de la isla de Tenerife y comentarios sobre algunos meloideos de la citada isla. *Eos, Revista Española de Entomología* **27**: 249-255.

Pardo Alcaide A. 1952: Los generos de Meloidae de la fauna hespérica, *Graellsia* **8** [1950]: 39-79.

Pardo Alcaide A. 1953: Contribución al conocimiento de la fauna entomólogica marroquí V. Una nueva especie de Stenoria Muls. Del sahara marroquí (col. Meloidae). *Tamuda* **1**: 288-293.

Pardo Alcaide A. 1954a: Études sur les Meloidae. V. Les Mylabrini du Maroc et du Sahara occidental espagnol (Col. Meloidae). *Bulletin de la Société de Sciences Naturelles et Physiques du Maroc* **34**: 55-88.

Pardo Alcaide A. 1954b: Estudios sobre Meloidae IV. Sobre algunos Mylabris Fabr. Del Norteafricano. *Eos, Revista Española de Entomología* **30**: 331-344.

Pardo Alcaide A. 1955a: Estudios sobre Meloidae Vi. Sobre algunos Mylabris (s. lat.) de la región etiópica. *Bulletin de l'Institut Royal de Sciences Naturelles du Belgique* **31** (49): 1-32.

Pardo Alcaide A. 1955b: Contribución al conocimiento de la fauna entomólogica marroquí VI. Coleopteros del valle inferior del Uad Muluya. *Tamuda* **3**: 39-74.

Pardo Alcaide A. 1956: Analectas entomologicas V. Los Zonitis de la fauna ibérica (Col. Meloidae). *Graellsia* **14**: 13-19.

Pardo Alcaide A. 1958a: Analectas entomologicas. VII. Los Sitarini de la Península Ibérica (Col. Meloidae). *Graellsia* **16**: 13-22.

Pardo Alcaide A. 1958b: Études sur les Meloidae (Col.) IX. Observations sur quelques Mylabris des régions éthiopienne et orientale. *Bulletin de l'Institut Royal de Sciences Naturelles de Belgique* **34** (13): 1-39.

Pardo Alcaide A. 1961a: Estudios sobre "Meloidae". XIII. Meloideos del Sahara occidental español y de la región du Draa (Marueccos) (Coleoptera). *Eos, Revista Española de Entomología* **37**: 91-111.

Pardo Alcaide A. 1961b: Études sur les Meloidae. XV. Sur deux Mylabris du Maroc méridional saharien. *Bulletin de la Société de Sciences Naturelles et Physiques du Maroc* **42**: 215-220.

Pardo Alcaide A. 1963: Études sur les Meloidae. XIV. Trois nouvelles espèces de Mylabris Fabricius (Coleoptera). *Eos, Revista Española de Entomología* **39**: 295-306.

Pardo Alcaide A. 1968: Contribution à la connaisance de la faune entomologique Marocaine X. Un nouveau type d'Oedemeridae (Col.) du Maroc Saharien: Gilotia perroti nov. gen. et nov. sp. *Arquivos do Museu Bocage* **2**: 41-49.

Pardo Alcaide A. 1969: Études sur les Meloidae. XXI. Matériaux pour une révision des mylabrides de l'Afrique du Nord et du Moyen orient (2e partie). *Eos, Revista Española de Entomología* **54**: 367-376.

Pardo Alcaide A. 1972: Études sur Oedemeridae II. Une nouvelle et singulière espèce de Chrysanthia Schmidt d'Asie Mineure (Coleoptera). *Fragmenta Entomologica* **8**: 213-221.

Pardo Alcaide A. 1975: Estudios sobre Meloidae (Coleoptera) XXIV. Una nueva especie de Mylabris Fabricius (Mylabris platai n.sp.) de Sierra nevada. *Quadernos Cientificas de Biologia* **4**: 45-48.

Pardo Alcaide A. 1977: Études sur les Meloidae (Col.). XXIII. Trois nouvelles taxa de Cerocoma Geoffroy d'Iran et Turquie d'Asie. *Nouvelle Revue d'Entomologie* **7**: 59-66.

Pardo Alcaide A. 1978: Un nuevo Alleculidae de Marruecos: Hymenorus candeli nov. sp. (Ins. Col.). *Butlleti de la Institució Catalana d'Historie Natural* (*Sec. Zool. 2*) **42**: 89-91.

Pascoe F. P. 1860a: Notices of new or little-known genera and species of Coleoptera. *Journal of Entomology* **1**: 36-64, pl. 2, 3.

Pascoe F. P. 1860b: Notices of new or little-known genera and species of Coleoptera. *Journal of Entomology* **1**: 98-132, pls 5-8.

Pascoe F. P. 1862: Notices of new or little-known genera and species of Coleoptera. Part III. *Journal of Entomology* **1**: 319-370, pls 16, 17.

Pascoe F. P. 1863a: Notices of new or little-known genera and species of Coleoptera. *Journal of Entomology* **2**: 26-56.

Pascoe F. P. 1863b: List of Colydiidae collected in the Amazons Valley by H.W. Bates, Esq., and descriptions of new species. *Journal of Entomology* **2**: 79-99.

Pascoe F. P. 1863c: List of Colydiidae collected in the Indian Islands by Alfred R. Wallace, Esq., and descriptions of new species. *Journal of Entomology* **2**: 121-143, pl. 8.

Pascoe F. P. 1866: Notices of new or little-known genera and species of Coleoptera. *Journal of Entomology* **2**: 443-493, pls. 17, 19.

Pascoe F. P. 1868: Contribution to the knowledge of the Coleoptera. *Transactions of the Entomological Society of London* **1868**: xi-xiii.

Pascoe F. P. 1869: Descriptions of new genera and species of Tenebrionidae from Australia and Tasmania. *The Annals and Magazine of Natural History* (4) **3**: 132-153, pl. 11, 277-296, pl. 12.

Pascoe F. P. 1870: Addtitions to the Tenebrionidae of Australia &c. *The Annals and Magazine of Natural History* (4) **5**: 94-107.

Pascoe F. P. 1871: Notes on Coleoptera, with descriptions of new genera and species. Part I. *The Annals and Magazine of Natural History* (4) **8**: 345-361, pl. 14.

Pascoe F. P. 1882: Notes on Coleoptera, with descriptions of new genera and species. Part IV. *The Annals and Magazine of Natural History* (5) **9**: 25-37.

Pascoe F. P. 1887: Notes on Coleoptera, with descriptions of new genera and species. Part VI. *The Annals and Magazine of Natural History* (5) **20**: 8-20, pl. 1.

Patti M. Z. 1840: Descrizione di un nuovo coleoptero. *Giornale di Scienze, Lettere ed Arti Sicilia* **72**: 292-294.

Paulus H. F. 1972: Neue Pyrochroidae aus Nepal (Coleoptera, Heteromera), mit einer Diskussion der verwandtschaftlichen Verhältnisse der Familie, 1. *Zeitschrift der Arbeitsgemeinschaft Österreichischer Entomologen* **23** [1971]: 75-85.

Paykull G. von. 1798: *Fauna Suecica. Insecta. Tomus I.* Uppsala: Joh. F. Edman, [8] + 358 + [2] pp.

Paykull G. von. 1799: *Fauna Suecica. Insecta. Tomus II.* Upsaliae: Joh. F. Edman, 234 pp.

Paykull G. von. 1800: *Fauna Suecica. Insecta. Tomus III.* Upsaliae: Joh. F. Edman, 459 pp.

Peacock E. R. 1982: Tydessa n. gen. from Japan (Coleoptera: Pyrochroidae). *Entomologica Scandinavica* **13**: 361-365.

Pecchioli V. 1839: Description d'une nouvelle espèce de Sitaris. *Annales de la Société Entomologique de France* **8**: 527-530, 1 pl.

Penecke K. A. 1912: Zwei neue Käferarten aus Bulgarien. *Wiener Entomologische Zeitung* **31**: 240-244.

Penrith M.-L. 1977: The Zophosini (Coleoptera: Tenebrionidae) of Western Southern Africa. *Cimbesiana, Memoir* **3**: 1-291.

Penrith M.-L. 1982a: Revision of the Zophosini (Coleoptera: Tenebrionidae). Part 5. A derived subgenus from Northern Africa. *Cimbebasia* (A) **6**: 165-226.

Penrith M.-L. 1982b: Revision of the Zophosini (Coleoptera: Tenebrionidae). Part 6. The subgenus Hologenosis Deyrolle. *Cimbebasia* (A) **6**: 227-289.

Penrith M.-L. 1983a: Revision of the Zophosini (Coleoptera: Tenebrionidae). Part 7. The African species of the subgenus Oculosis Penrith. *Cimbebasia* (A) **6**: 291-367.

Penrith M.-L. 1983b: Revision of the Zophosini (Coleoptera: Tenebrionidae). Part 8. The palaearctic species of the subgenus Oculosis Penrith, the subgenus Cheirosis Deyrolle, and a monotypical subgenus from Socotra. *Cimbebasia* (A) **6**: 369-384.

Penrith M.-L. 1984: Revision of the Zophosini (Coleoptera: Tenebrionidae). Part 9. The remaining species of the subgenus Septentriophosis Penrith. *Cimbebasia* (A) **6**: 385-416.

Penrith M.-L. 1986: Revision of the Zophosini (Coleoptera: Tenebrionidae). Part 10. Key to the subgenera, supplement, evolution and biogegraphy of the tribe, and catalogue. *Cimbebasia* (A) **6**: 417-502.

Perez Arcas L. 1865a: Insectos nuevos ó poco conocidos de la fauna española. Primera parte. *Revista de los Progresos de las Ciencias Exactas, Fisicas y Naturales* **15**: 166-186 [issued also separately, pp. 1-23].

Perez Arcas L. 1865b: Insectos nuevos ó poco conocidos de la fauna española. Secunda parte. *Revista de los Progresos de las Ciencias Exactas, Fisicas y Naturales* **15**: 413-444 [issued also separately, pp. 25-56].

Perez Arcas L. 1868: Insectos nuevos ó poco conocidos de la fauna española. Tercera parte. *Revista de los Progresos de las Ciencias Exactas, Fisicas y Naturales* **18**: 490-520 [issued also separately, pp. 57-86].

Perez Arcas L. 1872: Especies nuevas ó críticas de la fauna española. *Anales de la Sociedad Española de Historia Natural* **1**: 89-137, pls. I-II.

Perez Arcas L. 1874: Especies nuevas ó críticas de la fauna española. Tercera parte. *Anales de la Sociedad Española de Historia Natural* **3**: 111-155, pls I-III.

Péringuey L. 1909: Descriptive Catalogue of the Coleoptera of South Africa. Family Meloidae. *Transactions of the Royal Society of South Africa* **1**: 165-297, 4 pls.

Perris E. 1852: Seconde excursion dans les Grandes-Landes. *Annales de la Société Linnéenne de Bordeaux* **1852**: 1-72.

Perris E. 1864: Description de quelques espèces nouvelles de coléoptères et notes diverses. *Annales de la Société Entomologique de France* (4) **4**: 275-310.

Perris E. 1865: Descriptions de quelques nouvelles espèces de coléoptères, rectifications et notes. *Annales de la Société Entomologique de France* (4) **5**: 505-512.

Perris E. 1869: Descriptions de quelques coléoptères nouveaux. Rectifications et notes. *L'Abeille, Mémoires d'Entomologie* **7** [1869-1870]: 3-33.

Perris E. 1874: [new taxon]. In: Abeille de Perrin E. 1874a: [Diagnoses de Cisides nouveaux]. *Bulletin de la Société Entomologique de France* **1874**: lii-liii.

Perris E. 1875: Descriptions de quelques insectes jugés nouveaux. *L'Abeille, Journal d'Entomologie* **13**: 1-14.

Perroud B.-P. 1864: [new taxon]. In: Peroud B.-P. & Montrousier X.: La Faune entomologique de Kanala (Nouvelle-Calédonie) et description de quelques espèces nouvelles ou peu connues. *Annales de la Société Linnéenne de Lyon* **11**: 44-256, pl. 50.

Perroud B.-P. & Mulsant E. 1856: Description de deux nouvelles espèces de coléoptères constituant un genre nouveau dans la famille des ulomiens. *Opuscules Entomologiques* **7**:160-165.

Perty J. A. M. 1831: *Observationes nonnullae in Coleoptera Indiae orientalis. Dissertatio philosophico-entomologica, quam unacum praemissis thesibus auctoritate et consensus illustris philosophorum ordinis in Academia Ludovico-Maximilianea facultatem legendi rite adepturus.* Monachii: Mich. Lindauer, xxxxiv pp., 1 pl.

Petagna L. 1819: Memoria su di alcuni insetti del Regno di Napoli. *Memorie della Accademia delle Scienze. Classe de Fisica e di Storia Naturale* (Napoli) **1**: 19-40, 4 pls.

Petagna V. 1786: *Specimen Insectorum ulterioris Calabriae*. Neapoli: Typis Petri Perger, vii + 46 pp., 1 pl.

Petagna V. 1787: *Specimen Insectorum ulterioris Calabriae*. Francofurti et Moguntiae: apud Varrentrapp et Wenner, vi +46 pp., 1pl.

Petri K. 1912: *Siebenbürgens Käferfauna auf Grund ihrer Erforschung bis zum Jahre 1911*. Hermannstadt: Jos. Drotleff, xi + 374 columns.

Petri K. 1926: Ergänzungen und Berichtigungen zur Käferfauna Siebenbürgens, 1912. Zusammengestellt unter Mitwirkung der Herren Friedrich Deubel, Kronstadt, Oberstabsarzt Dr. Eugen Worell und Dr. Arnold Müller, Hermannstadt. *Verhandlungen und Mitteilungen des Siebenbürgischen Vereins für Naturwissenschaften zu Hemannstadt* **75-76**: 165-206.

Peyerimhoff P. de. 1907a: Liste de coléoptères du Sinaï. *L'Abeille, Journal d'Entomologie* **31**: 1-48.

Peyerimhoff P. de. 1907b: Recherches sur la synonymie des coléoptères du Nord-Africain (première note). *L'Abeille, Journal d'Entomologie* **31**: 185-187.

Peyerimhoff P. de. 1911: Nouveaux coléoptères du Nord-Africain (treizième note) (Tenebrionidae) récoltes du Dr. Edm. Sergent dans l'Extrême-Sud-Oranais. *Bulletin de la Société Entomologique de France* **1911**: 346-349.

Peyerimhoff P. de. 1913a: Nouveaux coléoptères du Nord-Africain (seizième note, faune du Djurdjura). *Bulletin de la Société Entomologique de France* **1913**: 253-255.

Peyerimhoff P. de. 1913b: Nouveaux coléoptères du Nord-Africain (quizième note: faune marocaine). *Annales de la Société Entomologique de France* **81**: 515-521.

Peyerimhoff P. de. 1915: Notes sur la biologie de quelques coléoptères phytophages du Nord-Africain (deuxième série). *Annales de la Société Entomologique de France* **84**: 19-61.

Peyerimhoff P. de. 1916a: Nouveaux coléoptères du Nord-Africain (vingt-deuxième note): Récoltes de M. le Dr. H. Foley dans l'extrême Sud-Oranais. *Bulletin de la Société Entomologique de France* **1916**: 71-76.

Peyerimhoff P. de. 1916b: Nouveaux coléoptères du Nord-Africain (vingt-troisième note): Récoltes de M. le Lt Nicloux dans le Sud-Ouest Marocain. *Bulletin de la Société Entomologique de France* **1916**: 82-84.

Peyerimhoff P. de. 1917a: Nouveaux coléoptères du Nord-African. Vingt-cinquième note avec des renseignements synonymique sur les Silphidae et les Histeridae. *Annales de la Société Entomologique de France* **86**: 117-148.

Peyerimhoff P. de. 1917b: Nouveaux coléoptères du Nord-Africain. Vingt-quatrième note: Faune de Djurdjura. *Bulletin de la Société Entomologique de France* **1916**: 318-319.

Peyerimhoff P. de. 1917c: Nouveaux coléoptères du Nord-African. (Vingt-sixième note: Faune du cèdre et du sapin de Numidie).*Bulletin de la Société Entomologique de France* **18**: 329-332.

Peyerimhoff P. de. 1918a: Nouveaux coléoptères du Nord-African. (Vingt-septième note: Faune du massif de Mouzaïa). *Bulletin de la Société Entomologique de France* **18** [1917]: 371-374.

Peyerimhoff P. de. 1918b: Nouveaux coléoptères du Nord-Africain (Vingt-huitième note: Faune du Pin d'Alep). *Bulletin de la Société Entomologique de France* **1918**: 141-144.

Peyerimhoff P. de. 1918c: Nouveaux coléoptères du Nord-Africain: (Vingt-neuvième note: Faune de Djurdjura. *Bulletin de la Société Entomologique de France* **1918**: 257-260.

Peyerimhoff P. de. 1919a: Notes sur la biologie de quelques coléoptères phytophages du Nord-Africain (troisième série). *Annales de la Société Entomologique de France* **88**: 169-258.

Peyerimhoff P. de. 1919b: Nouveaux coléoptères du Nord-Africain. Trentième note: Faune de l'Aurès. *Bulletin de la Société Entomologique de France* **1918**: 276-278.

Peyerimhoff P. de. 1920a: Nouveaux coléoptères du Nord-Africain. Trente-deuxième note: Faune saharienne. *Bulletin de la Société Entomologique de France* **1919**: 325-328.

Peyerimhoff P. de. 1920b: Nouveaux coléoptères du Nord-Africain. Trente-troisième note: Récoltes de M. le capitaine Surcouf dans le Grand Erg occidental. *Bulletin de la Société Entomologique de France* **1920**: 21-23.

Peyerimhoff P. de. 1920c: Nouveaux coléoptères du Nord-Africain. Trente-cinquième note: Faune del'Aurès. *Bulletin de la Société Entomologique de France* **1920**: 243-246.

Peyerimhoff P. de. 1921a: Sur quelques Tenebrionidae d'Orient (Col.). Description d'un Pimelia nouveau. *Bulletin de la Société Entomologique de France* **1920**: 327-329.

Peyerimhoff P. de. 1921b: Nouveaux coléoptères du Nord Africain. Trente-huitième note: Tenebrionidae. *Bulletin de la Société Entomologique de France* **1921**: 62-66.

Peyerimhoff P. de. 1921c: Nouveaux coléoptères du Nord Africain. Trente-neuvième note: Faune saharienne. *Bulletin de la Société Entomologique de France* **1921**: 231-236.

Peyerimhoff P. de. 1923: Nouveaux coléoptères du Nord-Africain. Quarante-troisième note: Récoltes de M. le Dr Heimsch dans l'Erg Iguidi (Sahara occidental). *Bulletin de la Société Entomologique de France* **1922**: 318-321.

Peyerimhoff P. de. 1924: Nouveaux coléoptères du Nord-Africain. Quarante-septième note: Récoltes de M. F. Peltier dans le Grand-erg occidental, le Gourara et le Touat. *Bulletin de la Société Entomologique de France* **1924**: 74-78.

Peyerimhoff P. de. 1925: Nouveaux coléoptères du Nord-Africain. Cinquantième note (avec table). *Annales de la Société Entomologique de France* **94**: 1-29.

Peyerimhoff P. de. 1926a: [new taxa]. In: Sainte Claire Deville J. & Peyerimhoff P. de: Nouveaux coléoptères du Nord Africain. Cinquante-cinquième note. Faune du Grand-Atlas marocain (suite). *Bulletin de la Société Entomologique de France* **1926**: 93-96.

Peyerimhoff P. de. 1926b: Nouveaux coléoptères du Nord-Africain. Cinquante-septième note. Faune de la Zousfana et du Guir. *Bulletin de la Société Entomologique de France* **1926**: 193-196.

Peyerimhoff P. de. 1927a: Études sur la systématique des coléoptères du Nord-Africain. I. - Les Pachychila (Tenebrionidae). *L'Abeille, Journal d'Entomologie* **34**: 1-57.

Peyerimhoff P. de. 1927b: Nouveaux coléoptères du Nord-Africain. Cinquante-huitième note. Récoltes de M. le Dr. A. Sicard dans le Maroc oriental. *Bulletin de la Société Entomologique de France* **1926**: 218-221.

Peyerimhoff O. de. 1928: Nouveaux coléoptères du Nord-African soixante-deuxième note. Un type nouveau de Pimeliinae (Tenebrionidae). *Bulletin de la Société Entomologique de France* **1928**: 61-63.

Peyerimhoff P. de. 1929a: Nouveaux coléoptères du Nord Africain. Soixante-septième note. Faune du Hoggar et des massifs voisins. *Bulletin de la Société Entomologique de France* **1929**: 123-126.

Peyerimhoff P. de. 1929b: Nouveaux coléoptères du Nord-Africain. Soixante-neuvième note. Faune du Hoggar et des massifs voisins. *Bulletin de la Société Entomologique de France* **1929**: 153-156.

Peyerimhoff P. de. 1929c: Nouveaux coléoptères du Nord-Africain. Soixante-treizième note. Faune du Hoggar et des massifs voisins. *Bulletin de la Société Entomologique de France* **1929**: 282-286.

Peyerimhoff P. de. 1931a: Nouveaux coléoptères du Nord-Africain. Soixante-quizième note (avec table et résumé). *Annales de la Société Entomologique de France* **100**: 1-20.

Peyerimhoff P. de. 1931b: Mission scientifique du Hoggar. Coléoptères. *Mémoires de la Société d'Histoire Naturelle de l'Afrique du Nord* **2**: 5-173, 3 pls, 2 maps.

Peyerimhoff P. de. 1933: Description de quatre Tenebrionidae (Col.) du Sahara central. *Bulletin de la Société d'Histoire Naturelle de l'Afrique du Nord* **24**: 344-356.

Peyerimhoff P. de. 1934: Mission transsaharienne de S. A. R. le Prince Sixte de Bourbon-Parme (1932). Insectes coléoptères. *Revue Française d'Entomologie* **1**: 104-108.

Peyerimhoff P. de. 1935a: Coléoptères (avec la description d'un genre et de deux espèces nouvelles). Mission au Tibesti (1930-1931) dirigée par M. Dalloni professeur à la faculté des Sciences d'Alger. *Mémoires de l'Académie des Sciences de l'Institut de France* **62**: 1-7, pl. 1.

Peyerimhoff P. de. 1935b: Coléoptères marocains récoltés par M. Ch. Rungs. *Bulletin de la Société des Sciences Narutelles du Maroc* **15**: 19-25.

Peyerimhoff P. de. 1935c: Coléoptères nouveaux ou mal connus de Berbérie. I. Le genre Storthocnemis Karsch (Tenebrionidae). *Bulletin de la Société Entomologique de France* **40**: 89-91.

Peyerimhoff P. de. 1936: Coléoptères (avec la description d'un nouveau genre et de deux espèces nouvelles). In: Dalloni M. M.: Mission au Tibesti (1930-31). *Mémoires de l'Academie des Sciences de l'Institut de France* **62**: 74-81.

Peyerimhoff P. de. 1938: Coléoptères nouveaux ou mal connus de Berbérie. XI. - Le genre Euthriptera Reitt. (Tenebrionidae). *Bulletin de la Société Entomologique de France* **43**: 178-180.

Peyerimhoff P. de. 1939a: Coléoptères nouveaux ou mal connus de Berbérie XII. - Le genre Belopus Gebien (Tenebrionidae). *Bulletin de la Société Entomologique de France* **44**: 25-28.

Peyerimhoff P. de. 1939b: Description d´un nouvel Heliotaurus Algérien (Col. Alleculidae). *Annales de la Société Entomologique de France* **108**: 120.

Peyerimhoff P. de. 1942a: Description d'un nouveau Crypticus (Col. Tenebrionidae) des Canaries. *Bulletin de la Société Entomologique de France* **47**: 11-12.

Peyerimhoff P. de. 1942b: Récoltes entomologiques faites par L. Berland à Villa-Cisneros (Rio de Oro). Coléoptères. *Bulletin du Muséum National d'Histore Naturelle, Paris* (2) **14**: 215-222.

Peyerimhoff P. de. 1942c: Sur un coléoptère nouveau du Rio de Oro. *Bulletin du Muséum National d'Histore Naturelle, Paris* (2) **14**: 276-277.

Peyerimhoff P. de. 1943: Matériaux pour un catalogue des coléoptères sahariens. II. – Descriptions d'espèces nouvelles. *Bulletin de la Société d'Histoire Naturelle de l'Afrique du Nord* **34**: 7-35.

Peyerimhoff P. de. 1945: Genres de coléoptères figurant à tort dans la faune euro-méditerranéenne. *Revue Française d'Entomologie* **12**: 1-5.

Peyerimhoff P. de. 1946: Coléopterès du Sahara marocain et du Sahara occidental. *Bulletin de la Société des Sciences Naturelles du Maroc* **24** [1944]: 90-110.

Peyerimhoff P. de. 1948a: Insectes coléoptères. Mission scientifique du Fezzan (1944-1945). *Mission Scientifique du Fezzan*: 5-84.

Peyerimhoff P. de. 1948b: Études sur la systematique des coléoptères du Nord Africain. II. Les Scaurus (Tenebrionidae). *Revue Française d'Entomologie* **14** [1947]: 157-193.

Peyerimhoff P. de. 1949a: Coléoptères nouveaux ou mal connus de Berbérie XXI. – Découverte dans le Grand Erg occidental du genre tauranien Sphenaria Mén. (Col. Tenebrionidae). *Bulletin de la Société Entomologie de France* **54**: 70-71.

Peyerimhoff P. de. 1949b: Études et descriptions de coléoptères marocains II. *Bulletin de la Société des Sciences Naturelles et Physiques du Maroc* **25-27** [1945-47]: 248-308.

Peyron E. 1854: Description d'une nouvelle espèces du genre Cossyphus. *Revue et Magasin de Zoologie Pure et Appliquée* (2) **6**: 224-225.

Pfaff C. H. 1824: Scharfer Stoff der Canthariden. In: *System der Materia medica nach chenmischen Principien mit Rücksicht auf die sinnlichen Merkmale und die Heilverhältnisse der Arzneymittel.* 7 (Zweyter Supplemenband). Leipzig: Vogel, xiv + 389 pp.

Philippi R. A. 1864: [new taxa]. In: Philippi R. A. & Philippi F.: Beschreibung einiger neuen chilenischen Käfer. *Entomologische Zeitung* **25**: 313-406.

Pic M. 1891a: Sur Crioceris tibialis Villa et Allecula morio Fabr. *L'Échange, Revue Linnéenne* **7**: 51.

Pic M. 1891b: Notes coléoptérologiques. *L'Échange, Revue Linnéenne* **7**: 117-118.

Pic M. 1892a: Descriptions et corrections. *L'Échange, Revue Linnéenne* **8**: 4.

Pic M. 1892b: Descriptions. - Beschreibungen. *Miscellanea Entomologica* **1**: 17-19, 25-27, 53-54.

Pic M. 1892c: Coléoptérologie descriptive. *L'Échange, Revue Linnéenne* **8**: 32-33.

Pic M. 1892d: Anthicus algériens. *Miscellanea Entomologica* **1**: 43-44.

Pic M. 1892e: Quelques mots sur les anthicides. *L'Échange, Revue Linnéenne* **8**: 43-44.
Pic M. 1892f: A corriger. *L'Échange, Revue Linnéenne* **8**: 66.
Pic M. 1892g: [new taxa]. *Bulletin de la Société Entomologiqe de France* **1892**: 75-77.
Pic M. 1892h: Contribution à l'étude des anthicides d'Algérie. *L'Échange, Revue Linnéenne* **8**: 102-103.
Pic M. 1892i: Notes entomologiques. *L'Échange, Revue Linnéenne* **8**: 114-115.
Pic M. 1892j: Contribution à l'étude des coléoptères. *L'Échange, Revue Linnéenne* **8**: 139.
Pic M. 1892k: Variétés. *L'Échange, Revue Linnéenne* **8**: 139-140.
Pic M. 1892l: Anthicides recueillis en Algérie en avril et mai 1892. *Revue Scientifique du Bourbonnais et du Centre de la France* **5**: 212-220.
Pic M. 1892m: Sur les Tomoderus Laf. Coléoptères de la famille des anthicides. *Revue Scientifique du Bourbonnais et du Centre de la France* **5**: 240-242.
Pic M. 1892n: [new taxa]. *Bulletin de la Société Entomologique de France* **1892**: cxcvi-cxcviii.
Pic M. 1892o: Deux coléoptères nouveaux. *Revue d'Entomologie* **11**: 313-314.
Pic M. 1893a: Anthicides nouveaux. *L'Échange, Revue Linnéenne* **9**: 15-16.
Pic M. 1893b: Descriptions de coléoptères. *L'Échange, Revue Linnéenne* **9**: 26-27.
Pic M. 1893c: Anthicides nouveaux. *L'Échange, Revue Linnéenne* **9**: 27-28.
Pic M. 1893d: Examen des anthicides de la collection Leprieur. *L'Échange, Revue Linnéenne* **9**: 38.
Pic M. 1893e: Étude sur les anthicides de la collection Leprieur. *L'Échange, Revue Linnéenne* **9**: 51-52.
Pic M. 1893f: Descriptions - Beschreibungen. *Miscellanea Entomologica* **1**: 53-54.
Pic M. 1893g: Rectifications. *Miscellanea Entomologica* **1**: 54.
Pic M. 1893h: Examen des anthicides de la collection Leprieur (Suite). *L'Échange, Revue Linnéenne* **9**: 64-65.
Pic M. 1893i: Examen des anthicides de la collection Leprieur (Suite et fin.). *L'Échange, Revue Linnéenne* **9**: 74-75.
Pic M. 1893j: Description d'espèces nouvelles de la faune européenne et circa - Anthicus nouveaux (Coléoptères). *Feuille des Jeunes Naturalistes* (3) **23**: 78-79.
Pic M. 1893k: Rectification. *L'Échange, Revue Linnéenne* **9**: 88.
Pic M. 1893l: Coléoptères nouveaux de la faune circa-européenne. *Feuille des Jeunes Naturalistes* (3) **23**: 111.
Pic M. 1893m: Essai d'une étude sur les Anthicus (lagenicolles) Cyclodinus Muls. *L'Échange, Revue Linnéenne* **9**: 113-116.
Pic M. 1893n: Anthicides tunisiens. *L'Échange, Revue Linnéenne* **9**: 122-123.
Pic M. 1893o: [Essai d'une étude sur les Anthicus (lagenicolles) Cyclodinus Muls, continuation]. *L'Échange, Revue Linnéenne* **9**: 125-128.
Pic M. 1893p: Anthicides nouveaux. *Miscellanea Entomologica* **1**: 126-127.
Pic M. 1893q: Description de deux coléoptères nouveaux. *L'Échange, Revue Linnéenne* **9**: 135.
Pic M. 1893r: Anthicides nouveaux. *Revue d'Entomologie* **12**: 155-156.
Pic M. 1893s: Xylophilides et anthicides recueillis en Algérie en Avril et Mai, 1893. *Revue Scientifique du Bourbonnais et du Centre de la France* **6**: 155-162.
Pic M. 1893t: Anthicides nouveaux de l'Ancien monde. *Feuille des Jeunes Naturalistes* (3) **23**: 175-176.
Pic M. 1893u: [Une espèce et une variété nouvelles de coléoptères]. *Bulletin de la Société Entomologique de France* **1893**: xxxiii-xxxiv.
Pic M. 1893v: [Description de trois Anthicus nouveaux (Coléoptères)]. *Bulletin de la Société Entomologique de France* **1893**: liii-lv.
Pic M. 1893w: Descriptions d'espèces nouvelles de coléoptères et notes synonymiques. *Bulletin de la Société Entomologique de France* **1893**: lxxxv-lxxxviii.
Pic M. 1893x: [Deux coléoptères nouveaux]. *Bulletin de la Société Entomologique de France* **1893**: clxi.
Pic M. 1893y: [Quelques notes synonymiques sur certains Anthicus (Coléoptères)]. *Bulletin de la Société Entomologique de France* **1892**: ccxi-ccxii.
Pic M. 1893z: [Descriptions suivantes de nouveaux Anthicus (Coléoptères)]. *Bulletin de la Société Entomologique de France* **1892**: cclviii-cclxxi.
Pic M. 1894a: Descriptions de coléoptères hétéromères africains. *Bulletin de la Société Zoologique de France* **1894**: 15-17.
Pic M. 1894b: Catalogue géographique des anthicides de France, Corse, Algérie et Tunisie (Suite). *Revue Scientifique du Bourbonnais et du Centre du France* **7**: 19-25.
Pic M. 1894c: Catalogue géographique des anthicides de France, Corse, Algérie et Tunisie (Suite). *Revue*

Scientifique du Bourbonnais et du Centre du France **7**: 40-49.

Pic M. 1894d: Liste des anthicides décrits postérieurement au Catalogus de MM. Gemminger et Harold (1870-1893). *Annales de la Société Entomologique de Belgique* **38**: 43-58.

Pic M. 1894e: Description de coléoptères de la faune circa-européenne. *Feuille des Jeunes Naturalistes* (3) **24**: 60-61.

Pic M. 1894f: Examen des anthicides de la collection H. Tournier. *L'Échange, Revue Linnéenne* **10**: 64-65.

Pic M. 1894g: Catalogue geographique des anthicides de France, Corse, Algérie et Tunisie (Suite). *Revue Scientifique du Bourbonnais et du Centre du France* **7**: 69-79.

Pic M. 1894h: Quelques notes en passant sur plusieurs coléoptères de la collection Henri Tournier. *L'Échange, Revue Linnéenne* **10**: 71-72.

Pic M. 1894i: Extrait de la Revue Scientifique du Bourbonnais et du Centre de la France. Xylophilides & Anthicides recueillis en Algérie en Mai et Juin 1894. *L'Échange, Revue Linnéenne* **10**: 84-88.

Pic M. 1894j: Descriptions de trois coléoptères hétéromères nouveaux. *Le Naturaliste* (2) **16**: 93.

Pic M. 1894k: Deux Coléoptères nouveaux. *Feuille des Jeunes Naturalistes* (3) **24**: 140.

Pic M. 1894l: Xylophilides et anthicides recueillis en Algérie en Mai et Juin 1894. *Revue Scientifique du Bourbonnais et du Centre du France* **7**: 140-146.

Pic M. 1894m: Contribution à la faune coléoptérologique d'Algérie. *Le Naturaliste* (2) **16**: 180.

Pic M. 1894n: Anthicides du Bengale. *Annales de la Société Entomologique de Belgique* **38**: 181-184.

Pic M. 1894o: Descriptions de coléoptères. *Mémoires de la Société Zoologique de France* **7**: 203-207.

Pic M. 1894p: Contribution à la faune coléoptérique d'Algérie. Description d'espèces nouvelles. *Le Naturaliste* **16**: 247-248.

Pic M. 1894q: Liste complète des xylophilides décrits jusqu'en 1894, avec descriptions d'espèces nouvelles. *Mémoires de la Société Zoologique de France* **7**: 427-436.

Pic M. 1894r: Nouvelles espèces d'anthicides appartenant au Musée Civique de Gênes. *Annali del Museo Civico di Storia Naturale di Genova* **14** [1894-1895]: 582-587.

Pic M. 1894s: [new taxa]. *Bulletin de la Société Entomologique de France* **1893**: cclxxvi-cclxxix.

Pic M. 1894t: [Descriptions de deux coléoptères nouveaux de la Chine orientale]. *Bulletin de la Société Entomologique de France* **1894**: x-xi.

Pic M. 1894u: [Trois Anthicus nouveaux et observations sur quelques anthicides]. *Bulletin de la Société Entomologique de France* **1894**: lxxvii-lxxix.

Pic M. 1895a: Examen des anthicides de la collection Lethierry. *L'Échange, Revue Linnéenne* **11**: 18-20.

Pic M. 1895b: Observations et renseignements divers. *L'Échange, Revue Linnéenne* **11**: 51-53.

Pic M. 1895c: Description de coléoptères nouveaux de la famille des anthicides. *Le Naturaliste* (2) **17**: 59-60.

Pic M. 1895d: Anthicides récoltés dans les tabacs. *Bulletin de la Société Zoologique de France* **20**: 61-64.

Pic M. 1895e: Descriptions de coléoptères d'Algérie. *L'Échange, Revue Linnéenne* **11**: 78-82.

Pic M. 1895f: Un nouveau Tomoderus de la Russie d'Asie (Col.). *Bulletin de la Société Entomologique de France* **1895**: 174.

Pic M. 1895g: Description de trois Anthicus méditerranéens. *L'Abeille, Journal d'Entomologie* **28** [1892-1896]: 195-196.

Pic M. 1895h: Descriptions de coléoptères. *Le Naturaliste* (2) **9**: 243.

Pic M. 1895i: Anthicides d'Afrique de la collection de M. Charles Alluaud. *Annales de la Société Entomologique de France* **63** [1894]: 665-668.

Pic M. 1895j: Anthicus nouveaux du Musée Civique de Gênes. *Annali del Museo Civico di Storia Naturale di Genova* **34**: 784-786.

Pic M. 1895l: Notes sur des coléoptères rares ou nouveaux d'Algérie. *Bulletin de la Société Entomologique de France* **1895**: cxxvi-cxxx.

Pic M. 1895m: [Description de deux coléoptères nouveaux et une note sur Anthicus dromioides Pic]. *Bulletin de la Société Entomologique de France* **1894**: cxci-cxcii.

Pic M. 1895n: [Diagnoses de plusiers coléoptères rapportés de Syrie par M. C. Delagrange]. *Bulletin de la Société Entomologique de France* **1894**: cclxxxiv-cclxxxv.

Pic M. 1896a: Descriptions de deux Anthicus syriens. *Miscellanea Entomologica* **4**: 5-6.

Pic M. 1896b: Excursion entomologique dans la province d'Oran (Algérie). *Revue Scientifique du Bourbonnais et du Centre de France* **8**: 10-13.

Pic M. 1896c: Xylophilides et Anthicides recueillis en Algérie Avril-Juin 1895. *Revue Scientifique du*

Bourbonnais et du Centre de France **9**: 37-41.

Pic M. 1896d: Descriptions et notes sur divers coléoptères. *Miscellanea Entomologica* **4**: 41-43.

Pic M. 1896e: Notes et renseignements sur les xylophilides (Coléoptères hétéromères). *Bulletin de la Société Zoologique de France* **21**: 49-53.

Pic M. 1896f: Descriptions et notes diverses. *L'Échange, Revue Linnéenne* **12**: 61-62.

Pic M. 1896g: Notes et diagnoses. *L'Échange, Revue Linnéenne* **12**: 87-88.

Pic M. 1896h: Coléoptères du Nord de L´Afrique. *Miscellanea Entomologica* **4**: 93-94.

Pic M. 1896i: Notes et renseignements descriptifs sur Anthicus insignis, Luc. et races voisines. *L'Échange, Revue Linnéenne* **12**: 131-132.

Pic M. 1896j: Descriptions de coléoptères d'Algérie et de Syrie. *Miscellanea Entomologica* **4**: 140-142.

Pic M. 1896l: Xylophilides et anthicides recueillis en Algérie (Mai-Juin 1896). *Revue Scientifique du Bourbonnais et du Centre de France* **9**: 162-169.

Pic M. 1896m: Examen des anthicides de la collection Reitter. *La Feuille des Jeunes Naturalistes* (3) **26**: 178-181.

Pic M. 1896n: Description d'un Anthicus d'Algérie (Col.). *Bulletin de la Société Entomologique de France* **1896**: 422-423.

Pic M. 1897a: Excursion entomologique dans le sud de l'Algérie. *Miscellanea Entomologica* **5**: 2-3, 35-38.

Pic M. 1897b: *Variétés II*. Lyon: L. Jacquet.

Pic M. 1897c: Coléoptères nouveaux. *Miscellanea Entomologica* **5**: 26-29.

Pic M. 1897d: Description de coléoptères nouveaux, d'Europe et circa. *Le Frelon* **6** [1897-1898]: 35-36.

Pic M. 1897e: Descriptions de coléoptères. *Miscellanea Entomologica* **5**: 5-6, 41-43.

Pic M. 1897f: Sur le groupe Liparoderus, Laf. dans le genre Anthicus, Payk. *L'Échange, Revue Linnéenne* **13**: 50-51.

Pic M. 1897g: Descriptions de coléoptères. *Miscellanea Entomologica* **5**: 61-62.

Pic M. 1897h: Description d'un Corticus nouveau d'Algérie (Col.). *Bulletin de la Société Entomologique de France* **1897**: 78-79.

Pic M. 1897i: Note sur les xylophilides (coléoptères hétéromères). *Bulletin de la Société Zoologique de France* **22**: 80-82.

Pic M. 1897j: Descriptions de coléoptères. *La Feuille des Jeunes Naturalistes* (3) **27**: 119-120.

Pic M. 1897l: Notes sur quelques Zonabris Har. (Mylabris F.). *Revue Scientifique du Bourbonnais et du Centre de France* **10**: 123-126.

Pic M. 1897m: En route pour la Mzab.- Bou Saada (fin). *Miscellanea Entomologica* **10**: 124-127.

Pic M. 1897n: Anthicides exotiques nouveaux. *Le Naturaliste* (2) **19**: 134.

Pic M. 1897o: [new taxon]. In: Ragusa E.: Catalogo ragionato dei coleotteri di Sicilia. Meloidae. *Il Naturalista Siciliano* (N. S.) **2**: 212-222.

Pic M. 1897r: Descriptions de coléoptères. *Comptes Rendus des Séances et des Excursions de la Société d'Histoire Naturelle d'Autun* **10**: 295-300.

Pic M. 1898a: Descriptions de coléoptères d´Anatolie & de Syrie. *Revue Scientifique du Bourbonais et du Centre de la France* **11**: 1-4.

Pic M. 1898b: Variétés et nouvelles espèces de coléoptères. *Miscellanea Entomologica* **6**: 2-4.

Pic M. 1898c: Description de coléoptères nouveaux, d'Europe et circa. *Le Frelon* **6**: 35-38.

Pic M. 1898d: Description de coléoptères nouveaux. *Le Naturaliste* (2) **20**: 63.

Pic M. 1898e: Anthicides (Col. hétéromères) africains nouveaux des collections du Muséum de Paris. *Bulletin de la Museum d'Histoire Naturelle, Paris* **4**: 67-72.

Pic M. 1898f: Notes descriptives sur plusieurs coléoptères et sur un Ichneumon (Hyménoptère). *Miscellanea Entomologica* **6**: 73-75.

Pic M. 1898g: Descriptions d'espèces ou variétés de coléoptères de la faune d'Eupope [sic] et circa. *Miscellanea Entomologica* **6**: 97-99.

Pic M. 1898h: Description de coléoptères. *Bulletin de la Société d'Histoire Naturelle d'Autun* **11**: 116-125.

Pic M. 1898i: Description de trois Anthicidae du Turkestan. *Revue d'Entomologie* **17**: 122-124.

Pic M. 1898j: Un Amblyderus (Col. Hétéromères) nouveau d'Abyssinie. *Bulletin du Museum d'Histoire Naturelle, Paris* **1898**: 181.

Pic M. 1898k: Description de trois coléoptères nouveaux de Madagascar et de l'île Maurice. *Bulletin de la Société Entomologique de France* **1898**: 182-183.

Pic M. 1898l: Répertoire des publications zoologiques (1889-1897). *Annales de la Société Entomologique de France* **67**: 191-224.

Pic M. 1898m: [new taxa]. In: Ragusa E.: Catalogo ragionato dei coleotteri di Sicilia. *Il Naturalista Siciliano* **2** [1897]: 197-237.

Pic M. 1899a: Contribution à l´étude du genre Chrysanthia Schm. *Feuille des Jeunes Naturalistes* **30**: 14-16.

Pic M. 1899b: Notes sur les anthicides d'Europe avec diagnoses. *Miscellanea Entomologica* **7**: 18-21.

Pic M. 1899c: Diagnoses de ptinides et anthicides (Col.) des collections du Muséum de Paris. *Bulletin du Muséum d'Histoire Naturelle, Paris* **5**: 28-32.

Pic M. 1899d: Liste de coléoptères de mes chasses en 1898 dans les Alpes. *Miscellanea Entomologica* **1**: 50-51.

Pic M. 1899e: Descriptions d'Anthicidae. *Revue de Entomologie* **18**: 73-76.

Pic M. 1899f: Anthicide et pedilide (Coléoptères hétéromères) recueillis au Sikkim par M. Harmand, et offerts par lui au Muséum d'Histoire Naturelle. *Bulletin du Muséum d'Histoire Naturelle, Paris* **5**: 76-79.

Pic M. 1899g: Diagnoses de coléoptères communiqués par M. L. Bleuse. *Bulletin de la Société Entomologique de France* **1899**: 85-86.

Pic M. 1899h: Sur divers coléoptères de la faune paléarctique. *Miscellanea Entomologica* **7**: 113-116.

Pic M. 1899i: Xylophides et anthicides recueillis en Orient, en mars, avril et mai 1899. *Revue Scientifique du Bourbonnais et du Centre de la France* **12**: 170-178.

Pic M. 1899j: Description de coléoptères. *Le Naturaliste* (2) **21**: 189.

Pic M. 1899k: Les coléoptères anthicides du centre de la France. *Bulletin Trimestriel de la Société d'Histoire Naturelle de Macon* **1**: 189-196.

Pic M. 1899l: Diagnoses de coléoptères d'Orient. *Bulletin de la Société d'Histoire Naturelle d'Autun* **12** (2): 204-210.

Pic M. 1899m: Description de coléoptères nouveaux. *Le Naturaliste* (2) **21**: 264.

Pic M. 1899n: Diagnoses de deux coléoptères nouveaux de Palestine. *Bulletin de la Société Entomologique de France* **1899**: 278-279.

Pic M. 1899o: Description d´une variété nouvelle d´Alleculidae (Col.) de l´Espagne centrale. *Bulletin de la Société Entomologique de France* **1899**: 300.

Pic M. 1899p: Diagnoses préliminaires d'Helopidae d'Asie Mineure (Col.). *Bulletin de la Société Entomologique de France* **1899**: 411.

Pic M. 1899q: Notes sur trois Anthicus de la région méditerranéenne. *Bulletin de la Société Entomologique de France* **1899**: 412.

Pic M. 1900a: *Bestimmungs-Tabellen der europäischen Coleopteren. XL. Heft. Hylophilidae (früher Euglenini und Xylophilini).* Paskau: Edm. Reitter, 21 pp.

Pic M. 1900b: [new taxa]. In: Bodemeyer E. von.: *Quer durch Klein-Asien in der Bulghar-Dagh. Eine naturwissenschaftliche Studien-Reise*. Emmendingen: Verlangs-Aktiengesellschaft vormals Dölter, v + 169 pp.

Pic M. 1900c: Neue Coleopteren des Hamburger Museums. *Mitteilungen des Zoologischen Staatsinstitut und Zoologischen Museum in Hamburg* **17** (2): 9-10.

Pic M. 1900d: Diagnoses de coléoptères d'Orient récoltés en 1899. *Bulletin de la Socété Naturelle d'Autun* **1900**: 24-31.

Pic M. 1900e: Notes sur divers coléoptères. *L'Échange, Revue Linnéenne* **16**: 37-38.

Pic M. 1900f: Descriptions de coléoptères algériens et tunisiens. *L'Èchange, Revue Linnéenne* **16**: 53-56.

Pic M. 1900g: Notes et diagnoses. *L'Échange, Revue Linnéenne* **16**: 61-63.

Pic M. 1900h: Contribution à l'étude des Notoxus (Coléoptères) d'Europe et de régions avoisinantes. *Feuille des Jeunes Naturalistes* **30**: 64-68.

Pic M. 1900i: Notes et diagnoses. 2e article. *L'Èchange, Revue Linnéenne* **16**: 77-80.

Pic M. 1900j: Diagnoses de coléoptères du globe. *L'Èchange, Revue Linnéenne* **16**: 89-91.

Pic M. 1900k: Contribution a l'étude des Notoxus (Coléoptères) d'Europe et des régions avoisinantes. *Feuille des Jeunes Naturalistes* **30**: 89-91.

Pic M. 1900l: Descriptions et habitats nouveaux de divers coléoptères d´Algérie et d´Orient. *Bulletin de la Société Entomologique de France* **1900**: 123-125.

Pic M. 1900m: Contribution à l'étude des coléoptères de la Tripolitaine et de la Tunisie. *Bulletin de la Société Entomologique de France* **1900**: 164-166.

Pic M. 1900n: Anthicidae de l'Erythrée. *Annali del Museo Civico di Storia Naturale di Genova* (2) **20**: 575-576.

Pic M. 1900o: Diagnoses d'Anthicidae de la Nouvelle Guinée. *Annali del Museo Civico di Storia Naturale di Genova* **20**: 602-608.

Pic M. 1901a: Notes diverses et diagnoses. *L'Échange, Revue Linnéenne* **17**: 2-4.

Pic M. 1901b: Notes diverses et diagnoses (2e article). *L'Échange, Revue Linnéenne* **17**: 9-12.

Pic M. 1901c: Notes diverses et diagnoses (5e article). *L'Échange, Revue Linnéenne* **17**: 33-35.

Pic M. 1901d: Coléoptères d'Orient. *Comptes Rendus des Séances et des Excursions de la Société d'Histoire Naturelle d'Autun* **14**: 33-39.

Pic M. 1901e: Descriptions abrégées de coléoptères provenant de Grèce. *L'Échange, Revue Linnéenne* **17**: 49-52.

Pic M. 1901f: Notes diverses et diagnoses (6e Article). *L'Échange, Revue Linnéenne* **17**: 57-59.

Pic M. 1901g: Notes diverses et diagnoses (7e article). *L'Échange, Revue Linnéenne* **17**: 65-68.

Pic M. 1901h: Notes diverses sur le genre Zonabris Harold. *Bulletin de la Société Zoologique de France* **26**: 77-78.

Pic M. 1901i: Communications diverses. 1. Espèces et variétés nouvelles de coléoptères paléarctiques. *L'Échange, Revue Linnéenne* **17**: 79-80.

Pic M. 1901j: Diagnoses d'Anthicidae exotiques. *Annales de la Société Entomologique de Belgique* **45**: 89-90.

Pic M. 1901k: Coléoptères nouveaux d'Orient, Arménie et Tunisie. *L'Échange, Revue Linnéenne* **17**: 89-91.

Pic M. 1901l: Sur quelques variétés de Zonabris, du Turkestan (Col.). *Bulletin de la Société Entomologique de France* **1901**: 110-111.

Pic M. 1901m: Note sur les Anthicus terminatus Schmt. et terminatus Laf. *Miscellanea Entomologica* **9**: 129-131.

Pic M. 1901n: Xylophilides et anthicides capturés en Italie et Grèce, du 17 avril au 14 juin 1901. *Revue Scientifique du Bourbonnais et du Centre de la France* **14**: 173-182.

Pic M. 1901o: Liste de quelques coléoptères recueillis en Grèce en 1901. *Bulletin de la Société Entomologique de France* **1901**: 253-255.

Pic M. 1901p: Diagnoses de Macratria et d'Anthicidae de la Malaysie. *Annali del Museo Civico di Storia Naturale di Genova* (2) **20** [1900]: 740-742, 791-803.

Pic M. 1902a: Coléoptères asiatiques nouveaux. *L'Échange, Revue Linnéenne* [18 sic] **17**: 1-3.

Pic M. 1902b: Descriptions et notes diverses. *L'Échange, Revue Linnéenne* [18 sic] **17**: 9-10.

Pic M. 1902c: Descriptions et notes diverses (2e article). *L'Échange, Revue Linnéenne* [18 sic] **17**: 17-19.

Pic M. 1902d: Diagnoses de coléoptères de l'ancien et du nouveau monde. *L'Échange, Revue Linnéenne* **17**: 23-26.

Pic M. 1902e: Nouvelles espèces et variétés de coléoptères paléarctiques. *L'Échange, Revue Linnéenne* **17**: 31-38.

Pic M. 1902f: Deux notes entomologiques avec synopsis. *L'Échange, Revue Linnéenne* [18 sic] **17**: 41-42.

Pic M. 1902g: Descriptions et notes diverses (4e article). *L'Échange, Revue Linnéenne* [18 sic] **17**: 55-57.

Pic M. 1902h: Coléoptères présumés nouveaux de la faune circaméditerranéenne. *L'Échange, Revue Linnéenne* [18 sic] **17**: 63-65.

Pic M. 1902i: Descriptions et notes diverses. *L'Échange, Revue Linnéenne* [18 sic] **17**: 72.

Pic M. 1902j: Nouvelles espèces et variétés de coléoptères paléarctiques. *L'Échange, Revue Linnéenne* [18 sic] **17**: 79-80.

Pic M. 1902k: Coléoptères nouveaux de l'Ile de Ténériffe. *L'Échange, Revue Linnéenne* [18 sic] **17**: 80-81.

Pic M. 1902l: Anthicides recueillis en Orient par M. Paul de Peyerimhoff (Col.). *Bulletin de la Société Entomologique de France* **1902**: 185-187.

Pic M. 1903a: Anthicides nouveaux des collections du Muséum de Paris (Col.). *Bulletin du Muséum d'Histoire Naturelle, Paris* **9**: 20-21.

Pic M. 1903b: Notes entomologiques. *Bulletin de la Société Entomologique de France* **1903**: 79-83.

Pic M. 1903c: Notes entomologiques. VI. Description d'un Notoxus de Djibouti. *Bulletin de la Société Entomologique de France* **1903**: 83.

Pic M. 1903d: Nouveaux anthicides des collections du Musée Zoologique de l'Académie Impériale des Sciences à St. Pétersbourg. (Coleoptera, Anthicidae). *Annuaire du Musée Zoologique de l'Académie Impériale des Sciences de St. Pétersbourg* **8**: 88-89.

Pic M. 1903e: Descriptions et notes complémentaires sur divers coléoptères paléarctiques. *L'Échange, Revue Linnéenne* **19**: 97-98.

Pic M. 1903f: Coléoptères exotiques nouveaux (2e article). *L'Échange, Revue Linnéenne* **19**: 105-107.

Pic M. 1903g: Descriptions et habitats de divers coléoptères du Nord de l'Afrique. *L'Échange, Revue Linnéenne* **19**: 113-115.

Pic M. 1903h: Coléoptères nouveaux d'Asie et d'Afrique. *L'Échange, Revue Linnéenne* **19**: 119-120.

Pic M. 1903i: Notes entomologiques et descriptions. *L'Échange, Revue Linnéenne* **19**: 121-125.

Pic M. 1903j: Coléoptères européens et Clytus syrien nouveaux. *L'Échange, Revue Linnéenne* **19**: 138-139.

Pic M. 1903k: Contribution à l'étude du genre «Salpingus» Gyllh. *L'Échange, Revue Linnéenne* **19**: 140-141.

Pic M. 1903l: Deux captures intéressantes. Diagnoses de divers coléoptères. *L'Échange, Revue Linnéenne* **19**: 145-147.

Pic M. 1903m: Espèces et variétés nouvelles de coléoptères. *L'Échange, Revue Linnéenne* **19**: 161-163.

Pic M. 1903n: Espèces et variétés nouvelles de coléoptères (2me article). *L'Échange, Revue Linnéenne* **19**: 169-171.

Pic M. 1903o: Espèces et variétés nouvelles de coléoptères (3me article). *L'Échange, Revue Linnéenne* **19**: 177-178.

Pic M. 1903p: Coléoptères rares ou nouveaux pour le département, recueilles en 1903. *Bulletin de la Société Histoire Naturelle d'Autun* **16**: 187-192.

Pic M. 1903q: Note synonymique (Col.). *Bulletin de la Société Entomologique de France* **1903**: 304.

Pic M. 1903r: Anthicides de Manille. *Annales de la Société Entomologique de France* **71** [1902]: 643-648.

Pic M. 1903s: Voyage de M. Maurice Maindron dans l'Inde méridionale (Mai à Novembre 1901). 3e mémoire. *Annales de la Société Entomologique de France* **72**: 348-352.

Pic M. 1904a: Notes et descriptions. *L'Échange, Revue Linnéenne* **20**: 2-5.

Pic M. 1904b: Notes et descriptions (Suite). *L'Échange, Revue Linnéenne* **20**: 9-11.

Pic M. 1904c: Diagnoses de seize coléoptères exotiques appartenant à diverses familles. *L'Échange, Revue Linnéenne* **20**: 18-20.

Pic M. 1904d: Diagnoses de coléoptères asiatiques provenant surtout de Sibéria. *L'Échange, Revue Linnéenne* **20**: 25-27.

Pic M. 1904e: Diagnoses de divers coléoptères d'Europe et Turquie d'Asie. *L'Échange, Revue Linnéenne* **20**: 57-58.

Pic M. 1904f: Coléoptères nouveaux de la Turquie d'Asie. *L'Échange, Revue Linnéenne* **20**: 73-74.

Pic M. 1904g: Nouvelles espèces et variétés de coléoptères paléarctiques. *L'Échange, Revue Linnéenne* **20**: 81-82.

Pic M. 1904h: Diagnoses ou descriptions abrégées de coléoptères paléarctiques. *L'Échange, Revue Linnéenne* **20**: 89-94.

Pic M. 1904i: Espèces et variétés nouvelles de coléoptères paléarctiques. *L'Échange, Revue Linnéenne* **21**: 97-100.

Pic M. 1905a: Notes entomologiques diverses (Col.). *Bulletin de la Société Entomologique de France* **1905**: 47-48.

Pic M. 1905b: Espèces et variétés nouvelles de coléoptères paléarctiques. *L'Échange, Revue Linnéenne* **21**: 97-110.

Pic M. 1905c: Description de deux Anthicus nouveaux (Col.). *Bulletin de la Société Entomologique de France* **1905**: 99-100.

Pic M. 1905d: Descriptions abrégées et notes diverses. *L'Échange, Revue Linnéenne* **21**: 105-107.

Pic M. 1905e: Descriptions abrégées et notes diverses (2e article). *L'Échange, Revue Linnéenne* **21**: 113-115.

Pic M. 1905f: Descriptions abrégées et notes diverses. *L'Échange, Revue Linnéenne* **21**: 129-131.

Pic M. 1905g: Coléoptères nouveaux provenant de France, Grèce, Algérie et Turquie d'Asie. *L'Échange, Revue Linnéenne* **21**: 161-163.

Pic M. 1905h: Petite contribution à l'étude des Myiodes ou Myodites Latr. paléarctiques. *L'Échange, Revue Linnéenne* **21**: 165-166.

Pic M. 1905i: Sur les Zonabris Wagneri et curta Chevrolat. *Revue Scientifique du Bourbonnais et du Centre de France* **1905**: 167-170.

Pic M. 1905j: Espèces et variétés nouvelles de coléoptères de France et Turquie d'Asie. *L'Échange, Revue Linnéenne* **21**: 177-179.

Pic M. 1905k: Notes entomologiques et descriptions. *L'Échange, Revue Linnéenne* **21**: 185-187.

Pic M. 1905l: Coléoptères asiatiques nouveaux. *Bulletin de la Société Entomologique de France* **1904**: 287-289.

Pic M. 1905m: Notes sur diverses espèces d'anthicides de Motschulsky et descriptions de plusiers nouveautés du même groupe. *Annuaire du Musée Zoologique de l'Académie Impériale des Sciences de St.Pétersbourg* **9** [1904]: 490-494.

Pic M. 1906a: Deux nouveaux coléoptères d'Espagne. *L'Échange, Revue Linnéenne* **22**: 12-13.

Pic M. 1906b: Coléoptères européens, algériens et asiatiques nouveaux. *L'Échange, Revue Linnéenne* **22**: 17-20.

Pic M. 1906c: Habitats et descriptions de divers coléoptères paléarctiques. *L'Échange, Revue Linnéenne* **22**: 25-27.

Pic M. 1906d: Contribution à l'étude des pyrochroides. *L'Échange, Revue Linnéenne* **22**: 28-30.

Pic M. 1906e: Habitats et descriptions de divers coléoptères paléarctiques. *L'Échange, Revue Linnéenne* **22**: 33-35.

Pic M. 1906f: Coléoptères exotiques nouveaux ou peu connus. *L'Échange, Revue Linnéenne* **22**: 43-45.

Pic M. 1906g: Un nouvel Anthicus du Turkestan (Col.). *Bulletin de la Société Entomologique* de *France* **1906**: 46-47.

Pic M. 1906h: Coléoptères en partie nouveaux provenant de la Perse et des régions avoisinantes. *L'Échange, Revue Linnéenne* **22**: 49-51.

Pic M. 1906i: Deux nouveaux Anthicus du Kashmir (Col.). *Bulletin de la Société Entomologique de France* **1906**: 78-79.

Pic M. 1906j: Anthicides et ptinides d'Algérie et de Tunisie (Col.) récoltés par M. le Capitaine Vibert. *Bulletin de la Société Entomologique de France* **1906**: 282-285.

Pic M. 1907a: Anthicides recueillis en Erythrée. *Annali del Museo Civico di Storia Naturale, Genova* **43**: 63-65.

Pic M. 1907b: Coléoptères exotiques nouveaux ou peu connus (Suite). *L'Échange, Revue Linnéenne* **23**: 119-120.

Pic M. 1907c: Notes entomologiques diverses. *L'Échange, Revue Linnéenne* **23**: 121-123.

Pic M. 1907d: Coléoptères exotiques nouveaux ou peu connus (Suite). *L'Échange, Revue Linnéenne* **23**: 134-135.

Pic M. 1907e: Notes entomologiques diverses (Suite). *L'Échange, Revue Linnéenne* **23**: 137-139.

Pic M. 1907f: Notes entomologiques diverses. *L'Échange, Revue Linnéenne* **23**: 153-154.

Pic M. 1907g: Coléoptères exotiques nouveaux ou peu connus (Suite). *L'Échange, Revue Linnéenne* **23**: 158.

Pic M. 1907h: Notes entomologiques diverses. *L'Échange, Revue Linnéenne* **23**: 161-162.

Pic M. 1907i: Coléoptères exotiques nouveaux ou peu connus (Suite). *L'Échange, Revue Linnéenne* **23**: 165-166.

Pic M. 1907j: Diagnoses de coléoptères asiatiques nouveaux. *L'Échange, Revue Linnéenne* **23**: 171-174.

Pic M. 1907k: Coléoptères exotiques nouveaux ou peu connus (Suite). *L'Échange, Revue Linnéenne* **23**: 175-176.

Pic M. 1907l: Notes entomologiques diverses. *L'Échange, Revue Linnéenne* **23**: 177-179.

Pic M. 1907m: Coléoptères exotiques nouveaux ou peu connus. (Suite.) *L'Échange, Revue Linnéenne* **23**: 181-183.

Pic M. 1907n: Coléoptères exotiques nouveaux ou peu connus (Suite). *L'Échange, Revue Linnéenne* **23**: 190-191.

Pic M. 1907o: Un Silis et deux anthicides nouveaux provenant des Indes (Col.). *Bulletin de la Société Entomologique de France* **1907**: 195-196.

Pic M. 1907p: Anthicides recueillis par M. Ch. Alluaud dans le Soudan Égyptien oriental (Col.) (Novembre 1905-Mai 1906). *Bulletin de la Société Entomologique de France* **1907**: 321-323.

Pic M. 1907q: Ueber neue oder wenig bekannte Käfer aus Asien, Amerika und Deutsch-Ostafrika. *Stettiner Entomologische Zeitung* **68**: 337-342.

Pic M. 1907r: Macratria et Anthicidae recueillis par feu L. Fea en Birmanie et regions avoisinantes. *Annali del Museo Civico di Storia Naturale di Genova* **43** [1907-1908]: 339-342.

Pic M. 1908a: Contribution à l'étude des Osphya paléarctiques. *L'Échange, Revue Linnéenne* **24**: 31-32.

Pic M. 1908b: Descriptions ou diagnoses et notes diverses (Suite). *L'Échange, Revue Linnéenne* **24**: 33-35.

Pic M. 1908c: Étude synoptique sur divers „Cteniopini" (Col. hétéromères). *L'Échange, Revue Linnéenne* **24**: 38-40.

Pic M. 1908d: Descriptions ou diagnoses et notes diverses (Suite). *L'Échange, Revue Linnéenne* **24**: 41-46.

Pic M. 1908e: Étude synoptique sur divers „Cteniopini" (Col. hétéromères). *L'Échange, Revue Linnéenne* **24**: 47-48.

Pic M. 1908f: Descriptions ou diagnoses et notes diverses (Suite). *L'Échange, Revue Linnéenne* **24**: 57-58.

Pic M. 1908g: Coléoptères exotiques nouveaux ou peu connus. *L'Échange, Revue Linnéenne* **24**: 59-61.
Pic M. 1908h: Étude synoptique sur divers „Cteniopini" (Col. hétéromères). *L'Échange, Revue Linnéenne* **24**: 61-62.
Pic M. 1908i: Descriptions ou diagnoses et notes diverses (Suite). *L'Échange, Revue Linnéenne* **24**: 65-66.
Pic M. 1908j: Captures intéressantes. Moeurs et nouvelles variétés de coléoptères français. *L'Échange, Revue Linnéenne* **24**: 71-72.
Pic M. 1908k: Descriptions ou diagnoses et notes diverses (Suite). *L'Échange, Revue Linnéenne* **24**: 73-75.
Pic M. 1908l: Deux nouveaux Pyrochroa Geoff. de Chine. *Bulletin de la Société Entomologique de France* **1908**: 75.
Pic M. 1908m: Description d'un Anthicus nouveau d'Égypte. *Bulletin de la Société Entomologique d'Egypte* **1** [1908-1909]: 143-144.
Pic M. 1908n: Deux nouveaux hétéromères exotiques (Col.). *Bulletin de la Société Entomologique de France* **1908**: 274-275.
Pic M. 1909a: Divers coléoptères exotiques mouveaux. *Le Naturaliste* **31**: 19.
Pic M. 1909b: Nouveaux coléoptères (malacodermes et hétéromères) des Indes. *Bulletin de la Société Entomologique de France* **1909**: 37-39.
Pic M. 1909c: Deux nouveaux Anthicus de l'Himalaya (Col.). *Bulletin de la Société Entomologique de France* **1909**: 56-57.
Pic M. 1909d: Descriptions ou diagnoses et notes diverses (Suite). *L'Échange, Revue Linnéenne* **25**: 105-106.
Pic M. 1909e: Coléoptères exotiques nouveaux ou peu connus (Suite). *L'Échange, Revue Linnéenne* **25**: 108-110.
Pic M. 1909f: Descriptions ou diagnoses et notes diverses (Suite). *L'Échange, Revue Linnéenne* **25**: 113-115.
Pic M. 1909g: Coléoptère hétéromère pedilide nouveau du Bengale. *Bulletin du Muséum Nationale d'Histoire Naturelle, Paris* **15**: 109-110.
Pic M. 1909h: Coléoptères exotiques nouveaux ou peu connus (Suite). *L'Échange, Revue Linnéenne* **25**: 117-118.
Pic M. 1909i: Descriptions ou diagnoses et notes diverses (Suite). *L'Échange, Revue Linnéenne* 25: 121-123
Pic M. 1909j: Descriptions ou diagnoses et notes diverses (Suite). *L'Échange, Revue Linnéenne* **25**: 129-131.
Pic M. 1909k: Coléoptères exotiques nouveaux ou peu connus (Suite). *L'Échange, Revue Linnéenne* **25**: 133-134.
Pic M. 1909l: Coléoptères exotiques nouveaux ou peu connus (Suite). *L'Échange, Revue Linnéenne* **25**: 142-143.
Pic M. 1909m: Coléoptères exotiques nouveaux ou peu connus (Suite). *L'Échange, Revue Linnéenne* **25**: 146-148, 160, 166-167.
Pic M. 1909n: Descriptions ou diagnoses et notes diverses (Suite). *L'Échange, Revue Linnéenne* **25**: 161-162, 169-171.
Pic M. 1909o: Coléoptères exotiques nouveaux ou peu connus (Suite). *L'Échange, Revue Linnéenne* **25**: 173-175.
Pic M. 1909p: Descriptions ou diagnoses et notes diverses (Suite). *L'Échange, Revue Linnéenne* **25**: 177-179.
Pic M. 1909q: Descriptions ou diagnoses et notes diverses (Suite). *L'Échange, Revue Linnéenne* **25**: 185-186.
Pic M. 1909r: Pyrochroidae nouveaux. *Bulletin de la Société Entomologique de France* **1909**: 330-331.
Pic M. 1910a: [new taxa]. In: Sainte-Claire Deville J.: Catalogue critique des coléoptères de la Corse. *Revue d'Entomologie* (Caen), *Supplément* **25-28** [1906-1910]: 1-400.
Pic M. 1910b: Description ou diagnoses et notes diverses (Suite). *L'Échange, Revue Linnéenne* **26**: 1.
Pic M. 1910c: Descriptions ou diagnoses et notes diverses (Suite). *L'Échange, Revue Linnéenne* **26**: 9-11.
Pic M. 1910d: Descriptions ou diagnoses et notes diverses. *L'Échange, Revue Linnéenne* **26**: 12-14.
Pic M. 1910e: Descriptions, diagnoses et notes diverses (Suite). *L'Échange, Revue Linnéenne* **26**: 17-18.
Pic M. 1910f: Coléoptères: pédilides et anthicides recueillis au Japon, par M.M. J. Harmand et E. Gallois. *Bulletin du Muséum National d'Histoire Naturelle, Paris* **16**: 19-21.
Pic M. 1910g: Coléoptères exotiques nouveaux ou peu connus (Suite.). *L'Échange, Revue Linnéenne* **26**: 20-22.
Pic M. 1910h: Coléoptères: hylophilides nouveaux originaires d'Abyssinie et du Japon (Col.). *Bulletin du Museum d'Histoire Naturelle, Paris* **16**: 21-24.
Pic M. 1910i: Description d'un nouveau Mecynotarsus d'Egypte. *Bulletin de la Société Entomologique d'Egypte* **3**: 26-27.
Pic M. 1910j: Descriptions, diagnoses et notes diverses. (Suite). *L'Échange, Revue Linnéenne* **26**: 33-34.

Pic M. 1910k: Description d´un nouveau Trotomma d´Algérie. *Bulletin de la Société Entomologique de France* **1910**: 36.

Pic M. 1910l: Coléoptères malacodermes et hétéromères en partie nouveaux. *Bulletin de la Société d'Histoire Naturelle d'Autun* **23**: 37-39.

Pic M. 1910m: Descriptions ou diagnoses et notes diverses. (Suite). *L'Échange, Revue Linnéenne* **26**: 41-43.

Pic M. 1910n: Descriptions ou diagnoses et notes diverses (Suite). *L'Échange, Revue Linnéenne* **26**: 57-58.

Pic M. 1910o: Descriptions ou diagnoses et notes diverses (Suite). *L'Échange, Revue Linnéenne* **26**: 65-66.

Pic M. 1910p: Coléoptères exotiques nouveaux ou peu connus (Suite). *L'Échange, Revue Linnéenne* **26**: 69-71.

Pic M. 1910q: Coléoptères exotiques nouveaux ou peu connus (Suite). *L'Échange, Revue Linnéenne* **26**: 74-78.

Pic M. 1910r: Description d'un Hylophilus du Japon (Col. Heteromera). *Bulletin de la Société Entomologique de France* **1910**: 75.

Pic M. 1910s: Coléoptères exotiques nouveaux ou peu connus (Suite.) *L'Échange, Revue Linnéenne* **26**: 86-87.

Pic M. 1910t: Hétéromères nouveaux du group des Zonitini. *Bulletin de la Société Entomologique de France* **1910**: 90-91.

Pic M. 1910u: Coléoptères exotiques nouveaux ou peu connus (Suite). *L'Échange, Revue Linnéenne* **26**: 94-95.

Pic M. 1910v: Malacodermes et hétéromères nouveaux d'Afrique et d'Asie. *Le Naturaliste* **32**: 259.

Pic M. 1910w: Malacodermes et hétéromères nouveaux d´Afrique et d´Asie. *Le Naturaliste* **32**: 271-272.

Pic M. 1911a: Anthicidae. Pars 36. In: Schenkling S. (ed.): *Coleopterorum Catalogus*. Berlin: W. Junk, 102 pp.

Pic M. 1911b: Descriptions bréves de vingt-huit hétéromères exotiques. *Mélanges Exotico-Entomologiques* **1**: 5-13.

Pic M. 1911c: Mission Niger-Tchad dirigée par le Capitaine Tilho. Collections recueillies par le Dr Gaillard. Coléoptères: Anthicides nouveaux (Hétéromères). *Bulletin du Muséum National d'Histoire Naturelle, Paris* **17**: 58-63.

Pic M. 1911d: Coléoptères égyptiens nouveaux ou plus ou moins rares. *Bulletin de la Société Entomologique d'Egypte* **4**: 71-76.

Pic M. 1911e: Descriptions ou diagnoses et notes diverses (Suite). *L'Échange, Revue Linnéenne* **27**: 97-98.

Pic M. 1911f: Coléoptères exotiques nouveaux ou peu connus (Suite). *L'Échange, Revue Linnéenne* **27**: 99-101.

Pic M. 1911g: Coléoptères exotiques nouveaux ou peu connus (Suite). *L'Échange, Revue Linnéenne* **27**: 117-118.

Pic M. 1911h: [new taxon]. In: Andres A.: Note sur quelques espèces de Zonabris de la région de Mariout (environs d'Alexandrie). *Bulletin de la Société Entomologique d'Egypte* **4**: 129-135.

Pic M. 1911i: Coléoptères exotiques nouveaux ou peu connus (Suite). *L'Échange, Revue Linnéenne* **27**: 132-134, 138.

Pic M. 1911j: Descriptions ou diagnoses et notes diverses (Suite). *L'Échange, Revue Linnéenne* **27**: 145-146.

Pic M. 1911k: Coléoptères exotiques nouveaux ou peu connus (Suite). *L'Échange, Revue Linnéenne* **27**: 155-159.

Pic M. 1911l: Notes pour aider à l'étude de divers coléoptères paléarctiques principalement gallo-rhénans. *L'Échange, Revue Linnéenne* **27**: 161-162.

Pic M. 1911m: Descriptions ou diagnoses et notes diverses (Suite). *L'Échange, Revue Linnéenne* **27**: 169-170.

Pic M. 1911n: Descriptions ou diagnoses et notes diverses. *L'Échange, Revue Linnéenne* **27**: 177-178.

Pic M. 1911o: Descriptions ou diagnoses et notes diverses (Suite). *L'Échange, Revue Linnéenne* **27**: 185-186.

Pic M. 1911p: H. Sauter's Formosa-Ausbeute. Cantharidae, Lampyridae, Mordellidae. (Col.). *Deutsche Entomologische National-Bibliothek* **24**: 188-189.

Pic M. 1911q: Coléoptères exotiques nouveaux ou peu connus (Suite.) *L'Échange, Revue Linnéenne* **27**: 190-191.

Pic M. 1912a: Descriptions d'espèces & variétés appartenant a diverses familles. *Mélanges Exotico-Entomologiques* **2**: 2-11.

Pic M. 1912b: Descriptions d'espèces et variétés diverses. *Mélanges Exotico-Entomologiques* **3**: 6-20.

Pic M. 1912c: Descriptions ou diagnoses et notes diverses (Suite). *L'Échange, Revue Linnéenne* **28**: 1-3, 9-10.

Pic M. 1912d: Synopsis pour aider à l'étude des Pyrochroa Geof. des Indes. *L'Échange, Revue Linnéenne* **28**: 18.

Pic M. 1912e: Descriptions ou diagnoses et notes diverses (Suite). *L'Échange, Revue Linnéenne* **28**: 25-26.

Pic M. 1912f: Coléoptères exotiques nouveaux ou peu connus. (Suite). *L'Échange, Revue Linnéenne* **28**: 29-32.

Pic M. 1912g: Descriptions ou diagnoses et notes diverses (Suite). *L'Échange, Revue Linnéenne* **28**: 33-35.

Pic M. 1912h: Descriptions ou diagnoses et notes diverses (Suite). *L'Échange, Revue Linnéenne*, **28**: 41-43.

Pic M. 1912i: Coléoptères exotiques nouveaux ou peu connus (Suite). *L'Échange, Revue Linnéenne* **28**: 46-47.
Pic M. 1912j: Descriptions ou diagnoses et notes diverses (Suite). *L'Échange, Revue Linnéenne* **28**: 49-51.
Pic M. 1912k: Descriptions ou diagnoses et notes diverses (Suite). *L'Échange, Revue Linnéenne* **28**: 57-58.
Pic M. 1912l: Coléoptères exotiques nouveaux ou peu connus (Suite). *L'Échange, Revue Linnéenne* **28**: 63-64.
Pic M. 1912m: Coléoptères exotiques nouveaux ou peu connus (Suite). *L'Échange, Revue Linnéenne* **28**: 68-69.
Pic M. 1912n: Descriptions ou diagnoses et notes diverses. *L'Échange, Revue Linnéenne* **28**: 73-74.
Pic M. 1912o: Nouveaux coléoptères paléarctiques. *L'Échange, Revue Linnéenne* **28**: 89-91.
Pic M. 1912p: Sur les Pyrochroidae du Muséum de Paris (Hétéromères). *Bulletin du Muséum National d'Histoire Naturelle, Paris* **18**: 142-144.
Pic M. 1912q: Notes sur quelques vésicants coléoptères. *Bulletin de la Société Entomologique de France* **1912**: 176-177.
Pic M. 1912r: Coléoptères du Maroc. *Bulletin du Muséum National d'Histoire Naturelle, Paris* **18**: 219-221.
Pic M. 1913a: Descriptions de 29 espèces et de plusiers variétés. *Mélanges Exotico- Entomologiques* **5**: 7-20.
Pic M. 1913b: Espèces et variétés nouvelles appartenant à diverses familles. *Mélanges Exotico- Entomologiques* **6**: 8-16.
Pic M. 1913c: Nouvelles espèces de "Toxicum" Latreille. *Mélanges Exotico-Entomologiques* **6**: 17-20.
Pic M. 1913d: Espèces et variétés nouvelles, ou peu connues, rentrant dans diverses familles. *Mélanges Exotico-Entomologiques* **7**: 10-20.
Pic M. 1913e: Coléoptères divers du Tonkin et de l'Indo-Chine. *Mélanges Exotico-Entomologiques* **9**: 2-20.
Pic M. 1913f: "Coryna denticulata" Mars. et "C. suturifera" Pic (Col. vésicants, hétéromères). *Revue Scientifique du Bourbonnais* **26**: 110-112.
Pic M. 1913g: Note diverses, descriptions et diagnoses (Suite.). *L'Échange, Revue Linnéenne* **29**: 113-115.
Pic M. 1913h: Notes diverses, descriptions et diagnoses(Suite.). *L'Échange, Revue Linnéenne* **29**: 121-122.
Pic M. 1913i: Notes diverses, descriptions et diagnoses (Suite). *L'Échange, Revue Linnéenne* **29**: 129-130.
Pic M. 1913j: H. Sauter's Formosa-Ausbeute. Anthicidae, Pedilidae und Hylophilidae. *Archiv für Naturgeschichte* (A) **79**: 129-135.
Pic M. 1913k: Notes diverses, descriptions et diagnoses (Suite). *L'Échange, Revue Linnéenne* **29**: 137-139.
Pic M. 1913l: Nouveaux coléoptères anthicides. *Bulletin de la Société Zoologique de France* **38**: 137-139.
Pic M. 1913m: Coléoptères exotiques en partie nouveaux (Suite.). *L'Échange, Revue Linnéenne* **29**: 142.
Pic M. 1913n: Notes diverses, descriptions et diagnoses (Suite). *L'Échange, Revue Linnéenne* **29**: 145-147.
Pic M. 1913o: Notes diverses, descriptions et diagnoses (Suite). *L'Échange, Revue Linnéenne* **29**: 153-154.
Pic M. 1913p: Notes diverses, descriptions et diagnoses (Suite). *L'Échange, Revue Linnéenne* **29**: 161-162.
Pic M. 1913q: Coléoptères exotiques en partie nouveau (Suite). *L'Échange, Revue Linnéenne* **29**: 163-166.
Pic M. 1913r: Notes diverses, descriptions et diagnoses (Suite.). *L'Échange, Revue Linnéenne* **29**: 169-171, 185-187.
Pic M. 1913s: Coléoptères exotiques en partie nouveaux (Suite). *L'Échange, Revue Linnéenne* **29**: 175-176, 185-186.
Pic M. 1913t: Deux nouveaux Formicomus Laf., de l'Inde (Col. Heteromera). *Bulletin de la Société Entomologique de France* **1913**: 204-205.
Pic M. 1913u: Deux vésicants africains nouveaux (Col. Heteromera). *Bulletin de la Société Entomologique de France* **1913**: 456-457.
Pic M. 1914a: Notes diverses, descriptions et diagnoses (Suite). *L'Échange, Revue Linnéenne* **30**: 1-2.
Pic M. 1914b: Coléoptères divers du Tonkin et de l'Indo-Chine. *Mélanges Exotico-Entomologiques* **9**: 2-20.
Pic M. 1914c: Nouveau genre, espèces et variétés nouvelles. *Mélanges Exotico-Entomologiques* **11**: 2-8.
Pic M. 1914d: Descriptions abrégées de malacodermes et hétéromères. *Mélanges Exotico-Entomologiques* **11**: 8-20.
Pic M. 1914e: Notes diverses, descriptions et diagnoses (Suite). *L'Échange, Revue Linnéenne* **30**: 10-13.
Pic M. 1914f: Descriptions abrégées d'anthicides exotiques. *L'Échange, Revue Linnéenne* **30**: 34-36.
Pic M. 1914g: Notes diverses, descriptions et diagnoses (Suite). *L'Échange, Revue Linnéenne* **30**: 41-42.
Pic M. 1914h: Coléoptères exotiques en partie nouveaux (Suite). *L'Échange, Revue Linnéenne* **30**: 44.
Pic M. 1914i: Coléoptères exotiques en partie nouveaux (Suite). *L'Échange, Revue Linnéenne* **30**: 54-56.
Pic M. 1914j: Deux nouvelles espèces d'Oedemerides du Kashmir (Col. Heteromera). *Bulletin de la Société Entomologique de France* **1914**: 57-58.
Pic M. 1914k: Coléoptères exotiques en partie nouveaux (Suite). *L'Échange, Revue Linnéenne* **30**: 61-62.

Pic M. 1914l: Coléoptères exotiques en partie nouveaux (Suite). *L'Échange, Revue Linnéenne* **30**: 67-68.
Pic M. 1914m: Notes diverses, descriptions et diagnoses (Suite). *L'Échange, Revue Linnéenne* **30**: 73-75.
Pic M. 1914n: Coléoptères exotiques en partie nouveaux (Suite). *L'Échange, Revue Linnéenne* **30**: 75-76.
Pic M. 1914o: Notes diverses, descriptions et diagnoses (Suite). *L'Échange, Revue Linnéenne* **30**: 77-79.
Pic M. 1914p: Notes sur divers vésicants des collections du Musée Civique de Gênes avec descriptions des nouveautés. *Annali del Museo Civico di Storia Naturale di Genova* (3) **6**: 103-115.
Pic M. 1914q: Anthicides exotiques nouveaux ou peu connus (Col.). *Bulletin de la Société Zoologique de France* **39**: 181-184.
Pic M. 1914r: Nouveaux Nemostira Fairm. asiatiques. *Bulletin de la Société Entomologique de France* **1914**: 304-305.
Pic M. 1914s: Description de trois nouveaux Hylophilus Berthold asiatiques (Col. Hétéromères). *Bulletin de la Société Entomologique de France* **1914**: 448-450.
Pic M. 1915a: Notes diverses, descriptions et diagnoses (Suite.). *L'Échange, Revue Linnéenne* **31**: 1-3.
Pic M. 1915b: Descriptions abrégées diverses. *Mélanges Exotico-Entomologiques* **12**: 1-20.
Pic M. 1915c: Nouveautés de diverses familles. *Mélanges Exotico-Entomologiques* **13**: 2-13.
Pic M. 1915d: Nouveautés rentrant dans diverses familles. *Mélanges Exotico-Entomologiques* **14**: 2-20.
Pic M. 1915e: Nouvelles espèces de diverses familles. *Mélanges Exotico-Entomologiques* **15**: 2-24.
Pic M. 1915f: Nouveau genre, espèces et variétés nouvelles. *Mélanges Exotico-Entomologiques* **16**: 2-13.
Pic M. 1915g: Diagnoses d'hétéromères. *Mélanges Exotico-Entomologiques* **16**: 14-24.
Pic M. 1915h: Notes diverses, descriptions et diagnoses (Suite). *L'Échange, Revue Linnéenne* **31**: 9-10.
Pic M. 1915i: Coléoptères exotiques en partie nouveaux (Suite). *L'Échange, Revue Linnéenne* **31**: 10-11.
Pic M. 1915j: Notes diverses, descriptions et diagnoses (Suite.). *L'Échange, Revue Linnéenne* **31**: 13-14.
Pic M. 1915k: Notes diverses, descriptions et diagnoses (Suite.). *L'Échange, Revue Linnéenne* **31**: 21-22.
Pic M. 1915l: Notes diverses, descriptions et diagnoses (Suite). *L'Échange, Revue Linnéenne* **31**: 37-38.
Pic M. 1915m: Notes diverses, descriptions et diagnoses (Suite). *L'Échange, Revue Linnéenne* **31**: 41-42.
Pic M. 1915n: Anthicides et Hylophilides des chasses de E. Gallois au Japon. *Bulletin du Muséum National d'Histoire Naturelle, Paris* **21**: 152-153.
Pic M. 1916a: Notes diverses, descriptions et diagnoses (Suite). *L'Échange, Revue Linnéenne* **32**: 5-7.
Pic M. 1916b: Coléoptères exotiques en partie nouveaux. *L'Échange, Revue Linnéenne* **32**: 7-8.
Pic M. 1916c: Diagnoses spécifiques. *Mélanges Exotico-Entomologiques* **17**: 8-20.
Pic M. 1916d: Espèces et variétés nouvelles. *Mélanges Exotico-Entomologiques* **19**: 6-20.
Pic M. 1916e: Descriptions abrégées diverses. *Mélanges Exotico-Entomologiques* **20**: 1-20.
Pic M. 1916f: Descriptions abrégées diverses. *Mélanges Exotico-Entomologiques* **21**: 2-20.
Pic M. 1916g: Notes diverses, descriptions et diagnoses (Suite). *L'Échange, Revue Linnéenne* **32**: 13-14.
Pic M. 1916h: Notes diverses, descriptions et diagnoses (Suite). *L'Échange, Revue Linnéenne* **32**: 21-23.
Pic M. 1916i: Nouveaux Zonabris de l'Inde et de Cochinchine (Col. Lyttidae). *Bulletin de la Société Entomologique de France* **1916**: 125-126.
Pic M. 1917a: Notes diverses, descriptions et diagnoses. *L'Échange, Revue Linnéenne* **33**: 1-3.
Pic M. 1917b: Descriptions abrégées diverses. *Mélanges Exotico-Entomologiques* **22**: 1-20.
Pic M. 1917c: Descriptions abrégées diverses. *Mélanges Exotico-Entomologiques* **23**: 2-20.
Pic M. 1917d: Descriptions abrégées diverses. *Mélanges Exotico-Entomologiques* **24**: 2-24.
Pic M. 1917e: Descriptions abrégées diverses. *Mélanges Exotico-Entomologiques* **25**: 2-24.
Pic M. 1917f: Descriptions abrégées diverses. *Mélanges Exotico-Entomologiques* **26**: 2-24.
Pic M. 1917g: Notes diverses, descriptions et diagnoses (Suite). *L'Échange, Revue Linnéenne* **33**: 5-6.
Pic M. 1917h: Coléoptères exotiques en partie nouveaux (Suite). *L'Échange, Revue Linnéenne* **33**: 11-12.
Pic M. 1917i: Anthicides de Mongolie (Col. hétéromères). *Bulletin du Muséum National d'Histoire Naturelle, Paris* **23**: 256-257.
Pic M. 1918a: Courtes descriptions diverses. *Mélanges Exotico-Entomologiques* **27**: 1-24.
Pic M. 1918b: Courtes descriptions diverses. *Mélanges Exotico-Entomologiques* **28**: 1-24.
Pic M. 1918c: Notes diverses, descriptions et diagnoses (Suite). *L'Échange, Revue Linnéenne* **34**: 13-15.
Pic M. 1918d: Coléoptères exotiques en partie nouveaux (Suite). *L'Échange, Revue Linnéenne* **34**: 15-16.
Pic M. 1919a: Étude détaillé de diverses insectes. *L'Échange, Revue Linnéenne* **35** [hors texte]: 1-20.
Pic M. 1919b: Nouveautés diverses. *Mélanges Exotico-Entomologiques* **30**: 1-24.
Pic M. 1919c: Notes diverses, descriptions et diagnoses (Suite). *L'Échange, Revue Linnéenne* **35**: 5-6.

Pic M. 1919d: Notes diverses, descriptions et diagnoses (Suite). *L'Échange, Revue Linnéenne* **35**: 9-11.
Pic M. 1919e: Notes diverses, descriptions et diagnoses (Suite). *L'Échange, Revue Linnéenne* **35**: 13-14.
Pic M. 1919f: Notes diverses, descriptions et diagnoses (Suite). *L'Échange, Revue Linnéenne* **35**: 17-20, 21-22.
Pic M. 1919g: Note sur divers coléoptères. *Miscellanea Entomologica* **24**: 66-69.
Pic M. 1920a: Nouveautés diverses. *Mélanges Exotico-Entomologiques* **32**: 1-32.
Pic M. 1920b: Coléoptères exotiques en partie nouveaux (Suite). *L'Échange, Revue Linnéenne* **36**: 2-3.
Pic M. 1920c: Notes diverses, descriptions et diagnoses (Suite). *L'Échange, Revue Linnéenne* **36**: 5-8.
Pic M. 1920d: Notes diverses, descriptions et diagnoses (Suite). *L'Échange, Revue Linnéenne* **36**: 9-12.
Pic M. 1920e: Notes diverses, descriptions et diagnoses (Suite). *L'Échange, Revue Linnéenne* **36**: 13.
Pic M. 1920f: Diagnoses de coléoptères exotiques. *L'Échange, Revue Linnéenne* **36**: 15-16.
Pic M. 1920g: Notes diverses, descriptions et diagnoses (Suite). *L'Échange, Revue Linnéenne* **36**: 17-18.
Pic M. 1921a: Notes diverses, descriptions et diagnoses (Suite). *L'Échange, Revue Linnéenne* **37**: 1-4, 5-6.
Pic M. 1921b: Nouveautés diverses. *Mélanges Exotico-Entomologiques* **33**: 1-32.
Pic M. 1921c: Nouveautés diverses. *Mélanges Exotico-Entomologiques* **34**: 1-33.
Pic M. 1921d: Notes diverses, descriptions et diagnoses (Suite). *L'Échange, Revue Linnéenne* **37**: 9-10.
Pic M. 1921e: Diagnoses de coléoptères exotiques (Suite). *L'Échange, Revue Linnéenne* **37**: 10-12, 13-15.
Pic M. 1921f: Étude détaillée de diverses insectes. *L'Échange, Revue Linnéenne* **37** [hors texte] (403): 21-24.
Pic M. 1921g: [new taxa]. In: Alfieri A.: Contributions à la faune entomologique du nord-est du Sinai. *Bulletin de la Société Entomologique d'Egypte* **6** [1919-1920]: 40-53.
Pic M. 1921h: Coléoptères nouveaux de la famille des hylophilides. *Bulletin du Muséum National d'Histoire Naturelle, Paris* **27**: 415-418.
Pic M. 1922a: Nouveautés diverses. *Mélanges Exotico-Entomologiques* **35**: 1-32.
Pic M. 1922b: Nouveautés diverses. *Mélanges Exotico-Entomologiques* **36**: 1-36.
Pic M. 1922c: Nouveautés diverses. *Mélanges Exotico-Entomologiques* **37**: 1-32.
Pic M. 1922d: Notes diverses, descriptions et diagnoses (Suite). *L'Échange, Revue Linnéenne* **38**: 17-19.
Pic M. 1922e: Notes diverses, descriptions et diagnoses (Suite). *L'Échange, Revue Linnéenne* **38**: 21-23.
Pic M. 1922f: Coléoptères exotiques en partie nouveaux. *L'Échange, Revue Linnéenne* **38**: 23-24.
Pic M. 1922g: Notes diverses, descriptions et diagnoses (Suite). *L'Échange, Revue Linnéenne* **38**: 29-30.
Pic M. 1922h: Coléoptères hétéromères exotiques nouveaux. *Bulletin de la Société Zoologique de France* **47**: 100-103.
Pic M. 1922i: Diagnoses d'hétéromères (Col.) de l'Indo-Chine. *Bulletin de la Société Entomologique de France* **1922**: 208-210.
Pic M. 1922j: Sur les hétéromères Amarygminae (Col.). *Bulletin de la Société Zoologique de France* **47**: 303-306.
Pic M. 1923a: Notes diverses, descriptions et diagnoses (Suite). *L'Échange, Revue Linnéenne* **39**: 1-3.
Pic M. 1923b: Nouveautés diverses. *Mélanges Exotico-Entomologiques* **38**: 1-32.
Pic M. 1923c: Nouveautés diverses. *Mélanges Exotico-Entomologiques* **39**: 3-32.
Pic M. 1923d: Nouveautés diverses. *Mélanges Exotico-Entomologiques* **40**: 3-32.
Pic M. 1923e: Coléoptères exotiques en partie nouveaux (Suite). *L'Échange, Revue Linnéenne* **39**: 4.
Pic M. 1923f: Notes diverses, descriptions et diagnoses (Suite). *L'Échange, Revue Linnéenne* **39**: 5-7.
Pic M. 1923g: Notes diverses, descriptions et diagnoses (Suite). *L'Échange, Revue Linnéenne* **39**: 9-11.
Pic M. 1923h: Coléoptères exotiques en partie nouveaux (Suite). *L'Échange, Revue Linnéenne* **39**: 11-12.
Pic M. 1923i: Mutations, synonymies, nouveautés. *L'Échange, Revue Linnéenne* **39**: 13-14.
Pic M. 1923j: Nouveaux coléoptères exotiques. *Annales de la Société Linnéenne de Lyon* (N. S.) **69** [1922]: 73-76.
Pic M. 1923k: Nouveaux coléoptères du Tonkin. *Bulletin de la Société Zoologique de France* **48**: 269-271.
Pic M. 1924a: Nouveautés diverses. *Mélanges Exotico-Entomologiques* **41**: 1-32.
Pic M. 1924b: Nouveautés diverses. *Mélanges Exotico-Entomologiques* **42**: 1-32.
Pic M. 1924c: Notes diverses, descriptions et diagnoses (Suite). *L'Échange, Revue Linnéenne* **39**: 21-23.
Pic M. 1924d: Notes diverses, descriptions et diagnoses (Suite). *L'Échange, Revue Linnéenne* **39**: 25-27.
Pic M. 1924e: Coléoptères exotiques en partie nouveaux (Suite.). *L'Échange, Revue Linnéenne* **39**: 27-28.
Pic M. 1924f: Notes diverses, descriptions et diagnoses (Suite.). *L'Échange, Revue Linnéenne* **39**: 29-30.
Pic M. 1924g: Vésicants en partie originaires du Congo. *Revue de Zoologie et Botanique Africaines* **12**: 450-454.
Pic M. 1925a: Notes diverses, descriptions et diagnoses (Suite). *L'Échange, Revue Linnéenne* **41**: 1-3.

Pic M. 1925b: Nouveautés diverses. *Mélanges Exotico-Entomologiques* **44**: 1-32.
Pic M. 1925c: Notes diverses, descriptions et diagnoses (Suite). *L'Échange, Revue Linnéenne* **41**: 5-6.
Pic M. 1925d: Notes diverses, descriptions et diagnoses (Suite). *L'Échange, Revue Linnéenne* **41**: 9-11.
Pic M. 1925ea: Coléoptères exotiques en partie nouveaux (Suite). *L'Échange, Revue Linnéenne* **41**: 11-12.
Pic M. 1925f: Notes diverses, descriptions et diagnoses (Suite). *L'Échange, Revue Linnéenne* **41**: 13-14.
Pic M. 1925g: Coléoptères exotiques en partie nouveaux (Suite). *L'Échange, Revue Linnéenne* **41** 15-16.
Pic M. 1925h: Coléoptères exotiques nouveaux ou peu connus. *Annales de la Société Linnéenne de Lyon* (N. S.) **72**: 73-77.
Pic M. 1925i: Coléoptères nouveaux. *Bulletin de la Société Royale Entomologique d'Egypte* **18**: 227-231.
Pic M. 1925j: Contribution à l'étude des ténébrionides du genre Basides Mots. *Bulletin du Muséum National d'Histoire Naturelle, Paris* **32**: 431-438.
Pic M. 1926a: Entomologische Ergebnisse der Schwedischen Kamtchatka-Expedition 1920-1922. 7. Coleoptera. *Arkiv för Zoologi* **18 B** (3): 1-5.
Pic M. 1926b: Nouveautés diverses. *Mélanges Exotico-Entomologiques* **45**: 1-32.
Pic M. 1926c: Nouveautés diverses. *Mélanges Exotico-Entomologiques* **46**: 1-32.
Pic M. 1926d: Nouveautés diverses. *Mélanges Exotico-Entomologiques* **47**: 1-32.
Pic M. 1926e: Notes diverses, descriptions et diagnoses (Suite). *L'Échange, Revue Linnéenne* **42**: 5-6.
Pic M. 1926f: Notes diverses, descriptions et diagnoses. *L'Échange, Revue Linnéenne* **42**: 9-10.
Pic M. 1926g: Coléoptères exotiques en partie nouveaux. *L'Échange, Revue Linnéenne* **42**: 15-16.
Pic M. 1926h: Nouveaux coléoptères du Tonkin (2e article). *Bulletin de la Société Zoologique de France* **51**: 45-48.
Pic M. 1926i: H. Sauter´s Formosa Ausbeute. *Entomologische Mitteillungen* **15**: 67-69.
Pic M. 1926j: Coléoptères exotiques nouveaux ou peu connus. *Annales de la Société Linnéenne de Lyon* (N. S.) **72** [1925]: 73-77.
Pic M. 1926k: Coléoptères nouveaux. *Bulletin de la Société Entomologique d'Egypte* **9** [1925-1926]: 227-231.
Pic M. 1926l: Coléoptères asiatiques nouveaux. *Bulletin de la Société Entomologique de France* **1925**: 301-303.
Pic M. 1926m: Nouveaux coléoptères exotiques. *Bulletin du Muséum National d'Histoire Naturelle, Paris* **32**: 354-359.
Pic M. 1927a: Entomologische Ergebnisse der Schwedischen Kamtchatka expedition 1920-1922. 14. Coleoptera 2. *Arkiv för Zoology* **19 B** (3): 1-3.
Pic M. 1927b: Nouveautés diverses. *Mélanges Exotico-Entomologiques* **48**: 1-32.
Pic M. 1927c: Coléoptères de l'Indochine. *Mélanges Exotico-Entomologiques* **49**: 1-36.
Pic M. 1927d: Coléoptères du globe. *Mélanges Exotico-Entomologiques* **50**: 1-36.
Pic M. 1927e: Notes diverses, descriptions et diagnoses (Suite). *L'Échange, Revue Linnéenne* **43**: 5-7.
Pic M. 1927f: Diagnoses de coléoptères exotiques. *Bulletin Bi-Mensuel de la Société Linnéenne de Lyon* **6**: 5-6.
Pic M. 1927g: Coléoptères exotiques en partie nouveaux (Suite). *L'Échange, Revue Linnéenne* **43**: 7-8.
Pic M. 1927h: Notes diverses, descriptions et diagnoses (Suite). *L'Échange, Revue Linnéenne* **43**: 9-10.
Pic M. 1927i: H. Sauter´s Formosa-Ausbeute: Heteromera [ex parte] (Col.). *Entomologische Mitteilungen* **16**: 48-49.
Pic M. 1927k: Coléoptères asiatiques nouveaux. *Bulletin Bi-Mensuel de la Société Linnéenne de Lyon* **6**: 132-133.
Pic M. 1927j: Coléoptéres nouveaux de Chine et du Japon. *Bulletin de la Société Entomologique de France* **1927**: 152-154.
Pic M. 1927l: Nouveaux coléoptères exotiques. *Bulletin de la Société Entomologique de France* **1927**: 183-184.
Pic M. 1927m: Nouveau coléoptères du Globe (IV). *Bulletin de la Société Zoologique de France* **52**: 185-189.
Pic M. 1928a: Notes diverses, descriptions et diagnoses (Suite). *L'Échange, Revue Linnéenne* **44**: 1-2.
Pic M. 1928b: Notes et descriptions. *Mélanges Exotico-Entomologiques* **51**: 1-36.
Pic M. 1928c: Nouveautés diverses. *Mélanges Exotico-Entomologiques* **52**: 1-32.
Pic M. 1928d: Coléoptères exotiques en partie nouveaux (Suite). *L'Échange, Revue Linnéenne* **44**: 4.
Pic M. 1928e: Notes diverses, descriptions et diagnoses (Suite). *L'Échange, Revue Linnéenne* **44**: 5-6.
Pic M. 1928f: Notes diverses, descriptions et diagnoses (Suite). *L'Échange, Revue Linnéenne* **44**: 9-10.
Pic M. 1928g: Coléoptères asiatiques nouveaux. *Bulletin de la Société Entomologique de France* **1928**: 247-248.
Pic M. 1928h: Nouveaux coléoptères exotiques. *Bulletin du Muséum National d'histoire Naturelle, Paris* **34**: 257-258.

Pic M. 1928i: Nouveaux coléoptères surtout tonkinois. *Bulletin de la Société Zoologique de France* **53**: 377-380.
Pic M. 1929a: Notes diverses, nouveautés. *L'Échange, Revue Linnéenne* **45**: 1-2.
Pic M. 1929b: Nouveautés diverses. *Mélanges Exotico-Entomologiques* **53**: 1-36.
Pic M. 1929c: Nouveautés diverses. *Mélanges Exotico-Entomologiques* **54**: 1-36.
Pic M. 1929d: Notes diverses, nouveautés. *L'Échange, Revue Linnéenne* **45**: 5-7.
Pic M. 1929e: Notes diverses, nouveautés. *L'Échange, Revue Linnéenne* **45**: 13-14.
Pic M. 1929f: Coléoptères exotiques nouveaux ou peu connus. *Bulletin de la Société Zoologique de France* **54**: 43-46.
Pic M. 1929g: Coléoptères d'Egypte et Sinaï. *Bulletin de la Société Royal Entomologique d'Egypte* (N. S.) **13**: 108-145.
Pic M. 1929c: Coléoptères nouveaux du Hoggar. *Bulletin de la Société Entomologique de France* **1929**: 126-128.
Pic M. 1930a: Notes diverses, nouveautés. *L'Échange, Revue Linnéenne* **46**: 1-3.
Pic M. 1930b: Nouveautés diverses. *Mélanges Exotico-Entomologiques* **56**: 1-36.
Pic M. 1930c: Notes diverses, nouveautés. *L'Échange, Revue Linnéenne* **46**: 5-7.
Pic M. 1930d: Notes diverses, nouveautés. *L'Échange, Revue Linnéenne* **46**: 9-11.
Pic M. 1930e: Trois nouveaux coléoptéres. *Bulletin de la Société Entomologique de France* **45**: 76-77.
Pic M. 1931a: Nouveautés diverses. *Mélanges Exotico-Entomologiques* **57**: 1-36.
Pic M. 1931b: [new taxa]. In: Miwa Y.: A systematic catalogue of Formosan Coleoptera. *Reports of the Department of Agriculture of the Governement Research Institute, Taihoku* **55**: 1-359.
Pic M. 1931c: Notes diverses, nouveautés. *L'Échange, Revue Linnéenne* **47**: 5-6.
Pic M. 1931d: Notes diverses, nouveautés (Suite). *L'Échange, Revue Linnéenne* **47**: 9-10.
Pic M. 1931e: Notes diverses, nouveautés (Suite). *L'Échange, Revue Linnéenne* **47**: 13-14.
Pic M. 1931f: [new taxa]. In: Koch C.: Risultati scientifici delle spedizioni entomologiche di S. A. S. il Principe Alessandro Della Torre e Tasso nell'Africa settentrionale e in Sicilia. *Memorie della Società Entomologica Italiana* **10**: 66-79.
Pic M. 1932a: Diagnoses préliminaires. *Mélanges Exotico-Entomologiques* **59**: 1-9.
Pic M. 1932b: [new taxa]. In: Pic M. & Lindberg Håk.: Inventa entomologica, itineris Hispanici et Maroccani, quod a. 1926 fecerunt Harald et Håkan Lindberg. XII. Anobiidae, Cleridae, Malacodermata, Heteromera (ex parte). *Commentationes Biologicae* **3** (18): 1-37.
Pic M. 1932c: Contributions à l'étude de la faune du Mozambique. Voyage de M. P. Lesne (1928-1929). 9.e note. - Coléoptères, Anthicidae, Hylophilidae, Pedilidae, Scraptiidae. *Memórias e Estudos do Museu Zoológico da Universidade de Coimbra* (1) **60**: 5-26.
Pic M. 1932d: Nouveautés diverses. *Mélanges Exotico-Entomologiques* **59**: 10-36.
Pic M. 1932e: Notes diverses, nouveautés (Suite.). *L'Échange, Revue Linnéenne* **47**: 17-18.
Pic M. 1932f: Notes diverses, nouveautés (Suite). *L'Échange, Revue Linnéenne* **47**: 25-26.
Pic M. 1932g: Diagnoses préliminaires. *L'Échange, Revue Linnéenne* **47**: 26-28.
Pic M. 1932h: Notes diverses, nouveautés (Suite). *L'Échange, Revue Linnéenne* **48**: 29-31.
Pic M. 1932i: Coléoptères exotiques en partie nouveaux (Suite). *L'Échange, Revue Linnéenne* **48**: 32.
Pic M. 1933a: Notes diverses, nouveautés. *L'Échange, Revue Linnéenne* **49**: 1-2.
Pic M. 1933b: Schwedisch-chinesische wissenschaftliche Expedition nach den nordwestlichen Provinzien Chinas unter Leitung von Dr. Sven Hedin und Prof. Sü Ping-chang. Insekten gesammelt vom schwedischen Arzt der Expedition Dr. David Hummel 1927-1930. 16. Coleoptera. 2. Helmidae, Dermestidae, Anobiidae, Cleridae, Malacodermata, Dascillidae, Heteromera (ex p.), Bruchidae, Cerambycidae, Phytophaga (ex p.). *Arkiv för Zoologi* **27 A** (2) [1935]: 1-14.
Pic M. 1933c: Nouveautés diverses. *Mélanges Exotico-Entomologiques* **61**: 3-36.
Pic M. 1933d: Nouveautés diverses. *Mélanges Exotico-Entomologiques* **62**: 1-36.
Pic M. 1933e: Notes diverses, nouveautés (Suite). *L'Échange, Revue Linnéenne* **49**: 9-11, 13-15.
Pic M. 1933f: Coléoptères du Japon recueillis par M. E. Gallois. *Bulletin du Muséum National d'Histoire Naturelle, Paris* (2) **5**: 318-321.
Pic M. 1934a: Nouveautés diverses. *Mélanges Exotico-Entomologiques* **63**: 1-36.
Pic M. 1934b: Nouveautés diverses. *Mélanges Exotico-Entomologiques* **64**: 1-36.
Pic M. 1934c: Coléoptères exotiques en partie nouveaux. *L'Échange, Revue Linnéenne* **50**: 22-23.
Pic M. 1934d: Notes diverses, nouveautés (Suite). *L'Échange, Revue Linnéenne* **50**: 21-22, 25-27.

Pic M. 1934e: Coléoptères exotiques en partie nouveaux. *L'Échange, Revue Linnéenne* **50**: 28, 31-32.
Pic M. 1934f: Nouveaux coléoptères de Chine. *Entomologisches Nachrichtenblatt* **8**: 84-87.
Pic M. 1935a: Nouveautés diverses. *Mélanges Exotico-Entomologiques* **65**: 1-36.
Pic M. 1935b: Nouveautés diverses. *Mélanges Exotico-Entomologiques* **66**: 1-36.
Pic M. 1935c: Notes diverses, nouveautés (Suite). *L'Échange, Revue Linnéenne* **51**: 3-4, 5-8.
Pic M. 1935d: Notes diverses, nouveautés (Suite). *L'Échange, Revue Linnéenne* **51**: 11-12.
Pic M. 1935e: Coléoptères oedémérides nouveaux. *Revue Française d'Entomologie* **2**: 98-102.
Pic M. 1935f: Sur diverses insectes paléarctiques. *Bulletin de la Société Entomologique de France* **40**: 108-111.
Pic M. 1935g: Descriptions de coléoptères paléarctiques. *Bulletin de la Société Entomologique de France* **40**: 158-160.
Pic M. 1936a: Nouveautés diverses. *Mélanges Exotico-Entomologiques* **67**: 1-36.
Pic M. 1936b: Nouveautés diverses. *Mélanges Exotico-Entomologiques* **68**: 10-36.
Pic M. 1936c: Notes diverses, nouveautés (Suite). *L'Échange, Revue Linnéenne* **52**: 21-23.
Pic M. 1936d: Notes diverses, nouveautés (Suite). *L'Échange, Revue Linnéenne* **52**: 25-27.
Pic M. 1936e: Notes diverses, nouveautés (Suite). *L'Échange, Revue Linnéenne* **52**: 29-30.
Pic M. 1936f: [new taxon]. In: Normand H.: Contribution au catalogue des coléoptères de la Tunisie (9e fascicule). *Bulletin de la Société d'Histoire Naturelle de l'Afrique du Nord* **27**: 144-164.
Pic M. 1936g: Descriptions de coléoptères. *Bulletin de la Société Entomologique de France* **41**: 213-216.
Pic M. 1936h: Nouveau coléoptères exotiques. *Bulletin de la Société Zoologique de France* **61**: 298-300.
Pic M. 1937a: Notes diverses, nouveautés (Suite). *L'Échange, Revue Linnéenne* **53**: 1.
Pic M. 1937b: Nouveautés diverses. *Mélanges Exotico-Entomologiques* **67**: 1-36.
Pic M. 1937c: Dasytidae: Dasytinae. Pars 155. In: Schenkling S. (ed.): *Coleopterorum Catalogus. Vol. 19.* Berlin and s'Gravenhage: W. Junk, 130 pp.
Pic M. 1937d: Notes diverses, nouveautés (Suite). *L'Échange, Revue Linnéenne* **53**: 1. 9-10.
Pic M. 1937e: Nouveautés diverses. *Mélanges Exotico-Entomologiques* **68**: 10-36.
Pic M. 1937f: Nouveautés diverses. *Mélanges Exotico-Entomologiques* **69**: 1-36.
Pic M. 1937g: Cis Latr. et Anobiides nouveaux des Indes (Coléoptères). *Indian Forest Records, Entomology* **3**: 123-126.
Pic M. 1937h: Nouveaux coléoptères d'Afrique et de Chine. *Revue Française d'Entomologique* **4**: 148-149.
Pic M. 1937i: Coléoptères nouveaux de Chine. *Notes d'Entomologie Chinoise, Musée Heude* (Shanghai) **4**: 161-176.
Pic M. 1938a: Nouveautés diverses, Mutations. *Mélanges Exotico-Entomologiques* **70**: 1-36.
Pic M. 1938b: Notes diverses, nouveautés (Suite). *L'Échange, Revue Linnéenne* **54**: 5-6.
Pic M. 1938c: Notes diverses, nouveautés (Suite). *L'Échange, Revue Linnéenne* **54**: 9-11.
Pic M. 1938d: Notes diverses, nouveautés. (Suite). *L'Échange, Revue Linnéenne* **54**: 13-14.
Pic M. 1938e: Mutations et coléoptères nouveaux. *L'Échange, Revue Linnéenne* **54**: 14-16.
Pic M. 1938f: Nouveaux coléoptères. *Bulletin de la Société Entomologique de France* **43**: 71-72.
Pic M. 1938g: Coléoptères asiatiques nouveaux. *Bulletin de la Société Entomologique de France* **43**: 220-221.
Pic M. 1939a: Mutations et nouveautés diverses. *Mélanges Exotico-Entomologiques* **71**: 1-36.
Pic M. 1939b: Notes diverses, nouveautés (Suite). *L'Échange, Revue Linnéenne* **55**: 17-19, 21-22.
Pic M. 1939c: Coléoptères exotiques en partie nouveaux (Suite). *L'Échange, Revue Linnéenne* **55**: 24.
Pic M. 1939d: Deux nouveaux coléoptères de Chine. *Notes d'Entomologie Chinoise, Musée Heude* (Shanghai) **6**: 135-136.
Pic M. 1939e: Tenthredid (Hymén.) et coléoptères nouveaux. *L'Échange, Revue Linnéenne* **55**: 29-30.
Pic M. 1939f: Coléoptères nouveaux d'Egypte et du Sinaï. *Bulletin de la Société Fouad 1er d'Entomologie* **23**: 143-149.
Pic M. 1939g: Nouveautés et observations sur divers coléoptères. *VII. Internationaler Kongress für Entomologie, Verhandlungen* **1**: 342-344.
Pic M. 1939h: Nouveaux coléoptères de la presqu'île Malaise (VI.). *Journal of the Federal Malay State Museums* **18**: 372-376.
Pic M. 1940a: Tenthredides (Hymén.) et coléoptères nouveaux. *L'Échange, Revue Linnéenne* **56**: 1-2.
Pic M. 1940b: Diagnoses de coléoptères exotiques. *L'Échange, Revue Linnéenne* **56**: 13-16.
Pic M. 1941a: Notes diverses, nouveautés. *L'Échange, Revue Linnéenne* **57**: 1, 5.
Pic M. 1941b: Opuscula martialis. *L'Échange, Revue Linnéenne. Numéro Spécial* **2**: 1-16.

Pic M. 1941c: Opuscula martialis. *L'Échange, Revue Linnéenne. Numéro Spécial* **3**: 1-16.
Pic M. 1941d: Opuscula martialis. *L'Échange, Revue Linnéenne. Numéro Spécial* **5**: 1-16.
Pic M. 1942a: Opuscula martialis. *L'Échange, Revue Linnéenne. Numéro Spécial* **7**: 1-16.
Pic M. 1942b: Opuscula martialis. *L'Échange, Revue Linnéenne. Numéro Spécial* **8**: 1-16.
Pic M. 1942c: Coléoptères du globe (Suite). *L'Échange, Revue Linnéenne* **58**: 3-12.
Pic M. 1942d: Coléoptères nouveaux de l'Adrar du Iforas. *Bulletin de la Société Entomologique de France* **47**: 77-79.
Pic M. 1943: Opuscula martialis. *L'Échange, Revue Linnéenne. Numéro Spécial* **9**: 1-16.
Pic M. 1944a: Opuscula martialis. *L'Échange, Revue Linnéenne. Numéro Spécial* **12**: 1-16.
Pic M. 1944b: Coléoptères du globe (Suite). *L'Échange, Revue Linnéenne* **60**: 5-8.
Pic M. 1945: Coléoptères du globe (Suite). *L'Échange, Revue Linnéenne* **61**: 1-4, 7-8.
Pic M. 1946: Coléoptères du globe (Suite). *L'Échange, Revue Linnéenne* **62**: 1-4.
Pic M. 1947a: Diversités entomologiques. *Diversités Entomologiques* **1**: 1-16.
Pic M. 1947b: Coléoptères du globe (Suite.). *L'Échange, Revue Linnéenne* **63**: 9-16.
Pic M. 1948a: Coléoptères du globe (Suite.). *L'Échange, Revue Linnéenne* **64**: 1-4.
Pic M. 1948b: Étude variétiste de Lytta vesicatoria L. *Diversités Entomologiques* **4**: 1-4.
Pic M. 1948c: Coléoptères du globe (Suite). *L'Échange, Revue Linnéenne* **64**: 6-8.
Pic M. 1949a: Coléoptères du globe. *L'Échange, Revue Linnéenne* **65**: 1-4, 9-12.
Pic M. 1949b: Notes et descriptions. *Diversités Entomologiques* **5**: 9-16.
Pic M. 1949c: Nouveaux coléoptères du Maroc. *Bulletin de la Société des Sciences Naturelles du Maroc* **29**: 73-75.
Pic M. 1950a: Descriptions et notes variées. *Diversités Entomologiques* **7**: 1-16.
Pic M. 1950b: Coléoptères du globe (Suite). *L'Échange, Revue Linnéenne* **66**: 5-8.
Pic M. 1951a: Coléoptères du globe (Suite). *L'Échange, Revue Linnéenne* **67**: 1-4, 10-18.
Pic M. 1951b: Descriptions diverses et notes. *Diversités Entomologiques* **8**: 7-16.
Pic M. 1951c: Notes et descriptions. *Diversités Entomologiques* **9**: 4-16.
Pic M. 1951d: Deux nouveaux Anthicus Payk. du Sahara marocain (Col. Hétéromères). *Comptes Rendus de la Société des Sciences Naturelles du Maroc* **17**: 37-38.
Pic M. 1951e: Un Alleculide nouveau du Maroc (Col. hétéromère). *Comptes Rendus de la Société des Sciences Naturalles du Maroc* **17**: 75.
Pic M. 1951f: Description de deux Anthicus nouveaux du Maroc. *Bulletin de la Société des Sciences Naturelles du Maroc* **29** [1949]: 81-82.
Pic M. 1951g: Entomological expeditions to Abyssinia, 1926-27 and 1948-49. Coleoptera. Anthicidae. *The Annals and Magazine of Natural History* (12) **4**: 125-136.
Pic M. 1951h: Résultats de la Mission Zoologique Suisse au Cameroun. Malacodermes, hétéromères vésicants et phytophages (p. p.). *Mémoires de l'Institut Français d'Afrique Noire, Centre du Cameroun, Duala, Sciences Naturelles* **1**: 217-222.
Pic M. 1952a: Coléoptères du globe (Suite). *L'Échange, Revue Linnéenne* **68**: 1-4.
Pic M. 1952b: Coléoptères du globe (Suite.). *L'Échange, Revue Linnéenne* **68**: 5-8.
Pic M. 1952c: Contribution à l'étude du peuplement de la Mauritanie. Coléoptères divers (deuxième note). *Bulletin de l'Institut Française d'Afrique Noire* **14**: 489-494.
Pic M. 1953a: Notes et descriptions. *Diversités Entomologiques* **12**: 5-16.
Pic M. 1953b: Coléoptères nouveaux de la région marocaine. *Comptes Rendus de la Société des Sciences Naturalles du Maroc* **19**: 61-63.
Pic M. 1953c: Melandryidae (Col. Hétéromères) nouveaux. *The Annals and Magazine of Natural History* (12) **6**: 333-336.
Pic M. 1953d: Nouveaux coléoptères Africains. *Bulletin de l'Institut Français d'Afrique Noire* **15**: 495-506.
Pic M. 1954a: Coléoptères nouveaux de Chine. *Bulletin de la Société Entomologique de Mulhouse* **1954**: 53-59.
Pic M. 1954b: Coléoptères nouveaux de Chine (Suite). *Bulletin de la Société Entomologique de Mulhouse* **1954**: 60-64.
Pic M. 1954c: Lagriidae et Alleculidae nouveaux du Musée royal du Congo belge (Coléoptères). *Revue de Zoologie et de Botanique Africaines* **49**: 225-264.
Pic M. 1955a: Coléoptères du globe (Suite). *L'Échange, Revue Linnéenne* **71**: 1-3.
Pic M. 1955b: Descriptions et notes. *Diversités Entomologiques* **15**: 9-16.

Pic M. 1955c: Coléoptères nouveaux de Chine. *Bulletin de la Société Entomologique de Mulhouse* **1955**: 29-32.

Pic M. 1955d: Contributions à l'étude de la faune entomologique du Ruanda-Urundi (Mission P. Basilewsky 1953). XII. Coleoptera Rhipiphoridae, Oedemeridae, Scraptiidae, Pedilidae, Aderidae, Anthicidae, Dermestidae, Byrrhidae, Anobiidae, Ptinidae, Cleridae, Lymexylonidae et Helodidae. *Annales du Musée Royal du Congo Belge, Tervuren* (*Série 8° Sciences Zoologiques*) **34**: 125-135.

Pic M. 1955e: Contribution à l'étude du peuplement de la Mauritanie (3e note). *Bulletin de l'Institut Français d'Afrique Noire, Série A: Sciences Naturelles* **67**: 1041-1050.

Pic M. 1956: Anobiidae, Ptinidae et Aderidae (Coleoptera) de l'expédition J. Klapperich en Afghanistan (1952-1953). *Bulletin de la Société Entomologique de Mulhouse* **1956**: 5-7.

Pic M. 1957: [new taxa]. In: Pic M. & Hawkins C. N.: 29. Coleoptera: Anthicidae, pp. 435-450. In: *British Museum* (*Natural History*) *Expedition to South-West Arabia 1937-8*. [1941-1957] *Volume 1*. London: Trustees of the British Museum, 504 pp., 1 pl., 1 map.

Pic T. 1900: Deux variétés de Notoxus d'Algérie. *L'Échange, Revue Linnéenne* **16**: 32.

Picka J. 1979: Eine neue mitteleuropäische Art der Gattung Blaps Fabricius, 1775 (Coleoptera, Tenebrionidae). *Reichenbachia* **17**: 99-103.

Picka J. 1983a: Eine neue Scaphidema-Art aus dem Kaukasus: Scaphidema rousi sp. n. (Coleoptera, Tenebrionidae). *Acta Entomologica Musei Nationalis Pragae* **41**: 261-263.

Picka J. 1983b: Eine neue Melanimon-Art aus Bulgarien: Melanimon inermus sp.n. (Coleoptera, Tenebrionidae). *Acta Entomologica Musei Nationalis Pragae* **41**: 265-268.

Picka J. 1984: Zur Faunistik und Taxonomie der Tenebrionidae (Coleoptera) der Insel Kreta. *Türkiye Bitki Koruma Dergisi* **7-8**: 17-31.

Picka J. 1987: Aphaleria brodskyi new species from Soviet Middle Asia (Coleoptera, Tenebrionidae). *Elytron* **1**: 33-35.

Picka J. 1988: Revision of the genus Brachyestes Fairmaire from North Africa (Coleoptera, Tenebrionidae). *Elytron* **2**: 107-112.

Picka J. 1990: Revision of the Plamius quadrinotatus species group (Coleoptera, Tenebrionidae). *Annales Historico-Naturales Musei Nationalis Hungarici* **81**: 109-114.

Pierre F. 1954: Le genre Leptonychus Chevrolat (Col. Tenebrionidae). *Annales de la Société Entomologique de France* **123**: 17-33.

Pierre F. 1958: Coléoptères ténébrionides du Tasili n'Ajjer (Mission F. Bernard). In: Bernard F. (ed.): *Mission scientifique au Tassili n'Ajjer* **3**: 133-145. [Publication de l'Institut de Recherches Sahariennes de l'Université d'Alger].

Pierre F. 1961: Expédition française au Jannu (Népal oriental) (Coleoptera, Tenebrionidae). *Bulletin de la Société Entomologique de France* **66**: 212-214.

Pierre F. 1964a: Notes sur quelque Blaps de Syrie et de Palestine (Tenebrionidae). *Livre Jubilaire dédié à George Petit, Supplément 17 Vie et Milieu* **1964**: 407-409.

Pierre F. 1964b: Storthocnemis nouveaux de la zone sahélienne et du Sahara. Remarques concernant les Leucolaephini, trib. nov. (Col. Tenebrionidae). *Bulletin de l'Institut Français d'Afrique Noire* (A) **26**: 866-874.

Pierre F. 1964c: Contribution à la connaissance de la faune du Moyen-Orient (Missions G. Remaudière 1955 et 1959). II. Coléoptères tenebrionides. *Vie et Milieu* **15** (4): 1043-1055.

Pierre F. 1968: Contribution à la faune de l'Iran 12. Étude écologique et biogéographique des ténébrionides Pimeliinae s. nov. des régions désertiques et semi-désertiques de l'Iran. *Annales de la Société Entomologique de France* (*N. S.*) **4**: 997-1036.

Pierre F. 1974: Contribution à la faune de l'Iran. Insectes coléoptères du désert du Lut. *Annales de la Société Entomologique de France* (N. S.) **10**: 229-235.

Pierre F. 1976: Une nouvelle espèce de Pimelia du Sahara nord-occidental et remarques sur les possibilités d'hybridation de quelques formes (Col. Tenebrionidae). *Bulletin de la Société Entomologigue de France* **81**: 16-21.

Pierre F. 1978: Note sur la distribution et la variation géographique des Pimelia du Maroc méridional, avec descriptions de nouvelles sous-espèces (Col. Tenebrionidae). *Bulletin de la Société Entomologigue de France* **83**: 197-206.

Pierre F. 1982: Variation spécifique des Pimelia du groupe d'interstitialis dans le Nord de l'Afrique et au Sahara, avec description de nouveaux taxa (Col. Tenebrionidae). *Bulletin de la Société Entomologigue de France* **87**:

2-13.

Pierre F. 1985: Sur quelques Pimelia recueillis par le Pr. A. Hollande dans le Haut Atlas Marocain (Coleoptera, Tenebrionidae). *Nouvelle Revue d'Entomologie* (N. S.) **1** [1984]: 287-297.

Piller M. & Mitterpacher L. 1783: *Iter per Poseganam Slavoniae provincian mensibus Junio, et Julio MDCCLXXXII susceptum*. Budae: Regiae Universitatis, 147 pp., 16 pls.

Piño M. 1985: Introduction al conocimiento de los Mordellidae (Coleoptera) de Chile, clave para generos y lista especies. *Revista Chilena de Entomologia* **12**: 101-103.

Piochard de la Brûlerie C. J. 1869: Descriptions de nouvelles espèces espagnoles du groupe des Pandarites. *Annales de la Société Entomologique de France* (4) **9**: 31-38.

Plavilshchikov N. N. 1927: Pseudocistela pamirica n. sp. (Col., Allecul.). *Entomologische Mitteilungen* **16**: 133-134.

Plaza E. 1977: Mordellidae de las Islas Canarias. *Nouvelle Revue d'Entomologie* **7**: 171-181.

Plaza E. 1986a: Contributión al conocimiento del género Mediimorda Meq., 1946; Mediimorda batteni nov. sp. de la fauna española. *Eos, Revista Española de Entomología* **61** [1985]: 265-273.

Plaza E. 1986b: Las especies españolas de Mordellistena Costa del grupo episternalis. *Eos, Revista Española de Entomología* **61**[1985]: 265-273.

Plaza E. & Compte A. 1980: Contributión al conocimiento de los Mordellidae de España. *Eos, Revista Española de Entomología* **56**: 195-214.

Pliginskij [= Pliginsky] V. G. 1910: Deux espèces nouvelles du genre Meloë Linn. (Coleoptera, Meloidae). *Revue Russe d'Entomologie* **10**: 170-172.

Pliginskij V. G. 1911: Notice sur un caractère chez les representant du genre Meloë. *Revue Russe d'Entomologie* **11**: 44-47.

Pliginskij V. G. 1914a: Les meloïnes (Coleoptera, Meloidae) de la collection de V. Motschulsky. *Revue Russe d'Entomologie* **14**: 254-261.

Pliginskij V. G. 1914b: Eine neue Meloë-Art (Coleoptera, Meloidae). *Revue Russe d'Entomologie* **14**: 264-265.

Pliginskij V. G. 1914c: Eine neue Zonitis-Art aus der Krim (Coleoptera, Meloidae). *Revue Russe d'Entomologie* **14**: 429-430.

Pliginskij V. G. 1923a: Novyy vid roda Mycetophagus Hellw. iz Kryma (Coleoptera, Mycetophagidae) – Mycetophagus tauricus sp. n. *Bulletin de la Société Entomologique de Moscou* **2** (2): 78-79.

Pliginskij V. G. 1923b: Espèces nouvelles des coléoptères paléarctiques. *Revue Russe d'Entomologie* **18**: 141-143.

Pliginskij V. G. 1930: Zwei neue Meloiden-Formen (Coleoptera). *Revue Russe d'Entomologie* **24**: 109-110.

Pliginskij V. G. 1935: Notizen über Meloiden (Coleoptera, Meloidae), III. *Revue Russe d'Entomologie* **25**: 320-323.

Pomel A. 1871: Le Sahara (Géologie, Géographie et Biologie). *Bulletin de la Société Algérienne de Climatologie Science Physique et Naturelle* **8**: 133-265.

Pool C. J. C. 1917: The Coleoptera of the family Cissidae found in Britain, with descriptions of two new species. *Proceedings of the Zoological Society of London* **1917**: 83-93.

Poppius B. 1904: Neue paläarktische Coleopteren. *Öfversigt af Finska Vetenskaps Forhandlingar* **46** (16)[1903-1904]: 1-14.

Pripisnova M. G. 1987a: Novye podrod i vid roda Meloe L. (Coleoptera, Meloidae) iz Tadzhikskoy S.S.R [New subgenus and species of the genus Meloe L. (Coleoptera, Meloidae) from the Tadzhik S.S.R.]. *Izvestia Akademii Nauk Respubliki Tadzikistan, Otdelenie Biologiskikh Nauk* **3**: 41-45.

Pripisnova M. G. 1987b: Fauna i ekologia zhukov-naryvnikov (Coleoptera, Meloidae) Tadzhikistana [Fauna and ecology of the meloid beetles (Coleoptera, Meloidae) of Tazhik SSR. *Entomologicheskoe Obozrenie* **66**: 555-571.

Prochazka J. 1892: Revision der Arten der Coleopteren-Gattung Hapalus Fabr. Aus der palaärctischen Fauna. *Wiener Entomologische Zeitung* **11**: 263-270.

Puel L. 1907: Tableau sur les variétés du "Zonabris geminata " F. *L'Échange, Revue Linnéenne* **23**: 115-119.

Puel L. 1919: Sur les Anthicus français du groupe de l'humilis Germ. *Miscellanea Entomologica* **24**: 49-60.

Quedenfeldt G. 1887: Ein neues Helopiden-Genus von Marokko. *Entomologische Nachrichten* **13**: 257-259.

Quedenfeldt G. 1889: Zwei neue afrikanische Tenebrioniden. *Entomologische Nachrichten* **15**: 353-356.

Quedenfeldt G. 1890: Drei neue Tenebrioniden aus Tripolitanien. *Berliner Entomologische Zeitschrift* **33** [1899]: 395-400.

Quedenfeldt G. 1891: Brachycryptus n. gen. Cistelidarum, prope Omophlus. *Entomologische Nachrichten* **17**: 129-130.

Quedenfeldt M. 1890: Akis Schweinfurthi n. sp. *Berliner Entomologische Zeitschrift* **35**: 139.

Quensel K. 1790: *Dissertatio historico-naturalis, ignotas insectorum species continents. 4.* Lundae: Resp. C. P. Lundgard, 20 pp.

Quensel K. 1806: In: Schoenherr C. J.: *Synonymia Insectorum, oder: Versuch einer Synonymie aller bisher bekannten Insecten; nach Fabricii Systema Eleutheratorum geordnet. Erster Band. Eleutherata oder Käfer.* Stockholm: Heinr. A. Nordström, xxii + 293 pp., 3 pls.

Raffray A. 1873: [new taxa]. In: Fairmaire L. & Raffray A.: Coléoptères du Nord de l'Afrique. *Revue et Magasin de Zoologie Pure et Appliquée* (3) **1**: 331-385.

Rafinesque C. S. 1815: *Analyse de la nature ou tableau de l'univers et des corps organisés*. Palermo, 224 pp.

Ragusa E. 1875: Gita entomologica all'isola di Pantelleria. *Bollettino della Società Entomologica Italiana* **7**: 238-256.

Ragusa E. 1881: Coleotteri nuovi o poco conosciuti della Sicilia. *Il Naturalista Siciliano* **1** [1881-1882]: 42-44.

Ragusa E. 1882: Coleotteri nuovi o poco conosciuti della Sicilia. *Il Naturalista Siciliano* **1** [1881-1882]: 251-252.

Ragusa A. 1892a: Catalogo ragionale dei coleotteri di Sicilia. *Il Naturalista Siciliano* **11** [1892-1893]: 122-143.

Ragusa E. 1892b: Coleotteri nuovi o poco conosciuti della Sicilia. *Il Naturalista Siciliano* **12** [1892-1893]: 26-31.

Ragusa E. 1898: Catalogo ragionato dei coleotteri di Sicilia. *Il Naturalista Siciliano* (N. S.) **2** [1897-1897]: 105-130, 197-237

Ragusa E. 1906: Coleotteri nuovi o poco conosciuti della Sicilia. *Il Naturalista Siciliano* (N. S.) **19** [1906-1907]: 4-8.

Ragusa E. 1923: Coleotteri nuovi o poco conosciuti nella Sicilia. *Atti della Reale Accademia delle Scienze e Lettere e Belle Arti di Palermo* (3) **12** [1920-1922]: 1-18.

Rallo G. 1974: Annotazioni sugli Oochrotus italiani (Coleoptera, Heteromera: IV contributo). *Bollettino del Museo Civico di Storia Naturale di Venezia* **25**: 61-63.

Rambur J. P. 1838: Pls. XIX, XX. In: *Faune Entomologique de l'Andalousie. Volume 1.* [1837-1838]. Paris: Arthus Bertrand, 144 pp., pls. i, ii, ixx, xx. [Note: livraison 1 published in 1837, livraison 2 in 1838; the plates XIX and XX issued presumedly in 1838].

Razzauti A. 1919: *Contributi alla conoscenza faunistica delle isole toscane II. Forme insulare tirreniche dell'Opatrum sculpturatum Fairmaire. (Coleoptera-Tenebrionidae).* Pisa: Succ. FF. Nistri, 7 pp

Redtenbacher L. 1843: Illustrationes et descriptiones Coleopterorum novorum Syriae. Pp. 971-990. In: Russegger J. von: *Reisen in Europa, Asien und Afrika, mit besonderer Rücksicht auf die naturwissenschaftliche Verhältnisse der betreffenden Länder unternommen in den Jahren 1835 bis 1841. Band 1 (Teil 2).* Stuttgart: Schweizerbart'sche Verlagshandlung.

Redtenbacher L. 1844: [new taxa]. In: Kollar V. & Redtenbacher L.: Aufzählung und Beschreibung der von Freiherr Carl von Huegel auf seiner Reise durch Kaschmir u. das Himalayagebirge gesammelten Insecten. Pp. 393-564. In: Hügel K. F. von. (ed.): *Kaschmir und das Reich der Siek. Vierter Band. Zweite Abtheilung.* Stuttgart: Hallbergerischer Verlag, pp. 244-586 pp. [1844]; 587-865 + [6] pp., 31 pls, 1 map [1848].

Redtenbacher L. 1845: *Die Gattungen der deutschen Käfer-Fauna nach der analytischen Methode bearbeitet, nebst einem kurzgefassten Leitfaden zum Studium dieses Zweiges der Entomologie.* Wien: Carl Überreuter, [14] +177 + [1] pp., 2 pls.

Redtenbacher L. 1847: [Part 3, pp. 321-480]. In: *Fauna Austriaca. Die Käfer. Nach der analytischen Methode bearbeitet* [1849]. Wien: C. Gerold, xxvii + 883 pp. (issued in parts).

Redtenbacher L. 1849: *Fauna Austriaca. Die Käfer. Nach der analytischen Methode bearbeitet.* Wien: C. Gerold, xxvii + 883 pp. (issued in parts).

Redtenbacher L. 1850: [new taxa]. In: Kollar V. & Redtenbacher L.: Ueber den Character der Insecten-Fauna von Südpersien. *Denkschriften der Kaiserlichen Akademie der Wissenschaften. Mathematisch-Naturwissenschaftliche Classe* **1**: 42-53.

Redtenbacher L. 1857: [Pp. 129-976]. In: *Fauna Austriaca. Die Käfer. Nach der analytischen Methode bearbeitet. Zweite, gänzlich umgearbeitete, mit mehreren Hunderten von Arten und mit der Charakteristik sämmtlicher europäischen Käfergattungen vermehrte Auflage* [1858]. Wien: Carl Gerold's Sohn, cxxxvi +1017 pp., 2 pls. (issued in parts).

Redtenbacher L. 1868: Zoologischer Theil. Zweiter Band. I. Abtheilung A. 2. Coleopteren. In: *Reise der Öesterreichischen Fregatte Novara um die Erde in den Jahren 1857, 1858, 1859 unter den Befehlen des Commodore B. von Wüllerstorf-Urbair.* [1867]. Wien: Karl Gerold's Sohn, iv + 249 pp., 5 pls.

Redtenbacher L. 1874: *Fauna Austriaca. Die Käfer. Nach der analytischen Methode bearbeitet. Dritte, gänzlich umgearbeitete und bedeutend vermehrte Auflage. Zweiter Band.* Wien: Carl Gerold's Sohn, 571 + viii pp.

Redtenbacher W. 1842: *Quaedam genera et species Coleopterorum Archiducatus Austriae nondum descriptiorum.* Vindobonae: Carolus Ueberreuter, 31 pp.

Reichardt A. N. 1934: Opredelitel i spisok zhukov naryvnikov v Turkmenii. *Trudy Sovieta po Izutcheniu Proizvoditelnykh Sil. Seriya Turkmenskaya* **6**: 205-239.

Reichardt A. N. 1935: Novye formy chernotelok triby Opatrini iz Tadzhikistana. *Trudy Tadzhikistanskoy Bazy ANSSR* **5**: 251-256.

Reichardt A. N. 1936a: Zhuki-chernotelki triby Opatrini palearkticheskoy oblasti. Révision des opatrines (Coleoptera Tenebrionidae) de la région paléarctique. *Tableaux analytiques de la Faune de l'URSS* **19**, Moskva, Leningrad: Nauka, 1-224 pp.

Reichardt A. N. 1936b: Obzor vidov roda Pedinus Latr. SSSR i Dalnego Vostoka (Coleoptera, Tenebrionidae). *Trudy Zoologicheskogo Instituta Akademii Nauk SSSR* **3**: 669-692.

Reiche L. J. 1850: Entomologie. Pp. 257-532. In: Ferret P. V. A. & Galinier J. G. (eds): *Voyage en Abyssinie dans les provinces du Tigre, du Samen et de l'Amhara. Tome Troisième.* [1847-1848]. Paris: Paulin, 536 pp.

Reiche L. J. 1860: Coléoptères de Sicile recueillis par M. E. Bellier de la Chavignerie et déterminés par M. L. Reiche et description de dix espèces nouvelles. *Annales de la Société Entomologique de France* (3) **8**: 717-737.

Reiche L. J. 1860: Description d'une nouvelle espèce de coléoptères du genre Microtelus. *Annales de la Société Entomologique de France* (3) **8**: 334.

Reiche L. J. 1861a: Species novae Coleopterorum descriptae, quae in Syria invenit Dom. Kindermann. *Wiener Entomologische Monatschrift* **5**: 1-8.

Reiche L. J. 1861b: Sur quelques espèces de coléoptères du Nord de l'Afrique. *Annales de la Société Entomologique de France* (4) **1**: 87-92.

Reiche L. J. 1861c: Coléoptères nouveaux recueillis en Corse par M. E. Bellier de la Chavignerie. *Annales de la Société Entomologique de France* (4) **1**: 201-210.

Reiche L. J. 1861d: Notes synonymiques. *Annales de la Société Entomologique de France* (4) **1**: 211-212.

Reiche L. J. 1862a: Espèces nouvelles de coléoptères appartenant à la faune circa-méditerranéenne. *Annales de la Société Entomologique de France* (4) **1** [1861]: 361-374.

Reiche L. J. 1862b: Espèces nouvelles de coléoptères découvertes en Corse par M. E. Bellier de la Chavignerie, en 1861. *Annales de la Société Entomologique de France* (4) **2**: 293-300.

Reiche L. J. 1862c: Espèces nouvelles de coléoptères appartenant à la faune circa-méditerranéenne. (Suite) (1). *Annales de la Société Entomologique de France* (4) **2**: 539-546.

Reiche L. J. 1864: Espèces nouvelles de coléoptères d'Algérie. *Annales de la Société Entomologique de France* (4) **4**: 233-246.

Reiche L. J. 1866: Étude sur les epèces de mylabrides de la collection de L. Reiche, suivie d'une note sur le genre Trigonurus Mulsant et description d'une espèce nouvelle. *Annales de la Société Entomologique de France* (4) **5** [1865]: 627-642.

Reiche L. J. 1867: [new taxon]. In: Marseul S.-A. de: Catalogue des coléoptères d'Europe et des pays limitrophes. *L'Abeille, Mémoires d'Entomologie* **4**: 1-131.

Reiche L. J. 1878: [new taxon]. *Bulletin de la Société Entomologique de France* **1878**: lxxiii-lxxiv.

Reiche L. J. & Saulcy F. 1857: Espèces nouvelles ou peu connues de coléoptères, recueillis par M. F. de Saulcy, membre de l'Institut, dans son voyage en Orient. *Annales de la Société Entomologique de France* (3) **5**: 169-276, 1 pl.

Reinig W. F. 1931: Entomologische Ergebnisse der Deutsch-Russischen Alai-Pamir-Expedition 1928 (II). 5. Coleoptera II. Tenebrionidae. *Mitteilungen aus dem Zoologischen Museum Berlin* **16** [1930]: 865-912.

Reinig W. F. 1934: 6. Coleoptera VII. (Tenebrionidae, Nachtrag). P. 162. In: Entomologische Ergebnisse der Deutsch-Russischen Alai-Pamir-Expedition 1928 (II). *Deutsche Entomologische Zeitschrift* **1933**: 129-176.

Reitter E. 1872: Neue Käferarten von Oran gesammelt von Hans Leder. *Berliner Entomologische Zeitschrift* **16**: 167-186.

Reitter E. 1874: Beschreibungen neuer Käfer-Arten nebst synonymischen Notizen. *Verhandlungen der Kaiserlich-Königlichen Zoologisch-Botanischen Gesellschaft in Wien* **24**: 509-528.

Reitter E. 1875: [new taxa]. In: Weise J.: Neue Käferarten aus Ungarn. *Deutsche Entomologische Zeitschrift* **19** [1930]: 355-364.

Reitter E. 1876a: Neue Gattungen und Arten aus der Familie der Cucujidae. *Coleopterologische Hefte* **15**: 37-64.

Reitter E. 1876b: Neue transcaucasische Coleopteren, gesammelt von H. Leder. *Berliner Entomologische Zeitschrift* **20**: 289-294.

Reitter E. 1877a: Coleopterologische Ergebnisse einer Reise nach Südungarn und die Transsylvanischen Alpen. *Verhandlungen des Naturforschenden Vereines in Brünn* **15** [1876]: 3-30, pl. 1.

Reitter E. 1877b: Neue Arten aus der Gattung Sitophagus Muls. *Mittheilungen des Münchener Entomologischen Vereins* **1**: 8-11.

Reitter E. 1877c: Neue Arten aus den Familien der Cucujidae, Nitidulidae, Colydiidae und Cryptophagidae. *Mittheilungen des Münchener Entomologischen Vereins* **1**: 22-28.

Reitter E. 1877d: Neuer Mycetophagus aus Japan. *Deutsche Entomologische Zeitschrift* **21**: 116.

Reitter E. 1877e: Neue kaukasische Coleoptereren, gesammelt von Hans Leder (Zweites Stück). *Deutsche Entomologische Zeitschrift* **21**: 289-296.

Reitter E. 1877f: Beitrag zur Kenntniss der Colydier. *Entomologische Zeitung* (Stettin) **38**: 323-355.

Reitter E. 1877g: Beiträge zur Käferfauna von Japan. (Drittes Stück). *Deutsche Entomologische Zeitschrift* **21**: 369-384.

Reitter E. 1877h: Atritomus nov. gen. Tritomidarum. *Deutsche Entomologische Zeitschrift* **21**: 384.

Reitter E. 1878a: [new taxa]. In: Schneider O. & Leder H.: Beiträge zur Kenntnis der kaukasischen Käferfauna. *Verhandlungen des Naturforschen Vereines in Brünn* **16** [1887]: 3-259, 3 pls.

Reitter E. 1878b: Nachträge und Ergänzungen zur Bearbeitung der Cioiden von H. v. Kiesenwetter. *Deutsche Entomologische Zeitschrift* **22**: 21-30.

Reitter E. 1878c: Neue Cioidae. *Mittheilungen des Münchener Entomologischen Vereins* **2**: 32-37.

Reitter E. 1878d: Beiträge zur Coleopteren-Fauna der Carpathen. *Deutsche Entomologische Zeitschrift* **22**: 33-64.

Reitter E. 1878e: Neue Colydiidae des Berliner Museums. *Deutsche Entomologische Zeitschrift* **22**: 113-125.

Reitter E. 1879: Verzeichniss der von H. Christoph in Ost-Sibirien gesammelten Clavicornier etc. *Deutsche Entomologische Zeitschrift* **23**: 208-226.

Reitter E. 1880a: Bestimmungs-Tabellen der europäischen Coleopteren. I. Enthaltend die Familien: Cucujidae, Telmatophilidae, Tritomidae, Mycetaeidae, Endomychidae, Lyctidae und Sphindidae. *Verhandlungen der Kaiserlich-Königlichen Zoologisch-Botanischen Gesellschaft in Wien* **29** [1879]: 71-100.

Reitter E. 1880b: Neue Lagria-Arten aus Africa und Ostindien. *Deutsche Entomologische Zeitschrift* **24**: 253-257.

Reitter E. 1880c: [new taxa]. In: Leder H.: Beitrag zur kaukasischen Käfer-Fauna. Unter Mitwirkung herrvoragender Fachgenossen. *Verhandlungen der Kaiserlich-Königlichen Zoologisch-Botanischen Gesellschaft in Wien* **29** [1879]: 451-488.

Reitter E. 1880d: Neue Coleopteren aus dem südöstlichen Russland aufgefunden von Herrn J. Faust aus Helsingfors. *Verhandlungen der Kaiserlich-Königlichen Zoologisch-Botanischen Gesellschaft in Wien* **29** [1879]: 543-546.

Reitter E. 1881a: Neue und seltene Coleopteren, im Jahre 1880 in Süddalmatien und Montenegro gesammelt und beschrieben. *Deutsche Entomologische Zeitschrift* **25**: 177-229, pls. 6-7.

Reitter E. 1881b: Coleopterologische Ergebnisse einer Reise nach Croatien, Dalmatien und der Herzegowina im Jahre 1879. *Verhandlungen der Kaiserlich-Königlichen Zoologisch-Botanischen Gesellschaft in Wien* **30** [1880]: 201-228.

Reitter E. 1881c: Neue Käferarten aus Asturien. *Deutsche Entomologische Zeitschrift* **25**: 231-232.

Reitter E. 1882a: Bestimmungs-Tabellen der europäischen Coleopteren. VI. Enhaltend die Familien: Colydiidae, Rhysodidae, Trogositidae. *Verhandlungen des Naturforschenden Vereines in Brünn* **22** [1881]: 113-149.

Reitter E. 1882b: Neuere Literatur. Histoire naturelle des coléoptères de France. (Par E. Mulsant.) Famille des Lathridiens, par le R. P. Fr. Marie-Joseph Belon. Lyon 1881. *Deutsche Entomologische Zeitschrift* **26**: 161-165.

Reitter E. 1883a: Neue Coleopteren aus Russland und Bemerkungen über bekannte Arten. *Revue Mensuele d'Entomologie* **1**: 111-118.

Reitter E. 1883b: Zwei neue Heteromeren-Genera aus Europa. *Wiener Entomologische Zeitung* **2**: 307-310, pl. 4.

Reitter E. 1884a: Diagnosen neuer Coleopteren aus Lenkoran. *Verhandlungen des Naturforschenden Vereines in Brünn* **22** [1883]: 3-10.

Reitter E. 1884b: [new taxa]. In: Brenske E. & Reitter E.: Neuer Beitrag zur Käferfauna Griechenlands. *Deutsche Entomologische Zeitschrift* **28**: 17-100, pls 1, 2.

Reitter E. 1884c: Resultate einer coleopterologischen Sammel-Campagne während den Monaten Februar bis April 1883 auf den jonischen Inseln. *Deutsche Entomologische Zeitschrift* **28**: 101-122.

Reitter E. 1884d: Revision der europäischen Mycetochares-Arten. *Deutsche Entomologische Zeitschrift* **28**: 241-250.

Reitter E. 1884e: Einige neue Coleopteren aus Süd-Europa. *Deutsche Entomologische Zeitschrift* **28**: 255-259.

Reitter E. 1884f: Ueber die syrischen Arten der Gattung Anemia Lap. *Deutsche Entomologische Zeitschrift* **28**: 259-260.

Reitter E. 1885a: Uebersicht der Arten der Gattung Cerocoma Geoff. *Deutsche Entomologische Zeitschrift* **29**: 1-14.

Reitter E. 1885b: Coleopterologische Ergebnisse einer Excursion nach Bosnien im Mai 1884. *Deutsche Entomologische Zeitschrift* **29**: 193-216.

Reitter E. 1885c: Coleopterologische Notizen. XIV. *Wiener Entomologische Zeitung* **4**: 313-317.

Reitter E. 1885d: Neue Coleopteren aus Europa und den angrenzenden Ländern, mit Bemerkungen über bekannte Arten. *Deutsche Entomologische Zeitschrift* **29**: 353-397.

Reitter E. 1885e: Uebersicht der bekannten Laena-Arten. *Deutsche Entomologische Zeitschrift* **29**: 398-402.

Reitter E. 1886a: Revision der mit Stenosis verwandten Coleopteren der alten Welt. *Deutsche Entomologische Zeitschrift* **30**: 97-144.

Reitter E. 1886b: Coleopterologische Notizen. XV. *Wiener Entomologische Zeitung.* **5**: 99-100.

Reitter E. 1886c: Ueber eine neue europäische Allecula. *Wiener Entomologische Zeitung* **5**: 140.

Reitter E. 1886d: Coleopterologische Notizen. XVI. *Wiener Entomologische Zeitung* **5**: 151-152.

Reitter E. 1886e: Coleopterologische Notizen. XIX. *Wiener Entomologische Zeitung* **5**: 347-352.

Reitter E. 1887a: Coleopterologische Notizen. XXI. *Wiener Entomologische Zeitung* **6**: 76-77.

Reitter E. 1887b: Sulle specie del genere di Coleotteri Tetratoma Fabricius. *Il Naturalista Siciliano* **6** [1886-1887]: 84-85.

Reitter E. 1887c: Neue Coleopteren aus Europa, den angrenzenden Ländern und Sibirien, mit Bemerkungen über bekannte Arten. *Deutsche Entomologische Zeitschrift* **31**: 241-288.

Reitter E. 1887d: Insecta in itinere Cl. N. Przewalskii in Asia Centrali novissime lecta. IX. Tenebrionidae. *Horae Societatis Entomologicae Rossicae* **21**: 355-389.

Reitter E. 1887e: Neue Coleopteren aus Europa, den angrenzenden Ländern und Sibirien, mit Bemerkungen über bekannte Arten. *Deutsche Entomologische Zeitschrift* **31**: 497-528.

Reitter E. 1888a: Coleopteren aus Circassien, gesammelt von Hans Leder in Jahr 1887. IV. Theil. *Wiener Entomologische Zeitung* **7**: 169-180.

Reitter E. 1888b: Coleopteren aus Circassien, gesammelt von Hans Leder im Jahre 1887. V. Theil. *Wiener Entomologische Zeitung* **7**: 207-214.

Reitter E. 1888c: Ueber einige mit Pimelia nahe verwandte Coleopteren-Gattungen. *Deutsche Entomologische Zeitschrift* **1888**: 329-331.

Reitter E. 1888d: Neue Coleopteren aus Europa, den angrenzenden Ländern und Sibirien, mit Bemerkungen über andere Arten. Fünfter Theil. *Deutsche Entomologische Zeitschrift* **1888**: 417-432.

Reitter E. 1889a: Neue Coleopteren aus Europa, den angrenzenden Ländern und Sibirien, mit Bemerkungen über bekannte Arten. Sechster Theil. *Deutsche Entomologische Zeitschrift* **1889**: 17-44.

Reitter E. 1889b: Coleopterologische Ergebnisse der im Jahre 1886 und 1887 in Transcaspien von Dr. G. Radde, Dr. A. Walter und A. Konchin ausgeführten Expedition. *Verhandlungen des Naturforschenden Vereines in Brünn* **27** [1888]: 95-133.

Reitter E. 1889c: Zwei neue Meloë-Arten aus Syrien und Armenien. *Wiener Entomologische Zeitung* **8**: 106-107.

Reitter E. 1889d: Beschreibungen der bekannten Tritomiden Japans mit Berücksichtigung der neuer Sammelergebnisse der Herrn George Lewis in den Jahren 1880 und 1881. *Wiener Entomologische Zeitung* **8**: 245-249.

Reitter E. 1889e: Berichte über die von E. v. Oertzen in Jahre 1887 in Griechenland u. Klein-Asien gesammelten Coleopteren. IX. Neue Arten aus verschiedenen Familien. *Deutsche Entomologische Zeitschrift* **1889**: 251-259.

Reitter E. 1889f: Uebersicht der palaearctischen Chrysanthia-Arten. *Deutsche Entomologische Zeitschrift* **1889**: 266.

Reitter E. 1889g: Uebersicht der bekannten Scraptia-Arten der palaearctischen Fauna. *Deutsche Entomologische Zeitschrift* **1889**: 267-268.

Reitter E. 1889h: Neue Coleopteren aus Europa, den angrenzenden Ländern und Sibirien, mit Bemerkungen über bekannte Arten. *Deutsche Entomologische Zeitschrift* **1889**: 273-288.

Reitter E. 1889i: Verzeichniss der Cucujiden Japans mit Beschreibungen neuer Arten. *Wiener Entomologische Zeitung* **8**: 313-320.

Reitter E. 1889j: Neue Coleopteren aus Europa, den angrenzenden Ländern und Sibirien, mit Bemerkungen über bekannte Arten. *Deutsche Entomologische Zeitschrift* **1889**: 369-376.

Reitter E. 1889k: Insecta, a cl. G. N. Potanin in China et in Mongolia novissime lecta. XIII. Tenebrionidae. *Horae Societatis Entomologicae Rossicae* **23**: 678-710.

Reitter E. 1890a: Coleopterologische Notizen. *Wiener Entomologische Zeitung* **9**: 13-15.

Reitter E. 1890b: Neue analytische Uebersicht der bekannten Arten der Coleopteren-Gattung Omophlus. *Deutsche Entomologische Zeitschrift* **1890**: 33-52.

Reitter E. 1890c: Eine neue behaarte Meloë-Art aus Turkestan. *Wiener Entomologische Zeitung* **9**: 152.

Reitter E. 1890d: Neue Coleopteren aus Europa, den angrenzenden Ländern und Sibirien, mit Bemerkungen über bekannte Arten. Neuenter Theil. *Deutsche Entomologische Zeitschrift* **1890**: 145-164.

Reitter E. 1890e: Neue Coleopteren aus Europa, den angrenzenden Ländern und Sibirien, mit Bemerkungen über bekannte Arten. Zehnter Theil. *Deutsche Entomologische Zeitschrift* **1890**: 165-176.

Reitter E. 1890f: Coleopterologische Notizen. XXXVIII. *Wiener Entomologische Zeitung* **9**: 210-213.

Reitter E. 1890g: Uebersicht der Cteniopus-Arten aus Europa und den angrenzenden Ländern, soweit mir dieselben in Natura bekannt sind. *Wiener Entomologische Zeitung* **9**: 256-258.

Reitter E. 1890h: [new taxa]. In: Heyden L. F. J. D. von.: XIII. Beitrag zur Coleopteren-Fauna von Turkestan. Unter Mitwirkung der Herren Reitter und Weise. *Deutsche Entomologische Zeitschrift* **1890**: 353-367.

Reitter E. 1890i: In: Heyden L. F. J. D. von: Beiträge zur Coleopteren-Fauna von Amasia und Samsoun in Nord-Kleinasien. *Deutsche Entomologische Zeitschrift* **1890**: 369-374.

Reitter E. 1890j: Neue Coleopteren aus Europa, den angrenzenden Ländern und Sibirien, mit Bemerkungen über bekannte Arten. Elfter Theil. *Deutsche Entomologische Zeitschrift* **1890**: 383-396.

Reitter E. 1891a: Neue Coleopteren aus Europa, den angrenzenden Ländern und Sibirien, mit Bemerkungen über bekannte Arten. *Deutsche Entomologische Zeitschrift* **1891**: 17-36.

Reitter E. 1891b: Drei neue Coleopteren. *Wiener Entomologische Zeitung* **10**: 33-34.

Reitter E. 1891c: Zweiter Beitrag zur Coleopteren-Fauna des russischen Reiches. *Wiener Entomologische Zeitung* **10**: 195-199.

Reitter E. 1891d: Dritter Beitrag zur Coleopterenfauna des russischen Reiches. *Wiener Entomologische Zeitung* **10**: 221-224.

Reitter E. 1891e: Coleopterologische Notizen. XLI. *Wiener Entomologische Zeitung* **10**: 226-228.

Reitter E. 1891f: Erster Beitrag zur Coleopteren-Fauna von Europa und den angrenzenden Ländern. *Wiener Entomologische Zeitung* **10**: 246-249.

Reitter E. 1891g: Zweiter Beitrag zur Coleopteren-Fauna von Europa und den angrenzenden Ländern. *Wiener Entomologische Zeitung* **10**: 259-262.

Reitter E. 1892a: Fünfter Beitrag zur Coleopterenfauna des russischen Reiches. *Wiener Entomologische Zeitung* **11**: 59-68.

Reitter E. 1892b: Ueber die Arten der Coleopteren-Gattung Leptodes Sol. *Wiener Entomologische Zeitung* **11**: 93-96.

Reitter E. 1892c: Sechster Beitrag zur Coleopteren-Fauna des russischen Reiches. *Wiener Entomologische Zeitung* **11**: 133-136.

Reitter E. 1893a: Achter Beitrag zur Coleopteren-Fauna des russischen Reiches. *Wiener Entomologische Zeitung* **12**: 109-114.

Reitter E. 1893b: Bestimmungs-Tabelle der unechten Pimeliden aus der palaearctischen Fauna. *Verhandlungen des Naturforschenden Vereines in Brünn* **31** [1892]: 201-250.

Reitter E. 1893c: Revision der Coleopteren-Gattung Prosodes Esch. *Deutsche Entomologische Zeitschrift* **1893**: 261-312.

Reitter E. 1893d: Ueber die Genus-Charaktere der Gattungen Blaps Fr., Prosodes Eschsch. und Verwandte. *Deutsche Entomologische Zeitschrift* **1893**: 313-316.

Reitter E. 1893e: Uebersicht der mir bekannten Ammobius-Arten. *Wiener Entomologische Zeitung* **12**: 315.

Reitter E. 1894a: [new taxa]. In: Hauser F.: Beitrag zur Coleopteren-Fauna von Transcaspien und Turkestan. *Deutsche Entomologische Zeitschrift* **1894**: 17-74.

Reitter E. 1894b: Einige neue Coleopteren von der Insel Malta. *Wiener Entomologische Zeitung* **13**: 81-83.

Reitter E. 1894c: Vierter Beitrag zur Coleopteren-Fauna von Europa und den angrenzenden Ländern. *Wiener Entomologische Zeitung* **13**: 101-107.

Reitter E. 1894d: Elfter Beitrag zur Coleopteren-Fauna des russischen Reiches. *Wiener Entomologische Zeitung* **13**: 238-248.

Reitter E. 1894e: Fünfter Beitrag zur Coleopteren-Fauna von Europa und den angrenzenden Ländern. *Wiener Entomologische Zeitung* **13**: 299-306.

Reitter E. 1895a: *Bestimmungs-Tabellen der europäischen Coleopteren Meloidae. 1. Theil: Meloini. 32.* Paskau: Verfasser, 13 pp.

Reitter E. 1895b: Beschreibung neuer oder wenig gekannter Coleopteren aus der Umgebung von Akbes in Syrien. *Wiener Entomologische Zeitung* **14**: 79-88.

Reitter E. 1895c: Uebersicht der mir bekannten Arten der Coleopteren-Gattung Phthora Germ. (Cataphronetis Lucas). *Entomologische Nachrichten* **21**: 147-149.

Reitter E. 1895d: Zwölfter Beitrag zur Coleopteren-Fauna des russischen Reiches. *Wiener Entomologische Zeitung* **14**: 149-162.

Reitter E. 1895e: Sechster Beitrag zur Coleopteren-Fauna von Europa und den angrenzenden Ländern. *Wiener Entomologische Zeitung* **14**: 199-204.

Reitter E. 1895f: Beschreibungen mit Abbildungen neuer Coleopteren, gesammelt von Herrn Hans Leder bei Urga in der nördlichen Mongolei. *Wiener Entomlogische Zeitung* **14**: 280-286.

Reitter E. 1896a: Beschreibungen neuer Coleopteren aus dem russischen Reich. *Deutsche Entomologische Zeitschrift* **1896**: 33-48.

Reitter E. 1896b: Dreizehnter Beitrag zur Coleopteren-Fauna des russischen Reiches. *Wiener Entomologische Zeitung* **15**: 64-77.

Reitter E. 1896c: Dichotomische Uebersicht der bekannten Arten der Coleopteren Gattung Capnisa Lac. = Gnathosia Fisch. *Entomologische Nachrichten* **22**: 129-135.

Reitter E. 1896d: Uebersicht der mir bekannten, palaearctischen Arten der Coleopteren-Gattung Crypticus Latreille. *Entomologische Nachrichten* **22**: 145-151.

Reitter E. 1896e: Uebersicht der Arten der Coleopteren-Gattung Heterophylus Muls. *Deutsche Entomologische Zeitschrift* **1896**: 159-160.

Reitter E. 1896f: Uebersicht der mir bekannten, mit Penthicus Fald. verwandten Coleopteren-Gattungen und Arten aus der paläarktischen Fauna. *Deutsche Entomologische Zeitschrift* **1896**: 161-172.

Reitter E. 1896g: Beitrag zur Kenntnis der Platysceliden. *Deutsche Entomologische Zeitschrift* **1896**: 173-176.

Reitter E. 1896h: Namensänderung. *Wiener Entomologische Zeitung* **15**: 191-192.

Reitter E. 1896i: Abbildungen mit Beschreibungen zehn neuer Coleopteren aus der palaearctischen Fauna. *Wiener Entomologische Zeitung* **15**: 233-240.

Reitter E. 1896j: Achter Beitrag zur Coleopteren-Fauna von Europa und den angrenzenden Ländern. *Wiener Entomologische Zeitung* **15**: 265-272.

Reitter E. 1896k: Vierzehnter Beitrag zur Coleopteren-Fauna des russisches Reiches. *Wiener Entomologische Zeitung* **15**: 285-291.

Reitter E. 1897a: Uebersicht der Arten der Coleopteren-Gattung Asclera Schm. In: *Fest-Schrift zur Feier des Fünfzigjährigen Bestehens des Vereins für Schlesische Insektenkunde in Breslau 1847-1897*. Breslau: Maruschke & Berendt, 141 pp.

Reitter E. 1897b: Ueber die Arten der Coleopteren-Gattung Anemia Cast., aus der palaearctischen Fauna. In: *Fest-Schrift zur Feier des Fünfzigjährigen Bestehens des Vereins für Schlesische Insektenkunde in Breslau 1847-1897*. Breslau: Maruschke & Berendt. 141 pp.

Reitter E. 1897c: [new taxon]. In: Escherich K.: Revision der palaearktischen Zonitiden, einer Unterfamilie der Meloiden. *Verhandlungen des Naturforschenden Vereines in Brünn* **35** [1896]: 96-133.

Reitter E. 1897d: Fünfzehnter Beitrag zur Coleopteren-fauna des russischen Reiches. *Wiener Entomologische Zeitung* **16**: 121-127.

Reitter E. 1897e: Dreissig neue Coleopteren aus russisch Asien und der Mongolei. *Deutsche Entomologische Zeitschrift* **1897**: 209-228.

Reitter E. 1897f: Analytische Revision der Coleopteren-Gattung Microdera Esch. *Deutsche Entomologische Zeitschrift* **1897**: 229-235.

Reitter E. 1897g: Dircaea subtilis n. sp. *Deutsche Entomologische Zeitschrift* **23**: 246.

Reitter E. 1897h: Dichotomische Uebersicht der mir bekannten Gattungen aus der Tenebrioniden-Abtheilung: Tentyrini. *Deutsche Entomologische Zeitschrift* **1896**: 297-303.

Reitter E. 1897i: Uebersicht der Arten der Coleopteren-Gattung Calyptopsis Sol. *Deutsche Entomologische Zeitschrift* **1896**: 305-311.

Reitter E. 1897j: Psammocryptus rugiceps Reitter n. sp. aus Turkestan. *Deutsche Entomologische Zeitschrift* **1896**: 311.

Reitter E. 1897k: Uebersicht der mir bekannten Arten der Coleopteren-Gattung: Dailognatha Stev. *Deutsche Entomologische Zeitschrift* **1896**: 313-317.

Reitter E. 1898a: Eine Decade neuer Coleopteren aus der Buchara. *Wiener Entomologische Zeitung* **17**: 10-16.

Reitter E. 1898b: Uebersicht der bekannten Arten der Coleopterengattung Scleropatrum Seidl. aus der palaearctischen Fauna. *Wiener Entomologische Zeitung* **17**: 36-39.

Reitter E. 1898c: Drei neue Mycetocharina-Arten. (Coleoptera, Alleculidae). *Entomologische Nachrichten* **24**: 65-66.

Reitter E. 1898d: Über die Gattungen: Apolites Duv. und Anisocerus Fald. (Coleopt. Tenebrionidae). *Wiener Entomologische Zeitung* **17**: 105-106.

Reitter E. 1898e: Siebzehnter Beitrag zur Coleopteren-Fauna des russischen Reiches. *Wiener Entomologische Zeitung* **17**: 109-127.

Reitter E. 1898f: Neue Coleopteren aus Europa und den angrenzenden Ländern. *Deutsche Entomologische Zeitschrift* **1898**: 337-360.

Reitter E. 1899a: Uebersicht der mir bekannten, palaearctischen Arten der Coleopteren-Gattung Leichenum Blanch. *Entomologische Nachrichten* **25**: 83-86.

Reitter E. 1899b: Zur weiteren Kenntniss der Coleopteren-Gattung Mycetochara Berthold. *Entomologische Nachrichten* **25**: 155-159.

Reitter E. 1899c: Elfter Beitrag zur Coleopteren-Fauna von Europa und den angrenzenden Ländern. *Wiener Entomologische Zeitung* **18**: 155-161.

Reitter E. 1899d: Beitrag zur Coleopteren-Fauna des russischen Reiches und der angrenzenden Ländern. *Deutsche Entomologische Zeitschrift* **1899**: 193-209.

Reitter E. 1899e: Ueber einige Coleopteren aus der palaearctischen Fauna und aus Japan. *Entomologische Nachrichten* **25**: 216-220.

Reitter E. 1900a: [new taxa]. In: Bodemeyer E. von.: *Quer durch Klein-Asien in den Bulghar-Dagh. Eine naturwissenschaftliche Studien-Reise*. Emmendingen: Verlangs-Aktiengesellschaft vormals Dölter, v + 169 pp.

Reitter E. 1900b: Beitrag zur Coleopteren-Fauna des russischen Reiches. *Deutsche Entomologische Zeitschrift* **1900**: 49-59.

Reitter E. 1900c: Beitrag zur Coleopteren-Fauna von Europa und den angrenzenden Ländern. *Deutsche Entomologische Zeitschrift* **1900**: 81-88.

Reitter E. 1900d: *Bestimmungs-Tabelle der Tenebrioniden-Abtheilungen: Tentyrini und Adelostomini aus Europa und den angrenzenden Ländern*. Paskau: Edmund Reitter, [1] + 82-197. [note: same issued in 1901 in *Verhandlungen des Naturforschendes Vereines in Brünn* **39** [1900]: 82-197.

Reitter E. 1900e: Coleoptera, gesammelt im Jahre 1898 in chinesisch Central-Asien von Dr. Holderer in Lahr. *Wiener Entomologische Zeitung* **19**: 152-166.

Reitter E. 1900f: Neue, von Herrn Dr. John Sahlberg auf seinen Reisen in Corfu, Palästina und Centralasien gesammelte Coleopteren. *Wiener Entomologische Zeitung* **19**: 217-220.

Reitter E. 1900g: Neue Coleopteren aus Europa, den angrenzenden Ländern, russisch Asien und der Mongolei. *Deutsche Entomologische Zeitschrift* **1899**: 273-281.

Reitter E. 1900h: Weitere Beiträge zur Kenntniss der Coleopteren-Gattung Laena Latr. *Deutsche Entomologische Zeitschrift* **1899**: 282-286.

Reitter E. 1900i: Uebersicht der mir bekannten Arten der Coleopteren-Gattung Dila Fisch. *Entomologische Nachrichten* **26**: 295-296.

Reitter E. 1900j: Uebersicht der mit Erodius verwandten palaearktischen Coleopteren-Genera. *Entomologische Nachrichten* **26**: 298-300.

Reitter E. 1900k: Uebersicht der bekannten Arten der Coleopteren-Gattung Ammozoum Sem. *Entomologische Nachrichten* **26**: 300-301.

Reitter E. 1900l: Uebersicht der bekannten zwei Arten der neuen Coleopteren-Gattung Diaphanidius Reitt. *Entomologische Nachrichten* **26**: 301-302.

Reitter E. 1901a: *Byrrhidae (Anobiidae) und Cioidae. Bestimmungs-Tabelle der europäischen Coleopteren. XLVII.* (*Heft 47*). Brünn: Edm. Reitter, 64 pp. [separate issue of Reitter E. 1902: Analytische Uebersicht der palaearctischen Gattungen und Arten der Coleopteren-Familien: Byrrhidae (Anobiidae) und Cioidae. *Verhandlungen des Naturforschenden Vereines in Brünn* **40** [1901]: 3-64].

Reitter E. 1901b: Ueber die turkestanische Arten der Coleopteren-Gattung Laena Latr. *Wiener Entomologische Zeitung* **20**: 61-63.

Reitter E. 1901c: Weitere Beiträge zur Coleopteren-Fauna des russischen Reiches. *Deutsche Entomologische Zeitschrift* **1901**: 65-84.

Reitter E. 1901d: Uebersicht der Coleopteren-Gattung Pedilus Fisch. der palaearctischen Fauna. *Wiener Entomologische Zeitung* **20**: 114-116.

Reitter E. 1901e: Abbildungen und Beschreibungen neuer Coleopteren aus der palaearctischen Fauna. *Wiener Entomologische Zeitung* **20**: 157-164.

Reitter E. 1902a: Neue und seltene Coleopteren, gesammelt im Jahre 1901, in der Herzegowina, in Dalmatien und Bosnien. *Wiener Entomologische Zeitung* **21**: 1-9, pl. 1.

Reitter E. 1902b: Eine Serie neuer Coleopteren aus dem russischen Reiche. *Deutsche Entomologische Zeitschrift* **1901**: 177-186.

Reitter E. 1902c: Beschreibung neuer Coleopteren aus dem russischen Reiche. *Wiener Entomologische Zeitung* **21**: 185-190.

Reitter E. 1902d: Dichotomische Übersicht der Coleopteren-Gattung Pterocoma Sol. *Deutsche Entomologische Zeitschrift* **1901**: 189-192.

Reitter E. 1902e: Fünfzehnter Beitrag zur Coleopteren-Fauna von Europa und den angrenzenden Ländern. *Wiener Entomologische Zeitung* **21**: 191-196.

Reitter E. 1902f: Coleopterologische Studien. *Wiener Entomologische Zeitung* **21**: 203-217.

Reitter E. 1902g: Verschiedenes über die Coleopteren der Tenebrioniden-Abtheilung Helopina. *Deutsche Entomologische Zeitschrift* **1901**: 209-224.

Reitter E. 1902h: Coleopterologische Notizen. *Wiener Entomologische Zeitung* **21**: 221-222.

Reitter E. 1903a: Uebersicht der Arten der Coleopteren-Gattung Entomogonus Sol. *Wiener Entomologische Zeitung* **22**: 18-20.

Reitter E. 1903b: Sechzehnter Beitrag zur Coleopteren-Fauna von Europa und den angrenzenden Ländern. *Wiener Entomologische Zeitung* **22**: 43-46.

Reitter E. 1903c: Die Arten der Gattung Cephalostenus Sol. *Wiener Entomologische Zeitung* **22**: 132.

Reitter E. 1903d: Siebzehnter Beitrag zur Coleopterenfauna von Europa und den angrenzenden Ländern. *Wiener Entomologische Zeitung* **22**: 173-178.

Reitter E. 1903e: Neue, von Hauptmann E. von Bodemeyer im Jahre 1903 in Kleinasien gesammelte Coleopteren, vorzüglich Pselaphiden und Scydmaeniden. *Wiener Entomologische Zeitung* **22**: 221-224.

Reitter E. 1903f: Übersicht der mit Zonabris variabilis Pall. zunächst verwandten Coleopteren-Arten. *Wiener Entomologische Zeitung* **22**: 230.

Reitter E. 1903g: Übersicht der palaearctischen Arten der Coleopteren-Gattung Evaniocera Guer. *Wiener Entomologische Zeitung* **22**: 237-238.

Reitter E. 1903h: Zwei neue Ctenioplus-Arten aus Kleiasien. *Wiener Entomologische Zeitung* **22**: 257-258.

Reitter E. 1904a: Bestimmungs-Tabelle der Tenebrioniden-Unterfamilien: Lachnogyini, Akidini, Pedinini, Opatrini und Trachyscelini aus Europa und den angrenzenden Ländern. *Verhandlungen des Naturforschenden Vereines in Brünn* **42** [1903]: 25-189.

Reitter E. 1904b: Una nuova varietà della Akis spinosa L. *Il Naturalista Siciliano* **17**: 97.

Reitter E. 1904c: Übersicht über die mit M. piceus F. zunächst verwandten Mycetophagus-Arten aus Europa und den angrenzenden Ländern. *Wiener Entomologische Zeitung* **23**: 165-166.

Reitter E. 1904d: Sechs neue Coleopteren aus der palaearktischen Region. *Wiener Entomologische Zeitung* **23**: 255-258.

Reitter E. 1905a: Acht neue Coleopteren aus der palaearktischen Fauna. *Wiener Entomologische Zeitung* **24**: 90-94.

Reitter E. 1905b: Übersicht der mir bekannten Arten der Coleopteren-Gattung Epicauta Redtb. aus der palaearktischen Fauna. *Wiener Entomologische Zeitung* **24**: 194-196.

Reitter E. 1905c: Neun neue Coleopteren aus der palaearktischen Fauna. *Wiener Entomologische Zeitung* **24**: 201-206.

Reitter E. 1905d: Sechzehn neue Coleopteren aus der palaearktischen Fauna. *Wiener Entomologische Zeitung* **24**: 241-251.

Reitter E. 1905e: [new taxa]. In: Ganglbauer L.: Coleoptera. Ergebnisse einer naturwissenschaftlichen Reise zum Erdschias-Dagh (Kleinasien). Ausgeführt von Dr. Arnold Penther und Dr. Emerich Zederbauer auf Kosten der "Gesellschaft zur Förderung des naturhistotrischen Erforschung des Orients in Wien" im Jahre 1902. *Annalen des k. k. Naturhistorischen Hofmuseums* **20**: 246-290.

Reitter E. 1905f: Drei neue Coleopteren aus der palaearktischen Fauna. *Wiener Entomologische Zeitung* **24**: 311-312.

Reitter E. 1906a: [new names]. In: Heyden L. F. J. D., Reitter E. & Weise J: *Catalogus Coleopterorum Europae, Caucasi et Armeniae Rossicae. Editio secunda.* Berlin, Paskau, Caen, 4 pp. + 774 columns.

Reitter E. 1906b: Neue Coleopteren aus der paläarktischen Fauna. *Wiener Entomologische Zeitung* **25**: 31-37.

Reitter E. 1906c: Drei neue im Quellgebiet des Indus von Professor Dr. Koken gesammlte Coleopteren. *Wiener Entomologische Zeitung* **25**: 40-42.

Reitter E. 1906d: Uebersicht der Coleopteren-Unterfamilie: Omophlini der Alleculidae aus Europa und den angrenzenden Ländern. *Verhandlungen des Naturforschenden Vereins in Brünn* **44**: 115-175.

Reitter E. 1906e: Vier neue von Herrn Paganetti-Hummler in Calabrien gesammelte Coleopteren. *Wiener Entomologische Zeitung* **25**: 129-130.

Reitter E. 1906f: Dreizehn neue Coleopteren aus der palaearktischen Fauna. *Wiener Entomologische Zeitung* **25**: 237-244.

Reitter E. 1906g: Neue Coleopteren aus Spanien. *Boletín de la Real Sociedad Española de Historia Natural* **6**: 375-380.

Reitter E. 1906h: Neue Uebersicht der mir bekannten Arten der Coleopterengattung Laena Latr. aus Turkestan und Kaschmir, nebst Beschreibung von 2 neuen Arten aus Osteuropa. *Deutsche Entomologische Zeitschrift* **1906**: 442-448.

Reitter E. 1907a: Coleopterologische Notizen. *Wiener Entomologische Zeitung* **26**: 27-30.

Reitter E. 1907b: Nachträge zur Bestimmungtabelle der unechten Pimeliden aus der palaearktischen Fauna. *Wiener Entomologische Zeitung* **26**: 81-92.

Reitter E. 1907c: Ein neuer Microtelus (Col.) aus Aegypten. *Wiener Entomologische Zeitung* **26**: 115.

Reitter E. 1907d: Nacerdes melanura var. nov. Zoufali. *Societatis Entomologica* **21**: 161.

Reitter E. 1907e: Parablops subchalybaeus n. sp. *Wiener Entomologische Zeitung* **26**: 204.

Reitter E. 1907f: Sechs neue Coleopteren aus Turkestan. *Wiener Entomologische Zeitung* **26**: 205-208.

Reitter E. 1907g: Zwei neue Meloë aus der palaearktischen Fauna. *Wiener Entomologische Zeitung* **26**: 214-215.

Reitter E. 1907h: Übersicht der Arten der Pimeliden-Gattung Podhomala Sol. (Col.). *Deutsche Entomologische Zeitschrift* **1907**: 412-414.

Reitter E. 1907i: Einige neue paläarktische Coleopteren. *Deutsche Entomologische Zeitschift* **51**: 484-487.

Reitter E. 1908a: Siebzehn Coleopteren-Neuheiten aus der palaearktischen Fauna. *Wiener Entomologische Zeitung* **27**: 17-24.

Reitter E. 1908b: Übersicht der Entomogonus-Aten aus der Verwandschaft des E. Barthelemyi Sol. *Wiener Entomologische Zeitung* **27**: 30.

Reitter E. 1908c: Beschreibung einiger neuen Käferarten von Egypten. *Bulletin de la Société Entomologique d'Egypte* **1** [1908-1909]: 39-40, 41-56.

Reitter E. 1908d: Omophlus subcostatus (♀) n. sp. *Wiener Entomologische Zeitung* **27**: 56.

Reitter E. 1908e: Neue paläarktische Käfer. *Entomologische Blätter* **4**: 129.

Reitter E. 1908f: Zwei neue Coleopteren aus Adana in Kleinasien. *Wiener Entomologische Zeitung* **26**: 133-136.

Reitter E. 1908g: Sieben neue Coleopteren aus Europa und den angrenzenden Ländern. *Wiener Entomologische Zeitung* **27**: 213-216.

Reitter E. 1908h: Übersicht über die Arten der Gattung Boromorphus Woll. mit Ausschluss der Spezies von Madeira. *Wiener Entomologische Zeitung* **27**: 217.

Reitter E. 1908i: Acht neue Coleopteren aus Europa, Kleinasien und Turkestan. *Wiener Entomologische Zeitung* **27**: 243-247.

Reitter E. 1908j: Eine auffällige Zonabris-Art aus der Herzegowina. *Wiener Entomologische Zeitung* **27**: 272.

Reitter E. 1908k: Die Arten der Gattung Laena Latr. (Tenebrionidae) aus dem westlichen Himalaya, gesammelt auf seinen Reisen von Herrn Karl Rost (Col.). *Deutsche Entomologische Zeitschrift* **1908**: 275-280.

Reitter E. 1909a: Conspectus specierum genere Tentyrina Reitt. (Tenebrionidae). *Bulletin de la Société Entomologique d'Egypte* **2**: 28-29.

Reitter E. 1909b: Übersicht der Arten der Coleopterengattung Hypulus Payk. *Entomologische Blätter* **5**: 51.

Reitter E. 1909c: Neun neue Coleopterenarten und Varietäten aus der paläarktischen Fauna. *Wiener Entomologische Zeitung* **27**: 99-103.

Reitter E. 1909d: Neue Revision der Arten der Coleopterengattung Prosodes Esch. *Wiener Entomologische Zeitung* **28**: 113-168.

Reitter E. 1909e: Dila leptoscelis n. sp. *Entomologische Blätter* **5**: 239.

Reitter E. 1909f: Solskia concavidorsis n. sp. *Entomologische Blätter* **5**: 296.

Reitter E. 1909g: Neue Coleopteren aus der paläarktischen Fauna. *Wiener Entomologische Zeitung* **28**: 303-312.

Reitter E. 1910a: Eine neue Gattung der Coleopterenfamilie der Tenebrionidae, zugleich ein Vertreter einer neuen Tribus bei den Lachnogiini. *Entomologische Blätter* 6: 20-21.

Reitter E. 1910b: Due nuovi coleotteri di Sardegna. *Rivista Coleotterologica Italiana* **8**: 115-116.

Reitter E. 1910c: Siagonadia, nov. subgen. von Macrosiagon. *Wiener Entomologische Zeitung* **29**: 131.

Reitter E. 1910d: Psilachnopus Aharonii. *Wiener Entomologische Zeitung* **29**: 197.

Reitter E. 1911a: *Fauna Germanica. Die Käfer des Deutschen Reiches. Nach der analytischen Methode bearbeitet. III. Band.* Stuttgart: K. G. Lutz, 436 pp., pls. 81-128.

Reitter E. 1911b: Paläarktische Coleopterennovitäten. *Wiener Entomologische Zeitung* **30**: 47-55.

Reitter E. 1911c: Vier neue Koleopteren-Arten aus der palaearctischen Fauna. *Wiener Entomologische Zeitung* **30**: 234-236.

Reitter E. 1911d: Übersicht der palaearctischen Alphitophagus-Arten (Col., Tenebrionidae). *Wiener Entomologische Zeitung* **30**: 268.

Reitter E. 1911e: Zwei neue Coleopteren-Varietäten. *Wiener Entomologische Zeitung* **30**: 291.

Reitter E. 1912a: Über die Myiodes-Arten aus der paläarktischen Fauna. *Wiener Entomologische Zeitung* **31**: 43.

Reitter E. 1912b: Rhinosimus (Cariderus) tapiroides n. sp. *Entomologische Blätter* **8**: 196-197.

Reitter E. 1912c: Übersicht der bekannten Arten der Coleopteren-Gattung Langelandia Aubé. *Wiener Entomologische Zeitung* **31**: 279-285.

Reitter E. 1912d: Tenebrio molitor L. subsp. nov. Codinae. *Wiener Entomologische Zeitung* **31**: 310.

Reitter E. 1913a: Zwei neue Coleopteren von der Balkan-Halbinsel. *Entomologische Blätter* **9**: 26-27.

Reitter E. 1913b: Coleopterologische Novitäten der palaearktischen Fauna. *Coleopterologische Rundschau* **2**: 121-125.

Reitter E. 1913c: Neue paläarktische Coleopteren. *Entomologische Blätter* **9**: 289-292.

Reitter E. 1913d: Eine Serie neuer Coleopteren aus der palaearktischen Fauna. *Deutsche Entomologische Zeitschrift* **1913**: 649-666.

Reitter E. 1914a: Die Microtelus-Arten aus der Verwandtschaft des M. careniceps Rche. (Col. Tenebrionidae). *Entomologische Blätter* **10**: 13-14.

Reitter E. 1914b: Bestimmungs-Tabelle für die Unterfamilie: Erodini der Tenebrionidae aus Europa und den angrenzenden Ländern. *Deutsche Entomologische Zeitschrift* **1914**: 43-85.

Reitter E. 1914c: Schellia nov. gen. Oedemeridarum. *Entomologische Blätter* **10**: 49-50.

Reitter E. 1914d: Neue Übersicht der palaearktischen Alphitobius-Arten. *Coleopterologische Rundschau* **3**: 81-82.

Reitter E. 1914e: Sechs neue Arten der Coleopteren-Gattung Hedyphanes Fischer. *Berliner Entomologische Zeitschrift* **58** [1913]: 184-187.

Reitter E. 1914f: Zweite Übersicht der Arten der Gattung Cerocoma Geoffr. (Col. Meloidae.). *Berliner Entomologische Zeitschrift* **58** [1913]: 190-193.

Reitter E. 1914g: Bestimmungs-Tabelle der Tenebrioniden-Abteilung der Scaurini. *Deutsche Entomologische Zeitschrift* **1914**: 369-380.

Reitter E. 1914h: Bestimmungs-Tabelle der Tenebrioniden-Abteilung der Sepidiini. *Deutsche Entomologische Zeitschrift* **1914**: 381-392.

Reitter E. 1915a: Bestimmungs-Tabelle der echten Pimeliiden aus der paläarktischen Fauna. *Wiener Entomologische Zeitung* **34**: 1-63.

Reitter E. 1915b: Über die Arten der Coleopterengattung Gnathosia Firch. (Capnisa Lac.) mit ganz gerandeter Basis der Flügeldecken. *Coleopterologische Rundschau* **4**: 60-62.

Reitter E. 1915c: Cis (Eridaulus) Mell. v. Matchanus nov. *Wiener Entomologische Zeitung* **34**: 66.

Reitter E. 1915d: Dichotomische Übersicht der Tenebrinoden-Gattung Scythis Schaum. *Entomologische Blätter* **11**: 66-72.

Reitter E. 1915e: Eine Serie neuer Coleopteren aus der Familie der Tenebrionidae der paläarktischen Fauna. *Wiener Entomologische Zeitung* **34**: 83-93.

Reitter E. 1915f: Neue Arten aus der Tenebrioniden-Gattung Microdera Eschsch. *Entomologische Blätter* **11**: 92-96.

Reitter E. 1915g: Coleopterologische Notizen. *Wiener Entomologische Zeitung* **34**: 124.

Reitter E. 1915h: Die Arten der Tenebrioniden-Gattung Kokeniella Reitt. *Wiener Entomologische Zeitung* **34**: 127-128.

Reitter E. 1915i: Über Arthrodosis-Arten aus der Buchara. *Wiener Entomologische Zeitung* **34**: 130-131.

Reitter E. 1916a: Bestimmungstabelle der Arten der Gattung Adesmia Fisch. aus der paläarktischen Fauna (Col. Tenebrionidae). *Wiener Entomologische Zeitung* **35**: 1-31.

Reitter E. 1916b: Bestimmungstabelle der Tenebrioniden-Gruppe der Phaleriini, aus der palaearktischen Fauna. *Entomologische Blätter* **12**: 3-10.

Reitter E. 1916c: Bestimmungstabelle der Tenebrioniden-Unterfamilie Zophosini aus der paläarktischen Fauna. *Wiener Entomologische Zeitung* **35**: 81-99.

Reitter E. 1916d: Bestimmungstabelle der Tenebrioniden, enthaltend die Zopherini, Elenophorini, Leptodini, Stenosini und Lachnogyini aus der paläarktischen Fauna. *Wiener Entomologische Zeitung* **35**: 129-171.

Reitter E. 1916e: Bestimmungs-Tabelle der Tenebrioniden-Abteilung der palaearctischen Epitragini. *Entomologische Blätter* **12**: 139-149.

Reitter E. 1916f: Lederia seidlitzi n. sp. *Wiener Entomologische Zeitung* **35**: 290.

Reitter E. 1916g: Coleopterologische Notizen. *Wiener Entomologische Zeitung* **35**: 294.

Reitter E. 1917a: Bestimmungs-Tabelle der palaearctischen Arten der Tenebrioniden-Abteilung Asidini. *Verhandlungen des Naturforschenden Vereins in Brünn* **55**: 1-74.

Reitter E. 1917b: Bestimmungstabelle der Cossyphini und Misolampini (Tribus der Tenebrionidae). *Wiener Entomologische Zeitung* **36**: 129-150.

Reitter E. 1918: Vier neue Coleopteren aus Albanien. *Wiener Entomologische Zeitung* **37**: 153-155.

Reitter E. 1920a: *Bestimmungstabelle der Unterfamilien: Belopinae, Borinae, Tenebrioninae und Coelometopinae der Tenebrioniden (Col.). Bestimmungs-Tabellen der europäischen Coleopteren 87*. Paskau: E. Reitter, 24 pp.

Reitter E. 1920b: Coleopterologische Notizen. *Entomologische Blätter* **15** [1919]: 220-221.

Reitter E. 1922a: *Bestimmungs-Tabellen der europaeischen Coleopteren. VI. Heft. Enhaltend die Familien: Colydiidae, Rhysodidae, Ostomidae. Zweite, gänzlich umgearbeitete und auf die palaearktische Fauna ausgedehnte Auflage*. Troppau: E. Reitter, 73 pp.

Reitter E. 1922b: Bestimmungstabelle der palaearktischen Helopinae (Col. Tenebrionidae). *Wiener Entomologische Zeitung* **39**: 1-44, 113-171.

Ren G.-D. 1992: Two new species of genus Melanesthes Lacord. from desert of N. W. China. *Zoological Research* **13**: 229-332 (in Chinese with English summary).

Ren G.-D. 1993a: [new taxa]. In: Ren G.-D., Yu Y., Ma F. & Jiang H.: Study on the genus Melanesthes Lacord. and description of its three new species and five new records species [sic] (Coleoptera: Tenebrionidae) in China. *Journal of the Ningxia Agricultural College* **14** (Supplement): 24-33 (in Chinese with English summary).

Ren G.-D. 1993b: Three new species of the genus Melanesthes Lacord. from China (Coleoptera: Tenebrionidae). *Acta Entomologica Sinica* **36**: 486-489 (in Chinese with English summary).

Ren G.-D. 1995a: Notes on Chinese species of Caedius Mulsant et Rey (Coleoptera: Tenebrionidae) with description of a new species. *Entomotaxonomia* **17** (Supplement): 22-24.

Ren G.-D. 1995b: Coleoptera: Tenebrionidae. Pp. 233-236. In: Wu H. (ed.): *Insects of Baishanzu Mountain,*

Eastern China. China Forestry Publishing House, xiii + 586 pp. (in Chinese with English summary).

Ren G.-D. 1998: Coleoptera: Tenebrionidae. Pp. 108-114. In: Wu H. (ed.): *Insects of Longwangshan Nature Reserve*. Beijing: China Forestry Publishing house, ix + 404 pp.

Ren G.-D. 1999: [new taxa]. In: Ren G.-D. & Yu Y.: *The darkling beetles from deserts and semideserts of China*. Hebei University Publishing House, [3] + 395 pp. (in Chinese with English summary).

Ren G.-D. & Dong S. 2001: The beetles of the tribe Platyopini in China and description of two new species (Coleoptera: Tenebrionidae). *Journal of Hebei University* (*Natural Science Edition*) **21**: 296-302.

Ren G.-D. & Hua H.-R. 2006: Tenebrionidae. Pp. 265-275. In: *Fanjingshan jingguan kunchong*. Guiyang: Guizhou Science and Technology Publishing House (in Chinese with English summary).

Ren G.-D. & Li Z. 2001: A new genus and two new species of the tribe Blaptini in China (Coleoptera: Tenebrionidae). *Zoological Research* **4**: 310-314.

Ren G.-D. & Liu S.-S. 2004: Six new species of the genus Uloma from Gaoligong Mountain in China (Coleoptera, Tenebrionidae). *Acta Zootaxonomica Sinica* **29**: 294-304.

Ren G.-D. & Luo Z. 1995: One new record genus [sic] and species of Blaptini (Coleoptera: Tenebrionidae) from China. *Entomotaxonomia* **17**: 27-30 (in Chinese with English summary).

Ren G.-D. & Meng L. 1995: Fauna of the Tenebrionidae (Coleoptera) from Xingjiang: I. *Acta Agriculturae Boreali-Occidentalis Sinica* **4** (3): 52-57 (in Chinese with English summary).

Ren G.-D. & Shi A.-M. 2006a: A taxonomic review of the Chinese species of the genus Ascelosodes Redtenbacher (Coleoptera: Tenebrionidae), with descriptions of three new species. *Zootaxa* **1228**: 1-24.

Ren G.-D. & Shi A.-M. 2006b: Two new record [sic] genera of Stenosini (Coleoptera, Tenebrionidae) from China, with descriptions of two new species. *Acta Zootaxonomica Sinica* **31**: 180-184.

Ren G.-D. & Wang X.-P. 1993a: Three new species of the genus Adesmia Fisch. from China (Coleoptera, Tenebrionidae, Adesmini). *Journal of the Ningxia Agricultural College* **14** (*Supplement*): 1-8 (in Chinese with English summary).

Ren G.-D. & Wang X.-P. 1993b: In: Ren G.-D., Wang X.-P. & Ma F.: Two new species and one new record of Platyope F.-W. from China. (Coleoptera: Tenebrionidae). *Journal of the Ningxia Agricultural College* **14** (Supplement): 44-49 (in Chinese with English summary).

Ren G.-D. & Wang X.-P. 2001: Eight new species of the genus Blaps Fabricius (Coleoptera: Tenebrionidae: Blaptini) of China. *Entomotaxonomia* **23**: 15-27 (in Chinese with English summary).

Ren G.-D., Wang X.-P. & Yu Y.-Z. 2000: Nine records and one new species of the tribe Blaptini from China (Coleoptera: Tenebrionidae). *Journal of Hebei University* (*Natural Science Edition*) **20**: Supplement: 26-33.

Ren G.-D. & Wu 1994: A new species of Belopus Gebin [sic] from China (Coleoptera: Tenebrionidae). *Acta Zootaxonomica Sinica* **19**: 351-353 (in Chinese with English summary).

Ren G.-D. & Yin H. 2004: Tenebrionidae. In: Yang X.-K. (ed.): *Insects of the Great Yarlung Zangbo Canyon of Xizang*. Beijing: China Science and Technology Press (in Chinese with English summary).

Ren G.-D. & Yu Y.-Z. 2000: On the darkling beetle tribe Akidini from China (Coleoptera: Tenebrionidae). *Acta Zootaxonomica Sinica* **25**: 320-329 (in Chinese with English summary).

Ren G.-D. & Yuan C.-X. 2005: Taxonomic study on the arboreal darkling beetles of the genus Ainu Lewis, 1894 from China (Coleoptera, Tenebrionidae). *Acta Zootaxonomica Sinica* **30**: 98-103.

Ren G.-D. & Zheng Z. 1993a: Six new species of the genus Crypticus Latr. from northwest of China (Coleoptera: Tenebrionidae). *Journal of the Ningxia Agricultural College* **14** (Supplement): 9-23 (in Chinese with English summary).

Ren G.-D. & Zheng Z. 1993b: Six new species of the genus Scytosoma Fisch. 1895 (Coleoptera, Tenebrionidae, Tentyriini). *Journal of the Ningxia Agricultural College* **14** (Supplement): 34-43 (in Chinese with English summary).

Ren G.-D. & Zheng Z. 1993c: Four new species of the tribe Epitragini from desert in Northwest of China (Coleoptera, Tenebrionidae). *Journal of the Ningxia Agricultural College* **14** (Supplement): 50-58 (in Chinese with English summary).

Ren G.-D. & Zheng Z. 1993d: One new species of genus Belopus GB. from Inner Mongolia (Coleoptera: Tenebrionidae: Belopini). *Journal Inner Mongolia Normal University* (*Natural Science Edition*) **4**: 52-54. (in Chinese with English summary).

Ren G.-D. & Zheng Z. 1993e: One new record tribe [sic] and genus from China with one new species of Phthora (Coleoptera: Tenebrionidae). *Journal of Ningxia University* (*Natural Science Edition*) **14** (4): 77-80 (in Chinese with English summary).

Rey C. 1857: [new taxon]. In: Mulsant E. & Rey C.: Description de quelques coléoptères de la tribu des longipèdes. *Annales de la Société Linnéenne de Lyon* (2) **4**: 410-413.
Rey C. 1889: Remarques en passant. *L'Échange, Revue Linnéenne* **5**: 23, 55.
Rey C. 1890: Note sur le genre Phaleria. *Revue d'Entomologie* **9**: 327-329.
Rey C. 1891a: Nouvelle note sur le genre Phaleria. *Revue d'Entomologie* **10**: 83-86.
Rey C. 1891b: Troixième note sur le genre Phaleria et description d'une espèce nouvelle ou peu connue. *Revue d'Entomologie* **10**: 236-239.
Rey C. 1892: Remarques en passant. *L'Échange, Revue Linnéenne* **8**: 2-3, 17, 30-31, 41-42, 53-54, 65, 77, 89-90, 101, 113-114, 137-138.
Reymond A. 1933: Description d'une espèce nouvelle de Meloë paléarctique. *Bulletin de la Société Entomologique de France* **38**: 14-16.
Reymond A. 1937: Un nouvel Earophanta d'Asie Centrale (Col. Tenebrionidae). *Bulletin de la Société Entomologique de France* **42**: 247-248.
Reymond A. 1938: Résultats scientifiques d'un voyage en Asie centrale (Mission Haardt – Audoin-Dubreuil) 1931. Sur le peuplement entomologique de l'Asie centrale. *Revue de Géographie Physique* **10**: 1-285, 4 maps, 20 pls.
Reymond A. 1956a: Présentation d'insectes (Coléoptères) nouveaux de la faune marocaine. *Comptes Rendus de la Société des Sciences Naturelles et Physiques du Maroc* **22**: 40-42.
Reymond A. 1956b: Description d'un nouveau coléoptère troglophile du Moyen-Atlas septentrional: Scaurus harbuloti nov. sp. (Col. Tenebrionidae). *Bulletin de la Société des Sciences Naturelles et Physiques du Maroc* **36**: 265-269.
Richter A. A. 1945: Novyy vid Scaurus iz doliny Araksa (Coleoptera, Tenebrionidae). *Doklady Akademii Nauk Armyanskoy SSR* **3**: 103-106.
Richter A. A. 1949: Novyy paraziticheskyj zhuk iz Zakavkazia (Coleoptera, Rhipiphoridae). *Entomologicheskoe Obozrenie* **30**: 359-361.
Říha P. 1958: Beitrag zur Kenntnis der Colydiiden von Adsharistan nebst einer Monographie der Gattung Dechomus Jacq. du Val (Coleoptera, Colydiidae). *Acta Entomologica Musei Nationalis Pragae* **32**: 465-69.
Rochebrune A. T. de. 1883: [new taxon]. In: *Bulletin de la Société des Philomatique de Paris* (7) **7**: 182-188, pl. 3.
Rosenhauer W. G. 1847: *Beiträge zur Insekten-Fauna Europas. Erstes Bändchen; enthält die Beschreibung von sechzig neuen Käfern aus Bayern, Tyrol, Ungarn, etc., so wie die Käfer Tyrols, nach dem Ergebnisse von vier Reisen.* Erlangen: Theodor Blaesing, x + 160 pp., 1 pl.
Rosenhauer W. G. 1856: *Thiere Andalusiens nach dem Resultate einer Reise zusammengestellt, nebst den Beschreibungen von 249 neuen oder bis jetzt unbeschriebenen Gattungen und Arten.* Erlangen: Theodor Blaesing, viii + 429 pp., 3 pls.
Rossi P. 1790: *Fauna Etrusca sistens insecta quae in provinciis Florentina et Pisana praesertim collegit Petrus Rossius. Tomus Primus.* Liburni: T. Masi & Sociorum, xxii + 272 pp., 10 pls.
Rossi P. 1792: *Mantissa insectorum exhibens species nuper in Etruria collectas a Petro Rossio adiectis faunae Etruscae illustrationibus ac emendationibus.* [Tomus primus]. Pisis: Polloni, 148 pp.
Rossi P. 1794: *Mantissa insectorum exhibens species nuper in Etruria collectas a Petro Rossio adiectis faunae Etruscae illustrationibus ac emendationibus. Tomus secundus.* Pisis: Polloni, 154 pp., 8 pls.
Rotrou M. 1938: Première contribution à l'étude des Planasida du groupe lanceocollis Esc. *Bulletin de la Société des Sciences Naturelles du Maroc* **18**: 119.
Rotrou M. 1943: Description d'une nouvelle espèce de Sepidium de l'Algérie. *Bulletin de la Société d'Histoire Naturelle de l'Afrique du Nord* **34**: 235-236.
Rottenberg A. von. 1871: Beiträge zur Coleopteren-Fauna von Sicilien. *Berliner Entomologische Zeitschrift* **14** [1870]: 235-260, pl. 2.
Rottenberg A. von. 1873: Synonymische Bemerkungen 9. *Berliner Entomologische Zeitschrift* **17**: 213-217.
Roubal J. 1910: Beitrag zur Kenntnis der Mordelliden von Bulgarien. *Entomologische Blätter* **6**: 153-157.
Roubal J. 1911a: Neuheiten der palaearktischen Koleopterenfauna. *Wiener Entomologische Zeitung* **30**: 133-134.
Roubal J. 1911b: Neue Käfer der palaearktischen Fauna. *Wiener Entomologische Zeitung* **30**: 272.
Roubal J. 1912a: Koleopterologické výsledky mé cesty na Kavkaz v červenci 1910. Quid novi de Coleopterorum Caucasi ad occidentem vergentis fauna in meo itinere Julio mense anni 1910 suscepto cognoverim. V. *Časopis České Společnosti Entomologické* **9**: 25-28.

Roubal J. 1912b: Drei palaearktische Nova. *Coleopterologische Rundschau* **1**: 70-72.

Roubal J. 1912c: Deux nouveaux coléoptères de la région européenne. *Bulletin de la Société Entomologique de France* **1911**: 401-402.

Roubal J. 1914: Eine neue Zonabris aus dem Kaukasus. *Coleopterologische Rundschau* **3**: 145-146.

Roubal J. 1916: Cis ragusai sp. n. *Časopis České Společnosti Entomologické.* **13**: 89-90.

Roubal J. 1917a: Eine neue Oedemera. *Neue Beiträge zur Systematischen Insektenkunde* **1**: 44.

Roubal J. 1917b: Cteniopus expulsus nova species. *Archiv für Naturgeschichte* A **82** (4): 47-48.

Roubal J. 1919: Drei coleopterologische Nova. *Deutsche Entomologische Zeitschrift* **1919**: 63-64.

Roubal J. 1921: Description d´un Mordella nouveau de Carniole. *Bulletin de la Société Entomologique de France* **1921**: 248-249.

Roubal J. 1924: Zwei neue Scraptia-Arten aus der paläarktischen Region. *Wiener Entomologische Zeitung* **41**: 181-182.

Roubal J. 1926: Sieben neue Coleopteren aus der paläarktischen Region. *Coleopterologisches Centralblatt* **1**: 244-249.

Roubal J. 1927: Ein Beitrag zur Kenntnis der Coleopteren-Fauna der westrumänischen Ebene. *Coleopterologisches Centralblatt* **2**: 112-118.

Roubal J. 1929a: Vier neue Coleopteren aus SSSR. *Entomologische Blätter* **25**: 46-48.

Roubal J. 1929b: Coleoptera nova asiatica. *Bollettino della Società Entomologica Italiana* **59**: 96-98.

Roubal J. 1929c: Descriptiones duorum Coleopterorum novorum regionis palaearcticae. *Entomologische Blätter* **25**: 192-193.

Roubal J. 1930: Coleoptera nova ex Europa descripta. *Časopis Československé Společnosti Entomologické* **27**: 75-76.

Roubal J. 1931a: Beschreibungen von vier neuen Coleopteren. *Časopis Československé Společnosti Entomologické* **28**: 131-132.

Roubal J. 1931b: Coleoptera nova pro Bohemia. *Časopis Československé Společnosti Entomologické* **28**: 132.

Roubal J. 1932a: Descriptions de quelques nouveaux coléoptères paléarctiques. *Miscellanea Entomologica* **34**: 1.

Roubal J. 1932b: Fragmente zur Koleopterenfaunistik des balkanischen Festlandes. *Entomologischer Anzeiger* **12**: 129-130.

Roubal J. 1933a: Beschreibungen von zwei neuen Cis-Arten aus Ostasien. *Stylops* **2**: 72.

Roubal J. 1933b: Beschreibungen einiger Coleopteren-Neuheiten. *Entomologisches Nachrichtenblatt* **7**: 77-78.

Roubal J. 1933c: Descriptions de deux mélandryides (Col.) d'Europe centrale. *Bulletin de la Société Entomologique de France* **38**: 128.

Roubal J. 1933d: Beschreibung einiger neuen Coleopterenformen aus der Slovakei. *Časopis Československé Společnosti Entomologické* **30**: 139.

Roubal J. 1934a: Inventa entomologica itineris Hispanici et Maroccani, quod a 1926 fecerunt Harald et Håkan Lindberg. XIX. Scydmaenidae, Pselaphidae, Mordellidae. *Commentationes Biologicae* **4** (11): 1-8.

Roubal J. 1934b: Eine neue Mycetochara Berth. aus Mitteleuropa: M. melandryina n. *Entomologische Blätter* **30**: 35-36.

Roubal J. 1935a: Descriptio novae speciei generis Mycetochara Berth. (Alleculidae) et adnotationes diversae de speciebus aliis. *Časopis Československé Společnosti Entomologické* **32**: 24-25.

Roubal J. 1935b: Quelques notes sur les mordellides paléarctiques avec la description de cinq formes nouvelles. *Bulletin de la Société Entomologique de France* **40**: 106-108.

Roubal J. 1936a: *Katalog coleopter (brouků) Slovenska a Podkarpatské Rusi na základě bionomickém a zoogeografickém a spolu systematický doplněk Ganglbauerových "Die Käfer von Mitteleuropa" a Reitterovy "Fauna germanica" Díl II.* Bratislava: Učená společnost Šafaříkova v Bratislavě, 434 pp.

Roubal J. 1936b: Beschreibung einer neuen Untergattung, einer Varietät und einer Form der Coleopteren. *Festschrift zum 60. Geburtstage von Professor Dr. Embrik Strand* (Riga) **1**: 53-58.

Roubal J. 1937a: Koleopterologische Notizen. *Entomologisches Nachrichtenblatt* **11**: 38-40.

Roubal J. 1937b: Dvě nové broučí variety z ČSR. Zwei neue Käfervarietäten aus ČSR. *Časopis Československé Společnosti Entomologické* **34**: 67-68.

Roubal J. 1938: Anidorus Lokvenci, n. sp. aderide (Col.) nouveau de l'Europe centrale. *Bulletin de la Société Entomologique de France* **43**: 40-41.

Ruiz J. L. & García-París M. 2004: Una specie nueva de Mylabris Fabricius, 1775 (Coleoptera, Meloidae) del sureste de la Península Ibérica. *Graellsia* **60**: 185-196.

Ruiz J. L. & García-París M. 2007: Nombre de reemplazo en el género Mylabris Fabricius, 1775 (Coleoptera, Meloidae). *Graellsia* **63**: 151-152.

Rye E. C. 1872: Additions, &c., to the list of British Coleoptera, with descriptions of three new species. *The Entomologist's Monthly Magazine* **9**: 5-11.

Rye E. C. 1873: Insecta. Coleoptera. Pp. 222-329. In: Rye E. C. (eds.) *The Zoological Record for 1871; being the volume eighth of the Record of Zoological literature.* London: John Van Voost.

Rye E. C. 1878: Insecta. 240pp. In: Rye E. C. (eds.) *The Zoological Record for 1876; being the volume thirteenth of the Record of Zoological literature.* London: John Van Voost.

Saha G. N. 1972: Studies on blister beetles (Coleoptera: Meloidae) from India. I. Two new species of the genus Eletica Lac. *Zoologischer Anzeiger* **189**: 98-101.

Saha G. N. 1979: Revision of Indian blister beetles (Coleoptera: Meloidae: Meloinae). *Records of the Zoological Survey of India* **74**: 1-146.

Saha G. N. 1988: Discovery of genus Platolenes (Coleoptera: Tenebrionidae) from India with description of two new species. *Records of the Zoological Survey of India* **85**: 433-437.

Sahlberg C. R. 1823: *Periculi Entomographici, species insectorum nondum descriptas propposituri, Fasciculus.* Aboae: Typis Frenckelliorum, 82 + (1), 4 pls.

Sahlberg C. R. 1833: Pars XXIX. Pp. 441-456. In: *Insecta Fennica enumerans, dissertationibus academicis. A. 1817-1834 editis. Tomus I.* Aboae: J. C. Frenckel [note: all parts issued as book in 1834,with viii + 520 pp.].

Sahlberg C. R. 1836: *Dissertatio entomologica insecta Fennica enumerans, cujus particulam decimam. Tomus II.* Aboae: J. C. Frenckel, 145-160 pp.

Sahlberg C. R. 1837: *Dissertatio entomologica Insecta Fennica enumerans, cujus particulam undecimam. Tomus II.* Aboae: J. C. Frenckel, 161-176 pp.

Sahlberg J. R. 1871: Othismopteryx ett nytt genus bland Finlands Coleoptera af familijen Colydiidae. *Notiser ur Sällskapets pro Fauna et Flora Fennica Förhandlingar* **11**: 441-444, pl. 1.

Sahlberg J. R. 1875: Ueber die finnischen Arten der Käfer-Gattung Pytho und deren Larven. *Deutsche Entomologische Zeitschrift* **19**: 219-224.

Sahlberg J. R. 1881: En ny art af Coleopter-slägtet Zilora Muls. af melandryidernas familij funne i Finland och beschrifven. *Meddelanden af Societas pro Fauna et Flora Fennica* **7**: 132-134.

Sahlberg J. R. 1901: Coleoptera nova vel minus cognita faunae Fennicae. *Acta Societatis pro Fauna et Flora Fennica* **19** (3):1-23.

Sahlberg J. R. 1903a: Coleoptera mediterranea et rosso-asiatica nova et minus cognita. II. *Öfversigt af Finska Vetenskaps-Societetens Förhandlingar* **15** (10): 1-40.

Sahlberg J. R. 1903b: Coleoptera mediterranea et rosso-asiatica nova et minus cognita, maxima ex parte itineribus annis 1895-1896 et 1898-1899 collecta. *Öfversigt af Finska Vetenskaps-Societetens Förhandlingar* **45** (10): 1-40.

Sahlberg J. R. 1903c: Ad cognitionem faunae Coleopterorum graecae fragmenta. *Öfversigt af Finska Vetenskaps-Societetens Förhandlingar* **45** (12): 1-9.

Sahlberg J. R. 1903d: Ad cognitionem faunae Coleopterorum italicae fragmenta. *Öfversigt af Finska Vetenskaps-Societetens Förhandlingar* **45** (13): 1-14.

Sahlberg J. R. 1903e: Coleoptera levantina mensibus februario et martio 1896 in Palaestina et Aegypto inferiore collecta. *Öfversigt af Finska Vetenskaps-Societetens Förhandlingar* **45** (18): 1-36.

Sahlberg J. R. 1903f: Coleoptera numido-punica mensibus martio, aprili et majo 1899 in Tunisia et Algeria orientali collecta. *Öfversigt af Finska Vetenskaps-Societetens Förhandlingar* **45** (19): 1-70.

Sahlberg J. R. 1908: Coleoptera mediterranea et rosso-asiatica nova et minus cognita, maxima ex parte itineribus annis 1895-1896, 1898-1899 et 1903-1904 collecta. *Öfversigt af Finska Vetenskaps-Societetens Förhandlingar* **50** (7) [1907-1908]: 1-94.

Sahlberg J. R. 1913a: Coleoptera mediterranea et rosso-asiatica nova et minus cognita maxima ex parte itineribus annis *1895-1896, 1898-1899* et 1903-1904 collecta. *Öfversigt af Finska Vetenskaps-Societetens Förhandlingar* **55** [1912-1913] A (8): 1-88.

Sahlberg J. R. 1913b: Coleoptera mediterranea orientalis quae in Aegypto, Palestina, Syria, Caramania collegerunt John Sahlberg et Unio Saalas. *Öfversigt af Finska Vetenskaps-Societetens Förhandlingar* **55** [1912-1913] A (19): 1-282.

Sahlberg R. F. 1834: Novae Coleopterorum fennicorum species. *Bulletin de la Société Impériale des Naturalistes de Moscou* **7**: 267-280.

Saint-Albin E. de. 1952: Note sur les anthicides (Col. Hétéromères). *Bulletin de la Société Entomologique de France* **57**: 20-23.

Saitô M. 1990: A new subspecies of pyrochroid beetle (Coleoptera, Pyrochroidae) from Okinawa-hontô, southwest Japan. *Elytra* **18**: 105-107.

Saitô M. 1995: Notes on Taiwanese species of the genus Arthromacra (Coleoptera: Tenebrionidae). *Elytra* **23**: 177-185.

Saitô M. 1999: Notes on the Japanese species of the genus Usechus (Coleoptera, Zopheridae). *Elytra* **27**: 103-111.

Saitô M. 2001: A new species of the genus Hymenalia (Coleoptera: Tenebrionidae: Alleculinae) from the Yaeyama Islands, Japan with notes of Japanese species. *Special Publication of the Japan Coleopterological Society of Osaka* **1**: 347-350.

Saitô M. 2003a: A new species of the genus Ischalia (Coleoptera, Anthicidae, Ischaliinae) from Hokkaido, Japan. *Elytra* **31**: 55-60.

Saitô M. 2003b: A new species of the genus Tomoderus (Coleoptera: Anthicidae) from the Ryukyu Islands, Southwest Japan. *Elytra* **31**: 321-323.

Sakai M. & Telnov D. 2001: Leptaleus sasajii, a new species and the first representative of the genus Leptaleus from Japan (Coleoptera, Anthicidae). *Sukunahikona, Special Publication of the Japan Coleopterological Society, Osaka* **1**: 327-332.

Salvaña J. M. 1870: *Apuntes para la geografía y fauna entomologica de Mataró*. Madrid: Junte, 44 pp.

Sasaji H. 1971: Description of a new Cicones-species in Japan (Coleoptera: Colydiidae). *The Life Study* (*Fukui*) **15**: 43-45.

Sasaji H. 1974: Notes on the Japanese Melandryidae, with descriptions of new species (Coleoptera). *Memoirs of the Faculty of Education, Fukui University Series II* (*Natural Science*) **24**: 1-14.

Sasaji H. 1983: Contribution to the taxonomy of the superfamily Cucujoidea (Coleoptera) of Japan and her adjacent districts, I. *Memoirs of the Faculty of Education, Fukui University Series II* (*Natural Science*) **33**: 17-52.

Sasaji H. 1984: Contribution to the taxonomy of the superfamily Cucujoidea (Coleoptera) of Japan and her adjacent districts, II. *Memoirs of the Faculty of Education, Fukui University Series II* (*Natural Science*) **34**: 21-63.

Sasaji H. 1985: Discovery of the subfamily Calopodinae (Coleoptera: Oedemeridae) in Formosa, with descriptions of a new genus and a new species. *Gekkan-Mushi* **177**: 7-12. (in Japanese, English title and descriptions).

Sasaji H. 1986: The Cucujoidea (Insecta: Coleoptera) collected by the Nagoya University Scientific Expedition to Formosa in 1984. *The Memoirs of the Faculty of Education, Fukui University, Series II* (*Natural Science*) **36**: 1-14.

Sasaji H. 1986: Notes on the Colydiidae (Coleoptera) of Japan and Formosa. Pp. 243-249. *Entomological papers presented to Yoshihiko Kurosawa on the occasion of his retirement.* Tokyo: The Coleopterists' Association of Japan, v + 342 pp.

Sasaji H. 1987: Contribution to the taxonomy of the superfamily Cucujoidea (Coleoptera) of Japan and her adjacent districts, III. *Memoirs of the Faculty of Education, Fukui University Series II* (*Natural Science*) **37**: 23-55.

Sasaji H. 1988: Contribution to the taxonomy of the superfamily Cucujoidea (Coleoptera) of Japan and her adjacent districts, IV. *Memoirs of the Faculty of Education, Fukui University Series II* (*Natural Science*) **38**: 13-48.

Sasaji H. 1995: On the adaptive characteristics of the genus Lederia (Coleoptera, Melandryidae), with description of a new species from Japan. *Special Bulletin of the Japanese Society of Coleopterology* **4**: 425-431.

Satô M. 1990: The genus Ischalia (Coleoptera, Pyrochroidae) from Taiwan. *Elytra* **18**: 101-104.

Satô M. & Hatta K. 1988: A new inopeplid beetle (Coleoptera, Inopeplidae) from the Ryukyu Islands. *Elytra* **16**: 85-88.

Say T. 1824a: Descriptions of coleopterous insects collected in the late expeditions to the Rocky Mountains, performed by order of Mr. Calhoun, Secretary of War, under the command of Major Long. *Journal of the Academy of Natural Sciences of Philadelphia* **3**: 238-282.

Say T. 1824b: Appendix. Part 1. - Natural History. 1. Zoology: E. Class Insecta. Coleoptera. Pp. 268-348. In: *Narrative of an expedition of the source of St. Peter's river, lake Winnepeek, lake of the Woods, etc. etc. performed in the year 1823, by order of the Hon. J. C. Calhoun, secretary of war, under the command of Stephen H. Long, Major U.S.T.E. compiled from the notes of major Long, Messrs. Say, Keating, and Calhoun by William H. Keating, A.M. etc. Vol. 2.* Philadelphia.

Say T. 1827: Descriptions of new species of coleopterous insects inhabiting the United States. *Journal of the Philadelphia Academy of Natural Sciences* **5**: 150-204, 237-284, 293-304.

Say T. 1835: Descriptions of new North American coleopterous insects, and observations on some already described. *Boston Journal of Natural History* **1** (2): 151-203.

Saylor L. W. 1940: Two new generic names for South American Coleoptera. *Proceedings of the Entomological Society of Washington* **42**: 46.

Schaller J. G. 1783: Neue Insecten. *Schriften der Naturforschenden Gesellschaft zu Halle* **1**: 217-328.

Schatzmayr A. 1926: Bestimmungstabelle der Gattung Nacerda nebst Bemerkungen über die bisher bekannten europäischen Arten. *Coleopterologisches Centralblatt* **1**: 51-65.

Schatzmayr A. 1938: Coleotteri raccolti dal Dott. Edgardo Moltoni durante le sue missioni ornitologiche in Libia. *Atti della Società Italiana di Scienze Naturali e del Museo Civico di Storia Naturale in Milano* **77**: 159-190.

Schatzmayr A. 1941a: Coleotteri raccolti nella Grande Sila del Dott. E. Moltoni. *Atti della Società Italiana di Scienze Naturali e del Museo Civico di Storia Naturale in Milano* **80**: 49-102.

Schatzmayr A. 1941b: Nuovi contributi alla conoscenza della fauna delle isole Italiane dell'Egeo. XIV: Cacce coleotterologiche nel Dodecaneso. *Bollettino del Laboratorio di Zoologia Generale ed Agraria della Facoltà di Agraria di Portici* **31**: 342-368.

Schatzmayr A. & Koch C. 1934a: Wissenschaftliche Ergebnisse der entomologischen Expeditionen seiner Durchlaucht des Fürsten Alexander C. della Torre e Tasso nach Aegypten und auf die Halbinsel Sinai. *Bulletin de la Société Royale Entomologique d'Egypte* **17** [1933]: 204-242, pl. 12.

Schatzmayr A. & Koch C. 1934b: Wissenschaftliche Ergebnisse der entomologischen Expedition Seiner Durchlaucht des Fürsten Alexander C. Della Torre e Tasso nach Aegypten und auf der Halbinsel Sinai. VII. Tenebrionidae. *Bulletin de la Société Royal Entomologique d'Egypte* **18**: 17-27.

Schaufuss L. W. 1861: Ueber 22 zum Theil neue Käferarten. *Sitzungsberichte der Naturwissenschaftlichen Gesellschaft Isis zu Dresden* **1821**: 47-51.

Schaufuss L. W. 1862a: [new taxa]. *Sitzungsberichte der Naturwissenschaftlichen Gesellschaft Isis zu Dresden* **1861**: 91-92.

Schaufuss L. W. 1862b: Diagnoses de coléoptères nouveaux. *Annales de la Société Entomologique de France* (4) **2**: 309-312.

Schaufuss L. W. 1869: [new taxa]. In: Ludwig Salvador & L. W. Schaufuss: *Beitrag zur Kenntniss der Coleopteren-Fauna der Balearen.* Prag: Selbstverlag, 31 pp.

Schaufuss L. W. 1870: Coleopterologisches. *Nunquam Otiosus* **1** [1870-1872]: 29-49.

Schaufuss L. W. 1872: Beschreibung einiger Käfer. *Nunquam Otiosus* **2** [1872-1877]: 275-278.

Schaufuss L. W. 1879: Zoologische Ergebnisse von Excursionen auf den Balearen. III. (Addenda und Fortsetzung). *Nunqaam Otiosus* **3** [1879-1890]: 527-552.

Schaufuss L. W. 1882a: Coléoptères aveugles de la famille des Colydiidae. *Annales de la Société Entomologique de France* (6) **2**: 46-48.

Schaufuss L. W. 1882b: Zoologische Ergebnisse von Excursionen auf den Balearen. III. (Addenda und Fortsetzung). *Nunquam Otiosus* **3** [1879-1890]: 527-552.

Schaufuss L. W. 1882c: Zoologische Ergebnisse von Excursionen auf den Balearen. *Verhandlungen der Kaiserlich-Königlichen Zoologisch-Botanischen Gesellschaft in Wien* **31** [1881]: 619-624.

Schaufuss L. W. 1887: Beitrag zur Fauna der Niederländischen Besitzungen auf den Sunda-Inseln. II. *Horae Societatis Entomologicae Rossicae* **21**: 103-147.

Schaum H. R. 1859: Beiträge zur europäischen Käferfauna. *Berliner Entomologische Zeitschrift* **3**: 42-59.

Schaum H. R. 1864: Discussion critique sur la synonymie de plusieurs espèces de coléoptères. Notes sur des observations synonymiques de M. Reiche. *Annales de la Société Entomologique de France* (4) **3** [1863]: 669-672.

Schaum H. R. 1865: [new taxon]. In: Kraatz G.: *Revision der Tenebrioniden der alten Welt.* Berlin: Nicolaische Verlagsbuchhandlung, 393 pp.

Schawaller W. 1987: Revision westpaläarktischer Tenebrionidae (Coleoptera). Teil 1. Die Arten der Gattung Akis Herbst. *Stuttgarter Beiträge zur Naturkunde, Serie A* (*Biologie*) **403**: 1-21.

Schawaller W. 1990a: Revision of the Western Palaearctic Tenebrionidae (Coleoptera). Part 2. The genus Ammogiton in Arabia with description of a new species. *Fauna of Saudi Arabia* **11**: 40-48.

Schawaller W. 1990b: Revision of the Western Palaearctic Tenebrionidae (Coleoptera). Part 3. A new genus of Erodiini from Arabia. *Fauna of Saudi Arabia* **11**: 49-54.

Schawaller W. 1991a: Prostomidae (Coleoptera) aus dem Himalaya mit einem Beitrag zur Larvalmorphologie. *Stuttgarter Beiträge zur Naturkunde Serie A* (*Biologie*) **461**: 1-17.

Schawaller W. 1991b: The genus Diphyrrhynchus in Arabia, with description of a new species (Coleoptera, Tenebrionidae). *Fauna of Saudi Arabia* **12**: 289-294.

Schawaller W. 1992: Leiochrini (Coleoptera: Tenebrionidae) aus dem Nepal-Himalaya. *Stuttgarter Beiträge zur Naturkunde Serie A* (*Biologie*) **482**: 1-15.

Schawaller W. 1993a: New and little known Tenebrionidae (Coleoptera) from the Arabian Peninsula. *Fauna of Saudi Arabia* **13**: 102-109.

Schawaller W. 1993b: Leiochrini (Coleoptera: Tenebrionidae) aus dem kontinentalen Südost-Asien. *Entomologia Basiliensia* **16**: 349-358.

Schawaller W. 1995a: Revision der Laena Mittelasiens (Insecta, Coleoptera, Tenebrionidae). *Spixiana* **18**: 65-73.

Schawaller W. 1995b: The genus Basanus Lacordaire (Coleoptera: Tenebrionidae) in continental south-eastern Asia. *Acta Zoologica Academiae Scientiarum Hungaricae* **41**: 327-334.

Schawaller W. 1996a: The genus Uloma Dejean (Coleoptera: Tenebrionidae) in the Himalayas. *Acta Zoologica Academiae Scietiarum Hungariae* **42**: 111-125.

Schawaller W. 1996b: Stenoria thakkola sp. n., the first species of the tribe Sitarini from the Himalayas (Coleoptera: Meloidae). *Elytron* **10**: 147-152.

Schawaller W. 1996c: Spinolyprops pakistanicus sp. n. (Coleoptera, Tenebrionidae), an Oriental element in the fauna of Pakistan. *Revue Suisse de Zoologie* **103**: 691-695.

Schawaller W. 1997: Luprops kaszabi n.sp. aus dem Himalaya, mit einer Bestimmungstabelle der Luprops-Arten Nepals (Coleoptera: Tenebrionidae). *Entomologische Zeitschrift* **107**: 295-298.

Schawaller W. 1998: New taxonomic and faunistic data on the genus Cneocnemis Gebien (Coleoptera: Tenebrionidae) with description of two new species from Nepal. *Linzer Biologische Beiträge* **30**: 463-468.

Schawaller W. 1999: Notes on Palaearctic and Oriental Phrenapatini (Coleoptera: Tenebrionidae), with description of four new species. *Revue Suisse de Zoologie* **106**: 419-428.

Schawaller W. 2000a: The genus Mesomorphus Seidlitz in Nepal with description of a new species (Insecta: Coleoptera: Tenebrionidae). *Faunistische Abhandlungen* **22**: 39-48.

Schawaller W. 2000b: The genera Chariophenus Blair, 1929, and Foochounus Pic, 1921, in the Himalayas, with description of a new species from Nepal (Coleoptera: Tenebrionidae). *Entomologische Zeitschrift* (Stuttgart) **100**: 354-357.

Schawaller W. 2001a: The genus Laena Latreille (Coleoptera: Tenebrionidae) in China, with descriptions of 47 new species. *Stuttgarter Beiträge zur Naturkunde Serie A* (*Biologie*) **632**: 1-62.

Schawaller W. 2001b: Probaticus kalavriticus n. sp., der Untergattung Helopotrichus Reitter, 1922, aus Griechenland (Coleoptera, Tenebrionidae). *Entomologische Blätter* **97**: 43-48.

Schawaller W. 2001c: Pseudethas jaegeri sp. n. from Tibet (Insecta: Coleoptera: Tenebrionidae: Stenosini). *Reichenbachia* **34**: 143-146.

Schawaller W. 2001d: Nepalolaena kira gen. nov., sp. nov. from the Nepal Himalayas (Coleoptera, Tenebrionidae). *Entomologia Basiliensia* **23**: 277-282.

Schawaller W. 2002: The genus Laena Latreille in Nepal, with taxonomic and faunistic notes on species from the adjacent Himalayas (Coleoptera: Tenebrionidae). *Stuttgarter Beiträge zur Naturkunde Serie A* (*Biologie*) **641**: 1-69.

Schawaller W. 2003a: Two new species of Scaphidema Redtenbacher (Coleoptera: Tenebrionidae) from Nepal and China. *Stuttgarter Beiträge zur Naturkunde Serie A* (*Biologie*) **653**: 1-7.

Schawaller W. 2003b: The genus Platydema Laporte & Brullé in the Himalayas and adjacent regions. With descriptions of five new species. Pp. 269-277. In: Hartmann M. & Baumbach H. (eds.): *Biodiversität und Naturausstattung im Himalaya – Biodiversity and Natural Heritage of the Himalaya*. Erfurt: Verein der Freunde & Forscher des Naturkundemuseum Erfurt e. V., 390 pp., XIV pls.

Schawaller W. 2004a: New species and records of the genus Spiloscapha Bates (Coleoptera: Tenebrionidae) from the Oriental and Papuan regions. *Stuttgarter Beiträge zur Naturkunde Serie A (Biologie)* **661**: 1-10.

Schawaller W. 2004b: The Oriental species of Platydema Laporte & Brullé, with descriptions of 16 new species (Coleoptera: Tenebrionidae). *Stuttgarter Beiträge zur Naturkunde Serie A (Biologie)* **671**: 1-49.

Schawaller W. 2004c: New species and records of Tenebrionidae (Coleoptera) from the Socotra Archipelago. *Fauna of Arabia* **20**: 439-458.

Schawaller W. 2005: New species and records of Leiochrini (Coleoptera, Tenebrionidae) from continental south-eastern Asia. *Entomologia Basiliensia et Collectionis Frey* **27**: 209-226.

Schawaller W. 2006a: Revision of the Oriental species of the genus Bradymerus Perroud with descriptions of 29 new species. *Stuttgarter Beiträge zur Naturkunde Serie A (Biologie)* **694**: 1-64.

Schawaller W. 2006b: First record of the subfamily Cossyphodinae (Coleoptera, Tenebrionidae) from the Socotra Archipelago. *Fauna of Arabia* **21**: 247-250.

Schawaller W. 2006c: New species, new records and new synonyms of Tenebrionidae (Insecta: Coleptera) from Nepal. *Veröffentlichungen des Naturkundemuseums Erfurt* **25**: 475-482, pls 9, 10.

Schellenberg J. R. 1798: *Helvetische Entomologie oder Verzeichniss der schweizerischen Insekten nach einer neuen Methode geordnet.* Erster Theil. Zürich: Orell, Füssli und Companie, 149 + [4] pp., 14 pls.

Schellenberg J. R. 1802: *Entomologische Beyträge.* Winterthur: Steiner, 24 pp., 10 pls.

Schenkling S. 1915: Oedemeridae. Pars 65. In: Schenkling S. (ed.): *Coleopterorum Catalogus.* Berlin: W. Junk, 82 pp.

Schenkling S. 1922: *Nomenclator Coleopterologicus. Eine etymologische Erklärung sämtlicher Gattungs- und Artnamen der Käfer der Deutschen Fauna sowie der Angrenzenden Gebiete. Zweite Auflage.* Jena: Verlag von Gustav Fischer, iv + 249 pp.

Schilder F. A. 1923a: Rhipiphoriden-Studien. IV. Bemerkungen zur Literatur und Nomenklatur. *Entomologische Blätter* **19**: 167-169.

Schilder F. A. 1923b: Sauter's Formosa-Ausbeute: Rhipiphoridae (Col.). (Zugleich: Rhipiphoriden-Studien III.). *Entomologische Mitteilungen* **12**: 202-204.

Schilder F. A. 1924: Rhipiphoriden-Studien. VI. Revision des Genus Metoecus Dej. (1833). *Deutsche Entomologische Zeitschrift* **66**: 237-244.

Schilsky J. 1888: Beitrag zur Kenntniss der deutschen Käferfauna. *Deutsche Entomologische Zeitschrift* **32**: 177-190.

Schilsky J. 1889: Synonymische und andere Bemerkungen zu Dr. Carl W. V. Dalla Torre´s „Synopsis der Inseckten Oberösterreichs" und „Die Käferfauna von Oberösterreich". *Deutsche Entomologische Zeitschrift* **1889**: 345-356.

Schilsky: 1892: VII. Beitrag zur Kenntnits der deutschen Käferfauna. *Deutsche Entomologische Zeitschrift* **1892**: 193-208.

Schilsky J. 1894a: *Die Käfer Europa´s. Nach der Natur Beschrieben von Dr. H. C. Küster und Dr. G. Kraatz Fortgesetzt von J. Schilsky.* Nürnberg: Bauer und Raspe (Emil Küster) **30**: viii + no. taxa 1-100.

Schilsky J. 1894b: IX. Beitrag zur deutschen Käferfauna. *Deutsche Entomologische Zeitschrift* **1894**: 329-332.

Schilsky J. 1895: *Die Käfer Europa´s. Nach der Natur Beschrieben von Dr. H. C. Küster und Dr. G. Kraatz Fortgesetzt von J. Schilsky.* Nürnberg: Bauer und Raspe (Emil Küster) **31**: viii + no. taxa 1-100 + [1] pp.

Schilsky J. 1899: *Die Käfer Europa´s. Nach der Natur Beschrieben von Dr. H. C. Küster und Dr. G. Kraatz Fortgesetzt von J. Schilsky.* Nürnberg: Bauer und Raspe (Emil Küster) **35**: viii + no. taxa 1-100 + 41 pp. from A to SS [= 268] pp.

Schilsky J. 1900: *Die Käfer Europa's. Nach der Natur Beschrieben von Dr. H. C. Küster und Dr. G. Kraatz Fortgesetzt von J. Schilsky.* Nürnberg: Bauer und Raspe (Emil Küster) **37**: [vi] + no. taxa 1-100 + A-W pp.

Schilsky J. 1903: *Die Käfer Europa´s. Nach der Natur Beschrieben von Dr. H. C. Küster und Dr. G. Kraatz Fortgesetzt von J. Schilsky.* Nürnberg: von Bauer und Raspe (Emil Küster) **40**: viii no. taxa 1-100 + A-PP pp.

Schilsky J. 1906: *Die Käfer Europa´s. Nach der Natur Beschrieben von Dr. H. C. Küster und Dr. G. Kraatz Fortgesetzt von J. Schilsky.* Nürnberg: von Bauer und Raspe (Emil Küster) **43**: viii + i-cxix pp. + no. taxa 1-100 + [1] pp.

Schilsky J. 1908a: *Die Käfer Europa´s. Nach der Natur Beschrieben von Dr. H. C. Küster und Dr. G. Kraatz Fortgesetzt von J. Schilsky.* Nürnberg: von Bauer und Raspe (Emil Küster) **45**: iv + no. taxa 1-100 + [2] pp.

Schilsky J. 1908b: Zoologische und Anthropologische Ergebnisse einer Forschungsreise in westlichen und zentralen Südafrika, Bd. 1, Lfg. 1, Mordellidae. *Denkschriften der Medicinisch-Naturwissenschaftlichen Gesellschaft* (Jena) **13**: 137-138.

Schilsky J. 1908c: XII. Beitrag zur deutschen Käferfauna. *Deutsche Entomologische Zeitschrift* **1908**: 604-608.

Schilsky J. 1910: Mordellistena engelharti. Eine neue Mordellide aus Dänemark. *Entomologiske Meddelelser* **3**: 255-256.

Schiödte J. C. 1893: [new taxa]. In: Bergroth E.: Bemerkungen zum "Catalogus Coleopterorum Europae, Caucasi et Armeniae rossicae. Ed. E. Reitter." *Entomologische Nachrichten* **19**: 305-311.

Schmidt K. 1913: Zur Kenntis der äthiopisch-afrikanischen Meloeformen (Coleopt.). *Stettiner Entomologische Zeitung* **74**: 327-339.

Schmidt W. L. E. 1842a: Die europäischen Arten der Gattung Anthicus Fbr. *Entomologische Zeitung* (Stettin) **3**: 74-88.

Schmidt W. L. E. 1842b: Die europäischen Arten der Gattung Anthicus Fbr. *Entomologische Zeitung* (Stettin) **3**: 122-135.

Schmidt W. L. E. 1842c: Die europäischen Arten der Gattung Anthicus Fbr. *Entomologische Zeitung* (Stettin) **3**: 170-189.

Schmidt W. L. E. 1842d: Die europäischen Arten der Gattung Anthicus Fbr. *Entomologische Zeitung* (Stettin) **3**: 193-200.

Schmidt W. L. E. 1844: [new taxa]. In: Schmidt O. G. C. F. 1844: *Catalogus Coleopterorum Europae*. Stettin: 76 + vi pp.

Schmidt W. L. E. 1846: Revision der europäischen Oedemeriden. *Linnaea Entomologica* **1**: 1-146.

Schneider D. H. 1792: Verzeichniss und Beschreibung der in der Sammlung des Herausgebers befindlichen zur Gattung Coccinella gehörigen europäischen Käfer (Sonnen-Käfer oder Blattlaus-Käfer). *Neuestes Magazin für die Liebhaber der Entomologie* **2**: 128-256.

Schneider O. 1902: Phaleria ligurica n.sp. nebst Bemerkungen über Ph. Reyi und Ph. Cava. *Wiener Entomologische Zeitung* **21**: 10-12.

Schneider W. 1991: New records of Meloidae (Insecta: Coleoptera) from Arabia with description of a new species. *Fauna of Saudi Arabia* **12**: 273-288.

Schoenherr C. J. 1806: *Synonymia Insectorum, oder: Versuch einer Synonymie aller bisher bekannten Insecten; nach Fabricii Systema Eleutheratorum geordnet. Erster Band. Eleutherata oder Käfer. Erster Theil. Lethrus --- Scolytes*. Stockholm: Heinr. A. Nordström, xxii + 293 pp., 3 pls.

Schoenherr C. J. 1817: *Synonymia Insectorum, oder: Versuch einer Synonymie aller bisher bekannten Insecten; nach Fabricii Systema Eleutheratorum geordnet. Erster Band. Eleutherata oder Käfer. Dritter Theil. Hispa ---Molorchus*. Uppsala: Em. Bruzelius, xi + 506 pp.

Schrank von Paula F. 1776: *Beyträge zur Naturgeschichte. Mit sieben von Verfasser gezeichneten und in Kupfer gestochenen Tabellen*. Augsburg: Veith Bros, [6] + 137 + [3] pp., 7 pls.

Schrank von Paula F. 1781: *Enumeratio insectorum Austriae indigenorum*. Augustae Vindelicorum: Eberhardi Klett et Franck, [xxiv] + 548 + [4] pp., 4 pls.

Schrank von Paula F. 1786: Erstes Brief. Pp. 1-23. In: *Baierische Reise*. München: Baptist Strobl, [14] + 418 + [10] pp., 2 pls.

Schrank von Paula F. 1789: Entomologische Beobachtungen. *Naturforscher* **24**: 60-90.

Schrank von Paula F. 1798: *Fauna Boica. Durchgedachte Geschichte der in Baiern einheimischen und zahmen Thiere. Erster Band*. Nürnberg: Stein'schen Buchhandlung, xii + 720 pp.

Schuh R. 1998a: Franzorphius franzi gen. et sp. n. from China (Coleoptera: Zopheridae: Colydiinae). *Koleopterologische Rundschau* **68**: 227-232.

Schuh R. 1998b: Revision of the Synchita variegata species group (Coleoptera: Zopheridae, Colydiinae). *Annales Zoologici* **48**: 313-324.

Schuh R. 1999: Revision of the Oriental species of the genus Namunaria Reitter, 1882 (Coleoptera: Zopheridae, Colydiinae). *Coleopterologische Rundschau* **69**: 137-152.

Schuh R. & Mifsud D. 2000: The cylindrical bark beetles of Malta (Insecta: Coleoptera: Zopheridae, Colydiinae). *Annalen des Naturhistorischen Museums Wien* **102 B**: 259-267.

Schulze P. 1912: Wanachia n. n. pro Cardinia Seidlitz (non M.-E.). *Internationale Entomologische Zeitschrift* **6** (2): 11.

Schuster A. 1914a: Itagonia Ganglbaueri nov. spec. (Col., Tenebr.). *Entomologische Mitteilungen* **3**: 58-59.

Schuster A. 1914b: Blaps kolbei nov. spec. (Col., Tenebr.). *Entomologische Blätter* **10**: 142-143.

Schuster A. 1914c: Dila Kuntzeni nov. spec. (Col., Tenebr.). *Entomologische Blätter* **10**: 182-183.

Schuster A. 1915a: Bemerkungen über die Tenebrionidenausbeute Paganettis auf Creta. *Entomologische Blätter* **11**: 1-6.

Schuster A. 1915b: Neue paläarktische Tenebrionidae (Col.) I. *Entomologische Blätter* **11**: 86-92.

Schuster A. 1915c: Neue Laena-Arten. *Wiener Entomologische Zeitung* **34**: 347-361.

Schuster A. 1916: Monographie der Coleopterengattung Laena Latreille. *Verhandlungen der Zoologisch-Botanischen Gesellschaft in Wien* **66**: 495-629.

Schuster A. 1919a: Neue paläarktische Tenebrioniden (Col.) II. *Entomologische Blätter* **15**: 27-33.

Schuster A. 1919b: Revision der Tenebrioniden-Gattungen Pachychilina Reitter und Pachychile Eschscholz. *Koleopterologische Rundschau* **8** [1919-1920]: 27-43.

Schuster A. 1920a: Neue paläarktische Tenebrioniden (Col.) III. *Entomologische Blätter* **16**: 199-201.

Schuster A. 1920b: Zwei neue Hionthis-Arten (Col., Tenebr.). *Entomologische Blätter* **16**: 234-236.

Schuster A. 1921a: Revision der Tenebrioniden-Gattungen Pachychilina Reitter und Pachychile Eschscholtz (Schluss). *Koleopterologische Rundschau* **9**: 1-27.

Schuster A. 1921b: Die paläarktischen Tenebrioniden des Deutschen Entomologischen Institutes. I. (Col.). *Entomologische Mitteilungen* **10**: 41-47.

Schuster A. 1922a: [new taxon]. In: Dodero A.: Missione zoologica del Dr. E. Festa in Cirenaica. *Bollettino di Musei di Zoologica ed Anatomia Comparata della R. Università di Torino* **37** N. 743: 1-7.

Schuster A. 1922b: Tenebrioniden aus Cyrenaica (Col.). *Memorie della Società Entomologica Italiana* **1**: 14-23.

Schuster A. 1922c: Die paläarktischen Tenebrioniden des Deutschen Entomologischen Institutes. No. II. (Col.). *Entomologische Mitteilungen* **11**: 47-50.

Schuster A. 1922d: Insecta, ex Sibiria meridionali et Mongolia in itinere Ørjan Olsen 1914 collecta. A. Coleoptera, a Fritz Jensen lecta. V. Tenebrionidae. *Norsk Entomologisk Tidsskrift* **1** [1922-1923]: 163-164.

Schuster A. 1922e: Die paläarktischen Tenebrioniden des Deutschen Entomologischen Institutes III. (Col.). *Entomologische Mitteilungen* **11**: 212-213.

Schuster A. 1923a: Neue chinesische Blaps (Col. Tenebrionidae). *Wiener Entomologische Zeitung* **40**: 30-32.

Schuster A. 1923b: Neue paläarktische Tenebrioniden (Coleopt.). *Wiener Entomologische Zeitung* **40**: 156-162.

Schuster A. 1925a: Neue Tenebrioniden aus der Cyrenaica. *Bollettino della Società Entomologica Italiana* **57**: 25-31.

Schuster A. 1925b: Zwei neue Helopini aus Libien (Coleopt.). *Bollettino della Società Entomologica Italiana* **57**: 88-90.

Schuster A. 1926a: Bestimmungstabelle der Laena-Arten aus dem Himalaya und den angrenzenden Gebieten. Mit Beschreibungen neuer Arten. *Koleopterologische Rundschau* **12**: 31-54.

Schuster A. 1926b: Neue Tenebrioniden aus der Cyrenaica II. (Col.). *Bollettino della Società Entomologica Italiana* **58**: 130-135.

Schuster A. 1926c: Heliophilus (s. str.) Liebei n. sp. (Col., Tenebr.). *Coleopterologisches Centralblatt* **1**: 243-244.

Schuster A. 1928a: Eine neue Pimelia (Col., Tenebr.) aus Tripolis. *Bollettino della Società Entomologica Italiana* **60**: 26-28.

Schuster A. 1928b: Neue Tenebrioniden aus Cyrenaica (IV). *Bollettino della Società Entomologica Italiana* **60**: 122-124.

Schuster A. 1928c: Die Gattung Caenoblaps König. (Col. Tenebr.). *Koleopterologische Rundschau* **14**: 122-125.

Schuster A. 1928d: Beitrag zur Kenntnis der Verbreitung der Tenebrioniden in Libien. *Bollettino della Società Entomologica Italiana* **60**: 149-154.

Schuster A. 1928e: Neue Tenebrioniden aus der Cyrenaica. III. *Annali del Museo Civico di Storia Naturale Giacomo Doria* **52**: 373-378.

Schuster A. 1928f: Tenebrionidae. Pp. 914-1024. In: Winkler A. (ed.): *Catalogus Coleopterorum regionis palaearcticae*. Vol. II [1927-1932]. Wien: Albert Winkler, 817-1698 pp.

Schuster A. 1929: Neue Pimelien (Col., Tenebr.) aus Marokko. *Koleopterologische Rundschau* **15** [1929-1930]: 38-41.

Schuster A. 1930a: Zwei neue Tenebrioniden aus dem Mittelmeergebiet. *Bollettino della Società Entomologica Italiana* **62**: 65-67.

Schuster A. 1930b: Neue Tenebrioniden (Coleopt.) aus Belutschistan. *Koleopterologische Rundschau* **15** [1929-

1930]: 235-239.

Schuster A. 1931: [new taxa]. In: Reinig W.F. 1931: Entomologische Ergebnisse der Deutsch-Russischen Alai-Pamir-Expedition 1928 (II). 5. Coleoptera II. Tenebrionidae. *Mitteilungen aus dem Zoologischen Museum Berlin* **16**: 865-912.

Schuster A. 1933a: [new taxon]. In: Lindberg Håk.: Inventa entomologica Hispanici et Maroccani, quod a. 1926 fererunt Harald et Håkan Lindberg. XVI. Tenebrionidae. *Commentationes Biologicae* **4** (5): 1-31, 1 pl.

Schuster A. 1933b: Die Gattung Myladina Rtt. (Coleoptera, Tenebrionidae). *Sborník Entomologického Oddělení Národního Musea v Praze* **11**: 96-98.

Schuster A. 1934a: Zur Nomenklatur des Subgenus Aulonoscelis Rtt. (Col., Tenebr.). *Koleopterologische Rundschau* **20**: 75.

Schuster A. 1934b: Tenebrioniden vom Toten Meer. *Entomologische Blätter* **30**: 180-182.

Schuster A. 1935a: Neue Tenebrioniden (Col.) aus der Cyrenaica. V. *Bollettino de la Società Italiana di Entomologica* **67**: 13-21.

Schuster A. 1935b: Die Tenebrioniden-Ausbeute Dr. Gabriels in Persien. *Koleopterologische Rundschau* **21**: 20-29.

Schuster A. 1935c: Die Gattung Asidoblaps,Fairm. (Col. Tenebr.). *Stylops* **4**: 161-165.

Schuster A. 1935d: Neue Laena-Arten aus dem Himalaya (Col., Fam. Tenebrionidae). *The Annals and Magazine of Natural History* (10) **14**: 437-466.

Schuster A. 1936a: Schwedish-chinesische wissenschaftliche Expedition nach den nordwestlichen Provinzen Chinas, unter Leitung von Dr. Hedin Svin und Prof. Sü Ping-chang. Insekten gesammelt vom schwedischen Arzt der Expedition Dr. David Hummel 1927-1930. 44. Coleoptera. 8. Tenebrionidae. *Arkiv für Zoologie* **27A** (24): 1-6.

Schuster A. 1936b: Die Tenebrioniden (Col.)-Ausbeute von Prof. Dr. Franz Werner und Kustos Dr. Otto Wettstein von den Griechischen, Ägäischen Inseln und dem italienischen Dodekanes in den Jahren 1934 und 1935. *Sitzberichte der Akademie der Wissenschaften in Wien, Mathematisch-naturwissenschaftliche Klasse, Abteilung I* **145**: 65-80.

Schuster A. 1936c: Tenebrionidae. Entomologische Sammelergebnisse der Deutschen Hindukusch-Expedition 1935 der Deutschen Forschungsgemeinschaft. *Arbeiten über Morphologische und Taxonomische Entomologie aus Berlin-Dahlem* **3**: 192-201.

Schuster A. 1937a: Eine Tenebrionides-Ausbeute von Mesopotamien, Palästina, Syrien und Aegypten. *Koleopterologische Rundschau* **23**: 44-52.

Schuster A. 1937b: Zwei neue palaearktische Tenebrioniden (Col.) und Bemerkungen zu einer ungedeuteten Art. *Entomologische Blätter* **33**: 79-83.

Schuster A. 1938a: [new taxa]. In: Schuster A: & Gebien H.: Tenebrioniden (Col.) aus Arabien. *Entomologische Blätter* **34**: 49-62.

Schuster A. 1938b: Neue Tenebrioniden (Col.) von Persich-Belutschistan. *Koleopterologische Rundschau* **24**: 77-90.

Schuster A. 1940a: Die Tenebrioniden (Col.) des Museums Hoang ho-Pei ho in Tientsin. *Koleopterologische Rundschau* **26**: 15-24.

Schuster A. 1940b: Über drei neue asiatische Tenebrioniden (Col. Tenebr.). *Mitteilungen der Münchener Entomologischen Gesellschaft* **30**: 861-866.

Schuster A. & Reymond A. 1937: Quatre nouveaux ténébrionides provenant de la Mission Citroën-Centre-Asie (Col.). *Bulletin de la Société Entomologique de France* **42**: 234-238.

Schwarz C. 1808: [new taxa]. In Schoenherr C. J.: *Synonymia Insectorum, oder: Versuch einer Synonymie aller bisher bekannten Insecten; nach Fabricii Systema Eleutheratorum geordnet. Erster Band. Eleutherata oder Käfer. Zweiter Theil. Spercheus --- Cryptocephalus.* Stockholm: Carl Friedr. Merquardsix + [1] + 424 pp., pl. 4.

Scopoli J. A. 1763: *Entomologia Carniolica exhibens insecta Carnioliae indigena et distributa in ordines, genera, species, varietates, methodo linnaeana.* Vindobonae: Ioannis Thomae Trattner, xxxvi + 420 pp., 3 pls.

Scriba L. G. 1790: *Journal für die Liebhaber der Entomologie.* Frankfurt: Varrentrapp und Werner, vol. **1**: 1-96.

Scriba W. 1865: Die Käfer im Grossherzogthum Hessen und seiner nächsten Umgebung. *Bericht der Obershessischen Gesellschaft für Natur- und Heilkunde* **11**: 1-59.

Scupola A. 1984: Una nuova Idastrandiella Strand di Grecia (Coleoptera Tenebrionidae). *Bollettino di Museo*

Civico di Storia Naturale di Verona **11**: 69-75.

Scupola A. 1994: Una nuova specie di Mesostena Esch. dello Yemen (Col. Tenebrionidae). (III. Contributo alla conoscenza dei Tenebrionidi). *Frustula Entomologica N. S.* **17**: 25-28.

Scupola A. 2000: Revisione della tribu Cossyphini Latreille, 1802. Parte I. Introduzione e genere Cossyphus Olivier, 1791 (Coleoptera Tenebrionidae). *Atti del Museo Civico di Storia Naturale di Trieste* **48**: 185-249.

Scupola A. 2002: A proposito di Ulomina carinata Baudi di Selve, 1876 (Coleoptera, Tenebrionidae). *Bollettino del Museo Regionale di Scienze Naturali di Torino* **19**: 185-189.

Scupola A. & Mifsud D. 2001: Descrizione di una nuova sottospecie di Heliopathes avarus Mulsant & Rey, 1854 dell'Isola di Gozo (Arcipelago Maltese) (Coleoptera, Tenebrionidae). *Doriana* **7** (324): 1-7.

Seidlitz G. C. M. von. 1889: Ueber die Coleopteren-Gattung Esarcus Reiche. *Deutsche Entomologische Zeitschrift* **1889**: 147-149.

Seidlitz G. C. M. von. 1890: *Fauna Baltica. Die Kaefer* (*Coleoptera*) *der deutschen Ostseeprovinzen Russlands. Zweite neu bearbeitete Auflage.* Königsberg: Hartungsche Verlagsdruckerei, [10] + lvi + 192 + 818 pp., 1 pl. [issued in parts: pp. i-xl, Gattungen 1-16, Arten 1-96 in 1887; pp. xli-xlviii, Gattungen 17-80, Arten 97-336 in 1888; pp. Gattungen 81-128, Arten 337-512 in 1889; pp. Gattungen 129-160, Arten 513-608 in 1890; pp. xlix-lvi, Gattungen 161-192, Arten 609-818 in 1891].

Seidlitz G. C. M. von. 1891: Gattungen pp. 129-160, Arten pp. 549-914. In: *Fauna Transsylvanica. Die Kaefer* (*Coleoptera*) *Siebenbürgens.* Königsberg: Hartungsche Verlagsdruckerei, [12] + lvi + 192 + 914 pp., 1 pl. [issued in parts: pp. i-xl, Gattungen 1-48, Arten 1-240 in 1888; pp. xli-lvi, Gattungen 49-128, Arten 241-544 in 1889; pp. xlix-lvi, Gattungen 129-192, Arten 545-914 in 1891].

Seidlitz G. C. M. von. 1893: Tenebrionidae. Pp. 201-400. In: Kiesenwetter H. von. & Seidlitz G. C. M. von.: *Naturgeschichte der Insecten Deutschlands. Begonnen von Dr. W. F. Erichson, fortgesetzt von Prof. Dr. H. Schaum, Dr. G. Kraatz, H. v. Kiesenwetter, Julius Weise, Edm. Reitter und Dr. G. Seidlitz. Erste Abteilung Coleoptera. Fünfter Band. Erste Hälfte.* Berlin, Nicolaische Verlags-Buchhandlung, xxviii + 877 + [1] pp. [issued in parts: pp. 201-400 in March 1893, pp. 401-608 in May 1894, pp. 609-800 in September 1896, pp. i-xxviiii + 801-877 in September 1898].

Seidlitz G. C. M. von. 1894: Tenebrionidae. Pp. 401-608. In: Kiesenwetter H. von. & Seidlitz G. C. M. von.: *Naturgeschichte der Insecten Deutschlands. Begonnen von Dr. W. F. Erichson, fortgesetzt von Prof. Dr. H. Schaum, Dr. G. Kraatz, H. v. Kiesenwetter, Julius Weise, Edm. Reitter und Dr. G. Seidlitz. Erste Abteilung Coleoptera. Fünfter Band. Erste Hälfte.* Berlin, Nicolaische Verlags-Buchhandlung, xxviii + 877 + [1] pp. [issued in parts: pp. 201-400 in 1893, pp. 401-608 in 1894, pp. 609-800 in 1896, pp. i-xxviiii + 801-877 1898].

Seidlitz G. C. M. von. 1896a: Tenebrionidae. Pp. 609-800. In: Kiesenwetter H. von. & Seidlitz G. C. M. von.: *Naturgeschichte der Insecten Deutschlands. Begonnen von Dr. W. F. Erichson, fortgesetzt von Prof. Dr. H. Schaum, Dr. G. Kraatz, H. v. Kiesenwetter, Julius Weise, Edm. Reitter und Dr. G. Seidlitz. Erste Abteilung Coleoptera. Fünfter Band. Erste Hälfte.* Berlin, Nicolaische Verlags-Buchhandlung, xxviii + 877 + [1] pp. [issued in parts: pp. 201-400 in 1893, pp. 401-608 in 1894, pp. 609-800 in 1896, pp. i-xxviiii + 801-877 in September 1898].

Seidlitz G. C. M. von. 1896b: Alleculidae. Pp. 1-305. In: *Naturgeschichte der Insecten Deutschlands. Begonnen von Dr. W. F. Erichson, fortgesetzt von Prof. Dr. H. Schaum, Dr. G. Kraatz, H. v. Kiesenwetter, Julius Weise, Edm. Reitter und Dr. G. Seidlitz. Erste Abtheilung Coleoptera. Fünfter Band. Zweite Hälfte. Lieferungen 1-3.* Berlin: Nicolaische Verlags-Buchhandlung, 968 pp.

Seidlitz G. C. M. von. 1898a: Tenebrionidae. Pp. 801-877. In: Kiesenwetter H. von. & Seidlitz G. C. M. von.: *Naturgeschichte der Insecten Deutschlands. Begonnen von Dr. W. F. Erichson, fortgesetzt von Prof. Dr. H. Schaum, Dr. G. Kraatz, H. v. Kiesenwetter, Julius Weise, Edm. Reitter und Dr. G. Seidlitz. Erste Abteilung Coleoptera. Fünfter Band. Erste Hälfte.* Berlin, Nicolaische Verlags-Buchhandlung, xxviii + 877 + [1] pp. [issued in parts: pp. 201-400 in 1893, pp. 401-608 in 1894, pp. 609-800 in 1896, pp. i-xxviiii + 801-877 in 1898].

Seidlitz G. C. M. von. 1898b: Lagriidae, pp. 306-364; Melandryidae, pp. 365-680. In: *Naturgeschichte der Insecten Deutschlands. Begonnen von Dr. W. F. Erichson, fortgesetzt von Prof. Dr. H. Schaum, Dr. G. Kraatz, H. v. Kiesenwetter, Julius Weise, Edm. Reitter und Dr. G. Seidlitz. Erste Abtheilung Coleoptera. Fünfter Band. Zweite Hälfte. Lieferungen 1-3.* Berlin: Nicolaische Verlags-Buchhandlung, 968 pp.

Seidlitz G. C. M. von. 1899: Oedemeridae. Pp. 681-968. In: *Naturgeschichte der Insecten Deutschlands. Begonnen von Dr. W. F. Erichson, fortgesetzt von Prof. Dr. H. Schaum, Dr. G. Kraatz, H. v. Kiesenwetter,*

Julius Weise, Edm. Reitter und Dr. G. Seidlitz. Erste Abtheilung Coleoptera. Fünfter Band. Zweite Hälfte. Lieferungen 1-3. Berlin: Nicolaische Verlags-Buchhandlung, 968 pp.

Seidlitz G. C. M. von. 1916: Die letzten Familien der Heteromeren. *Deutsche Entomologische Zeitschrift* **1916**: 113-128, 397-498.

Seidlitz G. C. M. von. 1917: Die letzten Familien der Heteromeren. *Deutsche Entomologische Zeitschrift* **1917**: 65-116.

Seidlitz G. C. M. von. 1920: *Naturgeschichte der Insecten Deutschlands. Begonnen von Dr. W. F. Erichson, fortgesetzt von Prof. Dr. H. Schaum, Dr. G. Kraatz, H. v. Kiesenwetter, Julius Weise, Edm. Reitter und Dr. G. Seidlitz. Erste Abtheilung. Coleoptera. Fünfter Band. Zweite Hälfte. Vierte Lieferung und Register.* Leipzig: K. F. Koehlers Antiquarium, 969-1206 pp.

Selander R. B. 1960: Bionomics, systematics, and phylogeny of Lytta, a genus of blister beetles (Coleoptera, Meloidae). *Illinois Biological Monographs* **28**: i-vi, 1-295.

Selander R. B. 1988: An annotated catalog and summary of bionomics of blister beetles of the genus Cylindrothorax (Coleoptera, Meloidae). *Transactions of the American Entomological Society* **114**: 15-70.

Selander R. B. 1991: On the nomenclature and classification of the Meloidae (Coleoptera). *Insecta Mundi* **5**: 65-94.

Semenov A. P. 1889a: Diagnoses Coleopterorum novorum ex Asia centrali et orientali. *Horae Societatis Entomologicae Rossicae* **24** [1889-1890]: 193-226.

Semenov A. P. 1889b: Ueber einige Arten der Gattung Sphenaria Mén. und Asphena Sem. *Wiener Entomologische Zeitung* **8**: 309-310.

Semenov A. P. 1891a: Diagnoses coleopterorum novorum ex Asia centrali et orientali. *Horae Societatis Entomologicae Rossicae* **25**: 262-382.

Semenov A. P. 1891b: Ueber die Arten der Coleopteren-Gattung Leptodes Sol. *Wiener Entomologische Zeitung* **10**: 268-272.

Semenov A. P. 1893a: Revisio synoptica meloidarum generis Ctenopus Fisch. *Bulletin de l'Académie de St. Petersburg* **13**: 11-20.

Semenov A. P. 1893b: Revisio synoptica Tenebrionidarum generis Leptodes Sol. *Horae Societatis Entomologicae Rossicae* **27** [1892-1893]: 225-235.

Semenov A. P. 1893c: Symbolae ad cognitionem pimeliidarum. I-III. *Horae Societatis Entomologicae Rossicae* **27** [1892-1893]: 249-264.

Semenov A. P. 1893d: Meloidarum species novae. *Horae Societatis Entomologicae Rossicae* **27** [1892-1893]: 265-281.

Semenov A. P. 1893e: Symbolae ad cognitionem pedilidarum. *Horae Societatis Entomologicae Rossicae* **27** [1892-1893]: 326-334.

Semenov A. P. 1893f: De coleopterorum familia nova. *Mélanges Biologiques tirés du Bulletin Physico-Mathématique de l'Académie Impériale des Sciences de St.-Pétersbourg* **13**: 359-366.

Semenov A. P. 1893g: Coleoptera asiatica nova. *Horae Societatis Entomologicae Rossicae* **27** [1892-1893]: 493-507.

Semenov A. P. 1894a: Supplementum ad Edm. Reitteri «Revisionem» Tenebrinidarum generis Prosodes Eschsch. *Horae Societatis Entomologicae Rossicae* **28** [1893-1894]: 179-223.

Semenov A. P. 1894b: Symbolae ad cognitionen oedemeridarum. *Horae Societatis Entomologicae Rossicae* **28** [1893-1894]: 449-474.

Semenov A. P. 1894c: Coleoptera asiatica nova. II. *Horae Societatis Entomologicae Rossicae* **28** [1893-1894]: 526-547.

Semenov A. P. 1894d: De coleopterorum familia nova. *Bulletin de l'Académie des Sciences de St.-Pétersbourg.* (N. S.) (35) **3**: 607-614.

Semenov A. P. 1895a: Coleoptera asiatica nova. III. *Horae Societatis Entomologicae Rossicae* **29** [1894-1895]: 125-144.

Semenov A. P. 1895b: Coleoptera nova rossiae europae caucasique I. *Horae Societatis Entomologicae Rossicae* **29** [1894-1895]: 242-250.

Semenov A. P. 1896a: Revisio specierum generis Petria Sem. *Annuaire du Musée Zoologique de l'Académie Impériale des Sciences de St.-Pétersbourg* **1**: 26-30.

Semenov A. P. 1896b: Coleopterorum genera faunae turanicae endemica vel praecipue peculiaria eorumque species. *Annuaire du Musée Zoologique de l'Académie Impériale des Sciences de St.-Pétersburg* **1**: 42-61.

Semenov A. P. 1896c: Coleoptera asiatica nova. VI. *Horae Societatis Entomologicae Rossicae* **29** [1894-1895]: 336-362.

Semenov A. P. 1896d: Insectorum quondam novorum faunae transcaspiae, diagnoses. *Annuaire du Muséum Zoologique de l'Académie Impériale des Sciences de St.-Pétersburg* **1**: 378-386.

Semenov A. P. 1896e: De genere Rhampholyssa Krtz. *Horae Societatis Entomologicae Rossicae* **29** [1894-1895]: 515-519.

Semenov A. P. 1898a: Zametki o zhestkokrylich europeyskoy Rossii i Kaukaza. *Bulletin de la Société Impériales des Naturalistes de Moscou* **71**: 68-115.

Semenov A. P. 1898b: Dua nova oedemeridarum genera. *Wiener Entomologische Zeitung* **17**: 175-178.

Semenov A. P. 1898c: Coleoptera nova Rossiae europaeae caucasique. III. *Horae Societatis Entomologicae Rossicae* **31** [1896-1897]: 543-554.

Semenov A. 1898d: Coleoptera nova Rossiae europaeae caucasique. *Horae Societatis Entomologicae Rossicae* **32** [1898-1899]: 280-290.

Semenov A. P. 1898e: Specierum palaearcticarum ad subgenus Zimioma Gozis generis Ostoma Laichart. spectancium synopsis. *Horae Societatis Entomologicae Rossicae* **32** [1898-1899]: 286-290.

Semenov A. P. 1899a: Coleoptera nova Rossiae europaeae caucasique. VI. *Horae Societatis Entomologicae Rossicae* **32** [1898-1899]: 604-615.

Semenov A. P. 1899b: Coleoptera asiatica nova. VI. *Horae Societatis Entomologicae Rossicae* **32** [1898-1899]: 632-656.

Semenov A. P. 1900a: Coleoptera asiatica nova. IX. *Horae Societatis Entomologicae Rossicae* **34**: 303-334.

Semenov A. P. 1900b: Commentarii de meloidis. *Horae Societatis Entomologicae Rossicae* **34**: 479-497.

Semenov A. P. 1900c: Coleoptera asiatica nova. X. *Horae Societatis Entomologicae Rossicae* **34**: 577-605.

Semenov A. P. 1900d: De nonnullis oedemeridarum generibus. *Horae Societatis Entomologicae Rossicae* **34**: 643-655.

Semenov A. P. 1900e: Coleoptera asiatica nova. XI. *Horae Societatis Entomologicae Rossicae* **34**: 676-688.

Semenov A. P. 1901a: K litoralnoy fauny Kryma: I. Phaleria pontica, sp. n. (Coleoptera, Tenebrionidae). *Revue Russe d'Entomologie* **1**: 90-97.

Semenov A. P. 1901b: Pervyy predstavitel′ v Rossii roda Hymenorus Muls. (Coleoptera, Alleculidae) i zoogeograficheskoe znachenie etoy pribavki k russkoy faune. *Revue Russe d'Entomologie* **4-5**: 167-172.

Semenov A. 1902a: Coleoptera nova Rossiae europeae Caucasique. VIII. *Horae Societatis Entomologicae Rossicae* **35** [1901-1902]: 253-265.

Semenov A. P. 1902b: Generica quaedam nomina mutanda vel emendanda. *Revue Russe d'Entomologie* **2**: 353.

Semenov A. P. 1903a: Coleoptera novae faunae turanicae. *Revue Russe d'Entomologie* **3**: 8-13.

Semenov A. P. 1903b: Analecta coleopterologica. III. *Revue Russe d'Entomologie* **3**: 169-173.

Semenov A. P. 1903c: Coleoptera novae faunae turanicae. II. *Revue Russe d'Entomologie* **3**: 353-357.

Semenov A. P. 1903d: Analecta coleopterologica. *Revue Russe d'Entomologie* **6**: 390-392.

Semenov A. P. 1905: K litoralnoy faune Kryma: II. Novyy predstavitel roda Ammobius Guér. = Ammophthorus Lac. (Coleoptera, Tenebrionidae). *Revue Russe d'Entomologie* **5**: 123-125.

Semenov A. P. 1907a: Synopsis generum tribus Platyopinorum (Coleoptera, Tenebrionidae Pimeliini). *Horae Societatis Entomologicae Rossicae* **38** [1907-1908]: 175-184.

Semenov A. P. 1907b: De novo Pimeliinorum genere, quod tribum peculiarem representat (Coleoptera, Tenebrionidae). *Revue Russe d'Entomologie* **6** [1906]: 257-260.

Semenov A. P. 1907c: Coleoptera nova heptapotamica. II. *Revue Russe d'Entomologie* **6** [1906]: 261-265.

Semenov A. P. 1908: Coleoptera nova faunae dzhungaro-tianshanicae. I. *Revue Russe d'Entomologie* **8**: 115-125.

Semenov A. P. 1909: Coleoptera nova faunae turanicae. III. *Revue Russe d'Entomologie* **8** [1908]: 295-304.

Semenov A. P. 1910a: Analecta coleopterologica. XV. *Revue Russe d'Entomologie* **9**: 24-35.

Semenov A. P. 1910b: Analecta coleopterologica. *Revue Russe d'Entomologie* **9**: 433-435.

Semenov A. P. 1910c: Symbolae ad faunam desertorum mesasiaticorum. I. Synopsis specierum generis Argyrophana Sem. 1889 (Coleoptera, Tenebrionidae). *Revue Russe d'Entomologie* **10**: 42-44.

Semenov A. P. 1910d: Coleoptera nova faunae kirgisicae. I. *Revue Russe d'Entomologie* **10**: 150-153.

Semenov A. P. 1914: Ob odnom novom vid roda Sympiezocnemis (Coleoptera, Tenebrionidae) i o nekotorykh ego osobennostyakh. *Revue Russe d'Entomologie* **13** [1913]: 506-509.

Semenov A. P. & Arnoldi C. 1937: Trois formes nouvelles de coléoptères, caractéristiques pour la faune hyémale des dèserts de la Transcaspienne. *Bulletin de la Société Entomologique de France* **42**: 221-224.

Semenov A. P. & Bogatchev A. V. 1936: Supplément à la révision du genre Blaps F. (Coleoptera, Tenebrionidae) de G. Seidlitz, 1893. *Festschrift zum 60. Geburtstage von Professor Dr. Embrik Strand* (Riga) **1**: 553-568.

Semenov A. P. & Bogatchev A. V. 1947: Novyy rod tribu Eurychorini (Col., Tenebrionidae). *Doklady Akademii Nauk Azerbaydzhanskoy SSR* **3**: 175-177.

Semenov A. P. & Skopin N. G. 1951: Novyy predstavitel roda Stalagmoptera Sols. (Coleoptera, Tenebrionidae) iz pustiny Betpak-Dala. *Uchenye Zapiski Kazaskhskogo Gos. Universiteta* (Alma Ata) **13** (1): 127-128.

Semenov A. P. & Ter-Minasjan M. E. 1937: Sur un nouveau genre paléarctique de la tribu des Nacerdini (Col. Oedemeridae). *Bulletin de la Société Entomologique de France* **1937**: 227-229.

Sénac H. 1880: Description de quatre nouvelles espèces appartenant au genre Pimelia. *Annales de la Société Entomologique de France* (5) **10**: 261-268.

Sénac H. 1881: [new taxa]. *Bulletin de la Société Entomologique de France* **1881**: xix-xx.

Sénac H. 1882a: [new taxa]. *Bulletin de la Société Entomologique de France* **1882**: xxx-xxxi.

Sénac H. 1882b: [new taxa]. *Bulletin de la Société Entomologique de France* **1882**: lvi-lviii.

Sénac H. 1884a: *Essai monographique sur le genre Pimelia (Fabricius). Première partie. Espèces à tarses postérieurs et intermédiaires comprimés.* Paris: Jacques Chevalier, xix + 106 pp.

Sénac H. 1884b: [new taxa]. *Bulletin de la Société Entomologique de France* **1884**: x-xi.

Sénac H. 1884c: [new taxa]. *Bulletin de la Société Entomologique de France* **1884**: xxiv-xxv.

Sénac H. 1885: [new taxa]. *Bulletin de la Société Entomologique de France* **1885**: lxxxi-lxxxii.

Sénac H. 1886: [new taxa]. *Bulletin de la Société Entomologique de France* **1886**: xlv-xlvi.

Sénac H. 1887a: *Essai monographique sur le genre Pimelia (Fabricius). Deuxième et dernière partie. Espèces à tarses postérieurs et intermédiaires non comprimés (2e Division de Solier).* Paris: Jacques Lechevalier, xv + 160 pp.

Sénac H.. 1887b: Sous-genre Pachyscelodes (subdivision nouvelle du genre Pachyscelis Sol.). *Annales de la Société Entomologique de France* (6) **7**: 187-194.

Sénac H. 1892: Voyage de M. Ch. Alluaud aux îles Canaries (novembre 1889 – juin 1890). *Annales de la Société Entomologique de France* **61**: 103-108.

Sengupta T. 1977: Discovery of the family Elacatidae (=Othniidae), Coleoptera from India with description of a new species. *Oriental Insect* **11**: 537-539.

Sengupta T. 1980: Ergebnisse der Bhutan-Expedition 1972 des Naturhistorischen Museums in Basel. Coleoptera: Fam. Elacatidae (= Othniidae). *Entomologica Basiliensia* **5**: 28-30.

Sengupta T., Pal T. K. & Mukhopadhyay P. 1977: On the family Inopeplidae (Coleoptera) from India. *Oriental Insects* **11**: 395-407.

Sharp D. S. 1876: Descriptions of some new genera and species of new Zealand Coleoptera. *The Entomologist's Monthly Magazine* **13** [1876-1877]: 70-79.

Sharp D. S. 1879: On some Coleoptera from Hawaiian Islands. *Transaction of the Entomological Society of London* **1879**: 77-105.

Sharp D. S. 1885a: On the Colydiidae collected by Mr. G. Lewis in Japan. *The Journal of the Linnaean Society. Zoology* **19**: 58-84, pl. III.

Sharp D. S. 1885b: On some Colydiidae obtained by Mr. Lewis in Ceylon. *The Journal of the Linnaean Society. Zoology* **19**: 117-131, pl. VI.

Sharp D. S. 1885: [new taxa]. In: Blackburn T. & Sharp D. S.: Memoirs on the Coleoptera of the Hawaiian Islands. *The Scientific Transactions of the Royal Dublin Society* (2) **3**: 119-300, pls. 4-5.

Sharp D. S. 1891: Descriptions of two new species and a new genus of Japanese Coleoptera. *The Entomologist* **24** (*Supplement*): 6-7.

Sharp D. S. 1894: Fam. Colydiidae. Pp. 441-488. In: Godman F. & Salvin O. (eds): *Biologia-Centrali-Americana. Insecta. Coleoptera. Vol. II. Part I.* London: Dulau & Co., 717 pp., 19 pls.

Sharp D. S. 1902: Mycetophagidae. Pp. 637-642. [1887-1905]. In: Godman F. & Salvin O. (eds): *Biologia Centrali-Americana. Insecta. Coleoptera. Vol. II. Part. I.* [1887-1905]. London: Dulau & Co. 717 pp., 19 pls.

Sharp D. S. 1909: Some critical remarks on the genus Rabocerus, Mulsant with descriptions of two new species. *The Entomologist's Monthly Magazine* (2) **20**: 245-247.

Shestoperov E. L. 1930: Materialy po Coleoptera ekspeditsii po obsledovaniyu saranchyovykh nasekomykh v Persii. Pp. 99-107. In: *Otchet o deyatelnosti 1926-1927, 1927-1928 i 1928-1929 operatsionnye goda.* Askhabad: Izdanie Narkomzema Turkmenskoy SSR, 107 + (10) pp.

Shi A.-M. & Ren G.-D. 2005: Study on the genus Agnaptoria Reitter (Coleoptera, Tenebrionidae) from China, with the description of a new species. *Entomotaxonomia* **27**: 282. (in Chinese with English summary).

Shi A.-M., Ren G.-D. & Merkl O. 2005: Two new species of Pseudognaptorina Kaszab, 1977 (Coleoptera, Tenebrionidae: Blaptini) from the Tibet Plateau. *Acta Zoologica Academiae Scientarum Hungaricae* **51**: 163-170.

Shibata T. 1978: Notes on Tenebrionidae from Taiwan and Japan, I. (Coleoptera). *Entomological Review of Japan* **32**: 19-27.

Shibata T. 1979: Notes on Tenebrionidae from Taiwan and Japan, II. (Coleoptera). *Entomological Review of Japan* **33**: 67-73.

Shibata T. 1980: Notes on Tenebrionidae from Taiwan and Japan, III. (Coleoptera). *Entomological Review of Japan* **34**: 63-74.

Shiyake S. 1996: Redescription of Falsomordellistena auromaculata (Kôno) with description of an allied new species from Japan (Coleoptera: Mordellidae). *Bulletin of the Osaka Museum of Natural History* **50**: 9-15.

Shiyake S. 2000: Mordellidae from East Asia with description of a new species. *Bulletin of the Osaka Museum of Natural History* **54**: 25-30.

Shiyake S. 2001: A new species of Glipostenoda from Shikoku, Japan. *Special Publication of the Japan Coleopterologal Society, Osaka* **1**: 333-336.

Shuckard W. E. 1838: Description of some new genera of Coleoptera in the author's collection. *The Entomological Magazine* **5** (5): 505-513 pp., pl. xviii.

Shuckard W. E. 1839: *Elements of British Entomology. Containing a general introduction to the science, a systematic description of all the genera, and a list of all the species of British insects, with a history of their transformation, habits, economy, and distribution, with outline figures of the families and their larvae and pupae, an explanation of the technical terms, and full directions. Vol. 1.* London: Hippolyte Baillière, 240 pp.

Shuckard W. E. 1840: *The British Coleoptera delineated, consisting of figures of all the genera of British Beetles, drawn in outline by W. Spry, M.E.S.* London, 83 pp., 94 pls.

Silfverberg H. 1984: The Coleopteran genera of Dejean 1821. II. Polyphaga 1. *Annales Entomologici Fennici* **50**: 58-60.

Silfverberg H. 1991: Nomenclatorial corrections in North European Coleoptera. *Entomologica Fennica* **2**: 21-22.

Silvestri F. 1905: Descrizione di uno nuovo genere di Rhipiphoridae. *Redia* **3**: 315-324, pl. 20.

Skopin N. G. 1951a: Novye i maloizvetnye vidy saranchovykh i zhestkokrylykh iz pustini Betpak-Dala. *Uchenye Zapiski Kazaskhskogo Gos. Universiteta* (Alma Ata) **13** (1): 121-126.

Skopin N. G. 1958: Novyy vid zhukov-chernotelok triby Platyscelini iz yugo-vostochnogo Kazakhastana. *Trudy Instituta Zoologii Akademii Nauk Kazakhskoy SSR* **8**: 146-150.

Skopin N. G. 1960: Neue Tenebrioniden aus Zentralasien. I. *Annales Historico-Naturales Musei Nationalis Hungarici* **52**: 295-311.

Skopin N. G. 1961a: Materialy po faune i ekologii chernotelok (Coleoptera, Tenebrionidae) yugo-vostochnogo Kazakhstana. *Trudy Nauchno-Isledovatelskogo Instituta Zashchtity Rasteniy Kazakhstanskoy Akademii Selskokhozyastvennykh Nauk* **6**: 172-208.

Skopin N. G. 1961b: Neue Tenebrioniden aus Zentralasien II, nebst einigen systematischen sowie synonymischen Bemerkungen. *Annales Historico-Naturales Musei Nationalis* **53**: 381-406.

Skopin N. G. 1962: Lichinki podsemeystva Pimeliinae (Coleoptera, Tenebrionidae). *Trudy Nauchno-Issledovatelskogo Instituta Zashchity Rastenii Kazakhstanskoy Akademii Selskokhozyastvennykh Nauk* **7**: 191-298.

Skopin N. G. 1964a: Materialy po faune i ekologii chernotelok (Coleoptera, Tenebrionidae) Mangyshlaka i severo-zapadnogo Ustyurta. *Trudy Nauchno-Issledovatelskogo Institute Zashchity Rastenii Kazakhstanskoy Akademii Selskokhozyastvennykh Nauk* **8**: 277-297.

Skopin N. G. 1964b: Novye vidy chernotelok (Coleoptera, Tenebrionidae) iz smezhnykh s Kazakhtanom rayonov Centralnoy Azii. *Trudy Nauchno-Issledovatelskogo Institute Zashchity Rastenii Kazakhstanskoy Akademii Selskokhozyastvennykh Nauk* **8**: 371-388.

Skopin N. G. 1964c: Neue Tenebrioniden aus Zentralasien. III, nebst einigen systematischen sowie synonymischen Bemerkungen. *Annales Historico-Naturales Musei Nationalis* **56**: 389-412.

Skopin N. G. 1965: Zhuki-chernotelki (Coleoptera, Tenebrionidae) triby Platyscilini Kazakhstana.*Trudy Kazakhskogo Nauchno-Issledovatelskogo Instituta Zashchtity Rasteniy* **9**: 51-74.

Skopin N. G. 1966a: Die Arten der Gattung Tentyria Latreille aus dem westlichen Zentralasien (Coleoptera, Tenebrionidae). *Entomologische Arbeiten aus dem Museum G. Frey* **17**: 1-6.

Skopin N. G. 1966b: Neue Tenebrioniden aus Zentralasien IV., nebst einigen systematischen so wie synonymischen Bemerkungen. *Annales Historico-Naturales Musei Nationalis Hungarici* **58**: 325-342.

Skopin N. G. 1966c: Eine neue Microdera-Art aus dem Balchashseegebiet (Coleoptera, Tenebrionidae). *Reichenbachia* **8**: 69-70.

Skopin N. G. 1967: K poznaniyu zhukov-chernotelok roda Melanesthes Lac. (Coleoptera, Tenebrionidae) Centralno-Kazakhstanskoy Fauny. *Entomologicheskoe Obozrenie* **46** (1): 206-211.

Skopin N. G. 1968: Eine neue Tenebrionidengattung und –Art aus Zentralasien (Coleoptera-Tenebrionidae-Pimeliini). *Entomologische Arbeiten aus dem Museum G. Frey* **19**: 300-304.

Skopin N. G. 1968b: Zhuki-chernotelki (Coleoptera, Tenebrionidae) Yuzhnogo Kazakhstana i ikh khozyaystvennoe znachenie. *Trudy Kazakhckogo Nauchno-Issledovatelskogo Institute Zashchity Rastenii* **10**: 73-114.

Skopin N. G. 1969: Dopolnenie k faune zhukov-chernotelok (Coleoptera, Tenebrionidae) khrebeta Kutitang-Tay v Turkmenii. *Entomologicheskoe Obozrenie* **48** (4): 881-887.

Skopin N. G. 1970: Revision der Tenebrioniden-Gattung Tagona Fisch.-W., 1822 (Coleoptera, Tenebrionidae). *Reichenbachia* **13**: 157-168.

Skopin N. G. 1971: Über einige Prosodes-Arten aus Mittelasien (Coleoptera: Tenebrionidae). *Beiträge zur Entomologie* **21**: 495-501.

Skopin N. G. 1971: Monographische Übersicht der Arten der Gattungen Diesia Fisch.-W. und Platyesia gen. nov. *Entomologische Abhandlungen des Staatlichen Museum für Tierkunde in Dresden* **38**: 325-349.

Skopin N. G. 1972: Zhuki-chernotelki (Coleoptera, Tenebrionidae) rodov Diesia Fisch.-W. i Platyesia, gen. n. *Entomologicheskoe Obozrenie* **51** (1): 114-124.

Skopin N. G. 1972: Revision der Tenebrioniden-Gattung Stalagmoptera Solsky, 1875 (Coleoptera). *Entomologische Arbeiten aus dem Museum G. Frey* **23**: 298-313.

Skopin N. G. 1973: Zhuki-chernotelki (Coleoptera, Tenebrionidae) kollekcii Balliona iz Kuldzhi i blizkikh rayonov Kazakhstana i Kirgizii. [Tenebrionid beetles (Coleoptera, Tenebrionidae) from Ballion's collection collected in Kuldzha and adjacent regions of Kazakhstan and Kirgizia]. *Entomologicheskoe Obozrenie* **52** (4): 866-875.

Skopin N. G. 1973: Revision der Tenebrioniden-Gattungsgruppe Trigonoscelis-Sternoplax (Coleoptera). *Entomologische Arbeiten aus dem Museum G. Frey* **24**: 104-185.

Skopin N. G. 1974a: Zur Revision der eurasiatischen Arten der Gattung Belopus Gb. *Entomologische Abhandlungen des Staatlichen Museum für Tierkunde in Dresden* **40**: 65-103.

Skopin N. G. 1974b: Revision der Gattung Pterocoma Dejean; Solier, 1836 (Coleoptera, Tenebrionidae). *Entomologische Abhandlungen des Staatlichen Museum für Tierkunde in Dresden* **40**: 127-164.

Skopin N. G. 1974c: Nachtrag zur "Revision der Tenebrioniden-Gattungsgruppe Trigonoscelis-Sternoplax (Coleoptera)". *Entomologische Arbeiten aus dem Museum G. Frey* **25**: 259-260.

Skopin N. G. 1977: Über Blaps transversimsulcata Ball. und ihre nähere Verwandten (Coleoptera, Tenebrionidae). *Annales Historico-Naturales Musei Nationalis Hungarici* **69**: 149-151.

Skopin N. G. 1978: In: Kaszab Z., Schawaller W. & Skopin N. G.: Systematik und Ökologie einiger Tenebrioniden aus Kashmir und Ladakh. *Senckenbergiana Biologica* **59**: 215-234.

Skopin N. G. 1979: Systematische Stellung der Gattung Scythis Schaum, 1865, sowie Revision der Arten (Coleoptera, Tenebrionidae). *Annales Historico-Naturales Musei Nationalis Hungarici* **71**: 169-183.

Skopin N. G. & Kaszab Z. 1978: Über die Arten der Gattung Blaps F. (Coleoptera, Tenebrionidae), gesammelt von Herrn Dr. W. Wittmer im Jahre 1976 in Kashmir. *Folia Entomologica Hungarica* (N. S.) **31**: 207-212.

Ślipiński S. A. 1981a: Notes on the Oriental Colydiidae (Coleoptera). I. *Entomologica Basiliensia* **6**: 422-426.

Ślipiński S. A. 1984: Insects of Saudi Arabia: Fam. Colydiidae. *Fauna of Saudi Arabia* **6**: 251-254.

Ślipiński S. A. 1985: New and little known species of Colydiidae (Coleoptera) from Asia, Madagascar and Comoro Islands. *Annales Zoologici* (Warszawa) **39**: 181-195.

Smith B. 1851: *Nomenclature of coleopterous insects in the British Museum. Volume 5. Cucujidae*. London, iv + 25 pp.

Snellen van Vollenhoven S. C. 1854: [new taxa]. In: Herklots J. A.: *Bouwstoffen voor eene Fauna fon Niederland, twende deel, stück 1: 88 pp. Korte beschrijving der nieuwe soorten van Schildvleugelige Insecten in deze naamlijst opgenoem.* Leiden: E. J. Brill, 70 p.

Soldati F. & Leo P. 2005: Revision of French and Corsican species of the genus Asida Latreille, 1802 (Insecta: Coleoptera: Tenebrionidae). *Annales Zoologici* **55**: 335-373.

Soldati L. & Soldati F. 1999: A new Blaps species from Turkey (Coleoptera: Tenebrionidae). *Bulletin de la Société Linnéenne de Bordeaux* **27**: 29-32.

Soldati F. & Soldati L. 2000: A new species of Dendarus Latreille of the subgenus Pandarus Mulsant (Coleoptera: Tenebrionidae: Dendarini). *Bulletin de la Société Linnéenne de Bordeaux* **28**: 67-71.

Soldati F. & Soldati L. 2001a: Asida (Polasida) poneli n. sp., nouvelle espèce de la Riviera, et remarques taxonomiques sur les Polasida de France et d'Italie (Coleoptera, Tenebrionidae, Asidini). *Bulletin de la Société Linnéenne de Bordeaux* **29**: 103-108.

Soldati F. & Soldati L. 2001b: Une nouvelle Asida de Corse, Asida ascoense n. sp. (Coleoptera: Tenebrionidae: Asidini). *Bulletin de la Société Linnéenne de Bordeaux* **29**: 179-184.

Soldati F. & Soldati L. 2001c: Le genre Asida en Grèce (Coleoptera, Tenebrionidae, Asidini). *L'Entomologiste* **57**: 185-198.

Soldati F. & Soldati L. 2002a: Un nouveau Stenohelops du sud-est de l'Espagne (Coleoptera, Tenebrionidae, Helopini). (31. Contribution à la connaissance des Tenebrionidae). *Bulletin de la Société Linnéenne de Bordeaux* **30**: 137-142.

Soldati F. & Soldati L. 2002b: Compte-rendu d'une expedition scientifique en Grèce et description de nouveaux taxons (Coleoptera: Tenebrionidae). *Biocosme Mésogéen* **18** [2001]: 167-185.

Soldati F. & Soldati L. 2003a: Une espèce nouvelle de Phthora appartenant à un sous-genre inédit (Coleoptera, Tenebrionidae, Phalerini). (33.Contribution à la connaissance des Tenebrionidae). *Bulletin de la Société Linnéenne de Bordeaux* **31**: 1-8.

Soldati F. & Soldati L. 2003b: Une nouvelle espèce du genre Asida de l'île de Skyros, Grèce (Coleoptera, Tenebrionidae, Asidini). *Revue de l'Association Roussillonnaise d'Entomologie (RARE)* **12** (2): 43-45.

Soldati F. & Leo P. 2005: Revision of French and Corsican species of the genus Asida Latreille, 1802 (Insecta, Coleoptera, Tenebrionidae). *Annales Zoologici* **55** (3): 335-373.

Solier A. J. J. 1834: Essai d'une division des coléoptères hétéromères, et d'une monographie de la famille des collaptèrides. *Annales de la Société Entomologique de France* **3**: 479-636, pls XII-XVI.

Solier M. 1835a: Prodrome de la famille des xystropides. *Annales de la Société Entomologiqué de France* **4**: 229-248.

Solier A. J. J. 1835b: Essai sur les collaptèrides (suite). 2e tribu. Tentyrites. *Annales de la Société Entomologique de France* **4**: 249-419, pls VI-IX.

Solier A. J. J. 1835c: Essai sur les collaptèrides (suite). 3e Tribu. Macropodites. *Annales de la Société Entomologique de France* **4**: 509-574, pls XIV, XV.

Solier A. J. J. 1836a: Essai sur les collaptèrides (suite). 4e Tribu. Pimélites. *Annales de la Société Entomologique de France* **5** [1835-1836]: 5-200, pls I-IV.

Solier A. J. J. 1836b: Essai sur les collaptèrides (suite). 5e Tribu. Nyctélites. *Annales de la Société Entomologique de France* **5** [1835-1836]: 303-355, pls VI, VII.

Solier A. J. J. 1836c: Essai sur les collaptèrides (suite). 6e Tribu. Asidites. *Annales de la Société Entomologique de France* **5** [1835-1836]: 403-512, pls XI-XIII.

Solier A. J. J. 1837a: Essai sur les collaptèrides (suite). 7e Tribu. Akisites. *Annales de la Société Entomologique de France* **5** [1835-1836]: 635-684, pl. XXIII.

Solier A. J. J. 1837b: Essai sur les collaptèrides (suite). 8e Tribu. - Adélostomites. *Annales de la Société Entomologique de France* **6** [1836-1837]: 151-172, pl. VII.

Solier A. J. J. 1838a: Essai sur les collaptèrides (suite). Deuxième division. *Annales de la Société Entomologique de France* **7**: 5-73, pls I-III.

Solier A. J. J. 1838b: Essai sur les collaptèrides (suite). 10e Tribu. Scaurites. *Annales de la Société Entomologique de France* **8**: 159-199, pls VII, VIII.

Solier A. J. J. 1838c: [new taxa]. In: Laporte F. L. N. de Caumont de Castelnau: *Histoire naturelle des insectes coléoptères; avec une introduction renfermant l'anatomie et la physiologie des animaux articulés, par M. Brullé. Tome deuxième.* Paris: P. Duménil, 563 + [1] pp., pls 20-37.

Solier A. J. J. 1844: Essai sur les collaptèrides de la tribu des Molurites. *Memorie della Reale Accademia delle Scienze di Torino, Serie II*, **6**: 213-339, 4 pls.

Solier A. J. J. 1848: Essai sur les collaptérides. 14e Tribu. Blapsites. Pp. 149-370, pls iv – xv. In: Baudi di Selve: *Studi Entomologici. Publ. Per di Flamino Baudi e di Eugenio Truqui. Tom. 1.* Torino: Stamporia Degli Artisti Tipografi, 376 pp.

Solier A. J. J. 1851: Insectos. Coleopteros. Pp. 1-285. In: Gay C.: *Historia Fisica y Politica de Chile. Fauna Chilena. Zoologie Tomo. 5.* Paris: Maulde et Renou, 563 pp.

Solsky [= Solskij] S. M. 1866: Matériaux pour servir à l'étude des insectes de la Russie. I. Notes sur quelques coléoptères nouveaux ou peu connus. *Horae Societatis Entomologicae Rossicae* **4** [1866-1867]: 79-96.

Solsky [= Solskij] S. M. 1871: Coléoptères de la Sibirie orientale. *Horae Societatis Entomologicae Rossicae* **7** [1870-1871]: 334-406.

Solsky [= Solskij] S. M. 1876: Matériaux pour l'entomographie des provinces asiatiques de la Russie. *Horae Societatis Entomologicae Rossicae* **11** [1875-1876]: 253-272, 273-299.

Solsky [= Solskij] S. M. 1881: Novyya ili maloizve'stnyya zhestkokrylyya okrain Rossiyskoy Imperii i prilegayushchikh k ney stran. *Trudy Russkago Entomologicheskago Obshchestva* **12** [1879-1881]: 231-265.

Solsky [= Solskij] S. M. 1882: Novyya ili maloizve'stnyya zhestkokrylyya okrain Rossiyskoy Imperii i prilegayushchikh k ney stran. *Trudy Russkago Entomologicheskago Obshchestva* **13** [1881-1882]: 31-84.

Spilman T. J. 1954: Generic names of the Salpingidae and their type species (Coleoptera). *Journal of the Washington Academy of Sciences* **44** (3): 85-94.

Spilman T. J. 1973: Nomenclatural problems in six genera of Tenebrionidae (Coleoptera). *Proceedings of the Entomological Society of Washington* **75**: 39-44.

Starck A. 1889: Coleoptera nova Imperii Rossici. *Wiener Entomologische Zeitung* **8**: 53-55.

Stchegoleva-Barovskaya T. 1927: De mordellidis (Coleoptera) palaearcticis tribus novis. *Annuaire du Musée Zoologique de l'Académie des Sciences de l'URSS* **28**: 158-161.

Stchegoleva-Barovskaya T. 1930a: De duabus novis mordellidarum speciebus e ponticis. *Revue Russe d'Entomologie* **24**: 56-58.

Stchegoleva-Barovskaya T. 1930b: De mordellidis (Coleoptera) novis in collectione Musei Zoologici Academiae Scientiarum. *Comptes Rendus de l'Académie des Sciences de l'URSS* **1930**: 750-752.

Stchegoleva-Barovskaya T. 1931: Aperçu des coléoptères de la fam. Mordellidae de la région nord-ouest de l'URSS (Ingrie). *Annuaire du Musée Zoologique de l'Académie des Sciences de l'URSS* **32**: 51-65.

Stchegoleva-Barovskaya T. 1932: Nouvelles espèces asiatiques de la fam. Mordellidae. *Comptes Rendus de l'Académie des Sciences de l'URSS* **1932**: 219-222.

Stephens J. F. 1829a: *The nomenclature of British insects; being a compendious list of such species as are contained in the systematic catalogue of the British insects, and forming a guide to their classification, &c. &c.* London: Baldwin & Craddock, (2) + 68 pp.

Stephens J. F. 1829b: *A systematic catalogue of British insects: being an attempt to arrange all the hitherto discovered indigenous insects in accordance with their natural affinities. Containing also the references to every English writer on entomology, and to the principal foreign authors. With all the published British genera to the present time.* London: Baldwin & Cradock, xxxiv + [2]+ 416 + 388 pp.

Stephens J. F. 1830: *Illustrations of British Entomology; or, a synopsis of indigenous insects: containing their generic and specific distinctions; with an account of their metamorphoses, times of appearance, localities, food, and economy, as far as practicable. Mandibulata. Volume III.* London: Baldwin and Craddock, 447 pp., pls. XVI-XIX.

Stephens J. F. 1831: *Illustrations of British Entomology; or, a synopsis of indigenous insects: containing their generic and specific distinctions; with an account of their metamorphoses, times of appearance, localities, food, and economy, as far as practicable. Mandibulata. Volume IV.* London: Baldwin & Cradock, 413 + [1] pp., pls XX-XXIII. [issued in parts, p. 1-366 iun 1831; pp. 367-413 in 1832].

Stephens J. F. 1832: Pp. 1-240, pls. XXIV-XXVI. In: *Illustrations of British Entomology; or, a synopsis of indigenous insects: containing their generic and specific distinctions; with an account of their metamorphoses, times of appearance, localities, food, and economy, as far as practicable. Mandibulata* [1832-1835] *Volume V.* London: Baldwin and Cradock, 447 pp. + list of illustrations, pls. XXIV-XXVII. [issued in parts: pp. 1-240, 1832; pp. 241-304, 1833; pp. 305-368, 1934; pp. 369-448, 1835].

Stephens J. F. 1835: Pp. 369-448. In: *Illustrations of British Entomology; or, a synopsis of indigenous insects containing their generic and specific distinctions; with an account of their metamorphoses, times of appearance, localities, food, and economy, as far as practicable. Mandibulata. Volume V.* London: Baldwin and Cradock, 447 pp. + list of illustrations, pls. XXIV-XXVII. [issued in parts: pp. 1-240, 1832; pp. 241-304, 1833; pp. 305-368, 1934; pp. 369-448, 1835].

Stephens J. F. 1839: *Manual of British Coleoptera, or beetles; containing a brief description of all the species of beetles hitherto ascertained to inhabit Great Britain and Ireland; together with a notice of their chief localities, times and places of appearances etc.* London: Longman, Orme Brown, Green and Longmans, xii + 443 pp.

Steven C. von. 1806: Déscription de dix insectes de la Russie méridionale. *Mémoires de la Société des Naturalistes de l'Université Impériale de Moscou* **1**: 155-167, pl. 10.

Steven C. von. 1811: Description de dix insectes de la Russie méridionale. *Mémoires de la Société des Naturalistes de l'Université Impériale de Moscou* **1**: 118-130, pl. 10.

Steven C. von. 1829: *Museum Historiae Naturalis. Universitatis Caesareae Mosquensis. Pars 2. Insecta.* Mosquae: Typis Universitatis Caesareae, 147 pp.

Steven C. von. 1829: Tentyriae et Opatra collectionis Stevenianae nunc Musei Universitatis Mosquensis. *Nouvelles Mémoires de la Société Impérial des Naturalistes de Moscou* **1**: 83-100.

Steven C. von. 1829: [new taxa]. In: Krynicki I.: Enumeratio Coleopterorum Rossiae meridionalis et praecipue in Universitatis Caesareae Charkoviensis circulo obvenientium, quae annorum 1827 - 1831 spatio observavit. *Bulletin de la Société Impériale des Naturalistes de Moscou* **5**: 65-179, 2 pls.

Stierlin G. 1867: Beschreibung einiger neuen Käfer-Arten. *Mitteilungen der Schweizerischen Entomologischen Gesellschaft* **2**: 218-228.

Stierlin G. 1876: Beschreibung einiger neuer Käferarten. *Mitteilungen der Schweizerischen Entomologischen Gesellschaft* **4**: 473-493.

Stierlin G. 1902: Beschreibung einiger neuen Species von Coleopteren. *Mitteilungen der Schweizerischen Entomologischen Gesellschaft* **10**: 425-427.

Stolz H. 1915: Eine neue Oedemera von der Insel Kreta. *Coleopterologische Rundschau* **4**: 91-93.

Strand A. 1965: Über die nordischen Arten der Gattung Cis Latr., Untergattung Eridaulus Thoms., mit Beschreibung einer neuen Art, Cis hanseni n. sp. (Col., Cisidae). *Norsk Entomologisk Tidsskrift* **13**: 61-65.

Strand E. 1917: Übersicht der in Gistel's "Achthundert und zwanzig neue oder unbeschriebene wirbellose Thiere" (1857) behandelten Insekten. *Archiv für Naturgeschichte* (A) **82** (5) [1916]: 75-101.

Strand E. 1929: Zoological and palaeontological nomenclatorial notes. *Acta Universitatis Latviensis* **20**: 1-29.

Strand E. 1935: Revision von Gattungsnamen palaearktischer Coleoptera. *Folia Zoologica et Hydrobiologica* **7**: 282-299.

Strassen R. zur. 1957: Zur Kenntnis der Arten-Gruppe Akis spinosa Linnaeus, genei Solier und granulifera Sahlberg (Ins. Col. Tenebr.). *Senckenbergiana Biologica* **38**: 41-59.

Sturm J. 1807: *Deutschlands Fauna in Abbildungen nach der Natur mit Beschreibungen. V. Abtheilung. Die Insecten. Zweites Bändchen. Käfer.* Nürnberg: J. Sturm, 279 pp., pls. 21-52.

Sturm J. 1826: *Catalog meiner Insecten-Sammlung. Erster Theil. Käfer.* Nürnberg: Verfasser, viii +207pp., 4 pls.

Sturm J. 1843: *Catalog der Käfersammlung von J. Sturm.* Nürnberg: Verfasser, 12 + 386 pp.

Sturm J. H. C. F. 1849: *Dr. Jacob Sturm's Deutschlands Fauna in Abbildungen nach der Natur mit Beschreibungen. V Abtheilung. Die Insekten. Zwanzigstes Bändchen. Käfer.* Nürnberg: Author, vi + 103 pp. pls. 361-376.

Suffrian C. W. L. E. 1848: Entomologische Bemerkungen. *Entomologische Zeitung* (Stettin) **9**: 98-108, 165-171.

Suffrian C. W. L. E. 1853: Synonymische Miscellaneen. *Entomologische Zeitschrift* (Stettin) **13**: 232-237.

Sulzer J. H. 1761: *Die Kennzeichen der Insecten.* Zürich: Heiddeger, 28 + 204 + 67 pp., 24 pls.

Sulzer J. H. 1776: *Abgekürzte Geschichte der Insecten nach dem Linneischen System.* Winterthur: H. Steiner & Co., Teil 1: 27 + 274 pp.; Teil 2: 71 pp., 32 pls.

Sumakov G. G. 1902: Materialy dlya fauny zhestokrylykh Zakaspitskoy oblasti. Beiträge zur Fauna der Coleopteren des Transkaspischen Gebietes. *Protokoly Obshchestva Estestvoispytateley pri Imperatskom Yuryevskom Universitete* **13** (1) [1901]: 21-48.

Sumakov G. G. 1908a: Novyy vid roda Steropes Stev. (Coleoptera, Anthicidae). *Revue Russe d'Entomologie* **8**: 133-134.

Sumakov G. G. 1908b: Novye sredneaziatskie zhestkokrylye (Col. Tenebrionidae i Carabidae). *Russkoe Entomologicheskoe Obozrenie* **8**: 245-247.

Sumakov G. G. 1909: Beiträge zur Fauna der Coleopteren von Turkestan und dem transkaspischen Gebiet. *Sitzungsberichte der Naturforschenden Gesellschaft, Universität Dorpat* **17**: 209-224.

Sumakov G. G. 1913a: Nouvelles variations de dessin sur les élytres des espèces asiatiques de Mylabris F. (Coleoptera, Meloidae). *Revue Russe d'Entomologie* **13**: 304-307.

Sumakov G. G. 1913b: Deux nouvelles espèces du genre Mylabris F. (Coleoptera Meloidae). *Revue Russe d'Entomologie* **13**: 473-475.

Sumakov G. G. 1915: Les espèces paléarctiques du genre Mylabris Fabr. (Coleoptera, Meloidae). *Horae Societatis Entomologicae Rossicae* **42**: 1-71.

Sumakov G. G. 1924: Corrections synonymiques sur quelques espèces du genre Mylabris F. *Sitzungsberichte der Naturforschenden Gesellschaft, Universität Dorpat* **31**: 80-90.

Sumakov G. G. 1929a: Nouvelles espèces, sous-espèces et morphes paléarctiques du genre Mylabris F. et quelques corrections synonymiques et systématiques. *L'Échange, Revue Linnéenne* **45** [hors texte] (435): 1-7.

Sumakov G. G. 1929b: Diagnosis einer neuen Art der Gattung Zonitis aus dem Kaukasus. *Entomologisches Nachrichtenblatt* **3**: 113-114.

Sumakov G. G. 1930a: Catalogue des espèces paléarctiques de tribu Mylabrina (Coleoptera, Meloidae). *Acta Instituti et Musei Zoologici Universitatis Tartuensis* **37**: 1-114.

Sumakov G. G. 1930b: [new name]. In: Pic M.: Annonces et avis divers. *L'Échange, Revue Linnéenne* 46: [9].

Sumakov G. G. 1934: Inventa entomologica itineris Hispanici et Maroccani quod a 1926 fecerunt Harald et Håkan Lindberg. XXI Méloides du basin Méditérranéen et du Maroc. *Commentationes Biologicae* **4** (13): 1-6.

Sundevall J. C. 1831: Beschreibung einer neuen Coleopteren-Gattung, Symbius Blattarum. *Isis von Oken* **11**: 1222-1228, pl. VIII.

Švihla V. 1978: Two new Oedemera species from Europe (Coleoptera, Oedemeridae). *Acta Entomologica Bohemoslovaca* **78**: 35-41.

Švihla V. 1980: Ergebnisse der Bhutan-Expedition des Naturhistorischen Museums in Basel. Coleoptera. Fam. Oedemeridae. *Entomologica Basiliensia* **5**: 45-58.

Švihla V. 1982: New species of the family Oedemeridae (Coleoptera) from the Palaearctic Region. *Entomologica Basiliensia* **7**: 381-386.

Švihla V. 1983a: Results of the Czechoslovak-Iranian entomological expeditions to Iran 1970, 1973 and 1977. Coleoptera, Oedemeridae. *Acta Entomologica Musei Nationalis Pragae* **41**: 107-132.

Švihla V. 1983b: Contribution to the knowledge of the palaearctic Oedemeridae (Coleoptera). *Entomologica Basiliensia* **8**: 334-341.

Švihla V. 1984: Insects of Saudi Arabia. Coleoptera: Fam. Oedemeridae (Part 2). *Fauna of Saudi Arabia* **6**: 260-266.

Švihla V. 1986a: Revision of the generic classification of the Old World Oedemeridae (Coleoptera). *Sborník Národniho Muzea v Praze* **41 B**: 141-238.

Švihla V. 1986b: Revision of Ischnomera species from Central Asia and the western Himalayas (Coleoptera, Oedemeridae). *Acta Entomologica Bohemoslovaca* **83**: 213-220.

Švihla V. 1987: Contribution to the knowledge of the Old World Oedemeridae (Coleoptera). *Annotationes Zoologicae et Botanicae* (Bratislava) **181**: 1-27.

Švihla V. 1988a: Revision of Ischnomera species from the western Palaearctic (Coleoptera, Oedemeridae). *Acta Entomologica Bohemoslovaca* **85**: 205-222.

Švihla V. 1988b: Review of the family Oedemeridae (Coleoptera) from Macaronesia. *Acta Entomologica Bohemoslovaca* **85**: 373-379.

Švihla V. 1991: Contribution to the knowledge of the Old World Oedemeridae (Coleoptera). *Annotationes Zoologicae et Botanicae* (Bratislava) **202**: 1-14.

Švihla V. 1992: Revision of Ischnomera species from the eastern Palaearctic (Coleoptera, Oedemeridae). *Acta Entomologica Bohemoslovaca* **89**: 35-46.

Švihla V. 1993a: Contribution to the knowledge of the Palaearctic Oedemeridae (Coleoptera). *Folia Heyrovskyana* **1**: 52-55.

Švihla V. 1993b: Revision of the genus Oedemera subgenus Oncomera (Coleoptera: Oedemeridae). *European Journal of Entomology* **90**: 189-208.

Švihla V. 1994: Revision of the subgenus Dryopomera s. str. (Coleoptera: Oedemeridae). *European Journal of Entomology* **91**: 237-254.

Švihla V. 1995: Contribution to the knowledge of the family Oedemeridae (Coleoptera) of arid regions of the Palaearctic. *Folia Heyrovskyana* **3**: 9-23.

Švihla V. 1996: Contribution to the knowledge of the Old World Oedemeridae (Coleoptera). *Acta Musei Nationalis Pragae, Series B, Historia Naturalis* **51**: 83-88.

Švihla V. 1997a: Contribution to the knowledge of the Palaearctic and Oriental Oedemeridae (Coleoptera). *Folia Heyrovskyana* **5**: 83-96.

Švihla V. 1997b: Revision of the genus Ascleropsis Seidlitz and related genera (Coleoptera, Oedemeridae). *Entomologica Basiliensia* **20**: 417-466.

Švihla V. 1998a: Revision of Nacerdes subgenera Xanthochroa and Asiochroa subgen. n. (Coleoptera: Oedemeridae) from western and southern China and adjacent regions. *Folia Heyrovskyana* **6**: 49-71.

Švihla V. 1998b: Revision of the genera Ascleranoncodes, Pseudonerdanus and Falsonerdanus (Coleoptera: Oedemeridae). *Folia Heyrovskyana* **6**: 147-166.

Švihla V. 1998c: Notes on some species of the genus Ischnomera (Coleoptera: Oedemeridae) from Turkey. *Klapalekiana* **34**: 239-241.

Švihla V. 1999a: Revision of the Old World Diplectrus with notes on the other genera (Coleoptera: Oedemeridae). *Folia Heyrovskyana* **7**: 73-86.

Švihla V. 1999b: Revision of the subgenera Stenaxis and Oedemera s. str. of the genus Oedemera (Coleoptera: Oedemeridae). *Folia Heyrovskyana*, Supplementum **4**: 1-117.

Švihla V. 2001a: New taxa of the Old World Ditylini (Coleoptera: Oedemeridae). *Folia Heyrovskyana* **9**: 89-98.

Švihla V. 2001b: Contribution to the knowledge of the genus Dryopomera (Coleoptera: Oedemeridae). *Acta Societatis Zoologicae Bohemicae* **65**: 127-134.

Švihla V. 2001c: Revision of the Nacerdes subgenus Xanthochroa (Coleoptera: Oedemeridae) from Malaysia and Indonesia with notes on species from other regions. *Folia Heyrovskyana* **9**: 217-234.

Švihla V. 2002: New species of the genus Indasclera Švihla, 1980 (Insecta: Coleoptera: Oedemeridae). *Reichenbachia* **34**: 333-339.

Švihla V. 2003: New taxa of the genus Oedemera (Coleoptera, Oedemeridae) from China. *Satonius, Special Bulletin of the Japanese Society of Coleopterology, Tokyo* **6**: 339-344.

Švihla V. 2004: Contribution to the knowledge of the Old World Oedemeridae (Coleoptera). *Acta Societatis Zoologicae Bohemiae* **68**: 61-78.

Švihla V. 2005a: On the taxonomy and nomenclature of Stenostoma rostratum (Coleoptera: Oedemeridae). *Časopis Národního Muzea, Řada Přírodovědná* **174**: 71-73.

Švihla V. 2005b: New species and synonymies in Oedemeridae (Coleoptera) from the Palaearctic and Oriental regions. *Folia Heyrovskyana* **13**: 71-80.

Švihla V. 2006a: A revision of the geus Sparedrus (Coleoptera: Oedemeridae) from central part of the Palaearctic Region and from western part of the Oriental Region. *Folia Heyrovskyana, Series A* **14**: 1-35.

Švihla V. 2006b: New species of the genus Sparedrus (Coleoptera: Oedemeridae) from Bulgaria. *Acta Entomologica Musei Nationalis Pragae* **46**: 123-125.

Švihla V. 2006c: Revision of Chitona species (Coleoptera: Oedemeridae) from eastern Mediterranean. *Acta Entomologica Musei Nationalis Pragae* **46**: 107-121.

Švihla V. 2006d: On the taxonomy of the Palaearctic Oedemeridae (Coleoptera). *Časopis Národního Muzea v Praze, Řada Přírodovědná* **175**: 1-12.

Švihla V. 2007: Preliminary revision of the genus Sparedrus (Coleoptera: Oedemeridae) from eastern and southeastern Asia. *Acta Entomologica Musei Nationalis Pragae* **47**: 153-168.

Szallies A. 1994: Drei neue Coleopteren-Arten aus der Türkei (Coleoptera: Prostomidae, Cerambycidae). *Entomologische Zeitschrift* **104**: 259-263.

Takakuwa M. 1977: A new species of the genus Glipa LeConte from the Ryukyu Islands. *Elytra* **5**: 9-11.

Takakuwa M. 1978: Occurence of a new species of the genus Yukahananomia Kôno (Mordellidae) in Central Honshu, Japan. *Elytra* **6**: 5-8.

Takakuwa M. 1985: Notes on the tribe Mordellini from the Ogasawara Islands with descriptions of three new species. *Gekkan-Mushi* **176**: 4-11.

Takakuwa M. 1985a: Two new species of the mordellid beetles from the Ryukyus. *Elytra* **13**: 71-74.

Takakuwa M. 1986: The group of Glipa formosana (Coleoptera, Mordellidae) from Amami-Oshima Island of the Ryukyus. *Entomological Papers presented to Yoshihiko Kurosawa on the Occasion of his Retirement* **1986**: 257-263.

Takakuwa M. 1986a: The group of Hoshihananomia hananomi (Coleoptera, Mordellidae) in Japan. *Papers on Entomology presented to Prof. Takehiko Nakane in Commemoration of his Retirement* **1986**: 237-244.

Takakuwa M. 1993: A peculiar new species of the mordellid genus Glipa (Coleoptera) from Yakushima Island, Southwest Japan. *Bulletin of the Kanagawa Prefectural Museum* (*Nat. Sci.*) **22**: 91-94.

Takakuwa M. 1995: A new species of the genus Yukahananomia (Coleoptera, Mordellidae) from Taiwan. *Special Bulletin of the Japannese Society of Coleopterology, Tokyo* **4**: 433-436.

Takakuwa M. 1999: A new mordellid species allied to Glipa oshimana (Coleoptera, Mordellidae) from the Ryukyu Islands. *Elytra* **27**: 65-69.

Takakuwa M. 2000: A taxonomic study of the mordellid subgenus Stenoglipa. *Bulletin Kanagawa Prefectural Museum* (*Nat. Sci.*) **29**: 53-105.

Takakuwa M. 2001: A remarkable new species of the genus Mordella (Coleoptera: Mordellidae) from the Central Ryukyus. *Special Publication of the Japan Coleopterologal Society, Osaka* **1**: 293-296.

Takakuwa M. 2001a: A beautiful new species of the genus Variimorda (Coleoptera, Mordellidae) from the Southern Ryukyus. *Elytra* **29**: 310-314.

Takakuwa M. 2002: Glipa watanabeorum (Coleoptera, Mordellidae) a new mordellid beetle from Southern Ryukyus. *Special Bulletin of the Japannese Society of Coleopterology, Tokyo* **5**: 357-361.

Tan J. J. 1958: On the genus Epicauta Redtb. of China (Meloidae, Coleoptera). *Acta Entomologica Sinica* **8**: 152-167 (in Chinese with English summary).

Tan J. J. 1981: Coleoptera Meloidae. In: Zhonggue & al. (ed.): *The series of the comprehensive scientific expedition to the Qhinghai-Xizang Plateau. Insects of Xizang. Vol. 1.* Peking: Sciences Press, xi + 600 pp. (in Chinese with English summary).

Tan J. J. 1988: [Meloidae]. Pp. 289–292. In: Climbing Expedition of Chinese Academy of Sciences (ed.): *Insects of Mt. Narjabarwa Region of Xizang*. Beijing: Science Press.

Tauscher A. M. 1812a: Tentyriae ruthenicae descriptionibus iconibusque illustratae. *Mémoires de la Société Impériale des Naturalistes de Moscou* **3**: 22-42, 313-316, pls 2, 3.

Tauscher A. M. 1812b: Enumeratio et descriptio insectorum e familia Cantharidarum quae in Russia observavit. *Memories de la Société Impériale des Naturalistes de Moscou* **3**: 129-164, 2 pls.

Telnov D. 1998: Anthicidae (Coleoptera) der Sammlung Sergej Kurbatov, mit Beschreibung von sechs neuen Arten aus der Orientalis. *Bulletin et Annales Société Royale Belge Entomologique* **134**: 81-100.

Telnov D. 2000: Neue Anthicidae (Insecta: Coleoptera) aus der Sammlung des Naturkundemuseums Erfurt. *Veröffentlichungen des Naturkundemuseums Erfurt* **19**: 173-184.

Telnov D. 2001: Anthicidae (Coleoptera) aus der Sammlung des Naturhistorischen Museums in Basel, Teil 1: Einleitung, Steropinae, Tomoderinae, Ischaliinae. *Entomologische Zeitschrift, Stuttgart* **111**: 117-123.

Telnov D. 2002a: Vorläufige Auflistung der Anthicidae (Coleoptera) von Afghanistan mit Beschreibung einer neuen Art. *Latvijas Entomologs* **39**: 20-29.

Telnov D. 2002b: Yunnanomonticola nanzhao - eine neue Gattung und Art der orientalischen Anthicini (Coleoptera: Anthicidae, Anthicini). *Entomologische Zeitschrift, Stuttgart* **112**: 47-50.

Telnov D. 2003a: Neue Arten der Gattung Anthelephila Hope, 1833 (Coleoptera: Anthicidae). *Entomologische Zeitschrift* **113**: 130-137.

Telnov D. 2003b: Trictenotomidae und Anthicidae (Insecta: Coleoptera) des Himalayas und angrenzender Regionen, Teil 1. Systematik, Faunistik, Zoogeographie, pp. 279-303. In: Hartmann M. & Baumbach H. (eds.): *Biodiversität und Naturausstattung im Himalaya.* Erfurt: Verein der Freunde und Förderer des Naturkundemuseums Erfurt e. V., 389 pp. + 16 pls.

Telnov D. 2004: [new names]. In: Chandler D. S., Nardi G., & Telnov D.: Nomenclatural notes on the Palaearctic Anthicidae (Coleoptera). *Mitteilungen der Internationalen Entomologischen Vereins e. V.* (Frankfurt am Main) **29**: 109-173.

Telnov D. 2005: Anthicidae (Coleoptera) aus der Sammlung des Naturhistorischen Museums in Basel. Teil II: Bemerkenswerte und neue Notoxini, Endomiini und Anthicini (Anthicinae) aus der paläarktischen und orientalischen Region. *Entomologica Basiliensia et Collectionis Frey* **27**: 161-180.

Telnov D. 2007: Redefinition of Pseudoleptaleus Pic, 1900 (Coleoptera: Anthicidae, Anthicinae). *Entomologische Zeitschrift* (Stuttgart) **117**: 71-82.

Tenenbaum S. 1915: Neue Käferarten von den Balearen. *Bulletin de l'Académie des Sciences de Cracovie* **1914**: 837-843, pls 63, 64.

Théry A. 1892: [new taxon]. *Bulletin de la Société Entomologique de France* **1892**: clxxx-clxxxi.

Théry A. 1921: Les Stenosini du Maroc (Coléoptères Tenebrionidae). *Bulletin de la Société des Sciences Naturelles du Maroc* **1**: 51-54.

Théry A. 1924: Description d'un Stenosis nouveau du Maroc (Coléoptère Tenebrionidae). *Bulletin de la Société des Sciences Naturelles du Maroc* **4**: 149-150.

Théry A. 1931a: Notes d'entomologie marocaine et nord-africaine (Huitième note). *Bulletin de la Société des Sciences Naturelles du Maroc* **11**: 26, pl. 1.

Théry A. 1931b: Notes d'entomologie marocaine et nord-africaine (Dixième note). *Bulletin de la Société des Sciences Naturelles du Maroc* **12**: 119-127.

Théry A. 1932a: Notes d'entomologie marocaine et nord-africaine (Onzième note). *Bulletin de la Société des Sciences Naturelles du Maroc* **12** [1931]: 189-198.

Théry A. 1932b: Notes d´entomologie marocaine et nord-africaine (Douxième note). *Bulletin de la Société des Sciences Naturelles du Maroc* **13**: 129-138.

Théry A. 1934: Notes d'entomologie marocaine et nord-africaine (Treizième note). *Bulletin de la Société des Sciences Naturelles du Maroc* **14**: 47.

Thomas M. K. 1897a: Note on Deridea Westwood (Lyttidae) with the description of a new species. *The Annals and Magazine of Natural History* (6) **19**: 389-390.

Thomas M. K. 1897b: Description of five new species of Mylabrina (Lyttidae) in the collection of the British Museum. *The Annals and Magazine of Natural History* (6) **19**: 501-503.

Thomson C. G. 1858: Insectes. I. Ordre des Coléoptères. Pp. 30-343. In: Voyage au Gabon. Histoire naturelle des insectes et des arachnides recueillis pendant un voyage fait au Gabon en 1856 et en 1857 par M. Henry C. Deyrolle sous les auspices de MM. le Comte de Mniszech et Jamen Thomson, précédée de l'hisotire du voyage par J. Thomson; Arachnides par H. Lucas. *Archives Entomologiques* **2**: 1-469 + 14 pp., XIV pls.

Thomson C. G. 1859: *Skandinaviens Coleoptera, synoptiskt bearbetade. 1. Tom.* Lund: Berlinska Boktryckeriet, 290 pp.

Thomson C. G. 1863: *Skandinaviens Coleoptera, synoptiskt bearbetade. 5. Tom.* Lund: Lundbergska Boktryckeriet, 340 pp.

Thomson C. G. 1864: *Skandinaviens Coleoptera, synoptiskt bearbetate. 6. Tom.* Lund: Lundbergska Boktryckeriet, 385 pp.

Thomson C. G. 1868: *Skandinaviens Coleoptera, synoptiskt bearbetade. 10. Tom.* Lund: Lundbergska Boktryckeriet, 420 pp.

Thomson C. G. 1886: [new taxa]. *Bulletin de la Société Entomologique de France* **1886**: ix-xi.

Thomson J. 1860a: *Musée scientifique ou recueil d'histoire naturelle.* [1860-1868]. Paris: J. Thomson, 96 pp., 9 pls.

Thomson J. 1860b: Monographie de la famille des Monommides. *Annales de la Société Entomologique de France* (3) **8**: 5-38, pls 1-3.

Thunberg C. P. 1784: Novae insectorum species descriptae. *Nova Acta Regiae Societatis Scientarum Upsaliensis* **4**: 1-28, pl. 1.

Thunberg C. P. 1784: *Dissertatio entomologica novas insectorum species. Pars III.* Upsaliae: Johan Edman, 53-69 pp., 1 pl.

Thunberg C. P. 1787: [new species]. In: Donation. Thunbergianae 1785. Continuat. II, pp. 43-58, 1 pl. In: Publico examini subjicit, Petrus A. Bjerkén. Cujus partem quartam. *D. D. Museum Naturalium Academiae Upsaliensis, parts 1-8.* Upsaliae: Joh. Edman, 106 pp., 1 pl.

Thunberg C. P. 1791: *Dissertatio entomologica novas insectorum species. Pars VI. Resp. A. J. Lagus.* Upsaliae: Johan Edman, [4] + 108-130 pp., 1 pl.

Thunberg C. P. 1806: Tvänne nya insect-slägten, Ptyocerus och Ripidius. *Kongliga Vetenskaps Academiens nya Handlingar* **27**: 1-6, 1 pl.

Thunberg C. P. 1814: Beskifning på tvänne nya insect-slågten, Gnathocerus och Taumacera. *Kongliga Vetenskaps Academiens Handlingar* **1814**: 46-50, pl. IV.

Thunberg C. P. 1821: Coleoptera capensia, antennis fusiformibus. *Nova Acta Regiae Societatis Scientiarum Upsaliensis* **8**: 157-193.

Tokeji M. 1953: Some notes on the mordelline-beetles from Japan and Saipan Island. *The Chûjô* **8**: 1-21.

Tournier H. 1868: Coléoptères nouveaux d'Europe, d'Algérie et d'Egypte. *L'Abeille, Mémoires d'Entomologie* **5**: 141-148.

Tournier H. 1874: Diagnoses de quelques coléoptères européens et circumeuropéens. *Comptes-Rendus des Séances de la Société Entomologique de Belgique* **1874**: lxxxvi-lxxxix.

Tôyama M. 1986: A new rhipiphorid Beetle (Coleoptera, Rhipiphoridae) from Malaysia. *Elytra* **14**: 43-45.

Tôyama M. 1987: A new rhipiphorid beetle (Coleoptera, Rhipiphoridae) from Japan. *Elytra* **15**: 33-35.

Tôyama M. & Hatayama T. 1985: Occurence of the Rhipiphorid subfamily Hemirhipidiinae in Japan and Taiwan (Coleoptera, Rhipiphoridae). *Gekkan-Mushi* **176**: 18-24.

Toyoshima R. 1999: A new species of the genus Phloeotrya (Coleoptera, Melandryidae) from Central Japan. *Elytra* **27**: 97-101.

Toyoshima R. & Ishikawa Y. 2000: A new Taiwanese species of the Tetratomidae and three new Japanese species of the Melandryidae (Coleoptera, Heteromera). *Elytra* **28**: 173-191.

Trost P. 1801: *Kleiner Beitrag zur Entomologie in einem Verzeichniss der Eichstettischen bekannten und neu entdecken Insecten mit Anmerkungen fuer Kenner und Liebhaber*. Erlangen: Palm, 71 pp.

Truqui E. 1855: Anthicini Insulae Cypri et Syriae. *Memorie della Reale Accademia delle Scienze di Torino* (2) **16**: 339-371, pl. 1.

Truqui E. 1857: Generis Iphthimi characteres. *Entomologische Zeitung* (Stettin) **18**: 92-94.

Tshernyshev S. E. & Mordkovich V. G. 2002: Novye nakhodki chernotelok roda Blaps Fabricius, 1775 (Coleoptera: Tenebrionidae) v Sibiri. *Russian Entomological Journal* **11**: 383-385.

Tsuru T. 2002: A new species of the genusMordellina (Coleoptera, Mordellidae) from the southern Ryukyus. *Special Bulletin of the Japanese Society of Coleopterology* **5**: 351-355.

Tsuru T. 2004: A taxonomic study of the genus Mordellistenoda (Coleoptera, Mordellidae) from the Ryukyu Islands, southwest Japan, with description of a new species. *Elytra* **32**: 405-413.

Tsuru T. & Takakuwa M. 2005: Two new subspecies of Variimorda miyarabi Nomura (Coleoptera, Mordellidae) from Yonaguni-jima Island of the southwesternmost Ryukyus and Taiwan. *Elytra* **33**: 79-85.

Uhmann G. 1978: Ein neuer Formicomus aus China (Coleopt., Anthicidae). *Entomologische Blätter* **74**: 81-83.

Uhmann G. 1980a: Eine bisher unbekannte Notoxus-Art aus dem Sudan, *Notoxus tiganii* nov. spec. (Coleoptera, Anthicidae). *Entomologische Blätter* **75**: 138-140.

Uhmann G. 1980b: Anthicidae aus Nepal, Kashmir und Ladakh (Insecta: Coleoptera). *Senckenbergiana Biologica* **60** [1979]: 163-169.

Uhmann G. 1982: Anthicidae aus Nepal (Insecta: Coleoptera) *Senckenbergiana Biologica* **62** [1981]: 233-236.

Uhmann G. 1983: Anthiciden der orientalischen Region (Coleoptera, Anthicidae). *Annales Historico-Naturales Musei Nationalis Hungarici* **75**: 185-206.

Uhmann G. 1988a: Die von Hans Mühle 1987 im Yemen gesammelten Anthicidae (Coleoptera). *Acta Coleopterologica* **4**: 1-4.

Uhmann G. 1988b: Die Anthicidae der Sammlung Johann Klapperich (Coleoptera). *Bonner Zoologische Beiträge* **39**: 395-407.

Uhmann G. 1988c: Anthiciden der orientalischen Region aus dem Museum in Genf (Coleoptera, Anthicidae). *Revue Suisse de Zoologie* **94** [1987]: 687-701.

Uhmann G. 1989: Anthicidae (Coleoptera) der orientalischen Region aus dem Naturhistorischen Museum in Genf, II. *Revue Suisse de Zoologie* **96**: 243-252.

Uhmann G. 1990a: Weitere Anthiciden verschiedener Regionen aus dem Museum in Genf (Coleoptera, Anthicidae). *Revue Suisse de Zoologie* **97**: 139-151.

Uhmann G. 1990b: Anthiciden aus Nepal III (Coleoptera). *Entomofauna* **11**: 333-347.

Uhmann G. 1994a: Die von Rudolf Schuh in Asien gefundenen Anthiciden (Insecta, Coleoptera, Anthicidae). *Entomofauna* **15**: 405-413.

Uhmann G. 1994b: Südostasiatische Anthiciden aus dem Naturhistorischen Museum in Genf, 4 (Coleoptera, Anthicidae). *Revue Suisse de Zoologie* **101**: 655-676.

Uhmann G. 1995: Anthicidae aus dem Sultanat von Oman (Coleoptera, Anthicidae). *Acta Coleopterologica* **11**: 38-40.

Uhmann G. 1996: Anthiciden aus dem Naturhistorischen Museum in Wien (Coleoptera, Anthicidae). *Entomologische Blätter* **92**: 19-36.

Uhmann G. 1997: Anthiciden aus China (Coleoptera, Anthicidae). *Acta Coleopterologica* **13**: 21-22.

Uhmann G. 1998a: Anthicidae (Insecta: Coleoptera) from Saudi Arabia with the description of a new species. *Fauna of Arabia* **17**: 93-105.

Uhmann G. 1998b: Eine neue Anthiciden-Art aus der Mongolei (Col., Anthicidae). *Entomologische Nachrichten und Berichte* **42**: 153-154.

Uhmann G. 2004: A new Mecynotarsus from Pakistan (Coleoptera, Anthicidae). *Bonner Zoologische Beiträge* **52**: 1-2.

Uhmann G. 2005: Eine eigenartige, neue Clavicomus-Art aus Japan (Coleoptera, Anthicidae). *Entomologische Blätter* **100** [2004]: 205-208.

Uhmann G., Chikatunov V. & Pavlíček T. 2005: Catalogue of the beetles (Coleoptera) in Israel and adjacent areas: 4. Anthicidae. *Biocosme Mésogéen* **22**: 1-64.

Uyttenboogaart D. L. 1929: Contributions to the knowledge of the fauna of the Canary Islands. IX. Description of new Tenebrionidae (Col.). *Tijdschrift voor Entomologie* **72**: 341-350.

Uyttenboogaart D. L. 1930: Contribution to the knowledge of the fauna of the Canary-Island. Synopsis of the results of the collecting-excursions 1925-1927. *Tijdschrift voor Entomologie* **73**: 211-235.

Uyttenboogaart D. L. 1933: [in Verslag]. *Tijdschrift voor Entomologie* **76**: xli-xlii.

Uyttenboogaart D. L. 1934: Revision des Genus Tribolium (Col. Ten.). *Entomologische Blätter* **30**: 20-31.

Uyttenboogaart D. L. 1937: Contributions to the knowledge of the fauna of the Canary Islands XIX. *Tijdschrift voor Entomologie* **40**: 75-118.

Vauloger de Beaupré M. 1895: Description de deux Notoxus nouveaux du nord de l'Afrique. *L'Abeille, Journal d'Entomologie* **28**: 193.

Vauloger de Beaupré M. 1897: Descriptions de deux erodiides nouveaux (Col.). *Bulletin de la Société Entomologique de France* **1897**: 294.

Vauloger de Beaupré M. 1897: Description de trois Erodius nouveaux du nord de l'Afrique (Col.). *Bulletin de la Société Entomologique de France* **1897**: 307-308.

Vauloger de Beaupré M. 1900: Contribution au catalogue des coléoptères du Nord de l'Afrique. *Annales de la Société Entomologique de France* **68** [1899]: 669-722.

Vázquez X. A. 1988: A new species of Oedemeridae from the Iberian Peninsula. *Nouvelle Revue d'Entomologie* **5**: 259-261.

Vázquez X. A. 1989: El género Chrysanthia Schmidt en la Península Ibérica (Coleoptera, Oedemeridae). *Elytron* **3**: 125-136.

Vázquez X. A. 1991: Contribución al conociemiento de los Oedemeridae de Marruecos, con descripción de una nueva especie (Coleoptera). *Elytron* **5**: 107-114.

Vázquez X. A. 1993: New data on northern African Alloxantha Seidlitz with description of a new species (Coleoptera, Oedemeridae). *Nouvelle Revue d'Entomologie* **10**: 127-132.

Vázquez X. A. 2002: Una nueva especie de Alloxantha de las Canarias (Coleoptera: Oedemeridae). *Boletín de la Asociación Española de Entomología* **26**: 23-29.

Vázquez X. A. 2006: Una nueva subespecie de Nacerdes (Xanthochroa) raymondi (Mulsant & Godart, 1860) de la isla de Mallorca (Coleoptera: Oedemeridae). *Bolletin de la Societat d'Història Natural de les Balears* **49**: 39-44.

Viedma M. G. de. 1959: Revisión del género Proctenius Reitt. (Coleoptera). *Eos, Revista Española de Entomología* **35**: 257-266.

Viedma M. G. de. 1962: El género Rhipidius Thunberg en España (Coleoptera). *Eos, Revista Española de Entomología* **38**: 47-49.

Villa A. & Villa J. B. 1833: *Coleoptera Europae dupleta in Collectione Villa quae pro mutua commutatione offerri possunt*. Mediolani: pp. 1-36.

Villa A. & Villa J. B. 1835: *Coleopterorum species novae in catalogo dupletorum extantes diagnosibus, adumbrationibus atque observationibus illustratae*. Mediolani, pp. 32-36.

Villa A. & Villa J. B. 1838: *Alterum supplementum Coleopterorum Europae sive additio ad Catalogum et Supplementum I. Dupletorum collectionis Villa continens species alias, nunc pro mutua commutatione adhunc offerrendas; nec non aliquarum specierum emendationes et synonima quae in catalogo anni 1833 et supplemento anni 1835 oblatae fuerunt*. Mediolani, pp. 51-66.

Villers C. J. de. 1789: *Caroli Linnaei entomologia, faunae Sueciae descriptionis aucta; DD. Scopoli, Geoffroy, de Geer, Fabricii, Schrank, &c. speciebus vel in systemate non enumeratis, vel nuperrime detectis, vel speciebus Galliae australis locupletana, generum specierumque rariorum iconibus ornata, curante &*

augente Carolo de Villiers, Acad. Lugd. Massil. Villa-Fr. Rhotom. necnom geometriae regio professore. Tomus primus. Lugduni: Piestre et Delamolliere, xiv + [4] + 4 + 765 + [1] pp., 3 pls.

Viñolas A. 1987: Révision de la Asida (Granulasida) ricoi (Martinez) (Col. Tenebrionidae). *Sessió Conjunta d'Entomologia Instució Catalana d'Historia Natural - Sociedad Catalana de Lepidopterologia* **5**: 31-34.

Viñolas A. 1990: Nueva ordenación de los géneros Phylan Stephens, 1832, y Micrositus Mulsant & Rey, 1854, de la tribu Dendarini (Coleoptera: Tenebrionidae). *VI. Sessió Conjunta d'Entomologia Instució Catalana d'Historia Natural - Societat Catalana de Lepidopterologia* **6** [1989]: 53-68.

Viñolas A. 1994: El género Pimelia Fabricius, 1775 en la Península Ibérica y Baleares, con nota sistemática sobre una especie de Canarias (Coleoptera, Tenebrionidae, Pimeliinae). *Sessió d'Entomologia Instució Catalana d'Historia Natural* **8** [1993]: 125-140.

Viñolas A. 1997a: Phylan (Litoboriolus) companyae sp. n. de Zamora, Península Ibérica (Coleoptera: Tenebrionidae, Dendarini). *Zapateri Revista Aragonesa de Entomología* **6** [1996]: 97-102.

Viñolas A. 1997b: Pseudoseriscius munyozi sp. n. d'Almería, Península Ibèrica (Coleoptera: Tenebrionidae, Crypticini). *Butlletí de la Institució Catalana d'Historia Natural* **65**: 97-104.

Viñolas A. & Cartagena M. C. 2003: *Revisión del género Phylan Stephens, 1837 (Coleoptera: Tenebrionidae: Dendarini)*. Entomological monographs of Arganio editio 1, Barcelona: 93 pp.

Viñolas A. & Cartagena M. C. 2005: *Fauna de Tenebrionidae de la Península Ibérica y Baleares*. Barcelona: Agrania editio, 428 pp.

Vít S. 1986: Révision du genre Prostheca Wollaston (Coleoptera, Colydiidae). *Mittheilungen der Schweizerischen Entomologischen Gesellschaft* **59**: 447-457.

Voet J. E. 1806: *Catalogus systematicus Coleopterorum - Catalogue systématique des coléoptères — Systematische Naamlyst van dat geslacht van Insecten dat men Torren noemt. Volume 2*. La Haye: Bakhuysen, 82 + 85 + 87 pp.

Voigts H. 1902: Zur Synonymie der Meloiden-Gattung Zonabris. *Wiener Entomologische Zeitung* **21**: 177-178.

Vuillet A. 1910: Description d'une nouvelle espèce du genre Autocrates Thoms. (Col. Trictenotomidae). *Bulletin de la Société Entomologique de France* **1910**: 347-348.

Vuillet A. 1912: Description d'une nouvelle espèce du genre Autocrates Thoms. *Insecta* 2: 297-299.

Wagner G. 2003: Zwei neue Arten der Gattung Mesostena Eschscholtz, 1831 aus Jemen mit einer Bestimmungstabelle der Untergattung Mesostenopa Kraatz, 1865 für die arabische Halbinsel (Coleoptera: Tenebrionidae: Tentyriini). *Mitteilungen des Internationalen Entomologischen Vereins* **28**: 111-125.

Wagner G. 2005: Zwei neue Arten der Gattung Capnisiceps Chatanay, 1914 aus dem Oman (Coleoptera: Tenebrionidae: Tentyriini). *Mitteilungen des Internationalen Entomologischen Vereins* **30**: 177-183.

Wagner H. 1917: Ueber Isomira semiflava Küst. und icteropa Küst. (Col.). *Entomologische Mitteilungen* **6**: 341-346.

Wagner H. 1928: Beschreibung neuer Coleopteren der europäischen Fauna nebst kritischen Bemerkungen zu bekannten Arten. *Coleopterologisches Centralblatt* **3**: 111-125.

Walker F. 1837: Characters of two undescribed British Coleoptera. *The Entomologists Monthly Magazine* **4**: 83.

Walker F. 1858: Characters of some apparently undescribed Ceylon insects. *The Annals and Magazine of Natural History* (3) **2**: 202-209, 280-286.

Walker F. 1859a: Characters of some apparently undescribed Ceylon insects. *The Annals and Magazine of Natural History* (3) **3**: 258-265.

Walker F. 1859b: Characters of some apparently undescribed Ceylon insects. *The Annals and Magazine of Natural History* (3) **4**: 217-224.

Walker F. 1871: *List of Coleoptera collected by J. K. Lord, Esq. in Egypt, Arabia and near the African shore of the Red Sea. With characters of the undescribed species*. London: E. W. Janson, 19 pp.

Waltl J. 1835: *Reise durch Tyrol, Oberitalien und Piemont nach dem südlichen Spanien*. Passau: Pustet'sche Buchhandlung, [8] + 247 pp. (Erster Theil) + 120 pp. (Zweiter Theil: Ueber die Thiere Andalusien).

Waltl J. 1838: Beiträge zur Kenntniss der Coleopteren der Türken. *Isis Encyclopädische Zeitschrift vorzüglich für Naturgeschichte, vergleichende Anatomie und Physiologie, von Oken* [**31**] 1838 (6): 449-472.

Waltl J. 1839: Käfer um Passau (Fortsetzung von Heft IV S. 273. 1838). *Isis Encyclopädische Zeitschrift vorzüglich für Naturgeschichte, vergleichende Anatomie und Physiologie, von Oken* [**32**] **1839** (3): 221-227.

Wankowicz [= Wankowiez] J. 1867: Descriptions de trois coléoptères nouveaux trouvés en Lithuanie. *Annales de la Société Entomologique de France* (4) 7: 249-254.

Wankowicz J. 1869: Notices entomologiques VI. Descriptions de nouvelles espèces de coléoptères. *Annales de la Société Entomologique de France* (4) **9**: 418-422.

Wasmann E. 1896a: Neue Termitophilen und Termiten aus Indien. *Annali del Museo Civico di Storia Naturale di Genova* **37**: 613-630, pl. II.

Wasmann E. 1896b: Neue Termitophilen und Termiten aus Indien. Nachtrag. *Annali del Museo Civico di Storia Naturale di Genova* **37** (2) [1896-1897]: 149-152.

Wasmann E. 1897: Zoologische Ergebnisse einer von Dr. K. Escherich und Dr. L. Kathariner nach Central-Kleinasien unternommenen Reise. – Myrmecophilen. *Deutsche Entomlogische Zeitschrift* **1896**: 234-241.

Wasmann E. 1899: Neue Termitophilen und Myrmecophilen aus Indien. *Deutsche Entomlogische Zeitschrift* **1899**: 145-169.

Waterhouse C. O. 1871: On some black species of Cantharis with red heads and filiform antennae. *Transactions of the Entomological Society of London* (4) **3**: 405-408.

Waterhouse C. O. 1874: Description of a new species of Heteromerous Coleoptera belonging to the genus Toxicum. *The Entomologist's Monthly Magazine* **11** [1874-5]: 126.

Waterhouse C. O. 1875: Descriptions of some new coleopterous insects belonging to the genera Myodites and Rhipidius (Rhipiphoridae). *Cistula Entomologica* **1**: 369-371.

Waterhouse O. C. 1877: Descriptions of new Coleoptera from various localities. *The Entomologist's Monthly Magazine* **14** [1877-78]: 72-75.

Waterhouse C. O. 1880: Description of new genus and species of Heteromerous Coleoptera. *The Annals and Magazine of Natural History* (5) **5**: 147-148.

Waterhouse C. O. 1881: On the coleopterous insects collected by Prof. I. Bailey Balfour in the island of Socotra. *Proceedings of the Zoological Society of London* **1881**: 469-478, pl. XLIII.

Waterhouse C. O. 1882: [new taxon]. *Proceedings of the Entomological Society of London* **1882**: iv-v.

Waterhouse C. O. 1886: A new genus of heteromerous Coleoptera allied to Notoxus. *The Annals and Magazine of Natural History* (5) **17**: 39.

Waterhouse C. O. 1889: Coleoptera. Pp. 123-131, pl. 14. In: Aitchison J. E. T.: On the Zoology of the Afghan Delimination Commission. *Transactions of the Linnean Society* (*Zoology*) (2) **5** [1888-1894]: 53-142, pls 6-14.

Waterhouse C. O. 1894: Coleoptera (partim). Pp. 64-71. In: Walker J.: A visit to Damma Island, East Indian Archipelago. *The Annals and Magazine of Natural History* (6) **14**: 49-71.

Weber F. 1801: *Observationes entomologicae, continentes novorum quae condidit generum characteres, et nuper detectarum specierum descriptiones*. Kiliae: Bibliopolii Academici, xii + 116 +[1] pp.

Weber F. & Obst M. H. 1804: *Naturhistorische Reise durch einen Theil Schwedens*. Göttingen: Heinrich Dieterich, xii + 13-207 + [1] pp., 3 pls.

Weise E. 1974: Die Isomira-Arten (Col., Alleculidae) Mitteleuropas und des Mittelmeer-Raums. *Entomologische Blätter* **70**: 65-128.

Weise J. 1877: [new taxa]. In Reitter E.: Coleopterologische Ergebnisse einer Reise nach Südungarn und die Transsylvanischen Alpen. *Verhandlungen des Naturforschendes Vereines in Brünn* **15** [1876]: 3-30, pl. I.

Weise J. 1878: [new taxa]. In: Schneider O. & Leder H.: Beiträge zur Kenntniss der kaukasischen Käferfauna. *Verhandlungen des Naturforschendes Vereines in Brünn* **16** [1877]: 3-262.

Weise J. 1880a: Laena Heydeni Weise n. sp. *Entomologische Monatsblätter* **2**: 102-103.

Weise J. 1880b: [new taxa]. In: Leder H.: Beitrag zur kaukasischen Käfer-Fauna. *Verhandlungen der Kaiserlich-Königlichen Zoologisch-Botanischen Gesellschaft in Wien* **29** 1879]: 451-488.

Weise J. 1883: [new taxa]. In: Heyden L. F. J. D. von. & Weise J.: Neue südost-europäische und klein-asiatische Käfer. *Deutsche Entomologische Zeitschrift* **27**: 310-314.

Wellman F. C. 1910a: On the synonymy of the Meloidae (Coleoptera). *Deutsche Entomologische Zeitschrift* **54**: 22-26.

Wellman F. C. 1910b: The generic and subgeneric types of the Lyttidae (Meloidae s. Cantharidae auct.), (Col.). *Canadian Entomologist* **41**: 389-396.

Wellman F. C. 1912: New species of Lyttidae, with notes on described species (Col.). *Entomological News* **23**: 29-38.

Westhoff F. 1881: *Die Käfer Westfalens*. Bonn: Max Cohen & Sohn, xxiii + 323 pp.

Westhoff F. 1882: Die Käfer Westfalens, II. Abtheilung. Pp. 141-323. *Verhandlungen des Naturhistorischen Vereins der Preussischen Rheinlande und Westfalens, Supplement* **38** [1881-1882]: xxviii + 323 pp.

Westwood J. O. 1830: Observations upon the Notoxidae, a family of coleopterous insects, with characters of two new British genera separated therein. *Zoological Journal* **5** [1829]: 57-61, pl. 41.

Westwood J. O. 1838: *An introduction to the modern classification of insects; founded on the natural habits and corresponding organisation of the different families. Vol. 1.* London: Longman, Orme, Brown, Green and Longmans: xii + 462 pp.

Westwood J. O. 1839: Descriptions of insects figured in plates 9 and 10. In: Royle J. F.: *Illustrations of the botany and other branches of the natural history of the Himalayan mountains, and of the flora of Cashmere. Volume I.* London: W. H. Allen and Co., xxxviii-lvii pp., pls 9-10.

Westwood J. O. 1841: Descriptions of some new genera of Australian heteromerous beetles in the collection of the Rev. F. W. Hope. *Arcana Entomologica* **1**: 43-44.

Westwood J. O. 1846: [new taxon]. In: Entomological society. *The Annals and Magazine of Natural History* **18**: 353.

Westwood J. O. 1851: Descriptions of three new genera of exotic Coleoptera. *Transactions of the Entomological Society of London, New Series* **1**: 167-172, pl. XIII.

Westwood J. O. 1883: Descriptions of some new exotic Coleoptera. *Tijdschrift voor Entomologie* **26** [1882-83]: 61-78, pl. III.

White A. 1846: Insects. Pp. 1-24, pls. i-vi. In: J. Richardson and E. Gray (eds): *Zoology of the voyage of H.M.S. Erebus & Terror, under the command of Captain Sir James Clark Ross, R.N., F.R.S, during the years 1839 to 1843. Part XI. – April. Insects of New Zealand.* London: Longman, Brown, Green, and Longmans, 24 pp., 6 pls.

Wiedemann C. R. W. 1819: Neue Käfer aus Bengalen und Java. *Zoologisches Magazin* **1** (3): 157-183.

Wiedemann C. R. W. 1821: In: Wiedemann C. R. W. & Germar E. F.: Neue exotische Käfer. *Magazin der Entomologie* **4**: 107-183.

Wiedemann C. R. W. 1823: Zweihundert neue Käfer von Java, Bengalen und dem Vorgebirge der Guten Hoffnung. *Zoologisches Magazin* **2** (1): 3-133.

Wilke S. 1921: Beiträge zur Systematik und geographischen Verbreitung ungeflügelten Tenebrioniden (Unterfam. Asidinae). *Archiv für Naturgeschichte* A **87**: 248-312, pls I-IV.

Winkler A. 1915: Ein neuer blinder Tenebrionide aus der Krim. *Wiener Entomologische Zeitung* **24**: 331-335.

Winkler A. 1927: Pp. 753-880. In: *Catalogus coleopterorum regionis palaearcticae.* Wien [1927-1932]: Albert Winkler, 1698 pp. [single column/page edition].

Wissmann O. L. 1848: Entomologische Notizen. *Entomologische Zeitung* (Stettin) **9**: 76-80.

Wittmer W. 1936: Neue Koleopteren aus Aegypten und dem Sinaï aus der Sammlung des Museums "Pietro Rossi", Duino. *Bulletin de la Société Royale Entomologique d'Egypte* **20**: 16-18.

Wollaston T. V. 1854: *Insecta Maderensia; being an account of the insects of the islands of the Madeiran group.* London: John van Voorst, xliii + 634 pp., 13 pls.

Wollaston T. V. 1857a: *Catalogue of the coleopterous insects of Madeira in the Collection of the British Museum.* London: Taylor and Francis, xvi + 234 pp.

Wollaston T. V. 1857b: Description of an uncharacterized British Anthicus. *The Zoologist: a Popular Miscellany of Natural History* **15**: 5721-5722.

Wollaston T. V. 1860a: On additions to the Madeiran Coleoptera. *The Annals and Magazine of Natural History* (3) **5**: 252-267, 358-365.

Wollaston T. V. 1860b: On additions to the Madeiran Coleoptera. *The Annals and Magazine of Natural History* (3) **6**: 48-54.

Wollaston T. V. 1861a: On certain Coleoptera from the Island of St. Vincent. *The Annals and Magazine of Natural History* (3) **7**: 197-206, 246-253.

Wollaston T. V. 1861b: An additions to the Madeiran Coleoptera. *The Annals and Magazine of Natural History* (3) **8**: 99-111.

Wollaston T. V. 1862a: On the Euphorbia-infesting Coleoptera of the Canary Islands. *Transactions of the Entomological Society of London* (3) **1**: 136-189.

Wollaston T. V. 1862b: Notes on Tarphii; with the description of an allied genus. *Journal of Entomology* **1**: 371-387, pls. XVIII, XIX.

Wollaston T. V. 1862c: Brief diagnostic characters of new Canarian Coleoptera. *The Annals and Magazine of Natural History* (3) **9**: 437-442.

Wollaston T. V. 1862d: On additions to the Madeiran Coleoptera. *The Annals and Magazine of Natural History* (3) **10**: 287-293, 331-342.

Wollaston T. V. 1864: *Catalogue of the Coleopterous insects of the Canaries in the Collection of the British Museum*. London: Taylor and Francis, xiii + 648 pp.

Wollaston T. V. 1865: *Coleoptera Atlantidum, being an enumeration of the Coleopterous Insects of the Madeiras, Salvages, and Canaries*. London: John van Voorst, xlvii + 526 + 140 (Appendix) pp., 1 map.

Wollaston T. V. 1867: *Coleoptera Hesperidum, being an enumeration of the Coleopterous insects of the Cape Verde Archipelago*. London: John van Voorst, xxxix + 285 pp.

Wollaston T. V. 1871: On additions to the Atlantic Coleoptera. *Transactions of the Entomological Society of London* **1871**: 203-314.

Wollaston T. V. 1873: On a new genus of Colydiidae, from Japan. *Transactions of the Entomological Society of London* **1873**: 1-4.

Wollaston T. V. 1873: On new Coleoptera from Japan. *Entomologists Monthly Magazine* **10** [1873-1874]: 167-172.

Wollaston T. V. 1874: On new Coleoptera from Japan. *Entomologists Monthly Magazine* **10** [1873-1874]: 200-203.

Wollaston T. V. 1875: [new taxa]. In: Bates H. W. & Wollaston T. V.: Descriptions of new species referred to in Mr. Blackmore's notes on Coleoptera from South Morocco. *The Entomologist's Monthly Magazine* **11** [1874-1875]: 217-220.

Wood T. 1884: A new species of Cis. *The Entomologist's Monthly Magazine* **21** [1884-1885]: 130-131, 147 [footnote].

Wu W. & Huang R.-X. 2005: [new taxa]. In: Wu W., Fan Z.-t. & Huang R.-x.: Three new species of Platyope (Coleoptera: Tenebrionidae) from Xinjiang, China. *Entomotaxonomia* **27**: 283-288. (in Chinese with English summary).

Yamazaki H. 1964: Plesiophthalmus nigrocyaneus Motschulsky and its allied species from Japan. *Entomological Review of Japan* **17**: 17-27.

Yamazaki H. 1968: A new species of the genus Plesiophthalmus from the Ryukyu Isls (Coleoptera: Tenebrionidae). *Tohoku Konchu Kenkyu* **3**: 27-28.

Yamazaki H. 2002: A new species of Plesiophthalmus (Coleoptera, Tenebrionidae) from Japan. *Special Bulletin of the Japanese Society of Coleopterology* **5**: 377-379.

Yang C. & Fan X. 1988: A new species of Glipa from Fanjingshan mountain. *Entomological investigations Fanxingshan. Institute of Biology, Guizhou Academy of Science* **1988**: 109-110. (in Chinese with English summary).

Yang X.-J. & Ren G.-D. 2004: A new species and twelve new records of the tribe Opatrini in China (Coleoptera, Tenebrionidae). *Acta Zootaxonomica Sinica* **29**: 305-309.

Yang Y.-X. & Ren G.-D. 2006: A new species of the genus Epicauta from Yunnan (Coleoptera: Meloidae). *Entomotaxonomia* **28**: 271-274 (in Chinese with English summary).

Yawata H. 1944: Description d'une espèce nouvelle appartenant au genre Anthicus du Japon (Col. Anthicidae). *Transactions of the Kansai Entomological Society* **14**: 1-2.

Young D. K. 1996: Status of Pseudopyrochroa types in the Museum National d'Histoire Naturelle, Paris (Coleoptera: Pyrochroidae: Pyrochroinae). *Oriental Insects* **30**: 213-235.

Young D. K. 2000: Five new species of Pseudopyrochroa (Coleoptera: Pyrochroidae: Pyrochroinae) from Taiwan. *Oriental Insects* **34**: 245-262.

Young D. K. 2001: A new species of Pseudopyrochroa (Coleoptera: Pyrochroidae: Pyrochroinae) from China. *Oriental Insects* **35**: 273-278.

Young D. K. 2003: A new species of Pseudopyrochroa (Coleoptera: Pyrochroidae: Pyrochroinae) from China and Korea. *Special Bulletin of the Japanese Society of Coleopterology* (Tokyo) **6**: 345-351.

Young D. K. 2004a: Himalapyrochroa gibbosa, a new genus and species of fire-colored beetle (Coleoptera: Pyrochroidae: Pyrochroinae) from Darjeeling, India. *Oriental Insects* **38**: 189-196.

Young D. K. 2004b: Dendroidopsis and Frontodendroidopsis: two new Asian genera of fire-colored beetles (Coleoptera: Pyrochroidae: Pyrochroinae), with descriptions of two new species of Dendroidopsis, and lectotype designations for Dendroides assamensis Blair and Pseudopyrochroa binhana Pic. *Oriental Insects* **38**: 197-217.

Young D. K. 2004c: Pyrochroa daglariensis: replacement name for Dendroides cyanipennis Pic 1896, primary junior homonym of Dendroides cyanipennis Latreille 1817 (Coleoptera: Pyrochroidae: Pyrochroinae). *Proceedings of the Washington Entomological Society* **106**(4): 933-934.

Young D. K. 2005a: Sinodendroides chinensis, a new genus and species of fire-colored beetle (Coleoptera: Pyrochroidae: Pyrochroinae) from China. *Oriental Insects* **39**: 93-100.

Young D. K. 2005b: A new species of Phyllocladus (Coleoptera: Pyrochroidae: Pyrochroinae) from China, with a key to males of the two known Chinese species. *Oriental Insects* **39**: 101-107.

Yu Y. & Ren G.-D. 1997: Five new species of the genus Cteniopinus from China (Tenebrionidae: Alleculinae). *Sichuan Journal of Zoology* **16** (1): 8-12 (in Chinese with English summary).

Yuan C.-X. & Ren G.-D. 2005: Study of the arboreal darkling beetles of the genus Strongylium from China (Coleoptera, Tenebrionidae). *Acta Zootaxonomica Sinica* **30**: 399-406.

Zaitzev P. A. 1941: Rasprostranenie vidov sem. uzkotelok v Zakavkaz"e (Coleoptera, Oedemeridae). *Trudy Zoologicheskogo Sektora* **3**: 229-243.

Zerche L. 2004: Revision der Gattung Aegialites Mannerheim (Coleoptera: Salpingidae: Aegialitinae). *Stuttgarter Beiträge zu Naturkunde Serie A* (*Biologie*) **666**: 1-116 pp.

Zetterstedt J. W. 1828: *Fauna insectorum Lapponica. Pars 1.* Hammone: Libraria Schulziana, xx + 563 pp.

Zetterstedt J. W. 1838: Coleoptera. Columns 7-868. In: *Insecta Lapponica descripta. Sectio prima. Coleoptera, Orthoptera et Hemiptera.* Lipsiae: Leopoldi Voss, vi pp. + 1140 columns [issued in parts: 1838-1840].

Znojko D. V. 1936: Ueber die systematische Stellung der Familie Petriidae (Coleoptera). *Travaux de l'Institut Zoologique de l'Académie des Sciences de l'URSS* **1936** (3): 67-77.

Znojko D. V. 1950: [new taxa]. In: Ogloblin D. A. & Znojko D. V.: *Fauna SSSR, Zhestkokrylye, Tom 18, vyp. 8. Pylceedy (Cem. Alleculidae), ch. 2, podsem. Omophlinae.* Moskva, Leningrad: Akademiya Nauk SSSR, 133 pp.

Zubkov [= Zoubkoff] B. 1829: Sur un nouveau genre et quelques nouvelles espèces de coléoptères. *Bulletin de la Société Impériale des Naturalistes de Moscou* **1**: 147-170, pls. 4, 5.

Zubkov [= Zoubkoff] B. 1833: Nouveaux coléoptères recueillis en Turcménie. *Bulletin de la Société Impériale des Naturalistes de Moscou* **6**: 310-340.

Zubkov [= Zoubkoff] B. 1837: Description de quelques coléoptères nouveaux. *Bulletin de la Société Impériale des Naturalistes de Moscou* **10** (6): 59-72, pls 3, 4.

Zoufal V. 1892: Uebersicht der mir bekannten Arten der Coleopteren-Gattung Tenebrio L. *Wiener Entomologische Zeitung* **12**: 283-286.

Zoufal V. 1893: Revision der Gattungen Centorus und Calcar aus Europa und den angrenzenden Ländern. *Wiener Entomologische Zeitung* **12**: 115-119.

Zürcher C. 1911: Colpotus Reitteri n. sp. (Col. Tenebr.). *Wiener Entomologische Zeitung* **30**: 292.

INDEX

Family-group names

INDEX

Genus-group names
(synonyms are in *italics*)

www.ingramcontent.com/pod-product-compliance
Lightning Source LLC
Chambersburg PA
CBHW071405230926
51846CB00001B/15
* 9 7 8 8 7 8 8 7 5 7 6 9 9 *

ISBN 8788757692

9 788788 757699